Flexible Multibody Dynamics

SOLID MECHANICS AND ITS APPLICATIONS
Volume 176

Series Editor: G.M.L. GLADWELL
Department of Civil Engineering
University of Waterloo
Waterloo, Ontario, Canada N2L 3GI

Aims and Scope of the Series

The fundamental questions arising in mechanics are: *Why?*, *How?*, and *How much?*
The aim of this series is to provide lucid accounts written by authoritative researchers giving vision and insight in answering these questions on the subject of mechanics as it relates to solids.

The scope of the series covers the entire spectrum of solid mechanics. Thus it includes the foundation of mechanics; variational formulations; computational mechanics; statics, kinematics and dynamics of rigid and elastic bodies: vibrations of
solids and structures; dynamical systems and chaos; the theories of elasticity, plasticity and viscoelasticity; composite materials; rods, beams, shells and membranes;
structural control and stability; soils, rocks and geomechanics; fracture; tribology;
experimental mechanics; biomechanics and machine design.

The median level of presentation is the first year graduate student. Some texts are
monographs defining the current state of the field; others are accessible to final year
undergraduates; but essentially the emphasis is on readability and clarity.

For other titles published in this series, go to
www.springer.com/series/6557

O. A. Bauchau

Flexible Multibody Dynamics

 Springer

O. A. Bauchau
Georgia Institute of Technology
School of Aerospace Engineering
Ferst Dr. 270
30332-0150 Atlanta Georgia
USA
olivier.bauchau@aerospace.gatech.edu

ISSN 0925-0042
ISBN 978-94-007-0334-6   e-ISBN 978-94-007-0335-3
DOI 10.1007/978-94-007-0335-3
Springer Dordrecht Heidelberg London New York

Library of Congress Control Number: 2010938509

Cover Design: CREST

Printed on acid-free paper

Springer is part of Springer Science+Business Media (www.springer.com)

To my wife, Yi-Ling, and my family

Preface

Multibody dynamics analysis was originally developed as a tool for modeling rigid multibody systems with simple tree-like topologies, but has considerably evolved to the point where it can handle linearly and nonlinearly elastic multibody systems with arbitrary topologies. It is now used widely as a fundamental design tool in many areas of engineering.

This textbook has emerged over the past two decades from efforts to teach graduate courses in advanced dynamics and flexible multibody dynamics to engineering students. Although this book reviews the basic principles of dynamics, it is assumed that students enrolling in these graduate courses have completed a comprehensive set of undergraduate courses in statics, dynamics, deformable bodies, energy methods, and numerical analysis. The advanced dynamics course is, of course, a prerequisite for the flexible multibody dynamics course.

The book is divided into six parts. The first part presents the basic tools and concepts that form the foundation for the other parts. It begins with a review of basic operations on vectors and tensors. The second chapter deals with coordinate systems. The differential geometry of both curves and surfaces is presented and leads to path and surface coordinates. Chapter 3 reviews the basic principles of dynamics, starting with Newton's laws. The important concept of conservative forces is discussed. Systems of particles are then treated, leading to Euler's first and second laws.

Chapter 4 concludes the first part of the book with a detailed description of three-dimensional rotation. For most graduate students, this chapter is not really a review. Indeed, many undergraduate dynamics courses focus primarily on two-dimensional systems. Problems involving three-dimensional rotation, if treated at all, are often rushed in the last few weeks of the semester, leaving most students with insufficient time to absorb this difficult material.

Part 2 develops rigid body dynamics, the foundation of multibody dynamics. The analysis of the kinematics of rigid bodies is the focus of chapter 5. It starts with the analysis of the general displacement and velocity fields of a rigid body. The classical topics of relative velocities and accelerations are also addressed. The motion tensor and its properties are given an in-depth treatment.

Kinetics of rigid bodies is the focus of chapter 6. The various forms of Euler's law governing the rotational motion of rigid bodies are presented. While the emphasis of the chapter is on three-dimensional problems, planar motion is also treated in details.

Part 3 presents the fundamental concepts of analytical dynamics. Chapter 7 introduces the concepts of virtual displacement, virtual rotation, and virtual work. The principle of virtual work for static problems is given extensive coverage as this is an indispensable topic for the study of the variational and energy principles of dynamics presented in chapter 8. D'Alembert's principle, Hamilton's principle, and Lagrange's formulation are derived and their use illustrated with numerous examples.

Multibody systems are characterized by two distinguishing features: system components undergo finite relative rotations and these components are connected by mechanical joints that impose restrictions on their relative motion. The first distinguishing feature is of a purely kinematic nature: in multibody systems, overall and relative motions are finite, leading to inherently nonlinear problems. The second distinguishing feature is the main culprit for the complexity of many multibody formulations. Each component of a flexible multibody system is a *constrained dynamical system* because of the restrictions imposed on it by the mechanical joints connecting it to others.

The first three parts of the book present background material on *unconstrained dynamical systems*, *i.e.*, systems for which the number of generalized coordinated used to describe the system equals the number of degrees of freedom. In contrast, part 4 focuses on *constrained dynamical systems*. Chapter 9 presents Lagrange's multiplier technique and the distinction between holonomic and nonholonomic constraints. The combination of the principle of virtual work with Lagrange's multiplier technique is shown to provide a powerful tool for the analysis of constrained static problems.

Chapter 10 reviews the classical formulations for constrained dynamical systems. D'Alembert's principle, Hamilton's principle, and Lagrange's formulation are updated to accommodate the presence of both holonomic and nonholonomic constraints. The kinematic constraints associated with the lower pair joints are described in details.

The advanced formulations presented in chapter 11 form the theoretical basis for the practical approaches to numerical solutions of multibody systems. Maggi's, the index-1, the null space, and Udwadia and Kalaba's formulations are presented and the chapter concludes with the geometric interpretation of constraints and Gauss' principle.

Finally, chapter 12 describes in a cursory manner the many numerical approaches used to treat constrained dynamical systems, most of which are rooted in the formulations presented in chapter 11. Chapter 12 is in fact a comprehensive review of the literature on methods of constrained dynamics applied to the solution of multibody systems. It is clearly impossible to treat each approach in detail. Rather, the salient features of each approach are given, and the relationships between them are underlined. The chapter concludes with a detailed description of scaling methods for Lagrange's equations of the first kind.

Part 5 presents a comprehensive overview of the many approaches used to parameterize rotation and motion. The vectorial parameterizations of rotation and motion are given special emphasis as they provide a unified approach to this complex topic. Specific parameterizations widely used in multibody formulations are reviewed in details, whereas other are presented in a more cursory manner.

The last part of the book focuses on flexible multibody dynamics problems, which are categorized into three groups: rigid multibody systems, linearly elastic multibody systems, and nonlinearly elastic multibody systems. The last three chapters of the book focus on the latter category, nonlinearly elastic multibody systems. Chapter 15 presents background material. The basic equations of linear elastodynamics are presented first. Next, finite displacement kinematics are studied, with special emphasis on small strain problems.

Chapter 16 develops the governing equations of flexible joints, cables, beams, and plates and shells. All formulations are *geometrically exact*, *i.e.*, all structural components are allowed to undergo arbitrarily large displacements and rotations, although strains are assumed to remain small. Finally, chapter 17 presents details of the implementation of these elements within the framework of finite element formulations. For instance, interpolation of the rotation fields is an issue that requires special attention.

The topics covered in the first three parts of the book form the basis for a three-credit hour, graduate level course in advanced dynamics, typically taken by first year graduate students. Topics selected from the last three parts provide an ample material for a follow-on, three-credit hour, graduate level course in flexible multibody dynamics. The advanced dynamics course is, of course, a prerequisite for the flexible multibody dynamics course.

Typically, engineering students generally grasp concepts more quickly when presented first with practical examples, which then lead to broader generalizations. Consequently, most concepts are first introduced by means of simple examples; more formal and abstract statements are presented later, when the student has a better grasp of the significance of the concepts.

Numerous homework problems are included throughout the book. Some are straightforward applications of basic concepts, others are small projects that require the use of computers and mathematical software, and others involve conceptual questions that are more appropriate for quizzes and exams. The text also provides many examples that treat practical problems in great details. Some of the examples are re-examined in successive chapters to illustrate alternative or more versatile solution methods.

Notation is a challenging issue in dynamics. Given the limitations of the Latin and Greek alphabets, the same symbol is sometimes used to indicate different quantities, but mostly in different contexts. Consequently, no attempt has been made to provide a comprehensive nomenclature, which would lead to even more confusion.

It is traditional to use a bold typeface to represent vectors and tensors, but this is very difficult to reproduce in handwriting, whether on a board or in personal notes. A notation that is more suitable for hand-written notes has been adopted here. Vectors and arrays are denoted using an underline, such as $\underline{u}$ or $\underline{F}$. Unit vectors are used frequently and are assigned a special notation using a single overbar, such as $\bar{\imath}_1$, which denotes the first Cartesian coordinate axis. The overbar notation also indicates non-dimensional scalar quantities, *i.e.*, $\bar{k}$ is a non-dimensional stiffness coefficient. This is inconsistent, but the two uses are in such different contexts that it should not lead to confusion. Second-order tensors and matrices are indicated using a double-underline, *i.e.*, $\underline{\underline{R}}$ indicates a 3×3 rotation tensor or $\underline{\underline{M}}$ a $n \times n$ mass matrix.

Notations $\underline{a}^T\underline{b}$, $\widetilde{a}\underline{b}$, and $\underline{a}\,\underline{b}^T$ indicate the scalar, vector, and tensor products, respectively, of two vectors, $\underline{a}$ and $\underline{b}$. While many students voice their displeasure with this mnemonic convention that departs from the classical "dot" and "cross product" notations, they very rapidly recognize and appreciate its power and conciseness.

Finally, I am indebted to the many students at Georgia Tech who have given me helpful and constructive feedback over the past decade as I developed the course notes that are the precursors of this book. The constructive use of their many questions and confusion has helped shape this book, and the treatment of many topics was modified numerous times before finding their final form.

Atlanta, Georgia, July 2010   *Olivier Bauchau*

Contents

Part III Concepts of analytical dynamics

Part IV Constrained dynamical systems

Part V Parameterization of rotation and motion

Part I

Basic tools and concepts

1

Vectors and tensors

Vectors and tensors are basic tools for the formulation of kinematics and dynamics problems. This chapter introduces notations and the fundamental operations on vectors and tensors that will be used throughout this book.

1.1 Free vectors

Consider two points, denoted **A** and **B**, in three-dimensional space, as shown in fig. 1.1. The line that connects point **A** to point **B** is called an *oriented line segment*, and denoted **AB**. In the following, the word "segment" will often be used to indicate an oriented line segment. Next, consider two points, **A'** and **B'**, such that **ABB'A'** forms a parallelogram. Segments **AB** and **A'B'** are then of identical length and are parallel to each other. Similarly, if two other points, **A''** and **B''**, are such that **ABB''A''** also forms a parallelogram, segments **AB** and **A''B''** are then of identical length and parallel to each other. Segments **AB**, **A'B'**, and **A''B''** are said to be *equivalent*. The ensemble of all equivalent segments define the *free vector* $\underline{a}$: a given oriented line segment defines a free vector.

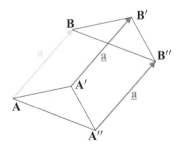

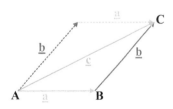

Fig. 1.1. A free vector.   **Fig. 1.2.** Sum of two vectors $\underline{c} = \underline{a} + \underline{b}$.

1.1.1 Vector sum

The addition of two free vectors, $\underline{a}$ and $\underline{b}$, is described in fig. 1.2. Let segments **AB** and **BC** define vectors $\underline{a}$ and $\underline{b}$, respectively, and point **B** is both the end of segment **AB** and the origin of segment **BC**. The *vector sum* of free vectors, $\underline{a}$ and $\underline{b}$, is then free vector $\underline{c}$ defined by all segments equivalent to **AC**.

As fig. 1.2 indicates, the vector sum is commutative, *i.e.*,

$$\underline{c} = \underline{a} + \underline{b} = \underline{b} + \underline{a}. \tag{1.1}$$

1.1.2 Scalar multiplication

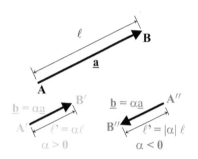

Fig. 1.3. Multiplying a vector by a scalar: $\underline{b} = \alpha \underline{a}$.

Figure 1.3 shows segment **AB** that defines vector $\underline{a}$; the length of vector $\underline{a}$ is defined as the distance, ℓ, between points **A** and **B**. The multiplication of a free vector, $\underline{a}$, by a scalar, α, is depicted in fig. 1.3 and is denoted as

$$\underline{b} = \alpha \underline{a}. \tag{1.2}$$

If α is positive, vector $\underline{b}$ is defined by segment $\mathbf{A'B'}$ parallel to **AB**, oriented in the same direction, and of length $\ell' = \alpha \ell$. If α is negative, vector $\underline{b}$ is defined by segment $\mathbf{A''B''}$ parallel to **AB**, oriented in the opposite direction, and of length $\ell' = |\alpha| \ell$.

1.1.3 Norm of a free vector

Segment **AB** shown in fig. 1.3 defines a free vector, $\underline{a}$. The *norm* of free vector, $\underline{a}$, is defined as the length, ℓ, of any segment defining it. Notation $\|\underline{a}\|$ is used to express the norm of a vector,

$$\|\underline{a}\| = a = \ell. \tag{1.3}$$

A *null vector* is a vector of vanishing length, *i.e.*, $\underline{a} = 0$ implies $\|\underline{a}\| = a = \ell = 0$.

From these definitions, it follows that

$$\|\alpha \underline{a}\| = |\alpha| \, \|\underline{a}\|, \tag{1.4}$$

and the triangular inequality implies

$$\|\underline{a} + \underline{b}\| \leq \|\underline{a}\| + \|\underline{b}\|. \tag{1.5}$$

A *unit vector* is a vector of unit norm and is indicated by an overbar, $(\bar{\cdot})$. A unit vector can be constructed from any vector, $\underline{a}$, by dividing it by its norm,

$$\bar{a} = \frac{\underline{a}}{\|\underline{a}\|}. \tag{1.6}$$

By construction, $\|\bar{a}\| = 1$.

1.1.4 Angle between two vectors

Figure 1.4 defines the angle, Φ, between two free vectors, $\underline{a}$ and $\underline{b}$. Let segments **AB** and **AC** define the free vectors $\underline{a}$ and $\underline{b}$, respectively. Angle Φ is that between segments **AB** and **AC** when these two segments share a common point **A**. The angle between two free vectors is denoted as

$$\Phi = (\underline{a}, \underline{b}). \tag{1.7}$$

Note that $(\underline{a}, \underline{b}) = (\underline{b}, \underline{a}) = \Phi$. The angle between two vectors is such that $0 \le \Phi \le \pi$, i.e., $\sin \Phi \ge 0$.

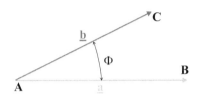

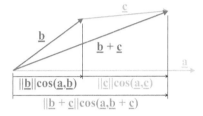

Fig. 1.4. Angle between two vectors.   **Fig. 1.5.** The scalar product is distributive.

1.1.5 The scalar product

The *scalar product*, σ, of two vectors, often called the *dot product*, is defined as

$$\sigma = \underline{a}^T \underline{b} = \|\underline{a}\| \|\underline{b}\| \cos(\underline{a}, \underline{b}). \tag{1.8}$$

Because $\cos(\underline{a}, \underline{b}) = \cos(\underline{b}, \underline{a})$, the scalar product is a commutative operation

$$\sigma = \underline{a}^T \underline{b} = \underline{b}^T \underline{a}. \tag{1.9}$$

Furthermore, it is a distributive operation

$$\sigma = \underline{a}^T (\underline{b} + \underline{c}) = \underline{a}^T \underline{b} + \underline{a}^T \underline{c}, \quad \sigma = (\underline{a} + \underline{b})^T \underline{c} = \underline{a}^T \underline{c} + \underline{b}^T \underline{c}. \tag{1.10}$$

This property follows from the fact that $\|\underline{b} + \underline{c}\| \cos(\underline{a}, \underline{b} + \underline{c}) = \|\underline{b}\| \cos(\underline{a}, \underline{b}) + \|\underline{c}\| \cos(\underline{a}, \underline{c})$, as illustrated in fig. 1.5.

The scalar product of a vector by itself yields the square of its norm, $\underline{a}^T \underline{a} = \|\underline{a}\|^2 = a^2$. Statement $\underline{a}^T \underline{b} = 0$ implies that either $\underline{a} = 0$ or $\underline{b} = 0$, or $\underline{a}$ is orthogonal to $\underline{b}$. The condition for the *orthogonality of two vectors* is

$$\underline{a}^T \underline{b} = 0, \tag{1.11}$$

provided that neither vector is null.

The projection ρ of a vector $\underline{a}$ along a unit direction $\bar{n}$ is readily expressed in terms of the scalar product as

$$\rho = \underline{a}^T \bar{n} = \|\underline{a}\| \cos(\underline{a}, \bar{n}). \tag{1.12}$$

1.1.6 Orthonormal bases

Figure 1.6 depicts an *orthonormal basis*, $\mathcal{I}$, defined by a set of three mutually orthogonal free unit vectors, $\bar{\imath}_1$, $\bar{\imath}_2$, and $\bar{\imath}_3$. Orthonormal bases, also called Cartesian bases, will be indicated with the following notation, $\mathcal{I} = (\bar{\imath}_1, \bar{\imath}_2, \bar{\imath}_3)$. In view of this definition, it is clear that $\bar{\imath}_1^T \bar{\imath}_1 = \bar{\imath}_2^T \bar{\imath}_2 = \bar{\imath}_3^T \bar{\imath}_3 = 1$ and $\bar{\imath}_1^T \bar{\imath}_2 = \bar{\imath}_2^T \bar{\imath}_3 = \bar{\imath}_3^T \bar{\imath}_1 = 0$.

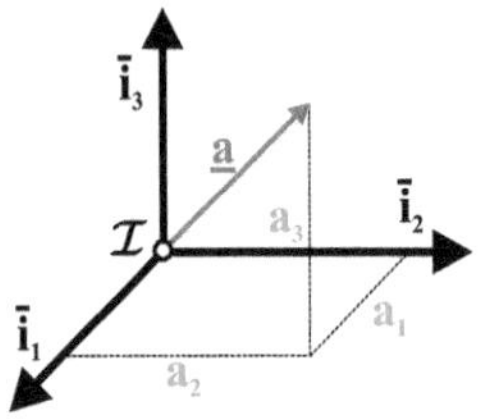

These relationships can be summarized as

$$\bar{\imath}_i^T \bar{\imath}_j = \delta_{ij}, \tag{1.13}$$

where δ_{ij} is the *Kronecker's symbol* defined as

$$\delta_{ij} = \begin{cases} 1, & i = j, \\ 0, & i \neq j. \end{cases} \tag{1.14}$$

Fig. 1.6. An orthonormal basis $\mathcal{I}$.

As shown in fig. 1.6, an arbitrary vector, $\underline{a}$, can be decomposed in the following manner

$$\underline{a} = (\underline{a}^T \bar{\imath}_1)\bar{\imath}_1 + (\underline{a}^T \bar{\imath}_2)\bar{\imath}_2 + (\underline{a}^T \bar{\imath}_3)\bar{\imath}_3 = a_1 \bar{\imath}_1 + a_2 \bar{\imath}_2 + a_3 \bar{\imath}_3, \tag{1.15}$$

where a_1, a_2, and a_3 are the projections, eq. (1.12), of vector $\underline{a}$ along unit vectors $\bar{\imath}_1$, $\bar{\imath}_2$, and $\bar{\imath}_3$, respectively.

The *components of vector* $\underline{a}$ resolved in orthonormal basis $\mathcal{I}$ are the projection of the vector along the unit vectors of the basis, $a_i = \underline{a}^T \bar{\imath}_i$, $i = 1, 2, 3$. The following notation is used

$$\underline{a}^{[\mathcal{I}]} = \begin{Bmatrix} a_1 \\ a_2 \\ a_3 \end{Bmatrix}. \tag{1.16}$$

Notation $\underline{a}$ is used to indicate a free vector, and notation $\underline{a}^{[\mathcal{I}]}$ indicates the components of vector $\underline{a}$ resolved in basis $\mathcal{I}$. The components of a vector consist of a set of three number, which are arranged in a column array, as shown in eq. (1.16). Braces are used to indicate a column array.

The transpose of the column array is a row array and is denoted with a superscript, $(\cdot)^T$. The following notation will be used

$$\underline{a}^{[\mathcal{I}]T} = \{a_1 \ a_2 \ a_3\}, \quad \text{or} \quad \underline{a}^{[\mathcal{I}]} = \{a_1 \ a_2 \ a_3\}^T. \tag{1.17}$$

The components of the unit vectors $\bar{\imath}_1$, $\bar{\imath}_2$, and $\bar{\imath}_3$ resolved in basis $\mathcal{I}$ are

$$\bar{\imath}_1^{[\mathcal{I}]} = \begin{Bmatrix} 1 \\ 0 \\ 0 \end{Bmatrix}, \quad \bar{\imath}_2^{[\mathcal{I}]} = \begin{Bmatrix} 0 \\ 1 \\ 0 \end{Bmatrix}, \quad \bar{\imath}_3^{[\mathcal{I}]} = \begin{Bmatrix} 0 \\ 0 \\ 1 \end{Bmatrix}. \tag{1.18}$$

Using the properties of an orthonormal basis, eq. (1.13), the scalar product of two vectors becomes

$$\underline{a}^T \underline{b} = a_1 b_1 + a_2 b_2 + a_3 b_3 = \{a_1, a_2, a_3\} \begin{Bmatrix} b_1 \\ b_2 \\ b_3 \end{Bmatrix} \tag{1.19}$$

where b_i, $i = 1, 2, 3$, are the components of $\underline{b}$ resolved in basis $\mathcal{I}$. For eq. (1.19) to hold, the components of vectors $\underline{a}$ and $\underline{b}$ must be evaluated *in the same orthonormal basis*.

The notation for the scalar product, $\underline{a}^T \underline{b}$, is a mnemonic notion for the result expressed by eq. (1.19): the scalar product is obtained by multiplying the row array of the components of vector $\underline{a}$ resolved in basis $\mathcal{I}$ by the column array of the components of vector $\underline{b}$ resolved in the same basis. The operation of computing the components of a vector in a given basis is a fundamental operation. The following expressions are used interchangeably: "computing the components of vector $\underline{a}$ in basis $\mathcal{I}$," or "expressing vector $\underline{a}$ in basis $\mathcal{I}$,"or "resolving vector $\underline{a}$ in basis $\mathcal{I}$." For sake of brevity, expressions such as "the components of $\underline{a}$ in $\mathcal{I}$," or "expressing $\underline{a}$ in $\mathcal{I}$" will also be used.

1.1.7 The vector product

The *vector product*, $\underline{c}$, of two vectors, $\underline{a}$ and $\underline{b}$, often called the *cross product*, is defined as

$$\underline{c} = \widetilde{a}\underline{b} = \|\underline{a}\|\|\underline{b}\| \, \sin(\underline{a}, \underline{b}) \, \bar{n}, \tag{1.20}$$

where $\bar{n}$ is a unit vector normal to both $\underline{a}$ and $\underline{b}$, and oriented according to the right-hand rule, as depicted in fig. 1.7. Note that $\mathcal{A} = \|\underline{a}\|\|\underline{b}\| \, \sin(\underline{a}, \underline{b})$ represents the area of the parallelogram spanned by vectors $\underline{a}$ and $\underline{b}$; hence, the norm of the vector product equals this area.

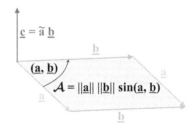

Fig. 1.7. The vector product of vectors $\underline{a}$ and $\underline{b}$.

The vector product is anti-commutative,

$$\widetilde{a}\underline{b} = -\widetilde{b}\underline{a}. \tag{1.21}$$

Indeed, the norms of the two vectors are equal, $\|\widetilde{a}\underline{b}\| = \|\widetilde{b}\underline{a}\| = \mathcal{A}$, but according to the right-hand rule, unit vector $\bar{n}$ will point in opposite directions when the order of vectors $\underline{a}$ and $\underline{b}$ is reversed.

Furthermore, the vector product is a distributive operation

$$(\widetilde{a} + \widetilde{b})\underline{c} = \widetilde{a}\underline{c} + \widetilde{b}\underline{c}, \quad \widetilde{a}(\underline{b} + \underline{c}) = \widetilde{a}\underline{b} + \widetilde{a}\underline{c}. \tag{1.22}$$

This property follows from geometric considerations detailed in fig. 1.8. Note that vectors $\widetilde{a}\underline{c}$, $\widetilde{b}\underline{c}$, and $(\widetilde{a} + \widetilde{b})\underline{c}$ are all in the plane normal to $\underline{c}$. Furthermore, triangles **OAB** and **OA′B′** are similar.

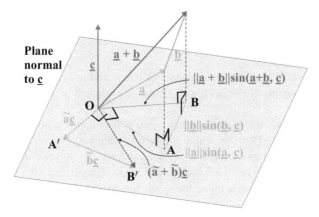

Fig. 1.8. The vector product is distributive.

Statement $\widetilde{a}\underline{b} = 0$ implies that either $\underline{a} = 0$ or $\underline{b} = 0$, or $\underline{a}$ is parallel to $\underline{b}$. The condition for the parallelism of two vectors is

$$\widetilde{a}\underline{b} = 0, \tag{1.23}$$

provided that neither vector is null.

The vector products of the unit vectors defining an orthonormal basis are readily obtained from the definition of the vector product, eq. (1.20), to find $\widetilde{\imath}_1\bar{\imath}_2 = \bar{\imath}_3$, $\widetilde{\imath}_3\bar{\imath}_1 = \bar{\imath}_2$, $\widetilde{\imath}_2\bar{\imath}_3 = \bar{\imath}_1$, $\widetilde{\imath}_2\bar{\imath}_1 = -\bar{\imath}_3$, $\widetilde{\imath}_1\bar{\imath}_3 = -\bar{\imath}_2$, and $\widetilde{\imath}_3\bar{\imath}_2 = -\bar{\imath}_1$. Of course, the cross product of a vector by itself vanishes. These relationships can be summarized as follows

$$\widetilde{\imath}_i\bar{\imath}_j = \epsilon_{ijk}\bar{\imath}_k, \tag{1.24}$$

where summation is implied over the repeated indices, and ϵ_{ijk} is the *Levi-Civita symbol* or *permutation symbol*

$$\epsilon_{ijk} = \begin{cases} +1, & \text{for a cyclic permutation of the indices,} \\ -1, & \text{for an acyclic permutation of the indices,} \\ 0, & \text{for all other cases.} \end{cases} \tag{1.25}$$

The above relationships implicitly assume that vector $\bar{\imath}_1$, $\bar{\imath}_2$, and $\bar{\imath}_3$ have been ordered in such a manner that eqs. (1.24) hold. Such bases are call *right-hand bases* and will be used exclusively in this book.

If a_i, b_i, and c_i, $i = 1, 2, 3$, are the components of vectors $\underline{a}$, $\underline{b}$, and $\underline{c}$, respectively, resolved in a common basis $\mathcal{I}$, the following relationship holds

$$\underline{c} = c_1\bar{\imath}_1 + c_2\bar{\imath}_2 + c_3\bar{\imath}_3 = \widetilde{a}\underline{b} = (a_2b_3 - a_3b_2)\bar{\imath}_1 + (a_3b_1 - a_1b_3)\bar{\imath}_2 + (a_1b_2 - a_2b_1)\bar{\imath}_3,$$

where eqs. (1.22) and (1.24) are used. Taking the scalar product of this expression by $\bar{\imath}_1$, $\bar{\imath}_2$, and $\bar{\imath}_3$ then yields

$$\underline{c}^{[\mathcal{I}]} = \left\{ \begin{array}{c} c_1 \\ c_2 \\ c_3 \end{array} \right\} = \left\{ \begin{array}{c} a_2 b_3 - a_3 b_2 \\ a_3 b_1 - a_1 b_3 \\ a_1 b_2 - a_2 b_1 \end{array} \right\} = \left[\begin{array}{ccc} 0 & -a_3 & a_2 \\ a_3 & 0 & -a_1 \\ -a_2 & a_1 & 0 \end{array} \right] \left\{ \begin{array}{c} b_1 \\ b_2 \\ b_3 \end{array} \right\} = \widetilde{a}^{[\mathcal{I}]} \underline{b}^{[\mathcal{I}]}. \quad (1.26)$$

It is now clear that $\widetilde{a}$ is a second-order, skew-symmetric tensor whose components in basis $\mathcal{I}$ are

$$\widetilde{a}^{[\mathcal{I}]} = \left[\begin{array}{ccc} 0 & -a_3 & a_2 \\ a_3 & 0 & -a_1 \\ -a_2 & a_1 & 0 \end{array} \right]. \quad (1.27)$$

The notation for the vector product, $\widetilde{a}\underline{b}$, is a mnemonic notion for the result expressed by eq. (1.26): the vector product is obtained by multiplying the components of the skew-symmetric tensor $\widetilde{a}$ resolved in basis $\mathcal{I}$ by the column array of the components of vector $\underline{b}$ resolved in the same basis.

1.1.8 The tensor product

The *tensor product* $\underline{\underline{T}}$ of two vectors is a second-order tensor defined as

$$\underline{\underline{T}} = \underline{a}\,\underline{b}^T. \quad (1.28)$$

The fundamental property of tensor $\underline{\underline{T}}$ is

$$\underline{\underline{T}}\,\underline{c} = (\underline{b}^T \underline{c})\underline{a}, \quad (1.29)$$

for any arbitrary vector $\underline{c}$. By letting $\underline{a} = \overline{\imath}_i$ and $\underline{b} = \overline{\imath}_j$, eq. (1.29) then implies

$$\overline{\imath}_1^{[\mathcal{I}]}\overline{\imath}_1^{[\mathcal{I}]T} = \left[\begin{array}{ccc} 1 & 0 & 0 \\ 0 & 0 & 0 \\ 0 & 0 & 0 \end{array} \right], \quad \overline{\imath}_2^{[\mathcal{I}]}\overline{\imath}_2^{[\mathcal{I}]T} = \left[\begin{array}{ccc} 0 & 0 & 0 \\ 0 & 1 & 0 \\ 0 & 0 & 0 \end{array} \right], \quad \overline{\imath}_3^{[\mathcal{I}]}\overline{\imath}_3^{[\mathcal{I}]T} = \left[\begin{array}{ccc} 0 & 0 & 0 \\ 0 & 0 & 0 \\ 0 & 0 & 1 \end{array} \right],$$

$$\overline{\imath}_1^{[\mathcal{I}]}\overline{\imath}_2^{[\mathcal{I}]T} = \left[\begin{array}{ccc} 0 & 1 & 0 \\ 0 & 0 & 0 \\ 0 & 0 & 0 \end{array} \right], \quad \overline{\imath}_1^{[\mathcal{I}]}\overline{\imath}_3^{[\mathcal{I}]T} = \left[\begin{array}{ccc} 0 & 0 & 1 \\ 0 & 0 & 0 \\ 0 & 0 & 0 \end{array} \right], \quad \overline{\imath}_2^{[\mathcal{I}]}\overline{\imath}_3^{[\mathcal{I}]T} = \left[\begin{array}{ccc} 0 & 0 & 0 \\ 0 & 0 & 1 \\ 0 & 0 & 0 \end{array} \right],$$

$$\overline{\imath}_2^{[\mathcal{I}]}\overline{\imath}_1^{[\mathcal{I}]T} = \left[\begin{array}{ccc} 0 & 0 & 0 \\ 1 & 0 & 0 \\ 0 & 0 & 0 \end{array} \right], \quad \overline{\imath}_3^{[\mathcal{I}]}\overline{\imath}_1^{[\mathcal{I}]T} = \left[\begin{array}{ccc} 0 & 0 & 0 \\ 0 & 0 & 0 \\ 1 & 0 & 0 \end{array} \right], \quad \overline{\imath}_3^{[\mathcal{I}]}\overline{\imath}_2^{[\mathcal{I}]T} = \left[\begin{array}{ccc} 0 & 0 & 0 \\ 0 & 0 & 0 \\ 0 & 1 & 0 \end{array} \right].$$

Letting $\underline{\underline{T}}^{[\mathcal{I}]}_{kl}$ represent the components of tensor $\underline{\underline{T}}^{[\mathcal{I}]} = \overline{\imath}_i^{[\mathcal{I}]}\overline{\imath}_j^{[\mathcal{I}]T}$, these relationships can be summarized as

$$\underline{\underline{T}}^{[\mathcal{I}]}_{k\ell} = \delta_{ki}\delta_{\ell j}. \quad (1.30)$$

If a_i and b_i, $i = 1, 2, 3$, are the components of vectors $\underline{a}$ and $\underline{b}$, respectively, in a common basis $\mathcal{I}$, the following relationship holds

$$\underline{\underline{T}}^{[\mathcal{I}]} = \left[\begin{array}{ccc} a_1 b_1 & a_1 b_2 & a_1 b_3 \\ a_2 b_1 & a_2 b_2 & a_2 b_3 \\ a_3 b_1 & a_3 b_2 & a_3 b_3 \end{array} \right] = \underline{a}^{[\mathcal{I}]}\underline{b}^{[\mathcal{I}]T}, \quad (1.31)$$

where eq. (1.30) was used. The notation for the tensor product, $\underline{a}\,\underline{b}^T$, is a mnemonic notion for the result expressed by eq. (1.31): the tensor product is obtained by multiplying the column array of components of vector $\underline{a}$ in basis $\mathcal{I}$ by the row array of the components of vector $\underline{b}$ in the same basis.

1.1.9 The mixed product

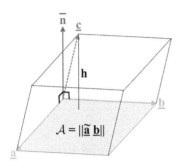

Fig. 1.9. The mixed product of vectors $\underline{a}$, $\underline{b}$, and $\underline{c}$.

Let $\underline{a}$, $\underline{b}$, and $\underline{c}$ be three arbitrary vectors. The scalar $\underline{c}^T \widetilde{a}\underline{b}$ is called the *mixed product* of these vectors. The geometric interpretation of this operation is illustrated in fig. 1.9. The vector product $\widetilde{a}\underline{b} = \mathcal{A}\bar{n}$ is defined by eq. (1.20), where $\mathcal{A}$ represents the area spanned by vectors $\underline{a}$ and $\underline{b}$ and the orientation of unit vector $\bar{n}$ is selected according to the right-hand rule. The mixed product then becomes $\underline{c}^T \widetilde{a}\underline{b} = \|\underline{c}\| \mathcal{A}\cos(\bar{n}, \underline{c})$, where $\|\underline{c}\| \cos(\bar{n}, \underline{c}) = h$ is the projection of vector $\underline{c}$ along the unit vector $\bar{n}$. It then follows that $\underline{c}^T \widetilde{a}\underline{b} = \mathcal{A}h$, where $\mathcal{A}$ is the area of the parallelogram spanned by vectors $\underline{a}$ and $\underline{b}$ and h the height of the parallelepiped defined by vectors $\underline{a}$, $\underline{b}$, and $\underline{c}$. Clearly, the mixed product represents the volume of this parallelepiped.

The above interpretation assumes that vectors $\underline{a}$, $\underline{b}$, and $\underline{c}$ are ordered according to the right-hand rule. If this is not the case, it is easily verified that the mixed product yields the negative of the volume spanned by the three vectors.

If a_i, b_i, and c_i are the components of vectors $\underline{a}$, $\underline{b}$, and $\underline{c}$, respectively, resolved in basis $\mathcal{I}$, the mixed product can be written as

$$\underline{c}^T \widetilde{a}\underline{b} = \det \begin{bmatrix} a_1 & a_2 & a_3 \\ b_1 & b_2 & b_3 \\ c_1 & c_2 & c_3 \end{bmatrix}, \qquad (1.32)$$

where eqs. (1.19) and (1.26) were used. It is now clear that $\underline{c}^T \widetilde{a}\underline{b} = \underline{b}^T \widetilde{c}\underline{a} = \underline{a}^T \widetilde{b}\underline{c}$, since these operations correspond to permutations of lines of the determinant. Of course, due to the anti-commutativity property of the vector product, eq. (1.21), $\underline{c}^T \widetilde{b}\underline{a} = \underline{b}^T \widetilde{a}\underline{c} = \underline{a}^T \widetilde{c}\underline{b}$.

1.1.10 Tensor identities

Important tensor identities will be used throughout this book. If $\underline{a}$, $\underline{b}$, and $\underline{c}$ are three arbitrary vectors, the following identities can be readily verified by painstakingly expanding the various products,

$$\widetilde{(\widetilde{a}\,\underline{b})} = \widetilde{a}\,\widetilde{b} - \widetilde{b}\,\widetilde{a}, \tag{1.33a}$$

$$\widetilde{a}\,\widetilde{b} = \underline{b}\,\underline{a}^T - (\underline{a}^T\underline{b})\underline{I}, \tag{1.33b}$$

$$\widetilde{a}\,\widetilde{b} - \widetilde{b}\,\widetilde{a} = \underline{b}\,\underline{a}^T - \underline{a}\,\underline{b}^T, \tag{1.33c}$$

$$\widetilde{a}\,\widetilde{b} - \underline{a}\,\underline{b}^T = \widetilde{(\widetilde{a}\,\underline{b})} - (\underline{a}^T\underline{b})\underline{I}, \tag{1.33d}$$

$$\widetilde{a}\,\widetilde{b}\,\underline{c} = (\underline{a}^T\underline{c})\underline{b} - (\underline{b}^T\underline{c})\underline{a}, \tag{1.33e}$$

$$\widetilde{a}\,\widetilde{b}\,\underline{c} = (\underline{a}^T\underline{c})\underline{b} - (\underline{a}^T\underline{b})\underline{c}, \tag{1.33f}$$

$$\underline{a}\,\underline{b}^T\underline{c} = (\underline{b}^T\underline{c})\underline{a}, \tag{1.33g}$$

$$\underline{a}^T\widetilde{b}\,\underline{c} = \underline{b}^T\widetilde{c}\,\underline{a} = \underline{c}^T\widetilde{a}\,\underline{b}. \tag{1.33h}$$

If $\bar{n}$ is a unit vector and $\underline{a}$ an arbitrary vector, the following identities also hold

$$(\underline{a}^T\bar{n})\bar{n} = \underline{a} + \widetilde{n}\widetilde{n}\underline{a}, \tag{1.34a}$$

$$\widetilde{n}\widetilde{n}\widetilde{n} = -\widetilde{n}, \tag{1.34b}$$

$$\widetilde{n}\dot{\widetilde{n}}\widetilde{n} = 0, \tag{1.34c}$$

where notation $(\cdot)^{\cdot}$ indicates a derivative with respect to time.

1.1.11 Solution of the vector product equation

Let $\underline{a}$, $\underline{b}$, and $\underline{x}$ be three vectors such that $\widetilde{a}\underline{x} = \underline{b}$. If $\underline{a}$ and $\underline{b}$ are known vectors, is it possible to solve for $\underline{x}$? Equation $\widetilde{a}\underline{x} = \underline{b}$ can be viewed as a set of three linear equations for the components of $\underline{x}$. Unfortunately, the matrix of the system of equations is singular because $\det(\widetilde{a}) = 0$; in fact, the null space of $\widetilde{a}$ is $\underline{a}$ since $\widetilde{a}\underline{a} = 0$. Hence, a solution only exists if the right-hand side of the system of equations is orthogonal to the the null space of $\widetilde{a}$, *i.e.*, if $\underline{a}^T\underline{b} = 0$.

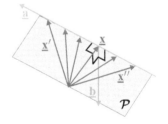

Fig. 1.10. The solution of the vector product equation.

Figure 1.10 gives a graphical illustration of the problem. The cross product equation, $\widetilde{a}\underline{x} = \underline{b}$, implies that $\underline{b}$ is orthogonal to both $\underline{a}$ and $\underline{x}$. Let plane $\mathcal{P}$ be normal to vector $\underline{b}$. Because $\underline{b}$ is orthogonal to $\underline{a}$, plane $\mathcal{P}$ contains vector $\underline{a}$. Any vector in plane $\mathcal{P}$ will be normal to vector $\underline{b}$.

The solution of the problem must be in plane $\mathcal{P}$ and hence, can be written as $\underline{x} = \mu\underline{a} + \alpha\,\widetilde{a}\underline{b}$, where μ and α are arbitrary scalars. Introducing this solution into the equation yields $\widetilde{a}\underline{x} = \widetilde{a}(\mu\underline{a} + \alpha\,\widetilde{a}\underline{b}) = \underline{b}$, or $\alpha\,\widetilde{a}\widetilde{a}\underline{b} = \underline{b}$. With the help of identity (1.33b), this becomes $\alpha(\underline{a}\,\underline{a}^T - \|\underline{a}\|^2\underline{I})\underline{b} = \underline{b}$. Because $\underline{a}^T\underline{b} = 0$, the equation then reduces to $-\alpha\|\underline{a}\|^2\underline{b} = \underline{b}$, and finally, $\alpha = -1/\|\underline{a}\|^2$.

The solution of the vector product equation is

$$\underline{x} = \mu\underline{a} - \frac{\widetilde{a}\underline{b}}{\|\underline{a}\|^2}, \tag{1.35}$$

where coefficient μ remains undetermined. Clearly, the vector product equation possesses an infinite number of solutions, because μ is arbitrary. Graphically, this corresponds to the various solution labeled as $\underline{x}'$ or $\underline{x}''$ in fig. 1.10.

To obtain a unique solution, an additional constraint must be enforced. For instance, the solution with the smallest norm is found by imposing the solution to be normal to vector $\underline{a}$, leading to $\mu = 0$ and finally $\underline{x} = -\widetilde{a}\underline{b}/\|\underline{a}\|^2$.

1.1.12 Problems

Problem 1.1. Lagrange's identity
Prove Lagrange's identity: $\|\widetilde{a}\underline{b}\|^2 + (\underline{a}^T\underline{b})^2 = \|\underline{a}\|^2\|\underline{b}\|^2$.

Problem 1.2. Geometric interpretation of identity
Prove identity (1.34a) and provide a geometric interpretation.

Problem 1.3. Geometric interpretation of identity
Prove the following identity $\underline{c}^T\widetilde{a}\underline{b} = \underline{a}^T\widetilde{b}\underline{c} = \underline{b}^T\widetilde{c}\underline{a}$, based on *(1)* geometric arguments, and *(2)* algebraic developments.

Problem 1.4. Jacobi's identity
With the help of the identities of section 1.1.10, prove Jacobi's identity $\widetilde{\widetilde{a}\,\underline{b}}\,\underline{c} + \widetilde{\widetilde{b}\,\underline{c}}\,\underline{a} + \widetilde{\widetilde{c}\,\underline{a}}\,\underline{b} = 0$.

Problem 1.5. Prove identity
Prove the following identity $\widetilde{a}\,\widetilde{\widetilde{b}\,\underline{b}}\,\underline{a} = \widetilde{b}\,\widetilde{a}\,\widetilde{a}^T\underline{b}$.

Problem 1.6. Prove identity
If $\bar{n}$ is a unit vector and $\underline{m}$ an arbitrary vector such that $\lambda = \bar{n}^T\underline{m}$, prove the following identity

$$\widetilde{n}\,\widetilde{n}\,\widetilde{m} + \widetilde{n}\,\widetilde{m}\,\widetilde{n} + \widetilde{m}\,\widetilde{n}\,\widetilde{n} = -\widetilde{m} - 2\lambda\widetilde{n}. \tag{1.36}$$

Problem 1.7. Criterion for linear independence
Show that three vectors $\underline{a}$, $\underline{b}$, and $\underline{c}$ are linearly independent if and only if their mixed product does not vanish.

Problem 1.8. Criterion for parallelism
Find the vector equation that expresses the fact that vectors $\underline{a}$ and $\underline{b}$ are parallel.

Problem 1.9. Criterion for orthogonality
Find the vector equation that expresses the fact that vectors $\underline{a}$ and $\underline{b}$ are orthogonal.

Problem 1.10. Criterion for coplanarity
Find the vector equation that expresses the fact that vectors $\underline{a}$, $\underline{b}$, and $\underline{c}$ are coplanar.

Problem 1.11. The projection tensor
Consider a plane, $\mathcal{P}$, defined by its unit normal, $\bar{n}$, and a free vector $\underline{a}$, as depicted in fig. 1.11. Vector $\underline{a}$ is decomposed as $\underline{a} = \underline{a}' + \underline{a}''$, where $\underline{a}'$ is in plane $\mathcal{P}$ and $\underline{a}''$ normal to $\mathcal{P}$. *(1)* Find the expression for the *projection tensor*, $\underline{\underline{P}}$, such that $\underline{a}' = \underline{\underline{P}}\,\underline{a}$. *(2)* Find tensor $\underline{\underline{Q}}$ such that $\underline{a}'' = \underline{\underline{Q}}\,\underline{a}$.

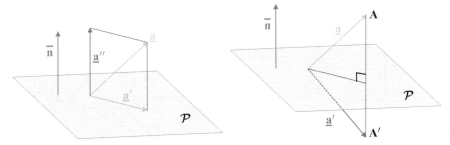

Fig. 1.11. Projection of vector $\underline{a}$ onto plane $\bar{n}$.

Fig. 1.12. Reflection of vector $\underline{a}$ onto plane $\bar{n}$.

Problem 1.12. The reflection tensor
Figure 1.12 depicts plane $\mathcal{P}$ defined by its unit normal $\bar{n}$ and a free vector, $\underline{a}$. Find the expression for the *reflection tensor* $\underline{\underline{R}}$ such that $\underline{a}' = \underline{\underline{R}}\,\underline{a}$, where $\underline{a}'$ is the reflection of $\underline{a}$ with respect the plane $\mathcal{P}$. Note that point $\mathbf{A}'$ is the mirror image of point $\mathbf{A}$ with respect the plane $\mathcal{P}$.

Problem 1.13. The covariant and contravariant components of a vector
Consider three non-coplanar vectors $\underline{a}_1$, $\underline{a}_2$, $\underline{a}_3$ such that $V = \underline{a}_1^T \widetilde{\underline{a}}_2 \underline{a}_3 \neq 0$. Define the following three vectors, $V\underline{a}^1 = \widetilde{\underline{a}}_2\underline{a}_3$, $V\underline{a}^2 = \widetilde{\underline{a}}_3\underline{a}_1$, and $V\underline{a}^3 = \widetilde{\underline{a}}_1\underline{a}_2$, called the *reciprocal vectors*. Prove that *(1)* $\underline{a}_i^T\underline{a}^j = \delta_{ij}$, *(2)* $\underline{a}^{1T}\widetilde{\underline{a}}^2\underline{a}^3 = 1/V$, and *(3)* $V\widetilde{\underline{a}}^2\underline{a}^3 = \underline{a}_1$, $V\widetilde{\underline{a}}^3\underline{a}^1 = \underline{a}_2$, and $V\widetilde{\underline{a}}^1\underline{a}^2 = \underline{a}_3$. Two arbitrary vectors, $\underline{u}$ and $\underline{v}$, are now resolved in the following manner

$$\underline{u} = u^1\underline{a}_1 + u^2\underline{a}_2 + u^3\underline{a}_3 = u_1\underline{a}^1 + u_2\underline{a}^2 + u_3\underline{a}^3;$$

$$\underline{v} = v^1\underline{a}_1 + v^2\underline{a}_2 + v^3\underline{a}_3 = v_1\underline{a}^1 + v_2\underline{a}^2 + v_3\underline{a}^3.$$

The components u^i and v^i are called the *contravariant components* of vectors $\underline{u}$ and $\underline{v}$, respectively, whereas the components u_i and v_i are called the *covariant components* of vectors $\underline{u}$ and $\underline{v}$, respectively. Prove that *(4)* $\underline{u}^T\underline{v} = u_1v^1 + u_2v^2 + u_3v^3 = u^1v_1 + u^2v_2 + u^3v_3$. *(5)* $\widetilde{\underline{u}}\underline{v}/V = (u^2v^3 - u^3v^2)\underline{a}^1 + (u^3v^1 - u^1v^3)\underline{a}^2 + (u^1v^2 - u^2v^3)\underline{a}^3$. *(6)* $V\widetilde{\underline{u}}\underline{v} = (u_2v_3 - u_3v_2)\underline{a}_1 + (u_3v_1 - u_1v_3)\underline{a}_2 + (u_1v_2 - u_2v_1)\underline{a}_3$.

1.2 Bound vectors

In section 1.1, free vectors were introduced as the ensemble of all segments equivalent to a given segment. In many practical applications, vectors are associated with a specific point in space; in that case they are called *bound vectors*. For instance, the description of a force applied to a rigid body requires knowledge of the *force vector*, $\underline{f}$, (magnitude and orientation of the applied force), and the *point of application* of the force, $\underline{x}_A$.

Figure 1.13 depicts a force vector, $\underline{f}$, applied to a rigid body at point $\mathbf{A}$; the force vector is a bound vector. On the other hand, a moment, $\underline{m}$, applied to a rigid body is not attached to a specific point of the body; it is a free vector. Similarly, the angular velocity vector, $\underline{\Omega}$, is a property of the rigid body. It is not associated with a specific point of the body, it is a free vector. The velocity vector, $\underline{v}$, describes the velocity at a specific point of the body; it is a bound vector.

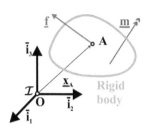

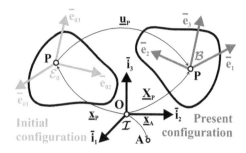

Fig. 1.13. A bound vector $\underline{f}$, a free vector $\underline{m}$, and the position vector $\underline{x}_A$.

Fig. 1.14. A reference frame defining the configuration of a rigid body.

1.2.1 The position vector

The position vector, $\underline{x}_A$, specifies the position of point **A** in three-dimensional space with respect to a reference point **O**, as depicted in fig. 1.13. The components of vector $\underline{x}_A$ resolved in basis $\mathcal{I}$ and denoted $\underline{x}_A^{[\mathcal{I}]}$, are the coordinates of point **A** in Cartesian basis $\mathcal{I}$.

1.2.2 Reference frames

Orthonormal or Cartesian bases were introduced in section 1.1.6 as a set of three mutually orthogonal unit vectors, $\mathcal{I} = (\bar{\imath}_1, \bar{\imath}_2, \bar{\imath}_3)$. The origin of this orthonormal basis, however, is not defined because it consists of three free vectors. Let point **O** be the common origin of the three unit vectors of the basis. It is now possible to define a *reference frame*, denoted $\mathcal{F} = [\mathbf{O}, \mathcal{I}]$, consisting of an orthonormal basis, $\mathcal{I}$, with its origin at point **O**, see fig. 1.14.

In dynamic problems, an *inertial reference frame* is always defined; the origin and orientation of such frame are invariant in time. Reference frames are conveniently used to define position vectors. The position of an arbitrary point **A** is given by its position vector, $\underline{x}_A$, with respect to the origin of reference frame $\mathcal{F}$, and the components of this vector, $\underline{x}_A^{[\mathcal{I}]}$, are resolved in basis $\mathcal{I}$.

Reference frames are closely related to the configuration of rigid bodies: let point **P** be a material point of the rigid body, and orthonormal basis $\mathcal{E}_0 = (\bar{e}_{01}, \bar{e}_{02}, \bar{e}_{03})$ a body attached basis defining its orientation. Clearly, the initial configuration of the rigid body is then completely defined by reference frame $\mathcal{F}_0 = [\mathbf{P}, \mathcal{E}_0]$, see fig. 1.14.

If the rigid body tumbles in space, it will move to its present configuration; the position vector of its reference point **P** is now $\underline{X}_P$, and its orientation is given by a new basis $\mathcal{E} = (\bar{e}_1, \bar{e}_2, \bar{e}_3)$. Reference frame $\mathcal{F} = [\mathbf{P}, \mathcal{E}]$ now defines the present configuration of the rigid body. The displacement vector, $\underline{u}_P$, of point **P** is such that $\underline{X}_P = \underline{x}_P + \underline{u}_P$. Clearly, a one to one correspondence exists between a reference frame and the configuration of a rigid body.

1.3 Geometric entities

Geometric problems can be conveniently formulated using a vector formalism. Lines, planes, circles, and spheres are briefly described in the following sections.

1.3.1 Lines

Figure 1.15 depicts a straight line is defined by the position vector, $\underline{x}_P$, of an arbitrary point **P** on the line, and the unit vector, $\bar{\ell}$, along the direction of the line. A straight line, $\mathcal{L}$, is denoted $\mathcal{L} = (\underline{x}_P, \bar{\ell})$. An arbitrary point **Q** on the line has a position vector, $\underline{x}_Q$, given by

$$\underline{x}_Q = \underline{x}_P + \lambda\bar{\ell}, \tag{1.37}$$

where λ is an arbitrary scalar.

An alternative definition of the line is in terms of its *Plücker coordinates* [1] defined as follows

$$\underline{\mathcal{Q}} = \left\{ \begin{matrix} \tilde{x}_P\bar{\ell} \\ \bar{\ell} \end{matrix} \right\} = \left\{ \begin{matrix} \underline{k} \\ \bar{\ell} \end{matrix} \right\}. \tag{1.38}$$

The first part of the Plücker coordinates, $\underline{k}$, defines a point of the line, and the second part, $\bar{\ell}$, its orientation.[1] Indeed, it is readily shown that $\underline{x}_P = \tilde{\ell}\underline{k}$. The two vectors forming the Plücker coordinates must be orthogonal, *i.e.*,

$$\underline{k}^T\bar{\ell} = 0. \tag{1.39}$$

Clearly, $\alpha\underline{\mathcal{Q}}$, where α is an arbitrary scalar such that $\alpha \neq 0$, defines the same line, $\underline{\mathcal{Q}}$.

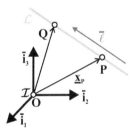

Fig. 1.15. The definition of a straight line.

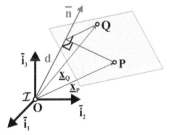

Fig. 1.16. The definition of a plane.

---

[1] Some authors define the Plücker coordinates as $\underline{\mathcal{Q}}^T = \{\bar{\ell}^T, \underline{k}^T\}$.

1.3.2 Planes

Similarly, a plane is defined by the position vector, $\underline{x}_P$, of an arbitrary point **P** of the plane, and the unit vector, $\bar{n}$, normal to the plane, see fig. 1.16. Plane $\mathcal{P}$ is denoted $\mathcal{P} = (\underline{x}_P, \bar{n})$. An arbitrary point **Q** of the plane has a position vector, $\underline{x}_Q$, satisfying the following relationship

$$\bar{n}^T(\underline{x}_Q - \underline{x}_P) = 0. \tag{1.40}$$

This equation expresses the condition that vector $\underline{x}_Q - \underline{x}_P$ must lie in plane $\mathcal{P}$, and is therefore normal to $\bar{n}$. The distance between point **O** and the plane is $d = \bar{n}^T \underline{x}_P$, and hence, the equation of a plane becomes

$$\bar{n}^T \underline{x}_Q = d. \tag{1.41}$$

1.3.3 Circles

A circle is defined by the position vector, $\underline{x}_C$, of its center, the unit vector, $\bar{n}$, normal to the plane of the circle, and its radius, ρ. Circle $\mathcal{C}$ is denoted $\mathcal{C} = (\underline{x}_C, \bar{n}, \rho)$. An arbitrary point **Q** of the circle has a position vector, $\underline{x}_Q$, satisfying the following relationships

$$\bar{n}^T(\underline{x}_Q - \underline{x}_C) = 0, \quad \|\underline{x}_Q - \underline{x}_C\| = \rho, \tag{1.42}$$

where the first equation expresses the fact that point **Q** is in plane $(\underline{x}_C, \bar{n})$ and the second that is it at a distance ρ from the center of the circle.

1.3.4 Spheres

A sphere is defined by the position vector, $\underline{x}_C$, of its center, and its radius, ρ. Sphere $\mathcal{S}$ is denoted $\mathcal{S} = (\underline{x}_C, \rho)$. An arbitrary point **Q** of the sphere has a position vector, $\underline{x}_Q$, satisfying the following relationship

$$\|\underline{x}_Q - \underline{x}_C\| = \rho. \tag{1.43}$$

Example 1.1. Intersection between two lines
Find the point at the intersection of two lines, $\mathcal{L}_1 = (\underline{x}_1, \bar{\ell}_1)$ and $\mathcal{L}_2 = (\underline{x}_2, \bar{\ell}_2)$. What is the condition for this intersection to exist? Figure 1.17 shows the two lines and their intersection at point **I**, assuming, of course, that this intersection exists.

Arbitrary points on lines $\mathcal{L}_1$ and $\mathcal{L}_2$, denoted $\underline{y}_1$ and $\underline{y}_2$, respectively, are given by eq. (1.37) as $\underline{y}_1 = \underline{x}_1 + \lambda_1 \bar{\ell}_1$ and $\underline{y}_2 = \underline{x}_2 + \lambda_2 \bar{\ell}_2$, respectively. If an intersection point exists, it must be on both lines, which implies the existence of scalars λ_1 and λ_2 such that

$$\underline{x}_I = \underline{x}_1 + \lambda_1 \bar{\ell}_1 = \underline{x}_2 + \lambda_2 \bar{\ell}_2, \tag{1.44}$$

where $\underline{x}_I$ is the position vector of the intersection point.

Let $\underline{x}_{21} = \underline{x}_2 - \underline{x}_1$ be the position vector of the reference point of line $\mathcal{L}_2$ with respect to that of line $\mathcal{L}_1$. Multiplying eq. (1.44) by $\underline{x}_{21}^T \widetilde{\ell}_1$ and $\underline{x}_{21}^T \widetilde{\ell}_2$ yields the two following conditions that must be satisfied for the intersection to exist,

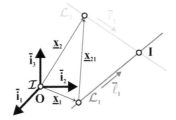

Fig. 1.17. Intersection between two lines.

$$\lambda_1(\bar{\ell}_1^T \tilde{x}_{21} \bar{\ell}_2) = 0, \quad \lambda_2(\bar{\ell}_1^T \tilde{x}_{21} \bar{\ell}_2) = 0.$$

The first solution of these equations is $\lambda_1 = \lambda_2 = 0$, which implies $\underline{x}_I = \underline{x}_1 = \underline{x}_2$: the reference points of the two lines are identical and this common point is the intersection of the two lines. The second solution is $\bar{\ell}_1^T \tilde{x}_{21} \bar{\ell}_2 = 0$, the vanishing of the mixed product of vectors $\bar{\ell}_1$, $\underline{x}_{21}$, and $\bar{\ell}_2$. Because the mixed product represent the volume spanned by these three vectors, the vanishing of the mixed product implies the coplanarity of the three vectors. As illustrated by fig. 1.17, the existence of an intersection point of the two lines does indeed require the coplanarity of vectors $\bar{\ell}_1$, $\underline{x}_{21}$, and $\bar{\ell}_2$.

To determine the location of the intersection point, scalars λ_1 and λ_2 must be determined. Multiplying eq. (1.44) by $\underline{x}_1^T \bar{\ell}_1$ and $\underline{x}_2^T \bar{\ell}_2$ yields $\lambda_2 = (\underline{x}_1^T \tilde{\ell}_1 \underline{x}_2)/(\bar{\ell}_1^T \tilde{x}_1 \bar{\ell}_2)$ and $\lambda_1 = (\underline{x}_2^T \tilde{\ell}_2 \underline{x}_1)/(\bar{\ell}_2^T \tilde{x}_2 \bar{\ell}_1)$, respectively. Point **I** is now found as

$$\underline{x}_I = \underline{x}_1 + \frac{\underline{x}_2^T \tilde{\ell}_2 \underline{x}_1}{\bar{\ell}_2^T \tilde{x}_2 \bar{\ell}_1} \bar{\ell}_1 = \underline{x}_2 + \frac{\underline{x}_1^T \tilde{\ell}_1 \underline{x}_2}{\bar{\ell}_1^T \tilde{x}_1 \bar{\ell}_2} \bar{\ell}_2.$$

Because the mixed product, $\bar{\ell}_1^T \tilde{x}_{21} \bar{\ell}_2$, must vanish for the intersection to exist, it follows that $\bar{\ell}_1^T \tilde{x}_1 \bar{\ell}_2 = \bar{\ell}_1^T \tilde{x}_2 \bar{\ell}_2$. If $\bar{\ell}_1^T \tilde{x}_1 \bar{\ell}_2 = \bar{\ell}_1^T \tilde{x}_2 \bar{\ell}_2 = 0$, the denominators in the above expressions vanish and the intersection does not exist because the two lines are parallel, $\bar{\ell}_1 \bar{\ell}_2 = 0$.

In summary, an intersection exists if $\bar{\ell}_1^T \tilde{x}_{21} \bar{\ell}_2 = 0$, implying the coplanarity of vectors $\bar{\ell}_1$, $\underline{x}_{21}$, and $\bar{\ell}_2$, and $\bar{\ell}_1 \bar{\ell}_2 \neq 0$, implying that the two lines are not parallel. A special case occurs $\bar{\ell}_1 = \bar{\ell}_2 = \bar{\ell}$ and $\tilde{\ell} \underline{x}_{12} = 0$: the two lines are coincident and all points on the line are intersection points.

Example 1.2. Intersection between two lines
Find the point at the intersection of two lines defined by their Plücker coordinates, $\mathcal{L}_1 = (\underline{k}_1, \bar{\ell}_1)$ and $\mathcal{L}_2 = (\underline{k}_2, \bar{\ell}_2)$. What is the condition for this intersection to exist? Figure 1.17 shows the two lines and their intersection at point **I**, assuming, of course, that this intersection exists.

The intersection point must be a point of both lines, and hence, by definition of the Plücker coordinates, $\underline{k}_1 = \tilde{x}_I \bar{\ell}_1$ and $\underline{k}_2 = \tilde{x}_I \bar{\ell}_2$. Multiplying the first equation by $\bar{\ell}_2^T$ and the second by $\bar{\ell}_1^T$, yields $\bar{\ell}_2^T \underline{k}_1 = \bar{\ell}_2^T \tilde{x}_I \bar{\ell}_1$ and $\bar{\ell}_1^T \underline{k}_2 = \bar{\ell}_1^T \tilde{x}_I \bar{\ell}_2$. Subtracting these two expressions then leads to

$$\underline{k}_1^T \bar{\ell}_2 + \underline{k}_2^T \bar{\ell}_1 = 0.$$

This is the condition that must be satisfied if an intersection exists. It is left to the reader to verify that the above condition is equivalent to that developed in example 1.1.

Next, the location of the intersection point must be evaluated. By definition of the Plücker coordinates, $\underline{k}_1 = \widetilde{x}_I \bar{\ell}_1$, which can be recast as $\widetilde{\ell}_1 \underline{x}_I = -\underline{k}_1$. This implies that the position vector of the intersection point is the solution of a vector product equation, see section 1.1.11. Because the solvability condition is satisfied, $\bar{\ell}_1^T \underline{k}_1 = 0$, the solution can be written as $\underline{x}_I = \alpha \bar{\ell}_1 + \widetilde{\ell}_1 \underline{k}_1$, where α is an arbitrary scalar. Multiplying this equation by $\underline{k}_2^T$ leads to $\underline{k}_2^T \underline{x}_I = (\underline{k}_2^T \bar{\ell}_1)\alpha + \underline{k}_2^T \widetilde{\ell}_1 \underline{k}_1 = 0$, because $\underline{k}_2$ must be orthogonal to $\underline{x}_I$. Coefficient α now becomes $\alpha = -(\underline{k}_2^T \widetilde{\ell}_1 \underline{k}_1)/(\underline{k}_2^T \bar{\ell}_1)$.

The intersection point is now

$$\underline{x}_I = -\frac{\bar{\ell}_1 \underline{k}_2^T \widetilde{\ell}_1 \underline{k}_1}{\underline{k}_2^T \bar{\ell}_1} + \widetilde{\ell}_1 \underline{k}_1 = \left[\underline{I} - \frac{\bar{\ell}_1 \underline{k}_2^T}{\underline{k}_2^T \bar{\ell}_1} \right] \widetilde{\ell}_1 \underline{k}_1 = -\frac{\widetilde{k}_2 \widetilde{\ell}_1 \underline{\ell}_1 \underline{k}_1}{\bar{\ell}_1^T \underline{k}_2},$$

where identity (1.33b) was used to evaluate the bracketed term. Using this same identity once more leads to

$$\underline{x}_I = \frac{\widetilde{k}_2(\underline{I} - \bar{\ell}_1 \bar{\ell}_1^T)\underline{k}_1}{\bar{\ell}_1^T \underline{k}_2} = \frac{\widetilde{k}_2 \underline{k}_1}{\bar{\ell}_1^T \underline{k}_2} = \frac{\widetilde{k}_1 \underline{k}_2}{\bar{\ell}_2^T \underline{k}_1},$$

where the second equality follows from the orthogonality of the Plücker coordinates, $\bar{\ell}_1^T \underline{k}_1 = 0$. Of course, the existence of the intersection point requires $\bar{\ell}_1^T \underline{k}_2 \neq 0$ or equivalently, $\bar{\ell}_2^T \underline{k}_1 \neq 0$, which imply that the two lines are not parallel.

It is left to the reader to verify that the solution found here is identical to that found in example 1.1. Comparing the solution obtained here with that found in example 1.1, it is clear that the use of the Plücker coordinates provides an elegant and compact solution of the problem.

Example 1.3. Intersection between two planes
Find the equation of the line at the intersection of two planes, $\mathcal{P}_1 = (\underline{x}_1, \bar{n}_1)$ and $\mathcal{P}_2 = (\underline{x}_2, \bar{n}_2)$. Does this line always exist? Under what conditions do the two planes coincide? Figure 1.18 shows the two planes and their intersection line, $\mathcal{L} = (\underline{x}_P, \bar{\ell})$, assuming, of course, that this intersection exists.

Line $\mathcal{L}$ must be entirely in both planes $\mathcal{P}_1$ and $\mathcal{P}_2$, and hence, must be normal to both $\bar{n}_1$ and $\bar{n}_2$, which implies

$$\bar{\ell} = \frac{\widetilde{n}_1 \bar{n}_2}{\|\widetilde{n}_1 \bar{n}_2\|}. \tag{1.45}$$

To fully define the intersection line, it is also necessary to find one of its point, say point **P**, as illustrated in fig. 1.18. This point must belong to both planes, *i.e.*, $\bar{n}_1^T \underline{x}_P = p_1$ and $\bar{n}_2^T \underline{x}_P = p_2$, where p_1 and p_2 are the distances from point **O** to planes $\mathcal{P}_1$ and $\mathcal{P}_2$, respectively. These two scalar equations are not sufficient to

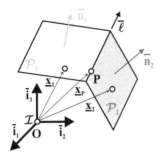

Fig. 1.18. Intersection between two planes.

determine the position of point **P** unequivocally. A third condition can be added, for instance, imposing that point **P** be at the shortest distance from point **O**, implying $\bar{\ell}^T \underline{x}_P = 0$.

The position vector of point **P** will be written as $\underline{x}_P = \alpha \bar{n}_1 + \beta \bar{n}_2 + \gamma \bar{\ell}$, where α, β, and γ are three unknown scalars. Multiplying this equation by $\bar{\ell}^T$ and using the shortest distance condition leads to $\gamma = 0$ and hence, $\underline{x}_P = \alpha \bar{n}_1 + \beta \bar{n}_2$. Imposing the remaining two conditions leads to a set of two algebraic equations for coefficients α and β, which are found as $\alpha = \left[p_1 - (\bar{n}_1^T \bar{n}_2)p_2 \right] / \|\tilde{n}_1 \bar{n}_2\|^2$ and $\beta = \left[p_2 - (\bar{n}_1^T \bar{n}_2)p_1 \right] / \|\tilde{n}_1 \bar{n}_2\|^2$. The position vector of point **P** now becomes

$$\underline{x}_P = \frac{p_1 - (\bar{n}_1^T \bar{n}_2)p_2}{\|\tilde{n}_1 \bar{n}_2\|^2} \bar{n}_1 + \frac{p_2 - (\bar{n}_1^T \bar{n}_2)p_1}{\|\tilde{n}_1 \bar{n}_2\|^2} \bar{n}_2.$$

If the two planes are parallel, $\tilde{n}_1 \bar{n}_2 = 0$, and the intersection line does not exist. Of course, if the two planes are coincident, *i.e.*, $\tilde{n}_1 \bar{n}_2 = 0$ and $p_1 = p_2$, all lines in the plane are intersection lines.

Example 1.4. Intersection between two planes
Find the Plücker coordinates of the line at the intersection of two planes. Does this line always exist? Under what conditions do the two planes coincide? Figure 1.18 shows the two planes and their intersection, $\mathcal{L} = (\underline{k}, \bar{\ell})$, assuming, of course, that this intersection exists.

As discussed in example 1.3, the orientation of the intersection line is given by eq. (1.45). By definition of the Plücker coordinates, $\underline{k} = \tilde{x}_P \bar{\ell}$, where $\underline{x}_P$ is any point on the line. It follows that $\underline{k}$ must be normal to $\bar{\ell}$, and hence, contained in the plane defined by unit vectors $\bar{n}_1$ and $\bar{n}_2$, *i.e.*, $\underline{k} = \alpha_1 \bar{n}_1 + \alpha_2 \bar{n}_2 = \tilde{x}_P \bar{\ell}$.

This equation can be recast as $\tilde{\ell} \underline{x}_P = -(\alpha_1 \bar{n}_1 + \alpha_2 \bar{n}_2)$. This implies that the position vector of the intersection point is the solution of a vector product equation, see section 1.1.11. Because the solvability condition is satisfied, $\bar{\ell}^T (\alpha_1 \bar{n}_1 + \alpha_2 \bar{n}_2) = 0$, the solution can be written as $\underline{x}_P = \mu \bar{\ell} + \tilde{\ell}(\alpha_1 \bar{n}_1 + \alpha_2 \bar{n}_2)$, where μ is an arbitrary scalar.

Multiplying this equation by $\bar{n}_1^T$ yields

$$\bar{n}_1^T \underline{x}_P = p_1 = \bar{n}_1^T \tilde{\ell}(\alpha_1 \bar{n}_1 + \alpha_2 \bar{n}_2) = \alpha_2 \bar{n}_1^T \tilde{\ell} \bar{n}_2 = -\alpha_2 \bar{\ell}^T \tilde{n}_1 \bar{n}_2 = -\alpha_2 \|\tilde{n}_1 \bar{n}_2\|.$$

It then follows that $\alpha_2 = -p_1/\|\tilde{n}_1\bar{n}_2\|$. Multiplying the equation by $\bar{n}_2^T$ and proceeding similarly yields $\alpha_1 = p_2/\|\tilde{n}_1\bar{n}_2\|$, and finally $\underline{k} = (p_2\bar{n}_1 - p_1\bar{n}_2)/\|\tilde{n}_1\bar{n}_2\|$. If the two planes are parallel, $\tilde{n}_1\bar{n}_2 = 0$, and the intersection line does not exist. Of course, if the two planes are coincident, *i.e.*, $\tilde{n}_1\bar{n}_2 = 0$ and $p_1 = p_2$, all lines in the plane are intersection lines.

Because the Plücker coordinates, $\underline{Q}$, of a line are defined within a constant, the Plücker coordinates of the intersection line can be written as

$$\underline{Q} = \left\{ \begin{matrix} p_2\bar{n}_1 - p_1\bar{n}_2 \\ \tilde{n}_1\bar{n}_2 \end{matrix} \right\}.$$

Comparing the solution obtained here with that found in example 1.3, it is clear that the use of the Plücker coordinates provides an elegant and compact solution of the problem.

1.3.5 Problems

Problem 1.14. Position vector of a point on a line
A line is defined by its Plücker coordinates, $\mathcal{L} = (\underline{k}, \bar{\ell})$. Find the position vector of an arbitrary point on line $\mathcal{L}$.

Problem 1.15. Line defined by two points
Find the equation of a line passing through two points $\mathbf{P}_1$ and $\mathbf{P}_2$. Does a solution always exist? Under what conditions do multiple solutions arise?

Problem 1.16. Distance from a point to a line
Find the distance between an arbitrary point $\mathbf{Q}$ (of position vector $\underline{x}_Q$) and a line $\mathcal{L} = (\underline{x}_P, \bar{\ell})$. Find the location of point $\mathbf{R}$ on line $\mathcal{L}$ that is at the shortest distance of point $\mathbf{Q}$.

Problem 1.17. Distance from a point to a line
Find the distance between an arbitrary point $\mathbf{Q}$ (of position vector $\underline{x}_Q$) and a line $\mathcal{L} = (\underline{k}, \bar{\ell})$. Find the location of point $\mathbf{R}$ on line $\mathcal{L}$ that is at the shortest distance of point $\mathbf{Q}$.

Problem 1.18. Distance from a point to a plane
Find the distance between an arbitrary point $\mathbf{Q}$ (of position vector $\underline{x}_Q$) and a plane $\mathcal{P} = (\underline{x}_P, \bar{n})$. Find the location of point $\mathbf{R}$ on plane $\mathcal{P}$ that is at the shortest distance of point $\mathbf{Q}$.

Problem 1.19. Intersection of a line and a plane
Find the point at the intersection of a line $\mathcal{L} = (\underline{x}_Q, \bar{\ell})$ and a plane $\mathcal{P} = (\underline{x}_P, \bar{n})$. Does this point always exist? Under what conditions does the line lie in the plane?

Problem 1.20. Intersection of a line and a pane
Find the point at the intersection of a line, $\mathcal{L} = (\underline{k}, \bar{\ell})$, expressed in terms of Plücker coordinates, and a plane, $\mathcal{P} = (\underline{x}_P, \bar{n})$. Does this point always exist? Under what conditions does the line lie in the plane?

Problem 1.21. Distance between two lines
Find the distance between two lines $\mathcal{L}_1 = (\underline{x}_1, \bar{\ell}_1)$ and $\mathcal{L}_2 = (\underline{x}_2, \bar{\ell}_2)$. Find the locations of points $\mathbf{R}_1$ and $\mathbf{R}_2$, on lines $\mathcal{L}_1$ and $\mathcal{L}_2$, respectively that are at the shortest distance from each other.

Problem 1.22. Distance between two lines

Find the distance between two lines $\mathcal{L}_1 = (\underline{k}_1, \bar{\ell}_1)$ and $\mathcal{L}_2 = (\underline{k}_2, \bar{\ell}_2)$, expressed in terms of Plücker coordinates. Find the locations of points $\mathbf{R}_1$ and $\mathbf{R}_2$, on lines $\mathcal{L}_1$ and $\mathcal{L}_2$, respectively that are at the shortest distance from each other.

Problem 1.23. Plane defined by three points

Find the equation of a plane passing through three points $\mathbf{P}_1$, $\mathbf{P}_2$, and $\mathbf{P}_3$ with position vectors $\underline{x}_1$, $\underline{x}_2$, and $\underline{x}_3$, respectively. Does a solution always exist? Under what conditions do multiple solutions arise?

Problem 1.24. Plane defined by the intersection of two planes

Let point $\mathbf{P}$ with position vector $\underline{x}_P$ be on the intersection of two given planes $\mathcal{P}_1 = (\underline{x}_P, \bar{n}_1)$ and $\mathcal{P}_2 = (\underline{x}_P, \bar{n}_2)$. Find the equation of plane $\mathcal{P}_3 = (\underline{x}_P, \bar{n}_3)$ passing through a given point $\mathbf{Q}$ with position vectors $\underline{x}_Q$ and the intersection of planes $\mathcal{P}_1$ and $\mathcal{P}_2$.

Problem 1.25. Circle defined by three points

Find circle $\mathcal{C} = (\underline{x}_C, \bar{n}, \rho)$ defined by three points $\mathbf{P}_1$, $\mathbf{P}_2$, and $\mathbf{P}_3$ with position vectors $\underline{x}_1$, $\underline{x}_2$, and $\underline{x}_3$, respectively. Does a solution always exist?

Problem 1.26. Tangent to a circle

Find the position vector $\underline{x}_P$ of point $\mathbf{P}$ such that the tangent to circle $\mathcal{C} = (\underline{x}_C, \bar{n}, \rho)$ at $\mathbf{P}$ passes through a given point $\mathbf{Q}$ with position vector $\underline{x}_Q$. Find the conditions for a solution to exist. Is the solution unique?

Problem 1.27. Distance between two circles

Find the shortest distance d between two arbitrary circles $\mathcal{C}_1 = (\underline{x}_{C1}, \bar{n}_1, \rho_1)$ and $\mathcal{C}_2 = (\underline{x}_{C2}, \bar{n}_2, \rho_2)$. Hint: let $\underline{x}_{Q1}$ and $\underline{x}_{Q2}$ be the position vectors of points $\mathbf{Q}_1$ and $\mathbf{Q}_2$ belonging to circles $\mathcal{C}_1$ and $\mathcal{C}_2$, respectively. If $\mathbf{Q}_1$ and $\mathbf{Q}_2$ are at the shortest distance, vector $\underline{x}_{Q2} - \underline{x}_{Q1}$ is then normal to the tangent to $\mathcal{C}_1$ at point $\mathbf{Q}_1$ and to the tangent to $\mathcal{C}_2$ at point $\mathbf{Q}_2$.

Problem 1.28. Intersection of a line and a sphere

Find the intersections between line $\mathcal{L} = (\underline{x}, \bar{\ell})$ and sphere $\mathcal{S} = (\underline{x}_C, \rho)$.

Problem 1.29. Distance from a disk to a plane

Consider plane $\mathcal{P} = (\underline{x}_P, \bar{n})$ and circle $\mathcal{C} = (\underline{x}_C, \bar{k}, \rho)$, as depicted in fig. 1.19. Find the shortest algebraic distance d between the disk and the plane. (A positive distance is defined when the disk is in the direction of $\bar{n}$).

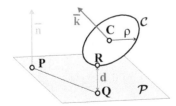

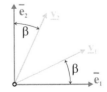

Fig. 1.19. The distance between a disk and a plane.

Fig. 1.20. Constructing an orthonormal basis from two vectors.

Problem 1.30. Orthonormal basis constructed from two vectors

Figure 1.20 shows two arbitrary vectors, $\underline{v}_1$ and $\underline{v}_2$. Construct a set of three mutually orthogonal unit vectors, $\bar{e}_1, \bar{e}_2, \bar{e}_3$, such that $\bar{e}_1, \bar{e}_2, \underline{v}_1$, and $\underline{v}_2$ are four coplanar vectors, and angle β between $\bar{e}_1$ and $\underline{v}_1$ is identical to that between $\bar{e}_2$ and $\underline{v}_2$. Find an expression for angle β.

1.4 Second-order tensors

Second-order tensors were encountered in previous sections: the tensor product of two vectors, eq. (1.28), yields a second-order tensor, and the vector product of two vector is conveniently expressed in terms of the second-order, skew-symmetric tensor defined by eq. (1.27).

In general, the *components of a second-order tensor*, $\underline{A}$, are denoted a_{ij}, where the indices $i = 1, 2, 3$ and $j = 1, 2, 3$. A second-order tensor is said to be a *symmetric tensor* if $a_{ij} = a_{ji}$. For instance, it is readily verified that the tensor product of a vector by itself, $\underline{T} = \underline{a}\,\underline{a}^T$, forms a symmetric tensor.

A second-order tensor is said to be a *skew-symmetric tensor* if $a_{ij} = -a_{ji}$. This implies that the diagonal terms vanish, $a_{ii} = 0$, $i = 1, 2, 3$. Skew-symmetric tensors were encountered when dealing with the vector product, see eq. (1.27). The superscript $(\cdot)^T$ is used to denote the transposition operation. If the components of $\underline{A}$ are a_{ij}, the components of $\underline{A}^T$ are a_{ji}.

1.4.1 Basic operations

The *trace of a second-order tensor* is a scalar defined as

$$\mathrm{tr}(\underline{A}) = a_{11} + a_{22} + a_{33}. \tag{1.46}$$

The *determinant of a second-order tensor* is also a scalar quantity defined as

$$\det(\underline{A}) = \; a_{11}a_{22}a_{33} + a_{12}a_{23}a_{31} + a_{13}a_{21}a_{32}$$
$$- a_{31}a_{22}a_{13} - a_{12}a_{21}a_{33} - a_{11}a_{23}a_{32}. \tag{1.47}$$

An arbitrary tensor can always be decomposed into its *symmetric part* and *skew-symmetric part*

$$\underline{A} = \frac{\underline{A} + \underline{A}^T}{2} + \frac{\underline{A} - \underline{A}^T}{2} = \mathrm{symm}(\underline{A}) + \mathrm{skew}(\underline{A}). \tag{1.48}$$

In this equation, $\mathrm{symm}(\underline{A})$ denoted the symmetric part of the tensor

$$\mathrm{symm}(\underline{A}) = \frac{\underline{A} + \underline{A}^T}{2} = \frac{1}{2} \begin{bmatrix} 2a_{11} & a_{12} + a_{21} & a_{13} + a_{31} \\ a_{12} + a_{21} & 2a_{22} & a_{23} + a_{32} \\ a_{13} + a_{31} & a_{23} + a_{32} & 2a_{33} \end{bmatrix}, \tag{1.49}$$

and $\mathrm{skew}(\underline{A})$ its skew-symmetric part

$$\text{skew}(\underline{\underline{A}}) = \frac{\underline{\underline{A}} - \underline{\underline{A}}^T}{2} = \frac{1}{2} \begin{bmatrix} 0 & (a_{12} - a_{21}) & (a_{13} - a_{31}) \\ -(a_{12} - a_{21}) & 0 & (a_{23} - a_{32}) \\ -(a_{13} - a_{31}) & -(a_{23} - a_{32}) & 0 \end{bmatrix}. \quad (1.50)$$

The *axial vector*, $\underline{a}$, associated with a second-order tensor, $\underline{\underline{A}}$, is denoted $\underline{a} = \text{axial}(\underline{\underline{A}})$. It is defined as follows

$$\underline{a} = \text{axial}(\underline{\underline{A}}) \iff \widetilde{\underline{a}} = \frac{\underline{\underline{A}} - \underline{\underline{A}}^T}{2}. \quad (1.51)$$

It is readily verified that

$$\underline{a} = \text{axial}(\underline{\underline{A}}) = \begin{Bmatrix} a_1 \\ a_2 \\ a_3 \end{Bmatrix} = \frac{1}{2} \begin{Bmatrix} a_{32} - a_{23} \\ a_{13} - a_{31} \\ a_{21} - a_{12} \end{Bmatrix}. \quad (1.52)$$

A second-order tensor, $\underline{\underline{T}}$, is *positive-definite* if and only if

$$\underline{u}^T \underline{\underline{T}} \underline{u} > 0, \quad (1.53)$$

for any arbitrary vector $\underline{u} \neq 0$. It is *semi positive-definite* if $\underline{u}^T \underline{\underline{T}} \underline{u} \geq 0$ for any vector $\underline{u} \neq 0$. For instance, consider the tensor corresponding to the tensor product of a vector by itself, $\underline{\underline{T}} = \underline{a}\,\underline{a}^T$, $\underline{a} \neq 0$. This tensor is semi positive-definite because $\underline{u}^T \underline{\underline{T}} \underline{u} = (\underline{a}^T \underline{u})^2 \geq 0$ for any choice of $\underline{u} \neq 0$; the equality hold when $\underline{a}$ is normal to $\underline{u}$.

1.4.2 Eigenvalue analysis

More often than not, the complete analysis of a second-order tensor will require the evaluation of its eigenvalues and eigenvectors. The following relationship

$$\underline{\underline{A}} \underline{u} = \lambda \underline{u}, \quad (1.54)$$

is satisfied by *eigenvector* $\underline{u}$ corresponding to *eigenvalue* λ. This relationship can be recast as $(\underline{\underline{A}} - \lambda \underline{\underline{I}})\underline{u} = 0$, where $\underline{\underline{I}}$ is the *identity tensor, i.e.*, $I_{ij} = \delta_{ij}$. This means that the eigenvector is the solution of a homogeneous system of algebraic equations. In general, this solution is the trivial solution, $\underline{u} = 0$. For a non-trivial solution to exist, the determinant of the set of linear equations must vanish, $\det(\underline{\underline{A}} - \lambda \underline{\underline{I}}) = 0$. This equation is called the *characteristic equation* satisfied by the eigenvalue of $\underline{\underline{A}}$; with the help of eq. (1.47), it expands to

$$-\lambda^3 + I_1 \lambda^2 - I_2 \lambda + I_3 = 0. \quad (1.55)$$

I_1, I_2, and I_3 are the *invariants* of the tensor

$$I_1 = \text{tr}(\underline{\underline{A}}), \quad (1.56a)$$

$$I_2 = a_{22}a_{33} + a_{33}a_{11} + a_{11}a_{22} - a_{23}a_{32} - a_{13}a_{31} - a_{12}a_{21}, \quad (1.56b)$$

$$I_3 = \det(\underline{\underline{A}}). \quad (1.56c)$$

Equation (1.55) will yield three solutions λ_1, λ_2, and λ_3, called the eigenvalues of $\underline{A}$. One eigenvalue is always real, the other two could be real, or a complex conjugate pair. To each eigenvalue corresponds an eigenvector, $\underline{u}$; the *eigenpairs* are denoted $(\lambda_1, \underline{u}_1)$, $(\lambda_2, \underline{u}_2)$, and $(\lambda_3, \underline{u}_3)$. Because the eigenvectors are the solution of a homogeneous, linear system, they are defined within a multiplicative constant. If an eigenvalue is real, the corresponding eigenvector is also real. A complex conjugate pair of eigenvectors is associated with complex conjugate eigenvalues.

Symmetric, positive-definite tensors

In general, a real second-order tensor will have one real eigenvalue and the remaining two could be real, or form a complex conjugate pair. If the tensor is symmetric and positive-definite, however, all three eigenvalues must be real and positive.

Indeed, assume λ is a complex eigenvalue and $\underline{u} = \underline{v} + i\underline{w}$ the corresponding eigenvector, where $i = \sqrt{-1}$. The eigenproblem, eq. (1.54), now becomes $\underline{A}(\underline{v} + i\underline{w}) = \lambda(\underline{v} + i\underline{w})$. Pre-multiplying by vector $(\underline{v} - i\underline{w})^T$ leads to

$$(\underline{v}^T \underline{\underline{A}} \underline{v} + \underline{w}^T \underline{\underline{A}} \underline{w}) + i(\underline{v}^T \underline{\underline{A}} \underline{w} - \underline{w}^T \underline{\underline{A}} \underline{v}) = \lambda(\underline{v}^T \underline{v} + \underline{w}^T \underline{w}). \tag{1.57}$$

If tensor $\underline{\underline{A}}$ is symmetric, $\underline{v}^T \underline{\underline{A}} \underline{w} = \underline{w}^T \underline{\underline{A}} \underline{v}$, and the term in the second parenthesis vanishes. It then follows that

$$\lambda = \frac{\underline{v}^T \underline{\underline{A}} \underline{v} + \underline{w}^T \underline{\underline{A}} \underline{w}}{\underline{v}^T \underline{v} + \underline{w}^T \underline{w}}. \tag{1.58}$$

Because $\underline{\underline{A}}$, $\underline{v}$, and $\underline{w}$ are real quantities, λ is also a real quantity. It follows that *the eigenvalues of a real, symmetric tensor are all real.*

The original eigenproblem, $\underline{A}(\underline{v} + i\underline{w}) = \lambda(\underline{v} + i\underline{w})$, splits into its real and imaginary parts, $\underline{A}\underline{v} = \lambda\underline{v}$ and $i\underline{A}\underline{w} = i\lambda\underline{w}$, respectively. Clearly the two problems are identical and the imaginary part is redundant; nothing is lost by setting $\underline{w} = 0$. The eigenvalue now becomes

$$\lambda = \frac{\underline{v}^T \underline{\underline{A}} \underline{v}}{\underline{v}^T \underline{v}}. \tag{1.59}$$

If $\underline{\underline{A}}$ is a positive-definite tensor, the numerator is a positive number, see eq. (1.53). On the other hand, the denominator is always a positive number. This proves that *the eigenvalues of a real, symmetric, positive-definite tensor are all real and positive.* If the tensor is symmetric and semi positive-definite, its eigenvalues are null or positive.

Similarity transformations

Consider now a linear transformation of the form $\underline{u} = \underline{\underline{Q}} \, \bar{\underline{u}}$, where $\underline{\underline{Q}}$ is an orthogonal tensor, *i.e.*, $\underline{\underline{Q}}^T \underline{\underline{Q}} = \underline{\underline{I}}$. This transformation is applied to the eigenproblem $\underline{A}\underline{u} = \lambda\underline{u}$ to yield $\underline{\underline{A}} \, \underline{\underline{Q}} \, \bar{\underline{u}} = \lambda \underline{\underline{Q}} \bar{\underline{u}}$. Pre-multiplying by $\underline{\underline{Q}}^T$ then leads to $\underline{\underline{Q}}^T \underline{\underline{A}} \, \underline{\underline{Q}} \, \bar{\underline{u}} = \lambda \underline{\underline{Q}}^T \underline{\underline{Q}} \, \bar{\underline{u}}$. Because $\underline{\underline{Q}}$ is an orthogonal transformation, this becomes

$$\underline{\underline{\bar{A}}}\,\bar{\underline{u}} = \lambda\bar{\underline{u}}, \tag{1.60}$$

where

$$\underline{\underline{\bar{A}}} = \underline{\underline{Q}}^T \underline{\underline{A}}\, \underline{\underline{Q}}, \tag{1.61}$$

is a *similarity transformation* of the original tensor. The transformed eigenproblem has the same form as the original problem, and the relationships between the eigenvalues of the two problems is sought.

The eigenvalues of the original and transformed problems are the solutions of the characteristic equations $\det(\underline{\underline{A}} - \lambda\underline{\underline{I}}) = 0$ and $\det(\underline{\underline{\bar{A}}} - \lambda\underline{\underline{I}}) = 0$, respectively. Because $\underline{\underline{Q}}$ is an orthogonal tensor, the latter equation can be stated as $\det\left[\underline{\underline{Q}}^T(\underline{\underline{A}} - \lambda\underline{\underline{I}})\underline{\underline{Q}}\right] = 0$. Since the determinant of a product equals the product of the determinant, this becomes $\det(\underline{\underline{Q}}^T)\det(\underline{\underline{A}} - \lambda\underline{\underline{I}})\det(\underline{\underline{Q}}) = 0$. The orthogonality of $\underline{\underline{Q}}$ implies $\det(\underline{\underline{Q}}) = 1$; indeed $\det(\underline{\underline{Q}}^T\underline{\underline{Q}}) = \det^2(\underline{\underline{Q}}) = 1$. Finally, the characteristic equation of the transformed problem becomes $\det(\underline{\underline{A}} - \lambda\underline{\underline{I}}) = 0$, the same as that of the original problem. Consequently, the eigenvalues of the two problems are identical; *similarity transformations preserve the spectrum of eigenvalues* and the corresponding eigenvectors are related as $\underline{u}_i = \underline{\underline{Q}}\,\bar{\underline{u}}_i$.

Orthogonality of the eigenvectors

Consider a symmetric tensor, $\underline{\underline{A}}$, and two of its eigenpairs, $(\lambda_i, \underline{u}_i)$ and $(\lambda_j, \underline{u}_j)$, satisfying relationships $\underline{\underline{A}}\,\underline{u}_i = \lambda_i\underline{u}_i$, and $\underline{\underline{A}}\,\underline{u}_j = \lambda_j\underline{u}_j$, respectively. Pre-multiplying the first statement by $\underline{u}_j$ and the second by $\underline{u}_i$ leads to $\underline{u}_j^T\underline{\underline{A}}\,\underline{u}_i = \lambda_i\underline{u}_j^T\underline{u}_i$ and $\underline{u}_i^T\underline{\underline{A}}\,\underline{u}_j = \lambda_j\underline{u}_i^T\underline{u}_j$, respectively. Subtracting these two equations results in $(\lambda_i - \lambda_j)\underline{u}_j^T\underline{u}_i = 0$, where the symmetry of tensor $\underline{\underline{A}}$ was invoked. If $\lambda_i \neq \lambda_j$, $\underline{u}_j^T\underline{u}_i = 0$: the *eigenvectors of a symmetric tensor associated with distinct eigenvalues are orthogonal to each other*. The orthogonality of the eigenvectors also implies their orthogonality in the space of tensor $\underline{\underline{A}}$, $\underline{u}_j^T\underline{\underline{A}}\,\underline{u}_i = 0$.

If the symmetric tensor possesses three distinct eigenvalues, the corresponding eigenvectors form an orthogonal triad: $\underline{\underline{P}} = [\underline{u}_1, \underline{u}_2, \underline{u}_3]$. Because the eigenvectors are defined within a multiplicative constant, it is possible to normalize this orthogonal triad and impose $\underline{\underline{P}}^T\underline{\underline{P}} = \underline{\underline{I}}$. This does not completely remove the indeterminacy of the eigenvectors that could still be multiplied by ± 1. It is customary to order the eigenvectors in such a way that they form a right-hand basis. With this normalization of the eigenvector, it follows that

$$\underline{\underline{P}}^T \underline{\underline{A}}\, \underline{\underline{P}} = \underline{\underline{P}}^T \left[\lambda_1\underline{u}_1, \lambda_2\underline{u}_2, \lambda_3\underline{u}_3\right] = \begin{bmatrix} \lambda_1 & 0 & 0 \\ 0 & \lambda_2 & 0 \\ 0 & 0 & \lambda_3 \end{bmatrix}. \tag{1.62}$$

The orthogonality of the eigenvectors is a very important property that has been proved, thus far, for distinct eigenvalues only. What happens if a tensor features repeated eigenvalues, a common occurrence? To be precise, let eigenvalue λ_1 have a multiplicity of 2. First, an eigenvector, $\underline{u}_1$, associated with this eigenvalue is evaluated; next, the following linear transformation is constructed $\underline{\underline{Q}} = [\underline{u}_1, \underline{\underline{\hat{q}}}]$, where $\underline{\underline{Q}}$ is an orthogonal matrix; this implies $\underline{u}_1^T \underline{\hat{q}} = 0$. Since $\underline{u}_1$ is an eigenvector, it is also true that $\underline{\hat{q}}^T \underline{u}_1 = \underline{\hat{q}}^T \underline{\underline{A}} \underline{u}_1 = 0$.

A similarity transformation, see eq. (1.61), of the original problem is performed to find

$$
\underline{\underline{\bar{A}}} = \begin{bmatrix} \underline{u}_1^T \\ \underline{\hat{q}}^T \end{bmatrix} \underline{\underline{A}} \begin{bmatrix} \underline{u}_1 & \underline{\hat{q}} \end{bmatrix} = \begin{bmatrix} \underline{u}_1^T \underline{\underline{A}} \underline{u}_1 & \underline{u}_1^T \underline{\underline{A}} \underline{\hat{q}} \\ \underline{\hat{q}}^T \underline{\underline{A}} \underline{u}_1 & \underline{\hat{q}}^T \underline{\underline{A}} \underline{\hat{q}} \end{bmatrix} = \begin{bmatrix} \lambda_1 & 0 \\ 0 & \underline{\hat{q}}^T \underline{\underline{A}} \underline{\hat{q}} \end{bmatrix}. \tag{1.63}
$$

By construction, the eigenvalues of $\underline{\underline{\bar{A}}}$ are identical to those of $\underline{\underline{A}}$. Hence, λ_1 is an eigenvalue of $\underline{\underline{\bar{A}}}$ with a multiplicity of 2. The first eigenpair is $(\lambda_1, \underline{\bar{u}}_1^T = \{1, 0, 0\})$ and the second is $(\lambda_1, \underline{\bar{u}}_2^T = \{0, \underline{\bar{u}}_2'^T\})$, where $\underline{\bar{u}}_2'$ is the eigenvector of the reduced tensor $\underline{\hat{q}}^T \underline{\underline{A}} \underline{\hat{q}}$ associated with its single eigenvalue λ_1; note that $\underline{\bar{u}}_1^T \underline{\bar{u}}_2 = 0$. Two eigenvectors of the original problem are now $\underline{u}_1 = \underline{\underline{Q}} \underline{\bar{u}}_1 = \underline{u}_1$, by construction, and $\underline{u}_2 = \underline{\underline{Q}} \underline{\bar{u}}_2 = \underline{\hat{q}} \underline{\bar{u}}_2'$. Finally, the orthogonality of the eigenvectors of the transformed problem implies that of their counterparts for the original problem: $0 = \underline{\bar{u}}_1^T \underline{\bar{u}}_2 = \underline{\bar{u}}_1^T \underline{\underline{Q}}^T \underline{\underline{Q}} \underline{\bar{u}}_2 = \underline{u}_1^T \underline{u}_2$.

In summary, in the presence of repeated eigenvalues, orthogonal eigenvectors can be always extracted. For eigenvalues of multiplicity 3, the above development could be recursively applied to extract three orthogonal eigenvectors of the symmetric tensor. The orthogonal tensor, $\underline{\underline{P}} = [\underline{u}_1, \underline{u}_2, \underline{u}_3]$, always exists and presents the important property of diagonalizing tensor $\underline{\underline{A}}$

$$
\underline{\underline{P}}^T \underline{\underline{A}} \underline{\underline{P}} = \text{diag}(\lambda_i) \tag{1.64}
$$

Example 1.5. Eigen analysis of the projection tensor
Figure 1.11 depicts an arbitrary vector $\underline{a}$ and a plane $\mathcal{P}$ defined by its unit normal $\bar{n}$. The *projection tensor* $\underline{\underline{P}}$ is such that $\underline{a}' = \underline{\underline{P}} \underline{a}$, where $\underline{a}'$ is the projection of vector $\underline{a}$ onto plane $\mathcal{P}$. Find the three eigenvalues of $\underline{\underline{P}}$ and the corresponding eigenvectors.

Inspection of fig. 1.11 reveals that $\underline{a} = (\bar{n}^T \underline{a})\bar{n} + \underline{a}'$. It then follows that $\underline{a}' = \underline{a} - \bar{n}\bar{n}^T \underline{a} = (\underline{\underline{I}} - \bar{n}\bar{n}^T)\underline{a}$, and hence, the projection tensor is

$$
\underline{\underline{P}} = \underline{\underline{I}} - \bar{n}\bar{n}^T = \begin{bmatrix} 1 - n_1^2 & -n_1 n_2 & -n_1 n_3 \\ -n_1 n_2 & 1 - n_2^2 & -n_2 n_3 \\ -n_1 n_3 & -n_2 n_3 & 1 - n_3^2 \end{bmatrix}. \tag{1.65}
$$

The projection tensor is symmetric and semi positive-definite. Indeed, $\underline{a}^T (\underline{\underline{I}} - \bar{n}\bar{n}^T)\underline{a} = \underline{a}^T \underline{a} - (\bar{n}^T \underline{a})^2 = \|\underline{a}\|^2 - \|\underline{a}\|^2 \cos^2 \alpha$, where α is the angle between vectors $\bar{n}$ and $\underline{a}$. It follows that $\underline{a}^T \underline{\underline{P}} \underline{a} = \|\underline{a}\|^2 \sin^2 \alpha \geq 0$ for any arbitrary vector $\underline{a} \neq 0$. Note that $\underline{a}^T \underline{\underline{P}} \underline{a} = 0$ when $\alpha = 0$, *i.e.*, when vector $\underline{a}$ is parallel to $\bar{n}$. It follows that the eigenvalues of $\underline{\underline{P}}$ must be real and greater or equal to zero.

One eigenvector of $\underline{\underline{P}}$ can be found by inspection: $\underline{\underline{P}} \bar{n} = (\underline{\underline{I}} - \bar{n}\bar{n}^T)\bar{n} = \bar{n} - \bar{n} = 0$. This implies that vector $\bar{n}$ is an eigenvector of the projection tensor corresponding to an eigenvalue $\lambda = 0$. The invariants of $\underline{\underline{P}}$ are readily found as $I_1 = 2$, $I_2 = 1$ and $I_3 = 0$. The eigenvalues then are the solutions of the characteristic equation $-\lambda^3 + 2\lambda^2 - \lambda = 0$ or $\lambda(\lambda - 1)^2 = 0$. The eigenvalues are $\lambda_1 = 0$, $\lambda_2 = +1$ and $\lambda_3 = +1$; note the multiplicity of two of the unit eigenvalue.

As discussed above, the eigenvector corresponding to the null eigenvalue $\lambda_1 = 0$ is the unit vector $\underline{u}_1 = \bar{n}$. The eigenvectors corresponding to the double unit eigenvalue are the solution of the homogeneous linear problem $(\underline{\underline{I}} - \bar{n}\bar{n}^T - \underline{\underline{I}})\underline{x} = 0$ or $\bar{n}\bar{n}^T \underline{x} = 0$. Clearly, any vector orthogonal to $\bar{n}$ will satisfy this equation. In other words, any vector $\underline{u}_2$ in plane $\mathcal{P}$, i.e., such that $\bar{n}^T \underline{u}_2 = 0$, is an eigenvector. This implies $\underline{\underline{P}}\,\underline{u}_2 = \underline{u}_2$, a result that is readily verified: $(\underline{\underline{I}} - \bar{n}\bar{n}^T)\underline{u}_2 = \underline{u}_2$ if $\bar{n}^T \underline{u}_2 = 0$. In geometric terms, this result is obvious: if vector $\underline{u}_2$ lie in plane $\mathcal{P}$, the projection of that vector onto the plane is the vector itself.

In view of the multiplicity of two of the unit eigenvalue, the eigenvector corresponding to λ_3 is identical to that corresponding to λ_2: an arbitrary vector in plane $\mathcal{P}$. It is, however, always possible to find an orthogonal vector by selecting $\underline{u}_3 = \tilde{n}\underline{u}_2$.

In summary, the three eigenvectors of the projection tensor are $\underline{u}_1 = \underline{n}$, the normal to plane $\mathcal{P}$, $\underline{u}_2$ an arbitrary vector in $\mathcal{P}$, and $\underline{u}_3 = \tilde{n}\underline{u}_2$. Clearly, the eigenvectors capture the essence of the projection tensor: $\bar{u}_1$ is the direction normal to the plane, and $\underline{u}_2$ and $\underline{u}_3$ are two orthogonal directions within this plane. The multiplicity of two of the unit eigenvector results in the fact that $\underline{u}_2$ can be chosen arbitrarily within plane $\mathcal{P}$. Geometrically, this is related to the isotropy of the projection tensor: it behaves in the same manner in all direction within plane $\mathcal{P}$. Finally, note that $\underline{\underline{P}} = \underline{\underline{P}}\,\underline{\underline{P}}$: once a vector has been projected onto the plane, any subsequent application of the projection tensor will leave the vector unchanged.

1.4.3 Problems

Problem 1.31. Solve linear system
Solve the following equation for $\underline{x}$, $\tilde{x}\underline{a} = \underline{b} - \underline{x}$.

Problem 1.32. Compute inverse
Show that $(\underline{\underline{I}} + \tilde{a})^{-1} = (\underline{\underline{I}} + \underline{a}\,\underline{a}^T - \tilde{a})/(1 + a^2)$

Problem 1.33. Eigenvalues of the reflection tensor
Figure 1.12 depicts an arbitrary vector $\underline{a}$ and plane $\mathcal{P}$, defined by its unit normal $\bar{n}$. *(1)* Find the expression for the *reflection tensor*, $\underline{\underline{R}}$, such that $\underline{a}' = \underline{\underline{R}}\,\underline{a}$, where vector $\underline{a}'$ is the reflection of vector $\underline{a}$ with respect the plane $\mathcal{P}$. Note that point $\mathbf{A}'$ is the mirror image of point $\mathbf{A}$ with respect the plane $\mathcal{P}$. *(2)* Is the reflection tensor positive-definite? *(3)* By inspection of $\underline{\underline{R}}$ find one of its eigenvectors and the corresponding eigenvalue. *(4)* Compute the three invariants of $\underline{\underline{R}}$. *(5)* Find the three eigenvalues of $\underline{\underline{R}}$ and the corresponding eigenvectors.

1.5 Tensor calculus

The derivative of a scalar function $s(t)$ of a single variable, t, say time, is defined in calculus textbooks (see, for instance, [2]), as

$$\frac{ds}{dt} = \dot{s} = \lim_{\Delta t \to 0} \frac{s(t + \Delta t) - s(t)}{\Delta t}. \tag{1.66}$$

The notation $(\dot{\cdot})$ will be used throughout this book to represent a derivative with respect to time. The derivative of a vector $\underline{u}(t)$ is defined in a similar manner as

$$\frac{d\underline{u}}{dt} = \dot{\underline{u}} = \lim_{\Delta t \to 0} \frac{\underline{u}(t + \Delta t) - \underline{u}(t)}{\Delta t}. \tag{1.67}$$

The following results stem from elementary rules for derivatives.

Derivative of a sum

If $\underline{u}(t)$ and $\underline{v}(t)$ are two arbitrary vectors

$$\frac{d}{dt}(\underline{u} + \underline{v}) = \dot{\underline{u}} + \dot{\underline{v}}. \tag{1.68}$$

Derivative of a product

If $s(t)$ is a scalar function of time,

$$\frac{d}{dt}(s\underline{u}) = \dot{s}\underline{u} + s\dot{\underline{u}}. \tag{1.69}$$

The derivative of the scalar product becomes

$$\frac{d}{dt}(\underline{u}^T \underline{v}) = \underline{v}^T \dot{\underline{u}} + \underline{u}^T \dot{\underline{v}}, \tag{1.70}$$

and that of the vector product

$$\frac{d}{dt}(\tilde{\underline{u}}\underline{v}) = \dot{\tilde{\underline{u}}}\underline{v} + \tilde{\underline{u}}\dot{\underline{v}} = \tilde{\underline{u}}\dot{\underline{v}} - \tilde{\underline{v}}\dot{\underline{u}}. \tag{1.71}$$

Chain rule for differentiation

If vector $\underline{u}$ is a function of a scalar function $s(t)$, the time derivative of this vector becomes

$$\frac{d}{dt}(\underline{u}(s(t))) = \frac{d\underline{u}}{ds}\frac{ds}{dt} = \dot{s}\frac{d\underline{u}}{ds}. \tag{1.72}$$

As an application of the above rules, consider the derivative of a unit vector, *i.e.*, vector $\underline{u}$ such that $\underline{u}^T \underline{u} = 1$. Equation (1.70) then implies

$$\frac{d}{dt}(\underline{u}^T \underline{u}) = 2\underline{u}^T \dot{\underline{u}} = 0. \tag{1.73}$$

In other words: the *derivative of a unit vector is orthogonal to the vector itself.* Next, consider two mutually orthogonal vector $\underline{u}$ and $\underline{v}$, $\underline{u}^T \underline{v} = 0$. A derivative of this expression then yields

$$\underline{u}^T \dot{\underline{v}} = -\underline{v}^T \dot{\underline{u}}. \tag{1.74}$$

1.6 Notational conventions

Several notational conventions are used in the literature to denote vectors and tensors. Three widely used notations, the *geometric notation*, the *matrix notation*, and the *index notation* [3] are presented in table 1.1. The *geometric notation* is widely used in the literature, sometimes the boldface notation for vectors is replaced by a specific "vector" superscript: $\vec{a}$. The index notation is frequently used, specially when higher-order tensors must be manipulated such as in the theory of elasticity. It is, however, less often used in kinematics and dynamics.

The matrix notation is a convenient mnemonic notation and will be used exclusively in this book. Vectors are denoted with an underline, $\underline{u}$, but unit vectors are simply denoted $\bar{n}$, rather than the more cumbersome $\underline{\bar{n}}$. Tensors are denoted by a double underline, $\underline{\underline{A}}$, but skew-symmetric tensors are denoted $\widetilde{a}$, rather than the more cumbersome $\underline{\underline{\widetilde{a}}}$. Note that the tensor product, $\underline{u}\,\underline{v}^T$, also yields a tensor.

Table 1.1. The geometric, matrix, and index notations for vectors and tensors.

	Geometric notation	Matrix notation	Index notation
vector	**a**	$\underline{a}$	a_i
tensor	$\underline{\underline{A}}$	$\underline{\underline{A}}$	A_{ij}
scalar product	$\mathbf{u} \cdot \mathbf{v}$	$\underline{u}^T\underline{v}$	$u_i v_i$
vector product	$\mathbf{u} \times \mathbf{v}$	$\widetilde{u}\underline{v}$	$u_i v_j \epsilon_{ijk}$
tensor product	$\mathbf{u} \otimes \mathbf{v}$	$\underline{u}\,\underline{v}^T$	$u_i v_j$

In practical situations, such computer implementations, it will be necessary to work with the components of specific tensors resolved in various bases. In such cases, the following notation will be used

$$\underline{a}^{[\mathcal{I}]} = \left\{ \begin{array}{c} a_1 \\ a_2 \\ a_3 \end{array} \right\},$$

where a_1, a_2, and a_3 are the components of vector $\underline{a}$ resolved in basis $\mathcal{I}$. Because the notation $\underline{a}^{[\mathcal{I}]}$ is rather cumbersome, it will be used only when necessary; for instance, when the components of a vector in two different bases are used in the same context. When there is no possible confusion, the notation $\underline{a}^{[\mathcal{I}]}$ will be simplified as $\underline{a}$, thereby blurring the distinction between a vector and its components in a given basis.

2

Coordinate systems

The practical description of dynamical systems involves a variety of coordinates systems. While the Cartesian coordinates discussed in section 2.1 are probably the most commonly used, many problems are more easily treated with special coordinate systems. The differential geometry of curves is studied in section 2.2 and leads to the concept of path coordinates, treated in section 2.3. Similarly, the differential geometry of surfaces is investigated in section 2.4 and leads to the concept of surface coordinates, treated in section 2.5. Finally, the differential geometry of three-dimensional maps is studied in section 2.6 and leads to orthogonal curvilinear coordinates developed in section 2.7.

2.1 Cartesian coordinates

The simplest way to represent the location of a point in three-dimensional space is to make use of a reference frame, $\mathcal{F} = [\mathbf{O}, \mathcal{I} = (\bar{\imath}_1, \bar{\imath}_2, \bar{\imath}_3)]$, consisting of an orthonormal basis $\mathcal{I}$ with its origin and point $\mathbf{O}$, as described in section 1.2.2. The time-dependent position vector of point $\mathbf{P}$ is represented by its *Cartesian coordinates*, $x_1(t)$, $x_2(t)$, and $x_3(t)$, resolved along unit vectors, $\bar{\imath}_1$, $\bar{\imath}_2$, and $\bar{\imath}_3$, respectively,

$$\underline{r}(t) = x_1(t)\bar{\imath}_1 + x_2(t)\bar{\imath}_2 + x_3(t)\bar{\imath}_3, \quad (2.1)$$

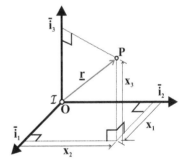

Fig. 2.1. Cartesian coordinate system.

where t denotes time. Figure 2.1 depicts the situation: Cartesian coordinate $x_1 = \bar{\imath}_1^T \underline{r}$ is the projection of the position vector of point $\mathbf{P}$ along unit vector $\bar{\imath}_1$. Similarly, Cartesian coordinates x_1 and x_2 are the projections of the same position vector along unit vectors $\bar{\imath}_2$ and $\bar{\imath}_3$, respectively.

The components of the velocity vector are readily obtained by differentiating the expression for the position vector, eq. (2.1), to find

$$\underline{v}(t) = \dot{x}_1(t)\bar{\imath}_1 + \dot{x}_2(t)\bar{\imath}_2 + \dot{x}_3(t)\bar{\imath}_3 = v_1(t)\bar{\imath}_1 + v_2(t)\bar{\imath}_2 + v_3(t)\bar{\imath}_3. \qquad (2.2)$$

The Cartesian components of the velocity vector are simply the time derivatives of the corresponding Cartesian components of the position vector: $v_1(t) = \dot{x}_1(t)$, $v_2(t) = \dot{x}_2(t)$, and $v_3(t) = \dot{x}_3(t)$.

Finally, the acceleration vector is obtained by taking a time derivative of the velocity vector to find

$$\underline{a}(t) = \ddot{x}_1(t)\bar{\imath}_1 + \ddot{x}_2(t)\bar{\imath}_2 + \ddot{x}_3(t)\bar{\imath}_3 = a_1(t)\bar{\imath}_1 + a_2(t)\bar{\imath}_2 + a_3(t)\bar{\imath}_3. \qquad (2.3)$$

Here again, the Cartesian components of the acceleration vector are simply the derivatives of the corresponding Cartesian components of the velocity vector, or the second derivatives of the position components: $a_1(t) = \dot{v}_1(t) = \ddot{x}_1(t)$, $a_2(t) = \dot{v}_2(t) = \ddot{x}_2(t)$, and $a_3(t) = \dot{v}_3(t) = \ddot{x}_3(t)$.

Cartesian coordinates are simple to manipulate and are the most commonly used coordinate system in computational applications that deal with problems presenting arbitrary topologies. On the other hand, several other coordinate systems, such as those discussed in the rest of this chapter, are often used because they can ease the solution process for specific problems. In such cases, a specific coordinate system is used solve a specific problem. For instance, polar coordinates are very efficient to describe the behavior of a particle constrained to move along a circular path.

2.2 Differential geometry of a curve

This section investigates the differential geometry of a curve, leading to the concept of path coordinates. Both intrinsic and arbitrary parameterizations will be considered. Frenet's triad is defined and its derivatives evaluated.

2.2.1 Intrinsic parameterization

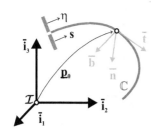

Fig. 2.2. Configuration of a curve in space.

Figure 2.2 depicts a curve, denoted $\mathbb{C}$, in three-dimensional space. A curve is the locus of the points generated by a single parameter, such that the position vector, $\underline{p}_0$, of such points can be written as

$$\underline{p}_0 = \underline{p}_0(s), \qquad (2.4)$$

where s is the parameter that generates the curve. If parameter s is the *curvilinear coordinate* that measures length along the curve, it is said to define the *intrinsic parameterization* or *natural parameterization* of the curve.

Frenet's triad

A differential element of length, $\mathrm{d}s$, along the curve is written as $\mathrm{d}s^2 = \mathrm{d}\underline{p}_0^T \mathrm{d}\underline{p}_0$, and in follows that $(\mathrm{d}\underline{p}_0/\mathrm{d}s)^T (\mathrm{d}\underline{p}_0/\mathrm{d}s) = 1$. The *unit tangent vector* to the curve is defined as

$$\bar{t} = \frac{d\underline{p}_0}{ds}.\tag{2.5}$$

By construction, this is a unit vector because $\bar{t}^T \bar{t} = 1$.

Taking a derivative of this relationship with respect to the curvilinear coordinate leads to $\bar{t}^T d\bar{t}/ds = 0$. Vector $d\bar{t}/ds$ is normal to the tangent vector. The *unit normal vector* to the curve is defined as

$$\bar{n} = \rho \frac{d\bar{t}}{ds},\tag{2.6}$$

where ρ is the *radius of curvature* of the curve, such that

$$\frac{1}{\rho} = \left\| \frac{d\bar{t}}{ds} \right\|.\tag{2.7}$$

The quantity $1/\rho$ is the *curvature* of the curve, and ρ its radius of curvature. The two unit vector, $\bar{t}$ and $\bar{n}$, are said to form the *osculating plane of the curve*.

An orthonormal triad is now constructed by defining the *binormal vector, $\bar{b}$*, as the cross product of the tangent by the normal vectors,

$$\bar{b} = \tilde{t}\,\bar{n}.\tag{2.8}$$

The unit tangent, normal, and binormal vectors form an orthonormal triad, called *Frenet's triad*, depicted in fig. 2.2.

Derivatives of Frenet's triad

First, the derivative of the normal vector is resolved in Frenet's triad as $d\bar{n}/ds = \alpha \bar{t} + \beta \bar{n} + \gamma \bar{b}$, where α, β, and γ are unknown coefficients. Pre-multiplying this relationship by $\bar{n}^T$ yields $\beta = \bar{n}^T d\bar{n}/ds = 0$, because $\bar{n}$ is a unit vector. Pre-multiplying by $\bar{t}^T$ yields $\alpha = \bar{t}^T d\bar{n}/ds = -\bar{n}^T d\bar{t}/ds = -1/\rho$, where eq. (2.6) was used. Finally, pre-multiplying by $\bar{b}^T$ yields $\gamma = \bar{b}^T d\bar{n}/ds = 1/\tau$. Combining all these results yields

$$\frac{d\bar{n}}{ds} = -\frac{1}{\rho}\bar{t} + \frac{1}{\tau}\bar{b},\tag{2.9}$$

where τ is the *radius of twist* of the curve, defined as

$$\frac{1}{\tau} = \bar{b}^T \frac{d\bar{n}}{ds}.\tag{2.10}$$

Next, the derivative of the binormal vector is resolved in Frenet's triad as $d\bar{b}/ds = \alpha \bar{t} + \beta \bar{n} + \gamma \bar{b}$, where α, β, and γ are unknown coefficients. Pre-multiplying this relationship by $\bar{b}^T$ yields $\gamma = \bar{b}^T d\bar{b}/ds = 0$, because $\bar{b}$ is a unit vector. Pre-multiplying by $\bar{t}^T$ yields $\alpha = \bar{t}^T d\bar{b}/ds = -\bar{b}^T d\bar{t}/ds = -\bar{b}^T \bar{n}/\rho = 0$. Finally, pre-multiplying by $\bar{n}^T$ yields $\beta = \bar{n}^T d\bar{b}/ds = -\bar{b}^T d\bar{n}/ds = -1/\tau$, where eq. (2.10) was used. Combining all these results yields

$$\frac{d\bar{b}}{ds} = -\frac{1}{\tau}\bar{n}.\tag{2.11}$$

It follows that the *twist of the curve* can also be written as

$$\frac{1}{\tau} = \left\| \frac{d\bar{b}}{ds} \right\|. \tag{2.12}$$

If the binormal vector has a constant direction at all points along the curve, $d\bar{b}/ds = 0$, and the curve entirely lies in the plane defined by vectors $\bar{t}$ and $\bar{n}$, *i.e.*, the osculating plane is the same at all points of the curve. The curve is then a *planar curve*, and eq. (2.12) implies that $1/\tau = 0$, *i.e.*, the twist of the curve vanishes.

The derivatives of Frenet's triad can be expressed in a compact manner by combining eqs. (2.6), (2.9), and (2.11),

$$\frac{d}{ds} \left\{ \begin{array}{c} \bar{t} \\ \bar{n} \\ \bar{b} \end{array} \right\} = \left[\begin{array}{ccc} 0 & 1/\rho & 0 \\ -1/\rho & 0 & 1/\tau \\ 0 & -1/\tau & 0 \end{array} \right] \left\{ \begin{array}{c} \bar{t} \\ \bar{n} \\ \bar{b} \end{array} \right\}. \tag{2.13}$$

2.2.2 Arbitrary parameterization

The previous section has developed a representation of a curve based on its natural or intrinsic parameterization. In many instances, however, this parameterization is difficult to obtain; instead, the curve is defined in terms of a single parameter, η, that does not measure length along the curve, see fig. 2.2. The position vector of a point on the curve is now $\underline{p}_0 = \underline{p}_0(\eta)$. The derivatives of the position vector with respect to parameter η will be denoted as

$$\underline{p}_1 = \frac{d\underline{p}_0}{d\eta}, \quad \underline{p}_2 = \frac{d^2\underline{p}_0}{d\eta^2}, \quad \underline{p}_3 = \frac{d^3\underline{p}_0}{d\eta^3}, \quad \underline{p}_4 = \frac{d^4\underline{p}_0}{d\eta^4}.$$

A similar notation will be used for the tangent and normal vectors,

$$\bar{t}_i = \frac{d^i\bar{t}}{d\eta^i}, \quad \bar{n}_i = \frac{d^i\bar{n}}{d\eta^i}.$$

The differential element of length along the curve can be written as $ds^2 = (d\underline{p}_0/d\eta)^T (d\underline{p}_0/d\eta) \, d\eta^2$. The ratio of the increment in length along the curve, ds, to the increment in parameter value, $d\eta$, is then

$$\frac{ds}{d\eta} = \sqrt{\underline{p}_1^T \underline{p}_1} = p_1. \tag{2.14}$$

Notation $(\cdot)'$ will be used to indicate a derivative with respect to η, and hence, $d/ds = (\cdot)'/p_1$. The unit tangent vector to the curve is evaluated with the help of eq. (2.5) as

$$\bar{t} = \frac{\underline{p}_1}{p_1} \tag{2.15}$$

Next, the derivative of the tangent vector is found as

$$\bar{t}_1 = \frac{p_1\, \underline{p}_2 - \bar{p}_1\, (\underline{p}_1^T \underline{p}_2)/p_1}{p_1^2} = \frac{1}{p_1}(1 - \bar{t}\,\bar{t}^T)\underline{p}_2 = \frac{1}{p_1}\left[\underline{p}_2 - (\bar{t}^T \underline{p}_2)\bar{t}\right]. \qquad (2.16)$$

From eq. (2.7), the radius of curvature now becomes

$$\frac{1}{\rho} = \left\|\frac{d\bar{t}}{ds}\right\| = \frac{1}{p_1}\|\bar{t}_1\|.$$

It follows that $\|\bar{t}_1\| = t_1 = p_1/\rho$. For a straight line, the tangent vector has a fixed direction in space, $\bar{t}_1 = 0$. It follows that for a straight line $1/\rho = 0$, *i.e.*, its radius of curvature is infinite. The curve's curvature is found to be

$$\frac{1}{\rho} = \frac{\sqrt{p_1^2 p_2^2 - (\underline{p}_2^T \underline{p}_1)^2}}{p_1^3} \qquad (2.17)$$

Higher-order derivatives of the tangent vector are found in a similar manner

$$\bar{t}_2 = \frac{1}{p_1}\left[\underline{p}_3 - (\bar{t}^T \underline{p}_3 + \bar{t}_1^T \underline{p}_2)\bar{t} - 2(\bar{t}^T \underline{p}_2)\bar{t}_1\right],$$

and

$$\bar{t}_3 = \frac{1}{p_1}\left[\underline{p}_4 - (\bar{t}^T \underline{p}_4 + 2\bar{t}_1^T \underline{p}_3 + \bar{t}_2^T \underline{p}_2)\bar{t} - 3(\bar{t}^T \underline{p}_3 + \bar{t}_1^T \underline{p}_2)\bar{t}_1 - 3(\bar{t}^T \underline{p}_2)\bar{t}_2\right].$$

Next, the normal vector defined in eq. (2.6) becomes

$$\bar{n} = \frac{\bar{t}_1}{\|\bar{t}_1\|} = \frac{1}{t_1}\bar{t}_1. \qquad (2.18)$$

For a straight line, $\bar{t}_1 = 0$, and hence, the normal vector is not defined. In fact, any vector normal to a straight line is a normal vector. The derivative of the normal vector with respect to η then follows as

$$\bar{n}_1 = \frac{1}{t_1}\left[\bar{t}_2 - (\bar{n}^T \bar{t}_2)\bar{n}\right]. \qquad (2.19)$$

The second-order derivative is then

$$\bar{n}_2 = \frac{1}{t_1}\left[\bar{t}_3 - (\bar{n}^T \bar{t}_3 + \bar{n}_1^T \bar{t}_2)\bar{n} - 2(\bar{n}^T \bar{t}_2)\bar{n}_1\right]. \qquad (2.20)$$

The binormal vector is readily expressed as

$$\bar{b} = \tilde{t}\,\bar{n} = \frac{1}{t_1}\tilde{t}\,\bar{t}_1 = \frac{\rho}{p_1^3}\tilde{p}_1\, \underline{p}_2. \qquad (2.21)$$

Because the normal vector is not defined for a straight line, the binormal vector is not defined in that case. In fact, any vector normal to a straight line is a binormal vector.
    The derivative of the binormal vector becomes

$$\bar{b}_1 = \left(\frac{\rho}{p_1^3}\right)' \tilde{p}_1 p_2 + \frac{\rho}{p_1^3} \tilde{p}_1 p_3. \tag{2.22}$$

Using eq. (2.10), the twist of the curve is found to be

$$\frac{1}{\tau} = -\frac{1}{p_1} \bar{n}^T \bar{b}_1 = -\frac{\rho}{p_1^5} \left[p_1^2 p_2^T - (p_1^T p_2) p_1^T \right] \bar{b}_1.$$

Finally, introducing eq. (2.22) leads to

$$\frac{1}{\tau} = -\frac{\rho^2}{p_1^6} p_2^T \tilde{p}_1 p_3. \tag{2.23}$$

The twist of the curve is closely related to the volume defined by vectors p_1, p_2, and p_3. Note that a straight line has a vanishing twist, $1/\tau = 0$.

Derivatives of the binormal vector are more easily expressed as $\bar{b}_1 = \tilde{t}_1 \bar{n} + \tilde{t} \bar{n}_1 = \tilde{t} \bar{n}_1$, and $\bar{b}_2 = \tilde{t}_1 \bar{n}_1 + \tilde{t} \bar{n}_2 = \tilde{n} t_2 + \tilde{t} \bar{n}_2$, where eqs. (2.18) and (2.19) were used.

Example 2.1. The helix

Figure 2.3 depicts a helix, which is a three-dimensional curve defined by the following position vector

$$p_0(\eta) = a \cos \eta \, \bar{\imath}_1 + a \sin \eta \, \bar{\imath}_2 + k\eta \, \bar{\imath}_3, \tag{2.24}$$

where a and k are two parameters defining the shape of the curve. The derivatives of the position vector are $p_1 = -a \sin \eta \, \bar{\imath}_1 + a \cos \eta \, \bar{\imath}_2 + k \, \bar{\imath}_3$, $p_2 = -a \cos \eta \, \bar{\imath}_1 - a \sin \eta \, \bar{\imath}_2$, and $p_3 = a \sin \eta \, \bar{\imath}_1 - a \cos \eta \, \bar{\imath}_2$. The curvature and twist of the helix are found with the help of eqs. (2.17) and (2.23), respectively, as

$$\frac{1}{\rho} = \frac{a}{a^2 + k^2}, \quad \frac{1}{\tau} = \frac{k}{a^2 + k^2}.$$

Note that both curvature and twist are constant along the helix. The unit tangent vector is evaluated with the help of eq. (2.15) as

$$\bar{t} = \frac{1}{\sqrt{a^2 + k^2}} p_1 = \frac{1}{\sqrt{a^2 + k^2}} (-a \sin \eta \bar{\imath}_1 + a \cos \eta \bar{\imath}_2 + k \bar{\imath}_3). \tag{2.25}$$

The ratio between an increment in length along the curve and the increment in the parameter value is then $ds = \sqrt{a^2 + k^2} \, d\eta$, see eq. (2.14). Next, the derivative of the tangent vector is computed with the help of eq. (2.16) as $\bar{t}_1 = p_2/p_1$ and the normal vector then follows as

$$\bar{n} = -\cos \eta \, \bar{\imath}_1 - \sin \eta \, \bar{\imath}_2.$$

Finally, the binormal vector found from eq. (2.21)

$$\bar{b} = \frac{1}{\sqrt{a^2 + k^2}} \left[k \sin \eta \, \bar{\imath}_1 - k \cos \eta \, \bar{\imath}_2 + a \, \bar{\imath}_3 \right].$$

The derivatives of Frenet's triad are found with the help of eq. (2.13) as

$$\frac{d\bar{t}}{ds} = \frac{a}{a^2 + k^2} \bar{n}, \quad \frac{d\bar{n}}{ds} = -\frac{a}{a^2 + k^2} \bar{t} + \frac{k}{a^2 + k^2} \bar{b}, \quad \frac{d\bar{b}}{ds} = -\frac{k}{a^2 + k^2} \bar{n}.$$

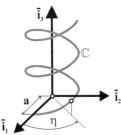

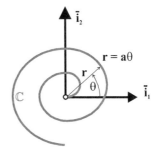

Fig. 2.3. Configuration of a helix in three-dimensional space.

Fig. 2.4. Configuration of a planar linear spiral.

Example 2.2. *The linear spiral*

Figure 2.4 depicts a linear spiral, which is a planar curve defined by the following position vector

$$\underline{p}_0 = a\theta \cos\theta\, \bar{\imath}_1 + a\theta \sin\theta\, \bar{\imath}_2, \tag{2.26}$$

where a is a parameter defining the shape of the curve. The derivatives of the position vector are $\underline{p}_1 = a\left[(\cos\theta - \theta\sin\theta)\bar{\imath}_1 + (\sin\theta + \theta\cos\theta)\bar{\imath}_2\right]$, $\underline{p}_2 = a\left[-(2\sin\theta + \theta\cos\theta)\bar{\imath}_1 + (2\cos\theta - \theta\sin\theta)\bar{\imath}_2\right]$. It is readily verified that $p_1^2 = a^2(1 + \theta^2)$, $p_2^2 = a^2(4 + \theta^2)$ and $\underline{p}_1^T\underline{p}_2 = a^2\theta$. The curvature of the linear spiral is found with the help of eq. (2.17)

$$\frac{a}{\rho} = \frac{2 + \theta^2}{(1 + \theta^2)^{3/2}}.$$

Note that the curvature varies along the spiral. Of course, the twist is zero since the curve is planar. The unit tangent vector is evaluated with the help of eq. (2.15) as

$$\bar{t} = \frac{(\cos\theta - \theta\sin\theta)\bar{\imath}_1 + (\sin\theta + \theta\cos\theta)\bar{\imath}_2}{\sqrt{1 + \theta^2}}.$$

Finally, the normal vector becomes

$$\bar{n} = \frac{-\left[2\sin\theta + \theta\cos\theta(2 + \theta^2)\right]\bar{\imath}_1 + \left[2\cos\theta - \theta\sin\theta(2 + \theta^2)\right]\bar{\imath}_2}{\sqrt{4 + \theta^2(2 + \theta^2)^2}}.$$

Example 2.3. *Using polar coordinates to represent curves*

Cams play an important role in numerous mechanical systems: cam-follower pairs typically transform the rotary motion of the cam into a desirable motion of the follower. Figure 2.5 depicts a typical cam whose outer shape is defined by a curve. It is convenient to define this curve using the polar coordinate system indicated on the figure: for each angle α, the distance from point **O** to point **P** is denoted r. The complete curve is then defined by function $r = r(\alpha)$; angle α provides an arbitrary parameterization of the curve. If $r(\alpha)$ is a periodic function of angle α, the curve will be a closed curve, as expected for a cam.

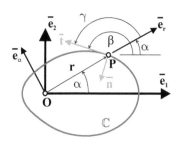

Fig. 2.5. Configuration of a cam.

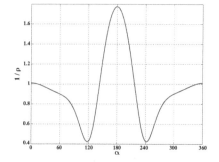

Fig. 2.6. Curvature distribution for the cam.

Vectors $\underline{p}_0$, $\underline{p}_1$, and $\underline{p}_2$ now become

$$\underline{p}_0 = rC_\alpha\, \bar{e}_1 + rS_\alpha\, \bar{e}_2, \tag{2.27a}$$

$$\underline{p}_1 = (r'C_\alpha - rS_\alpha)\, \bar{e}_1 + (r'S_\alpha + rC_\alpha)\, \bar{e}_2, \tag{2.27b}$$

$$\underline{p}_2 = (r''C_\alpha - 2r'S_\alpha - rC_\alpha)\, \bar{e}_1 + (r''S_\alpha + 2r'C_\alpha - rS_\alpha)\, \bar{e}_2, \tag{2.27c}$$

where the notation $(\cdot)'$ indicates a derivative with respect to α, $S_\alpha = \sin\alpha$, and $C_\alpha = \cos\alpha$. It then follows that $p_1^2 = r^2 + r'^2$ and $p_2^2 = (r'' - r)^2 + 4r'^2$. The various properties of the curve can then be evaluated; for instance, eqs. (2.15) and (2.17) yield the tangent vector and curvature along the curve, respectively.

The curve depicted in fig. 2.5 is defined by the following equation, $r(\alpha) = 1.0 + 0.5\cos\alpha + 0.15\cos 2\alpha$ and fig. 2.6 shows the curvature distribution as a function of angle α.

Figure 2.5 shows the unit tangent vector, $\bar{t}$, at point **P** of the curve and defines angles $\beta = (\bar{e}_1, \bar{t})$ and $\gamma = (\bar{e}_r, \bar{t})$; note that $\gamma = \beta - \alpha$. The unit tangent vector can now be written as $\bar{t} = C_\beta \bar{e}_1 + S_\beta \bar{e}_2 = \underline{p}_1/p_1$, where the second equality follows from eq. (2.15). Pre-multiplying this relationship by $\bar{e}_1^T$ and $\bar{e}_2^T$ yields $p_1 C_\beta = r'C_\alpha - rS_\alpha$ and $p_1 S_\beta = r'S_\alpha + rC_\alpha$, respectively. Solving these two equations for r and r' and using elementary trigonometric identities then leads to

$$r = p_1 \sin(\beta - \alpha) = p_1 S_\gamma, \tag{2.28a}$$

$$r' = p_1 \cos(\beta - \alpha) = p_1 C_\gamma, \tag{2.28b}$$

where $S_\gamma = \sin\gamma$, and $C_\gamma = \cos\gamma$. The quotient of these two equations then yields the following relationship

$$d\alpha = \tan\gamma \frac{dr}{r}. \tag{2.29}$$

The derivative of the unit tangent vector with respect to the curvilinear coordinate along the curve is $d\bar{t}/ds = (-S_\gamma \bar{e}_1 + C_\gamma \bar{e}_2) d\beta/ds$, and the curvature is then $1/\rho = |d\beta/ds|$. If the curve is convex, which is generally the case for cams, angle β is a monotonically increasing function of s, and hence, $1/\rho = d\beta/ds$. The chain rule for derivatives implies $d\beta = (1/\rho)(ds/d\alpha)(d\alpha/dr)dr$ and introducing eqs. (2.14), (2.28a), and (2.29) then yields

$$d\beta = \frac{dr}{\rho C_\gamma}. \tag{2.30}$$

It is left to the reader to verify that eq. (2.30) yields an alternative, simplified expression for the curvature of the cam

$$\frac{1}{\rho} = \frac{2r'^2 - rr'' + r^2}{p_1^3}. \tag{2.31}$$

Finally, an increment in angle γ can be expressed as $d\gamma = d\beta - d\alpha$ and introducing eqs. (2.30) and (2.29) yields

$$d\gamma = \left(\frac{1}{\rho C_\gamma} - \frac{\tan\gamma}{r}\right) dr. \tag{2.32}$$

2.3 Path coordinates

Consider a particle moving along a curve such that its position, $s(t)$, is a given function of time. The velocity vector, $\underline{v}$, of the particle is then

$$\underline{v} = \frac{d\underline{p}_0}{dt} = \frac{d\underline{p}_0}{ds}\frac{ds}{dt} = v\bar{t}, \tag{2.33}$$

where $v = ds/dt$ is the *speed of the particle*, Clearly, the velocity vector of the particle is along the tangent to the curve.

Next, the particle acceleration vector, $\underline{a}$, becomes

$$\underline{a} = \frac{d\underline{v}}{dt} = \frac{dv}{dt}\bar{t} + v\frac{d\bar{t}}{ds}\frac{ds}{dt} = \dot{v}\bar{t} + \frac{v^2}{\rho}\bar{n}. \tag{2.34}$$

The acceleration vector is contained in the osculating plane, and can be written as $\underline{a} = a_t\bar{t} + a_n\bar{n}$, where a_t and a_n are the tangential and normal components of acceleration, respectively. The tangential component of acceleration, $a_t = \dot{v}$, simply measures the change in particle speed. The normal component, $a_n = v^2/\rho$, is always directed towards the center of curvature since v^2/ρ is a positive number. This normal acceleration is clearly related to the curvature of the path; in fact, when the path is a straight line, $1/\rho = 0$, and the normal acceleration vanishes.

2.3.1 Problems

Problem 2.1. Prove identity
Prove that $1/\rho = p_2/p_1^2 \, |\sin\alpha|$, where $p_2 = \|\underline{p}_2\|$ and α is the angle between vectors $\underline{p}_1$ and $\underline{p}_2$.

Problem 2.2. Study of a curve

Consider the following spatial curve: $\underline{p}_0 = a(\eta + \sin \eta)\bar{\imath}_1 + a(1 + \cos \eta)\bar{\imath}_2 + a(1 - \cos \eta)\bar{\imath}_3$, where $a > 0$ is a given parameter. *(1)* Find the tangent, normal, and binormal vectors for this curve. *(2)* Determine the curvature, radius of curvature, and twist of the curve. Is this a planar curve? Is the tangent vector defined at all points of the curve?

Problem 2.3. Study of a curve

Consider the following spatial curve: $\underline{p}_0 = \rho(\cos \alpha\eta)(\cos \eta)\bar{\imath}_1 + \rho(\cos \alpha\eta)(\sin \eta)\bar{\imath}_2 + \rho(\sin \alpha\eta)\bar{\imath}_3$, where $\rho > 0$ and α are given parameters. *(1)* Find the tangent, normal, and binormal vectors for this curve. *(2)* Determine the curvature, radius of curvature, and twist of the curve.

Problem 2.4. Short questions

(1) A particle of mass m is sliding along a planar curve. Find the component of the particle's acceleration vector along the binormal vector of Frenet's triad. *(2)* A particle of mass m is sliding along a three-dimensional curve. Find the component of the particle's acceleration vector along the binormal vector of Frenet's triad. *(3)* State the criterion used to ascertain whether a curve is planar or three-dimensional.

Problem 2.5. Study of a curve defined in polar coordinates

The outer surface of a cam is specified by the following curve defined in polar coordinates, $r(\alpha) = 1.0 - 0.5 \cos \alpha + 0.18 \cos 2\alpha$. *(1)* Plot the curve. *(2)* Plot the curvature distribution for $\alpha \in [0, 2\pi]$.

2.4 Differential geometry of a surface

This section investigates the differential geometry of surfaces, leading to the concept of surface coordinates. The differential geometry of surfaces is more complex than that of curves. The first and second metric tensors of surfaces are introduced first, and the analysis of the curvature of surfaces leads to the concept of lines of curvatures and associated principal radii of curvature. Finally, the base vectors and their derivatives are evaluated, leading to Gauss' and Weingarten's formulæ.

2.4.1 The first metric tensor of a surface

Figure 2.7 depicts a surface, denoted $\mathbb{S}$, in three-dimensional space. A surface is the locus of the points generated by two parameters, η_1 and η_2, such that the position vector, $\underline{p}_0$, of such points can be written as

$$\underline{p}_0 = \underline{p}_0(\eta_1, \eta_2). \tag{2.35}$$

If η_2 is kept constant, $\eta_2 = c_2$, $\underline{p}_0 = \underline{p}_0(\eta_1, c_2)$ defines a curve embedded into the surface; such curve is called an "η_1 curve." Figure 2.7 shows a grid of such curves for various values of c_2. Similarly, "η_2 curves" can be defined, corresponding to $\underline{p}_0 = \underline{p}_0(c_1, \eta_2)$; a grid of η_2 curves obtained for different constant c_1 is also shown on the figure. In general, parameters η_1 and η_2 do not measure length along these embedded curves, and hence, they do not define intrinsic parameterizations of the curves.

The *surface base vectors* are defined as follows

$$\underline{a}_1 = \frac{\partial \underline{p}_0}{\partial \eta_1}, \quad \underline{a}_2 = \frac{\partial \underline{p}_0}{\partial \eta_2}, \quad (2.36)$$

and are shown in fig. 2.7. Clearly, vectors $\underline{a}_1$ and $\underline{a}_2$ are tangent to the η_1 and η_2 curves that intersect at point **P**, respectively.

Consequently, they lie in the plane tangent to the surface at this point. Since η_1 and η_2 do not form an intrinsic parameterization, vectors $\underline{a}_1$ and $\underline{a}_2$ are not unit tangent vectors. Furthermore, these two vectors are not, in general, orthogonal to each other.

The *first metric tensor of the surface*, $\underline{\underline{A}}$, is defined as

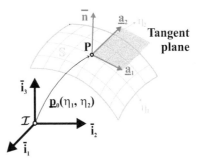

Fig. 2.7. The base vectors of a surface.

$$\underline{\underline{A}} = \begin{bmatrix} \underline{a}_1^T \underline{a}_1 & \underline{a}_1^T \underline{a}_2 \\ \underline{a}_2^T \underline{a}_1 & \underline{a}_2^T \underline{a}_2 \end{bmatrix} = \begin{bmatrix} a_{11} & a_{12} \\ a_{12} & a_{22} \end{bmatrix}, \quad (2.37)$$

and its determinant is denoted $a = \det(\underline{\underline{A}})$. A differential element of length on the surface is found as

$$\mathrm{d}s^2 = \mathrm{d}\underline{p}_0^T \mathrm{d}\underline{p}_0 = (\underline{a}_1^T \mathrm{d}\eta_1 + \underline{a}_2^T \mathrm{d}\eta_2)(\underline{a}_1 \mathrm{d}\eta_1 + \underline{a}_2 \mathrm{d}\eta_2) = \mathrm{d}\underline{\eta}^T \underline{\underline{A}} \, \mathrm{d}\underline{\eta}. \quad (2.38)$$

where $\mathrm{d}\underline{\eta}^T = \{\mathrm{d}\eta_1, \mathrm{d}\eta_2\}$. Clearly, the first metric tensor is closely related to length measurements on the surface.

Because the base vectors define the plane tangent to the surface, the unit vector, $\bar{n}$, normal to the surface is readily found as

$$\bar{n} = \frac{\widetilde{a}_1 \underline{a}_2}{\|\widetilde{a}_1 \underline{a}_2\|} = \frac{\widetilde{a}_1 \underline{a}_2}{\sqrt{a}}. \quad (2.39)$$

The area of a differential element of the surface then becomes

$$\mathrm{d}a = \|\widetilde{a}_1 \underline{a}_2 \, \mathrm{d}\eta_1 \mathrm{d}\eta_2\| = \|\widetilde{a}_1 \underline{a}_2\| \, \mathrm{d}\eta_1 \mathrm{d}\eta_2 = \sqrt{a} \, \mathrm{d}\eta_1 \mathrm{d}\eta_2. \quad (2.40)$$

2.4.2 Curve on a surface

Figure 2.8 depicts a curve, $\mathbb{C}$, entirely contained within surface $\mathbb{S}$. Let the curve be defined by its intrinsic parameter, s, the curvilinear variable along curve $\mathbb{C}$. The tangent vector, $\bar{t}$, to curve $\mathbb{C}$ is defined by eq. (2.5). This unit tangent vector clearly lies in the plane tangent to $\mathbb{S}$, and hence, it can be resolved along the base vectors, $\bar{t} = \lambda_1 \underline{a}_1 + \lambda_2 \underline{a}_2$.

Because $\bar{t}$ is a unit vector, it follows that

$$\vec{t}^T \vec{t} = \underline{\lambda}^T \underline{\underline{A}} \underline{\lambda} = 1, \qquad (2.41)$$

where $\underline{\lambda}^T = \{\lambda_1, \lambda_2\}$. On the other hand, eq. (2.38) can be recast as

$$\frac{d\underline{\eta}^T}{ds} \underline{\underline{A}} \frac{d\underline{\eta}}{ds} = 1. \qquad (2.42)$$

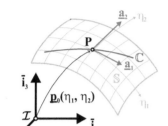

Because eqs. (2.41) and (2.42) must be identical for all curves on the surface,

$$\underline{\lambda} = \frac{d\underline{\eta}}{ds}. \qquad (2.43)$$

This result is expected since ds is an increment of length along $\mathbb{C}$, and $\vec{t}$ is tangent to $\mathbb{C}$. Angles $\theta_1 = (\vec{t}, \underline{a}_1)$ and $\theta_2 = (\vec{t}, \underline{a}_2)$ can be obtained by expanding the dot products $\vec{t}^T \underline{a}_1$ and $\vec{t}^T \underline{a}_2$, respectively, to find

Fig. 2.8. A curve, $\mathbb{C}$, entirely contained within surface, $\mathbb{S}$

$$\left\{ \begin{matrix} \sqrt{a_{11}} \cos \theta_1 \\ \sqrt{a_{22}} \cos \theta_2 \end{matrix} \right\} = \underline{\underline{A}} \underline{\lambda}. \qquad (2.44)$$

2.4.3 The second metric tensor of a surface

Consider once again a curve, $\mathbb{C}$, entirely contained within surface $\mathbb{S}$, as depicted in fig. 2.8. The unit tangent vector clearly lies in the plane tangent to the surface, but the curvature vector $d\vec{t}/ds$ will have components in and out of this tangent plane,

$$\frac{d\vec{t}}{ds} = \kappa_n \bar{n} + \kappa_g \bar{\rho}, \qquad (2.45)$$

where κ_n is the *normal curvature*, κ_g the *geodesic curvature*, and $\bar{\rho}$ a unit vector belonging to the plane tangent to $\mathbb{S}$. The normal curvature can be evaluated as

$$\kappa_n = \bar{n}^T \frac{d\vec{t}}{ds} = -\vec{t}^T \frac{d\bar{n}}{ds} = -\frac{d\underline{p}_0^T d\bar{n}}{ds^2}, \qquad (2.46)$$

where the normality condition, $\vec{t}^T \bar{n} = 0$, was used. The numerator can be written as

$$-d\underline{p}_0^T d\bar{n} = -\left(\underline{a}_1^T d\eta_1 + \underline{a}_2^T d\eta_2\right) \left(\frac{\partial \bar{n}}{\partial \eta_1} d\eta_1 + \frac{\partial \bar{n}}{\partial \eta_2} d\eta_2\right)$$

$$= -\left[\underline{a}_1^T \frac{\partial \bar{n}}{\partial \eta_1} d\eta_1^2 + \underline{a}_2^T \frac{\partial \bar{n}}{\partial \eta_2} d\eta_2^2 + \left(\underline{a}_1^T \frac{\partial \bar{n}}{\partial \eta_2} + \underline{a}_2^T \frac{\partial \bar{n}}{\partial \eta_1}\right) d\eta_1 d\eta_2\right],$$

$$= \left[\bar{n}^T \frac{\partial \underline{a}_1}{\partial \eta_1} d\eta_1^2 + \bar{n}^T \frac{\partial \underline{a}_2}{\partial \eta_2} d\eta_2^2 + \left(\bar{n}^T \frac{\partial \underline{a}_1}{\partial \eta_2} + \bar{n}^T \frac{\partial \underline{a}_2}{\partial \eta_1}\right) d\eta_1 d\eta_2\right],$$

where the orthogonality conditions, $\bar{n}^T \underline{a}_1 = 0$ and $\bar{n}^T \underline{a}_2 = 0$, were used to obtain the last equality.

The *second metric tensor of the surface* is defined as

$$
\underline{\underline{B}} = \begin{bmatrix} \bar{n}^T \dfrac{\partial \underline{a}_1}{\partial \eta_1} & \bar{n}^T \dfrac{\partial \underline{a}_1}{\partial \eta_2} \\[2mm] \bar{n}^T \dfrac{\partial \underline{a}_2}{\partial \eta_1} & \bar{n}^T \dfrac{\partial \underline{a}_2}{\partial \eta_2} \end{bmatrix} = \begin{bmatrix} \bar{n}^T \dfrac{\partial^2 \underline{p}_0}{\partial \eta_1^2} & \bar{n}^T \dfrac{\partial^2 \underline{p}_0}{\partial \eta_1 \partial \eta_2} \\[2mm] \bar{n}^T \dfrac{\partial^2 \underline{p}_0}{\partial \eta_1 \partial \eta_2} & \bar{n}^T \dfrac{\partial^2 \underline{p}_0}{\partial \eta_2^2} \end{bmatrix} = \begin{bmatrix} b_{11} & b_{12} \\ b_{12} & b_{22} \end{bmatrix},
\qquad (2.47)
$$

and its determinant is denoted $b = \det(\underline{\underline{B}})$. The second equality shows that the second metric tensor is a symmetric tensor. It follows that $-d\underline{p}_0^T \, d\bar{n} = d\underline{\eta}^T \underline{\underline{B}} d\underline{\eta}$, and the normal curvature, eq. (2.46), becomes

$$
\kappa_n = \frac{d\underline{\eta}^T \underline{\underline{B}} d\underline{\eta}}{ds^2} = \frac{d\underline{\eta}^T}{ds} \underline{\underline{B}} \frac{d\underline{\eta}}{ds} = \underline{\lambda}^T \underline{\underline{B}} \, \underline{\lambda}.
\qquad (2.48)
$$

2.4.4 Analysis of curvatures

Figure 2.9 shows a plane, $\mathcal{P}$, containing the normal, $\bar{n}$, to surface $\mathbb{S}$. Let curve $\mathbb{C}_n$ be at the intersection of plane $\mathcal{P}$ and surface $\mathbb{S}$. Because curve $\mathbb{C}_n$ is a planar curve, its curvature vector is in plane $\mathcal{P}$.

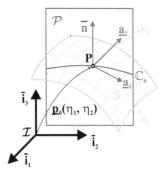

Fig. 2.9. Intersection of surface, $\mathbb{S}$, with plane, $\mathcal{P}$, that contains the normal to the surface.

Next, let plane $\mathcal{P}$ rotate about $\bar{n}$. For each new orientation of the plane, a new curve, $\mathbb{C}_n$, is generated with its own normal curvature κ_n. The following problem will be investigated: what is the orientation of plane $\mathcal{P}$ that maximizes the normal curvature κ_n? In mathematical terms, the maximum value of $\kappa_n = \underline{\lambda}^T \underline{\underline{B}} \, \underline{\lambda}$ is sought, under the normality constraint, $\underline{\lambda}^T \underline{\underline{A}} \, \underline{\lambda} = 1$.

This constrained maximization problem will be solved with the help of Lagrange's multiplier technique

$$
\max_{\underline{\lambda}, \mu} \left[\underline{\lambda}^T \underline{\underline{B}} \, \underline{\lambda} - \mu (\underline{\lambda}^T \underline{\underline{A}} \, \underline{\lambda} - 1) \right],
$$

where μ is the Lagrange multiplier used to enforce the constraint. The solution of this problem implies $(\underline{\underline{B}} - \mu \underline{\underline{A}}) \underline{\lambda} = 0$, and the normality condition $\underline{\lambda}^T \underline{\underline{A}} \, \underline{\lambda} = 1$. Pre-multiplying this equation by $\underline{\lambda}^T$ yields the physical interpretation of the Lagrange multiplier: $\underline{\lambda}^T \underline{\underline{B}} \, \underline{\lambda} - \mu \, \underline{\lambda}^T \underline{\underline{A}} \, \underline{\lambda} = 0$ or, in view of the normality constraint, $\mu = \underline{\lambda}^T \underline{\underline{B}} \, \underline{\lambda} = \kappa_n$, Hence, Lagrange's multiplier can be interpreted as the normal curvature itself.

The condition for maximum normal curvature can now be written as $(\underline{\underline{B}} - \kappa_n \underline{\underline{A}}) \underline{\lambda} = 0$. This set of homogeneous algebraic equations admits the trivial solution $\underline{\lambda} = 0$, but this solution violates the normality constraint. Non-trivial solutions correspond to the eigenpairs of the generalized eigenproblem $\underline{\underline{B}} \, \underline{\lambda} = \kappa_n \underline{\underline{A}} \, \underline{\lambda}$. Because $\underline{\underline{A}}$ and $\underline{\underline{B}}$ are symmetric and $\underline{\underline{A}}$ is positive-definite, the eigenvalues are always real, and mutually orthogonal eigenvectors can be constructed.

The eigenvalues are the solution of the quadratic equation $\det(\underline{\underline{B}} - \kappa_n \underline{\underline{A}}) = 0$, or

$$\kappa_n^2 - 2\kappa_m\kappa_n + \frac{b}{a} = 0, \tag{2.49}$$

where $\kappa_m = (a_{11}b_{22} + a_{22}b_{11} - 2a_{12}b_{12})/2a$. The solutions of this quadratic equation are called the *principal curvatures*

$$\kappa_n^I, \kappa_n^{II} = \kappa_m \pm \sqrt{\kappa_m^2 - b/a}. \tag{2.50}$$

The *mean curvature* is defined as

$$\kappa_m = \frac{\kappa_n^I + \kappa_n^{II}}{2} = \frac{a_{11}b_{22} + a_{22}b_{11} - 2a_{12}b_{12}}{2a}, \tag{2.51}$$

and the *Gaussian curvature* as

$$\kappa_n^I \kappa_n^{II} = \frac{b}{a}. \tag{2.52}$$

When $b/a > 0$, the principal curvatures have the same sign, corresponding to a convex shape; when $b/a < 0$, the principal curvatures are of opposite sign, corresponding to a saddle shape; finally, when $b/a = 0$, one of the principal curvatures is zero, the surface $\mathbb{S}$ has zero curvature in one of the principal curvature directions.

2.4.5 Lines of curvature

A *line of curvature* of a surface is defined as a curve whose tangent vector always points along the principal curvature directions of the surface. Consider now a set of coordinates, η_1 and η_2, such that $a_{12} = b_{12} = 0$. It follows that $a = a_{11}a_{22}$, $b = b_{11}b_{22}$ and $\kappa_m = (b_{11}/a_{11} + b_{22}/a_{22})/2$. The principal curvatures then simply become

$$\kappa_n^I = \frac{b_{11}}{a_{11}}, \quad \kappa_n^{II} = \frac{b_{22}}{a_{22}}. \tag{2.53}$$

On the other hand, in view of eq. (2.41), η_1 or η_2 curves are characterized by $\underline{\lambda}^T = \{1/\sqrt{a_{11}}, 0\}$ or $\underline{\lambda}^T = \{0, 1/\sqrt{a_{22}}\}$, respectively. Their normal curvature then follows from eq. (2.48) as $\kappa_n = b_{11}/a_{11}$ and $\kappa_n = b_{22}/a_{22}$, respectively. It is now clear that when $a_{12} = b_{12} = 0$, the η_1 and η_2 curves are indeed the lines of curvatures. It is customary to introduce the *principal radii of curvature*, R_1 and R_2, defined as

$$\kappa_n^I = \frac{b_{11}}{a_{11}} = \frac{1}{R_1}, \quad \kappa_n^{II} = \frac{b_{22}}{a_{22}} = \frac{1}{R_2}. \tag{2.54}$$

2.4.6 Derivatives of the base vectors

At this point, the discussion will focus exclusively on surface parameterizations defining lines of curvatures. In this case, vectors $\underline{a}_1$, $\underline{a}_2$ and $\bar{n}$ form a set of mutually orthogonal vectors, although the first two are not necessarily unit vectors. An orthonormal triad can be constructed as follows

$$\bar{e}_1 = \frac{a_1}{\|a_1\|}, \quad \bar{e}_2 = \frac{a_2}{\|a_2\|}, \quad \bar{e}_3 = \bar{n}. \tag{2.55}$$

To interpret the meaning of these unit vectors, the chain rule for derivatives is used to write

$$a_1 = \frac{\partial \underline{p}_0}{\partial \eta_1} = \frac{\partial \underline{p}_0}{\partial s_1} \frac{ds_1}{d\eta_1} = \frac{ds_1}{d\eta_1} \bar{e}_1,$$

where s_1 is the arc length measured along the η_1 curve. Because $\partial \underline{p}_0 / \partial s_1 = \bar{e}_1$ is the unit tangent vector to the η_1 curve, see eq. (2.5), it follows that

$$\|a_1\| = h_1 = \frac{ds_1}{d\eta_1}, \quad \|a_2\| = h_2 = \frac{ds_2}{d\eta_2}. \tag{2.56}$$

Notation $h_1 = \|a_1\|$ was introduced to simplify the writing. Clearly, h_1 is a *scale factor*, the ratio of the infinitesimal increment in length, ds_1, to the infinitesimal increment in parameter η_1, $d\eta_1$, along the curve.

It is interesting to compute the derivatives of the base vectors. To that effect, the following expression is considered

$$\frac{\partial^2 \underline{p}_0}{\partial \eta_1 \partial \eta_2} = \frac{\partial a_1}{\partial \eta_2} = \frac{\partial a_2}{\partial \eta_1} = \frac{\partial (h_1 \bar{e}_1)}{\partial \eta_2} = \frac{\partial (h_2 \bar{e}_2)}{\partial \eta_1}.$$

Expanding the derivatives leads to

$$\frac{\partial h_1}{\partial \eta_2} \bar{e}_1 + h_1 \frac{\partial \bar{e}_1}{\partial \eta_2} = \frac{\partial h_2}{\partial \eta_1} \bar{e}_2 + h_2 \frac{\partial \bar{e}_2}{\partial \eta_1}. \tag{2.57}$$

Pre-multiplying this relationship by $\bar{e}_1^T$ yields the following identity

$$\bar{e}_1^T \frac{\partial \bar{e}_2}{\partial \eta_1} = \frac{1}{h_2} \frac{\partial h_1}{\partial \eta_2}.$$

To obtain this result, the orthogonality of the base vectors, $\bar{e}_1^T \bar{e}_2 = 0$, was used; furthermore, $\bar{e}_1^T \partial \bar{e}_1 / \partial \eta_2 = 0$, since $\bar{e}_1$ is a unit vector. In terms of intrinsic parameterization, this expression becomes

$$\bar{e}_1^T \frac{\partial \bar{e}_2}{\partial s_1} = -\bar{e}_2^T \frac{\partial \bar{e}_1}{\partial s_1} = \frac{1}{h_1} \frac{\partial h_1}{\partial s_2} = \frac{1}{T_1}, \tag{2.58}$$

where T_1 is the first *radius of twist* of the surface.

Next, eq. (2.57) is pre-multiplied $\bar{e}_2^T$ to yield

$$\bar{e}_2^T \frac{\partial \bar{e}_1}{\partial s_2} = -\bar{e}_1^T \frac{\partial \bar{e}_2}{\partial s_2} = \frac{1}{h_2} \frac{\partial h_2}{\partial s_1} = \frac{1}{T_2}, \tag{2.59}$$

where T_2 is the second *radius of twist* of the surface. Since the parameterization defines lines of curvatures, $b_{12} = 0$, and eq. (2.47) then implies

$$\bar{e}_2^T \frac{\partial \bar{n}}{\partial s_1} = \bar{n}^T \frac{\partial \bar{e}_2}{\partial s_1} = 0, \quad \bar{e}_1^T \frac{\partial \bar{n}}{\partial s_2} = \bar{n}^T \frac{\partial \bar{e}_1}{\partial s_2} = 0.$$

The definitions of the diagonal terms, b_{11} and b_{22}, of the second metric tensor, eq. (2.47), lead to

$$\bar{e}_1^T \frac{\partial \bar{n}}{\partial s_1} = -\bar{n}^T \frac{\partial \bar{e}_1}{\partial s_1} = -\frac{1}{R_1}, \quad \bar{e}_2^T \frac{\partial \bar{n}}{\partial s_2} = -\bar{n}^T \frac{\partial \bar{e}_2}{\partial s_2} = -\frac{1}{R_2},$$

where the principal radii of curvature, R_1 and R_2, were defined in eq. (2.54).

The derivatives of the surface base vector $\bar{e}_1$ can be resolved in the following manner

$$\frac{\partial \bar{e}_1}{\partial s_1} = c_1 \underline{e}_1 + c_2 \underline{e}_2 + c_3 \bar{n}, \tag{2.60}$$

where the unknown coefficients c_1, c_2, and c_3 are readily found by pre-multiplying the above relationship by $\bar{e}_1^T$, $\bar{e}_2^T$, and $\bar{n}^T$ to find

$$\frac{\partial \bar{e}_1}{\partial s_1} = -\frac{1}{T_1} \bar{e}_2 + \frac{1}{R_1} \bar{n}. \tag{2.61}$$

A similar development leads to

$$\frac{\partial \bar{e}_1}{\partial s_2} = \frac{1}{T_2} \bar{e}_2. \tag{2.62}$$

The derivatives of the surface base vector $\bar{e}_2$ are found in a similar manner

$$\frac{\partial \bar{e}_2}{\partial s_1} = \frac{1}{T_1} \bar{e}_1, \quad \frac{\partial \bar{e}_2}{\partial s_2} = -\frac{1}{T_2} \underline{e}_1 + \frac{1}{R_2} \bar{n}. \tag{2.63}$$

These results are known as *Gauss' formulæ*.

Proceeding in a similar fashion, the derivatives of the normal vector are resolved in the following manner

$$\frac{\partial \bar{n}}{\partial s_1} = -\frac{1}{R_1} \bar{e}_1, \quad \frac{\partial \bar{n}}{\partial s_2} = -\frac{1}{R_2} \bar{e}_2. \tag{2.64}$$

These results are known as *Weingarten's formulæ*.

Gauss' and Weingarten's formulæ can be combined to yield the derivatives of the base vectors in a compact manner as

$$\frac{\partial}{\partial s_1} \begin{Bmatrix} \bar{e}_1 \\ \bar{e}_2 \\ \bar{n} \end{Bmatrix} = \begin{bmatrix} 0 & -1/T_1 & 1/R_1 \\ 1/T_1 & 0 & 0 \\ -1/R_1 & 0 & 0 \end{bmatrix} \begin{Bmatrix} \bar{e}_1 \\ \bar{e}_2 \\ \bar{n} \end{Bmatrix}, \tag{2.65a}$$

$$\frac{\partial}{\partial s_2} \begin{Bmatrix} \bar{e}_1 \\ \bar{e}_2 \\ \bar{n} \end{Bmatrix} = \begin{bmatrix} 0 & 1/T_2 & 0 \\ -1/T_2 & 0 & 1/R_2 \\ 0 & -1/R_2 & 0 \end{bmatrix} \begin{Bmatrix} \bar{e}_1 \\ \bar{e}_2 \\ \bar{n} \end{Bmatrix}. \tag{2.65b}$$

These equations should be compared to the derivatives of Frenet's triad, eq. (2.13).

Example 2.4. The spherical surface

The spherical surface in three-dimensional space depicted in fig. 2.10 is defined by following position vector $\underline{p}_0 = R\,(\sin\eta_1\cos\eta_2\,\bar{\imath}_1 + \sin\eta_1\sin\eta_2\,\bar{\imath}_2 + \cos\eta_1\,\bar{\imath}_3)$, where R is the radius of the sphere. The surface base vectors are readily evaluated as $\underline{a}_1 = \partial\underline{p}_0/\partial\eta_1 = R\,(\cos\eta_1\cos\eta_2\,\bar{\imath}_1 + \cos\eta_1\sin\eta_2\,\bar{\imath}_2 - \sin\eta_1\,\bar{\imath}_3)$, and $\underline{a}_2 = \partial\underline{p}_0/\partial\eta_2 = R\,(-\sin\eta_1\sin\eta_2\,\bar{\imath}_1 + \sin\eta_1\cos\eta_2\,\bar{\imath}_2)$.

The first metric tensor of the sphere now becomes

$$\underline{\underline{A}} = \begin{bmatrix} R^2 & 0 \\ 0 & R^2\sin^2\eta_1 \end{bmatrix}.$$

Clearly, $h_1 = R$, $h_2 = R\sin\eta_1$, and $\sqrt{a} = R^2\sin\eta_1$. The normal vector is then evaluated with the help of eq. (2.39), to find

$$\bar{n} = \frac{\tilde{a}_1\tilde{a}_2}{\|\tilde{a}_1\tilde{a}_2\|} = \sin\eta_1\cos\eta_2\bar{\imath}_1 + \sin\eta_1\sin\eta_2\bar{\imath}_2 + \cos\eta_1\bar{\imath}_3.$$

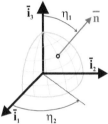

Fig. 2.10. Spherical surface configuration.   **Fig. 2.11.** Parabolic surface of revolution.

The second metric tensor of the spherical surface now follows from eq. (2.47)

$$\underline{\underline{B}} = \begin{bmatrix} -R & 0 \\ 0 & -R\sin^2\eta_1 \end{bmatrix}.$$

Note that since $a_{12} = 0$ and $b_{12} = 0$, the coordinates used here are lines of curvature for the spherical surface. The orthonormal triad to the surface is

$$\bar{e}_1 = \cos\eta_1\cos\eta_2\,\bar{\imath}_1 + \cos\eta_1\sin\eta_2\,\bar{\imath}_2 - \sin\eta_1\bar{\imath}_3,$$
$$\bar{e}_2 = -\sin\eta_2\,\bar{\imath}_1 + \cos\eta_2\,\bar{\imath}_2,$$
$$\bar{n} = \sin\eta_1\cos\eta_2\,\bar{\imath}_1 + \sin\eta_1\sin\eta_2\,\bar{\imath}_2 + \cos\eta_1\,\bar{\imath}_3.$$

These expressions are readily inverted to find

$$\bar{\imath}_1 = \cos\eta_1\cos\eta_2\,\bar{e}_1 - \sin\eta_2\,\bar{e}_2 + \sin\eta_1\cos\eta_2\,\bar{n},$$
$$\bar{\imath}_2 = \cos\eta_1\sin\eta_2\,\bar{e}_1 + \cos\eta_2\,\bar{e}_2 + \sin\eta_1\sin\eta_2\,\bar{n},$$
$$\bar{\imath}_3 = -\sin\eta_1\,\bar{e}_1 + \cos\eta_1\,\bar{n}.$$

The mean curvature, eq. (2.51), and Gaussian curvature, eq. (2.52), are

$$\kappa_m = \frac{1}{2}\left(-\frac{R}{R^2} - \frac{R\sin^2\eta_1}{R^2\sin^2\eta_1}\right) = -\frac{1}{R}, \quad \kappa_n^I\,\kappa_n^{II} = \frac{R^2\sin^2\eta_1}{R^4\sin^2\eta_1} = \frac{1}{R^2}.$$

Finally, the principal curvatures, eq. (2.53), become

$$\kappa_n^I = -\frac{1}{R}, \quad \kappa_n^{II} = -\frac{1}{R}.$$

As expected, the principal radii of curvature $R_1 = R_2 = -R$ are equal to the radius of sphere. The twists of the surface now follow from eqs. (2.58) and (2.59)

$$\frac{1}{T_1} = \frac{1}{h_1 h_2}\frac{\partial h_1}{\partial \eta_2} = 0, \quad \frac{1}{T_2} = \frac{1}{h_1 h_2}\frac{\partial h_2}{\partial \eta_1} = \frac{\cos\eta_1}{R\sin\eta_1}. \tag{2.66}$$

2.4.7 Problems

Problem 2.6. The parabola of revolution
Figure 2.11 depicts a parabolic surface of revolution. It is defined by the following position vector $\underline{p}_0 = r\cos\phi\,\bar{\imath}_1 + r\sin\phi\,\bar{\imath}_2 + ar^2\bar{\imath}_3$, where $r \geq 0$ and $0 \leq \phi \leq 2\pi$. The following notation was used $\eta_1 = r$ and $\eta_2 = \phi$. (1) Find the first and second metric tensors of the surface. (2) Find the orthonormal triad $\bar{e}_1$, $\bar{e}_2$, and $\bar{n}$. (3) Find the mean curvature, the Gaussian curvature, and the principal radii of curvature of the surface. (4) Find the twists of the surface.

Problem 2.7. Jacobian of the transformation
Consider two parameterizations of a surface defined by coordinates (η_1, η_2) and $(\hat{\eta}_1, \hat{\eta}_2)$. Show that the base vectors in the two parameterizations are related as follows $\hat{\underline{a}}_1 = J_{11}\underline{a}_1 + J_{12}\underline{a}_2$ and $\hat{\underline{a}}_2 = J_{21}\underline{a}_1 + J_{22}\underline{a}_2$, where $\underline{\underline{J}}$ is the Jacobian of the coordinate transformation

$$\underline{\underline{J}} = \begin{bmatrix} J_{11} & J_{12} \\ J_{21} & J_{22} \end{bmatrix} = \begin{bmatrix} \dfrac{\partial\eta_1}{\partial\hat{\eta}_1} & \dfrac{\partial\eta_2}{\partial\hat{\eta}_1} \\ \dfrac{\partial\eta_1}{\partial\hat{\eta}_2} & \dfrac{\partial\eta_2}{\partial\hat{\eta}_2} \end{bmatrix}.$$

If $\underline{\underline{A}}$ and $\underline{\underline{B}}$ are the first and second metric tensors in coordinate system (η_1, η_2) and $\underline{\underline{\hat{A}}}$ and $\underline{\underline{\hat{B}}}$ the corresponding quantities in coordinate system $(\hat{\eta}_1, \hat{\eta}_2)$, show that $\underline{\underline{\hat{A}}} = \underline{\underline{J}}\,\underline{\underline{A}}\,\underline{\underline{J}}^T$ and $\underline{\underline{\hat{B}}} = \underline{\underline{J}}\,\underline{\underline{B}}\,\underline{\underline{J}}^T$.

Problem 2.8. Finding the line of curvature system
Using the notations defined in problem 2.7, let (η_1, η_2) be a known coordinate system and $(\hat{\eta}_1, \hat{\eta}_2)$ the unknown line of curvature system. Find the Jacobian of the coordinate transformation that will bring (η_1, η_2) to the desired line of curvature system $(\hat{\eta}_1, \hat{\eta}_2)$. Show that the principal radii of curvature are

$$\frac{1}{R_1} = \frac{b_{11} + \gamma(2b_{12} + \gamma b_{22})}{a_{11} + \gamma(2a_{12} + \gamma a_{22})}, \quad \frac{1}{R_2} = \frac{b_{11} + \alpha(2b_{12} + \alpha b_{22})}{a_{11} + \alpha(2a_{12} + \alpha a_{22})}.$$

Hint: write the Jacobian as

$$\underline{\underline{J}} = \begin{bmatrix} 1 & \gamma \\ \alpha & 1 \end{bmatrix},$$

and compute the coefficients α and γ so as to enforce $\hat{a}_{12} = \hat{b}_{12} = 0$. The solution of the problem is $\alpha = C_\alpha/[\Delta/(1+\alpha\gamma)]$ and $\gamma = -C_\gamma/[\Delta/(1+\alpha\gamma)]$ where $C_\alpha = a_{22}b_{12} - b_{22}a_{12}$, $C_\gamma = a_{11}b_{12} - b_{11}a_{12}$, $\Delta = a_{11}b_{22} - b_{11}a_{22}$, and $\Delta/(1+\alpha\gamma) = \Delta/2 \pm \sqrt{(\Delta/2)^2 + C_\alpha C_\gamma}$.

2.5 Surface coordinates

A particle is moving on a surface and its position is given by the lines of curvature coordinates, $\eta_1(t)$ and $\eta_2(t)$. The velocity vector is computed with the help of the chain rule for derivatives

$$\underline{v} = \frac{d\underline{p}_0}{dt} = \frac{\partial \underline{p}_0}{\partial \eta_1}\dot{\eta}_1 + \frac{\partial \underline{p}_0}{\partial \eta_2}\dot{\eta}_2 = \dot{s}_1\bar{e}_1 + \dot{s}_2\bar{e}_2. \tag{2.67}$$

Note the close similarity between this expression and that obtained for path coordinates, eq. (2.33). The velocity vector is in the plane tangent to the surface, and the speed of the particle is $v = \sqrt{\dot{s}_1^2 + \dot{s}_2^2}$.

Next, the acceleration vector is computed as

$$\underline{a} = \ddot{s}_1\bar{e}_1 + \dot{s}_1\dot{\bar{e}}_1 + \ddot{s}_2\bar{e}_2 + \dot{s}_2\dot{\bar{e}}_2$$

$$= \ddot{s}_1\bar{e}_1 + \ddot{s}_2\bar{e}_2 + \dot{s}_1\left(\frac{\partial\bar{e}_1}{\partial s_1}\dot{s}_1 + \frac{\partial\bar{e}_1}{\partial s_2}\dot{s}_2\right) + \dot{s}_2\left(\frac{\partial\bar{e}_2}{\partial s_1}\dot{s}_1 + \frac{\partial\bar{e}_2}{\partial s_2}\dot{s}_2\right).$$

Introducing Gauss' formulae, eq. (2.61) to (2.63), then yields

$$\underline{a} = \left(\ddot{s}_1 + \frac{\dot{s}_1\dot{s}_2}{T_1} - \frac{\dot{s}_2^2}{T_2}\right)\bar{e}_1 + \left(\ddot{s}_2 + \frac{\dot{s}_1\dot{s}_2}{T_2} - \frac{\dot{s}_1^2}{T_1}\right)\bar{e}_2 + \left(\frac{\dot{s}_1^2}{R_1} + \frac{\dot{s}_2^2}{R_2}\right)\bar{n}. \tag{2.68}$$

Note here again the similarity between this expression and that obtained for path coordinates, eq. (2.34). The acceleration component along the normal to the surface is related to the principal radii of curvatures, R_1 and R_2. For a curve, the radius of curvature is always positive, see eq. (2.7), whereas for a surface, the radii of curvatures could be positive or negative, see eq. (2.54). Hence, the normal component of acceleration is not necessarily oriented along the normal to the surface.

The components of acceleration in the plane tangent to the surface are related to the second time derivative of the intrinsic parameters, as expected. Additional terms, however, associated with the surface radii of twist also appear. Clearly, the acceleration of a particle moving on the surface is affected by the surface radii of curvature and twist; the particle "feels" the curvatures and twists of the surface as it moves.

2.6 Differential geometry of a three-dimensional mapping

This section investigates the differential geometry of mappings of the three-dimensional space onto itself. The differential geometry of such mappings is more complex than that of curves or surfaces. For simplicity, the analysis focuses on orthogonal mappings, leading to the definition of the curvatures of the coordinate system and orthogonal curvilinear coordinates. Two orthogonal curvilinear coordinate systems of great practical importance, the cylindrical and spherical coordinate systems are reviewed.

2.6.1 Arbitrary parameterization

Consider the following mapping of the three-dimensional space onto itself in terms of three parameters, η_1, η_2, and η_3,

$$\underline{p}_0(\eta_1, \eta_2, \eta_3) = x_1(\eta_1, \eta_2, \eta_3)\bar{\imath}_1 + x_2(\eta_1, \eta_2, \eta_3)\bar{\imath}_2 + x_3(\eta_1, \eta_2, \eta_3)\bar{\imath}_3. \quad (2.69)$$

This relationship defines a mapping between the parameters and the Cartesian coordinates

$$x_1 = x_1(\eta_1, \eta_2, \eta_3), \quad x_2 = x_2(\eta_1, \eta_2, \eta_3), \quad x_3 = x_3(\eta_1, \eta_2, \eta_3). \quad (2.70)$$

Let η_2 and η_3 be constants whereas η_1 only is allowed to vary: a general curve in three-dimensional space is generated. The analysis of section 2.2 would readily apply to this curve, called an "η_1 curve." Similarly, η_2 and η_3 curves could be defined.

Next, let η_1 be a constant, whereas η_2 and η_3 are allowed to vary: a general surface in three-dimensional space is generated. The analysis of section 2.4 would readily apply to this surface, called an "η_1 surface." Here again, η_2 and η_3 surfaces could be similarly defined.

A point in space with parameters (η_1, η_2, η_3) is at the intersection of three η_1, η_2, and η_3 curves, or at the intersection of three η_1, η_2, and η_3 surfaces. Furthermore, an η_1 curve forms the intersection of η_2 and η_3 surfaces.

The inverse mapping defines the parameters as functions of the Cartesian coordinates

$$\eta_1 = \eta_1(x_1, x_2, x_3), \quad \eta_2 = \eta_2(x_1, x_2, x_3), \quad \eta_3 = \eta_3(x_1, x_2, x_3). \quad (2.71)$$

It is assumed here that eqs. (2.70) and (2.71) define a *one to one* mapping, which implies that the *Jacobian* of the transformation,

$$\underline{\underline{J}} = \begin{bmatrix} \dfrac{\partial x_1}{\partial \eta_1} & \dfrac{\partial x_1}{\partial \eta_2} & \dfrac{\partial x_1}{\partial \eta_3} \\[2mm] \dfrac{\partial x_2}{\partial \eta_1} & \dfrac{\partial x_2}{\partial \eta_2} & \dfrac{\partial x_2}{\partial \eta_3} \\[2mm] \dfrac{\partial x_3}{\partial \eta_1} & \dfrac{\partial x_3}{\partial \eta_2} & \dfrac{\partial x_3}{\partial \eta_3} \end{bmatrix}, \quad (2.72)$$

has a non vanishing determinant at all points in space. Next, the *base vectors* associated with the parameters are defined as

$$\underline{g}_1 = \frac{\partial \underline{p}_0}{\partial \eta_1}, \quad \underline{g}_2 = \frac{\partial \underline{p}_0}{\partial \eta_2}, \quad \underline{g}_3 = \frac{\partial \underline{p}_0}{\partial \eta_3}. \quad (2.73)$$

For an arbitrary parameterization, the base vectors will not be unit vectors, nor will they be mutually orthogonal.

Consider the example of the cylindrical coordinate system defined by the following parameterization

$$x_1 = r \cos \theta, \; x_2 = r \sin \theta, \; x_3 = z,$$

where $r \geq 0$ and $0 \leq \theta < 2\pi$. The following notation was used: $\eta_1 = r$, $\eta_2 = \theta$ and $\eta_3 = z$. The inverse mapping is readily found as

$$r = \sqrt{x_1^2 + x_2^2}, \; \theta = \tan^{-1} \frac{x_2}{x_1}, \; z = x_3.$$

Figure 2.12 depicts this mapping; clearly, the familiar polar coordinates are used in the $(\bar{\imath}_1, \bar{\imath}_2)$ plane and z is the distance point **P** is above this plane. The Jacobian of the transformation becomes

$$J = \begin{bmatrix} \cos \theta & -r \sin \theta & 0 \\ \sin \theta & r \cos \theta & 0 \\ 0 & 0 & 1 \end{bmatrix}.$$

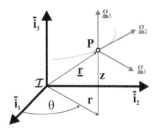

Note that $\det J = r$, and hence, vanishes at $r = 0$. Indeed, cylindrical coordinates are not defined at the origin since when $r = 0$, any angle θ maps to the same point, the origin.

Fig. 2.12. The cylindrical coordinate system.

The base vectors of this coordinate system are $g_1 = \cos \theta \, \bar{\imath}_1 + \sin \theta \, \bar{\imath}_2$, $g_2 = -r \sin \theta \, \bar{\imath}_1 + r \cos \theta \, \bar{\imath}_2$, and $g_3 = \bar{\imath}_3$. Note that g_1 is a unit vector, since $\|g_1\| = 1$, but g_2 is not, $\|g_2\| = r$. Also note that for cylindrical coordinates, $g_2^T g_3 = g_1^T g_3 = g_1^T g_2 = 0$, the base vectors are mutually orthogonal, as shown in fig. 2.12.

2.6.2 Orthogonal parameterization

When the base vectors associated with the parameterization are mutually orthogonal, the parameters define an *orthogonal parameterization* of the three-dimensional space. The rest of this section will be restricted to such parameterization. In this case, it is advantageous to define a set of orthonormal vectors

$$\bar{e}_1 = \frac{1}{\|g_1\|} g_1, \quad \bar{e}_2 = \frac{1}{\|g_2\|} g_2, \quad \bar{e}_3 = \frac{1}{\|g_3\|} g_3. \tag{2.74}$$

To interpret the meaning of these unit vectors, the chain rule for derivatives is used to write

$$g_1 = \frac{\partial p_0}{\partial \eta_1} = \frac{\partial p_0}{\partial s_1} \frac{ds_1}{d\eta_1} = \bar{e}_1 \frac{ds_1}{d\eta_1}, \tag{2.75}$$

where s_1 is the arc length measured along the η_1 curve. Because $\partial p_0/\partial s_1 = \bar{e}_1$ is the unit tangent to the η_1 curve, see eq. (2.5), it follows that

$$\|g_1\| = h_1 = \frac{ds_1}{d\eta_1}, \quad \|g_2\| = h_2 = \frac{ds_2}{d\eta_2}, \quad \|g_3\| = h_3 = \frac{ds_3}{d\eta_3}. \tag{2.76}$$

Notation $h_1 = \|g_1\|$ is introduced to simplify the notation. Clearly, h_1 is a *scale factor*, the ratio of the infinitesimal increment in length, ds_1, to the infinitesimal increment in parameter η_1, $d\eta_1$, along the curve.

2.6.3 Derivatives of the base vectors

Here again, the derivatives of the base vectors will be evaluated. To that effect, the following expression is considered

$$\frac{\partial^2 \underline{p}_0}{\partial \eta_1 \eta_2} = \frac{\partial \underline{g}_1}{\partial \eta_2} = \frac{\partial \underline{g}_2}{\partial \eta_1} = \frac{\partial(h_1 \bar{e}_1)}{\partial \eta_2} = \frac{\partial(h_2 \bar{e}_2)}{\partial \eta_1}. \tag{2.77}$$

Expanding the derivatives leads to

$$\frac{\partial h_1}{\partial \eta_2}\bar{e}_1 + h_1 \frac{\partial \bar{e}_1}{\partial \eta_2} = \frac{\partial h_2}{\partial \eta_1}\bar{e}_2 + h_2 \frac{\partial \bar{e}_2}{\partial \eta_1}. \tag{2.78}$$

Pre-multiplying this relationship by $\bar{e}_1^T$ yields the following identity

$$\bar{e}_1^T \frac{\partial \bar{e}_2}{\partial \eta_1} = \frac{1}{h_2}\frac{\partial h_1}{\partial \eta_2}. \tag{2.79}$$

To obtain this result, the orthogonality of the base vectors, $\bar{e}_1^T \bar{e}_2 = 0$, was used; furthermore, $\bar{e}_1^T \partial \bar{e}_1/\partial \eta_2 = 0$, since $\bar{e}_1$ is a unit vector. Next, eq. (2.78) is pre-multiplied $\bar{e}_2^T$ to yield

$$\bar{e}_2^T \frac{\partial \bar{e}_1}{\partial \eta_2} = -\bar{e}_1^T \frac{\partial \bar{e}_2}{\partial \eta_2} = \frac{1}{h_1}\frac{\partial h_2}{\partial \eta_1}. \tag{2.80}$$

Finally, pre-multiplication by $\bar{e}_3^T$ leads to

$$h_1 \bar{e}_3^T \frac{\partial \bar{e}_1}{\partial \eta_2} = h_2 \bar{e}_3^T \frac{\partial \bar{e}_2}{\partial \eta_1}. \tag{2.81}$$

Since $\bar{e}_3^T \partial \bar{e}_2/\partial \eta_1 = -\bar{e}_2^T \partial \bar{e}_3/\partial \eta_1$, this result can be manipulated as follows

$$h_1 \bar{e}_3^T \frac{\partial \bar{e}_1}{\partial \eta_2} = -h_2 \bar{e}_2^T \frac{\partial \bar{e}_3}{\partial \eta_1} = -\frac{h_1 h_2}{h_3} \bar{e}_2^T \frac{\partial \bar{e}_1}{\partial \eta_3}, \tag{2.82}$$

where identity (2.81) was used with a permutation of the indices. Using the same identities once again leads to

$$h_1 \bar{e}_3^T \frac{\partial \bar{e}_1}{\partial \eta_2} = \frac{h_1 h_2}{h_3} \bar{e}_3^T \frac{\partial \bar{e}_2}{\partial \eta_3} = h_1 \bar{e}_1^T \frac{\partial \bar{e}_3}{\partial \eta_2} = -h_1 \bar{e}_3^T \frac{\partial \bar{e}_1}{\partial \eta_2}.$$

This result clearly implies

$$\bar{e}_3^T \frac{\partial \bar{e}_1}{\partial \eta_2} = 0. \tag{2.83}$$

The derivatives of the base vector can be resolved as

$$\frac{\partial \bar{e}_1}{\partial \eta_1} = c_1 \bar{e}_1 + c_2 \bar{e}_2 + c_3 \bar{e}_3,$$

where the unknown coefficients c_1, c_2, and c_3 are found by pre-multiplying this expression by $\bar{e}_1$, $\bar{e}_2$, and $\bar{e}_3$, respectively, and using identities (2.79), (2.80) and (2.83) to find

$$\frac{\partial \bar{e}_1}{\partial \eta_1} = -\frac{1}{h_2}\frac{\partial h_1}{\partial \eta_2}\bar{e}_2 - \frac{1}{h_3}\frac{\partial h_1}{\partial \eta_3}\bar{e}_3.$$

Proceeding in a similar manner, the derivatives of base vector $\bar{e}_1$ with respect to η_2 and η_3 are found as

$$\frac{\partial \bar{e}_1}{\partial \eta_2} = \frac{1}{h_1}\frac{\partial h_2}{\partial \eta_1}\bar{e}_2, \qquad \frac{\partial \bar{e}_1}{\partial \eta_3} = \frac{1}{h_1}\frac{\partial h_3}{\partial \eta_1}\bar{e}_3.$$

Similar expression are readily found for the derivatives of the unit base vectors $\bar{e}_2$ and $\bar{e}_3$ through index permutations and are summarized as

$$\frac{\partial}{\partial s_1}\begin{Bmatrix}\bar{e}_1\\\bar{e}_2\\\bar{e}_3\end{Bmatrix} = \begin{bmatrix} 0 & 1/R_{13} & -1/R_{12}\\ -1/R_{13} & 0 & 0\\ 1/R_{12} & 0 & 0 \end{bmatrix}\begin{Bmatrix}\bar{e}_1\\\bar{e}_2\\\bar{e}_3\end{Bmatrix}, \qquad (2.84a)$$

$$\frac{\partial}{\partial s_2}\begin{Bmatrix}\bar{e}_1\\\bar{e}_2\\\bar{e}_3\end{Bmatrix} = \begin{bmatrix} 0 & 1/R_{23} & 0\\ -1/R_{23} & 0 & 1/R_{21}\\ 0 & -1/R_{21} & 0 \end{bmatrix}\begin{Bmatrix}\bar{e}_1\\\bar{e}_2\\\bar{e}_3\end{Bmatrix}, \qquad (2.84b)$$

$$\frac{\partial}{\partial s_3}\begin{Bmatrix}\bar{e}_1\\\bar{e}_2\\\bar{e}_3\end{Bmatrix} = \begin{bmatrix} 0 & 0 & -1/R_{32}\\ 0 & 0 & 1/R_{31}\\ 1/R_{32} & -1/R_{31} & 0 \end{bmatrix}\begin{Bmatrix}\bar{e}_1\\\bar{e}_2\\\bar{e}_3\end{Bmatrix}, \qquad (2.84c)$$

where the curvatures of the system were defined as

$$\frac{1}{R_{12}} = \frac{1}{h_1}\frac{\partial h_1}{\partial s_3}, \qquad \frac{1}{R_{13}} = -\frac{1}{h_1}\frac{\partial h_1}{\partial s_2}, \qquad (2.85a)$$

$$\frac{1}{R_{21}} = -\frac{1}{h_2}\frac{\partial h_2}{\partial s_3}, \qquad \frac{1}{R_{23}} = \frac{1}{h_2}\frac{\partial h_2}{\partial s_1}, \qquad (2.85b)$$

$$\frac{1}{R_{31}} = \frac{1}{h_3}\frac{\partial h_3}{\partial s_2}, \qquad \frac{1}{R_{32}} = -\frac{1}{h_3}\frac{\partial h_3}{\partial s_1}. \qquad (2.85c)$$

2.7 Orthogonal curvilinear coordinates

Consider a particle moving in three-dimension space. The position of this particle can be defined by eq. (2.69) in terms of an orthogonal parameterization of space. These parameter define a set of *orthogonal curvilinear coordinates* for the particle. The velocity vector is computed with the help of the chain rule for derivatives

$$\underline{v} = \frac{\mathrm{d}\underline{p}_0}{\mathrm{d}t} = \frac{\partial \underline{p}_0}{\partial s_1}\dot{s}_1 + \frac{\partial \underline{p}_0}{\partial s_2}\dot{s}_2 + \frac{\partial \underline{p}_0}{\partial s_3}\dot{s}_3 = \dot{s}_1\bar{e}_1 + \dot{s}_2\bar{e}_2 + \dot{s}_3\bar{e}_3. \qquad (2.86)$$

The expression for the acceleration vector will involve term in $\ddot{s}_1\bar{e}_1$ and $\dot{s}_1\dot{\bar{e}}_1$, and similar terms for the other two indices. The latter term is further expanded using the chain rule for derivatives, and expressing the derivatives of the base vectors using eqs. (2.84) then yields

$$\underline{a} = \left[\ddot{s}_1 - \dot{s}_2^2/R_{23} + \dot{s}_3^2/R_{32} - \dot{s}_1\dot{s}_2/R_{13} + \dot{s}_1\dot{s}_3/R_{12}\right]\bar{e}_1$$
$$+ \left[\ddot{s}_2 + \dot{s}_1^2/R_{13} - \dot{s}_3^2/R_{31} + \dot{s}_1\dot{s}_2/R_{23} - \dot{s}_2\dot{s}_3/R_{21}\right]\bar{e}_2 \qquad (2.87)$$
$$+ \left[\ddot{s}_3 - \dot{s}_1^2/R_{12} + \dot{s}_2^2/R_{21} - \dot{s}_1\dot{s}_3/R_{32} + \dot{s}_2\dot{s}_3/R_{31}\right]\bar{e}_3.$$

Note here again the similarity between this expression and that obtained for path or surface coordinates, eqs. (2.34) or (2.68), respectively. The acceleration components in each direction involve the second time derivative of the intrinsic parameters, as expected. Additional terms, however, associated with the radii of curvature of the curvilinear coordinate system also appear.

2.7.1 Cylindrical coordinates

The *cylindrical coordinate system*, depicted in fig. 2.13, is an orthogonal curvilinear coordinate system defined as follows

$$\underline{p}_0 = r\cos\theta\,\bar{\imath}_1 + r\sin\theta\,\bar{\imath}_2 + z\,\bar{\imath}_3, \qquad (2.88)$$

where $r \geq 0$ and $0 \leq \theta < 2\pi$. The following notation was used: $\eta_1 = r$, $\eta_2 = \theta$, and $\eta_3 = z$. Note that if $z = 0$, the cylindrical coordinate system reduces to coordinates r and θ in plane $(\bar{\imath}_1, \bar{\imath}_2)$ and are then often called polar coordinates.

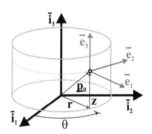
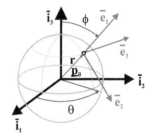

Fig. 2.13. The cylindrical coordinate system.   **Fig. 2.14.** The spherical coordinate system.

The following summarizes important formulæ in cylindrical coordinates. The scale factors are $h_1 = 1$, $h_2 = r$, and $h_3 = 1$. The curvatures of the cylindrical coordinate system all vanish, except that $R_{23} = r$. The base vectors expressed in terms of the Cartesian system are

$$\bar{e}_1 = \quad\cos\theta\,\bar{\imath}_1 + \sin\theta\,\bar{\imath}_2, \qquad (2.89a)$$
$$\bar{e}_2 = -\sin\theta\,\bar{\imath}_1 + \cos\theta\,\bar{\imath}_2, \qquad (2.89b)$$
$$\bar{e}_3 = \qquad\qquad\qquad \bar{\imath}_3. \qquad (2.89c)$$

The time derivatives of the based vectors resolved along this triad are

$$\dot{\bar{e}}_1 = \quad\dot{\theta}\,\bar{e}_2, \qquad (2.90a)$$
$$\dot{\bar{e}}_2 = -\dot{\theta}\,\bar{e}_1, \qquad (2.90b)$$
$$\dot{\bar{e}}_3 = \quad 0. \qquad (2.90c)$$

Finally, the position, velocity, and acceleration vectors, resolved along the base vectors of the cylindrical coordinate system are

$$\underline{p}_0 = r\,\bar{e}_1 + z\,\bar{e}_3, \tag{2.91a}$$

$$\underline{v} = \dot{r}\,\bar{e}_1 + r\dot{\theta}\,\bar{e}_2 + \dot{z}\,\bar{e}_3, \tag{2.91b}$$

$$\underline{a} = (\ddot{r} - r\dot{\theta}^2)\,\bar{e}_1 + (r\ddot{\theta} + 2\dot{r}\dot{\theta})\,\bar{e}_2 + \ddot{z}\,\bar{e}_3. \tag{2.91c}$$

respectively.

2.7.2 Spherical coordinates

The *spherical coordinate system*, depicted in fig. 2.14, is an orthogonal curvilinear coordinate system defined as follows

$$\underline{p}_0 = r\sin\phi\cos\theta\,\bar{\imath}_1 + r\sin\phi\sin\theta\,\bar{\imath}_2 + r\cos\phi\,\bar{\imath}_3, \tag{2.92}$$

where $r \geq 0, 0 \leq \phi \leq \pi$, and $0 \leq \theta < 2\pi$. The following notation was used: $\eta_1 = r$, $\eta_2 = \phi$, and $\eta_3 = \theta$.

The following summarizes important formulæ in spherical coordinates. The scale factors are $h_1 = 1$, $h_2 = r$, and $h_3 = r\sin\phi$. The curvatures of the spherical coordinate system all vanish, except that $R_{23} = r$, $R_{31} = r\tan\phi$ and $R_{32} = -r$.

The base vectors expressed in terms of the Cartesian system are

$$\bar{e}_1 = \sin\phi\cos\theta\,\bar{\imath}_1 + \sin\phi\sin\theta\,\bar{\imath}_2 + \cos\phi\,\bar{\imath}_3, \tag{2.93a}$$

$$\bar{e}_2 = \cos\phi\cos\theta\,\bar{\imath}_1 + \cos\phi\sin\theta\,\bar{\imath}_2 - \sin\phi\,\bar{\imath}_3, \tag{2.93b}$$

$$\bar{e}_3 = - \sin\theta\,\bar{\imath}_1 + \cos\theta\,\bar{\imath}_2. \tag{2.93c}$$

The time derivatives of the based vectors resolved along this triad are

$$\dot{\bar{e}}_1 = \dot{\phi}\,\bar{e}_2 + \dot{\theta}\sin\phi\,\bar{e}_3, \tag{2.94a}$$

$$\dot{\bar{e}}_2 = -\dot{\phi}\,\bar{e}_1 + \dot{\theta}\cos\phi\,\bar{e}_3, \tag{2.94b}$$

$$\dot{\bar{e}}_3 = -\dot{\theta}(\sin\phi\,\bar{e}_1 + \cos\phi\,\bar{e}_2). \tag{2.94c}$$

Finally, the position, velocity, and acceleration vectors, resolved along the base vectors of the spherical coordinate system are

$$\underline{p}_0 = r\,\bar{e}_1, \tag{2.95a}$$

$$\underline{v} = \dot{r}\,\bar{e}_1 + r\dot{\phi}\,\bar{e}_2 + r\dot{\theta}\sin\phi\,\bar{e}_3, \tag{2.95b}$$

$$\underline{a} = (\ddot{r} - r\dot{\phi}^2 - r\dot{\theta}^2\sin^2\phi)\,\bar{e}_1 + (r\ddot{\phi} + 2\dot{r}\dot{\phi} - r\dot{\theta}^2\sin\phi\cos\phi)\,\bar{e}_2$$
$$+ (r\ddot{\theta}\sin\phi + 2\dot{r}\dot{\theta}\sin\phi + 2r\dot{\phi}\dot{\theta}\cos\phi)\,\bar{e}_3. \tag{2.95c}$$

3

Basic principles

This chapter reviews the basic principles of dynamics. Newton's laws are the foundation of mechanics and dynamics and deal with the behavior of particles subjected to forces. Section 3.1 presents Newton's three laws and the principle of work and energy. Section 3.2 introduces the concept of conservative forces that play a fundamental role in dynamics. The principle of conservation of energy is discussed in section 3.2.1.

The potentials of common conservative forces are given in section 3.2.2, which also introduces the concept of strain energy for rectilinear and torsional springs. The principle of impulse and momentum is discussed in section 3.2.4. Section 3.3 presents basic facts about contact forces because they play an important role in dynamics.

Newton's law only apply to a single particle; section 3.4 introduces Euler's first and second laws, which are applicable to very general systems of particles.

3.1 Newtonian mechanics for a particle

Newton's laws deal with the motion of a *particle*, *i.e.*, a body of mass m that presents no physical dimension. This abstraction can be visualized by considering a body of mass m and finite dimensions. Next, the dimensions of the body are allowed to shrink, while the mass remains constant; at the limit, a particle of mass m is obtained that occupies a single point in space. As the particle moves, the locus of all positions it occupies in time describes a curve in three-dimensional space called the *path of the particle.*

3.1.1 Kinematics of a particle

The position vector of particle **P** with respect to an inertial frame will be denoted as $\underline{x}_{P/O}$, meaning "position vector of particle **P** with respect to point **O**," which is the origin of the inertial frame. Newton's laws assume the existence of an *inertial* *frame*, that is, a *frame that is stationary with respect to the distant stars.* In many practical applications, a frame attached to the earth may be used as an inertial frame. For instance, when studying the dynamics of a jet engine on a test bench, a frame of reference attached to the test bench is appropriate. If the same engine is mounted on an aircraft wing, a frame attached to the wing would not be inertial, because the aircraft is itself moving; for such a problem, a frame attached to the surface of the earth could be considered to be inertial. Finally, when studying the motion of satellites, it becomes necessary to select an inertial frame attached to the sun.

The *inertial velocity* vector or *absolute velocity* vector of the particle is the time derivative of its position vector with respect to the origin of the inertial frame

$$\underline{v} = \frac{d\underline{x}_{P/O}}{dt} = \dot{\underline{x}}_{P/O},\tag{3.1}$$

where t indicates time. More often than not, the term "velocity vector" will be used instead of "inertial velocity vector." The norm of the velocity vector is called the speed, v, of the particle

$$v = \|\underline{v}\|.\tag{3.2}$$

Finally, the particle *inertial acceleration* vector or *absolute acceleration* vector is defined as the derivative of the absolute velocity vector

$$\underline{a} = \frac{d\underline{v}}{dt} = \frac{d^2\underline{x}_{P/O}}{dt^2}.\tag{3.3}$$

3.1.2 Newton's laws

This section presents Newton's three laws and Newton's law of gravitation. These laws provide the foundation of dynamics and mechanics.

Newton's first law

Newton's first law of motion states that *every object in a state of uniform motion tends to remain in that state of motion unless an external force is applied to it.* The expression "state of uniform motion" means that the object moves at a constant velocity. If several forces are applied to the object, the "external force" is, in fact, the resultant, *i.e.*, the vector sum, of all externally applied forces. Finally, the "object" mentioned in the law is to be understood as a particle, as defined in the previous section.

With all these clarifications, Newton's first law can be restated: *a particle moves at a constant velocity unless the sum of the externally applied forces does not vanish.* This also implies that if the sum of the externally applied forces does not vanish, the particle no longer moves at a constant velocity. A more mathematical statement of Newton's first law is

Law 1 (Newton's first law) *A particle moves at a constant velocity if and only if the sum of the externally applied forces vanishes.*

The expression "if and only if" is included in the statement because the vanishing of the externally applied forces is both a necessary and sufficient condition for the particle to move at a constant velocity.

For statics problems, is is customary to focus on particles at rest rather than moving at a constant velocity. Within this framework, Newton's first law becomes: *a particle is at rest if and only if the sum of the externally applied forces vanishes.* This statement provides the definition of *static equilibrium* and is the foundation of statics and structural mechanics.

Newton's second law

Newton's second law states that *if a force is acting on a particle, its acceleration is proportional to this force; the constant of proportionality is the mass of the particle.* Here again, the force acting on the particle is the vector sum of all externally applied forces. Both externally applied force and resulting acceleration must be understood as vector quantities, and furthermore, the acceleration vector is the inertial acceleration vector as defined by eq. (3.3). Newton's second law then states

Law 2 (Newton's second law) *The inertial acceleration vector of a particle is proportional to the vector sum of the externally applied forces; the constant of proportionality is the mass of the particle.*

In mathematical terms, Newton's second law becomes

$$\underline{F} = m\underline{a}, \tag{3.4}$$

where $\underline{F}$ is the sum of the externally applied forces acting on the particle, $\underline{a}$ its inertial acceleration vector, and m its mass.

Clearly, the Newton's first law is implied by the second. Newton's second law provides the *equations of motion* for a particle; it relates the motion of the particle to the externally applied forces.

Newton's third law

Newton's third law is also of fundamental importance to dynamics. It states: *if particle **A** exerts a force on particle **B**, particle **B** simultaneously exerts on particle **A** a force of identical magnitude and opposite direction.* It is also postulated that *these two forces share a common line of action.* In a more compact manner, Newton's third law states that

Law 3 (Newton's third law) *Two interacting particles exert on each other forces of equal magnitude, opposite directions, and sharing a common line of action.*

Newton's third law is most useful when dealing with systems of particles: it enables the appropriate modeling of the interaction forces among the particles. It also allows "isolating" or "disconnecting" a particle from its surroundings and replacing the connection by a set of forces of equal magnitudes, opposite directions, and sharing a common line of action. This technique is the basis for drawing free body diagrams of a particle or system of particles.

Newton's law of gravitation

Newton's law of gravitation also plays an important role in dynamics. It states that

Law 4 (Newton's law of gravitation) *Two particles attract each other in proportion to their masses and in inverse proportion to the square of their relative distance. The line of action of this attractive force joins the two particles.*

This implies

$$F = G\frac{m_1 m_2}{r^2}, \tag{3.5}$$

where F is the magnitude of the attractive force, m_1 and m_2 the masses of the two particles, r their relative distance, and G the constant of proportionality know as the *universal constant of gravitation*.

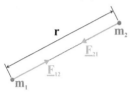

Fig. 3.1. Gravitation force acting between two particles.

Figure 3.1 shows the force, $\underline{F}_{12}$, that the second particle exerts on the first, and the force, $\underline{F}_{21}$, that the first exerts on the second. Forces $\underline{F}_{12}$ and $\underline{F}_{21}$ have the same magnitude $F = \|\underline{F}_{12}\| = \|\underline{F}_{21}\|$, opposite directions $\underline{F}_{12} + \underline{F}_{21} = 0$, and share a common line of action that joins the two particles. Clearly, these two forces present an important example of Newton's third law.

3.1.3 Systems of units

The quantities involved in Newton's three laws are length, mass, time, and force, denoted L, M, T, and F, respectively. In view of Newton's second law, eq. (3.4), these three quantities are not independent, rather $F = ML/T^2$.

This text uses the SI system of units exclusively. In this system of units, the three basic units are length, mass, and time, measured in meters, denoted "m," kilograms, denoted "kg," and seconds, denoted "s," respectively. Force is then a derived unit measured in Newtons, denoted "N." A force of 1 N imparts an acceleration of 1 m/s^2 to a mass of 1 kg. Systems of units where mass is a basic unit are said to be absolute: the SI system is an *absolute system of units*.

In this set of units, the universal constant of gravitation is

$$G = 6.6732 \; 10^{-11} \, \mathrm{m}^3/(\mathrm{kg} \cdot \mathrm{s}^2). \tag{3.6}$$

In view of the small value of this constant, the attractive force acting between objects of small masses is very small. The attractive force between particles, however, is large if one of the particles has a large mass.

The weight, w, of a particle at the surface of the earth is defined as the gravitational force applied by the earth to the particle,

$$w = \frac{GM}{r_e^2} m, \tag{3.7}$$

where $M = 5.976\ 10^{24}$ kg is the mass of the earth, $r_e = 6,378$ km its radius, and m the mass of a particle located at the surface of the earth. Using these constants, it follows that the weight of a particle is $w = 9.803\ m = gm$, where $g = 9.803$ m/s² is the *gravitational constant at the surface of the earth*.

Because the earth is not a perfect sphere and its mass distribution is not uniform, small variations of the gravitational constant should be expected from point to point. For most dynamics problems, $g = 9.81$ m/s² will be a sufficiently accurate value of the gravitational constant. At the surface of the earth, the weight of an 80 kg person is $w = 9.81 \times 80 = 785$ N.

In the US customary system of units, the three basic unit are length, time, and force, measured in feet, denoted "ft," seconds, denoted "s," and pounds, denoted "lbs," respectively. In this system, mass is then a derived unit measured in slugs, denoted "slug." A mass of 1 slug weighs 1 lb when subjected to a gravitational acceleration of 1 ft/s². Systems of units where force is a basic unit are said to be *gravitational*: the US customary system is a gravitational system. In the US customary system, $g = 32.17$ ft/s², and the mass of a particle at the surface of the earth is then found as $m = w/g$. It should be noted that in the US customary system, length is sometimes measured in inches rather than feet; in this case, $g = 386$ in/s².

3.1.4 The principle of work and energy

Figure 3.2 depicts a particle of mass m whose position is described by position vector $\underline{r}(t)$ with respect to an inertial frame, $\mathcal{F}^I = [\mathbf{O}, \mathcal{I} = (\bar{\imath}_1, \bar{\imath}_2, \bar{\imath}_3)]$. While moving along its path, the particle is acted upon by forces, the resultant of which is $\underline{F}(t)$. These forces are called *externally applied forces*, or *impressed forces*.

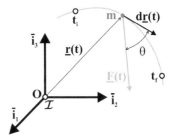

The *differential work*, $\mathrm{d}W$, the resultant force performs on the particle as it moves by an differential distance, $\mathrm{d}\underline{r}$, is defined as the scalar

Fig. 3.2. Force acting on a particle.

product of the force vector by the differential displacement vector of its point of application

$$\mathrm{d}W = \underline{F}^T \mathrm{d}\underline{r}. \tag{3.8}$$

In view of the definition of the scalar product, this differential work can be written as $\mathrm{d}W = \|\underline{F}\|\|\mathrm{d}\underline{r}\| \cos\theta$, where θ is the angle between the force and the differential displacement vectors, see fig. 3.2. If the force is normal to the differential displacement, the differential work vanishes, although the force is of finite magnitude. The notation $\mathrm{d}W$ is used to indicate the differential work, but it does not imply the existence of a work function, W, such that $\mathrm{d}(W)$ is the differential work.

Introducing Newton's second law, eq. (3.4), into the definition of the differential work leads to

$$\mathrm{d}W = \underline{F}^T \mathrm{d}\underline{r} = m\underline{a}^T \mathrm{d}\underline{r} = m\frac{\mathrm{d}\underline{v}^T}{\mathrm{d}t}\frac{\mathrm{d}\underline{r}}{\mathrm{d}t}\,\mathrm{d}t = m\frac{\mathrm{d}\underline{v}^T}{\mathrm{d}t}\underline{v}\,\mathrm{d}t = m\,\underline{v}^T\mathrm{d}\underline{v}. \tag{3.9}$$

The *kinetic energy*, K, of the particle is defined as

$$K = \frac{1}{2} m \underline{v}^T \underline{v}. \tag{3.10}$$

The differential change in kinetic energy is $\mathrm{d}(K) = m\,\underline{v}^T\,\mathrm{d}\underline{v}$, and it follows that

$$\mathrm{d}W = \mathrm{d}(K). \tag{3.11}$$

Consider now two arbitrary instants during the motion of the particle, say times t_i and t_f, as illustrated in fig. 3.2. The work done by the force over this period is denoted $W_{t_i \to t_f}$ and can be evaluated as follows

$$W_{t_i \to t_f} = \int_{t_i}^{t_f} \underline{F}^T \mathrm{d}\underline{r} = \int_{t_i}^{t_f} \mathrm{d}(K) = K(t_f) - K(t_i) = K_f - K_i = \Delta K. \tag{3.12}$$

This result is known as the principle of work and energy.

Principle 1 (Principle of work and energy for a particle) *The work done by the external forces acting on a particle equals the change in the particle's kinetic energy.*

3.2 Conservative forces

Figure 3.2 depicts a particle of mass m whose position is described by position vector $\underline{r}(t)$ with respect to an inertial frame, $\mathcal{F}^I = [\mathbf{O}, \mathcal{I} = (\bar{\imath}_1, \bar{\imath}_2, \bar{\imath}_3)]$. Conservative forces are a class of forces that depend only upon the position of the particles on which they act, $\underline{F} = \underline{F}(\underline{r})$. Although these forces may vary with time as the particle moves, they do not depend explicitly on time or velocity. Figure 3.3 shows two arbitrary paths, denoted **ACB** and **ADB**, along which the particle moves in space from point **A** to point **B**.

Definition

By definition, force $\underline{F}$ is conservative if and only if the work it performs along any path joining the same initial and final points is identical. This is expressed by the following equation

$$W_{A \to B} = \int_{\text{Path ACB}} \underline{F}^T \mathrm{d}\underline{r} = \int_{\text{Path ADB}} \underline{F}^T \mathrm{d}\underline{r}. \tag{3.13}$$

Since reversing the limits of integration simply changes the sign of the integral, the work done by the force along path **ADB** is equal in magnitude and opposite in sign to that along path **BDA**. Equation (3.13) then implies the vanishing of the work done by the force over the closed path **ACBDA**. Because path **ACB** and **ADB** are arbitrary paths joining points **A** and **B**, it follows that a force is conservative if and only if the work it performs vanishes over any arbitrary closed path,

$$W = \oint_{\text{Any path}} \underline{F}^T \mathrm{d}\underline{r} = \oint_{\mathbb{C}} \underline{F}^T \mathrm{d}\underline{r} = 0, \tag{3.14}$$

where $\mathbb{C}$ is an arbitrary closed curve.

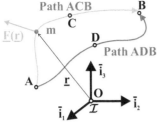

Fig. 3.3. Paths **ACB** and **ADB** join the same two points, **A** and **B**.

Fig. 3.4. Path enclosing a surface of area $\mathbb{S}$ with a normal $\bar{n}$.

Potential of a conservative force

Based on the definition of conservative forces, eq. (3.14), Stokes' theorem [2] then implies that

$$\oint_{\mathbb{C}} \underline{F}^T \mathrm{d}\underline{r} = \int_{\mathbb{S}} \bar{n}^T \widetilde{\underline{\nabla}} \underline{F} \, \mathrm{d}\mathbb{S} = 0, \tag{3.15}$$

where $\mathbb{S}$ is a surface bounded by curve $\mathbb{C}$, $\bar{n}$ the outward normal to surface $\mathbb{S}$, as shown in fig. 3.4, and $\widetilde{\underline{\nabla}} \underline{F} = \mathrm{curl}(\underline{F})$. If the force is conservative, the surface integral must vanish for any surface, $\mathbb{S}$, and this can only occur if the integrand vanishes, leading to $\widetilde{\underline{\nabla}} \underline{F} = 0$ for any curve, $\mathbb{C}$, and surface, $\mathbb{S}$. Textbooks on vector algebra [2], prove the following identity: $\widetilde{\underline{\nabla}} \underline{\nabla} V = 0$, where V is an arbitrary scalar function and $\underline{\nabla} V = \mathrm{grad}(V)$. It can then be shown that the solution of equation $\widetilde{\underline{\nabla}} \underline{F} = 0$ is simply

$$\underline{F} = -\underline{\nabla} V, \tag{3.16}$$

where $\underline{\nabla}$ is the gradient operator.

If a vector field, $\underline{F}$, can be derived from a scalar function, V, this function is called a *potential*, and the vector function is said to "be derived from a potential." Because the potential is an arbitrary scalar function, the minus sign is redundant, but is, however, a convention that will be justified later.

It has now been established that if a force is conservative, it can be "derived from a potential." In more mathematical terms, a conservative force must be the gradient a scalar function, called the *potential of the force*. If $\mathcal{I} = (\bar{\imath}_1, \bar{\imath}_2, \bar{\imath}_3)$ is an orthonormal basis, conservative forces can be expressed as

$$\underline{F} = -\underline{\nabla} V = -\frac{\partial V}{\partial x_1}\bar{\imath}_1 - \frac{\partial V}{\partial x_2}\bar{\imath}_2 - \frac{\partial V}{\partial x_3}\bar{\imath}_3. \tag{3.17}$$

The work done by a conservative force over an arbitrary path joining point **1** to point **2**, with position vectors $\underline{r}_1$ and $\underline{r}_2$, respectively, is then

$$W_{1\to2} = \int_{\underline{r}_1}^{\underline{r}_2} \underline{F}^T \mathrm{d}\underline{r} = -\int_{\underline{r}_1}^{\underline{r}_2} \underline{\nabla}^T V \mathrm{d}\underline{r} = -\int_{\underline{r}_1}^{\underline{r}_2}\left[\frac{\partial V}{\partial x_1}\mathrm{d}x_1 + \frac{\partial V}{\partial x_2}\mathrm{d}x_2 + \frac{\partial V}{\partial x_3}\mathrm{d}x_3\right]$$

$$= -\int_{\underline{r}_1}^{\underline{r}_2} \mathrm{d}V = V(\underline{r}_1) - V(\underline{r}_2).$$

Thus the work done by a conservative force *along any path* joining point **1** to point **2** depends only on the positions of these points and can be evaluated as the difference between the values of the potential function expressed at these two points,

$$W_{1\to2} = V(\underline{r}_1) - V(\underline{r}_2) = -\Delta V. \tag{3.18}$$

If point **1** and **2** are an infinitesimal distance apart,

$$dW = V(\underline{r}_1) - V(\underline{r}_1 + d\underline{r}) = -d(V). \tag{3.19}$$

The differential work is now the true derivative of the potential function.

Summary

Conservative forces enjoy a number of remarkable properties. Initially, conservative forces are defined as forces that perform the same work along any path joining the same initial and final points, as expressed by eq. (3.13). Simple calculus reasoning is then used to prove that a force is conservative if and only if the work it performs vanishes over any arbitrary closed path, see eq. (3.14). Finally, conservative forces are shown to be derivable from a potential, as expressed by eq. (3.16). Consequently, the work done by a conservative force *along any path* joining two points can be evaluated as the difference between the potential function evaluated at these two points, see eq. (3.18).

Examples of conservative forces

To illustrate these concepts, consider the gravity force acting on a particle of mass m located at the surface of the earth. It can easily be shown that this force is conservative. Therefore, the scalar potential, V, of the gravity forces is $V = mg\,\underline{r}\cdot\bar{\imath}_3 = mgx_3$, where $\underline{r} = x_1\bar{\imath}_1 + x_2\bar{\imath}_2 + x_3\bar{\imath}_3$ is the position vector of the particle. The gravity force, $\underline{F}_g$, acting on the particle can be obtained from this potential using eq. (3.17) to find $\underline{F}_g = -\nabla V = -\partial V/\partial x_3\,\bar{\imath}_3 = -mg\bar{\imath}_3$, and the gravity forces is said to be "derived from a potential."

The work done by the gravity force as the particle moves from elevation x_{3a} to x_{3b} then becomes $W = \int_{x_{3a}}^{x_{3b}} \underline{F}_g \cdot d\underline{r} = -\int_{x_{3a}}^{x_{3b}} \partial V/\partial x_3 \, dx_3 = V(x_{3a}) - V(x_{3b})$. Clearly, this work depends on the initial and final elevations only, but not on the particular path followed by the particle as it moved from the initial to the final elevation. If the particle moves along a closed path starting and ending at the same elevation, the work done by the gravity force vanishes.

As another example, consider the restoring force of an elastic spring of stiffness constant k. If the spring is stretched by an amount u, the restoring force is $-ku$, and can be derived from a potential of the form $V(u) = 1/2\,ku^2$. Indeed, using eq. (3.17), the elastic force in the spring becomes $F_s = -\partial V/\partial u = -ku$. This relationship is the constitutive law for the spring because it relates the force in the spring to its elongation.

Quantity $V(u)$ is called the *strain energy* and it can be viewed as a "potential of the elastic forces" in the spring. Hence, the strain energy function implicitly defines the constitutive behavior of the component. Finally, the work done by the elastic restoring force as the spring stretches from u_a to u_b is $W = \int_{u_a}^{u_b} F_s \, du = -\int_{u_a}^{u_b} \partial V/\partial u \, du = V(u_a) - V(u_b)$. Here again, the work depends only on the initial and final positions.

At first glance, the potential of a gravity force and the strain energy of an elastic spring seem to be distinct, unrelated concepts. Both quantities, however, share a common property: forces can be derived from these scalar potentials. Consider a particle of mass m connected to an elastic spring of stiffness constant k and subjected to a gravity force acting in the direction of the spring. The downward displacement, u, of the mass measures both the spring stretch and the elevation of the particle. The externally applied gravity force can be derived from the potential, $V = mgu$, as $F_g = -\partial V/\partial u = -mg$; the restoring force in the spring can be derived from the strain energy, $V = 1/2 \, ku^2$, which can also be viewed as the potential of the internal forces, as $F_s = -\partial V/\partial u = -ku$. The two forces acting on the particle can therefore be derived from a potential.

3.2.1 Principle of conservation of energy

The forces applied to a particle can be divided into two categories: the conservative forces, which can be derived from a potential, and the *non-conservative forces*, for which no potential function exists. The principle of work and energy, eq. (3.11), now becomes

$$dW = dW_c + dW_{nc} = -d(V) + d\underline{r}^T \underline{F}_{nc} = d(K), \tag{3.20}$$

where dW_c and dW_{nc} indicate the differential work done by the conservative and non-conservatives forces, respectively, and $\underline{F}_{nc}$ denotes the non-conservative forces. The work done by these forces over the period from time t_i to t_f now becomes

$$\int_{t_i}^{t_f} -d(V) + \int_{t_i}^{t_f} \underline{F}_{nc}^T d\underline{r} = K_f - K_i. \tag{3.21}$$

The first term of this expression readily integrates to yield

$$\int_{t_i}^{t_f} \underline{F}_{nc}^T d\underline{r} = (K_f + V_f) - (K_i + V_i), \tag{3.22}$$

where $V_i = V(t_i)$ and $V_f = V(t_f)$ are the values of the potential function at the initial and final times, respectively.

The *total mechanical energy*, E, is defined as the sum of the kinetic energy and potential function,

$$E = K + V. \tag{3.23}$$

The principle of work an energy principle now becomes

$$\int_{t_i}^{t_f} \underline{F}_{nc}^T d\underline{r} = E_f - E_i. \tag{3.24}$$

If the particle is acted upon by conservative forces only, the principle of work and energy reduces to

$$E_f = E_i. \tag{3.25}$$

This statement is known as the principle of conservation of energy.

Principle 2 (Principle of conservation of energy for a particle) *If a particle is subjected to conservative forces only, the total mechanical energy is preserved.*

Clearly, the term "conservative forces" stems from the fact that in the sole presence of such forces, the total mechanical energy of the particle is conserved.

In view of the principle of work and energy, work, kinetic energy, potential energy, and total mechanical energy all share the same units, force times distance, N·m. A *Joule* is defined as 1 J = 1 N·m. Although the moment of a force has the same units, N·m, Joules are used only when dealing with energy; in other words, a 10 N·m moment should not be referred to as a 10 J moment.

The work done by force over a period of time from t_i to t_f, see eq. (3.12), can be written as

$$W_{t_i \to t_f} = \int_{t_i}^{t_f} \underline{F}^T d\underline{r} = \int_{t_i}^{t_f} \underline{F}^T \frac{d\underline{r}}{dt} dt = \int_{t_i}^{t_f} \underline{F}^T \underline{v} \, dt. \tag{3.26}$$

The last integrand, $\underline{F}^T \underline{v}$, is the *power* of the externally applied forces; it is a measure of the work done by the forces per unit time. Power has units of work divided by time, J/s. A *Watt* is defined as 1 W = 1 J/s = 1 N·m/s.

3.2.2 Potential of common conservative forces

In the previous section, it was shown that conservative forces are associated with special functions called *potential functions*, from which they can be derived. A few commonly used potential functions will be derived in this section.

Work done by a central force

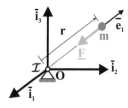

First, the work done by a *central force* will be evaluated. A central force is such that its line of action passes through a fixed inertial point in space and its magnitude depends on the sole distance r between the particle and the fixed point. Figure 3.5 shows a particle of mass m subjected to a central force $\underline{F}$ whose line of action passes through point **O**, the origin of an inertial frame.

Fig. 3.5. A central force.

Because distance r between the origin and the particle is inherent to the definition of the central force, it seems natural to use the spherical coordinate system defined in section 2.7.2 to express the position of the particle.

The velocity of the particle expressed in spherical coordinates, see eq. (2.95b), is $\underline{v} = \dot{r}\,\bar{e}_1 + r\dot{\phi}\,\bar{e}_2 + r\dot{\theta}\sin\phi\,\bar{e}_3$. Multiplying this relationship by dt reveals the relationship between an increment in particle position, $d\underline{r}$, and increments in coordinates dr, $d\theta$, and $d\phi$, as $d\underline{r} = dr\,\bar{e}_1 + rd\phi\,\bar{e}_2 + rd\theta\sin\phi\,\bar{e}_3$. On the other hand, the central force is expressed as $\underline{F} = -f(r)\bar{e}_1$, where $f(r)$ is its magnitude that depends on r only, and $\bar{e}_1$ its line of action, always passing through point **O**.

The differential work done by the central force now becomes $dW = \underline{F}^T d\underline{r} = -f(r)\bar{e}_1^T(dr\,\bar{e}_1 + rd\phi\,\bar{e}_2 + rd\theta\sin\phi\,\bar{e}_3) = -f(r)dr$. The potential, V, of the central force is defined as

$$f(r) = \frac{dV}{dr}. \tag{3.27}$$

With this definition, the differential work done by the central force becomes

$$dW = -\frac{dV}{dr}dr = -d(V).$$

Because the differential work can be expressed as an exact differential, the central force is a conservative force, and its potential is the integral of the magnitude of the central force. The potential is defined within a constant: adding a constant to the potential does not alter the magnitude of the central force.

The potential of gravity forces

An important example of central forces are gravitational forces, as described by Newton's gravitation law. The magnitude of the gravitational force is given by eq. (3.5) as $f(r) = GMm/r^2$. The gravitational force acts on an particle of mass m due to the presence of another particle of mass M assumed to be fixed with respect to an inertial frame; fig. 3.1 shows that such force is a central force. The potential function for the gravity forces then follows from eq. (3.27) as

$$V(r) = -G\frac{Mm}{r}. \tag{3.28}$$

This potential is called the *potential of gravity forces*.

Consider now a particle located at a height h above the surface of the earth; this implies $r = r_e + h$, where r_e is the radius of the earth. If the particle is close to the surface of the earth, $h \ll r_e$ and $1/r = 1/[r_e(1 + h/r_e)] \approx (1 - h/r_e)/r_e$. The potential function now becomes: $V(r) = -GMm/r_e + GMmh/r_e^2$. Because the first term of this expression is a constant, it can be omitted to yield the potential function as

$$V(r) = G\frac{Mmh}{r_e^2} = mgh. \tag{3.29}$$

This potential function is the potential of gravity forces for particles located near the surface of the earth. The height, h, of the particle is measured from a reference elevation, called the *datum*, which is selected in an arbitrary manner. Indeed, changing the datum is equivalent to adding a constant to the potential function, leaving the gravitation forces unchanged.

The strain energy function of an elastic spring

Consider now a particle of mass m connected to a rectilinear spring; the other end of the spring is attached to inertial point **O**, as depicted in fig. 3.6. The spring can stretch elastically, but is massless; it practice, this means that the mass of the spring is negligible with respect to that of the particle.

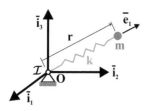

Clearly, the situation is similar to that shown in fig. 3.5: the particle is subjected to a central force $\underline{F}(r) = -f(r)\bar{e}_1$. The magnitude of the central force is related to the *stretch of the spring*, $\Delta = r - r_0$, where r_0 is the *un-stretched length of the spring*. For a *linearly elastic spring*, the force in the spring is proportional to its stretch, $f(r) = k(r - r_0) = k\Delta$, where k is the *spring stiffness constant*. The units of the spring stiffness constant are N/m.

Fig. 3.6. Particle connected to an elastic spring.

The potential function of the elastic forces in the spring then follows from eq. (3.27) as

$$V(r) = \frac{1}{2}k\,\Delta^2. \tag{3.30}$$

This work function is often called the *strain energy function* for the elastic spring.

The present formulation is not limited to linearly elastic springs: the magnitude of the elastic force in the spring could be a nonlinear function of the stretch, such as $f(r) = k_1\Delta + k_3\Delta^3$. In this case, the strain energy function of the nonlinearly elastic spring is $V = 1/2\,k_1\Delta^2 + 1/4\,k_3\Delta^4$.

The principle of work and energy affords a description of the kinetics of a particle in terms of energies rather than displacements and accelerations. Consider the system depicted in fig. 3.6, at time t_0, the particle is at rest and the spring is un-stretched: the velocity of the particle vanishes, implying $K_0 = 0$, and $V_0 = 0$, because the spring is un-stretched. External forces are applied to the particle that bring it to a new rest configuration at time t_1, hence $K_1 = 0$. Because the system is conservative, the work done by the external forces is $W_{0\rightarrow1}^{\text{ext}} = E_1 - E_0 = V_1$. For this simple case, the principle of work and energy implies that the work done by the externally applied force equals the strain energy in spring. This work is stored in the system in the form of strain energy: no energy has been lost, but its nature has changed from potential to strain energy.

In this description, the trajectory of the particle from time t_0 to time t_1 is irrelevant; the only important quantity is the stretch, Δ_1, of the spring at time t_1, which determines the strain energy, V_1. This is a characteristic of conservative forces: the work they perform does not depend on the particular path followed from time t_0 to t_1, but only on the initial and final configurations of the system that determine the initial and final stretch of the spring.

Next, the set of external forces that maintained the steady deformation Δ_1 of the spring is released; the particle evolves along a certain trajectory and at time t_2, the stretch of the spring vanishes, $\Delta_2 = 0$. Because no external forces are applied between time t_1 and t_2, the principle of work and energy implies $W_{1\to 2}^{\text{ext}} = 0 = E_2 - E_1 = K_2 - V_1$, where K_2 is the kinetic energy of the particle at time t_2. No energy has been lost: the energy transformed from strain to kinetic energy, $K_2 = V_1$. The speed v_2 of the particle at time t_2 is $v_2 = \sqrt{k/m}\ \Delta_1$. Here again, the specific trajectory followed by the particle is not relevant.

Both kinetic and strain energy functions are *positive-definite functions*, i.e., $K = 1/2\ mv^2 > 0$ for any arbitrary speed of the particle $v \neq 0$ and $V = 1/2\ k\Delta^2 > 0$ for any stretch of the elastic spring $\Delta \neq 0$. Consider a strain energy function of the form $V = 1/2\ k_0\Delta^2 + 1/3\ k_1\Delta^3$; this strain energy function vanishes for $\Delta_{\text{cr}} = -3/2\ k_0/k_1$. For stretches $\Delta < \Delta_{\text{cr}}$, the strain energy becomes negative, hence this strain energy function is invalid because it is not positive-definite. For $\Delta < \Delta_{\text{cr}}$, the spring will add energy to the system; energy is being created, a physical impossibility for a passive device.

The strain energy function of a torsional spring

Consider the planar problem depicted in fig. 3.7: a particle of mass m is connected to a rigid rod of length ℓ. The rod pivots about inertial point $\mathbf{O}$, where a torsional spring of stiffness constant k is located. The torsional spring applies a moment to the rigid rod about point $\mathbf{O}$, which is then transmitted to the particle in the form of a force $\underline{F}$, acting in the direction normal to the rod; this force is clearly not a central force. The position of the particle will be represented by polar coordinates, r and θ, see sec-

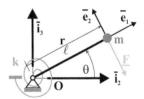

Fig. 3.7. Particle subjected to a force generated by a torsional spring.

tion 2.7.1. The velocity of the particle is $\underline{v} = \dot{r}\ \bar{e}_1 + r\dot{\theta}\ \bar{e}_2$, see eq. (2.91b). Because the rod is rigid, $\dot{r} = 0$, and multiplying the velocity relationship by dt implies $d\underline{r} = \ell d\theta\ \bar{e}_2$. The force vector, $\underline{F}$, has a line of action along $\bar{e}_2$ and its magnitude is a function of the sole angle θ: $\underline{F} = -f(\theta)\bar{e}_2$. The differential work done by this force now becomes

$$dW = \underline{F}^T d\underline{r} = -f(\theta)\bar{e}_2^T \ell d\theta\bar{e}_2 = -\ell f(\theta)d\theta. \tag{3.31}$$

Clearly, $M(\theta) = \ell f(\theta)$ is the moment the torsional spring applies to the rigid rod and hence, $dW = -M(\theta)d\theta$. For a linearly elastic torsional spring, $M(\theta) = k(\theta - \theta_0)$, where θ_0 is the angular position of the rigid rod for which the torsional spring is unstretched. The units for the stiffness constant k are N·m/rad. The potential function for the torsional spring now becomes

$$V(\theta) = \frac{1}{2}\ k(\theta - \theta_0)^2. \tag{3.32}$$

This potential function is called the strain energy function of the torsional spring. It is also possible to define nonlinearly elastic torsional springs, for which the elastic moment is a nonlinear function of angle θ; for instance, if $M(\theta) = k_1(\theta - \theta_0) + k_3(\theta - \theta_0)^3$, the strain energy function is then

$V(\theta) = k_1(\theta - \theta_0)^2/2 + k_3(\theta - \theta_0)^4/4.$

3.2.3 Non-conservative forces

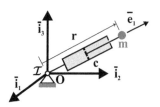

Fig. 3.8. Particle connected to a dashpot.

Consider now a particle of mass m connected to a rectilinear dashpot; the other end of the dashpot is attached to an inertial point **O**, as depicted in fig. 3.8. The dashpot can slide axially and it is massless; in practice, this means that its mass is negligible compared to that of the particle.

For a linear dashpot, the magnitude of viscous force it generates is proportional to the time rate of change of its length, *i.e.*, $f(\dot{r}) = c\dot{r}$. The coefficient c is called the *dashpot constant* and it units are N·s/m. Although the line of action of the force generated by the dashpot passes through an inertial point, it is not a central force because its magnitude does not depend on the sole distance between the particle and the inertial point.

If the position of the particle is expressed in terms of spherical coordinates, differential displacements are then $d\underline{r} = dr\,\bar{e}_1 + rd\phi\,\bar{e}_2 + rd\theta \sin\phi\,\bar{e}_3$. The differential work done by the dashpot force now becomes $dW = \underline{F}^T d\underline{r} = -f(\dot{r})\bar{e}_1^T(dr\,\bar{e}_1 + rd\phi\,\bar{e}_2 + rd\theta \sin\phi\,\bar{e}_3) = -c\dot{r}\,dr$. Because of the $\dot{r}$ dependency of the viscous force, the differential work cannot be cast in the form of an exact differential; there exist no potential function, $V(r)$, such that $dV/dr = -c\dot{r}dr$. The force in the dashpot is a *non-conservative force*.

The work done by the viscous forces in the dashpot is

$$W_{t_i \to t_f} = -\int_{t_i}^{t_f} c\dot{r}\,dr = -\int_{t_i}^{t_f} c\dot{r}\frac{dr}{dt}\,dt = -\int_{t_i}^{t_f} c\dot{r}^2\,dt < 0. \qquad (3.33)$$

The presence of the $\dot{r}^2$ term implies that the work done by the viscous forces is always negative, *i.e.*, they are *dissipative forces*. For the system depicted in fig. 3.8, the principle of work and energy implies that $W_{t_i \to t_f} = E_f - E_i$, or $E_f = E_i + W_{t_i \to t_f}$. Because the work is a negative quantity, the total mechanical energy of the system monotonically decreases in time; furthermore, the change in total mechanical energy exactly equals the work done by the viscous forces in the dashpot. This result explains the term "dissipative forces" or "non-conservative forces" used to qualify the viscous forces in the dashpot.

Of course, dashpots are not always linear; the magnitude of the viscous force could be a nonlinear function of velocity, such as $f(\dot{r}) = c_1\dot{r} + c_3\dot{r}^3$, for instance. Function $f(\dot{r})\dot{r}$, however, must be a positive-definite function of $\dot{r}$ to guarantee the dissipative nature of the resulting viscous force.

Finally, it is also possible to encounter *torsional dashpots*; in fig. 3.7, the torsional spring would be replaced by a dashpot that applies to the rigid bar a moment whose magnitude is a function of the time rate of change of angle θ. The differential work done by the viscous forces in the torsional is then $dW = -\ell f(\dot{\theta})d\theta = -M(\dot{\theta})d\theta$; for a linear torsional dashpot, $M(\dot{\theta}) = c\dot{\theta}$, where the dashpot constant now has units of N·m·s.

The energy closure equation

Consider the work and energy principle given by eq. (3.24), written as $\int_{t_i}^{t_f} \underline{F}_{nc}^T d\underline{r} = E_f - E_i$. In this expression, the initial and final time instants can be selected arbitrarily; in particular, let the final time be an arbitrary time, $t_f = t$, during the evolution of the system. The principle of work and energy now becomes

$$E(t) - \int_{t_i}^{t} \underline{F}_{nc}^T d\underline{r} = E_i. \tag{3.34}$$

In the absence of non-conservative forces, this equation reduces to $E(t) = E_i$, the statement of conservation of the total mechanical energy of the system. Even in the presence of non-conservative forces, however, equation (3.34) implies the conservation a scalar quantity, the difference between the total mechanical energy and the cumulative work done by the non-conservative forces, must remain constant. This relationship is known as the *energy closure equation*.

Example 3.1. bungee jumping
A man of mass m is jumping off a bridge while attached to a bungee cord of unstretched length d_0. An inertial frame, $\mathcal{F} = [\mathbf{O}, \mathcal{I} = (\bar{\imath}_1, \bar{\imath}_2)]$, is attached to the bridge. The man is jumping from point $\mathbf{O}$ with an initial velocity, v_0, oriented along horizontal axis $\bar{\imath}_2$, and the acceleration of gravity is acting along vertical axis $\bar{\imath}_1$.

During the first part of his fall, the man is in free flight under the effect of gravity, and at some instant in time, the bungee becomes taut. During the second portion of his fall, the man is subjected to the combined effects of gravity and the elastic force of the bungee. The potential of the bungee is of the following form: $V_b = 1/2\, k_0 d_0^2 \ln^2(1 + \bar{\Delta})$, where $\bar{\Delta} = (d - d_0)/d_0 = \Delta/d_0$ is the non-dimensional stretch of the bungee, and d the distance from point $\mathbf{O}$ to the man. The magnitude of the force the bungee applied to the man is $F_b = dV/d\Delta = k_0 d_0 \ln(1 + \bar{\Delta})/(1 + \bar{\Delta})$. Determine the trajectory of the fall.

Free fall

Let the man's trajectory be denoted $\underline{r}(t) = x_1(t)\bar{\imath}_1 + x_2(t)\bar{\imath}_2$. During free fall, Newton's second law writes $m\ddot{\underline{r}} = mg\bar{\imath}_1$, where g is the acceleration of gravity. Integration yields

$$\bar{\underline{v}} = \bar{g}\tau\bar{\imath}_1 + \bar{\imath}_2, \quad \bar{\underline{r}} = \frac{1}{2}\bar{g}\tau^2\bar{\imath}_1 + \tau\bar{\imath}_2. \tag{3.35}$$

The following non-dimensional quantities were introduced: $\bar{\underline{r}} = \underline{r}/d_0$, $\bar{\underline{v}} = \dot{\underline{r}}/v_0$, $\tau = v_0 t/d_0$, and $\bar{g} = g d_0/v_0^2$.

The bungee cord becomes taut when $\|\underline{r}\| = d_0$, or $\|\bar{\underline{r}}(\tau_t)\|^2 = 1$, where τ_t denotes the instant at which the bungee becomes taut. Introducing this condition in eq. (3.35) and solving for τ_t yields

$$\tau_t = \frac{\sqrt{2}}{\bar{g}}\sqrt{\sqrt{1 + \bar{g}^2} - 1}. \tag{3.36}$$

Trajectory when the bungee cord is taut

Once the bungee is taut, Newton's second law implies $m\ddot{\underline{r}} = mg\bar{\imath}_1 + F_b\bar{u}$, were F_b is the magnitude of the elastic force the bungee applies on the man and $\bar{u}$ the unit vector pointing from the man to point **O**. The distance from the man to point **O** is $d = d_0 + \Delta = \sqrt{x_1^2 + x_2^2}$, and $\bar{u} = (x_1 \bar{\imath}_1 + x_2 \bar{\imath}_2)/d$. In non-dimensional form, the equation of motion becomes

$$\bar{\underline{r}}'' = \bar{g}\bar{\imath}_1 - \bar{k}_0 \frac{\ln(1 + \bar{\Delta})}{(1 + \bar{\Delta})^2}(\bar{x}_1 \bar{\imath}_1 + \bar{x}_2 \bar{\imath}_2), \tag{3.37}$$

where $(\cdot)'$ indicates a derivative with respect to the non-dimensional time τ, $\bar{\Delta} = \sqrt{\bar{x}_1^2 + \bar{x}_2^2} - 1$, $\bar{k}_0 = k_0 d_0^2/(m v_0^2)$, $\bar{x}_1 = x_1/d_0$, and $\bar{x}_2 = x_2/d_0$.

Because the equations of motion are nonlinear, their solution can only be obtained by means of numerical methods, which often require recasting the governing equations in first-order form. In the present case, the first-order form of the equations is

$$\left\{\begin{array}{c} \bar{x}_1 \\ \bar{x}_2 \\ \bar{v}_1 \\ \bar{v}_2 \end{array}\right\}' = \left\{\begin{array}{c} \bar{v}_1 \\ \bar{v}_2 \\ \bar{g} - \bar{k}_0 \dfrac{\ln(1 + \bar{\Delta})}{(1 + \bar{\Delta})^2} \bar{x}_1 \\ - \bar{k}_0 \dfrac{\ln(1 + \bar{\Delta})}{(1 + \bar{\Delta})^2} \bar{x}_2 \end{array}\right\},$$

where $\bar{v}_1 = v_1/v_0$ and $\bar{v}_2 = v_2/v_0$ are the non-dimensional components of vertical and horizontal velocity, respectively. The first two equations, $\bar{x}_1' = \bar{v}_1$ and $\bar{x}_2' = \bar{v}_2$, simply define the velocity components, $\bar{v}_1$ and $\bar{v}_2$, and the last two equations are the actual equations of motion. In this form, many standard time integration methods such as Runge-Kutta integrators, among many others, can be used. Extensive discussion of these integrators can be found in many textbook on numerical analysis, see refs. [4, 5], for instance.

The following non-dimensional parameters are used for the simulation: $\bar{g} = 12$, and $\bar{k}_0 = 50$. The end of the free fall phase occurs at time $\tau_t = 0.4254$. Figure 3.9 show the man's trajectory during free fall and when the bungee cord is taut. For all times $\tau < \tau_t$ the bungee cord is slack and its stretch vanishes; fig. 3.10 shows the bungee's non-dimensional stretch, $\bar{\Delta}$, for $\tau \geq \tau_t$.

At time $\tau = 1.5018$, the bungee becomes slack again, and equation of motion, eq. (3.37), is no longer valid because it include the force stemming from the bungee cord. To continue the simulation past that time, the equation of motion for free fall under gravity, $m\ddot{\underline{r}} = mg\bar{\imath}_1$, would be used again, with initial conditions corresponding to the man's position and and velocity at the end of the previous phase, *i.e.*, at time $\tau = 1.5018$.

Figure 3.11 depicts the bungee non-dimensional force, $\bar{F}_b = F_b/(k_0 d_0)$, versus its non-dimensional stretch, $\bar{\Delta}$. The apparent stiffness, k, of the bungee cord is the tangent to the force-stretch curve,

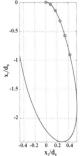
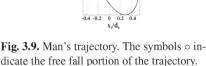

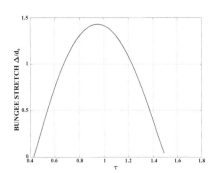

Fig. 3.9. Man's trajectory. The symbols ∘ indicate the free fall portion of the trajectory.

Fig. 3.10. Non-dimensional stretch of the bungee, $\bar{\Delta}$, versus time τ.

$$k = \frac{\mathrm{d}F_b}{\mathrm{d}\Delta} = k_0 \frac{1 - \ln(1 + \bar{\Delta})}{(1 + \bar{\Delta})^2}. \tag{3.38}$$

As the stretch of the cord increases, its stiffness decreases and vanishes when $\ln(1 + \bar{\Delta}) = 1$, or $\bar{\Delta} \approx 1.718$. Clearly, the parameters selected for the present simulation result in a very large stretching of the bungee cord, which would threaten the safety of the jumper.

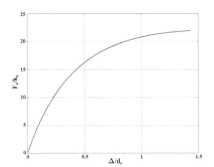

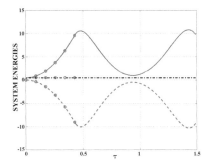

Fig. 3.12. System energies: kinetic energy, solid line; potential energy, dashed line; total mechanical energy, dashed-dotted line.

Fig. 3.11. Magnitude of the non-dimensional force, F_b, in the bungee versus stretch, $\bar{\Delta}$.

The kinetic energy of the system is $K = 1/2\, m\dot{r}^2$. The potential of the gravity forces is $V_g = -mgx_1$, and the potential of the elastic bungee cord $V_b = 1/2\, k_0 d_0^2\, \ln^2(1 + \bar{\Delta})$. In non-dimensional form, the total mechanical energy of the system becomes

$$\bar{E} = \frac{E}{mv_0^2} = \frac{K}{mv_0^2} + \frac{V}{mv_0^2} = \frac{1}{2}\bar{r}'^2 - \bar{g}\bar{x}_1 + \frac{1}{2}\bar{k}_0 \ln^2(1 + \bar{\Delta}).$$

Figure 3.12 depicts the evolution of the system's energies versus time τ. Because the forces acting on the system are conservative forces, the total mechanical energy remains constant during the simulation. This observation provides a validation of the derivation of the equation of motion and of its numerical solution.

Effect of drag forces.

The developments presented in the previous paragraphs have ignored the effect of air friction on the man's trajectory. These forces can be taken into account in an approximate manner applying to the man a drag force, $\underline{F}_d = -1/2\, C_d \rho \mathcal{A} \|\underline{v}\| \underline{v}$, where C_d is the non-dimensional drag coefficient, ρ the air density, and $\mathcal{A}$ the man's cross-sectional area. This drag force is at all times proportional to the square of the speed, aligned with the velocity vector, and oriented in the direction opposite to this vector. During free fall, the equation of motion is $m\ddot{\underline{r}} = mg\bar{\imath}_1 - 1/2\, C_d \rho \mathcal{A} \|\underline{v}\| \underline{v}$; as before, the bungee cord will become taut when $\|\underline{r}(\tau_t)\| = 1$. Because the governing differential equation is now a nonlinear differential equation, a numerical process must be used for its solution and time τ_t must be determined numerically. A closed form analytical solution such as that given by eq. (3.36) no longer exists.

When the bungee cord is taut, the differential equation governing the problem becomes

$$\underline{r}'' = \bar{g}\,\bar{\imath}_1 - \bar{k}_0 \frac{\ln(1+\bar{\Delta})}{(1+\bar{\Delta})^2}(\bar{x}_1\,\bar{\imath}_1 + \bar{x}_2\,\bar{\imath}_2) - \frac{1}{2}\bar{\mu} C_d \sqrt{\bar{v}_1^2 + \bar{v}_2^2}\,(\bar{v}_1\,\bar{\imath}_1 + \bar{v}_2\,\bar{\imath}_2),$$

where $\bar{\mu} = \rho \mathcal{A} d_0 / m$. Here again, the equation of motion is nonlinear, and its solution can be obtained only by means of numerical methods

3.2.4 The principle of impulse and momentum

The principle of impulse and momentum involves two sets of new quantities. First, the linear and angular momentum vectors of a particle are introduced; the angular momentum is the moment of the linear momentum vector. Next, the linear and angular impulse vectors of the externally applied forces are introduced.

Principle of linear impulse and momentum

Figure 3.13 shows a particle of mass m in motion with respect to an inertial frame $\mathcal{F}^I = [\mathbf{O}, \mathcal{I} = (\bar{\imath}_1, \bar{\imath}_2, \bar{\imath}_3)]$. The inertial velocity vector of the particle is denoted $\underline{v}$. The *linear momentum* vector of a particle is defined as the product of its mass by its inertial velocity vector

$$\underline{p} = m\underline{v}. \tag{3.39}$$

Taking a time derivative of the linear momentum vector yields $\dot{\underline{p}} = m\underline{a}$. Comparing this result with Newton's second law, eq. (3.4), leads to

$$\underline{F} = \dot{\underline{p}}. \tag{3.40}$$

This result implies that *the time derivative of the linear momentum vector of a particle equals the sum of the externally applied forces.* Clearly, this result is a direct corollary of Newton's second law.

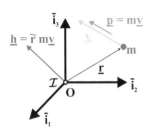

Fig. 3.13. Linear and angular momenta vectors of a particle.

It is interesting to integrate the above equation in time, between an initial and a final time, denoted t_i and t_f, respectively. These two instants are chosen arbitrarily, but $t_i < t_f$,

$$\int_{t_i}^{t_f} \underline{F}(t)\, dt = \int_{t_i}^{t_f} \underline{\dot{p}}\, dt = \underline{p}(t_f) - \underline{p}(t_i). \quad (3.41)$$

The term on the left-hand side is called the *linear impulse of the externally applied forces,* and has units of mass times velocity, or N·s. Equation (3.41) expresses the principle of linear impulse and momentum for a particle.

Principle 3 (Principle of linear impulse and momentum for a particle) *The linear impulse of the externally applied forces equals the change in linear momentum.*

In the absence of external forces, this principle implies $\underline{p}(t_f) = \underline{p}(t_i)$, *i.e.*, the linear momentum remains constant at all times, since t_i and t_f are instants chosen arbitrarily. In other words, the linear momentum vector of a particle remains a constant when the externally applied forces vanish.

Principle of angular impulse and momentum

Next, the moment of the particle's linear momentum vector is computed with respect to point **O**. This quantity if more often called the *angular momentum* vector of the particle, $\underline{h}_O$, where the subscript, $(\cdot)_O$ indicates that the angular momentum is computed with respect to point **O**. As illustrated in fig. 3.13, the moment of the linear momentum vector is expressed as the cross product of the particle's inertial position vector, $\underline{r}$, by its linear momentum vector, $m\underline{v}$, to find

$$\underline{h}_O = \tilde{r}\, m\underline{v}. \quad (3.42)$$

Taking a time derivative of the angular momentum vector yields $\underline{\dot{h}}_O = \dot{\tilde{r}} m\underline{v} + \tilde{r}m\underline{a}$. The time derivative of the inertial position vector, $\underline{\dot{r}}$, equals the inertial velocity vector, $\underline{v}$, eq. (3.1); it then follows that $\underline{\dot{h}}_O = \tilde{v}m\underline{v} + \tilde{r}m\underline{a}$. Finally, since $\tilde{v}m\underline{v} = \underline{0}$, the time derivative of the angular momentum vector reduces to $\underline{\dot{h}}_O = \tilde{r}m\underline{a}$

The moment of Newton's second law computed with respect to the origin of the inertial frame implies $\tilde{r}\underline{F} = \tilde{r}m\underline{a}$. Comparing these two results then leads to $\tilde{r}\underline{F} = \underline{\dot{h}}_O$, where the left-hand side term can be interpreted as the moment of the externally applied forces evaluated with respect to point **O**, denoted $\underline{M}_O$. In summary,

$$\underline{M}_O = \underline{\dot{h}}_O \quad (3.43)$$

This result implies that *the time derivative of the angular momentum vector of a particle computed with respect to an inertial point equals the sum of the externally applied moments computed with respect to the same point.* Here again, this result is a direct corollary of Newton's second law.

As in the case of the linear momentum, the above equation can be integrated in time between two arbitrary instants to yield

$$\int_{t_i}^{t_f} \underline{M}_O(t)\, dt = \int_{t_i}^{t_f} \underline{\dot{h}}_O\, dt = \underline{h}_O(t_f) - \underline{h}_O(t_i). \tag{3.44}$$

The term on the left-hand side is called the *angular impulse of the externally applied forces,* and has units of N·m·s. Equation (3.44) expresses the principle of angular impulse and momentum for a particle.

Principle 4 (Principle of angular impulse and momentum for a particle) *The angular impulse of the externally applied forces equals the change in angular momentum when both angular impulse and momentum are computed with respect to the same inertial point.*

In the absence of external moments, this principle implies $\underline{h}_O(t_f) = \underline{h}_O(t_i)$, *i.e.,* the angular momentum remains constant at all times. In other words, the angular momentum vector of a particle remains a constant when the externally applied moments vanish.

Example 3.2. Particle in a pinned tube
Figure 3.14 depicts a particle of mass m connected to inertial point **A** by means of a spring of stiffness k and dashpot of constant c. At the initial time, the particle is located at $\theta = 0$, $\phi = \pi/2$, and $r = r_0$, which corresponds to the un-stretched configuration of the spring; r, ϕ, and θ form a spherical coordinate system, see section 2.7.2. The initial velocity vector of the particle is $\underline{v}_0$. Derive the equations of motion of the system.

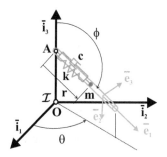

Fig. 3.14. Particle subjected to a central force due to a spring and dashpot.

First, Newton's second law is used to obtain the desired equations of motion as

$$m\left[(\ddot{r} - r\dot{\phi}^2 - r\dot{\theta}^2\sin^2\phi)\bar{e}_1 + (r\ddot{\phi} + 2\dot{r}\dot{\phi} - r\dot{\theta}^2\sin\phi\cos\phi)\bar{e}_2\right.$$
$$\left. + (r\ddot{\theta}\sin\phi + 2\dot{r}\dot{\theta}\sin\phi + 2r\dot{\phi}\dot{\theta}\cos\phi)\bar{e}_3\right] = -[k(r - r_0) + c\dot{r}]\,\bar{e}_1,$$

where the components of the acceleration vector in the spherical coordinate system are given by eq. (2.95c). Projecting this equation along the unit vectors $\bar{e}_1$, $\bar{e}_2$, and $\bar{e}_3$, then yields $m(\ddot{r} - r\dot{\phi}^2 - r\dot{\theta}^2\sin^2\phi) = -k(r - r_0) - c\dot{r}$, $r\ddot{\phi} + 2\dot{r}\dot{\phi} - r\dot{\theta}^2\sin\phi\cos\phi = 0$, and $r\ddot{\theta}\sin\phi + 2\dot{r}\dot{\theta}\sin\phi + 2r\dot{\phi}\dot{\theta}\cos\phi = 0$, respectively. These three nonlinear differential equations can be solved for the coordinates of the particle, r, ϕ, and θ. Although this approach will indeed yield the solution of the problem, much information about the nature of the particle's motion can be obtained from the application the principle of angular impulse and momentum.

Because the line of action of the forces applied to the particle passes through point $\mathbf{A}$, the moment of these forces with respect to point $\mathbf{A}$ vanishes. The principle of angular impulse and momentum, eq. (3.44), then implies that the angular momentum must remain constant, $\underline{H}_A = \underline{H}_{A0} = r_0\bar{\imath}_1\,m\underline{v}_0$.

It follows that $r\bar{e}_1\,m\underline{v} = \underline{H}_{A0}$. This vector product equation, see section 1.1.11, affords a solution if and only if the particle's position vector, $r\bar{e}_1$, and velocity vector, $\underline{v}$, are both contained in the plane normal to the initial angular momentum vector, $\underline{H}_{A0}$. Because the particle's position and velocity vectors are contained in the same plane, the particle's motion is contained entirely in the plane normal to the angular momentum vector.

This result is quite general: *if a particle is subjected to forces with a line of action passing through a fixed inertial point, its trajectory is contained in the plane normal to the initial angular momentum vector.* In particular, if a particle is subjected to a central force, its trajectory lies in the plane normal to the initial angular momentum vector. In the present example, the force associated with the elastic spring is a central force, whereas that associated with the dashpot is not.

The solution of the problem is now considerably simplified. Without loss generality, axis $\bar{\imath}_3$ is selected to be along $\underline{H}_{A0}$ and hence, $\phi = \pi/2$, $\dot{\phi} = 0$, $\underline{v}_0 = \dot{r}_0\bar{\imath}_1 + r_0\dot{\theta}_0\bar{\imath}_2$, and $\underline{H}_{A0} = mr_0^2\dot{\theta}_0\bar{\imath}_3$. The constancy of the angular momentum then implies $\underline{H}_A = mr^2\dot{\theta}\bar{\imath}_3 = mr_0^2\dot{\theta}_0\bar{\imath}_3$, or $r^2\dot{\theta} = r_0^2\dot{\theta}_0$. The first equation of motion now becomes $m(\ddot{r} - r\dot{\theta}^2) = -k(r - r_0) - c\dot{r}$, whereas the last two are identically satisfied. It is convenient to introduce the following parameters: $\Omega = \sqrt{k/m}$, the frequency of the spring mass system, $\zeta = cm/(2\Omega)$, the damping ratio of the dashpot and $\tau = \Omega t$, the non-dimensional time.

The equations of motion then reduce to $r'' - r\theta'^2 = -(r - r_0) - 2\zeta r'$, and $r^2\theta' = r_0^2\theta_0'$, where notation $(\cdot)'$ indicates a derivative with respect to τ. Finally, the non-dimensional position of the particle is introduced, $\bar{r} = r/r_0$, and the equations of motion simply become $\bar{r}'' = \theta_0'^2/\bar{r}^3 - (\bar{r} - 1) - 2\zeta\bar{r}'$ and $\theta' = \theta_0'/\bar{r}^2$, respectively; the initial conditions are $\bar{r}(t = 0) = 1$, $\bar{r}'(t = 0) = (\bar{\imath}_1^T\underline{v}_0)/(\Omega r_0)$, and $\theta_0' = (\bar{\imath}_2^T\underline{v}_0)/(\Omega r_0)$.

Example 3.3. Particle sliding on a helix

Consider the motion of a particle sliding without friction along the helix depicted in fig. 2.3. Gravity acts down, in the opposite direction of axis $\bar{\imath}_3$. Since the particle is constrained to move along the helix, a constraint force is applied to the particle. This force acts in the plane normal to the curve, *i.e.*, it has components along the normal and binormal vectors, but not along the tangent vector.

Newton's second law then implies

$$-mg\bar{\imath}_3 + F_n\bar{n} + F_b\bar{b} = m(\ddot{s}\bar{t} + \frac{\dot{s}^2}{\rho}\bar{n}),$$

where F_n and F_b are the components of the reaction force in the normal and binormal directions, respectively, and the components of the acceleration vector are expressed in terms of path coordinates, eq. (2.34). The application of Newton's second law requires the consideration of all externally applied forces acting on the particle, including the reaction forces.

Projecting this equation along the tangent direction and using eq. (2.25) yields

$$\ddot{s} = -g\bar{t}^T\bar{\imath}_3 = -\frac{kg}{\sqrt{a^2 + k^2}},$$

The particle slides down the helix acted upon by an "apparent gravity," $kg/\sqrt{a^2 + k^2}$. The equation of motion is expressed in terms of the curvilinear coordinate s; it can be readily modified to be expressed in terms of parameter η defined in fig. 2.3. Indeed, using the results established in example 2.1, $\dot{s} = \dot{\eta}\sqrt{a^2 + k^2}$ and $\ddot{s} = \ddot{\eta}\sqrt{a^2 + k^2}$. It follows that $\ddot{\eta} = -kg/(a^2 + k^2)$.

Projecting Newton's law along the normal and binormal vectors yields

$$F_n = \frac{ma\dot{s}^2}{a^2 + k^2}, \quad F_b = \frac{mga}{\sqrt{a^2 + k^2}},$$

respectively. The normal component of the constraint force stems from the normal component of acceleration. Because the component of acceleration in the binormal direction vanishes, the corresponding component of the constraint force is solely due to the gravity component in that direction.

Example 3.4. Particle sliding on a spherical surface

Consider the motion of a particle sliding on the spherical surface depicted in fig. 2.10. Gravity acts down, in the opposite direction of axis $\bar{\imath}_3$. Since the particle is constrained to move on the spherical surface, a constraint force is applied to the particle. This force acts in the direction normal to the surface, *i.e.*, it has a single component along the surface normal.

Using the surface coordinates introduced in section 2.5 for a sphere, see example 2.4, Newton's second law states that

$$-mg\bar{\imath}_3 + F_n\bar{n} = m\left[(\ddot{s}_1 - \frac{\dot{s}_2^2}{T_2})\bar{e}_1 + (\ddot{s}_2 + \frac{\dot{s}_1\dot{s}_2}{T_2})\bar{e}_2 - \frac{\dot{s}_1^2 + \dot{s}_2^2}{R}\bar{n}\right], \quad (3.45)$$

where F_n is the magnitude of the reaction force in the normal direction. For a sphere, the following results were derived in example 2.4, $1/R_1 = 1/R_2 = -1/R$ and

$1/T_1 = 0$. Projecting this equation along unit vectors $\bar{e}_1$ and $\bar{e}_2$ yields $\ddot{s}_1 - \dot{s}_2^2/T_2 = g \sin \eta_1$ and $\ddot{s}_2 + \dot{s}_1 \dot{s}_2/T_2 = 0$, respectively.

The equations of motion are expressed in terms of the curvilinear coordinates, s_1 and s_2, but can be readily modified to be expressed in terms of the surface coordinates η_1 and η_2. Indeed, $\dot{s}_1 = R \dot{\eta}_1$ and $\dot{s}_2 = R \dot{\eta}_2 \sin \eta_1$. Introducing the expression for the twist of the spherical surface, eq. (2.66), then yields

$$\ddot{\eta}_1 - \dot{\eta}_2^2 \sin \eta_1 \cos \eta_1 = \frac{g}{R} \sin \eta_1, \quad \ddot{\eta}_2 \sin \eta_1 + 2 \dot{\eta}_1 \dot{\eta}_2 \cos \eta_1 = 0. \tag{3.46}$$

Projecting Newton's law along the normal vector yields

$$F_n = m \left(g \cos \eta_1 - \frac{\dot{s}_1^2 + \dot{s}_2^2}{R} \right). \tag{3.47}$$

The magnitude of the constraint force stems from the normal component of acceleration and from the component of the gravity force in that direction.

For small motions of the particle near the lowest point on the sphere, *i.e* for $\eta_1 = \pi + \hat{\eta}_1$, $\hat{\eta}_1 \ll 1$. The equations of motion can be linearized as $\ddot{\hat{\eta}}_1 + g \hat{\eta}_1/R = 0$, the well known equation governing the small amplitude motion of a pendulum under gravity.

The same results could have been obtained using spherical coordinates, see section 2.7.2, instead of surface coordinates; the fact that the particle is moving on the surface of a sphere then implies $\dot{r} = 0$.

3.2.5 Problems

Problem 3.1. Simple spring mass system
Consider a simple spring mass system: a particle of mass m is connected to a spring of stiffness k and a gravity field with an acceleration g is acting on the system. At time t_0, the system is at rest and the spring is un-stretched. Consider the following two scenarios. *Scenario 1*: the mass is released from rest and oscillates freely thereafter. *Scenario 2*: the mass is slowly brought to its static equilibrium position. *(1)* Find the maximum displacement of the particle for *scenario 1*. *(2)* Find the maximum displacement of the particle for *scenario 2*. *(3)* If there exist any difference in the maximum displacements for *scenarios 1* and *2*, give work and energy arguments to justify the discrepancy.

Problem 3.2. Work done by conservative forces
Prove that the work done by a conservative force applied to a particle between times t_i and t_f is independent of the path of the particle during that time.

Problem 3.3. Is a constant force a conservative force?
Is a constant force a conservative force? If yes, find the potential of this force.

Problem 3.4. Particle subjected to friction forces
Consider a particle of mass $m = 1$ kg subjected to a friction force $\underline{F}_f = -kv\underline{v}$, where k is the friction coefficient and $v = \|\underline{v}\|$. The particle is also subjected to gravity forces ($g = 9.81$ m/s^2), see fig. 3.15. At time $t = 0$, the particle is launched with an initial speed $v_0 = 100$ m/s with an angle $\theta = 30$ deg with respect to the horizontal. *(1)* Write the equations of motion for

the particle. *(2)* Solve these equations for $k = 0, 0.001$, and 0.002 kg/m. *(3)* Plot the trajectory of the particle for the three cases on the same graph. *(4)* Determine the distance d and the maximum height h from the computed trajectory. *(5)* Plot d and h as a function of the friction coefficient $k \in [0, 0.003]$ kg/m.

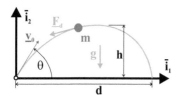

Fig. 3.15. Particle subjected to friction.

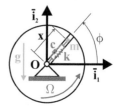

Fig. 3.16. Particle in a slot on a rotating disk.

Problem 3.5. Particle in a slot on a rotating disk

Figure 3.16 depicts a disk rotating in a vertical plane at a constant angular speed, $\dot{\phi} = \Omega$, around inertial point **O**. Mass m is free to slide in a radial slot on the disk and is connected to the center of the disk by means of a spring of stiffness constant k and a dashpot of constant c. The system is subjected to gravity and a torque, Q, is applied to the disk. The spring's unstretched length is denoted x_0. *(1)* Derive the equation of motion of the system in terms of distance x from point **O** to the particle. *(2)* Find the horizontal and vertical components of the reaction force at point O. *(3)* Find the applied torque, Q, required to maintain this constant angular speed.

Problem 3.6. Free falling parachute

Figure 3.17 shows a payload of mass $m = 120$ kg attached to a parachute. The payload is dropped from an altitude $h = 1000$ m with a horizontal velocity of magnitude $v_0 = 100$ m/s. The payload is subjected to a drag force $\underline{F}_d = -1/2\ C_d \rho \mathcal{A}\ v\underline{v}$, where $C_d = 1.42$ is the drag coefficient, $\rho = 1.23$ kg/m^3 the air density, $\mathcal{A} = \pi D^2/4$ the cross-sectional area of the parachute, and D its diameter. The velocity vector is denoted $\underline{v}$ and the speed is $v = \|\underline{v}\|$. The payload is also subjected to gravity forces ($g = 9.81$ m/s^2), see fig. 3.17. *(1)* Write the equations of motion for the payload. *(2)* Solve these equations numerically for parachutes of diameter $D = 3, 4$, and 6 m. *(3)* Plot the horizontal position of the payload for the three cases on the same graph. *(4)* Plot the vertical position of the payload for the three cases on the same graph. *(5)* Plot the horizontal velocity of the payload for the three cases on the same graph. *(6)* Plot the vertical velocity of the payload for the three cases on the same graph. *(7)* Find an **analytical expression** for the **constant** horizontal velocity that is eventually reached by the payload. *(8)* Find an **analytical expression** for the **constant** vertical velocity that is eventually reached by the payload. *(9)* Based on this **constant** vertical velocity, find an **analytical expression** for the time it takes for the payload to reach the ground. *(10)* Compute the time to reach the ground as a function of parachute diameter $D \in [3, 6]$ m. On the same graph, plot the numerical and analytical solutions. *(11)* Compute the final vertical velocity as a function of parachute diameter $D \in [3, 6]$ m. On the same graph, plot the numerical and analytical solutions

Problem 3.7. Pendulum under gravity forces

Consider a pendulum with a bob of mass $m = 1.5$ kg, length $\ell = 0.75$ m and subjected to gravity forces ($g = 9.81$ m/s^2). The pendulum is released from rest with $\theta = 0$, see fig. 3.18.

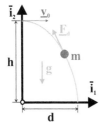

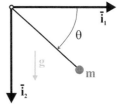

Fig. 3.17. Parachute subject to drag force.   **Fig. 3.18.** Pendulum under gravity forces.

(1) Write the equations of motion for the pendulum. *(2)* Solve these equations numerically. *(3)*Plot the angular motion θ and angular velocity $\dot{\theta}$ as functions of time on two separate graphs. *(4)* Compute and plot the tension in the cord as a function of time. *(5)* On one graph, plot the kinetic energy, potential energy, and total mechanical energy of the system versus time. *(6)* Plot the total mechanical energy of the system versus time. Does it remain constant? Comment on your results

Problem 3.8. Inverted pendulum

Consider an inverted pendulum with a bob of mass $m = 1$ kg. A massless, rigid bar of length $\ell = 1$ m supports the bob. Gravity, $g = 9.81$ m/s^2, acts in the direction indicated on fig. 3.19. A torsional spring of stiffness $k = 10$ N·m/rad is located at point **O** and applies a moment $M = -k\theta$ on the rigid bar. The pendulum is released from $\theta = 0$, with an initial speed $v_0 = 2$ m/s to the right, see fig. 3.19. *(1)* Write the equations of motion for the system. *(2)* Solve these equations numerically. *(3)* Plot the angular motion θ as a function of time. *(4)* Plot the angular velocity $\dot{\theta}$ as a function of time. *(5)* Plot the load in the rigid bar as a function of time. *(6)* On one graph, plot the kinetic energy, potential energy, strain energy, and total mechanical energy of the system versus time. *(7)* On one graph, plot the total mechanical energy of the system versus time. Does it remain constant? Comment on your results. *(8)* Consider two states ot the system: the initial configuration, $(\theta = 0)$, and a final configuration, $(\theta = \theta_f)$, where the angle θ_f is maximum. Find the maximum angular deflection, θ_f. Check your answer against the numerical simulation.

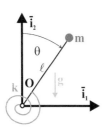

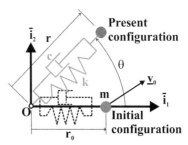

Fig. 3.19. Inverted pendulum under gravity forces.   **Fig. 3.20.** Particle connected to the ground by a spring and damper.

Problem 3.9. Particle connected to the ground with spring and damper

A particle of mass m is connected to the ground by a spring of stiffness k and a damper of constant c. The initial configuration of the system is indicated on fig. 3.20, and the initial velocity vector is $\underline{v}_0$. The following quantities are defined: $\Omega^2 = k/m$ and $c = 2m\Omega\zeta$. For this problem, it is convenient to use the polar coordinate system indicated on the figure. (1) Set up the equations of motion of the system. (2) Plot $\bar{r} = r/r_0$ as a function of the non-dimensional time $\tau = \Omega t$ for $\tau \in [0, 20\pi]$. (3) Plot $\theta(\tau)$. (4) Plot the trajectory of the particle in space. (5) Plot the history of the non-dimensional angular velocity $\bar{\Omega}(\tau) = \dot{\theta}/\Omega$. (6) Plot the history of the components of the velocity vector in the inertial frame, $\bar{v}_x = v_x/(\Omega r_0)$ and $\bar{v}_y = v_y/(\Omega r_0)$. (7) Plot the history of the non-dimensional total mechanical energy of the system $\bar{E}(\tau) = E/(m\Omega^2 r_0^2)$; comment your result. (8) Compute the non-dimensional cumulative energy dissipated in the damper $\bar{W}(\tau) = W/(m\Omega^2 r_0^2)$. (9) Plot the history of the quantity $\bar{E}(\tau) + \bar{W}(\tau)$; comment your result. Use the following data: $\zeta = 0.05$; at time $\tau = 0$, $\underline{v}_0/(\Omega r_0) = 1.2\,\bar{\imath}_1 + 0.8\,\bar{\imath}_2$, $\theta_0 = 0$ and $r(t=0)/r_0 = 1$; the un-stretched length of the spring is r_0.

Problem 3.10. Particle sliding along a curve

A particle of mass m freely slides along a given curve $\mathbb{C}$ in three-dimensional space. A point on the curve has a position vector $\underline{p}_0(s)$. Find the equation of motion for the particle if it is subjected to externally applied forces $\underline{F}(t)$. What are the components of the constraint force acting on the particle.

Problem 3.11. Particle sliding along a helix

A particle slides along a helix and is subjected to a gravity force acting along the $\bar{\imath}_1$ direction, see fig. 2.3. Find the equation of motion for the particle in terms of the parameter η. If the initial condition at $t = 0$ are $\eta = 0$ and $\dot{\eta} = v_0/\sqrt{a^2 + k^2}$, find the minimum value of v_0 such that the particle proceeds along the helix with $\dot{\eta} > 0$ at all time.

Problem 3.12. Particle sliding along a circular ring

Figure 3.21 depicts a particle of mass m sliding along a circular ring under the effect of gravity. The ring rotates on two bearing about an axis parallel to $\bar{\imath}_3$; a torque $Q(t)$, acting about axis $\bar{\imath}_3$, is applied to the ring. (1) Find the equations of motion for the particle. (2) Write the expression for the potential of the gravity forces. (3) Write the expression for the kinetic energy of the particle. (4) Write the expression for the work done by the applied torque $Q(t)$.

Problem 3.13. Particle sliding along a circular ring

Figure 3.21 depicts a particle of mass m sliding along a circular ring under the effect of gravity. The ring rotates on two bearing about an axis parallel to $\bar{\imath}_3$; a torque $Q(t)$, acting about axis $\bar{\imath}_3$, is applied to the ring. (1) Find the equations of motion for the particle based on the principle of impulse and momentum.

Problem 3.14. Particle in a massless tube

Figure 3.22 shows a particle of mass m sliding in a massless tube is connected to a spring of stiffness k_r and a damper of constant c_r. The un-stretched length of the spring is r_0. A spherical coordinate system r, ϕ and θ with corresponding unit vectors $\bar{e}_1$, $\bar{e}_2$ and $\bar{e}_3$ will be convenient to use. The spring/damper assembly is attached to the ground at point $\mathbf{A}$ by means of a joint that allows rotation about axis $\bar{e}_3$. This joint features a torsional spring of stiffness k_ϕ and a torsional damper of constant c_ϕ. The torsional spring is un-stretched when $\phi = \pi/2$. The angle θ has a prescribed schedule $\theta(t) = \omega t$. The following quantities are defined: the non-dimensional time $\tau = \omega t$, the axial spring frequency $\Omega_r = \sqrt{k_r/m}$, and

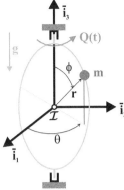

Fig. 3.21. Particle sliding on a circular ring.

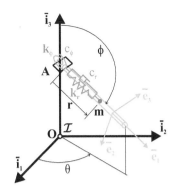

Fig. 3.22. Particle connected to the ground by a spring, damper and revolute joint.

its critical damping ratio $\zeta_r = c_r/(2m\Omega_r)$, the torsional spring frequency $\Omega_\phi = \sqrt{k_\phi/mr_0^2}$ and its critical damping ratio $\zeta_\phi = c_\phi/(2mr_0^2\Omega_\phi)$. *(1)* Set up the equations of motion of the system. *(2)* Plot $\bar{r} = r/r_0$ as a function of the non-dimensional time $\tau \in [0, 20\pi]$. *(3)* Plot $\phi(\tau)$. *(4)* Plot the trajectory of the particle in three-dimensional space. *(5)* Plot the history of the non-dimensional force $\bar{F}_3 = F_3/(mr_0\omega^2)$ that the massless tube applies on the particle. *(6)* Plot the history of the non-dimensional total mechanical energy of the system $\bar{E}(\tau) = E/(mr_0^2\omega^2)$; comment your result. *(7)* Plot the history of the cumulative non-dimensional energy dissipated in the dampers $\bar{W}^d(\tau) = W^d/(mr_0^2\omega^2)$. *(8)* Plot the history of cumulative non-dimensional work $\bar{W}^M = W^M/(mr_0^2\omega^2)$ done by the torque required to prescribe $\theta(t) = \omega t$. *(9)* Plot the history of the quantity $\bar{E}(\tau) + \bar{W}^d(\tau) - \bar{W}^M(\tau)$; comment your result. Use the following data: $\bar{\Omega}_r = \Omega_r/\omega = 5$, $\zeta_r = 0.05$; $\bar{\Omega}_\phi = \Omega_\phi/\omega = 1.5$, $\zeta_\phi = 0.05$; at time $\tau = 0$, $\underline{v}_0/(\omega r_0) = 0.6\,\bar{\imath}_1 + 1\,\bar{\imath}_2 + 0.75\,\bar{\imath}_3$, $\phi_0 = \pi/2$ and $r/r_0 = 1$.

Problem 3.15. Particle moving on a track

Figure 3.23 shows particle of mass m moving on a track defined by a curve $\mathbb{C}$ while constrained to remain within a slot inside a massless arm. The massless arm is prescribed to move at a constant angular speed, $\dot{\theta} = \Omega$. *(1)* Plot the radial location, r/R, of the particle as a function of θ. *(2)* Plot the moment $\bar{m} = M/(mR^2\Omega^2)$ necessary to drive the system at a constant angular speed. *(3)* Plot the non-dimensional normal force $\bar{F}_n = F_n/(mR\Omega^2)$ the curved track applies on the particle. *(4)* Determine the minimum stiffness of the spring, i.e., the minimum non-dimensional frequency $\bar{\Omega}$, for which the particle remains on the track at all times. The curve is defined in the polar coordinate system as $p_0(\theta) = r(\theta)\bar{e}_1$, where $r(\theta) = R - b\cos N\theta$. It will be convenient to define the normal to the curve as $\bar{n} = \tilde{\imath}_3\bar{t}$, where $\bar{t}$ is the tangent to the curve. Use the following data: $\bar{b} = b/R = 0.25$; $N = 6$; $\omega^2 = k/m$; $\bar{\Omega} = \omega/\Omega = 3$. The spring is un-stretched when $r = 0$. At time $t = 0$, $\theta = 0$.

Problem 3.16. Particle moving on a track

Figure 3.23 shows particle of mass m moving on a track defined by a curve $\mathbb{C}$ while constrained to remain within a slot inside a massless arm. A moment, M, is applied to the arm at point **O**. *(1)* Plot the time history of the angle θ. *(2)* Plot the angular speed $\dot{\theta}/\omega$. *(3)* Plot the normal force $\bar{F}_n = F_n/(mR\omega^2)$ the curved track applies on the particle. *(4)* Plot the total mechanical energy of the system, $\bar{E} = E/(mR^2\omega^2)$. Discuss your results. *(5)* Compare the responses of the system at $\bar{M}_0 = 0.75$ and 0.80. Explain your results. The curve is defined in the polar coordinate system as $p_0(\theta) = r(\theta)\bar{e}_1$, where $r(\theta) = R - b\cos N\theta$. It will be convenient to define the normal to the curve as $\bar{n} = \tilde{i}_3 \bar{t}$, where $\bar{t}$ is the tangent to the curve. Use the following data: $\bar{b} = b/R = 0.25$; $N = 6$; $\omega^2 = k/m$. The spring is un-stretched when $r = 0$. At time $t = 0$, $\theta = 0$ and $\dot{\theta} = 0$. The applied moment is given as $\bar{M} = M/(mR^2\omega^2) = \bar{M}_0(1 - \cos\tau)$ for $\tau \leq 2\pi$ and $\bar{m} = 0$ for $\tau > 2\pi$, where $\tau = \omega t$ is the non-dimensional time and $\bar{M}_0 = 0.75$. Simulate the system for $\tau \in [0, 6\pi]$.

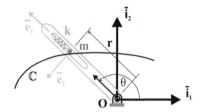

Fig. 3.23. Particle moving on a track.

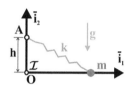

Fig. 3.24. Particle connected to a spring with unilateral contact to a horizontal plane.

Problem 3.17. Particle with unilateral contact

The particle of mass m depicted in fig. 3.24 is subjected to a gravity field of acceleration g and is connected to a spring of stiffness k and un-stretched length h. The spring is attached to inertial point **A**, located a distance h above point **O**. A unilateral contact condition is imposed on the particle by the horizontal plane $\mathcal{P} = (\mathbf{O}, \bar{i}_2)$; this means that the particle can only move in the half-space above this plane. At the initial time, the particle is at point **O** and has a velocity $\underline{v}(t = 0) = v_0\bar{i}_1$. *(1)* Write the equation of motion for the particle while it is in contact with the plane. *(2)* Plot the non-dimensional position of the particle as a function of non-dimensional time. *(3)* Plot the non-dimensional velocity of the particle as a function of non-dimensional time. *(4)* Find the time at which the particle leaves the plane and its corresponding position and velocity. *(5)* Under what condition will the particle always remain on the plane for any magnitude of the initial velocity v_0? *(6)* Find the time at which the particle will first hit the plane after leaving it. *(7)* Plot the trajectory of the particle during its free flight. Use the following data: $\bar{v}_0 = v_0/(\omega h) = 1$; $\bar{g} = mg/(kh) = 0.25$; $\omega^2 = k/m$. Use the following non-dimensional time $\tau = \omega t$. All lengths are non-dimensionalized by h, velocities by ωh.

Problem 3.18. Particle moving on a parabolic surface of revolution

Figure 2.11 shows a particle sliding on a parabolic surface of revolution and subjected to a gravity force acting along the negative $\bar{i}_3$ direction. This surface is defined by the position vector of one of its points, $\underline{p}_0 = r\cos\phi\,\bar{i}_1 + r\sin\phi\,\bar{i}_2 + ar^2\bar{i}_3$, where $r \geq 0$ and $0 \leq \phi \leq 2\pi$. The following notation was used $\eta_1 = r$ and $\eta_2 = \phi$. *(1)* Find the equation of motion for the particle in terms of the surface coordinates r and ϕ. *(2)* Find the constraint force acting on the particle.

Problem 3.19. Particle sliding on a linear spiral

A particle of mass m is sliding along a linear spiral, as defined in example 2.2, under the effect of gravity acting down along the $\bar{i}_2$ axis, see fig. 2.4. *(1)* Derive the governing equation of motion using Newton's law. *(2)* Plot the angle θ as a function of time. *(3)* Plot the $\dot{\theta}$ as a function of time. *(4)* Plot the time history of the magnitude of the normal reaction force that the spiral applies to the particle. *(5)* On one graph, plot the time history of the kinetic energy of the particle, its potential energy, and its total mechanical energy. *(6)* Derive the governing equation of motion of the particle from the principle of work and energy. Use the following data: $m = 2.5$ kg; $a = 0.2$ m; $g = 9.81$ m/s^2. At time $t = 0$, $\theta = 0$ and $\dot{\theta} = 50$ rad/s. Present all your results for $t \in [0, 15]$ s.

Problem 3.20. Bungee jumping

A man of mass m is jumping off a bridge while attached to a bungee cord of un-stretched length d_0, as described in example 3.1. An inertial frame, $\mathcal{F} = [\mathbf{O}, \mathcal{I} = (\bar{\imath}_1, \bar{\imath}_2)]$, is attached to the bridge. The man is jumping from point $\mathbf{O}$ with an initial velocity, v_0, oriented along horizontal axis $\bar{\imath}_2$, and the acceleration of gravity is acting along vertical axis $\bar{\imath}_1$. In the developments presented in example 3.1, the effect of air friction on the man was ignored. In this problem, these forces will be taken into account in an approximate manner applying to the man a drag force, $\underline{F}_d = -1/2\,C_d\rho\mathcal{A}\,\|\underline{v}\|\underline{v}$, where C_d is the non-dimensional drag coefficient, ρ the air density, and A the man's cross-sectional area. This drag force is at all times proportional to the square of the speed, aligned with the velocity vector, and oriented in the direction opposite to the velocity vector. *(1)* Derive the equation of motion for the free fall portion of the man's trajectory. Solve the equations numerically to find the time τ_t at which the bungee becomes taught. *(2)* Derive the equations of motion once the bungee is taut. Solve the equations numerically. *(3)* On one graph, plot the components $\bar{x}_1$ and $\bar{x}_2$ of the man's position vector as functions of τ. *(4)* Plot the trajectory of the man. *(5)* On one graph, plot the components $\bar{v}_1$ and $\bar{v}_2$ of the velocity vector as functions of τ. *(6)* Plot the stretch of the bungee as a function of τ. *(7)* On one graph, plot the non-dimensional kinetic energy, $\bar{K} = K/(mv_0^2)$, potential energy, $\bar{V} = V/(mv_0^2)$, and total potential energy $\bar{E} = \bar{K} + \bar{V}$. *(8)* Determine the non-dimensional time at which the bungee becomes slack again. Use the following non-dimensional quantities: $\bar{x}_1 = x_1/d_0$, $\bar{x}_2 = x_2/d_0$; $\bar{v}_1 = v_1/v_0$, $\bar{v}_2 = v_2/v_0$; use the non-dimensional time $\tau = v_0 t/d_0$. Use the following data: $\bar{g} = gd_0/v_0^2 = 10$, $\bar{k}_0 = k_0 d_0^2/(mv_0^2) = 60$, $C_d = 0.47$ and $\bar{\mu} = \rho\mathcal{A}d_0/m = 0.03$. Present all your results for $\tau \in [0, 3.5\tau_t]$.

3.3 Contact forces

When dealing with particle dynamics, it is often the case that the particle is in contact with another body. Contact can be of a continuous nature; for instance, a particle is moving while in continuous contact with a curve or a surface, see example 3.3 or 3.4, respectively. Contact could also be of an intermittent nature, such as, for instance, the impact of a particle on an obstacle. These contact forces are forces acting on the particle, which must therefore be included in the statement of Newton's second law when studying the dynamic response of the particle. Both magnitude and direction of these forces must be studied to properly state Newton's second law.

The kinematics of contact of a particle with a surface and a curve will be studied first in sections 3.3.1 and 3.3.2, respectively; contact forces are categorized into normal and tangential contact forces. Next, the magnitudes of these forces will be studied in section 3.3.3. Typically, constitutive laws are postulated that relate the magnitude of the contact forces to contact parameters. For instance, Coulomb's friction law relates the friction force to both normal force and relative velocity of the particle with respect to surface it is in contact with.

3.3.1 Kinematics of particles in contact with a surface

Figure 3.25 depicts a particle of mass m in continuous contact with surface $\mathbb{S}$ at point **P**. The differential geometry of a surface was studied in details in section 2.4, and it is assumed here that parameters η_1 and η_2 define the lines of curvature presented section 2.4.5. Unit vectors $\bar{e}_1$ and $\bar{e}_2$ given by eq. (2.55) define the plane tangent to the surface at point **P**. The normal to the surface is now defined as $\bar{n} = \tilde{e}_1 \bar{e}_2$ and $\mathcal{E} = (\bar{e}_1, \bar{e}_2, \bar{n})$ forms an orthonormal basis.

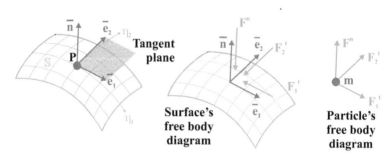

Fig. 3.25. Particle moving on a surface.

The contact force, $\underline{F}^c$, between the particle and the surface is conveniently divided into two components, the *normal contact force*, $\underline{F}^n = F^n \bar{n}$, which acts along the normal to the surface, and the *tangential contact force*, $\underline{F}^t = F_1^t \bar{e}_1 + F_2^t \bar{e}_2$, which acts in the plane tangent to the surface. Hence, the contact force is written as

$$\underline{F}^c = \underline{F}^n + \underline{F}^t = F^n \bar{n} + (F_1^t \bar{e}_1 + F_2^t \bar{e}_2). \tag{3.48}$$

Imagine first that the particle slides over the surface without any friction: the tangential contact forces vanish in eq. (3.48). Further assume that the surface on which the particle slides is a plane. If the particle slides on this plane under the effect of externally applied forces acting in the same plane, $\underline{F}^a = F_1^a \bar{e}_1 + F_2^a \bar{e}_2$, the normal contact force also vanishes. Indeed, Newton's law now reduces to $F_1^a \bar{e}_1 + F_2^a \bar{e}_2 + 0\bar{n} = m(\ddot{x}_1 \bar{e}_1 + \ddot{x}_2 \bar{e}_2 + 0\bar{n})$: both externally applied forces and accelerations vanish along the normal direction.

Consider now the same particle sliding on a curved surface under the effect of external forces applied in the plane tangent to the surface; Newton's law now becomes $F_1^a \bar{e}_1 + F_2^a \bar{e}_2 + F^n \bar{n} = m(a_1 \bar{e}_1 + a_2 \bar{e}_2 + a_n \bar{n})$, where the acceleration components are given by eq. (2.68). Since the particle has to follow the curvature of the surface, the acceleration component in the normal direction, a_n, does not vanish, and the normal contact force, F^n, is necessary to equilibrate the corresponding inertial forces. The normal contact force can be interpreted as the constraint force that constrains the particle to remain on the surface.

If the interface between the particle and the surface is rough, friction forces acting in the plane tangent to the surface will appear in addition to the normal contact force. Figure 3.25 also gives free body diagrams for the particle and surface. Force components F^n, F_1^t, and F_2^t are the forces the surface applies to the particle. Note that according to Newton's third law, the particle applies equal and opposite forces to the surface.

3.3.2 Kinematics of particles in contact with a curve

Figure 3.26 shows a common situation where a particle moves along a curve. This would be, for instance, the case of a train moving along its rails, or of a roller coaster car moving along its track. The differential geometry of a curve was studied in details in section 2.2. At point **P** of the curve, it is possible to define Frenet's triad consisting of the tangent, normal, and binormal vectors, denoted $\bar{t}$, $\bar{n}$, and $\bar{b}$, respectively.

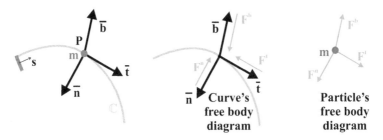

Fig. 3.26. Particle moving on a curve.

The contact force, $\underline{F}^c$, between the particle and the curve is conveniently divided into two components, the *normal contact force*, $\underline{F}^n = F^n\bar{n} + F^b\bar{b}$, which acts in the plane normal to the curve, and the *tangential contact force*, $\underline{F}^t = F^t\bar{t}$, which acts along the tangent to the curve. Hence, the contact force is written as

$$\underline{F}^c = F^t\bar{t} + (F^n\bar{n} + F^b\bar{b}). \tag{3.49}$$

Note the different expressions for the contact forces between a particle and a surface, eq. (3.48), and those between a particle and a curve, eq. (3.49). For a surface, the normal contact force has a single component along the normal to the surface, whereas for a curve, the normal contact force has two components along the normal and binormal to the curve. On the other hand, for a surface, the tangential contact force has two components in the plane tangent to the surface, whereas for a curve, the tangential contact force has a single component along the tangent to the curve.

It is important to distinguish the difference between *unilateral* and *bilateral contact*. For instance, a train is in unilateral contact with its rails: the train cannot go through the rails, nothing, however, prevents the train from moving off the rails in the upward direction. Of course, gravity forces are, in general, sufficient to keep the train on its rails. This contrasts with roller coasters: in this case, cars are connected to the track by a set of wheels that prevent them from running off track in any direction.

When dealing with unilateral contact, it is often important to determine when a particle will loose contact with the surface or curve. Consider, at first, the case of a particle on a surface and assume the particle can freely move in the direction of the normal to the surface. In that case, F^n is positive when the particle is on the surface and the unilateral contact condition cannot support a negative normal force. Clearly, the particle is about to leave the surface when $F^n = 0$.

In the case of a particle on a curve, the normal to the curve is always pointing to the concave side of the curve: the normal flips direction at an inflection point of the curve. Due to this discontinuity, the condition $F^n = 0$ must be applied with care when dealing with particle moving along a curve. For more details about the complex problems associated with unilateral contact conditions can be found in the textbook by Pfeiffer [6].

3.3.3 Constitutive laws for tangential contact forces

The tangential contact forces are friction forces between the particle and surface or curve it moves on. Coulomb's friction law is commonly used to evaluate the friction forces, and sometimes, friction forces are assumed to be of a viscous type.

Coulomb's friction law

Coulomb's friction law has been extensively used to model friction forces. It postulates that the friction force between the particle and surface is proportional to the absolute value of the normal contact force.

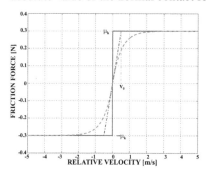

The empirical coefficient of proportionality, μ_k, is called the *coefficient of kinetic friction*. The friction force always acts in the direction opposing the relative velocity of the particle with respect to the surface,

$$\underline{F}^f = -\mu_k \|\underline{F}^n\| \frac{\underline{v}_{rel}}{\|\underline{v}_{rel}\|}, \qquad (3.50)$$

where $\underline{v}_{rel}$ is the relative velocity of the particle with respect to the surface and $\|\underline{F}^n\| = |F^n|$, see eq. (3.48).

Fig. 3.27. Coulomb's friction law: solid line; continuous friction law: dashed line.

In the case of a particle moving along a curve, the relative velocity of the particle with respect to the curve is along the tangent to the curve, *i.e.*, $\underline{F}^f = F^f \bar{t}$, where

$$F^f = -\mu_k \|\underline{F}^n\| \, \text{sign}(v_{rel}), \qquad (3.51)$$

where v_{rel} is the speed of the particle with respect to the curve, $\underline{v}_{rel} = v_{rel}\bar{t}$, and $\|\underline{F}^n\|^2 = F^{n2} + F^{b2}$, see eq. (3.49).

Sliding gives way to sticking when the relative velocity vanishes. In that case, the magnitude of the friction force must be smaller than that of the normal contact force times an empirical coefficient μ_s, the *coefficient of static friction*, or

$$\|\underline{F}^f\| \le \mu_s \,\|\underline{F}^n\|. \tag{3.52}$$

Figure 3.27 depicts the friction force as a function of the relative velocity, for the simple case of a constant friction coefficient $\mu_k = 0.3$ and unit normal force $F^n = 1$.

Coulomb's friction law presents a discontinuity of the friction force at zero relative velocity, as shown in fig. 3.27. This discontinuity causes numerical difficulties in computer simulations and hence, various approximations to Coulomb's law have been proposed in the literature [7, 8, 9, 10, 11]. These various approximations can be viewed as *continuous friction laws* that replace the discontinuity at zero relative velocity by a smooth, rapidly varying function of the relative velocity. A typical expression for continuous friction laws is

$$F^f = -\mu_k \|\underline{F}^n\| \, \mathrm{sign}(v_{\mathrm{rel}}) \, (1 - e^{-|v_{\mathrm{rel}}|/v_0}), \tag{3.53}$$

where v_0 is a characteristic relative velocity typically chosen to be small compared to the maximum relative velocity encountered during the simulation. Figure 3.27 shows the friction force corresponding to the continuous friction law for $v_0 = 0.5$ m/s. The continuous friction law replaces both kinetic and static friction laws.

Viscous friction law

It is sometimes assumed that friction forces are of a viscous type, *i.e.*, the friction forces are proportional to the relative velocity of the particle with respect to the surface or curve it moves on. The coefficient of proportionality, c, is called the *coefficient of viscous friction*, and hence

$$\underline{F}^f = -c\underline{v}_{\mathrm{rel}}. \tag{3.54}$$

Note that the normal force does not appear in this expression.

Coulomb's friction is sometimes called "dry friction," as opposed to the present "viscous friction" phenomenon. For cases of friction between a particle and a lubricated surface, a combination of dry and viscous frictions forces is often observed. Both dry and viscous friction laws are approximations to the experimentally observed friction forces. In fact, friction is a very complex phenomenon that involves many, often poorly understood physical processes; the following references give detailed descriptions of the friction process and a wealth of experimental observations Rabinowicz [12], or Oden and Martins [13].

Example 3.5. Particle elastically suspended to a straight track
Consider a particle of mass m suspended to a straight track by means of a spring in parallel with a dashpot of constant c, as depicted in fig. 3.28. The magnitude of the force in the spring, F_s, is a nonlinear function of its stretch, $F_s = k_1\Delta + k_3\Delta^3$, where $\Delta = r - r_0$ is the stretch of the spring, r the distance from the particle to point **A**, and r_0 the un-stretched length of the spring.

The spring-dashpot system is connected at point **A** to a massless slider that moves along a straight track, which makes an angle α with respect to the horizontal. The motion of the slider along the track is prescribed as $s(t) = s_0 \sin \omega t$. Considering

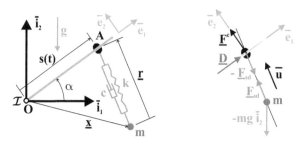

Fig. 3.28. Particle on a straight track.

the free body diagram for the particle illustrated on the right portion of fig. 3.28, the equation of motion of the system is found as

$$-mg\bar{\imath}_2 + \underline{F}_{sd} = m\underline{\ddot{x}},$$

where $\underline{x}^T = \{x_1, x_2\}$ are the components of the position vector of the particle resolved in the inertial frame $\mathcal{F}^I = [\mathbf{O}, \mathcal{I} = (\bar{\imath}_1, \bar{\imath}_2)]$, and $\underline{F}_{sd}$ those of the the force vector applied to the particle by the spring-dashpot system.

The force in the spring-dashpot system has a line of action that joins the particle to point **A**; the unit vector along this line of action is $\bar{u} = [s(t)\bar{e}_1 - \underline{x}]/r$, where $r = \|s(t)\bar{e}_1 - \underline{x}\|$ and $\bar{e}_1 = \cos \alpha\, \bar{\imath}_1 + \sin \alpha\, \bar{\imath}_2$ is the unit vector along the track. The force vector generated by the spring-dashpot system is now readily expressed as $\underline{F}_{sd} = [k_1\Delta + k_3\Delta^3 + c\dot{\Delta}]\, \bar{u}$, where $\dot{\Delta} = \dot{r} = \bar{u}^T(\dot{s}\bar{e}_1 - \underline{\dot{x}})$ is the stretch rate, *i.e.*, the projection of the relative velocity of the two ends of the dashpot, $(\dot{s}\bar{e}_1 - \underline{\dot{x}})$, along unit vector $\bar{u}$ pointing from one end of the damper to the other.

Because the time history of the slider motion is given, the equation of motion for the particle can be integrated numerically to yield the response of the system. The following physical parameters are used: $\alpha = \pi/6$; $s_0 = 0.45$ m; $\omega = 2$ rad/s; $m = 1.5$ kg; $r_0 = 0.25$ m; $k_1 = 50$ N/m, $k_3 = 20$ kN/m^3; $c = 2.6$ N·s/m; and $g = 9.81$ m/s^2. At the initial time, the particle is at rest at the following position: $\underline{x}^T = \{0, -r_0\}$. Figure 3.29 shows the inertial position and velocity of the particle as a function of time. The spring stretch and force, and the damper stretch rate and force are depicted in fig. 3.31 and 3.32, respectively.

The kinetic energy of the system is readily evaluated as $K = 1/2\, m\, \underline{\dot{x}}^T \underline{\dot{x}}$. The potential energy of the system consists of two terms, the potential of the gravity forces, mgx_2, and the strain energy in the spring, leading to

$$V = mgx_2 + \frac{1}{2}k_1\Delta^2 + \frac{1}{4}k_3\Delta^4.$$

The second term of the strain energy expression, $1/4\, k_3\Delta^4$, is associated with the nonlinear force term in the spring, $k_3\Delta^3$. Figure 3.30 shows the time histories of the kinetic, potential and total mechanical energies of the system. As expected, the total mechanical energy does not remain constant, because the present system is not conservative.

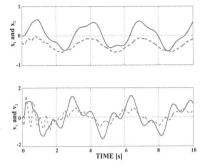

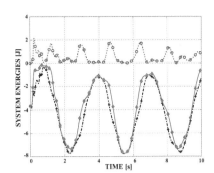

Fig. 3.29. Top figure. Particle position: x_1, solid line and x_2, dashed line. Bottom figure. Particle velocity: $v_1 = \dot{x}_1$, solid line and $v_2 = \dot{x}_2$, dashed line.

Fig. 3.30. System energies: kinetic energy: ($\circ$); potential energy: ($+$); total mechanical energy: ($\diamond$).

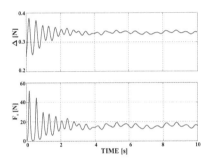

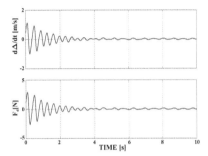

Fig. 3.31. Top plot: spring stretch, Δ. Bottom plot: spring force, F_s.

Fig. 3.32. Top plot: damper stretch rate, $\dot{\Delta}$. Bottom plot: damper force, F_d.

To verify the energy closure condition for this system, the work done by two additional forces, the driving force and the damper force, must be brought into the picture. The statement of the problem specifies that "The motion of the slider along the track is prescribed as $s(t) = s_0 \sin \omega t$." Clearly a *driving force* must be applied to the slider, if this desired motion is to be achieved. It is implicitly assumed that a device applies this force and is sufficiently powerful to instantaneously generate the required force that achieves the desired motion of the slider. The right portion of fig. 3.28 shows the free body diagram for the slider, leading to the following equilibrium equation

$$D\bar{e}_1 + F^c\bar{e}_2 - \underline{F}_{sd} = 0,$$

where $D(t)$ is the magnitude of the driving force acting along the direction of axis $\bar{e}_1$, and F^c the magnitude of the force that the track applies to the slider along the direction normal to the track, $\bar{e}_2$. Because the slider is massless, the right-hand side of the equation vanishes, leading to a static equilibrium condition.

Pre-multiplying this equation by $\bar{e}_1^T$ yields the magnitude of the driving force: $D = \bar{e}_1^T \underline{F}_{sd}$. The cumulative work done by this force is now found as

$$W_D = \int_0^t (D\bar{e}_1)^T (\mathrm{d}s\bar{e}_1) = \int_0^t D\dot{s}\ \mathrm{d}t = \int_0^t (\bar{e}_1^T \underline{F}_{sd})\dot{s}\ \mathrm{d}t.$$

Figure. 3.33 shows the history of the required driving force together with the work it performs.

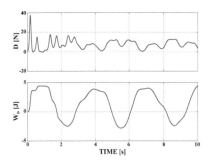

Fig. 3.33. Top plot: history of the driving force, $D(t)$. Bottom plot: cumulative work done by the driving force, W_D.

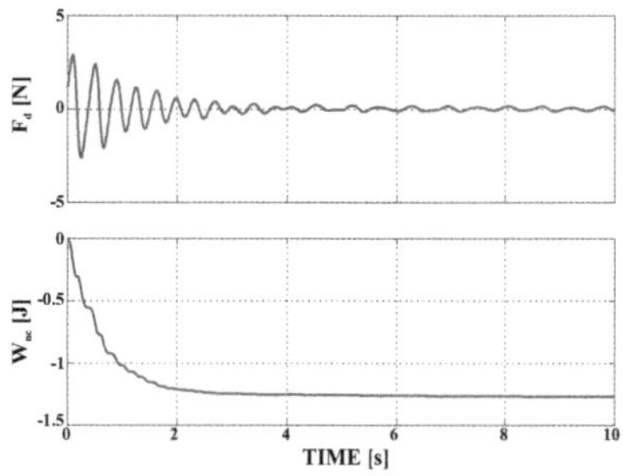

Fig. 3.34. Top plot: history of the damper force, F_d. Bottom plot: cumulative work done by the damper force, W_{nc}.

As discussed in section 3.2.3, the cumulative work done by the damper force is

$$W_{nc} = \int_0^t (-c\dot{\Delta}\ \bar{u})^T (\bar{u}\ \mathrm{d}\Delta) = -\int_0^t c\dot{\Delta}^2\ \mathrm{d}t.$$

This quantity represents the energy dissipated in the dashpot. Figure. 3.34 depicts the history of the damper force and the work it performs. As expected, this is a monotonically decreasing function of time, which represents the amount of energy dissipated in the damper in the form of heat. Finally, the energy closure equation for this problem writes

$$E(t) - W_D - W_{nc} = E_i.$$

Figure 3.35 show the various quantities in this equation: the total mechanical energy, the work done by the driving force, and the energy dissipated in the damper. As expected, the sum of these three energies remains constant during the evolution of the system, expressing the energy closure equation.

3.3.4 Problems

Problem 3.21. Particle sliding along a slot in a rotating disk

Figure 3.36 shows a particle of mass m sliding along a slot in a rotating disk. The disk rotates at a constant angular velocity, Ω, while the time-dependent position of the particle along the slot

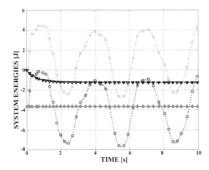

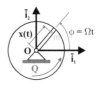

Fig. 3.35. Total mechanical energy ($\circ$), work done by the driving force ($\square$); energy dissipated in the damper ($\triangledown$); energy closure equation ($\diamond$).

Fig. 3.36. Particle sliding along a slot in a rotating disk.

is prescribed as $x(t)$. *(1)* Find the driving torque, $Q(t)$, required to keep the angular velocity of the disk constant. *(2)* Find the driving force, F^d, required to prescribe to position of the particle along the slot. *(3)* Assume now that a friction force, F^f, acts between the particle and the slot; repeat questions *(1)* and *(2)*.

Problem 3.22. Particle sliding along a helix with friction

A particle slides along a helix as defined by eq. (2.24) and is subjected to a gravity force acting along the $-\bar{\imath}_3$ direction, see fig. 2.3. The particle is also subjected to a friction force of magnitude $F_f = \mu_k \sqrt{F_n^2 + F_b^2}$, where F_n and F_b are the components of the constraint force acting on the particle in the normal and binormal directions, respectively, and μ_k the kinetic coefficient of friction. *(1)* Find the equation of motion for the particle. *(2)* Plot the time history of the particle curvilinear coordinate. *(3)* Plot the speed of the particle versus time. *(4)* Find an analytical expression for the limit velocity of the particle, *i.e.*, the velocity reached by the particle after it has been sliding along the helix for a long time. *(5)* Find an analytical expression for the limit magnitude of the acceleration vector. *(6)* What condition must the satisfied by the kinetic coefficient of friction is the particle does not remain stuck. Use the following parameters: non-dimensional time $\tau = \sqrt{a/g}\,t$; lengths are non-dimensionalized by a, velocities by $\sqrt{ag}$, accelerations by g. $\bar{k} = k/a = 0.35$, $\mu_k = 0.3$.

Problem 3.23. Motion of a particle on a track

Figure 3.37 depicts a particle sliding along a planar track under the effect of gravity forces. The constraint force between the particle and the track is unilateral, *i.e.*, the particle cannot go through the track, but it can leave it moving upwards. *(1)* Find the condition that must be satisfied by the kinetic energy of the particle if it is about to leave the track. *(2)* Could this condition be satisfied at point **A**? or at point **B**? *(3)* If a friction force (friction coefficient μ) is present between the particle and the track, what is the condition that must be satisfied by the kinetic energy of the particle if it is about to leave the track.

Problem 3.24. Particle on circular track

The particle of mass m is sliding on a circular track under the effect of gravity forces, as depicted in fig. 3.38. The particle is connected to fixed point **A** by means of a spring of stiffness constant k in parallel with a dashpot of constant c. The spring has an un-stretched

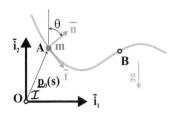

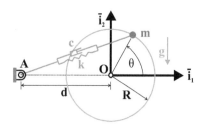

Fig. 3.37. Particle sliding on a track under the effect of gravity.

Fig. 3.38. Particle on a circular track.

length Δ_0. A viscous friction force $F^f = \mu\dot{s}$ is acting between the particle and the track. *(1)* Derive the equation of motion of the system. *(2)* Plot angle θ as a function of time. *(3)* Plot the history of the angular rate $\theta' = \dot{\theta}/\omega$. *(4)* Plot the history of the normal force, F^c, the track applies on the particle, as $\bar{F}^c = F^c/(mR\omega^2)$. *(5)* Plot the non-dimensional kinetic energy, $\bar{K} = K/(mR^2\omega^2)$, potential energy, $\bar{V} = V/(mR^2\omega^2)$, energy dissipated in the damper, $\bar{W}^c = W^c/(mR^2\omega^2)$, and energy dissipated at the viscous friction interface, $\bar{W}^f = W^f/(mR^2\omega^2)$. Verify that the energy closure equation is satisfied. Use the following non-dimensional quantities: non-dimensional time $\tau = \omega t$, where $\omega^2 = k/m$. Use the following data: $\bar{d} = d/R = 2$, $\zeta = c/(2m\omega) = 0.1$; $\bar{\Delta}_0 = \Delta_0/R = 1$; $\bar{g} = g/(R\omega^2) = 2.5$; $\bar{\mu} = \mu/(m\omega) = 0.1$. At the initial time, the system is at rest and $\theta = \pi/2$. Present all your results for $\tau \in [0, 20]$.

3.4 Newtonian mechanics for a system of particles

Newton's laws, as presented in section 3.1.2, are concerned with a single particle. For many practical engineering applications, these laws must be extended to deal with systems of particles, rather than a single particle.

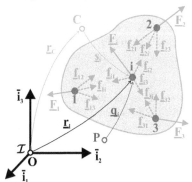

Fig. 3.39. A system of particles.

Figure 3.39 depicts a system of N particles. The particles are of mass m_i, $i = 1, 2, \ldots, N$. Each particle is subjected to forces that can be divided into two categories: the *externally applied forces* and the *internal forces*. The words "internal" and "external" should be understood with respect to the system of particles. *Internal forces* act and are reacted within the system, whereas *external forces* act on the system but are reacted outside the system.

The externally applied force acting on particle i, also called *impressed force*, is denoted $\underline{F}_i$. The origin of these forces is external to the set of particles; for instance, if the system is subjected to a gravity field or an electromagnetic field, the resulting gravity or electromagnetic forces, respectively, would be external forces. Interaction forces between the particles of the system and other material particles would also give rise to external forces.

The internal forces, denoted $\underline{f}_{ij}$, correspond to the forces that the various particles of the system apply on each other. According to Newton's third law, these forces appear in pairs of forces of equal magnitude and opposite sign, sharing a common line of action. Force $\underline{f}_{ij}$ is the force exerted by particle j on particle i, and the companion force, $\underline{f}_{ji}$, is that applied by particle i on particle j. Newton's third law then implies $\underline{f}_{ij} + \underline{f}_{ji} = 0$.

The system of particles under scrutiny is very general; it could be a rigid body, a flexible body, or a large number of sand particles. For the rigid body, the internal forces are the cohesion forces that make the body a "rigid body." Of course, there exist no truly rigid body; all bodies will exhibit some amount of elastic deformation under load. For an elastic body, the internal forces are the stresses acting between the particles of the body. The deformable body could also exhibit internal energy dissipation mechanisms; such would be the case of two deformable bodies connected by dashpots, or a single body with internal material damping. If two contacting bodies are taken to form a single system of particles, the contact forces between the bodies are internal forces. If one of the two bodies, however, constitute the system of particles, the contact forces applied on that body will be external forces.

3.4.1 The center of mass

As shown in fig. 3.39, the inertial position vector of particle i is denoted $\underline{r}_i$ and its mass m_i. The total mass of the system, denoted m, is then found by summing up the masses of individual particles

$$m = \sum_{i=1}^{N} m_i. \tag{3.55}$$

The *center of mass* of the system of particles will play an important role in the analysis. The location of the center of mass is defined as follows

$$\underline{r}_C = \frac{1}{m} \sum_{i=1}^{N} m_i \underline{r}_i. \tag{3.56}$$

Let $\underline{s}_i$ denote the relative position vector of particle i with respect to the center of mass, see fig. 3.39. It follows that $\underline{r}_i = \underline{r}_C + \underline{s}_i$, and hence,

$$\underline{r}_C = \frac{1}{m} \sum_{i=1}^{N} m_i (\underline{r}_C + \underline{s}_i) = \underline{r}_C + \frac{1}{m} \sum_{i=1}^{N} m_i \underline{s}_i.$$

This result reveals an important property of the center of mass: $\sum_{i=1}^{N} m_i \underline{s}_i = 0$. Successive time derivatives then yield

$$\sum_{i=1}^{N} m_i \underline{s}_i = \underline{0}, \tag{3.57a}$$

$$\sum_{i=1}^{N} m_i \dot{\underline{s}}_i = \underline{0}, \tag{3.57b}$$

$$\sum_{i=1}^{N} m_i \ddot{\underline{s}}_i = \underline{0}. \tag{3.57c}$$

3.4.2 The forces and moments

The forces applied on each particle can be divided into two categories: the externally applied forces and the internal forces. The sum of all the forces externally applied on the system is

$$\underline{F} = \sum_{i=1}^{N} \underline{F}_i. \tag{3.58}$$

On the other hand, the sum of all internal forces applied on the system vanishes because all the internal forces can be grouped in pairs $(\underline{f}_{ij} + \underline{f}_{ji})$ that individually vanish due to Newton's third law; hence

$$\sum_{i=1}^{N} \sum_{j=1, \ j \neq i}^{N} \underline{f}_{ij} = \underline{0}. \tag{3.59}$$

The sum of the moments of all the forces externally applied to the system evaluated with respect to the origin of the inertial frame, point **O**, is

$$\underline{M}_O = \sum_{i=1}^{N} \widetilde{r}_i \underline{F}_i. \tag{3.60}$$

The subscript, $(\cdot)_O$, indicates the point about which the moments are evaluated. Indeed, the moments could have been evaluated with respect to any arbitrary point. For instance, the moments evaluated with respect to the center of mass would write $\underline{M}_C = \sum_{i=1}^{N} \widetilde{s}_i \underline{F}_i$, and those evaluated with respect to point **P** are $\underline{M}_P = \sum_{i=1}^{N} \widetilde{q}_i \underline{F}_i$. These various quantities are not independent; indeed,

$$\underline{M}_P = \sum_{i=1}^{N} \widetilde{q}_i \underline{F}_i = \sum_{i=1}^{N} (\widetilde{r}_{PC} + \widetilde{s}_i) \underline{F}_i = \widetilde{r}_{PC} \underline{F} + \sum_{i=1}^{N} \widetilde{s}_i \underline{F}_i,$$

where $\underline{r}_{PC}$ is the relative position vector of the center of mass with respect to point **P**. It the follows that

$$\underline{M}_P = \underline{M}_C + \widetilde{r}_{PC} \underline{F}. \tag{3.61}$$

Finally, the sum of the moments of all the internal forces of the system evaluated with respect to point **O** vanishes. First, the moments of all internal forces are grouped in pairs $(\tilde{r}_i \underline{f}_{ij} + \tilde{r}_j \underline{f}_{ji})$. Next, it is clear that $\tilde{r}_i \underline{f}_{ij} = \tilde{r}_\perp \underline{f}_{ij}$ and $\tilde{r}_j \underline{f}_{ji} = \tilde{r}_\perp \underline{f}_{ji}$, where $\tilde{r}_\perp$ is the vector that joins point **O** to the point on the common line of action of the internal force pair that is at the shortest distance from point **O**. It then follows that $\tilde{r}_i \underline{f}_{ij} + \tilde{r}_j \underline{f}_{ji} = \tilde{r}_\perp (\underline{f}_{ij} + \underline{f}_{ji}) = \underline{0}$, by virtue on Newton's third law. In summary,

$$\sum_{i=1}^{N} \sum_{j=1, \, j\neq i}^{N} \tilde{r}_i \underline{f}_{ij} = \underline{0}. \tag{3.62}$$

Note that the sum of these moments vanishes when computed with respect to any arbitrary point.

3.4.3 Linear and angular momenta

The *linear momentum* of the system, $\underline{P}$, is the sum of the linear momenta of the individual particles

$$\underline{P} = \sum_{i=1}^{N} m_i \underline{v}_i. \tag{3.63}$$

A time derivative of expression $\underline{r}_i = \underline{r}_C + \underline{s}_i$ leads to $\underline{\dot{r}}_i = \underline{\dot{r}}_C + \underline{\dot{s}}_i = \underline{v}_C + \underline{\dot{s}}_i$, where $\underline{v}_C = \underline{\dot{r}}_C$ is the velocity vector of the center of mass. The system's linear momentum now becomes $\underline{P} = \sum_{i=1}^{N} m_i (\underline{\dot{r}}_C + \underline{\dot{s}}_i) = m \underline{\dot{r}}_C + \sum_{i=1}^{N} m_i \underline{\dot{s}}_i = m \underline{v}_C$, where the second property of the center mass, eq. (3.57b), was used. The linear momentum of the system then simply becomes

$$\underline{P} = m \underline{v}_C. \tag{3.64}$$

The angular momentum of the system computed with respect to the origin of the inertial frame, denoted $\underline{H}_O$, is the sum of the corresponding angular momenta of all particles the system

$$\underline{H}_O = \sum_{i=1}^{N} \tilde{r}_i \, m_i \underline{v}_i. \tag{3.65}$$

The subscript, $(\cdot)_O$, indicates the point about which the angular momentum is evaluated. The angular momentum vector can be computed with respect to any point; for instance, $\underline{H}_C = \sum_{i=1}^{N} \tilde{s}_i \, m_i \underline{v}_i$ is the angular momentum vector computed with respect to the center of mass and $\underline{H}_P = \sum_{i=1}^{N} \tilde{q}_i \, m_i \underline{v}_i$ the corresponding quantity evaluated with respect to an arbitrary point **P**, see fig. 3.39. These various quantities are not independent of each other; indeed

$$\underline{H}_P = \sum_{i=1}^{N} \tilde{q}_i \, m_i \underline{v}_i = \sum_{i=1}^{N} (\tilde{r}_{PC} + \tilde{s}_i) \, m_i \underline{v}_i = \tilde{r}_{PC} \underline{P} + \sum_{i=1}^{N} \tilde{s}_i \, m_i \underline{v}_i. \tag{3.66}$$

It follows that

$$\underline{H}_P = \underline{H}_C + \tilde{r}_{PC} \underline{P}. \tag{3.67}$$

3.4.4 Euler's laws for a system of particles

Euler's first law

Newton's second law applied to each of the N particles writes

$$\underline{F}_i + \sum_{j=1,\ j\neq i}^{N} \underline{f}_{ij} = m_i \underline{a}_i, \quad i = 1, 2, 3, \ldots N. \tag{3.68}$$

Although these equations are all correct, they are difficult to manipulate because, in general, the system comprises a very large number of particles.

To circumvent this problem, the equations of motion of all particles are added together to yield

$$\sum_{i=1}^{N} \underline{F}_i + \sum_{i=1}^{N} \sum_{j=1,\ j\neq i}^{N} \underline{f}_{ij} = \sum_{i=1}^{N} m_i \underline{a}_i. \tag{3.69}$$

The first term represents the sum of all externally applied forces on the system, see eq. (3.58). The second term vanishes in view of eq. (3.59). The last term is simplified by introducing the expression for the center of mass: $\sum_{i=1}^{N} m_i \underline{a}_i = \sum_{i=1}^{N} m_i (\underline{\ddot{r}}_C + \underline{\ddot{s}}_i) = \sum_{i=1}^{N} m_i \underline{\ddot{r}}_C = m \underline{\ddot{r}}_C = m \underline{a}_C$, where the property of the center mass, eq. (3.57c), was used. It follows that

$$\underline{F} = m \underline{a}_C. \tag{3.70}$$

This result is known as *Euler's first law [14, 15]*.

Law 5 (Euler's first law) *The inertial acceleration vector of the center of mass of a system of particles is proportional to the vector sum of all externally applied forces; the constant of proportionality is the total mass of the system.*

Note the striking resemblance between eq. (3.70) and Newton's second law for a single particle, eq. (3.4). It appears that Newton's second law can be applied to a fictitious particle of mass m located at the center of mass of the system and subjected to all the forces externally applied on the system.

Equation (3.70) is much more convenient to use than the N equations of motion for each individual particle; it gives information about the overall response of the system in terms of the motion of its center of mass. Much information, however, has been lost: the N individual vector equations, eqs. (3.68), gave rise to a single vector equation of motion for the system, eq. (3.70). In fact, this latter equation cannot predict the motion of individual particles, nor does it allow to predict the internal forces in the system. In view of eq. (3.64), the time derivative of the linear momentum is $\underline{\dot{P}} = m \underline{a}_C$, and hence

$$\underline{F} = \underline{\dot{P}}. \tag{3.71}$$

Clearly, this equation is identical to Euler's first law, eq. (3.70).

Law 6 (Alternative statement of Euler's first law) *The time derivative of the linear momentum vector of a system of particles equals the sum of all externally applied forces.*

Euler's second law

To extract additional information about the response of the system, the moments of the equations of motion for individual particles, eqs. (3.68), with respect to the origin of the inertial frame are evaluated and summed up for all particles to yield

$$\sum_{i=1}^{N} \tilde{r}_i \underline{F}_i + \sum_{i=1}^{N} \sum_{j=1, \, j\neq i}^{N} \tilde{r}_i \underline{f}_{ij} = \sum_{i=1}^{N} m_i \tilde{r}_i \underline{a}_i. \qquad (3.72)$$

The first term represents the moment of the externally applied forces computed with respect to point **O**, see eq. (3.60). The second term vanishes in view of eq. (3.62). Equation (3.72) now reduces to

$$\underline{M}_O = \sum_{i=1}^{N} m_i \tilde{r}_i \underline{a}_i. \qquad (3.73)$$

The right-hand side of this equation can be expressed in a simpler manner in terms of the angular momentum vector; indeed, a time derivative of eq. (3.65) yields

$$\dot{\underline{H}}_O = \sum_{i=1}^{N} m_i \tilde{v}_i \underline{v}_i + \sum_{i=1}^{N} m_i \tilde{r}_i \underline{a}_i = \sum_{i=1}^{N} m_i \tilde{r}_i \underline{a}_i. \qquad (3.74)$$

Comparing the last two equations then leads to

$$\underline{M}_O = \dot{\underline{H}}_O. \qquad (3.75)$$

This result is known as *Euler's second law [14, 15].*

Law 7 (Euler's second law) *The time derivative of the angular momentum vector of a system of particles equals the sum of all moments externally applied to the system, when these quantities are evaluated with respect to a common inertial point.*

Introducing eqs. (3.61) and (3.67) into eq. (3.75) leads to

$$\underline{M}_C + \tilde{r}_C \underline{F} = \dot{\underline{H}}_C + \tilde{v}_C \underline{P} + \tilde{r}_C \dot{\underline{P}} = \dot{\underline{H}}_C + \tilde{r}_C \underline{F},$$

which reduces to

$$\underline{M}_C = \dot{\underline{H}}_C. \qquad (3.76)$$

This is another form of Euler's second law for a system of particles.

Law 8 (Alternative statement of Euler's second law) *The time derivative of the angular momentum vector of a system of particles equals the sum of all moments externally applied to the system, when these quantities are evaluated with respect to the system's center of mass.*

It would be erroneous to believe that this statement holds when moments and angular momentum vectors are evaluated with respect to an arbitrary point **P**. Indeed, introducing eqs. (3.61) and (3.67) into eq. (3.75) leads to $\underline{M}_P - \tilde{r}_{PC} \underline{F} = \dot{\underline{H}}_P - (\tilde{v}_C - \tilde{v}_P) \underline{P} - \tilde{r}_{PC} \dot{\underline{P}} = \dot{\underline{H}}_P + \tilde{v}_P \underline{P} - \tilde{r}_{PC} \underline{F}$, and finally

$$\underline{M}_P = \dot{\underline{H}}_P + \tilde{v}_P \underline{P}. \qquad (3.77)$$

3.4.5 The principle of work and energy

In section 3.1.4, the principle of work and energy was developed for a single particle, see eq. (3.12). When dealing with a system of particles, this principle can be applied to each individual particle, leading to

$$\int_{t_i}^{t_f} (\underline{F}_i^T + \sum_{j=1,\ j\neq i}^{N} \underline{f}_{ij}^T)\, d\underline{r}_i = K_i(t_f) - K_i(t_i). \tag{3.78}$$

In this expression, $K_i = 1/2\, m_i v_i^2$ represents the kinetic energy of particle i. As was done in the previous sections, the equations for each individual particle are added together to find

$$\int_{t_i}^{t_f} \sum_{i=1}^{N} (\underline{F}_i^T\, d\underline{r}_i) + \int_{t_i}^{t_f} \sum_{i=1}^{N} \sum_{j=1,\ j\neq i}^{N} (\underline{f}_{ij}^T\, d\underline{r}_i) = \sum_{i=1}^{N} K_i(t_f) - \sum_{i=1}^{N} K_i(t_i). \tag{3.79}$$

This first term clearly represents the work done by all externally applied forces. The second term is a complex double summation over the work done by all internal forces. It would be erroneous to believe that this term vanishes; indeed, consider two internal forces that obey Newton's third law: $\underline{f}_{ij} + \underline{f}_{ji} = 0$. The differential work done by these two forces is $\underline{f}_{ij}^T d\underline{r}_i + \underline{f}_{ji}^T d\underline{r}_j = \underline{f}_{ij}^T(d\underline{r}_i - d\underline{r}_j) \neq 0$, since the two particles have two distinct differential displacements along their distinct paths. Finally, the terms on the right-hand side represent the difference between the total kinetic energies of the system at the final and initial times. The total kinetic energy of the system, K, is found by summing up the contributions of each individual particle, $K = \sum_{i=1}^{N} K_i$.

The *principle of work and energy for a system of particles* now becomes

$$W_{t_i \to t_f} = \int_{t_i}^{t_f} \sum_{i=1}^{N} (\underline{F}_i^T\, d\underline{r}_i) + \int_{t_i}^{t_f} \sum_{i=1}^{N} \sum_{j=1,\ j\neq i}^{N} (\underline{f}_{ij}^T\, d\underline{r}_i) = K(t_f) - K(t_i). \tag{3.80}$$

Although this statement is correct, it is of limited practical use because it requires the evaluation of the work done by all internal forces. This contrasts with the equations of motion derived in the previous section, eqs. (3.71) and (3.75), which do not involve the internal forces.

For specific systems of particles it will be possible to prove that the term involving the work done by all internal forces does indeed vanish; this is the case for a rigid body, for instance. In such case, the internal forces do not appear in the statement of the principle of work and energy that can then be used conveniently.

3.4.6 The principle of impulse and momentum

It is interesting to integrate Euler's first law, eq. (3.71), over a period of time from t_i to t_f, to find

$$\int_{t_i}^{t_f} \underline{F}(t) \, dt = \int_{t_i}^{t_f} \underline{\dot{P}} \, dt = \underline{P}(t_f) - \underline{P}(t_i). \tag{3.81}$$

The term on the left-hand side is called the *linear impulse of all externally applied forces*. Equation (3.81) expresses the principle of linear impulse and momentum for a system of particles.

Principle 5 (Principle of linear impulse and momentum for a system) *The lin-ear impulse of all externally applied forces equals the change in linear momentum of the system of particles.*

In the absence of external forces, this principle implies $\underline{P}(t_f) = \underline{P}(t_i)$, *i.e.*, the system's linear momentum remains constant at all times, since t_i and t_f are instants chosen arbitrarily.

A similar treatment of Euler's second law, eq. (3.75), leads to

$$\int_{t_i}^{t_f} \underline{M}_O(t) \, dt = \int_{t_i}^{t_f} \underline{\dot{H}}_O \, dt = \underline{H}_O(t_f) - \underline{H}_O(t_i). \tag{3.82}$$

The term on the left-hand side is called the *angular impulse of all externally applied forces*. Equation (3.82) expresses the principle of angular impulse and momentum for a system of particles.

Principle 6 (Principle of angular impulse and momentum for a system) *The angular impulse of all externally applied forces equals the change in angular momentum of the system when both angular impulse and momentum are computed with respect to the same inertial point.*

In the absence of external moments with respect to point **O**, this principle implies $\underline{H}_O(t_f) = \underline{H}_O(t_i)$, *i.e.*, the system's angular momentum remains constant at all times.

Of course, a similar principle can be derived from eq. (3.76); in this case, both angular impulse and momentum must be evaluated with respect to the center of mass of the system of particles.

3.4.7 Problems

Problem 3.25. Particles interconnected by a massless link
Consider the dumbbell consisting of two particles of mass m_1 and m_2, respectively, connected by a massless arm of constant length ℓ, as depicted in fig. 3.40. *(1)* Show that the work done by all internal forces in the system vanishes. *(2)* Write the principle of work and energy for the system. Hint: the differential displacements of the two particles $d\underline{r}_1$ and $d\underline{r}_2$, respectively, are not independent; they must satisfy the constraint imposed by the constant length bar.

Problem 3.26. Particles linked by an inextensible cable
The system depicted in fig. 3.41 consists of two particles of mass m_1 and m_2, respectively, linked by an inextensible cable. *(1)* Show that the work done by all internal forces in the system vanishes. *(2)* Write the principle of work and energy for the system.

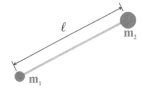

Fig. 3.40. Particles interconnected by a mass-less link.

Fig. 3.41. Two particles linked by an inexten-sible cable.

Problem 3.27. System of three rigidly connected particles

Figure 3.42 depicts three particles of masses m_1, m_2, and m_3, respectively, located at the vertices of an equilateral triangle with sides of length ℓ. Particle m_1 touches a fixed plane at point **A** at all times and the system is subjected to a gravity field as indicated on the figure. (1) Derive the equation of motion of the system based on Euler's second law. (2) Show that the same equation can be obtained from the principle of conservation of total mechanical energy. (3) Plot the time history of angle θ. (4) Plot the time history of angular velocity θ'. (5) Find the reaction forces at point **A**. (6) On one graph, plot the time histories of the non-dimensional normal contact force, $\bar{F}^n$, and friction force, $\bar{F}^f$, at point **A**. (7) If the static friction coefficient at point **A** is $\mu_s = 0.5$, for what value of angle θ will particle m_1 start sliding? Use the following data: $m_1 = 10$, $m_2 = 2$, and $m_3 = 10$ kg. Use the non-dimensional time $\tau = t\sqrt{g/\ell}$, and non-dimensional forces $\bar{F} = F/(mg)$, where $m = m_1 + m_2 + m_3$. At time $t = 0$, $\theta = 2\pi/3$ radians and $\theta' = -1$, where $(\cdot)'$ indicates a derivative with respect to the non-dimensional time τ. Present all your results for $\tau \in [0, 2]$.

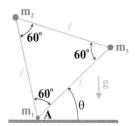

Fig. 3.42. Three interconnected particles touching a plane at point **A**.

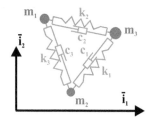

Fig. 3.43. Three interconnected particles.

Problem 3.28. System of three interconnected particles

Figure 3.43 shows a system of three particles of masses m_1, m_2, and m_3, respectively. The particles are linked by springs of stiffness constants k_1, k_2, and k_3, respectively, and dashpots of constants c_1, c_2, and c_3, respectively. The un-stretched lengths of the springs are ℓ_1, ℓ_2, and ℓ_3 respectively. (1) Draw a free body diagram of each particle. (2) Derive the equations of motion of the system. (3) Solve these equations numerically for a period of 50 s. (4) On one graph, plot the coordinates of particle 1 relative to the center of mass as a function of time. (5) Plot the relative coordinates of particle 2 versus time. (6) Plot the relative coordinates of particle 3 versus time. (7) Plot the magnitude of the forces in the three spring/dashpot systems. (8) Plot the components of the linear momentum vector of the system. Comment on your results. *(9)* Plot the components of the angular momentum vector of the system computed with respect to the center of mass. Comment on your results. *(10)* Plot the kinetic, and strain energies of the system. *(11)* Compute the energy dissipated in the dashpots. *(12)* Demonstrate that the principle of work and energy is verified by your solution. Use the following data: $m_1 = 5$, $m_2 = 3$ and $m_3 = 7$ kg; $k_1 = 12$, $k_2 = 25$ and $k_3 = 6$ N/m; $c_1 = 0.25$, $c_2 = 0.12$ and $c_3 = 0.16$ N·s/m; $\ell_1 = 0.6$, $\ell_2 = 0.4$ and $\ell_3 = 0.9$ m. At the initial time, the position vector of the particles are $\underline{r}_1 = \{0, 0, 0\}$, $\underline{r}_2 = \{0.9, 0, 0\}$ and $\underline{r}_3 = \{0, 0.4, 0\}$ m, respectively. The initial velocities of the particles are $\underline{v}_1 = \{-25, 25, 0\}$ and $\underline{v}_2 = \underline{v}_3 = \{0, 0, 0\}$.

Problem 3.29. System of three interconnected particles

Figure 3.43 shows a system of three particles of masses m_1, m_2, and m_3, respectively. The particles are linked by springs of stiffness constants k_1, k_2, and k_3, respectively, and dashpots of constants c_1, c_2, and c_3, respectively. The un-stretched lengths of the springs are ℓ_1, ℓ_2, and ℓ_3 respectively. The system of particle evolves freely in two-dimensional space. *(1)* Is the linear momentum of the system preserved? *(2)* Is the angular momentum of the system preserved. *(3)* Is the total mechanical energy of the system preserved. *(4)* Write an energy related quantity that is preserved during the evolution of the system.

Problem 3.30. Particles interconnected by a spring and damper

Figure 3.44 shows two particles of mass m_1 and m_2 connected together by a spring of stiffness k and a damper of constant c. The initial configuration of the system is indicated on the figure and the initial velocity vectors of the two particles are $\underline{v}_{10}$ and $\underline{v}_{20}$, respectively. The following quantities are defined: $\Omega^2 = k/m$ and $c = 2m\Omega\zeta$, where $m = m_1 + m_2$. In the present configuration, r_1 and r_2 measure the distance from the center of mass to particles m_1 and m_2, respectively. For this problem, it is convenient to use the polar coordinate system indicated on the figure with its origin at the center of mass of the system. *(1)* When applying Newton's second law to this problem, can the accelerations of the particles with respect to the center of mass be used? Justify your answer. *(2)* Are r_1 and r_2 independent variables? *(3)* Set up the equations of motion of the system. *(4)* Plot r_1 as a function of the non-dimensional time $\tau = \Omega t$ for $\tau \in [0, 10\pi]$. *(5)* Plot $\theta(\tau)$. *(6)* Plot the trajectory of the particle in space. *(7)* Plot the history of the non-dimensional angular velocity $\bar{\Omega}(\tau) = \dot{\theta}/\Omega$. *(8)* Plot the history of the components of the velocity vector of particle m_1 in the inertial frame, $\bar{v}_{1x} = v_{1x}/(\Omega r_{10})$ and $\bar{v}_{1y} = v_{1y}/(\Omega r_{10})$. Plot the corresponding quantities for the velocity vector of the second particle. *(9)* Plot the history of the non-dimensional total mechanical energy of the system $\bar{E}(\tau) = E/(m\Omega^2 r_{10}^2)$; comment your result. *(10)* Compute the non-dimensional cumulative energy dissipated in the damper $\bar{W}(\tau) = W/(m\Omega^2 r_{10}^2)$ as a function of τ. *(11)* Plot the history of the quantity $\bar{E}(\tau) + \bar{W}(\tau)$; comment your result. Use the following data: $\mu_1 = m_1/m = 0.3$; $\zeta = 0.02$; $\underline{v}_1(t = 0)/(\Omega r_{10}) = -0.1\bar{\imath}_1 - 0.5\,\bar{\imath}_2$; $\underline{v}_2(t = 0)/(\Omega r_{10}) = 2.2\,\bar{\imath}_1 + 0.6\,\bar{\imath}_2$; $\theta_0 = 0$; $r_1(t = 0)/r_{10} = 1$;

Problem 3.31. Particle suspended from a circular track

Figure 3.45 shows a particle of mass M sliding along a track defined by a curve $\underline{p}_0(s)$ under the effect of gravity. A particle of mass m is suspended from the first particle by means of a spring of stiffness constant k in parallel with a dashpot of constant c. The un-stretched length of the spring is Δ_0. A viscous friction force $F^f = \mu\dot{s}$ is acting between particle M and the track. *(1)* Derive the three equations of motion of the system for a curve of arbitrary shape. *(2)* Particularize the equations of motion to the case where the curve is a circle of radius R, as depicted in the right portion of fig. 3.45. *(3)* Solve these equations numerically. *(4)* On one graph, plot the coordinates, $\bar{x}_1 = x_1/R$ and $\bar{x}_2 = x_2/R$, of particle m. *(5)* Plot the history of

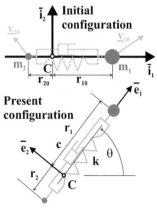

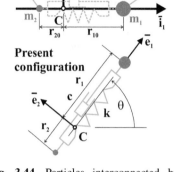

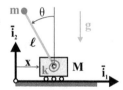

Fig. 3.44. Particles interconnected by a spring and damper.

Fig. 3.45. Particle suspended from a circular track.

angle θ. *(6)* On one graph, plot the velocity components, $\bar{v}_1 = v_1/(R\omega)$ and $\bar{v}_2 = v_2/(R\omega)$, of particle m. *(7)* Plot the history of the angular rate $\theta' = \dot{\theta}/\omega$. *(8)* Plot the history of the normal force, F^c, the track applies on the particle, as $\bar{F}^c = F^c/(mR\omega^2)$. *(9)* Plot the non-dimensional kinetic energy, $\bar{K} = K/mR^2\omega^2$, potential energy, $\bar{V} = V/(mR^2\omega^2)$, energy dissipated in the damper, $\bar{W}^c = W^c/(mR^2\omega^2)$, and energy dissipated at the viscous friction interface, $\bar{W}^f = W^f/(mR^2\omega^2)$. Verify that the energy closure equation is satisfied. Use the following non-dimensional quantities: $\theta = s/R$, and non-dimensional time $\tau = \omega t$, where $\omega^2 = k/m$. Use the following data: $\zeta = c/(2m\omega) = 0.2$; $\bar{\Delta}_0 = \Delta_0/R = 0.5$; $\bar{g} = g/(R\omega^2) = 2.5$; $\bar{\mu} = \mu/(M\omega) = 0.2$; $\bar{m} = m/M = 0.25$. At the initial time, the system is at rest, the position vector of particle m is $\underline{x} = \{-(1 + \bar{\Delta}_0), 0\}$ and $\theta = 0$. Present all your results for $\tau \in [0, 20]$

Problem 3.32. Two particles linked by an elastic spring
Consider the system depicted in fig. 3.46 that consists of two particles of mass m_1 and m_2, respectively, connected by a massless spring of stiffness k. *(1)* Show that the work done by the force in the elastic spring, can be derived from a potential. *(2)* What is the expression of the strain energy function of the spring if it is a linearly elastic spring of stiffness constant k.

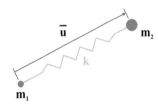

Fig. 3.46. Two particles linked by an elastic spring.

Fig. 3.47. Inverted pendulum mounted on a track.

Problem 3.33. Inverted pendulum mounted on a track

Figure 3.47 shows an inverted pendulum of length ℓ with a tip mass m. The pendulum is mounted on a cart of mass M free to translate along a horizontal track. A torsional spring of stiffness constant k restrains the pendulum at its attachment point. The spring is un-stretched when angle $\theta = 0$. *(1)* Derive the two equations of motion of the system. *(2)* Solve these equations numerically. *(3)* Plot the cart's position, $\bar{x} = x/\ell$ versus τ. *(4)* Plot angle θ. *(5)* Plot the cart's velocity, $\bar{x}'$. *(6)* Plot θ'. *(7)* Plot the cart's acceleration, $\bar{x}''$. *(8)* Plot θ''. *(9)* Plot the system's kinetic, $\bar{K} = K/m\ell^2\omega^2$, potential, $\bar{V} = V/(m\ell^2\omega^2)$, and total mechanical energies. Use the following data: $\mu = M/m = 1.5$. Use non-dimensional time $\tau = \omega t$, where $\omega^2 = k/(m\ell^2)$ and $(\cdot)'$ denotes a derivative with respect to τ. At the initial time, $\bar{x} = 0$, $\bar{x}' = 1$, $\theta = \pi/4$, $\theta' = 0$. Present all your results for $\tau \in [0, 20]$. Study two cases, $\bar{g} = g/(\ell\omega^2) = 0.8$ and $\bar{g} = 4$, and comment on the differences.

Problem 3.34. Flexible pendulum on a slider

Figure 3.48 depicts a slider of mass M constrained to move along a horizontal track. A bob of mass m is attached to the slider at point **A** by means of a spring of stiffness constant k and un-stretched length r_0. The displacement of the slider is denoted x, and the position of the bob is expressed by its polar coordinates r and θ. Gravity acts on the system as indicated in the figure. The bob is subjected to a drag force $\underline{F}_d = -\varrho A c_d \|\underline{v}_m\|\underline{v}_m$, where ϱ is the fluid mass density, A the cross-sectional area of the bob,

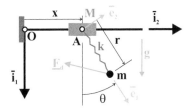

Fig. 3.48. Flexible pendulum mounted on a slider.

c_d is the drag coefficient, and $\underline{v}_m$ the velocity vector of the bob. *(1)* Derive the equations of motion of the system using the coordinates x, r, and θ. *(2)* Solve these equations numerically. *(3)* Plot the cart's position, $\bar{x} = x/r_0$ versus τ. *(4)* Plot distance $\bar{r} = r/r_0$. *(5)* Plot angle θ. *(6)* Plot the cart's velocity, $\bar{x}'$. *(7)* Plot $\bar{r}'$. *(8)* Plot θ'. *(9)* Plot the cart's acceleration, $\bar{x}''$. *(10)* Plot $\bar{r}''$. *(11)* Plot θ''. *(12)* Plot the cumulative dissipated energy, $\bar{W}_d = W_d/(mr_0^2\omega^2)$. *(13)* Plot the system's kinetic, $\bar{K} = K/mr_0^2\omega^2$, potential, $\bar{V} = V/(mr_0^2\omega^2)$, and total mechanical energies. Check the energy closure equations. Use the following data: $\mu = M/m = 1.5$, $\bar{g} = g/(r_0\omega^2) = 0.2$, and $\zeta = \bar{\varrho}\bar{A}c_d = 0.01$, where $\bar{\varrho} = r_0^3\varrho/m$ and $\bar{A} = A/r_0^2$. Use non-dimensional time $\tau = \omega t$, where $\omega^2 = k/m$ and $(\cdot)'$ denotes a derivative with respect to τ. At the initial time, $\bar{x} = 0$, $\bar{r} = 2$, $\theta = \pi/4$, $\bar{x}' = 1$, $\bar{r}' = 1$, $\theta' = 0$. Present all your results for $\tau \in [0, 20]$.

4

The geometric description of rotation

The most natural way of describing rotations is rooted in their *geometric represen-tation*, which is the focus of this chapter. More abstract approaches, however, also exist and will be presented in chapter 13.

Consider an orthonormal basis $\mathcal{I} = (\bar{\imath}_1, \bar{\imath}_2, \bar{\imath}_3)$. The *rotation operation* brings orthonormal basis $\mathcal{I}$ to a new orthonormal basis $\mathcal{E} = (\bar{e}_1, \bar{e}_2, \bar{e}_3)$. In section 4.1, the rotation operation is characterized by expressing the unit vectors of basis $\mathcal{E}$ in terms of those of basis $\mathcal{I}$. This leads to the concept of *direction cosine matrix*. The simplest rotation operation consists of a rotation of basis $\mathcal{I}$ about one of its unit vectors. This operation, called a *planar rotation*, is discussed in section 4.2. The fact that successive planar rotations in distinct planes do not commute is emphasized in section 4.3, and leads to the representation of arbitrary rotations in terms of three successive planar rotations. The resulting *Euler angle representation* is described in section 4.4.

Euler's theorem on rotations presented in section 4.5 states that any arbitrary rota-tion that leave a point fixed can be viewed as a single rotation about a unit vector. This fundamental result leads to the concept of *rotation tensor* presented in section 4.6; a formal definition of tensors follows. Important rotation operations are examined in details: the composition of rotations is presented in section 4.9, and time and space derivatives of rotations in sections 4.10 and 4.12, respectively. Applications to parti-cle dynamics are presented in section 4.13.

4.1 The direction cosine matrix

Consider the two orthonormal bases $\mathcal{I} = (\bar{\imath}_1, \bar{\imath}_2, \bar{\imath}_3)$ and $\mathcal{E} = (\bar{e}_1, \bar{e}_2, \bar{e}_3)$ shown in fig. 4.1. A *rotation* is defined as the operation that brings basis $\mathcal{I}$ to basis $\mathcal{E}$. Unit vector $\bar{e}_1$ can be expressed as a linear combination of the vectors of basis $\mathcal{I}$

$$\bar{e}_1 = D_{11}\bar{\imath}_1 + D_{21}\bar{\imath}_2 + D_{31}\bar{\imath}_3. \tag{4.1}$$

The coefficients of this linear combination are readily expressed as $D_{k1} = \bar{\imath}_k^T \bar{e}_1$. Proceeding similarly with the three unit vectors defining basis $\mathcal{E}$ yields the terms of the *direction cosine matrix, $\underline{\underline{D}}$,* as

$$D_{k\ell} = \bar{\imath}_k^T \bar{e}_\ell. \tag{4.2}$$

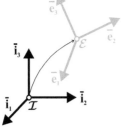

Using eq. (1.8), and observing that vectors $\bar{\imath}_k$ and $\bar{e}_\ell$ are unit vectors yields an alternative expression for the direction cosine matrix is obtained

$$D_{kl} = \cos\left(\bar{\imath}_k, \bar{e}_\ell\right). \tag{4.3}$$

This expression gives its name to the direction cosine matrix: its entries are the cosine of the angle between $\bar{\imath}_k$ and $\bar{e}_\ell$, the unit vectors defining bases $\mathcal{I}$ and $\mathcal{E}$, respectively. Each component of the direction cosine matrix is a scalar quantity. The direction cosine matrix, however, is not a second-order tensor, see section 4.8.2.

Fig. 4.1. rotation from basis $\mathcal{I}$ to $\mathcal{E}$.

The matrix of direction cosines provides a simple description of rotations. Each term of the direction cosine matrix is a scalar quantity representing the cosine of the angle between two vectors, eq. (4.2). As will be shown in the following sections, rotations can be represented by as few as three parameters. This basic property of rotation is not apparent in this description.

4.2 Planar rotations

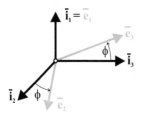

Fig. 4.2. Planar rotation of magnitude ϕ about axis $\bar{\imath}_1$.

A simple example of a rotation is a *planar rotation* defined as a rotation of angular magnitude ϕ about one of the axes defining basis $\mathcal{I}$, say $\bar{\imath}_1$, as depicted in fig. 4.2. The direction cosine matrix corresponding to this planar rotation can be readily obtained from eq. (4.3) and inspection of fig. 4.2. The coefficients of the direction cosine matrix are obtained from elementary trigonometry as

$$\begin{cases} \bar{e}_1 = \bar{\imath}_1 \\ \bar{e}_2 = \quad \cos\phi\,\bar{\imath}_2 + \sin\phi\,\bar{\imath}_3 \\ \bar{e}_3 = \quad -\sin\phi\,\bar{\imath}_2 + \cos\phi\,\bar{\imath}_3 \end{cases} \Longleftrightarrow \underline{\underline{D}}_1(\phi) = \begin{bmatrix} 1 & 0 & 0 \\ 0 & \cos\phi & -\sin\phi \\ 0 & \sin\phi & \cos\phi \end{bmatrix}. \tag{4.4}$$

The direction cosine matrix corresponding to planar rotation of magnitude ϕ about axis $\bar{\imath}_2$ is found in a similar manner as

$$\begin{cases} \bar{e}_1 = \cos\phi\,\bar{\imath}_1 \quad -\sin\phi\,\bar{\imath}_3 \\ \bar{e}_2 = \quad \bar{\imath}_2 \\ \bar{e}_3 = \sin\phi\,\bar{\imath}_1 \quad +\cos\phi\,\bar{\imath}_3 \end{cases} \Longleftrightarrow \underline{\underline{D}}_2(\phi) = \begin{bmatrix} \cos\phi & 0 & \sin\phi \\ 0 & 1 & 0 \\ -\sin\phi & 0 & \cos\phi \end{bmatrix}. \tag{4.5}$$

The corresponding matrix for a planar rotation of magnitude ϕ about axis $\bar{\imath}_3$ becomes

$$\begin{cases} \bar{e}_1 = \quad \cos\phi\,\bar{\imath}_1 + \sin\phi\,\bar{\imath}_2 \\ \bar{e}_2 = -\sin\phi\,\bar{\imath}_1 + \cos\phi\,\bar{\imath}_2 \\ \bar{e}_3 = \qquad\qquad\qquad \bar{\imath}_3 \end{cases} \iff \underline{\underline{D}}_3(\phi) = \begin{bmatrix} \cos\phi & -\sin\phi & 0 \\ \sin\phi & \cos\phi & 0 \\ 0 & 0 & 1 \end{bmatrix}. \quad (4.6)$$

4.3 Non-commutativity of rotations

Rotation operations do not commute. This means that the order in which successive rotations are performed is important. This point is most easily understood by looking at the simple example depicted in fig. 4.3.

A rigid block is rotated by 90° about $\bar{\imath}_2$, then by 90° about $\bar{\imath}_3$. The final configuration of the block is shown in the top portion of fig. 4.3. The same rigid block is now rotated by 90° about $\bar{\imath}_3$, then by 90° about $\bar{\imath}_2$. The final con-

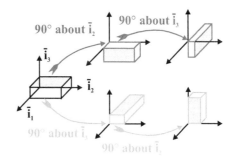

Fig. 4.3. Two successive planar rotations do not commute.

figuration, depicted in the bottom portion of fig. 4.3, is clearly different from that obtained when the two successive rotations were performed in the reverse order.

In this example, the two successive rotations are performed about axes fixed in space. If the two rotations are performed about body fixed axes, the same conclusion is reached: the final configuration depends on the order of the rotation operations.

In the next section, it will be shown that an arbitrary rotation can be viewed as a succession of three planar rotations. The fact that rotation operations about distinct axes do not commute implies that the order in which these three successive planar rotations are performed is important. More generally, when several rotations are involved in a problem, the order of application of these rotations must be carefully specified.

4.4 Euler angles

An arbitrary rotation from $\mathcal{I} = (\bar{\imath}_1, \bar{\imath}_2, \bar{\imath}_3)$ to $\mathcal{E} = (\bar{e}_1, \bar{e}_2, \bar{e}_3)$ can be viewed as a succession of three planar rotations about three different axes [16].

Figure 4.4 shows one possible set of three planar rotations, which can be described as follows.

1. A planar rotation of magnitude ϕ, called *precession*, about axis $\bar{\imath}_3$ brings basis $\mathcal{I}$ to basis $\mathcal{A} = (\bar{a}_1, \bar{a}_2, \bar{a}_3)$. Equation (4.6) gives the corresponding direction cosine matrix

$$\begin{cases} \bar{a}_1 = \quad \cos\phi\,\bar{\imath}_1 + \sin\phi\,\bar{\imath}_2, \\ \bar{a}_2 = -\sin\phi\,\bar{\imath}_1 + \cos\phi\,\bar{\imath}_2, \\ \bar{a}_3 = \qquad\qquad\qquad \bar{\imath}_3. \end{cases} \iff \underline{\underline{D}}_3(\phi) = \begin{bmatrix} \cos\phi & -\sin\phi & 0 \\ \sin\phi & \cos\phi & 0 \\ 0 & 0 & 1 \end{bmatrix}. \quad (4.7)$$

2. A planar rotation of magnitude θ, called *nutation*, about axis $\bar{a}_1$ brings basis $\mathcal{A}$ to basis $\mathcal{B} = (\bar{b}_1, \bar{b}_2, \bar{b}_3)$. Equation (4.4) gives the corresponding direction cosine matrix

$$
\begin{cases} \bar{b}_1 = \bar{a}_1, \\ \bar{b}_2 = \cos\theta\,\bar{a}_2 + \sin\theta\,\bar{a}_3, \\ \bar{b}_3 = -\sin\theta\,\bar{a}_2 + \cos\theta\,\bar{a}_3. \end{cases} \iff \underline{\underline{D}}_1(\theta) = \begin{bmatrix} 1 & 0 & 0 \\ 0 & \cos\theta & -\sin\theta \\ 0 & \sin\theta & \cos\theta \end{bmatrix}. \quad (4.8)
$$

3. A planar rotation of magnitude ψ, called *spin*, about axis $\bar{b}_3$ brings basis $\mathcal{B}$ to basis $\mathcal{E}$. Once again, eq. (4.6) gives the corresponding direction cosine matrix

$$
\begin{cases} \bar{e}_1 = \cos\psi\,\bar{b}_1 + \sin\psi\,\bar{b}_2, \\ \bar{e}_2 = -\sin\psi\,\bar{b}_1 + \cos\psi\,\bar{b}_2, \\ \bar{e}_3 = \bar{b}_3. \end{cases} \iff \underline{\underline{D}}_3(\psi) = \begin{bmatrix} \cos\psi & -\sin\psi & 0 \\ \sin\psi & \cos\psi & 0 \\ 0 & 0 & 1 \end{bmatrix}. \quad (4.9)
$$

The relationship between bases $\mathcal{I}$ and $\mathcal{E}$ is obtained by combining the three successive rotations described by eqs. (4.7) to (4.9) to find

$$
\begin{cases} \bar{e}_1 = (C_\phi C_\psi - S_\phi C_\theta S_\psi)\,\bar{\imath}_1 + (S_\phi C_\psi + C_\phi C_\theta S_\psi)\,\bar{\imath}_2 + S_\theta S_\psi\,\bar{\imath}_3, \\ \bar{e}_2 = (-C_\phi S_\psi - S_\phi C_\theta C_\psi)\,\bar{\imath}_1 + (-S_\phi S_\psi + C_\phi C_\theta C_\psi)\,\bar{\imath}_2 + S_\theta C_\psi\,\bar{\imath}_3, \\ \bar{e}_3 = S_\phi S_\theta\,\bar{\imath}_1 - C_\phi S_\theta\,\bar{\imath}_2 + C_\theta\,\bar{\imath}_3, \end{cases} \quad (4.10)
$$

where the following short-hand notations were used: $C_\phi = \cos\phi$, $S_\phi = \sin\phi$, etc.

The three rotation angles, ϕ, θ, and ψ, are called the *Euler angles*. The direction cosine matrix expressed in terms of Euler angles becomes

$$
\underline{\underline{D}}_{3\text{-}1\text{-}3} = \begin{bmatrix} C_\phi C_\psi - S_\phi C_\theta S_\psi & -C_\phi S_\psi - S_\phi C_\theta C_\psi & S_\phi S_\theta \\ S_\phi C_\psi + C_\phi C_\theta S_\psi & -S_\phi S_\psi + C_\phi C_\theta C_\psi & -C_\phi S_\theta \\ S_\theta S_\psi & S_\theta C_\psi & C_\theta \end{bmatrix}. \quad (4.11)
$$

It is often important to perform the inverse operation: given a direction cosine matrix, find the corresponding Euler angles. The following process will yield the desired angles. Assuming $D_{32} \neq 0$,

$$\tan\psi = D_{31}/D_{32}, \quad (4.12a)$$

$$\sin\theta = D_{31}\sin\psi + D_{32}\cos\psi, \quad \cos\theta = D_{33}, \quad (4.12b)$$

$$\sin\phi = D_{21}\cos\psi - D_{22}\sin\psi, \quad \cos\phi = D_{11}\cos\psi - D_{12}\sin\psi. \quad (4.12c)$$

To remove the ambiguity associated with inverse trigonometric functions, both sine and cosines of the angles are derived, leading to a definition of each angle in the range $[-\pi, +\pi]$.[1]

When $\theta = 0$ or π, a singularity occurs. In fact, the process then reduces to a single rotation of magnitude $(\phi + \psi)$ or $(\phi - \psi)$ for $\theta = 0$ or π, respectively, because the direction cosine matrix reduces to

---

[1] In computer implementations, these operations are conveniently performed with the help of the function $\text{atan2}(y, x) = \tan^{-1}(y/x)$, yielding an angle in the range $[-\pi, +\pi]$.

$$\underline{\underline{D}} = \begin{bmatrix} \cos(\phi \pm \psi) & -\sin(\phi \pm \psi) & 0 \\ \sin(\phi \pm \psi) & \cos(\phi \pm \psi) & 0 \\ 0 & 0 & 1 \end{bmatrix}. \tag{4.13}$$

Clearly, angles ϕ and ψ cannot be determined individually, the sole combination $\phi \pm \psi$ can be evaluated.

The Euler angles introduced above correspond to the following sequence of planar rotations: a rotation of magnitude ϕ, about axis $\bar{\imath}_3$, then, a rotation of magnitude θ about axis $\bar{a}_1$, and finally, a rotation of magnitude ψ about axis $\bar{b}_3$. This sequence will be called the "*3-1-3* sequence" to indicate the sequence of body axes about which the three successive rotations are taking place.

Clearly, Euler angles could be defined in several different manners: the first rotation could occur about either of the three axes, $\bar{\imath}_1$, $\bar{\imath}_2$, or $\bar{\imath}_3$, offering three choices. Because two consecutive rotations cannot take place about the same axis, two alternatives are possible for the second rotation. Two choices are again possible for the last rotation.

In all, $3 \times 2 \times 2 = 12$ possible choices exist, corresponding to sequences labeled *1-2-1*, *1-2-3*, *1-3-1*, *1-3-2*, *2-1-2*, *2-1-3*, *2-3-1*, *2-3-2*, *3-1-2*, *3-1-3*, *3-2-1* and *3-2-3*. Three of these sequences, *3-2-3*, *3-2-1* and *3-1-2* will be the focus of problems below. A summary of expressions and formulæ involving Euler angles appears in section 4.11.

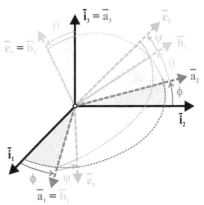

Fig. 4.4. An arbitrary rotation viewed as three successive planar rotations.

The representation of rotation in terms of three Euler angles shows that the direction cosine matrix can be expressed in terms of three parameters only. This representation, however, presents several drawbacks. First, Euler angles can be defined in several different manners, and the choice of the rotation sequence is entirely arbitrary. Furthermore, the expression for the direction cosine matrix, as seen for this example in eq. (4.11), is rather complicated and involves the evaluation of numerous trigonometric functions. Finally, singularities will occur in the evaluation of Euler angles from a direction cosine matrix for all 12 possible sequences.

4.4.1 Problems

Problem 4.1. Euler angles, sequence *3-2-3*
A popular choice of Euler angles is the *3-2-3* sequence that corresponds to the following sequence of planar rotations: a rotation of magnitude ψ, called *precession*, about axis $\bar{\imath}_3$, then, a rotation of magnitude θ, called *nutation*, about axis $\bar{a}_2$, and finally, a rotation of magnitude ϕ, called *spin*, about axis $\bar{b}_3$. *(1)* Find the rotation matrix in terms of this Euler angle sequence. *(2)* Determine the singularities associated with this choice of Euler angles.

Problem 4.2. Euler angles, sequence 3-2-1
A popular choice of Euler angles for airplane flight mechanics is the *3-2-1* sequence that corresponds to the following sequence of planar rotations: a rotation of magnitude ψ, called *heading*, about axis $\bar{i}_3$, then, a rotation of magnitude θ, called *attitude*, about axis $\bar{a}_2$, and finally, a rotation of magnitude ϕ, called *bank*, about axis $\bar{b}_1$. *(1)* Find the rotation matrix in terms of this Euler angle sequence. *(2)* Determine the singularities associated with this choice of Euler angles.

Problem 4.3. Euler angles, sequence 3-1-2
A possible choice of Euler angles is the *3-1-2* sequence that corresponds to the following sequence of planar rotations: a rotation of magnitude ϕ, about axis $\bar{i}_3$, then, a rotation of magnitude θ about axis $\bar{a}_1$, then finally, a rotation of magnitude ψ about axis $\bar{b}_2$. *(1)* Find the rotation matrix in terms of this Euler angle sequence. *(2)* Determine the singularities associated with this choice of Euler angles.

4.5 Euler's theorem on rotations

Euler's theorem [17] on rotations states the following.

Theorem 4.1 (Euler's theorem on rotations). *Any arbitrary rotation of a rigid body that leaves on of its point fixed can be viewed as a single rotation of magnitude ϕ about a unit vector $\bar{n}$.*

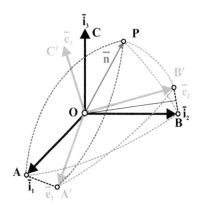

Fig. 4.5. An arbitrary rotation viewed as a single rotation about axis $\bar{n}$.

To prove this statement, consider two frames, $\mathcal{F}_1 = [\mathbf{O}, \mathcal{I} = (\bar{i}_1, \bar{i}_2, \bar{i}_3)]$ and $\mathcal{F}_2 = [\mathbf{O}, \mathcal{E} = (\bar{e}_1, \bar{e}_2, \bar{e}_3)]$, shown in fig. 4.5, and associated with two configurations of a rigid body that its material point $\mathbf{O}$ fixed. Because the vectors defining bases $\mathcal{I}$ and $\mathcal{E}$ are unit vectors, they all are radii of a sphere of unit radius and center $\mathbf{O}$.

Vector $\bar{i}_1$ can be brought to vector $\bar{e}_1$ by a single rotation about axis $\bar{n}_1$. This axis passes through point $\mathbf{O}$ and belongs to plane $\mathcal{P}_1$ that is normal to segment $\mathbf{AA'}$ and passes through point $\mathbf{O}$, as shown in fig. 4.5.

On the other hand, vector $\bar{i}_2$ can be brought to vector $\bar{e}_2$ by a single rotation about axis $\bar{n}_2$, passing through point $\mathbf{O}$ and belonging to plane $\mathcal{P}_2$, which is normal to the segment $\mathbf{BB'}$ and passes through point $\mathbf{O}$. If both operations must be achieved by a single rotation, axis $\bar{n}$ about which this common rotation takes place must be at the intersection of planes $\mathcal{P}_1$ and $\mathcal{P}_2$. Let point $\mathbf{P}$ be the intersection of axis $\bar{n}$ with the unit sphere.

Figure 4.5 shows the great circle segments $\mathbf{PA}$, $\mathbf{PA'}$, $\mathbf{PB}$, $\mathbf{PB'}$, $\mathbf{AB}$, and $\mathbf{A'B'}$. By construction, $\mathbf{PA} = \mathbf{PA'}$ because point $\mathbf{A}$ can be brought to point $\mathbf{A'}$ by a rotation about $\bar{n}$. Similarly, **PB** = **PB**′. Finally, **AB** = **A B**′ since both segments correspond to a 90 degree rotation.

Consequently, the spherical triangles **APB** and **A′PB**′ are equal. This, in turn, implies the equality of angles ∠**APB** and ∠**A′PB**′. Subtracting from these two angles their common part, ∠**A′PB**, yields the following result: ∠**APA**′ = ∠**BPB**′ = ϕ, where ϕ is now the magnitude of the rotation about axis $\bar{n}$ that simultaneously brings $\bar{\imath}_1$ to $\bar{e}_1$ and $\bar{\imath}_2$ to $\bar{e}_2$.

A rotation of magnitude ϕ about axis $\bar{n}$ has been shown to bring $\bar{\imath}_1$ to $\bar{e}_1$, and $\bar{\imath}_2$ to $\bar{e}_2$ simultaneously. It remains to prove that a rotation of the same magnitude will bring $\bar{\imath}_3$ to $\bar{e}_3$. Let the rotation of magnitude ϕ about axis $\bar{n}$ bring $\bar{\imath}_3$ to vector $\bar{\imath}_3'$. Reasoning as before, it is clear that **PC** = **PC**′, **PA** = **PA**′, and by construction ∠**APC** = ∠**A′PC**′ = ϕ + ∠**APC**′. This shows the equality of spherical triangles **APC** and **A′PC**′. This in turns implies the equality of segments **AC** and **A′C**′. Since segment **AC** corresponds to a 90 degree rotation, so does segment **A′C**′, implying the orthogonality of $\bar{\imath}_3'$ and $\bar{e}_1$. A similar reasoning on spherical triangles **BPC** and **B′PC**′ leads to the orthogonality of $\bar{\imath}_3'$ and $\bar{e}_2$. Finally, since $\bar{\imath}_3'$ is orthogonal to both $\bar{e}_1$ and $\bar{e}_2$, it is clear that $\bar{\imath}_3 = \bar{e}_3$.

In summary, basis $\mathcal{I}$ can be brought to basis $\mathcal{E}$ by a single rotation of magnitude ϕ about axis $\bar{n}$, which proves Euler theorem on rotations.

4.6 The rotation tensor

Euler's theorem on rotations leads a compact expression for the rotation tensor. Consider an arbitrary vector $\underline{a}$ and let the rotation of magnitude ϕ about unit vector $\bar{n}$ bring this vector to $\underline{b}$. The rotation tensor, $\underline{R}$, relates these two vectors, $\underline{b} = \underline{R}\,\underline{a}$

Basic expression for the rotation tensor

Figure 4.6 depicts the configuration of the problem. Vector $\underline{b}$ is the sum of segments **OC** and **CB**, and elementary geometry then yields $\underline{b} =$ **OC** + **CB** $= \|\underline{b}\|\cos\alpha\,\bar{n} + \|\underline{b}\|\sin\alpha\,[\bar{s}\cos\phi + \bar{t}\sin\phi]$. Unit vector $\bar{t}$ is along the vector product of vectors $\bar{n}$ and $\underline{a}$, $\bar{t} = \widetilde{n}\underline{a}/\|\widetilde{n}\underline{a}\|$, and unit vector $\bar{s}$ is $\bar{s} = \bar{t}\,\bar{n} = (\widetilde{n}\underline{a})\bar{n}/\|\widetilde{n}\underline{a}\|$.

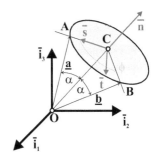

Fig. 4.6. A rotation of magnitude ϕ about axis $\bar{n}$.

The fundamental property of rotation is to preserve length, *i.e.*, the norms of vectors $\underline{a}$ and $\underline{b}$ must be identical, leading to $\bar{n}^T\underline{a} = \|\underline{a}\|\cos\alpha = \|\underline{b}\|\cos\alpha$ and $\|\widetilde{n}\underline{a}\| = \|\underline{a}\|\sin\alpha = \|\underline{b}\|\sin\alpha$. With the help of these relationships, vector $\underline{b}$ becomes

$\underline{b} = (\bar{n}^T\underline{a})\,\bar{n} + (\widetilde{n}\underline{a})\bar{n}\cos\phi + (\widetilde{n}\underline{a})\sin\phi$. Applying identity (1.34a) then leads to

$$\underline{b} = \underline{a} + \sin\phi\,(\widetilde{n}\underline{a}) + (1 - \cos\phi)\,\widetilde{n}\widetilde{n}\underline{a} = \underline{R}\,\underline{a}, \tag{4.14}$$

where the *rotation tensor* [15], $\underline{\underline{R}}$, is defined as

$$\underline{\underline{R}} = \underline{\underline{I}} + \sin\phi\,\widetilde{n} + (1 - \cos\phi)\widetilde{n}\widetilde{n}. \tag{4.15}$$

This result is known as *Rodrigues' rotation formula*.

This fundamental result expresses the rotation tensor in terms of a unit vector $\bar{n}$, and a rotation of magnitude ϕ about this unit vector. It is a direct consequence of Euler's theorem on rotations, theorem 4.1.

In view of eq. (4.15), vector $\bar{n}$ can be expressed as

$$\bar{n} = \frac{1}{2\sin\phi}\begin{Bmatrix} R_{32} - R_{23} \\ R_{13} - R_{31} \\ R_{21} - R_{12} \end{Bmatrix} = n_1 n_2 n_3 (1 - \cos\phi)\begin{Bmatrix} 1/(R_{32} + R_{23}) \\ 1/(R_{13} + R_{31}) \\ 1/(R_{21} + R_{12}) \end{Bmatrix}.$$

Hence, the orientation of this vector is

$$\bar{n} \parallel \begin{Bmatrix} R_{32} - R_{23} \\ R_{13} - R_{31} \\ R_{21} - R_{12} \end{Bmatrix}, \quad \bar{n} \parallel \begin{Bmatrix} 1/(R_{32} + R_{23}) \\ 1/(R_{13} + R_{31}) \\ 1/(R_{21} + R_{12}) \end{Bmatrix},$$

where symbol $\parallel$ indicates the parallelism of two vectors.

Relating the rotation tensor to the matrix of direction cosines

The rotation tensor and matrix of direction cosines are closely related to each other. Consider a rotation that brings basis $\mathcal{I} = (\bar{\imath}_1, \bar{\imath}_2, \bar{\imath}_3)$ to basis $\mathcal{E} = (\bar{e}_1, \bar{e}_2, \bar{e}_3)$ and let the matrix of direction cosines, $\underline{\underline{D}}$, eq. (4.1), define this rotation. Resolving the vector quantities in basis $\mathcal{I}$ then yields

$$\bar{e}_1^{[\mathcal{I}]} = \underline{\underline{D}}\,\bar{\imath}_1^{[\mathcal{I}]}, \tag{4.16}$$

where the following identities were used: $\bar{\imath}_1^{[\mathcal{I}]T} = \{1, 0, 0\}$, $\bar{\imath}_2^{[\mathcal{I}]T} = \{0, 1, 0\}$, and $\bar{\imath}_3^{[\mathcal{I}]T} = \{0, 0, 1\}$.

On the other hand, if rotation tensor $\underline{\underline{R}}$ rotates vector $\bar{\imath}_1$ to $\bar{e}_1$, eq. (4.14) implies $\bar{e}_1 = \underline{\underline{R}}\,\bar{\imath}_1$. Resolving this tensor relationship in basis $\mathcal{I}$ then yields

$$\bar{e}_1^{[\mathcal{I}]} = \underline{\underline{R}}^{[\mathcal{I}]}\bar{\imath}_1^{[\mathcal{I}]}. \tag{4.17}$$

Identifying eqs. (4.16) and (4.17) yields the relationship between the direction cosine matrix and the rotation tensor as

$$\underline{\underline{D}} = \underline{\underline{R}}^{[\mathcal{I}]}. \tag{4.18}$$

The entries of the direction cosine matrix describing the rotation from $\mathcal{I}$ to $\mathcal{E}$ are identical to the components of the rotation tensor describing the same rotation and resolved in basis $\mathcal{I}$.

Multiplicative decompositions of the rotation tensor

Two multiplicative decompositions of the rotation tensor are now presented. The first is the "square root," $\underline{\underline{G}}$, of the rotation tensor, $\underline{\underline{R}}$, defined as

$$\underline{\underline{R}} = \underline{\underline{G}}\,\underline{\underline{G}}. \tag{4.19}$$

It is readily verified that

$$\underline{\underline{G}} = \underline{\underline{I}} + \sin\frac{\phi}{2}\tilde{n} + \left(1 - \cos\frac{\phi}{2}\right)\tilde{n}\tilde{n}. \tag{4.20}$$

It is interesting to note that the "square root" of the rotation tensor corresponds to a rotation of $\phi/2$ about axis $\bar{n}$, and hence, is itself an orthogonal tensor, $\underline{\underline{G}}\,\underline{\underline{G}}^T = \underline{\underline{I}}$. The following results then follow

$$\underline{\underline{R}} - \underline{\underline{I}} = \underline{\underline{G}}\,\underline{\underline{G}} - \underline{\underline{G}}\,\underline{\underline{G}}^T = \underline{\underline{G}}(\underline{\underline{G}} - \underline{\underline{G}}^T) = 2\sin\frac{\phi}{2}\,\underline{\underline{G}}\tilde{n} = 2\sin\frac{\phi}{2}\,\tilde{n}\,\underline{\underline{G}}. \tag{4.21}$$

The second multiplicative decomposition of the rotation tensor is

$$\begin{aligned}
\underline{\underline{R}} &= \left(\underline{\underline{I}} + \tan\frac{\phi}{2}\,\tilde{n}\right)\left(\underline{\underline{I}} + \tan\frac{\phi}{2}\,\tilde{n}\right)^{-T} \\
&= \left(\underline{\underline{I}} + \tan\frac{\phi}{2}\,\tilde{n}\right)^{-T}\left(\underline{\underline{I}} + \tan\frac{\phi}{2}\,\tilde{n}\right).
\end{aligned} \tag{4.22}$$

Note that

$$\left(\underline{\underline{I}} + \tan\frac{\phi}{2}\,\tilde{n}\right)^{-T} = \frac{1}{2}\left(\underline{\underline{R}} + \underline{\underline{I}}\right). \tag{4.23}$$

4.7 Properties of the rotation tensor

Inspection of equation (4.15) reveals that $\mathrm{symm}(\underline{\underline{R}}) = \underline{\underline{I}}\cos\phi + (1 - \cos\phi)\bar{n}\bar{n}^T$, and $\mathrm{skew}(\underline{\underline{R}}) = \tilde{n}\sin\phi$. It then follows that $\mathrm{axial}(\underline{\underline{R}}) = \bar{n}\sin\phi$.

The invariants of the rotation tensor can also be directly evaluated from eq. (4.15) as $I_1 = \mathrm{tr}(\underline{\underline{R}}) = 1 + 2\cos\phi$, $I_2 = 1 + 2\cos\phi$, and $I_3 = \det(\underline{\underline{R}}) = 1$. The characteristic equations, eq. (1.55), now becomes $-\lambda^3 + (1 + 2\cos\phi)\lambda^2 - (1 + 2\cos\phi)\lambda + 1 = 0$.

Eigenvalues and eigenvectors of the rotation tensor

The first fundamental property of the rotation tensor is that *it possesses a unit eigenvalue, $\lambda = +1$, associated with eigenvector $\bar{n}$*. Indeed, it follows from eq. (4.15) that

$$\underline{\underline{R}}\,\bar{n} = \bar{n}. \tag{4.24}$$

This indicates that $(\lambda - 1)$ should be a factor of the characteristic equation, which can indeed be written as $(\lambda - 1)(\lambda^2 - 2\lambda \cos\phi + 1) = 0$.

The other two eigenvalues of the rotation tensor are complex conjugate roots, $\cos\phi \pm i\sin\phi$, where $i = \sqrt{-1}$. In summary, the eigenvalues of the rotation tensor are

$$\lambda_1 = 1, \quad \lambda_{2,3} = \cos\phi \pm i\sin\phi = e^{\pm i\phi}. \tag{4.25}$$

Consider now two mutually orthogonal, unit vectors $\bar{u}$ and $\bar{v}$, which lie in the plane normal to $\bar{n}$, such that $\mathcal{E} = (\bar{n}, \bar{u}, \bar{v})$ forms an orthonormal basis. It is easily verified that

$$\underline{\underline{R}}\,\bar{u} = \cos\phi\,\bar{u} + \sin\phi\,\bar{v}, \quad \underline{\underline{R}}\,\bar{v} = -\sin\phi\,\bar{u} + \cos\phi\,\bar{v}.$$

Clearly, vectors $\bar{u}$ and $\bar{v}$ undergo a planar rotation of magnitude ϕ in the plane normal to the axis of rotation, $\bar{n}$. Linear combinations of these two equations then leads to

$$\underline{\underline{R}}(\bar{u} - i\underline{v}) = (\cos\phi + i\sin\phi)(\bar{u} - i\bar{v}), \quad \underline{\underline{R}}(\bar{u} + i\underline{v}) = (\cos\phi - i\sin\phi)(\bar{u} + i\bar{v}).$$

This reveals that $(\bar{u} \mp i\bar{v})$ are the complex conjugate eigenvectors associated with the complex conjugate eigenvalues $\cos\phi \pm i\sin\phi$, respectively.

Orthogonality of the rotation tensor

The second fundamental property of the rotation tensor is that it is an *orthogonal tensor*. Using eq. (4.15), it is readily verified that

$$\underline{\underline{R}}\,\underline{\underline{R}}^T = \underline{\underline{R}}^T\underline{\underline{R}} = \underline{\underline{I}}, \tag{4.26}$$

which implies $\det(\underline{\underline{R}}) = \pm 1$. in general, orthogonal tensors have a determinant of ± 1. Equation (4.25) shows, however, that $\det(\underline{\underline{R}}) = \lambda_1\lambda_2\lambda_3 = +1$: the rotation tensor belongs to the class of *proper orthogonal tensors* for which $\det(\underline{\underline{R}}) = +1$.

4.8 Change of basis operations

4.8.1 Vector components in various orthonormal bases

Consider an orthonormal basis $\mathcal{B}^1 = (\bar{\imath}_1^1, \bar{\imath}_2^1, \bar{\imath}_3^1)$, and an arbitrary vector $\underline{a}^1$. Next, consider a rotation of magnitude ϕ about a unit vector $\bar{n}$. The corresponding rotation tensor is denoted $\underline{\underline{R}}$. Let vectors $\bar{\imath}_1^2, \bar{\imath}_2^2, \bar{\imath}_3^2$, and $\underline{a}^2$ be the vectors resulting from the application of rotation $\underline{\underline{R}}$ to vectors $\bar{\imath}_1^1, \bar{\imath}_2^1, \bar{\imath}_3^1$, and $\underline{a}^1$, respectively. It is clear that vectors $\bar{\imath}_1^2, \bar{\imath}_2^2$, and $\bar{\imath}_3^2$ define a new orthonormal basis $\mathcal{B}^2$. Vectors $\underline{a}^1$ and $\underline{a}^2$ are related by eq. (4.14), *i.e.*, $\underline{a}^2 = \underline{\underline{R}}\,\underline{a}^1$. This tensor relationship is now resolved in basis $\mathcal{B}^1$ to find $\underline{a}^{2[\mathcal{B}^1]} = \underline{\underline{R}}^{[\mathcal{B}^1]}\underline{a}^{1[\mathcal{B}^1]}$.

By construction, the components of vector $\underline{a}^1$ resolved in basis $\mathcal{B}^1$ are identical those of $\underline{a}^2$ resolved in basis $\mathcal{B}^2$, $\underline{a}^{1[\mathcal{B}^1]} = \underline{a}^{2[\mathcal{B}^2]}$. It follows that the relationship between the components of vector $\underline{a}^2$ expressed in bases $\mathcal{B}^1$ and $\mathcal{B}^2$ is $\underline{a}^{2[\mathcal{B}^1]} = \underline{\underline{R}}^{[\mathcal{B}^1]} \underline{a}^{2[\mathcal{B}^2]}$. Because vector $\underline{a}^1$ is arbitrary, this relationship holds for any vector $\underline{v}$, *i.e.*,

$$\underline{v}^{[\mathcal{B}^1]} = \underline{\underline{R}}^{[\mathcal{B}^1]} \underline{v}^{[\mathcal{B}^2]}, \qquad (4.27a)$$

$$\underline{v}^{[\mathcal{B}^2]} = \underline{\underline{R}}^{[\mathcal{B}^1]T} \underline{v}^{[\mathcal{B}^1]}. \qquad (4.27b)$$

These equations express the transformation laws for the components of a first-order tensor. In fact, the rigorous definition of a first-order tensor is as follows.

Definition 4.1 (first-order tensor). *A first-order tensor is a mathematical entity whose components resolved in two bases are related by eqs. (4.27).*

The component of vector $\bar{n}$ resolved in bases $\mathcal{B}^1$ and $\mathcal{B}^2$ are identical, $\bar{n}^{[\mathcal{B}^1]} = \bar{n}^{[\mathcal{B}^2]}$. Consequently, in view of eq. (4.15), the components of the rotation tensor that bring basis $\mathcal{B}^1$ to $\mathcal{B}^2$, resolved in those two bases, are also identical

$$\underline{\underline{R}}^{[\mathcal{B}^1]} = \underline{\underline{R}}^{[\mathcal{B}^2]}. \qquad (4.28)$$

4.8.2 Second-order tensor components in various orthonormal bases

Consider now a second-order tensor such as $\underline{\underline{T}} = \underline{a}\,\underline{b}^T$, where $\underline{a}$ and $\underline{b}$ are two arbitrary vectors. The components of this tensor in two distinct bases are $\underline{\underline{T}}^{[\mathcal{B}^1]} = \underline{a}^{[\mathcal{B}^1]}\underline{b}^{[\mathcal{B}^1]T}$ and $\underline{\underline{T}}^{[\mathcal{B}^2]} = \underline{a}^{[\mathcal{B}^2]}\underline{b}^{[\mathcal{B}^2]T}$. Using the transformation law for first-order tensors, eq. (4.27), the transformation laws for the components of second-order tensors are found to be

$$\underline{\underline{T}}^{[\mathcal{B}^2]} = \underline{\underline{R}}^{[\mathcal{B}^1]T} \underline{\underline{T}}^{[\mathcal{B}^1]} \underline{\underline{R}}^{[\mathcal{B}^1]}, \qquad (4.29a)$$

$$\underline{\underline{T}}^{[\mathcal{B}^1]} = \underline{\underline{R}}^{[\mathcal{B}^1]} \underline{\underline{T}}^{[\mathcal{B}^2]} \underline{\underline{R}}^{T[\mathcal{B}^1]}. \qquad (4.29b)$$

The rigorous definition of second-order tensors is as follows.

Definition 4.2 (second-order tensor). *A second-order tensor is a mathematical entity whose components resolved in two bases are related by eqs. (4.29).*

Example 4.1. First- and second-order tensors
The previous sections have given precise definitions of first- and second-order tensors as mathematical entities whose components resolved in two bases are related by eqs. (4.27) and (4.29), respectively.

Consider a vector (first-order tensor) whose components in two bases, $\mathcal{B}$ and $\mathcal{B}^*$, are denoted $\underline{a}$ and $\underline{a}^*$, respectively. In the notation of the above sections, $\underline{a} = \underline{a}^{[\mathcal{B}]}$ and $\underline{a}^* = \underline{a}^{[\mathcal{B}^*]}$. The simplified notation, $\underline{a}$ and $\underline{a}^*$, is clearly much simpler, provided that all symbols are clearly defined. If the components of the rotation tensor that brings basis $\mathcal{B}$ to $\mathcal{B}^*$, resolved in $\mathcal{B}$, are denoted $\underline{\underline{R}}$, eq. (4.27) implies that $\underline{a}^* = \underline{\underline{R}}^T \underline{a}$. Consider now the skew symmetric operators $\widetilde{a}$ and $\widetilde{a}^*$ formed with the components of vector $\underline{a}$ resolved in bases $\mathcal{B}$ and $\mathcal{B}^*$, respectively. Prove that $\tilde{a}$ and $\tilde{a}^*$ are the components of a skew-symmetric, second-order tensor, $\tilde{\underline{a}}$.

If $\tilde{\underline{a}}$ is a second-order tensor, its components resolved in two bases must be related by eq. (4.29). Hence, the tensorial nature of $\tilde{\underline{a}}$ will be established if and only if

$$\tilde{\underline{a}}^* = \underline{\underline{R}}^T \tilde{\underline{a}} \underline{\underline{R}} \iff \underline{a}^* = \underline{\underline{R}}^T \underline{a}. \tag{4.30}$$

This statement can be proved based on simple, but tedious algebraic manipulations, taking into account the fact that $\underline{\underline{R}}$ is an orthogonal tensor.

Example 4.2. The rotation tensor

The rotation tensor was introduced in section 4.6 and was called a "tensor." Prove that the rotation tensor is indeed a second-order tensor.

Euler's theorem defines the rotation tensor in terms of a unit vector $\bar{n}$ about which a rotation of magnitude ϕ is taking place. Let $\bar{n}$ be the components of this unit vector resolved in basis $\mathcal{B}$; the components of the rotation tensor, $\underline{\underline{S}}$, resolved in the same basis are then given by eq. (4.15) as $\underline{\underline{S}} = \underline{\underline{I}} + \sin\phi\,\tilde{n} + (1 - \cos\phi)\,\tilde{n}\tilde{n}$.

Consider now an arbitrary basis $\mathcal{B}^*$ and let the components of unit vector $\bar{n}$, resolved in this basis, be denoted $\bar{n}^* = \underline{\underline{R}}^T \bar{n}$, where $\underline{\underline{R}}$ are the components of the rotation tensor that brings basis $\mathcal{B}$ to $\mathcal{B}^*$, resolved in basis $\mathcal{B}$. It then follows that

$$\underline{\underline{S}} = \underline{\underline{I}} + \sin\phi\,\underline{\underline{R}}\tilde{n}^*\underline{\underline{R}}^T + (1 - \cos\phi)\,\underline{\underline{R}}\tilde{n}^*\underline{\underline{R}}\,\underline{\underline{R}}^T\tilde{n}^*\underline{\underline{R}}^T$$
$$= \underline{\underline{R}}\left[\underline{\underline{I}} + \sin\phi\,\tilde{n}^* + (1 - \cos\phi)\,\tilde{n}^*\tilde{n}^*\right]\underline{\underline{R}}^T = \underline{\underline{R}}\,\underline{\underline{S}}^*\underline{\underline{R}}^T,$$

where the orthogonality property of the rotation tensor was used together with eq. (4.30). Clearly, $\underline{\underline{S}}^* = \underline{\underline{I}} + \sin\phi\,\tilde{n}^* + (1 - \cos\phi)\,\tilde{n}^*\tilde{n}^*$ are the components of the rotation tensor resolved in basis $\mathcal{B}^*$, and the above result then provides the transformation rule for the components of the rotation tensor. This transformation rule is, as expected, the rule that characterizes the transformation of components of second-order tensors, see eq. (4.29). Hence, the tensorial nature of the rotation tensor is established.

The proof of the tensorial nature of the rotation tensor rests on the definition of the rotation tensor provided by Rodrigues' rotation formula, eq. (4.15), and on the tensorial nature of the unit vector about which the rotation is taking place, expressed as $\bar{n} = \underline{\underline{R}}\,\bar{n}^*$. Consequently, the definition of the rotation tensor by Rodrigues' rotation formula guarantees the following equivalence

$$\underline{\underline{S}} = \underline{\underline{R}}\,\underline{\underline{S}}^*\underline{\underline{R}}^T \iff \bar{n} = \underline{\underline{R}}\,\bar{n}^*. \tag{4.31}$$

The components of the first-order tensor, unit vector $\bar{n}$, transform according to eqs. (4.27), and the components of the second-order tensor, $\underline{\underline{S}}$, transform according to eqs. (4.29).

Example 4.3. Canonical basis for the rotation tensor

In section 4.7, the following orthonormal basis was introduced

$$\mathcal{E} = (\bar{n}, \bar{u}, \bar{v}), \tag{4.32}$$

where $\bar{u}$ and $\bar{v}$ are mutually orthogonal unit vectors in the plane normal to $\bar{n}$. Such basis is called a *canonical basis* for the rotation tensor, $\underline{\underline{R}}$.

Let $\underline{\underline{S}}$ be the rotation tensor that brings basis $\mathcal{I}$ to basis $\mathcal{E}$; the components of this tensor resolved in basis $\mathcal{I}$ are $S^{[\mathcal{I}]} = (\bar{n}^{[\mathcal{I}]}, \bar{u}^{[\mathcal{I}]}, \bar{v}^{[\mathcal{I}]})$. Because the rotation tensor is a second-order tensor, eq. (4.15) yields its components in basis $\mathcal{E}$ as

$$
\begin{aligned}
\underline{\underline{R}}^{[\mathcal{E}]} &= \underline{\underline{S}}^{[\mathcal{I}]T} \underline{\underline{R}}^{[\mathcal{I}]} \underline{\underline{S}}^{[\mathcal{I}]} \\
&= \underline{\underline{S}}^{[\mathcal{I}]T} \left[\bar{n}^{[\mathcal{I}]}, \cos\phi\, \bar{u}^{[\mathcal{I}]} + \sin\phi\, \bar{v}^{[\mathcal{I}]}, -\sin\phi\, \bar{u}^{[\mathcal{I}]} + \cos\phi\, \bar{v}^{[\mathcal{I}]} \right].
\end{aligned}
$$

Finally, the components of the rotation tensor in the canonical basis become

$$
\underline{\underline{R}}^{[\mathcal{E}]} = \begin{bmatrix} 1 & 0 & 0 \\ 0 & \cos\phi & -\sin\phi \\ 0 & \sin\phi & \cos\phi \end{bmatrix}. \tag{4.33}
$$

In this canonical form, the rotation tensor takes the expected form of the direction cosine matrix for a planar rotation, see eq. (4.4).

When resolved in the same canonical basis, the components of rotation tensor $\underline{\underline{G}}$ defined by eq. (4.20) become

$$
\underline{\underline{G}}^{[\mathcal{E}]} = \begin{bmatrix} 1 & 0 & 0 \\ 0 & \cos\phi/2 & -\sin\phi/2 \\ 0 & \sin\phi/2 & \cos\phi/2 \end{bmatrix}. \tag{4.34}
$$

In this canonical form, rotation tensor $\underline{\underline{G}}$ takes the expected form of the direction cosine matrix for a planar rotation of half angle, $\phi/2$.

4.8.3 Tensor operations

Sections 4.8.1 and 4.8.2 give formal definitions of first- and second-order tensors. For completeness of the discussion, a formal definition of zeroth order tensors is also given.

Definition 4.3 (Zeroth order tensor). *A zeroth order tensor is a mathematical entity that remains invariant under a change of basis operation.*

Take, for instance, the mass of a particle. This scalar quantity is invariant under a change of basis operation and hence, is a zeroth order tensor. The length of a vector or the angle between two vectors are two other examples of scalar quantities that remain invariant under a change of basis and hence, are also zeroth order tensor.

Chapter 1 defines a number of operations between vectors: the scalar product, the vector product and the tensor product, among others. A *tensor operation* is an operation using two or more tensors and resulting in another tensor.

As a first example of a tensor operation, consider the differential work defined in eq. (3.8) as the scalar product of the force vector by the differential displacement of its point of application, $dW = \underline{F}^T d\underline{r}$. Two analysts working with two different bases, $\mathcal{B}$ and $\mathcal{B}^*$, will write this differential work as $dW = \underline{F}^T d\underline{r}$ and $dW^* = \underline{F}^{*T} d\underline{r}^*$, respectively. The three numbers, $\underline{F}$, representing the force vector in basis $\mathcal{B}$ are different from the three numbers, $\underline{F}^*$, representing the same force vector in basis $\mathcal{B}^*$. Similarly, the numbers representing the components of the differential displacement vector resolved in the two bases, $d\underline{r}$ and $d\underline{r}^*$, differ.

Because the force and differential displacement vectors are first-order tensors, their components in the two bases are related by eq. (4.27), i.e., $\underline{F}^* = \underline{\underline{R}}^T \underline{F}$ and $d\underline{r}^* = \underline{\underline{R}}^T d\underline{r}$, respectively. It then follows that

$$dW^* = \underline{F}^{*T} d\underline{r}^* = \underline{F}^T \underline{\underline{R}}\,\underline{\underline{R}}^T d\underline{r} = \underline{F}^T d\underline{r} = dW. \tag{4.35}$$

This well known results stems from the orthogonality of the rotation tensor, eq. (4.26). Because $dW^* = dW$, the differential work is a zeroth order tensor, i.e., a quantity that remains invariant under a change of basis.

The same conclusion can be reached by looking at the definition of the scalar product, eq. (1.8), $dW = \|\underline{F}\|\,\|d\underline{r}\|\,\cos(\underline{F}, d\underline{r}) = \|\underline{F}^*\|\,\|d\underline{r}^*\|\,\cos(\underline{F}^*, d\underline{r}^*) = dW^*$. In this case, the invariance of the differential work under a change of basis stems from the fact that the length of a vector and the angle between two vectors are zeroth order tensors. In summary, the scalar product is an operation based on two first-order tensors, which produces a zeroth order tensor. This proves that the scalar product is a tensor operation.

While this proof seems rather technical, it has fundamental physical implications. Because the differential work is obtained from a scalar product, i.e., from a tensor operation, it is invariant under a change of basis and hence, is a physically meaningful quantity. Indeed, if the value of the differential work were to depend on the basis in which the force and differential displacement vectors are resolved, this quantity would have no physical meaning because two analysts using two different bases to represent the same vectors would find two different values of the differential work.

A second example of tensor operation is the moment of a force, defined as the vector product of the position vector of the point of application of a force by the force vector itself, $\underline{M} = \widetilde{r}\underline{F}$. Two analysts working with two different bases, $\mathcal{B}$ and $\mathcal{B}^*$, will write this moment as $\underline{M} = \widetilde{r}\underline{F}$ and $\underline{M}^* = \widetilde{r}^*\underline{F}^*$, respectively. Here again, because the position and force vectors are first-order tensors, their components in the two bases are related by eq. (4.27), i.e., $\underline{r}^* = \underline{\underline{R}}^T \underline{r}$ and $\underline{F}^* = \underline{\underline{R}}^T \underline{F}$. The components of the moment are as follows,

$$\underline{M}^* = \widetilde{r}^*\underline{F}^* = \widetilde{\underline{\underline{R}}^T \underline{r}}\,\underline{\underline{R}}^T \underline{F} = \underline{\underline{R}}^T \widetilde{r}\underline{\underline{R}}\,\underline{\underline{R}}^T \underline{F} = \underline{\underline{R}}^T \widetilde{r}\underline{F} = \underline{\underline{R}}^T \underline{M}, \tag{4.36}$$

where eq. (4.30) and the property of orthogonality of the rotation tensor were used. The result of the vector product operation is a quantity, $\underline{M}$, whose components obey the rules of transformation for first-order tensors, eq. (4.27), $\underline{M}^* = \underline{\underline{R}}^T \underline{M}$; hence, the vector product is a tensor operation.

The same conclusion can be reached by looking at the definition of the vector product, eq. (1.20), $\underline{M} = \|\underline{r}\|\,\|\underline{F}\|\,\sin(\underline{r}, \underline{F})\,\bar{n} = \|\underline{r}^*\|\,\|\underline{F}^*\|\,\sin(\underline{r}^*, \underline{F}^*)\,\underline{\underline{R}}\bar{n}^* = \underline{\underline{R}}\,\underline{M}^*$. This is a tensor operation because the length of a vector and the angle between two vectors are invariant under a change of basis. Furthermore, unit vector $\bar{n}$, normal to vectors $\underline{r}$ and $\underline{F}$, is a first-order tensor, implying the following transformation rule for its components, $\bar{n} = \underline{\underline{R}}\,\bar{n}^*$.

In summary, the vector product is an operation based on two first-order tensors, which produces a first-order tensor. This proves that the vector product is a tensor operation. As a corollary, the moment of a force, the vector product of the position vector of the point of application of a force by the force vector itself, is a physically meaningful quantity because its a first-order tensor.

It is left to the reader to verify that the various operations defined in chapter 1 are indeed tensor operations, *i.e.*, operations that are invariant under a change of basis, see problem 4.7. As a last example, consider the product of a zeroth order by a first-order tensor, which defines the linear momentum vector, $\underline{p} = m\underline{v}$. The mass of the particle is a zeroth order tensor and its inertial velocity a first-order tensor, implying $\underline{v}^* = \underline{\underline{R}}^T\underline{v}$. It then follows that

$$\underline{p}^* = m\underline{v}^* = m\underline{\underline{R}}^T\underline{v} = \underline{\underline{R}}^T m\underline{v} = \underline{\underline{R}}^T\underline{p}. \tag{4.37}$$

Because the components of the linear momentum obey the rules of transformation for first-order tensors, eq. (4.27), it is a first-order tensor and hence, the product of a zeroth order by a first-order tensor is a tensor operation. It follows that the linear momentum is a physically meaningful quantity.

4.8.4 The concept of tensor analysis

Zeroth-, first-, and second-order tensors are mathematical entities whose components resolved in different bases transform according to strict rules. Manipulation of tensors through tensor operations lead to new tensors. For instance, the vector product of the position vector of the point of application of a force by the force vector itself produces a new vector, the moment of the force. These is a rather abstract mathematical concepts have important physical implications. In fact, the use of tensors expresses the invariance of the laws of physics with respect to change of basis operations [3].

Consider, for instance, Newton's second law, which states that the force and acceleration vectors must be parallel to each other and the ratio of their lengths must equal the mass of the particle. Clearly, Newton's second law is invariant under a change of basis. Indeed, the condition of parallelism between the force and acceleration vectors is invariant under a change of basis. Furthermore, because the mass of the particle and the length of the force and acceleration vectors are three invariant quantities, the equality of the length ratio with the particle's mass is also invariant under a change of basis.

Using the vector formalism, Newton's second law is written as $\underline{F} = m\underline{a}$. This law involves three tensors: a zeroth order tensor, the particle's mass, and two first-order tensors, the externally applied force vector and the particle's acceleration vector. Furthermore, Newton's second law uses tensor operations only: the product of the mass by the acceleration vector is indeed a tensor operation, the product of a zeroth order by a first-order tensor, see eq. (4.37). The combined use of tensor quantities and tensor operations guarantees the invariance of Newton's second law under change of basis operations.

Two analysts working with two different bases, $\mathcal{B}$ and $\mathcal{B}^*$, will write Newton's second law as $\underline{F} = m\underline{a}$ and $\underline{F}^* = m\underline{a}^*$, respectively. Yet both analysts express the same physical law: the force and acceleration vectors must be parallel and the ratio of their lengths must equal the mass of the particle.

In summary, the laws of physics should be expressed in terms of tensors exclusively and should only involve tensor operations. When these two conditions are met, the invariance of the laws of physics under a change of basis is achieved.

4.8.5 Problems

Problem 4.4. Geometric interpretation of tensor
Prove eq. (4.19), where $\underline{G}$ is given by eq. (4.20). Give the geometric interpretation of this result.

Problem 4.5. Orthogonality of the rotation tensor
Prove the orthogonality of the rotation tensor, eq. (4.26), based on its expression based on Euler theorem, eq. (4.15).

Problem 4.6. Base transformation for skew symmetric tensor
Prove eq. (4.30). Hint: remember that the rotation tensor is orthogonal, $\underline{\underline{R}}^{-1} = \underline{\underline{R}}^T$.

Problem 4.7. Tensor operations
(1) Prove that the product of a zeroth order tensor by a first-order tensor is a tensor operation. (2) Prove that the product of a zeroth order tensor by a second-order tensor is a tensor operation. (3) Prove that the tensor product of two vectors, eq. (1.28), is a tensor operation. (4) Prove that the mixed product of three vectors is a tensor operation. (5) Let $\underline{\underline{A}}$ be a second-order tensor and let $\underline{\underline{A}}$ and $\underline{\underline{A}}^*$ its components in two bases, $\mathcal{B}$ and $\mathcal{B}^*$. Prove that the eigenvalues of $\underline{\underline{A}}$ and $\underline{\underline{A}}^*$ are identical and that the eigenvectors of $\underline{\underline{A}}$ are first-order tensors.

Problem 4.8. Tensors $\underline{\underline{R}}$ and $\underline{\underline{G}}$
The components of rotation tensor $\underline{\underline{R}}$ resolved in basis $\mathcal{I}$ are given as follows,

$$\underline{\underline{R}}^{[\mathcal{I}]} = \begin{bmatrix} 0.6272 & -0.7305 & 0.2700 \\ -0.1268 & -0.4379 & -0.8900 \\ 0.7684 & 0.5240 & -0.3673 \end{bmatrix}.$$

(1) Find the components of tensor $\underline{\underline{G}}$ in the same basis such that $\underline{\underline{R}}^{[\mathcal{I}]} = \underline{\underline{G}}^{[\mathcal{I}]}\underline{\underline{G}}^{[\mathcal{I}]}$. (2) Verify that your answer is correct by evaluating the product $\underline{\underline{G}}^{[\mathcal{I}]}\underline{\underline{G}}^{[\mathcal{I}]}$.

Problem 4.9. Components of a vector in two bases
Rotation tensor $\underline{\underline{R}}$ brings basis $\mathcal{I}$ to basis $\mathcal{E}$. The components of tensor $\underline{\underline{R}}$ and vector $\underline{a}$, both resolved in basis $\mathcal{I}$, are given as follows,

$$\underline{\underline{R}} = \begin{bmatrix} 0.2944 & 0.9433 & -0.1536 \\ -0.9005 & 0.2199 & -0.3751 \\ -0.3200 & 0.2488 & 0.9142 \end{bmatrix}, \quad \underline{a} = \begin{Bmatrix} 7.54 \\ -3.44 \\ 1.77 \end{Bmatrix}.$$

(1) Find the components of vector $\underline{a}$ in basis $\mathcal{E}$, denoted $\underline{a}^*$, as $\underline{a}^* = \underline{\underline{R}}^T\underline{a}$. (2) Verify that $\widetilde{a}^* = \underline{\underline{R}}^T\widetilde{a}\underline{\underline{R}}$.

Problem 4.10. Relationship among unit vectors of a basis

The rotation tensor can be written as $\underline{\underline{R}} = [\bar{e}_1, \bar{e}_2, \bar{e}_3]$, where $\bar{e}_1$, $\bar{e}_2$, and $\bar{e}_3$ form an orthonormal basis. Show that

$$\tilde{e}_1 = \bar{e}_3 \bar{e}_2^T - \bar{e}_2 \bar{e}_3^T, \quad \tilde{e}_2 = \bar{e}_1 \bar{e}_3^T - \bar{e}_3 \bar{e}_1^T, \quad \tilde{e}_3 = \bar{e}_2 \bar{e}_1^T - \bar{e}_1 \bar{e}_2^T, \tag{4.38a}$$

$$\bar{e}_1 \bar{e}_1^T + \bar{e}_2 \bar{e}_2^T + \bar{e}_3 \bar{e}_3^T = \underline{\underline{I}}. \tag{4.38b}$$

Problem 4.11. Analysis of the projection operator

Prove that the projection operator defined in example 1.5 is a second-order tensor.

Problem 4.12. Analysis of the reflection operator

Prove that the reflection operator defined in problem 1.12 is a second-order tensor.

Problem 4.13. Rotation tensor in canonical form

(1) Compute the components of tensor $\underline{\underline{R}}^{[\mathcal{E}]}$ in the canonical basis, $\mathcal{E}$, defined by eq. (4.32), *i.e.*, verify eq. (4.33). *(2)* Compute the eigenvalues of $\underline{\underline{R}}^{[\mathcal{E}]}$, *i.e.*, verify eq. (4.25).

Problem 4.14. Square root of rotation tensor

(1) Compute the components of tensors $\underline{\underline{R}}$ and $\underline{\underline{G}}$ in the canonical basis, $\mathcal{E}$, defined by eq. (4.32), denoted $\underline{\underline{R}}^{[\mathcal{E}]}$ and $\underline{\underline{G}}^{[\mathcal{E}]}$, respectively. *(2)* Verify eq. (4.19) by checking that $\underline{\underline{R}}^{[\mathcal{E}]} = \underline{\underline{G}}^{[\mathcal{E}]} \underline{\underline{G}}^{[\mathcal{E}]}$. *(3)* Find the kth root of the rotation tensor $\underline{\underline{R}}$, denoted $\underline{\underline{G}}_k$. Discuss the geometric meaning of this tensor.

Problem 4.15. Orthogonality in canonical form

Verify the orthogonality property of the rotation tensor, $\underline{\underline{R}}\,\underline{\underline{R}}^T = \underline{\underline{I}}$, by first computing the components of $\underline{\underline{R}}$ in the canonical basis, $\mathcal{E}$, defined by eq. (4.32), denoted $\underline{\underline{R}}^{[\mathcal{E}]}$, then checking that $\underline{\underline{R}}^{[\mathcal{E}]} \underline{\underline{R}}^{[\mathcal{E}]T} = \underline{\underline{I}}$.

Problem 4.16. Multiplicative decomposition of the rotation tensor

Consider three rotation tensors, $\underline{\underline{R}}$, $\underline{\underline{R}}_1$, and $\underline{\underline{R}}_2$, corresponding to rotations of magnitude ϕ, $\eta\phi$, and $(1 - \eta)\phi$, respectively, about the same unit vector, $\bar{n}$, where $\eta \in [0, 1]$. Prove that $\underline{\underline{R}} = \underline{\underline{R}}_1 \underline{\underline{R}}_2 = \underline{\underline{R}}_2 \underline{\underline{R}}_1$. Hint: write the three rotation tensors in their common canonical basis.

Problem 4.17. Properties of rotation tensors $\underline{\underline{R}}$ and $\underline{\underline{G}}$

Prove the following relationships.

$$(\underline{\underline{R}} - \underline{\underline{I}})(\underline{\underline{R}} + \underline{\underline{I}})^{-1} = (\underline{\underline{R}} + \underline{\underline{I}})^{-1}(\underline{\underline{R}} - \underline{\underline{I}}) = \tilde{n} \tan \phi/2, \tag{4.39a}$$

$$(\underline{\underline{G}} - \underline{\underline{I}})(\underline{\underline{G}} + \underline{\underline{I}})^{-1} = (\underline{\underline{G}} + \underline{\underline{I}})^{-1}(\underline{\underline{G}} - \underline{\underline{I}}) = \tilde{n} \tan \phi/4, \tag{4.39b}$$

$$(\underline{\underline{I}} - \underline{\underline{R}}^T)(\underline{\underline{I}} + \underline{\underline{R}}^T)^{-1} = (\underline{\underline{I}} + \underline{\underline{R}}^T)^{-1}(\underline{\underline{I}} - \underline{\underline{R}}^T) = \tilde{n} \tan \phi/2, \tag{4.39c}$$

$$(\underline{\underline{I}} - \underline{\underline{G}}^T)(\underline{\underline{I}} + \underline{\underline{G}}^T)^{-1} = (\underline{\underline{I}} + \underline{\underline{G}}^T)^{-1}(\underline{\underline{I}} - \underline{\underline{G}}^T) = \tilde{n} \tan \phi/4. \tag{4.39d}$$

Problem 4.18. Multiplicative decomposition of rotation tensors $\underline{\underline{R}}$ and $\underline{\underline{G}}$

Prove the following relationships.

$$\underline{\underline{R}} = (\underline{\underline{I}} - \alpha\tilde{n})^{-1}(\underline{\underline{I}} + \alpha\tilde{n}) = (\underline{\underline{I}} + \alpha\tilde{n})(\underline{\underline{I}} - \alpha\tilde{n})^{-1}, \quad \alpha = \tan \phi/2, \tag{4.40a}$$

$$\underline{\underline{G}} = (\underline{\underline{I}} - \beta\tilde{n})^{-1}(\underline{\underline{I}} + \beta\tilde{n}) = (\underline{\underline{I}} + \beta\tilde{n})(\underline{\underline{I}} - \beta\tilde{n})^{-1}, \quad \beta = \tan \phi/4. \tag{4.40b}$$

Problem 4.19. Properties of rotation tensors $\underline{\underline{R}}$ and $\underline{\underline{G}}$

Prove the following relationships.

$$(\underline{\underline{R}} + \underline{\underline{I}})(\underline{\underline{I}} - \alpha \widetilde{n}) = 2\underline{\underline{I}}, \quad \alpha = \tan\phi/2, \tag{4.41a}$$

$$(\underline{\underline{G}} + \underline{\underline{I}})(\underline{\underline{I}} - \beta \widetilde{n}) = 2\underline{\underline{I}}, \quad \beta = \tan\phi/4. \tag{4.41b}$$

Problem 4.20. Properties of rotation tensors $\underline{\underline{R}}$ and $\underline{\underline{G}}$

Prove the following relationships.

$$(\underline{\underline{R}} + \underline{\underline{I}})^{-1} + (\underline{\underline{R}}^T + \underline{\underline{I}})^{-1} = \underline{\underline{I}}, \tag{4.42a}$$

$$(\underline{\underline{G}} + \underline{\underline{I}})^{-1} + (\underline{\underline{G}}^T + \underline{\underline{I}})^{-1} = \underline{\underline{I}}. \tag{4.42b}$$

4.9 Composition of rotations

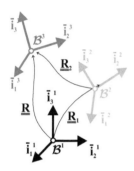

Fig. 4.7. Composition of rotations.

Figure 4.7 shows three orthonormal bases, $\mathcal{B}^1 = (\bar{\imath}_1^1, \bar{\imath}_2^1, \bar{\imath}_3^1)$, $\mathcal{B}^2 = (\bar{\imath}_1^2, \bar{\imath}_2^2, \bar{\imath}_3^2)$, and $\mathcal{B}^3 = (\bar{\imath}_1^3, \bar{\imath}_2^3, \bar{\imath}_3^3)$. Let the rotations of magnitude ϕ_1 about a unit vector $\bar{n}_1$ and of magnitude ϕ_2 about a unit vector $\bar{n}_2$, represented by tensors $\underline{\underline{R}}_1$ and $\underline{\underline{R}}_2$, respectively, express the rotations from basis $\mathcal{B}^1$ to $\mathcal{B}^2$ and from basis $\mathcal{B}^2$ to $\mathcal{B}^3$, respectively.

Application of the rotation operation, eq. (4.14), yields $\bar{\imath}_1^2 = \underline{\underline{R}}_1 \bar{\imath}_1^1$ and $\bar{\imath}_1^3 = \underline{\underline{R}}_2 \bar{\imath}_1^2$. Eliminating $\bar{\imath}_1^2$ from these two expressions leads to $\bar{\imath}_1^3 = \underline{\underline{R}}_2 \underline{\underline{R}}_1 \bar{\imath}_1^1 = \underline{\underline{R}} \bar{\imath}_1^1$, where $\underline{\underline{R}} = \underline{\underline{R}}_2 \underline{\underline{R}}_1$ is the rotation tensor that brings basis $\mathcal{B}^1$ to $\mathcal{B}^3$. The operation that combines two rotations, that from basis $\mathcal{B}^1$ to $\mathcal{B}^2$ and that from basis $\mathcal{B}^2$ to $\mathcal{B}^3$, into a single rotation from basis $\mathcal{B}^1$ to $\mathcal{B}^3$ is called *composition of rotations*, a concept that was first addressed by Rodrigues [18]. Mathematically, the composition of two rotations is expressed by the multiplication of the corresponding rotation tensors. Finite rotations do not form a linear space: the expression "composition of finite rotation" is used to underline the fact these quantities are not additive.

The tensor relationship, $\underline{\underline{R}} = \underline{\underline{R}}_2 \underline{\underline{R}}_1$, can be resolved in any basis, in particular bases $\mathcal{B}^1$ and $\mathcal{B}^3$, to find

$$\underline{\underline{R}}^{[\mathcal{B}^1]} = \underline{\underline{R}}^{[\mathcal{B}^3]} = \underline{\underline{R}}_2^{[\mathcal{B}^1]} \underline{\underline{R}}_1^{[\mathcal{B}^1]} = \underline{\underline{R}}_2^{[\mathcal{B}^3]} \underline{\underline{R}}_1^{[\mathcal{B}^3]}. \tag{4.43}$$

where eq. (4.28) was used to obtain the first equality.

It is often convenient to resolve rotation tensor $\underline{\underline{R}}_2$ in basis $\mathcal{B}^2$. The second-order tensor component transformation law, eq. (4.29), relates the components of this tensor resolved in the two bases as $\underline{\underline{R}}_2^{[\mathcal{B}^1]} = \underline{\underline{R}}_1^{[\mathcal{B}^1]} \underline{\underline{R}}_2^{[\mathcal{B}^2]} \underline{\underline{R}}_1^{[\mathcal{B}^1]T}$. Introducing this transformation into eq. (4.43) yields the additional result

$$\underline{\underline{R}}^{[\mathcal{B}^1]} = \underline{\underline{R}}^{[\mathcal{B}^3]} = \underline{\underline{R}}_1^{[\mathcal{B}^1]} \underline{\underline{R}}_2^{[\mathcal{B}^2]}. \tag{4.44}$$

Hence, the total rotation $\underline{\underline{R}}$ from $\mathcal{B}^1$ to $\mathcal{B}^3$ can be expressed in two alternative ways

$$\underline{\underline{R}}^{[\mathcal{B}^1]} = \underline{\underline{R}}^{[\mathcal{B}^3]} = \underline{\underline{R}}_2^{[\mathcal{B}^1]}\underline{\underline{R}}_1^{[\mathcal{B}^1]} = \underline{\underline{R}}_1^{[\mathcal{B}^1]}\underline{\underline{R}}_2^{[\mathcal{B}^2]}. \tag{4.45}$$

Note that *the order in which the rotation tensors appear depends on the basis in which they are resolved.*

Example 4.4. Euler angles

In section 4.4, Euler angles were defined as the magnitudes of three successive planar rotations describing an arbitrary rotation, as illustrated in fig 4.4. Considering the *3-1-3* sequence, rotation tensor $\underline{\underline{R}}_\phi$ rotates basis $\mathcal{I}$ to basis $\mathcal{A}$, next, tensor $\underline{\underline{R}}_\theta$ brings basis $\mathcal{A}$ to basis $\mathcal{B}$, and finally, tensor $\underline{\underline{R}}_\psi$ rotates basis $\mathcal{B}$ to $\mathcal{E}$. The operations can be summarized as: $\bar{a}_1 = \underline{\underline{R}}_\phi \bar{\imath}_1$, $\bar{b}_1 = \underline{\underline{R}}_\theta \bar{a}_1$, and $\bar{e}_1 = \underline{\underline{R}}_\psi \bar{b}_1$. Eliminating the intermediate bases then yields $\bar{e}_1 = \underline{\underline{R}}_\psi \underline{\underline{R}}_\theta \underline{\underline{R}}_\phi \bar{\imath}_1 = \underline{\underline{R}} \bar{\imath}_1$, where $\underline{\underline{R}}$ is the tensor that brings basis $\mathcal{I}$ to $\mathcal{E}$.

The statement $\underline{\underline{R}} = \underline{\underline{R}}_\psi \underline{\underline{R}}_\theta \underline{\underline{R}}_\phi$ is a tensor relationship that is true when expressed in any basis, provided that all tensors are resolved in the same basis; for instance, one could write $\underline{\underline{R}}^{[\mathcal{I}]} = \underline{\underline{R}}_\psi^{[\mathcal{I}]}\underline{\underline{R}}_\theta^{[\mathcal{I}]}\underline{\underline{R}}_\phi^{[\mathcal{I}]}$. In this expression, $\underline{\underline{R}}_\phi$ represents a planar rotation and the components of this tensor resolved in basis $\mathcal{I}$, denoted $\underline{\underline{R}}_\phi^{[\mathcal{I}]}$, are in the form of the direction cosine matrix given by eq. (4.6). Tensor $\underline{\underline{R}}_\theta$ also represents a planar rotation, but its components resolved in basis $\mathcal{I}$, denoted $\underline{\underline{R}}_\theta^{[\mathcal{I}]}$, *are not of the form of a direction cosine matrix* as given in eq. (4.4). However, the components of this tensor resolved in basis $\mathcal{A}$, denoted $\underline{\underline{R}}_\theta^{[\mathcal{A}]}$, would be of the form given in eq. (4.4). The same remarks can be made about tensor $\underline{\underline{R}}_\psi$: its components in basis $\mathcal{B}$, denoted $\underline{\underline{R}}_\psi^{[\mathcal{B}]}$, are of the form of the direction cosine matrix for a planar rotation as given by eq. (4.6), whereas its components in basis $\mathcal{I}$, denoted $\underline{\underline{R}}_\psi^{[\mathcal{I}]}$, are not.

The above discussion indicates that the evaluation of rotation tensor $\underline{\underline{R}}$ will be easier if the tensor relationship $\underline{\underline{R}} = \underline{\underline{R}}_\psi \underline{\underline{R}}_\theta \underline{\underline{R}}_\phi$ is expressed in component form as $\underline{\underline{R}}^{[\mathcal{I}]} = \underline{\underline{R}}_\phi^{[\mathcal{I}]}\underline{\underline{R}}_\theta^{[\mathcal{A}]}\underline{\underline{R}}_\psi^{[\mathcal{B}]}$, a recursive application of eq. (4.44). This yields

$$\underline{\underline{R}}^{[\mathcal{I}]} = \begin{bmatrix} \cos\phi & -\sin\phi & 0 \\ \sin\phi & \cos\phi & 0 \\ 0 & 0 & 1 \end{bmatrix} \begin{bmatrix} 1 & 0 & 0 \\ 0 & \cos\theta & -\sin\theta \\ 0 & \sin\theta & \cos\theta \end{bmatrix} \begin{bmatrix} \cos\psi & -\sin\psi & 0 \\ \sin\psi & \cos\psi & 0 \\ 0 & 0 & 1 \end{bmatrix}.$$

Performing the triple matrix multiplication yields the components of the rotation tensor in basis $\mathcal{I}$; the result yields the entries of the direction cosine matrix defined in eq. (4.74), as expected from eq. (4.18).

Example 4.5. Time-dependent motion of a rigid body

Consider a rigid body moving in three-dimensional space. In many computational schemes, it is necessary to track down the motion of the body by determining its actual position and orientation in space at various instants in time, as depicted in fig. 4.8.

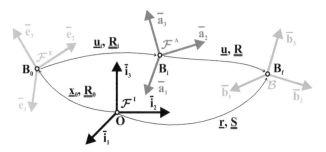

Fig. 4.8. Time-dependent motion of a rigid body.

The following frames will be used in this problem: $\mathcal{F}^I = [\mathbf{O}, \mathcal{I} = (\bar{\imath}_1, \bar{\imath}_2, \bar{\imath}_3)]$, is an inertial frame, $\mathcal{F}^E = [\mathbf{B_0}, \mathcal{E} = (\bar{e}_1, \bar{e}_2, \bar{e}_3)]$, a body attached frame that defines the configuration of the rigid body in its reference configuration (say at time $t = 0$), $\mathcal{F}^A = [\mathbf{B_i}, \mathcal{A} = (\bar{a}_1, \bar{a}_2, \bar{a}_3)]$, a body attached frame that defines the configuration of the body at t_i, and finally $\mathcal{F}^B = [\mathbf{B_f}, \mathcal{B} = (\bar{b}_1, \bar{b}_2, \bar{b}_3)]$, a body attached frame that defines the configuration of the body at t_f. Typically, t_i and t_f would be the initial and final times, respectively, for a time step of the computation that proceeds in an incremental manner.

The position vector of point $\mathbf{B_0}$ of the rigid body with respect to point $\mathbf{O}$ is denoted $\underline{x}_0$, and the orientation of the body is determined by rotation tensor $\underline{\underline{R}}_0$ that brings basis $\mathcal{I}$ to basis $\mathcal{E}$. Next, the position vector of point $\mathbf{B_i}$ with respect to point $\mathbf{B}_0$ is denoted $\underline{u}_i$ and the corresponding orientation of the body is determined by rotation tensor $\underline{\underline{R}}_i$ that brings basis $\mathcal{E}$ to basis $\mathcal{A}$. Note that $\underline{u}_i$ and $\underline{\underline{R}}_i$ define the configuration of the rigid body at time t_i relative to that at time $t = 0$. The configuration of the body with respect to the inertial frame would have to be obtained from a composition of the partial displacements and rotations. Finally, the incremental motion of the body from time t_i to t_f is defined by position vector $\underline{u}$ of point $\mathbf{B}_f$ with respect to point $\mathbf{B}_i$ and rotation tensor $\underline{\underline{R}}$ that brings basis $\mathcal{A}$ to basis $\mathcal{B}$. Determine the inertial position and orientation of the body at time t_f.

The inertial position of the body is readily found by adding the various displacements to find $\underline{r} = \underline{x}_0 + \underline{u}_i + \underline{u}$. This vector equation can be resolved in any basis, for instance the inertial basis.

The various bases are related to each other through the corresponding rotation tensors: $\bar{e}_1 = \underline{\underline{R}}_0 \bar{\imath}_1$, $\bar{a}_1 = \underline{\underline{R}}_i \bar{e}_1$, and $\bar{b}_1 = \underline{\underline{R}} \bar{a}_1$. Eliminating the intermediate bases yields $\bar{b}_1 = \underline{\underline{R}} \, \underline{\underline{R}}_i \underline{\underline{R}}_0 \bar{\imath}_1 = \underline{\underline{S}} \, \bar{\imath}_1$, where $\underline{\underline{S}} = \underline{\underline{R}} \, \underline{\underline{R}}_i \underline{\underline{R}}_0$ is the rotation tensor that brings basis $\mathcal{I}$ to basis $\mathcal{B}$. This tensor relationship can be expressed in component form as follows: $\underline{\underline{S}}^{[\mathcal{I}]} = \underline{\underline{R}}^{[\mathcal{I}]} \underline{\underline{R}}_i^{[\mathcal{I}]} \underline{\underline{R}}_0^{[\mathcal{I}]}$, where all tensors have been expressed in a common basis $\mathcal{I}$, see eq. (4.43).

It is sometimes more convenient to express each rotation tensor in the local basis; in that case eq. (4.44) yields: $\underline{\underline{S}}^{[\mathcal{I}]} = \underline{\underline{R}}_0^{[\mathcal{I}]} \underline{\underline{R}}_i^{[\mathcal{E}]} \underline{\underline{R}}^{[\mathcal{A}]}$. Note the reversing of the order of the individual rotations depending on the basis in which the tensors are expressed. This behavior is a consequence of the nonlinear nature of rotation operations.

The operation of *composing* displacements corresponds to a simple addition of vectors. In contrast, the corresponding operation for rotations is far more complex: rotations cannot be added by simply adding "rotation vectors." Rather, the components of the corresponding rotation tensors are multiplied, and the order in which the tensors appear depends on the bases in which their components are resolved. This fundamental difference is reflected in the vocabulary: *displacement vectors are added, rotations are composed.*

4.9.1 Problems

Problem 4.21. Sequence of rotations

Consider a sequence of n orthonormal bases denoted $\mathcal{B}^1, \mathcal{B}^2, \ldots \mathcal{B}^k, \ldots \mathcal{B}^n$. Let rotation tensor $\underline{\underline{R}}_k$ define the rotation from $\mathcal{B}^k$ to $\mathcal{B}^{k+1}$. Rotation tensors $\underline{\underline{R}}_k$, $k = 1, 2, \ldots n - 1$, then define the successive rotations between these bases. *(1)* Prove the following tensor relationship $\underline{\underline{R}} = \underline{\underline{R}}_{n-1} \underline{\underline{R}}_{n-2} \cdots \underline{\underline{R}}_2 \underline{\underline{R}}_1$, where rotation tensor $\underline{\underline{R}}$ defines the rotation from basis $\mathcal{B}^1$ to basis $\mathcal{B}^n$. *(2)* Prove that $\underline{\underline{R}}^{[\mathcal{B}^1]} = \underline{\underline{R}}^{[\mathcal{B}^n]} = \underline{\underline{R}}_{n-1}^{[\mathcal{B}^1]} \underline{\underline{R}}_{n-2}^{[\mathcal{B}^1]} \cdots \underline{\underline{R}}_2^{[\mathcal{B}^1]} \underline{\underline{R}}_1^{[\mathcal{B}^1]}$, and $\underline{\underline{R}}^{[\mathcal{B}^1]} = \underline{\underline{R}}^{[\mathcal{B}^n]} = \underline{\underline{R}}_1^{[\mathcal{B}^1]} \underline{\underline{R}}_2^{[\mathcal{B}^2]} \cdots \underline{\underline{R}}_{n-2}^{[\mathcal{B}^{n-2}]} \underline{\underline{R}}_{n-1}^{[\mathcal{B}^{n-1}]}$.

Problem 4.22. Composition of rotations

Consider three orthonormal bases $\mathcal{B}$, $\mathcal{B}_0$, and $\mathcal{B}^*$. Let rotation tensor $\underline{\underline{R}}_0$ describe the rotation from basis $\mathcal{B}$ to $\mathcal{B}_0$ and $\underline{\underline{R}}$ that from basis $\mathcal{B}_0$ to $\mathcal{B}^*$. The components of tensors $\underline{\underline{R}}_0$ and $\underline{\underline{R}}$ resolved in basis $\mathcal{B}$ are

$$
\underline{\underline{R}}_0^{[\mathcal{B}]} = \begin{bmatrix} 0.3258 & -0.9377 & -0.1212 \\ 0.8683 & 0.3474 & -0.3540 \\ 0.3740 & 0.0101 & 0.9274 \end{bmatrix}, \quad
\underline{\underline{R}}^{[\mathcal{B}]} = \begin{bmatrix} 0.2944 & 0.9433 & -0.1536 \\ -0.9005 & 0.2199 & -0.3751 \\ -0.3200 & 0.2488 & 0.9142 \end{bmatrix}.
$$

Let $\underline{\underline{R}}_i = \underline{\underline{R}}\,\underline{\underline{R}}_0$. Prove the following relationships: *(1)* $\underline{\underline{R}}_i^{[\mathcal{B}]} = \underline{\underline{R}}_i^{[\mathcal{B}^*]}$. *(2)* $\underline{\underline{R}}_i^{[\mathcal{B}]} = \underline{\underline{R}}^{[\mathcal{B}]} \underline{\underline{R}}_0^{[\mathcal{B}]}$. *(3)* $\underline{\underline{R}}_i^{[\mathcal{B}]} = \underline{\underline{R}}_0^{[\mathcal{B}]} \underline{\underline{R}}^{[\mathcal{B}_0]}$. *(4)* Verify each relationship numerically by performing the matrix multiplications.

Problem 4.23. Robotic system with spinning disk

The system depicted in fig. 4.9 consists of a shaft of height h rigidly connected to an arm of length L_a and of a spinning disk of radius R mounted at the free end of the arm. Frame $\mathcal{F}^S = [\mathbf{S}, \mathcal{S}^+ = (\bar{s}_1, \bar{s}_2, \bar{s}_3)]$ is attached to the shaft at point $\mathbf{S}$, and frame $\mathcal{F}^D = [\mathbf{C}, \mathcal{B}^* = (\bar{b}_1, \bar{b}_2, \bar{b}_3)]$ is attached to the disk at point $\mathbf{C}$. Superscripts $(.)^+$ and $(.)^*$ indicate components of tensors resolved in bases $\mathcal{S}^+$ and $\mathcal{B}^*$, respectively. Angle $\alpha(t)$ and $\beta(t)$ are the magnitudes of the planar rotations about axis $\bar{\imath}_3$ and $\bar{s}_1$, respectively, that bring basis $\mathcal{I}$ to $\mathcal{S}^+$ and basis $\mathcal{S}^+$ to $\mathcal{B}^*$, respectively. Tensors $\underline{\underline{R}}_\alpha$ and $\underline{\underline{R}}_\beta$ are the rotation tensors associated with those two rotations, and $\underline{\underline{R}}$ is the rotation tensor that brings basis $\mathcal{I}$ to $\mathcal{B}^*$. If angles $\alpha(t)$ and $\beta(t)$ are given, write compact expressions for the following tensor components: *(1)* $\underline{\underline{R}}_\alpha$, $\underline{\underline{R}}_\alpha^+$, and $\underline{\underline{R}}_\alpha^*$; *(2)* $\underline{\underline{R}}_\beta$, $\underline{\underline{R}}_\beta^+$, and $\underline{\underline{R}}_\beta^*$. *(3)* Express $\underline{\underline{R}}$ in terms of $\underline{\underline{R}}_\alpha$ and $\underline{\underline{R}}_\beta$. *(4)* express $\underline{\underline{R}}$ in terms of $\underline{\underline{R}}_\alpha$ and $\underline{\underline{R}}_\beta^+$.

Problem 4.24. Relative rotation at a revolute joint

Figure 4.10 depicts two rigid bodies denoted with superscripts $(\cdot)^k$ and $(\cdot)^\ell$, respectively, linked together by a revolute joint. In the reference configuration, the orientation of the rigid

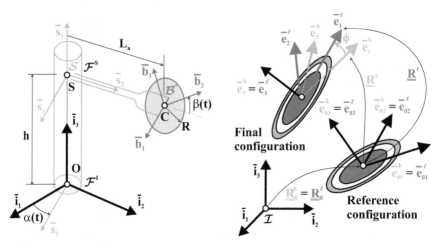

Fig. 4.9. Spinning disk mounted on a rotating arm.

Fig. 4.10. Revolute joint in the reference and final configurations.

bodies is defined by coincident bases $\mathcal{B}_0^k = \mathcal{B}_0^\ell$. In the deformed configuration, the orientations of the bodies are defined by two distinct bases $\mathcal{B}^k$ and $\mathcal{B}^\ell$, respectively. No relative displacement is permitted between the bodies that are allowed to rotate with respect to each other in such a way that $\bar{e}_3^k = \bar{e}_3^\ell$. Rotation tensor $\underline{\underline{R}}_0^k = \underline{\underline{R}}_0^\ell$ describes the rotation from $\mathcal{I}$ to $\mathcal{B}_0^k = \mathcal{B}_0^\ell$, tensor $\underline{\underline{R}}^k$ that from $\mathcal{B}_0^k$ to $\mathcal{B}^k$, and tensor $\underline{\underline{R}}^\ell$ that from $\mathcal{B}_0^\ell$ to $\mathcal{B}^\ell$. If

$$\underline{\underline{R}}_0^{[\mathcal{I}]} = \begin{bmatrix} 0.1043 & 0.5561 & 0.8245 \\ -0.1873 & 0.8252 & -0.5329 \\ -0.9767 & -0.0989 & 0.1902 \end{bmatrix} ;$$

$$\underline{\underline{R}}^{k[\mathcal{I}]} = \begin{bmatrix} 0.6311 & -0.7492 & 0.2010 \\ 0.0140 & -0.2480 & -0.9687 \\ 0.7756 & 0.6141 & -0.1460 \end{bmatrix}, \quad \underline{\underline{R}}^{\ell[\mathcal{I}]} = \begin{bmatrix} 0.6272 & -0.7305 & 0.2700 \\ -0.1268 & -0.4379 & -0.8900 \\ 0.7684 & 0.5240 & -0.3673 \end{bmatrix},$$

find the relative rotation, ϕ, of the revolute joint.

Problem 4.25. Rigid bodies connected by torsional springs

Figure 4.11 shows two rigid bodies denoted with superscripts $(\cdot)^k$ and $(\cdot)^\ell$, respectively, linked together by torsional springs at a point. In the reference configuration, the orientation of the rigid bodies is defined by coincident bases $\mathcal{B}_0^k = \mathcal{B}_0^\ell$. In the deformed configuration, the orientations of the bodies are defined by two distinct bases $\mathcal{B}^k$ and $\mathcal{B}^\ell$, respectively. No relative displacement is permitted between the bodies that are allowed to rotate with respect to each other in an arbitrary manner. Rotation tensor $\underline{\underline{R}}_0^k = \underline{\underline{R}}_0^\ell$ describes the rotation from $\mathcal{I}$ to $\mathcal{B}_0^k = \mathcal{B}_0^\ell$, tensor $\underline{\underline{R}}^k$ that from $\mathcal{B}_0^k$ to $\mathcal{B}^k$, and tensor $\underline{\underline{R}}^\ell$ that from $\mathcal{B}_0^\ell$ to $\mathcal{B}^\ell$. Let $\underline{\underline{R}}$ be the rotation tensor from $\mathcal{B}^k$ to $\mathcal{B}^\ell$. The deformation of the torsional springs will be measured by the following vector

$$\underline{s} = \frac{1}{2} \left\{ \begin{array}{l} \bar{e}_3^{kT} \bar{e}_2^\ell - \bar{e}_2^{kT} \bar{e}_3^\ell \\ \bar{e}_1^{kT} \bar{e}_3^\ell - \bar{e}_3^{kT} \bar{e}_1^\ell \\ \bar{e}_2^{kT} \bar{e}_1^\ell - \bar{e}_1^{kT} \bar{e}_2^\ell \end{array} \right\} .$$

(1) Find the relationship between components $\underline{s}^{[\mathcal{B}^k]}$ and the components of tensor $\underline{\underline{R}}$ resolved in an appropriate basis. Clearly define this appropriate basis, and give the components of $\underline{\underline{R}}$ in that basis. *(2)* Find the relationship between components $\underline{s}^{[\mathcal{B}^k]}$ and the magnitude ϕ and unit axis $\bar{n}$ characterizing the rotation tensor $\underline{\underline{R}}$ expressed in the previously defined basis.

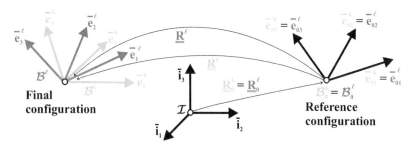

Fig. 4.11. Rigid bodies linked by torsional springs. For clarity of the figure, the reference and final configurations have been translated with respect to each other.

4.10 Time derivatives of rotation operations

Consider a fixed orthonormal basis $\mathcal{I} = (\bar{\imath}_1, \bar{\imath}_2, \bar{\imath}_3)$ and a time-dependent orthonormal basis $\mathcal{E} = (\bar{e}_1, \bar{e}_2, \bar{e}_3)$. It is often the case that the orientation of this moving orthonormal basis depend on a scalar variable, say time t. If $\underline{\underline{R}}(t)$ is the time-dependent rotation tensor that bring $\mathcal{I}$ to $\mathcal{E}$, $\bar{e}_1(t) = \underline{\underline{R}}(t)\bar{\imath}_1$. The time derivative of this expression is $\dot{\bar{e}}_1(t) = \underline{\underline{\dot{R}}}(t)\bar{\imath}_1$, where notation $(\dot{\cdot})$ indicates a time derivative. Clearly, $\underline{\underline{\dot{R}}}(t)$ can be evaluated directly by taking a time derivative of the rotation tensor, eq. (4.15). The concept of *angular velocity vector,* however, considerably simplifies this operation and will be explored in the next sections.

4.10.1 The angular velocity vector: an intuitive approach

Consider a constant norm, time-dependent vector, $\underline{b}(t)$, and a rotation operation characterized by an instantaneous unit vector, $\bar{n}(t)$, and an infinitesimal rotation, $\Delta\phi$. Figure 4.12 shows the effect of this infinitesimal rotation on the orientation of vector $\underline{b}$: at time t, its orientation is $\underline{b}(t)$, at time $t + \Delta t$, its orientation is $\underline{b}(t + \Delta t)$. Because vector $\underline{b}$ is of constant norm, it sweeps the outer surface of a cone, whose summit is at the origin and its basis is a circle in a plane normal to $\bar{n}$. The radius of the circle is $r = \|\underline{b}\| \sin \alpha$, where α is the angle between vectors $\underline{b}$ and $\bar{n}$. The increment in $\underline{b}$ is $\Delta\underline{b} = \underline{b}(t + \Delta t) - \underline{b}(t)$, a vector that lies in a plane normal to $\bar{n}$.

If $\Delta\phi \to 0$, vector $\Delta\underline{b}$ becomes tangent to the circle, and hence, normal to $\underline{b}(t)$. In this case, $\Delta\underline{b}$ is normal to both $\bar{n}$ and $\underline{b}(t)$, and hence, $\Delta\underline{b} = c\, \tilde{n}\underline{b}$. The unknown constant, c, can be determined by taking the norm of both sides of this equation to find $\|\Delta\underline{b}\| = c\|\underline{b}\| \sin\alpha$. Because $\|\Delta\underline{b}\| = r\Delta\phi$, it follows that $r\Delta\phi = cr$, and finally $c = \Delta\phi$. The incremental change in vector $\underline{b}$ now becomes $\Delta\underline{b} = \Delta\phi\,\widetilde{n}\underline{b}$.

By definition, the time derivative of vector $\underline{b}$ is

$$\dot{\underline{b}} = \lim_{\Delta t\to 0} \frac{\underline{b}(t+\Delta t) - \underline{b}(t)}{\Delta t} = \lim_{\Delta t\to 0}\frac{\Delta\underline{b}}{\Delta t} = \lim_{\Delta t\to 0}\frac{\Delta\phi}{\Delta t}\,\widetilde{n}\underline{b} = \dot{\phi}\,\widetilde{n}\underline{b}.$$

Vector $\underline{\omega} = \dot{\phi}\bar{n}$ is the angular velocity vector; the time derivative of vector $\underline{b}$ now becomes

$$\dot{\underline{b}} = \widetilde{\omega}\underline{b}. \tag{4.46}$$

This important relationship implies that *the time derivative of a constant norm vector equals the vector product of the angular velocity vector by the constant norm vector itself.* Clearly, the angular velocity vector will play a fundamental role in computing the time derivatives of vectors.

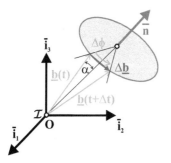
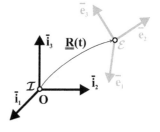

Fig. 4.12. Rotation of a constant norm vector $\underline{b}$.

Fig. 4.13. Fixed and rotating orthonormal bases.

The above result can also be obtained from eq. (4.14) that describes the rotation an arbitrary vector $\underline{a}$. Let $\phi = 0$ and $\phi = \Delta\phi$ at time $t = 0$ and $t = \Delta t$, respectively. It follows that $\underline{b}(0) = \underline{a}$ and $\underline{b}(\Delta t) = \underline{a} + \Delta\phi\,\widetilde{n}\underline{a}$, where higher order terms were neglected. The increment in vector $\underline{b}$ is then $\Delta\underline{b} = \Delta\phi\,\widetilde{n}\underline{b}$, a result identical to that obtained above. Clearly, this result is valid for infinitesimal rotations about $\phi = 0$.

If the axis $\bar{n}$ about which the rotation is taking place has a constant direction in time, the angular velocity vector can be written as $\underline{\omega} = \dot{\phi}\bar{n} = \mathrm{d}(\phi\bar{n})/\mathrm{dt}$, *i.e.*, the angular velocity is the time derivative of vector $\phi\bar{n}$. The results obtained above, however, are not limited to the case where axis $\bar{n}$ is of constant direction. In the general case, $\underline{\omega} = \dot{\phi}\bar{n}(t)$, and because $\phi(t)$ and $\bar{n}(t)$ are independent functions of time, there exist no vector such that its derivative equals the angular velocity vector.

The angular velocity vector is a *nonholonomic vector*, *i.e.*, a vector that cannot be integrated. This contrasts with the expression of the velocity vector of a particle: the position vector is $\underline{u}(t)$ and the velocity $\underline{v}(t) = \dot{\underline{u}}$. In this case, the integral of the velocity vector is the position vector. When it comes to the angular velocity, there exist no vector $\underline{x}$ such that $\underline{\omega} = \dot{\underline{x}}$.

4.10.2 The angular velocity vector: a rigorous approach

The development presented in the previous section is limited to rotations about $\phi = 0$; consequently, the resulting expressions are not general. In the present section, a rigorous definition of the angular velocity vector is derived by considering the time-dependent orthonormal basis, $\mathcal{E}(t) = (\bar{e}_1, \bar{e}_2, \bar{e}_3)$, depicted in fig. 4.13.

Definition of the angular velocity vector

Because $\bar{e}_1$ is a unit vector $\bar{e}_1^T \bar{e}_1 = 1$ and a time derivative of this equation yields $\bar{e}_1^T \dot{\bar{e}}_1 = 0$, *i.e.*, vectors $\dot{\bar{e}}_1$ and $\bar{e}_1$ must be perpendicular to each other. This implies the existence of vector $\underline{a}_1$, such that

$$\dot{\bar{e}}_1 = \widetilde{a}_1 \bar{e}_1. \tag{4.47}$$

To determine $\underline{a}_1$, this equation is recast as $\widetilde{e}_1 \underline{a}_1 = -\dot{\bar{e}}_1$, a vector product equation for unknown vector $\underline{a}_1$. In view of eq. (1.35), the solution of this equation is $\underline{a}_1 = w_1 \bar{e}_1 + \widetilde{e}_1 \dot{\bar{e}}_1$, where w_1 is an arbitrary constant; this solution exists because the right-hand side of the equation, $-\dot{\bar{e}}_1$, is orthogonal to the the null space of $\widetilde{e}_1$, *i.e.*, $\bar{e}_1^T \dot{\bar{e}}_1 = 0$.

Vectors $\bar{e}_2$ an $\bar{e}_3$ are also unit vectors and a reasoning similar to that developed above leads to $\underline{a}_1 = w_1 \bar{e}_1 + \widetilde{e}_1 \dot{\bar{e}}_1$, $\underline{a}_2 = w_2 \bar{e}_2 + \widetilde{e}_2 \dot{\bar{e}}_2$, and $\underline{a}_3 = w_3 \bar{e}_3 + \widetilde{e}_3 \dot{\bar{e}}_3$, where w_1, w_2, and w_3 are arbitrary constants.

The components of vector $\underline{a}_1$ in the rotating basis $\mathcal{E}$ are readily found as $\bar{e}_1^T \underline{a}_1 = w_1$, $\bar{e}_2^T \underline{a}_1 = -\dot{\bar{e}}_1^T \bar{e}_3$ and $\bar{e}_3^T \underline{a}_1 = \dot{\bar{e}}_1^T \bar{e}_2$. A similar reasoning applied to vectors $\underline{a}_2$ and $\underline{a}_3$ then leads to

$$\underline{a}_1 = w_1\, \bar{e}_1 - (\dot{\bar{e}}_1^T \bar{e}_3)\, \bar{e}_2 + (\dot{\bar{e}}_1^T \bar{e}_2)\, \bar{e}_3, \tag{4.48a}$$

$$\underline{a}_2 = (\dot{\bar{e}}_2^T \bar{e}_3)\, \bar{e}_1 + w_2\, \bar{e}_2 - (\dot{\bar{e}}_2^T \bar{e}_1)\, \bar{e}_3, \tag{4.48b}$$

$$\underline{a}_3 = -(\dot{\bar{e}}_3^T \bar{e}_2)\, \bar{e}_1 + (\dot{\bar{e}}_3^T \bar{e}_1)\, \bar{e}_2 + w_3\, \bar{e}_3, \tag{4.48c}$$

where the last two equations were obtained by evaluating the components of vectors $\underline{a}_2$ and $\underline{a}_3$ in rotating basis $\mathcal{E}$.

Unit vectors $\bar{e}_2$ and $\bar{e}_3$ are mutually orthogonal, *i.e.*, $\bar{e}_2^T \bar{e}_3 = 0$; a time derivative of this orthogonality condition implies $\dot{\bar{e}}_2^T \bar{e}_3 = -\dot{\bar{e}}_3^T \bar{e}_2$. Since w_1 is arbitrary, it is possible to select $w_1 = \dot{\bar{e}}_2^T \bar{e}_3 = -\dot{\bar{e}}_3^T \bar{e}_2$. The three arbitrary constants are selected as follows,

$$w_1 = \dot{\bar{e}}_2^T \bar{e}_3 = -\dot{\bar{e}}_3^T \bar{e}_2, \quad w_2 = \dot{\bar{e}}_3^T \bar{e}_1 = -\dot{\bar{e}}_1^T \bar{e}_3, \quad w_3 = \dot{\bar{e}}_1^T \bar{e}_2 = -\dot{\bar{e}}_2^T \bar{e}_1, \tag{4.49}$$

where the last two equations stem from the orthogonality conditions, $\bar{e}_1^T \bar{e}_3 = 0$ and $\bar{e}_1^T \bar{e}_2 = 0$, respectively. Inspection of eqs. 4.48 and 4.49 then reveals that the three vectors $\underline{a}_1$, $\underline{a}_2$, and $\underline{a}_3$ are equal to each other, *i.e.*, $\underline{a}_1 = \underline{a}_2 = \underline{a}_3 = \underline{\omega}$, where

$$\underline{\omega} = (\dot{\bar{e}}_2^T \bar{e}_3)\, \bar{e}_1 + (\dot{\bar{e}}_3^T \bar{e}_1)\, \bar{e}_2 + (\dot{\bar{e}}_1^T \bar{e}_2)\, \bar{e}_3. \tag{4.50}$$

This relationship provides a formal definition of the angular velocity vector. Since the quantities in parentheses are scalar products of vector, the angular velocity vector is indeed a first-order tensor, because it is a linear combination of first-order tensors.

Properties of the angular velocity vector

The fundamental property of the angular velocity vector is its relationship to the time derivative of an orthonormal basis; indeed, eq. (4.47) now becomes

$$\dot{\bar{e}}_1 = \widetilde{w}\bar{e}_1, \quad \dot{\bar{e}}_2 = \widetilde{w}\bar{e}_2, \quad \dot{\bar{e}}_3 = \widetilde{w}\bar{e}_3. \tag{4.51}$$

Clearly, these results are identical to those obtained in the previous section, see eq. (4.46): the vector product of the angular velocity vector by a constant norm vector yields the time derivative of the vector itself.

The concept of angular velocity is associated with the time derivative of orthonormal basis $\mathcal{E}$. Let $\underline{b}$ be an arbitrary vector attached to basis $\mathcal{E}$, *i.e.*, $\underline{b} = \alpha_1 \bar{e}_1 + \alpha_2 \bar{e}_2 + \alpha_3 \bar{e}_3$, where α_1, α_2, and α_3 are time independent constants. The time derivative of this vector then becomes $\dot{\underline{b}} = \alpha_1 \dot{\bar{e}}_1 + \alpha_2 \dot{\bar{e}}_2 + \alpha_3 \dot{\bar{e}}_3 = \widetilde{w}(\alpha_1 \bar{e}_1 + \alpha_2 \bar{e}_2 + \alpha_3 \bar{e}_3) = \widetilde{w}\underline{b}$.

The time derivative of any vector attached to basis $\mathcal{E}$ is $\dot{\underline{b}} = \widetilde{w}\underline{b}$, *i.e.*, the angular velocity vector characterizes the time derivative of the angular motion of the basis, not just that of a single unit vector. Because a one to one correspondence exists between the angular motion of an orthonormal basis and that of a rigid body, the angular velocity vector characterizes the time derivative of the angular motion of a rigid body.

The following alternative expression, which presents a higher symmetry in the indices, can also be used to define the angular velocity vector

$$\underline{w} = \frac{1}{2}\left[(\bar{e}_3^T \dot{\bar{e}}_2 - \bar{e}_2^T \dot{\bar{e}}_3)\,\bar{e}_1 + (\bar{e}_1^T \dot{\bar{e}}_3 - \bar{e}_3^T \dot{\bar{e}}_1)\,\bar{e}_2 + (\bar{e}_2^T \dot{\bar{e}}_1 - \bar{e}_1^T \dot{\bar{e}}_2)\,\bar{e}_3\right]. \tag{4.52}$$

The components of the angular velocity vector resolved in the rotating basis, $\mathcal{E}$, denoted $\underline{w}^*$ are

$$\underline{w}^* = \frac{1}{2} \left\{ \begin{matrix} \bar{e}_3^T \dot{\bar{e}}_2 - \bar{e}_2^T \dot{\bar{e}}_3 \\ \bar{e}_1^T \dot{\bar{e}}_3 - \bar{e}_3^T \dot{\bar{e}}_1 \\ \bar{e}_2^T \dot{\bar{e}}_1 - \bar{e}_1^T \dot{\bar{e}}_2 \end{matrix} \right\}. \tag{4.53}$$

Relating the angular velocity vector to the rotation tensor

Let $\underline{\underline{R}}$ be the rotation tensor that bring basis $\mathcal{I}$ to basis $\mathcal{E}$; the components of this tensor in basis $\mathcal{I}$ are $\underline{\underline{R}} = [\bar{e}_1, \bar{e}_2, \bar{e}_3]$ and it then follows that

$$\begin{aligned}
\underline{\underline{R}}^T \dot{\underline{\underline{R}}} &= \begin{bmatrix} 0 & \bar{e}_1^T \dot{\bar{e}}_2 & \bar{e}_1^T \dot{\bar{e}}_3 \\ \bar{e}_2^T \dot{\bar{e}}_1 & 0 & \bar{e}_2^T \dot{\bar{e}}_3 \\ \bar{e}_3^T \dot{\bar{e}}_1 & \bar{e}_3^T \dot{\bar{e}}_2 & 0 \end{bmatrix} \\
&= \frac{1}{2} \begin{bmatrix} 0 & -(\bar{e}_2^T \dot{\bar{e}}_1 - \bar{e}_1^T \dot{\bar{e}}_2) & (\bar{e}_1^T \dot{\bar{e}}_3 - \bar{e}_3^T \dot{\bar{e}}_1) \\ (\bar{e}_2^T \dot{\bar{e}}_1 - \bar{e}_1^T \dot{\bar{e}}_2) & 0 & -(\bar{e}_3^T \dot{\bar{e}}_2 - \bar{e}_2^T \dot{\bar{e}}_3) \\ -(\bar{e}_1^T \dot{\bar{e}}_3 - \bar{e}_3^T \dot{\bar{e}}_1) & (\bar{e}_3^T \dot{\bar{e}}_2 - \bar{e}_2^T \dot{\bar{e}}_3) & 0 \end{bmatrix}.
\end{aligned} \tag{4.54}$$

Comparing eq. (4.53) and (4.54) then yields

$$\widetilde{w}^* = \underline{\underline{R}}^T \dot{\underline{\underline{R}}}. \tag{4.55}$$

Because the angular velocity vector is a tensor, its components in the fixed basis $\mathcal{I}$ are then obtained from eq. (4.29) as

$$\underline{\widetilde{\omega}} = \underline{\underline{R}}\,\underline{\widetilde{\omega}}^*\underline{\underline{R}}^T = \underline{\underline{R}}(\underline{\underline{R}}^T\underline{\dot{\underline{R}}})\underline{\underline{R}}^T = \underline{\dot{\underline{R}}}\,\underline{\underline{R}}^T. \tag{4.56}$$

This equation defines the angular velocity in terms of the rotation tensor and its time derivative. Since it is a tensor relationship, it is true in all bases, and could be taken as the definition of the angular velocity vector, although it more abstract and algebraic than the definition given by eq. (4.50), which is rooted in more geometric arguments.

The results derived above can be recovered from purely algebraic manipulation. Let $\underline{\underline{R}}(t)$ be the time-dependent rotation tensor that brings basis $\mathcal{I}$ to basis $\mathcal{E}(t)$, $\bar{e}_1(t) = \underline{\underline{R}}(t)\bar{\imath}_1$. A time derivative of this expression yields $\dot{\bar{e}}_1 = \underline{\dot{\underline{R}}}\,\bar{\imath}_1(t) = \underline{\dot{\underline{R}}}\,\underline{\underline{R}}^T\,\bar{e}_1(t)$. A time derivative of the orthogonality property of the rotation tensor, eq. (4.26), leads to $\underline{\dot{\underline{R}}}\,\underline{\underline{R}}^T = -(\underline{\dot{\underline{R}}}\,\underline{\underline{R}}^T)^T$, which shows that tensor $\underline{\dot{\underline{R}}}\,\underline{\underline{R}}^T$ is skew symmetric, as implied by eq. (4.56), which is taken to be the definition of the angular velocity vector. The time derivative of unit vector $\bar{e}_1$ then becomes $\dot{\bar{e}}_1(t) = \underline{\widetilde{\omega}}\bar{e}_1(t)$, as expected from earlier developments. The components of this vector in basis $\mathcal{E}$ are now $\underline{\underline{R}}^T\,\dot{\bar{e}}_1(t) = \underline{\underline{R}}^T\underline{\widetilde{\omega}}\bar{e}_1 = \underline{\underline{R}}^T\underline{\widetilde{\omega}}\underline{\underline{R}}\,\underline{\underline{R}}^T\bar{e}_1 = \underline{\widetilde{\omega}}^*\bar{\imath}_1$.

Explicit expression of the angular velocity vector

The angular velocity vector can be expressed in terms of quantities $\phi(t)$ and $\bar{n}(t)$ that characterize the rotation. Introducing the rotation tensor, eq. (4.15), into eq. (4.56) and using identity (1.34c) yields

$$\underline{\widetilde{\omega}} = \dot{\phi}\widetilde{n} + \sin\phi\,\dot{\widetilde{n}} + (1 - \cos\phi)(\widetilde{n}\dot{\widetilde{n}} - \dot{\widetilde{n}}\widetilde{n}). \tag{4.57}$$

The angular velocity vector now becomes

$$\underline{\omega} = \dot{\phi}\,\bar{n} + \sin\phi\,\dot{\bar{n}} + (1 - \cos\phi)\widetilde{n}\dot{\bar{n}} \tag{4.58}$$

Note that for $\phi = 0$, $\underline{\omega} = \dot{\phi}\bar{n}$, the result obtained with the simplified approach of the previous section. The time derivative of unit vector $\bar{n}$ about which the rotation takes place explicitly appears in the rigorous expression of the angular velocity vector. Because eq. (4.58) cannot be integrated in general, the angular velocity vector is a nonholonomic vector.

Example 4.6. Angular velocity in terms of Euler angles
Find the angular velocity vector expressed in terms of Euler angles and their time derivatives; use the *3-1-3* sequence to define the Euler angles.

The components of the angular velocity vector resolved in the moving basis will be evaluated first. In section 4.4, the rotation tensor expressed in terms of Euler angles was found to be $\underline{\underline{R}}^{[\mathcal{I}]} = \underline{\underline{R}}_\phi^{[\mathcal{I}]}\underline{\underline{R}}_\theta^{[\mathcal{A}]}\underline{\underline{R}}_\psi^{[\mathcal{B}]}$. In the following, the superscripts will be dropped to simplify the writing. Equation (4.55) then yields

$$\underline{\widetilde{\omega}}^* = \underline{\underline{R}}^T\underline{\dot{\underline{R}}} = (\underline{\underline{R}}_\phi\underline{\underline{R}}_\theta\underline{\underline{R}}_\psi)^T(\underline{\dot{\underline{R}}}_\phi\underline{\underline{R}}_\theta\underline{\underline{R}}_\psi + \underline{\underline{R}}_\phi\underline{\dot{\underline{R}}}_\theta\underline{\underline{R}}_\psi + \underline{\underline{R}}_\phi\underline{\underline{R}}_\theta\underline{\dot{\underline{R}}}_\psi)$$

$$= (\underline{\underline{R}}_\theta\underline{\underline{R}}_\psi)^T(\underline{\underline{R}}_\phi^T\underline{\dot{\underline{R}}}_\phi)(\underline{\underline{R}}_\theta\underline{\underline{R}}_\psi) + \underline{\underline{R}}_\psi^T(\underline{\underline{R}}_\theta^T\underline{\dot{\underline{R}}}_\theta)\underline{\underline{R}}_\psi + (\underline{\underline{R}}_\psi^T\underline{\dot{\underline{R}}}_\psi).$$

$\underline{\underline{R}}_\phi^T \underline{\dot{R}}_\phi$ represents the angular velocity associated with the first planar rotation of the 3-1-3 Euler angle sequence. It can be readily evaluated using elementary trigonometric formulæ to find $\underline{\underline{R}}_\phi^T \underline{\dot{R}}_\phi = \dot\phi\, \bar{\imath}_3^{[\mathcal{I}]}$.

Proceeding in a similar manner with the other terms leads to $\widetilde{\omega}^* = (\underline{\underline{R}}_\theta \underline{\underline{R}}_\psi)^T (\dot\phi \widetilde{\imath}_3^{[\mathcal{I}]})(\underline{\underline{R}}_\theta \underline{\underline{R}}_\psi) + \underline{\underline{R}}_\psi^T (\dot\theta \widetilde{\bar{a}}_1^{[\mathcal{A}]})\underline{\underline{R}}_\psi + (\dot\psi \widetilde{\bar{b}}_3^{[\mathcal{B}]})$, and finally

$$\underline{\omega}^* = (\underline{\underline{R}}_\theta^{[\mathcal{A}]} \underline{\underline{R}}_\psi^{[\mathcal{B}]})^T (\dot\phi\, \bar{\imath}_3^{[\mathcal{I}]}) + \underline{\underline{R}}_\psi^{[\mathcal{B}]T} (\dot\theta\, \bar{a}_1^{[\mathcal{A}]}) + (\dot\psi\, \bar{b}_3^{[\mathcal{B}]}),$$

where $\bar{\imath}_3^{[\mathcal{I}]T} = \bar{b}_3^{[\mathcal{B}]T} = \{0, 0, 1\}$ and $\bar{a}_1^{[\mathcal{A}]T} = \{1, 0, 0\}$. Performing the matrix multiplications and casting the result in a matrix form leads to

$$\underline{\omega}^* = \underline{\underline{H}}^*_{3\text{-}1\text{-}3} \begin{Bmatrix} \dot\phi \\ \dot\theta \\ \dot\psi \end{Bmatrix}, \quad \text{with} \quad \underline{\underline{H}}^*_{3\text{-}1\text{-}3} = \begin{bmatrix} S_\theta S_\psi & C_\psi & 0 \\ S_\theta C_\psi & -S_\psi & 0 \\ C_\theta & 0 & 1 \end{bmatrix}. \tag{4.59}$$

Operator $\underline{\underline{H}}^*_{3\text{-}1\text{-}3}$ is called the *tangent operator* because it is tangent to the rotation manifold.

Of course, the components of the angular velocity vector resolved in the fixed basis could also be evaluated. Starting from eq. (4.56), the desired components are found as $\underline{\omega} = \dot\phi\, \bar{\imath}_3^{[\mathcal{I}]} + \dot\theta \underline{\underline{R}}_\phi\, \bar{\imath}_1^{[\mathcal{I}]} + \dot\psi \underline{\underline{R}}_\phi \underline{\underline{R}}_\theta\, \bar{\imath}_3^{[\mathcal{I}]}$; in matrix form, this becomes

$$\underline{\omega} = \underline{\underline{H}}_{3\text{-}1\text{-}3} \begin{Bmatrix} \dot\phi \\ \dot\theta \\ \dot\psi \end{Bmatrix}, \quad \text{with} \quad \underline{\underline{H}}_{3\text{-}1\text{-}3} = \begin{bmatrix} 0 & C_\phi & S_\phi S_\theta \\ 0 & S_\phi & -C_\phi S_\theta \\ 1 & 0 & C_\theta \end{bmatrix}. \tag{4.60}$$

Sometimes, the angular velocity components of a rigid body are known, or have been computed from dynamical equations of motion. The orientation of the rigid body is then obtained by integration the following kinematical equations

$$\begin{Bmatrix} \dot\phi \\ \dot\theta \\ \dot\psi \end{Bmatrix} = \underline{\underline{H}}^{*-1}_{3\text{-}1\text{-}3}\, \underline{\omega}^*, \quad \text{with} \quad \underline{\underline{H}}^{*-1}_{3\text{-}1\text{-}3} = \frac{1}{S_\theta} \begin{bmatrix} S_\psi & C_\psi & 0 \\ S_\theta C_\psi & -S_\theta S_\psi & 0 \\ -C_\theta S_\psi & -C_\theta C_\psi & S_\theta \end{bmatrix}.$$

$$\begin{Bmatrix} \dot\phi \\ \dot\theta \\ \dot\psi \end{Bmatrix} = \underline{\underline{H}}^{-1}_{3\text{-}1\text{-}3}\, \underline{\omega}, \quad \text{with} \quad \underline{\underline{H}}^{-1}_{3\text{-}1\text{-}3} = \frac{1}{S_\theta} \begin{bmatrix} -S_\phi C_\theta & C_\phi C_\theta & S_\theta \\ C_\phi S_\theta & S_\phi S_\theta & 0 \\ S_\phi & -C_\phi & 0 \end{bmatrix}.$$

These relationships become singular when $S_\theta = 0$; as was noted in section 4.4, singularities occurs when using Euler angles to represent rotations, for all possible sequence choices. Because $\underline{\omega}^* = \underline{\underline{R}}^T \underline{\omega}$, it follows that $\underline{\underline{R}}\, \underline{\underline{H}}^*_{3\text{-}1\text{-}3} = \underline{\underline{H}}_{3\text{-}1\text{-}3}$, or $\underline{\underline{R}} = \underline{\underline{H}}_{3\text{-}1\text{-}3} \underline{\underline{H}}^{*-1}_{3\text{-}1\text{-}3}$.

4.10.3 The addition theorem

Consider now the problem of two time-dependent bases rotating with respect to an inertial frame of reference, $\mathcal{I}$. The first basis is denoted $\mathcal{E}^* = (\bar{e}_1, \bar{e}_2, \bar{e}_3)$ and the second $\mathcal{B} = (\bar{b}_1, \bar{b}_2, \bar{b}_3)$, as depicted in fig. 4.14. Rotation tensor $\underline{\underline{R}}_1$ brings basis $\mathcal{I}$ to basis $\mathcal{E}^*$, and tensor $\underline{\underline{R}}_2$ brings $\mathcal{E}^*$ to $\mathcal{B}$.

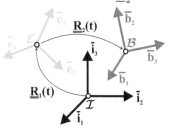

Fig. 4.14. Two time-dependent bases, $\mathcal{E}$ and $\mathcal{B}$.

Clearly, the instantaneous orientation of basis $\mathcal{B}$ with respect to basis $\mathcal{I}$ is a function of both tensors $\underline{\underline{R}}_1$ and $\underline{\underline{R}}_2$. Similarly, the angular velocity of basis $\mathcal{B}$ with respect to $\mathcal{I}$ depends on the angular velocities of both bases $\mathcal{E}^*$ and $\mathcal{B}$.

If the components of tensors $\underline{\underline{R}}_1$ and $\underline{\underline{R}}_2$ are resolved in basis $\mathcal{I}$,

$$\bar{b}_1 = \underline{\underline{R}}_2 \underline{\underline{R}}_1 \bar{\imath}_1 = \underline{\underline{R}}_1 \underline{\underline{R}}_1^T \underline{\underline{R}}_2 \underline{\underline{R}}_1 \bar{\imath}_1 = \underline{\underline{R}}_1 \underline{\underline{R}}_2^* \bar{\imath}_1,$$

where $\underline{\underline{R}}_2^*$ are the components of tensor $\underline{\underline{R}}_2$ resolved in basis $\mathcal{E}^*$, see eq. (4.29). Superscript $(\cdot)^*$ is used here to indicate tensor components resolved in basis $\mathcal{E}^*$.

The time derivative of unit vector $\bar{b}_1$ now becomes

$$\dot{\bar{b}}_1 = (\dot{\underline{\underline{R}}}_1 \underline{\underline{R}}_2^* + \underline{\underline{R}}_1 \dot{\underline{\underline{R}}}_2^*)(\underline{\underline{R}}_1 \underline{\underline{R}}_2^*)^T \bar{b}_1 = (\dot{\underline{\underline{R}}}_1 \underline{\underline{R}}_1^T + \underline{\underline{R}}_1 \dot{\underline{\underline{R}}}_2^* \underline{\underline{R}}_2^{*T} \underline{\underline{R}}_1^T)\bar{b}_1. \tag{4.61}$$

The first term of the last equality is the angular velocity of basis $\mathcal{E}^*$ with respect to basis $\mathcal{I}$, denoted $\underline{\omega}_1 = \text{axial}(\dot{\underline{\underline{R}}}_1 \underline{\underline{R}}_1^T)$. The components of this angular velocity vector are resolved in basis $\mathcal{I}$. Next, $\underline{\omega}_2^* = \text{axial}(\dot{\underline{\underline{R}}}_2^* \underline{\underline{R}}_2^{*T})$ are the components of the angular velocity vector of basis $\mathcal{B}$ with respect to basis $\mathcal{E}^*$, resolved in $\mathcal{E}^*$. The second term of the last equality involves the components of the angular velocity vector, $\underline{\omega}_2 = \underline{\underline{R}}_1 \underline{\omega}_2^*$, of basis $\mathcal{B}$ with respect to basis $\mathcal{E}^*$, resolved in basis $\mathcal{I}$, because

$$\underline{\underline{R}}_1 \dot{\underline{\underline{R}}}_2^* \underline{\underline{R}}_2^{*T} \underline{\underline{R}}_1^T = \underline{\underline{R}}_1 \widetilde{\omega}_2^* \underline{\underline{R}}_1^T = \widetilde{\underline{\underline{R}}_1 \underline{\omega}_2^*} = \widetilde{\omega}_2.$$

The derivative of the unit vector, eq. (4.61), now reduces to $\dot{\bar{b}}_1 = (\widetilde{\omega}_1 + \widetilde{\omega}_2)\bar{b}_1 = \widetilde{\omega}\bar{b}_1$, where

$$\underline{\omega} = \underline{\omega}_1 + \underline{\omega}_2. \tag{4.62}$$

Vector $\underline{\omega}$ is the angular velocity of basis $\mathcal{B}$ with respect to $\mathcal{I}$. This result is know as the addition theorem.

Theorem 4.2 (Addition theorem). *The angular velocity of basis $\mathcal{B}$ with respect to basis $\mathcal{I}$ is the sum of the angular velocities of basis $\mathcal{A}$ with respect to $\mathcal{I}$ and of basis $\mathcal{B}$ with respect to $\mathcal{A}$, where $\mathcal{A}$ is an arbitrary basis.*

The angular velocity of basis $\mathcal{B}$ with respect to $\mathcal{I}$ is $\underline{\omega} = \underline{\omega}_1 + \underline{\omega}_2$, where $\underline{\omega}_1$ is the angular velocity of basis $\mathcal{E}^*$ with respect to basis $\mathcal{I}$, and $\underline{\omega}_2$ is the angular velocity of basis $\mathcal{B}$ with respect to basis $\mathcal{E}^*$. This is a tensor relationship can be expressed in any basis; for instance, $\underline{\omega}^{[\mathcal{I}]} = \underline{\omega}_1^{[\mathcal{I}]} + \underline{\omega}_2^{[\mathcal{I}]}$ or $\underline{\omega}^{[\mathcal{B}]} = \underline{\omega}_1^{[\mathcal{B}]} + \underline{\omega}_2^{[\mathcal{B}]}$.

It would appear that angular velocity vector $\underline{\omega}_2$ is more naturally resolved in basis $\mathcal{E}^*$; its components are then denoted $\underline{\omega}_2^*$. It is clearly incorrect, however, to write

$\underline{\omega} = \underline{\omega}_1 + \underline{\omega}_2^*$, because it is *wrong to add the components of two vectors resolved in different bases*. In many applications, the angular velocity of the second basis will be defined by its components in that basis, $\underline{\omega}_2^*$; in that case, the addition theorem states $\underline{\omega} = \underline{\omega}_1 + \underline{R}_1 \underline{\omega}_2^*$.

In applications of the addition theorem, it is important to correctly identify and evaluate the angular velocities of the two bases at hand. For instance, it would be incorrect to believe that the angular velocity of basis $\mathcal{B}$ with respect to $\mathcal{E}^*$ is $\underline{\dot{R}}_2 \underline{R}_2^T$. Indeed, because $\underline{R}_2 = \underline{R}_1 \underline{R}_2^* \underline{R}_1^T$, it follows that

$$
\begin{aligned}
\underline{\dot{R}}_2 \underline{R}_2^T &= (\underline{\dot{R}}_1 \underline{R}_2^* \underline{R}_1^T + \underline{R}_1 \underline{\dot{R}}_2^* \underline{R}_1^T + \underline{R}_1 \underline{R}_2^* \underline{\dot{R}}_1^T)(\underline{R}_1 \underline{R}_2^* \underline{R}_1^T)^T \\
&= \underline{\dot{R}}_1 \underline{R}_1^T + \underline{R}_1 \underline{\dot{R}}_2^* \underline{R}_2^{*T} \underline{R}_1^T + (\underline{R}_1 \underline{R}_2^*) \underline{\dot{R}}_1^T \underline{R}_1 (\underline{R}_1 \underline{R}_2^*)^T \\
&= \tilde{\omega}_1 + \underline{R}_1 \tilde{\omega}_2^* \underline{R}_1^T + (\underline{R}_2 \underline{R}_1) \underline{\dot{R}}_1^T \underline{R}_1 (\underline{R}_2 \underline{R}_1)^T \\
&= \tilde{\omega}_1 + \tilde{\omega}_2 + \underline{R}_2 \underline{R}_1 \underline{\dot{R}}_1^T \underline{R}_2^T = \tilde{\omega}_1 + \tilde{\omega}_2 + \underline{R}_2 \tilde{\omega}_1^T \underline{R}_2^T \\
&= \tilde{\omega}_2 + \widetilde{(\underline{I} - \underline{R}_2)\omega_1}.
\end{aligned}
$$

Clearly, $\underline{\dot{R}}_2 \underline{R}_2^T \neq \tilde{\omega}_2$. This is due to the fact that although $\underline{R}_2$ is the rotation tensor that rotates basis $\mathcal{E}^*$ to basis $\mathcal{B}$, the components of this tensor are resolved in basis $\mathcal{I}$. The components of the angular velocity of basis $\mathcal{B}$ with respect to basis $\mathcal{E}^*$, resolved in $\mathcal{E}^*$, are $\underline{\omega}_2^* = \text{axial}(\underline{\dot{R}}_2^* \underline{R}_2^{*T})$, because $\underline{R}_2^*$ are the components of the rotation tensor that rotates basis $\mathcal{E}^*$ to basis $\mathcal{B}$, resolved in basis $\mathcal{E}^*$.

Example 4.7. Angular velocity in terms of Euler angles

In section 4.4, Euler angles were defined as the magnitudes of three successive planar rotations that produce an arbitrary rotation, as shown in fig 4.4. In example 4.6, expressions were derived for the components of the angular velocity vector in terms of Euler angles and their time derivatives. Derive these expressions using the addition theorem.

According to this theorem, the angular velocity of basis $\mathcal{E}$ with respect to basis $\mathcal{I}$ is simply $\underline{\omega} = \underline{\omega}_\phi + \underline{\omega}_\theta + \underline{\omega}_\psi$, where $\underline{\omega}_\phi$ is the angular velocity associated with the planar rotation that brings basis $\mathcal{I}$ to basis $\mathcal{A}$, $\underline{\omega}_\theta$ that associated with the planar rotation from basis $\mathcal{A}$ to basis $\mathcal{B}$, and $\underline{\omega}_\psi$ that associated with the planar rotation from basis $\mathcal{B}$ to $\mathcal{E}$. It follows that $\underline{\omega}^* = \underline{\omega}^{[\mathcal{E}]} = \underline{\omega}_\phi^{[\mathcal{E}]} + \underline{\omega}_\theta^{[\mathcal{E}]} + \underline{\omega}_\psi^{[\mathcal{E}]}$; while correct, this expression is not convenient to use because the partial angular velocities are all expressed in the same basis, $\mathcal{E}$.

Using the rules of transformation for the components of first-order tensors, eq. (4.27), yields $\underline{\omega}^* = (\underline{R}_\theta^{[\mathcal{A}]} \underline{R}_\psi^{[\mathcal{B}]})^T \underline{\omega}_\phi^{[\mathcal{A}]} + \underline{R}_\psi^{[\mathcal{B}]T} \underline{\omega}_\theta^{[\mathcal{B}]} + \underline{\omega}_\psi^{[\mathcal{E}]}$. Because each partial rotation is a planar rotation, it is clear that $\underline{\omega}_\phi^{[\mathcal{A}]} = \underline{\omega}_\phi^{[\mathcal{I}]} = \dot{\phi} \bar{\imath}_3^{[\mathcal{I}]}$, $\underline{\omega}_\theta^{[\mathcal{B}]} = \underline{\omega}_\theta^{[\mathcal{A}]} = \dot{\theta} \bar{a}_1^{[\mathcal{A}]}$, $\underline{\omega}_\psi^{[\mathcal{E}]} = \underline{\omega}_\psi^{[\mathcal{B}]} = \dot{\psi} \bar{b}_3^{[\mathcal{B}]}$, and finally, $\underline{\omega}^* = (\underline{R}_\theta^{[\mathcal{A}]} \underline{R}_\psi^{[\mathcal{B}]})^T (\dot{\phi} \bar{\imath}_3^{[\mathcal{I}]}) + \underline{R}_\psi^{[\mathcal{B}]T} (\dot{\theta} \bar{a}_1^{[\mathcal{A}]}) + (\dot{\psi} \bar{b}_3^{[\mathcal{B}]})$. This expression is identical to that found in example 4.6.

4.10.4 Angular acceleration

The angular velocity vector enables the evaluation of the derivative of a unit vector, see eq. (4.51). The second derivative of a unit vector then becomes

$$\ddot{\bar{e}}_1 = \dot{\tilde{\omega}}\bar{e}_1 + \tilde{\omega}\dot{\bar{e}}_1 = \dot{\tilde{\omega}}\bar{e}_1 + \tilde{\omega}\tilde{\omega}\bar{e}_1 = (\dot{\tilde{\omega}} + \tilde{\omega}\tilde{\omega})\bar{e}_1. \tag{4.63}$$

The *angular acceleration vector* is defined as $\underline{\alpha} = \dot{\underline{\omega}}$; it then follows that

$$\ddot{\bar{e}}_1 = (\tilde{\alpha} + \tilde{\omega}\tilde{\omega})\bar{e}_1. \tag{4.64}$$

Consider now the problem of two time-dependent bases rotating with respect to an inertial frame of reference, $\mathcal{I}$, as depicted in fig. 4.14. The addition theorem, eq. (4.62), implies $\dot{\bar{b}}_1 = \tilde{\omega}\bar{b}_1$, where $\underline{\omega} = \underline{\omega}_1 + \underline{\omega}_2$. The second derivative of the unit vector now becomes $\ddot{\bar{b}}_1 = (\tilde{\alpha} + \tilde{\omega}\tilde{\omega})\bar{b}_1$, where the angular acceleration of basis $\mathcal{B}$ with respect to basis $\mathcal{I}$ is

$$\underline{\alpha} = \dot{\underline{\omega}}_1 + \dot{\underline{\omega}}_2. \tag{4.65}$$

This result corresponds to the addition theorem for angular acceleration, and echoes the corresponding result for angular velocities, eq. (4.62).

The angular acceleration of basis $\mathcal{B}$ with respect to $\mathcal{I}$ is $\underline{\alpha} = \dot{\underline{\omega}} = \dot{\underline{\omega}}_1 + \dot{\underline{\omega}}_2$, where $\underline{\omega}_1$ is the angular velocity of basis $\mathcal{E}^*$ with respect to basis $\mathcal{I}$ and $\underline{\omega}_2$ is the angular velocity of basis $\mathcal{B}$ with respect to basis $\mathcal{E}^*$. This tensor relationship is true in any basis, $\underline{\alpha}^{[\mathcal{I}]} = \dot{\underline{\omega}}_1^{[\mathcal{I}]} + \dot{\underline{\omega}}_2^{[\mathcal{I}]}$ or $\underline{\alpha}^{[\mathcal{B}]} = \dot{\underline{\omega}}_1^{[\mathcal{B}]} + \dot{\underline{\omega}}_2^{[\mathcal{B}]}$. Of course, it would be incorrect to write $\underline{\alpha} = \dot{\underline{\omega}}_1 + \dot{\underline{\omega}}_2^*$, where notation $(\cdot)^*$ indicates tensor components resolved in basis $\mathcal{E}^*$, because it is wrong to add the components of two vectors resolved in different bases. If the angular acceleration is to be written in terms of $\underline{\omega}_2^*$, the following expression should be used instead $\underline{\alpha} = \dot{\underline{\omega}}_1 + (\underline{\underline{R}}_1\underline{\omega}_2^*)^{\cdot} = \dot{\underline{\omega}}_1 + \underline{\underline{R}}_1\dot{\underline{\omega}}_2^* + \tilde{\omega}_1\underline{\underline{R}}_1\underline{\omega}_2^*$.

4.11 Euler angle formulas

This section gives a summary of formulas used for the manipulation of rotation operations expressed in terms of Euler angles. As mentioned in section 4.4, twelve different sequences of planar rotations can be used to express an arbitrary rotation; of those twelve possible sequences, the four sequences starting with a rotation about the third axis will be detailed in this section. Arbitrary rotations from basis $\mathcal{I} = (\bar{\imath}_1, \bar{\imath}_2, \bar{\imath}_3)$ to basis $\mathcal{E} = (\bar{e}_1, \bar{e}_2, \bar{e}_3)$ will be considered, with two intermediate bases, $\mathcal{A} = (\bar{a}_1, \bar{a}_2, \bar{a}_3)$ and $\mathcal{B} = (\bar{b}_1, \bar{b}_2, \bar{b}_3)$. The complete sequence of bases is as follows: $\mathcal{I} \rightarrow \mathcal{A} \rightarrow \mathcal{B} \rightarrow \mathcal{E}$. The three Euler angles are denoted q_1, q_2, and q_3, and the array of Euler angles is then $\underline{q}^T = \{q_1, q_2, q_3\}$.

For each sequence, the following information is given.

1. The matrix of direction cosines expressed in terms of the Euler angles, $\underline{\underline{D}} = \underline{\underline{D}}(\underline{q})$. See section 4.4 and eq. (4.11) for the *3-1-3* sequence.

2. The Euler angles expressed in terms of the components of the direction cosine matrix, $q_1 = q_1(\underline{\underline{D}})$, $q_2 = q_2(\underline{\underline{D}})$, and $q_3 = q_3(\underline{\underline{D}})$. See section 4.4 and eq. (4.12) for the *3-1-3* sequence. For computer implementation, it is convenient to use the function $\phi = \text{atan2}(\sin q, \cos q)$ that defines angle $\phi \in [-\pi, \pi]$.
3. The tangent operators, $\underline{\underline{H}}(\underline{q})$ and $\underline{\underline{H}}^*(\underline{q})$, which express the components of the angular velocity vector resolved in basis $\mathcal{I}$ and $\mathcal{E}$, respectively, in terms of the time derivatives of the Euler angles, *i.e.*, $\underline{\omega} = \underline{\underline{H}}(\underline{q})\dot{\underline{q}}$ and $\underline{\omega}^* = \underline{\underline{H}}^*(\underline{q})\dot{\underline{q}}$, respectively. See example 4.6 and eqs. (4.60) and (4.59), respectively, for the *3-1-3* sequence.
4. The inverses of the tangent operators are also given.

4.11.1 Euler angles: sequence *3-1-3*

Euler angles with the *3-1-3* sequence are defined as follows.

1. A planar rotation of magnitude ϕ, called *precession*, about axis $\bar{\imath}_3$ brings $\mathcal{I}$ to $\mathcal{A}$.
2. A planar rotation of magnitude θ, called *nutation*, about axis $\bar{a}_1$ brings $\mathcal{A}$ to $\mathcal{B}$.
3. A planar rotation of magnitude ψ, called *spin*, about axis $\bar{b}_3$ brings $\mathcal{B}$ to $\mathcal{E}$.

The array of Euler angles is now $\underline{q}^T = \{\phi, \theta, \psi\}$.
1) The direction cosine matrix is

$$\underline{\underline{D}}_{3\text{-}1\text{-}3} = \begin{bmatrix} C_\phi C_\psi - S_\phi C_\theta S_\psi & -C_\phi S_\psi - S_\phi C_\theta C_\psi & S_\phi S_\theta \\ S_\phi C_\psi + C_\phi C_\theta S_\psi & -S_\phi S_\psi + C_\phi C_\theta C_\psi & -C_\phi S_\theta \\ S_\theta S_\psi & S_\theta C_\psi & C_\theta \end{bmatrix}. \tag{4.66}$$

2) Euler angle expressed in terms of the direction cosine matrix components are

$$\begin{aligned} \psi &= \text{atan2}(D_{31}, D_{32}), \quad \text{if } D_{32} \neq 0, \\ \theta &= \text{atan2}(D_{31}\sin\psi + D_{32}\cos\psi, D_{33}), \\ \phi &= \text{atan2}(D_{13}, -D_{23}). \end{aligned} \tag{4.67}$$

A singularity occurs when $\theta = 0$ or π, .
3) The tangent operators are

$$\underline{\underline{H}}_{3\text{-}1\text{-}3} = \begin{bmatrix} 0 & C_\phi & S_\phi S_\theta \\ 0 & S_\phi & -C_\phi S_\theta \\ 1 & 0 & C_\theta \end{bmatrix}, \quad \underline{\underline{H}}^*_{3\text{-}1\text{-}3} = \begin{bmatrix} S_\theta S_\psi & C_\psi & 0 \\ S_\theta C_\psi & -S_\psi & 0 \\ C_\theta & 0 & 1 \end{bmatrix}, \tag{4.68}$$

respectively.
4) The inverses of the tangent operators are

$$\underline{\underline{H}}^{-1}_{3\text{-}1\text{-}3} = \frac{1}{S_\theta}\begin{bmatrix} -S_\phi C_\theta & C_\phi C_\theta & S_\theta \\ C_\phi S_\theta & S_\phi S_\theta & 0 \\ S_\phi & -C_\phi & 0 \end{bmatrix}, \quad \underline{\underline{H}}^{*-1}_{3\text{-}1\text{-}3} = \frac{1}{S_\theta}\begin{bmatrix} S_\psi & C_\psi & 0 \\ S_\theta C_\psi & -S_\theta S_\psi & 0 \\ -C_\theta S_\psi & -C_\theta C_\psi & S_\theta \end{bmatrix}. \tag{4.69}$$

4.11.2 Euler angles: sequence *3-2-3*

Euler angles with the *3-2-3* sequence are defined as follows.

1. A planar rotation of magnitude ψ, called *precession*, about axis $\bar{\imath}_3$ brings $\mathcal{I}$ to $\mathcal{A}$.
2. A planar rotation of magnitude θ, called *nutation*, about axis $\bar{a}_2$ brings $\mathcal{A}$ to $\mathcal{B}$.
3. A planar rotation of magnitude ϕ, called *spin*, about axis $\bar{b}_3$ brings $\mathcal{B}$ to $\mathcal{E}$.

The array of Euler angles is now $\underline{q}^T = \{\psi, \theta, \phi\}$.
   1) The direction cosine matrix is

$$\underline{\underline{D}}_{3\text{-}2\text{-}3} = \begin{bmatrix} C_\psi C_\theta C_\phi - S_\psi S_\phi & -C_\psi C_\theta S_\phi - S_\psi C_\phi & C_\psi S_\theta \\ S_\psi C_\theta C_\phi + C_\psi S_\phi & -S_\psi C_\theta S_\phi + C_\psi C_\phi & S_\psi S_\theta \\ -S_\theta C_\phi & S_\theta S_\phi & C_\theta \end{bmatrix}. \qquad (4.70)$$

   2) Euler angle expressed in terms of the direction cosine matrix components are

$$\begin{aligned} \phi &= \text{atan2}(\ D_{32}, -D_{31}), \quad \text{if } D_{31} \neq 0, \\ \theta &= \text{atan2}(-D_{31}\cos\phi + D_{32}\sin\phi, D_{33}), \\ \psi &= \text{atan2}(\ D_{23}, D_{13}). \end{aligned} \qquad (4.71)$$

It is clear that when $\theta = 0$ or π, a singularity occurs.
   3) The tangent operators are

$$\underline{\underline{H}}_{3\text{-}2\text{-}3} = \begin{bmatrix} 0 & -S_\psi & C_\psi S_\theta \\ 0 & C_\psi & S_\psi S_\theta \\ 1 & 0 & C_\theta \end{bmatrix}, \ \underline{\underline{H}}^*_{3\text{-}2\text{-}3} = \begin{bmatrix} -S_\theta C_\phi & S_\phi & 0 \\ S_\theta S_\phi & C_\phi & 0 \\ C_\theta & 0 & 1 \end{bmatrix}, \qquad (4.72)$$

respectively.
   4) The inverses of the tangent operators are

$$\underline{\underline{H}}^{-1}_{3\text{-}2\text{-}3} = \frac{1}{S_\theta} \begin{bmatrix} -C_\psi C_\theta & -S_\psi C_\theta & S_\theta \\ -S_\psi S_\theta & C_\psi S_\theta & 0 \\ C_\psi & S_\psi & 0 \end{bmatrix}, \ \underline{\underline{H}}^{*-1}_{3\text{-}2\text{-}3} = \frac{1}{S_\theta} \begin{bmatrix} -C_\phi & S_\phi & 0 \\ S_\theta S_\phi & S_\theta C_\phi & 0 \\ C_\theta C_\phi & -C_\theta S_\phi & S_\theta \end{bmatrix}. \qquad (4.73)$$

4.11.3 Euler angles: sequence *3-2-1*

Euler angles with the *3-2-1* sequence are commonly used in airplane flight mechanics formulations and are defined as follows.

1. A planar rotation of magnitude ψ, called *heading*, about axis $\bar{\imath}_3$ brings $\mathcal{I}$ to $\mathcal{A}$.
2. A planar rotation of magnitude θ, called *attitude*, about axis $\bar{a}_2$ brings $\mathcal{A}$ to $\mathcal{B}$.
3. A planar rotation of magnitude ϕ, called *bank*, about axis $\bar{b}_1$ brings $\mathcal{B}$ to $\mathcal{E}$.

The array of Euler angles is now $\underline{q}^T = \{\psi, \theta, \phi\}$.
   1) The direction cosine matrix is

$$\underline{\underline{D}}_{3\text{-}2\text{-}1} = \begin{bmatrix} C_\psi C_\theta & -S_\psi C_\phi + C_\psi S_\theta S_\phi & S_\psi S_\phi + C_\psi S_\theta C_\phi \\ S_\psi C_\theta & C_\psi C_\phi + S_\psi S_\theta S_\phi & -C_\psi S_\phi + S_\psi S_\theta C_\phi \\ -S_\theta & C_\theta S_\phi & C_\theta C_\phi \end{bmatrix}. \qquad (4.74)$$

2) Euler angle expressed in terms of the direction cosine matrix components are

$$\phi = \text{atan2}(\ D_{32}, D_{33}), \quad \text{if } D_{33} \neq 0,$$
$$\theta = \text{atan2}(-D_{31}, D_{32} \sin \phi + D_{33} \cos \phi), \tag{4.75}$$
$$\psi = \text{atan2}(\ D_{21}, D_{11}).$$

It is clear that when $\theta = \pi/2$ or $3\pi/2$, a singularity occurs.

3) The tangent operators are

$$\underline{\underline{H}}_{3\text{-}2\text{-}1} = \begin{bmatrix} 0 & -S_\psi & C_\psi C_\theta \\ 0 & C_\psi & S_\psi C_\theta \\ 1 & 0 & -S_\theta \end{bmatrix}, \quad \underline{\underline{H}}^*_{3\text{-}2\text{-}1} = \begin{bmatrix} -S_\theta & 0 & 1 \\ C_\theta S_\phi & C_\phi & 0 \\ C_\theta C_\phi & -S_\phi & 0 \end{bmatrix}, \tag{4.76}$$

respectively.

4) The inverses of the tangent operators are

$$\underline{\underline{H}}^{-1}_{3\text{-}2\text{-}1} = \frac{1}{C_\theta} \begin{bmatrix} C_\psi S_\theta & S_\psi S_\theta & C_\theta \\ -S_\psi C_\theta & C_\psi C_\theta & 0 \\ C_\psi & S_\psi & 0 \end{bmatrix}, \quad \underline{\underline{H}}^{*-1}_{3\text{-}2\text{-}1} = \frac{1}{C_\theta} \begin{bmatrix} 0 & S_\phi & C_\phi \\ 0 & C_\theta C_\phi & -C_\theta S_\phi \\ C_\theta & S_\theta S_\phi & S_\theta C_\phi \end{bmatrix}. \tag{4.77}$$

4.11.4 Euler angles: sequence 3-1-2

Euler angles with the *3-1-2* sequence are defined as follows.

1. A planar rotation of magnitude ϕ about axis $\bar{\imath}_3$ brings $\mathcal{I}$ to $\mathcal{A}$.
2. A planar rotation of magnitude θ about axis $\bar{a}_1$ brings $\mathcal{A}$ to $\mathcal{B}$.
3. A planar rotation of magnitude ψ about axis $\bar{b}_2$ brings $\mathcal{B}$ to $\mathcal{E}$.

The array of Euler angles is now $\underline{q}^T = \{\phi, \theta, \psi\}$.

1) The direction cosine matrix is

$$\underline{\underline{D}}_{3\text{-}1\text{-}2} = \begin{bmatrix} C_\phi C_\psi - S_\phi S_\theta S_\psi & -S_\phi C_\theta & C_\phi S_\psi + S_\phi S_\theta C_\psi \\ S_\phi C_\psi + C_\phi S_\theta S_\psi & C_\phi C_\theta & S_\phi S_\psi - C_\phi S_\theta C_\psi \\ -C_\theta S_\psi & S_\theta & C_\theta C_\psi \end{bmatrix} \tag{4.78}$$

2) Euler angle expressed in terms of the direction cosine matrix components are

$$\psi = \text{atan2}(-D_{31}, D_{33}), \quad \text{if } D_{33} \neq 0;$$
$$\theta = \text{atan2}(\ D_{32}, -D_{31} \sin \psi + D_{33} \cos \psi); \tag{4.79}$$
$$\phi = \text{atan2}(-D_{12}, D_{22}).$$

It is clear that when $\theta = \pi/2$ or $3\pi/2$, a singularity occurs.

3) The tangent operators are

$$\underline{\underline{H}}_{3\text{-}1\text{-}2} = \begin{bmatrix} 0 & C_\phi & -S_\phi C_\theta \\ 0 & S_\phi & C_\phi C_\theta \\ 1 & 0 & S_\theta \end{bmatrix}, \quad \underline{\underline{H}}^*_{3\text{-}1\text{-}2} = \begin{bmatrix} -C_\theta S_\psi & C_\psi & 0 \\ S_\theta & 0 & 1 \\ C_\theta C_\psi & S_\psi & 0 \end{bmatrix}, \tag{4.80}$$

respectively.

4) The inverses of the tangent operators are

$$\underline{\underline{H}}_{3\text{-}1\text{-}2}^{-1} = \frac{1}{C_\theta} \begin{bmatrix} S_\phi S_\theta & -C_\phi S_\theta & C_\theta \\ C_\phi C_\theta & S_\phi C_\theta & 0 \\ -S_\phi & C_\phi & 0 \end{bmatrix}, \quad \underline{\underline{H}}_{3\text{-}1\text{-}2}^{*\,-1} = \frac{1}{C_\theta} \begin{bmatrix} -S_\psi & 0 & C_\psi \\ C_\theta C_\psi & 0 & C_\theta S_\psi \\ S_\theta S_\psi & C_\theta & -S_\theta C_\psi \end{bmatrix}. \quad (4.81)$$

4.11.5 Problems

Problem 4.26. Angular velocity for *3-2-3* Euler angles

A popular choice of Euler angles is the *3-2-3* sequence that corresponds to the following sequence of planar rotations. First, a rotation of magnitude ψ about axis $\bar{\imath}_3$, called *precession*, brings basis $\mathcal{I}$ to $\mathcal{A}$. Second, a rotation of magnitude θ about axis $\bar{a}_2$, called *nutation*, brings basis $\mathcal{A}$ to $\mathcal{B}$. Finally, a rotation of magnitude ϕ about axis $\bar{b}_3$, called *spin*, brings basis $\mathcal{B}$ to $\mathcal{E}$. *(1)* Find the angular velocity vector associated with this rotation. *(2)* Determine the components of this vector in the fixed and moving bases. *(3)* Discuss the occurrence of singularities.

Problem 4.27. Angular velocity for *3-2-1* Euler angles

A popular choice of Euler angles for airplane flight mechanics is the *3-2-1* sequence that corresponds to the following sequence of planar rotations. First, a rotation of magnitude ψ about axis $\bar{\imath}_3$, called *heading*, brings basis $\mathcal{I}$ to $\mathcal{A}$. Second, a rotation of magnitude θ about axis $\bar{a}_2$, called *attitude*, brings basis $\mathcal{A}$ to $\mathcal{B}$. Finally, a rotation of magnitude ϕ about axis $\bar{b}_1$, called *bank*, brings basis $\mathcal{B}$ to $\mathcal{E}$. *(1)* Find the angular velocity vector associated with this rotation. *(2)* Determine the components of this vector in the fixed and moving bases. *(3)* Discuss the occurrence of singularities.

Problem 4.28. Angular velocity for *3-1-2* Euler angles

A choice of Euler angles is the *3-1-2* sequence that corresponds to the following sequence of planar rotations. First, a rotation of magnitude ϕ about axis $\bar{\imath}_3$ brings basis $\mathcal{I}$ to $\mathcal{A}$. Second, a rotation of magnitude θ about axis $\bar{a}_1$ brings basis $\mathcal{A}$ to $\mathcal{B}$. Finally, a rotation of magnitude ψ about axis $\bar{b}_2$ brings basis $\mathcal{B}$ to $\mathcal{E}$. *(1)* Find the angular velocity vector associated with this rotation. *(2)* Determine the components of this vector in the fixed and moving bases. *(3)* Discuss the occurrence of singularities.

Problem 4.29. Spinning disk on a rotating arm

The system depicted in fig. 4.9 consists of a shaft of height h rigidly connected to an arm of length L_a and of a spinning disk of radius R mounted at the free end of the arm. Frame $\mathcal{F}^S = [\mathbf{S}, \mathcal{S}^+ = (\bar{s}_1, \bar{s}_2, \bar{s}_3)]$ is attached to the shaft at point $\mathbf{S}$, whereas frame $\mathcal{F}^D = [\mathbf{C}, \mathcal{B}^* = (\bar{b}_1, \bar{b}_2, \bar{b}_3)]$ is attached to the disk at point $\mathbf{C}$. Superscripts $(.)^+$ and $(.)^*$ will be used to denote tensor components in bases $\mathcal{S}^+$ and $\mathcal{B}^*$, respectively. Angle $\alpha(t)$ and $\beta(t)$ are the magnitudes of the planar rotations about axis $\bar{\imath}_3$ and $\bar{s}_1$, respectively, that bring basis $\mathcal{I}$ to $\mathcal{S}^+$ and basis $\mathcal{S}^+$ to $\mathcal{B}^*$, respectively. *(1)* Find the angular velocity vector of basis $\mathcal{B}^*$ with respect to basis $\mathcal{I}$. *(2)* Find the components of this vector resolved in basis $\mathcal{I}$, then resolved in basis $\mathcal{B}^*$. *(3)* Find the angular acceleration vector of basis $\mathcal{B}$ with respect to basis $\mathcal{I}$. *(4)* Find the components of this vector resolved in basis $\mathcal{I}$, then in basis $\mathcal{B}^*$.

Problem 4.30. Alternative expression of the angular velocity vector

Show that the angular velocity vector can be written as $\underline{\omega} = \left[\tilde{e}_1 \dot{\bar{e}}_1 + \tilde{e}_2 \dot{\bar{e}}_2 + \tilde{e}_3 \dot{\bar{e}}_3 \right] / 2$. Give a geometric interpretation of this result.

Problem 4.31. Components of the angular velocity vector in the rotating basis

Based on eqs. (4.55) and (4.15), show that the components of the angular velocity vector in the rotating basis can be written as $\underline{\omega}^* = \dot{\phi}\,\bar{n} + \sin\phi\,\dot{\bar{n}} + (1 - \cos\phi)\widetilde{\bar{n}}\bar{n}$. Compare your result with eq. (4.58).

Problem 4.32. Derivatives with respect to the rotation parameters

This section has focused on the time derivatives of the rotation tensor. In some applications, derivatives of the rotation tensor with respect to the rotation parameters are needed. Let $\underline{R} = \underline{R}(\underline{q})$, where $\underline{q}^T = \{q_1, q_2, q_3\}$ are the rotation parameters that could be, for instance, Euler angles with a specific sequence of planar rotations, as discussed in section 4.11. Let

$$\frac{\partial\underline{R}}{\partial q_i} = \underline{R}_i, \quad i = 1, 2, 3.$$

(1) Show that $\underline{R}_i\underline{R}^T = \widetilde{h}_i$, $i = 1, 2, 3$, where $\underline{h}_i$ are the columns of the tangent operator $\underline{H}$, i.e., $\underline{H} = [\underline{h}_1, \underline{h}_2, \underline{h}_3]$, $\underline{\omega} = \underline{H}\dot{\underline{q}}$ and $\widetilde{\omega} = \dot{\underline{R}}\underline{R}^T$. *(2)* Show that $\underline{R}^T\underline{R}_i = \widetilde{h}_i^*$, $i = 1, 2, 3$, where $\underline{h}_i^*$ are the columns of the tangent operator $\underline{H}^*$, i.e., $\underline{H}^* = [\underline{h}_1^*, \underline{h}_2^*, \underline{h}_3^*]$, $\underline{\omega}^* = \underline{H}^*\dot{\underline{q}}$ and $\widetilde{\omega}^* = \underline{R}^T\dot{\underline{R}}$. *(3)* If vector $\underline{u}$ is not a function of $\underline{q}$ and $\underline{u}^* = \underline{R}^T\underline{u}$, show that

$$\frac{\partial\underline{u}^*}{\partial\underline{q}} = \widetilde{u}^*\underline{H}^* = \underline{R}^T\widetilde{u}\underline{H}. \tag{4.82}$$

(4) If vector $\underline{u}^*$ is not a function of $\underline{q}$ and $\underline{u} = \underline{R}\,\underline{u}^*$, show that

$$\frac{\partial\underline{u}}{\partial\underline{q}} = \widetilde{u}^T\underline{H} = \underline{R}\,\widetilde{u}^{*T}\underline{H}^*. \tag{4.83}$$

Problem 4.33. Derivatives of angular velocity with respect to the rotation parameters

Prove the following two identities

$$\dot{\underline{H}} = \frac{\partial\underline{\omega}}{\partial\underline{q}} + \widetilde{\omega}\underline{H} = \underline{R}\frac{\partial\underline{\omega}^*}{\partial\underline{q}}, \tag{4.84}$$

$$\dot{\underline{H}}^* = \frac{\partial\underline{\omega}^*}{\partial\underline{q}} - \widetilde{\omega}^*\underline{H}^* = \underline{R}^T\frac{\partial\underline{\omega}}{\partial\underline{q}}. \tag{4.85}$$

Hint: be familiar with the results of the previous problem. First show that $\widetilde{h_1 h_2} = \underline{R}_2\underline{R}_1^T - \underline{R}_1\underline{R}_2^T$, and because $\underline{R}_{12} = \underline{R}_{21}$, show that $\widetilde{h_1 h_2} = (\underline{R}_{21}\underline{R}^T + \underline{R}_2\underline{R}_1^T) - (\underline{R}_{12}\underline{R}^T + \underline{R}_1\underline{R}_2^T)$. The following relationships result

$$\widetilde{h_1 h_2} = \frac{\partial\underline{h}_2}{\partial q_1} - \frac{\partial\underline{h}_1}{\partial q_2}, \quad \widetilde{h_2 h_3} = \frac{\partial\underline{h}_3}{\partial q_2} - \frac{\partial\underline{h}_2}{\partial q_3}, \quad \widetilde{h_3 h_1} = \frac{\partial\underline{h}_1}{\partial q_3} - \frac{\partial\underline{h}_3}{\partial q_1}.$$

Combining these equations then yields

$$\frac{\partial\underline{h}_1}{\partial\underline{q}} = \underline{H}_1 - \widetilde{h}_1\underline{H}, \quad \frac{\partial\underline{h}_2}{\partial\underline{q}} = \underline{H}_2 - \widetilde{h}_2\underline{H}, \quad \frac{\partial\underline{h}_3}{\partial\underline{q}} = \underline{H}_3 - \widetilde{h}_3\underline{H},$$

where notation $\underline{H}_i = \partial\underline{H}/\partial q_i$ was introduced. Application of the chain rule for derivatives then leads to the desired identities.

4.12 Spatial derivatives of rotation operations

In the previous sections, time-dependent rotations were treated. Space-dependent rotations will be treated in a similar manner in this section. In fact, space-dependent rotations were encountered in chapter 2 when dealing with path coordinates, surface coordinates, and orthogonal curvilinear coordinates, as discussed in sections 2.3, 2.5, and 2.7, respectively.

Consider now a space-dependent orthonormal basis $\mathcal{E}(s) = (\bar{e}_1, \bar{e}_2, \bar{e}_3)$. If $\underline{\underline{R}}(s)$ is the space-dependent rotation tensor that bring $\mathcal{I}$ to $\mathcal{E}(s)$, $\bar{e}_1(s) = \underline{\underline{R}}(s)\bar{\imath}_1$. A spatial derivative of this expression yields

$$\bar{e}_1'(s) = \underline{\underline{R}}'\bar{\imath}_1(t) = \underline{\underline{R}}'\underline{\underline{R}}^T\bar{e}_1(s) = \widetilde{\kappa}\bar{e}_1(s),$$

where $\widetilde{\kappa} = \underline{\underline{R}}'\underline{\underline{R}}^T$ and notation $(\cdot)'$ indicates a derivative with respect to the spatial variable s. $\underline{\kappa}$ is the *curvature vector*, and by analogy with eq. (4.58), is expressed as

$$\underline{\kappa} = \phi'\bar{n} + \sin\phi\,\bar{n}' + (1 - \cos\phi)\widetilde{n}\bar{n}'. \tag{4.86}$$

This result is similar to that obtained for the angular velocity vector, eq. (4.58): the time derivative, $(\dot{\cdot})$, is replaced by the spatial derivative, $(\cdot)'$. The curvature vector resolved in basis $\mathcal{E}$ is $\underline{\kappa}^* = \underline{\underline{R}}^T\underline{\kappa}$, and $\widetilde{\kappa}^* = \underline{\underline{R}}^T\underline{\underline{R}}'$.

4.12.1 Path coordinates

Section 2.2 studies the differential geometry of curves in three-dimensional space. Unit vector $\bar{t}$ was shown to define the tangent to the curve at a point, vector $\bar{n}$ to be normal to the curve at the same point, and the binormal vector $\bar{b}$ was selected to be orthogonal to the two other vectors, see fig. 2.2. In section 2.2.1, vectors $\bar{t}$, $\bar{n}$, and $\bar{b}$ were shown to form an orthonormal basis, $\mathcal{F} = (\bar{t}, \bar{n}, \bar{b})$, called *Frenet's triad*.

The following orthogonal tensor is now defined

$$\underline{\underline{F}}(s) = \left[\bar{t}, \bar{n}, \bar{b}\right]. \tag{4.87}$$

This tensor can be interpreted as the space-dependent rotation tensor that bring the reference triad $\mathcal{I}$, to Frenet's triad $\mathcal{F}$. The *curvature tensor*, $\widetilde{\kappa}^*$, of the curve is defined as

$$\underline{\underline{F}}^T\frac{d\underline{\underline{F}}}{ds} = \widetilde{\kappa}^*. \tag{4.88}$$

With the help of eqs. (2.13), the curvature vector becomes

$$\underline{\kappa}^{*T} = \left\{\frac{1}{\tau}, 0, \frac{1}{\rho}\right\}. \tag{4.89}$$

Clearly, the twist and curvature of the curve, defined in eqs. (2.12) and (2.7), respectively, are the two non-vanishing components of the curvature vector resolved in Frenet's triad. The components of the curvature vector resolved in the reference frame $\mathcal{I}$ are then $\underline{\kappa} = \underline{\underline{F}}\,\underline{\kappa}^*$.

Finally, let curvilinear variable s be a function of time. Frenet's triad now becomes an implicit function of time, $\underline{\underline{F}}(t) = \underline{\underline{F}}(s(t))$. Using the chain rule for derivatives, the angular velocity of Frenet's triad, resolved in basis $\mathcal{F}$, is now

$$\widetilde{\omega}^* = \underline{\underline{F}}^T \dot{\underline{\underline{F}}} = \underline{\underline{F}}^T \frac{\mathrm{d}\underline{\underline{F}}}{\mathrm{d}s} \dot{s} = \dot{s}\widetilde{\kappa}^*. \tag{4.90}$$

This implies $\underline{\omega}^* = \dot{s}\underline{\kappa}^*$: the angular velocity vector is parallel to the curvature vector.

4.12.2 Surface coordinates

In the study of the differential geometry of surfaces in three-dimensional space, see section 2.4, unit vectors $\bar{e}_1$ and $\bar{e}_2$ were shown to define the plane tangent to the surface at a point and vector $\bar{n}$ to be normal to the surface at the same point, see fig. 2.7. In section 2.4.6, vectors $\bar{e}_1$, $\bar{e}_2$, and $\bar{n}$ were shown to form an orthonormal basis $\mathcal{B} = (\bar{e}_1, \bar{e}_2, \bar{n})$, when using lines of curvature.

The following orthogonal tensor is now defined

$$\underline{\underline{F}}(\eta_1, \eta_2) = [\bar{e}_1, \bar{e}_2, \bar{n}]. \tag{4.91}$$

This tensor can be interpreted as the space-dependent rotation tensor that brings the reference triad $\mathcal{I}$ to triad $\mathcal{B}$. The *curvature tensors* of the surface are now defined as

$$\underline{\underline{F}}^T \frac{\partial \underline{\underline{F}}}{\partial s_1} = \widetilde{\kappa}_1^*, \quad \underline{\underline{F}}^T \frac{\partial \underline{\underline{F}}}{\partial s_2} = \widetilde{\kappa}_2^*. \tag{4.92}$$

With the help of Gauss' and Weingarten's formulæ, eqs. (2.65), the curvature vectors are found as

$$\underline{\kappa}_1^{*T} = -\left\{0, \frac{1}{R_1}, \frac{1}{T_1}\right\}, \quad \underline{\kappa}_2^{*T} = \left\{\frac{1}{R_2}, 0, \frac{1}{T_2}\right\}. \tag{4.93}$$

The principal radii of curvature, see eqs. (2.54), and the twists, eqs. (2.58) and (2.59), of the surface, are the components of the curvature vectors resolved in frame $\mathcal{B}$. The components of the curvature vectors resolved in frame $\mathcal{I}$ are then $\underline{\kappa}_1 = \underline{\underline{F}}\,\underline{\kappa}_1^*$ and $\underline{\kappa}_2 = \underline{\underline{F}}\,\underline{\kappa}_2^*$.

If curvilinear coordinates s_1 and s_2 are functions of time, tensor $\underline{\underline{F}}$ becomes an implicit function of time, $\underline{\underline{F}}(t) = \underline{\underline{F}}(s_1(t), s_2(t))$. The angular velocity of tensor $\underline{\underline{F}}$, resolved in basis $\mathcal{B}$, is now

$$\widetilde{\omega}^* = \underline{\underline{F}}^T \dot{\underline{\underline{F}}} = \underline{\underline{F}}^T \left[\frac{\partial \underline{\underline{F}}}{\partial s_1} \dot{s}_1 + \frac{\partial \underline{\underline{F}}}{\partial s_2} \dot{s}_2\right] = \dot{s}_1 \widetilde{\kappa}_1^* + \dot{s}_2 \widetilde{\kappa}_2^*, \tag{4.94}$$

where the chain rule for derivatives was used. This implies $\underline{\omega}^* = \dot{s}_1\underline{\kappa}_1^* + \dot{s}_2\underline{\kappa}_2^*$: both curvature vectors contribute to the total angular velocity of basis $\mathcal{B}$.

In terms of the surface coordinates, eqs. (4.92) imply $\partial\underline{\underline{F}}/\partial\eta_1 = \underline{\underline{F}}\,h_1\widetilde{\kappa}_1^*$, and $\partial\underline{\underline{F}}/\partial\eta_2 = \underline{\underline{F}}\,h_2\widetilde{\kappa}_2^*$, where h_1 and h_2 are the scale factors introduced in eq. (2.56).

Taking partial derivatives with respect to η_2 and η_1 of the first and second equations, respectively, leads to

$$h_1 h_2 \, \tilde{\kappa}_2^* \tilde{\kappa}_1^* + \frac{\partial}{\partial \eta_2}(h_1 \tilde{\kappa}_1^*) = h_1 h_2 \, \tilde{\kappa}_1^* \tilde{\kappa}_2^* + \frac{\partial}{\partial \eta_1}(h_2 \tilde{\kappa}_2^*),$$

because $\partial^2 \underline{\underline{F}}/\partial \eta_1 \partial \eta_2 = \partial^2 \underline{\underline{F}}/\partial \eta_2 \partial \eta_1$.

With the help of identity (1.33a), the *Gauss-Codazzi conditions* are then obtained

$$\widetilde{\tilde{\kappa}_1^* \tilde{\kappa}_2^*} = \frac{1}{h_1}\frac{\partial}{\partial s_2}(h_1 \tilde{\kappa}_1^*) - \frac{1}{h_2}\frac{\partial}{\partial s_1}(h_2 \tilde{\kappa}_2^*). \tag{4.95}$$

More explicitly, this vector condition gives rise to three scalar conditions

$$\frac{\partial}{\partial s_1}\left(\frac{h_2}{R_2}\right) = \frac{h_2}{R_1 T_2}, \tag{4.96a}$$

$$\frac{\partial}{\partial s_2}\left(\frac{h_1}{R_1}\right) = \frac{h_1}{R_2 T_1}, \tag{4.96b}$$

$$\frac{1}{h_1}\frac{\partial}{\partial s_2}\left(\frac{h_1}{T_1}\right) + \frac{1}{h_2}\frac{\partial}{\partial s_1}\left(\frac{h_2}{T_2}\right) + \frac{1}{R_1 R_2} = 0. \tag{4.96c}$$

These equations express three conditions that must be satisfied by the radii of curvature, twists, and their spatial derivatives.

4.12.3 Orthogonal curvilinear coordinates

In the study of the differential geometry of a mapping of the three-dimensional space onto itself, see section 2.6, vectors $\bar{e}_1$, $\bar{e}_2$, and $\bar{e}_3$ were defined along the base vectors of the mapping. In section 2.6.2, these vectors were shown to form an orthonormal triad, $\mathcal{E}$, in the case of orthogonal curvilinear coordinate systems. The following orthogonal tensor is now defined

$$\underline{\underline{F}}(\eta_1, \eta_2, \eta_3) = [\bar{e}_1, \bar{e}_2, \bar{e}_3]. \tag{4.97}$$

This tensor can be interpreted as the rotation tensor that brings the reference triad, $\mathcal{I}$, to orthonormal triad $\mathcal{E}$.

The *curvature tensors* of the orthogonal curvilinear coordinate system are now defined as

$$\underline{\underline{F}}^T \frac{\partial \underline{\underline{F}}}{\partial s_1} = \tilde{\kappa}_1^*, \quad F^T \frac{\partial F}{\partial s_2} = \tilde{\kappa}_2^*, \quad \underline{\underline{F}}^T \frac{\partial \underline{\underline{F}}}{\partial s_3} = \tilde{\kappa}_3^*. \tag{4.98}$$

With the help of eqs. (2.84), these curvature vectors are found to be

$$\underline{\kappa}_1^* = \left\{0, \frac{1}{R_{12}}, \frac{1}{R_{13}}\right\}, \; \underline{\kappa}_2^* = \left\{\frac{1}{R_{21}}, 0, \frac{1}{R_{23}}\right\}, \; \underline{\kappa}_3^* = \left\{\frac{1}{R_{31}}, \frac{1}{R_{32}}, 0\right\}, \tag{4.99}$$

where the radii of curvatures of the coordinate system were defined in eqs. (2.85). The various derivatives of the scale factors are the components of the curvature vectors resolved in frame $\mathcal{B}$. The components of the curvature vectors resolved in frame $\mathcal{I}$ are then $\underline{\kappa}_1 = \underline{\underline{F}}\,\underline{\kappa}^*$, $\underline{\kappa}_2 = \underline{\underline{F}}\,\underline{\kappa}_2^*$, and $\underline{\kappa}_3 = \underline{\underline{F}}\,\underline{\kappa}_3^*$.

If the curvilinear coordinates s_1, s_2, and s_3 are functions of time, tensor $\underline{\underline{F}}$ becomes an implicit function of time, $\underline{\underline{F}}(t) = \underline{\underline{F}}(s_1(t), s_2(t), s_3(t))$. The angular velocity of tensor F, resolved in basis $\mathcal{B}$, is now

$$\widetilde{\underline{\underline{\omega}}}^* = \underline{\underline{F}}^T \underline{\underline{\dot{F}}} = \underline{\underline{F}}^T \left[\frac{\partial \underline{\underline{F}}}{\partial s_1} \dot{s}_1 + \frac{\partial \underline{\underline{F}}}{\partial s_2} \dot{s}_2 + \frac{\partial \underline{\underline{F}}}{\partial s_3} \dot{s}_3 \right] = \dot{s}_1 \widetilde{\underline{\kappa}}_1^* + \dot{s}_2 \widetilde{\underline{\kappa}}_2^* + \dot{s}_3 \widetilde{\underline{\kappa}}_3^*, \quad (4.100)$$

where the chain rule for derivatives was used. This implies $\underline{\omega}^* = \dot{s}_1 \underline{\kappa}_1^* + \dot{s}_2 \underline{\kappa}_2^* + \dot{s}_3 \underline{\kappa}_3^*$: the three curvature vectors contribute to the total angular velocity of basis $\mathcal{B}$.

The various components of curvature are not independent of each other; following the process outlined in the previous section, relationships similar to the Gauss-Codazzi conditions can be readily derived.

Example 4.8. Motion of a particle on a curve

Figure 4.15 depicts a particle sliding along curve $\mathbb{C}$ embedded in a rigid body. The curvilinear variable along the curve is denoted s. The rigid body is moving with respect to an inertial frame of reference, $\mathcal{F}^I = [\mathbf{O}, \mathcal{I} = (\bar{\imath}_1, \bar{\imath}_2, \bar{\imath}_3)]$. The configuration of the rigid body is defined by the body attached frame, $\mathcal{F}^B = [\mathbf{B}, \mathcal{B}^* = (\bar{b}_1, \bar{b}_2, \bar{b}_3)]$. Superscript $(\cdot)^*$ indicates the components of tensors resolved in basis $\mathcal{B}^*$. The components of the position vector of point $\mathbf{B}$ with respect to point $\mathbf{O}$, resolved in basis $\mathcal{I}$, are denoted $\underline{r}_B$, and $\underline{\underline{R}}(t)$ are the components of the rotation tensor that brings basis $\mathcal{I}$ to $\mathcal{B}^*$, resolved in basis $\mathcal{I}$.

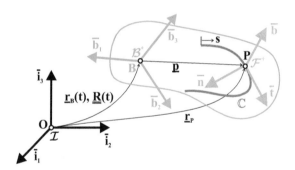

Fig. 4.15. Rigid body with an embedded curve.

Let point $\mathbf{P}$ be a point along curve $\mathbb{C}$; the curvilinear variable at the location of point $\mathbf{P}$ is denoted s. The position vector of point $\mathbf{P}$ with respect to point $\mathbf{B}$ is denoted $\underline{p}$, and the components of this vector resolved in basis $\mathcal{B}^*$ are denoted $\underline{p}^*$. Because curve $\mathbb{C}$ is embedded in the rigid body, its shape is defined by the position vector of a point on the curve, $\underline{p}^* = \underline{p}^*(s)$; clearly, the components of this position vector are most naturally resolved in the body attached basis, $\mathcal{B}^*$. Find the position, velocity, and acceleration vector of point $\mathbf{P}$.

The inertial position vector of point $\mathbf{P}$ is $\underline{r}_P = \underline{r}_B + \underline{p} = \underline{r}_B + \underline{\underline{R}} \underline{p}^*$, where $\underline{r}_P$ are the components of the position vector of point $\mathbf{P}$ with respect to point $\mathbf{O}$, resolved in basis $\mathcal{I}$.

The component of the inertial velocity vector of point **P** resolved in basis $\mathcal{I}$ are readily obtained by taking a time derivative of the position vector to find

$$\underline{v}_P = \underline{v}_B + \underline{\dot{R}}\,\underline{p}^* + \underline{R}\,\underline{\dot{p}}_0^* = \underline{v}_B + \tilde{\omega}\,\underline{R}\,\underline{p}^* + \dot{s}\underline{R}\,\bar{t}^*,$$

where $\underline{v}_B = \underline{\dot{r}}_B$ is the inertial velocity of point **B**, $\underline{\omega} = \mathrm{axial}(\underline{\dot{R}}\,\underline{R}^T)$ are the components of the angular velocity vector of the rigid body resolved in basis $\mathcal{I}$, and $\bar{t}^*$ the component of the unit tangent vector to the curve resolved in basis $\mathcal{B}^*$. The first term of this expression stems from the translation of the rigid body, and the second from its rotation. The last term describes the velocity associated with the sliding of the particle along the curve.

The components of the inertial velocity vector resolved in the body attached basis, $\mathcal{B}^*$, now become

$$\underline{R}^T\underline{v}_P = \underline{R}^T\,\underline{v}_B + \tilde{\omega}^*\underline{p}^* + \dot{s}\bar{t}^*.$$

Note that the expression for the components of the inertial velocity of point **P** is simpler when resolved in the body attached basis, $\mathcal{B}^*$, than when resolved in the inertial basis, $\mathcal{I}$. This is expected, because the quantities associated with curve $\mathbb{C}$, $\underline{p}^*$ and $\bar{t}^*$, are most naturally expressed in basis $\mathcal{B}^*$.

Next, the components of the inertial acceleration of point **P** resolved in basis $\mathcal{I}$ are obtained by taking a derivative of the velocity components in the same basis to find

$$\underline{a}_P = \underline{a}_B + \dot{\tilde{\omega}}\,\underline{R}\,\underline{p}^* + \tilde{\omega}\,\underline{\dot{R}}\,\underline{p}^* + \tilde{\omega}\,\underline{R}\,\underline{\dot{p}}^* + \ddot{s}\underline{R}\,\bar{t}^* + \dot{s}\underline{\dot{R}}\,\bar{t}^* + \dot{s}\underline{R}\,\dot{\bar{t}}^*$$

$$= \underline{a}_B + (\dot{\tilde{\omega}} + \tilde{\omega}\tilde{\omega})\underline{R}\,\underline{p}^* + 2\dot{s}\,\tilde{\omega}\underline{R}\,\bar{t}^* + \ddot{s}\underline{R}\,\bar{t}^* + \frac{\dot{s}^2}{\rho}\,\underline{R}\,\bar{n}^*,$$

where $\underline{a}_B = \underline{\dot{v}}_B$ is the inertial acceleration of point **B** and $\bar{n}^*$ are the component of the unit normal vector to the curve resolved in basis $\mathcal{B}^*$.

Here again, the components of the inertial acceleration vector are simpler when expressed in the body attached basis,

$$\underline{R}^T\underline{a}_P = \underline{R}^T\,\underline{a}_B + (\widetilde{\underline{R}^T\dot{\omega}} + \tilde{\omega}^*\tilde{\omega}^*)\underline{p}^* + 2\dot{s}\,\tilde{\omega}^*\bar{t}^* + \ddot{s}\bar{t}^* + \frac{\dot{s}^2}{\rho}\,\bar{n}^*.$$

The first two terms represent the contributions of the translation and rotation of the rigid body, respectively. The third term is the Coriolis acceleration. Finally, the last two terms are the acceleration of the particle with respect to the rigid body, which in this case, are the acceleration of the particle obtained using path coordinates, see eq. (2.34).

The components of the angular acceleration vector expressed in basis $\mathcal{B}^*$, $\underline{R}^T\dot{\omega}$, are easily evaluated. Indeed, $\underline{R}^T\dot{\omega} = \underline{R}^T(\underline{R}\omega^*)^{\cdot} = \underline{R}^T\underline{\dot{R}}\,\omega^* + \dot{\omega}^* = \tilde{\omega}^*\omega^* + \dot{\omega}^* = \dot{\omega}^*$. A simplified expression for the components of the inertial acceleration vector expressed in the body attached basis is then

$$\underline{R}^T\underline{a}_P = \underline{R}^T\,\underline{a}_B + (\dot{\tilde{\omega}}^* + \tilde{\omega}^*\tilde{\omega}^*)\underline{p}^* + 2\dot{s}\,\tilde{\omega}^*\bar{t}^* + \ddot{s}\bar{t}^* + \frac{\dot{s}^2}{\rho}\,\bar{n}^*.$$

It is possible to define a frame associated with the particle, $\mathcal{F}^F = [\mathbf{P}, \mathcal{F}^+ = (\bar{t}, \bar{n}, \bar{b})]$, where $\mathcal{F}^+$ is the curve's Frenet's triad at point $\mathbf{P}$. Superscript $(\cdot)^+$ indicates tensor components resolved in this basis. The components of the rotation tensor that brings basis $\mathcal{B}^*$ to basis $\mathcal{F}^+$, resolved in basis $\mathcal{B}^*$, are denoted $\underline{\underline{F}}^*$. This tensor, like all other characteristics of the curve, is most naturally expressed in basis $\mathcal{B}^*$, a basis attached to the body in which the curve is embedded.

The components of the unit vector tangent to curve $\mathbb{C}$, resolved in basis $\mathcal{B}^*$, are $\bar{t}^* = \underline{\underline{F}}^* \bar{b}_1^* = \underline{\underline{F}}^* \bar{\imath}_1$, and hence, its components resolved in basis $\mathcal{I}$ become $\bar{t} = \underline{\underline{R}}\,\underline{\underline{F}}^* \bar{\imath}_1$. The angular velocity of basis $\mathcal{F}^+$ with respect to basis $\mathcal{I}$ is now evaluated by taking a time derivative of this tangent vector to find

$$\dot{\bar{t}} = (\underline{\dot{\underline{R}}}\,\underline{\underline{F}}^* + \underline{\underline{R}}\,\underline{\dot{\underline{F}}}^*)(\underline{\underline{R}}\,\underline{\underline{F}}^*)^T \bar{t}$$
$$= \left[\widetilde{\omega} + \dot{s}(\underline{\underline{R}}\,\underline{\underline{F}}^*)\underline{\underline{F}}^{*T}\underline{\underline{F}}^{*\prime}(\underline{\underline{R}}\,\underline{\underline{F}}^*)^T\right] \bar{t} = (\widetilde{\omega} + \dot{s}\widetilde{\underline{\underline{R}}\,\underline{\underline{F}}^*\kappa^+})\bar{t},$$

where $\underline{\kappa}^+ = \underline{\underline{F}}^{*T}\underline{\underline{F}}^{*\prime}$ are the components of the curvature vector of curve $\mathbb{C}$ resolved in basis $\mathcal{F}^+$, see eq. (4.88); it follows that $\underline{\kappa} = \underline{\underline{R}}\,\underline{\underline{F}}^*\underline{\kappa}^+$ are its components resolved in basis $\mathcal{I}$. The angular velocity of basis $\mathcal{F}^+$ with respect to basis $\mathcal{I}$, denoted $\underline{\Omega}$, is now

$$\underline{\Omega} = \underline{\omega} + \dot{s}\underline{\underline{R}}\,\underline{\underline{F}}^*\underline{\kappa}^+ = \underline{\omega} + \dot{s}\underline{\kappa}.$$

The first term represents the contribution of the angular velocity of the rigid body; the second term stems from the change in orientation of Frenet's triad as the particle moves along the curve. Note that the above result could have been established more expeditiously with the help of the addition theorem.

Finally, the angular acceleration of basis $\mathcal{F}^+$ is the time derivative of the angular velocity

$$\underline{\dot{\Omega}} = \underline{\dot{\omega}} + \ddot{s}\underline{\kappa} + \dot{s}(\underline{\dot{\underline{R}}}\,\underline{\underline{F}}^*\underline{\kappa}^+ + \dot{s}\underline{\underline{R}}\,\underline{\underline{F}}^{*\prime}\underline{\kappa}^+ + \dot{s}\underline{\underline{R}}\,\underline{\underline{F}}^*\underline{\kappa}^{+\prime})$$
$$= \underline{\dot{\omega}} + \ddot{s}\underline{\kappa} + \dot{s}(\widetilde{\omega}\underline{\kappa} + \dot{s}\underline{\underline{R}}\,\underline{\underline{F}}^*\widetilde{\kappa}^+\underline{\kappa}^+ + \dot{s}\underline{\underline{R}}\,\underline{\underline{F}}^*\underline{\kappa}^{+\prime}).$$

The second term inside the parentheses vanishes because $\widetilde{\kappa}^+\underline{\kappa}^+ = 0$. The final expression for the angular acceleration is

$$\underline{\dot{\Omega}} = \underline{\dot{\omega}} + \ddot{s}\underline{\kappa} + \dot{s}\,\widetilde{\omega}\underline{\kappa} + \dot{s}^2\,\underline{\underline{R}}\,\underline{\underline{F}}^*\underline{\kappa}^{+\prime}.$$

The inertial acceleration of Frenet's triad depends on the curvature vector $\underline{\kappa} = \underline{\underline{R}}\,\underline{\underline{F}}^*\underline{\kappa}^+$, but also on its derivative along the curve, $\underline{\kappa}^{+\prime}$. Both quantities, $\underline{\kappa}^+$ and $\underline{\kappa}^{+\prime}$, are intrinsic properties of curve $\mathbb{C}$ because they are components of the curvature vector and its spatial derivative resolved in Frenet's triad, $\mathcal{F}^+$.

4.12.4 The differential rotation vector

Let time-dependent rotation tensor $\underline{\underline{R}}(t)$ describe the rotation from basis $\mathcal{I} = (\bar{\imath}_1, \bar{\imath}_2, \bar{\imath}_3)$, called the fixed basis, to basis $\mathcal{B} = (\bar{b}_1, \bar{b}_2, \bar{b}_3)$, called the rotating basis. The differential rotation vector is defined by analogy to the angular velocity vector, see eq. (4.56), as

$$\underline{d\psi} = \text{axial}(d\underline{\underline{R}}\ \underline{\underline{R}}^T),\tag{4.101}$$

where $\underline{d\psi}$ is called the *differential rotation vector*. Note that there exist no "rotation vector," ψ, such that $d(\underline{\psi})$ gives the differential rotation vector. To emphasize this important fact, notation $\underline{d\psi}$ is used to indicate the differential rotation vector, rather than $d\underline{\psi}$.

Taking a differential of eq. (4.56) and a time derivative of eq. (4.101) then yields $d\underline{\underline{\omega}} = d\underline{\dot{\underline{R}}}\ \underline{\underline{R}}^T + \underline{\dot{\underline{R}}}d\underline{\underline{R}}^T$ and $\dot{\widetilde{d\psi}} = d\underline{\dot{\underline{R}}}\ \underline{\underline{R}}^T + d\underline{\underline{R}}\ \underline{\dot{\underline{R}}}^T$, respectively. Subtracting these two equations, and using the orthogonality of the rotation tensor, eq. (4.26), then leads to

$$d\widetilde{\underline{\omega}} = \dot{\widetilde{d\psi}} + \underline{\dot{\underline{R}}}\,d\underline{\underline{R}}^T - d\underline{\underline{R}}\,\underline{\dot{\underline{R}}}^T = \dot{\widetilde{d\psi}} + (\underline{\dot{\underline{R}}}\,\underline{\underline{R}}^T)(\underline{\underline{R}}\,d\underline{\underline{R}}^T) - (d\underline{\underline{R}}\,\underline{\underline{R}}^T)(\underline{\underline{R}}\,\underline{\dot{\underline{R}}}^T).$$

With the help of the definition of the angular velocity vector, eq. (4.56), and of the differential rotation vector, eq. (4.101), this reduces to

$$d\widetilde{\underline{\omega}} = \dot{\widetilde{d\psi}} + \widetilde{d\psi}\widetilde{\underline{\omega}} - \widetilde{\underline{\omega}}\widetilde{d\psi} = \dot{\widetilde{d\psi}} + \widetilde{(\widetilde{d\psi}\,\underline{\omega})}$$

where identity (1.33a) was used. Finally, a differential in the angular velocity vector becomes $d\underline{\omega} = \dot{\underline{d\psi}} - \widetilde{\underline{\omega}}\,\underline{d\psi}$. This important result relates differentials in the angular velocity vector to the differential rotation vector and its derivatives.

Differentials of the components of the angular velocity vector expressed in the rotating frame can also be obtained in a similar manner

$$d\underline{\omega} = \dot{\underline{d\psi}} - \widetilde{\underline{\omega}}\underline{d\psi}, \qquad d\underline{\omega} = \underline{\underline{R}}\,\dot{\underline{d\psi}}^*, \tag{4.102a}$$

$$d\underline{\omega}^* = \dot{\underline{d\psi}}^* + \widetilde{\underline{\omega}}^*\underline{d\psi}^*, \qquad d\underline{\omega}^* = \underline{\underline{R}}^T\,\dot{\underline{d\psi}}. \tag{4.102b}$$

4.13 Applications to particle dynamics

The geometric description of rotation presented in the previous sections is used extensively when analyzing the dynamic behavior of systems of particles when rotations are required to describe the kinematics of the system. Furthermore, Newton's laws will be expressed in various bases to ease the analysis and help understand the physical interpretation of the various quantities involved in the problem.

Example 4.9. Pendulum with rotating mass
Figure 4.16 depicts a pendulum of length ℓ and tip mass M featuring an additional rotating mass m located at a fixed distance d from the tip mass. Frame $\mathcal{F}^I = [\mathbf{O}, \mathcal{I} = (\bar{\imath}_1, \bar{\imath}_2, \bar{\imath}_3)]$ is inertial and the pendulum is attached to the ground at point $\mathbf{O}$ where a bearing allows rotation about axis $\bar{\imath}_3$; gravity acts along axis $\bar{\imath}_1$.

A second frame, $\mathcal{F}^E = [\mathbf{O}, \mathcal{E}^+ = (\bar{e}_1, \bar{e}_2, \bar{e}_3)]$, is defined; tensor components resolved in basis $\mathcal{E}^+$ are denoted with a superscript $(\cdot)^+$. A planar rotation of magnitude ϕ about axis $\bar{\imath}_3$ brings basis $\mathcal{I}$ to basis $\mathcal{E}^+$. Axis $\bar{e}_1$ is aligned with the massless rigid arm $\mathbf{OA}$ of the pendulum.

A third frame, $\mathcal{F}^B = \left[\mathbf{A}, \mathcal{B}^* = (\bar{b}_1, \bar{b}_2, \bar{b}_3)\right]$, is also defined; tensor components resolved in basis $\mathcal{B}^*$ are denoted with a superscript $(\cdot)^*$. At point $\mathbf{A}$, a bearing allows rotation of the massless rigid bar $\mathbf{AT}$ about axis $\bar{e}_1$. A planar rotation of magnitude θ about axis $\bar{e}_1$ brings basis $\mathcal{E}^+$ to basis $\mathcal{B}^*$. Axis $\bar{b}_2$ passes through the rotating mass m. Derive the equation of motions of the system using Newton's second law.

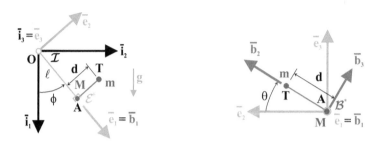

Fig. 4.16. Configuration of a pendulum with a rotating mass.

Let $\underline{\underline{R}}_\phi$ and $\underline{\underline{R}}_\theta^+$ be the components of the rotation tensors that bring basis $\mathcal{I}$ to $\mathcal{E}^+$ and basis $\mathcal{E}^+$ to $\mathcal{B}^*$, respectively, resolved in basis $\mathcal{I}$ and $\mathcal{E}^+$, respectively. The components of the tensor that brings basis $\mathcal{I}$ to $\mathcal{B}^*$, resolved in basis $\mathcal{I}$, are then $\underline{\underline{R}} = \underline{\underline{R}}_\phi \underline{\underline{R}}_\theta^+$, and hence

$$\underline{\underline{R}} = \begin{bmatrix} C_\phi & -S_\phi & 0 \\ S_\phi & C_\phi & 0 \\ 0 & 0 & 1 \end{bmatrix} \begin{bmatrix} 1 & 0 & 0 \\ 0 & C_\theta & -S_\theta \\ 0 & S_\theta & C_\theta \end{bmatrix} = \begin{bmatrix} C_\phi & -S_\phi C_\theta & S_\phi S_\theta \\ S_\phi & C_\phi C_\theta & -C_\phi S_\theta \\ 0 & S_\theta & C_\theta \end{bmatrix},$$

where the short-hand notation, $S_\phi = \sin\phi$, $C_\phi = \cos\phi$ was used, with similar conventions for angle θ.

The angular velocity vector of basis $\mathcal{B}^*$ with respect to basis $\mathcal{I}$ is found with the help of the addition theorem to be $\underline{\omega} = \dot{\phi}\,\bar{e}_3 + \dot{\theta}\,\bar{b}_1 = \dot{\theta}\,\bar{b}_1 + \dot{\phi}S_\theta\,\bar{b}_2 + \dot{\phi}C_\theta\,\bar{b}_3$. The components of the angular and acceleration vectors, resolved in basis $\mathcal{B}^*$, now become

$$\underline{\omega}^* = \left\{ \begin{array}{c} \dot{\theta} \\ \dot{\phi}S_\theta \\ \dot{\phi}C_\theta \end{array} \right\}, \quad \dot{\underline{\omega}}^* = \left\{ \begin{array}{c} \ddot{\theta} \\ \ddot{\phi}S_\theta + \dot{\phi}\dot{\theta}C_\theta \\ \ddot{\phi}C_\theta - \dot{\phi}\dot{\theta}S_\theta \end{array} \right\}.$$

The position vector of particle m with respect to inertial point $\mathbf{O}$ is $\underline{x}_m = \ell\,\bar{b}_1 + d\,\bar{b}_2$. The inertial velocity vector then becomes $\dot{\underline{x}}_m = \ell\widetilde{\omega}\,\bar{b}_1 + d\widetilde{\omega}\,\bar{b}_2$ and finally, the acceleration vector is $\ddot{\underline{x}}_m = \ell(\dot{\widetilde{\omega}} + \widetilde{\omega}\widetilde{\omega})\,\bar{b}_1 + d(\dot{\widetilde{\omega}} + \widetilde{\omega}\widetilde{\omega})\,\bar{b}_2$. The components of this vector, resolved in basis $\mathcal{B}$, then become

$$\underline{\underline{R}}^T \ddot{\underline{x}}_m = \ell(\dot{\widetilde{\omega}}^* + \widetilde{\omega}^* \widetilde{\omega}^*)\bar{b}_1^* + d(\dot{\widetilde{\omega}}^* + \widetilde{\omega}^* \widetilde{\omega}^*)\bar{b}_2^* = \left\{ \begin{array}{c} -\ell\dot{\phi}^2 - d\ddot{\phi}C_\theta + 2d\dot{\phi}\dot{\theta}S_\theta \\ \ell\ddot{\phi}C_\theta - d\dot{\phi}^2 C_\theta^2 - d\dot{\theta}^2 \\ -\ell\ddot{\phi}S_\theta + d\ddot{\theta} + d\dot{\phi}^2 S_\theta C_\theta \end{array} \right\},$$

where $\bar{b}_1^{*T} = \{1, 0, 0\}$ and $\bar{b}_2^{*T} = \{0, 1, 0\}$ are the components of vectors $\bar{b}_1$ and $\bar{b}_2$, respectively, resolved in basis $\mathcal{B}^*$.

The position vector of particle M with respect to inertial point $\mathbf{O}$ is $\underline{x}_M = \ell\,\bar{b}_1$. The velocity and acceleration vectors of mass M can be found by letting $d = 0$ in the corresponding expressions for the velocity and acceleration vectors of mass m.

The left portion of fig. 4.17 shows a free body diagram of the two mass particles; $\underline{F}_M$ and $\underline{F}_m$ are the reaction forces exerted by the rigid bars onto particles M and m, respectively. For particle M, the applied forces are $\underline{F}_M$ and the gravity force, $Mg\bar{\iota}_1$; Newton's second law, resolved in basis $\mathcal{B}^*$, then yields

$$ M \left\{ \begin{array}{c} -\ell\dot{\phi}^2 \\ \ell\ddot{\phi}C_\theta \\ -\ell\ddot{\phi}S_\theta \end{array} \right\} = \left\{ \begin{array}{c} F_{M1}^* + MgC_\phi \\ F_{M2}^* - MgS_\phi C_\theta \\ F_{M3}^* + MgS_\phi S_\theta \end{array} \right\}, \tag{4.103} $$

where $\underline{F}_M^{*T} = \{F_{M1}^*, F_{M2}^*, F_{M3}^*\}$ are the components of vector $\underline{F}_M$ in basis $\mathcal{B}^*$. The components of the gravity force vector, resolved in basis $\mathcal{B}^*$, are $Mg\underline{R}^T\bar{\iota}_1$.

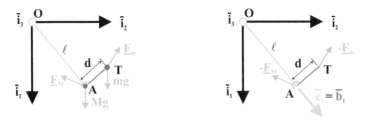

Fig. 4.17. Left portion: free body diagram of the two masses. Right portion: free body diagram of rigid bars **OA** and **AT**.

For particle m, the applied forces are $\underline{F}_m$ and the gravity force $mg\bar{\iota}_1$; Newton's second law, resolved in basis $\mathcal{B}^*$, then yields

$$ m \left\{ \begin{array}{c} -\ell\dot{\phi}^2 - d\ddot{\phi}C_\theta + 2d\dot{\phi}\dot{\theta}S_\theta \\ \ell\ddot{\phi}C_\theta - d(\dot{\phi}^2 C_\theta^2 + \dot{\theta}^2) \\ -\ell\ddot{\phi}S_\theta + d(\ddot{\theta} + \dot{\phi}^2 S_\theta C_\theta) \end{array} \right\} = \left\{ \begin{array}{c} F_{m1}^* + mgC_\phi \\ F_{m2}^* - mgS_\phi C_\theta \\ F_{m3}^* + mgS_\phi S_\theta \end{array} \right\}, \tag{4.104} $$

where $\underline{F}_m^{*T} = \{F_{m1}^*, F_{m2}^*, F_{m3}^*\}$ are the components of vector $\underline{F}_m$ in basis $\mathcal{B}^*$.

Equations (4.103) and (4.104) are the equations of motion of the system. They involve two kinematic unknowns, ϕ and θ, and six force unknowns, the components of the reaction forces $\underline{F}_M$ and $\underline{F}_m$, for a total of eight unknowns. Since the system comprises two particles, Newton's equations yield a total of six equations. Consequently, two additional relationships are required to solve the problem.

The right portion of fig. 4.17 illustrates the two massless bars **OA** and **TA**. In view of the presence of a bearing at point **A**, the moment of the forces applied to bar **TA** must vanish about axis $\bar{b}_1$, *i.e.*, $-\bar{b}_1^T d\tilde{b}_2 \underline{F}_m = 0$; this condition reduces to

$-d\bar{b}_3^T \underline{F}_m = 0$ and finally, $F_{m3}^* = 0$. The third line of eq. (4.104) the yields the following equation

$$m \left[-\ell\ddot{\phi}S_\theta + d(\ddot{\theta} + \dot{\phi}^2 S_\theta C_\theta) \right] = mgS_\phi S_\theta. \tag{4.105}$$

Similarly, the moment of the forces applied to bars **OA** and **TA** must vanish about axis $\bar{\imath}_3$, because a bearing is present at point **O**. This condition implies $-\bar{\imath}_3^T [\widetilde{\ell b}_1 \underline{F}_M + (\widetilde{\ell b}_1 + \widetilde{db}_2)\underline{F}_m] = 0$, or, expressed in basis $\mathcal{B}^*$, $-\bar{\imath}_3^T \underline{\underline{R}}[\widetilde{\ell b}_1^* \underline{F}_M^* + (\widetilde{\ell b}_1^* + \widetilde{db}_2^*)\underline{F}_m^*] = 0$. Expanding this relationship leads to

$$\{0, S_\theta, C_\theta\} \left\{ \begin{array}{c} 0 \\ \ell F_{M3}^* \\ dF_{m1}^* - \ell(F_{M2}^* + F_{m2}^*) \end{array} \right\} = 0.$$

Eliminating the reaction force components with the help of eqs. (4.103) and (4.104) leads to the following equation

$$\left[M\ell^2 + m(\ell^2 + d^2)C_\theta^2\right] \ddot{\phi} - m\ell dC_\theta(\dot{\phi}^2 C_\theta^2 + \dot{\theta}^2)$$
$$+ m\ell d\dot{\phi}^2 C_\theta - 2md^2\dot{\phi}\dot{\theta}S_\theta C_\theta = -Mg\ell S_\phi - mgC_\theta(\ell S_\phi C_\theta + dC_\phi). \tag{4.106}$$

Equations (4.105) and (4.106) are two nonlinear, coupled, ordinary differential equations for the two kinematic variables, ϕ and θ. Once these equations have been solved, eqs. (4.103) and (4.104) will yield the reaction forces, thereby completing the solution of the problem.

4.13.1 Problems

Problem 4.34. Relationships between angular velocity and curvature
Consider a rotation field that is a function of both space and time, *i.e.*, $\underline{\underline{R}} = \underline{\underline{R}}(s, t)$. It is now possible to define the components of the angular velocity vector as $\underline{\omega}(s, t) = \text{axial}(\dot{\underline{R}}\,\underline{\underline{R}}^T)$ and $\underline{\omega}^*(s, t) = \text{axial}(\underline{\underline{R}}^T \dot{\underline{R}})$ resolved in the inertial and rotating frames, respectively. Similarly, the components of the curvature vector are $\underline{\kappa}(s, t) = \text{axial}(\underline{\underline{R}}'\underline{\underline{R}}^T)$ and $\underline{\kappa}^*(s, t) = \text{axial}(\underline{\underline{R}}^T \underline{\underline{R}}')$ in the inertial and rotating frames, respectively. Based on the developments presented in section (4.12.4), prove the following results $\underline{\omega}' = \dot{\underline{\kappa}} + \widetilde{\kappa}\underline{\omega}$, $\underline{\omega}^{*\prime} = \dot{\underline{\kappa}}^* + \widetilde{\omega}^*\underline{\kappa}^*$, $\underline{\omega}' = \underline{\underline{R}}\,\dot{\underline{\kappa}}^*$, and $\underline{\omega}^{*\prime} = \underline{\underline{R}}^T \dot{\underline{\kappa}}$.

Problem 4.35. Rigid body with a slot
Figure 4.18 depicts a rigid body with a slot in its reference configuration as defined by frame $\mathcal{F}_0 = [\mathbf{B}, \mathcal{B}_0 = (\bar{b}_{01}, \bar{b}_{02}, \bar{b}_{03})]$, where basis $\mathcal{B}_0$ determines the orientation of the body. Position vector $\underline{x}_O$ determines the location of a reference point **O** on the rigid body with respect to inertial frame $\mathcal{I}$. In the final configuration, the rigid body is defined by frame $\mathcal{F} = [\mathbf{B}, \mathcal{B} = (\bar{b}_1, \bar{b}_2, \bar{b}_3)]$, where its orientation is determined by basis $\mathcal{B}$. The displacement of point **O** is denoted $\underline{u}_O$. Point **P** moves along a slot fixed with respect to the rigid body in such a way that the distance from point **B** to point **P** is a given function of time $d(t)$. The unit vectors aligned with the slot in the reference and final configurations are denoted $\bar{s}$ and $\bar{S}$, respectively. Let $\underline{\underline{R}}_0$ and $\underline{\underline{R}}$ be the rotation tensors that bring basis $\mathcal{I}$ to $\mathcal{B}_0$ and basis $\mathcal{B}_0$ to $\mathcal{B}$,

respectively. *(1)* Find the inertial velocity and acceleration of point **P** in terms of the velocity and acceleration of point **B**, the angular velocity and acceleration of the rigid body, and function $d(t)$. Express this vector in inertial basis $\mathcal{I}$. All tensor components should be resolved in basis $\mathcal{I}$. *(2)* Express these inertial velocity and acceleration vectors in the body attached basis $\mathcal{B}$. All tensor components should be resolved in basis $\mathcal{B}$.

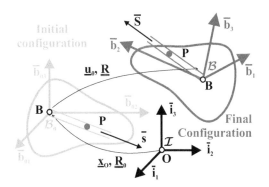
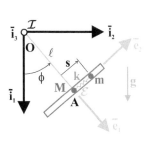

Fig. 4.18. Rigid body with a slot in the reference and final configurations.

Fig. 4.19. Pendulum with a sliding mass connected to a spring.

Problem 4.36. Pendulum with a sliding mass

Figure 4.19 depicts a pendulum of mass M and length ℓ connected at point **A** to a massless tube in which a point mass m is sliding while restrained by a spring of stiffness constant k. Frame $\mathcal{F}^I = [\mathbf{O}, \mathcal{I} = (\bar{\imath}_1, \bar{\imath}_2, \bar{\imath}_3)]$ is inertial and the pendulum is attached to the ground at point **O** where a bearing allows rotation about axis $\bar{\imath}_3$; gravity acts along axis $\bar{\imath}_1$. A second frame $\mathcal{F}^E = [\mathbf{A}, \mathcal{E} = (\bar{e}_1, \bar{e}_2, \bar{e}_3)]$ is defined. A planar rotation of magnitude ϕ about axis $\bar{\imath}_3$ brings basis $\mathcal{I}$ to basis $\mathcal{E}$; axis $\bar{e}_2$ is aligned with the massless tube. The position of mass m with respect to point **A** is denoted s. *(1)* Using Newton's second law, derive the equations of motion of the system for $\phi(\tau)$ and $s(\tau)$. *(2)* Plot the time history of $\phi(\tau)$. *(3)* Plot the time history of $\bar{s}(\tau)$. *(4)* On one graph, plot the kinetic, potential and total mechanical energies of the system. *(5)* Plot the normalized components of the reaction force vector acting on particle M. *(6)* Plot the normalized components of the reaction force vector acting on particle m. *(7)* Compute the angular momentum vector of the system evaluated with respect to point **O**. From this expression, derive a differential equation that must be satisfied by ϕ and s. Show that this relationship can be derived from the equations of motion obtained in step *1*. Use the following data: $\mu = M/m = 1$, $\bar{g} = g/(\ell\omega^2) = 0.6$, $\bar{s} = s/\ell$. The following non-dimensional time is defined: $\tau = \omega t$, where $\omega^2 = k/m$; derivatives with respect to τ are denoted $(\cdot)'$. Forces are normalized by $m\ell\omega^2$. At time $t = 0$, $\phi = \pi/2$, $\bar{s} = 0$, $\phi' = 0$ and $\bar{s}' = 2$. For all plot, $\tau \in [0, 200]$.

Problem 4.37. Mass particle moving in a tube

Figure 4.20 shows a particle of mass m moving in a rigid slot under the effect of an actuator. The actuator is connected to the particle at point **P** and to the slot at point **A**; for clarity, the actuator is not shown on the figure. A rigid bar **OA** of length r rotates in plane $\mathcal{P} = (\bar{\imath}_1, \bar{\imath}_2)$ at a constant angular velocity, Ω. A planar rotation of magnitude $\psi = \Omega t$ about axis $\bar{\imath}_3$ brings basis $\mathcal{I} = (\bar{\imath}_1, \bar{\imath}_2, \bar{\imath}_3)$ to axis $\mathcal{E} = (\bar{e}_1, \bar{e}_2, \bar{e}_3)$; axis $\bar{e}_1$ is along rigid bar **OA**. The rigid slot

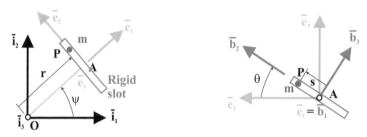

Fig. 4.20. Mass particle moving in a tube.

is connected to bar **OA** at point **A** and is allowed to rotate with respect the bar about axis $\bar{e}_1$. A planar rotation of magnitude θ about axis $\bar{e}_1$ brings basis $\mathcal{E} = (\bar{e}_1, \bar{e}_2, \bar{e}_3)$ to axis $\mathcal{B} = (\bar{b}_1, \bar{b}_2, \bar{b}_3)$; axis $\bar{b}_2$ is along the rigid slot. The position of the mass in the slot is defined by the curvilinear variable s that is positive along axis $\bar{b}_2$. *(1)* Find the position, velocity and acceleration vectors of point **P** with respect to inertial point **O**. *(2)* Find the components of the acceleration vector resolved in basis $\mathcal{B}$. *(3)* Write Newton's second law for the particle, resolved in basis $\mathcal{B}$. *(4)* Identify the nature of the forces applied on the particle. *(5)* On one graph, plot the components of the force vector acting on the particle, resolved in basis $\mathcal{B}$. *(6)* Find the moment of the forces that the particle and actuator apply on the slot with respect to point **O**. *(7)* On one graph, plot the components of this moment vector resolved in basis $\mathcal{E}^+$. Use the following data: $\theta(t) = \theta_0 + \theta_s \sin \Omega t + \theta_c \cos \Omega t$, where $\theta_0 = 15$, $\theta_s = -6$ and $\theta_c = 8$ degrees; $s(t) = s_s \sin \Omega t + s_c \cos \Omega t$, where $s_s = 0.05$ and $s_c = -0.03$ m. $m = 10$ kg, $\Omega = 27.02$ rad/s, $r = 0.8$ m.

Problem 4.38. Pendulum with a rotating mass

Figure 4.16 depicts a pendulum of length ℓ and tip mass M featuring an additional rotating mass m located at a fixed distance d from the tip mass, as treated in example 4.9 on page 149. Frame $\mathcal{F}^I = [\mathbf{O}, \mathcal{I} = (\bar{\imath}_1, \bar{\imath}_2, \bar{\imath}_3)]$ is inertial and the pendulum is attached to the ground at point **O** where a bearing allows rotation about axis $\bar{\imath}_3$; gravity acts along axis $\bar{\imath}_1$. A second frame, $\mathcal{F}^E = [\mathbf{O}, \mathcal{E}^+ = (\bar{e}_1, \bar{e}_2, \bar{e}_3)]$, is defined. A planar rotation of magnitude ϕ about axis $\bar{\imath}_3$ brings basis $\mathcal{I}$ to basis $\mathcal{E}^+$. Axis $\bar{e}_1$ is aligned with the massless rigid arm **OA** of the pendulum. A third frame, $\mathcal{F}^B = [\mathbf{A}, \mathcal{B}^* = (\bar{b}_1, \bar{b}_2, \bar{b}_3)]$, is also defined. At point **A**, a bearing allows rotation of the massless rigid bar **AT** about axis $\bar{e}_1$. A planar rotation of magnitude θ about axis $\bar{e}_1$ brings basis $\mathcal{E}^+$ to basis $\mathcal{B}^*$. Axis $\bar{b}_2$ passes through the rotating mass m. Derive the equation of motions of the system using Newton's second law. *(1)* Using Newton's second law, derive the equations of motion of the system for $\phi(\tau)$ and $\theta(\tau)$. *(2)* On one graph, plot the time history of $\phi(\tau)$ and $\theta(\tau)$. *(3)* On one graph, plot the time history of ϕ' and θ'. *(4)* On one graph, plot the kinetic, potential and total mechanical energies of the system. *(5)* Plot the normalized components of the reaction force vector acting on particle M, resolved in basis $\mathcal{B}$. *(6)* Plot the normalized components of the reaction force vector acting on particle m, in basis $\mathcal{B}$. *(7)* Plot the normalized components of the reaction force vector at point **O**, in basis $\mathcal{I}$. *(8)* Plot the normalized components of the reaction moment vector at point **O**, in basis $\mathcal{I}$. *(9)* Compute the angular momentum vector of the system evaluated with respect to point **O**. From this expression, derive a differential equation that must be satisfied by ϕ and θ. Show that this relationship can derived from the equations of motion obtained in step *1*. Use the following data: $\mu = m/M = 1$, $\bar{d} = d/\ell = 0.2$. The following non-dimensional time is defined: $\tau = \omega t$, where $\omega^2 = g/\ell$; a derivative with respect to τ is denoted $(\cdot)'$. Forces are normalized by $M\ell\omega^2$, moments by $M\ell^2\omega^2$. At time $t = 0$, $\phi = \pi/2$, $\theta = 0$, $\phi' = 0$ and $\theta' = 0.1$. For all plot, $\tau \in [0, 50]$.

4.14 Change of reference frame operations

Figure 4.21 depicts the configuration of a rigid body characterized by frame $\mathcal{F}_0^A = \left[\mathbf{A}, \mathcal{E}_0^A = (\bar{e}_{01}^A, \bar{e}_{02}^A, \bar{e}_{03}^A)\right]$. In the reference configuration, the position vector of point $\mathbf{P}$ with respect to inertial point $\mathbf{O}$ is $\underline{u}_0^A$ and rotation tensor $\underline{\underline{R}}_0^A$ bring basis $\mathcal{I}$ to basis $\mathcal{E}_0^A$.

In its final configuration, the rigid body is characterized by reference frame $\mathcal{F}^A = \left[\mathbf{A}, \mathcal{E}^A = (\bar{e}_1^A, \bar{e}_2^A, \bar{e}_3^A)\right]$. The displacement vector of point $\mathbf{A}$ from the reference to the final configuration of the rigid body is $\underline{u}^A$ and rotation tensor $\underline{\underline{R}}^A$ bring basis $\mathcal{E}_0^A$ to basis $\mathcal{E}^A$. The position vector of point $\mathbf{A}$ in the reference configuration with respect to point $\mathbf{O}$ is $\underline{u}_0^A + \underline{u}^A$ and rotation tensor $\underline{\underline{R}}^A\underline{\underline{R}}_0^A$ brings basis $\mathcal{I}$ to basis $\mathcal{E}^A$.

Figure 4.21 also shows a second rigid body in its reference and final configurations. All quantities belonging this second rigid body are denoted with superscript $(\cdot)^B$. All vectors and

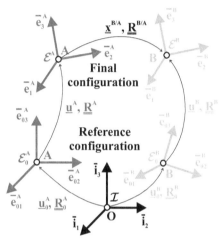

Fig. 4.21. Change of reference frame

tensor are expressed by their components in the inertial frame $\mathcal{I}$; *i.e.*, all quantities are "viewed by an inertial observer."

The position vector of point $\mathbf{B}$ with respect to point $\mathbf{A}$, denoted $\underline{r}^{B/A}$, is

$$\underline{r}^{B/A} = \left(\underline{u}_0^B + \underline{u}^B\right) - \left(\underline{u}_0^A + \underline{u}^A\right).$$

The components of this vector resolved in basis $\mathcal{E}^A$, denoted $\underline{r}^{*B/A}$, are

$$\underline{r}^{*B/A} = \left(\underline{\underline{R}}^A\underline{\underline{R}}_0^A\right)^T\left[\left(\underline{u}_0^B + \underline{u}^B\right) - \left(\underline{u}_0^A + \underline{u}^A\right)\right].$$

These components are often called the "components of the position vector of point $\mathbf{B}$ as viewed by an observer on frame $\mathcal{F}^A$."

Similarly, the rotation tensor that bring basis $\mathcal{E}^A$ to basis $\mathcal{E}^B$, denoted $\underline{\underline{R}}^{B/A}$, can be evaluated as follows

$$\bar{e}_\alpha^B = \left(\underline{\underline{R}}^B\underline{\underline{R}}_0^B\right)\left(\underline{\underline{R}}^A\underline{\underline{R}}_0^A\right)^T\bar{e}_\alpha^A = \underline{\underline{R}}^{B/A}\bar{e}_\alpha^A.$$

The components of this tensor resolved in $\mathcal{E}^A$, denoted $\underline{\underline{R}}^{*B/A}$, are

$$\underline{\underline{R}}^{*B/A} = \left(\underline{\underline{R}}^A \underline{\underline{R}}_0^A\right)^T \left(\underline{\underline{R}}^B \underline{\underline{R}}_0^B\right).$$

These are often called the "components of the rotation tensor of basis $\mathcal{E}^B$ as viewed by an observer on frame $\mathcal{F}^A$."

The position vector of point **B** with respect to point **A** in the reference configuration is $\underline{r}_0^{B/A} = \underline{u}_0^B - \underline{u}_0^A$. The components of this vector resolved in basis $\mathcal{E}_0^A$ are $\underline{r}_0^{*B/A} = \underline{\underline{R}}_0^{AT}(\underline{u}_0^B - \underline{u}_0^A)$.

If body **B** is rigidly connected to body **A**, the components of the position vector of point **B** as viewed by an observer on frame $\mathcal{F}^A$ are still $\underline{x}_0^{*B/A}$ in the final configuration.

The components of *displacement vector* of point **B** with respect to point **A** as viewed by an observer on frame $\mathcal{F}^A$ in the final configuration are

$$\underline{u}^{*B/A} = \left(\underline{\underline{R}}^A \underline{\underline{R}}_0^A\right)^T \left[(\underline{u}_0^B + \underline{u}^B) - (\underline{u}_0^A + \underline{u}^A)\right] - \underline{\underline{R}}_0^{AT}(\underline{u}_0^B - \underline{u}_0^A). \quad (4.107)$$

The rotation tensor that brings basis $\mathcal{E}_0^A$ to basis $\mathcal{E}_0^B$ is $\underline{\underline{R}}_0^{B/A} = \underline{\underline{R}}_0^B \underline{\underline{R}}_0^{AT}$. The components of this tensor resolved in $\mathcal{E}_0^A$ are $\underline{\underline{R}}_0^{*B/A} = \underline{\underline{R}}_0^{AT} \underline{\underline{R}}_0^B$.

If body **B** is rigidly connected to body **A**, the components of the rotation tensor that bring basis $\mathcal{E}^A$ to basis $\mathcal{E}^B$, resolved in basis $\mathcal{E}^A$, are still $\underline{\underline{R}}_0^{*B/A}$. Let $\underline{\underline{Q}}^{B/A}$ be the rotation tensor that measures the *change in orientation* of basis $\mathcal{E}^B$ with respect to basis $\mathcal{E}^A$, between the reference and final configurations. The components of this change in orientation of basis $\mathcal{E}^B$ as viewed by an observer on frame $\mathcal{F}^A$ in the final configuration are

$$\underline{\underline{Q}}^{*B/A} = \underline{\underline{R}}^{*B/A} \underline{\underline{R}}_0^{*B/AT} = \underline{\underline{R}}_0^{AT}\left(\underline{\underline{R}}^{AT}\underline{\underline{R}}^B\right)\underline{\underline{R}}_0^A. \quad (4.108)$$

Let the final configuration of the system be time-dependent. The inertial velocities of points **A** and **B**, denoted $\underline{v}^A$ and $\underline{v}^B$, respectively, are easily found as $\underline{v}^A = \underline{\dot{u}}^A$ and $\underline{v}^B = \underline{\dot{u}}^B$, respectively. Similarly, the angular velocity vectors of bases $\mathcal{E}^A$ and $\mathcal{E}^B$, denoted $\underline{\omega}^A$ and $\underline{\omega}^B$, respectively, are easily found as $\underline{\omega}^A = \text{axial}(\underline{\dot{\underline{R}}}^A \underline{\underline{R}}^{AT})$ and $\underline{\omega}^B = \text{axial}(\underline{\dot{\underline{R}}}^B \underline{\underline{R}}^{BT})$, respectively. All these vectors are expressed by their components in the inertial frame $\mathcal{I}$; *i.e.*, all quantities are "viewed by an inertial observer."

The components of the velocity vector of point **B** as viewed by an observer on frame $\mathcal{F}^A$, denoted $\underline{v}^{*B/A}$, are readily found by taking a time derivative of eq. (4.107) to find

$$\underline{v}^{*B/A} = \left(\underline{\underline{R}}^A \underline{\underline{R}}_0^A\right)^T \left\{\widetilde{\omega}^{AT}\left[(\underline{u}_0^B + \underline{u}^B) - (\underline{u}_0^A + \underline{u}^A)\right] + (\underline{v}^B - \underline{v}^A)\right\}. \quad (4.109)$$

Similarly, the components of the angular velocity vector of basis $\mathcal{E}^B$ as viewed by an observer on frame $\mathcal{F}^A$ are $\underline{\omega}^{*B/A} = \text{axial}(\underline{\dot{\underline{Q}}}^{*B/A}\underline{\underline{Q}}^{*B/AT})$, and it the follows from eq. (4.108) that

$$\underline{\omega}^{*B/A} = \left(\underline{\underline{R}}^A \underline{\underline{R}}_0^A\right)^T (\underline{\omega}^B - \underline{\omega}^A). \quad (4.110)$$

4.15 Orientation of a unit vector

Consider a unit vector, $\bar{\imath}_3$, called a *director*, that rotates to a final orientation $\bar{e}_3$, as depicted in fig. 4.22. For convenience, this director is considered to be the third unit vector of a basis $\mathcal{I} = (\bar{\imath}_1, \bar{\imath}_2, \bar{\imath}_3)$, rotating to a basis $\mathcal{E} = (\bar{e}_1, \bar{e}_2, \bar{e}_3)$.

The relationship between these two bases is

$$\bar{e}_\alpha = \underline{\underline{R}}\,\bar{\imath}_\alpha, \tag{4.111}$$

where $\underline{\underline{R}}$ is an orthogonal rotation tensor. If attention solely focuses on the director, this rotation tensor is not uniquely defined, because any rotation about the director leaves its orientation unchanged.

A differential change in the director's orientation is $\mathrm{d}\bar{e}_3 = \widetilde{\bar{e}_3}^T\,\mathrm{d}\underline{\psi}$, where $\mathrm{d}\underline{\psi}$ is the differential rotation vector defined by eq. (4.101). The components of the differential change in director orientation resolved in basis $\mathcal{E}$ become

$$\underline{\underline{R}}^T\,\mathrm{d}\bar{e}_3 = \underline{\underline{R}}^T\,\widetilde{e}_3^T\,\mathrm{d}\underline{\psi} = \widetilde{\imath}_3^T\,\underline{\underline{R}}^T\,\mathrm{d}\underline{\psi} = \left\{ \begin{array}{c} \mathrm{d}\psi_2^* \\ -\mathrm{d}\psi_1^* \\ 0 \end{array} \right\},$$

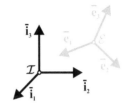

Fig. 4.22. Change of orientation of the director $\bar{\imath}_3$.

where $\mathrm{d}\psi^*$ are the components of the differential rotation vector resolved in basis $\mathcal{E}$.

This relationship demonstrates that differential changes in the orientation vector only depend on two components, $\mathrm{d}\psi_1^*$ and $\mathrm{d}\psi_2^*$, of the differential rotation vector. Arbitrary values of $\mathrm{d}\psi_3^*$, corresponding to differential rotations of the director about its own orientation, will not affect differential changes in the director orientation, and hence, setting $\mathrm{d}\psi_3^* = 0$ is a valid choice.

The following notation is adopted $\underline{\mathrm{d}\psi}^* = \bar{\imath}_1\mathrm{d}\alpha_1^* + \bar{\imath}_2\mathrm{d}\alpha_2^* = \underline{b}\,\underline{\mathrm{d}\alpha}^*$, where

$$\underline{b} = [\bar{\imath}_1, \bar{\imath}_2] = \begin{bmatrix} 1 & 0 \\ 0 & 1 \\ 0 & 0 \end{bmatrix}. \tag{4.112}$$

The "two parameter" differential rotation vector is denoted $\underline{\mathrm{d}\alpha}^*$. Array $\underline{b}$ simply expands this two parameter differential rotation vector, $\underline{\mathrm{d}\alpha}^*$, to the differential rotation vector, $\underline{\mathrm{d}\psi}^*$, by imposing the condition $\mathrm{d}\psi_3^* = 0$. It follows that $\underline{\mathrm{d}\psi} = \underline{\underline{R}}\,\underline{\mathrm{d}\psi}^* = \underline{\underline{R}}\,\underline{b}\,\underline{\mathrm{d}\alpha}^*$, and finally, differential changes in the orientation of the triad become

$$\mathrm{d}\bar{e}_\alpha = \underline{\underline{R}}\,\widetilde{\imath}_\alpha^T\,\underline{b}\,\underline{\mathrm{d}\alpha}^*. \tag{4.113}$$

Part II

Rigid body dynamics

5

Kinematics of rigid bodies

Newton's laws deal with the dynamic behavior of a single particle and Euler's laws generalize the analysis to the case of a system of particles. Rigid bodies form a special case of "systems of particles," and their dynamic behavior is studied in depth in chapter 6. This chapter focuses on the kinematics of rigid bodies, *i.e.*, the description of the motion of rigid bodies without consideration of the forces that create this motion.

Sections 5.1 and 5.2 study the displacement and velocity fields, respectively, of rigid bodies undergoing arbitrary, time-dependent motion. The concept of relative velocity and acceleration is treated in section 5.3, while section 5.4 addresses the problem of contact between two rigid bodies. The chapter concludes with the analysis of the motion tensor.

5.1 General motion of a rigid body

Figure 5.1 depicts a rigid body defined in its reference configuration by frame $\mathcal{F}_0 = [\mathbf{A}, \mathcal{E}_0 = (\bar{e}_{01}, \bar{e}_{02}, \bar{e}_{03})]$. The position vector of point $\mathbf{A}$ with respect to point $\mathbf{O}$ is denoted $\underline{r}_0$. Let $\underline{r}_P$ be the position vector of a material point $\mathbf{P}$ of the rigid body with respect to inertial frame $\mathcal{F}^I = [\mathbf{O}, \mathcal{I} = (\bar{\imath}_1, \bar{\imath}_2, \bar{\imath}_3)]$. The position vector of the same material point with respect to point $\mathbf{A}$ is denoted $\underline{s}_P$. Hence, $\underline{r}_P = \underline{r}_0 + \underline{s}_P$.

The rigid body now undergoes an arbitrary motion that brings it to a final configuration defined by frame $\mathcal{F} = [\mathbf{A}, \mathcal{E} = (\bar{e}_1, \bar{e}_2, \bar{e}_3)]$. Let $\underline{\underline{R}}_0$ and $\underline{\underline{R}}$ be the rotation tensors that bring basis $\mathcal{I}$ to $\mathcal{E}_0$ and basis $\mathcal{E}_0$ to $\mathcal{E}$, respectively. Considering fig. 5.1, the following vector relationship is easily established,

$$\underline{u}_P = \underline{u} + \underline{S}_P - \underline{s}_P, \tag{5.1}$$

where $\underline{S}_P$ is the position vector of material point $\mathbf{P}$ with respect to point $\mathbf{A}$ in the final configuration. Let $\underline{s}_P^* = \underline{\underline{R}}_0^T \underline{s}_P$ and $\underline{S}_P^+ = (\underline{\underline{R}}\,\underline{\underline{R}}_0)^T \underline{S}_P$ denote the components of vector $\underline{s}_P$ in basis $\mathcal{E}_0$ and of vector $\underline{S}_P$ in basis $\mathcal{E}$, respectively.

Because the body is assumed to be rigid, the components of vector $\underline{s}_P$ in $\mathcal{E}_0$ are identical to those of $\underline{S}_P$ in $\mathcal{E}$, i.e., $\underline{s}_P^* = \underline{S}_P^+$, and hence, $\underline{S}_P = \underline{\underline{R}}\,\underline{s}_P$. Equation (5.1) now becomes

$$\underline{u}_P = \underline{u} + \left(\underline{\underline{R}} - \underline{\underline{I}}\right)\underline{s}_P. \tag{5.2}$$

This relationship describes the displacement of a material point **P** of the rigid body in terms of $\underline{u}$, the displacement of its reference point, and tensor $\underline{\underline{R}}$ that defines its orientation. Note that the choice of reference point **A** is arbitrary, and hence, eq. (5.2) is not an intrinsic relationship.

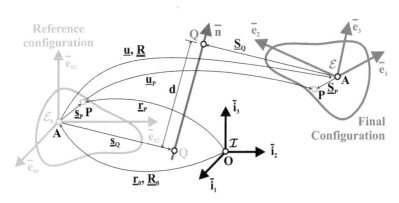

Fig. 5.1. General motion of a rigid body.

To obtain a more general expression of the displacement field, the following question can be asked: is it possible to find a material point of the rigid body, say point **Q**, whose displacement is parallel to $\bar{n}$, the axis defining rotation tensor $\underline{\underline{R}}$? If point **Q** exist, its relative position vector, $\underline{s}_Q$, must satisfy the following relationship

$$\underline{u}_Q = \underline{u} + \left(\underline{\underline{R}} - \underline{\underline{I}}\right)\underline{s}_Q = d\bar{n}. \tag{5.3}$$

Constant d can be evaluated by taking the scalar product this equation by $\bar{n}^T$ to find $d = \bar{n}^T\underline{u}$. It then follows that

$$\left(\underline{\underline{R}} - \underline{\underline{I}}\right)\underline{s}_Q = d\bar{n} - \underline{u} = \left(\bar{n}\bar{n}^T - \underline{\underline{I}}\right)\underline{u}. \tag{5.4}$$

In view of eq. (4.21) and identity (1.33b), this equation can be written as $\tilde{n}\left[2\sin\phi/2\,\underline{\underline{G}}\,\underline{s}_Q - \tilde{n}\underline{u}\right] = 0$. The bracketed must be parallel to unit vector $\bar{n}$, which implies $2\sin\phi/2\,\underline{\underline{G}}\,\underline{s}_Q - \tilde{n}\underline{u} = \beta\bar{n}$, where β is an arbitrary constant. The location of point **Q** is now readily found as

$$\underline{s}_Q = \frac{\tilde{n}\underline{\underline{G}}^T}{2\sin\phi/2}\underline{u} + \frac{\beta}{2\sin\phi/2}\bar{n}.$$

This represents the equation of a line passing through point **Q** and parallel to $\bar{n}$. The displacements of all points on this line are along $\bar{n}$.

Point **Q** can be defined uniquely by requiring $\underline{s}_Q$ to be orthogonal to $\bar{n}$, *i.e.*, $\bar{n}^T \underline{s}_Q = 0$, and hence, $\beta = 0$. The location of point **Q** [19] now becomes

$$\underline{s}_Q = \frac{\widetilde{n}\underline{\underline{G}}^T}{2\sin\phi/2}\underline{u}. \tag{5.5}$$

By construction, the displacement of point **Q** is parallel to $\bar{n}$, see eq. (5.3). Combining eqs. (5.2) and (5.3) now yields

$$\underline{u}_P = d\bar{n} + (\underline{\underline{R}} - \underline{\underline{I}})(\underline{s}_P - \underline{s}_Q). \tag{5.6}$$

This relationship expresses the displacement of a material point **P** of the rigid body as a translation, $d\bar{n}$, parallel to axis $\bar{n}$, followed by a rotation about that same axis. The displacement

$$d = \bar{n}^T \underline{u}, \tag{5.7}$$

is the *intrinsic displacement of the rigid body*: all points of the rigid body undergo the same displacement, d, followed by a rotation.

If the rigid body undergoes a general planar motion, $\underline{u}$ lies in the plane of the motion, and $\bar{n}$ is perpendicular this plane. Hence, $d = \bar{n}^T \underline{u} = 0$, the intrinsic displacement, d, of a rigid body in general planar motion always vanishes. If the rigid body undergoes a pure translation, axis $\bar{n}$ is along the displacement $\underline{u}$ of all the points of the body. The motion is then decomposed into a translation, $d\bar{n}$, followed by a rotation of vanishing magnitude about the same axis.

Equation (5.6) expresses the general motion of a rigid body as *screw motion* about axis $\bar{n}$. The *pitch of the screw*, ϖ, is defined as

$$\varpi = \frac{2\pi d}{\phi}. \tag{5.8}$$

Mozzi-Chasles' theorem [20, 21] states the results obtained here in a compact manner.

Theorem 5.1 (Mozzi-Chasles' theorem). *The most general motion of a rigid body consists of a translation along an axis followed by a rotation about the same axis.*

The Mozzi-Chasles axis is defined by its orientation, $\bar{n}$, and the position of one of its points, $\underline{s}_Q$, given by eq. (5.5). Alternatively, this axis can be defined by its Plücker coordinates [19, 22]

$$\underline{\mathcal{Q}}_{MC} = \left\{ \begin{array}{c} -\dfrac{\widetilde{n}\widetilde{n}\underline{\underline{G}}^T}{2\sin\phi/2}\underline{u} \\ \bar{n} \end{array} \right\} \tag{5.9}$$

5.2 Velocity field of a rigid body

The time-dependent motion of a rigid body, as depicted in fig. 5.2, will now be investigated. The structure of the velocity field of the entire rigid body is the focus of the analysis.

The inertial velocity of an arbitrary point **P** is obtained from a time derivative of eq. (5.2), $\underline{v}_P = \underline{v} + \underline{\dot{R}}\underline{s}_P = \underline{v} + \underline{\dot{R}}\underline{R}^T\underline{S}_P$, where $\underline{v}_P = \underline{\dot{u}}_P$ and $\underline{v} = \underline{\dot{u}}$ are the inertial velocity vectors of point **P** and **A**, respectively. This equation becomes

$$\underline{v}_P = \underline{v} + \tilde{\omega}\underline{S}_P, \tag{5.10}$$

where $\underline{\omega} = \text{axial}(\underline{\dot{R}}\,\underline{R}^T)$ is the angular velocity vector of the rigid body. This relationship describes the velocity of an arbitrary point **P** of the rigid body in terms of $\underline{v}$, the velocity of a reference point, and $\underline{\omega}$, the angular velocity vector of the rigid body. Here again, the choice of reference point **A** is arbitrary, and hence, eq. (5.10) is not an intrinsic relationship.

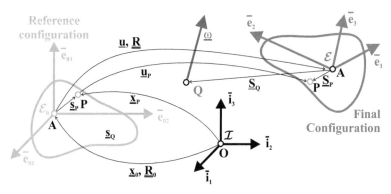

Fig. 5.2. Time-dependent motion of a rigid body.

To obtain a more general description of the velocity field, the following question can be asked: is it possible to find a material point of the rigid body, say point **Q**, whose velocity vector is parallel to the angular velocity vector? If such a point exists, the following relationship must hold

$$\underline{v}_Q = \underline{v} + \tilde{\omega}\underline{S}_Q = \mu\underline{\omega}, \tag{5.11}$$

where μ is an arbitrary scalar that can be found by taking the scalar product of this equation by $\underline{\omega}^T$ to find $\mu = (\underline{\omega}^T\underline{v})/\omega^2$.

Equation (5.11) now becomes $\tilde{\omega}\underline{S}_Q = (\underline{\omega}\,\underline{\omega}^T/\omega^2 - \underline{I})\underline{v} = \tilde{\omega}\tilde{\omega}\underline{v}/\omega^2$, where identity (1.33b) was used. This equation can be recast as $\tilde{\omega}\left[\underline{S}_Q - \tilde{\omega}\underline{v}/\omega^2\right] = 0$. The bracketed term is parallel to the angular velocity vector, which implies $\underline{S}_Q - \tilde{\omega}\underline{v}/\omega^2 = \alpha\underline{\omega}$, where α is an arbitrary constant. The location of point **Q** is now found as

$$\underline{S}_Q = \alpha\underline{\omega} + \frac{\tilde{\omega}}{\omega^2}\underline{v}.$$

The solution is the locus of points along a straight line parallel to $\underline{\omega}$, and hence, no unique solution exists for the location of point **Q**.

To remove this ambiguity, point **Q** will be selected as that at the shortest distance from point **A**, *i.e.*, $\underline{\omega}^T\underline{S}_Q = 0$. It follows that $\alpha = 0$, and

$$\underline{S}_Q = \frac{\widetilde{\omega}}{\omega^2}\underline{v}. \tag{5.12}$$

In summary, material point **Q** of the rigid body exists whose velocity vector is parallel to the angular velocity vector. The location of this point is given by eq. (5.12). Combining eqs. (5.10) and (5.11) now yields

$$\underline{v}_P = \frac{\underline{\omega}^T \underline{v}}{\omega^2}\underline{\omega} + \widetilde{\omega}(\underline{S}_P - \underline{S}_Q) = \underline{v}_Q + \widetilde{\omega}(\underline{S}_P - \underline{S}_Q) \tag{5.13}$$

This relationship expresses the velocity of material point **P** of the rigid body as the velocity of point **Q**, $\underline{v}_Q$, which is parallel to angular velocity vector $\underline{\omega}$, followed by a rotation about that same axis. This is referred to as *screw motion* about axis $\underline{\omega}$. The *screw axis* is defined as the line passing through point **Q** and parallel to $\underline{\omega}$. The Plücker coordinates, $\underline{Q}$, of the screw axis are

$$\underline{Q}_{SA} = \left\{ \begin{array}{c} -\dfrac{\widetilde{\omega}\widetilde{\omega}}{\omega^2}\underline{v} \\ \underline{\omega} \end{array} \right\} \tag{5.14}$$

5.2.1 Problems

Problem 5.1. General motion of a rigid body
Figure 5.1 depicts the general motion of a rigid body. Find material point **Q** of the rigid body whose displacement vector is of minimum norm. Is this point unique? Hint: The condition for minimization of the displacement norm is $||\underline{u}_Q||^2 = \min_{\underline{s}_q} [\underline{u} + (\underline{R} - \underline{I})\underline{s}_q]^T[\underline{u} + (\underline{R} - \underline{I})\underline{s}_q]$. The minimum displacement norm is found when $(\underline{R} - \underline{I})^T[\underline{u} + (\underline{R} - \underline{I})\underline{s}_q] = (\underline{R} - \underline{I})^T\underline{u}_Q = 0$. The solution of this system then $\underline{u}_Q = \underline{u} + (\underline{R} - \underline{I})\underline{s}_q = d\bar{n}$.

Problem 5.2. Time-dependent motion of a rigid body
Figure 5.2 shows the time-dependent motion of a rigid body. Find material point **Q** of the rigid body whose velocity vector is of minimum norm. Is this point unique? Hint: The condition for minimization of the velocity norm is $||\underline{v}_Q||^2 = \min_{\underline{S}_Q} [\underline{v} + \widetilde{\omega}\underline{S}_Q]^T[\underline{v} + \widetilde{\omega}\underline{S}_Q]$. The minimum velocity norm is found when $\widetilde{\omega}^T[\underline{v} + \widetilde{\omega}\underline{S}_Q] = \widetilde{\omega}^T\underline{v}_Q = 0$.

Problem 5.3. Location of the average velocity point
Consider three material points, **P**, **Q**, and **R**, of a rigid body with position vectors $\underline{x}_P$, $\underline{x}_Q$, and $\underline{x}_R$, respectively, and velocity vectors $\underline{v}_P$, $\underline{v}_Q$, and $\underline{v}_R$, respectively. Find the location of point **C** of the rigid body whose velocity is $\underline{v}_C = (\underline{v}_P + \underline{v}_Q + \underline{v}_R)/3$.

Problem 5.4. Relating the velocity vectors of three points of a rigid body
Consider two material points **P** and **Q** of a rigid body and their velocity vectors, $\underline{v}_P$ and $\underline{v}_Q$, respectively. *(1)* Find the velocity vector of point **R** of the rigid body, assuming that points **P**, **Q**, and **R** are not collinear. *(2)* Is the velocity of point **R** fully determined?

Problem 5.5. Relating the velocity vectors of three points of a rigid body
Consider two material points **P** and **Q** of a rigid body and their velocity vectors, $\underline{v}_P$ and $\underline{v}_Q$, respectively. *(1)* Find the velocity vector of point **R** of the rigid body, assuming that points **P**, **Q**, and **R** are collinear. *(2)* Is the velocity of point **R** fully determined?

Problem 5.6. Computing the angular velocity of a rigid body

The velocity vectors of material points $\mathbf{P}$, $\mathbf{Q}$, and $\mathbf{R}$ of a rigid body are given as $\underline{v}_P$, $\underline{v}_Q$, and $\underline{v}_R$, respectively. *(1)* Find the angular velocity vector of the rigid body. *(2)* State the three scalar constraints that the given velocity vectors must satisfy.

Problem 5.7. Relating the velocity vectors of four points of a rigid body

Consider three material points $\mathbf{P}$, $\mathbf{Q}$, and $\mathbf{R}$ of a rigid body and their velocities, $\underline{v}_P$, $\underline{v}_Q$, and $\underline{v}_R$, respectively. *(1)* Find the velocity of point $\mathbf{S}$ of the rigid body, assuming that points $\mathbf{P}$, $\mathbf{Q}$, and $\mathbf{R}$ are not collinear. *(2)* State the three scalar constraints that the given velocity vectors must satisfy.

Problem 5.8. Determination of Mozzi-Chasles axis

Figure 5.3 depicts a cube of unit size. Point $\mathbf{A}$ is selected as the reference point of the body; its displacement vector is denoted $\underline{u}$. The rotation of the rigid body is defined as a rotation of magnitude ϕ about unit vector $\bar{n}$. *(1)* Determine the coordinates of a point on the Mozzi-Chasles axis characterizing the motion of the rigid body. *(2)* Find the Plücker coordinates of the Mozzi-Chasles axis. *(3)* Compute the intrinsic displacement of the rigid body. *(4)* Using eq. (5.6), compute the displacements of points $\mathbf{A}$, $\mathbf{B}$, $\mathbf{C}$, and $\mathbf{D}$. Use the following data: $\underline{u}^T = \{3.2, 4.5, 0.76\}$ m, $\phi = 1.25$ rad, $\underline{n}^T = \{0.20, -0.26, 0.95\}$ (normalize this vector to make it a unit vector).

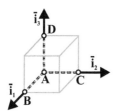

Fig. 5.3. Arbitrary motion of a rigid body.

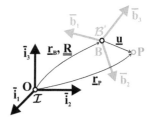

Fig. 5.4. Motion of a point defined with respect to frame $\mathcal{F}^B$.

5.3 Relative velocity and acceleration

Figure 5.4 depicts a practical situation that occurs in many engineering problem. The motion of a point, $\mathbf{P}$, is defined with respect to a rigid body associated with frame $\mathcal{F}^B = [\mathbf{B}, \mathcal{B}^* = (\bar{b}_1, \bar{b}_2, \bar{b}_3)]$. Tensor components resolved in basis $\mathcal{B}^*$ are denoted with superscript $(\cdot)^*$. The motion of the rigid body is defined with respect to an inertial frame $\mathcal{F}^I = [\mathbf{O}, \mathcal{I} = (\bar{\imath}_1, \bar{\imath}_2, \bar{\imath}_3)]$. The components of the position vector of point $\mathbf{P}$ with respect to point $\mathbf{B}$, resolved in basis $\mathcal{B}^*$, are denoted $\underline{u}^*$.

The motion of frame $\mathcal{F}^B$ is defined by the components of the position vector of point $\mathbf{B}$ with respect to point $\mathbf{O}$, denoted $\underline{r}_B$, and the components of the rotation tensor that bring basis $\mathcal{I}$ to basis $\mathcal{B}^*$, denoted $\underline{\underline{R}}$, both resolved in basis $\mathcal{I}$.

Consider, for instance, a piece of rotating machinery such as a jet engine or a helicopter rotor. It makes sense to attach a frame at the hub of the rotor: point **B** is at the hub center point and basis $\mathcal{B}^*$ rotates with the rotor. The position vector of a point on the blade, say point **P**, is then most naturally expressed in terms of its components in the hub attached basis $\mathcal{B}^*$; in fact, if the rotor blade is rigid, the components of the position vector of a point on the blade resolved in basis $\mathcal{B}^*$ are constant. The motion of the hub attached frame, defined by position vector $\underline{r}_B$ and rotation tensor $\underline{\underline{R}}$, characterize the motion of the hub with respect to an inertial frame.

Two important cases will be considered. In the most general case, point **P** is in relative motion with respect to the rigid body, *i.e.*, the components of the position vector of point **P** with respect to point **B**, resolved in the body attached basis, are a function of time, $\underline{u}^* = \underline{u}^*(t)$. In the second case, point **P** is a fixed, or *material point of the rigid body*, which implies that the components of the position vector of point **P** with respect to point **B**, resolved in the body attached basis, are constant in time, $\underline{u}^* \neq \underline{u}^*(t)$.

5.3.1 Point P is in motion with respect to the rigid body

The inertial velocity and acceleration vectors of point **P** will now be evaluated assuming that this point is in motion with respect to the rigid body. The inertial position of point **P**, denoted $\underline{r}_P$, is expanded as

$$\underline{r}_P = \underline{r}_B + \underline{u} = \underline{r}_B + \underline{\underline{R}}\,\underline{u}^*, \tag{5.15}$$

where $\underline{u} = \underline{\underline{R}}\,\underline{u}^*$ are the components of the position vector of point **P** with respect to point **B** resolved in basis $\mathcal{I}$.

The inertial velocity vector of point **P**, denoted $\underline{v}_P$, now becomes

$$\begin{aligned}\underline{v}_P &= \underline{v}_B + \underline{\dot{\underline{R}}}\,\underline{u}^* + \underline{\underline{R}}\,\underline{\dot{u}}^* = \underline{v}_B + \underline{\dot{\underline{R}}}\,\underline{\underline{R}}^T(\underline{r}_P - \underline{r}_B) + \underline{\underline{R}}\,\underline{\dot{u}}^* \\ &= \underline{v}_B + \widetilde{\omega}(\underline{r}_P - \underline{r}_B) + \underline{\underline{R}}\,\underline{\dot{u}}^*,\end{aligned} \tag{5.16}$$

where $\underline{v}_B = \underline{\dot{r}}_B$ is the inertial velocity of point **B** and, eq. (5.15), written as $\underline{u}^* = \underline{\underline{R}}^T(\underline{r}_P - \underline{r}_B)$, is used to eliminate $\underline{u}^*$.

The first term of eq. (5.16) represents the inertial velocity of the origin of the body attached frame, $\mathcal{F}^B$, and the second term accounts for the effects of its angular velocity. The last term is the *relative velocity* of point **P** with respect to point **B**, resolved in inertial basis $\mathcal{I}$. Of course, the inertial velocity vector of point **P** could also be resolved in the body attached basis $\mathcal{B}^*$; multiplication by $\underline{\underline{R}}^T$ yields

$$\underline{\underline{R}}^T\underline{v}_P = \underline{\underline{R}}^T\underline{v}_B + \widetilde{\omega}^*\underline{u}^* + \underline{\dot{u}}^*. \tag{5.17}$$

Next, the inertial acceleration of point **P**, denoted $\underline{a}_P$, is obtained by taking a time derivative of the inertial velocity, eq. (5.16), to find

$$\underline{a}_P = \underline{\dot{v}}_B + \widetilde{\dot{\omega}}(\underline{r}_P - \underline{r}_B) + \widetilde{\omega}(\underline{v}_P - \underline{v}_B) + \underline{\dot{\underline{R}}}\,\underline{\dot{u}}^* + \underline{\underline{R}}\,\underline{\ddot{u}}^*.$$

The velocities appearing in the third term are eliminated using eq. (5.16), to find

$$
\begin{aligned}
\underline{a}_P &= \underline{a}_B + \dot{\tilde{\omega}}(\underline{r}_P - \underline{r}_B) + \tilde{\omega}\left[\tilde{\omega}(\underline{r}_P - \underline{r}_B) + \underline{R}\,\dot{\underline{u}}^*\right] + \dot{\underline{R}}\,\underline{R}^T\underline{R}\,\dot{\underline{u}}^* + \underline{R}\,\ddot{\underline{u}}^* \\
&= \underline{a}_B + (\dot{\tilde{\omega}} + \tilde{\omega}\tilde{\omega})(\underline{r}_P - \underline{r}_B) + 2\tilde{\omega}\underline{R}\,\dot{\underline{u}}^* + \underline{R}\,\ddot{\underline{u}}^*,
\end{aligned} \tag{5.18}
$$

where $\underline{a}_B = \dot{\underline{v}}_B$ is the inertial acceleration of point **B**.

The first term of this expression represents the inertial acceleration of the origin of the body attached frame $\mathcal{F}^B$, and the second term accounts for the effects of its angular acceleration and velocity. The third term is known as the *Coriolis acceleration*. Finally, the last term is the *relative acceleration* of point **P** with respect to point **B**, resolved in the inertial basis $\mathcal{I}$.

Here again, the inertial acceleration vector of point **P** could also be resolved in the body attached basis $\mathcal{B}^*$; multiplication by $\underline{R}^T$ yields

$$
\underline{R}^T\underline{a}_P = \underline{R}^T\underline{a}_B + \left[\widetilde{(\underline{R}^T\dot{\omega})} + \tilde{\omega}^*\tilde{\omega}^*\right]\underline{u}^* + 2\tilde{\omega}^*\dot{\underline{u}}^* + \ddot{\underline{u}}^*. \tag{5.19}
$$

Term $\underline{R}^T\dot{\omega}$ represents the angular acceleration of the rigid body, resolved in body attached frame $\mathcal{B}^*$; this quantity is readily evaluated as $\underline{R}^T\dot{\omega} = \underline{R}^T(\underline{R}\,\omega^*)^{\cdot} = \underline{R}^T\dot{\underline{R}}\,\omega^* + \dot{\omega}^* = \tilde{\omega}^*\omega^* + \dot{\omega}^*$, and finally

$$
\underline{R}^T\dot{\omega} = \dot{\omega}^*. \tag{5.20}
$$

With this result at hand, the components of the inertial acceleration vector of point **P**, resolved in basis $\mathcal{B}^*$, become

$$
\underline{R}^T\underline{a}_P = \underline{R}^T\underline{a}_B + (\dot{\tilde{\omega}}^* + \tilde{\omega}^*\tilde{\omega}^*)\underline{u}^* + 2\tilde{\omega}^*\dot{\underline{u}}^* + \ddot{\underline{u}}^*. \tag{5.21}
$$

5.3.2 Point P is a material point of the rigid body

If point **P** is a material point of the rigid body, the components of its position vector with respect to point **B**, resolved in the body attached basis, are constant in time, $\underline{u}^* \neq \underline{u}^*(t)$. The velocity vector of point **P**, eq. (5.16), now reduces to

$$
\underline{v}_P = \underline{v}_B + \tilde{\omega}(\underline{r}_P - \underline{r}_B). \tag{5.22}
$$

Points **B** and **P** are two arbitrary material points of the rigid body. This means that eq. (5.22) relates the *velocity vectors of two arbitrary points of the same rigid body*. Of course, this relationship is identical to that found earlier, see (5.13), using a different reasoning. When expressed in the body attached basis, the same relationship becomes

$$
\underline{R}^T\underline{v}_P = \underline{R}^T\underline{v}_B + \tilde{\omega}^*\underline{u}^*. \tag{5.23}
$$

The acceleration vector of point **P**, eq. (5.18), reduces to

$$
\underline{a}_P = \underline{a}_B + (\dot{\tilde{\omega}} + \tilde{\omega}\tilde{\omega})(\underline{r}_P - \underline{r}_B). \tag{5.24}
$$

This equation relates the *acceleration vectors of two arbitrary points of the same rigid body*. When expressed in the body attached basis, the same relationship becomes

$$\underline{R}^T \underline{a}_P = \underline{R}^T \underline{a}_B + (\dot{\widetilde{\omega}}^* + \widetilde{\omega}^* \widetilde{\omega}^*) \underline{u}^*. \tag{5.25}$$

Example 5.1. Velocities and acceleration of a robotic arm

Figure 5.5 depicts a robotic system. The shaft is allowed to rotate about axis $\bar{\imath}_3$ with respect to inertial frame $\mathcal{F}^I = [\mathbf{O}, \mathcal{I} = (\bar{\imath}_1, \bar{\imath}_2, \bar{\imath}_3)]$. The time-dependent rotation angle of unit vector $\bar{s}_1$ with respect to axis $\bar{\imath}_1$ is denoted $\alpha(t)$. Frame $\mathcal{F}^S = [\mathbf{S}, \mathcal{S}^+ = (\bar{s}_1, \bar{s}_2, \bar{s}_3)]$ is attached to the shaft at a distance h from the origin of the inertial frame, as indicated on the figure; tensor quantities resolved in basis $\mathcal{S}^+$ are denoted with superscript $(\cdot)^+$. An arm of length L_a extends along the direction of axis $\bar{s}_2$ and is attached to the shaft at point $\mathbf{S}$.

Finally, a rigid manipulator of length L_b is connected to the arm at point $\mathbf{B}$. The manipulator is allowed to rotate with respect to frame $\mathcal{F}^S$, about axis $\bar{s}_1$. The time-dependent rotation angle of unit vector $\bar{b}_2$ with respect to axis $\bar{s}_2$ is denoted $\beta(t)$. Frame $\mathcal{F}^B = [\mathbf{B}, \mathcal{B}^* = (\bar{b}_1, \bar{b}_2, \bar{b}_3)]$ is attached to the manipulator; tensor quantities resolved in basis $\mathcal{B}^*$ will be denoted with superscript $(\cdot)^*$. Determine the velocity and acceleration vectors of point $\mathbf{P}$, located at the tip of the manipulator, at a distance L_b from point $\mathbf{B}$.

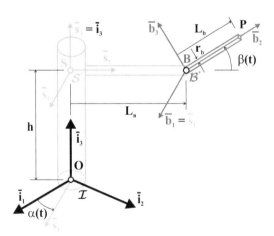

Fig. 5.5. Robotic arm configuration.

Let $\underline{R}_\alpha$ and $\underline{R}_\beta$ be the components of the rotation tensors that bring basis $\mathcal{I}$ to basis $\mathcal{S}^+$ and basis $\mathcal{S}^+$ to basis $\mathcal{B}^*$, respectively, both resolved in basis $\mathcal{I}$. This implies $\bar{s}_1 = \underline{R}_\alpha \bar{\imath}_1$ and $\bar{b}_1 = \underline{R}_\beta \bar{s}_1$; it follows that $\bar{b}_1 = \underline{R}_\beta \underline{R}_\alpha \bar{\imath}_1 = \underline{R}\bar{\imath}_1$, where $\underline{R} = \underline{R}_\beta \underline{R}_\alpha$ are the components of the rotation tensor that brings basis $\mathcal{I}$ to basis $\mathcal{B}^*$, resolved in basis $\mathcal{I}$. It is more natural to work with the components of rotation tensor $\underline{R}_\beta$ resolved in basis $\mathcal{S}^+$, $\underline{R}_\beta^+ = \underline{R}_\alpha^T \underline{R}_\beta \underline{R}_\alpha$, see eq. (4.29). The components of rotation tensor $\underline{R}$ now become

$$\underline{R} = \underline{R}_\alpha \underline{R}_\beta^+ = \begin{bmatrix} C_\alpha & -S_\alpha & 0 \\ S_\alpha & C_\alpha & 0 \\ 0 & 0 & 1 \end{bmatrix} \begin{bmatrix} 1 & 0 & 0 \\ 0 & C_\beta & -S_\beta \\ 0 & S_\beta & C_\beta \end{bmatrix} = \begin{bmatrix} C_\alpha & -S_\alpha C_\beta & S_\alpha S_\beta \\ S_\alpha & C_\alpha C_\beta & -C_\alpha S_\beta \\ 0 & S_\beta & C_\beta \end{bmatrix},$$

because $\underline{R}_\alpha$ and $\underline{R}_\beta^+$ are planar rotations about axes $\bar{\imath}_3$ and $\bar{s}_1$, respectively, see eqs. (4.6) and (4.4), respectively. Notations $S_\alpha = \sin\alpha$ and $C_\alpha = \cos\alpha$ were used to simplify the writing; similar expressions are used for angle β.

The angular velocity of basis $\mathcal{B}^*$ with respect to basis $\mathcal{I}$, denoted ω, is readily found by using the addition theorem, eq. (4.62), as $\underline{\omega} = \dot\alpha \bar{s}_3 + \dot\beta \bar{b}_1 = \dot\alpha(S_\beta \bar{b}_2 + C_\beta \bar{b}_3) + \dot\beta \bar{b}_1 = \dot\beta \bar{b}_1 + \dot\alpha S_\beta \bar{b}_2 + \dot\alpha C_\beta \bar{b}_3$. Expressing this tensor relationship in basis $\mathcal{B}^*$ yields $\underline{\omega}^* = \dot\beta \bar{b}_1^* + \dot\alpha S_\beta \bar{b}_2^* + \dot\alpha C_\beta \bar{b}_3^*$, and hence, the components of the angular velocity vector expressed in basis $\mathcal{B}^*$ are $\underline{\omega}^{*T} = \{\dot\beta, \dot\alpha S_\beta, \dot\alpha C_\beta\}$. The components of this vector in basis $\mathcal{I}$ are then evaluated as $\underline{\omega} = \underline{R}\,\underline{\omega}^*$, to find $\underline{\omega}^T = \{\dot\beta C_\alpha, \dot\beta S_\alpha, \dot\alpha\}$.

The position vector of point $\mathbf{P}$ with respect to point $\mathbf{O}$ is $\underline{r}_P = h\bar{\imath}_3 + L_a\bar{s}_2 + L_b\bar{b}_2 = h\bar{\imath}_3 + (L_b + L_aC_\beta)\bar{b}_2 - L_aS_\beta\bar{b}_3$. The inertial velocity of point $\mathbf{P}$, denoted $\underline{v}_P = \dot{\underline{r}}_P$, is then obtained from time differentiation $\underline{v}_P = -L_a\dot\beta S_\beta\bar{b}_2 - L_a\dot\beta C_\beta\bar{b}_3 + (L_b + L_aC_\beta)\dot{\bar{b}}_2 - L_aS_\beta\dot{\bar{b}}_3$.

Time derivatives of unit vectors $\bar{b}_2$ and $\bar{b}_3$ are readily evaluated as $\dot{\bar{b}}_2 = \widetilde{\omega}\bar{b}_2$ and $\dot{\bar{b}}_3 = \widetilde{\omega}\bar{b}_3$, respectively, see eq. (4.51). Regrouping the terms then yields $\underline{v}_P = -(L_a + L_bC_\beta)\dot\alpha\bar{b}_1 + L_b\dot\beta\bar{b}_3$. Expressing this tensor relationship in basis $\mathcal{B}^*$ yields $\underline{R}^T\underline{v}_P = -(L_a + L_bC_\beta)\dot\alpha\bar{b}_1^* + L_b\dot\beta\bar{b}_3^*$, and hence, the components of the velocity vector expressed in basis $\mathcal{B}^*$ are $(\underline{R}^T\underline{v}_P)^T = \{-(L_a + L_bC_\beta)\dot\alpha, 0, L_b\dot\beta\}$. The components of this vector in basis $\mathcal{I}$ are then readily obtained as

$$\underline{R}^T\underline{v}_P = \left\{ \begin{array}{c} -(L_a + L_bC_\beta)\dot\alpha \\ 0 \\ L_b\dot\beta \end{array} \right\}, \quad \underline{v}_P = \left\{ \begin{array}{c} -(L_a + L_bC_\beta)C_\alpha\dot\alpha + L_bS_\alpha S_\beta\dot\beta \\ -(L_a + L_bC_\beta)S_\alpha\dot\alpha - L_bC_\alpha S_\beta\dot\beta \\ L_bC_\beta\dot\beta \end{array} \right\}.$$

Next, the inertial acceleration vector of point $\mathbf{P}$ is obtained from a time derivative of its inertial velocity vector

$$\underline{a}_P = L_bS_\beta\dot\alpha\dot\beta\bar{b}_1 - (L_a + L_bC_\beta)\ddot\alpha\bar{b}_1 - (L_a + L_bC_\beta)\dot\alpha\dot{\bar{b}}_1 + L_b\ddot\beta\bar{b}_3 + L_b\dot\beta\dot{\bar{b}}_3.$$

Here again, the time derivatives of the unit vectors defining basis $\mathcal{B}^*$ are evaluated with the help of eq. (4.51), to yield

$$\underline{a}_P = \left[-(L_a + L_bC_\beta)\ddot\alpha + 2L_b\dot\alpha\dot\beta S_\beta\right]\bar{b}_1 - \left[(L_a + L_bC_\beta)\dot\alpha^2 C_\beta + L_b\dot\beta^2\right]\bar{b}_2$$
$$+ \left[(L_a + L_bC_\beta)\dot\alpha^2 S_\beta + L_b\ddot\beta\right]\bar{b}_3.$$

This expression reveals the components of the inertial acceleration vector, resolved in basis $\mathcal{B}^*$, as

$$a_P^* = \left\{ \begin{array}{c} -(L_a + L_b C_\beta)\ddot{\alpha} + 2L_b \dot{\alpha}\dot{\beta}S_\beta \\ -(L_a + L_b C_\beta)\dot{\alpha}^2 C_\beta - L_b \dot{\beta}^2 \\ (L_a + L_b C_\beta)\dot{\alpha}^2 S_\beta + L_b \ddot{\beta} \end{array} \right\},$$

and the corresponding components in basis $\mathcal{I}$ are $a_P = \underline{\underline{R}}\, a_P^*$,

$$a_P = \left\{ \begin{array}{c} (L_a + L_b C_\beta)(\dot{\alpha}^2 S_\alpha - \ddot{\alpha}C_\alpha) - L_b S_\alpha(\dot{\beta}^2 C_\beta - \ddot{\beta}S_\beta) + 2L_b \dot{\alpha}\dot{\beta}C_\alpha S_\beta \\ -(L_a + L_b C_\beta)(\dot{\alpha}^2 C_\alpha + \ddot{\alpha}S_\alpha) + L_b C_\alpha(\dot{\beta}^2 C_\beta - \ddot{\beta}S_\beta) + 2L_b \dot{\alpha}\dot{\beta}S_\alpha S_\beta \\ L_b(\dot{\beta}^2 S_\beta + \ddot{\beta}C_\beta) \end{array} \right\}.$$

In this example, the components of various vectors were derived in both bases $\mathcal{B}^*$ and $\mathcal{I}$. Of course, it is possible to work with the components of vectors in any basis, and hence, the choice of a specific basis is just a matter of convenience. For this problem, the body attached basis $\mathcal{B}^*$ is a good choice because the expressions for the components of the velocity and acceleration vectors appear to be simpler in that basis as compared to the corresponding expressions in basis $\mathcal{I}$.

Example 5.2. Velocities and acceleration of a spatial mechanism

The spatial mechanism depicted in fig. 5.6 consists of an arm of length L_a attached to the ground at point **S** and rotating about axis $\bar{\imath}_1$ of inertial frame $\mathcal{F}^I = [\mathbf{O}, \mathcal{I} = (\bar{\imath}_1, \bar{\imath}_2, \bar{\imath}_3)]$; the time-dependent rotation angle of unit vector $\bar{s}_2$ with respect to axis $\bar{\imath}_2$ is denoted $\theta(t)$. Frame $\mathcal{F}^S = [\mathbf{S}, \mathcal{S}^+ = (\bar{s}_1, \bar{s}_2, \bar{s}_3)]$, is attached to the arm; tensor quantities resolved in basis $\mathcal{S}^+$ will be denoted with superscript $(\cdot)^+$.

A rigid link connects point **P**, at the tip of the arm, to point **Q** that is free to slide along axis $\bar{\imath}_1$. The link is of length L_b and the distance from point **O** to point **Q** is denoted x. Find the inertial velocity and acceleration of point **Q** and the angular velocity of the link.

The inertial position vectors of points **P** and **Q** are readily found as $r_P = L_a \cos\theta\, \bar{\imath}_2 + (h + L_a \sin\theta)\bar{\imath}_3$ and $r_Q = x\,\bar{\imath}_1$, respectively. Vector $\underline{s}_{PQ}$, extending from point **P** to point **Q**, then becomes

$$\underline{s}_{PQ} = x\bar{\imath}_1 - L_a \cos\theta\bar{\imath}_2 - (h + L_a \sin\theta)\bar{\imath}_3.$$

The link is of length L_b, and hence, $L_b^2 = \|\underline{s}_{PQ}\|^2$. Expressing the norm of vector $\underline{s}_{PQ}$ implies that $L_b^2 = x^2 + L_a^2 + h^2 + 2hL_a \sin\theta$, which yields the position of point **Q** along axis $\bar{\imath}_1$ as $x = [L_b^2 - L_a^2 - h^2 - 2hL_a \sin\theta]^{1/2}$. A first derivative of this expression yields $x\dot{x} = -hL_a\dot{\theta}\cos\theta$, and a second derivative leads to $x\ddot{x} = -hL_a\ddot{\theta}\cos\theta + hL_a\dot{\theta}^2\sin\theta - \dot{x}^2$. The inertial velocity and acceleration of point **Q** are then $v_Q = \dot{x}\,\bar{\imath}_1$ and $a_Q = \ddot{x}\,\bar{\imath}_1$, respectively.

Because points **P** and **Q** are two material points of the same rigid body, link **PQ**, eq. (5.22) implies $v_Q = v_P + \tilde{w}(r_Q - r_P)$, where v_P is the inertial velocity of point **P**, and w the angular velocity of the link. This equation can be cast as $\tilde{s}_{PQ}\, w = -\dot{s}_{PQ}$, where $\dot{s}_{PQ} = v_Q - v_P$. In view of eq. (1.35), this vector product equation admits the following solution

$$w = \mu \underline{s}_{PQ} + \frac{\tilde{s}_{PQ}\dot{s}_{PQ}}{L_b^2}, \tag{5.26}$$

where μ is an arbitrary constant. This solution exists if and only if $\underline{s}_{PQ}^T \dot{\underline{s}}_{PQ}$, a condition that is always satisfied because vector $\underline{s}_{PQ}$ is of constant length. The indeterminacy of the solution is due to the fact that the link is free to rotate about its own axis, because its end points rotate freely.

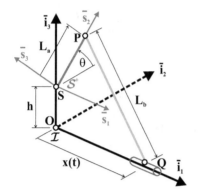

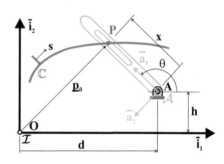

Fig. 5.6. Configuration of the spatial mechanism.

Fig. 5.7. Particle **P** sliding along curve $\mathbb{C}$ and through a slotted arm.

The first term of eq. (5.26), $\mu \underline{s}_{PQ}$, reflects this indeterminacy, which can be removed by assuming that the component of the angular velocity vector along the link vanishes, $\underline{s}_{PQ}^T \underline{\omega} = 0$, *i.e.*, the link is not allowed to rotate about its own axis. This condition leads to $\mu = 0$, and hence, $L_b^2 \underline{\omega} = \tilde{s}_{PQ} \dot{\underline{s}}_{PQ}$. Expanding the vector product then leads to

$$L_b^2 \underline{\omega} = L_a \dot{\theta}(L_a + h\sin\theta)\bar{\imath}_1 + \left[xL_a\dot{\theta}\cos\theta - \dot{x}(h + L_a\sin\theta) \right] \bar{\imath}_2$$
$$+ L_a(x\dot{\theta}\sin\theta + \dot{x}\cos\theta)\bar{\imath}_3.$$

This expression gives the components of the angular velocity vector of the link in basis $\mathcal{I}$.

Example 5.3. Particle sliding on a curve

Figure 5.7 shows particle **P** sliding along a planar curve fixed with respect to an inertial frame $\mathcal{F}^I = [\mathbf{O}, \mathcal{I} = (\bar{\imath}_1, \bar{\imath}_2, \bar{\imath}_3)]$. A slotted arm pivots about point **A** whose position vector is $\underline{r}_A = d\bar{\imath}_1 + h\bar{\imath}_2$. Frame $\mathcal{F}^A = [\mathbf{A}, \mathcal{A} = (\bar{a}_1, \bar{a}_2, \bar{a}_3)]$ is attached to the arm. The rotation angle between unit vector $\bar{\imath}_1$ and axis $\bar{a}_1$, denoted $\theta(t)$, is a given function of time. The particle slides along the curve and through the slot in the arm. Find the velocity and acceleration of the particle along the curve and the relative velocity and acceleration of the particle with respect to the arm.

Let x denote the distance between the particle and point **A**. The inertial position vector of the particle then becomes $\underline{p}_0(s) = d\bar{\imath}_1 + h\bar{\imath}_2 + x\bar{a}_1$. This equation involves two unknowns: the position of the particle along the curve, s, and the position of the particle along the arm, x. Projecting this equation along unit vectors $\bar{a}_2$ and $\bar{a}_1$ yields two scalar equations $\bar{a}_2^T \underline{p}_0(s) = -d\sin\theta + h\cos\theta$, and $x = \bar{a}_1^T \underline{p}_0 - (d\cos\theta + h\sin\theta)$, respectively, that can be solved for s and x, respectively, as functions of angle θ. The first equation is a nonlinear scalar equation; in general, several solutions might exist. The given initial configuration of the system should remove any ambiguity in the solution process at the initial time; for subsequent times, the requirement of a continuous solution for s should remove any further ambiguities.

Because particle **P** slides along the curve, its inertial velocity vector is $\underline{v}_P = \dot{s}\bar{t}$, see eq. (2.33). On the other hand, the velocity vector of point **P** on the arm is $\underline{v}_P = \dot{x}\bar{a}_1 + x\dot{\theta}\bar{a}_2$. Equating these two expressions yields

$$\dot{s}\bar{t} = \dot{x}\bar{a}_1 + x\dot{\theta}\bar{a}_2. \tag{5.27}$$

Let $\underline{\underline{R}}_\alpha$ be the components of Frenet's triad, see eq. (4.87), at point **P** of curve $\mathbb{C}$; hence, $\bar{t} = \underline{\underline{R}}_\alpha \bar{\imath}_1$. Let $\underline{\underline{R}}_\theta$ be the components of the rotation tensor that bring basis $\mathcal{I}$ to basis $\mathcal{A}$; hence, $\bar{a}_1 = \underline{\underline{R}}_\theta \bar{\imath}_1$. With the help of these definitions, eq. (5.27) becomes

$$\left\{ \begin{matrix} \dot{s} \\ 0 \\ 0 \end{matrix} \right\} = \underline{\underline{R}}_\alpha^T \underline{\underline{R}}_\theta \left\{ \begin{matrix} \dot{x} \\ x\dot{\theta} \\ 0 \end{matrix} \right\} = \begin{bmatrix} \cos(\alpha - \theta) & \sin(\alpha - \theta) & 0 \\ -\sin(\alpha - \theta) & \cos(\alpha - \theta) & 0 \\ 0 & 0 & 1 \end{bmatrix} \left\{ \begin{matrix} \dot{x} \\ x\dot{\theta} \\ 0 \end{matrix} \right\}. \tag{5.28}$$

The first two scalar equations are readily solved to find $\dot{x}$ and $\dot{s}$ as

$$\dot{x} = x\dot{\theta}\, \frac{\cos(\alpha - \theta)}{\sin(\alpha - \theta)}, \quad \dot{s} = x\dot{\theta}\, \frac{1}{\sin(\alpha - \theta)}.$$

The relative velocity of the particle with respect to the arm is $\dot{x}$, and $\dot{s}$ is the speed of the particle along the curve. Both results depend on the angle $(\alpha - \theta)$, which represents the relative rotation of Frenet's triad with respect to basis $\mathcal{A}$. When those two bases are parallel to each other, $\alpha = \theta$ and the tangent to the curve is parallel to the arm. Clearly, the mechanism "locks" in such a case, as implied by the infinite velocities $\dot{x}$ and $\dot{s} \to \infty$.

The accelerations of the system are obtained by taking a time derivative of eq. (5.28) to find

$$\left\{ \begin{matrix} \ddot{s} \\ 0 \\ 0 \end{matrix} \right\} = (\dot{\alpha} - \dot{\theta}) \begin{bmatrix} 0 & 1 & 0 \\ -1 & 0 & 0 \\ 0 & 0 & 0 \end{bmatrix} \left\{ \begin{matrix} \dot{s} \\ 0 \\ 0 \end{matrix} \right\} + \begin{bmatrix} \cos(\alpha - \theta) & \sin(\alpha - \theta) & 0 \\ -\sin(\alpha - \theta) & \cos(\alpha - \theta) & 0 \\ 0 & 0 & 1 \end{bmatrix} \left\{ \begin{matrix} \ddot{x} \\ \dot{x}\dot{\theta} + x\ddot{\theta} \\ 0 \end{matrix} \right\}.$$

Because $\underline{\underline{R}}_\alpha^T \dot{\underline{\underline{R}}}_\alpha = \dot{s}\widetilde{\kappa}^*$, see eq. (4.88), $\dot{\alpha} = \dot{s}/\rho$, where ρ is the radius of curvature of curve $\mathbb{C}$. Here again, the first two scalar equations are readily solved to find the desired accelerations, $\ddot{x}$ and $\ddot{s}$, leading to

$$\ddot{x} = \frac{(\dot{x}\dot{\theta} + x\ddot{\theta})\cos(\alpha - \theta) - (\dot{\alpha} - \dot{\theta})\dot{s}}{\sin(\alpha - \theta)}, \quad \ddot{s} = \frac{(\dot{x}\dot{\theta} + x\ddot{\theta}) - (\dot{\alpha} - \dot{\theta})\dot{s}}{\sin(\alpha - \theta)}.$$

Several observation can be made concerning this example. This problem involves several bases: the inertial basis, $\mathcal{I}$, the arm attached basis, $\mathcal{A}$, and Frenet's triad for curve $\mathbb{C}$. The components of the velocity vector of the particle are most easily expressed in the arm attached basis, $\mathcal{A}$. Because the particle slides along curve $\mathbb{C}$, it is natural to use Frenet's triad of the curve, since its properties are also expressed naturally in this triad. The analyst should always use the most appropriate basis to express the various kinematic characteristics of the system; typically, this implies selecting the basis that leads to the simplest, or most natural, analytical expressions.

Once kinematics conditions have been expressed in different bases, it is often necessary to "reconcile" the various equations, *i.e.*, express them in a common basis. This operation is most effectively achieved with the help of rotation tensors and the systematic use of their orthogonality property: the inverse of the rotation tensor equals its transpose, and the time derivative of the rotation tensor calls for the use of the angular velocity vector.

5.3.3 Problems

Problem 5.9. Retraction of a landing gear
Figure 5.8 depicts the extension of a simple landing gear. It consists of a link of length $L = 1.2$ m and of a wheel. The length $\ell(t)$ of the hydraulic actuator is a given function of time: $\ell(t) = h + g[1 - \cos \pi t/T]$, where $g = [\sqrt{(L^2/2 + hL + h^2)} - h]/2$, $h = 0.6$ m and $T = 1.5$ s is the total time required for extending the landing gear. (1) Compute and plot the angular velocity of the link as a function of time. (2) Compute and plot the angular acceleration of the link as a function of time. (3) Compute the inertial velocity vector of point **P** at the tip of the link. Plot the components of this vector resolved in $\mathcal{I}$. (4) Compute the inertial acceleration vector of point **P**. Plot the components of this vector resolved in $\mathcal{I}$.

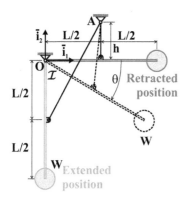
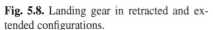
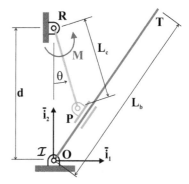

Fig. 5.8. Landing gear in retracted and extended configurations.

Fig. 5.9. Quick return mechanism configuration.

Problem 5.10. Quick return mechanism
The quick return mechanism shown in fig. 5.9 consists of a crank of length $L_c = 0.30$ m and of a bar of length $L_b = 1.6$ m. The crank is pinned at point **R** and the bar is pinned at point **O**. The distance between these two points is $d = 0.35$ m. At point **P**, a slider allows the tip

of the crank to slide along the bar. The time history of angle θ is $\theta(t) = \omega t$, where $\omega = 1.25$ rad/s. (1) On the same graph, plot the time history of the angular velocities of the crank and bar for two revolutions of the crank. (2) On the same graph, plot the angular accelerations of the crank and bar. (3) Compute the inertial velocity vector of point **T** at the tip of the bar. Plot the components of this vector resolved in $\mathcal{I}$. (4) Compute the inertial acceleration vector of point **T**. Plot the components of this vector resolved in $\mathcal{I}$.

Problem 5.11. Crank-slider mechanism

Figure 5.10 depicts a crank-slider mechanism. The crank of length $\ell_1 = 0.20$ m rotates counterclockwise at a constant angular velocity $\omega_1 = 200$ rad/s and is connected to the ground at point **O**. At point **A** the crank connects to a linkage of length $\ell_2 = 0.6$ m. Finally, at point **B**, this linkage connects of a piston that is constrained to move in the horizontal direction. The angular position of the crank is $\theta(t) = \omega_1 t$. (1) Compute the angular velocity ω_2 of the linkage and the velocity $v_p = \dot{x}$ of the piston. (2) Plot the horizontal position x of the piston as a function of time. (3) On one graph, plot the angles θ and ϕ as a function of time. (4) On one graph, plot the angular velocities ω_1 and ω_2 of the two bodies as a function of time. (5) Plot the velocity v_p of the piston as a function of time. (6) Compute the angular acceleration α_2 of the linkage and the acceleration $a_p = \ddot{x}$ of the piston. (7) On one graph, plot the angular acceleration α_1 and α_2 of the two bodies as a function of time. (8) Plot the acceleration a_p of the piston as a function of time. For all plots, the time scale should cover a complete revolution of the crank, i.e., $t \in [0, 2\pi/\omega_1]$.

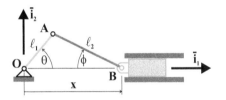

Fig. 5.10. Crank-slider mechanism rotating at a constant angular velocity.

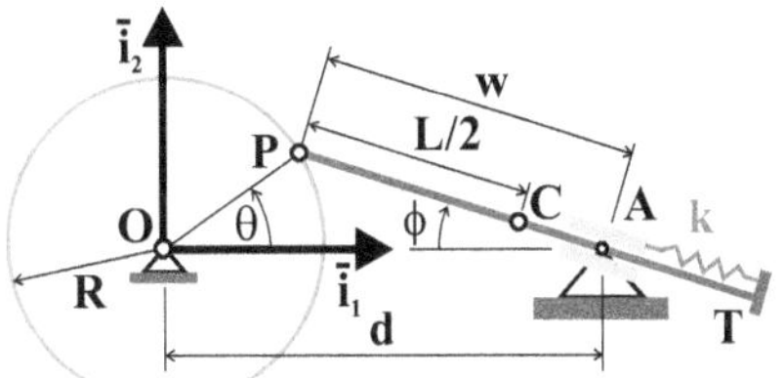

Fig. 5.11. Locking mechanism configuration.

Problem 5.12. Locking mechanism

Figure 5.11 shows a locking mechanism used in the deployment of large space structures. When the homogeneous disk of radius R rotates about its fixed point **O**, bar **PT** of length L slides at point **A** through a collar that is allowed to swivel about the pin at point **A**. The bar has a length L, and $w(t)$ denotes the part of the bar between point **P** and **A**. The time history of angle θ is prescribed as $\theta(t) = \pi(1 + \cos \pi t/T)/4$. (1) On the same graph, plot angles θ and ϕ as a function of time. (2) Plot $\dot{\theta}$ and $\dot{\phi}$. (3) Plot $\ddot{\theta}$ and $\ddot{\phi}$. (4) Plot w. (5) Plot $\dot{w}$. (6) Plot $\ddot{w}$. Use the following data: $R = 0.15$ m; $d = 0.2$ m; $L = 0.4$ m; $T = 2$ s. Present the response of the system for a duration of 2 s.

Problem 5.13. Crank-slider mechanism

Figure 5.12 depicts a crank-slider mechanism. The crank of length $\ell = 0.30$ m rotates counterclockwise at a constant angular velocity $\omega_1 = 200$ rad/s and is connected to the ground at point **O**. At point **B**, the crank connects to a linkage that slides along point **P**, a fixed point in space, located at a distance $d = 0.6$ m from point **O**. The angular position of the crank is

$\theta(t) = \omega_1 t$. Let w denote the distance from point **B** to point **P**. Point **T** is located at the tip of the linkage, at a distance $b = 1.2$ m from point **B**. *(1)* On one graph, plot the angles θ and ϕ as a function of time. *(2)* Plot the distance w as a function of time. *(3)* Compute the angular velocity ω_2 of the linkage and the relative velocity, $\dot{w}$, of point **P** with respect to the linkage. *(4)* On one graph, plot the angular velocities ω_1 and ω_2 of the two bodies as a function of time. *(5)* Plot the relative velocity, $\dot{w}$, of point **P** with respect to the linkage as a function of time. *(6)* Compute the angular acceleration, α_2, of the linkage and the relative acceleration, $\ddot{w}$, of point **P** with respect to the linkage. *(7)* On one graph, plot the angular acceleration α_1 and α_2 of the two bodies as a function of time. *(8)* Plot the relative acceleration, $\ddot{w}$, of point **P** with respect to the linkage as a function of time. *(9)* On one graph, plot the horizontal and vertical components of the inertial velocity vector of point **T**. *(10)* On one graph, plot the horizontal and vertical components of the inertial acceleration vector of point **T**. For all plots, the time scale should cover a complete revolution of the crank, *i.e.*, $t \in [0, 2\pi/\omega_1]$.

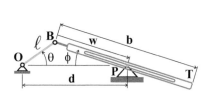

Fig. 5.12. Crank-slider mechanism rotating at a constant angular velocity.

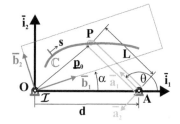

Fig. 5.13. Rotating curve connected to link.

Problem 5.14. Rotating curve connected to link

Planar curve $\mathbb{C}$ is embedded into a rigid body that rotates with respect to an inertial frame $\mathcal{F}^I = [\mathbf{O}, \mathcal{I} = (\bar{\imath}_1, \bar{\imath}_2, \bar{\imath}_3)]$, as depicted in fig. 5.13. A frame $\mathcal{F}^B = [\mathbf{O}, \mathcal{B} = (\bar{b}_1, \bar{b}_2, \bar{b}_3)]$, is attached to the body and the rotation angle is a known quantity $\alpha(t)$. A link is attached to inertial point **A** whose position vector is $\underline{r}_A = d\bar{\imath}_1$. The other end of the link is connected to a particle that slides along curve $\mathbb{C}$. A frame $\mathcal{F}^A = [\mathbf{A}, \mathcal{A} = (\bar{a}_1, \bar{a}_2, \bar{a}_3)]$, is attached to the link; the rotation angle for frame $\mathcal{F}^A$ is denoted θ. Let $F_\beta^*(s)$ be the components of Frenet's triad of curve $\mathbb{C}$ resolved in basis $\mathcal{B}$ and β the angle that brings basis $\mathcal{B}$ to Frenet's triad. *(1)* Find a scalar equation to determine the location s of point **P** along curve $\mathbb{C}$. Is the solution uniquely defined? *(2)* Find a scalar equation to determine angle θ. *(3)* Determine the angular velocity of the link. *(4)* Determine the speed $\dot{s}$ of the particle along curve $\mathbb{C}$. *(5)* Determine the angular acceleration of the link. *(6)* Determine the acceleration $\ddot{s}$ of the particle along curve $\mathbb{C}$. *(7)* Under what condition does the mechanism lock? Explain your answer in geometric terms. Express your answers in terms of the angles α, θ and β.

Problem 5.15. Spinning disk mounted on rotating arm

The system depicted in fig. 4.9 consists of a shaft of height h rigidly connected to an arm of length L_a and of a spinning disk of radius R mounted at the free end of the arm. Frame $\mathcal{F}^S = [\mathbf{S}, \mathcal{S}^+ = (\bar{s}_1, \bar{s}_2, \bar{s}_3)]$ is attached to the shaft at point **S** and frame $\mathcal{F}^D = [\mathbf{C}, \mathcal{B}^* = (\bar{b}_1, \bar{b}_2, \bar{b}_3)]$ is attached to the disk at point **C**. Superscripts $(\cdot)^+$ and $(\cdot)^*$ will be used to denote tensor components resolved in bases $\mathcal{S}^+$ and $\mathcal{B}^*$, respectively. Angle $\alpha(t)$ and $\beta(t)$ are the magnitudes of the planar rotations about axis $\bar{\imath}_3$ and $\bar{s}_1$, respectively, that bring basis $\mathcal{I}$ to $\mathcal{S}^+$ and basis $\mathcal{S}^+$ to $\mathcal{B}^*$, respectively. *(1)* Find the angular velocity vector of basis $\mathcal{B}^*$ with respect to basis $\mathcal{I}$. *(2)* Find the components of this vector in basis $\mathcal{I}$. *(3)* Find the components of this vector in basis $\mathcal{B}^*$. *(4)* Find the angular acceleration vector of basis $\mathcal{B}^*$ with respect to basis $\mathcal{I}$. *(5)* Find the components of this vector in basis $\mathcal{I}$. *(6)* Find the components of this vector in basis $\mathcal{B}^*$.

Problem 5.16. Robotic system

Figure 5.5 depicts a robotic system. The shaft is allowed to rotate with respect to an inertial frame $\mathcal{F}^I$, about axis $\bar{\imath}_3$; the time-dependent angle of rotation is denoted $\alpha(t)$. A frame $\mathcal{F}^S = [\mathbf{S}, \mathcal{S} = (\bar{s}_1, \bar{s}_2, \bar{s}_3)]$ is attached to the shaft at a distance h from the origin of the inertial frame, as indicated on the figure. An arm of length $L_a = 1.2$ m, extending along the direction of axis $\bar{s}_2$, is attached to the shaft at point $\mathbf{S}$. Finally, a rigid manipulator of length $L_b = 0.5$ m is connected to the arm at point $\mathbf{B}$. The manipulator is allowed to rotate with respect to frame $\mathcal{F}^S$, about axis $\bar{s}_1$; the time-dependent angle of rotation is denoted $\beta(t)$. Frame $\mathcal{F}^B = [\mathbf{B}, \mathcal{B} = (\bar{b}_1, \bar{b}_2, \bar{b}_3)]$ is attached to the manipulator. Angles $\alpha(t)$ and $\beta(t)$ are prescribed as $\alpha(t) = \pi(1 - \cos \pi t/T)/2$ and $\beta(t) = 2\pi \left(1 - \cos \pi t/T\right)$, where $T = 2$ s. *(1)* Compute the angular velocity vector $\underline{\omega}$ of the manipulator with respect to the inertial system. On one graph, plot the components of this vector in basis $\mathcal{B}$. *(2)* On one graph, plot the components of this vector in basis $\mathcal{I}$. *(3)* Compute the position vector $\underline{r}_P$ of point $\mathbf{P}$ with respect to point $\mathbf{O}$. On one graph, plot the components of this vector in basis $\mathcal{I}$. *(4)* Evaluate the inertial velocity vector of point $\mathbf{P}$. On one graph, plot the components of this vector in basis $\mathcal{B}$. *(5)* On one graph, plot the components of this vector in basis $\mathcal{I}$. *(6)* Compute the inertial acceleration of point $\mathbf{P}$. On one graph, plot the components of this vector in basis $\mathcal{B}$. *(7)* On one graph, plot the components of this vector in basis $\mathcal{I}$.

Problem 5.17. Swiveling plate

Figure 5.14 depicts a homogeneous, rectangular plate of height a, width b and mass m connected to the ground by a rigid, massless link of length d. At point $\mathbf{O}$, a bearing allows the link to rotate with respect to axis $\bar{\imath}_3$, whereas at point $\mathbf{B}$, the plate is free to rotate with respect to the link about axis $\bar{a}_1$. Three frames will be used in this problem: the inertial frame, $\mathcal{F}^I = [\mathbf{O}, \mathcal{I} = (\bar{\imath}_1, \bar{\imath}_2, \bar{\imath}_3)]$, a frame attached to the link, $\mathcal{F}^A = [\mathbf{O}, \mathcal{A} = (\bar{a}_1, \bar{a}_2, \bar{a}_3)]$, and finally, a frame attached to the plate at its center of mass, $\mathcal{F}^B = [\mathbf{C}, \mathcal{B} = (\bar{b}_1, \bar{b}_2, \bar{b}_3)]$. A planar rotation of magnitude α about axis $\bar{\imath}_3$ brings basis $\mathcal{I}$ to basis $\mathcal{A}$, whereas a planar rotation of magnitude β about axis $\bar{a}_1$ brings basis $\mathcal{A}$ to basis $\mathcal{B}$. *(1)* Find the components of the inertial position vector of point $\mathbf{P}$ in basis $\mathcal{B}$. *(2)* Find the components of the inertial velocity vector of point $\mathbf{P}$ in basis $\mathcal{B}$. *(3)* Find the components of the inertial acceleration vector of point $\mathbf{P}$ in basis $\mathcal{B}$.

Problem 5.18. Robotic system with a sliding manipulator

Figure 5.15 depicts a robotic system with a sliding manipulator. The shaft is allowed to rotate with respect to an inertial frame $\mathcal{F}^I$, about axis $\bar{\imath}_3$; the time-dependent angle of rotation is denoted $\alpha(t)$. A frame $\mathcal{F}^S = [\mathbf{S}, \mathcal{S} = (\bar{s}_1, \bar{s}_2, \bar{s}_3)]$ is attached to the shaft at a distance h from the origin of the inertial frame, as indicated on the figure. An arm of length L_b, connected to the shaft at point $\mathbf{S}$ is allowed to pivot with respect to the shaft about axis $\bar{s}_1$; the time-dependent angle of rotation is denoted $\beta(t)$. A frame $\mathcal{F}^B = [\mathbf{B}, \mathcal{B} = (\bar{b}_1, \bar{b}_2, \bar{b}_3)]$ is attached to the arm. Finally, a rigid manipulator slides with respect to the arm along axis $\bar{b}_2$; the displacement of the manipulator is denoted $u(t)$. Angles $\alpha(t)$ and $\beta(t)$ and displacement $u(t)$ are known, prescribed functions of time. *(1)* Compute the angular velocity vector $\underline{\omega}$ of the manipulator with respect to the inertial system. *(2)* Give the components of this vector in

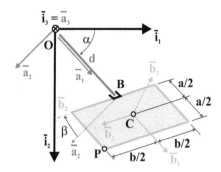

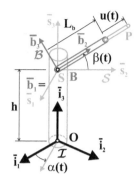

Fig. 5.14. Configuration of the swiveling plate.

Fig. 5.15. Robotic arm with a sliding manipulator.

basis $\mathcal{B}$. *(3)* Compute the position vector $\underline{r}_P$ of point **P** with respect to point **O**. *(4)* Evaluate the inertial velocity vector of point **P**. *(5)* Compute the inertial acceleration of point **P**.

Problem 5.19. Robotic system with a sliding manipulator

Figure 5.15 depicts a robotic system with a sliding manipulator. The shaft is allowed to rotate with respect to an inertial frame $\mathcal{F}^I$, about axis $\bar{\imath}_3$; the time-dependent angle of rotation is denoted $\alpha(t)$. Frame $\mathcal{F}^S = [\mathbf{S}, \mathcal{S} = (\bar{s}_1, \bar{s}_2, \bar{s}_3)]$ is attached to the shaft at a distance $h = 1$ m from the origin of the inertial frame, as indicated on the figure. An arm of length $L_b = 0.75$ m, connected to the shaft at point **S** is allowed to pivot with respect to the shaft about axis $\bar{s}_1$; the time-dependent angle of rotation is denoted $\beta(t)$. A frame $\mathcal{F}^B = [\mathbf{B}, \mathcal{B} = (\bar{b}_1, \bar{b}_2, \bar{b}_3)]$ is attached to the arm. Finally, a rigid manipulator slides with respect to the arm along axis $\bar{b}_2$; the displacement of the manipulator is denoted $u(t)$. Angles $\alpha(t)$ and $\beta(t)$ are prescribed as $\alpha(t) = \pi(1 - \cos \pi t/T)/2$ and $\beta(t) = \pi(1 - \cos \pi t/T)/6$, where $T = 2$ s. The sliding motion is prescribed as $u(t) = 0.5\,(1 - \exp(-5t/T))$ m. *(1)* Compute the angular velocity vector $\underline{\omega}$ of the manipulator with respect to the inertial system. On one graph, plot the components of this vector in basis $\mathcal{B}$. *(2)* On one graph, plot the components of this vector in basis $\mathcal{I}$. *(3)* Compute the position vector $\underline{r}_P$ of point **P** with respect to point **O**. On one graph, plot the components of this vector in basis $\mathcal{I}$. *(4)* Evaluate the inertial velocity vector of point **P**. On one graph, plot the components of this vector in basis $\mathcal{B}$. *(5)* On one graph, plot the components of this vector in basis $\mathcal{I}$. *(6)* Compute the inertial acceleration of point **P**. On one graph, plot the components of this vector in basis $\mathcal{B}$. *(7)* On one graph, plot the components of this vector in basis $\mathcal{I}$. For all plots, the time scale should cover $t \in [0, 4T]$ s.

Problem 5.20. Robotic system with a manipulator on screw joint

Consider the robotic system with a manipulator mounted on a screw joint depicted in fig. 5.16. The shaft is allowed to rotate with respect to an inertial frame $\mathcal{F}^I$, about axis $\bar{\imath}_3$; the time-dependent angle of rotation is denoted $\alpha(t)$. Frame $\mathcal{F}^S = [\mathbf{S}, \mathcal{S} = (\bar{s}_1, \bar{s}_2, \bar{s}_3)]$ is attached to the shaft at a distance $h = 0.5$ m from the origin of the inertial frame, as indicated on the figure. An arm of length $L_a = 0.6$ m, extending along the direction of axis $\bar{s}_2$, is attached to the shaft at point **S**. Finally, a rigid manipulator is connected to the arm by means of a screw joint. Frame $\mathcal{F}^B = [\mathbf{B}, \mathcal{B} = (\bar{b}_1, \bar{b}_2, \bar{b}_3)]$ is attached to the manipulator. The manipulator slides and rotates with respect to the arm; the sliding distance is denoted $u(t)$ and the rotation angle is $\beta(t)$. The screw joint implies the following relationship between these two motions: $u(t) = \varpi\,\beta(t)/(2\pi)$, where $\varpi = 0.5$ m is the pitch of the screw. Angles $\alpha(t)$ and $\beta(t)$ are prescribed as $\alpha(t) = \pi/2 \, (1 - \cos \pi t/T)$ and $\beta(t) = 5\pi \, t/T$, where $T = 2$ s. *(1)* Compute the angular velocity vector $\underline{\omega}$ of the manipulator with respect to the inertial system. On one graph, plot the components of this vector in basis $\mathcal{B}$. *(2)* On one graph, plot the components of this vector in basis $\mathcal{I}$. *(3)* Compute the position vector $\underline{r}_P$ of point **P** with respect to point **O**; point **P** is located at a distance $L_b = 0.25$ m from the manipulator elbow. On one graph, plot the components of this vector in basis $\mathcal{I}$. *(4)* Evaluate the inertial velocity vector of point **P**. On one graph, plot the components of this vector in basis $\mathcal{B}$. *(5)* On one graph, plot the components of this vector in basis $\mathcal{I}$. *(6)* Compute the inertial acceleration of point **P**. On one graph, plot the components of this vector in basis $\mathcal{B}$. *(7)* On one graph, plot the components of this vector in basis $\mathcal{I}$. For all plots, the time scale should cover $t \in [0, 2T]$ s.

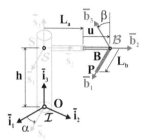

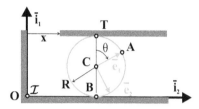

Fig. 5.16. Robotic arm with manipulator.   **Fig. 5.17.** Wheel rolling between two plates.

Problem 5.21. Wheel rolling between two plates
Figure 5.17 depicts a wheel of radius R rolling without sliding between two horizontal plates. The top plate moves horizontally and is at a distance x from axis $\bar{\imath}_1$. Frame $\mathcal{F} = [\mathbf{C}, \mathcal{E} = (\bar{e}_1, \bar{e}_2)]$ rotates with the wheel. A planar rotation of magnitude $\theta(t)$ about unit vector $\bar{\imath}_3$ brings basis $\mathcal{I}$ to basis $\mathcal{E}$. *(1)* Find the velocity vector of material point **A** of the wheel. *(2)* Resolve this vector in basis $\mathcal{I}$ then in basis $\mathcal{E}$. *(3)* Find the acceleration vector of material point **A** of the wheel. *(2)* Resolve this vector in basis $\mathcal{I}$ then in basis $\mathcal{E}$.

5.4 Contact between rigid bodies

Many commonly used mechanical systems involve contacting rigid bodies. Figure 5.18 shows two rigid bodies, denoted body k and body ℓ, with outer shapes defined by two closed curves, denoted curves $\mathbb{C}^k$ and $\mathbb{C}^\ell$, respectively. Point **P** is the instantaneous point of contact between the two rigid bodies. For the purpose of illustration, the two bodies are assumed to undergo planar motion and rotate about fixed inertial points $\mathbf{O}^k$ and $\mathbf{O}^\ell$, respectively.

The mechanism shown in fig. 5.18 is generally called a *cam-follower pair*. The angular motion of body k, called the cam, is typically prescribed to be a constant angular speed, say Ω. As the cam rotates, body ℓ, called the follower, is assumed to remain in contact with the cam at all times at a single point. The cam-follower pair transforms the constant angular motion of the cam into a rocking motion of the follower. By tailoring the shapes of curves $\mathbb{C}^k$ and $\mathbb{C}^\ell$, desirable periodic schedules of the follower can be achieved.

The contact point between the two rigid bodies is not a material point of either bodies. At the cam rotates, the location of the contact point coincides with a different material point of the cam at each instant: the contact point slides along curve $\mathbb{C}^k$. Similarly, the contact point slides along curve $\mathbb{C}^\ell$ because the location of the contact point coincides with a different material point of the follower at each instant.

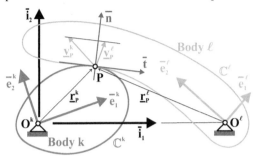

At instant t, let $\mathbf{K}$ and $\mathbf{L}$ be the material points of the cam and follower, respectively, that are *instantaneously coincident* with the contact point, $\mathbf{P}(t)$. To avoid confusion, points $\mathbf{K}$ and $\mathbf{L}$ are not shown in fig. 5.18 because their location is identical to that of point $\mathbf{P}$. Let $\underline{r}_K$ and $\underline{r}_L$ be the position vectors of mate-

Fig. 5.18. Body k and body ℓ in contact at point $\mathbf{P}$.

rial points $\mathbf{K}$ and $\mathbf{L}$ with respect to point $\mathbf{O}$, respectively. Furthermore, the position vector of the instantaneous point of contact with respect to point $\mathbf{O}$ is denoted $\underline{r}_P(t)$. Given these definitions, $\underline{r}_K = \underline{r}_L = \underline{r}_P(t)$.

At instant $t' > t$, $\mathbf{K}'$ and $\mathbf{L}'$ are the material points of the cam and follower, respectively, that are instantaneously coincident with the contact point, $\mathbf{P}(t')$. If $\underline{r}_{K'}$ and $\underline{r}_{L'}$ are the position vectors of material points $\mathbf{K}'$ and $\mathbf{L}'$ with respect to point $\mathbf{O}$, respectively, and if $\underline{r}_P(t')$ denotes the position vector of the instantaneous point of contact with respect to point $\mathbf{O}$, it is still true that $\underline{r}_{K'} = \underline{r}_{L'} = \underline{r}_P(t')$.

In general, however, $\underline{r}_K \neq \underline{r}_{K'}$ because points $\mathbf{K}$ and $\mathbf{K}'$ are two different material points of the cam and $\underline{r}_L \neq \underline{r}_{L'}$ because points $\mathbf{L}$ and $\mathbf{L}'$ are two different material points of the follower. Because the instantaneous point of contact slides over curves $\mathbb{C}^k$ and $\mathbb{C}^\ell$, $\underline{r}_P(t) \neq \underline{r}_P(t')$ and furthermore, $\underline{r}_K \neq \underline{r}_P(t')$ and $\underline{r}_L \neq \underline{r}_P(t')$.

Because point $\mathbf{K}$ is a material point of body k, its inertial velocity and acceleration vectors can be evaluated using eqs. (5.22) and (5.24), respectively. The relative position vectors of point $\mathbf{P}$ with respect to points $\mathbf{O}^k$ and $\mathbf{O}^\ell$ are denoted $\underline{r}_P^k$ and $\underline{r}_P^\ell$, respectively. Let point $\mathbf{O}^k$ be the reference point for body k; the velocity of material point $\mathbf{K}$, denoted $\underline{v}_P^k$, is given by eq. (5.22) as $\underline{v}_P^k = \widetilde{\omega}^k \underline{r}_P^k$, where $\underline{\omega}^k$ is the angular velocity of body k. Vectors $\underline{v}_P^k$ and $\underline{r}_P^k$ are perpendicular to each other, as illustrated in fig. 5.18. A similar expression holds for the velocity of point $\mathbf{L}$, denoted $\underline{v}_P^\ell$.

The components of the same velocity vectors in the body attached basis are given by eq. (5.23). Let $\mathcal{F}^k = \left[\mathbf{O}^k, \mathcal{B}^k = (\bar{e}_1^k, \bar{e}_2^k)\right]$ be a frame attached to the cam, as shown in fig. 5.18, and notation $(\cdot)^*$ indicates tensor components resolved in basis $\mathcal{B}^k$. The components of the inertial velocity vector of point $\mathbf{K}$ resolved in this basis are then $\underline{v}_P^{*k} = \widetilde{\omega}^{*k} \underline{r}_P^{*k}$; because array $\underline{r}_P^{*k}$ stores the components of the relative position vector of material point $\mathbf{K}$ of the cam resolved in a cam attached basis, this array is *time-independent*.

It is assumed that the two bodies are in contact at a single point, and the unit tangent vectors to curves $\mathbb{C}^k$ and $\mathbb{C}^\ell$ at point **P** are coincident and denoted $\bar{t}$. The unit vector perpendicular to this common tangent is the unit normal vector, denoted $\bar{n}$. As discussed in section 2.2, the unit vector tangent to curve $\mathbb{C}^k$ is given by eq. (2.5) and its orientation depends on the curvilinear variable used to parameterize the curve. If curves $\mathbb{C}^k$ and $\mathbb{C}^\ell$ are both parameterized in the counterclockwise direction, and if the unit vectors tangent to the two curves are denoted $\bar{t}^k$ and $\bar{t}^\ell$, respectively, and fig. 5.18 shows that at the instantaneous contact point, $\bar{t} = -\bar{t}^k = \bar{t}^\ell$. For the configuration illustrated in the figure, $\bar{n} = -\bar{n}^k = -\bar{n}^\ell$.

If the two bodies of the cam-follower pair remain in contact at a single point, the normal projections of the velocity vectors of the material points that are instantaneously coincident with the contact point must be identical,

$$\bar{n}^T \underline{v}_P^k = \bar{n}^T \underline{v}_P^\ell. \tag{5.29}$$

If this condition were not satisfied, the two bodies would either separate or interpenetrate and contact at a single point would not be maintained. The relative velocity of the material points that are instantaneously coincident with the contact point, denoted $\underline{v}_P^r$, is

$$\begin{aligned}
\underline{v}_P^r = \underline{v}_P^\ell - \underline{v}_P^k &= (\bar{n}^T \underline{v}_P^\ell)\bar{n} + (\bar{t}^T \underline{v}_P^\ell)\bar{t} - (\bar{n}^T \underline{v}_P^k)\bar{n} - (\bar{t}^T \underline{v}_P^k)\bar{t} \\
&= (\bar{t}^T \underline{v}_P^\ell)\bar{t} - (\bar{t}^T \underline{v}_P^k)\bar{t} = \left[(\bar{t}^T \underline{v}_P^\ell) - (\bar{t}^T \underline{v}_P^k) \right] \bar{t}.
\end{aligned} \tag{5.30}$$

where the third equality follows from the contact condition, eq. (5.29). As expected, the relative velocity of the material points that are instantaneously located at the contact point is oriented along to the common tangent vector at this point.

Smooth operation of cam-follower systems generally require a single point contact between the two rigid bodies. For arbitrary shapes of the bounding curves, contact could occur at two or more points simultaneously, or even along a line if portions of the outer curves are straight, for instance. Such situations rarely occur in mechanical systems. To guarantee single point contact, the bounding curves must satisfy specific conditions at the contact point. For instance, a sufficient condition for single point contact is for the cam and follower to be bounded by strictly convex curves. For the case illustrated in fig. 5.18, the cam and follower are convex and concave, respectively, at the point of contact. For single point contact to occur, the cam's radius of curvature must be smaller than that of the follower.

The discussion has focused thus far on contacting rigid bodies undergoing planar motion. If the problem is fully three-dimensional, it becomes necessary to define the external surfaces of bodies k and ℓ, denoted $\mathbb{S}^k$ and $\mathbb{S}^\ell$, respectively. If the contact between the two bodies occurs at a single point, the planes tangent to surfaces $\mathbb{S}^k$ and $\mathbb{S}^\ell$ at the instantaneous contact point must coincide and it is still possible to define a unit normal vector that is perpendicular to this common tangent plane. The contact condition expressed by eq. (5.29) still holds for this problem, but the relative velocity vector will have components along two directions within the common tangent plane.

Example 5.4. Cam-follower pair

Consider the planar cam-follower pair depicted in fig. 5.19. The cam rotates at a constant angular velocity, Ω, about fixed inertial point **O**. Frame $\mathcal{F}^I = [\mathbf{O}, \mathcal{I} = (\bar{\imath}_1, \bar{\imath}_2)]$ is inertial and frame $\mathcal{F}^E = [\mathbf{O}, \mathcal{E} = (\bar{e}_1, \bar{e}_2)]$ is attached to the cam. The external shape of the cam is defined by curve $\mathbb{C}$ and the follower slides over this curve; the contact point between the cam and follower is denoted **P**. The motion of the follower is constrained to be along axis $\bar{\imath}_2$ and its displacement is denoted x. Find the velocity and acceleration of the follower.

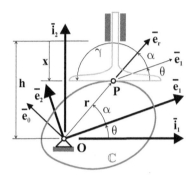

Fig. 5.19. Configuration of the cam-follower pair.

Let angle θ define the rotation of the cam; it follows that $\dot{\theta} = \Omega$. The geometry of curve $\mathbb{C}$ is defined in polar coordinates, as discussed in example 2.3 on page 37, and angle α defines an arbitrary parameterization of the curve. The shape of the curve is then defined by a given function, $r = r(\alpha)$, and notation $(\cdot)'$ indicates a derivative with respect to angle α.

Given the configuration of the system, the tangent to curve $\mathbb{C}$ at the point of contact must remain horizontal at all times; this implies

$$\theta + \alpha + \gamma = \pi. \tag{5.31}$$

It then follows that $\sin(\theta + \alpha) = S_\gamma = r/p_1$, where $S_\gamma = \sin\gamma$ and the second equality results from eq. (2.28a). Similarly, $\cos(\theta + \alpha) = -C_\gamma = -r'/p_1$, where $C_\gamma = \cos\gamma$ and the second equality results from eq. (2.28b). Eliminating p_1 from these two relationships leads to $r(\alpha)\cos(\theta + \alpha) + r'(\alpha)\sin(\theta + \alpha) = 0$.

For a given value of angle θ, this transcendental equation can be solved for angle α, which determines the location of the point of contact between the cam and the follower. It then becomes possible to evaluate $r(\alpha)$, $r'(\alpha)$, and angle γ then follows from eqs. (2.28a) and (2.28b).

A time derivative of eq. (5.31) yields $\Omega + \dot{\alpha} + \dot{\gamma} = 0$. Introducing eqs. (2.29) and (2.32) then yields

$$\dot{r} = -\Omega\rho C_\gamma.$$

The velocity of the material point of the cam located at the instantaneous point of contact between the cam and the follower is $\Omega r \bar{e}_\theta$. The velocity of the material point of the follower at the same location is $-\dot{x}\bar{\imath}_2$. The contact condition for these two bodies, eq. (5.29), then implies $\bar{\imath}_2^T \Omega r \bar{e}_\theta = -\bar{\imath}_2^T \dot{x} \bar{\imath}_2$, or

$$\dot{x} = \Omega r C_\gamma.$$

The acceleration of the follower is obtained by taking a time derivative of this expression to find $\ddot{x} = \Omega \dot{r} C_\gamma - \Omega r \dot{\gamma} S_\gamma$. Introducing eq. (2.32) then yields $\ddot{x} = (1 - S_\gamma r/\rho)\Omega \dot{r}/C_\gamma$ and finally

$$\ddot{x} = \Omega^2 (r S_\gamma - \rho). \tag{5.32}$$

5.4.1 Problems

Problem 5.22. Cam-follower pair questions
Figure 5.20 shows the instantaneous point of contact, **P**, between to rigid bodies, denoted body k and body ℓ. Let $\bar{t}$ be the unit vector tangent to the curves bounding the two bodies and $\bar{n}$ is perpendicular to this tangent vector. Vectors $\underline{r}_P^k$ and $\underline{r}_P^\ell$ are the relative position vectors of point **P** with respect to points $\mathbf{O}^k$ and $\mathbf{O}^\ell$, respectively. *(1)* Is $\dot{\theta}^k$ the angular velocity of body k? *(2)* Let $\underline{v}_P^k$ and $\underline{v}_P^\ell$ be the velocity vectors of the material points of body k and ℓ, respectively, that are coincident with the instantaneous point of contact, **P**. An analyst has evaluated these vectors, which are shown in fig. 5.20. Are his predictions correct? *(3)* Is the relative velocity of body ℓ with respect to body k parallel to unit vector $\bar{n}$? *(4)* If $\|\underline{v}_P^\ell\| = 4.5$ m/s in the upwards direction, determine $\|\underline{v}_P^k\|$. *(5)* Find the relative velocity vector of body ℓ with respect to body k. Justify all your answers; YES/NO answers are not sufficient.

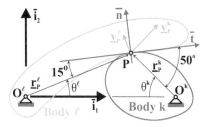

Fig. 5.20. Configuration of the cam-follower system.

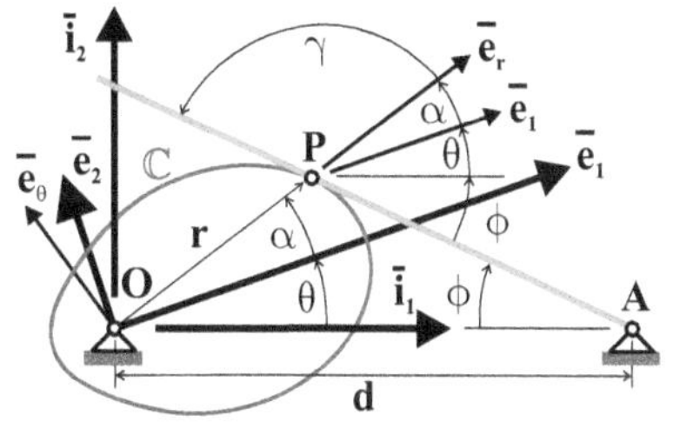

Fig. 5.21. Configuration of the cam-follower system.

Problem 5.23. Cam-rocking bar pair
Consider the planar cam-rocking bar pair depicted in fig. 5.21. The cam rotates at a constant angular velocity, Ω, about fixed inertial point **O**. Frame $\mathcal{F}^I = [\mathbf{O}, \mathcal{I} = (\bar{\imath}_1, \bar{\imath}_2)]$ is inertial and frame $\mathcal{F}^E = [\mathbf{O}, \mathcal{E} = (\bar{e}_1, \bar{e}_2)]$ is attached to the cam. The external shape of the cam is defined by curve $\mathbb{C}$ and the rocking bar slides over this curve; the contact point between the cam and bar is denoted **P**. The bar is pivoted about point **A** and the distance between point **O** and **A** is denoted d. *(1)* Plot the curve defining the outer shape of the cam. *(2)* Plot the curvature of curve $\mathbb{C}$ as a function of $\alpha \in [0, 360]$ deg. *(3)* Find the location of contact point **P** as a function of the cam rotation angle θ. This step requires the numerical solution of a transcendental equation

for each value of angle θ. (4) On one graph, plot angles α, γ, and ϕ at the point of contact for $\theta \in [0, 360]$ deg. (5) On one graph, plot r, r', and r'' at the point of contact for $\theta \in [0, 360]$ deg. (6) Plot the non-dimensional angular velocity of the bar, $\dot{\phi}/\Omega$, for $\theta \in [0, 360]$ deg. (7) Plot the non-dimensional angular acceleration of the bar, $\ddot{\phi}/\Omega^2$, for $\theta \in [0, 360]$ deg. Use the following data: $r(\alpha) = r_0 + r_{1c} \cos \alpha + r_{2c} \cos 2\alpha$, $r_0 = 1$, $r_{1c} = 0.54$, and $r_{2c} = 0.18$.

Problem 5.24. Cam-follower pair

Consider the planar cam-follower pair depicted in fig. 5.19. The cam rotates at a constant angular velocity, Ω, about fixed inertial point **O**. Frame $\mathcal{F}^I = [\mathbf{O}, \mathcal{I} = (\bar{\imath}_1, \bar{\imath}_2)]$ is inertial and frame $\mathcal{F}^E = [\mathbf{O}, \mathcal{E} = (\bar{e}_1, \bar{e}_2)]$ is attached to the cam. The external shape of the cam is defined by curve $\mathbb{C}$ and the follower slides over this curve; the contact point between the cam and follower is denoted **P**. The motion of the follower is constrained to be along axis $\bar{\imath}_2$ and its displacement is denoted x. (1) Plot the curve defining the outer shape of the cam. (2) Plot the curvature of curve $\mathbb{C}$ as a function of $\alpha \in [0, 360]$ deg. (3) Find the location of contact point **P** as a function of the cam rotation angle θ. This step requires the numerical solution of a transcendental equation for each value of angle θ. (4) On one graph, plot angles α and γ at the point of contact versus $\theta \in [0, 360]$ deg. (5) On one graph, plot r, r', and r'' at the point of contact versus θ. (6) Plot the non-dimensional velocity of the follower, $\dot{x}/(\Omega r_0)$, versus θ. (7) Plot the non-dimensional acceleration of the follower, $\ddot{x}/(\Omega^2 r_0)$, versus θ. Use the following data: $r(\alpha) = r_0 + r_{1c} \cos \alpha + r_{2c} \cos 2\alpha$, $r_0 = 1$, $r_{1c} = 0.5$, and $r_{2c} = 0.18$.

Problem 5.25. Cam-push rod pair

Figure 5.22 depicts a planar cam-push rod pair. The cam rotates at a constant angular velocity, Ω, about fixed inertial point **O**. Frame $\mathcal{F}^I = [\mathbf{O}, \mathcal{I} = (\bar{\imath}_1, \bar{\imath}_2)]$ is inertial and frame $\mathcal{F}^E = [\mathbf{O}, \mathcal{E} = (\bar{e}_1, \bar{e}_2)]$ is attached to the cam. The external shape of the cam is defined by curve $\mathbb{C}$ and the push rod slides over this curve; the contact point between the cam and push rod is denoted **P**. The push rod's axis is at a distance d from axis $\bar{\imath}_2$ and its support at a distance h from axis $\bar{\imath}_1$. (1) Plot the curve defining the outer shape of the cam. (2) Plot the curvature of curve $\mathbb{C}$ as a function of $\alpha \in [0, 360]$ deg. (3) Find the location of contact point **P** as a function of the cam rotation angle θ. This step requires the numerical solution of a transcendental equation for each value of angle θ. (4) On one graph, plot angles α, β, and γ at the point of contact versus $\theta \in [0, 360]$ deg. (5) On one graph, plot r, r', and r'' at the point of contact versus θ. (6) Plot the non-dimensional position of the push rod, x/r_0, versus θ. (7) Plot the non-dimensional angular velocity of the push rod, $\dot{x}/(\Omega r_0)$, versus θ. (8) Plot the non-dimensional angular acceleration of the push rod, $\ddot{x}/(\Omega^2 r_0)$, versus θ. Use the following data: $r(\alpha) = r_0 + r_{1c} \cos \alpha + r_{2c} \cos 2\alpha$, $r_0 = 1$, $r_{1c} = 0.50$, $r_{2c} = 0.18$, $\bar{d} = d/r_0 = 0.5$, $\bar{h} = h/r_0 = 1.8$.

Problem 5.26. Oscillating disk with sliding bar

Figure 5.23 shows an oscillating disk (body ℓ) pinned to the ground at its center point **C**, while a bar of length L, pinned at point **O**, slides in a radial track of the disk. The angular motion of the disk is prescribed as $\phi = \phi_0 \sin \omega t$. The distance between points **O** and **C** is denoted b. The angular position of the bar (body k) is denoted θ and the point of contact between the bar and the track is at a distance r from the center of the disk. (1) On one graph, plot angles ϕ and θ versus τ. (2) Plot $\bar{r} = r/L$ versus τ. (3) On one graph, plot angular velocities ϕ' and θ' versus τ. (4) Plot $\bar{r}'$ versus τ. (5) On one graph, plot angular accelerations ϕ'' and θ'' versus τ. (6) Plot $\bar{r}''$ versus τ. (7) Evaluate the velocities of the material points of body k and ℓ, denoted $\underline{v}_P^k$ and $\underline{v}_P^\ell$, respectively, that are instantaneously located at the point of contact of the two bodies. (8) Verify that eq. (5.29) is satisfied for your solution. (9) Evaluate the relative velocity of body ℓ with respect to body k, denoted $\underline{v}_P^r$. (10) Let $\underline{v}_P$ be the velocity vector of

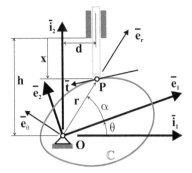

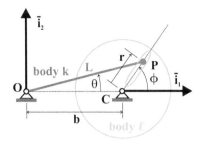

Fig. 5.22. Configuration of the cam-push rod pair.

Fig. 5.23. Configuration of the oscillating disk with sliding bar.

the material point at the tip of the bar. On one graph, plot the inertial components of $\underline{v}_P^r/(\omega L)$ and $\underline{v}_P/(\omega L)$. Use the following data: $\bar{b} = b/L = 0.75$, $\phi_0 = \pi/3$. Use non-dimensional time $\tau = \omega t$ and notation $(\cdot)'$ indicates a derivative with respect to τ. Present all the results for $\tau \in [0, 2\pi]$.

Problem 5.27. Piston with track and pin

Figure 5.24 depicts an oscillating piston with a track along which a vertical pin is sliding. The motion of the piston is prescribed as $x = x_0(1 - \cos \omega t)/2$. The shape of the track is defined by an arbitrary parameterization, $\underline{p}_0(\eta)$. *(1)* On one graph, plot $\bar{x} = x/L$ and $\bar{u} = u/L$ versus τ. *(2)* Plot η versus τ. *(3)* On one graph, plot $\bar{x}'$ and $\bar{u}'$ versus τ. *(4)* Plot η' versus τ. *(5)* On one graph, plot $\bar{x}''$ and $\bar{u}''$ versus τ. *(6)* Plot η'' versus τ. *(7)* Evaluate the relative velocity vector of body ℓ with respect to body k. *(8)* Evaluate the tangential and normal components of this relative velocity vector. *(9)* Plot the tangential components of the relative velocity vector versus τ. Use the following data: $\bar{a} = a/L = 1.5$, $\bar{b} = b/L = 3$, $\bar{x}_0 = x_0/L = 1$, $\bar{h} = h/L = 1$, and $\underline{p}_0(\eta) = a\eta\bar{e}_1 + b\eta^2\bar{e}_2$. Use non-dimensional time $\tau = \omega t$ and notation $(\cdot)'$ indicates a derivative with respect to τ. Present all the results for $\tau \in [0, 2\pi]$.

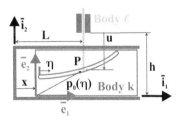

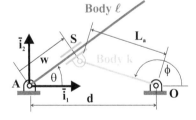

Fig. 5.24. Configuration of sliding piston with track and pin.

Fig. 5.25. Configuration of the two-bar mechanism.

Problem 5.28. Two-bar mechanism

Figure 5.25 shows a planar, two-bar mechanism. Crank **OS** is of length L_a and rotates at a constant angular velocity, Ω. Its tip slides along bar **AS**. The distance between points **A** and

O is denoted d and that between point **A** and **S** is denoted w. *(1)* On one graph, plot angles ϕ and θ versus τ. *(2)* Plot $\bar{w} = w/d$ versus τ. *(3)* On one graph, plot angular velocities ϕ' and θ' versus τ. *(4)* Plot $\bar{w}'$ versus τ. *(5)* On one graph, plot angular accelerations ϕ'' and θ'' versus τ. *(6)* Plot $\bar{w}''$ versus τ. *(7)* Evaluate the relative velocity vector of body ℓ with respect to body k. *(8)* Evaluate the tangential and normal components of this relative velocity vector. *(9)* On one graph, plot the magnitude of the relative velocity vector and that of the slider. Use the following data: $\bar{L}_a = L_a/d = 0.8$. Use non-dimensional time $\tau = \Omega t$ and notation $(\cdot)'$ indicates a derivative with respect to τ. Present all the results for $\tau \in [0, 2\pi]$.

Problem 5.29. Disk-follower mechanism

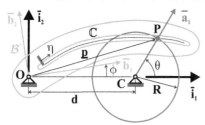

Figure 5.26 depicts a disk-follower mechanism. The disk of radius R rotates at a constant angular velocity, $\dot{\theta} = \Omega$, about point **C**. A pin is located at the rim of the disk. The slotted follower is hinged at point **O** and the pin slides inside the slot. Frame $\mathcal{F}^F = [\mathbf{O}, \mathcal{B}^+ = (\bar{b}_1, \bar{b}_2, \bar{b}_3)]$ is attached to the follower and the distance from point **O** to **C** is denoted d. The shape of the slot is defined by curve $\mathbb{C}$ and the position vector of a point on this curve with respect to point **O**, re-

Fig. 5.26. The disk-follower mechanism.

solved in basis $\mathcal{B}^+$, is denoted $\underline{p}^+ = (x_0+x_1\eta+x_2\eta^2+x_3\eta^3)\bar{b}_1+(y_0+y_1\eta+y_2\eta^2+y_3\eta^3)\bar{b}_2$, where η is an arbitrary parameterization of the curve. *(1)* Plot the shape of curve $\mathbb{C}$. *(2)* Plot angle ϕ versus τ. *(3)* Plot parameter η. *(4)* Plot angular velocity ϕ'. *(5)* Plot η'. *(6)* Plot angular acceleration ϕ''. *(7)* Plot η''. *(8)* Show that eq. (5.30) holds for your solution. *(8)* Plot the magnitude of the relative velocity vector. Use non-dimensional time $\tau = \Omega t$; notation $(\cdot)'$ denotes a derivative with respect to τ. Use the following data: $R = 1.2$, $d = 1.8$, $x_0 = 0$, $x_1 = 1$, $x_2 = 0$, $x_3 = 0.5$, $y_0 = y_1 = 0$, $y_2 = -1.4$, and $y_3 = 1$ m.

Problem 5.30. Geneva wheel mechanism

Figure 5.27 depicts the Geneva wheel mechanism, which consists of a disk and slotted arm. The disk of radius R rotates about inertial point **O** at a constant angular velocity, $\dot{\theta} = \Omega$. A pin is located at the rim of the disk at point **P**. The slotted arm is hinged at point **A** and the pin slides inside the slot. The distance from point **A** to the pin is denoted w. *(1)* On one graph, plot angle ϕ versus θ for one revolution of the disk. *(2)* Plot distance $\bar{w} = w/R$. *(3)* Plot angular velocity ϕ'. *(4)* Plot $\bar{w}'$. *(5)* Plot angular acceleration ϕ''. *(6)* Plot $\bar{w}''$. *(7)* Show that eq. (5.30) holds for your solution. Use non-dimensional time $\tau = \Omega t$; notation $(\cdot)'$ denotes a derivative with respect to τ. Use the following data: $\bar{L} = L/R = 1.5$.

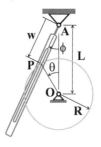

Fig. 5.27. Geneva wheel mechanism.

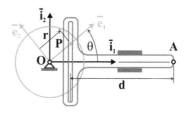

Fig. 5.28. Scotch yoke mechanism.

Problem 5.31. Scotch yoke mechanism

Figure 5.28 depicts the Scotch yoke mechanism, which consists of a disk and slotted yoke. The disk rotates about inertial point **O** at a constant angular velocity, $\dot\theta = \Omega$. A pin is located at a distance r from the center of the disk. The slotted yoke is allowed to move horizontally and the pin slides inside the slot. *(1)* Find the position of point **A** as a function of angle θ. *(2)* Find the velocity of point **A**. *(3)* Find the acceleration of point **A**. *(4)* Show that eq. (5.30) holds for your solution.

5.5 The motion tensor

In this section, the motion tensor is introduced as the tensor that relates the Plücker coordinates of a line of a rigid body in its initial and final configurations.

5.5.1 Transformation of a line of a rigid body

Figure 5.29 shows a rigid body in its reference configuration defined by frame $\mathcal{F}^I = [\mathbf{O}, \mathcal{I} = (\bar\imath_1, \bar\imath_2, \bar\imath_3)]$. Two points of this rigid body, denoted points **P** and **Q**, are defined by their position vectors with respect to point **O** given as $\underline{s}_P$ and $\underline{s}_Q$, respectively. In the final configuration, the rigid body is associated with frame $\mathcal{F} = [\mathbf{A}, \mathcal{B}^* = (\bar{b}_1, \bar{b}_2, \bar{b}_3)]$. Superscripts $(\cdot)^*$ indicate tensor components resolved in basis $\mathcal{B}^*$. The position vectors of material points **P** and **Q** with respect to point **A** are now $\underline{S}_P$ and $\underline{S}_Q$, respectively. Because points **P** and **Q** are material points of the rigid body, $\underline{S}_P = \underline{\underline{R}}\,\underline{S}_P^*$ and $\underline{S}_Q = \underline{\underline{R}}\,\underline{S}_Q^*$.

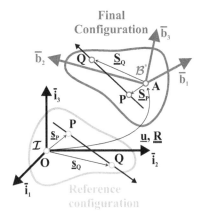

Fig. 5.29. A line of a rigid body in the reference and final configurations.

Consider now the line passing through these two points in the final configuration. Its orientation, resolved in basis $\mathcal{B}^*$, is $\bar\ell^* = (\underline{S}_Q^* - \underline{S}_P^*)/(\|\underline{S}_Q^* - \underline{S}_P^*\|)$. The Plücker coordinates of this line, eq. (1.38), evaluated with respect to point **A**, are

$$\underline{Q}^* = \left\{\begin{matrix}\widetilde{S}_P^*\,\bar\ell^* \\ \bar\ell^*\end{matrix}\right\} = \left\{\begin{matrix}k^* \\ \bar\ell^*\end{matrix}\right\}. \tag{5.33}$$

The Plücker coordinates of the same line with respect to point **O** will now be evaluated and resolved in basis $\mathcal{I}$. First, the orientation of the line is now

$$\bar\ell = \frac{(\underline{u} + \underline{S}_Q) - (\underline{u} + \underline{S}_P)}{\|(\underline{u} + \underline{S}_Q) - (\underline{u} + \underline{S}_P)\|} = \frac{\underline{S}_Q - \underline{S}_P}{\|\underline{S}_Q - \underline{S}_P\|} = \underline{\underline{R}}\frac{\underline{S}_Q^* - \underline{S}_P^*}{\|\underline{S}_Q^* - \underline{S}_P^*\|} = \underline{\underline{R}}\,\bar\ell^*.$$

Next, the Plücker coordinates of the same line become

$$\underline{\mathcal{Q}} = \left\{ \begin{matrix} (\tilde{u} + \tilde{S}_P)\bar{\ell} \\ \bar{\ell} \end{matrix} \right\} = \left\{ \begin{matrix} \tilde{u}\underline{R}\bar{\ell}^* + \underline{R}\tilde{S}_P^* \underline{R}^T \underline{R} \bar{\ell}^* \\ \underline{R}\bar{\ell}^* \end{matrix} \right\} = \begin{bmatrix} \underline{R} & \tilde{u}\underline{R} \\ 0 & \underline{R} \end{bmatrix} \left\{ \begin{matrix} \tilde{S}_P^* \bar{\ell}^* \\ \bar{\ell}^* \end{matrix} \right\}. \tag{5.34}$$

The *motion tensor* is defined as

$$\underline{\underline{\mathcal{C}}} = \begin{bmatrix} \underline{R} & \tilde{u}\underline{R} \\ 0 & \underline{R} \end{bmatrix}, \tag{5.35}$$

and eq. (5.34) can now be written in a compact form as

$$\underline{\mathcal{Q}} = \left\{ \begin{matrix} k \\ \bar{\ell} \end{matrix} \right\} = \underline{\underline{\mathcal{C}}} \, \underline{\mathcal{Q}}^* = \underline{\underline{\mathcal{C}}} \left\{ \begin{matrix} k^* \\ \bar{\ell}^* \end{matrix} \right\}. \tag{5.36}$$

Clearly, the motion tensor relates the Plücker coordinates of an arbitrary line of the rigid body resolved in two frames. This change of frame operation is more complex than the change in basis operation discussed in section 4.8: it involve both a change of basis and a change of reference point [1, 23, 22]. Equation (5.36) can be written in a more explicit manner as

$$\underline{\mathcal{Q}}^{[\mathcal{F}^I]} = \underline{\underline{\mathcal{C}}}^{[\mathcal{F}^I]} \underline{\mathcal{Q}}^{[\mathcal{F}]}.$$

In this form, the present change of frame operation mirrors the change of basis operation expressed by eq. (4.27).

5.5.2 Properties of the motion tensor

The motion tensor can be factorized in the following manner

$$\underline{\underline{\mathcal{C}}} = \begin{bmatrix} \underline{I} & \tilde{u} \\ 0 & \underline{I} \end{bmatrix} \begin{bmatrix} \underline{R} & 0 \\ 0 & \underline{R} \end{bmatrix} = \underline{\underline{\mathcal{T}}} \, \underline{\underline{\mathcal{R}}}, \tag{5.37}$$

where $\underline{\underline{\mathcal{R}}}$ is the *rotation tensor* and $\underline{\underline{\mathcal{T}}}$ the *translation tensor*. The eigenvalues of the motion tensor are now easily computed. Indeed, $\det(\underline{\underline{\mathcal{C}}}) = \det(\underline{\underline{\mathcal{T}}}) \det(\underline{\underline{\mathcal{R}}}) = \det(\underline{\underline{\mathcal{T}}}) \det^2(\underline{R})$ and because $\det(\underline{\underline{\mathcal{T}}}) = 1$, $\det(\underline{\underline{\mathcal{C}}}) = \det^2(\underline{R})$. Hence, the eigenvalues of the motion tensor are identical to those of the rotation tensor, but each with a multiplicity of two. The motion tensor, however, unlike the rotation tensor, is not an orthogonal tensor.

Two linearly independent eigenvectors of the motion tensor associated with its unit eigenvalues are found to be

$$\underline{\mathcal{N}}_1^\dagger = \left\{ \begin{matrix} \bar{n} \\ 0 \end{matrix} \right\}, \quad \text{and} \quad \underline{\mathcal{N}}_2^\dagger = \left\{ \begin{matrix} \dfrac{\underline{G}^T u}{2 \sin \phi/2} \\ \bar{n} \end{matrix} \right\}. \tag{5.38}$$

The fact that $\underline{\mathcal{N}}_1^\dagger$ is an eigenvector of the motion tensor stems from the corresponding property for the rotation tensor, $\underline{\underline{R}} \bar{n} = \bar{n}$. It is readily verified that $\underline{\mathcal{N}}_2^\dagger$

is also an eigenvector of the motion tensor, indeed, $\underline{\underline{R}} \, \underline{\underline{G}}^T \underline{u} / (2 \sin \phi / 2) + \tilde{u} \underline{\underline{R}} \, \bar{n} = (\underline{\underline{G}} - 2\tilde{n} \sin \phi / 2) \underline{u} / (2 \sin \phi / 2) = \underline{\underline{G}}^T \underline{u} / (2 \sin \phi / 2)$.

Any linear combination of eigenvectors $\underline{\mathcal{N}}_1^{\dagger}$ and $\underline{\mathcal{N}}_2^{\dagger}$ is still an eigenvector of the motion tensor. Consequently, the family of eigenvectors associated with the unit eigenvalue is expressed as follows

$$\underline{\mathcal{N}} = \left\{ \begin{array}{c} \underline{m} \\ \bar{n} \end{array} \right\} = \frac{(\alpha - 1)d}{2 \sin \phi / 2} \underline{\mathcal{N}}_1^{\dagger} + \underline{\mathcal{N}}_2^{\dagger}, \tag{5.39}$$

where α is an arbitrary scalar and d the intrinsic displacement of the rigid body. The displacement related part of the eigenvector is

$$\underline{m} = \frac{\underline{\underline{G}}^T \underline{u}}{2 \sin \phi / 2} + \frac{(\alpha - 1)d}{2 \sin \phi / 2} \bar{n}. \tag{5.40}$$

The scalar product of the two vectors forming the eigenvector is closely related to the intrinsic displacement of the rigid body

$$\lambda = \bar{n}^T \underline{m} = \frac{\alpha d}{2 \sin \phi / 2}. \tag{5.41}$$

5.5.3 Mozzi-Chasles' axis

In general, an arbitrary line of a rigid body is different in the reference and final configurations. The following question can then be asked: is it possible to find a line of the rigid body that is identical in the reference and final configurations? If such line exists, its Plücker coordinates in the reference and final configurations are identical, *i.e.*, $\underline{\mathcal{Q}} = \underline{\mathcal{Q}}^*$, or, using eq. (5.36), $\underline{\mathcal{Q}} = \underline{\underline{C}} \, \underline{\mathcal{Q}}$.

This implies that the Plücker coordinates of this line must form an eigenvector of the motion tensor, as given by eq. (5.39). Because the first three components of the Plücker coordinates of a line must be orthogonal to the last three, eq. (5.41) implies $\lambda = \alpha = 0$, and hence,

$$\underline{\mathcal{Q}}_{MC} = \underline{\mathcal{N}}_2^{\dagger} - \frac{d}{2 \sin \phi / 2} \underline{\mathcal{N}}_1^{\dagger} = \left\{ \begin{array}{c} -\dfrac{\underline{\underline{G}}^T \tilde{n} \tilde{n}}{2 \sin \phi / 2} \underline{u} \\ \bar{n} \end{array} \right\}. \tag{5.42}$$

In summary, the Plücker coordinates of the line of the rigid body that is identical in the reference and final configurations are given by eq. (5.42). These coordinates are those of Mozzi-Chasles' axis, see eq. (5.9). Hence, Mozzi-Chasles' axis is the line of the rigid body that is identical in the reference and final configurations. This can be written as $\underline{\mathcal{Q}}_{MC} = \underline{\underline{C}} \, \underline{\mathcal{Q}}_{MC}$: Mozzi-Chasles' axis is an eigenvector of the motion tensor corresponding to a unit eigenvalue.

5.5.4 Intrinsic expression of the motion tensor

The motion tensor was defined by eq. (5.35), which is not an intrinsic expression because the displacement vector of the reference point of the rigid body, $\underline{u}$, explicitly appears in this definition. In this section, an intrinsic expression of the motion tensor is sought, *i.e.*, an expression in which vector $\underline{u}$ does not appear explicitly.

Rodrigues' rotation formula, eq. (4.15), provides an intrinsic equation for the rotation tensor in terms of $\bar{n}$, the eigenvector of the rotation tensor associated with its unit eigenvalue, and ϕ, the magnitude of the rotation. A similar approach is followed here for the motion tensor, which should be expressed in terms of $\underline{\mathcal{N}}$, the eigenvector of the motion tensor associated with its unit eigenvalue, ϕ, the magnitude of the rotation, and d, the intrinsic displacement of the rigid body.

The motion tensor, eq. (5.35), is composed of two sub-matrices: the rotation tensor, repeated twice along the diagonal, and tensor $\widetilde{u}\underline{R}$, appearing as an off-diagonal term. The intrinsic expression of the rotation tensor is provided by Rodrigues' rotation formula, eq. (4.15). In contrast, the term $\widetilde{u}\underline{R}$ is not intrinsic because the displacement vector of the reference point, $\underline{u}$, appear explicitly.

Using the definition of the intrinsic displacement of the rigid body, eq. (5.7), the displacement vector is related to the eigenvector of the motion tensor, with the help of eq. (5.40) to find $\underline{m} = \left[\underline{G}^T \underline{u} + (\alpha - 1)\bar{n}\bar{n}^T \underline{u} \right] / (2 \sin \phi/2)$. Introducing the expression for the half-angle rotation tensor, eq. (4.20), then yields

$$\underline{m} = \underline{\underline{E}}\,\underline{u}, \tag{5.43}$$

where second-order tensor $\underline{\underline{E}}$ is defined as

$$\underline{\underline{E}} = \frac{\alpha}{2 \sin \phi/2}\underline{\underline{I}} - \frac{1}{2}\widetilde{n} + \left(\frac{\alpha}{2 \sin \phi/2} - \frac{1}{2 \tan \phi/2} \right)\widetilde{n}\widetilde{n}. \tag{5.44}$$

It now becomes possible to express the displacement vector in terms of the first part of the eigenvector of the motion tensor as

$$\underline{u} = \underline{\underline{J}}\,\underline{m}, \tag{5.45}$$

where tensor $\underline{\underline{J}} = \underline{\underline{E}}^{-1}$ is easily found as

$$\underline{\underline{J}} = \frac{2 \sin \phi/2}{\alpha}\underline{\underline{I}} + (1 - \cos \phi)\widetilde{n} + \left(\frac{2 \sin \phi/2}{\alpha} - \sin \phi \right)\widetilde{n}\widetilde{n}. \tag{5.46}$$

Equation (5.45) now yields an explicit expression of the displacement of the body's reference point

$$\widetilde{u} = \widetilde{\underline{J}\,\underline{m}} = \sin \phi\,\widetilde{m} + d(1 - \alpha \cos \frac{\phi}{2})\widetilde{n} + (1 - \cos \phi)(\widetilde{n}\widetilde{m} - \widetilde{m}\widetilde{n}). \tag{5.47}$$

Finally, tedious algebra reveals the following result,

$$\widetilde{u}\underline{R} = \widetilde{\underline{J}\,\underline{m}}\,\underline{R} = \sin \phi\,\widetilde{m} + dc_1\widetilde{n} + (1 - \cos \phi)\left(\widetilde{n}\widetilde{m} + \widetilde{m}\widetilde{n} \right) + dc_2\widetilde{n}\widetilde{n}, \tag{5.48}$$

where coefficients c_1 and c_2 are defined as

$$c_1 = \cos\phi - \alpha\cos\phi/2, \tag{5.49a}$$

$$c_2 = \sin\phi - 2\alpha\sin\phi/2. \tag{5.49b}$$

Combining Rodrigues' rotation formula, eq. (4.15), and eq. (5.48), the motion tensor, eq. (5.35), becomes

$$\underline{\underline{C}} = \underline{\underline{I}} + \begin{bmatrix} \sin\phi\,\underline{\underline{I}} & dc_1\underline{\underline{I}} \\ \underline{\underline{0}} & \sin\phi\,\underline{\underline{I}} \end{bmatrix}\begin{bmatrix} \widetilde{n} & \widetilde{m} \\ \underline{0} & \widetilde{n} \end{bmatrix} + \begin{bmatrix} (1-\cos\phi)\,\underline{\underline{I}} & dc_2\underline{\underline{I}} \\ \underline{\underline{0}} & (1-\cos\phi)\,\underline{\underline{I}} \end{bmatrix}\begin{bmatrix} \widetilde{n} & \widetilde{m} \\ \underline{0} & \widetilde{n} \end{bmatrix}\begin{bmatrix} \widetilde{n} & \widetilde{m} \\ \underline{0} & \widetilde{n} \end{bmatrix}. \tag{5.50}$$

To simplify the writing of this seemingly complicated expression, the following notation is introduced. First, tensor $\underline{\underline{Z}}$, a function of two scalars, α and β, is introduced

$$\underline{\underline{Z}}(\alpha,\beta) = \begin{bmatrix} \beta\underline{\underline{I}} & \alpha\underline{\underline{I}} \\ \underline{\underline{0}} & \beta\underline{\underline{I}} \end{bmatrix}. \tag{5.51}$$

Second, the *generalized vector product tensor* is defined

$$\widetilde{\mathcal{N}} = \begin{bmatrix} \widetilde{n} & \widetilde{m} \\ \underline{0} & \widetilde{n} \end{bmatrix}. \tag{5.52}$$

Notation $\widetilde{\mathcal{N}}$ does not indicate a 6×6 skew-symmetric tensor, but rather the above 6×6 tensor formed by three skew-symmetric sub-tensors.

Introducing these various notations into eq. (5.50) yields the desired intrinsic expression of the motion tensor and of its inverse

$$\underline{\underline{C}}(\underline{\mathcal{N}}) = \underline{\underline{I}} + \underline{\underline{Z}}(dc_1, \sin\phi)\widetilde{\mathcal{N}} + \underline{\underline{Z}}(dc_2, 1 - \cos\phi)\widetilde{\mathcal{N}}\widetilde{\mathcal{N}}, \tag{5.53a}$$

$$\underline{\underline{C}}^{-1}(\underline{\mathcal{N}}) = \underline{\underline{I}} - \underline{\underline{Z}}(dc_1, \sin\phi)\widetilde{\mathcal{N}} + \underline{\underline{Z}}(dc_2, 1 - \cos\phi)\widetilde{\mathcal{N}}\widetilde{\mathcal{N}}. \tag{5.53b}$$

The parallel between this intrinsic expression for the motion tensor and that for the rotation tensor given by Rodrigues' rotation formula, eq. (4.15), is striking. Clearly, the skew-symmetric tensor, $\widetilde{n}$, appearing in the expression for the rotation tensor is replaced by the generalized vector product tensor, $\widetilde{\mathcal{N}}$, appearing in that for the motion tensor. The two scalars, $\sin\phi$ and $(1-\cos\phi)$, appearing in the expression for the rotation tensor becomes the second arguments of tensor $\underline{\underline{Z}}$ appearing in that for the motion tensor.

Rodrigues' rotation formula, eq. (4.15), provides an intrinsic expression for the rotation tensor and is a direct consequence of Euler's theorem on rotations, theorem 4.1. Similarly, the intrinsic expression for the motion tensor is a direct consequence of the Mozzi-Chasles theorem 5.1. The parallel between the rotation and motion tensors will be further explored in section 5.6.3.

5.5.5 Properties of the generalized vector product tensor

The generalized vector product tensor defined by eq. (5.52) enjoys remarkable properties that generalize those of the skew-symmetric tensor. First, the skew-symmetric operator, $\widetilde{n}$, possesses a null eigenvalue, $\widetilde{n}\bar{n} = 0\bar{n}$. Similarly, the generalized vector product tensor also possesses a null eigenvalue, $\widetilde{\mathcal{N}}\underline{\mathcal{N}} = 0\underline{\mathcal{N}}$.

The second property of the generalized vector product tensor generalizes the behavior of the skew-symmetric tensor under a change of basis operation, eq. (4.30). Consider the following triple matrix product

$$\begin{bmatrix} \widetilde{n}_3 & \widetilde{m}_3 \\ \underline{0} & \widetilde{n}_3 \end{bmatrix} = \begin{bmatrix} \underline{R}_2^T & \underline{R}_2^T \widetilde{u}_2^T \\ \underline{0} & \underline{R}_2^T \end{bmatrix} \begin{bmatrix} \widetilde{n}_1 & \widetilde{m}_1 \\ \underline{0} & \widetilde{n}_1 \end{bmatrix} \begin{bmatrix} \underline{R}_2 & \widetilde{u}_2\underline{R}_2 \\ \underline{0} & \underline{R}_2 \end{bmatrix}.$$

This equality implies two conditions. The first condition is $\widetilde{n}_3 = \underline{R}_2^T \widetilde{n}_1 \underline{R}_2$, which, in view of eq. (4.30), implies $\bar{n}_3 = \underline{R}_2^T \bar{n}_1$. The second condition is $\widetilde{m}_3 = \underline{R}_2^T(\widetilde{m}_1 + \widetilde{n}_1\widetilde{u}_2 - \widetilde{u}_2\widetilde{n}_1)\underline{R}_2$, and tensor identities then lead to $\underline{m}_3 = \underline{R}_2^T(\underline{m}_1 + \widetilde{n}_1\underline{u}_2)$. These results can be summarized by the following equivalence,

$$\widetilde{\mathcal{N}}_3 = \underline{\mathcal{C}}^{-1}(\underline{\mathcal{N}}_2)\widetilde{\mathcal{N}}_1\underline{\mathcal{C}}(\underline{\mathcal{N}}_2) \iff \underline{\mathcal{N}}_3 = \underline{\mathcal{C}}^{-1}(\underline{\mathcal{N}}_2)\underline{\mathcal{N}}_1. \tag{5.54}$$

The third property of the generalized vector product tensor generalizes identity (1.34b), which holds for unit vectors and is rewritten here as $\widetilde{n}\widetilde{n}\widetilde{n} + \widetilde{n} = 0$.

$$\widetilde{\mathcal{N}}\widetilde{\mathcal{N}}\widetilde{\mathcal{N}} + \underline{\mathcal{Z}}(2\lambda, 1)\widetilde{\mathcal{N}} = \underline{0}. \tag{5.55}$$

The use of identities (1.34b) and (1.36) yields the above result, where $\lambda = \bar{n}^T \underline{m}$.

5.5.6 Change of frame operation for linear and angular velocities

Let the reference configuration of the rigid body shown in fig. 5.29 be the configuration of the body at time $t = 0$, and its final configuration is time-dependent. Consider now two vectors associated with the rigid body: the velocity vector of point **A**, denoted $\underline{v}_A$, a bound vector, see section 1.2, and the angular velocity vector of the body, denoted $\underline{\omega}_A$, a free vector, see section 1.1.

The components of these two vectors resolved in basis $\mathcal{B}^*$ are denoted $\underline{v}_A^*$ and $\underline{\omega}_A^*$, respectively, where the subscript on the latter symbol is, of course, superfluous because the angular velocity is identical for all points of the body. The following velocity vector is now defined

$$\underline{\mathcal{V}}^* = \begin{Bmatrix} \underline{v}_A^* \\ \underline{\omega}_A^* \end{Bmatrix}. \tag{5.56}$$

Strictly speaking, quantity $\underline{\mathcal{V}}^*$ should not be called a vector: it is, in fact, an array composed of two individual vectors, the linear and angular velocity vectors. The rules of transformation of first-order tensors, eq. (4.27), apply to these two vectors, but not to quantity $\underline{\mathcal{V}}^*$. It is convenient, however, to call quantity $\underline{\mathcal{V}}^*$ a vector to underline the tensorial nature of the two vectors it is composed of. Symbols in calligraphic type, such as $\underline{\mathcal{V}}^*$, are used to denote quantities composed of two vectors. For simplicity, these quantities will be referred to as vectors in the following.

In the previous section, the motion tensor was shown to transform the Plücker coordinates of a line from one frame to the other, and hence, it is interesting to consider the following transformation

$$\underline{\mathcal{V}} = \begin{Bmatrix} \underline{v}_O \\ \underline{\omega}_O \end{Bmatrix} = \underline{\underline{\mathcal{C}}}\,\underline{\mathcal{V}}^*. \tag{5.57}$$

To understand the physical meaning of this transformation, the physical interpretation of the velocity vectors $\underline{v}_O$ and $\underline{\omega}_O$ must be identified first. It is clear that $\underline{v}_A = \underline{\underline{R}}\,\underline{v}_A^*$ and $\underline{\omega}_A = \underline{\underline{R}}\,\underline{\omega}_A^*$ are the components of vectors $\underline{v}_A$ and $\underline{\omega}_A$, respectively, resolved in basis $\bar{\mathcal{I}}$. This corresponds to a *change of basis* operation, which establishes the relationship between the components of vectors in two bases.

Next, because $\underline{v}_O = \underline{\underline{R}}\,\underline{v}_A^* + \widetilde{\underline{u}}\,\underline{\underline{R}}\,\underline{\omega}_A^* = \underline{v}_A - \widetilde{\omega}_A\underline{u}$, velocity vector $\underline{v}_O$ is that of the point of the rigid body which instantaneously coincides with the origin of the reference frame, point **O**. Of course, $\underline{\omega}_O$ can also be interpreted as the angular velocity vector of the same point, because the angular velocity vector is the same for all points of a rigid body. Hence, this second operation corresponds to a *change of reference point* operation, which establishes the relationship between the velocities of two different points of the rigid body. In summary, the operation described by eq. (5.57) corresponds to a *change of frame* operation, which combines a change of basis operation and a change of reference point operation.

The factorized form of the motion tensor, eq. (5.37), clearly underlines the double effect of a frame change. It consists of two operations: first a change of basis operation characterized by the rotation operator, $\underline{\underline{R}}$, then a change of reference point operation characterized by the translation operator, $\underline{\underline{T}}$.

This change of frame operation can be inverted to yield

$$\underline{\mathcal{V}}^* = \underline{\underline{\mathcal{C}}}^{-1}\underline{\mathcal{V}}, \tag{5.58}$$

where the inverse of the motion tensor is

$$\underline{\underline{\mathcal{C}}}^{-1} = \underline{\underline{\mathcal{R}}}^{-1}\underline{\underline{\mathcal{T}}}^{-1} = \underline{\underline{\mathcal{R}}}^T\underline{\underline{\mathcal{T}}}^{-1} = \begin{bmatrix} \underline{\underline{R}}^T & \underline{\underline{0}} \\ \underline{\underline{0}} & \underline{\underline{R}}^T \end{bmatrix}\begin{bmatrix} \underline{\underline{I}} & \widetilde{\underline{u}}^T \\ \underline{\underline{0}} & \underline{\underline{I}} \end{bmatrix} = \begin{bmatrix} \underline{\underline{R}}^T & \underline{\underline{R}}^T\widetilde{\underline{u}}^T \\ \underline{\underline{0}} & \underline{\underline{R}}^T \end{bmatrix}. \tag{5.59}$$

An an intrinsic expression for this inverse is given by eq. (5.53b).

5.5.7 Change of frame operation for forces and moments

A similar study can be made concerning two other vectors associated with the rigid body: the force vector acting on the rigid body, $\underline{F}_A$, and the moment acting on the rigid body, $\underline{M}_A$, evaluated with respect to point **A**.

The components of these two vectors resolved in basis $\mathcal{B}^*$ are denoted $\underline{F}_A^*$ and $\underline{M}_A^*$, respectively, where the subscript on the former symbol is, of course, superfluous because the force vector can be applied at any point of the rigid body. The following applied load vector is defined

$$\underline{\mathcal{A}}^* = \begin{Bmatrix} \underline{F}_A^* \\ \underline{M}_A^* \end{Bmatrix}. \tag{5.60}$$

Consider now the effect of the following transformation

$$\underline{A} = \left\{ \begin{matrix} \underline{F}_O \\ \underline{M}_O \end{matrix} \right\} = \underline{\underline{C}}^{-T} \underline{A}^* = \begin{bmatrix} \underline{\underline{R}} & \underline{\underline{0}} \\ \tilde{u}\underline{\underline{R}} & \underline{\underline{R}} \end{bmatrix} \underline{A}^*. \tag{5.61}$$

Here again, it is clear that $\underline{F}_A = \underline{\underline{R}}\,\underline{F}_A^*$ and $\underline{M}_A = \underline{\underline{R}}\,\underline{M}_A^*$ are the components of the force and moment vectors, respectively, resolved in the inertial basis. This corresponds to a *change of basis* operation, which establishes the relationship between the components of vectors in two orthonormal bases.

Next, because $\underline{M}_O = \underline{\underline{R}}\,\underline{M}_A^* + \tilde{u}\underline{\underline{R}}\,\underline{F}_A^* = \underline{M}_A + \tilde{u}\underline{F}_A$, moment vector $\underline{M}_O$ is the applied moment computed with respect to the point of the rigid body that instantaneously coincides with point **O**. Of course, $\underline{F}_O$ can also be interpreted as the force applied on the rigid body at the same point, because this force is the same at all points of the rigid body.

In summary, the operation described by eq. (5.61) is a change of frame operation that combines a change of basis and a change of reference point. This change of frame operation can be inverted to yield

$$\underline{A}^* = \underline{\underline{C}}^T \underline{A}. \tag{5.62}$$

The components of velocity quantities and applied loads quantities transform differently under a frame change operation, as indicated in eqs. (5.57) and (5.61), respectively. Both transformations, however, are based on the motion tensor which appears to be a fundamental quantity associated with frame changes.

5.6 Derivatives of finite motion operations

The derivatives of finite rotation operations were discussed in section 4.10 and led to the concept of angular velocity vector. The present section focuses on the study of time derivatives of the motion tensor, which leads to both velocity and angular velocity vectors. Differential changes in motion are also investigated.

5.6.1 The velocity vector

The time-dependent motion of a rigid body is represented by the time-dependent motion of the body attached frame, $\mathcal{F} = \left[\mathbf{A}, \mathcal{B}^* = (\bar{b}_1, \bar{b}_3, \bar{b}_3) \right]$, depicted in fig. 5.29. Let $\underline{\underline{C}}$ be the motion tensor that brings reference frame $\mathcal{F}^I$ to frame $\mathcal{F}$, and eq. (5.36) then implies $\underline{\underline{Q}}(t) = \underline{\underline{C}}(t)\underline{\underline{Q}}^*$. Taking a time derivative of this equation leads to $\underline{\dot{\underline{Q}}} = \underline{\dot{\underline{C}}}\,\underline{\underline{Q}}^*$, and eliminating $\underline{\underline{Q}}^*$ then yields

$$\underline{\dot{\underline{Q}}} = \underline{\dot{\underline{C}}}\,\underline{\underline{C}}^{-1}\underline{\underline{Q}}. \tag{5.63}$$

Comparing this equation with eq. (4.56) reveals that expression $\underline{\dot{\underline{C}}}\,\underline{\underline{C}}^{-1}$, associated with the motion tensor, generalizes expression $\underline{\dot{\underline{R}}}\,\underline{\underline{R}}^T$, associated with the rotation tensor. The use of identity (1.33a) leads to

$$\underline{\dot{\underline{C}}}\,\underline{\underline{C}}^{-1} = \begin{bmatrix} \dot{\underline{R}} & \dot{\tilde{u}}\underline{R} + \tilde{u}\dot{\underline{R}} \\ \underline{0} & \dot{\underline{R}} \end{bmatrix} \begin{bmatrix} \underline{R}^T & \underline{R}^T \tilde{u}^T \\ \underline{0} & \underline{R}^T \end{bmatrix} = \begin{bmatrix} \tilde{\omega} & \widetilde{(\dot{u} + \tilde{u}\omega)} \\ \underline{0} & \tilde{\omega} \end{bmatrix} = \begin{bmatrix} \tilde{\omega} & \tilde{v} \\ \underline{0} & \tilde{\omega} \end{bmatrix}. \tag{5.64}$$

This expression gives rise to two quantities. First, the angular velocity of the rigid body emerges from the time derivative of the rotation tensor, $\omega = \text{axial}(\dot{\underline{R}}\,\underline{R}^T)$; as expected, this quantity is identical to that which arose for the study of time derivatives of time-dependent rotations, see section 4.10. Second, the velocity vector of the rigid body, $v = \dot{u} + \tilde{u}\omega$, also emerges from the time derivative of the motion tensor. This quantity can be interpreted as the linear velocity of the point of the rigid body that instantaneously coincides with the origin of the reference frame, point **O**, see section 5.5.6.

The velocity vector of the rigid body resolved in frame $\mathcal{F}^I$ is now defined as

$$\underline{V} = \left\{ \begin{array}{c} v \\ \omega \end{array} \right\}, \tag{5.65}$$

and eq. (5.63) becomes $\dot{\underline{Q}} = \tilde{\mathcal{V}}\underline{Q}$, where the generalized vector product tensor is given by eq. (5.52).

It is also possible to resolve the components of the velocity vector in the moving frame,

$$\underline{\underline{C}}^{-1}\dot{\underline{Q}} = \underline{\underline{C}}^{-1}\dot{\underline{\underline{C}}}\,\underline{Q}^*. \tag{5.66}$$

Comparing this equation with eq. (4.55) reveals that expression $\underline{\underline{C}}^{-1}\dot{\underline{\underline{C}}}$, associated with the motion tensor, generalizes expression $\underline{R}^T\dot{\underline{R}}$, associated with the rotation tensor. It is readily found that

$$\underline{\underline{C}}^{-1}\dot{\underline{\underline{C}}} = \begin{bmatrix} \underline{R}^T & \underline{R}^T \tilde{u}^T \\ \underline{0} & \underline{R}^T \end{bmatrix} \begin{bmatrix} \dot{\underline{R}} & \dot{\tilde{u}}\underline{R} + \tilde{u}\dot{\underline{R}} \\ \underline{0} & \dot{\underline{R}} \end{bmatrix} = \begin{bmatrix} \tilde{\omega}^* & \widetilde{\underline{R}^T \dot{u}} \\ \underline{0} & \tilde{\omega} \end{bmatrix} = \begin{bmatrix} \tilde{\omega}^* & \tilde{v}^* \\ \underline{0} & \tilde{\omega}^* \end{bmatrix}. \tag{5.67}$$

This expression gives rise to two quantities. First, the components of the angular velocity of the rigid body resolved in the rotating basis, $\omega^* = \text{axial}(\underline{R}^T\dot{\underline{R}})$. Second, the components of the velocity vector of the reference point of rigid body resolved in the rotating basis, $v^* = \underline{R}^T\dot{u}$.

The components of the velocity vector of the rigid body resolved in the material frame are now defined as

$$\underline{V}^* = \left\{ \begin{array}{c} v^* \\ \omega^* \end{array} \right\} = \underline{\underline{C}}^{-1}\underline{V}, \tag{5.68}$$

where the second equality follows from eq. (5.58). Equation (5.66) now becomes $\underline{\underline{C}}^{-1}\dot{\underline{Q}} = \tilde{\mathcal{V}}^*\underline{Q}^*$, where the generalized vector product operator is given by eq. (5.52).

The above developments are summarized in the following relationships

$$\dot{\underline{\underline{C}}}\,\underline{\underline{C}}^{-1} = \tilde{\mathcal{V}}, \qquad \underline{\underline{C}}\,\dot{\underline{\underline{C}}}^{-1} = -\tilde{\mathcal{V}}, \tag{5.69a}$$

$$\underline{\underline{C}}^{-1}\dot{\underline{\underline{C}}} = \tilde{\mathcal{V}}^*, \qquad \dot{\underline{\underline{C}}}^{-1}\underline{\underline{C}} = -\tilde{\mathcal{V}}^*. \tag{5.69b}$$

It is readily shown that

$$\tilde{\mathcal{V}}^* = \underline{\underline{C}}^{-1} \tilde{\mathcal{V}} \underline{\underline{C}}, \tag{5.70a}$$

$$\tilde{\mathcal{V}} = \underline{\underline{C}} \, \tilde{\mathcal{V}}^* \underline{\underline{C}}^{-1}, \tag{5.70b}$$

as can be expected from the transformation formulae for the velocity vectors, eqs. (5.57) and (5.58).

5.6.2 The differential motion vector

The concept of differential rotation vector was introduced in section 4.12.4 based on the rotation tensor, eq. (4.101). By analogy, the following expression is formed

$$d\underline{\underline{C}} \, \underline{\underline{C}}^{-1} = \begin{bmatrix} d\underline{R} \, \widetilde{du} \, \underline{R} + \tilde{u} \, d\underline{R} \\ \underline{0} & d\underline{R} \end{bmatrix} \begin{bmatrix} \underline{R}^T & \underline{R}^T \tilde{u}^T \\ \underline{0} & \underline{R}^T \end{bmatrix} = \begin{bmatrix} \widetilde{d\psi} & \widetilde{(du + \tilde{u} d\psi)} \\ \underline{0} & \widetilde{d\psi} \end{bmatrix} = \begin{bmatrix} \widetilde{d\psi} & \widetilde{du} \\ \underline{0} & \widetilde{d\psi} \end{bmatrix}.$$

This expression gives rise to two quantities. First, the differential rotation vector of the rigid body emerges from differential changes of the rotation tensor, $d\psi = \text{axial}(d\underline{R} \, \underline{R}^T)$; this quantity is identical to that defined by eq. (4.101). As discussed in section 4.12.4, no vector ψ exists such that $d(\psi)$ gives the differential rotation vector.

Second, the differential displacement vector of the rigid body, $\underline{du} = d\underline{u} + \tilde{u} \, d\psi$, also emerges from the differential of the motion tensor. $d\underline{u}$ is the differential displacement of point **A** and $\underline{du} = d\underline{u} + \tilde{u} \, d\psi$ the differential displacement of the material point of the rigid body that instantaneously coincides with point **O**. Of course, there exist no displacement vector, say $\underline{x}$, such $d(\underline{x}) = d\underline{u} + \tilde{u} \, d\psi$. Notations $\underline{du}$ and $d\psi$ will be used to denote the differential displacement and rotation vectors, respectively.

By analogy to eqs. (5.69a) and (5.69b), the following compact notation is adopted

$$d\underline{\underline{C}} \, \underline{\underline{C}}^{-1} = \widetilde{d\mathcal{U}}, \qquad \underline{\underline{C}} d\underline{\underline{C}}^{-1} = -\widetilde{d\mathcal{U}}, \tag{5.71a}$$

$$\underline{\underline{C}}^{-1} d\underline{\underline{C}} = \widetilde{d\mathcal{U}}^*, \qquad d\underline{\underline{C}}^{-1} \underline{\underline{C}} = -\widetilde{d\mathcal{U}}^*, \tag{5.71b}$$

where the components of the *differential motion vector* are defined as

$$d\mathcal{U} = \begin{Bmatrix} du \\ d\psi \end{Bmatrix} = \underline{\underline{C}} \, d\mathcal{U}^*, \tag{5.72a}$$

$$d\mathcal{U}^* = \begin{Bmatrix} du^* \\ d\psi^* \end{Bmatrix} = \underline{\underline{C}}^{-1} d\mathcal{U}, \tag{5.72b}$$

in the fixed and moving frames, respectively. The components of the differential rotation and displacement vectors, both resolved in the moving frame, are $d\psi^* = \text{axial}(\underline{R}^T d\underline{R})$ and $du^* = \underline{R}^T du$, respectively.

It is readily shown that

$$\widetilde{d\mathcal{U}}^* = \underline{\underline{C}}^{-1} \widetilde{d\mathcal{U}} \underline{\underline{C}}, \tag{5.73a}$$

$$\widetilde{d\mathcal{U}} = \underline{\underline{C}} \, \widetilde{d\mathcal{U}}^* \underline{\underline{C}}^{-1}. \tag{5.73b}$$

Taking a differential of eq. (5.69a) and a time derivative of eq. (5.71a) leads to $\mathrm{d}\widetilde{\mathcal{V}} = \mathrm{d}\dot{\underline{\mathcal{C}}}\,\underline{\mathcal{C}}^{-1} + \dot{\underline{\mathcal{C}}}\,\mathrm{d}\underline{\mathcal{C}}^{-1}$ and $\mathrm{d}\widetilde{\dot{\mathcal{U}}} = \mathrm{d}\dot{\underline{\mathcal{C}}}\,\underline{\mathcal{C}}^{-1} + \mathrm{d}\underline{\mathcal{C}}\,\dot{\underline{\mathcal{C}}}^{-1}$, respectively. Subtracting these two equations and using eqs. (5.69a) and (5.71a) then yields

$$\mathrm{d}\dot{\widetilde{\mathcal{V}}} - \mathrm{d}\dot{\widetilde{\mathcal{U}}} = -\widetilde{\mathcal{V}}\,\mathrm{d}\widetilde{\mathcal{U}} + \mathrm{d}\widetilde{\mathcal{U}}\,\widetilde{\mathcal{V}}.$$

Expanding these expressions and using identity (1.33a) then leads to this important result $\mathrm{d}\dot{\mathcal{V}} = \mathrm{d}\dot{\mathcal{U}} - \widetilde{\mathcal{V}}\mathrm{d}\mathcal{U}$, which relates differentials in the velocity vector to the differential motion vector and its time derivative. This equation generalizes eq. (4.102a) written for the sole angular velocity.

The following results are obtained in a similar manner

$$\mathrm{d}\underline{\mathcal{V}} = \mathrm{d}\dot{\underline{\mathcal{U}}} - \widetilde{\underline{\mathcal{V}}}\mathrm{d}\underline{\mathcal{U}}, \qquad \mathrm{d}\underline{\mathcal{V}} = \underline{\mathcal{C}}\,\mathrm{d}\dot{\underline{\mathcal{U}}}^{*}, \qquad (5.74a)$$

$$\mathrm{d}\underline{\mathcal{V}}^{*} = \mathrm{d}\dot{\underline{\mathcal{U}}}^{*} + \widetilde{\underline{\mathcal{V}}}^{*}\mathrm{d}\underline{\mathcal{U}}^{*}, \qquad \mathrm{d}\underline{\mathcal{V}}^{*} = \underline{\mathcal{C}}^{-1}\mathrm{d}\dot{\underline{\mathcal{U}}}. \qquad (5.74b)$$

5.6.3 Change of frame operations

Section 4.8.1 discussed change of basis operations. By definition 4.1, a vector, or first-order tensor, is a mathematical entity whose components resolved in two bases are related by eqs. (4.27). This definition applies equally to kinematic quantities such as displacement and rotation vectors, and load quantities such as force or moment vectors. For instance, the components of the velocity vector resolved in inertial and material bases, denoted $\underline{v}$ and $\underline{v}^{*}$, respectively, are such that $\underline{v} = \underline{R}\,\underline{v}^{*}$, if $\underline{R}$ are the components of the rotation tensor that brings the inertial to the material basis, resolved in the inertial basis. The components of the angular velocity vector resolved in the same bases are such that $\underline{\omega} = \underline{R}\,\underline{\omega}^{*}$. Similar relationships hold for the components of the force and moment vectors. In fact, according to definition 4.1, the components of all vectors follow the same transformation rule under a change of basis.

Section 5.5.6 presented the change of frame operation for the linear and angular velocity vectors. For instance, eq. (5.57) provides the relationship between the components of the linear and angular velocity vectors resolved in the inertial and material frames, denoted $\underline{\mathcal{V}}$ and $\underline{\mathcal{V}}^{*}$, respectively, as $\underline{\mathcal{V}} = \underline{\mathcal{C}}\,\underline{\mathcal{V}}^{*}$, if $\underline{\mathcal{C}}$ are the components of the motion tensor that brings the inertial frame to the material frame, resolved in the inertial frame. The change of frame transformation operates on the linear and angular velocity vectors *simultaneously*. The notational convention, $\underline{\mathcal{V}}^{T} = \{\underline{v}^{T}, \underline{\omega}^{T}\}$ and the use of a 6×6 motion tensor enable the simultaneous manipulation of the two vectors.

On the other hand, section 5.5.7 introduced the change of frame operation for forces and moments. For instance, eq. (5.61) provides the relationship between the components of the force and moment vectors resolved in the inertial and material frames, denoted $\underline{\mathcal{A}}$ and $\underline{\mathcal{A}}^{*}$, respectively, as $\underline{\mathcal{A}} = \underline{\mathcal{C}}^{-T}\underline{\mathcal{A}}^{*}$. The change of frame operation for kinematic quantities is based on the motion tensor, $\underline{\mathcal{C}}$, but the same change of frame operation for loads uses the transpose of its inverse, $\underline{\mathcal{C}}^{-T}$.

Energetically conjugate quantities

To understand the crucial difference between change of basis and change of frame operations, consider the differential work done by the force and moment vectors, denoted $\underline{F}$ and $\underline{M}$, respectively, applied at point **A** of the rigid body,

$$dW = \underline{F}^T d\underline{u} + \underline{M}^T d\underline{\psi}, \tag{5.75}$$

where $d\underline{u}$ and $d\underline{\psi}$ are the differential displacement vector of the point of application of the force and the differential rotation of the rigid body, respectively, see fig. 5.29. The force and differential displacement vectors are said to be *energetically conjugate* quantities because their scalar product yields the differential work. Similarly, the moment and differential rotation vectors are also energetically conjugate quantities.

   Because the scalar product is a tensor operation, the differential work can also be expressed as $dW = \underline{F}^{*T} d\underline{u}^* + \underline{M}^{*T} d\underline{\psi}^*$, where $\underline{F}^* = \underline{\underline{R}}^T \underline{F}$ and $\underline{M}^* = \underline{\underline{R}}^T \underline{M}$ are the components of the force and moment vectors, resolved in the body attached basis, and $d\underline{u}^* = \underline{\underline{R}}^T d\underline{u}$ and $d\underline{\psi}^* = \underline{\underline{R}}^T d\underline{\psi}$ the components of the differential displacement and rotation vectors resolved in the same basis. Energetically conjugate quantities, such as the moment and differential rotation vectors, follow the same rules of transformation under a change of basis.

   The following compact notation is introduced

$$dW = \underline{F}^{*T} d\underline{u}^* + \underline{M}^{*T} d\underline{\psi}^* = \underline{A}^{*T} d\underline{U}^*, \tag{5.76}$$

where $\underline{A}^{*T} = \left\{ \underline{F}^{*T}, \underline{M}^{*T} \right\}$ is the applied loading vector and $d\underline{U}^*$ the differential motion vector defined by eq. (5.72b). These two quantities, $\underline{A}^*$ and $d\underline{U}^*$, are energetically conjugate because their scalar product yields the differential work.

   To explore the effect of a change of frame, the following transformation is performed,

$$dW = \underline{A}^{*T} d\underline{U}^* = \underline{A}^{*T} \underline{\underline{C}}^{-1} \underline{\underline{C}} d\underline{U}^* = \underline{A}^T d\underline{U}, \tag{5.77}$$

where $\underline{A} = \underline{\underline{C}}^{-T} \underline{A}^*$, as expected from eq. (5.61), and $d\underline{U} = \underline{\underline{C}} d\underline{U}^*$, as expected from eq. (5.72a). Under a change of frame, the rules of transformations for energetically conjugate quantities differ. This difference stems from the fact that the motion tensor is not an orthogonal tensor, $\underline{\underline{C}}^{-1} \neq \underline{\underline{C}}^T$. In contrast, the rotation tensor is orthogonal, $\underline{\underline{R}}^{-1} = \underline{\underline{R}}^T$ and consequently, the rules of transformations for energetically conjugate quantities are identical for change of basis operations.

Generalization of the concept of tensor analysis

Section 4.8.4 introduced the concept of tensor analysis. The combined use of tensor quantities and tensor operations leads to a formulation of the laws of physics that guarantees their invariance with respect to change of basis operations. Intuitively, the laws of physics should be invariant under a change of basis operation because this operation simply corresponds to the selection of a different basis in which all tensor components are resolved, but does not change the physical behavior of the system.

Intuitively, the laws of physics should also be invariant with respect to a change in the location of the origin of the coordinate system, as long as this new origin is still an inertial point. Combining these two intuitive observations, the invariance of the laws of physics with respect to both basis and origin selection, leads to the natural conclusion that the laws of physics must be invariant with respect to a change of frame.

The generalization of the concept of tensor analysis to the invariance of the laws of physics to change of frame operations involves two parts, the use of generalized tensors quantities and of generalized tensor operations. These two concepts generalize the use of tensor quantities and tensor operations characteristics of tensor analysis, see sections 4.8.3 and 4.8.4.

When dealing with change of frame operation, linear and angular quantities becomes coupled. For instance, linear and angular velocity vectors are paired to form the generalized velocity vector defined by eq. (5.56), similarly, the force and moment vectors are paired to form the generalized loading vector defined by eq. (5.60). The generalized velocity vector is composed of two vectors, or first-order tensors

Consider now the change of frame operation expressed by eq. (5.57) and repeated here in more explicit details

$$
\left.\begin{aligned}
\underline{v}_O &= \underline{\underline{R}}\,\underline{v}_A^* + \widetilde{u}\underline{\underline{R}}\,\underline{\omega}_A^* \\
\underline{\omega}_O &= \phantom{\underline{\underline{R}}\,\underline{v}_A^* + \widetilde{u}} \underline{\underline{R}}\,\underline{\omega}_A^*
\end{aligned}\right\}
\iff
\underline{\mathcal{V}} = \begin{bmatrix} \underline{\underline{R}} & \widetilde{u}\underline{\underline{R}} \\ \underline{\underline{0}} & \underline{\underline{R}} \end{bmatrix} \underline{\mathcal{V}}^* = \underline{\underline{\mathcal{C}}}\,\underline{\mathcal{V}}^*.
\tag{5.78}
$$

The two equations on the left-hand side are basis invariant because they only involve tensor quantities and tensor operations; they satisfy all the rules of tensor analysis. Taken together, they express a change of frame operation, which is repeated on the right-hand side with a more compact notation. The generalized motion tensor, $\underline{\underline{\mathcal{C}}}$, of size 6×6, is composed of four sub-matrices, each of size 3×3, which are each second-order tensors.

Clearly, the right-hand side of eq. (5.78) generalizes the change of basis operation, $\underline{v} = \underline{\underline{R}}\,\underline{v}^*$, to the change of frame operation, $\underline{\mathcal{V}} = \underline{\underline{\mathcal{C}}}\,\underline{\mathcal{V}}^*$. In the following developments, quantities such as $\underline{\mathcal{V}}$ or $\underline{\mathcal{V}}^*$ will be called vectors or first-order tensors, and quantities such as the motion tensor will be called second-order tensors. This terminology is more convenient to use in place of the more awkward "generalized velocity vector" and "generalized motion tensor." Because generalized vectors and tensors are indicated in calligraphic type, their generalized nature is clearly implied.

Example 5.5. First- and second-order tensors

In the previous sections, first- and second-order tensors have already been encountered. The second property of the generalized vector product operator given by eq. (5.54) is repeated here for convenience

$$
\widetilde{\mathcal{N}}_3 = \underline{\underline{\mathcal{C}}}^{-1}(\underline{\mathcal{N}}_2)\widetilde{\mathcal{N}}_1\underline{\underline{\mathcal{C}}}(\underline{\mathcal{N}}_2) \iff \underline{\mathcal{N}}_3 = \underline{\underline{\mathcal{C}}}^{-1}(\underline{\mathcal{N}}_2)\underline{\mathcal{N}}_1.
$$

The right-hand side of this equivalence expresses the rules of transformation for the first-order tensor, $\underline{\mathcal{N}}$: its components in frames $\mathcal{F}_1$ and $\mathcal{F}_3$ are $\underline{\mathcal{N}}_1$ and $\underline{\mathcal{N}}_3$, respectively, and $\underline{\underline{C}}(\underline{N}_2)$ are the components of the motion tensor that brings frame $\mathcal{F}_1$ to $\mathcal{F}_3$, resolved in frame $\mathcal{F}_1$.

The left-hand side of this equivalence expresses the rules of transformation for the second-order tensor, $\widetilde{\mathcal{N}}$: its components in frames $\mathcal{F}_1$ and $\mathcal{F}_3$ are $\widetilde{\mathcal{N}}_1$ and $\widetilde{\mathcal{N}}_3$, respectively. This gives a formal proof that the generalized vector product operator defined by eq. (5.52) is in fact a second-order tensor. Clearly, these results generalize the corresponding results obtained for the skew-symmetric tensor in eq. (4.30).

Example 5.6. The motion tensor

The motion tensor was introduced in section 5.5.1 and was called a "tensor." Prove that the motion tensor is indeed a second-order tensor.

Consider the intrinsic expression of the motion tensor given by eq. (5.53a) as $\underline{\underline{S}}(\underline{N}) = \underline{\underline{I}} + \underline{\underline{Z}}(dc_1, \sin\phi)\widetilde{\mathcal{N}} + \underline{\underline{Z}}(dc_2, 1 - \cos\phi)\widetilde{\mathcal{N}}\widetilde{\mathcal{N}}$. The arguments of operator $\underline{\underline{Z}}$ are functions of two variables, rotation angle ϕ and the intrinsic displacement of the rigid body, d. Both quantities are zeroth-order tensor because they are unaffected by a change of frame operation.

Next, it is easily verified that operator $\underline{\underline{Z}}(\alpha, \beta)$, where α and β are zeroth-order tensors, is itself a zeroth-order tensor. Indeed,

$$\underline{\underline{Z}}(\alpha, \beta) = \underline{\underline{C}}^{-1}(\underline{N}_2)\underline{\underline{Z}}(\alpha, \beta)\underline{\underline{C}}(\underline{N}_2), \tag{5.79}$$

which implies the invariance of $\underline{\underline{Z}}(\alpha, \beta)$ under a change of frame operation,

$$\underline{\underline{C}}_2^{-1}\underline{\underline{S}}(\underline{N})\underline{\underline{C}}_2 = \underline{\underline{I}} + \underline{\underline{Z}}(dc_1, \sin\phi)\widetilde{\underline{\underline{C}}_2^{-1}\underline{N}} + \underline{\underline{Z}}(dc_2, 1 - \cos\phi)\widetilde{\underline{\underline{C}}_2^{-1}\underline{N}}\,\widetilde{\underline{\underline{C}}_2^{-1}\underline{N}}$$
$$= \underline{\underline{S}}(\underline{\underline{C}}_2^{-1}\underline{N}),$$

where the tensorial nature of the generalized vector product tensor, eq. (5.54), was taken into account. The tensorial nature of the motion tensor is now established.

A more formal expression of the tensorial nature of the motion tensor is

$$\underline{\underline{C}}(\underline{N}_3) = \underline{\underline{C}}^{-1}(\underline{N}_2)\underline{\underline{C}}(\underline{N}_1)\underline{\underline{C}}(\underline{N}_2) \iff \underline{N}_3 = \underline{\underline{C}}^{-1}(\underline{N}_2)\underline{N}_1. \tag{5.80}$$

This result for the motion tensor should be compared with the corresponding result for the rotation tensor, eq. (4.31).

6

Kinetics of rigid bodies

In section 3.4, the dynamic response of a system of particles subjected to both internally and externally applied loads is studied and leads to Euler's first and second laws [14, 15]. Rigid bodies can be viewed as systems of particles subjected to both internal and external forces. The former forces are those that maintain the shape of the rigid body. By definition, a rigid body is one for which the distance between any two of its particles remains constant at all times. The displacement field of the rigid body must satisfy the kinematic constraints developed in section 5.1 and its velocity field those described in section 5.2.

The configuration of the rigid body is defined by six parameters: the three coordinates describing the location of one of its points and three parameters describing its orientation. Similarly, the velocity field of the rigid body is determined by six parameters: the three components of the linear velocity vector of one of its points and the three components of its angular velocity vector.

Clearly, all the results derived in section 3.4 concerning the Newtonian mechanics of systems of particles are readily applicable to rigid bodies. In particular, the motion of the center of mass of the rigid body is governed by the following equation: $\underline{F} = \underline{\dot{P}}$, where $\underline{P} = m\underline{v}_C$ is the linear momentum of the body, m its total mass, $\underline{v}_C$ the velocity of its center of mass, and $\underline{F}$ the sum of all externally applied forces. Another vector equation that applies to systems of particles is $\underline{M}_C = \underline{\dot{H}}_C$, where $\underline{H}_C$ is the angular momentum vector of the rigid body and $\underline{M}_C$ the sum of the externally applied moments, both computed with respect to the center of mass of the rigid body. It is also true that $\underline{M}_O = \underline{\dot{H}}_O$, i.e., both angular momentum and externally applied moments can be evaluated with respect to an inertial point **O**.

These two differential vector equations in time provide the six scalar equations necessary to solve for the motion of the rigid body. The first equation is very similar to Newton's second law for a single particle, eq. (3.4). The mass of the entire rigid body multiplied by the acceleration of its center of mass equals the sum of all externally applied forces. The rigid body can be replaced by a fictitious particle of mass m located at its center of mass and subjected to all the forces externally applied to the body.

The second equation describes the motion of the rigid body *around its center of mass*. This equation is more complex than the first and does require the evaluation of the angular momentum vector of the rigid body, which will bring to light an additional inertial characteristics of the rigid body, the mass moment of inertia tensor.

The evaluations of the angular velocity vector and of the kinetic energy of a rigid body are presented in sections 6.1 and 6.2, respectively. The evaluation of these quantities gives rise to the tensor of mass moments of inertia whose properties are reviewed in section 6.3. The equations of motion of a rigid body are derived in section 6.5 and the principle of work and energy in section 6.6. A special case of particular interest for many applications is the planar motion of rigid bodies, which is the focus of section 6.7.

6.1 The angular momentum vector

The angular momentum vector of a system of particles, computed with respect to an arbitrary point $\mathbf{O}$, is defined in section 3.4 as $\underline{H}_O = \sum_{i=1}^{N} \tilde{r}_i \, m_i \underline{v}_i$, where m_i is the mass of a particle of the system, $\underline{v}_i$ its inertial velocity vector, $\underline{r}_i$ its position vector with respect to point $\mathbf{O}$, and N the total number of particles of the system.

When dealing with a rigid body, this definition is not easy to handle: the number of particles is very large while the mass of each one is very small. Each atom the rigid body could be considered to be a particle of very small mass and the total number of particles would be extremely large. Consequently, the familiar concepts of continuum mechanics are introduced: each differential volume element of the body, dV, is considered to be a particle of mass $m_i = \rho dV$, where ρ is the mass density of the body. The sum over all particles is then replaced by an integral over the entire volume of the body.

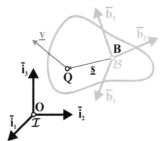

Figure 6.1 show the configuration of the rigid body; frame $\mathcal{F}^I = [\mathbf{O}, \mathcal{I} = (\bar{\imath}_1, \bar{\imath}_2, \bar{\imath}_3)]$ is an inertial frame. Point $\mathbf{B}$ is a reference point of the body while point $\mathbf{Q}$ is a material point of the body. The position vector of point $\mathbf{Q}$ with respect to point $\mathbf{B}$ is denoted $\underline{s}$. The angular momentum vector of the body, computed with respect to point $\mathbf{B}$, is then

Fig. 6.1. Configuration of a rigid body.

$$\underline{H}_B = \int_V \tilde{s}\underline{v}_Q \, \rho dV,$$

where $\underline{v}_Q$ is the inertial velocity of point $\mathbf{Q}$, and V the total volume of the body.

Because the body is rigid, its velocity field is described by eq. (5.10), *i.e.*, $\underline{v}_Q = \underline{v}_B + \tilde{\omega}\underline{s}$ and the angular momentum vector now becomes

$$\underline{H}_B = \int_V \tilde{s}(\underline{v}_B + \tilde{\omega}\underline{s}) \, \rho dV.$$

Because $\underline{v}_B$ is the velocity of reference point **B** and $\underline{\omega}$ the angular velocity of the rigid body, this expression can be recast as

$$\underline{H}_B = \left[\int_\mathcal{V} \widetilde{s}\,\rho\mathrm{d}\mathcal{V}\right]\underline{v}_B + \left[\int_\mathcal{V} \widetilde{s}\,\widetilde{s}^T\rho\mathrm{d}\mathcal{V}\right]\underline{\omega}.$$

The first bracketed term is related to the location of the center of mass of the rigid body, see eq. (3.56); $\int_\mathcal{V} \underline{s}\,\rho\mathrm{d}\mathcal{V} = m\underline{r}_{BC}$, where $\underline{r}_{BC}$ is the position vector of the rigid body center of mass with respect to point **B**.

The second bracketed term is the *tensor of mass moments of inertia*, evaluated with respect to point **B**. This second-order, symmetric tensor is defined as

$$\underline{\underline{I}}^B = \int_\mathcal{V} \widetilde{s}\,\widetilde{s}^T\rho\mathrm{d}\mathcal{V}, \tag{6.1}$$

With these definitions, the angular momentum vector takes the following form

$$\underline{H}_B = m\widetilde{r}_{BC}\underline{v}_B + \underline{\underline{I}}^B\underline{\omega}. \tag{6.2}$$

The sole inertial characteristic of a particle is its mass, but the characterization of the inertial properties of a rigid body is more complex. Ten quantities are required: the total mass of the body, m (a single scalar quantity), the location of the center of mass, $\underline{r}_{BC}$ (three components of this vector), and the mass moments of inertia tensor, $\underline{\underline{I}}^B$ (six independent components of this symmetric tensor). The units of the mass moments of inertia tensor are kg·m^2.

An arbitrary orientation of the inertial basis, $\mathcal{I}$, was selected. A different basis, say $\mathcal{I}'$, could have been selected. Let $\underline{\underline{R}}$ be the components of the rotation tensor that brings basis $\mathcal{I}$ to basis $\mathcal{I}'$, resolved in basis $\mathcal{I}$. If $\underline{s}$ and $\underline{s}'$ denote the components of vector $\underline{s}$ in bases $\mathcal{I}$ and $\mathcal{I}'$, respectively, eq. (4.27) implies $\underline{s}' = \underline{\underline{R}}^T\underline{s}$. It then follows that

$$(\underline{\underline{I}}^B)' = \int_\mathcal{V} \widetilde{s}'\,\widetilde{s}'^T\,\rho\mathrm{d}\mathcal{V} = \int_\mathcal{V} \underline{\underline{R}}^T\widetilde{s}\,\underline{\underline{R}}\,\underline{\underline{R}}^T\widetilde{s}^T\underline{\underline{R}}\,\rho\mathrm{d}\mathcal{V}$$
$$= \underline{\underline{R}}^T\left[\int_\mathcal{V} \widetilde{s}\,\widetilde{s}^T\,\rho\mathrm{d}\mathcal{V}\right]\underline{\underline{R}} = \underline{\underline{R}}^T\underline{\underline{I}}^B\underline{\underline{R}}. \tag{6.3}$$

This expression relates the components of tensor of mass moments of inertia resolved in two bases, $\mathcal{I}$ and $\mathcal{I}'$, denoted $\underline{\underline{I}}^B$ and $(\underline{\underline{I}}^B)'$, respectively. The fact that these components are related by the transformation rules for the components of second-order tensors, eq. (4.29), proves the tensorial nature of the mass moments of inertia tensor defined in eq. (6.1).

Often, it will be convenient to compute the angular momentum vector with respect to the center of mass of the rigid body. Indeed, selecting the center of mass as the reference point of the body implies $\underline{r}_{CC} = 0$, and eq. (6.2) reduces to

$$\underline{H}_C = \underline{\underline{I}}^C\underline{\omega}. \tag{6.4}$$

A similar simplification is achieved if the reference point on the rigid body happens to be an inertial point, *i.e.*, if $\underline{v}_B = 0$. In this case, eq. (6.2) reduces to

$$H_B = \underline{\underline{I}}^B \underline{\omega}. \tag{6.5}$$

Note that the angular momentum vector and the tensor of mass moments of inertia are quantities computed with respect to a specific point. Notations $\underline{\underline{I}}^C$ and $\underline{\underline{I}}^B$ denote the tensors of mass moments of inertia computed with respect to points **C** and **B**, respectively. Similarly, $\underline{H}_C$ and $\underline{H}_B$ indicate the angular momentum vectors evaluated with respect to points **C** and **B**, respectively.

6.2 The kinetic energy

The kinetic energy of a particle, eq. (3.10), is defined as $K = 1/2\, m\underline{v}^T\underline{v}$, where m is the mass of the particle and $\underline{v}$ its inertial velocity vector. The kinetic energy of a differential element of the rigid body located at point **Q** is now $K = 1/2\, \rho \mathrm{d}\mathcal{V}\, \underline{v}_Q^T \underline{v}_Q$, and that of the entire body becomes $K = 1/2 \int_\mathcal{V} \underline{v}_Q^T \underline{v}_Q\, \rho \mathrm{d}\mathcal{V}$. The velocity field of a rigid body is described by eq. (5.10) as $\underline{v}_Q = \underline{v}_B + \widetilde{\omega}\underline{s}$ and the kinetic energy now becomes

$$K = \frac{1}{2} \int_\mathcal{V} (\underline{v}_B^T \underline{v}_B + 2\underline{v}_B^T \widetilde{\omega}\underline{s} + \underline{s}^T \widetilde{\omega}^T \widetilde{\omega}\underline{s})\, \rho \mathrm{d}\mathcal{V}.$$

Because vector $\underline{v}_B$ is the velocity of reference point **B** and $\underline{\omega}$ the angular velocity vector of the rigid body, this expression is recast as

$$K = \frac{1}{2} \left\{ \left[\int_\mathcal{V} \rho \mathrm{d}\mathcal{V} \right] \underline{v}_B^T \underline{v}_B + 2\underline{v}_B^T \left[\int_\mathcal{V} \widetilde{s}^T\, \rho \mathrm{d}\mathcal{V} \right] \underline{\omega} + \underline{\omega}^T \left[\int_\mathcal{V} \widetilde{s}\,\widetilde{s}^T\, \rho \mathrm{d}\mathcal{V} \right] \underline{\omega} \right\}.$$

The first bracketed term simply represents the total mass of the rigid body. The second bracketed term is related to the location of the center of mass of the rigid body, eq. (3.56), $\int_\mathcal{V} \underline{s}\, \rho \mathrm{d}\mathcal{V} = m\underline{r}_{BC}$. Finally, the last bracketed term is the tensor of mass moments of inertia evaluated with respect to point **B** defined by eq. (6.1).

The kinetic energy expression now reduces to

$$K = \frac{1}{2} \left(m\, \underline{v}_B^T \underline{v}_B + 2m\underline{v}_B^T \widetilde{r}_{BC}^T \underline{\omega} + \underline{\omega}^T \underline{\underline{I}}^B \underline{\omega} \right). \tag{6.6}$$

Here again, it is possible to simplify this expression by selecting the center of mass of the rigid body as the reference point; this implies $\underline{r}_{CC} = 0$, and hence,

$$K = \frac{1}{2}\, m\, \underline{v}_C^T \underline{v}_C + \frac{1}{2}\, \underline{\omega}^T \underline{\underline{I}}^C \underline{\omega}. \tag{6.7}$$

The first term represents the kinetic energy associated with the translational motion of the rigid body, and the second represents that associated with the rotation of the body. The expression for the *translational kinetic energy* of the rigid body, $1/2\, m\, \underline{v}_C^T \underline{v}_C$, is identical to that of a particle of mass m moving at velocity $\underline{v}_C$.

The rotational motion of the body about its center of mass is associated with an additional amount of kinetic energy called *rotational kinetic energy*, $1/2\, \underline{\omega}^T \underline{\underline{I}}^C \underline{\omega}$, that is a quadratic function of the angular velocity of the rigid body.

Because the kinetic energy is a positive quantity, the translational energy must always be positive, *i.e.*, $1/2 \, m \, \underline{v}_C^T \underline{v}_C > 0$ for any vector $\underline{v}_C \neq 0$; it follows that the mass of the body, m, must be positive number, a forgone conclusion. The same argument applied to the rotational kinetic energy yields $1/2 \, \underline{\omega}^T \underline{\underline{I}}^C \underline{\omega} > 0$ for all angular velocity vectors $\underline{\omega} \neq 0$; this implies that the tensor of mass moments of inertia, $\underline{\underline{I}}^C$, is a *positive-definite tensor*, eq. (1.53).

6.3 Properties of the mass moment of inertia tensor

This section investigates the properties of the mass moment of inertia tensor. If the location of the reference point of the rigid body is changed, the components of the mass moment of inertia tensor change according to the parallel axis theorem studied in section 6.3.1. Furthermore, section 6.3.2 shows that the components of the mass moment of inertia tensor change according to the rules of transformation for second-order tensors if the orientation of the body attached basis is modified.

6.3.1 The parallel axis theorem

In the previous section, the mass moment of inertia tensor was evaluated with respect to an arbitrary reference point of the body and with respect to its center of mass. Figure 6.2 depicts the configuration of the rigid body: $\underline{s}$ is the position vector of a material point **Q** of the rigid body with respect to reference point **B**, and $\underline{q}$ is the position vector of the same point with respect to the center of mass **C**. Clearly, $\underline{s} = \underline{r}_{BC} + \underline{q}$, where $\underline{r}_{BC}$ is the position vector of the center of mass with respect to point **B**.

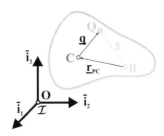

Fig. 6.2. Evaluating the mass moments of inertia with respect to a reference point **B** and the center of mass **C**.

The tensor of mass moments of inertia evaluated with respect to point **B** is now

$$\underline{\underline{I}}^B = \int_{\mathcal{V}} \widetilde{s}\widetilde{s}^T \, \rho \mathrm{d}\mathcal{V} = \int_{\mathcal{V}} (\widetilde{r}_{BC} + \widetilde{q})(\widetilde{r}_{BC}^T + \widetilde{q}^T) \, \rho \mathrm{d}\mathcal{V}.$$

Expanding the integrand and taking advantage of the fact that $\underline{r}_{BC}$ can be factored out of the integral sign leads to

$$\underline{\underline{I}}^B = m \, \widetilde{r}_{BC}\widetilde{r}_{BC}^T + \widetilde{r}_{BC}\left[\int_{\mathcal{V}} \widetilde{q}^T \, \rho \mathrm{d}\mathcal{V}\right] + \left[\int_{\mathcal{V}} \widetilde{q} \, \rho \mathrm{d}\mathcal{V}\right]\widetilde{r}_{BC}^T + \int_{\mathcal{V}} \widetilde{q}\widetilde{q}^T \, \rho \mathrm{d}\mathcal{V}.$$

The two middle terms vanish because $\int_{\mathcal{V}} \widetilde{q} \, \rho \mathrm{d}\mathcal{V} = m\widetilde{r}_{CC} = 0$. The last term is the mass moment of inertia tensor, $\underline{\underline{I}}^C$, evaluated with respect to the center of mass, and hence,

$$\underline{\underline{I}}^B = \underline{\underline{I}}^C + m \, \widetilde{r}_{BC}\widetilde{r}_{BC}^T. \tag{6.8}$$

Let I_{ij}^B and I_{ij}^C be the components of the mass moment of inertia tensors $\underline{\underline{I}}^B$ and $\underline{\underline{I}}^C$, respectively, and let $\left\{x_1, x_2, x_3\right\}^T$ be the components of vector $\underline{r}_{BC}$, all resolved in a given basis. The diagonal components of tensor $\underline{\underline{I}}^B$ now become

$$I_{11}^B = I_{11}^C + m(x_2^2 + x_3^2), \tag{6.9a}$$
$$I_{22}^B = I_{22}^C + m(x_1^2 + x_3^2), \tag{6.9b}$$
$$I_{33}^B = I_{33}^C + m(x_1^2 + x_2^2). \tag{6.9c}$$

Components I_{11}^B and I_{11}^C of the mass moment of inertia tensor are evaluated with respect to two different points, an arbitrary point **B** and the center of mass, respectively, but in the same basis, *i.e.*, with respect to parallel axis systems; hence, the name of *parallel axes theorem*.

The properties of the center of mass were used in the derivation of this theorem, hence, it is incorrect to write $I_{11}^B = I_{11}^R + m(x_2^2 + x_3^2)$ if points **B** and **R** are two arbitrary points of the rigid body.

Because the second term on the right-hand side of eqs. (6.9) is strictly positive, it follows $I_{11}^B > I_{11}^C$, that is, the moment of inertia always increases when moving away from the center of mass. In other words, the minimum value of I_{11} is obtained when it is computed with respect to the center of mass.

The off-diagonal terms of tensor of moments of inertia are called *products of inertia*; in view of eq. (6.8), they become

$$I_{23}^B = I_{23}^C - mx_2x_3, \tag{6.10a}$$
$$I_{13}^B = I_{13}^C - mx_1x_3, \tag{6.10b}$$
$$I_{12}^B = I_{12}^C - mx_1x_2. \tag{6.10c}$$

In this case, the second term on the right-hand side could be positive or negative; consequently, products of inertia could increase of decrease when moving away from the center of mass.

Theorem 6.1 (Parallel axis theorem). *The components of the mass moment of inertia tensor of a rigid body computed with respect to an arbitrary point* **B** *are related to their counterparts resolved in the same basis but computed with respect to the body's center of mass by eqs. (6.9) and (6.10).*

6.3.2 Change of basis

In the previous section, relationships were derived between the components of the tensor of mass moments of inertia evaluated with respect to two different points, but resolved in the same basis. In this section, relationships are sought between the components of this tensor resolved in two different bases, but evaluated with respect to the same point.

Consider two bases, $\mathcal{B}$ and $\mathcal{B}'$, and let $\underline{\underline{R}}$ be the components of the rotation tensor that brings basis $\mathcal{B}$ to basis $\mathcal{B}'$, resolved in basis $\mathcal{B}$. Equation (6.3) then implies

$$
\begin{bmatrix} I'_{11} & I'_{12} & I'_{13} \\ I'_{12} & I'_{22} & I'_{23} \\ I'_{13} & I'_{23} & I'_{33} \end{bmatrix} = \underline{\underline{R}}^T \begin{bmatrix} I_{11} & I_{12} & I_{13} \\ I_{12} & I_{22} & I_{23} \\ I_{13} & I_{23} & I_{33} \end{bmatrix} \underline{\underline{R}},
\tag{6.11}
$$

where $\underline{\underline{I}}'$ and $\underline{\underline{I}}$ are the components of the mass moment of inertia tensor in bases $\mathcal{B}'$ and $\mathcal{B}$, respectively.

It is instructive to look at the transformation laws for specific components. The rotation tensor will be represented by its direction cosines,

$$
\underline{\underline{R}} = \begin{bmatrix} \ell_1 & m_1 & n_1 \\ \ell_2 & m_2 & n_2 \\ \ell_3 & m_3 & n_3 \end{bmatrix}.
\tag{6.12}
$$

First, the diagonal terms of $\underline{\underline{I}}'$ are

$$
I'_{11} = \ell_1^2 I_{11} + \ell_2^2 I_{22} + \ell_3^2 I_{33} + 2\ell_2\ell_3 I_{23} + 2\ell_1\ell_3 I_{13} + 2\ell_1\ell_2 I_{12},
\tag{6.13a}
$$
$$
I'_{22} = m_1^2 I_{11} + m_2^2 I_{22} + m_3^2 I_{33} + 2m_2 m_3 I_{23} + 2m_1 m_3 I_{13} + 2m_1 m_2 I_{12},
\tag{6.13b}
$$
$$
I'_{33} = n_1^2 I_{11} + n_2^2 I_{22} + n_3^2 I_{33} + 2n_2 n_3 I_{23} + 2n_1 n_3 I_{13} + 2n_1 n_2 I_{12},
\tag{6.13c}
$$

Next, the off-diagonal terms of $\underline{\underline{I}}'$ are

$$
\begin{aligned}
I'_{23} &= n_1 m_1 I_{11} + n_2 m_2 I_{22} + n_3 m_3 I_{33} + (n_2 m_3 + n_3 m_2) I_{23} \\
&\quad + (n_1 m_3 + n_3 m_1) I_{13} + (n_1 m_2 + n_2 m_1) I_{12}.
\end{aligned}
\tag{6.14a}
$$
$$
\begin{aligned}
I'_{13} &= n_1 \ell_1 I_{11} + n_2 \ell_2 I_{22} + n_3 \ell_3 I_{33} + (n_2 \ell_3 + n_3 \ell_2) I_{23} \\
&\quad + (n_1 \ell_3 + n_3 \ell_1) I_{13} + (n_1 \ell_2 + n_2 \ell_1) I_{12}.
\end{aligned}
\tag{6.14b}
$$
$$
\begin{aligned}
I'_{12} &= \ell_1 m_1 I_{11} + \ell_2 m_2 I_{22} + \ell_3 m_3 I_{33} + (\ell_2 m_3 + \ell_3 m_2) I_{23} \\
&\quad + (\ell_1 m_3 + \ell_3 m_1) I_{13} + (\ell_1 m_2 + \ell_2 m_1) I_{12}.
\end{aligned}
\tag{6.14c}
$$

6.3.3 Principal axes of inertia

The tensor of mass moments of inertia was shown to be a symmetric, positive-definite tensor. In view of section 1.4.2, its eigenvalues must be real and positive. Furthermore, it is always possible to construct a set the orthogonal eigenvectors that will diagonalize this tensor, see eq. (1.64). Let $\underline{u}_1$, $\underline{u}_2$, and $\underline{u}_2$ be the eigenvector of the tensor of mass moments of inertia, and $\underline{\underline{P}} = [\underline{u}_1, \underline{u}_2, \underline{u}_3]$. It the follows that

$$
\underline{\underline{P}}^T \underline{\underline{I}}\, \underline{\underline{P}} = \mathrm{diag}(I_i^*) = \begin{bmatrix} I_1^* & 0 & 0 \\ 0 & I_2^* & 0 \\ 0 & 0 & I_3^* \end{bmatrix},
\tag{6.15}
$$

where I_1^*, I_2^*, and I_3^* are the eigenvalues of the tensor of mass moments of inertia. For symmetric, positive-definite tensors, the eigenvectors can be selected to form an orthogonal tensor, which itself, can be interpreted as a rotation tensor.

Let $\mathcal{B}^*$ be the basis defined by the eigenvectors; $\underline{\underline{P}}$ is then the rotation tensor that brings basis $\mathcal{B}$ to basis $\mathcal{B}^*$. In view of eq. (6.3), the statement $\mathrm{diag}(I_i^*) = \underline{\underline{P}}^T \underline{\underline{I}} \, \underline{\underline{P}}$ is a change of basis operation: $\underline{\underline{I}}$ and $\mathrm{diag}(I_i^*)$ are the components of the moment of inertia tensor in bases $\mathcal{B}$ and $\bar{\mathcal{B}}^*$, respectively. The transformation defined by the principal axes of inertia brings the components of the mass moments of inertia tensor to a diagonal form.

6.3.4 Problems

Problem 6.1. Kinetic energy for a rigid body undergoing rotational motion
A rigid body is in rotational motion about fixed inertial point **O**, which does not coincide with the center of mass of the rigid body. Starting from eq. (6.7), prove that the kinetic energy of the rigid body can be expressed in the following form

$$K = \frac{1}{2}\,\underline{\omega}^{*T} \underline{\underline{I}}^{O*} \underline{\omega}^*, \qquad (6.16)$$

where $\underline{\underline{I}}^{O*}$ is the mass moment of inertia tensor computed with respect to point **O**. Notation $(\cdot)^*$ indicates the components of vectors and tensors resolved in a body attached basis.

Problem 6.2. Two interconnected particles in planar motion
Figure 6.3 depicts a system of two rigidly interconnected particles undergoing planar motion and subjected to the acceleration of gravity, $-g\bar{\imath}_2$. Point **C** is the center of mass of the system and θ the angle the massless rigid bar connecting the particles makes with the horizontal. *(1)* Does the following relationship hold, $m_1\ddot{\underline{r}}_1 = -m_1 g\bar{\imath}_2$? (2) If d_1 and d_2 are the distances from the two particles to the center of mass, prove that $d_1 m_1 = d_2 m_2$. *(3)* Does the following relationship hold, $\ddot{\underline{r}}_C = -g\bar{\imath}_2$? *(4)* Evaluate the angular momentum vector of the system with respect to its center of mass in terms of $\dot{\theta}$. *(5)* Is this angular momentum of the system preserved? Justify all your answers. YES/NO answers are not valid.

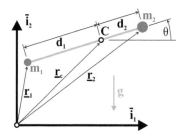
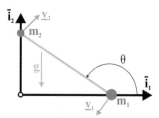

Fig. 6.3. Two interconnected particles in planar motion.

Fig. 6.4. Two interconnected particles in planar motion.

Problem 6.3. Two interconnected particles in planar motion

Figure 6.4 show a system of two rigidly interconnected particles undergoing planar motion and subjected to the acceleration of gravity, $-g\bar{\imath}_2$. The angle the massless rigid bar connecting the particles makes with the horizontal is denoted $\theta(t)$. (1) Determine the initial location of the center of mass, $\underline{r}_{C0}$, and its initial velocity, $\underline{v}_{C0}$. (2) Find the initial angular velocity vector of the system, $\underline{\omega}_0$. (3) Find the condition that must be satisfied by the initial velocity vectors of the two particles, $\underline{v}_{10}$ and $\underline{v}_{20}$. (4) Is the angular momentum of the system preserved? (5) Find the time history of the position vectors of the two particles, $\underline{r}_1(t)$ and $\underline{r}_2(t)$. Use the following data: $m_1 = 1.3$ kg, $m_2 = 5.2$ kg, $\underline{r}_1(t = 0) = \underline{r}_{10} = 5\,\bar{\imath}_1$ m, $\underline{r}_2(t = 0) = \underline{r}_{20} = 3\,\bar{\imath}_2$ m, $\underline{v}_1(t = 0) = \underline{v}_{10} = -2.2\,\bar{\imath}_1 - 3\,\bar{\imath}_2$ m/s, $\underline{v}_2(t = 0) = \underline{v}_{20} = 2.12\,\bar{\imath}_1 + 4.2\,\bar{\imath}_2$ m/s.

Problem 6.4. Moments of inertia of a rectangular plate with side bar

A homogeneous rectangular plate of mass M, length a, and width b is connected to a homogeneous rod of mass m and length $a/2$, as depicted in fig. 6.5. (1) Determine the mass moment of inertia tensor of the system evaluated with respect to point **O**, the plate's geometric center, and resolved in a set of axes parallel to the edges of the plate. (2) Determine the orientation of the principal axes of inertia at point **O** and the corresponding principal mass moments of inertia. (3) Find the location of the center of mass of the system, point **C**. (4) Determine the orientation of the principal axes of inertia at point **C** and the corresponding principal mass moments of inertia. Use the following data: $a = 0.48$ m, $b = 0.24$ m, $M = 0.5$ kg, and $m = 0.3$ kg.

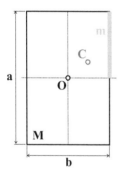
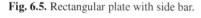

Fig. 6.5. Rectangular plate with side bar.

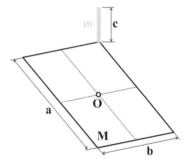

Fig. 6.6. Rectangular plate with corner normal bar.

Problem 6.5. Moments of inertia of a rectangular plate with corner normal bar

A homogeneous rod of mass m and length c is connected at the corner of a homogeneous rectangular plate of mass M, length a, and width b, as depicted in fig. 6.6. The rod is normal to the plate. (1) Determine the mass moment of inertia tensor of the system evaluated with respect to point **O**, the plate's geometric center, and resolved in a set of axes parallel to the edges of the plate. (2) Determine the orientation of the principal axes of inertia at point **O** and the corresponding principal mass moments of inertia. (3) Find the location of the center of mass of the system, point **C**. (4) Determine the orientation of the principal axes of inertia at point **C** and the corresponding principal mass moments of inertia. Use the following data: $a = 0.64$ m, $b = 0.36$ m, $c = 0.48$ m, $M = 0.5$ kg, and $m = 0.4$ kg.

Problem 6.6. Moments of inertia of the flywheel governor
Figure 6.7 shows a simplified configuration of the flywheel governor. Two particles of mass m are connected to four articulated bars of length L, which remain in a plane at all times. Angle θ is changing according to the following schedule: $\theta(t) = \pi/4 + \pi/6 \sin 2\pi t$. At time $t = 0$, the system has an angular velocity $\underline{\omega} = \Omega_0 \bar{\imath}_3$. (1) Is the angular momentum of the system preserved? (2) Is the angular velocity of the system preserved? (3) Find the time history of the angular velocity vector of the system, $\underline{\omega}(t)$. (4) Plot $\|\underline{\omega}(t)\|/\Omega_0$ for $t \in [0, 1]$ s.

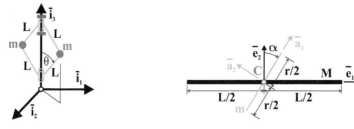

Fig. 6.7. Flywheel governor.   **Fig. 6.8.** Two rigidly connected bars.

Problem 6.7. Rigid bar connected to a shaft
Figure 6.8 shows rigid shaft of length L and mass M. Basis $\mathcal{E} = (\bar{e}_1, \bar{e}_2, \bar{e}_3)$ is attached to the shaft; unit vector $\bar{e}_1$ is aligned with the shaft. A rigid bar of length r and mass m is rigidly connected to the shaft. Basis $\mathcal{A} = (\bar{a}_1, \bar{a}_2, \bar{a}_3)$ is attached to the bar, $\bar{e}_3 = \bar{a}_3$ and $\alpha = (\bar{e}_2, \bar{a}_1)$; unit vector $\bar{a}_1$ is aligned with the bar. The shaft and bar are homogeneous slender rods, see fig. 6.42, and their centers of mass coincide at point **C**. (1) Determine the tensor of mass moments of inertia of the assembly in basis $\mathcal{E}$.

Problem 6.8. Rigid disk connected to a shaft
Figure 6.8 shows rigid shaft of length L, radius R, and mass M. Basis $\mathcal{E} = (\bar{e}_1, \bar{e}_2, \bar{e}_3)$ is attached to the shaft; unit vector $\bar{e}_1$ is aligned with the shaft. A rigid disk of radius r and mass m is rigidly connected to the shaft. Basis $\mathcal{A} = (\bar{a}_1, \bar{a}_2, \bar{a}_3)$ is attached to the disk, $\bar{e}_3 = \bar{a}_3$ and $\alpha = (\bar{e}_2, \bar{a}_1)$; unit vector $\bar{a}_2$ is normal the disk. The shaft is a homogeneous cylinder, see fig. 6.40, and the disk a homogeneous thin disk, see fig. 6.41. Their centers of mass coincide at point **C**. (1) Determine the tensor of mass moments of inertia of the assembly in basis $\mathcal{E}$.

6.4 Derivatives of the angular momentum vector

Because a rigid body is a system of particles, Euler's second law, $\underline{M}_C = \underline{\dot{H}}_C$, applies. Use of this equation calls for the evaluation of the time derivative of the angular momentum vector evaluated with respect to the center of mass of the rigid body, $\underline{H}_C = \underline{\underline{I}}^C \underline{\omega}$, see eq. (6.4). The time derivative of this quantity is

$$\underline{\dot{H}}_C = \underline{\underline{\dot{I}}}^C \underline{\omega} + \underline{\underline{I}}^C \underline{\dot{\omega}}.$$

Evaluation of the mass moment of inertia tensor is a cumbersome task. If the basis in which the components of this tensor are computed changes its orientation with respect to the rigid body, $\underline{\underline{I}}^C = \underline{\underline{I}}^C(t)$, and this evaluation must be repeated at each instant in time; furthermore, this implies $\underline{\dot{\underline{I}}}^C \neq 0$. Consequently, it is convenient to select *a body attached frame* for the evaluation of the mass moments of inertia, which then become *constant quantities in time,* and their time derivatives vanish.

Consider an inertial basis, $\mathcal{I}$, and a body attached basis, $\mathcal{B}^*$; superscript $(\cdot)^*$ indicates tensor components resolved in basis $\mathcal{B}^*$. Let $\underline{\underline{R}}$ be the components of the rotation tensor that brings basis $\mathcal{I}$ to basis $\mathcal{B}^*$, resolved in $\mathcal{I}$. Furthermore, let $\underline{H}_C$, $\underline{\underline{I}}^C$, and $\underline{\omega}$ be the components of the angular momentum vector, mass moment of inertia tensor, and angular velocity vector, respectively, all resolved in basis $\mathcal{I}$. It then follows that $\underline{\underline{R}}^T \underline{H}_C = \underline{\underline{R}}^T \underline{\underline{I}}^C \underline{\underline{R}}\, \underline{\underline{R}}^T \underline{\omega} = \underline{\underline{I}}^{C*} \underline{\omega}^*$, where $\underline{\underline{I}}^{C*}$ and $\underline{\omega}^*$ are the components of the mass moment of inertia tensor and angular velocity vector, respectively, resolved in basis $\mathcal{B}^*$. The orthogonality of the rotation tensor then implies $\underline{H}_C = \underline{\underline{R}}\, \underline{\underline{I}}^{C*} \underline{\omega}^*$. While the component of $\underline{\underline{I}}^C$ are time-dependent quantities, those of $\underline{\underline{I}}^{C*}$ are constants and their time derivatives vanish. The time derivative of the angular momentum vector now becomes

$$\underline{\dot{H}}_C = \underline{\dot{\underline{R}}}\,\underline{\underline{I}}^{C*}\underline{\omega}^* + \underline{\underline{R}}\,\underline{\underline{I}}^{C*}\underline{\dot{\omega}}^* = \underline{\underline{R}}\left(\underline{\underline{I}}^{C*}\underline{\dot{\omega}}^* + \widetilde{\omega}^*\underline{\underline{I}}^{C*}\underline{\omega}^*\right). \tag{6.17}$$

Euler's second law holds true when expressed about the rigid body's center of mass, but it is equally valid when expresses with respect to an inertial point $\mathbf{O}$, $\underline{M}_O = \underline{\dot{H}}_O$, see eq. (3.75). Here again, evaluation of the derivative of the angular momentum vector involves derivatives of the mass moment of inertia tensor.

To ease the evaluation of these derivatives, the first step is to work in a body attached basis $\mathcal{B}^*$, and the components of the mass moment of inertia tensor resolved in this basis are denoted $\underline{\underline{I}}^{O*}$. Unfortunately, this is not yet sufficient to guarantee $\underline{\dot{\underline{I}}}^{O*} = 0$. Indeed, point $\mathbf{O}$ is an inertial point that is not necessarily a material point of the body; consequently, the mass moments of inertia might still be time-dependent.

If inertial point $\mathbf{O}$ is a fixed point on the body, the mass moments of inertia resolved in the body attached basis become constant and $\underline{\dot{\underline{I}}}^{O*}$ vanishes. This happens only if the rigid body is undergoing pure rotation about inertial point $\mathbf{O}$. If $\underline{\dot{\underline{I}}}^{O*} = 0$, developments similar to those presented above lead to $\underline{\dot{H}}_O = \underline{\underline{R}}(\underline{\underline{I}}^{O*}\underline{\dot{\omega}}^* + \widetilde{\omega}^*\underline{\underline{I}}^{O*}\underline{\omega}^*)$.

In summary, the time derivative of the angular momentum vector is

$$\underline{\dot{H}}_A = \underline{\underline{R}}\left(\underline{\underline{I}}^{A*}\underline{\dot{\omega}}^* + \widetilde{\omega}^*\underline{\underline{I}}^{A*}\underline{\omega}^*\right), \tag{6.18}$$

when *(1)* point $\mathbf{A}$ is the center of mass of the rigid body, or *(2)* the rigid body is undergoing pure rotation about inertial point $\mathbf{A}$.

6.5 Equations of motion for a rigid body

The equations of motion for a rigid body are derived from the equations of motion for a general system of particles presented in section 3.4.4. The first equation of motion governs the motion of the center of mass of the rigid body, eq. (3.70), and the second equation governs its angular motion, eq. (3.76). The general form of the first equation is

$$\underline{F} = m\underline{a}_C, \tag{6.19}$$

where $\underline{a}_C$ is the inertial acceleration of the center of mass of the rigid body and $\underline{F}$ the sum of the externally applied forces.

The general form of the second equation is

$$\underline{M}_C = \underline{\dot{H}}_C, \tag{6.20}$$

where $\underline{H}_C$ is the angular momentum computed with respect to the center of mass and $\underline{M}_C$ the sum of the externally applied moments computed with respect to the center of mass. This second equation of motion can be written in several different manners depending on the point with respect to which the externally applied moments are computed. The following four cases will be considered.

1. The sum of the externally applied moments is computed with respect to *the center of mass of the rigid body*.
2. The sum of the externally applied moments is computed with respect to *a pivot point of the rigid body*. A pivot point is a point of the body that happens to be an inertial point; clearly, such a point does not always exist.
3. The sum of the externally applied moments is computed with respect to *a material point of the rigid body*.
4. The sum of the externally applied moments is computed with respect to *an arbitrary point*. This arbitrary point is not necessarily inertial and is not a material point of the body.

The choice among the various forms of the equations is purely a matter of convenience: for specific applications, one formulation might lead to simpler equations. The four approaches are detailed in the following sections.

6.5.1 Euler's equations

In view of eq. (6.17), the second equation of motion of the rigid body, eq. (6.20), becomes $\underline{M}_C = \underline{\underline{R}}(\underline{\underline{I}}^{C*}\underline{\dot{\omega}}^* + \widetilde{\omega}^*\underline{\underline{I}}^{C*}\underline{\omega}^*)$. Multiplying this equation by $\underline{\underline{R}}^T$ then leads to

$$\underline{M}_C^* = \underline{\underline{I}}^{C*}\underline{\dot{\omega}}^* + \widetilde{\omega}^*\underline{\underline{I}}^{C*}\underline{\omega}^*, \tag{6.21}$$

where $\underline{M}_C^* = \underline{\underline{R}}^T\underline{M}_C$ is the sum of the externally applied moments computed with respect to the center of mass and resolved in a body attached basis.

If this basis coincides with the principal axes of inertia, the mass moment of inertia tensor reduces to a diagonal form, see eq. (6.15), and the governing equations further simplify to

$$\begin{aligned}
M_{C1}^* &= I_1^{C*}\dot{\omega}_1^* - \left(I_2^{C*} - I_3^{C*}\right)\omega_2^*\omega_3^*, \\
M_{C2}^* &= I_2^{C*}\dot{\omega}_2^* - \left(I_3^{C*} - I_1^{C*}\right)\omega_3^*\omega_1^*, \\
M_{C3}^* &= I_3^{C*}\dot{\omega}_3^* - \left(I_1^{C*} - I_2^{C*}\right)\omega_1^*\omega_2^*,
\end{aligned} \tag{6.22}$$

where $\underline{M}_C^{*T} = \{M_{C1}^*, M_{C2}^*, M_{C3}^*\}$ and $\underline{\underline{I}}^{C*} = \mathrm{diag}(I_1^{C*}, I_2^{C*}, I_3^{C*})$. These equations are known as *Euler's equations* for the angular motion of a rigid body. The sum of the externally applied moments is computed with respect to the center of mass of the rigid body.

6.5.2 The pivot equations

The governing equations of a system of particles can also be written with respect to an inertial point, $\underline{M}_O = \dot{\underline{H}}_O$, see eq. (3.75). Furthermore, if this inertial point is also a material point of the rigid body, the time derivative of the angular momentum vector is given by eq. (6.18) and the equation of motion becomes $\underline{M}_O = \underline{\underline{R}}(\underline{\underline{I}}^{O*}\dot{\underline{\omega}}^* + \widetilde{\omega}^* \underline{\underline{I}}^{O*}\underline{\omega}^*)$. Multiplying this equation by $\underline{\underline{R}}^T$ then leads to

$$\underline{M}_O^* = \underline{\underline{I}}^{O*}\dot{\underline{\omega}}^* + \widetilde{\omega}^*\underline{\underline{I}}^{O*}\underline{\omega}^*, \tag{6.23}$$

where $\underline{M}_O^* = \underline{\underline{R}}^T \underline{M}_O$ is the sum of the externally applied moments computed with respect to point **O**, resolved in a body attached basis.

Equation (6.23) only holds if point **O** is *an inertial point that is also a material point of the rigid body*; this implies that the rigid body is undergoing pure rotational motion about inertial point **O**. Point **O** is then often called a *pivot point* of the rigid body, and hence, eqs. (6.23) are known as the *pivot equations* for the angular motion of a rigid body; the sum of the externally applied moments is computed with respect to this pivot point.

6.5.3 Equations of motion with respect to a material point of the rigid body

Let point **B** be *a material point of the rigid body*. The angular momentum vector computed with respect to this point is given by eq. (6.2) as $\underline{H}_B = \widetilde{r}_{BC}\, m\underline{v}_B + \underline{\underline{I}}^B\underline{\omega}$, and it can be related to the angular momentum computed with respect to the center of mass by eq. (3.67) as $\underline{H}_B = \underline{H}_C + \widetilde{r}_{BC}\, m\underline{v}_C$.

Equating these two expressions and taking a time derivative leads to

$$\dot{\underline{H}}_C + \widetilde{r}_{BC}\, m\underline{a}_C + \dot{\widetilde{r}}_{BC}\, m\underline{v}_C = \widetilde{r}_{BC}\, m\underline{v}_B + \widetilde{r}_{BC}\, m\underline{a}_B + (\underline{\underline{I}}^B\underline{\omega})^{\cdot}.$$

In view of eqs. (6.19) and (6.20), the first two terms on the left-hand side are expressed as $\dot{\underline{H}}_C + \widetilde{r}_{BC}\, m\underline{a}_C = \underline{M}_C + \widetilde{r}_{BC}\, \underline{F} = \underline{M}_B$, where the last equality follows from eq. (3.61), and the above expression becomes $\underline{M}_B = \dot{\widetilde{r}}_{BC}\, m(\underline{v}_B - \underline{v}_C) + \widetilde{r}_{BC}\, m\underline{a}_B + (\underline{\underline{I}}^B\underline{\omega})^{\cdot}$. The first term on the right-hand side vanishes because $\dot{\widetilde{r}}_{BC}\, m(\underline{v}_B - \underline{v}_C) = -\widetilde{v}_{BC}\, m\underline{v}_{BC} = 0$, and finally

$$\underline{M}_B = \widetilde{r}_{BC}\, m\underline{a}_B + (\underline{\underline{I}}^B\underline{\omega})^{\cdot}. \tag{6.24}$$

To evaluate the time derivative of the last term, it is convenient to express the moment of inertia tensor in the body attached basis: $(\underline{\underline{I}}^B\underline{\omega})^{\cdot} = (\underline{\underline{R}}\,\underline{\underline{I}}^{B*}\underline{\omega}^*)^{\cdot} =$

$\underline{R}(\underline{\underline{I}}^{B*}\underline{\dot{\omega}}^* + \widetilde{\omega}^*\underline{\underline{I}}^{B*}\underline{\omega}^*)$. Resolving all quantities in the body attached basis, eq. (6.24) becomes

$$M_B^* = \widetilde{r}_{BC}^* \, \underline{R}^T m \underline{a}_B + \underline{\underline{I}}^{B*}\underline{\dot{\omega}}^* + \widetilde{\omega}^*\underline{\underline{I}}^{B*}\underline{\omega}^*, \tag{6.25}$$

where $\underline{M}_B^*$ is the sum of the externally applied moments computed with respect to *material point B of the rigid body*, $\underline{r}_{BC}^*$ is the position vector of the center of mass with respect to point **B**, and $\underline{R}^T \underline{a}_B$ the components of the acceleration vector of point **B**, all resolved in the body attached frame.

6.5.4 Equations of motion with respect to an arbitrary point

Let point **P** be *an arbitrary point, i.e.*, point **P** is neither inertial, nor a material point of the rigid body. Using eq. (3.61), the moment computed with respect to point **P** is related to that computed with respect to the center of mass as $\underline{M}_P = \underline{M}_C + \widetilde{r}_{PC}\underline{F}$. Introducing eqs. (6.19) and (6.20) then leads to

$$\underline{M}_P = \underline{\dot{H}}_C + \widetilde{r}_{PC}\, m\underline{a}_C = \widetilde{r}_{PC}\, m\underline{a}_C + \underline{R}\left(\underline{\underline{I}}^{C*}\underline{\dot{\omega}}^* + \widetilde{\omega}^*\underline{\underline{I}}^{C*}\underline{\omega}^*\right), \tag{6.26}$$

where the last equality follows from eq. (6.17).

Resolving all quantities in the body attached basis, eq. (6.26) becomes

$$\underline{M}_P^* = \widetilde{r}_{PC}^* \, \underline{R}^T m\underline{a}_C + \underline{\underline{I}}^{C*}\underline{\dot{\omega}}^* + \widetilde{\omega}^*\underline{\underline{I}}^{C*}\underline{\omega}^*, \tag{6.27}$$

where $\underline{M}_P^*$ is the sum of the externally applied moments, computed with respect to *an arbitrary point P*, $\underline{r}_{PC}^*$ are the components of the position vector of the center of mass with respect to point **P**, and $R^T \underline{a}_C$ the components of the acceleration vector of the center of mass, all resolved in the body attached basis.

6.6 The principle of work and energy

In section 3.4.5, the principle of work and energy was derived for a system of particles, see eq. (3.80). For an arbitrary system of particles, the work done by the internal forces explicitly appears in the statement of the principle, which is, consequently, of little practical use. If the system of particles, however, is a rigid body, the work done by the internal forces can be eliminated from the statement of the principle of work and energy, making it a powerful, practical tool.

The work done by all external and internal forces acting of the rigid body is found by summing up the work done by all external and internal forces acting on each particle of the body

$$W_{t_i \to t_f} = \sum_{i=1}^N \int_{t_i}^{t_f} (\underline{F}_i^T + \sum_{j=1,\, j\neq i}^N \underline{f}_{ij}^T)\mathrm{d}\underline{r}_i. \tag{6.28}$$

The sum of the externally applied forces acting on particle i is denoted $\underline{F}_i$, $\underline{f}_{ij}$ denotes the internal forces resulting from the interaction of particles i and j, and $\mathrm{d}\underline{r}_i$ is the differential displacement of particle i.

Because the body is rigid, the differential displacements of one of its points can be expressed in terms of the differential displacement of reference point **B** of the body, $d\underline{r}_B$, and the differential rotation vector of the body, $d\underline{\psi}$, as $d\underline{r}_i = d\underline{r}_B + \widetilde{d\underline{\psi}}\underline{s}_i$, where $\underline{s}_i$ is the position vector of particle i with respect to reference point **B**. Introducing this expression for the differential displacement into eq. (6.28) leads to

$$W_{t_i \to t_f} = \sum_{i=1}^{N} \int_{t_i}^{t_f} \left(\underline{F}_i^T + \sum_{j=1, j\neq i}^{N} \underline{f}_{ij}^T \right) \left(d\underline{r}_B + \widetilde{d\underline{\psi}}\underline{s}_i \right).$$

Expanding the scalar products then yields

$$W_{t_i \to t_f} = \int_{t_i}^{t_f} \left\{ \sum_{i=1}^{N} \underline{F}_i^T \, d\underline{r}_B + \sum_{i=1}^{N} \underline{F}_i^T \widetilde{s}_i^T \, d\underline{\psi} \right.$$
$$\left. + \sum_{i=1}^{N} \sum_{j=1, j\neq i}^{N} \underline{f}_{ij}^T \, d\underline{r}_B + \sum_{i=1}^{N} \sum_{j=1, j\neq i}^{N} \underline{f}_{ij}^T \widetilde{s}_i^T \, d\underline{\psi} \right\}.$$

Because quantities $d\underline{r}_B$ and $d\underline{\psi}$ do not depend on the particle number, the various summations appearing in this expression can be regrouped in the following manner

$$W_{t_i \to t_f} = \int_{t_i}^{t_f} \left[\sum_{i=1}^{N} \underline{F}_i \right]^T d\underline{r}_B + \int_{t_i}^{t_f} \left[\sum_{i=1}^{N} \widetilde{s}_i \underline{F}_i \right]^T d\underline{\psi}$$
$$+ \int_{t_i}^{t_f} \left[\sum_{i=1}^{N} \sum_{j=1, j\neq i}^{N} \underline{f}_{ij} \right]^T d\underline{r}_B + \int_{t_i}^{t_f} \left[\sum_{i=1}^{N} \sum_{j=1, j\neq i}^{N} \widetilde{s}_i \underline{f}_{ij} \right]^T d\underline{\psi}. \tag{6.29}$$

In the first term, the bracketed expression represents the sum of all externally applied forces to the rigid body, $\underline{F} = \sum_{i=1}^{N} \underline{F}_i$. In the second term, the bracketed expression represents the sum of all moments externally applied to the rigid body, $\underline{M}_B = \sum_{i=1}^{N} \widetilde{s}_i \underline{F}_i$. The third term in this expression vanishes in view of eq. (3.59), and eq. (3.62) implies the vanishing of the last term. Equation (6.29) finally reduces to $W_{t_i \to t_f} = \int_{t_i}^{t_f} \underline{F}^T d\underline{r}_B + \int_{t_i}^{t_f} \underline{M}_B^T d\underline{\psi}$. The principle of work and energy, eq. (3.80), applied to a rigid body now becomes

$$\int_{t_i}^{t_f} \underline{F}^T d\underline{r}_B + \int_{t_i}^{t_f} \underline{M}_B^T d\underline{\psi} = K(t_f) - K(t_i).$$

This result is known as the principle of work and energy.

Principle 7 (Principle of work and energy for a rigid body) *The work done by the external forces and moments acting on a rigid body equals the change in the rigid body's kinetic energy.*

Example 6.1. Rotating disk on a bent arm

Figure 6.9 shows a rotating disk connected to a bent arm. Massless arm **OAB** features a bend of β rad at point **A**. At point **O**, a bearing allows the arm to rotate at a constant angular velocity, Ω, with respect to ground. A disk of mass m and radius r rotates at a constant angular velocity, ω, and is connected to the arm at point **B** by means of a massless shaft. The dimensions of the system are indicated on the figure. A planar rotation of magnitude Ωt about axis $\bar{\imath}_3$ brings inertial frame, $\mathcal{F}^I = [\mathbf{O}, \mathcal{I} = (\bar{\imath}_1, \bar{\imath}_2, \bar{\imath}_3)]$ to frame $\mathcal{F}^A = [\mathbf{O}, \mathcal{A}' = (\bar{a}_1, \bar{a}_2, \bar{a}_3)]$ that is attached to the arm; all tensor components resolved in basis $\mathcal{A}'$ are denoted with superscripts $(\cdot)'$.

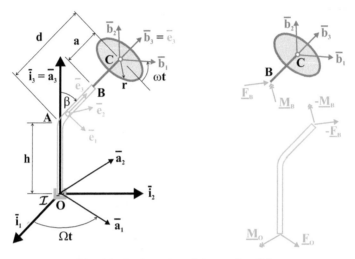

Fig. 6.9. Configuration of the rotating disk.

A second planar rotation of magnitude β about axis $\bar{a}_2$ brings basis $\mathcal{A}'$ to basis $\mathcal{E}^+$ and frame $\mathcal{F}^E = [\mathbf{A}, \mathcal{E}^+ = (\bar{e}_1, \bar{e}_2, \bar{e}_3)]$ is attached to the arm, with axis $\bar{e}_3$ pointing along the bent segment **AB**; all tensor components resolved in basis $\mathcal{E}^+$ are denoted with superscripts $(\cdot)^+$. Finally, a planar rotation of magnitude ωt about axis $\bar{e}_3$ brings basis $\mathcal{E}^+$ to basis $\mathcal{B}^*$ and frame $\mathcal{F}^B = [\mathbf{C}, \mathcal{B}^* = (\bar{b}_1, \bar{b}_2, \bar{b}_3)]$ is attached to the rotating disk; all tensor components resolved in basis $\mathcal{B}^*$ are denoted with superscripts $(\cdot)^*$.

The components of the rotation tensor that brings inertial basis $\mathcal{I}$ to basis $\mathcal{B}^*$, resolved in basis $\mathcal{I}$, will be constructed as $\underline{\underline{R}} = \underline{\underline{R}}_\Omega \underline{\underline{R}}'_\beta \underline{\underline{R}}^+_\omega$, where $\underline{\underline{R}}_\Omega$, $\underline{\underline{R}}'_\beta$, and $\underline{\underline{R}}^+_\omega$ are the components of the rotation tensors that bring basis $\mathcal{I}$ to basis $\mathcal{A}'$, $\mathcal{A}'$ to $\mathcal{E}^+$, and $\mathcal{E}^+$ to $\mathcal{B}^*$, respectively, resolved in bases $\mathcal{I}$, $\mathcal{A}'$, and $\mathcal{E}^+$, respectively. Compute the forces and moments acting in the shaft at point **B**, those acting in the arm at the same point, and finally, the forces and moments acting in the bearing at point **O**.

The angular velocity of the disk is readily obtained from the addition theorem as $\underline{\omega} = \Omega \bar{\imath}_3 + \omega \bar{b}_3$. The components of the angular velocity and acceleration vectors then become

$$\underline{\omega}^* = \Omega(R_\Omega R'_\beta R^+_\omega)^T \bar{\imath}_3 + \omega \underline{b}^*_3 = \begin{Bmatrix} -\Omega S_\beta C_\omega \\ \Omega S_\beta S_\omega \\ \Omega C_\beta + \omega \end{Bmatrix}, \text{ and } \underline{\dot{\omega}}^* = \begin{Bmatrix} \Omega\omega S_\beta S_\omega \\ \Omega\omega S_\beta C_\omega \\ 0 \end{Bmatrix},$$

where $C_\Omega = \cos\Omega t$, $C_\beta = \cos\beta$, and $C_\omega = \cos\omega t$, with similar expressions for the sine functions of the corresponding angles.

The inertial position of the center of mass of the disk is $\underline{r}_C = d\bar{b}_3$; its acceleration vector then becomes $\underline{a}_C = d(\tilde{\dot{\omega}} + \tilde{\omega}\tilde{\omega})\bar{b}_3$. The components of this acceleration vector in basis $\mathcal{B}^*$ then become $\underline{a}^*_C = d(\tilde{\dot{\omega}}^* + \tilde{\omega}^*\tilde{\omega}^*)\bar{b}^*_3$, or

$$\underline{a}^*_C = d \begin{Bmatrix} \dot{\omega}^*_2 + \omega^*_1\omega^*_3 \\ -\dot{\omega}^*_1 + \omega^*_2\omega^*_3 \\ -\omega^{*2}_1 - \omega^{*2}_2 \end{Bmatrix} = d \begin{Bmatrix} -\Omega^2 S_\beta C_\beta C_\omega \\ \Omega^2 S_\beta C_\beta S_\omega \\ -\Omega^2 S^2_\beta \end{Bmatrix}.$$

Body attached frame $\mathcal{F}^B$ is located at the center of mass of the disk and is aligned with its principal axes of inertia. Figure 6.41 gives the principal mass moments of inertia of the disk as $I^{C*}_{11} = I^{C*}_{22} = mr^2/4$ and $I^{C*}_{33} = mr^2/2$. With the help of the free body diagram shown in fig. 6.9, the equations of motion of the disk, see eqs. (6.19) and (6.21), then become

$$\underline{F}^*_B = md\Omega^2 \begin{Bmatrix} -S_\beta C_\beta C_\omega \\ S_\beta C_\beta S_\omega \\ -S^2_\beta \end{Bmatrix},$$

and

$$\underline{M}^*_B - a\tilde{b}^*_3\underline{F}^*_B = \frac{1}{4}mr^2 \begin{Bmatrix} \Omega\omega S_\beta S_\omega + \Omega S_\beta S_\omega(\Omega C_\beta + \omega) \\ \Omega\omega S_\beta C_\omega + \Omega S_\beta C_\omega(\Omega C_\beta + \omega) \\ 0 \end{Bmatrix}$$
$$= \frac{mr^2}{4}\Omega(2\omega + \Omega C_\beta)S_\beta \begin{Bmatrix} S_\omega \\ C_\omega \\ 0 \end{Bmatrix}, \tag{6.30}$$

respectively. The resultant of the externally applied moments was computed with respect to the center of mass, as required by eq. (6.21). In these equations, $\underline{F}_B$ and $\underline{M}_B$ are the externally applied force and moment vectors acting on the disk at point **B**.

Eliminating force $\underline{F}^*_B$ from the equations of motion yields an expression for moment $\underline{M}^*_B$,

$$\underline{M}^*_B = \left[\frac{mr^2}{4}\Omega(2\omega + \Omega C_\beta)S_\beta - mad\Omega^2 S_\beta C_\beta\right] \begin{Bmatrix} S_\omega \\ C_\omega \\ 0 \end{Bmatrix}.$$

The third component of moment, M^*_{B3}, vanishes; this implies that no moment needs to be applied to the disk about unit vector $\bar{b}_3$ to maintain its constant angular velocity, ω. Due to the presence of the trigonometric functions $S_\omega = \sin\omega t$ and $C_\omega = \cos\omega t$, the components of the moment vector, $\underline{M}_B$, acting on the disk are time-dependent when resolved in the disk attached basis $\mathcal{B}^*$. This implies that the shaft will be subjected to fatigue loading and as the angular speed, ω, of the disk increases, it will accumulate an increasing amount of loading cycles per unit time.

On the other hand, the components of the same moment *resolved in the arm attached basis*, $\mathcal{E}^+$, denoted $\underline{M}_B^+ = \underline{\underline{R}}_\omega^+ \underline{M}_B^*$, become

$$\underline{M}_B^+ = \left[\frac{mr^2}{4} \Omega(2\omega + \Omega C_\beta)S_\beta - mad\Omega^2 S_\beta C_\beta \right] \begin{Bmatrix} 0 \\ 1 \\ 0 \end{Bmatrix}. \qquad (6.31)$$

Unlike the shaft carrying the disk, the arm is subjected to a *constant bending moment*. It is easily verified that the components of force in the arm attached basis, $\underline{F}^+ = \underline{\underline{R}}_\omega^+ \underline{F}^*$, are also constant in time.

Figure 6.9 also shows a free body diagram of the arm; because this component is massless, the equations of statics apply: the sum of both forces and moments must vanish. This yields the following expressions for the externally applied force and moment vectors at point **O**, denoted $\underline{F}_O$ and $\underline{M}_O$, respectively: $\underline{F}_O = \underline{F}_B$ and $\underline{M}_O = \underline{M}_B + \tilde{r}_{OB}\underline{F}_B$. The components of these vectors, expressed in basis $\mathcal{A}'$, are $\underline{F}_O' = \underline{\underline{R}}_\beta' \underline{F}_B^+$ and $\underline{M}_O' = \underline{\underline{R}}_\beta'(\underline{M}_B^+ + \tilde{r}_{OB}^+ \underline{F}_B^+)$, respectively, and are, of course, constant in time.

Finally, the forces and moments acting on the bearing at point **O**, resolved in the inertial frame, are $\underline{F}_O = \underline{\underline{R}}_\Omega \underline{F}_O'$ and $\underline{M}_O = \underline{\underline{R}}_\Omega \underline{M}_O'$, respectively. The third component of moment, M_{O3}, vanishes: no moment needs to be applied to the arm about axis $\bar{\imath}_3$ to maintain the constant angular velocity, Ω, of the system. Because $S_\Omega = \sin\Omega t$ and $C_\Omega = \cos\Omega t$, the other loading components are time-dependent: as expected, the bearing will be subjected to cyclic loading. The bearing is subjected to loads oscillating at a frequency Ω, in contrast with those acting in the shaft, which have a frequency ω.

Point **A** is a fixed point, or pivot point, of the disk: rotation about the shaft and rotation of the bent arm leave point **A** at an inertial location. Furthermore, point **A** is a material point of the disk, consequently, the pivot equation, eq. (6.23), could have been used instead of Euler's equation, eq. (6.21). The mass moment of inertia tensor of the disk with respect to point **A** can be obtained from its counterpart about point **C** with the help of the parallel axis theorem, eq. (6.9), to find $I_1^{A*} = m(r^2/4 + d^2)$, $I_2^{A*} = m(r^2/4 + d^2)$, $I_3^{A*} = mr^2/2$. The pivot equation about point **A** now becomes

$$\underline{M}_B^* + (d-a)\tilde{b}_3^* \underline{F}_B^* = m\Omega S_\beta \begin{Bmatrix} (r^2/4 + d^2)\omega S_\omega + (r^2/4 - d^2)S_\omega(\Omega C_\beta + \omega) \\ (r^2/4 + d^2)\omega C_\omega + (r^2/4 - d^2)C_\omega(\Omega C_\beta + \omega) \\ 0 \end{Bmatrix}.$$

This equation is equivalent to that derived above. Indeed, introducing the expression for the force $\underline{F}_B^*$ from eq. (6.30), leads again to eq. (6.31) for the externally applied moment $\underline{M}_B^*$.

Example 6.2. Swiveling plate

Figure 6.10 shows a homogeneous, rectangular plate of height a, width b, and mass m connected to the ground by a rigid, massless link of length d. At point **O**, a bearing allows the link to rotate with respect to axis $\bar{\imath}_3$, and at point **B**, the plate is free to rotate with respect to the link about axis $\bar{a}_1$.

Three frames are used in this problem: the inertial frame, $\mathcal{F}^I = [\mathbf{O}, \mathcal{I} = (\bar{\imath}_1, \bar{\imath}_2, \bar{\imath}_3)]$, a frame attached to the link, $\mathcal{F}^A = [\mathbf{O}, \mathcal{A}^+ = (\bar{a}_1, \bar{a}_2, \bar{a}_3)]$, and finally, a frame attached to the plate's center of mass, $\mathcal{F}^B = [\mathbf{C}, \mathcal{B}^* = (\bar{b}_1, \bar{b}_2, \bar{b}_3)]$. Tensor components resolved in bases $\mathcal{A}^+$ and $\mathcal{B}^*$ are denoted with superscripts $(\cdot)^+$ and $(\cdot)^*$, respectively. A planar rotation of magnitude α about axis $\bar{\imath}_3$ brings basis $\mathcal{I}$ to basis $\mathcal{A}^+$, a planar rotation of magnitude β about axis $\bar{a}_1$ brings basis $\mathcal{A}^+$ to basis $\mathcal{B}^*$, and rotation tensors $\underline{\underline{R}}_\alpha$ and $\underline{\underline{R}}_\beta$ are associated with these two planar rotations, respectively

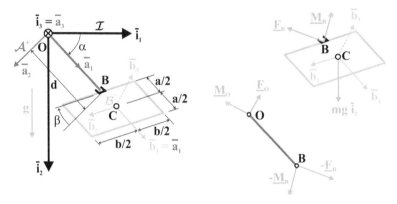

Fig. 6.10. Configuration of the swiveling plate.

The inertial angular velocity vector of the plate is readily found with the help of the addition theorem as $\underline{\omega} = \dot{\alpha}\bar{a}_3 + \dot{\beta}\bar{a}_1$. The components of the angular velocity and acceleration vectors resolved in basis $\mathcal{B}^*$ are then found to be

$$\underline{\omega}^* = \dot{\alpha}R_\beta^{+T}\bar{a}_3^+ + \dot{\beta}\bar{a}_1^+ = \left\{\begin{array}{c} \dot{\beta} \\ \dot{\alpha}S_\beta \\ \dot{\alpha}C_\beta \end{array}\right\}, \quad \text{and} \quad \dot{\underline{\omega}}^* = \left\{\begin{array}{c} \ddot{\beta} \\ \ddot{\alpha}S_\beta + \dot{\alpha}\dot{\beta}C_\beta \\ \ddot{\alpha}C_\beta - \dot{\alpha}\dot{\beta}S_\beta \end{array}\right\},$$

respectively. The inertial position of the center of mass of the plate is $\underline{r}_C = (d + a/2)\bar{a}_1$ and the acceleration vector $\underline{a}_C = (d + a/2)(\ddot{\alpha}\bar{a}_2 - \dot{\alpha}^2\bar{a}_1)$; the components of this vector resolved in basis $\mathcal{B}^*$ then become

$$\underline{a}_C^* = (d + \frac{a}{2})\left\{\begin{array}{c} -\dot{\alpha}^2 \\ \ddot{\alpha}C_\beta \\ -\ddot{\alpha}S_\beta \end{array}\right\}.$$

Body attached frame $\mathcal{F}^B$ is located at the plate's center of mass and is aligned with its principal axes of inertia, see fig. 6.45. The principal mass moments of inertia are $I_{11}^{C*} = mb^2/12$, $I_{22}^{C*} = ma^2/12$, $I_{33}^{C*} = m(a^2 + b^2)/12$. Figure 6.10 shows a free body diagram of the plate; $\underline{F}_B$ and $\underline{M}_B$ are the externally applied force and moment vectors acting on the plate at point **B**. The equations of motion for the plate, see eqs. (6.19) and (6.21), then become

$$\underline{F}_B^* + mg \left\{ \begin{matrix} S_\alpha \\ C_\alpha C_\beta \\ -C_\alpha S_\beta \end{matrix} \right\} = m\left(d + \frac{a}{2}\right) \left\{ \begin{matrix} -\dot\alpha^2 \\ \ddot\alpha C_\beta \\ -\ddot\alpha S_\beta \end{matrix} \right\}, \tag{6.32}$$

and

$$\underline{M}_B^* - \frac{a}{2}\tilde{b}_1^*\underline{F}_B^* = \frac{m}{12} \left\{ \begin{matrix} b^2\ddot\beta + b^2\dot\alpha^2 S_\beta C_\beta \\ a^2(\ddot\alpha S_\beta + \dot\alpha\dot\beta C_\beta) - a^2\dot\alpha\dot\beta C_\beta \\ (a^2 + b^2)(\ddot\alpha C_\beta - \dot\alpha\dot\beta S_\beta) - (b^2 - a^2)\dot\alpha\dot\beta S_\beta \end{matrix} \right\},$$

respectively. After simplification, this last equation becomes

$$\left\{ \begin{matrix} M_{B1}^* \\ M_{B2}^* + a/2\, F_{B3}^* \\ M_{B3}^* - a/2\, F_{B2}^* \end{matrix} \right\} = \frac{m}{12} \left\{ \begin{matrix} b^2(\ddot\beta + \dot\alpha^2 S_\beta C_\beta) \\ a^2\ddot\alpha S_\beta \\ (a^2 + b^2)\ddot\alpha C_\beta - 2b^2\dot\alpha\dot\beta S_\beta \end{matrix} \right\} = \underline{M}_I^*, \tag{6.33}$$

where $\underline{M}_I^*$ represents the right-hand side of this equation. The first component of moment, M_{B1}^*, must vanish because the plate is free to rotate with respect to the arm about axis $\bar{b}_1$; this reveals the first equation of motion of the problem, $\ddot\beta + \dot\alpha^2 S_\beta C_\beta = 0$.

Figure 6.10 also shows a free body diagram of the massless arm **OB**; $\underline{F}_O$ and $\underline{M}_O$ are the externally applied force and moment vectors acting on the arm at point **O**. The moment equilibrium equation about point **O**, expressed in basis $\mathcal{A}^+$, is $\underline{M}_O^+ - d\tilde{a}_1^+\underline{F}_B^+ - \underline{M}_B^+ = 0$. Introducing eq. (6.33) then yields $\underline{M}_O^+ = (d+a/2)\tilde{a}_1^+\underline{F}_B^+ + \underline{M}_I^+$. With the help of eqs. (6.32) and (6.33), this applied moment at point **O** becomes

$$\underline{M}_O^+ = m \left\{ \begin{matrix} 0 \\ -\frac{b^2}{12} S_\beta(\ddot\alpha C_\beta - 2\dot\alpha\dot\beta S_\beta) \\ \frac{a^2}{12}\ddot\alpha + \frac{b^2}{12} C_\beta(\ddot\alpha C_\beta - 2\dot\alpha\dot\beta S_\beta) + (d + a/2)^2\ddot\alpha - g(d + a/2)C_\alpha \end{matrix} \right\}.$$

Here again, the configuration of the system implies the vanishing of the third component of this moment, M_{O3}^+. The equations of motion of the system correspond to the vanishing of two components of moment, $M_{O3}^+ = 0$ and $M_{B1}^* = 0$, or

$$\left[a^2/12 + b^2/12\, C_\beta^2 + (d + a/2)^2\right]\ddot\alpha - b^2/6\, \dot\alpha\dot\beta S_\beta C_\beta = (d + a/2)gC_\alpha,$$

$$\ddot\beta + \dot\alpha^2 S_\beta C_\beta = 0,$$

respectively. The two conditions leading to the equations of motion of the problem can be expressed as scalar products: $\bar{b}_1^T \underline{M}_B = 0$ and $\bar{\imath}_3^T \underline{M}_O = 0$, which are easily evaluated when the vectors are expressed in an appropriate basis: $\bar{b}_1^{*T} \underline{M}_B^* = 0$ and $\bar{\imath}_3^{+T} \underline{M}_O^+ = 0$.

Example 6.3. *Rigid body connected to a spring and dashpot*

Figure 6.11 depicts a rigid body connected to the ground at point **B** by means of a spring of stiffness constant k and dashpot of constant c. The rigid body is of mass M and its moment of inertia tensor with respect to the center of mass is $\underline{\underline{I}}^C$. Vector η defines the position of the center of mass with respect to point **B**. Frame $\mathcal{F}^B = \left[\mathbf{B}, \mathcal{B}^* = (\bar{b}_1, \bar{b}_2, \bar{b}_3) \right]$ is attached to the rigid body; superscript $(\cdot)^*$ indicates tensor components resolved in basis $\mathcal{B}^*$. The components of the rotation tensor that brings basis $\mathcal{I}$ to basis $\mathcal{B}^*$, resolved in basis $\mathcal{I}$, are denoted $\underline{\underline{R}}$. Find the equations of motion of the system.

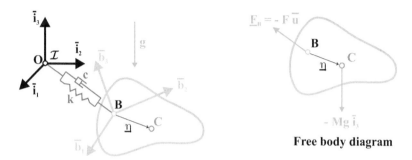

Fig. 6.11. Configuration of the rigid body connected to a spring and dashpot.

Let $\underline{x}_B$ and $\underline{x}_C$ be the position vectors of points **B** and **C** with respect to point **O**; it follows that $\underline{x}_B = \underline{x}_C - \eta$. The force vector, $\underline{F}_B$, applied to the rigid body at point **B** then acts in the direction of unit vector $\bar{u} = \underline{x}_B/\|\underline{x}_B\|$, or $\underline{F}_B = -F\bar{u}$, where F is the magnitude of the applied force. It then follows that $F = k\Delta + c\dot{\Delta}$, where $\Delta = \|\underline{x}_B\| - \ell_0$ is the stretch of the spring and ℓ_0 its un-stretched length. The time rate of change of the stretch is easily found as $\dot{\Delta} = \bar{u}^T \dot{\underline{x}}_B$. The equation of motion for the center of mass is $M\ddot{\underline{x}}_C = -F\bar{u} - Mg\,\bar{\imath}_3$. Euler's equation, eq. (6.21), implies $\underline{\underline{I}}^{C*}\dot{\underline{\omega}}^* + \tilde{\omega}^*\underline{\underline{I}}^{C*}\underline{\omega}^* = \underline{\underline{R}}^T \left[-\tilde{\eta}(-F\bar{u}) \right]$, where the right-hand side represents the moment of the externally applied forces, resolved in basis $\mathcal{B}^*$.

Although it is convenient to work with the components of all vectors and tensors resolved in the body attached basis, it is also possible to use the corresponding components resolved in the inertial basis. Multiplying by $\underline{\underline{R}}$ leads to $\underline{\underline{I}}^C\dot{\underline{\omega}} + \tilde{\omega}\underline{\underline{I}}^C\underline{\omega} = F\tilde{\eta}\bar{u}$, where all vectors and tensors are now resolved in the inertial basis.

For numerical solution of the equations of motion, it is convenient to recast them as a set of first-order equations by introducing the velocities of the center of mass, $\underline{v}_C$, and a set of parameters, $\underline{q}$, that represent the rotation of the rigid body. These parameters could be selected as Euler angles with a specific sequence of planar rotations, see section 4.11. The angular velocity of the body then becomes $\underline{\omega} = \underline{\underline{H}}(\underline{q})\dot{\underline{q}}$. For Euler angles with the sequence *3-1-3* defined in section 4.11.1, the tangent operator, $\underline{\underline{H}}$, is given by eq. (4.68). The complete set of first-order equations now becomes

$$\left\{\begin{matrix} \dot{x}_C \\ \dot{q} \\ \dot{v}_C \\ \dot{\omega} \end{matrix}\right\} = \left\{\begin{matrix} v_C \\ \underline{\underline{H}}^{-1}\omega \\ -F/M\,\bar{u} - g\bar{\imath}_3 \\ (\underline{\underline{I}}^C)^{-1}(F\tilde{\eta}\bar{u} - \tilde{\omega}\,\underline{\underline{I}}^C\omega) \end{matrix}\right\}.$$

(6.34)

This formulation requires the tangent operator, $\underline{\underline{H}}$, to be singularity free; as discussed in section 4.11.1, such is not the case for Euler angles with the sequence *3-1-3*, for which $\underline{\underline{H}}^{-1}$ is singular then angle $\theta = 0$.

6.6.1 Problems

Problem 6.9. Kinetic energy of a rigid body
Derive an expression for the rotational kinetic energy of a rigid body. Use a body attached frame with its origin at the center of mass and orientation that coincides with the principal axes of inertia. The orientation of the body attached frame with respect to an inertial frame will be determined by Euler angles with the *3-1-3* sequence, see section 4.11.

Problem 6.10. Rigid body connected to a fixed point
Point **C** is the center of mass of a rigid body of arbitrary shape. This point is connected to an inertial point **O** by means of a ball and socket joint. Point **C** and **O** are coincident. The only externally applied forces are the gravity forces and the reactions at point **O**. *(1)* Prove that the angular momentum vector $\underline{H}_O$ of the body is of constant magnitude and direction. *(2)* Prove that the kinetic energy of the body remains a constant. *(3)* Show that the magnitude of the projection of the angular velocity vector along the direction of the angular momentum vector is a constant; find this constant.

Problem 6.11. Rigid body moving along a curve
A rigid body freely slides along a given curve $\mathbb{C}$ in three-dimensional space. A point of the curve has a position vector $\underline{p}_0(s)$. The position of the reference point of the rigid body is $\underline{p}_0(s)$ and its orientation is determined by Frenet's triad $\underline{\underline{R}}(s) = [\bar{t}(s), \bar{n}(s), \bar{b}(s)]$. Find the equation of motion for the rigid body if it is subjected to externally applied forces and moments. Hint: the only degree of freedom of the problem is s, the position of the body along the curve.

Problem 6.12. Spinning rotor mounted on a rotating disk
Figure 6.12 depicts a homogeneous disk of mass M and radius R rotating about inertial axis $\bar{\imath}_3$. Frame $\mathcal{F}^D = [\mathbf{O}, \mathcal{E}^+ = (\bar{e}_1, \bar{e}_2, \bar{e}_3)]$ is attached to the disk. The disk rotates about unit vector $\bar{\imath}_3$ at a constant angular velocity, Ω. At the rim of the disk, a rigid massless shaft of length d extends in the radial direction and connects to a homogeneous disk of mass m and radius r spinning about unit vector $\bar{e}_1$ at a constant angular velocity, ω. *(1)* Find the three components of the reaction force in the bearing at point **B**, resolved in basis $\mathcal{E}^+$. *(2)* Find the three components of the reaction moment in the bearing at point **B**, resolved in basis $\mathcal{E}^+$.

Problem 6.13. Spinning rotor mounted on a rotating disk
Figure 6.12 depicts a homogeneous disk of mass M and radius R rotating about inertial axis $\bar{\imath}_3$. Frame $\mathcal{F}^D = [\mathbf{O}, \mathcal{E}^+ = (\bar{e}_1, \bar{e}_2, \bar{e}_3)]$ is attached to the disk. Torque T is applied to the disk and act about unit vector $\bar{\imath}_3$. At the rim of the disk, a rigid massless shaft of length d extends in the radial direction and connects to a homogeneous disk of mass m and radius r spinning about unit vector $\bar{e}_1$. Frame $\mathcal{F}^R = [\mathbf{R}, \mathcal{B}^* = (\bar{b}_1, \bar{b}_2, \bar{b}_3)]$ is attached to the rotor. Torque Q is applied to the rotor and act about unit vector $\bar{e}_1$. *(1)* Develop the equations of

motion of the system in terms of angles ϕ and θ, where θ is the rotation of the rotor about unit vector $\bar{e}_1$. *(2)* Find the three components of the reaction moment in the bearing at point **B**, resolved in basis $\mathcal{E}^+$. *(3)* Find the three components of the reaction force in the bearing at point **B**, resolved in basis $\mathcal{E}^+$. *(4)* Find the three components of the reaction moment in the bearing at point **O**, resolved in basis $\mathcal{E}^+$ and $\mathcal{I}$. *(5)* Find the three components of the reaction force in the bearing at point **O**, resolved in basis $\mathcal{E}^+$ and $\mathcal{I}$. *(6)* After an initial start-up phase, the disk and rotor spin at constant angular velocities, $\dot{\phi} = \Omega$ and $\dot{\theta} = \omega$, respectively. Determine the reaction forces and moments of questions *(2)* to *(5)*.

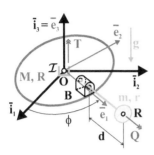

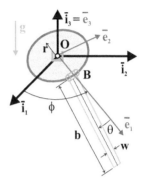

Fig. 6.12. Spinning rotor mounted on a rotating disk.

Fig. 6.13. Plate hinged at the rim of a rotating disk.

Problem 6.14. Plate hinged at the rim of a rotating disk

Figure 6.13 depicts a homogeneous disk of mass M and radius R rotating about inertial axis $\bar{i}_3$. Frame $\mathcal{F}^D = [\mathbf{O}, \mathcal{E}^+ = (\bar{e}_1, \bar{e}_2, \bar{e}_3)]$ is attached to the disk. At point **B**, a point on the rim of the disk, a homogeneous plate of mass m, length b, and width w is hinged to the disk. The hinge's axis is aligned with unit vector $\bar{e}_2$; a torsional spring of stiffness constant k and a torsional dashpot of constant c are located at the hinge. The torsional spring is un-stretched when $\theta = \theta_0$. The system is subjected to gravity acting in the direction indicated on the figure. *(1)* Develop the equations of motion of the system in terms of angles ϕ and θ indicated on the figure. *(2)* On one graph, plot angles ϕ and θ versus τ. *(3)* On one graph, plot angular speeds ϕ' and θ'. *(4)* On one graph, plot angular accelerations ϕ'' and θ''. *(5)* Plot the cumulative energy, $\bar{W}^d = W^d/k$, dissipated in the dashpot. *(6)* On one graph, plot the kinetic, $\bar{K} = K/k$ and potential, $\bar{V} = V/k$ energies of the system. Check that the energy closure equation is satisfied. *(7)* On one graph, plot the three components of the moment in the bearing at point **B**, resolved in basis $\mathcal{E}^+$. *(8)* On one graph, plot the three components of the force in the bearing at point **B**, resolved in basis $\mathcal{E}^+$. Use the following data: $\mu = M/m = 1.5$, $\bar{w} = w/b = 0.2$, $\bar{R} = R/b = 0.2$, $\zeta = wc/(2k) = 0.05$, $\bar{g} = g/(bw^2) = 2$, $\theta_0 = 0$. A non-dimensional time is defined, $\tau = \omega t$, where $\omega^2 = 3k/(mb^2)$; notation $(\cdot)'$ indicates a derivative with respect to τ. Use the following initial conditions, $\phi(\tau = 0) = 0$, $\theta = 0$, $\phi' = 1$, $\theta' = -1$. Present all your results for $\tau \in [0, 20]$.

Problem 6.15. Spinning Satellite

Frame $\mathcal{F} = [\mathbf{B}, \mathcal{B}^* = (\bar{b}_1, \bar{b}_2, \bar{b}_3)]$ is attached to a satellite. Point **B** is the satellite's center of mass and basis $\mathcal{B}^*$ is aligned with its principal axes of inertia. Tensor components resolved in basis $\mathcal{B}^*$ are denoted with a superscript $(\cdot)^*$. The components of the angular velocity vector of the satellite, resolved in $\mathcal{B}^*$, are denoted $\underline{\omega}^*$. The mass moments of inertia are $I_1^* = 12$, $I_2^* = 16$ and $I_3^* = 20$ kg.m^2. During a maneuver, thrusters apply a moment $\underline{M}(t)$ to the satellite. For $t \le T$, $\underline{M}^*(t) = Q^* \sin 2\pi t/T$, and for $t > T$, $\underline{M}^*(t) = \underline{0}$, where $T = 5$ s. The initial angular velocity of the satellite is $\underline{\omega}^{*T}(t = 0) = \{0, 0.5, 0\}$ rad/s. Moment vector Q^* is defined by its components in the body attached basis. Two cases will be considered here: for *case 1*, $Q^{*T} = \{0, 5, 0\}$ N·m, for *case 2*, $Q^{*T} = \{5, 0, 0\}$ N·m. *(1)* Solve Euler's equation for the time history of the angular velocity of the satellite. *(2)* On one graph, plot the three components of the angular velocity vector in the body attached frame as a function of time for $t \in [0, 30T]$. Present one graph for *case 1* and one for *case 2*. *(3)* At the end of the maneuver, will the orientation of the satellite remain fixed with respect to an inertial frame for *case 1*? What about *case 2*?

Problem 6.16. Double spatial pendulum

Figure 6.14 depicts a double spatial pendulum consisting of two bodies subjected to gravity. The first body, of mass m_a and mass moment of inertia tensor $\underline{\underline{I}}^{C_a}$, is connected to the ground at point $\mathbf{O}$ by means of a ball and socket joint. The position vectors of points $\mathbf{O}$ and $\mathbf{B}$ with respect to the center of mass, $\mathbf{C_a}$, of the body are denoted η_a and μ_a, respectively. The second body, of mass m_b and mass moment of inertia tensor $\underline{\underline{I}}^{C_b}$, is connected to the first body at point $\mathbf{B}$ though a ball and socket joint. The position vector of point $\mathbf{B}$ with respect to the center of mass, $\mathbf{C_b}$, of the body is denoted η_b. Two frames, $\mathcal{F}^A = [\mathbf{C_a}, \mathcal{A} = (\bar{a}_1, \bar{a}_2, \bar{a}_3)]$ and $\mathcal{F}^B = [\mathbf{C_b}, \mathcal{B} = (\bar{b}_1, \bar{b}_2, \bar{b}_3)]$, are attached to the first and second body, respectively. Let $\underline{\underline{R}}_a$ and $\underline{\underline{R}}_b$ be the rotation tensors that bring basis $\mathcal{I}$ to $\mathcal{A}$ and basis $\mathcal{I}$ to $\mathcal{B}$, respectively. Tensor $\underline{\underline{R}}_a$ will be represented with Euler angles ϕ_a, θ_a, and ψ_a using the *3-1-3* sequence, and Euler angles ϕ_b, θ_b, and ψ_b, also using the *3-1-3* sequence, represent tensor $\underline{\underline{R}}_b$. *(1)* Draw free body diagrams for each of the bodies. *(2)* Derive the equations of motion of the system. Carefully define all terms appearing in the equations.

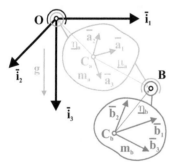 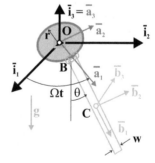

Fig. 6.14. Configuration of the double spatial pendulum.

Fig. 6.15. Configuration of the plate hinged at the rim of a disk.

Problem 6.17. Plate hinged at the rim of a disk

The system depicted in fig. 6.15 features a disk of radius r rotating at a constant angular velocity Ω about inertial axis $\bar{i}_3$. Frame $\mathcal{F}^A = [\mathbf{O}, \mathcal{A} = (\bar{a}_1, \bar{a}_2, \bar{a}_3)]$ is attached to the disk. At point $\mathbf{B}$, a point on the rim of the disk, a homogeneous plate of mass m, length b, and width

w is hinged to the disk; the plate is free to rotate with respect to the disk about axis $\bar{a}_2$. Frame $\mathcal{F}^B = \left[\mathbf{C}, \mathcal{B} = (\bar{b}_1, \bar{b}_2, \bar{b}_3) \right]$ is attached to the plate; point $\mathbf{C}$ is the center of mass of the plate. Line $\mathbf{BC}$ is at an angle θ with respect to the vertical. The system is subjected to gravity acting in the direction indicated on the figure. Determine the angular speed, Ω, required to maintain a given, constant angle θ.

Problem 6.18. Robotic arm

Figure 5.5 shows a robotic system. The shaft is allowed to rotate with respect to an inertial frame $\mathcal{F}^I$, about axis $\bar{\imath}_3$; the time-dependent angle of rotation is denoted $\alpha(t)$. Frame $\mathcal{F}^S = \left[\mathbf{S}, \mathcal{S}^+ = (\bar{s}_1, \bar{s}_2, \bar{s}_3) \right]$ is attached to the shaft at a distance $h = 0.5$ m from the origin of the inertial frame, as indicated on the figure. An arm of length $L_a = 1.2$ m, extending along the direction of axis $\bar{s}_2$, is attached to the shaft at point $\mathbf{S}$. Finally, a rigid manipulator of length $L_b = 0.5$ m, radius $r_b = 0.02$ m, and mass $m_b = 10$ kg is connected to the arm at point $\mathbf{B}$. The manipulator is allowed to rotate with respect to frame $\mathcal{F}^S$, about axis $\bar{s}_1$; the time-dependent angle of rotation is denoted $\beta(t)$. Frame $\mathcal{F}^B = \left[\mathbf{B}, \mathcal{B}^* = (\bar{b}_1, \bar{b}_2, \bar{b}_3) \right]$ is attached to the manipulator. Superscripts $(\cdot)^*$ and $(\cdot)^+$ denote tensor components resolved in bases $\mathcal{B}^*$ and $\mathcal{S}^+$, respectively. Angles $\alpha(t)$ and $\beta(t)$ are prescribed as $\alpha(t) = \pi/2 \, (1 - \cos \pi t/T)$, and $\beta(t) = 2\pi(1 - \cos \pi t/T)$, respectively, where $T = 2$ s. The acceleration of gravity is $g = 9.81$ m/s^2. (1) Compute the components of the force vector $\underline{F}^*$ and moment vector $\underline{M}^*$ applied to the manipulator at point $\mathbf{B}$. (2) On one graph, plot the components of the force vector $\underline{F}^*$. (3) On one graph, plot the components of the moment vector $\underline{M}^*$. (4) What is the moment required to rotate the manipulator?

Problem 6.19. Rotating disk on a bent arm

Figure 6.9 shows a rotating disk connected to a bent arm. Massless arm $\mathbf{OAB}$ features a bend of β rad at point $\mathbf{A}$. At point $\mathbf{O}$, a bearing allows the arm to rotate with an angular velocity Ω with respect to ground. A disk of mass m and radius r rotates with an angular velocity ω and is connected to the arm at point $\mathbf{B}$ by means of a massless shaft. A planar rotation about axis $\bar{\imath}_3$ brings inertial frame $\mathcal{F}^I = [\mathbf{O}, \mathcal{I} = (\bar{\imath}_1, \bar{\imath}_2, \bar{\imath}_3)]$ to frame $\mathcal{F}^A = [\mathbf{O}, \mathcal{A}' = (\bar{a}_1, \bar{a}_2, \bar{a}_3)]$ that is attached to the arm. A second planar rotation of magnitude β about axis $\bar{a}_2$ brings frame $\mathcal{A}'$ to frame $\mathcal{F}^E = [\mathbf{A}, \mathcal{E}^+ = (\bar{e}_1, \bar{e}_2, \bar{e}_3)]$ that is also attached to the arm, with axis $\bar{e}_3$ pointing along the bent segment $\mathbf{AB}$. Superscripts $(\cdot)'$ and $(\cdot)^+$ denote tensor components resolved in basis $\mathcal{A}'$ and $\mathcal{E}^+$, respectively. Finally, a planar rotation about axis $\bar{e}_3$ brings frame $\mathcal{E}^+$ to frame $\mathcal{F}^B = [\mathbf{C}, \mathcal{B}^* = (\bar{b}_1, \bar{b}_2, \bar{b}_3)]$ that is attached to the rotating disk; all tensor components resolved in basis $\mathcal{B}^*$ are denoted with superscripts $(\cdot)^*$. The components of the rotation tensor that brings inertial basis $\mathcal{I}$ to basis $\mathcal{B}^*$, resolved in $\mathcal{I}$, is constructed as $\underline{\underline{R}} = \underline{\underline{R}}_\Omega \underline{\underline{R}}'_\beta \underline{\underline{R}}^+_\omega$, where $\underline{\underline{R}}_\Omega$, $\underline{\underline{R}}'_\beta$, and $\underline{\underline{R}}^+_\omega$ are the components of the rotation tensors that bring basis $\mathcal{I}$ to basis $\mathcal{A}'$, $\mathcal{A}'$ to $\mathcal{E}^+$, and $\mathcal{E}^+$ to $\mathcal{B}^*$, respectively, resolved in bases $\mathcal{I}$, $\mathcal{A}'$, and $\mathcal{E}^+$, respectively. The angular velocities of the bent arm are prescribed to be $\Omega = \Omega_f(1 - \cos 2\pi t/T)/2$ for $0 \leq t \leq T/2$ and $\Omega = \Omega_f$ for $t > T/2$. The angular velocities of the disk are prescribed to be $\omega = \omega_f(1 - \cos 2\pi t/T)/2$ for $0 \leq t \leq T/2$, and $\omega = \omega_f$ for $t > T/2$. This represents the start-up sequence for the system from the rest condition to a nominal operating point where the angular velocities of the arm and disk are stabilized to their final values, Ω_f and ω_f, respectively. These angular velocity profiles are achieved by applying to the bent arm a torque $Q_O(t)$ about axis $\bar{\imath}_3$ at point $\mathbf{O}$ and to the shaft a torque $Q_B(t)$ about axis $\bar{b}_3$ at point $\mathbf{B}$. (1) On one graph, plot the time history of angular velocities Ω and ω. (2) On one graph, plot angular accelerations $\dot{\Omega}$ and $\dot{\omega}$. (3) Plot the three components of the angular velocity vector of the disk, $\underline{\omega}^*$. (4) Plot the three components of the angular acceleration vector of the disk, $\underline{\dot{\omega}}^*$. (5) Plot the three components of the moment vector applied to the shaft at point

B, $\underline{M}_B^*$. (6) Plot the three components of the moment vector applied to the arm at point **B**, $\underline{M}_B^+$. (7) Plot the three components of the force vector applied to the arm at point **O**, $\underline{F}$. (8) Plot the three components of the moment vector applied to the arm at point **O**, $\underline{M}_O^+$. (9) Plot the three components of the moment vector applied to the arm at point **O**, $\underline{M}_O$. (10) Plot the cumulative work done by torques $Q_O(t)$ and $Q_B(t)$, and the total kinetic energy of the system. (11) Demonstrate by a graph that your predictions satisfy the principle of work and energy. (12) Plot the instantaneous power required by the servomotors located at points **B** and **O**. (13) If the servomotors can deliver a maximum power of 50 Watts each, find the minimum time T required to bring the system to steady angular velocities. Use the following data: $\beta = \pi/6$ rad, $\omega_f = 50$ rad/s, $\Omega_f = 10$ rad/s, $r = 0.2$ m, $m = 10$ kg, $h = 0.6$ m, $d = 0.3$ m, $a = 0.1$ m, and $T = 15$ s.

Problem 6.20. Swiveling plate

Figure 6.10 shows a homogeneous, rectangular plate of height a, width b, and mass m connected to the ground by a rigid, massless link of length d. At point **O**, a bearing allows the link to rotate with respect to axis $\bar{\imath}_3$, and at point **B**, the plate is free to rotate with respect to the link about axis $\bar{a}_1$. Three frames will be used in this problem: inertial frame $\mathcal{F}^I = [\mathbf{O}, \mathcal{I} = (\bar{\imath}_1, \bar{\imath}_2, \bar{\imath}_3)]$, a frame attached to the link, $\mathcal{F}^A = [\mathbf{O}, \mathcal{A}^+ = (\bar{a}_1, \bar{a}_2, \bar{a}_3)]$, and finally, a frame attached to the plate at its center of mass, $\mathcal{F}^P = [\mathbf{C}, \mathcal{B}^* = (\bar{b}_1, \bar{b}_2, \bar{b}_3)]$. A planar rotation of magnitude α about axis $\bar{\imath}_3$ brings basis $\mathcal{I}$ to basis $\mathcal{A}^+$, and a planar rotation of magnitude β about axis $\bar{a}_1$ brings basis $\mathcal{A}^+$ to basis $\mathcal{B}^*$. Rotation tensors $\underline{\underline{R}}_\alpha$ and $\underline{\underline{R}}_\beta$ represent these two planar rotations, respectively; tensor components resolved in basis $\mathcal{A}^+$ and $\mathcal{B}^*$ are denoted with superscripts $(\cdot)^+$ and $(\cdot)^*$, respectively. (1) Derive the equations of motion of the problem. (2) On one graph, plot the time histories of angles α and β. (3) Plot $\dot{\alpha}$ and $\dot{\beta}$. (4) On one graph, plot the components of the angular velocity of the plate in basis $\mathcal{I}$. (5) Plot the components of the same vector in basis $\mathcal{B}^*$. (6) On one graph, plot the kinetic, potential, and total mechanical energies of the system. Comment on your results. (7) On one graph, plot the components of the force applied to the plate at point **B** resolved in basis $\mathcal{B}^*$. (8) Plot the components of the moment applied to the plate at point **B** in basis $\mathcal{B}^*$. (9) Plot the components of the moment applied to the link at point **O** in basis $\mathcal{I}$. Use the following data: $a = 0.2$ m, $b = 0.2$ m, $d = 0.5$ m, acceleration of gravity $g = 9.81$ m/s², and $m = 2$ kg. Present the response on the system for a period of 15 s. At first, use the following initial conditions: $\alpha = \pi/4$, $\beta = \pi/12$, and $\dot{\alpha} = \dot{\beta} = 0$. Next, consider a different set of initial conditions: $\alpha = 0$, $\beta = \pi/4$, and $\dot{\alpha} = \dot{\beta} = 0$. Comment on the response of the system for these two sets of initial conditions.

Problem 6.21. Rigid body connected to spring and dashpot

Figure 6.11 depicts a rigid body connected to the ground at point **B** by means of a spring of stiffness constant k and dashpot of constant c. The rigid body is of mass M and its moment of inertia tensor with respect to the center of mass is $\underline{\underline{I}}^C$. Vector η defines the position of the center of mass with respect to point **B**. Frame $\mathcal{F}^B = [\mathbf{B}, \mathcal{B} = (\bar{b}_1, \bar{b}_2, \bar{b}_3)]$ is attached to the rigid body. The components of the rotation tensor that brings inertial basis $\mathcal{I}$ to basis $\mathcal{B}$, resolved in basis $\mathcal{I}$, are denoted $\underline{\underline{R}}$. (1) Derive the equations of motion of the problem; resolve the components of all vectors and tensors in the inertial frame. (2) On one graph, plot the time histories of the three components of vector $\bar{x}_C$. (3) On one graph, plot Euler angles ψ, θ, and ϕ as a function of τ. (4) On one graph, plot the time histories of the three components of the velocity vector, $\bar{v}_C = v_C/(\Omega \ell_0)$. (5) On one graph, plot the time histories of the three components of the angular velocity vector, $\bar{\omega} = \omega/\Omega$. (6) On one graph, plot the time histories of the forces in the elastic spring and dashpot. (7) On one graph, plot the kinetic and potential energies of the system as well as the energy dissipated in the dashpot. Show that the energy closure equation is verified. Treat the problem using a non-dimensional scheme with $\tau = \Omega t$, $\Omega^2 = k/M$ and $\bar{x}_C = x_C/\ell_0$. Use the following data: $\bar{g} = g/(\ell_0\Omega^2) = 0.4$, $\zeta = c/(2M\Omega) = 0.1$, $\underline{\bar{\eta}}^* = \underline{\eta}^*/\ell_0 = [0.8, 1.25, -1.8]^T$, and $\underline{\bar{I}}^{C*} = \underline{I}^{C*}/(M\ell_0^2) =$ diag$(1, 2.3, 1.5)$. Use the following initial conditions: $\underline{\bar{x}}_C^T(\tau = 0) = [0, 1, 0]$, $\underline{q}^T(\tau = 0) = [0, 0, 0]$, and the system is at rest. Present the response on the system for $\tau \in [0, 100]$. Hint: to avoid singularities, use Euler angles, ψ, θ, and ϕ, with the *3-2-1* sequence, as defined in section 4.11.3, to represent the rotation of the rigid body.

6.7 Planar motion of rigid bodies

The previous sections have focused on the three-dimensional motion of rigid bodies. In some cases, the motion of the body is restricted to a *planar motion*: the center of mass of the body moves in an inertial plane and its angular velocity vector is at all time normal to this plane.

Let axes $\bar{\imath}_1$ and $\bar{\imath}_2$ defines the inertial plane in which the center of mass moves; the position vector of the center of mass then becomes $\underline{r}_C = x_{C1}\bar{\imath}_1 + x_{C2}\bar{\imath}_2$ and the angular velocity vector is $\underline{\omega} = \omega\bar{\imath}_3$. Next, a body attached frame, $\mathcal{F} = [\mathbf{C}, \mathcal{B} = (\bar{b}_1, \bar{b}_2, \bar{b}_3)]$, is defined, where point $\mathbf{C}$ is the body's center of mass. For convenience, axes $\bar{b}_1$ and $\bar{b}_2$ are selected to be in the plane of the motion whereas $\bar{b}_3$ is normal to the same plane.

It follows that the position vector of the center of mass becomes $\underline{r}_C = x_{C1}^*\bar{b}_1 + x_{C2}^*\bar{b}_2$ and the angular velocity vector is $\underline{\omega} = \omega\bar{b}_3$. The components of the position vector of the mass center resolved in the inertial basis are x_{C1} and x_{C2}, and x_{C1}^* and x_{C2}^* are their counterparts resolved in the body attached basis. The only non-vanishing component, ω, of angular velocity vector is the same in both frames: indeed, $\underline{\omega} = \omega\bar{\imath}_3 = \omega^*\bar{b}_3$ implies $\omega = \omega^*$, since $\bar{b}_3 = \bar{\imath}_3$ is an inertial direction.

The acceleration vector of the center of mass is now $\underline{a}_C = a_{C1}\bar{\imath}_1 + a_{C2}\bar{\imath}_2 = a_{C1}^*\bar{b}_1 + a_{C2}^*\bar{b}_2$, and the equations of motion for the center of mass, eq. (6.19), becomes $F_1 = ma_{C1}$, $F_2 = ma_{C2}$, and $F_3 = 0$. This last equation implies that the sum of the externally applied forces acting in the direction normal to the plane of motion must vanish if the motion is to remain planar. The following two equations of motion are sufficient to determine the motion of the center of mass

$$F_1 = ma_{C1}, \quad F_2 = ma_{C2}. \tag{6.35}$$

The second equation of motion can be written in several different manners depending on the point with respect to which the externally applied moments are computed, as discussed in section 6.5. The various options are detailed in the following sections.

6.7.1 Euler's equations

First, Euler's equations, see eqs. (6.21), specialized to the planar motion case become

$$M_{C1}^* = I_{13}^{C*}\dot{\omega} - I_{23}^{C*}\omega^2, \tag{6.36a}$$

$$M_{C2}^* = I_{23}^{C*}\dot{\omega} + I_{13}^{C*}\omega^2, \tag{6.36b}$$

$$M_{C3}^* = I_{33}^{C*}\dot{\omega}. \tag{6.36c}$$

Moment components M_{C1}^* and M_{C2}^* must be applied to sustain the planar motion; such moments are called *gyroscopic moments*. The sum of the externally applied moments is computed with respect to the center of mass of the rigid body.

If axes $\bar{b}_1$, $\bar{b}_2$, and $\bar{b}_3$ coincide with the principal axes of inertia of the body, the equations of motion further simplify since the cross products of inertia vanish, $I_{13}^{C*} = I_{23}^{C*} = 0$, leading to

$$M_{C3} = I_3^{C*}\dot{\omega}. \tag{6.37}$$

In this case, the two components of moment in the plane of motion must vanish, $M_{C2}^* = M_{C3}^* = 0$. The only non-vanishing force components are those in the plane of motion, F_1 and F_2. A single component of moment remains, M_{C3}; of course, $\underline{M} = M_{C3}\bar{i}_3 = M_{C3}^*\bar{b}_3$ implies $M_{C3} = M_{C3}^*$, because $\bar{b}_3 = \bar{i}_3$.

When a rigid body is in planar motion, its configuration is defined by three parameters only: two displacement components locate its center of mass, and a single rotation component determines its orientation. Equations of motion (6.35) and (6.37) provide the three equations necessary to solve the problem.

6.7.2 The pivot equations

When the rigid body undergoes *pure rotation about an inertial point O*, eqs. (6.23) were shown to hold. Specializing these equations to the case of planar motion leads to

$$M_{O1}^* = I_{13}^{O*}\dot{\omega} - I_{23}^{O*}\omega^2, \tag{6.38a}$$

$$M_{O2}^* = I_{23}^{O*}\dot{\omega} + I_{13}^{O*}\omega^2, \tag{6.38b}$$

$$M_{O3} = I_{33}^{O*}\dot{\omega}. \tag{6.38c}$$

where $\underline{M}_O^{*T} = \{M_{O1}^*, M_{O2}^*, M_{O3}^*\}$ is the sum of the externally applied moments computed with respect to a pivot point O, resolved in the body attached basis. If axes $\bar{b}_1$, $\bar{b}_2$, and $\bar{b}_3$ coincide with the principal axes of inertia, the equations of motion further simplify since the cross products of inertia vanish, leading to $M_{O3} = I_3^{O*}\dot{\omega}$.

6.7.3 Equations of motion with respect to a material point of the body

Let point **B** be *a material point of the rigid body*; eqs. (6.25) then govern the motion of the rigid body. Introducing the assumption of planar motion, these equations become

$$M_{B1}^* = I_{13}^{B*}\dot{\omega} - I_{23}^{B*}\omega^2, \tag{6.39a}$$

$$M_{B2}^* = I_{23}^{B*}\dot{\omega} + I_{13}^{B*}\omega^2, \tag{6.39b}$$

$$M_{B3} = [\tilde{r}_{BC}\, m\underline{a}_B]_3 + I_{33}^{B*}\dot{\omega}, \tag{6.39c}$$

where notation $[\underline{v}]_3$ indicates the third component of vector $\underline{v}$. Because vectors $\underline{r}^*_{BC}$ and $\underline{R}^T m\underline{a}_B$ both lie in the plane of the motion, the vector product $\widetilde{r}^*_{BC}\,\underline{R}^T m\underline{a}_B$ appearing in eq. (6.25) is normal to the plane of motion and it follows that $\left[\widetilde{r}^*_{BC}\,\underline{R}^T m\underline{a}_B\right]_1 = 0$ and $\left[\widetilde{r}^*_{BC}\,\underline{R}^T m\underline{a}_B\right]_2 = 0$. It is also easy to verify that $\left[\widetilde{r}^*_{BC}\,\underline{R}^T m\underline{a}_B\right]_3 = \left[\widetilde{r}_{BC}\,m\underline{a}_B\right]_3$. Moment $\underline{M}^{*T}_B = \{M^*_{B1}, M^*_{B2}, M^*_{B3}\}$ is the sum of the externally applied moments computed with respect to *material point B of the rigid body*. If axes $\bar{b}_1$, $\bar{b}_2$, and $\bar{b}_3$ coincide with the principal axes of inertia, the equations of motion further simplify because the cross products of inertia vanish.

6.7.4 Equations of motion with respect to an arbitrary point

Let point $\mathbf{P}$ be *an arbitrary point, i.e.*, point $\mathbf{P}$ is neither inertial, nor a material point of the rigid body; the motion of the body is then governed by eq. (6.27). Introducing the assumption of planar motion, these equations become

$$M^*_{P1} = I^{C*}_{13}\dot{\omega} - I^{C*}_{23}\omega^2, \tag{6.40a}$$

$$M^*_{P2} = I^{C*}_{23}\dot{\omega} + I^{C*}_{13}\omega^2, \tag{6.40b}$$

$$M_{P3} = [\widetilde{r}_{PC}\,m\underline{a}_C]_3 + I^{C*}_{33}\dot{\omega}, \tag{6.40c}$$

where $\underline{M}^*_P$ is the sum of the externally applied moments, computed with respect to *an arbitrary point $\mathbf{P}$*.

Example 6.4. Rolling disk with bar
A homogeneous cylinder of mass M and radius R rolls without sliding on a horizontal plane under the effect of gravity. A homogeneous bar of mass m and length ℓ is rigidly attached to the center of the cylinder, as shown in fig. 6.16. Angle θ denotes the orientation of the bar with respect to the vertical axis. At the tip of the bar, denoted point $\mathbf{T}$, a spring of stiffness constant k connects the bar to inertial point $\mathbf{A}$; the un-stretched length of the spring vanishes. Derive the equations of motion of the system in terms of angle θ.

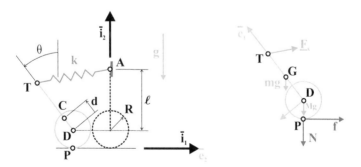

Fig. 6.16. Configuration of the rolling cylinder.

The center of mass of the system is located on the line joining the centers of the disk and bar, at a distance $d = m\ell/\left[2(M + m)\right]$ from the center of the disk. Let axes

$\bar{e}_1$ and $\bar{e}_2$ be a system of body attached axes, as indicated in fig. 6.16. The position of the center of mass of the system now becomes $\underline{r}_C = -R\theta\,\bar{\imath}_1 + R\,\bar{\imath}_2 + d\,\bar{e}_1$ and its acceleration is then $\underline{a}_C = (-R\ddot{\theta} - d\ddot{\theta}C_\theta + d\dot{\theta}^2 S_\theta)\bar{\imath}_1 + (-d\ddot{\theta}S_\theta - d\dot{\theta}^2 C_\theta)\bar{\imath}_2$, where $S_\theta = \sin\theta$ and $C_\theta = \cos\theta$. Equation (6.35) governing the motion of the center of mass of the system now becomes $f\bar{\imath}_1 + N\bar{\imath}_2 + \underline{F}_s - (M+m)g\bar{\imath}_2 = (M+m)\,\underline{a}_C$, where $\underline{F}_s = k[(R\theta + \ell S_\theta)\bar{\imath}_1 + \ell(1-C_\theta)\bar{\imath}_2]$ is the elastic force the spring applies at the tip of the bar, and N and f are the normal reaction and friction forces the plane applies to the disk, respectively.

The following two scalar equations of motion are obtained

$$f + k(R\theta + \ell S_\theta) = -(M+m)R\ddot{\theta} - \frac{m\ell}{2}(\ddot{\theta}C_\theta - \dot{\theta}^2 S_\theta), \qquad (6.41a)$$

$$N - (M+m)g + k\ell(1-C_\theta) = -\frac{m\ell}{2}(\ddot{\theta}S_\theta + \dot{\theta}^2 C_\theta). \qquad (6.41b)$$

Equation (6.40) will be used to derive the third equation of motion governing the angular behavior of the rigid body. It is convenient to compute the sum of the externally applied moments with respect to point **P**, the instantaneous point of contact of the cylinder with the ground because the normal reaction and friction forces will be eliminated from the equation, as their lines of action pass through point **P**,

$$[\tilde{r}_{PT}\underline{F}_s - \tilde{r}_{PG}mg\bar{\imath}_2]_3 = [\tilde{r}_{PC}\,(m+M)\underline{a}_C]_3 + I_{33}^{C*}\ddot{\theta},$$

where $\underline{r}_{PC} = R\bar{\imath}_2 + d\bar{e}_1$ is the position vector of the center of mass with respect to point **P**. The position vectors of points **T** and **G** with respect to point **P** are denoted $\underline{r}_{PT}$ and $\underline{r}_{PG}$, respectively.

The moment of inertia of the system with respect to the center of mass is found by adding the contributions of the cylinder and bar to find

$$I_{33}^{C*} = \left[\frac{MR^2}{2} + Md^2\right] + \left[\frac{m\ell^2}{12} + m(\frac{\ell}{2} - d)^2\right].$$

Note the use of the parallel axis theorem: the moment of inertial of the cylinder with respect to its own center of mass is $MR^2/2$, see fig. 6.40, and the transport term is Md^2. The rotational equation now becomes

$$mg\frac{\ell}{2}S_\theta - k\left[R^2\theta + R\ell(S_\theta + \theta C_\theta) + \ell^2 S_\theta\right]$$
$$= \left[\frac{3MR^2}{2} + m(R^2 + \frac{\ell^3}{3} + R\ell C_\theta)\right]\ddot{\theta} - m\frac{R\ell}{2}\dot{\theta}^2 S_\theta.$$

Given initial conditions, this differential equation can be solved to find the response of the system. Introducing θ into eqs. (6.41a) and (6.41b) then yields the friction and normal forces, respectively.

The derivation presented here assumes that at all times, the cylinder is rolling without slipping. To make sure the analysis is consistent, it is then important to check that $N > 0$ and $|f| \leq \mu_s N$ at all times, where μ_s is the static friction coefficient between the cylinder and the ground.

Example 6.5. The double pendulum with elastic joint

Figure 6.17 depicts a double pendulum comprising bar 1, of mass m_1 and length ℓ_1, and bar 2, of mass m_2 and length ℓ_2. Let frame $\mathcal{F}^A = [\mathbf{A}, \mathcal{A} = (\bar{a}_1, \bar{a}_2)]$ be attached to bar 1 and frame $\mathcal{F}^E = [\mathbf{E}, \mathcal{E} = (\bar{e}_1, \bar{e}_2)]$ be attached to bar 2. A massless tube allows bar 2 to slide in the direction of $\bar{a}_2$; the slider is of mass M and is connected to bar 1 at point $\mathbf{A}$ by means of a spring of stiffness constant k. The position of the slider is determined by its distance, x, from point $\mathbf{A}$, the tip of bar 1; the angular positions of the two bars with respect to the vertical are denoted θ_1 and θ_2, respectively. The system is subjected to gravity along the inertial $\bar{\imath}_1$ direction. Find the equations of motion of the system.

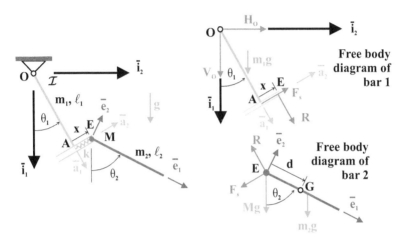

Fig. 6.17. Configuration of the double pendulum with elastic joint.

First, the equations of motion of bar 2, including the concentrated mass of the slider, M, will be derived. The center of mass of the combined body is at a distance d from point $\mathbf{E}$, where $(M + m_2)d = m_2 \ell_2/2$; for simplicity, the following notation is used, $\mu_2 = M + m_2$. Considering the free body diagram shown in fig. 6.17, eq. (6.39) gives the sum of the moments computed with respect to point $\mathbf{E}$ as

$$-\frac{m_2 \ell_2}{2} g S_2 = \frac{m_2 \ell_2^2}{3} \ddot{\theta}_2 + [d\tilde{e}_1 \mu_2 \underline{a}_E]_3,$$

where the following notation was introduced: $S_1 = \sin\theta_1$ and $C_1 = \cos\theta_1$, with similar conventions for angle θ_2. Point $\mathbf{E}$ was selected as the point about which moments were computed because this choice automatically eliminates the reaction force, R, and spring force, F_s, from the resulting equation of motion.

The acceleration, $\underline{a}_E$, of point $\mathbf{E}$ is readily computed as the second time derivative of the position vector of point $\mathbf{E}$, $\underline{r}_E = \ell_1 \bar{a}_1 + x \bar{a}_2$. The first equation of motion is now

$$\frac{m_2\ell_2^2}{3}\ddot{\theta}_2 + \frac{m_2\ell_2}{2}gS_2 + \frac{m_2\ell_2}{2}\left[(\ddot{x} + \ell_1\ddot{\theta}_1)C_{21} + (\dot{x} + \ell_1\dot{\theta}_1)\dot{\theta}_1 S_{21}\right.$$
$$\left. +(\dot{x}\dot{\theta}_1 + x\ddot{\theta}_1)S_{21} - x\dot{\theta}_1^2 C_{21}\right] = 0, \tag{6.42}$$

where the following notation was introduced: $S_{21} = \sin(\theta_2 - \theta_1)$ and $C_{21} = \cos(\theta_2 - \theta_1)$.

The acceleration of the center of mass of bar 2 is found by taking two derivatives of its position vector, $\underline{r}_G = \ell_1\bar{a}_1 + x\bar{a}_2 + d\bar{e}_1$; hence, the equations of motion for the center of mass becomes $-R\bar{a}_1 - F_s\bar{a}_2 + \mu_2 g\bar{\imath}_1 = \mu_2[-(2\dot{x}\dot{\theta}_1 + \ell_1\dot{\theta}_1^2 + x\dot{\theta}_1)\bar{a}_1 + (\ddot{x} + \ell_1\ddot{\theta}_1 - x\dot{\theta}_1^2)\bar{a}_2 + d\ddot{\theta}_2 - d\dot{\theta}_2^2\bar{e}_1]$. Taking a scalar product of this relationship by $\bar{a}_1$ and $\bar{a}_2$ yields the reaction force

$$R = \mu_2 gC_1 + \mu_2(2\dot{x}\dot{\theta}_1 + \ell_1\dot{\theta}_1^2 + x\ddot{\theta}_1) + \frac{m_2\ell_2}{2}(\ddot{\theta}_2 S_{21} + \dot{\theta}_2^2 C_{21}), \tag{6.43}$$

and the spring force

$$F_s = -\mu_2 gS_1 - \mu_2(\ddot{x} + \ell_1\ddot{\theta}_1 - x\dot{\theta}_1^2) - \frac{m_2\ell_2}{2}(\ddot{\theta}_2 C_{21} - \dot{\theta}_2^2 S_{21}), \tag{6.44}$$

respectively.

Next, the equations of motion for bar 1 are derived from the free body diagram shown in fig. 6.17. The pivot equation, eq. (6.38), is applied about point **O** to find

$$\frac{m_1\ell_1^2}{3}\ddot{\theta}_1 = -\frac{m_1\ell_1}{2}gS_1 + \ell_1 F_s - xR. \tag{6.45}$$

The three equations of motion of the problem can now be summarized. The first equation is eq. (6.45), where the reaction and spring forces are eliminated by means of eqs. (6.43) and (6.44), respectively; the second equation is eq. (6.42); finally, the last equation is the constitutive equation for the elastic spring, $F_s = kx$, where the elastic force is eliminated with the help of eq. (6.44). These three equations are recast in a matrix form, leading to

$$\begin{bmatrix} (\frac{m_1}{3} + \mu_2)\ell_1^2 + \mu_2 x^2 & \frac{m_2\ell_2}{2}(\ell_1 C_{21} + xS_{21}) & \mu_2\ell_1 \\ \frac{m_2\ell_2}{2}(\ell_1 C_{21} + xS_{21}) & \frac{m_2\ell_2^2}{3} & \frac{m_2\ell_2}{2}C_{21} \\ \mu_2\ell_1 & \frac{m_2\ell_2}{2}C_{21} & \mu_2 \end{bmatrix} \begin{Bmatrix} \ddot{\theta}_1 \\ \ddot{\theta}_2 \\ \ddot{x} \end{Bmatrix}$$

$$+ \begin{Bmatrix} 2\mu_2 x\dot{x}\dot{\theta}_1 - \frac{m_2\ell_2}{2}(\ell_1 S_{21} - xC_{21})\dot{\theta}_2^2 + (m_1/2 + \mu_2)g\ell_1 S_1 + \mu_2 gxC_1 \\ m_2\ell_2\dot{x}\dot{\theta}_1 S_{21} + \frac{m_2\ell_2}{2}(\ell_1 S_{21} - xC_{21})\dot{\theta}_1^2 + \frac{m_2\ell_2}{2}gS_2 \\ -\mu_2 x\dot{\theta}_1^2 - \frac{m_2\ell_2}{2}\dot{\theta}_2^2 S_{21} + kx + \mu_2 gS_1 \end{Bmatrix} = 0.$$

These equations form a set of coupled, nonlinear, second-order, ordinary differential equations in time for the three unknowns of the problem, θ_1, θ_2, and x.

The equations of motion for the center of mass of bar 1 are readily found as

$$\frac{m_1\ell_1}{2}(-\ddot{\theta}_1 S_1 - \dot{\theta}_1^2 C_1) = m_1 g \bar{\imath}_1 + R\bar{a}_1 + F_s a_2 + V_O \bar{\imath}_1 + H_O \bar{\imath}_2.$$

Projection of this relationship along unit vectors $\bar{\imath}_1$ and $\bar{\imath}_2$ yields the components of the reaction force at point O in the vertical and horizontal directions as $V_O = F_s S_1 - R C_1 - m_1 g - m_1 \ell_1 (\ddot{\theta}_1 S_1 + \dot{\theta}_1^2 C_1)/2$ and $H_O = -F_s C_1 - R S_1 + m_1 \ell_1 (\ddot{\theta}_1 C_1 - \dot{\theta}_1^2 S_1)/2$, respectively.

Example 6.6. Pendulum with sliding mass

Figure 6.18 shows a pendulum comprising a bar of mass m and length ℓ and a rigid body of mass M. Frame $\mathcal{F}^E = [O, \mathcal{E} = (\bar{e}_1, \bar{e}_2)]$ is attached to the bar. The rigid body is connected at point B to the tip of the bar at point A by means of a spring of stiffness constant k and a dashpot of constant c. The stretch of the spring is denoted x and its un-stretched length vanishes. The center of mass of the rigid body is located at point C and vector η defines the position of the center of mass with respect to point B; the moment of inertia of the body with respect to center of mass is denoted I^C. The angular position of the bar with respect to the vertical is defined by angle θ. The system is subjected to gravity along unit vector $\bar{\imath}_1$. Find the equations of motion of the system.

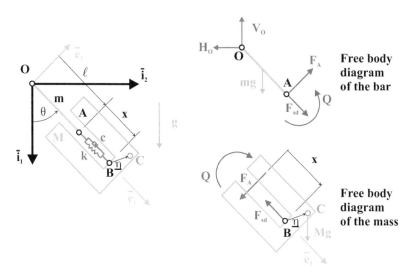

Fig. 6.18. Configuration of the pendulum with sliding mass.

First, the equations of motion of the bar will be derived based on the free body diagrams depicted in fig. 6.18. Because point O is a pivot point, eq. (6.38) yields

$$\frac{m\ell^2}{3}\ddot{\theta} = Q + \ell F_A - \frac{m\ell}{2}g S_\theta, \tag{6.46}$$

where the notation $S_\theta = \sin\theta$ and $C_\theta = \cos\theta$ was introduced. Because the angular orientations of the bar and rigid body must remain identical, equal and opposite moments of magnitude Q must be applied to the bar and rigid body, as shown in fig. 6.18. Similarly, equal and opposite forces of magnitude F_A must be applied to the tip of the bar at point **A** and to the rigid body, along a common line of action normal to the bar and passing through point **A**.

The position of the center mass of the bar is $\underset{\sim}{r} = \ell/2\,\bar{e}_1$, and hence, the equation of motion for the center of mass becomes $m\ell(-\dot{\bar\theta}^2\bar{e}_1 + \ddot\theta\bar{e}_2)/2 = -V_O\,\bar\imath_1 - H_O\,\bar\imath_2 + mg\,\bar\imath_1 + F_{sd}\,\bar{e}_1 + F_A\,\bar{e}_2$, where V_O and H_O are the vertical and horizontal components of the reaction force at point **O**, and F_{sd} the sum of the forces acting in the spring and dashpot. The equation of motion for the center of mass can be projected along axes $\bar\imath_1$ and $\bar\imath_2$ to obtain the reaction forces at point **O**

$$V_O = mg + (F_{sd} + \frac{m\ell}{2}\dot\theta^2)C_\theta - (F_A - \frac{m\ell}{2}\ddot\theta)S_\theta, \qquad (6.47a)$$

$$H_O = \qquad (F_{sd} + \frac{m\ell}{2}\dot\theta^2)S_\theta + (F_A - \frac{m\ell}{2}\ddot\theta)C_\theta, \qquad (6.47b)$$

respectively.

Next, the equations of motion of the rigid body will be derived. Because point **B** is a fixed point of the body, eq. (6.39) gives the sum of the moments computed with respect to point **B** as $-Q + xF_A + \tilde{\eta}\,Mg\bar\imath_1 = \tilde{\eta}\,M a_B + I^B\ddot\theta$. The moment of inertia of the body with respect to point **B** is found with the help of the parallel axis theorem, eq. (6.8), as $I^B = I^C + M(\eta_1^{*2} + \eta_2^{*2})$, where η_1^* and η_2^* are the components of vector $\underset{\sim}{\eta}$ resolved in basis $\mathcal{E}$. The position vector of point **B** is $\underset{\sim}{r}_B = (\ell + x)\bar{e}_1$ and its acceleration, $\underset{\sim}{a}_B$, is then readily obtained. Expanding the various terms then leads to

$$
\begin{aligned}
&- Q + xF_A - Mg\eta_1^* S_\theta - Mg\eta_2^* C_\theta \\
&= M\eta_1^*\left[2\dot{x}\dot\theta + (\ell + x)\ddot\theta\right] - M\eta_2^*\left[\ddot{x} - (\ell + x)\dot\theta^2\right] + I^B\ddot\theta.
\end{aligned} \qquad (6.48)
$$

The acceleration of the center of mass of the rigid body is found by taking two time derivatives of its position vector, $\underset{\sim}{r} = (\ell + x + \eta_1^*)\bar{e}_1 + \eta_2^*\bar{e}_2$; hence, the equations of motion for the center of mass become

$$
\begin{aligned}
&- F_A\bar{e}_2 - F_{sd}\bar{e}_1 + Mg\bar\imath_1 \\
&= M\left[(\ddot{x} - \eta_2^*\ddot\theta) - (\ell + x + \eta_1^*)\dot\theta^2\right]\bar{e}_1 + M\left[(2\dot{x}\dot\theta - \eta_2^*\dot\theta^2) + (\ell + x + \eta_1^*)\ddot\theta\right]\bar{e}_2.
\end{aligned}
$$

Projecting this equation along unit vectors $\bar{e}_2$ and $\bar{e}_1$ yields

$$F_A = MgS_\theta - M\left[(2\dot{x}\dot\theta - \eta_2^*\dot\theta^2) + (\ell + x + \eta_1^*)\ddot\theta\right], \qquad (6.49a)$$

$$M(\ddot{x} - \eta_2^*\ddot\theta) - M(\ell + x + \eta_1^*)\dot\theta^2 + kx + c\dot{x} - MgC_\theta = 0. \qquad (6.49b)$$

The first equation gives the interaction force acting at point **A**. The constitutive law for the spring dashpot assembly is $F_{sd} = kx + c\dot{x}$.

Equation (6.49b) is the first equation of motion; the second equation of motion is found by introducing into eq. (6.46) the expression for the interaction moments from eq. (6.48), and interaction force from eq. (6.49a) to find

$$
\left[\frac{m\ell^2}{3} + I^C + M\eta_2^{*2} + M(\ell + \eta_1^* + x)^2\right]\ddot{\theta} - M\eta_2^*\ddot{x}
$$
$$
+ 2M(\ell + \eta_1^* + x)\dot{x}\dot{\theta} + \left[\frac{m\ell}{2} + M(\ell + \eta_1^* + x)\right]gS_\theta + Mg\eta_2^*C_\theta = 0. \tag{6.50}
$$

In summary, the equations of motion can be recast as a system of coupled, ordinary differential equations by combining eqs. (6.50) and (6.49b) to find

$$
\begin{bmatrix} m\ell^2/3 + I^C + M\eta_2^{*2} + M(\ell + \eta_1^* + x)^2 & -M\eta_2^* \\ -M\eta_2^* & M \end{bmatrix} \begin{Bmatrix} \ddot{\theta} \\ \ddot{x} \end{Bmatrix}
$$
$$
+ \begin{Bmatrix} 2M(\ell + \eta_1^* + x)\dot{x}\dot{\theta} + [m\ell/2 + M(\ell + \eta_1^* + x)]gS_\theta + Mg\eta_2^*C_\theta \\ -M(\ell + x + \eta_1^*)\dot{\theta}^2 + kx + c\dot{x} - MgC_\theta \end{Bmatrix} = 0.
$$

Once the solution of these equations has been obtained, the vertical and horizontal components of the reaction force at point **O** can be obtained by eqs. (6.47a) and (6.47b), respectively; next, the interaction moment Q is obtained from eq. (6.48) and the interaction force at point **A** by eq. (6.49a).

Example 6.7. The unbalanced rotor
Figure 6.19 shows a rigid rotor of length L and mass M supported by two end bearings at points **B** and **D**. A torque, T, is applied to the rotor at point **D**. Let frame $\mathcal{F}^B = \left[\mathbf{B}, \mathcal{B}^* = (\bar{b}_1, \bar{b}_2, \bar{b}_3)\right]$ be attached to the body; superscript $(\cdot)^*$ denotes components resolved in basis $\mathcal{B}^*$. Point **G** is located at the intersection of the shaft's axis with the plane passing through the center of mass of the rotor and normal to the shaft's axis. The coordinates of the center of mass of the rotor, resolved in basis $\mathcal{B}^*$, are denoted x_{1c}^*, x_{2c}^*, and x_{3c}. At point **B**, three reaction forces, denoted B_1^*, B_2^*, and B_3 are applied to the shaft; at point **D**, two reaction forces, D_1^* and D_2^*, are applied to the shaft together with the torque T. Find the reactions forces applied to the bearings.

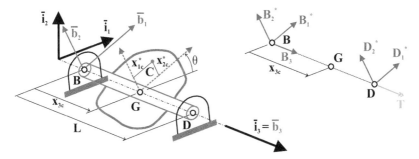

Fig. 6.19. Configuration of a rotor with an imbalance.

The components of tensor of moment of inertia of the rigid rotor with respect to point **B**, resolved in basis $\mathcal{B}^*$, are denoted $\underline{\underline{I}}^{B*}$. The position vector of the center of mass is $\underline{r}_C = x_{1c}^* \bar{b}_1 + x_{2c}^* \bar{b}_2 + x_{3c} \bar{b}_3$ and the corresponding acceleration is readily found. Equation (6.35) then yields the first three equations of motion of the rotor $B_1^* + D_1^* = -M(\omega^2 x_{1c}^* + \dot{\omega} x_{2c}^*)$, $B_2^* + D_2^* = -M(\omega^2 x_{2c}^* - \dot{\omega} x_{1c}^*)$, and $B_3 = 0$. Because point **B** is a material point of the body and is inertial, it is a pivot point for the rigid body and eq. (6.39) leads to $-LD_2^* = I_{13}^{B*} \dot{\omega} - I_{23}^{B*} \omega^2$, $LD_1^* = I_{23}^{B*} \dot{\omega} + I_{13}^{B*} \omega^2$, and $T = I_{33}^{B*} \dot{\omega}$.

Given the applied torque, this last equation can be integrated to find the angular velocity and acceleration of the system. The remaining equations then yield expressions for the reaction forces at the bearing

$$LB_1^* = -I_{23}^{B*} \dot{\omega} - I_{13}^{B*} \omega^2 - ML(\omega^2 x_{1c}^* + \dot{\omega} x_{2c}^*),$$
$$LB_2^* = +I_{13}^{B*} \dot{\omega} - I_{23}^{B*} \omega^2 - ML(\omega^2 x_{2c}^* - \dot{\omega} x_{1c}^*),$$
$$LD_1^* = I_{23}^{B*} \dot{\omega} + I_{13}^{B*} \omega^2,$$
$$LD_2^* = -I_{13}^{B*} \dot{\omega} + I_{23}^{B*} \omega^2.$$

If the rotor rotates at a constant angular speed, $\omega = \Omega$, $\dot{\omega} = 0$, the reaction forces at the bearing will be constant when resolved in the body attached basis. Of course, in the inertial system, these reaction forces will be harmonic forces at frequency Ω, as expected. It is often desirable to minimize or eliminate the reaction forces at the bearing. To eliminate these reaction forces, two conditions must be satisfied: *(1)* the rotor center of mass must be located on the axis of the shaft, *i.e.*, $x_{1c}^* = x_{2c}^* = 0$, and *(2)* axis $\bar{b}_3$ must be a principal axis of inertia, *i.e.*, $I_{13}^{B*} = I_{23}^{B*} = 0$.

Example 6.8. The cam-valve system

Figure 6.20 shows a planar cam-valve system. The cam rotates at a constant angular velocity, Ω, about fixed inertial point **O**. Frame $\mathcal{F}^I = [\mathbf{O}, \mathcal{I} = (\bar{\imath}_1, \bar{\imath}_2)]$ is inertial and frame $\mathcal{F}^E = [\mathbf{O}, \mathcal{E} = (\bar{e}_1, \bar{e}_2)]$ is attached to the cam. The external shape of the cam is defined by curve $\mathbb{C}$ and the valve of mass m slides over this curve; the contact point between the cam and valve is denoted **P**. The motion of the valve is constrained to be along axis $\bar{\imath}_2$ and its displacement is denoted x. A spring of stiffness constant k is connected to the valve and is pre-compressed by a distance d. The kinematics of this problem have been treated in example 5.4 on page 182. Find the contact force acting between the cam and the valve.

Assuming that the cam and valve are in contact at all times, the motion of the valve is known once the shape of curve $\mathbb{C}$ is given: this problem has no degrees of freedom. The right portion of fig 6.20 depicts a free body diagram of the valve and its equation of motion is $m\ddot{x} = k(d - x) - N$, where $k(d - x)$ is the force the spring applies on the valve and N the desired contact force. Solving for the contact force yields

$$N = k(d - x) - m\ddot{x} = k(d - h + rS_\gamma) - m\Omega^2(rS_\gamma - \rho).$$

To obtain the second equality, the valve's position was evaluated using elementary trigonometry as $x = h - r\sin(\theta + \alpha)$, its acceleration was found using eq. (5.32), and the following notation was defined, $S_\gamma = \sin(\theta + \alpha)$.

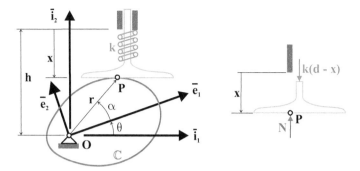

Fig. 6.20. Configuration of the cam-valve system.

In non-dimensional form, the contact force becomes

$$\bar{N} = \frac{N}{mr_0\Omega^2} = \bar{\rho} + (\frac{\omega^2}{\Omega^2} - 1)\bar{r}S_\gamma + \frac{\omega^2}{\Omega^2}(\bar{d} - \bar{h}),$$

where r_0 is a reference length, $\bar{r} = r/r_0$, $\bar{\rho} = \rho/r_0$, $\bar{d} = d/r_0$, and $\bar{h} = h/r_0$. The natural frequency of the spring-valve system was defined as $\omega^2 = k/m$.

6.7.5 Problems

Problem 6.22. Retraction of a landing gear
Figure 5.8 shows a simple landing gear system. It consists of a uniform link of mass m_L and length L, and a wheel of mass m_W (a point mass). The length $\ell(t)$ of the hydraulic actuator is given as function of time: $\ell(t) = h + g[1 - \cos(\pi t/T)]$, where $g = [\sqrt{(L^2/2 + hL + h^2)} - h]/2$. *(1)* Compute the magnitude F of the force that the actuator must apply to generate the desired motion. Plot F versus time $\tau = t/T$. *(2)* Compute the vertical and horizontal components of reaction at point $\mathbf{O}$, denoted V and H respectively. Plot V and H versus τ. Use the following data: $h = 0.6$ m, $L = 1.2$ m, $T = 1.5$ s, $m_L = 120$ kg, $m_W = 80$ kg, and gravity $= 9.81$ m/s^2.

Problem 6.23. Locking mechanism
Figure 5.11 shows a locking mechanism used in the deployment of large space structures. When the homogeneous disk of radius R and mass M rotates about its fixed point O, bar $\mathbf{PT}$ of length L and mass m_b slides at point $\mathbf{A}$ through a collar that is allowed to swivel about the pin at point A. The mechanism is spring loaded by connecting a spring of stiffness k between the tip of the bar at point T and the collar at point $\mathbf{A}$. The spring is un-stretched when $\theta = 90$ deg. The bar has a length L, and $w(t)$ denotes the portion of the bar between points $\mathbf{P}$ and $\mathbf{A}$. The time history of angle θ is prescribed as $\theta(t) = \pi/4 \, (1 + \cos \pi t/T)$. *(1)* Compute the reaction forces S and Q at point $\mathbf{A}$. Force S is oriented in the direction parallel to the bar, and Q is perpendicular to the bar. On the same graph, plot S and Q as a function of time $\tau = t/T$. *(2)* Compute the horizontal and vertical components of force, denoted H and V, respectively, at point $\mathbf{P}$. Plot H and V. *(3)* Compute and plot the torque required to rotate the disk. Use the following data: $R = 0.15$ m, $M = 1.2$ kg, $d = 0.2$ m, $k = 1.5$ kN/m, $L = 0.4$ m, $m_b = 0.5$ kg, $T = 2$ s, and $g = 10$ m/s^2.

Problem 6.24. Deployment of a satellite

The satellite depicted in fig. 6.21 is powered by solar panels. Initially, the three articulated solar panels are in the stowed configuration indicated on the figure. To become operational, these panels are deployed by means of motors located at points **A**, **B**, and **C**. These motors provide torques that will deploy the system in such a way that the time schedule of angle θ is $\theta(t) = \pi \left[1 - \cos(\pi t/T)\right]/4$, where T is the total time required for the deployment. Each panel of the solar array is uniform, has a mass $m_P = 120$ kg and a length $\ell_P = 5$ m. The total time to complete the deployment is $T = 5$ s. Let M_A, M_B, and M_C be the torques that the motors located at points **A**, **B**, and **C**, respectively, must apply to complete the desired schedule of deployment. Let H_A, H_B, and H_C be the horizontal components of force at the joint located at points **A**, **B**, and **C**, respectively; V_A, V_B, and V_C are the corresponding vertical force components. Finally, F_A, F_B, and F_C are the magnitudes of the force at each joint. (1) Plot θ, $\dot{\theta}$, and $\ddot{\theta}$ versus time. (2) Draw free body diagrams for each of the three panels and the corresponding dynamic equations of motion. (3) Plot M_A, M_B, and M_C. Find the instant at which each torque is maximum. Which motor will have to produce the highest torque? Why? (4) On one graph, plot H_A, H_B, and H_C versus time. (5) Plot V_A, V_B, and V_C. (6) Plot F_A, F_B, and F_C. Find the instant at which each force component is maximum. Which joint is the most heavily loaded? Why? (7) If the maximum torque the motors can produce is $M_{\text{MAX}} = 100$ N·m, what is the minimum time in which the deployment can be completed?

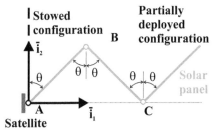

Fig. 6.21. Satellite in the stowed and partially deployed configurations.

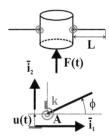

Fig. 6.22. Satellite release configuration.

Problem 6.25. Satellite release

A satellite is released from a launch vehicle, as depicted in fig. 6.22. The satellite is composed of a rigid body and of two solar panels of length $L = 5$ m. During release, force $\underline{F}(t)$ imparts to the satellite the following vertical motion (rectilinear motion along $\bar{\imath}_2$): $u(t) = \Delta_0(1 - \cos \pi t/T)/2$ for $t \leq T/2$ and $u(t) = \Delta_0 \left[\pi(2t/T - 1)/2 + 1\right]/2$ for $t > T/2$, where $\Delta_0 = 0.5$ m and $T = 0.5$ s is the characteristic release time. Due to the impulsive nature of the applied force, the solar panel will start to vibrate. Each panel is uniform and has a mass $m_p = 100$ kg. The elasticity of the panels will be represented by torsional springs of stiffness $k = 5$ kN·m/rad at their root. In view of the symmetry of the problem, the motions of the two solar panels will be identical. Consequently, the sole right panel will be investigated here. (1) Draw a free body diagram for the right panel. (2) Derive the differential equation of motion of the panel. (3) Solve this equation numerically. (4) On three separate graphs, plot ϕ, $\dot{\phi}$, and $\ddot{\phi}$ as a function of time, for $t \in [0, 5]$ s. (5) Plot the horizontal and vertical components of reaction at point **A**. (6) Plot the torque in the torsional spring as a function of time.

Problem 6.26. Quick return mechanism

The quick return mechanism shown in fig. 5.9 consists of a uniform crank of length $\ell_c = 0.30$ m and mass $m_c = 12$ kg, and of a uniform bar of length $\ell_b = 1.6$ m and mass $m_b = 60$ kg. The crank is pinned at point $\mathbf{R}$ and the bar is pinned at point $\mathbf{O}$. The distance between these two points is $d = 0.35$ m. At point $\mathbf{P}$, a slider allows the tip of the crank to slide along the bar. Gravity acts in the vertical direction, $g = 9.81$ m/s^2. The system is driven by a torque M applied to the crank at point $\mathbf{R}$. The time history of angle θ is: $\theta(t) = \pi(1 - \cos \pi t/T)/2$, where $T = 5$ s is the time required for the crank to rotate 180 degrees. *(1)* Draw free body diagrams of the bar and crank. Write the equations of motion of the system. *(2)* Plot the time history of the contact force at the slider. *(3)* On the same graph, plot the horizontal and vertical components of the reaction force at point $\mathbf{O}$. *(4)* Plot the time history of the torque M required to drive the system. *(5)* On the same graph, plot the horizontal and vertical components of the reaction force at point $\mathbf{R}$.

Problem 6.27. Bar hinged at rim of rotating disk

Figure 6.23 shows a homogeneous disk of radius R and mass M rotating in a vertical plane around inertial point $\mathbf{O}$. Frame $\mathcal{F}^D = \left[\mathbf{O}, \mathcal{E}^+ = (\bar{e}_1, \bar{e}_2, \bar{e}_3)\right]$ is attached to the disk. At point $\mathbf{B}$, a homogeneous bar of length ℓ and mass m is hinged to the disk. A torsional spring of stiffness constant k and a torsional dashpot of constant c are located at the hinge. The spring is un-stretched when $\phi = \phi_0$. *(1)* Derive the equations of motion of the system in terms of angles θ and ϕ. *(2)* On one graph, plot angles θ and ϕ versus

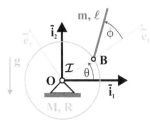

Fig. 6.23. Bar hinged at rim of rotating disk.

τ. *(3)* On one graph, plot angular velocities θ' and ϕ'. *(4)* On one graph, plot angular accelerations θ'' and ϕ''. *(5)* Plot the cumulative energy dissipated in the dashpot, $\bar{W}_d = W_d/k$. *(6)* Plot the system's kinetic, $\bar{K} = K/k$, potential, $\bar{V} = V/k$, energies. Check that the energy closure equation is satisfied. *(7)* Plot the components of the force, $\bar{F}_B = \underline{F}_B/(m\ell\omega^2)$, in the hinge, resolved in basis $\mathcal{E}^+$. Use the following data: $\mu = M/m = 3$, $\bar{R} = R/\ell = 1$, $\bar{g} = g/(\ell\omega^2) = 1.2$, $\zeta = \omega c/(2k) = 0.05$, and $\phi_0 = 0$. Use the following non-dimensional time $\tau = \omega t$, where $\omega^2 = 3k/(m\ell^2)$ and $(\cdot)'$ indicates a derivative with respect to τ. Plot all results for $\tau \in [0, 30]$. The initial conditions are $\theta = \phi = 0$, $\theta' = 1$, and $\phi' = -1$.

Problem 6.28. Robotic arm in space

Consider a robotic arm in space depicted in fig. 6.24. The flexibility of the arm will be represented in a crude manner by a mid-span torsional spring of stiffness $k_t = 1,500$ N·m/rad. The first segment of the robotic arm is of length $L_a = 2.4$ m and its orientation is prescribed as $\theta(t) = \pi(1 - \cos \pi t/T)/6$ for $t \leq T$ and $\theta(t) = \pi/3$ for $t > T$, where $T = 25$ s. The second segment of the robotic arm is of length $L_b = 2.4$ m and mass $m_b = 60$ kg. The system is used to manipulate a payload of mass $M_p = 1,000$ kg and moment of inertia $I_p = 250$ kg·m^2 connected to the tip of the second segment of the robotic arm. *(1)* Derive the equation of motion for the orientation angle ϕ of the second segment of the robotic arm. *(2)* Solve this differential equation numerically assuming initial conditions at rest. *(3)* On the same graph, plot θ and ϕ as a function of time. What is the maximum overshoot, $(\phi_{\max} - \theta_{\max})/\theta_{\max}$, of ϕ with respect to the command signal θ? *(4)* Plot the angular velocity $\dot{\phi}$ as a function of time. *(5)* Plot the torque, Q, in the torsional spring as a function of time. *(6)* Find the components of force through the pin at point $\mathbf{A}$. *(7)* On the same graph, plot the horizontal and vertical components of this force and its magnitude as a function of time. For all graphs, use $t \in [0, 100]$ s.

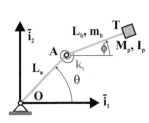

Fig. 6.24. Robotic arm configuration.

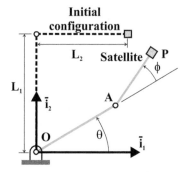

Fig. 6.25. Satellite capture configuration.

Problem 6.29. Satellite capture
A satellite is to be brought to the cargo bay of the space shuttle. Figure 6.25 shows the initial configuration of the system, with the satellite connected to the end of the shuttle robotic system. The first part of the robotic system, bar 1, is a uniform bar of length $L_1 = 4$ m and mass $m_1 = 100$ kg. The second part of the robotic system, bar 2, is a uniform bar of length $L_2 = 3$ m, and mass $m_2 = 65$ kg. The satellite has a mass $m_{pl} = 1,500$ kg and a moment of inertia $I_{pl} = 1,200$ kg·m². Torques T_O and T_A are applied at the joints located at points O and A, respectively, in such a way that the time histories of angles θ and ϕ are as follows: $\theta(t) = \pi(1 + \cos \pi t/T)$ and $\phi(t) = \pi(1 - 3 \cos \pi t/T)/4$, respectively, where $T = 10$ s is the total time needed bring the satellite into the cargo bay. (1) On the same graph, plot the time history of angles θ and ϕ. (2) Plot the angular velocities of bars 1 and 2. (3) Plot the angular accelerations of bars 1 and 2. (4) Plot the trajectory of the satellite as it is brought into the cargo bay. (5) Draw free body diagrams for bars 1 and 2. (6) Plot the horizontal and vertical components of the reaction force at point A, denoted H_A and V_A, respectively. (7) Plot the horizontal and vertical components of the reaction force at point O, denoted H_O and V_O, respectively. (8) Plot the torques T_O and T_A applied at points O and A, respectively. (9) If the actuators at points O and A can generate a maximum torque of 1,000 N·m, find the minimum maneuver time, T_{min}.

Problem 6.30. Rolling cylinder with bar
Figure 6.26 shows a homogeneous cylinder of mass M and radius r rolling without sliding on a horizontal plane under the effect of gravity. A homogeneous bar of mass m and length ℓ is rigidly attached to the center of the cylinder. Angle θ denotes the orientation of the bar with respect to the vertical axis. At the tip of the bar, denoted point T, a spring of stiffness constant k connects the bar to fixed point A; the un-stretched length of the spring vanishes. (1) Derive the equations of motion of the system in terms of angle θ. (2) Plot θ as a function of time. (3) Plot $\dot{\theta}$ as a function of time. (4) On one graph, plot the kinetic, potential and total mechanical energies of the system. (5) On one graph, plot the normal reaction and friction forces acting on the disk. (5) If the friction coefficient between the disk and the horizontal plane is $\mu = 0.3$, will the disk start sliding? Use the following data: $r = 0.25$, $\ell = 1.25$ m, $M = 5$, $m = 1.25$

kg, $k = 10$ N/m, and $g = 9.81$m/s^2. At time $t = 0$, $\theta = 4\pi/5$ rad, and $\dot{\theta} = 0$. Present all your results for $t \in [0, 5]$ s.

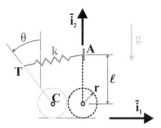

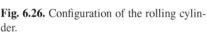

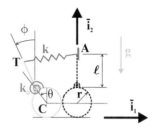

Fig. 6.26. Configuration of the rolling cylin-der.

Fig. 6.27. Configuration of the rolling cylin-der with an articulated bar.

Problem 6.31. Rolling cylinder with articulated bar

Figure 6.27 shows a homogeneous cylinder of mass M and radius r rolling without sliding on a horizontal plane under the effect of gravity. A homogeneous bar of mass m and length ℓ is articulated to the rim of the cylinder. Angle θ denotes the rolling angle of the cylinder and angle ϕ the orientation of the bar with respect to the vertical axis. At the tip of the bar, denoted point **T**, a spring of stiffness constant k connects the bar to inertial point **A**; the un-stretched length of the spring vanishes. A torsional spring of constant k_ϕ acts at the connection between the cylinder and the bar; the spring is un-stretched when the bar point radially outwards. *(1)* Derive the equations of motion of the system in terms of angles θ and ϕ. *(2)* On one graph, plot θ and ϕ as a function of time. *(3)* On one graph, plot $\dot{\theta}$ and $\dot{\phi}$ as a function of time. *(4)* On one graph, plot the kinetic, potential and total mechanical energies of the system. Use the following data: $r = 0.25$, $\ell = 1.25$ m, $M = 5$, $m = 1.25$ kg, $k = 10$ N/m, $k_\phi = 15$ N·m/rad, and $g = 9.81$m/s^2. At time $t = 0$, $\theta = \pi/2$, $\phi = 0$ rad, and $\dot{\theta} = \dot{\phi} = 0$. Present all your results for $t \in [0, 5]$ s.

Problem 6.32. Balancing a rotor

Consider the rigid rotor of length L and mass M supported by two end bearings at points **B** and **D**, as depicted in fig. 6.19 and discussed in example 6.7. Due to manufacturing imperfections, the rotor is not balanced, *i.e.*, the coordinates of the center of mass do not vanish, $x_{1c}^* \neq 0$ and $x_{2c}^* \neq 0$, and axis $\bar{b}_3$ is not a principal axis of inertia of the rotor, $I_{13}^{B*} \neq 0$ and $I_{23}^{B*} \neq 0$. To estimate these unknown parameters, the rotor is spun at a constant angular velocity, Ω, and the bearing reactions B_1^*, B_2^*, D_1^*, and D_2^* are measured. To balance the rotor, *i.e.*, to eliminate the reaction forces at the bearings, it is proposed to add two point masses to the rotor. The first point mass is located at point $(x_{1a}^* = R_a \cos\theta_a, x_{2a}^* = R_a \sin\theta_a, x_{3a})$ and is of mass m_a; similarly, the second point mass is located at point $(x_{1b}^* = R_b \cos\theta_b, x_{2b}^* = R_b \sin\theta_b, x_{3b})$ and is of mass m_b. This implies that the balancing masses m_a and m_b are located on circles of radii R_a and R_b, respectively, at angular locations θ_a and θ_b, respectively. *(1)* Based on the measured reactions, evaluate $\bar{I}_{13}^{B*} = I_{13}^{B*}/(ML^2)$ and $\bar{I}_{23}^{B*} = I_{23}^{B*}/(ML^2)$. *(2)* Based on the measured reactions, evaluate $\bar{x}_{1c}^* = x_{1c}^*/L$ and $\bar{x}_{2c}^* = x_{2c}^*/L$. *(3)* Find the magnitude of the balancing masses, m_a and m_b, and the angular locations, θ_a and θ_b. Use the following data: $\bar{B}_1^* = B_1^*/(M\Omega^2 L) = -0.0005$; $\bar{B}_2^* = B_2^*/(M\Omega^2 L) = .0008$; $\bar{D}_1^* = D_1^*/(M\Omega^2 L) = 0.0005$; $\bar{D}_2^* = D_2^*/(M\Omega^2 L) = -.0004$. $L = 0.5$ m; $M = 10$ kg; $x_{3a} = 0.2$ and $x_{3b} = 0.3$ m; $R_a = 0.2$ and $R_b = 0.1$ m.

Problem 6.33. Double pendulum with elastic joint

Figure 6.17 depicts a double pendulum comprising bar 1, of mass m_1 and length ℓ_1, and bar 2, of mass m_2 and length ℓ_2. Let frame $\mathcal{F}^A = [\mathbf{A}, \mathcal{A} = (\bar{a}_1, \bar{a}_2)]$ be attached to bar 1 and frame $\mathcal{F}^E = [\mathbf{E}, \mathcal{E} = (\bar{e}_1, \bar{e}_2)]$ be attached to bar 2. A massless tube allows bar 2 to slide in the direction of $\bar{a}_2$; the slider has a mass M and is connected to bar 1 at point $\mathbf{A}$ by means of a spring of stiffness constant k. The position of the slider is determined by its distance, x, from point $\mathbf{A}$, the tip of bar 1; the angular positions of the two bars with respect to the vertical are denoted θ_1 and θ_2, respectively. The system is subjected to gravity along the inertial $\bar{\imath}_1$ direction. *(1)* Derive the equations of motion of the system in terms of θ_1, θ_2 and x. *(2)* On one graph, plot θ_1 and θ_2 as a function of time. *(3)* Plot x as a function of time. *(4)* On one graph, plot the angular velocities of the two bars. *(5)* Plot $\dot{x}$ as a function of time. *(6)* On one graph, plot the kinetic, potential and total mechanical energies of the system. *(7)* On one graph, plot the reaction and elastic forces at the joint. *(8)* On one graph, plot the vertical and horizontal components of the reaction force at point $\mathbf{O}$. Use the following data: $M = 1$, $m_1 = 1$ and $m_2 = 1$ kg; $\ell_1 = 0.4$ and $\ell_2 = 0.5$ m; $k = 400$ N/m; $g = 9.81$ m/s^2. At the initial time $t = 0$, $\theta_1 = \theta_2 = \pi/2$ and $x = 0$. Present all the results of the simulation for $t \in [0, 15]$ s.

Problem 6.34. Pendulum with sliding mass

Figure 6.18 shows a pendulum comprising a bar of mass m and length ℓ and a rigid body of mass M, as discussed in example 6.6. Let frame $\mathcal{F}^E = [(\mathbf{O}, \mathcal{E} = (\bar{e}_1, \bar{e}_2)]$ be attached to the bar. The rigid body is connected at point $\mathbf{B}$ to the tip of the bar at point $\mathbf{A}$ by means of a spring of stiffness constant k and a dashpot of constant c. The stretch of the spring is denoted x and its un-stretched length vanishes. The center of mass of the rigid body is located at point $\mathbf{C}$ and vector η defines the position of the center of mass with respect to point $\mathbf{B}$; the moment of inertia of the body with respect to center of mass is denoted $\underline{\underline{I}}^C$. The angular position of the bar with respect to the vertical is denoted θ. The system is subjected to gravity along the inertial $\bar{\imath}_1$ direction. *(1)* Derive the equations of motion of the system in terms of θ and x. *(2)* Plot θ as a function of time. *(3)* Plot x as a function of time. *(4)* Plot the angular velocity of the bar. *(5)* Plot $\dot{x}$ as a function of time. *(6)* On one graph, plot the kinetic and potential energies of the system as well as the energy dissipated in the dashpot. Verify the energy closure equation. *(7)* On one graph, plot the interaction force at point $\mathbf{A}$ and the total force in the spring and dashpot assembly. *(8)* On one graph, plot the vertical and horizontal components of the reaction force at point $\mathbf{O}$. *(9)* Plot the interaction moment between the bar and the rigid body. Use the following data: $m = 0.4$, $M = 2.5$ kg, $\ell = 0.45$ m, $k = 10$ N/m, $c = 0.05$ N·s/m, $I^{C*} = 0.75$ kg·m^2, $\eta_1^* = 0.2$, and $\eta_2^* = 0.3$ m are the components of vector η in basis $\mathcal{E}$, and $g = 9.81$ m/s^2. At the initial time $t = 0$, $\theta = \pi/2$ and $x = 0$. Present all the results of the simulation for $t \in [0, 50]$ s.

Problem 6.35. Milling machine

Consider the simplified model of a milling machine as depicted in fig. 6.28. The tool support is a rigid body of mass m and moment of inertia I^O with respect to point $\mathbf{O}$ connected to the ground at point $\mathbf{O}$. Its center of mass is located at point $\mathbf{A}$, which is at a distance ℓ_1 from point $\mathbf{O}$. A torsional spring of stiffness constant k_θ, the un-stretched rotation of the spring is denoted θ_0, and a torsional dashpot of constant c_θ act at point $\mathbf{O}$. Let frame $\mathcal{F}^E = (\mathbf{O}, \mathcal{E})$, $\mathcal{E} = (\bar{e}_1, \bar{e}_2)$, be attached to the tool support; the angle between axes $\bar{\imath}_1$ and $\bar{e}_1$ is denoted θ. A massless, rigid bar $\mathbf{DB}$ of length ℓ_2 is free to slide inside the tool support. A spring of stiffness constant k_x, the un-stretched length of the spring is denoted x_0, and a dashpot of constant c_x connect the tool support at point $\mathbf{A}$ to the bar at point $\mathbf{D}$. At point $\mathbf{B}$, the bar connects to the milling machine tool, which is free to rotate about point **B**. Let frame $\mathcal{F}^B = (\mathbf{B}, \mathcal{B})$, $\mathcal{B} = (\bar{b}_1, \bar{b}_2)$, be attached to the tool, whose center of mass **C** is located a distance d along axis $\bar{b}_1$. The tool rotates at a constant angular velocity, Ω, with respect to the bar, such that the angle between axes $\bar{e}_1$ and $\bar{b}_1$ is $\phi = \Omega t$. The tool is of mass M and moment of inertia I^B with respect to point **B**. *(1)* Derive the equations of motion of the system in terms of angle θ and x, the distance from point **A** to **D**. *(2)* Plot θ as a function of time. *(3)* Plot x as a function of time. *(4)* Plot the angular velocity of the tool support. *(5)* Plot $\dot{x}$ as a function of time. *(6)* Plot the torque T_B applied to the tool at point **B**. *(7)* On one graph, plot the cumulative work dissipated in the two dashpots and that done by torque T_B. *(8)* On one graph, plot the kinetic and potential energies of the system. Verify the energy closure equation. Use the following data: $\ell_1 = 0.25$, $\ell_2 = 0.3$, $d = 0.002$ m, $m = 2$, $M = 4$ kg, $I^O = 0.2$, $I^B = 0.0125$ kg·m^2, $\Omega = 400$ rad/s, $k_x = 10$ kN/m, $k_\theta = 15$ kN·m/rad, $c_x = 10$ N·s/m, $c_\theta = 10$ N·m·s/rad, $\theta_0 = \pi/4$, $x_0 = 0.1$ m, and $g = 9.81$ m/s^2. At the initial time $t = 0$, $\theta = \pi/4$ and $x = 0.1$ m. Present all the results of the simulation for $t \in [0, 0.5]$ s.

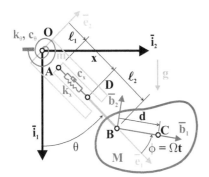

Fig. 6.28. Configuration of the milling machine.

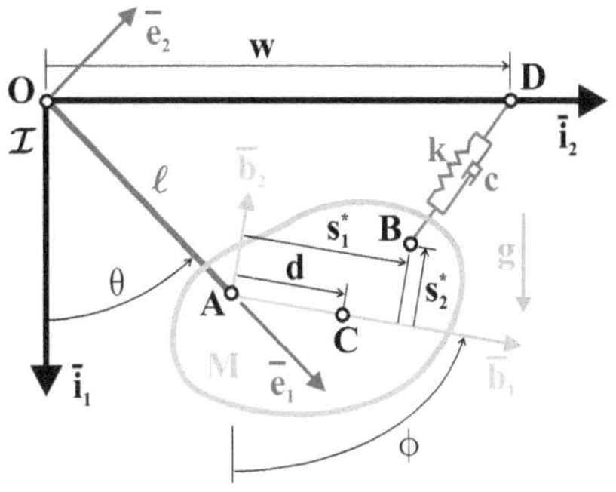

Fig. 6.29. Configuration of the simplified suspension system.

Problem 6.36. Suspension system
Figure 6.29 shows the configuration of a simplified planar suspension system. A rigid body of mass M is connected to the ground at point **A** by means of a massless rigid bar of length ℓ and at point **B** by means of a spring of stiffness constant k and dashpot of constant c. Reference frame $\mathcal{F}^B [\mathbf{A}, \mathcal{B} = (\bar{b}_1, \bar{b}_2)]$ is attached to the rigid body at point **A**; the center of mass of the rigid body is located at distance d from point **A**, along axis $\bar{b}_1$. The coordinates of point **B**, resolved in $\mathcal{B}$, are (s_1^*, s_2^*). Point **D** is located a distance w from point **O**. The configuration of the system is represented by angles θ and ϕ, as indicated in the figure. *(1)* Draw a free body diagram of the system. *(2)* Derive the two equations of motion of the system. *(3)* Find the load in the bar.

Problem 6.37. Bar rocking on top of a curve
A homogeneous bar a length L, thickness h, and mass M is rocking without sliding on top of a fixed curve, as depicted in fig. 6.30. At contact point **P**, a normal contact force, N, and a friction force, F, are acting on the bar. *(1)* Find the work done by the normal contact force, N. Under what condition will this force perform work? *(2)* Find the work done by the friction force, F. Under what condition will this force perform work? *(3)* By means of high-speed

cameras, an experimentalist is monitoring the elevation, d, of the bar's center of mass above the apex of the curve. At times t_1 and t_2, the elevations of the center of mass were measured to be d_1 and d_2, respectively. What can be said about the evolution of the bar's kinetic energy during that time. (4) Is the system's total mechanical energy preserved? (5) Does the bar's angular momentum remain constant? Justify all your answers.

Problem 6.38. Bar rocking atop a curve

Figure 6.30 depicts a homogeneous bar a length L, thickness h, and mass M rocking without sliding on top of a fixed curve. The curve is defined by its intrinsic parametrization, $\underline{r}(s)$, where s is the curvilinear variable measuring length along the curve. (1) Find the equation of motion of the system. (2) Evaluate the normal contact and friction force at point $\mathbf{P}$. (3) If the curve is a circle of radius R, what is the form of the equation of motion?

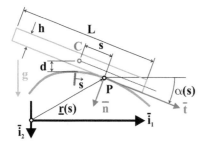

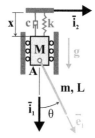

Fig. 6.30. Homogeneous bar rocking without sliding atop a curve.

Fig. 6.31. Pendulum connected to a plunging mass.

Problem 6.39. Pendulum connected to a plunging mass

A pendulum of mass m and length L is connected to a mass M that is allowed to slide vertically, as depicted in fig. 6.31. Mass M is connected to the ground be means of a spring of stiffness constant k and dashpot of constant c. The spring is un-stretched when $x = 0$. (1) Find the equations of motion of the system. (2) Plot the time history of the plunging motion, $\bar{x}(\tau)$. (3) Plot the time history of angle $\theta(\tau)$. (4) Plot the velocity of the plunging mass, $\bar{x}'(\tau)$. (5) Plot the angular velocity of the pendulum, $\theta'(\tau)$. (6) On one graph, plot the non-dimensional horizontal and vertical components of the force applied to the plunging mass at point $\mathbf{A}$, denoted $\bar{H}_A = H_A/(kL)$ and $\bar{V}_A = V_A/(kL)$, respectively. (7) Plot the cumulative energy dissipated in the damper, $\bar{W}^d = W^d/(kL^2)$. (8) On one graph, plot the system's kinetic energy, $\bar{K} = K/(kL^2)$, potential energy, $\bar{V} = V/kL^2$, and the energy closure equation. Use the following data: non-dimensional time, $\tau = \omega t$, where $\omega^2 = k/(M+m)$, $(\cdot)'$ indicates a derivative with respect to τ, $\bar{x} = x/L$, $\mu = m/(M+m) = 0.5$, $\zeta = cw/(2k) = 0.05$, and $\bar{g} = (M+m)g/(kL) = 1.5$. At the initial time ($\tau = 0$), $\bar{x} = 0.5$, $\bar{x}' = 0$, $\theta = \pi/3$, and $\theta' = 0$. Present all results for $\tau \in [0, 20]$.

Problem 6.40. Two-bar mechanism

The two bar mechanism shown in fig. 6.32 comprises bar $\mathbf{OB}$ of length L_1 and mass m_1, and bar $\mathbf{BAT}$ of length L_2 and mass m_2. Bar $\mathbf{BAT}$ passes through a slider located at fixed point $\mathbf{A}$ but free to swivel about that point. A spring of stiffness constant k connects the tip of the bar at point $\mathbf{T}$ to the slider at point $\mathbf{A}$ and is of vanishing un-stretched length. A viscous friction force, $F^f = -c\dot{w}$, acts at the interface between the bar and the slider. *(1)* Derive the equation of motion of the system using generalized coordinate θ_1. *(2)* On one graph, plot angles θ_1 and θ_2 as functions of the non-dimensional time τ. *(3)* On one graph, plot angular velocities θ_1' and θ_2'. *(4)* On one graph, plot angular accelerations θ_1'' and θ_2''. *(5)* Plot the spring stretch, $\bar{\Delta} = \Delta/L_1$. *(6)* On one graph, plot the friction force, $\bar{F}^f = F^f/(kL_1)$, and reaction force at the slider, $\bar{S} = S/(kL_1)$. *(7)* Plot the cumulative energy dissipated at the slider, $\bar{W}^d = W^d/(kL_1^2)$. *(8)* On one graph, plot the kinetic energy, $\bar{K} = K/(kL_1^2)$, potential energy, $\bar{V} = V/(kL_1^2)$, and the energy closure equation. Use the following data: $\mu_1 = m_1/(m_1 + m_2) = 0.6$, $\mu_2 = m_2/(m_1 + m_2) = 1 - \mu_1$, $\bar{d} = d/L_1 = 3$, $\bar{L}_2 = L_2/L_1 = 5$, and $\bar{g} = (m_1 + m_2)g/(kL_1) = 0.2$. Use the non-dimensional time $\tau = \omega t$, where $\omega^2 = k/(m_1 + m_2)$. The viscous friction coefficient is written as $c = 2(m_1 + m_2)\omega\zeta$, where $\zeta = 0.02$. At the initial time, $\theta_1 = 0$ and $\theta_1' = 2.4$, where $(\cdot)'$ indicates a derivative with respect to τ. Present all your results for $\tau \in [0, 40]$.

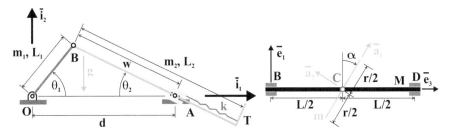

Fig. 6.32. Two-bar mechanism.   **Fig. 6.33.** Rotor with skewed bar.

Problem 6.41. Rigid bar connected to a rotor

Figure 6.33 shows rigid rotor of length L and mass M. Basis $\mathcal{E} = (\bar{e}_1, \bar{e}_2, \bar{e}_3)$ is attached to the rotor; unit vector $\bar{e}_3$ is aligned with the shaft. A rigid bar of length r and mass m is rigidly connected to the shaft. Basis $\mathcal{A} = (\bar{a}_1, \bar{a}_2, \bar{a}_3)$ is attached to the bar, $\bar{e}_3 = \bar{a}_3$ and $\alpha = (\bar{e}_2, \bar{a}_1)$; unit vector $\bar{a}_1$ is aligned with the bar. The shaft and bar are homogeneous slender rods, see fig. 6.42, and their centers of mass coincide at point **C**. The rotor rotates at a constant angular velocity, Ω, about axis $\bar{e}_3$, and is supported by bearings at points **B** and **D**. *(1)* Compute the components of the reaction forces at points **B** and **D** resolved in basis $\mathcal{E}$. *(2)* Compute the components of the reaction forces at points **B** and **D** resolved in the inertial basis.

Problem 6.42. Rigid disk connected to a rotor

Figure 6.33 shows rigid rotor of length L, radius R, and mass M. Basis $\mathcal{E} = (\bar{e}_1, \bar{e}_2, \bar{e}_3)$ is attached to the rotor; unit vector $\bar{e}_3$ is aligned with the shaft. A rigid disk of radius r and mass m is rigidly connected to the shaft. Basis $\mathcal{A} = (\bar{a}_1, \bar{a}_2, \bar{a}_3)$ is attached to the disk, $\bar{e}_3 = \bar{a}_3$ and $\alpha = (\bar{e}_2, \bar{a}_1)$; unit vector $\bar{a}_2$ is normal the disk. The shaft is a homogeneous cylinder, see fig. 6.40, and the disk a homogeneous thin disk, see fig. 6.41. Their centers of mass coincide at point **C**. The rotor rotates at a constant angular velocity, Ω, about axis $\bar{e}_3$, and is supported by bearings at points **B** and **D**. *(1)* Compute the components of the reaction forces at points **B** and **D** resolved in basis $\mathcal{E}$. *(2)* Compute the components of the reaction forces at points **B** and **D** resolved in the inertial basis.

Problem 6.43. Particle sliding in a rolling wheel

Figure 6.34 shows a homogeneous wheel of mass M and radius R rolling without sliding on a horizontal plane under the effect of gravity. A particle of mass m slides in a radial slot of the wheel and is connect to its center by means of a spring of stiffness constant k and dashpot of constant c. The un-stretched length of the spring is x_0. (1) Derive the system's equations of motion using the generalized coordinates x and θ indicated on the figure. (2) Plot the position of mass m, $\bar{x} = x/R$, versus τ. (3) Plot angle θ versus τ. (4) Plot the velocity of mass m, $\bar{x}'$. (5) Plot the wheel's angular velocity, θ'. (6) Plot the acceleration of mass m, $\bar{x}''$. (7) Plot the wheel's angular acceleration, θ''. (8) Plot the cumulative energy dissipated in the dashpot, $\bar{W}_d = W_d/(m\omega^2 R^2)$. (9) On one graph, plot the kinetic, $\bar{K} = K/(m\omega^2 R^2)$, and potential, $\bar{V} = V/(m\omega^2 R^2)$, energies of the system. Verify that the energy closure equation is satisfied. (10) On one graph, plot the force in the spring-dashpot system, $\bar{F}_{sd} = F_{sd}/(m\omega^2 R)$, and the contact force between the particle and slot, $\bar{F}^c = F^c/(m\omega^2 R)$. (11) On one graph, plot the normal and tangential force components at the point of contact of the wheel with the plane, $\bar{N} = N/(m\omega^2 R)$ and $\bar{F}^f = F^f/(m\omega^2 R)$, respectively. (12) What is the minimum required friction coefficient if the wheel is to roll without sliding. Use the following data: $\mu = M/m = 5$, $\zeta = c/(2m\omega) = 0.01$, $\bar{g} = g/(R\omega^2) = 0.2$, $\bar{x}_0 = x_0/R = 0.5$. Use the following non-dimensional time $\tau = \omega t$, where $\omega^2 = k/m$ and $(\cdot)'$ indicates a derivative with respect to τ. Plot all results for $\tau \in [0, 200]$. The initial conditions are $\bar{x} = \theta = \bar{x}' = 0$, and $\theta' = 0.1$.

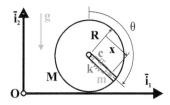
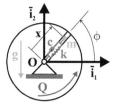

Fig. 6.34. Particle sliding in a rolling wheel.   **Fig. 6.35.** Particle in a slot on a rotating disk.

Problem 6.44. Particle in a slot on a rotating disk

Figure 6.35 depicts a homogeneous disk of mass M and radius R rotating in a vertical plane around inertial point $\mathbf{O}$. Mass m is free to slide in a radial slot on the disk and is connected to the center of the disk by means of a spring of stiffness constant k and a dashpot of constant c. The system is subjected to gravity and a torque, Q, is applied to the disk. The spring's un-stretched length is denoted x_0. (1) Derive the equations of motion of the system in terms of angle ϕ and distance x from point $\mathbf{O}$ to the particle. (2) Find the horizontal and vertical components of the reaction force at point O. (3) If the disk is to rotate at a constant angular velocity, $\dot{\phi} = \Omega$, find the equation of motion for the particle. (4) Find the applied torque, Q, required to maintain this constant angular speed.

Problem 6.45. Pendulum connected to horizontal piston

Figure 6.36 shows a pendulum of length ℓ with a tip mass m. A piston of mass M is rigidly connected to a horizontal rod sliding along the pendulum by means of a slider at point $\mathbf{S}$. A spring of stiffness constant k and dashpot of constant c connect the piston to the ground. The spring is un-stretched when angle $\theta = 0$. The distance from the vertical to point $\mathbf{S}$ is denoted x. (1) Derive the system's equation of motion in terms of angle θ indicated on the figure. *(2)* Plot angle θ versus time τ. *(3)* Plot the rod's angular velocity, θ'. *(4)* Plot the rod's angular acceleration, θ''. *(5)* Plot distance $\bar{x} = x/\ell$. *(6)* Plot the piston's speed $\bar{x}'$. *(7)* Plot the piston's acceleration $\bar{x}''$. *(8)* Plot the cumulative energy dissipated in the dashpot, $\bar{W}_d = W_d/(m\omega^2\ell^2)$. *(9)* On one graph, plot the kinetic, $\bar{K} = K/(m\omega^2\ell^2)$, and potential, $\bar{V} = V/(m\omega^2\ell^2)$, energies of the system. Verify that the energy closure equation is satisfied. *(10)* Plot the normal slider force, $\bar{S} = S/(m\omega^2\ell)$. Use the following data: $\mu = M/m = 2$, $\bar{h} = h/\ell = 0.25$, $\bar{g} = g/(\ell\omega^2) = 0.1$, $\zeta = c/(2m\omega) = 0.01$. Use the following non-dimensional time $\tau = \omega t$, where $\omega^2 = k/m$ and $(\cdot)'$ indicates a derivative with respect to τ. Plot all results for $\tau \in [0, 50]$. The initial conditions are $\theta = 0$, and $\theta' = 1$.

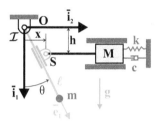

Fig. 6.36. Pendulum connected to horizontal piston.

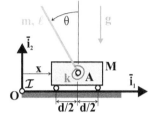

Fig. 6.37. Inverted pendulum mounted on a cart.

Problem 6.46. Inverted pendulum mounted on a cart

Figure 6.37 depicts an inverted homogeneous pendulum of mass m and length ℓ. The pendulum is mounted on a cart of mass M free to translate along a horizontal track. A torsional spring of stiffness constant k restrains the pendulum at its attachment point. The spring is un-stretched when angle $\theta = \theta_0$. *(1)* Derive the two equations of motion of the system. *(2)* Plot the cart's position, $\bar{x} = x/\ell$ versus τ. *(3)* Plot angle θ. *(4)* Plot the cart's velocity, $\bar{x}'$. *(5)* Plot θ'. *(6)* Plot the cart's acceleration, $\bar{x}''$. *(7)* Plot θ''. *(8)* Plot the system's kinetic, $\bar{K} = K/m\ell^2\omega^2$, potential, $\bar{V} = V/(m\ell^2\omega^2)$, and total mechanical energies. *(9)* Plot the horizontal and vertical components of the internal force at point **A**, denoted $\bar{H} = H/(m\ell\omega^2)$ and $\bar{V} = V/(m\ell\omega^2)$, respectively. *(10)* Plot the vertical force components in the front and rear wheels, denoted $\bar{F}_f = F_f/(m\ell\omega^2)$ and $\bar{F}_r = F_r/(m\ell\omega^2)$, respectively. Use the following data: $\mu = M/m = 1.5$, $\theta_0 = 0$, and $\bar{d} = d/\ell = 1$. Use non-dimensional time $\tau = \omega t$, where $\omega = k/(m\ell^2)$ and $(\cdot)'$ denotes a derivative with respect to τ. At the initial time, $\bar{x} = 0$, $\bar{x}' = 1$, $\theta = \pi/4$, $\theta' = 0$. Present all your results for $\tau \in [0, 20]$. Study two cases, $\bar{g} = g/(\ell\omega^2) = 0.8$ and $\bar{g} = 4$, and comment on the differences.

Problem 6.47. Geneva wheel mechanism

Figure 6.38 depicts the Geneva wheel mechanism, which consists of a disk and slotted arm. The disk of radius R and mass M is free to rotate about inertial point **O**. A pin is located at the rim of the disk at point **P**. The slotted arm of length ℓ and mass m is hinged at point **A** and the pin slides inside the slot. The distance from point **A** to the pin is denoted w. At point **A**, the arm is restrained by a torsional spring of stiffness constant k and a torsional dashpot of constant c_1. The spring is un-stretched when $\theta = 0$. A viscous friction force, $F^f = -c_2\dot{w}$, acts at the interface between the pin and the slot. *(1)* On one graph, plot angles ϕ and θ versus τ. *(2)* On one graph, plot angular velocities ϕ' and θ'. *(3)* On one graph, plot angular accelerations ϕ'' and θ''. *(4)* Plot the cumulative energy dissipated in the dashpot and friction mechanism,

$\bar{W}_d = W_d/k$. (5) Plot the system's kinetic, $\bar{K} = K/k$, potential, $\bar{V} = V/k$, energies. Check that the energy closure equation is satisfied. (6) On one graph, plot the normal contact force, $\bar{F}^c = F^c/(mR\omega^2)$, and viscous friction force, $\bar{F}^f = F^f/(mR\omega^2)$, at the pin. Use the following data: $\mu = M/m = 2$, $\bar{L} = L/R = 1.5$, $\bar{\ell} = \ell/R = 2$, $\bar{g} = g/(\ell\omega^2) = 0.2$, $\zeta_1 = \omega c_1/(2k) = 0.02$, and $\zeta_2 = \omega c_2 R^2/(2k) = 0.01$. Use the following non-dimensional time $\tau = \omega t$, where $\omega^2 = 3k/(m\ell^2)$ and $(\cdot)'$ indicates a derivative with respect to τ. Plot all results for $\tau \in [0, 5]$. The initial conditions are $\theta = 0$ and $\theta' = 1.5$.

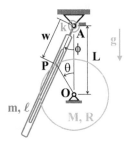

Fig. 6.38. Geneva wheel mechanism.

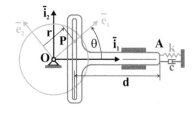

Fig. 6.39. Scotch yoke mechanism.

Problem 6.48. Scotch yoke mechanism

Figure 6.39 depicts the Scotch yoke mechanism, which consists of a disk and slotted yoke. The disk of radius R and mass M is free to rotate about inertial point **O**. A pin is located at a distance r from the center of the disk. The slotted yoke of length ℓ and mass m is allowed to move horizontally and the pin slides inside the slot. At point **A**, the yoke is restrained by a spring of stiffness constant k and a dashpot of constant c_1. The spring is un-stretched when $\theta = 0$. A viscous friction force, $F^f = -c_2 v_r$, acts at the interface between the pin and the slot; v_r is the relative velocity of the pin with respect to the slot. (1) Plot angle θ versus τ. (2) Plot angular velocity θ'. (3) Plot angular acceleration θ''. (4) Plot the cumulative energy dissipated in the dashpot and friction mechanism, $\bar{W}_d = W_d/(m\omega^2 r^2)$. (5) Plot the system's kinetic, $\bar{K} = K/(m\omega^2 r^2)$, potential, $\bar{V} = V/(m\omega^2 r^2)$, energies. Check that the energy closure equation is satisfied. (6) On one graph, plot the normal contact force, $\bar{F}^c = F^c/(mr\omega^2)$, and viscous friction force, $\bar{F}^f = F^f/(mr\omega^2)$, at the pin. Use the following data: $\mu = M/m = 2$, $\bar{R} = R/r = 0.8$, $\zeta_1 = \omega c_1/(2k) = 0.01$, and $\zeta_2 = \omega c_2/(2k) = 0.01$. Use the following non-dimensional time $\tau = \omega t$, where $\omega^2 = k/m$ and $(\cdot)'$ indicates a derivative with respect to τ. Plot all results for $\tau \in [0, 30]$. The initial conditions are $\theta = 0$ and $\theta' = 1.5$.

6.8 Inertial characteristics

The inertial characteristics of rigid bodies with simple shapes are presented below. For each rigid body, the volume, $\mathcal{V}$, of the body and the principal mass moments of inertia I_{11}^*, I_{22}^*, and I_{33}^* are given. The figures also indicate the location of the center of mass and the orientation of the principal axes of inertia.

- **Cylinder (Figure 6.40):** volume, $\mathcal{V} = \pi R^2 L$; principal mass moments of inertia, $I_{11}^* = I_{22}^* = mR^2/4 + mL^2/12$, $I_{33}^* = mR^2/2$.

- **Thin disk (Figure 6.41):** volume, $\mathcal{V} = \pi R^2 L$; principal mass moments of inertia, $I_{11}^* = I_{22}^* = mR^2/4$, $I_{33}^* = mR^2/2$. These results are obtained from their counterparts for a cylinder when $L/R \ll 1$.

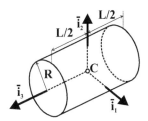

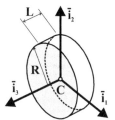

Fig. 6.40. Cylinder.   **Fig. 6.41.** Thin disk.

- **Slender rod (Figure 6.42):** volume, $\mathcal{V} = AL$; principal mass moments of inertia, $I_{11}^* = I_{22}^* = mL^2/12$, $I_{33}^* \approx 0$. These results are obtained from their counterparts for a cylinder when $d/R \ll 1$, where d is a representative dimension of area A.
- **Half cylinder (Figure 6.43):** volume, $\mathcal{V} = \pi R^2 L/2$; principal mass moments of inertia, $I_{11}^* = m(R^2/4 - d^2) + mL^2/12$, $I_{22}^* = mR^2/4 + mL^2/12$, $I_{33}^* = m(R^2/2 - d^2)$; center of mass location, $d = 4R/3\pi$.

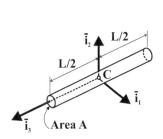

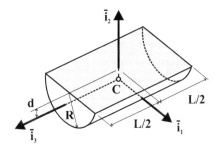

Fig. 6.42. Slender rod.   **Fig. 6.43.** Half cylinder.

- **Parallelepiped (Figure 6.44):** volume, $\mathcal{V} = abc$; principal mass moments of inertia, $I_{11}^* = m(b^2 + c^2)/12$, $I_{22}^* = m(a^2 + c^2)/12$, $I_{33}^* = m(a^2 + b^2)/12$.
- **Thin plate (Figure 6.45):** volume, $\mathcal{V} = abc$; principal mass moments of inertia, $I_{11}^* = mb^2/12$, $I_{22}^* = ma^2/12$, $I_{33}^* = m(a^2 + b^2)/12$; center of mass location, $d = 4R/3\pi$.
- **Sphere (Figure 6.46):** volume, $\mathcal{V} = 4\pi R^3/3$; principal mass moments of inertia, $I_{11}^* = I_{22}^* = I_{33}^* = 2mR^2/5$.
- **Half sphere (Figure 6.47):** volume, $\mathcal{V} = 2\pi R^3/3$; principal mass moments of inertia, $I_{11}^* = I_{22}^* = 83mR^2/320$, $I_{33}^* = 2mR^2/5$.

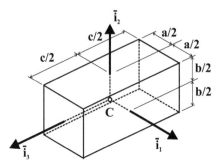

Fig. 6.44. Parallelepiped.

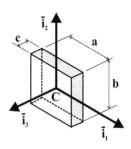

Fig. 6.45. Thin plate.

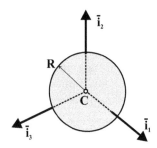

Fig. 6.46. Sphere.

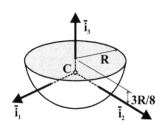

Fig. 6.47. Half sphere.

- **Ellipsoid (Figure 6.48):** volume, $\mathcal{V} = 4\pi abc/3$; principal mass moments of inertia, $I_{11}^* = 1/5\, m(b^2 + c^2)$, $I_{22}^* = 1/5\, m(a^2 + c^2)$, $I_{33}^* = 1/5\, m(a^2 + b^2)$.
- **Hollow cylinder (Figure 6.49):** volume, $\mathcal{V} = \pi(R_o^2 - R_i^2)L$; principal mass moments of inertia, $I_{11}^* = I_{22}^* = m(R_o^2 - R_i^2)/4 + mL^2/12$, $I_{33}^* = m(R_o^2 - R_i^2)/2$.

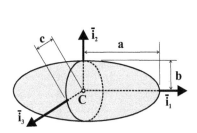

Fig. 6.48. Ellipsoid.

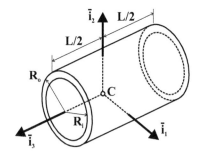

Fig. 6.49. Hollow cylinder.

Part III

Concepts of analytical dynamics

7

Basic concepts of analytical dynamics

Newtonian mechanics deals with the response of particles to externally applied loads and Euler generalized these concepts to systems of particles. For simple systems of particles, it is convenient to use Cartesian coordinates to represent the configuration of the system, but more often than not, other types of coordinates are used as well. For instance, path or surface coordinates were introduced in chapter 2. The manipulation of finite rotation also plays an important role in dynamics and was studied in depth in chapter 4.

In fact, the ability to use various types of coordinates considerably simplifies the description of dynamical systems and the analysis of their response to externally applied loads. The concepts of generalized coordinates, kinematic constraints, and degrees of freedom are introduced in section 7.2. Next, the important concepts of virtual displacements and rotations presented in section 7.3 lead to the definition of a scalar quantity of fundamental importance to dynamics, the virtual work presented in section 7.4.

The principle of virtual work for static problems is introduced in section 7.5 and is shown to be equivalent to Newton's first law. Examples of application of this important principle are presented using both arbitrary virtual displacements and kinematically admissible virtual displacements. Finally, in the presence of conservative forces, the statement of the principle of virtual work is shown to simplify remarkably. The first section introduces the mathematical tools required for the comprehension of this chapter.

7.1 Mathematical preliminaries

In this section, the stationarity conditions of a function of several variables are expressed as both differential and variational conditions. These concepts will play a fundamental role in the remainder of the chapter.

7.1.1 Stationary point of a function

Consider a function of n variables, $F = F(u_1, u_2, \ldots, u_n)$. By definition, the stationary points [2] of this function are defined as those for which

$$\frac{\partial F}{\partial u_i} = 0, \quad i = 1, 2, \ldots, n. \tag{7.1}$$

For a function of a single variable, this condition corresponds to a horizontal tangent to the graph of the function, as illustrated in fig. 7.1. At a stationary point, the function can present a minimum, a maximum, or a saddle point.

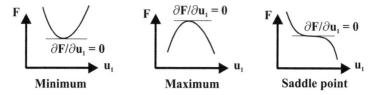

Fig. 7.1. Stationary points of a function.

If a function is stationary at a point, conditions (7.1) hold and the following statement is then true

$$\frac{\partial F}{\partial u_1} w_1 + \frac{\partial F}{\partial u_2} w_2 + \ldots + \frac{\partial F}{\partial u_n} w_n = 0,$$

where $w_1, w_2, \ldots, w_n$ are arbitrary quantities. It is convenient to use a special notation for these arbitrary quantities, $w_i = \delta u_i$, where δu_i are called *virtual changes* in u_i. The above statement now becomes

$$\frac{\partial F}{\partial u_1} \delta u_1 + \frac{\partial F}{\partial u_2} \delta u_2 + \ldots + \frac{\partial F}{\partial u_n} \delta u_n = 0.$$

Comparison of this result with a similar expression for the differential, dF, of the same function expanded using the chain rule for derivatives implies that virtual changes, δu_i, are similar to differentials in the variables, du_i. Consequently, the virtual change operator, denoted "δ," behaves in a manner similar to the differential operator, denoted "d". This relationship between the two operators will be further investigated in later sections.

The *variation in* function F, noted δF, is defined as

$$\delta F = \frac{\partial F}{\partial u_1} \delta u_1 + \frac{\partial F}{\partial u_2} \delta u_2 + \ldots + \frac{\partial F}{\partial u_n} \delta u_n. \tag{7.2}$$

If follows that the stationarity conditions, eq. (7.1), now become

$$\delta F = 0, \tag{7.3}$$

for all arbitrary variations $\delta u_1, \delta u_2, \ldots, \delta u_n$. The differential conditions, eq. (7.1), and the variational condition, eq. (7.3), both express the necessary and sufficient conditions for the stationarity of function F at a point. From the above developments, it is clear that eq. (7.1) implies eq. (7.3) and since the above reasoning can be reversed, it is simple to prove that eq. (7.3) implies eq. (7.1). Hence, the two conditions are equivalent.

The process defined by a "variation of function F" can be thought of as a "*mathematical experiment*," or "*what if?*" scenario. The condition $\delta F = 0$ for all arbitrary variations $\delta u_1, \delta u_2, \ldots, \delta u_n$ at a stationary point means "the change in function F would vanish if I were to change the values of all variables at the stationary point." Or, "if I were to experiment with changes in all variables about a stationary point, I would find no corresponding change in function F." Because the changes in variable values defined by such a mathematical experiment are not actual changes, the words "virtual change" are used. The symbol "δ" is associated with such *virtual changes* as opposed to the symbol "d" that refers to *actual, infinitesimal changes*.

To determine whether a stationary point is a minimum, a maximum, or a saddle point it is necessary to consider the second derivatives [2] of the function. If

$$\sum_{i,j=1}^{n} \frac{\partial^2 F}{\partial u_i \partial u_j} \, du_i du_j > 0 \tag{7.4}$$

at a stationary point for all differentials du_i and du_j, the function presents a minimum. If, on the other hand, the same quantity is negative for all du_i and du_j, the function presents a maximum. Finally, if the same quantity can be positive or negative depending on the choice of the differentials, the function presents a saddle point.

From the definition of the variation of a function, eq. (7.2), it follows that

$$\delta^2 F = \sum_{i,j=1}^{n} \frac{\partial^2 F}{\partial u_i \partial u_j} \, \delta u_i \delta u_j.$$

It is now clear that a stationary point is a minimum if

$$\delta^2 F > 0, \tag{7.5}$$

for all arbitrary variations δu_i and δu_j. It is a maximum if $\delta^2 F < 0$ for all variations, and a saddle point occurs if the sign of the second variation depends on the choice of the variations of the independent variables.

7.1.2 Stationary point of a definite integral

Next, consider the determination of the stationary point of the following definite integral

$$I = \int_a^b F(y, y', x) \, dx, \tag{7.6}$$

where notation $(\cdot)'$ indicates a derivative with respect to variable x. The integrand involves an unknown function, $y(x)$, which is subjected to boundary conditions $y(a) = \alpha$ and $y(b) = \beta$.

This problem seems to be of a completely different nature from that treated in the previous section. Indeed, integral I is a "function of a function," *i.e.*, the value of the definite integral, I, depends on the choice of the unknown function, $y(x)$. Because there are an infinite number of values of function $y(x)$ for $x \in [a, b]$, definite integral, I, is equivalent to a function of an infinite number of variables.

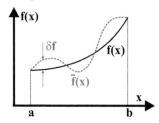

This problem will be treated using the variational formalism introduced in the previous section. The concept of variation of a variable, δu, is extended to the concept of *variation of a function*, denoted δf. Figure 7.2 shows two functions, $f(x)$ and $\bar{f}(x)$, such that

Fig. 7.2. The concept of variation of a function.

$$\delta f = \bar{f}(x) - f(x) = \phi(x), \qquad (7.7)$$

where $\phi(x)$ is a continuous and differentiable, but otherwise arbitrary function such that $\phi(a) = \phi(b) = 0$. In other words, δf is a virtual change that brings the function $f(x)$ to a new, arbitrary function $\bar{f}(x)$. Note that $\delta f(a) = \delta f(b) = 0$.

The stationarity of I requires

$$\delta I = \delta \int_a^b F(y, y', x) \, \mathrm{d}x = \int_a^b \delta F(y, y', x) \, \mathrm{d}x = 0.$$

With the help of eq. (7.2), this becomes

$$\delta I = \int_a^b \left[\frac{\partial F}{\partial y} \delta y + \frac{\partial F}{\partial y'} \delta y' \right] \mathrm{d}x = 0.$$

Functions $y(x)$ and $y'(x)$ are not independent of each other; hence, variations δy and $\delta y'$ are not independent, making it difficult to draw any conclusion from this statement. To eliminate the variation $\delta y'$, an integration by parts is performed on the second term in the square bracket

$$\int_a^b \frac{\partial F}{\partial y'} \delta\left(\frac{\mathrm{d}y}{\mathrm{d}x}\right) \mathrm{d}x = \int_a^b \frac{\partial F}{\partial y'} \frac{\mathrm{d}}{\mathrm{d}x}(\delta y) \, \mathrm{d}x = -\int_a^b \frac{\mathrm{d}}{\mathrm{d}x}\left(\frac{\partial F}{\partial y'}\right) \delta y \, \mathrm{d}x + \left[\frac{\partial F}{\partial y'} \delta y\right]_a^b.$$

The boundary terms vanish because $\delta y(a) = \delta y(b) = 0$, and the stationarity condition now becomes

$$\delta I = \int_a^b \left[\frac{\partial F}{\partial y} - \frac{\mathrm{d}}{\mathrm{d}x}\left(\frac{\partial F}{\partial y'}\right) \right] \delta y \, \mathrm{d}x = 0.$$

The bracketed term must vanish because the integral must vanish for *all arbitrary variations* δy. This yields

$$\frac{\partial F}{\partial y} - \frac{\mathrm{d}}{\mathrm{d}x}\left(\frac{\partial F}{\partial y'}\right) = 0. \tag{7.8}$$

Here again, the above reasoning can be reversed. Starting from eq. (7.8), and performing the integration by parts in the reverse order implies $\delta I = 0$. In summary, *the necessary and sufficient condition for the definite integral to be at a stationary point is that eq. (7.8) be satisfied.* This differential equation is called the *Euler-Lagrange equation* of the problem.

The variational formalism introduced in this section will be systematically applied to dynamics problem. It will be shown that the equations of motions of dynamics can be viewed as the Euler-Lagrange equations associated with the stationarity condition of definite integrals. Various forms of the equations of dynamics can be easily obtained by direct manipulations of these definite integrals. It is therefore important to understand the variational formalism and its implications.

A crucial difference exists between a differential, $\mathrm{d}f$, of function $f(x)$ and a variation, δf, of the same function, as depicted in fig. 7.3. A differential, $\mathrm{d}f$, is an infinitesimal change in $f(x)$ resulting from an infinitesimal change, $\mathrm{d}x$, in the independent variable; $\mathrm{d}f/\mathrm{d}x$ represents the tangent at the point. On the other hand, δf is an arbitrary virtual change that brings $f(x)$ to $\bar{f}(x)$. The two quantities, $\mathrm{d}f$ and δf, are clearly unrelated, the former is positive in fig. 7.3, but the latter is negative.

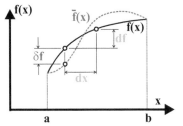

Fig. 7.3. The difference between a differential, $\mathrm{d}f$, and a variation, δf.

Although the concepts associated with a differential of a function, $\mathrm{d}f$, and a variation of the same function, δf, are clearly distinct, manipulations of the two symbols are quite similar. For instance, the order of application of the two operations can be interchanged. Indeed,

$$\frac{\mathrm{d}}{\mathrm{d}x}(\delta f) = \frac{\mathrm{d}}{\mathrm{d}x}(\bar{f} - f) = \frac{\mathrm{d}\bar{f}}{\mathrm{d}x} - \frac{\mathrm{d}f}{\mathrm{d}x} = \delta\left(\frac{\mathrm{d}f}{\mathrm{d}x}\right).$$

Similarly, the order of the integration and variation operations commutes

$$\delta\left(\int_a^b F \,\mathrm{d}x\right) = \int_a^b \bar{f} \,\mathrm{d}x - \int_a^b F \,\mathrm{d}x = \int_a^b (\bar{f} - F) \,\mathrm{d}x = \int_a^b \delta F \,\mathrm{d}x.$$

7.2 Generalized coordinates

Consider a system consisting of N particles that are free to move in three-dimensional space. The position vector of particle i will be expressed in terms of its Cartesian coordinates as $\underline{r}_i = x_i \bar{\imath}_1 + y_i \bar{\imath}_2 + z_i \bar{\imath}_3$. The total number of parameters required to define the configuration of the system is $3N$, three parameters for each of the N particles. The solution of the problem involves the determination of the time history of these $3N$ Cartesian coordinates when the system is subjected to a set of time-dependent forces.

Of course, Cartesian coordinates are not the only way to determine the position of a particle in space; for instance, the spherical coordinates introduced in section 2.7.2 could be used, and the Cartesian coordinates of particle i would then be expressed in terms of the spherical coordinates r_i, ϕ_i, and θ_i as $x_i = r_i \sin \phi_i \cos \theta_i$, $y_i = r_i \sin \phi_i \sin \theta_i$, and $z_i = r_i \cos \phi_i$. As discussed in section 2.6, this coordinate transformation corresponds to a mapping of the three-dimensional space onto itself.

Generalized coordinates

In general, the Cartesian coordinates of particle i could be expressed in terms of $n = 3N$ parameters, called *generalized coordinates*, as

$$x_i = x_i(q_1, q_2, \ldots q_n), \tag{7.9a}$$
$$y_i = y_i(q_1, q_2, \ldots q_n), \tag{7.9b}$$
$$z_i = z_i(q_1, q_2, \ldots q_n). \tag{7.9c}$$

The solution of the problem now involves the determination of the time history of the n generalized coordinates, q_i, $i = 1, 2, \ldots n$. Presumably, the choice of appropriate generalized coordinates will ease the solution of the problem. For instance, the solution of a problem involving spherical symmetry is often simplified by using spherical coordinates. It is assumed here that eqs. (7.9) define a *one to one* mapping between Cartesian and generalized coordinates; this implies that the Jacobian of the coordinate transformation, see eq. (2.72), has a non vanishing determinant at all points in space.

Cartesian coordinates determine the position of a particle in space: the three parameters x, y, and z are the projections of the position vector of the particle along the axes of an orthonormal basis in three-dimensional space. Let the position of the particle be determined by spherical coordinates, $q_1 = r$, $q_2 = \phi$, and $q_3 = \theta$. It now becomes possible to consider the three numbers, q_1, q_2, and q_3, to be the rectangular coordinates of a point in a three-dimensional space, called the *configuration space*.

Figure 7.4 depicts this concept: on the left, the particle is shown in the geometric space defined by Cartesian coordinates x, y, and z; on the right, is it shown in the configuration space defined by generalized coordinates q_1, q_2, and q_3. The geometry of the problem is distorted in the configuration space; if the particle is constrained to move on a spherical surface in the geometric space, it must remain in the shaded rectangular area shown on the right portion of fig. 7.4.

The concept of configuration space can be generalized to higher-dimensional problems. If the system is defined by n generalized coordinates, $q_1, q_2, \ldots, q_n$, these n numbers become the rectangular coordinates of a point in the n dimensional configuration space. The trajectory of a particle is defined by three functions, $x_i = x_i(t)$, $y_i = y_i(t)$, and $z_i = z_i(t)$, a curve in three-dimensional space. In the configuration space, the trajectories of all particles are defined by a single curve in the n dimensional configuration space, $q_1 = q_1(t)$, $q_2 = q_2(t)$, $\ldots$, $q_n = q_n(t)$.

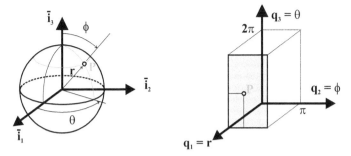

Fig. 7.4. Geometric and configuration spaces for $n = 3$.

Kinematic constraints

The concept of generalized coordinates is intimately linked to that of *kinematic constraints*. For instance, fig. 7.5 depicts a dumbbell consisting of two masses moving in two-dimensional space and linked by a rigid bar. At first, the configuration of the system will defined by four generalized coordinates consisting of the Cartesian coordinates of the two particles: $q_1 = x_1$ and $q_2 = y_1$ for the first particle, and $q_3 = x_2$ and $q_4 = y_2$ for the second.

This representation, however, ignores the fact that the rigid bar imposes a kinematic constraint on the system: at all times, the two particles must remain at a distance ℓ from each other, and hence, $(q_3 - q_1)^2 + (q_4 - q_2)^2 = \ell^2$. Of the four generalized coordinates, three only are independent.

Next, the configuration of the system will be defined by the Cartesian coordinates of its center of mass, point **C**, $q_1 = x_C$ and $q_2 = y_C$, and the orientation, $q_3 = \theta$, of the rigid bar with respect to axis $\bar{\imath}_1$, as shown in fig. 7.5. This second approach bypasses the need for kinematic constraints. Clearly, the number of generalized coordinates is not a characteristic of the system: the dumbbell system can be represented alternatively by three or four generalized coordinates.

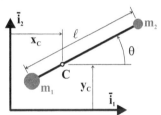

Fig. 7.5. Dumbbell in two-dimensional space.

Degrees of freedom

This discussion also leads to the concept of *degree of freedom*: the system depicted in fig. 7.5 presents three degrees of freedom because three parameters are required to uniquely define its configuration. Let n denote the number of generalized coordinates, m the number of kinematic constraints, and d the number of degrees of freedom; it then follows that

$$d = n - m. \tag{7.10}$$

The first approach discussed above involves 4 generalized coordinates - 1 kinematic constraint = 3 degrees of freedom. The second approach features 3 generalized coordinates - 0 kinematic constraint = 3 degrees of freedom.

The number of degrees of freedom is an intrinsic characteristic of the system. On the other hand, the choice of the number of generalized coordinates is left to analyst. If the number of generalized coordinates exceeds that of degrees of freedom, $m = n - d$ kinematic constraints must exist among the n generalized coordinates. If the number of generalized coordinates equals that of degrees of freedom, all generalized coordinates are independent and no kinematic constraints are involved. Finally, if the number of generalized coordinates is less than that of degrees of freedom, the configuration of the system cannot be fully defined.

The number of degrees of freedom is an *invariant characteristic* of a given mechanical system; it is defined as the minimum number of parameters necessary to determine the configuration of the system.

Here are a few sample mechanical systems involving various numbers of degrees of freedom.

1. *One degree of freedom:* a particle moving along a fixed curve in space, a rigid body rotating about a fixed axis in space while one of its points remains a fixed inertial point.
2. *Two degrees of freedom:* a particle moving on a surface, a planar double pendulum.
3. *Three degrees of freedom:* a particle moving in three-dimensional space, the planar motion of a rigid body, the three-dimensional motion of a rigid body rotating about a fixed inertial point.
4. *Four degrees of freedom:* a double pendulum moving in three-dimensional space.
5. *Five degrees of freedom:* two particles linked by a rigid bar and moving in three-dimensional space.
6. *Six degrees of freedom:* the arbitrary motion of a rigid body in three-dimensional space.

The time derivatives of the generalized coordinates are called the *generalized velocities*. The 2n dimensional space defined by the generalized coordinates and velocities is called the *state space*.

Example 7.1. The rigid body

Consider a rigid body consisting of N particles, where N is a very large number. In the first approach, the configuration of the rigid body will be defined by the $3N$ Cartesian coordinates of its N particles. This representation involves a large number of kinematic constraints that enforce the rigidity of the body: the distance between any two particles of the body must remain constant.

To evaluate the number of kinematic constraints, consider four particles of the body located at the vertices of a tetrahedron. This simplified configuration features $4 \times 3 = 12$ generalized coordinates, the positions of the four particles, linked by the six kinematic constraints enforcing to the constant length conditions for the six edges of the tetrahedron. The fifth particle of the system adds three new generalized coordinates, the Cartesian coordinates of the particle, and three new kinematic constraints, three constant length constraints linking the particle to the previous four. The complete rigid body is then constructed by adding the particles one at a time; each new particle adds three new generalized coordinates and three new constraints.

The complete system involves $n = 3N$ generalized coordinates and $m = 6 + 3(N - 4) = 3N - 6$ kinematic constraints, for a total of $d = 3N - (3N - 6) = 6$ degrees of freedom. This reasoning establishes the fact that a rigid body involves six degrees of freedom only, a very intuitive fact.

A second approach to the representation of a rigid body takes advantage of the of the fact that six parameters only are required to define the configuration of the body. Such a representation could use the Cartesian coordinates of one arbitrary reference point of the body, and three rotation components to define the orientation of the body; Euler angles, for instance, could be used for this purpose.

The last approach to be discussed here is one that involves 12 generalized coordinates, selected to be the $4 \times 3 = 12$ Cartesian coordinates of four points on the body forming a tetrahedron and 6 kinematic constraints, imposing the constant length constraint for the six edges of the tetrahedron. One advantage of this formulation is that it bypasses the need for the nonlinear kinematics associated with rotations: this is a *rotationless formulation*.

Example 7.2. The slider-arm mechanism

Figure 7.6 depicts a mechanism consisting of a slider free to move along unit vector $\bar{\imath}_1$ and connected to arm **AP** of length ℓ. The arm is free to rotate in the plane normal to $\bar{\imath}_1$. This mechanical system features two degrees of freedom: the position of the slider $q_1 = x_1$, and angle $q_2 = \theta$ between the arm and the horizontal plane, for instance. Indeed, the configuration of the system is unequivocally defined once these generalized coordinates are known.

Although the number of degrees of freedom, d, is an inherent property of the system, the choice of a specific set of generalized coordinates is far from being unique. Consider the following choice of generalized coordinates: $q_1 = x_1$ and $q_2 = x_2$. In this case, the number of generalized coordinates still equals the number of degrees of freedom and there are no kinematic constraints. This simple choice, however, might not be the most appropriate: for a given value of q_2, two configurations of the system are possible, corresponding to arm positions above and below the horizontal plane, respectively.

Alternatively, it is possible to select more generalized coordinates than strictly necessary. For instance, three generalized coordinates could be used to define this system, the Cartesian coordinates of point **P**: $q_1 = x_1$, $q_2 = x_2$, and $q_3 = x_3$. Clearly, this choice does not increase the number of degrees of freedom to three; rather, it implies that a single relationship or kinematic constraint must exist between the three generalized coordinates. Indeed, q_2 and q_3 must be such that $q_2^2 + q_3^2 = \ell^2$. Hence, the system presents two degrees of freedom: 3 generalized coordinates - 1 constraint = 2 degrees of freedom.

Example 7.3. The crank-slider mechanism

Figure 7.7 depicts a crank slider mechanism. An experienced analyst will correctly identify this system as presenting a single degree of freedom; indeed, selecting a single generalized coordinate, $q_1 = \theta_1$, unequivocally defines the configuration of the system.

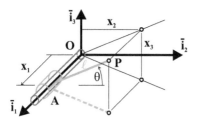

Fig. 7.6. Slider with arm mechanism.   **Fig. 7.7.** Crank slider mechanism.

A less experimented analyst might select four generalized coordinates, $q_1 = \theta_1$, $q_2 = \theta_2$, $q_3 = \phi$, and $q_4 = x$. A second look at the system, however, reveals that these generalized coordinates are linked by a number of constraints

$$q_3 = q_1 + q_2, \quad \ell_1 \cos q_1 + \ell_2 \cos q_2 = q_4, \quad \frac{\ell_1}{\sin q_1} = \frac{\ell_2}{\sin q_2} = \frac{q_4}{\sin q_3}. \quad (7.11)$$

The first constraint is an angle equality in triangle **OAB**; the second stems from the projection of segments **OA** and **AB** along unit vector $\bar{\imath}_1$; finally, the last two constraints express the laws of sines in triangle **OAB**. Consequently, the number of degrees of freedom is: 4 generalized coordinates - 4 constraints = 0 degrees of freedom.

This reasoning is erroneous because the four kinematic constraints are not independent. Indeed, the law of sine constraint implies $\ell_1 = q_4 \sin q_1 / \sin q_3$ and $\ell_2 = q_4 \sin q_2 / \sin q_3$; introducing these expressions in the second constraint leads to $\sin(q_1 + q_2) / \sin q_3 = 1$, a result that is implied by the first constraint. Consequently, three constraints only are independent, and hence, the system presents a single degree of freedom: 4 generalized coordinates - 3 independent constraints = 1 degree of freedom.

Here again, the choice of a specific set of generalized coordinates is far from being unique. Clearly, each one of the three angles θ_1, θ_2, or ϕ would be a valid choice for the generalized coordinate. Position x of the piston could be another possible choice for the generalized coordinate, although not a very desirable choice. Indeed, two possible configurations of the system are associated with the same value of x: the two configurations are mirror images about unit vector $\bar{\imath}_1$.

Furthermore, when x reaches its maximum value, $\ell_1 + \ell_2$, i.e., when the two linkages become collinear, the value of x does not accurately determine the position of the system. Let y be the position of point **A** above unit vector $\bar{\imath}_1$, see fig. 7.7; kinematic arguments yield the following relationship

$$\bar{x} = 1 - \frac{1}{2}\left[1 + \left(\frac{\ell_2}{\ell_1}\right)^3\right]\left[1 + \left(\frac{\ell_2}{\ell_1}\right)\right]\bar{y}^2 = 1 - c\bar{y}^2$$

where $\bar{x} = x/(\ell_1 + \ell_2)$ and $\bar{y} = y/(\ell_1 + \ell_2)$. It then follows that $d\bar{x}/d\bar{y} = -2c\bar{y}$. When $\bar{y} \to 0$, $d\bar{x}/d\bar{y} \to 0$; this means that when $\bar{y}$ becomes small, generalized coordinate $\bar{x}$ does not accurately define the configuration of the system.

7.3 The virtual displacement and rotation vectors

Consider a particle whose displacement vector is given as $\underline{r}(t) = x_1(t)\bar{\imath}_1 + x_2(t)\bar{\imath}_2 + x_3(t)\bar{\imath}_3$ in a Cartesian coordinate system. The *variation of the position vector* is then $\delta\underline{r} = \delta x_1\bar{\imath}_1 + \delta x_2\bar{\imath}_2 + \delta x_3\bar{\imath}_3$, where $\delta x_i(t)$, $i = 1, 2, 3$, are the variations of the corresponding Cartesian coordinates, as defined by eq. (7.7) for an arbitrary function.

The virtual displacement vector

Next, the Cartesian coordinates of the particle are assumed to be expressed in terms of generalized coordinates, $x_i = x_i(q_1, q_2, \ldots, q_n)$, $i = 1, 2, 3$. The variation of this Cartesian coordinate then follows from the definition of the variation of a function, eq. (7.2),

$$\delta x_i = \frac{\partial x_i}{\partial q_1}\delta q_1 + \frac{\partial x_i}{\partial q_2}\delta q_2 + \ldots + \frac{\partial x_i}{\partial q_n}\delta q_n.$$

Applying the same treatment to each coordinate leads to the following expression for the variation of the position vector,

$$\delta\underline{r} = \frac{\partial r}{\partial q_1}\delta q_1 + \frac{\partial r}{\partial q_2}\delta q_2 + \ldots + \frac{\partial r}{\partial q_n}\delta q_n. \tag{7.12}$$

The terms *variation of position vector, virtual change of the position vector* or *virtual displacement vector* are used interchangeably, because a virtual change in position is, in fact, a virtual change in displacement.

The study of constrained dynamical systems will be delayed up to chapter 10. For the remainder of this chapter, it is assumed that the number of generalized coordinates used to represent the system is equal to its number of degrees of freedom, hence, the systems is not subjected to any kinematic constraints. Under this restriction, the virtual displacements defined by eq. (7.12) are called *virtual displacements compatible with the constraints*, or *kinematically admissible virtual displacements*.

Comparing the differential and virtual displacement vectors

It is interesting to compare eq. (7.12) with its counterpart for the *differential position vector* or *differential displacement vector*

$$d\underline{r} = \frac{\partial r}{\partial q_1}dq_1 + \frac{\partial r}{\partial q_2}dq_2 + \ldots + \frac{\partial r}{\partial q_n}dq_n + \frac{\partial r}{\partial t}dt. \tag{7.13}$$

If the position vector is an explicit function of time, the last term, involving the partial derivative with respect to time, appears in the expression for the differential displacement. This contrasts with the expression for the virtual displacement vector that does not involve partial derivatives with respect to time. This important difference stems from the fact that a virtual displacement is an arbitrary change in displacement *at a given, fixed instant.* Consequently, when evaluating virtual displacements, time is held constant, and partial derivatives with respect to time vanish.

Dividing eq. (7.13) by a time increment, dt, yields the expression for the velocity vector

$$\dot{\underline{r}} = \frac{\partial \underline{r}}{\partial q_1}\dot{q}_1 + \frac{\partial \underline{r}}{\partial q_2}\dot{q}_2 + \ldots + \frac{\partial \underline{r}}{\partial q_n}\dot{q}_n + \frac{\partial \underline{r}}{\partial t}. \tag{7.14}$$

A comparison between eqs. (7.12), (7.13) and (7.14) for the virtual displacement, differential displacement and velocity vectors, respectively, reveals close similarities, but also important differences among these three concepts.

The velocity vector is simply the time derivative of the position vector, a familiar concept. The differential displacement vector is the infinitesimal change in position resulting from infinitesimal changes in the generalized coordinates and time. Finally, the virtual displacement vector corresponds to the change in displacement associated with virtual changes in the generalized coordinates at a fixed instant in time. The differential displacement is the actual displacement resulting from actual infinitesimal changes in generalized coordinates and time. In contrast, a virtual displacement is associated with an arbitrary virtual changes that bring the configuration of the system described by generalized coordinates q_i to a new configuration described by generalized coordinates $\bar{q}_i$, at a given, fixed instant in time.

While it is important to keep in mind the fundamental differences between these concepts, the similarities between eqs. (7.12) and (7.14) can be used to expeditiously evaluate virtual displacement vectors. Consider, for instance, the velocity vector expressed in cylindrical coordinates, see eq. (2.91b), and the corresponding expression for virtual displacements

$$\underline{v} = \dot{r}\,\bar{e}_1 + r\dot{\theta}\,\bar{e}_2 + \dot{z}\,\bar{e}_3 \iff \delta\underline{r} = \delta r\,\bar{e}_1 + r\delta\theta\,\bar{e}_2 + \delta z\,\bar{e}_3.$$

The velocity vector, $\underline{v}$, is replaced by the virtual displacement vector, $\delta\underline{r}$, and the time derivatives of the generalized coordinates, $\dot{r}$, $\dot{\theta}$, and $\dot{z}$ are replaced by the corresponding virtual changes in generalized coordinates, δr, $\delta\theta$, and δz, respectively.

Similar guidelines are used to obtain the expression for the the virtual displacement vector in spherical coordinates from the corresponding expression for the velocity vector, eq. (2.95b),

$$\underline{v} = \dot{r}\,\bar{e}_1 + r\dot{\phi}\,\bar{e}_2 + r\dot{\theta}\sin\phi\,\bar{e}_3 \iff \delta\underline{r} = \delta r\,\bar{e}_1 + r\delta\phi\,\bar{e}_2 + r\delta\theta\sin\phi\,\bar{e}_3.$$

The virtual rotation vector

A striking example of the analogy between velocity and virtual displacement vectors is the concept of virtual rotation vector. The angular velocity vector was defined by eq. (4.56) as $\underline{\omega} = \text{axial}(\underline{\dot{R}}\,\underline{R}^T)$; the *virtual rotation vector*, $\delta\underline{\psi}$, is defined in an analogous manner as

$$\delta\underline{\psi} = \text{axial}(\delta\underline{R}\,\underline{R}^T). \tag{7.15}$$

In section 4.12.4, the differential rotation vector itself was introduced by analogy to the angular velocity vector, underlining the close connection between the three concepts.

Here again, it is crucial to understand that there exist no vector $\underline{\psi}$ such that $\delta(\underline{\psi})$ is the virtual rotation vector; to emphasize is important fact, the notation $\delta\underline{\psi}$, rather than $\delta\underline{\psi}$, is used denote the virtual rotation vector. Note the parallel between the virtual rotation vector defined here and the differential rotation vector defined in section 4.12.4.

Developments identical to those presented in section 4.12.4 lead to the following important relationship between virtual changes in angular velocity and the virtual rotation vector

$$\delta\underline{\omega} = \dot{\overline{\delta\underline{\psi}}} - \widetilde{\omega}\delta\underline{\psi}. \tag{7.16}$$

Virtual changes in the components of the angular velocity vector expressed in the rotating frame can also be obtained in a similar manner

$$\delta\underline{\omega} = \dot{\overline{\delta\underline{\psi}}} - \widetilde{\omega}\delta\underline{\psi}, \qquad \delta\underline{\omega} = \underline{R}\,\dot{\overline{\delta\underline{\psi}}}^*, \tag{7.17a}$$

$$\delta\underline{\omega}^* = \dot{\overline{\delta\underline{\psi}}}^* + \widetilde{\omega}^*\delta\underline{\psi}^*, \qquad \delta\underline{\omega}^* = \underline{R}^T\,\dot{\overline{\delta\underline{\psi}}}. \tag{7.17b}$$

Example 7.4. The two-bar linkage with slider system
Figure 7.8 shows a single degree of freedom planar mechanism. The system is represented by a single generalized coordinate, θ. Determine the kinematically admissible virtual displacement vector at point **T** in terms of the virtual rotation component, $\delta\theta$.

The position vector of point **T** is $\underline{r}_T = L_b\,\bar{e}_1$, where $\mathcal{F}^A = [\mathbf{A}, \mathcal{E} = (\bar{e}_1, \bar{e}_2)]$ is a frame attached to bar **AT** at point **A**. A virtual change is the position vector of point **T** then becomes $\delta\underline{r}_T = L_b\,\delta\bar{e}_1 = L_b\delta\phi\,\bar{e}_2$. This relationship should be compared with its counterpart for velocities, $\dot{\underline{r}}_T = L_b\,\dot{\bar{e}}_1 = L_b\dot{\phi}\,\bar{e}_2$, where $\dot{\phi}$ is the angular velocity of bar **AT**.

The problem now reduces to finding a relationship between virtual rotations $\delta\phi$ and $\delta\theta$, or equivalently, between the angular velocities of bars **AT** and **OB**, denoted $\dot{\phi}$ and $\dot{\theta}$, respectively. To that effect, the position vector of point **B** is written in two alternative manners: $\underline{r}_B = w\,\bar{e}_1 = L_c\,\bar{a}_1$, where $\mathcal{F}^O = [\mathbf{O}, \mathcal{A} = (\bar{a}_1, \bar{a}_2)]$ is a frame attached to bar **OB** at point **O**. A virtual change in the position vector of point **B** then becomes $\delta\underline{r}_B = \delta w\,\bar{e}_1 + w\delta\phi\,\bar{e}_2 = L_c\delta\theta\,\bar{a}_2$. Here again, it is interesting to compare this expression with its counterpart relating velocities: $\dot{\underline{r}}_B = \dot{w}\bar{e}_1 + w\dot{\phi}\bar{e}_2 = L_c\dot{\theta}\bar{a}_2$.

The desired result is then obtained by evaluating the scalar product of the virtual displacement by $\bar{e}_2$ to find $w\,\delta\phi = L_c\delta\theta\,\bar{e}_2^T\bar{a}_2 = L_c\delta\theta\cos(\theta + \phi)$. The virtual displacement at point **T** then follows as

$$\delta\underline{r}_T = \frac{L_b L_c}{w}\cos(\theta + \phi)\delta\theta\,\bar{e}_2.$$

If so desired, the components of the virtual displacement vector could be evaluated in the fixed basis $\mathcal{I} = (\bar{\imath}_1, \bar{\imath}_2)$ by projecting vector $\bar{e}_2$ along that basis.

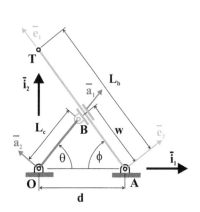

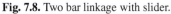

Fig. 7.8. Two bar linkage with slider.

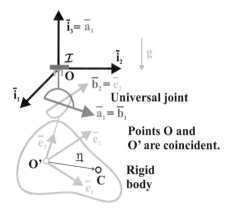

Fig. 7.9. Configuration of the rigid body connected to a universal joint.

Example 7.5. The rigid body/universal joint system

Figure 7.9 depicts a rigid body attached to the ground by means of a universal joint. This common mechanical joint, shown in detail in fig. 7.10, consists of a rigid cruciform articulated to two rigid components, denoted components k and ℓ. The cruciform consists of two orthogonal bars, and unit vectors $\bar{b}_1$ and $\bar{b}_2$ are aligned with those bars. Component k is articulated with respect to the cruciform and is allowed to rotate about unit vector $\bar{b}_1$. Similarly, component ℓ is also articulated to the cruciform and rotates about unit vector $\bar{b}_2$.

Component k of the universal joint is connected to the ground at point **O** by means of a bearing allowing rotation about axis $\bar{\imath}_3$. Component ℓ is connected to a rigid body at point **O′**. A first planar rotation about axis $\bar{\imath}_3$, of magnitude ϕ, brings inertial basis $\mathcal{I} = (\bar{\imath}_1, \bar{\imath}_2, \bar{\imath}_3)$ to basis $\mathcal{A} = (\bar{a}_1, \bar{a}_2, \bar{a}_3)$, where $\bar{a}_1$ is aligned with unit vector $\bar{b}_1$ of the cruciform. A second planar rotation about axis $\bar{a}_1$, of magnitude θ, brings basis $\mathcal{A}$ to basis $\mathcal{B} = (\bar{b}_1, \bar{b}_2, \bar{b}_3)$, where $\bar{b}_2$ is the second unit vector aligned with the cruciform. Finally, a third planar rotation about axis $\bar{b}_2$, of magnitude ψ, bring basis $\mathcal{B}$ to basis $\mathcal{E} = (\bar{e}_1, \bar{e}_2, \bar{e}_3)$ that is attached to the rigid body. Points **O** and **O′** are coincident. These three planar rotations describe the orientation of the rigid body using Euler angles with the *3-1-2* sequence, see eq. (4.78).

The first planar rotation is prescribed to be $\phi = \Omega t$. Compute the velocity of point **C**, the center of mass of the rigid body. The position vector of point **C** with respect to point **O** is denoted η. Because angle ϕ is a known function of time, the system features two degrees of freedom. Evaluate the kinematically admissible virtual displacement vector of the center of mass in terms of the virtual rotation components $\delta\theta$ and $\delta\psi$.

Let $\underline{R}$ denote the rotation tensor that brings basis $\mathcal{I}$ to basis $\mathcal{E}$. The components of the inertial position of point **C** in basis $\mathcal{I}$ are $\underline{r}_C = \underline{\underline{R}}\,\underline{\eta}^*$, where $\underline{\eta}^*$ are the components of the vector η in the body attached basis, $\mathcal{E}$. The components of the inertial velocity of point **C** now become $\underline{v}_C = \underline{\underline{R}}\,\widetilde{\underline{\omega}}^*\underline{\eta}^*$, where $\underline{\omega}^*$ are the components of the angular velocity vector of the rigid body resolved in basis $\mathcal{E}$. The relationship between the components of the angular velocity vector and the time derivatives of the Euler angles is given by eq. (4.80), and hence,

$$\underline{v}_C = \underline{\underline{R}}\,\widetilde{\eta}^{*T} \begin{bmatrix} -C_\theta S_\psi & C_\psi & 0 \\ S_\theta & 0 & 1 \\ C_\theta C_\psi & S_\psi & 0 \end{bmatrix} \begin{Bmatrix} \dot\phi \\ \dot\theta \\ \dot\psi \end{Bmatrix},$$

where $\dot\phi = \Omega$ and the components of the rotation tensor expressed in terms of Euler angles are given by eq. (4.78).

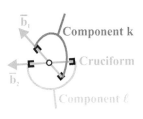

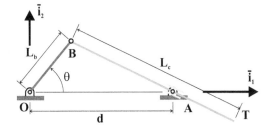

Fig. 7.10. Configuration of the universal joint.

Fig. 7.11. Two bar linkage.

Next, the components of the virtual displacement vector of point $\mathbf{C}$ are evaluated as $\delta\underline{r}_C = \underline{\underline{R}}\,\widetilde{\delta\psi}^{*}\,\underline{\eta}^{*}$, where $\delta\psi^*$ are the components of the virtual rotation vector of the rigid body resolved in basis $\mathcal{E}$. Using eq. (4.80) once again leads to

$$\delta\underline{r}_C = \underline{\underline{R}}\,\widetilde{\eta}^{*T} \begin{bmatrix} -C_\theta S_\psi & C_\psi & 0 \\ S_\theta & 0 & 1 \\ C_\theta C_\psi & S_\psi & 0 \end{bmatrix} \begin{Bmatrix} \delta\phi \\ \delta\theta \\ \delta\psi \end{Bmatrix}.$$

Because virtual changes are taken at a given, fixed instant in time, $\delta\phi = \delta(\Omega t) = 0$. The virtual displacement vector now becomes

$$\delta\underline{r}_C = \underline{\underline{R}}\,\widetilde{\eta}^{*T} \begin{bmatrix} C_\psi & 0 \\ 0 & 1 \\ S_\psi & 0 \end{bmatrix} \begin{Bmatrix} \delta\theta \\ \delta\psi \end{Bmatrix}.$$

The components of the velocity or virtual displacement vectors can be evaluated in any basis; for instance, their components in the body attached basis $\mathcal{E}$ are $\underline{\underline{R}}^T \underline{v}_C$ and $\underline{\underline{R}}^T \delta\underline{r}_C$, respectively. This example illustrates an important difference between the velocity and virtual displacement vectors. In contrast with the velocity vector that does depends on the prescribed angular velocity, $\dot\phi = \Omega$, the virtual displacement vector is independent of this quantity.

7.3.1 Problems

Problem 7.1. Virtual displacement of a two-bar linkage system
The two bar linkage shown in fig. 7.11 comprises bar **OB** of length L_b and bar **BAT** of length L_c. Bar **BAT** passes through fixed point **A** but is free to swivel about that point. *(1)* Compute the virtual displacement vector of point **T** in terms of the virtual rotation component, $\delta\theta$.

7.4 Virtual work and generalized forces

The differential work done by a force was defined as the scalar product of the force vector by the differential displacement vector of its point of application, see eq. (3.8). By analogy, the *virtual work* done by a force is defined in this section as the scalar product of the force vector by the virtual displacement vector of its point of application. The concept of virtual work then gives rise to that of generalized forces.

7.4.1 Virtual work

The *virtual work* done by the forces externally applied to a particle is defined as

$$\delta W = \underline{F}^T \delta \underline{r}. \tag{7.18}$$

Note the parallel between the definition of the virtual work and that of the differential work, see eq. (3.8). The virtual work corresponds to the work that would be performed by the externally applied forces if the particle were to undergo virtual displacement $\delta \underline{r}$. This contrasts with the differential work that corresponds to the work performed by the same forces when the particle undergoes an actual, infinitesimal displacement $d\underline{r}$.

Notation δW denotes the virtual work, but this does not imply the existence of a work function, W, such that $\delta(W)$ is the virtual work. In general, the virtual work is a *nonholonomic quantity*, *i.e.*, a quantity that cannot be integrated.

For a system of N particles, the virtual work is found by summing the contributions of all particles, each undergoing its own virtual displacement $\delta \underline{r}_i$: $\delta W = \sum_{i=1}^{N} F_i^T \delta \underline{r}_i$.

7.4.2 Generalized forces

As discussed in section 7.2, it is often convenient to represent the configuration of a system by a set of generalized coordinates, $\underline{q}^T = \{q_1, q_2, \ldots, q_n\}$. Let the position vector of a particle be a function of generalized coordinates: $\underline{r} = \underline{r}(\underline{q})$. The virtual work done by the externally applied forces now becomes

$$\delta W = \underline{F}^T \delta \underline{r} = \underline{F}^T \left(\frac{\partial \underline{r}}{\partial q_1} \delta q_1 + \frac{\partial \underline{r}}{\partial q_2} \delta q_2 + \ldots + \frac{\partial \underline{r}}{\partial q_n} \delta q_n \right) \tag{7.19}$$
$$= Q_1 \delta q_1 + Q_2 \delta q_2 + \ldots + Q_n \delta q_n,$$

where the quantities,

$$Q_i = \underline{F}^T \frac{\partial \underline{r}}{\partial q_i}, \tag{7.20}$$

are called the *generalized forces*.

In section 3.1.4, the forces applied to a particle were shown fall into two categories: conservative forces, *i.e.*, those that can be derived from a potential, and non-conservative forces, *i.e.*, those for which no potential function exists. Similarly, generalized forces that can be derived from a potential are called *conservative generalized forces*; in this case, a potential function, V, exists such that

$$Q_i^c = -\frac{\partial V}{\partial q_i}. \tag{7.21}$$

The virtual work done by a generalized conservative force, denoted δW_c, can now be computed as

$$\delta W_c = -\frac{\partial V}{\partial q_1} \delta q_1 - \frac{\partial V}{\partial q_2} \delta q_2 - \ldots - \frac{\partial V}{\partial q_n} \delta q_n = -\delta(V). \tag{7.22}$$

The virtual work done by a generalized conservative force can be evaluated as the variation of a potential function, V, and becomes an integrable expression.

7.4.3 Virtual work done by internal forces

It is of interest to compute the virtual work done by the internal forces of a system. Consider the single degree of freedom, planar mechanism shown in fig. 7.12; the system is represented by a single generalized coordinate, θ. At first, the virtual work done by the internal force at point **B** will be computed. To that effect, bar **OB** is separated from the slider at point **B** and the corresponding free body diagram is shown in fig. 7.12a, revealing the internal force vector, $\underline{F}^B$.

The virtual work done by this internal force is $\delta W^B = \delta \underline{r}_B^T \underline{F}^B + \delta \underline{r}_{B'}^T (-\underline{F}^B)$. In view of Newton's third law, the internal forces acting at points **B** and **B'** are of equal magnitudes, opposite directions, and share a common line of action; on the other hand, because the virtual displacements are kinematically admissible, they do not violate the kinematic constraints of the system, and hence, $\delta \underline{r}_B = \delta \underline{r}_{B'}$. The virtual work done by the internal force at point **B** now becomes

$$\delta W^B = \delta \underline{r}_B^T \underline{F}^B - \delta \underline{r}_B^T \underline{F}^B = 0. \tag{7.23}$$

The virtual work done by internal forces vanishes. This important result will be used extensively in many methods of analytical dynamics.

The evaluation of the work done by the internal force at point **B** will now be contrasted with that done by the friction force, $\underline{F}^f$, acting between the slider and bar **AT**. The virtual work done by the friction force is $\delta W^f = \delta \underline{r}_B^T \underline{F}^f + \delta \underline{r}_{B''}^T (-\underline{F}^f)$. Here again, in view of Newton's third law, the friction forces acting at points **B** and **B''** are of equal magnitudes, opposite directions, and share a common line of action;

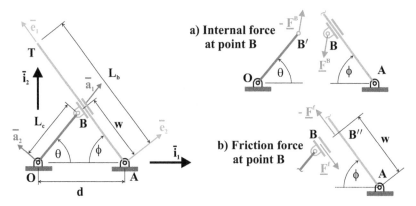

Fig. 7.12. Two bar linkage with slider. a) Internal force at point **B**. b) Friction force in the slider.

the work done by the friction force becomes $\delta W^f = F^f \bar{e}_1^T (-\delta \underline{r}_B + \delta \underline{r}_{B''})$, where F^f is the magnitude of the friction force.

On bar **AT**, the point of application of the friction force is point **B''**, the *material point* on bar **AT** that is at a distance w from point **A**. Point **B''** is the material point of bar **AT** located at the instantaneous point of contact between the slider and bar **AT**, see section 5.4. The position vector of this point is $\underline{r}_{B''} = w\,\bar{e}_1$ and the virtual displacement vector become $\delta \underline{r}_{B''} = w\,\delta \bar{e}_1 = w\delta\phi\,\bar{e}_2$, because w remains a constant for material point **B''**. With this result, the virtual work done by the friction force becomes $\delta W^f = F^f \bar{e}_1^T (-\delta \underline{r}_B + w\delta\phi\,\bar{e}_2) = -F^f \bar{e}_1^T \delta \underline{r}_B$.

The position vector of point **B** is $\underline{r}_B = w\,\bar{e}_1$, and the virtual displacement vector of this point is then $\delta \underline{r}_B = \delta w\,\bar{e}_1 + w\,\delta\phi\,\bar{e}_2$; the virtual work now becomes $\delta W^f = -F^f \bar{e}_1^T (\delta w\,\bar{e}_1 + w\,\delta\phi\,\bar{e}_2) = -F^f \delta w$. This result is rather intuitive: the virtual work done by the friction force equals the product of the magnitude of the friction force by the virtual displacement of its point of application. Since the friction force is directed along $\bar{e}_1$, any virtual displacement along the direction perpendicular to the bar, $\bar{e}_2$, does not contribute to the virtual work.

The law of cosines applied to triangle **OBA** reveals that $w^2 = d^2 + L_c^2 - 2dL_c \cos\theta$, and hence, $w\delta w = dL_c\delta\theta \sin\theta$. Finally, the virtual work done by the friction force becomes

$$\delta W^B = -F^f \delta w = -F^f \frac{dL_c}{w} \sin\theta\,\delta\theta. \tag{7.24}$$

The work done by the *internal force* at point **B** vanishes, see eq. (7.23), but the work done by the friction force does not, see eq. (7.24). The force at point **B** is a constraint force: it imposes the kinematic constraint that the displacement of the slider must equal that of the tip of bar **AB** at all times. The virtual work done by the constraint forces vanishes because the virtual displacements of the points of application of constraint forces, $\underline{F}^B$ and $-\underline{F}^B$, are *identical*.

The work done by the *friction force* does not vanish because the virtual displacements of the points of application of friction forces, $\underline{F}^f$ and $-\underline{F}^f$, are *different*.

Indeed, although the position vectors of points **B** and **B″** are identical, $\underline{r}_B = \underline{r}_{B''} = w\bar{e}_1$, the corresponding virtual displacement vectors are not, $\delta\underline{r}_B = \delta w\,\bar{e}_1 + w\,\delta\phi\,\bar{e}_2$ and $\delta\underline{r}_{B''} = w\,\delta\phi\,\bar{e}_2$. Point **B″** is the material point of bar **AT** that is at the location of the point of contact of the slider with the bar. Because point **B″** is a material point of bar **AT**, the value of w that defines its location remains constant, *i.e.*, $\delta w = 0$ when computing the virtual displacement $\delta\underline{r}_{B''}$. Here again, it is important to distinguish the contact point from the material points that instantaneously coincide with this contact point, see section 5.4.

7.4.4 Problems

Problem 7.2. Virtual work done by friction force
The two bar linkage shown in fig. 7.11 comprises bar **OB** of length L_b and bar **BAT** of length L_c. Bar **BAT** passes through fixed point **A** but is free to swivel about that point. *(1)* Assuming that a friction torque, M^f, is acting in the joint at point **B**, compute the virtual work done by this torque. *(2)* Assuming that a friction force, F^f, is acting in the sliding joint at point **A**, compute the virtual work done by this force. In both cases, express the virtual work in terms of the virtual rotation component, $\delta\theta$.

Problem 7.3. Two rigid bodies connected by an actuator
Figure 7.13 depicts two rigid bodies, denoted bodies k and ℓ, respectively, connected by an actuator. Frame $\mathcal{F}^k = [\mathbf{K}, \mathcal{E}^k = (\bar{e}_1^k, \bar{e}_2^k, \bar{e}_3^k)]$ is attached to body k and a similarly defined frame, $\mathcal{F}^\ell$, is attached to body ℓ. The configuration of frame $\mathcal{F}^k$ is determined by the position vector, $\underline{u}^k$, of its reference point **K** and rotation tensor $\underline{\underline{R}}^k$ that brings triad $\mathcal{I}$ to triad $\mathcal{E}^k$. The configuration of frame $\mathcal{F}^\ell$ is defined by corresponding quantities, $\underline{u}^\ell$ and $\underline{\underline{R}}^\ell$. The actuator is connected

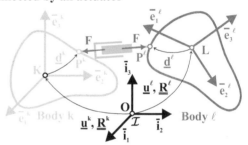

Fig. 7.13. Two rigid bodies connected connected by an actuator.

at points **P**^{k} and **P**$^\ell$ to bodies k and ℓ, respectively. Let $\underline{d}^k$ and $\underline{d}^\ell$ be the position vectors of points **P**^{k} and **P**$^\ell$ with respect to the reference points **K** and **L**, respectively. The actuator applies known forces F of equal magnitudes and opposite signs to bodies k and ℓ, respectively, as indicated on the figure. *(1)* Find the virtual work done by the actuator. *(2)* Find the generalized forces applied to body k and ℓ, respectively. *(3)* Discuss the physical interpretation of the various generalized force components.

7.5 The principle of virtual work for statics

As discussed in section 3.1.2, the static equilibrium condition for a particle, as stated by Newton's first law, is written as a vector equation that imposes the vanishing of the externally applied forces. In the present section, an alternative formulation will be developed, which results in the *principle of virtual work*. Although expressed in terms of work rather than force vectors, the principle of virtual work will be shown to be equivalent to Newton's first law. In this section, the principle of virtual work is develop *for static problems only*; applications of this principle to dynamical system will be treated in chapter 8. The principle will be developed first for a single particle; next, it will be generalized to systems of particles.

The principle of virtual work introduces the fundamental concept of "arbitrary virtual displacements" sometimes called "arbitrary test displacements," or also "arbitrary fictitious displacements," and all of these expressions will be used interchangeably. The word "arbitrary" is easily understood: it simply means that the displacements can be chosen in an arbitrary manner without any restriction imposed on their magnitudes or orientations. More difficult to understand are the words "virtual," "test," or "fictitious." All three imply that these are not real, actual displacements. More importantly, these fictitious displacements *do not affect the forces acting on the particle*. These important concepts will be explained in the following sections.

7.5.1 Principle of virtual work for a single particle

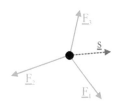

Consider a particle in static equilibrium under a set a externally applied loads, as depicted in fig. 7.14. According to Newton's first law, the sum of the externally applied load must vanish. Next, consider a fictitious displacement of arbitrary magnitude and orientation, denoted $\underline{s}$. Although the problem appears to be two-dimensional in the figure, both forces and fictitious displacements are three-dimensional quantities.

Fig. 7.14. A particle with applied forces subjected to a fictitious test displacement.

The virtual work done by the externally applied forces is now evaluated by computing the scalar product of the externally applied load by the fictitious displacement vector to find

$$W = \underline{s}^T \left[\sum \underline{F} \right] = 0. \tag{7.25}$$

Because the particle is in static equilibrium, Newton's first law implies the vanishing of the bracketed term. It follows that the scalar product vanishes *for any arbitrary fictitious displacement*.

This result sheds some light on the special nature of the fictitious, or virtual displacements. If the particle is in static equilibrium in a given configuration, the sum of the forces vanishes, *i.e.*, $\sum \underline{F} = 0$. Assume now that one of the externally applied forces, say $\underline{F}_1$, is the force acting in an elastic spring connected to the particle. If the particle undergoes a *real, but arbitrary displacement*, $\underline{d}$, the force in the spring will change to become $\underline{F}_1'$. All displacement-dependent forces applied to the particle will change, and the sum of the externally applied loads becomes $\sum \underline{F}'$. In the new configuration resulting from the application of the real displacement, $\underline{d}$, static equilibrium will not be satisfied, *i.e.*, $\sum \underline{F}' \neq 0$. Indeed, if the particle is in static equilibrium in the configuration resulting from the application of an *arbitrary displacement*, it would be in static equilibrium in *any* configuration, which makes little sense.

In contrast with real displacements, *virtual or fictitious displacements do not affect the forces applied to the particle*. This means that even in the presence of displacement-dependent loads such as those arising within an elastic spring, if the particle is in static equilibrium, it remains in static equilibrium when virtual or fictitious displacements are applied. This is the reason why eq. (7.25) remains true for *all arbitrary virtual displacements*. The discussion thus far has thus established that if the particle is in static equilibrium, eq. (7.25) holds for all arbitrary fictitious displacements.

Next, the following question is asked: if eq. (7.25) holds, is the particle in static equilibrium? Consider fig. 7.14, and let the components of the applied forces be $\underline{F}_1 = F_{11}\bar{\imath}_1 + F_{12}\bar{\imath}_2 + F_{13}\bar{\imath}_3$, $\underline{F}_2 = F_{21}\bar{\imath}_1 + F_{22}\bar{\imath}_2 + F_{23}\bar{\imath}_3$, and $\underline{F}_3 = F_{31}\bar{\imath}_1 + F_{32}\bar{\imath}_2 + F_{33}\bar{\imath}_3$, while the components of the virtual displacement are $\underline{s} = s_1\bar{\imath}_1 + s_2\bar{\imath}_2 + s_3\bar{\imath}_3$, where $\mathcal{I} = (\bar{\imath}_1, \bar{\imath}_2, \bar{\imath}_3)$ is an orthonormal basis. Equation (7.25) now states $(F_{11} + F_{21} + F_{31})s_1 + (F_{12} + F_{22} + F_{32})s_2 + (F_{13} + F_{23} + F_{33})s_3 = 0$.

At first, assume that the particle is not in static equilibrium, *i.e.*, $\sum \underline{F} \neq 0$. It is always possible to find *a particular virtual displacement* for which eq. (7.25) will be satisfied. Indeed, for a given set of forces, select s_1 and s_2 in an arbitrary manner, then solve eq. (7.25) for s_3 to find $s_3 = -[(F_{11} + F_{21} + F_{31})s_1 + (F_{12} + F_{22} + F_{32})s_2]/(F_{13} + F_{23} + F_{33})$. Consequently, the fact that eq. (7.25) is satisfied *for a particular virtual displacement* does not imply that it is in static equilibrium. In fact, even if it is satisfied *for many virtual displacements*, static equilibrium is still not guaranteed. Indeed, for each new arbitrary choice of s_1 and s_2, it is possible to compute an s_3 for which eq. (7.25) is satisfied.

Different conclusions are reached if eq. (7.25) is satisfied *for all arbitrary virtual displacements*. Indeed, if $(F_{11} + F_{21} + F_{31})s_1 + (F_{12} + F_{22} + F_{32})s_2 + (F_{13} + F_{23} + F_{33})s_3 = 0$ for all independently chosen quantities s_1, s_2, and s_3, it follows that $F_{11} + F_{21} + F_{31} = 0$, $F_{12} + F_{22} + F_{32} = 0$, and $F_{13} + F_{23} + F_{33} = 0$, is the only solution of eq. (7.25). In turn, this can be written as $(F_{11} + F_{21} + F_{31})\bar{\imath}_1 + (F_{12} + F_{22} + F_{32})\bar{\imath}_2 + (F_{13} + F_{23} + F_{33})\bar{\imath}_3 = 0$, and finally, $\sum \underline{F} = 0$. Thus, if eq. (7.25) is satisfied *for all arbitrary virtual displacements*, then $\sum \underline{F} = 0$, and the particle is in static equilibrium.

In conclusion, if a particle is in static equilibrium, the virtual work done by the externally applied forces vanishes for all arbitrary virtual displacements. Furthermore, it is also true that if the virtual work vanishes for all arbitrary fictitious test displacements, the sum of the externally applied forces vanishes, and hence, the particle is in static equilibrium. These two facts can be combined into the statement of the principle of virtual work for a particle.

Principle 8 (Principle of virtual work for a particle) *A particle is in static equilibrium if and only if the virtual work done by the externally applied forces vanishes for all arbitrary virtual displacements.*

Because the condition for static equilibrium is nothing but Newton's first law, it follows that the principle of virtual work, which states the condition for static equilibrium, is equivalent to Newton's first law, and either statement provides a fundamental definition of static equilibrium. Simple examples will now be used to illustrate the principle of virtual work.

Example 7.6. Equilibrium of a particle

Consider the particle depicted in fig. 7.15, which is subjected to two vertical forces $\underline{F}_1 = 1\bar{\imath}_1$ and $\underline{F}_2 = -3\bar{\imath}_1$. The following question is asked: is the particle in static equilibrium? Rather than relying on Newton's first law, the principle of virtual work will used to answer the question. Consider the following arbitrary virtual displacement, $\underline{s} = s_1\bar{\imath}_1 + s_2\bar{\imath}_2$, and the associated virtual work

$$ W = (1\bar{\imath}_1 - 3\bar{\imath}_1)^T (s_1\bar{\imath}_1 + s_2\bar{\imath}_2) = -2\bar{\imath}_1^T (s_1\bar{\imath}_1 + s_2\bar{\imath}_2) = -2s_1 \neq 0. $$

The fact that $\underline{s}$ is an arbitrary virtual displacement implies that s_1 and s_2 are arbitrary scalars, and hence, $W = -2s_1 \neq 0$. Because the virtual work done by the externally applied forces does not vanish for all virtual displacements, the principle of virtual work, principle 8, implies that the particle is not in static equilibrium.

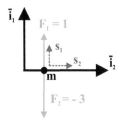

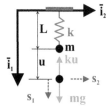

Fig. 7.15. A particle under the action of two forces.

Fig. 7.16. A particle suspended to an elastic spring.

It is important to understand the implications of the last part of the principle of virtual work, "for all arbitrary virtual displacements." Consider the following arbitrary virtual displacement, $\underline{s} = s_2\bar{\imath}_2$, and the associated virtual work

$$ W = (1\bar{\imath}_1 - 3\bar{\imath}_1)^T s_2\bar{\imath}_2 = -2\bar{\imath}_1^T s_2\bar{\imath}_2 = 0. $$

This result is due to the fact that the sum of the externally applied loads, $-2\bar{\imath}_1$, is orthogonal to the virtual displacement, $s_2\bar{\imath}_2$, and hence, the virtual work vanishes. One might be tempted to conclude from the above result that the particle is in static equilibrium because the virtual work vanishes. To satisfy the principle of virtual work, however, the virtual work must vanish *for all arbitrary virtual displacements*.

The above result shows that the virtual work may vanish for "a particular virtual displacement," but this is not a sufficient condition to guarantee static equilibrium. For the two-dimensional problem shown in fig. 7.15, an arbitrary fictitious displacement must span the plane of the problem, *i.e.*, must be of the form $\underline{s} = s_1\bar{\imath}_1 + s_2\bar{\imath}_2$. For three-dimensional problems, a three-dimensional virtual displacement must be selected, $\underline{s} = s_1\bar{\imath}_1 + s_2\bar{\imath}_2 + s_3\bar{\imath}_3$, where s_1, s_2, and s_3 are three arbitrary scalars, and $\mathcal{I} = (\bar{\imath}_1, \bar{\imath}_2, \bar{\imath}_3)$ a basis that spans the three-dimensional space.

Example 7.7. Equilibrium of a particle connected to an elastic spring
Consider next a particle in static equilibrium under the effect of gravity and the restoring force of an elastic spring of stiffness constant k, as depicted in fig. 7.16. Find the displacement of the particle in its actual static equilibrium configuration.

For this two-dimensional problem, assume that the particle is at position u. An arbitrary fictitious displacement is selected as $\underline{s} = s_1\bar{\imath}_1 + s_2\bar{\imath}_2$, where s_2 and s_2 are two arbitrary scalars. The virtual work done by the externally applied loads becomes

$$W = (mg\bar{\imath}_1 - ku\bar{\imath}_1)^T(s_1\bar{\imath}_1 + s_2\bar{\imath}_2) = [mg - ku]s_1.$$

The principle of virtual work now implies that the particle is in static equilibrium at position u if and only if the virtual work done by the externally applied loads vanishes for all arbitrary virtual displacements, *i.e.*, if and only if $[mg - ku]s_1 = 0$ for all values of s_1. Equation $[mg - ku]s_1 = 0$ possesses two solutions, $[mg - ku] = 0$ or $s_1 = 0$; the second solution, however, is not valid because, as implied by the principle of virtual work, s_1 is arbitrary.

In conclusion, the vanishing of the virtual work for all arbitrary virtual displacements implies that $mg - ku = 0$, and the equilibrium configuration of the system is found as $u = mg/k$. Of course, the same conclusion can be drawn more expeditiously from a direct application of Newton's first law, which requires the sum of the externally applied forces to vanish, *i.e.*, $mg\bar{\imath}_1 - ku\,\bar{\imath}_1 = 0$, or $(mg - ku)\bar{\imath}_1 = 0$, and finally, $mg - ku = 0$.

This example involves the restoring force of an elastic spring, a displacement-dependent force. Indeed, the elastic force in the spring is $-ku\bar{\imath}_1$, and if the particle undergoes a *real downward displacement* of magnitude d, the restoring force becomes $-k(u + d)\bar{\imath}_1$. In contrast, if the particle undergoes a *virtual downward displacement* of magnitude s_1, the restoring force *remains unchanged* as $-ku\bar{\imath}_1$. This difference has profound implications on the computation of work. First, consider the work done by the elastic force, $-ku\bar{\imath}_1^T du\,\bar{\imath}_1$, under a *virtual displacement*, s_1,

$$W = \int_u^{u+s_1} -ku\,du = -ku\int_u^{u+s_1} du = -ku\,[u]_u^{u+s_1} = -kus_1. \tag{7.26}$$

It is possible to factor out the elastic force, $-ku$, from the integral because this force remains unchanged by the virtual displacement, and hence, it can be treated as a constant.

In contrast, the work done by the same elastic force under a *real displacement*, d, is

$$W = \int_u^{u+d} -ku\,du = \left[-\frac{1}{2}ku^2\right]_u^{u+d} = -kud - \frac{1}{2}kd^2. \tag{7.27}$$

In this case, the real work includes an additional term that is quadratic in d and represents the work done by the change in force that develops due to the stretching of the spring. Even if the magnitude of the real displacement is equal to that of the virtual displacement, *i.e.*, even if $d = s_1$, the two expressions for the work done by the elastic restoring force differ.

These observations help explain the terminology used when dealing with the principle of virtual work. The concept of virtual displacement is key to the correct use of the principle of virtual work, which requires the virtual work done by displacement-dependent forces to be evaluated according to eq. (7.26) rather than eq. (7.27). Of course, the *real* work done by the elastic force as it undergoes a real displacement is correctly evaluated by eq. (7.27).

Clearly, it is important to keep in mind the crucial difference between "real displacements" and "virtual" or "fictitious displacements." The words "virtual" or "fictitious" are used to emphasize the fact the forces remain unaffected by these displacements. In practice, the term "real displacement" is rarely used; real displacements are simply called displacements. The terms "virtual," "fictitious," or "test displacements" all imply that the forces acting on the system remain unaffected by the application of such displacements. The term "virtual displacement" is the most widely used.

Example 7.8. Equilibrium of a particle sliding on a track

Figure 7.17 shows a particle of mass m sliding on a track. The externally applied horizontal force is resited by friction between the particle and track. Newton's first law expresses the condition for static equilibrium as $mg\,\bar{\imath}_1 - R\,\bar{\imath}_1 + P\,\bar{\imath}_2 - F\,\bar{\imath}_2 = 0$, where $-R\,\bar{\imath}_1$ is the reaction force the track exerts on the particle, and $-F\,\bar{\imath}_2$ the friction force applies to the particle.

The four forces applied to the particle are of different physical natures: $P\,\bar{\imath}_2$ is an externally applied force, $mg\bar{\imath}_1$ the force of gravity, $-R\,\bar{\imath}_1$ a reaction force, and $-F\,\bar{\imath}_2$ a friction force. Yet, all forces play an equal role in Newton's law, which states that the sum of all forces must vanish. The law simply states "all forces" without making any distinction among them. Newton's first law is readily solved to find $(mg-R)\,\bar{\imath}_1 + (P - F)\,\bar{\imath}_2 = 0$, and finally $R = mg$ and $F = P$, as expected.

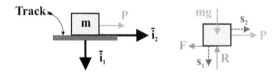

Fig. 7.17. A particle sliding on a track.

Next, the principle of virtual work will be used to solve the same problem. For this two dimensional problem, an arbitrary virtual displacement will be written as $\underline{s} = s_1\,\bar{\imath}_1 + s_2\,\bar{\imath}_2$, and the vanishing of the virtual work it performs implies

$$W = (mg\,\bar{\imath}_1 - R\,\bar{\imath}_1 + P\,\bar{\imath}_2 - F\,\bar{\imath}_2)^T (s_1\,\bar{\imath}_1 + s_2\,\bar{\imath}_2)$$
$$= [mg - R]s_1 + [P - F]s_2 = 0. \tag{7.28}$$

Following a reasoning similar to that developed in the previous example, it is easy to show that the vanishing of the virtual work for all arbitrary scalars s_1 and s_2 implies the vanishing of the two bracketed terms in the above equation: $mg - R = 0$

and $P - F = 0$. This result is identical to that obtained from Newton's first law, as expected, since the principle of virtual work and Newton's first law are identical.

This example illustrates a crucial relationship between Newton's first law and the principle of virtual work. The projection of Newton's law along unit vectors $\bar{\imath}_1$ and $\bar{\imath}_2$ yields two scalar equilibrium equations, $mg - R = 0$ and $P - F = 0$, respectively. The same two equilibrium equations are obtained by imposing the vanishing of the factors multiplying the arbitrary virtual displacement components, s_1 and s_2, resolved along the same unit vectors, $\bar{\imath}_1$ and $\bar{\imath}_2$, respectively.

The principle of virtual work yields scalar equilibrium equations which are the projections of Newton's first law along the directions associated with the virtual displacement components. Because it is based on a scalar quantity, the virtual work, the principle of virtual work yields scalar equations of equilibrium, rather than their vector counterparts inherent to the application of Newton's first law.

7.5.2 Kinematically admissible virtual displacements

Example 7.8 illustrates an important feature of virtual displacements, which are selected to have components in the horizontal direction, $s_2\bar{\imath}_2$, and the vertical direction, $s_1\bar{\imath}_1$. This raises a basic question: how could the particle move in the vertical direction when it is constrained to remain on the track? The answer to this question lies in the nature of the virtual displacements that are not real, but rather are virtual or fictitious displacements. Of course, the particle cannot possibly undergo *real displacements* in the vertical direction because it must remain on the track, but *virtual* or *fictitious displacements* in that same direction are allowed.

In the derivation of the principle of virtual work, it is necessary to use completely arbitrary virtual displacements to prove that the vanishing of the virtual work implies Newton's first law. The completely arbitrary nature of the virtual displacements is key to the successful use of the principle of virtual work. The expression, "arbitrary virtual displacements" means *any virtual displacements, including those that violate the kinematic constraints of the problem.*

In fig. 7.17, the particle is confined to remain on the track; it can move along the track, but not in the direction perpendicular to it. The direction along the track is called the *kinematically admissible direction*, and the direction normal to it is called the *kinematically inadmissible direction*, or the *infeasible direction*.

It is sometimes convenient to introduce the concept of *kinematically admissible virtual displacements*. These are virtual displacements that satisfy the kinematic constraints of the problem.

For the problem depicted in fig. 7.17, the kinematic constraint enforces the particle to remain on the track. Arbitrary virtual displacements are written as $\underline{s} = s_1\,\bar{\imath}_1 + s_2\,\bar{\imath}_2$, but since these include a component in the vertical direction, *i.e.*, in a kinematically inadmissible direction, these are not kinematically admissible virtual displacements. On the other hand, virtual displacements of the form $\underline{s} = s_2\,\bar{\imath}_2$, are kinematically admissible because these are oriented along the track.

At this point, the relationship between kinematic constraints and reaction forces should be clarified. Reaction forces are those forces arising from the enforcement of kinematic constraints. The particle depicted in fig. 7.17 is constrained to move along the track, and this kinematic constraint gives rise to a reaction force. Note that the reaction force acts along the kinematically inadmissible direction, *i.e.*, the direction normal to the track.

Consider now the virtual work done by the reaction force under arbitrary virtual displacements,

$$W = (-R\bar{\imath}_1)^T(s_1\bar{\imath}_1 + s_2\bar{\imath}_2) = -Rs_1 \neq 0.$$

Next, consider the virtual work done by the same reaction force under arbitrary *kinematically admissible virtual displacements,*

$$W = (-R\bar{\imath}_1)^T(s_2\,\bar{\imath}_2) = 0.$$

Because the reaction force acts along the infeasible direction and the kinematically admissible virtual displacement is along the admissible direction, these two vectors are normal to each other, and hence, the virtual work done by the reaction force vanishes. In contrast, the work done by the same reaction force under arbitrary virtual displacements does not.

The vanishing of the virtual work done by reaction forces under kinematically admissible virtual displacements has profound implications for applications of the principle of virtual work. The principle is repeated here: "a particle is in static equilibrium if and only if the virtual work done by the externally applied forces vanishes *for all arbitrary virtual displacements*". Because this principle calls for the use of arbitrary virtual displacements, it is of crucial importance to treat reaction forces as externally applied forces. For instance, in example 7.8, the virtual work done by the reaction force must be included in the statement of the principle, as is done in eq. (7.28), because completely arbitrary virtual displacements are used.

Consider now a modified version of the principle of virtual work: "a particle is in static equilibrium if and only if the virtual work done by the externally applied forces vanishes *for all arbitrary kinematically admissible virtual displacements*". Rather than considering completely arbitrary virtual displacements, only kinematically admissible virtual displacements are considered now. Because the virtual work done by the constraint forces vanishes for kinematically admissible virtual displacements, constraint forces are automatically eliminated from this statement of the principle of virtual work. This often simplifies the statement of the principle because fewer terms are involved. On the other hand, because the constraint forces are eliminated from the formulation, this modified principle will not yield the equations required to evaluate the reaction forces, which are often quantities of great interest.

As pointed out earlier, Newton's first law requires the sum of all forces to vanish for static equilibrium to be achieved. The "sum of all forces" involves all forces without distinction. While the principle of virtual work is shown to be identical to Newton's first law, this principle creates an important distinction between reaction forces stemming from kinematic constraints, and all other forces. Indeed, reaction forces, also called forces of constraint, can be completely eliminated from the formulation by using kinematically admissible virtual displacements.

All other forces, such as those generated by springs, gravity, friction, temperature, electric or magnetic fields, are of a physical origin. It is easy to recognize such forces because their description involves physical constants that can only be determined by experiment. For instance, the stiffness constant of a spring, the universal constant of gravitation appearing in gravity forces, or the friction coefficient appearing in Coulomb's friction law. All these forces are referred to as *natural forces*, which can be further differentiated into *internal* and *external* forces. *Internal forces* are natural forces arising from and reacted within the structural system under consideration, whereas *external forces* are natural forces that act on the system but stem from outside it; these forces are also called *externally applied loads*.

Example 7.9. Equilibrium of a particle sliding on a track

Consider once again the particle of mass m sliding on a track and shown in fig. 7.17. For this simple problem, the kinematically admissible direction is along axis $\bar{\imath}_2$, while the infeasible direction is along axis $\bar{\imath}_1$. The free body diagram in the right part of fig. 7.17 shows the forces acting on the particle. The reaction force, $-R\bar{\imath}_1$, acts in the infeasible direction, as expected.

In contrast with example 7.8, which uses completely arbitrary virtual displacements, kinematically admissible virtual displacements will be used here, and hence, $\underline{s} = s_2\bar{\imath}_2$. The vanishing of the virtual work then implies

$$W = (mg\,\bar{\imath}_1 - R\,\bar{\imath}_1 + P\,\bar{\imath}_2 - F\,\bar{\imath}_2)^T s_2\,\bar{\imath}_2 = [P - F]s_2 = 0.$$

Because s_2 is an arbitrary quantity, the bracketed term must vanish, leading to $F = P$.

First, reaction force R is eliminated from the formulation: the statement of the principle of virtual work becomes simply $(P - F)s_2 = 0$ for all values of s_2. The reaction force does not appear in this statement. It is also possible to apply external loads along the infeasible direction: for instance, in this problem, gravity loads act in the infeasible direction and are also eliminated from the formulation. Of course, if gravity acts along the kinematically admissible direction, *i.e.*, along the track, this force will appear in the statement of the principle. In contrast, reaction forces always act along the infeasible direction and hence, are always eliminated from the formulation.

Second, note that less information about the system is obtained. In example 7.8 that uses arbitrary virtual displacements, two equations are obtained: $F = P$ and $R = mg$. In contrast, the use of kinematically admissible virtual displacements yields a single equation, $F = P$. On the other hand, the solution process is simpler and involves one single equation; however, no information about the reaction force is available.

Finally, it is shown here that the modified version of the principle of virtual work stating "a particle is in static equilibrium if and only if the virtual work done by the externally applied forces vanishes *for all arbitrary kinematically admissible virtual displacements,*" is not entirely correct. The vanishing of the virtual work for all kinematically admissible virtual displacements is a necessary condition, but it is not sufficient, because it does not guarantee equilibrium of the particle in the infeasible direction. Indeed, this latter condition, $R = mg$, is not recovered by the modified principle.

Example 7.10. Equilibrium of a particle on a curved track

Figure 7.18 depicts a particle of mass m constrained to move on a semi-circular track of radius R under the combined effects of gravity, friction, and elastic forces. Determine the equilibrium position of the particle and the forces acting on it in the equilibrium state.

The spring of stiffness constant k is pinned at point $\mathbf{C}$ located at coordinates $x_1 = c_1 R$ and $x_2 = c_2 R$ and its un-stretched length vanishes. Force N is the reaction force acting on the particle due to its contact with the track and acts in direction $\bar{n}$, which is normal to the track. Force F is the force exerted by the track on the particle and acts in the tangential direction, $\bar{t}$; this force arises from friction between the particle and track.

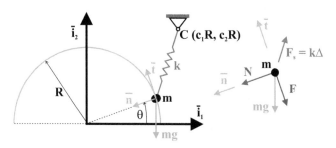

Fig. 7.18. Particle constrained to slide with friction on a circular track.

The position of the particle on the track is conveniently represented by angle θ. The unit vector tangent to the circular track is given by $\bar{t} = -\sin\theta\,\bar{\imath}_1 + \cos\theta\,\bar{\imath}_2$, and the normal to the track is $\bar{n} = -\cos\theta\,\bar{\imath}_1 - \sin\theta\,\bar{\imath}_2$. For this problem, the kinematically admissible direction is $\bar{t}$, and $\bar{n}$ the infeasible direction. In contrast with the previous example, the admissible direction is not a fixed direction in space, but instead, it depends on the position of the particle on the track, $\bar{t} = \bar{t}(\theta)$. The reaction force of magnitude N acts along the infeasible direction, as expected. The friction force of magnitude F acts in the admissible direction.

The force, $\underline{F}_s$, applied by the elastic spring to the particle is given by the spring stiffness constant times the distance between the particle and point $\mathbf{C}$ and is oriented in that same direction: $\underline{F}_s = kR[(c_1 - \cos\theta)\bar{\imath}_1 + (c_2 - \sin\theta)\bar{\imath}_2]$. This can be expressed in terms of admissible and infeasible directions, $\bar{t}$ and $\bar{n}$, respectively, as $\underline{F}_s = kR[(-c_1\sin\theta + c_2\cos\theta)\bar{t} + (1 - c_1\cos\theta - c_2\sin\theta)\bar{n}]$ where use is made of the following relationships: $\bar{\imath}_1 = -\sin\theta\,\bar{t} - \cos\theta\,\bar{n}$ and $\bar{\imath}_2 = \cos\theta\,\bar{t} - \sin\theta\,\bar{n}$.

An arbitrary virtual displacement of the form $\underline{s} = s_t\,\bar{t} + s_n\,\bar{n}$ is selected, where s_t and s_n are arbitrary quantities. The virtual work done by the forces acting on the particle now becomes

$$W = \{kR\left[(-c_1 \sin\theta + c_2 \cos\theta)\bar{t} + (1 - c_1\cos\theta - c_2\sin\theta)\bar{n}\right] + N\bar{n} - F\bar{t}$$
$$+ mg(-\cos\theta\bar{t} + \sin\theta\bar{n})\}^T (s_t\,\bar{t} + s_n\,\bar{n})$$
$$= [kR(-c_1\sin\theta + c_2\cos\theta) - F - mg\cos\theta]\,s_t$$
$$+ [kR(1 - c_1\cos\theta - c_2\sin\theta) + N + mg\sin\theta]\,s_n.$$

Because the virtual work must vanish for arbitrary s_t and s_n, the two bracketed terms must vanish, leading to the two equilibrium equations of the problem,

$$F = kR(-c_1\sin\theta + c_2\cos\theta) - mg\cos\theta, \tag{7.29a}$$
$$N = -kR(1 - c_1\cos\theta - c_2\sin\theta) - mg\sin\theta. \tag{7.29b}$$

This forms a set of two equations for the three unknowns of the problem: the reaction force, N, the friction force, F, and the equilibrium position of the particle, θ.

One additional equation is required to solve the problem. Coulomb's law of static friction requires the friction force to be smaller than the normal contact force multiplied by the static friction coefficient, μ_s, i.e., $|F| \leq \mu_s|N|$. Substituting the friction and normal forces from eqs. (7.29a) and (7.29b), respectively, leads to $kR(-c_1\sin\theta + c_2\cos\theta) - mg\cos\theta \leq \pm\mu_s\left[-kR(1 - c_1\cos\theta - c_2\sin\theta) - mg\sin\theta\right]$. This equation can be solved to find two solutions, θ_ℓ and θ_u: the particle is in equilibrium for all configurations, θ, such that $\theta_\ell \leq \theta \leq \theta_u$.

Next, kinematically admissible virtual displacements of the form $\underline{s} = s_t\bar{t}$ will be selected, where s_t is an arbitrary quantity. The virtual work done by the forces acting on the particle then becomes

$$W = \{kR\left[(-c_1\sin\theta + c_2\cos\theta)\bar{t} + (1 - c_1\cos\theta - c_2\sin\theta)\bar{n}\right] + N\bar{n} - F\bar{t}$$
$$+ mg(-\cos\theta\bar{t} + \sin\theta\bar{n})\}^T s_t\,\bar{t}$$
$$= [kR(-c_1\sin\theta + c_2\cos\theta) - F - mg\cos\theta]\,s_t.$$

Because the virtual work must vanish for all arbitrary s_t, the bracketed term must vanish, yielding a single equilibrium equation of the problem, which is the same as eq. (7.29a) above. As expected, the normal reaction force, N, is eliminated from the formulation. The problem still features three unknowns, N, F, and θ, and the addition of the static friction law provides a second equation for the problem. Clearly, the principle of virtual work with kinematically admissible virtual displacements does not provide enough equations to solve this problem. This is because the static friction law establishes a relationship between friction and normal forces. By eliminating the normal contact force from the formulation, the use of kinematically admissible virtual displacements yields too little information to solve the problem.

If friction is neglected, the friction force will vanish, $F = 0$, and the single equation stemming from the use of kinematically admissible virtual displacements yields the solution of the problem, $kR(-c_1\sin\theta + c_2\cos\theta) - mg\cos\theta = 0$, or $\tan\theta = (c_2 - mg/kR)/c_1$.

In summary, when using kinematically admissible virtual displacements, the principle of virtual work yields a reduced set of equilibrium equations from which the forces of constraints are eliminated. This often greatly simplifies and streamlines the solution process. In some cases, however, too few equations are obtained, giving the false impression that the problem cannot be solved. Arbitrary virtual displacements, *i.e.*, virtual displacements that violate the kinematic constraints must then be used to obtain the missing equations of equilibrium, which correspond to the projection of Newton's first law along the infeasible directions.

7.5.3 Use of infinitesimal displacements as virtual displacements

In the previous sections, three-dimensional virtual displacements are denoted $\underline{s} = s_1 \bar{\imath}_1 + s_2 \bar{\imath}_1 + s_3 \bar{\imath}_3$, where s_1, s_2, and s_3 are arbitrary quantities. In view of the fundamental role they play in energy and variational principles, a special notation is commonly used to denote virtual displacements,

$$\underline{s} = \delta\underline{u}. \tag{7.30}$$

The symbol "δ" is placed in front of the displacement vector, $\underline{u}$, to indicate that it should be understood as a virtual displacement. Similarly, the virtual work done by a force undergoing a virtual displacement will be denoted δW to distinguish it from the real work done by the same force undergoing real displacements. The new notation changes nothing to the special nature of virtual displacements, which are fictitious displacements that do not alter the applied forces.

In many applications of the principle of virtual work, it will also be convenient to use virtual displacements of infinitesimal magnitude. Because virtual displacements are of arbitrary magnitude, virtual displacements of infinitesimal magnitude qualify as valid virtual displacements. The infinitesimal magnitude of virtual displacements is a convenience that often simplifies algebraic developments, but is by no means a requirement.

Displacement-dependent forces

A key simplification arising from the use of virtual displacements of infinitesimal magnitude is that displacement-dependent forces automatically remain unaltered by their application, as illustrated in the following example.

Example 7.11. Equilibrium of a particle connected to an elastic spring
Consider a particle connected to an elastic spring, as illustrated in fig. 7.19. This is the same problem treated in example 7.7.

The principle of virtual work requires that

$$\delta W = (mg\,\bar{\imath}_1^T - ku\,\bar{\imath}_1^T)(\delta u\,\bar{\imath}_1 + \delta v\,\bar{\imath}_2) = [mg - ku]\delta u = 0,$$

for all virtual displacements, δu, where the virtual displacements must leave the forces applied to the particle unchanged. Consider now a virtual displacement of infinitesimal magnitude, $\delta u = du$. The virtual work done by this virtual displacement of infinitesimal magnitude is still given by eq (7.27) as

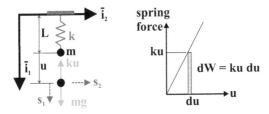

Fig. 7.19. Use of a differential displacement as a virtual displacement.

$$\int_{u}^{u+du} -ku\,du = \left[-\frac{1}{2}ku^2\right]_{u}^{u+du} = -kudu - \frac{1}{2}k(du)^2 = -ku\,du,$$

where the last equality follows from neglecting the higher-order differential quantity. The virtual work is now equal to the real work done by an infinitesimal displacement of magnitude $du = \delta u$. The right portion of fig. 7.19 illustrates the differential work, dW, for a displacement of infinitesimal magnitude.

Rigid bodies

Next, the close relationship between infinitesimal displacements and virtual displacements of infinitesimal magnitude will be explored further in the context of rigid bodies. Consider two material points, **P** and **Q**, of a rigid body. When the rigid body undergoes arbitrary motions, the velocities of these two points must satisfy eq. (5.22), $\underline{v}_P = \underline{v}_Q + \widetilde{\omega}\,\underline{r}_{QP}$, where $\underline{v}_P$ and $\underline{v}_Q$ are the velocities of points **P** and **Q**, respectively, $\underline{\omega}$ is the angular velocity of the rigid body, and $\underline{r}_{QP}$ the position vector of point **P** with respect to **Q**. This relationship is now written as $d\underline{u}_P/dt = d\underline{u}_Q/dt + (\widetilde{d\psi}/dt)\underline{r}_{QP}$, where $d\underline{u}_P$ and $d\underline{u}_Q$ are the infinitesimal displacement vectors of points **P** and **Q**, respectively, and $d\underline{\psi}$ is the differential rotation vector for the rigid body. After multiplication by dt, the differential displacements are found to satisfy the following equation, $d\underline{u}_P = d\underline{u}_Q + \widetilde{d\psi}\,\underline{r}_{QP}$.

Because virtual displacements can be of infinitesimal magnitude, it is possible to write

$$\delta\underline{u}_P = \delta\underline{u}_Q + \widetilde{\delta\psi}\,\underline{r}_{QP}. \tag{7.31}$$

where $\delta\underline{u}_P$ and $\delta\underline{u}_Q$ are the virtual displacement vectors of arbitrary points **P** and **Q**, respectively, and $\delta\underline{\psi}$ is the virtual rotation vector for the rigid body. Equation (7.31) describes the field of kinematically admissible virtual displacements for a rigid body. Indeed, these virtual displacements satisfy the kinematic constraints for two points belonging to the same rigid body.

The discussion of the previous paragraph underlines the close relationship between infinitesimal quantities, denoted with symbol "d," and virtual quantities, denoted with symbol "δ." To obtain eq. (7.31) symbol "d" is replaced by "δ" in the last step of the reasoning. While this approach is correct, it must be emphasized that virtual displacements remain fictitious displacements, whereas infinitesimal displacements are real displacements. Furthermore, virtual displacements leave the forces unchanged, whereas no such requirement applies for real infinitesimal displacements. Finally, virtual displacements are allowed to violate the kinematic constraints, whereas real displacement are not.

Using virtual displacements of infinitesimal magnitude greatly simplifies the treatment of many problems. In the mathematical treatment of virtual quantities, a branch of mathematics called *calculus of variations,* virtual quantities are systematically assumed to be of infinitesimal magnitude [24, 25].

7.5.4 Principle of virtual work for a system of particles

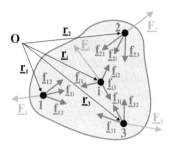

Fig. 7.20. A system of particles.

Figure 7.20 depicts a system of N particles. This problem is treated in section 3.4 using the classical Newtonian approach. Particle i is subjected to an external force, $\underline{F}_i$, and to $N - 1$ interaction forces, $\underline{f}_{ij}, j = 1, 2, \ldots, N, j \neq i$. For particle i, the virtual work, denoted δW_i, done by all applied forces when subjected to a virtual displacement, $\delta \underline{u}_i$, is

$$\delta W_i = (\underline{F}_i^T + \sum_{j=1, j \neq i}^{N} \underline{f}_{ij}^T) \delta \underline{u}_i. \qquad (7.32)$$

According to the principle of virtual work, this virtual work must vanish for all virtual displacements, $\delta \underline{u}_i$. The principle can be applied to each particle independently, leading to $\delta W_i = 0$, where δW_i is given by eq. (7.32), for $i = 1, 2, \ldots N$.

Because the virtual work must vanish for each particle independently, the sum of the virtual work for all particles must also vanish, leading to the following statement of the principle of virtual work for a system of N particles: a system of particle is in static equilibrium if and only if the virtual work,

$$\delta W = \sum_{i=1}^{N} \left\{ \left[\underline{F}_i^T + \sum_{j=1, j \neq i}^{N} \underline{f}_{ij}^T \right] \delta \underline{u}_i \right\}, \qquad (7.33)$$

vanishes for all virtual displacements, $\delta \underline{u}_i, i = 1, 2, \ldots, N$. Because the N virtual displacements are all arbitrary and independent, the bracketed term in eq. (7.33) must vanish for $i = 1, 2, \ldots, N$, leading to equilibrium equations that are identical to those obtained from Newton's first law.

Because each of the N virtual displacement vectors involves three scalar components, the principle of virtual work yields $3N$ scalar equations for a system of N particles; all must be satisfied for the system to be in static equilibrium. The system is said to present $3N$ *degrees of freedom.* For a two-dimensional, or planar system, the number of scalar equations would reduce to $2N$, *i.e.*, $2N$ degrees of freedom.

The above developments have shown, once again, that the principle of virtual work is equivalent to Newton's first law, and gives the necessary and sufficient conditions for the static equilibrium of the system. Equilibrium is the most fundamental requirement in structural analysis, and must always be satisfied. This means that Newton's first law and the principle of virtual work, because they are both equivalent, always apply. The system of particles considered above is very general; it could represent a rigid body, a flexible body deforming elastically or plastically, a fluid, or a planetary system. Yet, the same equilibrium requirements apply equally to all systems.

Internal and external virtual work

Equation (7.33) also affords another important interpretation. The forces acting on the system are separated into two groups, the externally applied forces, $\underline{F}_i$, and the internal forces, $\underline{f}_{ij}$. The words "internal" and "external" should be understood with respect to the system of particles. *Internal forces* act and are reacted within the system, and *external forces* act on the system but are reacted outside the system. The virtual work done by the external and internal forces, denoted δW_E and δW_I, respectively, are defined as

$$\delta W_E = \sum_{i=1}^{N} \underline{F}_i^T \, \delta \underline{u}_i, \tag{7.34a}$$

$$\delta W_I = \sum_{i=1}^{N} \left[\sum_{j=1, j \neq i}^{N} \underline{f}_{ij}^T \right] \delta \underline{u}_i, \tag{7.34b}$$

respectively. With these definitions, eq. (7.33) is becomes

$$\delta W = \delta W_E + \delta W_I = 0, \tag{7.35}$$

for all arbitrary virtual displacements. This leads to the principle of virtual work for a system of particles.

Principle 9 (Principle of virtual work for a system of particles) *A system of particles is in static equilibrium if and only if the sum of the virtual work done by the internal and external forces vanishes for all arbitrary virtual displacements.*

Finally, note that because the virtual displacements are arbitrary, it is possible to choose them to be the actual displacements, and eq. (7.35) then implies

$$W = W_E + W_I = 0, \tag{7.36}$$

where W_E and W_I are the actual work done by the external and internal forces, respectively. Equation (7.36) states that *if a system of particles is in static equilibrium, the sum of the work done by the internal and external forces vanishes.*

Euler's laws

The $3N$ scalar equations implied by the vanishing of the virtual work expressed in eq. (7.33) are often cumbersome to use because they all involve the interaction forces between the particles of the system. To obtain equations that are more convenient to use, a special set of virtual displacements will be selected.

Inspired by eq. (7.31), the virtual displacement of particle i is written as

$$\delta \underline{u}_i = \delta \underline{u}_B + \widetilde{\delta \psi}\, \underline{r}_i, \tag{7.37}$$

where $\delta \underline{u}_B$ is the virtual displacement of the reference point **B** of the rigid body, $\delta \psi$ the virtual rotation vector, and $\underline{r}_i$ the relative position vector of particle i with respect to point **B**. The virtual displacements of all particles are now expressed in terms of a virtual translation of the rigid body, $\delta \underline{u}_B$, and its virtual rotation, $\delta \psi$, both chosen to be of infinitesimal magnitude. This corresponds to 6 independent virtual displacement components, far fewer than the original $3N$. The virtual work done by all forces acting on the system under these virtual displacements is

$$\delta W = \sum_{i=1}^{N} \left\{ \left[\underline{F}_i + \sum_{j=1,j\neq i}^{N} \underline{f}_{ij} \right]^T (\delta \underline{u}_B + \widetilde{\delta \psi}\, \underline{r}_i) \right\} = \left(\sum_{i=1}^{N} \underline{F}_i^T \right) \delta \underline{u}_B$$

$$+ \left(\sum_{i=1}^{N} \sum_{j=1,j\neq i}^{N} \underline{f}_{ij}^T \right) \delta \underline{u}_B + \sum_{i=1}^{N} \underline{F}_i^T \widetilde{\delta \psi}\, \underline{r}_i + \sum_{i=1}^{N} \sum_{j=1,j\neq i}^{N} \underline{f}_{ij}^T \widetilde{\delta \psi}\, \underline{r}_i.$$

The last two terms of this expression can be simplified using identity (1.33h), and the above equation now becomes

$$\delta W = \delta \underline{u}_B^T \left(\sum_{i=1}^{N} \underline{F}_i \right) + \delta \underline{u}_B^T \left(\sum_{i=1}^{N} \sum_{j=1,j\neq i}^{N} \underline{f}_{ij} \right)$$

$$+ \delta \psi^T \left(\sum_{i=1}^{N} \widetilde{r}_i \underline{F}_i \right) + \delta \psi^T \left(\sum_{i=1}^{N} \sum_{j=1,j\neq i}^{N} \widetilde{r}_i \underline{f}_{ij} \right).$$

In view of eqs. (3.59) and (3.62), the terms in the second and last sets of parenthesis now vanish, reducing the expression to

$$\delta W = \delta \underline{u}_B^T \left[\sum_{i=1}^{N} \underline{F}_i \right] + \delta \psi^T \left[\sum_{i=1}^{N} \widetilde{r}_i\, \underline{F}_i \right] = \delta \underline{u}_B^T \underline{F} + \delta \psi^T \underline{M}_B,$$

where the last equality follows from eqs. (3.58) and (3.60). Because the virtual work must vanish for all virtual displacements and virtual rotations, the sum of the externally applied forces and moments must vanish, $\underline{F} = 0$ and $\underline{M}_B = 0$. Clearly, these two equations are identical to Euler's first and second laws obtained directly from Newtonian arguments, see eqs. (3.70) and (3.75). The present problem is a static problem, and hence, the time derivatives of the linear and angular momenta appearing in eqs. (3.70) and (3.75) are absent.

These two vector equations are necessary but not sufficient conditions to guarantee static equilibrium. Indeed, static equilibrium requires a total of N vector equations to be satisfied; eqs. (3.70) and (3.75) are two linear combinations of those N equations. Only two vector equations are obtained from the principle of virtual work because the virtual displacement field, eq. (7.37), selected for the rigid body involves a single virtual displacement vector, $\delta \underline{u}_B$, and a single virtual rotation vector, $\delta \underline{\psi}$.

7.5.5 The use of generalized coordinates

In the previous sections, the configuration of the system was represented by the Cartesian coordinates of the various particles. As discussed in section 7.2, it is often convenient to represent the configuration of the system by means of generalized coordinates, which give rise to the concept of the generalized forces defined by eq. (7.20).

When using generalized coordinates, the virtual work done by a force is expressed by eq. (7.19). This expression can be written for both internal and external forces, leading to

$$\delta W_I = \sum_{i=1}^{N} Q_i^I \delta q_i, \tag{7.38a}$$

$$\delta W_E = \sum_{i=1}^{N} Q_i^E \delta q_i, \tag{7.38b}$$

where Q_i^I and Q_i^E are the generalized forces associated with the internal forces and externally applied loads, respectively.

The principle of virtual work, expressed by eq. (7.35), now becomes

$$\delta W_I + \delta W_E = \sum_{i=1}^{N} Q_i^I \delta q_i + \sum_{i=1}^{N} Q_i^E \delta q_i = \sum_{i=1}^{N} \left[Q_i^I + Q_i^E \right] \delta q_i = 0,$$

for all virtual generalized displacements, δq_i. Because the virtual generalized displacements, δq_i, are arbitrary, each of the N bracketed terms under the summation sign must vanish, leading to

$$Q_i^I + Q_i^E = 0, \quad i = 1, 2, \ldots, N. \tag{7.39}$$

This equation represents yet another statement of the principle of virtual work.

As discussed in section 7.5.2, the principle of virtual work can be used with either arbitrary or kinematically admissible virtual displacements. Similarly, the present statement of the principle can be used with either arbitrary or kinematically admissible virtual changes in generalized coordinates. When using arbitrary virtual generalized coordinates, the virtual work done by the reaction forces must be included in the evaluation of the virtual work done by the external forces; this implies that the generalized forces associated with the reaction forces must be included in Q_i^E. If the virtual generalized coordinates are kinematically admissible, the reaction forces are eliminated from the formulation.

Example 7.12. Pendulum with torsional spring

A rigid arm of length R connects mass m to a pinned support point where a torsional spring of stiffness constant k acts between ground and the rod. The torsional spring is un-stretched when the arm is horizontal. The mass is subjected to gravity loading. The configuration of the system is conveniently represented by the angular position, ϕ, of the arm and is selected to be the single generalized coordinate for this single degree of freedom problem.

Consider first the virtual work done by the gravity load, $\delta W_E = -mg\bar{\imath}_2^T \delta \underline{u}_T$, where $\delta \underline{u}_T$ is the virtual displacement at point **T**. Since $\underline{u}_T = R(\cos \phi\, \bar{\imath}_1 + \sin \phi\, \bar{\imath}_2)$, an infinitesimal virtual displacement of the same quantity is $\delta \underline{u}_T = R(-\sin \phi\, \bar{\imath}_1 + \cos \phi\, \bar{\imath}_2)\delta\phi$. It now follows that $\delta W_E = -mgR\cos \phi\, \delta\phi$, and by defining the generalized force as $Q_\phi^E = -mgR\cos \phi$, the virtual work becomes $\delta W_E = Q_\phi^E \delta\phi$. The same result can be obtained in a more expeditious manner by using eq. (7.20) to find $Q_\phi^E = -mg\bar{\imath}_2^T \partial \underline{u}_T / \partial\phi = -mg\bar{\imath}_2^T R(-\sin \phi\, \bar{\imath}_1 + \cos \phi\, \bar{\imath}_2) = -mgR\cos \phi$.

An even simpler interpretation is as follows. Because the virtual displacement is a rotation, $\delta\phi$, it must be multiplied by a moment to yield a virtual work; hence, the generalized force is simply the moment of the gravity load, $-mgR\cos \phi$.

For this problem, the virtual work done by the internal forces reduces to the virtual work done by the restoring moment of the elastic spring, $\delta W_I = -k\phi\, \delta\phi = Q_\phi^I \delta\phi$, where $Q_\phi^I = -k\phi$ is the generalized internal force of the system. The generalized force is, in this case, a moment, and hence, the expression "generalized force" must be interpreted carefully.

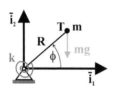
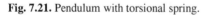
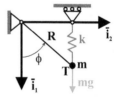

Fig. 7.21. Pendulum with torsional spring.   **Fig. 7.22.** Rotating mass with vertical spring.

The principle of virtual work, eq. (7.39), yields the equilibrium equation for the system as $Q_\phi^I + Q_\phi^E = -mgR\cos \phi - k\phi = 0$. This is a transcendental equation, but if the angular displacement of the pendulum remains small, $\cos \phi \approx 1$, and the equilibrium configuration becomes $\phi = -mgR/k$.

Example 7.13. Pendulum with rectilinear spring

Consider next the modified system shown in fig. 7.22 where a rigid arm of length R connects mass m to a pinned support at the ground. A linear spring of stiffness constant k supports the mass; this spring remains vertical because its support point is free to move horizontally on rollers. The spring is un-stretched when the arm is horizontal.

As in the previous example, the virtual work done by the gravity load is easily found as $\delta W_E = m g \bar{\imath}_1^T \delta \underline{u}_T$, where $\delta \underline{u}_T$ is the virtual displacement at point **T**. Because $\underline{u}_T = R(\cos\phi\,\bar{\imath}_1 + \sin\phi\,\bar{\imath}_2)$, an infinitesimal virtual displacement of the same quantity is $\delta\underline{u}_T = R(-\sin\phi\,\bar{\imath}_1 + \cos\phi\,\bar{\imath}_2)\delta\phi$. The virtual work done by the gravity load now becomes $\delta W_E = -m g R \sin\phi\,\delta\phi$, and the corresponding generalized force is $Q_\phi^E = -m g R \sin\phi$. Next, the virtual work done by the restoring force in the spring is $\delta W_I = -k R \cos\phi\,\bar{\imath}_1^T \delta\underline{u}_T$, which yields $Q_\phi^I = k R^2 \cos\phi \sin\phi$.

The principle of virtual work, as expressed by eq. (7.39), now implies

$$Q_\phi^I + Q_\phi^E = k R^2 \cos\phi \sin\phi - m g R \sin\phi = R\sin\phi(k R\cos\phi - mg) = 0.$$

Two solutions are possible. First, $\sin\phi = 0$: this leads to $\phi = 0$ or π, *i.e.*, the arm is in the down or up vertical position, respectively. The second solution is $\cos\phi = mg/(kR)$. For $mg/(kR) > 1$, however, this solution no longer exists, leaving the first solution as the only valid solution of the problem.

7.5.6 The principle of virtual work and conservative forces

The principle of virtual work was first developed for a single particle, then extended to a system of particles. In this latter case, a distinction was made between internal and external forces acting on the system. On the other hand, section 3.2 introduced the concept of conservative forces.

In this section, the internal and external forces applied to the system of particles will be divided into two groups, the conservative and the non-conservative forces. The principle of virtual work is now expressed as $\delta W = \delta W_c + \delta W_{nc} = 0$, where δW_c and δW_{nc} denote the virtual work done by the conservative and non-conservative forces, respectively. The virtual work done by the conservative forces can be evaluated with the help of eq. (7.22) to yield

$$\delta W = -\delta(V) + \delta W_{nc} = 0, \tag{7.40}$$

where V is the potential of the conservative forces. This leads to the following principle.

Principle 10 *A system of particles is in static equilibrium if and only if virtual changes in the potential of the conservative force equal the virtual work done by the non-conservative forces for all arbitrary virtual displacements.*

If all the forces applied to a system of particles are conservative, the system is called a *conservative system*. The virtual work done by the non-conservative forces is absent, and principle of virtual work, eq. (7.40), takes on a particularly simple form,

$$\delta W = -\delta(V) = 0. \tag{7.41}$$

The following principle follows.

Principle 11 *A conservative system of particles is in static equilibrium if and only if virtual changes in the potential is stationary for all arbitrary virtual displacements.*

Statements (7.35), (7.39), (7.40), or (7.41) all are statements of the principle of virtual work. In the first two statements, a distinction is made between internal and external forces. In the last two statements, a distinction is made between conservative and non-conservative forces. For conservative forces, the virtual work can be expressed as the variation of a potential function, whereas this is not possible for non-conservative forces.

It was shown that Newton's first law and the principle of virtual work are equivalent; indeed, the principle of virtual work was derived from Newton's first law. Newton's law does not distinguish among the various types of forces: it simply states that "the sum of *all forces* must vanish." On the other hand, the nature of the applied forces profoundly affects the statement of the principle of virtual work: conservative forces are derived from a potential, but non-conservative forces are not; this fundamental difference is reflected in the principle.

Example 7.14. Four particles on a single rigid bar

Consider the system depicted in fig. 7.23: four particles of masses m_a, m_b, m_c, and m_d, respectively, are connected to the ground by four springs of identical stiffness k. The un-stretched length of the springs are ℓ_a, ℓ_b, ℓ_c, and ℓ_d, respectively. The four particles are also connected to a rigid bar, as indicated on the figure.

The rotation of the rigid bar is assumed to remain small, and hence, the motion of the particles is purely vertical. This system could be represented by four generalized coordinates, the vertical motions of the four particles, subjected to two kinematic constraints imposed by the rigid bar, for a total of two degrees of freedom.

Another approach is to select two generalized coordinates only, the vertical motion of the bar's mid-span, u, and its rotation, θ. Using this latter approach, the potential of the forces associated with the elastic springs is

$$V^e = \frac{k}{2} \left[(u - L\theta/2 - \ell_a)^2 + (u - L\theta/6 - \ell_b)^2 \right.$$
$$\left. + (u + L\theta/6 - \ell_c)^2 + (u + L\theta/2 - \ell_d)^2 \right],$$

and the potential of the gravity forces is

$$V^m = g\left[m_a (u - L\theta/2) + m_b (u - L\theta/6) + m_c (u + L\theta/6) + m_d (u + L\theta/2) \right].$$

Because all forces acting on the system are conservative, the statement of the principle of virtual work based on kinematically admissible virtual displacements reduces to $\delta(V^e + V^m) = 0$, see eq. (7.41), and leads to

$$kL^2 \left[\ (\bar{u} - \theta/2 - \bar{\ell}_a)(\delta u - \delta\theta/2) + (\bar{u} - \theta/6 - \bar{\ell}_b)(\delta u - \delta\theta/6) \right.$$
$$+ (\bar{u} + \theta/6 - \bar{\ell}_c)(\delta u + \delta\theta/6) + (\bar{u} + \theta/2 - \bar{\ell}_d)(\delta u + \delta\theta/2) \big]$$
$$+ gL \left[\ m_a (\delta u - \delta\theta/2) + m_b (\delta u - \delta\theta/6) \right.$$
$$+ m_c (\delta u + \delta\theta/6) + m_d (\delta u + \delta\theta/2) \big] = 0,$$

where $\bar{u} = u/L$, $\bar{\ell}_a = \ell_a/L$, and similar notations are used for the un-stretched length of the other springs. Because the kinematically admissible virtual displacements are arbitrary, two equations are obtained; these are readily solved to find

$$\bar{u} = \bar{\ell} - m\bar{g}, \tag{7.42a}$$

$$\frac{5\theta}{9} = \left[\frac{\bar{\ell}_d - \bar{\ell}_a}{2} + \frac{\bar{\ell}_c - \bar{\ell}_b}{6} \right] - \left[\frac{m_d - m_a}{2} + \frac{m_c - m_b}{6} \right] \bar{g}. \tag{7.42b}$$

where $\bar{\ell} = (\bar{\ell}_a + \bar{\ell}_b + \bar{\ell}_c + \bar{\ell}_d)/4$ is the average non dimensional un-stretched length of the springs, $m = (m_a + m_b + m_c + m_d)/4$ the average mass of the particles and $\bar{g} = g/(kL)$.

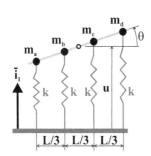

Fig. 7.23. Four spring supporting a rigid bar.

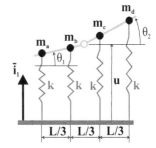

Fig. 7.24. Four spring supporting an articulated rigid bar.

The principle of virtual work, *per se*, does not provide any information about the loads acting in the rigid bar. To compute the bar mid-span bending moment, for instance, the basic methods of statics could be used: summing up the bending moments acting on a free body diagram of the right side of the beam yields

$$\bar{M} = \frac{2}{3}(\bar{\ell} - m\bar{g}) - \frac{1}{2}\left[\frac{\bar{\ell}_a + \bar{\ell}_d}{2} + \frac{\bar{\ell}_b + \bar{\ell}_c}{6} \right] + \left[\frac{m_a + m_d}{2} + \frac{m_b + m_c}{6} \right] \frac{\bar{g}}{2}, \tag{7.43}$$

where $\bar{M} = M/(kL^2)$ is the non dimensional mid-span bending moment.

Example 7.15. Four particles on two rigid bars

Consider now the system depicted in fig. 7.24: the four springs support two rigid bars connected at mid-span by means of a hinge. The system now presents three degrees of freedom that are conveniently chosen as the vertical displacement of the hinge, u, and the orientations, θ_1 and θ_2, of the left and right bars, respectively.

The potential of the elastic forces in the springs becomes

$$V^e = \frac{k}{2} \Big[(u - L\theta_1/2 - \ell_a)^2 + (u - L\theta_1/6 - \ell_b)^2$$
$$+ (u + L\theta_2/6 - \ell_c)^2 + (u + L\theta_2/2 - \ell_d)^2 \Big], \tag{7.44}$$

and the potential of the gravity forces is

$$V^m = m_a g \left(u - L\theta_1/2 \right) + m_b g \left(u - L\theta_1/6 \right)$$
$$+ m_c g \left(u + L\theta_2/6 \right) + m_d g \left(u + L\theta_2/2 \right),$$

where the rotations, θ_1 and θ_2, are assumed to remain small. Here again, all forces acting on the system are conservative, the statement of the principle of virtual work based on kinematically admissible virtual displacements reduces to $\delta(V^e + V^m) = 0$, see eq. (7.41), and leads to the following set of equations

$$\begin{bmatrix} 4 & -2/3 & 2/3 \\ -2/3 & 5/18 & 0 \\ 2/3 & 0 & 5/18 \end{bmatrix} \begin{Bmatrix} \bar{u} \\ \theta_1 \\ \theta_2 \end{Bmatrix} = \frac{1}{6} \begin{Bmatrix} 24\bar{\ell} - 24m\bar{g} \\ -(3\ell_a + \ell_b) + (3m_a + m_b)\bar{g} \\ (3\ell_d + \ell_c) - (3m_d + m_c)\bar{g} \end{Bmatrix}. \quad (7.45)$$

This solution is of course different from that of the previous problem; indeed, the mid-span hinge relieves the bending moment at the middle of the bar.

7.5.7 Problems

Problem 7.4. Rotating disk with spring restraint
A mechanism consists of the rotating circular disk pinned at its center as shown in fig. 7.25. A cable is wrapped around the outer edge and a force, P, is applied tangentially. The rotation is resisted by a spring of stiffness constant k attached to a pin on the disk's outer radius and fixed horizontally to a support that can move vertically, leaving the spring horizontal at all times. The spring is un-stretched when $\theta = 0$. Use the principle of virtual work to determine the force, P, required to keep the disk in static equilibrium as a function of angle θ.

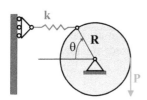

Fig. 7.25. Rotating disk with spring restraint.

Fig. 7.26. Crank-slider mechanism with a spring.

Problem 7.5. Crank-slider mechanism with a spring
Consider the crank-slider mechanism depicted in fig. 7.26. The crank of length R is actuated by a torque Q, and the link of length L transforms the rotary motion of the crank into a linear motion of the slider. A spring of stiffness constant k connects the slider to the ground and is un-stretched when $x = 0$. Use the principle of virtual work to find the static equilibrium configuration of the system.

Problem 7.6. Lever with sliding pivots
Bar **ABC** is of length $b + a$ and is constrained to move vertically at point **A** and horizontally at **B**, while a horizontal force, P, is applied at point **C**, as depicted in fig. 7.27. Point **A** is restrained by a vertical spring of stiffness constant k, which is relaxed when angle $\theta = 0$. Use the principle of virtual work to determine the static equilibrium configuration of the system.

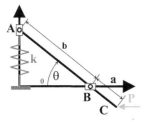

Fig. 7.27. Lever with spring-restrained sliding pivots.

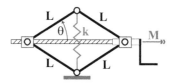

Fig. 7.28. Screw jack type of scissor lift.

Problem 7.7. Screw jack scissor lift

Consider the scissor lift with a spring of stiffness constant k linking the opposite joints shown
in fig. 7.28. The configuration of the system is represented by a single generalized coordinate,
θ, the angle between the jack legs and the horizontal. Using the principle of virtual work,
determine the crank moment, M, for which static equilibrium of the system is achieved. The
threaded screw has a pitch of N threads per unit length. All bars of the jack are articulated.

Problem 7.8. Lever mechanism

A bar of length $3b$ is pinned at its lower end, point **O**, and a spring of stiffness constant k
connects its tip point **T** to the ground a point **A**, as shown in fig. 7.29. A second bar, of length
b, is pinned to the first bar as shown and to a slider that is constrained to move vertically on
a frictionless rod. A force of magnitude F is acting on the slider. Use the principle of virtual
work to determine the static equilibrium configuration of the system.

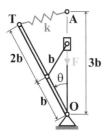

Fig. 7.29. Lever mechanism.

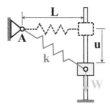

Fig. 7.30. Mechanism with nonlinear geometry.

Problem 7.9. Spring-mass problem with nonlinear geometry

A spring of stiffness constant k and un-stretched length L is fastened to a support at point
A and is connected to a weight, W, as shown in fig. 7.30. The weight slides on a frictionless vertical rod and the spring is un-stretched when horizontal. *(1)* Using the principle of
virtual work, determine the static equilibrium configuration of the system. *(2)* Plot the non-
dimensional weight, $\bar{W} = W/(kL)$, as function of the non-dimensional displacement of the
slider, $\bar{u} = u/L$.

Problem 7.10. Linked bars with lateral springs and forces

Figure 7.31 shows a mechanical system consisting of two articulated bars pinned together at
point **B** and to the ground at point **C**. Two springs of stiffness constants k_1 and k_2 support the

bars at their mid-span and two forces, P and Q, are applied at points **B** and **A**, respectively. Let q_A and q_B, the downward deflection of points **A** and **B**, be the two generalized coordinates of the system. Use the principle of virtual work to determine the two static equilibrium equations of the system. Assume small displacements: $|q_A| \ll L$ and $|q_B| \ll L$.

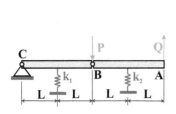

Fig. 7.31. Two articulated bars supported by springs.

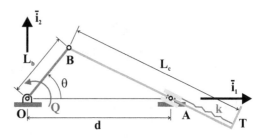

Fig. 7.32. Two articulated bars supported by springs.

Problem 7.11. Two-bar linkage system

The two bar linkage shown in fig. 7.32 comprises bar **OB** of length L_b and bar **BAT** of length L_c. Bar **BAT** passes through a slider located at fixed point **A** but free to swivel about that point. A spring of stiffness constant k connects the tip of the bar at point **T** to the slider at point **A** and is of vanishing un-stretched length. A torque of magnitude Q is applied to bar **OB**. Use the principle of virtual work to determine the static equilibrium configuration of the system.

8

Variational and energy principles

This chapter investigates applications of the principles of analytical mechanics developed in chapter 7 to dynamical systems. First, the principle of virtual work presented in section 7.5 for static problems will be generalized to dynamic problems, leading to d'Alembert's principle, see section 8.1. Next, Hamilton's principle is presented in section 8.2 as an integral version of d'Alembert's principle. Finally, Lagrange's formulation is presented in section 8.3, leading to Lagrange's equations of motion.

8.1 D'Alembert's principle

Newton's second law, eq. (3.4), states that if external forces, $\underline{F}^a$, are acting on a particle, its acceleration is proportional to the sum of these forces, $\underline{F}^a = m\underline{a}$. The product of the mass by the acceleration vector is a force vector, called the *inertial force vector*, $\underline{F}^I$, defined as

$$\underline{F}^I = -m\underline{a}. \tag{8.1}$$

The minus sign in the definition of the inertial force indicates that such force *always opposes motion*. With this definition, Newton's second law becomes

$$\underline{F}^I + \underline{F}^a = 0. \tag{8.2}$$

Of course, this equation looks like a trivial manipulation of Newton's law: inertial forces have been brought from the right- to the left-hand side of the equation. The importance of the above statement, however, is that it generalizes the concept of equilibrium, a concept of statics, to dynamics problems.

As mentioned in section 3.1.2, Newton's first law is generally stated as "a particle is in static equilibrium if and only if the sum of the externally applied forces vanishes" within the context of statics problems.

Equation (8.2) expresses the condition for *dynamic equilibrium:* the sum of the externally applied forces must vanish, provided that the *inertial forces are treated as externally applied forces*. Of course, the concept of dynamic equilibrium does not imply that the particle is at rest; indeed, the particle moves under the effect of the externally applied forces. Rather, dynamic equilibrium implies the vanish of the resultant of the set of forces acting on a particle in motion; this set of forces includes all externally applied forces *and* the inertial forces. The importance of the concept of inertial force is that the same law, "the sum of the forces must vanish," now applies to both statics and dynamics problems; *dynamics is reduced to statics*. D'Alembert's principle can now be stated as follows.

Principle 12 (D'Alembert's principle) *A system of particles is in dynamic equilibrium if and only if the sum of the externally applied forces and inertial forces vanishes.*

In section 7.5, the principle of virtual work for static problems was derived from Newton's first law and shown to imply that $\delta(V) = \delta W_{nc}$, for all arbitrary virtual displacements, see eq. (7.40). In this expression, V is the potential of the conservative forces acting on the system of particles, and δW_{nc} the virtual work done by the non-conservative forces.

For dynamic equilibrium, D'Alembert's principle requires the vanishing of the sum of the externally applied forces and inertial forces. Inertial forces are non-conservative force because they cannot be derived from a potential. It follows that the principle of virtual work, the condition for static equilibrium, can be generalized to becomes the condition for dynamic equilibrium, if the *virtual work done by the inertial forces*, denoted δW^I, is added to the virtual work done by the other non-conservative forces. In summary, a system of particles is in dynamic equilibrium if and only if

$$\delta(V) = \delta W^{nc} + \delta W^I, \tag{8.3}$$

for all arbitrary virtual displacements. D'Alembert's principle can also be stated as follows.

Principle 13 (D'Alembert's principle) *A system of particles is in dynamic equilibrium if and only if virtual changes in the potential of the conservative force equal the virtual work done by the non-conservative forces and inertial forces for all arbitrary virtual displacements.*

The principle of virtual work presented in section 7.5 is equivalent to Newton's first law. By treating inertial forces as "externally applied forces," dynamic problems are reduced to static problems and d'Alembert's principle becomes equivalent to Newton's second law. The two alternative statements of d'Alembert's principle given above are equivalent to Newton's second law, and hence, provide an alternative basis for dynamics.

For a system composed of N of particles, the virtual work done by the inertial forces is

$$\delta W^I = \sum_{i=1}^{N} \underline{F}_i^{IT} \delta \underline{r}_i = -\sum_{i=1}^{N} m_i \underline{a}_i^T \delta \underline{r}_i, \tag{8.4}$$

where $\underline{a}_i$ is the inertial acceleration vector of the ith particle, m_i its mass, and $\delta \underline{r}_i$ an arbitrary virtual displacement of the same particle.

Newton's formulation relates the sum of all externally applied forces to the acceleration of the system, but d'Alembert's principle only involves the virtual work done by the forces acting on the system. It follows that if the virtual work done by a specific force vanishes, this force will be automatically eliminated from the equations of motion obtained from d'Alembert's principle. In section 7.4.3, it was shown that the virtual work done by the forces that impose a kinematic constraint does vanish; hence, such forces will not appear in a formulation based on d'Alembert's principle but will explicitly appear when using Newton's formulation.

Example 8.1. Conservation of energy

Consider a system acted upon by conservative forces only. D'Alembert's principle then reduces to

$$\delta V + \sum_{i=1}^{N} m_i \, \ddot{\underline{r}}_i^T \, \delta \underline{r}_i = 0.$$

Because the virtual displacements are arbitrary, they can be selected to equal the actual, differential displacements of the system, i.e., $\delta \underline{r}_i = \mathrm{d}\underline{r}_i$.

This selection, however, is only possible for specific systems; indeed, virtual displacements are arbitrary virtual changes that bring the configuration of the system to a new configuration, at a given, fixed instant in time. Consequently, equating virtual displacements to differential displacements is only possible when dealing with time-independent potential functions.

Under this restriction, d'Alembert's principle now becomes

$$\mathrm{d}V + \sum_{i=1}^{N} m_i \, \ddot{\underline{r}}_i^T \dot{\underline{r}}_i \, \mathrm{d}t = \mathrm{d}V + \frac{\mathrm{d}}{\mathrm{d}t} \left(\frac{1}{2} \sum_{i=1}^{N} m_i \, \dot{\underline{r}}_i^T \dot{\underline{r}}_i \right) \mathrm{d}t = 0,$$

The term in parenthesis is the kinetic energy, K, of the system, and hence, $\mathrm{d}V/\mathrm{d}t + \mathrm{d}K/\mathrm{d}t = \mathrm{d}E/\mathrm{d}t = 0$, where E is the total mechanical energy of the system. This is the principle of conservation of energy, see eq. (3.25), previously derived directly from Newton's second law.

8.1.1 Equations of motion for a rigid body

Consider a rigid body with a body attached frame $\mathcal{F}^B = \left[\mathbf{B}, \mathcal{B}^* = (\bar{b}_1, \bar{b}_2, \bar{b}_3) \right]$, where point $\mathbf{B}$ is a reference point on the body and $\mathcal{B}^*$ a body attached basis, as depicted in fig. 8.1. The configuration of the body is described with respect to an inertial frame $\mathcal{F}^I = [\mathbf{O}, \mathcal{I} = (\bar{\imath}_1, \bar{\imath}_2, \bar{\imath}_3)]$. The position of reference point $\mathbf{B}$ of the body is $\underline{r}_B$ and its orientation is determined by rotation tensor $\underline{\underline{R}}$, which brings inertial basis $\mathcal{I}$ to basis $\mathcal{B}^*$.

The rigid body is composed of N particles each of mass m_i and located at point $\mathbf{P_i}$; the position vector of the ith particle is denoted $\underline{r}_i$ and its position with respect to reference point $\mathbf{B}$ is denoted $\underline{s}_i$. Superscript $(\cdot)^*$ indicates tensor components resolved in material basis $\mathcal{B}^*$.

The virtual work done by the inertial forces is

$$\delta W^I = \sum_{i=1}^{N} \underline{F}^{IT} \delta \underline{r}_i = \sum_{i=1}^{N} \underline{F}^{IT} \underline{\underline{R}} \, \underline{\underline{R}}^T \delta \underline{r}_i = \sum_{i=1}^{N} -m_i \underline{a}_i^T \underline{\underline{R}} \, \underline{\underline{R}}^T \delta \underline{r}_i.$$

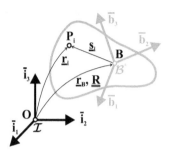

Fig. 8.1. Configuration of a rigid body.

The components of the virtual displacement vector in the body attached basis, $\underline{\underline{R}}^T \delta \underline{r}_i$, will be evaluated using the corresponding expression for the velocity, see eq. (5.23), as $\underline{\underline{R}}^T \delta \underline{r}_i = \underline{\underline{R}}^T \delta \underline{r}_B + \widetilde{\delta \psi}^* \underline{s}_i^*$, where $\delta \underline{r}_B$ is the virtual displacement of the reference point of the body and $\delta \underline{\psi}^*$ are the components of its virtual rotation vector resolved in the material basis. Similarly, the components of the inertial acceleration resolved in the same basis, $\underline{\underline{R}}^T \underline{a}_i$, are given by eq. (5.25) as $\underline{\underline{R}}^T \underline{a}_i = \underline{\underline{R}}^T \underline{a}_B + (\dot{\widetilde{\omega}}^* + \widetilde{\omega}^* \widetilde{\omega}^*) \underline{s}_i^*$, where $\underline{a}_B$ is the inertial acceleration vector of point **B** and $\underline{\omega}^*$ the components of the angular velocity vector resolved in the material basis. The virtual work done by the inertial forces now becomes

$$\delta W^I = -\sum_{i=1}^{N} \left[\delta \underline{r}_B^T \underline{\underline{R}} + \delta \underline{\psi}^{*T} \widetilde{s}_i^* \right] m_i \left[\underline{\underline{R}}^T \underline{a}_B + (\dot{\widetilde{\omega}}^* + \widetilde{\omega}^* \widetilde{\omega}^*) \underline{s}_i^* \right].$$

Expanding the products then leads to

$$\begin{aligned}
\delta W^I = -&\left[\sum_{i=1}^{N} m_i \right] \delta \underline{r}_B^T \underline{a}_B - \delta \underline{r}_B^T \underline{\underline{R}} (\dot{\widetilde{\omega}}^* + \widetilde{\omega}^* \widetilde{\omega}^*) \left[\sum_{i=1}^{N} m_i \underline{s}_i^* \right] \\
&- \delta \underline{\psi}^{*T} \left[\sum_{i=1}^{N} m_i \widetilde{s}_i^* \right] \underline{\underline{R}}^T \underline{a}_B - \delta \underline{\psi}^{*T} \left[\sum_{i=1}^{N} m_i \widetilde{s}_i^* \widetilde{s}_i^{*T} \right] \dot{\underline{\omega}}^* \\
&- \delta \underline{\psi}^{*T} \widetilde{\omega}^* \left[\sum_{i=1}^{N} m_i \widetilde{s}_i^* \widetilde{s}_i^{*T} \right] \underline{\omega}^*.
\end{aligned}$$

The first bracketed term is simply the total mass of the rigid body, and the components of the tensor of mass moments of inertia, see eq. (6.1), resolved in the material basis, appear in the last two bracketed terms. The second and third bracketed term are related to the location of the center of mass of the body

$$\underline{\eta}^* = \frac{1}{m} \sum_{i=1}^{N} m_i \underline{s}_i^*, \tag{8.5}$$

where $\underline{\eta}^*$ are the components of the position vector of the center of mass of the rigid body with respect to its reference point **B**, resolved in the material basis.

The virtual work done by the inertial forces now becomes

$$\delta W^I = - \delta \underline{r}_B^T R \left[m \underline{\underline{R}}^T \underline{a}_B + (\dot{\tilde{\omega}}^* + \tilde{\omega}^* \tilde{\omega}^*) m \underline{\eta}^* \right]$$
$$- \delta \underline{\psi}^{*T} \left[m \tilde{\underline{\eta}}^* \underline{\underline{R}}^T \underline{a}_B + \underline{\underline{I}}^{B*} \dot{\underline{\omega}}^* + \tilde{\omega}^* \underline{\underline{I}}^{B*} \underline{\omega}^* \right] . \tag{8.6}$$

Let $\underline{F}^*$ and $\underline{M}_B^*$ be the components, resolved in the material basis, of the force and moment vectors, respectively, applied to the rigid body. The virtual work done by these externally applied loads is then $\delta W^a = \delta \underline{r}_B^T \underline{\underline{R}} \underline{F}^* + \delta \underline{\psi}^{*T} \underline{M}_B^*$; note that $\underline{M}_B^*$ are the components of the applied moment computed with respect to the reference point **B**.

D'Alembert's principle states that $\delta W^I + \delta W^a = 0$, for all kinematically admissible virtual displacements $\underline{\underline{R}}^T \delta \underline{r}_B$ and $\delta \underline{\psi}^*$. Of course, the constraint forces that keep the rigid body rigid vanish from the formulation because kinematically admissible virtual displacements are used here. The equations of motion of the rigid body then follow as

$$m \underline{\underline{R}}^T \underline{a}_B + (\dot{\tilde{\omega}}^* + \tilde{\omega}^* \tilde{\omega}^*) m \underline{\eta}^* = \underline{F}^*, \tag{8.7a}$$
$$m \tilde{\underline{\eta}}^* \underline{\underline{R}}^T \underline{a}_B + \underline{\underline{I}}^{B*} \dot{\underline{\omega}}^* + \tilde{\omega}^* \underline{\underline{I}}^{B*} \underline{\omega}^* = \underline{M}_B^*. \tag{8.7b}$$

In this derivation, no assumptions were made concerning the location to the reference point **B** of the rigid body. Consequently, the two vector equations of motion become coupled: each equation involves both the acceleration of the reference point, $\underline{a}_B$, and the angular velocity, $\underline{\omega}^*$, and acceleration, $\dot{\underline{\omega}}^*$, of the rigid body. The relative position of the center of mass with respect to the chosen reference point **B**, $\underline{\eta}$, appears explicitly in the equations of motion.

Clearly, the center of mass of the rigid body could be chosen as the reference point; in this case, $\underline{\eta} = 0$, and the governing equations of motion simplify to $m \underline{\underline{R}}^T \underline{a}_C = \underline{F}^*$, and $\underline{\underline{I}}^{C*} \dot{\underline{\omega}}^* + \tilde{\omega}^* \underline{\underline{I}}^{C*} \underline{\omega}^* = \underline{M}_C^*$. The first equation describes the motion of the center of mass of the rigid body, and the second equation describes the motion of the body around this point. These equations are, of course, identical to those obtained earlier in section 6.5. If the orientation of the material frame is selected to coincide with that of the principal axes of inertia, the tensor of mass moments of inertia becomes diagonal, and the equations further simplify, see eqs. (6.21).

8.1.2 Equations of motion for the planar motion of a rigid body

When dealing with the planar motion of a rigid body, the equations of motion derived in the previous section simplify considerably. Let the planar motion take place in the plane defined by unit vectors $\bar{\imath}_1$ and $\bar{\imath}_2$; the angular velocity, angular acceleration, and virtual rotation vectors now become $\underline{\omega} = \dot{\theta} \bar{\imath}_3$, $\dot{\underline{\omega}} = \ddot{\theta} \bar{\imath}_3$ and $\delta \underline{\psi} = \delta \theta \bar{\imath}_3$, respectively, where θ is the rotation angle of the rigid body. It will be assumed here that unit vector $\bar{\imath}_3$, the normal to the plane in which the motion is taking place, is a principal axis of inertia, and hence, the mass moment of inertia tensor becomes

$$\underline{\underline{I}}^B = \begin{bmatrix} I_{11}^B & I_{12}^B & 0 \\ I_{12}^B & I_{22}^B & 0 \\ 0 & 0 & I_{33}^B \end{bmatrix} .$$

Introducing these expressions into eq. (8.6), the virtual work done by the inertial forces becomes

$$\delta W^I = -\delta \underline{r}_B^T \left[m\underline{a}_B + m\ddot{\theta}\tilde{\imath}_3\underline{\eta} - m\dot{\theta}^2\underline{\eta} \right] - \delta\theta \left[m\tilde{\imath}_3^T\tilde{\eta}\underline{a}_B + I_{33}^B\ddot{\theta} \right]. \tag{8.8}$$

D'Alembert's principle then yields the equations of motion of the rigid body undergoing planar motion as

$$m\,\underline{a}_B + m\,\ddot{\theta}\tilde{\imath}_3\underline{\eta} - m\,\dot{\theta}^2\underline{\eta} = \underline{F}, \tag{8.9a}$$

$$m\,\tilde{\imath}_3^T\tilde{\eta}\underline{a}_B + I_{33}^B\ddot{\theta} = M_B, \tag{8.9b}$$

where $\underline{F}$ and M_B are the components of the externally applied force vector and moment, respectively. These equations further simplify if the reference point is chosen to coincide with the center of mass of the rigid body.

Example 8.2. The double pendulum

Consider the double pendulum system depicted in fig. 8.2. The first bar is of length L_1, mass m_1, and is connected by hinges to the ground at point **O** and to the second bar at point **A**. The second bar is of length L_2 and mass m_2. The bars have orientation angles θ_1 and θ_2 with respect to the vertical, respectively. $\mathcal{I} = (\bar{\imath}_1, \bar{\imath}_2, \bar{\imath}_3)$ is an inertial basis, and bases $\mathcal{E} = (\bar{e}_1, \bar{e}_2, \bar{e}_3)$ and $\mathcal{A} = (\bar{a}_1, \bar{a}_2, \bar{a}_3)$ are material bases attached to the first and second bars, respectively. Derive the equations of motion of the system.

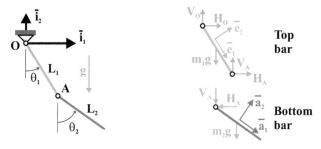

Fig. 8.2. Configuration of the double pendulum system.

Newtonian formulation

First, the equations of motion will be derived using the classical Newtonian approach. Figure 8.2 shows the free body diagrams for the two bars: V_O and H_O are the vertical and horizontal components of the reaction force at point **O** and V_A and H_A the components of the force vector transmitted through the hinge at point **A**. The equations of motion for the first bar are found using eq. (6.19) and the pivot equation, eq. (6.38), about inertial point **O**, to find

$$(H_O + H_A)\bar{\imath}_1 + (V_O + V_A)\bar{\imath}_2 - m_1 g \bar{\imath}_2 = \frac{m_1 L_1}{2}(\ddot{\theta}_1 \bar{e}_2 - \dot{\theta}_1^2 \bar{e}_1), \quad (8.10a)$$

$$\frac{m_1 L_1^2}{3}\ddot{\theta}_1 = -m_1 g \frac{L_1}{2}S_1 + V_A L_1 S_1 + H_A L_1 C_1. \quad (8.10b)$$

The short-hand notation $\sin\theta_1 = S_1$ and $\sin\theta_2 = S_2$ was used, with similar expressions for the cosine function. The equations of motion for the second bar are obtained in a similar manner as

$$-H_A \bar{\imath}_1 - V_A \bar{\imath}_2 - m_2 g \bar{\imath}_2 = m_2 L_1(\ddot{\theta}_1 \bar{e}_2 - \dot{\theta}_1^2 \bar{e}_1) + \frac{m_2 L_2}{2}(\ddot{\theta}_2 \bar{a}_2 - \dot{\theta}_2^2 \bar{a}_1), \quad (8.11a)$$

$$\frac{m_2 L_2^2}{12}\ddot{\theta}_2 = V_A \frac{L_2}{2}S_2 + H_A \frac{L_2}{2}C_2. \quad (8.11b)$$

In this case, the second equation was written about the center of mass of the bar, see eq. (6.37).

Newton's approach gives a total of six scalar equations, involving six unknowns: angles θ_1 and θ_2, two components of reaction force, H_O and V_O, and two components of internal force, H_A and V_A. Clearly, the system features two degrees of freedom only, and can be represented with two generalized coordinates, θ_1 and θ_2, for instance. The two equations of motion of the system would be obtained by eliminating the four components of reaction force, resulting in two coupled differential equations for θ_1 and θ_2.

After tedious algebra, the following equations are obtained

$$(\frac{m_1}{3} + m_2)L_1^2\ddot{\theta}_1 + m_2\frac{L_1 L_2}{2}C_{21}\ddot{\theta}_2 = -(\frac{m_1}{2} + m_2)gL_1 S_1 + m_2\frac{L_1 L_2}{2}\dot{\theta}_2^2 S_{21},$$

$$m_2\frac{L_1 L_2}{2}C_{21}\ddot{\theta}_1 + \frac{m_2 L_2^2}{3}\ddot{\theta}_2 = -m_2 g\frac{L_2}{2}S_2 - m_2\frac{L_1 L_2}{2}\dot{\theta}_1^2 S_{21}$$

where $C_{21} = \cos(\theta_2 - \theta_1)$ and $S_{21} = \sin(\theta_2 - \theta_1)$.

D'Alembert's formulation

The same problem will now be approached with d'Alembert's principle using kinematically admissible virtual displacements. For the first bar, the potential of the gravity forces is $V_1 = -m_1 g L_1 C_1/2$. The virtual work done by the inertial forces is obtained from eq. (8.8) as $\delta W_1^I = -\delta\theta_1 I_1^O \ddot{\theta}_1$, where I_1^O is the moment of inertia of the first bar with respect to point **O**. This point was chosen as the reference point on the body, and hence, $\underline{a}_O = 0$ and $\delta\underline{r}_O = 0$, because kinematically admissible virtual displacements are used. It follows that

$$\delta V_1 - \delta W_1^I = \delta\theta_1(m_1 g\frac{L_1}{2}S_1 + \frac{m_1 L_1^2}{3}\ddot{\theta}_1) \quad (8.12)$$

The potential of the gravity forces acting on the second bar is $V_2 = -m_2 g(C_1 L_1 + C_2 L_2/2)$. The virtual work done by the inertial forces is once again obtained from eq. (8.8), using point **A** as the reference point

$$\delta W_2^I = - \delta \underline{r}_A^T m_2 \left[\underline{a}_A + \ddot{\theta}_2 \widetilde{\imath}_3 \frac{L_2}{2} \bar{a}_1 - \dot{\theta}_2^2 \frac{L_2}{2} \bar{a}_1 \right]$$

$$- \delta \theta_2 \left[m_2 \widetilde{\imath}_3^T \frac{L_2}{2} \widetilde{a}_1 \underline{a}_A + I_2^A \ddot{\theta}_2 \right].$$

(8.13)

The position vector of point **A** is $\underline{r}_A = L_1 \bar{e}_1$; its velocity vector is then $\underline{v}_A = L_1 \dot{\theta}_1 \bar{e}_2$, and hence, $\delta \underline{r}_A = L_1 \delta \theta_1 \bar{e}_2$; finally, its acceleration vector is $\underline{a}_A = L_1 (\ddot{\theta}_1 \bar{e}_2 - \dot{\theta}_1^2 \bar{e}_1)$. Introducing all these results in the above equation leads to

$$\delta V_2 - \delta W_2^I = m_2 g (L_1 S_1 \delta \theta_1 + \frac{L_2}{2} S_2 \delta \theta_2)$$

$$+ m_2 L_1 \delta \theta_1 \bar{e}_2^T \left[L_1 (\ddot{\theta}_1 \bar{e}_2 - \dot{\theta}_1^2 \bar{e}_1) + \frac{L_2}{2} \ddot{\theta}_2 \bar{a}_2 - \frac{L_2}{2} \dot{\theta}_2^2 \bar{a}_1 \right]$$

$$+ \delta \theta_2 \left[m_2 \frac{L_1 L_2}{2} \bar{a}_1^T \widetilde{\imath}_3^T (\ddot{\theta}_1 \bar{e}_2 - \dot{\theta}_1^2 \bar{e}_1) + I_2^A \ddot{\theta}_2 \right].$$

D'Alembert's principle now implies $\delta(V_1 + V_2) - \delta(W_1^I + W_2^I) = 0$ for all arbitrary variations, $\delta \theta_1$ and $\delta \theta_2$; this directly leads to the equations of motion given above.

In contrast with Newton's approach, d'Alembert's principle yields the equations of motion of the system without reference to the reaction and internal forces that are eliminated from the onset of the formulation. Consequently, the equations of motion are obtained in a more direct and convenient manner, reducing the risk of errors. The reaction and internal forces do not appear in d'Alembert's formulation because the virtual work done by such forces vanishes, as discussed in section 7.4.3.

Of course, reaction and internal forces are quantities of primary interest that must often be evaluated as an integral part of the analysis. Newton's equations could be used to introduce reaction forces into the formulation; for instance, eqs. (8.11) could be used to evaluate H_A and V_A, then eqs. (8.10) would yield the other two components H_O and V_O.

Example 8.3. The rigid body/universal joint system

Figure 7.9 depicts a rigid body attached to the ground by means of a universal joint. This problem was treated in example 7.5 on page 266, where the configuration of the universal joint is described.

Component k of the universal joint is connected to the ground at point **O** by means of a bearing allowing rotation about axis $\bar{\imath}_3$. Component ℓ is connected to a rigid body at point **O'**. The orientation of the rigid body will be defined by Euler angles, using the 3-1-2 sequence. A first planar rotation about axis $\bar{\imath}_3$, of magnitude ϕ, brings inertial basis $\mathcal{I} = (\bar{\imath}_1, \bar{\imath}_2, \bar{\imath}_3)$ to basis $\mathcal{A} = (\bar{a}_1, \bar{a}_2, \bar{a}_3)$, where $\bar{a}_1$ is aligned with unit vector $\bar{b}_1$ of the cruciform. This rotation is associated with a constant angular speed $\dot{\phi} = \Omega$, implying $\bar{a}_1(t) = \cos(\Omega t) \bar{\imath}_1 + \sin(\Omega t) \bar{\imath}_2$. A second planar rotation about axis $\bar{a}_1$, of magnitude θ, brings basis $\mathcal{A}$ to basis $\mathcal{B} = (\bar{b}_1, \bar{b}_2, \bar{b}_3)$, where $\bar{b}_2$ is the second unit vector aligned with the cruciform. Finally, a third planar rotation about axis $\bar{b}_2$, of magnitude ψ, bring basis $\mathcal{B}$ to basis $\mathcal{E}^* = (\bar{e}_1, \bar{e}_2, \bar{e}_3)$ that is attached to the rigid body. Tensor components resolved in basis $\mathcal{E}^*$ will be denoted with the superscript $(\cdot)^*$. Points **O** and **O**$'$ are coincident. Establish the equations of motion of the system using d'Alembert's principle.

The potential of the gravity forces acting at the center of mass of the rigid body is $V = mg\bar{\imath}_3^T\underline{\eta}$, where m is the total mass of the rigid body and $\underline{\eta}$ the position vector of the center of mass with respect to point **O**. The variation of this potential is then $\delta V = \delta\underline{\psi}^{*T}mg\widetilde{\underline{\eta}}^*\underline{\underline{R}}^T\bar{\imath}_3$, where $\delta\underline{\psi}^*$ are the components of the virtual rotation vector resolved in the body attached frame. The virtual work done by the inertial forces is given by eq. (8.6), and d'Alembert's principle now implies

$$\delta\underline{r}_O^T R\left[m\underline{\underline{R}}^T\underline{a}_O + (\dot{\widetilde{\omega}}^* + \widetilde{\omega}^*\widetilde{\omega}^*)m\underline{\eta}^*\right]$$
$$+ \delta\underline{\psi}^{*T}\left[mg\widetilde{\underline{\eta}}^*\underline{\underline{R}}^T\bar{\imath}_3 + m\widetilde{\underline{\eta}}^*\underline{\underline{R}}^T\underline{a}_O + \underline{\underline{I}}^{O*}\dot{\underline{\omega}}^* + \widetilde{\omega}^*\underline{\underline{I}}^{O*}\underline{\omega}^*\right] = 0,$$

where inertial point **O** was used as the reference point on the rigid body; hence, $\underline{a}_O = 0$.

The system presents two degrees of freedom and two generalized coordinates, the two Euler angles, θ and ψ, will be selected here. Kinematically admissible virtual displacements will be used for this problem, hence $\delta\underline{r}_O = 0$ and, in view of eq. (4.80), the virtual rotation vector becomes

$$\delta\underline{\psi}^* = \underline{\underline{H}}^* \begin{Bmatrix} \delta\phi \\ \delta\theta \\ \delta\psi \end{Bmatrix} = \begin{Bmatrix} -C_\theta S_\psi \\ S_\theta \\ C_\theta C_\psi \end{Bmatrix}\delta\phi + \begin{bmatrix} C_\psi & 0 \\ 0 & 1 \\ S_\psi & 0 \end{bmatrix}\begin{Bmatrix} \delta\theta \\ \delta\psi \end{Bmatrix}$$

$$= \underline{h}^*\,\delta\phi + \underline{\underline{G}}^*\begin{Bmatrix} \delta\theta \\ \delta\psi \end{Bmatrix}, \tag{8.14}$$

where $\underline{h}^*$ stores the first column of the tangent operator, $\underline{\underline{H}}^*$, and $\underline{\underline{G}}^*$ its last two. Because the first Euler angle is prescribed to be $\phi = \Omega t$, the corresponding variation vanishes, $\delta\phi = 0$.

D'Alembert's principle now reduces to $\{\delta\theta, \delta\psi\}\underline{\underline{G}}^{*T}[mg\widetilde{\underline{\eta}}^*\underline{\underline{R}}^T\bar{\imath}_3 + \underline{\underline{I}}^{O*}\dot{\underline{\omega}}^* + \widetilde{\omega}^*\underline{\underline{I}}^{O*}\underline{\omega}^*] = 0$. Because the virtual changes in the two generalized coordinates are arbitrary, the equations of motion of the rigid body are

$$\underline{\underline{G}}^{*T}\left[mg\widetilde{\underline{\eta}}^*\underline{\underline{R}}^T\bar{\imath}_3 + \underline{\underline{I}}^{O*}\dot{\underline{\omega}}^* + \widetilde{\omega}^*\underline{\underline{I}}^{O*}\underline{\omega}^*\right] = 0. \tag{8.15}$$

D'Alembert's principle is a very powerful tool for the derivation of the equations of motion of the system. Two equations of motion are obtained for the two generalized coordinates of the problem. In contrast, Newton's method would generate six equations involving six unknown: two Euler angles θ and ψ, three components of the reaction force at point **O**, and the torque Q required to impose the constant angular velocity $\dot{\phi} = \Omega$. These latter four unknowns would need to be eliminated from the set of equations to obtain two equations of motion for θ and ψ. Derivation of the equations of motion based on Newton's approach is left to the reader as an exercise.

8.1.3 Problems

Problem 8.1. Euler's first and second laws
Prove Euler's first and second laws for a system of particles based on d'Alembert's principle. Hint: first read sections 3.4.4 and 7.5.4.

Problem 8.2. Two bar linkage system
The two bar linkage shown in fig. 7.11 comprises bar **OB** of length L_b and mass m_b, and bar **BAT** of length L_c and mass m_c. Bar **BAT** passes through fixed point **A** but is free to swivel about that point. (1) Derive the equation of motion of the system using d'Alembert's principle. Use angle θ as the generalized coordinate.

Problem 8.3. Homogeneous bar sliding on guides
Figure 8.3 depicts a homogeneous bar of length L and mass m sliding on two guides at its end points. At the left end, the bar is connected to a spring of stiffness constant k that is unstretched when the bar is horizontal. At the right end, the bar is connected to a point mass M. Gravity acts along unit vector $\bar{\imath}_2$. (1) Use d'Alembert's principle to derive the equation of motion of the system. Use a single generalized coordinate, θ.

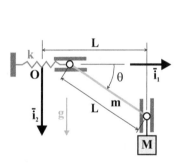
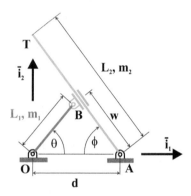

Fig. 8.3. Homogeneous bar sliding on guides at both ends.

Fig. 8.4. The two-bar linkage with slider system.

Problem 8.4. The two-bar linkage with slider
The two-bar linkage with slider system shown in fig. 8.4 is a planar mechanism. It consists of a uniform crank of length L_1 and mass m_1 connected to the ground at point **O**; let θ be the angle from the horizontal to the crank. At point **B**, the crank slides over a uniform linkage of length L_2 and mass m_2 that is connected to the ground at point **A**. Let w denote the distance from point **B** to point **A** and ϕ the angle from the horizontal to link **BA**. (1) Derive the equation of motion of the system using d'Alembert's principle. Use angle θ as the generalized coordinate.

Problem 8.5. Pendulum mounted on a cart
Figure 8.5 shows a pendulum of length L and mass m mounted on a cart of mass M that is connected to the ground by means of a spring of stiffness constant k and of a dashpot of constant c. The displacement of the cart is denoted x which is also the stretch of the spring, and angle θ measures the deflection of the pendulum with respect to the vertical. Gravity acts on

the system as indicated in the figure. *(1)* Based on d'Alembert's principle, derive the equations of motion of the system using the generalized coordinates x and θ. *(2)* Plot the time history of the cart displacement, x. *(3)* Plot the history of angle θ. *(4)* Plot the trajectory of the point at the tip of the pendulum. *(5)* Plot the cart velocity, $\dot{x}$. *(6)* Plot the angular velocity of the pendulum, $\dot{\theta}$. *(7)* Plot the system kinetic and potential energies and the energy dissipated in the damper. Check the energy closure equation. *(8)* Plot the components of the internal force at point **A**. Use the following data: $M = 5$ kg; $m = 2$ kg; $L = 0.4$ m; $k = 10$ N/m; acceleration of gravity $g = 9.81$ m/s^2; $c = 0.5$ N·s/m. Present all your results for a period of 10 s. Initial condition are at rest with $x(t = 0) = 0.2$ m and $\theta(t = 0) = \pi$.

Problem 8.6. Flexible pendulum on a slider
Solve problem 3.34 using d'Alembert's principle.

Problem 8.7. Pendulum with rotating mass
Solve problem 4.38 using d'Alembert's principle.

Problem 8.8. Plate hinged at the rim of a rotating disk
Solve problem 6.14 using d'Alembert's principle.

Problem 8.9. Pendulum with sliding mass
Solve problem 6.34 using d'Alembert's principle.

Problem 8.10. Bar rocking on top of a curve
Solve problem 6.38 using d'Alembert's principle.

Problem 8.11. Pendulum connected to a plunging mass
Solve problem 6.39 using d'Alembert's principle.

Problem 8.12. Particle sliding in a rolling wheel
Solve problem 6.43 using d'Alembert's principle.

Problem 8.13. Pendulum connected to horizontal piston
Solve problem 6.45 using d'Alembert's principle.

Problem 8.14. Inverted pendulum mounted on a cart
Solve problem 6.46 using d'Alembert's principle.

Problem 8.15. Bar hinged at rim of rotating disk
Solve problem 6.27 using d'Alembert's principle.

Problem 8.16. Geneva wheel mechanism
Solve problem 6.47 using d'Alembert's principle.

8.2 Hamilton's principle

When dealing d'Alembert's principle, the virtual work done by the inertial force acting on the kth particle, $-\delta \underline{r}_k^T m_k \underline{a}_k$, cannot not be derived from a "potential of the inertial forces." Consequently, accelerations appear explicitly in formulations based on d'Alembert's principle. It is possible to remedy this situation by further manipulation of this principle.

In general, dynamical systems must be studied over a period of interest, say from an initial time t_i to a final time t_f. At those times, boundary conditions will be enforced; at time t_i, the position of each particle could be known, $\underline{r}_k(t_i) = \hat{\underline{r}}_k(t_i)$, or the velocity of each particle could be known, $\underline{v}_k(t_i) = \hat{\underline{v}}_k(t_i)$, or both position and velocity could be known. In fact, it is even possible to imagine situations were positions could be known for some particles, and velocities for others. At time t_f, similar boundary conditions will exist.

Initial boundary value problems form an important class of problems in dynamics: the positions and velocities of all particles are known at time t_i; the solution of the dynamical equations of motion will reveal the corresponding quantities at time t_f. Another class of problems seeks to determine the configuration of the system at time t_i such that a desirable configuration is achieved at time t_f. In practice, more complex combinations of initial and final boundary conditions could be encountered.

To deal with all possible cases, the following notation will be used: at time t_i, $\underline{r}_k(t_i) = \hat{\underline{r}}_k(t_i)$, where $\hat{\underline{r}}_k(t_i)$ denotes the known or prescribed position of the kth particle if this position is known, or the resulting position if this quantity will result from the solution of the equations of motion. Similarly, $\hat{\underline{v}}_k(t_i)$ denotes the prescribed or resulting velocity at time t_i depending on the specific problem at hand. Of course, similar notations are used at the final time t_f.

8.2.1 Use of physical coordinates

If a system of N particles is in dynamic equilibrium, d'Alembert's principle, eq. (8.3), must be satisfied together with the boundary conditions at the initial and final times. Hence, the following statement must hold for all arbitrary virtual displacements

$$
\int_{t_i}^{t_f} \left(\delta W^I - \delta V + \delta W^{nc}\right) \, dt + \left[\sum_{k=1}^{N} \delta \underline{r}_k^T m_k (\underline{v}_k - \hat{\underline{v}}_k)\right]_{t_i}^{t_f}
$$
$$
+ \left[\sum_{k=1}^{N} m_k \delta \underline{v}_k^T (\underline{r}_k - \hat{\underline{r}}_k)\right]_{t_i}^{t_f} = 0. \tag{8.16}
$$

At each instant in time, the integrand must vanish because of d'Alembert's principle, eq. (8.3), and consequently, the first integral vanishes for all virtual displacements. The next two terms of the statement also vanish because they simply state the initial and final boundary conditions of the problem. The sum of the three terms must hence vanish for all arbitrary virtual displacements.

The expression for the integral of the virtual work done by the inertial forces acting on each particle, see eq. (8.4), is expressed in terms of velocity vectors, $\underline{v}_k$, and virtual displacement vectors, $\delta \underline{r}_k$. Integration by parts then yields

$$\int_{t_i}^{t_f} \delta W^I \, dt = - \int_{t_i}^{t_f} \sum_{k=1}^{N} \delta \underline{r}_k^T m_k \frac{d \underline{v}_k}{dt} \, dt$$

$$= \int_{t_i}^{t_f} \sum_{k=1}^{N} \delta \underline{v}_k^T m_k \underline{v}_k \, dt - \left[\sum_{k=1}^{N} \delta \underline{r}_k^T m_k \underline{v}_k \right]_{t_i}^{t_f}. \tag{8.17}$$

The first term of the last equality is directly related to the kinetic energy of the system; $\delta \underline{v}_k^T m_k \underline{v}_k = \delta(1/2 \, m_k \underline{v}_k^T \underline{v}_k) = \delta K_k$, where K_k is the kinetic energy of the kth particle. The sum then yields the total kinetic energy, K, of the system. The linear momentum vector, $\underline{p}_k$, of the kth particle equals the derivative of the total kinetic energy with respect to the corresponding velocity vector

$$\underline{p}_k = m_k \underline{v}_k = \frac{\partial K}{\partial \underline{v}_k}. \tag{8.18}$$

Equation (8.17) now simplifies to

$$\int_{t_i}^{t_f} \delta W^I \, dt = \int_{t_i}^{t_f} \delta K \, dt - \left[\sum_{k=1}^{N} \delta \underline{r}_k^T \underline{p}_k \right]_{t_i}^{t_f}. \tag{8.19}$$

Introducing these results into statement (8.16) now leads to

$$\int_{t_i}^{t_f} (\delta K - \delta V + \delta W^{nc}) \, dt = \left[\sum_{k=1}^{N} \delta \underline{r}_k^T \hat{\underline{p}}_k \right]_{t_i}^{t_f} - \left[\sum_{k=1}^{N} \delta \underline{p}_k^T (\underline{r}_k - \hat{\underline{r}}_k) \right]_{t_i}^{t_f}. \tag{8.20}$$

Hamilton's principle can be stated as follows.

Principle 14 (Hamilton's principle) *A system of particles is in dynamic equilibrium if and only if equation (8.20) holds for all arbitrary virtual displacements.*

If the system is subjected to conservative forces only, $\delta W^{nc} = 0$, and Hamilton's principle simplifies to

$$\int_{t_i}^{t_f} (\delta K - \delta V) \, dt = \left[\sum_{k=1}^{N} \delta \underline{r}_k^T \hat{\underline{p}}_k \right]_{t_i}^{t_f} - \left[\sum_{k=1}^{N} \delta \underline{p}_k^T (\underline{r}_k - \hat{\underline{r}}_k) \right]_{t_i}^{t_f}. \tag{8.21}$$

It is convenient to introduce here a scalar function, L, called the *Lagrangian of the system,*

$$L = K - V. \tag{8.22}$$

With the help of this function, Hamilton's principle applied to conservative systems further reduces to

$$\int_{t_i}^{t_f} \delta L \, dt = \left[\sum_{k=1}^{N} \delta \underline{r}_k^T \hat{\underline{p}}_k \right]_{t_i}^{t_f} - \left[\sum_{k=1}^{N} \delta \underline{p}_k^T (\underline{r}_k - \hat{\underline{r}}_k) \right]_{t_i}^{t_f}.$$

In this statement, the left-hand side takes care of the dynamics of the system, and the right-hand side deals with the boundary conditions at the initial and final times.

In some instances, the boundary conditions are rather simple, and Hamilton's principle is used to derive the sole equations of motion of the system. In such case, the boundary terms on the right-hand side are simply dropped from the statement that now reduces to the *principle of least action*.

$$\int_{t_i}^{t_f} \delta L \, \mathrm{d}t = 0. \tag{8.23}$$

Principle 15 (Principle of least action) *A system of particles is in dynamic equilibrium if and only if equation (8.23) holds for all arbitrary virtual displacements.*

A fundamental difference between d'Alembert's and Hamilton's principles is that in the latter principle, the virtual work done by the inertial forces is expressed in terms of the variation of the kinetic energy. As was the case for d'Alembert's principle, if the virtual work done by a specific force vanishes, this force will be automatically eliminated from the equations of motion obtained from Hamilton's principle. In section 7.4.3, it was shown that the virtual work done by forces that impose kinematic constraints does vanish; hence, such forces will not appear in formulations based on Hamilton's principle. Of course, if non-conservative forces are applied to the system, the virtual work of these forces must also be taken into account, see eq. (8.20).

8.2.2 Use of generalized coordinates

In the previous section, the system was represented by the Cartesian coordinates of each particles. As discussed in section 7.2, it is often convenient to use generalized, rather than Cartesian coordinates to describe dynamical systems. To that effect, eq. (8.16) is now recast as

$$\int_{t_i}^{t_f} \left(\delta W^I - \delta V + \delta W^{nc} \right) \, \mathrm{d}t + \left[\delta \underline{q}^T (\underline{p} - \hat{\underline{p}}) \right]_{t_i}^{t_f} + \left[\delta \underline{p}^T (\underline{q} - \hat{\underline{q}}) \right]_{t_i}^{t_f} = 0. \tag{8.24}$$

In this statement, the boundary conditions are written in terms of the array of generalized coordinates, $\underline{q}$, used to describe the system and the generalized momenta, $\underline{p}$, defined as

$$\underline{p} = \frac{\partial L}{\partial \dot{\underline{q}}}, \tag{8.25}$$

where $\dot{\underline{q}}$ are the system's generalized velocities.

The boundary conditions are as follows: at time t_i, $\underline{q}(t_i) = \hat{\underline{q}}(t_i)$, where $\hat{\underline{q}}(t_i)$ denotes the known, or prescribed generalized coordinates defining the configuration of the system if this configuration is known, or the configuration resulting from the solution of the equations of motion. Similarly, $\hat{\underline{p}}(t_i)$ denotes the prescribed or resulting generalized momentum at time t_i depending on the specific problem at hand. Of course, similar notations are used at the final time t_f.

If the configuration of the system is described in terms of generalized coordinates, the virtual work done by the inertial forces, eq. (8.19), is expressed as

$$
\int_{t_i}^{t_f} \delta W^I \, \mathrm{d}t = \int_{t_i}^{t_f} \delta K \, \mathrm{d}t - \left[\sum_{k=1}^{N} \delta \underline{r}_k^T \frac{\partial K}{\partial \underline{v}_k} \right]_{t_i}^{t_f}
$$
$$
= \int_{t_i}^{t_f} \delta K \, \mathrm{d}t - \left[\delta \underline{q}^T \sum_{k=1}^{N} \left[\frac{\partial \underline{r}_k}{\partial \underline{q}} \right]^T \frac{\partial K}{\partial \underline{v}_k} \right]_{t_i}^{t_f} .
\tag{8.26}
$$

In the last bracketed term, the virtual displacement vector of the kth particle was expressed in terms of the generalized coordinates using the chain rule for derivatives as $\delta \underline{r}_k = (\partial \underline{r}_k / \partial \underline{q}) \delta \underline{q}$. Similarly, the velocity vector of the kth particle can be evaluated in terms of the generalized velocities as $\underline{v}_k = (\partial \underline{r}_k / \partial \underline{q}) \dot{\underline{q}}$; it then follows that $\partial \underline{v}_k / \partial \dot{\underline{q}} = \partial \underline{r}_k / \partial \underline{q}$. The boundary term of eq. (8.26) now simplifies to

$$
\left[\delta \underline{q}^T \sum_{k=1}^{N} \left[\frac{\partial \underline{r}_k}{\partial \underline{q}} \right]^T \frac{\partial K}{\partial \underline{v}_k} \right]_{t_i}^{t_f} = \left[\delta \underline{q}^T \sum_{k=1}^{N} \left[\frac{\partial \underline{v}_k}{\partial \dot{\underline{q}}} \right]^T \frac{\partial K}{\partial \underline{v}_k} \right]_{t_i}^{t_f}
$$
$$
= \left[\delta \underline{q}^T \frac{\partial K}{\partial \dot{\underline{q}}} \right]_{t_i}^{t_f} = \left[\delta \underline{q}^T \underline{p} \right]_{t_i}^{t_f} ,
$$

where the definition of the generalized momenta, eq. (8.25), was introduced.

The virtual work done by the inertial forces, eq. (8.26), now becomes

$$
\int_{t_i}^{t_f} \delta W^I \, \mathrm{d}t = \int_{t_i}^{t_f} \delta K \, \mathrm{d}t - \left[\delta \underline{q}^T \underline{p} \right]_{t_i}^{t_f} ,
$$

where it is understood that variations should be taken with respect to the generalized coordinates.

Introducing all these results into eq. (8.24) now leads to

$$
\int_{t_i}^{t_f} (\delta K - \delta V + \delta W^{nc}) \, \mathrm{d}t = \left[\delta \underline{q}^T \hat{\underline{p}} \right]_{t_i}^{t_f} - \left[\delta \underline{p}^T (\underline{q} - \hat{\underline{q}}) \right]_{t_i}^{t_f} .
\tag{8.27}
$$

Hamilton's principle can be stated as follows.

Principle 16 (Hamilton's principle) *A system of particles is in dynamic equilibrium if and only if equation (8.27) holds for all arbitrary virtual changes in the generalized coordinates.*

This principle should be compared with principle 14: both statement express the condition for dynamic equilibrium of the system, the former when the configuration of the system in described in terms of Cartesian coordinates, the latter when the configuration is expressed in terms of generalized coordinates. Of course, the statement of Hamilton's principle can be further simplified by introducing the system's Lagrangian as defined by eq. (8.22).

Example 8.4. Pendulum mounted on a cart

Figure 8.5 depicts a pendulum of length L and mass m mounted on a cart of mass M that is connected to the ground by means of a spring of stiffness constant k and of a dashpot of constant c. This two degree of freedom problem will be represented by two generalized coordinates: the displacement of the cart, denoted x, which is also the stretch of the spring, and the angular deflection of the pendulum with respect to the vertical, denoted θ. Gravity acts on the system as indicated in the figure. Derive the equations of motion of the system.

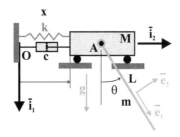

Fig. 8.5. Pendulum mounted on a cart.

The position of the center of mass of the pendulum is $\underline{r}_C = x\,\bar{\imath}_2 + L/2\,\bar{e}_1$ and its velocity $\underline{v}_C = \dot{x}\,\bar{\imath}_2 + L/2\dot{\theta}\,\bar{e}_2$. The total kinetic energy of the system is the sum of that of the cart, $1/2\,M\dot{x}^2$, and that of the pendulum, $1/2\,mL^2/12\,\dot{\theta}^2 + 1/2\,m\underline{v}_C^T\underline{v}_C$, i.e.,

$$K = \frac{1}{2}(M + m)\dot{x}^2 + \frac{1}{2}\frac{mL^2}{3}\dot{\theta}^2 + \frac{1}{2}mL\dot{x}\dot{\theta}C_\theta, \qquad (8.28)$$

where the short-hand notation $C_\theta = \cos\theta$ and $S_\theta = \sin\theta$ was used. The potential energy of the system consists of the strain energy of the spring, $1/2\,kx^2$, and the potential energy of the gravity forces, $mgL/2\,(1 - C_\theta)$. The potential of the system is now

$$V = \frac{1}{2}kx^2 + \frac{mgL}{2}(1 - C_\theta). \qquad (8.29)$$

Variation in the Lagrangian of the system becomes

$$\delta L = \delta K - \delta V = (M + m)\dot{x}\delta\dot{x} + \frac{mL^2}{3}\dot{\theta}\delta\dot{\theta}$$
$$+ \frac{mL}{2}(\dot{\theta}C_\theta\delta\dot{x} + \dot{x}C_\theta\delta\dot{\theta} - \dot{x}\dot{\theta}S_\theta\delta\theta) - kx\delta x - mg\frac{L}{2}S_\theta\delta\theta.$$

Finally, the virtual work done the non-conservative forces is $\delta W^{nc} = -c\dot{x}\delta x$. Introducing these results into Hamilton's principle leads to

$$\int_{t_i}^{t_f} \left\{ -\left[(M + m)\dot{x} + \frac{mL}{2}\dot{\theta}C_\theta\right]^{\cdot}\delta x - \left[\frac{mL^2}{3}\dot{\theta} + \frac{mL}{2}\dot{x}C_\theta\right]^{\cdot}\delta\theta \right.$$
$$\left. - \frac{mL}{2}\dot{x}\dot{\theta}S_\theta\delta\theta - kx\delta x - mg\frac{L}{2}S_\theta\delta\theta - c\dot{x}\delta x \right\} dt = 0.$$

To obtain this expression, integration by parts were performed for all terms involving time derivatives of virtual changes in the generalized coordinates, $\delta\dot{x}$ and $\delta\dot{\theta}$. Each integration by parts generates boundary terms at times t_i and t_f, which were ignored in this example.

Because the variations δx and $\delta\theta$ are arbitrary, the sum of their coefficients in the integrand must vanish, leading to the two governing equations of the system

$$(M+m)\ddot{x} + \frac{mL}{2}C_\theta\ddot{\theta} - \frac{mL}{2}S_\theta\dot{\theta}^2 + kx + c\dot{x} = 0,$$
$$\frac{mL}{2}C_\theta\ddot{x} + \frac{mL^2}{3}\ddot{\theta} + mg\frac{L}{2}S_\theta = 0. \tag{8.30}$$

Several distinguishing features of Hamilton's principle are apparent in this example. First, there is no need to compute accelerations because the Lagrangian of the system only involves velocities appearing in the expression for the kinetic energy. Both Newtonian approach or application of d'Alembert's principle would require the computation of accelerations, adding to the complexity of the kinematic analysis.

Second, internal and reaction forces are eliminated from the formulation. If this problem were treated using Newton's approach, the components of the internal force at point **A** would enter the formulation and derivation of the equations of motion would requires elimination of these forces.

Example 8.5. Kinetic energy for a rigid body
Figure 8.6 depicts a rigid body in its reference configuration, as defined by frame $\mathcal{F}^E = [\mathbf{B}, \mathcal{E} = (\bar{e}_1, \bar{e}_2, \bar{e}_3)]$, where **B** is a reference point on the body. Position vector $\underline{r}_{B0}$ determines the location of a reference point **B** on the rigid body with respect to inertial frame $\mathcal{F}^I = [\mathbf{O}, \mathcal{I} = (\bar{\imath}_1, \bar{\imath}_2, \bar{\imath}_3)]$. In the final configuration, the configuration of the rigid body is defined by frame $\mathcal{F}^B = [\mathbf{B}, \mathcal{B}^* = (\bar{b}_1, \bar{b}_2, \bar{b}_3)]$. The position vector of point **B** in the present configuration is denoted $\underline{r}_B = \underline{r}_{B0} + \underline{u}$, where $\underline{u}$ is the displacement vector of point **B**. Let $\underline{\underline{R}}_0$ and $\underline{\underline{R}}$ be the rotation tensors that bring bases $\mathcal{I}$ to $\mathcal{E}$ and $\mathcal{E}$ to $\mathcal{B}^*$, respectively, both resolved in the inertial basis, $\mathcal{I}$. Superscripts $(\cdot)^*$ denote tensor components resolved in the body attached basis. Find the kinetic energy of the rigid body.

In section 6.2, the kinetic energy of a rigid body was found to be given by eq. (6.6). Expressing all tensor components in the body attached basis leads to

$$K = \frac{1}{2}\left[m\underline{\dot{u}}^T(\underline{\underline{R}}\,\underline{\underline{R}}_0)(\underline{\underline{R}}\,\underline{\underline{R}}_0)^T\underline{\dot{u}} + 2m\underline{\dot{u}}^T(\underline{\underline{R}}\,\underline{\underline{R}}_0)\tilde{\eta}^{*T}\underline{\omega}^* + \underline{\omega}^{*T}\underline{\underline{I}}^{B*}\underline{\omega}^*\right].$$

The components of angular velocity vector of the rigid body resolved in the inertial basis are $\underline{\omega} = \text{axial}(\underline{\dot{R}}\,\underline{R}^T)$ and $\underline{\omega}^* = (\underline{\underline{R}}\,\underline{\underline{R}}_0)^T\underline{\omega}$ are the components of the same vector resolved in the material basis; $\underline{\eta}$ and $\underline{\eta}^* = (\underline{\underline{R}}\,\underline{\underline{R}}_0)^T\underline{\eta}$ are the components of the position vector of the center of mass of the body with respect to reference point **B** resolved in the inertial and material bases, respectively; finally, $\underline{\underline{I}}^{B*}$ and $\underline{\underline{I}}^B = (\underline{\underline{R}}\,\underline{\underline{R}}_0)\underline{\underline{I}}^{B*}(\underline{\underline{R}}\,\underline{\underline{R}}_0)^T$ are the components of the mass moment of inertia tensor of the body with respect to reference point **B**, resolved in material and inertial bases, respectively.

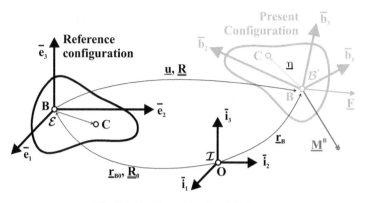

Fig. 8.6. Configuration of a rigid body.

The kinetic energy is now written in a compact form as

$$K = \frac{1}{2} \underline{\mathcal{V}}^{*T} \underline{\underline{\mathcal{M}}}_B^* \underline{\mathcal{V}}^*, \tag{8.31}$$

where the *mass matrix* of the rigid body, resolved in the body attached frame, is

$$\underline{\underline{\mathcal{M}}}_B^* = \begin{bmatrix} m\underline{\underline{I}} & m\widetilde{\eta}^{*T} \\ m\widetilde{\eta}^* & \underline{\underline{I}}^{B*} \end{bmatrix}, \tag{8.32}$$

and the *velocity vector*, resolved in the same frame, is

$$\underline{\mathcal{V}}^* = \left\{ \begin{matrix} (\underline{\underline{R}}\,\underline{\underline{R}}_0)^T \dot{\underline{u}} \\ \underline{\omega}^* \end{matrix} \right\}. \tag{8.33}$$

With this notation, the mass matrix is of size 6×6 and the velocity vector of size 6×1. The components of the linear and angular momentum vectors of the rigid body, resolved in the material basis, are denoted $\underline{p}^*$ and $\underline{h}^*$, respectively, and defined as

$$\underline{\mathcal{P}}^* = \left\{ \begin{matrix} \underline{p}^* \\ \underline{h}^* \end{matrix} \right\} = \underline{\underline{\mathcal{M}}}_B^* \underline{\mathcal{V}}^*. \tag{8.34}$$

As for the velocity vector, the momentum vector, $\underline{\mathcal{P}}^*$, is of size 6×1. Strictly speaking, quantities $\underline{\mathcal{V}}^*$ and $\underline{\mathcal{P}}^*$ are not vectors, but rather arrays, consisting of two separate vectors. To underline this fact, the script type is used to denote such quantities. The same remarks apply to the mass matrix, $\underline{\underline{\mathcal{M}}}_B^*$.

Example 8.6. Equations of motion for a rigid body with respect to a material point

Figure 8.6 depicts a rigid body in its reference and present configurations. The external loading applied to the rigid body consist of a force, $\underline{F}$, and a moment, $\underline{M}_B$, computed with respect to reference point **B**. The kinematics of the problem are described in the previous example. Find the equations of motion of the rigid body using Hamilton's principle.

The virtual work done by the externally applied loads is

$$\delta W^{nc} = \delta \underline{u}^T \underline{F} + \delta \underline{\psi}^T \underline{M}_B, \tag{8.35}$$

where $\delta \underline{u}$ and $\delta \underline{\psi}$ are the components of the virtual displacement and rotation vectors, respectively, resolved in the inertial basis.

For this problem, Hamilton's principle, eq. (8.20), reduces to

$$\int_{t_i}^{t_f} \left(\delta K + \delta W^{nc} \right) \, \mathrm{d}t = 0,$$

for all arbitrary virtual displacements, where the terms associated with the temporal boundary conditions are neglected. Introducing the expression for the kinetic energy of the rigid body, eq. (8.31), and virtual work done by the non-conservative forces, eq. (8.35), this principle becomes

$$\int_{t_i}^{t_f} \left(\delta \underline{\mathcal{V}}^{*T} \underline{\mathcal{P}}^* + \delta \underline{u}^T \underline{F} + \delta \underline{\psi}^T \underline{M}_B \right) \, \mathrm{d}t = 0. \tag{8.36}$$

With the help of eqs. (4.101) and (7.17b), virtual changes in the velocity vector are readily evaluated as

$$\delta \underline{\mathcal{V}}^* = \left\{ \begin{matrix} \delta \left[(\underline{\underline{R}}\, \underline{\underline{R}}_0)^T \dot{\underline{u}} \right] \\ \delta \underline{\omega}^* \end{matrix} \right\} = \left\{ \begin{matrix} (\underline{\underline{R}}\, \underline{\underline{R}}_0)^T (\tilde{\dot{u}}\, \delta \underline{\psi} + \delta \dot{\underline{u}}) \\ (\underline{\underline{R}}\, \underline{\underline{R}}_0)^T \delta \dot{\underline{\psi}} \end{matrix} \right\}.$$

Introducing these results into Hamilton's principle, eq. (8.36), then yields

$$\int_{t_i}^{t_f} \left\{ \left(\tilde{\dot{u}}\, \delta \underline{\psi} + \delta \dot{\underline{u}} \right)^T \left(\underline{\underline{R}}\, \underline{\underline{R}}_0 \right) \underline{p}^* + \delta \dot{\underline{\psi}}^T \left(\underline{\underline{R}}\, \underline{\underline{R}}_0 \right) \underline{h}^* + \delta \underline{u}^T \underline{F} + \delta \underline{\psi}^T \underline{M}_B \right\} \, \mathrm{d}t = 0.$$

In this expression, $\underline{p} = (\underline{\underline{R}}\, \underline{\underline{R}}_0)\underline{p}^*$ and $\underline{h} = (\underline{\underline{R}}\, \underline{\underline{R}}_0)\underline{h}^*$ are the components of linear and angular momentum vectors of the body, both resolved in the inertial basis.

After integration by parts, Hamilton's principle becomes

$$\int_{t_i}^{t_f} \left\{ \delta \underline{u}^T \left[-\dot{\underline{p}} + \underline{F} \right] + \delta \underline{\psi}^T \left[-\dot{\underline{h}} + \tilde{\dot{u}}^T \underline{p} + \underline{M}_B \right] \right\} \, \mathrm{d}t = 0.$$

Because the virtual displacements are arbitrary, the bracketed terms must vanish, leading to the governing equations of the problem.

In summary, the governing differential equations of motion for a rigid body are

$$\dot{\underline{p}} = \underline{F}, \tag{8.37a}$$

$$\dot{\underline{h}} + \tilde{\dot{u}}\underline{p} = \underline{M}_B, \tag{8.37b}$$

$$\left\{ \begin{matrix} \underline{p} \\ \underline{h} \end{matrix} \right\} = \underline{\underline{M}}_B \left\{ \begin{matrix} \dot{\underline{u}} \\ \underline{\omega} \end{matrix} \right\}. \tag{8.37c}$$

The mass matrix of the rigid body computed with respect to point **B** and resolved in the inertial frame is

$$\underline{\underline{M}}_B = \underline{\underline{R}} \, \underline{\underline{M}}^*_B \underline{\underline{R}}^T = \begin{bmatrix} m\underline{\underline{I}} & m\widetilde{\eta}^T \\ m\widetilde{\eta} & \underline{\underline{I}}^B \end{bmatrix}, \tag{8.38}$$

where

$$\underline{\underline{R}} = \begin{bmatrix} (\underline{\underline{R}}\,\underline{\underline{R}}_0) & \underline{\underline{0}} \\ \underline{\underline{0}} & (\underline{\underline{R}}\,\underline{\underline{R}}_0) \end{bmatrix}. \tag{8.39}$$

The twelve first-order differential equations in time, eqs. (8.37a) to (8.37c) should be solved to find the twelve unknowns of the problem: the three components of the displacement vector, $\underline{u}$, the three rotation components required to express the rotation tensor, $\underline{\underline{R}}$, and the three components for each linear and angular momentum vectors, $\underline{p}$ and $\underline{h}$, respectively.

Equations (8.37a) to (8.37c) present a high level of nonlinearity; indeed, the mass matrix involves a term, $\underline{\underline{I}}^B = (\underline{\underline{R}}\,\underline{\underline{R}}_0)\underline{\underline{I}}^{B*}(\underline{\underline{R}}\,\underline{\underline{R}}_0)^T$, containing a product of the unknown rotation tensor, $\underline{\underline{R}}$. The equations of motion can be recast in a manner that decreases the level of nonlinearity

$$(\underline{\underline{R}}\,\underline{\underline{R}}_0 \underline{p}^*)\dot{} = \underline{F}, \tag{8.40a}$$

$$(\underline{\underline{R}}\,\underline{\underline{R}}_0 \underline{h}^*)\dot{} + \widetilde{\dot{u}}(\underline{\underline{R}}\,\underline{\underline{R}}_0 \underline{p}^*) = \underline{M}_B, \tag{8.40b}$$

$$\begin{Bmatrix} \underline{p}^* \\ \underline{h}^* \end{Bmatrix} = \underline{\underline{M}}^*_B \begin{Bmatrix} (\underline{\underline{R}}\,\underline{\underline{R}}_0)^T \underline{\dot{u}} \\ \underline{\omega}^* \end{Bmatrix}. \tag{8.40c}$$

In this form, the mass matrix of the body is a constant, because it is expressed in the material frame.

Of course, the equations of motion can be written in many different manners. Eliminating the momenta will lead to six, second-order differential equations for the six displacement components, three displacements of reference point **B** and three rotation components. The equations of motion could also be recast in terms of displacements and velocities, rather than momenta. In each case, the equations might be resolved in the inertial basis or in the material basis.

The equations presented here are *intrinsic equations of motion*, i.e., they are independent of the variables used to represent rotations: the above equations could be used with any set of Euler angles, see section 4.11. In practice, however, a specific parametrization of finite rotations must be selected to solve the equations of motion. If the rigid body is tumbling in space, it is important to select a parametrization that is singularity free, as discussed in chapter 13.

Example 8.7. *Equations of motion for a rigid body with respect to an inertial point*

In the previous example, the applied moment was computed with respect to a material point, reference point **B**. Similarly, the linear and angular momenta defined by eq. (8.34) are the product of the mass matrix resolved in the material frame by the velocity vector. The mass matrix, eq. (8.32), is evaluated with respect to reference point **B**: indeed, η^* is the position vector of the body's center of mass with respect to point **B** and $\underline{\underline{I}}^{B*}$ the body's mass moment of inertia tensor computed with respect to the same point. Derive the equations of motion of the rigid body with respect to an inertial point.

Referring to fig. 8.6, the loading externally applied to the rigid body is

$$\underline{F}_O = \underline{F}, \tag{8.41a}$$
$$\underline{M}_O = \underline{M}_B + (\tilde{r}_{B0} + \tilde{u})\underline{F}. \tag{8.41b}$$

The externally applied force, $\underline{F}_O$, and moment, $\underline{M}_O$, are now computed with respect to inertial point **O**. Similarly, the linear and angular momenta of the rigid body are

$$\underline{p}_O = \underline{p}, \tag{8.42a}$$
$$\underline{h}_O = \underline{h} + (\tilde{r}_{B0} + \tilde{u})\underline{p}, \tag{8.42b}$$

where the linear momentum, $\underline{p}_O$, and angular momentum, $\underline{h}_O$, are now computed with respect to inertial point **O**.

The governing equations of the problem derived in the previous example are given by eqs. (8.37). A linear combination of eqs. (8.37a) and (8.37b) yields $\dot{\underline{h}} + \tilde{u}\underline{p} + (\tilde{r}_{B0} + \tilde{u})\dot{\underline{p}} = \underline{M}_B + (\tilde{r}_{B0} + \tilde{u})\underline{F}$; the left-hand side of this equation is an exact derivative, leading to $[\underline{h} + (\tilde{r}_{B0} + \tilde{u})\underline{p}]^{\cdot} = \underline{M}_O$, where eq. (8.41b) was used to simplify the right-hand side of the equation. The equations of motion of the problem, are now recast in a more compact manner as

$$\dot{\underline{p}}_O = \underline{F}_O, \tag{8.43a}$$
$$\dot{\underline{h}}_O = \underline{M}_O. \tag{8.43b}$$

In the absence of externally applied forces, these equations reduce to $\dot{\underline{p}}_O = 0$ and $\dot{\underline{h}}^O = 0$; this implies the conservation of linear and angular momenta, a general result that was derived for a system of particles, see eqs. (3.81) and (3.82), respectively.

To obtain a complete set of governing equations, rotation parameters must now be selected. Assuming that the rotation tensor is expressed in terms of three Euler angles stored in array q, the angular velocity vector becomes $\underline{\omega} = \underline{\underline{H}}(q)\dot{q}$, where expressions for the tangent operator, $\underline{\underline{H}}(q)$, are given in section 4.11 for various Euler angle sequences; see, for instance, eq. (4.68) for the 3-1-3 sequence. Equation (8.37c) now become

$$\left\{ \begin{matrix} u \\ q \end{matrix} \right\}^{\cdot} = \begin{bmatrix} \underline{\underline{I}} & \underline{\underline{0}} \\ \underline{\underline{0}} & \underline{\underline{H}}^{-1}(q) \end{bmatrix} \left(\underline{\underline{M}}_B\right)^{-1} \left\{ \begin{matrix} \underline{p}_O \\ \underline{h}_O - (\tilde{r}_{B0} + \tilde{u})\underline{p}_O \end{matrix} \right\}. \tag{8.44}$$

While the externally applied loads and momenta are now referred to inertial point **O**, reference to material point **B** has not been completely eliminated from the formulation; indeed, the mass matrix, $\underline{\underline{M}}_B$, is still computed with respect to point **B**. Example 8.9 will eliminate this restriction by using a more general formalism based on the use of the motion tensor.

Example 8.8. Rigid body connected to spring and dashpot

Consider a rigid body connected to the ground at point **B** by means of a spring of stiffness constant k and dashpot of constant c, as depicted in fig. 8.7 and discussed in example 6.3 on page 221. The spring is of un-stretched length ℓ_0. The rigid body is of mass M and its moment of inertia tensor with respect to the center of mass is $\underline{\underline{I}}^C$. Vector η defines the position of the center of mass with respect to point **B**. Frame $\mathcal{F}^B = \left[\mathbf{B}, \mathcal{B}^* = (\bar{b}_1, \bar{b}_2, \bar{b}_3)\right]$ is attached to the rigid body. The components of the rotation tensor that brings inertial basis $\mathcal{I}$ to material basis $\mathcal{B}^*$, resolved in basis $\mathcal{I}$, is denoted $\underline{\underline{R}}$. Find the equations of motion of the system.

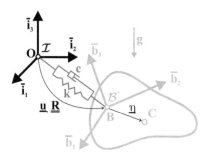

Fig. 8.7. Configuration of the rigid body connected to a spring and dashpot.

The force, $\underline{F}_B$, applied to the rigid body at point **B** acts in the direction opposite to unit vector $\bar{e} = \underline{u}/\|\underline{u}\|$, or $\underline{F}_B = -F\bar{e}$, where $\underline{u}$ is the position vector of point **B** with respect to point **O** and F the magnitude of the applied force. It then follows that $F = k\Delta + c\dot{\Delta}$, where $\Delta = \|\underline{u}\| - \ell_0$ is the stretch of the spring. The time rate of change of the stretch is easily found as $\dot{\Delta} = \bar{e}^T \underline{\dot{u}}$.

The formalism developed in example 8.6 will be used to solve this problem. In this case, eqs. (8.37a) and (8.37b) become $\dot{p} = -F\bar{e} - Mg\bar{\imath}_3$ and $\underline{\dot{h}} + \tilde{\underline{u}}\underline{p} = -Mg\tilde{\eta}\bar{\imath}_3$, respectively, and eq. (8.37c) is unchanged. The equations of motion are further simplified by expressing the linear and angular momenta with respect to inertial point **O**, as defined in eqs. (8.42). The governing equations now become

$$\left\{\begin{matrix} \underline{p}_O \\ \underline{h}_O \end{matrix}\right\}^{\cdot} = \left\{\begin{matrix} -F\bar{e} - Mg\,\bar{\imath}_3 \\ -Mg(\tilde{\eta} + \tilde{u})\bar{\imath}_3 \end{matrix}\right\},$$

and eqs. (8.44) remain unchanged.

Example 8.9. Equations of motion for a rigid body using the motion formalism

Consider the rigid body depicted in fig. 8.6; the kinetic energy of the body was evaluated in example 8.5. Use the motion formalism and the motion tensor to derive the equations of motion of the rigid body. Let $\underline{\underline{C}}_0$ and $\underline{\underline{C}}$ be the motion tensors that bring frame $\mathcal{F}^I$ to $\mathcal{F}^E$ and frame $\mathcal{F}^E$ to $\mathcal{F}^B$, respectively, both resolved in frame $\mathcal{F}^I$. Frame $\mathcal{F}^E$ is attached to the rigid body in its reference configuration.

Figure 8.6 shows the force, $\underline{F}$, and moment, $\underline{M}_B$, externally applied to the rigid body and computed with respect to reference point **B**. The virtual work done by these loads is $\delta W^{nc} = \underline{F}^T \delta \underline{u} + \underline{M}_B^T \delta \underline{\psi}$, where $\delta \underline{u}$ is the virtual displacement vector of point **B** and $\delta \underline{\psi}$ the virtual rotation vector of the rigid body. Resolving the loads in the body attached basis yields $\delta W^{nc} = \underline{F}^{*T} \underline{R}^T \delta \underline{u} + \underline{M}_B^{*T} \delta \underline{\psi}^*$, and the following compact notation is then used,

$$\delta W^{nc} = \underline{A}^{*T} \delta \underline{U}^* = \underline{A}^T \left(\underline{\underline{C}}\,\underline{\underline{C}}_0\right) \left(\underline{\underline{C}}\,\underline{\underline{C}}_0\right)^{-1} \delta \underline{U} = \underline{A}^T \delta \underline{U}, \qquad (8.45)$$

where $\underline{A}^{*T} = \{\underline{F}^{*T}, \underline{M}^{*T}\}$ is the applied load vector, and $\delta \underline{U}^*$ the virtual motion vector, both resolved in frame $\mathcal{F}^B$. The virtual motion vector is similar to the differential motion vector defined in eq. (5.72b). The second equality follows from the frame change operations expressed by eqs. (5.72b) and (5.62).

The velocity vector $\underline{V}^*$ defined in eq. (8.33) combines the linear and angular velocity vectors of the rigid body resolved in the body attached frame. It follows that $\underline{V} = (\underline{\underline{C}}\,\underline{\underline{C}}_0)\underline{V}^*$ is the velocity vector resolved in the inertial frame. Virtual changes in the velocity vector become

$$\delta \underline{V}^* = (\underline{\underline{C}}\,\underline{\underline{C}}_0)^{-1} \left[\delta \underline{V} + \underline{\underline{C}}\,\delta \underline{\underline{C}}^{-1} \underline{V}\right]$$
$$= (\underline{\underline{C}}\,\underline{\underline{C}}_0)^{-1} \left[\delta \underline{U} - \widetilde{\underline{V}}\delta \underline{U} - \widetilde{\delta \underline{U}}\,\underline{V}\right] = (\underline{\underline{C}}\,\underline{\underline{C}}_0)^{-1} \delta \underline{U}, \qquad (8.46)$$

where eqs. (5.74a) and (5.74b) were used.

Because virtual changes in the kinetic energy of the system are expressed as $\delta K = \delta \underline{V}^{*T} \underline{P}^*$, where the linear and angular momenta of the rigid body, $\underline{P}^*$, are defined in eq. (8.34), Hamilton's principle, eq. (8.20), becomes

$$\int_t \left[\delta \underline{U}^T (\underline{\underline{C}}\,\underline{\underline{C}}_0)^{-T} \underline{P}^* + \delta \underline{U}^T \underline{A}\right] \mathrm{d}t = 0,$$

Introducing eqs. (8.34) and (5.58) then leads to

$$\int_t \left[\delta \underline{U}^T (\underline{\underline{C}}\,\underline{\underline{C}}_0)^{-T} \underline{\underline{M}}_B^* (\underline{\underline{C}}\,\underline{\underline{C}}_0)^{-1} \underline{V} + \delta \underline{U}^T (\underline{\underline{C}}\,\underline{\underline{C}}_0)^{-T} \underline{A}^*\right] \mathrm{d}t = 0.$$

The mass matrix of the rigid body in the inertial frame is defined as

$$\underline{\underline{M}}_O = (\underline{\underline{C}}\,\underline{\underline{C}}_0)^{-T} \underline{\underline{M}}_B^* (\underline{\underline{C}}\,\underline{\underline{C}}_0)^{-1}. \qquad (8.47)$$

This change of frame operation performs two tasks: a change of reference point from **B** to **O** and a change of basis from $\mathcal{B}^*$ to $\mathcal{I}$.

Hamilton's principle now becomes $\int_t (\delta \underline{U}^T \underline{P} + \delta \underline{U}^T \underline{A}) \, \mathrm{d}t = 0$, where the momentum vector in the inertial frame is defined as $\underline{P} = \underline{\underline{M}}_O \underline{V}$. Integration by parts now yields $\int_t \delta \underline{U}^T [-\dot{\underline{P}} + \underline{A}] \, \mathrm{d}t = 0$. Because the virtual motions are arbitrary, the bracketed term must vanish.

In summary, the governing differential equations of motion for a rigid body are

$$\dot{\underline{P}} = \underline{A}, \tag{8.48a}$$

$$\underline{P} = \underline{\underline{M}}_o \underline{V}. \tag{8.48b}$$

These twelve first-order differential equations in time, should be solved to find the twelve unknowns of the problem: six motion components to express the motion tensor, $\underline{C}$, and six components of momentum, $\underline{P}$. Note that eqs. (8.48a) and (8.48b) present a high level of nonlinearity; indeed, the mass matrix, eq. (8.47), involves the unknown motion tensor $\underline{C}$ and its inverse.

Example 8.10. Numerical application
This example illustrates the use of Hamilton's principle for numerical applications. Consider a particle of mass m and position vector $\underline{r}(t)$ with respect to an inertial point, subjected to an arbitrary, time varying force, $\underline{F}(t)$. The kinetic energy of the system simply writes $K = 1/2\, m\, \underline{\dot{r}}^T \underline{\dot{r}}$ and the virtual work done by the externally applied force becomes $\delta W^{nc} = \delta \underline{r}^T \underline{F}(t)$.

Hamilton's principle could be used to derive the equations of motion of the system, which are $m \underline{\ddot{r}} = \underline{F}$ for this simple problem; integration of these equations would then yield the trajectory of the particle. The traditional approach to solving dynamics problems consists of two steps; first, one of the formulations of dynamics, such as Newton's formulation or Hamilton's principle, is used to derive the ordinary differential equations of motion of the system, and second, an approximate solution of these equations is obtained with the help of numerical methods. Because the equations of motion are, in general, nonlinear differential equations, numerical methods, such as Runge-Kutta integrators [5, 26], for instance, are often used to obtain numerical solutions.

A different approach is taken here: algebraic or discretized equations of motion will be directly obtained from Hamilton's principle, bypassing the derivation of the ordinary differential equations of motion. To that effect, the behavior of the particle is investigated during a small, but finite period of time between t_i and t_f, called a *time step*. The time step size is denoted $\Delta t = t_f - t_i$. For simplicity, a non-dimensional time, τ, is defined, which is related to the dimensional time as $t = 1/2\,(t_i + t_f) + \tau\,\Delta t/2$, i.e., $\tau = \pm 1$ at times t_i and t_f, respectively. To obtain algebraic equations, the trajectory of the particle over the time step is discretized by assuming its motion to be the straight line joining its positions at times t_i and t_f, denoted $\underline{r}_i$ and $\underline{r}_f$, respectively. It then follows that $\underline{r}(\tau) = \underline{r}_i\,(1 - \tau)/2 + \underline{r}_f\,(1 + \tau)/2$.

Within the time step, the velocity of the particle is constant: $\underline{v} = \mathrm{d}\underline{r}/\mathrm{d}t = (\mathrm{d}\underline{r}/\mathrm{d}\tau)(\mathrm{d}\tau/\mathrm{d}t) = (\underline{r}_f - \underline{r}_i)/\Delta t$. Introducing these approximations into the expression for the kinetic energy then yields $\delta K = \underline{p}_m^T (\delta \underline{r}_f - \delta \underline{r}_i)/\Delta t$, where $\underline{p}_m = m(\underline{r}_f - \underline{r}_i)/\Delta t$ is the constant linear momentum of the particle within the time step. The statement of Hamilton's principle governing the motion of the particle now becomes

$$\int_{t_i}^{t_f} \left\{ \frac{\delta \underline{r}_f^T - \delta \underline{r}_i^T}{\Delta t}\, \underline{p}_m + \left[\frac{1 - \tau}{2} \delta \underline{r}_i^T + \frac{1 + \tau}{2} \delta \underline{r}_f^T \right] \underline{F}(t) \right\} \mathrm{d}t$$

$$= \left[\delta \underline{r}^T \underline{\hat{p}} \right]_{t_i}^{t_f} - \left[\delta \underline{p}^T (\underline{r} - \underline{\hat{r}}) \right]_{t_i}^{t_f}.$$

Note that the boundary terms on the right-hand side of the principle are not ignored for this application and form an essential part of the solution.

Because the time history of the particle's trajectory was assumed, the left-hand side of the principle is readily integrated to yield

$$
\begin{aligned}
(\delta \underline{r}_f - \delta \underline{r}_i)^T \underline{p}_m + \delta \underline{r}_i^T \underline{G}_i + \delta \underline{r}_f^T \underline{G}_f \\
= \delta \underline{r}_f^T \hat{\underline{p}}_f - \delta \underline{r}_i^T \hat{\underline{p}}_i - \delta \underline{p}_f^T (\underline{r}_f - \hat{\underline{r}}_f) + \delta \underline{p}_i^T (\underline{r}_i - \hat{\underline{r}}_i),
\end{aligned}
\tag{8.49}
$$

where the following two quantities were defined

$$
\underline{G}_i = \frac{\Delta t}{2} \int_{-1}^{+1} \frac{1-\tau}{2} \underline{F}(\tau) \, d\tau, \quad \underline{G}_f = \frac{\Delta t}{2} \int_{-1}^{+1} \frac{1+\tau}{2} \underline{F}(\tau) \, d\tau.
$$

Equation (8.49) must be satisfied for all arbitrary variations $\delta \underline{p}_i$, $\delta \underline{p}_f$, $\delta \underline{r}_i$, and $\delta \underline{r}_f$, leading to the four equations of motion of the system

$$
\underline{r}_i = \hat{\underline{r}}_i, \quad \underline{r}_f = \hat{\underline{r}}_f,
\tag{8.50a}
$$

$$
-\underline{p}_m + \underline{G}_i + \hat{\underline{p}}_i = 0, \quad \underline{p}_m + \underline{G}_f - \hat{\underline{p}}_f = 0.
\tag{8.50b}
$$

These four equations involve six unknowns, $\hat{\underline{p}}_i$, $\hat{\underline{p}}_f$, $\hat{\underline{r}}_i$, $\hat{\underline{r}}_f$, $\underline{r}_i$, and $\underline{r}_f$; hence, two of these six quantities must be given if a solution is to be found.

Initial value problems are a common class of problems for which the initial position and velocity of the particle are given. In this example, it is assumed that the position vector, $\hat{\underline{r}}_i$, and velocity vector, $\hat{\underline{v}}_i = \hat{\underline{p}}_i/m$, of the particle are given at the beginning of the time step. Summing up the two eqs. (8.50b) yields $\underline{G}_i + \underline{G}_f + \hat{\underline{p}}_i - \hat{\underline{p}}_f = 0$; clearly,

$$
\underline{G} = \underline{G}_i + \underline{G}_f = \frac{\Delta t}{2} \int_{-1}^{+1} \underline{F}(\tau) \, d\tau,
$$

is the impulse of the externally applied force over the time step. It follows that $\underline{G} = \hat{\underline{p}}_f - \hat{\underline{p}}_i$: this equation expresses the principle of impulse and momentum applied to the particle.

Although the trajectory of the particle was approximated, this equation is exact, since the principle of impulse and momentum is a first integral of Newton's law. Because $\hat{\underline{p}}_i = m\hat{\underline{v}}_i$ and $\hat{\underline{p}}_f = m\hat{\underline{v}}_f$, this equation yields the velocity of the particle at the end of the time step as

$$
\hat{\underline{v}}_f = \hat{\underline{v}}_i + \underline{G}/m.
\tag{8.51}
$$

Next, subtracting the two eqs. (8.50b) yields $2\underline{p}_m + \underline{G}_f - \underline{G}_i - \hat{\underline{p}}_f - \hat{\underline{p}}_i = 0$. With the help of eq. (8.51), this relationship can be used to evaluate the final position of the particle as

$$
\hat{\underline{r}}_f = \hat{\underline{r}}_i + \Delta t \, \hat{\underline{v}}_i + \frac{\Delta t}{m} \underline{G}_i.
\tag{8.52}
$$

Finally, eqs. (8.52) and (8.51) can be recast as a matrix equation relating the initial position and velocity vectors of the particle to the corresponding quantities at the end of the time step

$$\left\{ \begin{matrix} \hat{r}_f \\ \Delta t\, \hat{\underline{v}}_f \end{matrix} \right\} = \begin{bmatrix} 1 & 1 \\ 0 & 1 \end{bmatrix} \left\{ \begin{matrix} \hat{r}_i \\ \Delta t\, \hat{\underline{v}}_i \end{matrix} \right\} + \frac{\Delta t}{m} \left\{ \begin{matrix} \underline{G}_i \\ \underline{G} \end{matrix} \right\}.$$

A recursive application of this formula for a number of consecutive time steps will yield the trajectory of the particle over any period of time. Of course, more accurate solutions will be obtained when smaller time steps are used, at the expense of increased computational cost.

More accurate, but also more complex, integration schemes can be obtained by assuming more complex trajectories of the particle within each time step. For instance, the particle's trajectory could be assumed to be parabolic: $\underline{r}(\tau) = -\tau(1 - \tau)\, \underline{r}_i/2 + (1 - \tau^2)\, \underline{r}_m + \tau(1 + \tau)\, \underline{r}_f/2$, where $\underline{r}_m$ is the position vector of the particle at $\tau = 0$, *i.e.*, at the mid-point of the time step for $t = (t_i + t_f)/2$. Following the procedure described above, discrete equations of motion will be found corresponding to this new approximation.

8.2.3 Problems

Problem 8.17. Equations of motion of rigid body
Prove that the equations of motion for a rigid body obtained from Hamilton's principle, eqs. (8.40), are identical to those obtained from d'Alembert's principle, eqs. (8.7).

Problem 8.18. Equations of motion of a rigid body
Consider the equations of motion of a rigid body, eqs. (8.37). *(1)* Show how these equations simplify if reference point **B** is chosen to be the center of mass of the rigid body, point **C**. Eliminate the momentum variables. Express the components of all tensors in the inertial basis. *(2)* Write the same equations with all tensors expressed in the body attached frame. *(3)* Let the orientation of the body attached frame coincide with the principal axes of inertia of the rigid body. In that case, the components the moment of inertia tensor in the material basis are denoted $\underline{\underline{I}}^{C*} = \mathrm{diag}(I_1^*, I_2^*, I_3^*)$. Show how the equations of motion expressed in the body attached frame further simplify to Euler's equations.

Problem 8.19. Rigid body subjected to gravity forces
Find the equations of motion of a rigid body subjected to gravity forces, as depicted in fig. 8.8. Vector $\underline{\eta}$ defines the position of the body's center of mass, point **C**, with respect to its reference point **B**. The potential of the gravity forces is $V = -mg^T \underline{r}_C$, where m is the total mass of the rigid body, $\underline{g}$ the components of the gravity acceleration vector, and $\underline{r}_C$ the components of the position vector of the center of mass in the present configuration, both resolved in $\mathcal{I}$.

Problem 8.20. Pendulum with sliding mass
Solve problem 6.34 using Hamilton's principle.

Problem 8.21. Pendulum with rotating mass
Solve problem 4.38 using Hamilton's principle.

Problem 8.22. Equations of motion of a rigid body in terms of Euler angles
The equations of motion for a rigid body, eqs. (8.40), are intrinsic equations, *i.e.*, no reference is made to a specific representation of finite rotations, the sole rotation tensor appears in the equations. Let the configuration of a rigid body be described with the following six generalized

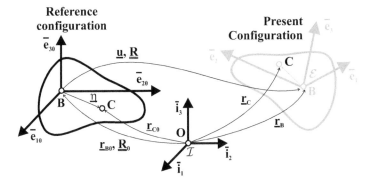

Fig. 8.8. Configuration of a rigid body with center of mass.

coordinates: the three components of the position vector, $\underline{u}$, of reference point **B**, and the three Euler angles ϕ, θ, and ψ using the *3-1-3* sequence that express the orientation of the body. *(1)* Express the kinetic energy of the body in terms of these generalized coordinates. *(2)* Write Hamilton's principle if the body is subjected to externally applied forces, $\underline{F}$, and moments, $\underline{M}_B$. *(3)* Derive the equations of motion of the system from Hamilton's principle. *(4)* Compare your results with eqs. (8.40); comment on the difference between the two sets of equations.

Problem 8.23. Rigid body connected to spring and dashpot

Consider a rigid body connected to the ground at point **B** by means of a spring of stiffness constant k and dashpot of constant c, as depicted in fig. 8.7 and discussed in example 8.8. The rigid body is of mass M and its moment of inertia tensor with respect to the center of mass is $\underline{I}^C$. Vector η defines the position of the center of mass with respect to point **B**. Frame $\mathcal{F}^B = \left[\mathbf{B}, \mathcal{B} = (\bar{b}_1, \bar{b}_2, \bar{b}_3)\right]$ is attached to the rigid body. The components of the rotation tensor that brings the inertial basis $\mathcal{I}$ to basis $\mathcal{B}$, resolved in basis $\mathcal{I}$, is denoted $\underline{R}$. Find the equations of motion of the system based on Hamilton's principle. *(1)* Solve the equations of motion of the problem using the momenta computed with respect to inertial point **O**. *(2)* On one graph, plot the time histories of the three components of vector $\bar{x}_C$. *(3)* On one graph, plot the Euler angles ψ, θ, and ϕ as a function of τ. *(4)* On one graph, plot the time histories of the three components of the velocity vector, $\bar{v}_C = v_C/(\Omega\ell_0)$. *(5)* On one graph, plot the time histories of the three components of the angular velocity vector, $\bar{\omega} = \omega/\Omega$. *(6)* On one graph, plot the time histories of the forces in the elastic spring and dashpot. *(7)* On one graph, plot the kinetic and potential energies of the system as well as the energy dissipated in the dashpot. Verify the energy closure equation. Treat the problem using a non-dimensional scheme with $\tau = \Omega t$, $\Omega^2 = k/M$, $\bar{x}_C = x_C/\ell_0$, and ℓ_0 the un-stretched length of the spring. Use the following data: $\bar{g} = g/(\ell_0\Omega^2) = 0.4$; $\zeta = c/(2M\Omega) = 0.1$; $\bar{\eta}^* = \eta^*/\ell_0 = [0.8, 1.25, -1.8]^T$; $\underline{I}^{C*} = \underline{I}^{C*}/(M\ell_0^2) = \text{diag}(1, 2.3, 1.5)$. Use the following initial conditions: $\bar{x}_B^T(\tau = 0) = [0, 1, 0]$; $q^T(\tau = 0) = [0, 0, 0]$ and the system is at rest. Present the response on the system for $\tau \in [0, 100]$. Hint: to avoid singularities, use Euler angles, ψ, θ, and ϕ, with the *3-2-1* sequence, as defined in section 4.11.3, to represent the rotation of the rigid body.

Problem 8.24. Change of frame for mass matrix

Consider two frames: an inertial frame $\mathcal{F}^I$ and a material frame $\mathcal{F}$. Let $\underline{M}^*$, as given by eq. (8.47), be the mass matrix of a rigid body expressed in the material frame. Show that mass matrix $\underline{M}$ defined as $\underline{M} = \underline{C}^{-T}\underline{M}^*\underline{C}^{-1}$, is the mass matrix of the rigid body in the inertial frame, if $\underline{\underline{C}}$ is the motion tensor that bring frame $\mathcal{F}^I$ to frame $\mathcal{F}$, resolved in $\mathcal{F}^I$. The above transformation performs two tasks: a change of reference point and a change of orthonormal basis. Note: it will be necessary to use the parallel axis theorem.

Problem 8.25. Higher order integration scheme
Generalize the time integration scheme derived in example 8.10 by approximating the trajectory of the particle as $\underline{r}(\tau) = -\tau(1 - \tau)\,\underline{r}_i/2 + (1 - \tau^2)\,\underline{r}_m + \tau(1 + \tau)\,\underline{r}_f/2$, where $\underline{r}_m$ is the position vector of the particle at $\tau = 0$, *i.e.*, at the mid-point of the time step for $t = (t_i + t_f)/2$.

8.3 Lagrange's formulation

As demonstrated in the previous sections, the equations of motions of dynamical systems can be derived from Hamilton's principle. The procedure involves integrations by parts of all the terms featuring variations of time derivatives of the generalized coordinates. Such derivation can become cumbersome when a large number of generalized coordinates is present; hence, it is desirable obtain the equations of motion of dynamical systems in a more systematic manner, bypassing many of the steps required by Hamilton's principle. Consider a system featuring n degrees of freedom and described by n generalized coordinates, q.

In general, the Lagrangian of the system will be a function of all generalized coordinates and their time derivatives, as well as time, $L = L(q, \dot{q}, t)$. Variation of the Lagrangian then becomes

$$\delta L = \frac{\partial L}{\partial q}^T \delta q + \frac{\partial L}{\partial \dot{q}}^T \delta \dot{q}. \tag{8.53}$$

Introducing this expression into Hamilton's principle, eq. (8.20), and ignoring the boundary terms leads to

$$\int_{t_i}^{t_f} \left(\frac{\partial L}{\partial q}^T \delta q + \frac{\partial L}{\partial \dot{q}}^T \delta \dot{q} + \delta W^{nc} \right) dt = 0. \tag{8.54}$$

The second term in the integrand involves variations of the generalized velocities, $\delta \dot{q}$, that are clearly not independent of the variations in the generalized coordinates, δq. To remedy this situation, this second term is integrated by parts

$$\int_{t_i}^{t_f} \frac{\partial L}{\partial \dot{q}}^T \delta \dot{q}\, dt = - \int_{t_i}^{t_f} \frac{d}{dt}\left(\frac{\partial L}{\partial \dot{q}}^T \right) \delta q\, dt + \left[\frac{\partial L}{\partial \dot{q}}^T \delta q \right]_{t_i}^{t_f}. \tag{8.55}$$

With the help of this result, Hamilton's principle becomes

$$\int_{t_i}^{t_f} \delta q^T \left[-\frac{d}{dt}\left(\frac{\partial L}{\partial \dot{q}} \right) + \frac{\partial L}{\partial q} + Q^{nc} \right] dt = 0, \tag{8.56}$$

for all arbitrary variations $\delta \underline{q}$. Here again, boundary terms are ignored; the virtual work done by the non-conservative forces was written as $\delta W^{nc} = \delta \underline{q}^T \underline{Q}^{nc}$, where $\underline{Q}^{nc}$ are the generalized, non-conservative forces acting on the system.

Because variations in generalized coordinates are arbitrary, the bracketed term must vanish, revealing *Lagrange's equations of motion*

$$\frac{d}{dt}\left(\frac{\partial L}{\partial \dot{\underline{q}}}\right) - \frac{\partial L}{\partial \underline{q}} = \underline{Q}^{nc}. \tag{8.57}$$

In the above developments, the partial derivatives of the Lagrangian with respect to the generalized velocities play a key role. These quantities, denoted $\underline{p}$, are called *generalized momenta* and are defined by eq. (8.25). With the help of these generalized momenta, Lagrange's equations of motion take the following form

$$\left(\dot{\underline{p}} - \frac{\partial K}{\partial \underline{q}}\right) + \frac{\partial V}{\partial \underline{q}} = \underline{Q}^{nc}.$$

The third term of this equation, $\partial V/\partial \underline{q}$, is associated with the conservative forces acting on the system: $\underline{Q}^c = -\partial V/\partial \underline{q}$.

Lagrange's equations now become

$$-\left(\dot{\underline{p}} - \frac{\partial K}{\partial \underline{q}}\right) + \underline{Q}^c + \underline{Q}^{nc} = 0.$$

The first term represents the inertial forces acting on the system; these forces are generalized forces, because they act in the configuration space. The following notation is introduced for the *generalized inertial forces*

$$\underline{Q}^I = -\left(\dot{\underline{p}} - \frac{\partial K}{\partial \underline{q}}\right). \tag{8.58}$$

Lagrange's equations are now cast in a simple form,

$$\underline{Q}^I + \underline{Q}^c + \underline{Q}^{nc} = 0,$$

and imply that the sum of all forces acting on the system must vanish at each instant in time. Clearly, Lagrange's equations are a *statement of dynamic equilibrium*; of course, inertial forces must be considered together with all externally applied forces for dynamic equilibrium conditions to be satisfied.

Example 8.11. Pendulum mounted on a cart

Figure 8.5 depicts a pendulum of length L and mass m mounted on a cart of mass M that is connected to the ground by means of a spring of stiffness constant k and of a dashpot of constant c. The displacement of the cart is denoted x, which is also the stretch of the spring, and angle θ measures the deflection of the pendulum with respect to the vertical. Gravity acts on the system as indicated in the figure. This problem was treated in example 8.4 on page 310 using Hamilton's principle. Derive the equations of motion of the system using Lagrange's formulation.

The kinetic and potential energies of the system are given by eqs. (8.28) and (8.29), respectively. The Lagrangian of the system then becomes

$$L = \frac{1}{2}(M+m)\dot{x}^2 + \frac{1}{2}\frac{mL^2}{3}\dot{\theta}^2 + \frac{mL}{2}\dot{x}\dot{\theta}C_\theta - \frac{1}{2}kx^2 - \frac{mgL}{2}(1 - C_\theta).$$

The system's generalized momenta are found as

$$p_x = \frac{\partial L}{\partial \dot{x}} = (M+m)\dot{x} + \frac{mL}{2}\dot{\theta}C_\theta, \quad p_\theta = \frac{\partial L}{\partial \dot{\theta}} = \frac{mL^2}{3}\dot{\theta} + \frac{mL}{2}\dot{x}C_\theta.$$

The derivatives of the Lagrangian with respect to the generalized coordinates are

$$\frac{\partial L}{\partial x} = -kx, \quad \frac{\partial L}{\partial \theta} = -\frac{mL}{2}\dot{x}\dot{\theta}S_\theta - \frac{mgL}{2}S_\theta.$$

Finally, the virtual work done the non-conservative forces is $\delta W^{nc} = -c\dot{x}\delta x$, and hence, the generalized forces are $Q_x = -c\dot{x}$ and $Q_\theta = 0$.

Lagrange's formulation then yield the equations of motion of the system

$$(M+m)\ddot{x} + \frac{mL}{2}(\ddot{\theta}C_\theta - \dot{\theta}^2 S_\theta) + kx = -c\dot{x};$$

$$\frac{mL^2}{3}\ddot{\theta} + \frac{mL}{2}(\ddot{x}C_\theta - \dot{x}\dot{\theta}S_\theta) + \frac{mL}{2}\dot{x}\dot{\theta}S_\theta + mg\frac{L}{2}S_\theta = 0.$$

These equations are, of course, identical to those obtained earlier, see eqs. (8.30), using Hamilton's principle. Clearly, Lagrange's formulation provides an effective procedure for deriving the system's equations of motion. The integrations by parts associated with the application of Hamilton's principle are completely bypassed in this approach.

Example 8.12. Swiveling plate

Figure 6.10 depicts a homogeneous, rectangular plate of height a, width b, and mass m connected to the ground by a rigid, massless link of length d, as discussed in example 6.2 on page 219. At point $\mathbf{O}$, a bearing allows the link to rotate with respect to unit vector $\bar{\imath}_3$, and at point $\mathbf{B}$, the plate is free to rotate with respect to the link about axis $\bar{a}_1$. Three frames will be used in this problem: the inertial frame, $\mathcal{F}^I = [\mathbf{O}, \mathcal{I} = (\bar{\imath}_1, \bar{\imath}_2, \bar{\imath}_3)]$, a frame attached to the link, $\mathcal{F}^A = [\mathbf{O}, \mathcal{A}^+ = (\bar{a}_1, \bar{a}_2, \bar{a}_3)]$, and finally, a frame attached to the plate at its center of mass, $\mathcal{F}^P = [\mathbf{C}, \mathcal{B}^* = (\bar{b}_1, \bar{b}_2, \bar{b}_3)]$. A planar rotation of magnitude α about axis $\bar{\imath}_3$ brings basis $\mathcal{I}$ to $\mathcal{A}^+$, and a planar rotation of magnitude β about axis $\bar{a}_1$ brings basis $\mathcal{A}^+$ to $\mathcal{B}^*$. Rotation tensors $\underline{\underline{R}}_\alpha$ and $\underline{\underline{R}}_\beta$ denotes these two planar rotations, respectively; tensor components resolved in bases $\mathcal{A}^+$ and $\mathcal{B}^*$ are denoted with superscripts $(\cdot)^+$ and $(\cdot)^*$, respectively.

The inertial angular velocity vector of the plate is readily found with the help of the addition theorem as $\underline{\omega} = \dot{\alpha}\,\bar{a}_3 + \dot{\beta}\,\bar{a}_1$; the components of this vector in basis $\mathcal{B}^*$ then become $\underline{\omega}^{*T} = \{\dot{\beta}, \dot{\alpha}S_\beta, \dot{\alpha}C_\beta\}$. The inertial position of the center of mass of the plate is $\underline{r}_C = (d + a/2)\,\bar{a}_1$ and the velocity vector $\underline{v}_C = (d + a/2)\dot{\alpha}\,\bar{a}_2$. Body attached frame $\mathcal{F}^P$ is located at the center of mass of the plate and is aligned with its principal axes of inertia. The principal moments of inertia are given in fig. 6.45 as $I_{11}^{*C} = mb^2/12$, $I_{22}^{*C} = ma^2/12$, $I_{33}^{*C} = m(a^2 + b^2)/12$. The kinetic energy of the system becomes

$$K = \frac{1}{2}m(d + \frac{a}{2})^2\dot{\alpha}^2 + \frac{1}{2}\left[m\frac{b^2}{12}\dot{\beta}^2 + m\frac{a^2}{12}\dot{\alpha}^2 S_\beta^2 + m\frac{a^2 + b^2}{12}\dot{\alpha}^2 C_\beta^2\right],$$

where $S_\beta = \sin\beta$ and $C_\beta = \cos\beta$.

The potential energy associated with the forces of gravity is easily found as $V = -mg(d + a/2)S_\alpha$, and the Lagrangian of the system becomes

$$L = \frac{1}{2}m(d + \frac{a}{2})^2\dot{\alpha}^2 + \frac{m}{2}\left[\frac{b^2}{12}\dot{\beta}^2 + \frac{a^2}{12}\dot{\alpha}^2 + \frac{b^2}{12}\dot{\alpha}^2 C_\beta^2\right] + mg(d + \frac{a}{2})S_\alpha.$$

The generalized momenta of the system are found by taking derivatives of the Lagrangian with respect to the generalized velocities $\dot{\alpha}$ and $\dot{\beta}$ to find

$$p_\alpha = \frac{\partial L}{\partial\dot{\alpha}} = m\left[(d + \frac{a}{2})^2 + \frac{a^2}{12} + \frac{b^2}{12}C_\beta^2\right]\dot{\alpha}, \quad \text{and} \quad p_\beta = \frac{\partial L}{\partial\dot{\beta}} = m\frac{b^2}{12}\dot{\beta},$$

respectively. The partial derivatives of the Lagrangian with respect to the generalized coordinates α and β are

$$\frac{\partial L}{\partial\alpha} = mg(d + \frac{a}{2})C_\alpha \quad \text{and} \quad \frac{\partial L}{\partial\beta} = -m\frac{b^2}{12}\dot{\alpha}S_\beta C_\beta,$$

respectively.

Lagrange's formulation then yield the equations of motion of the system

$$\frac{d}{dt}\left\{m\left[(d + \frac{a}{2})^2 + \frac{a^2}{12} + \frac{b^2}{12}C_\beta^2\right]\dot{\alpha}\right\} - mg(d + \frac{a}{2})C_\alpha = 0,$$

$$\frac{d}{dt}\left\{m\frac{b^2}{12}\dot{\beta}\right\} + m\frac{b^2}{12}\dot{\alpha}S_\beta C_\beta = 0.$$

It is readily verified that these equations of motion are identical to those obtained from Newton's approach, eqs. (6.34).

Example 8.13. The double pendulum with elastic joint

Figure 6.17 depicts a double pendulum comprising bar 1 of mass m_1 and length ℓ_1, and bar 2 of mass m_2 and length ℓ_2, as treated in example 6.5 on page 231. Let frame $\mathcal{F}^A = [\mathbf{A}, \mathcal{A} = (\bar{a}_1, \bar{a}_2)]$ be attached to bar 1 and frame $\mathcal{F}^E = [\mathbf{E}, \mathcal{E} = (\bar{e}_1, \bar{e}_2)]$ be attached to bar 2. A massless tube allows bar 2 to slide in the direction of $\bar{a}_2$; the slider has a mass M and is connected to bar 1 at point $\mathbf{A}$ by means of a spring of stiffness constant k. The position of the slider is determined by its distance, x, from point $\mathbf{A}$, the tip of bar 1; the angular positions of the two bars with respect to the vertical are denoted θ_1 and θ_2, respectively. The system is subjected to gravity along the inertial direction $\bar{\imath}_1$. Derive the equations of motion of the system.

This problem features three degrees of freedom and will be described by means of three generalized coordinates: angles θ_1 and θ_2 giving the angular positions of the two bars and the sliding distance, x. The angular velocity of bar 1 is $\dot{\theta}_1$ and hence, its kinetic energy is $K_1 = 1/2\, m_1 \ell_1^2 \dot{\theta}_1^2 / 3$. The position vector of mass M is $\underline{r}_M = \ell_1 \bar{a}_1 + x \bar{a}_2$ and its kinetic energy then follows as

$$
K_M = \frac{1}{2} M \left[(\ell_1^2 + x^2) \dot{\theta}_1^2 + \dot{x}^2 + 2\ell_1 \dot{x} \dot{\theta}_1 \right].
$$

Similarly, the position vector of the center of mass of bar 2 is $\underline{r}_2 = \ell_1 \bar{a}_1 + x \bar{a}_2 + \ell_2 / 2\, \bar{e}_1$ and its kinetic energy becomes

$$
K_2 = \frac{1}{2} m_2 \left[(\ell_1^2 + x^2) \dot{\theta}_1^2 + \dot{x}^2 + 2\ell_1 \dot{x} \dot{\theta}_1 + \frac{\ell_2^2}{4} \dot{\theta}_2^2 \right.
$$
$$
\left. + \ell_2 x \dot{\theta}_1 \dot{\theta}_2 S_{21} + \ell_2 (\dot{x} + \ell_1 \dot{\theta}_1) \dot{\theta}_2 S_{21} \right] + \frac{1}{2} \frac{m_2 \ell_2^2}{12} \dot{\theta}_2^2,
$$

where the following notation was introduced: $S_{21} = \sin(\theta_2 - \theta_1)$ and $C_{21} = \cos(\theta_2 - \theta_1)$. The last term of this expression represent the kinetic energy associated with the angular motion of the bar.

The strain energy for the elastic spring is simply $V_s = 1/2\, kx^2$. The potential energy of the gravity forces is

$$
V_g = -m_1 g \bar{\imath}_1^T (\frac{\ell_1}{2} \bar{a}_1) - M g \bar{\imath}_1^T (\ell_1 \bar{a}_1 + x \bar{a}_2) - m_2 g \bar{\imath}_1^T (\ell_1 \bar{a}_1 + x \bar{a}_2 + \frac{\ell_2}{2} \bar{e}_1).
$$

Combining kinetic, strain, and potential energies yields the Lagrangian of the system as

$$
L = \frac{1}{2} \left[(\frac{m_1}{3} + M_2) \ell_1^2 \dot{\theta}_1^2 + M_2 (\dot{x}^2 + x^2 \dot{\theta}_1^2) + \frac{m_2 \ell_2^2}{3} \dot{\theta}_2^2 \right.
$$
$$
\left. + 2M_2 \ell_1 \dot{x} \dot{\theta}_1 + m_2 \ell_2 \dot{x} \dot{\theta}_2 C_{21} + m_2 \ell_2 \dot{\theta}_1 \dot{\theta}_2 (\ell_1 C_{21} + x S_{21}) \right]
$$
$$
- \frac{1}{2} kx^2 + (\frac{m_1}{2} + M_2) g \ell_1 C_1 - M_2 gx S_1 + \frac{m_2 \ell_2}{2} g C_2,
$$

where the following notation was introduced: $S_1 = \sin\theta_1$ and $C_1 = \cos\theta_1$, with similar conventions for angle θ_2, and $M_2 = M + m_2$.

The generalized momenta of the system are found by taking derivatives of the Lagrangian with respect to the generalized velocities $\dot{\theta}_1$, $\dot{\theta}_2$, and $\dot{x}$ to find

$$
p_1 = \frac{\partial L}{\partial \dot{\theta}_1} = (\frac{m_1}{3} + M_2) \ell_1^2 \dot{\theta}_1 + M_2 x^2 \dot{\theta}_1 + M_2 \ell_1 \dot{x} + \frac{m_2 \ell_2}{2} \dot{\theta}_2 (\ell_1 C_{21} + x S_{21}),
$$

$$
p_2 = \frac{\partial L}{\partial \dot{\theta}_2} = \frac{m_2 \ell_2^2}{3} \dot{\theta}_2 + \frac{m_2 \ell_2}{2} \dot{x} C_{21} + \frac{m_2 \ell_2}{2} \dot{\theta}_1 (\ell_1 C_{21} + x S_{21}),
$$

$$
p_3 = \frac{\partial L}{\partial \dot{x}} = M_2 (\dot{x} + \ell_1 \dot{\theta}_1) + \frac{m_2 \ell_2}{2} \dot{\theta}_2 C_{21},
$$

respectively. The partial derivatives of the Lagrangian with respect to the generalized coordinates θ_1, θ_2, and x are

$$\frac{\partial L}{\partial \theta_1} = \frac{m_2\ell_2}{2}\dot{\theta}_2\left[\dot{x}S_{21} + \dot{\theta}_1(\ell_1 S_{21} - xC_{21})\right] - (\frac{m_1}{2} + \mathcal{M}_2)g\ell_1 S_1 + \mathcal{M}_2 gxC_1,$$

$$\frac{\partial L}{\partial \theta_2} = -\frac{m_2\ell_2}{2}\dot{x}\dot{\theta}_2 S_{21} - \frac{m_2\ell_2}{2}\dot{\theta}_1\dot{\theta}_2(\ell_1 S_{21} - xC_{21}) - \frac{m_2\ell_2}{2}gS_2,$$

$$\frac{\partial L}{\partial x} = \mathcal{M}_2 x\dot{\theta}_1^2 + \frac{m_2\ell_2}{2}\dot{\theta}_1\dot{\theta}_2 S_{21} - kx - \mathcal{M}_2 gS_1,$$

respectively.

Lagrange's formulation then yields the equations of motion of the problem as

$$
\begin{bmatrix}
(\frac{m_1}{3} + \mathcal{M}_2)\ell_1^2 + \mathcal{M}_2 x^2 & \frac{m_2\ell_2}{2}(\ell_1 C_{21} + xS_{21}) & \mathcal{M}_2\ell_1 \\
\frac{m_2\ell_2}{2}(\ell_1 C_{21} + xS_{21}) & \frac{m_2\ell_2^2}{3} & \frac{m_2\ell_2}{2}C_{21} \\
\mathcal{M}_2\ell_1 & \frac{m_2\ell_2}{2}C_{21} & \mathcal{M}_2
\end{bmatrix}
\begin{Bmatrix}
\ddot{\theta}_1 \\ \ddot{\theta}_2 \\ \ddot{x}
\end{Bmatrix}
$$

$$
+ \begin{Bmatrix}
2\mathcal{M}_2 x\dot{x}\dot{\theta}_1 - \frac{m_2\ell_2}{2}(\ell_1 S_{21} - xC_{21})\dot{\theta}_2^2 \\
m_2\ell_2\dot{x}\dot{\theta}_1 S_{21} + \frac{m_2\ell_2}{2}(\ell_1 S_{21} - xC_{21})\dot{\theta}_1^2 \\
-\mathcal{M}_2 x\dot{\theta}_1^2 - \frac{m_2\ell_2}{2}\dot{\theta}_2^2 S_{21} + kx
\end{Bmatrix}
$$

$$
+ \begin{Bmatrix}
(\frac{m_1}{2} + \mathcal{M}_2)g\ell_1 S_1 + \mathcal{M}_2 gxC_1 \\
\frac{m_2\ell_2}{2}gS_2 \\
\mathcal{M}_2 gS_1
\end{Bmatrix} = 0.
$$

This Lagrangian approach to the problem should be contrasted with Newton's formulation discussed in example 6.5. As shown in fig. 6.17, Newton's formulation involves the internal forces at the joint, F_s and R, and the reaction forces at point **O**, V_O and H_O. With Lagrange's formulation, the internal and reaction forces are eliminated because the work they perform vanishes and the elastic force in the spring is taken into account by the strain energy of the spring.

Example 8.14. The milling machine

Consider the simplified model of a milling machine depicted in fig. 6.28. The tool support is a rigid body of mass m and moment of inertia I^O with respect to point **O** connected to the ground at point **O**. Its center of mass is located at point **A**, which is at a distance ℓ_1 from point **O**. A torsional spring of stiffness constant k_θ and un-stretched rotation θ_0, and a torsional dashpot of constant c_θ act at point **O**. Let frame $\mathcal{F}^E = [\mathbf{O}, \mathcal{E} = (\bar{e}_1, \bar{e}_2)]$ be attached to the tool support; the angle between unit vectors $\bar{\imath}_1$ and $\bar{e}_1$ is denoted θ. A massless, rigid bar **DB** of length ℓ_2 is free to slide inside the tool support. A spring of stiffness constant k_x and the un-stretched length x_0, and a dashpot of constant c_x connect the tool support at point **A** to the bar at point **D**.

At point **B**, the bar connects to the milling machine tool, which is free to rotate about point **B**. Let frame $\mathcal{F}^B = \left[\mathbf{B}, \mathcal{B} = (\bar{b}_1, \bar{b}_2)\right]$ be attached to the tool, which center of mass **C** is located a distance d along axis $\bar{b}_1$. The tool rotates at a constant angular velocity, Ω, with respect to the bar, such that the angle between unit vectors $\bar{e}_1$ and $\bar{b}_1$ is $\phi = \Omega t$. The tool is of mass M and moment of inertia I^B with respect to point **B**. Derive the equations of motion of the system using Lagrange's formulation and the following generalized coordinates: θ and x, the distance from point **A** to **D**.

The kinetic energy of the system consists of two parts: the kinetic energy of the tool support and that of the tool. Since point **O** is an inertial point, the kinetic energy of the tool support is simply $I^O \dot{\theta}^2/2$. The kinetic energy of the tool is the sum of its translational and rotational components. The position vector of the center of mass of the tool is $\underline{r}_C = (\ell_1 + \ell_2 + x)\,\bar{e}_1 + d\,\bar{b}_1$ and its velocity then becomes $\underline{v}_C = \dot{x}\,\bar{e}_1 + (\ell_1 + \ell_2 + x)\dot{\theta}\,\bar{e}_2 + d(\dot{\theta} + \Omega)\,\bar{b}_2$. Because the angular velocity of the tool is $(\dot{\theta} + \Omega)$, the kinetic energy of the system becomes

$$
\begin{aligned}
K &= \frac{1}{2}I^O\dot{\theta}^2 + \frac{1}{2}M\|\underline{v}_C\|^2 + \frac{1}{2}I^C(\dot{\theta} + \Omega)^2 \\
&= \frac{1}{2}\left[I^O + M(\ell_{12} + x)^2 + 2Md(\ell_{12} + x)C_\phi\right]\dot{\theta}^2 + \frac{1}{2}M\dot{x}^2 + \frac{1}{2}I^B(\dot{\theta} + \Omega)^2 \\
&\quad - Md\dot{x}(\dot{\theta} + \Omega)S_\phi + Md(\ell_{12} + x)\Omega\dot{\theta}C_\phi,
\end{aligned}
$$

where $S_\phi = \sin\phi$, $C_\phi = \cos\phi$, and $\ell_{12} = \ell_1 + \ell_2$. The parallel axis theorem was used to define $I^B = I^C + Md^2$, where I^B is the moment of inertia of the tool evaluated with respect to point **B**.

The potential of the gravity forces is easily found as $V = -mg\ell_1 C_\theta - Mg\left[(\ell_{12} + x)C_\theta + dC_{\theta+\phi}\right]$, where $S_\theta = \sin\theta$, $C_\theta = \cos\theta$, $S_{\theta+\phi} = \sin(\theta + \phi)$, and $C_{\theta+\phi} = \cos(\theta + \phi)$. The Lagrangian of the system now becomes

$$
\begin{aligned}
L &= \frac{1}{2}\left[I^O + M(\ell_{12} + x)^2 + 2Md(\ell_{12} + x)C_\phi\right]\dot{\theta}^2 + \frac{1}{2}M\dot{x}^2 \\
&\quad + \frac{1}{2}I^B(\dot{\theta} + \Omega)^2 - Md\dot{x}(\dot{\theta} + \Omega)S_\phi + Md(\ell_{12} + x)\Omega\dot{\theta}C_\phi \\
&\quad + mg\ell_1 C_\theta + Mg\left[(\ell_{12} + x)C_\theta + dC_{\theta+\phi}\right].
\end{aligned}
$$

The generalized momenta of the system are found by taking the partial derivatives of the Lagrangian with respect to the generalized velocities to find

$$
\begin{aligned}
p_\theta = \frac{\partial L}{\partial \dot{\theta}} &= \left[I^O + M(\ell_{12} + x)^2 + 2Md(\ell_{12} + x)C_\phi\right]\dot{\theta} \\
&\quad + I^B(\dot{\theta} + \Omega) - Md\dot{x}S_\phi + Md(\ell_{12} + x)\Omega C_\phi,
\end{aligned}
$$

and

$$
p_x = \frac{\partial L}{\partial \dot{x}} = M\dot{x} - Md(\dot{\theta} + \Omega)S_\phi.
$$

Finally, the derivatives of the Lagrangian with respect to the generalized coordinates are found as

$$\frac{\partial L}{\partial \theta} = -mg\ell_1 S_\theta - Mg\left[(\ell_{12} + x)S_\theta + dS_{\theta+\phi}\right],$$

and

$$\frac{\partial L}{\partial x} = M(\ell_{12} + x)\dot\theta^2 + Md\dot\theta^2 C_\phi + Md\Omega\dot\theta C_\phi + MgC_\theta.$$

The governing equations of motion of the system then follow from Lagrange's equation as

$$\begin{bmatrix} I^O + I^B + M(\ell_{12} + x)^2 + 2Md(\ell_{12} + x)C_\phi & -MdS_\phi \\ -MdS_\phi & M \end{bmatrix} \begin{Bmatrix} \ddot\theta \\ \ddot x \end{Bmatrix}$$
$$+ \begin{Bmatrix} 2M\left[(\ell_{12} + x) + dC_\phi\right]\dot x\dot\theta - Md(\ell_{12} + x)\Omega S_\phi(2\dot\theta + \Omega) \\ -M(\ell_{12} + x)\dot\theta^2 - Md(\dot\theta + \Omega)^2 C_\phi \end{Bmatrix}$$
$$+ \begin{Bmatrix} Mg\left[(\ell_{12} + x)S_\theta + dS_{\theta+\phi}\right] + mg\ell_1 S_\theta - Q_\theta \\ -MgC_\theta - Q_x \end{Bmatrix} = 0.$$

To evaluate the generalized forces, Q_θ and Q_x, the virtual work done by the externally applied forces is computed. The linear spring/dashpot system applies forces of equal magnitude and opposite directions at point **A** and **D**; the virtual work is then $\delta W = \delta \underline{r}_A^T(F_{sd}\bar{e}_1) + \delta \underline{r}_D^T(-F_{sd}\bar{e}_1)$. Since $\underline{r}_A = \ell_1 \bar{e}_1$ and $\underline{r}_D = (\ell_1 + x)\bar{e}_1$, it follows that $\delta W = \ell_1 \delta\theta \bar{e}_2^T(F_{sd}\bar{e}_1) + \left[\delta x\bar{e}_1^T + (\ell_1 + x)\delta\theta\bar{e}_2^T\right](-F_{sd}\bar{e}_1) = -\delta x F_{sd}$. Hence, $Q_x = -F_{sd}$, where $F_{sd} = k_x(x - x_0) + c_x\dot x$. Similarly, the virtual work done by the torsional spring/dashpot is simply $\delta W = \delta\theta(-M_{sd})$, leading to the generalized force $Q_\theta = -M_{sd}$, where $M_{sd} = k_\theta(\theta - \theta_0) + c_\theta\dot\theta$. The elastic springs of stiffness constants k_θ and k_x could have been accounted for through strain energy expressions, $V = 1/2\, k_\theta(\theta - \theta_0)^2$ and $1/2\, k_x(x - x_0)^2$, respectively. In this particular case, however, it is convenient to treat the spring/dashpot components as single units and compute the virtual work done by both elastic and viscous forces simultaneously.

Example 8.15. The quick return mechanism

The quick return mechanism shown in fig. 8.9 consists of a uniform crank of length L_c and mass m_c, and of a uniform arm of length L_a and mass m_a. The crank is pinned at point **R** and the arm at point **O**; the distance between these two points is denoted d. At point **S**, a slider allows the tip of the crank to slide along the arm. A mass M is attached at point **T**, the tip of the arm. A spring of stiffness constant k connects point **T** to inertial point **A**; the spring is un-stretched when the arm is in the vertical position. Use Lagrange's formulation to derive the equations of motion of the problem.

This is a single degree of freedom problem, and angle θ will be selected as the single generalized coordinate. The kinematics of the problem are addressed first. Considering triangle **ORS**, it is clear that $\beta = \phi - \theta$, and the law of sines then yields $L_c \sin(\phi - \theta) = d\sin\phi$. Solving for angle ϕ leads to

$$\tan\phi = -\frac{\bar{L}_c S_\theta}{1 - \bar{L}_c C_\theta}, \quad C_\phi = -\frac{1 - \bar{L}_c C_\theta}{\sqrt{2(1 - \bar{L}_c C_\theta) - (1 - \bar{L}_c^2)}},$$

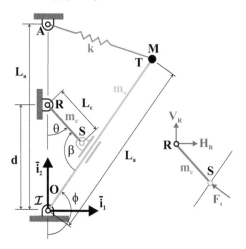

Fig. 8.9. Quick return mechanism.

where $\bar{L}_c = L_c/d$, $\sin\theta = S_\theta$, $\cos\theta = C_\theta$, and $\cos\phi = C_\phi$. A time derivative of the first equation yields the relationship between the angular velocities of the two bar,

$$\dot{\phi} = \frac{(1 - \bar{L}_c C_\theta) - (1 - \bar{L}_c^2)}{2(1 - \bar{L}_c C_\theta) - (1 - \bar{L}_c^2)}\dot{\theta} = h(\theta)\dot{\theta}.$$

The system's kinetic energy is $K = [m_c L_c^2 \dot{\theta}^2/3 + (M + m_a/3)L_a^2\dot{\phi}^2]/2$, where the first term represents the kinetic energy of the crank and the second that of the arm. The following non-dimensional time is introduced: $\tau = \omega t$, where $\omega^2 = k/M$, and notation $(\cdot)'$ indicates a derivative with respect to τ. The system's non-dimensional kinetic energy now becomes

$$\bar{K} = \frac{K}{kd^2} = \frac{1}{2}\left[\frac{\mu_c \bar{L}_c^2}{3} + (1 + \frac{\mu_a}{3})\bar{L}_a^2 h^2\right]\theta'^2,$$

where $\bar{L}_a = L_a/d$, $\mu_a = m_a/M$, and $\mu_c = m_c/M$.

The potential energy of the spring is $V = 1/2\ k\Delta^2$, where Δ is the stretch of the spring. The law of cosines applied to triangle **OMA** yields $\Delta^2 = L_a^2 + L_a^2 - 2L_a^2\cos(\pi - \phi) = 2L_a^2(1 + C_\phi)$. The non-dimensional potential of the spring then becomes

$$\bar{V} = \frac{V}{kd^2} = \bar{L}_a^2(1 + C_\phi).$$

The non-dimensional Lagrangian of the system is $\bar{L} = L/(kd^2) = \bar{K} - \bar{V}$, and the generalized momentum is

$$p_{\theta'} = \frac{\partial\bar{L}}{\partial\theta'} = \left[\frac{\mu_c \bar{L}_c^2}{3} + (1 + \frac{\mu_a}{3})\bar{L}_a^2 h^2\right]\theta'.$$

The derivative of the Lagrangian with respect to the generalized coordinate θ is

$$\frac{\partial \bar{L}}{\partial \theta} = \left(1 + \frac{\mu_a}{3}\right)\bar{L}_a^2 \theta'^2 h \frac{\partial h}{\partial \theta} - \bar{L}_a^2 \frac{\partial C_\phi}{\partial \theta},$$

where

$$\frac{\partial h}{\partial \theta} = \frac{(1 - \bar{L}_c^2)\bar{L}_c S_\theta}{\left[2(1 - \bar{L}_c C_\theta) - (1 - \bar{L}_c^2)\right]^2}, \quad \frac{\partial C_\phi}{\partial \theta} = -\frac{h\bar{L}_c S_\theta}{\sqrt{2(1 - \bar{L}_c C_\theta) - (1 - \bar{L}_c^2)}}.$$

Finally, the single equation of motion of the problem is found from Lagrange's formulation as

$$\left[\frac{\mu_c \bar{L}_c^2}{3\bar{L}_a^2} + \left(1 + \frac{\mu_a}{3}\right)h^2\right]\theta'' + \left(1 + \frac{\mu_a}{3}\right)\theta'^2 h \frac{\partial h}{\partial \theta} + \frac{\partial C_\phi}{\partial \theta} = 0. \tag{8.61}$$

Although the quick return mechanism is a simple mechanical system that features a single degree of freedom, the system's equation of motion is complex and extensive algebraic manipulations are required for its derivation.

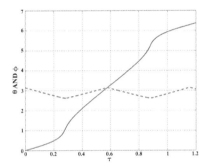

Fig. 8.10. Angular position of the two bars, θ, solid line, and ϕ, dashed line.

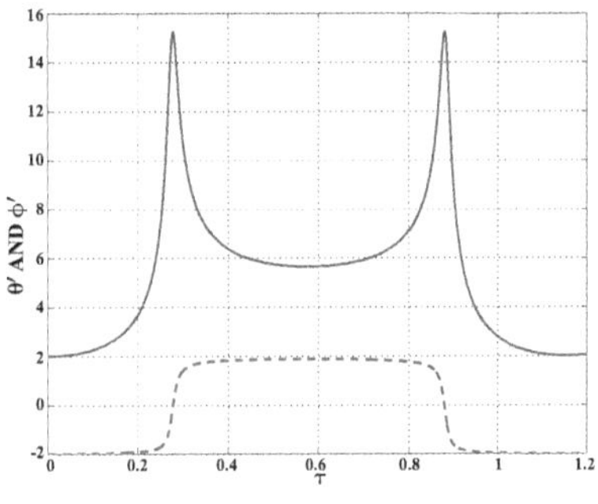

Fig. 8.11. Angular velocities of the two bar, θ', solid line, and ϕ', dashed line.

The system was simulated for the following values of the non-dimensional parameters of the system: $\bar{L}_a = 3$, $\bar{L}_c = 0.5$, $\mu_a = 1$, $\mu_c = 2.4$. At the initial time, the crank is in the vertical position, $\theta = 0$ and $\theta' = 2$. The system was simulated for $\tau \in [0, 1.2]$, which corresponds approximately to one complete revolution of the crank. Figure 8.10 shows the time histories of the angular positions of the two bars, θ and ϕ, and their angular velocities are depicted in fig. 8.11. At times $\tau = 0.2762$ and 0.8806, the angular velocity of the crank change very rapidly and that of the arm quickly changes sign. This corresponds to the configuration of the system when the crank is perpendicular to the arm, forcing a quick reversal of the arm's direction of motion.

Figure 8.12 shows the evolution of the system's kinetic and potential energies. Because the system is conservative, the total mechanical energy of the system remains constant, as also depicted in the figure.

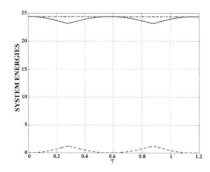

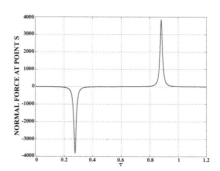

Fig. 8.12. System energies: kinetic energy, solid line; potential energy, dashed line; total mechanical energy, dashed-dotted line.

Fig. 8.13. Non-dimensional magnitude of the force at point **S**.

The reaction and internal forces of the system are automatically eliminated when using Lagrange's formulation. It is often important, however, to evaluate both reaction and internal force. Of particular interest in this case is the internal force in the slider. Because this force was eliminated from the formulation, Lagrange's equations provide no information concerning these forces.

Figure 8.9 shows a free body diagram of the crank; at its tip, the force exerted by the slider on the crank, denoted F_s, is acting in the direction normal to the arm. For this planar system, the pivot equation written with respect to point **R** yields $m_c L_c^2 \ddot{\theta}/3 = -F_s \cos(\pi - \phi + \theta)$. In non-dimensional form, this becomes

$$\bar{F}_s = \frac{F_s}{k L_c} = \frac{\mu_c \bar{L}_c}{3} \frac{\sqrt{2(1 - \bar{L}_c C_\theta) - (1 - \bar{L}_c^2)}}{(1 - \bar{L}_c C_\theta) - (1 - \bar{L}_c^2)} \theta''.$$

Figure 8.13 shows the time history of the force the slider applies on the crank. At times $\tau = 0.2762$ and 0.8806, large magnitudes of this force is observed. At these two instants, the crank is normal to the arm, which reverses the direction of its motion, leading to large accelerations and hence, large forces.

Example 8.16. Equations of motion of a rigid body

Figure 8.14 depicts a rigid body defined by frame $\mathcal{F}^B = \left[\mathbf{B}, \mathcal{B}^* = (\bar{b}_1, \bar{b}_2, \bar{b}_3)\right]$, where point **B** is a reference point on the body and $\mathcal{B}^*$ a body attached basis. The displacement vector, $\underline{u}$, defines the location of the reference point **B** of the rigid body with respect to inertial frame $\mathcal{F}^I = [\mathbf{O}, \mathcal{I} = (\bar{\imath}_1, \bar{\imath}_2, \bar{\imath}_3)]$. Let $\underline{R}$ be the rotation tensor that bring basis $\mathcal{I}$ to basis $\mathcal{B}^*$, resolved in basis $\mathcal{I}$. The body is subjected to external forces, $\underline{F}$, and external moments, $\underline{M}_B$, computed with respect to the reference point of the body. Find the equations of motion of the rigid body using Lagrange's formulation and contrast the resulting equations with those obtained in example 8.6 using Hamilton's principle.

In example 8.6, the kinetic energy of the rigid body was derived as eq. (8.31), where the 6×6 mass matrix, $\underline{\underline{M}}_B^*$, of the rigid body is given by eq. (8.32) and the velocity array, $\underline{V}^*$, by eq. (8.33). Finally, the components of the linear and angular

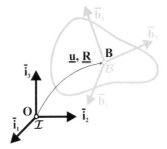

Fig. 8.14. Configuration of a rigid body.

momentum vectors of the body were defined by eq. (8.34). The generalized coordinates of the system are the three displacements, $\underline{u}$, of the reference point of the body and three parameters, $\underline{q}$, defining rotation tensor $\underline{\underline{R}}$. These parameters could be a set of Euler angles with a specific sequence, as defined in section 4.11. For this problem, the Lagrangian of the system simply equals its kinetic energy, $L = K$.

The generalized momenta of the system are computed with the help of the chain rule for derivatives

$$\frac{\partial L}{\partial \underline{u}} = \left[\frac{\partial \underline{\mathcal{V}}^*}{\partial \underline{u}}\right]^T \frac{\partial K}{\partial \underline{\mathcal{V}}^*} = \{\underline{\underline{R}}\ \underline{\underline{0}}\}\underline{\mathcal{P}}^* = \underline{\underline{R}}\,\underline{p}^* = \underline{p},$$

$$\frac{\partial L}{\partial \underline{\dot{q}}} = \left[\frac{\partial \underline{\mathcal{V}}^*}{\partial \underline{\dot{q}}}\right]^T \frac{\partial K}{\partial \underline{\mathcal{V}}^*} = \{\underline{\underline{0}}\ \underline{\underline{H}}^T\underline{\underline{R}}\}\underline{\mathcal{P}}^* = \underline{\underline{H}}^T\underline{\underline{R}}\,\underline{h}^* = \underline{\underline{H}}^T\underline{h},$$

where $\underline{p} = \underline{\underline{R}}\,\underline{p}^*$ and $\underline{h} = \underline{\underline{R}}\,\underline{h}^*$ are the components of linear and angular momentum vectors of the body, both resolved in basis $\mathcal{I}$. As expected, the generalized momentum associated with the generalized velocities $\underline{\dot{u}}$ are the components of the linear momentum vector of the rigid body in the inertial frame. To compute the generalized momenta associated with the generalized velocities $\underline{\dot{q}}$ it is important to recognize that the angular velocity vector is related to the time derivative of the rotation parameters, $\underline{\dot{q}}$. Indeed, $\underline{\omega}^* = \underline{\underline{R}}^T\underline{\omega} = \underline{\underline{R}}^T\underline{\underline{H}}\,\underline{\dot{q}}$, where the tangent operator, $\underline{\underline{H}}$, is defined in section 4.11 if Euler angles with various sequences are used to represent the orientation of the rigid body. It then follows that $\partial\underline{\omega}^*/\partial\underline{\dot{q}} = \underline{\underline{R}}^T\underline{\underline{H}}$. The generalized momentum associated with the generalized velocities $\underline{\dot{q}}$ is not the angular momentum, $\underline{h}$, of the body, but rather $\underline{\underline{H}}^T\underline{h}$.

The next step is to evaluate the derivatives of the Lagrangian with respect to he generalized coordinates. Clearly, $\partial L/\partial\underline{u} = 0$ since the displacement vector $\underline{u}$ does not appear in the Lagrangian. The derivative of the Lagrangian with respect to the rotation parameters now becomes

$$\frac{\partial L}{\partial \underline{q}} = \left[\frac{\partial \underline{\mathcal{V}}^*}{\partial \underline{q}}\right]^T \frac{\partial K}{\partial \underline{\mathcal{V}}^*} = \left\{\underline{\underline{H}}^T\widetilde{\dot{u}}^T\underline{\underline{R}},\ \left[\frac{\partial \underline{\omega}^*}{\partial \underline{q}}\right]^T\underline{\underline{R}}^T\underline{\underline{R}}\right\}\underline{\mathcal{P}}^*$$

$$= \underline{\underline{H}}^T\widetilde{\dot{u}}^T\underline{p} + \left[\frac{\partial \underline{\omega}^*}{\partial \underline{q}}\right]^T\underline{\underline{R}}^T\underline{h}.$$

To derive these results, eq. (4.82) was used to obtain $\partial(\underline{\underline{R}}^T \underline{\dot{u}})/\partial \underline{q} = \underline{\underline{R}}^T \tilde{\dot{u}} H$. Furthermore, $\partial\left[\underline{\underline{R}}^T \underline{\omega}\right]/\partial\underline{q} = \partial\underline{\omega}^*/\partial\underline{q} = \underline{\underline{R}}^T \underline{\underline{R}} \, \partial\underline{\omega}^*/\partial\underline{q}$. Finally, in view of identity (4.85), this simplifies to

$$\frac{\partial L}{\partial \underline{q}} = \underline{\underline{H}}^T \tilde{\dot{u}}^T \underline{p} + \underline{\dot{H}}^T \underline{h}.$$

The equations of motion of the system now follow from Lagrange's formulation as

$$\dot{\underline{p}} = \underline{F},$$

where $\underline{F}$ are the generalized forces associated with generalized coordinates $\underline{u}$ because the corresponding virtual work is $\delta W = \delta\underline{u}^T \underline{F}$, and

$$\left[\underline{\underline{H}}^T \underline{h}\right]' - \underline{\underline{H}}^T \tilde{\dot{u}}^T \underline{p} - \underline{\dot{H}}^T \underline{h} = \underline{\underline{H}}^T \underline{M}_B,$$

where $\underline{\underline{H}}^T \underline{M}_B$ are the generalized forces associated with generalized coordinates $\underline{q}$ because the corresponding virtual work is $\delta W = \delta\underline{\psi}^T \underline{M}_B = \delta\underline{q}^T \underline{\underline{H}}^T \underline{M}_B$.

The first equation is the familiar equation of motion for the center of mass and is identical to that obtained from Hamilton's principle, see eq. (8.37a). The second equation is not identical to its counterpart, eq. (8.37c), obtained from Hamilton's principle. Expanding the time derivative of the first term leads to

$$\underline{\underline{H}}^T \left[\underline{\dot{h}} + \tilde{\dot{u}}\underline{p}\right] = \underline{\underline{H}}^T \underline{M}_B.$$

In this form, it is apparent that the equations of motion obtained from Lagrange's formulation are a linear combination of those resulting from Hamilton's principle.

This example, which involves the three-dimensional rotation of a rigid body, calls for the following remarks concerning the application of Lagrange's method. From the onset, this approach describes the system in terms of a specific set of generalized coordinates. The formulation requires the choice of a specific set of parameters, $\underline{q}$, to represent the three-dimensional rotation because the derivation of the equations of motion will require the computation of the following derivatives: $\partial L/\partial\underline{\dot{q}}$ and $\partial L/\partial\underline{q}$. Consequently, the angular velocity vector, which the kinetic energy explicitly depends on, must be expressed in terms of these parameters as $\underline{\omega} = \underline{\underline{H}}(\underline{q})\underline{\dot{q}}$. The equations of motion depend on the tangent operator, $\underline{\underline{H}}$, which is specific to the parametrization selected to represent finite rotations.

This contrasts with d'Alembert's and Hamilton's formulations that do not require the selection of specific rotation parameters, see eqs. (8.7a) and (8.7b) for d'Alembert's principle and eqs. (8.37a) to (8.37c) for Hamilton's principle. These two approaches are based on the concept of virtual rotation and lead to *intrinsic equations of motion*, i.e., equations that are independent of the specific choice of rotation parameters. The use of Lagrange's formulation also complicates the algebra because derivatives such as $\partial\underline{\omega}/\partial\underline{\dot{q}}$ and $\partial\underline{\omega}/\partial\underline{q}$ are required obtain the equations of motion. Furthermore, identity (4.85) is required to simplify the equations of motion.

8.3.1 Problems

Problem 8.26. Particle sliding along a circular ring
Solve problem 3.12 using Lagrange's formulation.

Problem 8.27. Flexible pendulum on a slider
Solve problem 3.34 using Lagrange's formulation.

Problem 8.28. Pendulum with rotating mass
Solve problem 4.38 using Lagrange's formulation.

Problem 8.29. Plate hinged at the rim of a rotating disk
Solve problem 6.14 using Lagrange's formulation.

Problem 8.30. Bar hinged at rim of rotating disk
Solve problem 6.27 using Lagrange's formulation.

Problem 8.31. Rolling cylinder with bar
Solve problem 6.30 using Lagrange's formulation.

Problem 8.32. Rolling cylinder with articulated bar
Solve problem 6.31 using Lagrange's formulation.

Problem 8.33. Double pendulum with elastic joint
Solve problem 6.33 using Lagrange's formulation.

Problem 8.34. Pendulum with sliding mass
Solve problem 6.34 using Lagrange's formulation.

Problem 8.35. Suspension system
Solve problem 6.36 using Lagrange's formulation.

Problem 8.36. Bar rocking on top of a curve
Solve problem 6.38 using Lagrange's formulation.

Problem 8.37. Pendulum connected to a plunging mass
Solve problem 6.39 using Lagrange's formulation.

Problem 8.38. Two-bar mechanism
Solve problem 6.40 using Lagrange's formulation.

Problem 8.39. Particle sliding in a rolling wheel
Solve problem 6.43 using Lagrange's formulation.

Problem 8.40. Particle in a slot on a rotating disk
Solve problem 6.44 using Lagrange's formulation.

Problem 8.41. Pendulum connected to horizontal piston
Solve problem 6.45 using Lagrange's formulation.

Problem 8.42. Inverted pendulum mounted on a cart
Solve problem 6.46 using Lagrange's formulation.

Problem 8.43. Geneva wheel mechanism

Solve problem 6.47 using Lagrange's formulation.

Problem 8.44. Scotch yoke mechanism

Solve problem 6.48 using Lagrange's formulation.

Problem 8.45. Particle in a circular slot with guiding arm

Figure 8.15 shows a particle of mass M sliding along a circular slot of radius R. The particle also slides in a rectilinear slot in an arm of mass m and length L. The arm is pivoted to the ground at point O and is restrained by a torsional spring of stiffness constant k and a dashpot of constant c. The spring is un-stretched when angle $\phi = 0$. A viscous friction force, $F^f = -\mu \dot{w}$ is acting at the interface between the particle and the arm. (1) Derive the equation of motion of the system based on Lagrange's formulation using angle ϕ as generalized coordinate. (2) Find an expression for the normal contact force that the circular slot applies on the particle. (3) Find an expression for the normal contact force that the arm applies on the particle.

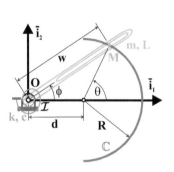

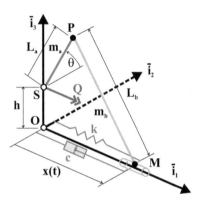

Fig. 8.15. Particle in a circular slot with guiding arm.

Fig. 8.16. Configuration of the spatial mechanism.

Problem 8.46. Spatial mechanism subjected to a torque

The spatial mechanism depicted in fig. 8.16 consists of an arm of length L_a and mass m_a attached to the ground at point S and rotating about the unit vector $\bar{\imath}_1$; the angle of rotation is denoted $\theta(t)$. A slender rigid link connects point P, at the tip of the arm, to point M that is free to slide along unit vector $\bar{\imath}_1$. The link is of length L_b and mass m_b and the distance from point O to point M is denoted x. A spring of stiffness constant k and a dashpot of constant c connect the slider to point O, the origin of the inertial system; the spring is un-stretched when $x = 0$. Torque $Q(t)$ is applied to the arm at point S, acting about an axis parallel to axis $\bar{\imath}_1$. (1) Use Lagrange's formulation to derive the equation of motion of the system.

8.4 Analysis of the motion

The many examples presented in this chapter demonstrate that the equations of motion of even the simplest mechanical systems are often highly nonlinear. Consequently, analytical solutions can rarely be developed and numerical procedures for the solution of ordinary differential equations are typically used to obtained approximate, although highly accurate solutions.

It is often required to obtain the dynamic response of a system subjected to externally applied, time-dependent loads. Such information can only be obtained through analytical or numerical integration of the equations of motion. On the other hand, information about the nature of the motion is sometimes equally important. Assume that for a given set of initial conditions, a dynamical system is at an equilibrium point in the configuration space. The following question then arises: if perturbed, does the system remain in the neighborhood of this equilibrium point for an extended period of time?

Let $\underline{q}_e$ be an equilibrium point of the system in the configuration space and consider a hyper-sphere of radius ε around this point, $\mathcal{S} = (\underline{q}_e, \varepsilon)$. The question raised in the previous paragraph is now rephrased in more precise terms: if the system is at the equilibrium point and a small perturbation is applied, will the response of the system remain in $\mathcal{S}$ for all subsequent times? The equilibrium point is said to be stable if the response of the system, $\underline{q}(t)$, is such that $\underline{q}(t) \in \mathcal{S}$ for all times $t \in [0, \infty]$. If this condition is not met, the equilibrium point is said to be unstable.

In most practical cases, it is desirable for dynamical systems to be stable about their nominal operating conditions. Consequently, assessing the stability characteristics of the system about its equilibrium points is an important task and a procedure for determining the linearized stability characteristics of dynamical systems is described below.

8.4.1 General procedure for the analysis of motion

Given the nonlinear equations of motion of a dynamical system, the determination of the linearized stability characteristics proceeds in the following steps.

1. *Determine the equilibrium points of the system.* An equilibrium point, $\underline{q}_e$, is a steady solution, $\underline{q}_e \neq \underline{q}_e(t)$, of the nonlinear governing equations of the system. Consequently, equilibrium points are the solutions of the nonlinear algebraic equations obtained by imposing the vanishing of all time derivatives appearing in the governing differential equations. Because these algebraic equations are nonlinear, multiple equilibrium points could exist.

2. *Linearize the governing equations of motion about the equilibrium points.* The dynamic response of the system, $\underline{q}(t)$, is assumed to take the following form,

$$\underline{q}(t) = \underline{q}_e + \hat{\underline{q}}(t), \tag{8.62}$$

where $\hat{\underline{q}}(t)$ are small perturbations in the generalized coordinates about an the equilibrium point, $\underline{q}_e$. Each term in the governing equations of motion is then expanded using Taylor series about the equilibrium point. For instance, if $f(\underline{q})$ is a term appearing in the equations,

$$f(\underline{q}(t)) = f(\underline{q}_e + \hat{\underline{q}}(t)) = f(\underline{q}_e) + \left.\frac{\partial f}{\partial \underline{q}}\right|_{\underline{q}_e} \hat{\underline{q}}(t) + \text{h.o.t.} \tag{8.63}$$

Because the perturbation, $\hat{q}(t)$, is very small, the higher-order terms can be neglected and $f(\underline{q})$ now becomes a linear function of the perturbation. A similar procedure is applied to all terms in the governing equations that now become linear, ordinary differential equations with constant coefficients.

3. *Study the stability of the motion.* The equations governing the behavior of small perturbations about an equilibrium point are now in the form of linear, second-order ordinary differential equations with constant coefficients. The solution of this type of equations is of the following form [27],

$$\hat{\underline{q}}(t) = \sum_{i=1}^{2n} A_i \exp(p_i t), \tag{8.64}$$

where n is the number of degrees of freedom of the system, A_i the integration constants, and p_i the *characteristic exponents*. Both integration constants and characteristic exponents are, in general, complex numbers.

It now becomes possible to asses the behavior of small perturbations about an equilibrium point. Each characteristic exponent is written as $p_i = \alpha_i + j\sigma_i$, where $j = \sqrt{-1}$, and consequently, if the real part of any characteristic exponent is positive, the magnitude of the small perturbation grows exponentially, eventually leaving sphere $\mathcal{S}$. The conditions for stability of perturbations about an equilibrium point become

$$\Re(p_i) < 0, i = 1, 2, \ldots, 2n. \tag{8.65}$$

This procedure is illustrated in the following examples.

Example 8.17. Particle sliding along a circular ring

Figure 8.17 depicts a particle of mass m sliding along a circular ring of radius ℓ under the effect of gravity. The ring rotates on two bearing about an unit vector $\bar{\imath}_3$; a torque $Q(t)$, acting about unit vector $\bar{\imath}_3$, is applied to the ring. Find the equilibrium points of the system and study the stability of the system at those equilibrium points.

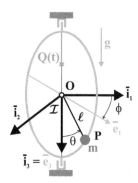

Fig. 8.17. Particle sliding along a circular ring.

Derivation of the equations of motion

The first step of the procedure is to develop the equations of motion of the system. The addition theorem gives the angular velocity of rod **OP** as $\omega = \dot{\phi}\bar{\imath}_3 + \dot{\theta}\bar{e}_2$. The velocity of the particle then becomes $\underline{v}_P = \ell(\dot{\theta}C_\theta\bar{e}_1 + \dot{\phi}S_\theta\bar{e}_2 - \dot{\theta}S_\theta\bar{e}_3)$, where $S_\theta = \sin\theta$ and $C_\theta = \cos\theta$. The kinetic energy of the system is now $K = 1/2\, m\ell^2(\dot{\theta}^2 + \dot{\phi}^2 S_\theta^2)$. The potential of the gravity forces is $V = mg\ell(1 - C_\theta)$ and the Lagrangian of the system becomes

$$L = \frac{1}{2}m\ell^2(\dot{\theta}^2 + \dot{\phi}^2 S_\theta^2) - mg\ell(1 - C_\theta).$$

Using Lagrange's formulation, the equations of motion of the system are then obtained

$$\ddot{\theta} + (\omega^2 - \dot{\phi}^2 C_\theta)S_\theta = 0, \tag{8.66a}$$

$$m\ell^2(\dot{\phi}S_\theta^2)^{\cdot} = Q, \tag{8.66b}$$

where $\omega^2 = g/\ell$.

Next, it is assumed that the circular ring rotates at a constant angular velocity, $\dot{\phi} = \Omega$. The system now features a single degree of freedom represented by generalized coordinate θ, and the associated governing equation, eq. (8.66a), now becomes $\ddot{\theta} + (\omega^2 - \Omega^2 C_\theta)S_\theta = 0$, and eq. (8.66b) yields the torque required to keep the ring rotating at a constant angular velocity, $Q = m\ell^2\Omega(S_\theta^2)^{\cdot}$.

It will be convenient to recast the governing equation of the problem in a non-dimensional form by using the non-dimensional time $\tau = \Omega t$ to find

$$\theta'' + (\bar{\omega}^2 - C_\theta)S_\theta = 0, \tag{8.67}$$

where $\bar{\omega} = \omega/\Omega$ and notation $(\cdot)'$ indicates a derivative with respect to τ.

Determination of the equilibrium points

Equilibrium points, denoted θ_e, are steady solutions of the governing differential equation of motion and are obtained by imposing the vanishing of all time derivatives appearing in eq. (8.67), leading to

$$(\bar{\omega}^2 - C_{\theta_e})S_{\theta_e} = 0. \tag{8.68}$$

The vanishing of the second factor yields two equilibrium points, $\theta_e = 0$ or π. The other solutions, $\theta_e = n\pi$ for $n = 2, 3, \ldots, \infty$ are indistinguishable from the first two and hence, need not be considered. The vanishing of the first factor yields additional equilibrium points, $\theta_e = \arccos\bar{\omega}^2$. Solutions $\pm\theta_e$ are symmetric with respect to the vertical axis of the ring and are physically indistinguishable. Of course, this solution only exists when $\bar{\omega}^2 \leq 1$. In summary, the system presents three distinct equilibrium points: $\theta_e = 0$, $\theta_e = \pi$, and $\theta_e = \arccos\bar{\omega}^2$ if $\bar{\omega}^2 \leq 1$.

At the equilibrium points, the particle is not necessarily at rest in space. For $\theta_e = \arccos\bar{\omega}^2$, the particle moves on a circular path of radius ℓS_{θ_e} in a plane normal to unit vector $\bar{\imath}_3$, but this trajectory corresponds to a constant value of generalized coordinate $\theta(t) = \theta_e$.

Linearization of the equations of motion

The next step of the procedure is to linearize the equations of motion about an equilibrium point. The solution is written as

$$\theta(\tau) = \theta_e + \hat{\theta}(\tau), \tag{8.69}$$

where θ_e is one of the equilibrium points determined in the previous section and $\hat{\theta}(\tau)$ a time-dependent perturbation of infinitesimal magnitude about the equilibrium point.

The governing equation of the problem, eq. (8.67), is nonlinear due to the presence of transcendental functions, $\cos\theta$ and $\sin\theta$. Introducing the assumed form of the solution, eq. (8.69), the cosine function is approximated using a Taylor series expansion, eq. (8.63), to find

$$\cos\theta = \cos(\theta_e + \hat{\theta}) = \cos\theta_e + \left.\frac{\partial\cos\theta}{\partial\theta}\right|_{\theta_e}\hat{\theta}(t) + \text{h.o.t.} \approx \cos\theta_e - \hat{\theta}\sin\theta_e.$$

This expression is now a linear function of the perturbation, $\hat{\theta}$. A similar treatment of the sine function yields

$$\sin\theta = \sin(\theta_e + \hat{\theta}) = \sin\theta_e + \left.\frac{\partial\sin\theta}{\partial\theta}\right|_{\theta_e}\hat{\theta}(t) + \text{h.o.t.} \approx \sin\theta_e + \hat{\theta}\cos\theta_e.$$

Introducing these expansions into the nonlinear governing equation of the problem, eq. (8.67), yields $\hat{\theta}'' + (\bar{\omega}^2 - \cos\theta_e + \hat{\theta}\sin\theta_e)(\sin\theta_e + \hat{\theta}\cos\theta_e) = 0$. Taking into account the definition of the equilibrium points, eq. (8.68), and neglecting higher-order terms leads to the desired linearized equation of motion for small perturbations about an equilibrium point,

$$\hat{\theta}'' + (\bar{\omega}^2\cos\theta_e - \cos 2\theta_e)\hat{\theta} = 0. \tag{8.70}$$

Analysis of motion

Because eq. (8.70) is a linear differential equation with constant coefficients, its solution is of the form $\hat{\theta} = A\exp(\bar{p}\tau)$. Introducing this solution into eq. (8.70) yields the characteristic exponent of the system as

$$\bar{p}^2 = \cos 2\theta_e - \bar{\omega}^2\cos\theta_e. \tag{8.71}$$

It now becomes possible to discuss the stability of the system about its three distinct equilibrium points.

1. *Equilibrium point* $\theta_e = 0$. For this point, $\cos\theta_e = 1$, $\cos 2\theta_e = 1$, and the characteristic exponent becomes $\bar{p}^2 = 1 - \bar{\omega}^2$. The system is stable if $1 - \bar{\omega}^2 < 0$, *i.e.* when $\bar{\omega} > 1$. In this case, the non-dimensional frequency of the motion is $\bar{\sigma} = \sigma/\Omega = \sqrt{\bar{\omega}^2 - 1}$.

2. *Equilibrium point,* $\theta_e = \pi$. For this point, $\cos\theta_e = -1$, $\cos 2\theta_e = 1$, and the characteristic exponent becomes $\bar{p}^2 = 1 + \bar{\omega}^2$. Because $1 + \bar{\omega}^2 > 0$, the system is always unstable about this equilibrium point.

3. *Equilibrium point* $\theta_e = \arccos\bar{\omega}^2$. For this point, $\cos\theta_e = \bar{\omega}^2$, $\cos 2\theta_e = 2\bar{\omega}^4 - 1$, and the characteristic exponent becomes $\bar{p}^2 = \bar{\omega}^4 - 1$. The system is stable if $\bar{\omega}^4 - 1 < 0$, *i.e.* when $\bar{\omega} < 1$. In this case, the non-dimensional frequency of the motion is $\bar{\sigma} = \sqrt{1 - \bar{\omega}^4}$.

In the discussion thus far, parameter $\bar{\omega}$ was used as a variable. To better understand the physical behavior of the system, it is easier to take $\omega = g/\ell$ to be constant, and evaluate the stability of the system as the non-dimensional angular speed of the ring, $\bar{\Omega} = \Omega/\omega = 1/\bar{\omega}$, increases. For equilibrium point $\theta_e = \pi$, the system is always unstable for any value of the ring's angular speed.

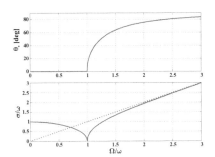

Fig. 8.18. Top figure: location of the stable equilibrium point. Bottom figure: non-dimensional frequency of the motion, $\bar{\sigma} = \sigma/\omega$. Horizontal axis is the non-dimensional angular speed of the ring, $\bar{\Omega} = \Omega/\omega$.

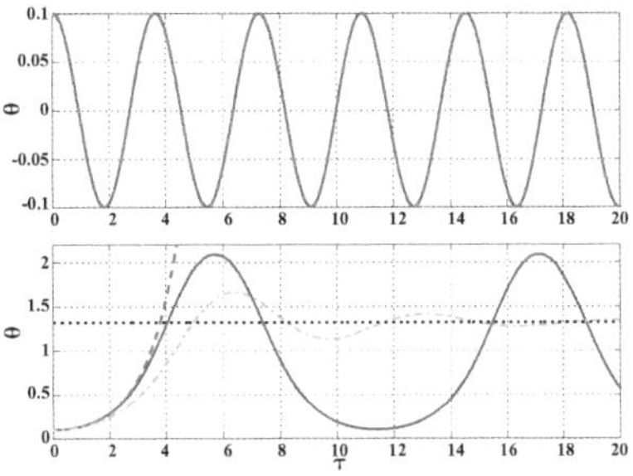

Fig. 8.19. System response for $\bar{\Omega} = 0.5$ (top figure) and $\bar{\Omega} = 2$ (bottom figure). Solid line: nonlinear solution; dashed line: linearized solution. The dashed-dotted line is the nonlinear solution with damping.

When the ring rotates slowly, $\bar{\Omega} < 1$, the only stable equilibrium point is $\theta_e = 0$. As the angular speed of the ring increases, the critical speed, $\bar{\Omega} = 1$, is reached. Above that speed, equilibrium point $\theta_e = 0$ becomes unstable, but a new equilibrium position arises, $\theta_e = \arccos\bar{\omega}^2$, which is stable. The top portion of fig. 8.18 shows the stable equilibrium point as function of the ring's angular speed.

At low speed, the non-dimensional frequency of the motion is $\bar{\sigma} = \sigma/\omega = \sqrt{1 - \bar{\Omega}^2}$, while above the critical speed, the frequency is $\bar{\sigma} = \sqrt{\bar{\Omega}^2 - 1/\bar{\Omega}^2}$. The bottom portion of fig. 8.18 shows this frequency as function of the ring's angular speed. For high angular speeds of the ring, the frequency of the motion is $\bar{\sigma} \approx \bar{\Omega}$, *i.e.*, the frequency of oscillation of the point mass is nearly identical to the ring's angular speed, as indicated by the asymptote shown in the figure.

To verify the predictions of the stability analysis presented above, the nonlinear equation of motion of the problem, eq. (8.67), was integrated numerically for two operating conditions, $\bar{\Omega} = 0.5$ and $\bar{\Omega} = 2$. In both cases, the initial conditions are

$\theta(\tau = 0) = 0.1$ and $\theta'(\tau = 0) = 0$, which put the particle in the neighborhood of equilibrium point $\theta_e = 0$.

For the low speed case, $\bar{\Omega} = 0.5$, the system is stable about equilibrium point $\theta_e = 0$. The top portion of fig. 8.19 shows the predicted small amplitude oscillations of the mass about the equilibrium point for both nonlinear and linearized equations of motion, eqs. (8.67) and (8.70), respectively. As expected, both equations predict nearly identical responses. Indeed, because the system is stable, the amplitude of the perturbation remains small, and the assumptions inherent to the linearization process are valid.

In contrast, for the high speed case, $\bar{\Omega} = 2$, the system is unstable about equilibrium point $\theta_e = 0$. The figure also shows the exponential growth of the response predicted by the linearized equation. As the magnitude of angle θ increases, the linearized and nonlinear solutions diverge. The linearized solution, however, correctly predicts that the particle does not remain in the neighborhood of $\theta_e = 0$. If a small amount of damping is added to the system, the solution of the nonlinear equation, shown in dashed-dotted line in the bottom portion of fig. 8.19, quickly settles to a new equilibrium point, which is correctly predicted by the linearized analysis as $\theta_e = \arccos \bar{\omega}^2 = \arccos(1/2)^2 = 1.32$ rad.

Example 8.18. Bar pivoted to a rigid frame

Figure 8.20 shows a homogeneous bar of length L and mass m connected at its mid point **M** to a rotating frame **ABCD** by means of two revolute joints at points $\mathbf{R}_1$ and $\mathbf{R}_2$. A torsional spring of stiffness k and un-stretched rotation angle β_0 is present in one of the revolute joints; torque $Q(t)$, acting about unit vector $\bar{\imath}_3$, is applied to the frame. Basis $\mathcal{A} = (\bar{a}_1, \bar{a}_2, \bar{a}_3)$ is attached to the rotating frame. A planar rotation about unit vector $\bar{\imath}_3$ of magnitude α bring the inertial basis $\mathcal{I}$ to $\mathcal{A}$. A planar rotation about axis $\bar{a}_1$ of magnitude β brings basis $\mathcal{A}$ to $\mathcal{B} = (\bar{b}_1, \bar{b}_2, \bar{b}_3)$, which is attached to the bar. The configuration of the system will be represented by two generalized coordinates, α and β. Find the equilibrium points of the system and study the stability characteristics of the system at those equilibrium points.

Derivation of the equations of motion

The first step of the procedure is to develop the equations of motion of the system. The addition theorem gives the angular velocity of the bar as $\omega = \dot{\beta}\bar{b}_1 + \dot{\alpha}S_\beta\bar{b}_2 + \dot{\alpha}C_\beta\bar{b}_3$, where $S_\beta = \sin\beta$ and $C_\beta = \cos\beta$, with similar notational conventions for the trigonometric functions of angle α. The mass moment of inertia tensor of the bar is $\underline{\underline{I}}^* = m\ell^2 \mathrm{diag}(1,0,1)/12$, see fig. 6.42. The kinetic energy of the system is now $K = 1/2\ m\ell^2(\dot{\beta}^2 + \dot{\alpha}^2 C_\beta^2)/12$. The potential of the torsional spring is $V = 1/2\ k(\beta - \beta_0)^2$ and the Lagrangian of the system becomes

$$L = \frac{1}{2}\frac{m\ell^2}{12}(\dot{\beta}^2 + \dot{\alpha}^2 C_\beta^2) - \frac{1}{2}k(\beta - \beta_0)^2.$$

Using Lagrange's formulation, the equations of motion of the system are then obtained

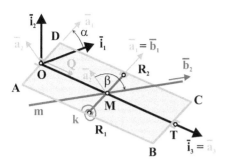

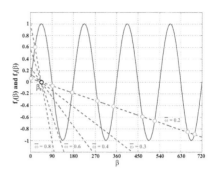

Fig. 8.20. Homogeneous bar connected to a rotating frame.

Fig. 8.21. Graphical solution of the transcendental equation for determining the equilibrium points. Solid line: function $f_1(\beta)$; dashed line: straight line $f_2(\beta)$ for $\beta_0 = \pi/4$ and $\bar{\omega} = 0.2, 0.3, 0.4, 0.6,$ and 0.8.

$$\ddot{\beta} + \dot{\alpha}^2 S_\beta C_\beta + \omega^2(\beta - \beta_0) = 0, \tag{8.72a}$$

$$m\ell^2(\dot{\alpha}C_\beta^2)^{\cdot}/12 = Q, \tag{8.72b}$$

where $\omega^2 = 12k/(m\ell^2)$.

Next, it is assumed that the rigid frame rotates at a constant angular velocity, $\dot{\alpha} = \Omega$. The system now features single degree of freedom represented by generalized coordinate, β, and the associated governing equation, eq. (8.72a), now becomes $\ddot{\beta} + \Omega^2 S_\beta C_\beta + \omega^2(\beta - \beta_0) = 0$, and eq. (8.72b) yields the torque required to keep the frame rotating at a constant angular velocity, $Q = m\ell^2\Omega(C_\beta^2)^{\cdot}/12$.

It will be convenient to recast the governing equation of the problem in a non-dimensional form by using the non-dimensional time $\tau = \Omega t$ to find

$$\beta'' + S_\beta C_\beta + \bar{\omega}^2(\beta - \beta_0) = 0, \tag{8.73}$$

where $\bar{\omega} = \omega/\Omega$ and notation $(\cdot)'$ indicates a derivative with respect to τ.

Determination of the equilibrium points

Equilibrium points, denoted β_e, are steady solutions of the governing differential equation of motion and are obtained by imposing the vanishing of all time derivatives appearing in eq. (8.73), leading to

$$\sin 2\beta_e + 2\bar{\omega}^2(\beta_e - \beta_0) = 0. \tag{8.74}$$

This transcendental equations could have a single solution, but for specific values of the two parameters, $\bar{\omega}$ and β_0, it could feature an infinite number of solutions. A graphical solution of eq. (8.74) can be obtained by defining two functions, a trigonometric function, $f_1(\beta_e) = \sin 2\beta_e$, and a straight line with a negative slope,

$f_2(\beta_e) = -2\bar{\omega}^2(\beta_e - \beta_0)$. The solutions of eq. (8.74) are at the intersections of these two curves.

Figure 8.21 illustrates the graphical solution process for the following values of the parameters, $\beta_0 = \pi/4$ and $\bar{\omega} = 0.2, 0.3, 0.4, 0.6$, and 0.8. Function $f_1(\beta_e)$ is independent of the parameters, while the straight line, $f_2(\beta_e)$, pivots about point $\beta = \beta_0$ for different values of $\bar{\omega}$. A single solution, $\beta_e \approx 0.32$ rad, is found for $\bar{\omega} = 0.8$. In contrast, for $\bar{\omega} = 0.2$, nine distinct solutions exist.

In a dimensional form, eq. (8.74) can be recast as $1/2\ (m\ell^2/12)\Omega^2 \sin 2\beta_e + k(\beta_e - \beta_0) = 0$. The first term represents the moment of the inertial forces applied to the bar, and the second term is the restoring force due to the elastic spring. Hence, the equation for the equilibrium point is a static moment equilibrium equation. As the bar rotates multiple turns around its axis, the restoring moment in the spring increases and can be equilibrated by inertial forces for different magnitudes of the frame's angular speed.

Linearization of the equations of motion

The linearization procedure described in section 8.4.1 yields the following linearized equations of motion for small perturbations about an equilibrium point,

$$\hat{\beta}'' + (\bar{\omega}^2 + \cos\theta_e)\hat{\beta} = 0. \tag{8.75}$$

Analysis of motion

The solution of eq. (8.75) is of the form $\hat{\beta} = A\exp(\bar{p}\tau)$, and the characteristic exponent of the system becomes

$$\bar{p}^2 = -(\bar{\omega}^2 + \cos 2\beta_e). \tag{8.76}$$

The stability condition becomes $\bar{\omega}^2 + \cos 2\beta_e > 0$, where $\beta_e = \beta_e(\bar{\omega}, \beta_0)$ because the equilibrium point is a solution of eq. (8.74).

As a first example, consider the following parameter values, $\beta_0 = \pi/4$ and $\bar{\omega} = 0.8$. Figure 8.21 shows that a single solution is possible and a numerical solution of eq. (8.74) yields $\beta_e = 0.319$. This equilibrium point is stable and the frequency of the motion is $\bar{\sigma} = \sigma/\Omega = 1.2$.

Next, consider the case where $\beta_0 = \pi/4$ and $\bar{\omega} = 0.2$. Figure 8.21 shows that nine solutions are possible and a numerical solution of eq. (8.74) yield $\beta_e = 0.030$, 1.60, 3.05, 4.88, 6.07, 8.17, 9.06, 11.51, and 12.01 rad. These solutions are alternatively stable and unstable. The frequencies of the stable motion are $\bar{\sigma} = 1.02$, 1.01, 0.973, 0.888, and 0.693, for equilibrium points $\beta_e = 0.030$, 3.05, 6.07, 9.06, and 12.01 rad, respectively. For $\beta_e = 1.60$, 4.88, 8.17, and 11.51 rad, the motion is unstable. It is left to the reader to verify that other equilibrium points exist for negative values of angle β.

8.4.2 Problems

Problem 8.47. Rotor blade with flap and lag motions

Figure 8.22 depicts a very simplified model of a helicopter blade of length L and mass m rotating at a constant angular velocity Ω about unit vector $\bar{\imath}_3 = \bar{a}_3$. At point $\mathbf{O}$, basis $\mathcal{A} = (\bar{a}_1, \bar{a}_2, \bar{a}_3)$ is attached to the hub represented by a massless bar of length e. Angular motion *out of the plane of rotation*, i.e., a planar rotation about axis $\bar{a}_2$ of magnitude ϕ, called the *flapping angle*, is shown on the left part of the figure; this motion is resisted by a torsional spring of stiffness k_ϕ. Angular motion *in the plane of rotation*, i.e., a planar rotation about axis $\bar{a}_3$ of magnitude θ, called the *lead-lag angle*, is shown on the right part of the figure; this motion is resisted by a torsional spring of stiffness k_θ. When both in- and out-of-plane motions are considered simultaneously, the configuration of the blade can be described by three successive planar rotations: first, a rotation of magnitude Ωt about axis $\bar{\imath}_3$ that brings basis $\mathcal{I}$ to $\mathcal{A}$, next, a rotation of magnitude θ about axis $\bar{a}_3$ that brings basis $\mathcal{A}$ to $\mathcal{E} = (\bar{e}_1, \bar{e}_2, \bar{e}_3)$, and finally, a rotation of magnitude ϕ about axis $\bar{e}_2$ that brings basis $\mathcal{E}$ to $\mathcal{B} = (\bar{b}_1, \bar{b}_2, \bar{b}_3)$, a blade attached basis. Use Lagrange's formulation to derive the equations of motion of the system in the following cases. (1) At first, assume that the sole flapping motion is allowed; derive the equation of motion. (2) Linearize the equations of motion. Find the natural frequency. (3) Next, assume that the sole lead-lag motion is allowed; derive the equation of motion. (4) Linearize the equation of motion. Find the natural frequency. (5) Finally, assume that both flap and lead-lag motions are allowed; derive the equations of motion. (6) Linearize the equations of motions. Find the natural frequencies.

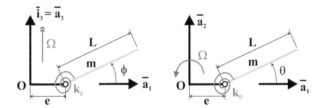

Fig. 8.22. Out-of-plane and in-plane motions of a uniform rotor blade.

Problem 8.48. Bar on two guides

A homogeneous bar of length L and mass m slides on two guides at its end points, as shown in fig. 8.3. At the left end, the bar is connected to a spring of stiffness constant k that is unstretched when the bar is horizontal. At the right end, the bar is connected to a point mass M. Gravity acts along axis $\bar{\imath}_2$. (1) Use Lagrange's formulation to derive the equation of motion of the system. Use a single generalized coordinate, θ. (2) Find the equilibrium configuration of the system. (3) Find the natural frequency of the system. (4) Is the system stable at the equilibrium point?

Problem 8.49. Spinning disk

The circular disk of mass m_1 and radius R spins at a constant angular velocity $\Omega \bar{b}_1$ about arm $\mathbf{OD}$, as depicted in fig. 8.23. This arm is of length L_2, mass m_2, and is connected to the ground at point $\mathbf{O}$ by means of two hinges. The orientation of this arm is determined by two planar rotations: first, a rotation of magnitude ψ about axis $\bar{\imath}_3$ that brings inertial basis $\mathcal{I} = (\bar{\imath}_1, \bar{\imath}_2, \bar{\imath}_3)$ to $\mathcal{A} = (\bar{a}_1, \bar{a}_2, \bar{a}_3)$, and second, a rotation of magnitude θ' about axis $\bar{b}_2$ that brings basis $\mathcal{A}$ to $\mathcal{B} = (\bar{b}_1, \bar{b}_2, \bar{b}_3)$. In the analysis, it will be more convenient to use the angle $\theta = \pi/2 - \theta'$. The configuration of the system is represented by the angles ψ and θ, shown in fig. 8.23. *(1)* Use Lagrange's formulation to derive the equations of motion of the system. *(2)* Linearize these equations of motion. *(3)* Show that the linearized equations are uncoupled and discuss the nature of the motion.

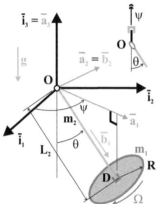

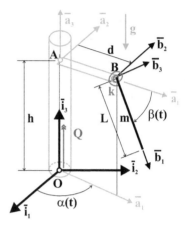

Fig. 8.23. Circular disk spinning at a constant angular velocity Ω.

Fig. 8.24. Bar of length L connected at point **B** to a torsional spring of stiffness constant k.

Problem 8.50. Spherical elastic pendulum

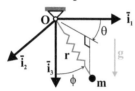

Fig. 8.25. Elastic spherical pendulum.

Figure 8.25 depicts an elastic spherical pendulum that consists of a bob of mass m connected to an inertial point by means of a spring of stiffness constant k and un-stretched length ℓ_0. The configuration of the system will be represented by the spherical coordinates r, ϕ, and θ. *(1)* Use Lagrange's approach to derive the equations of motion of the system. *(2)* Linearize the equations of motion. *(3)* Study the stability of the equilibrium points.

Problem 8.51. Spinning arm

A shaft of height h is fixed at point **O** and free to rotate about axis $\bar{i}_3$, as shown in fig. 8.24. An arm of length d, rigidly attached to the shaft at point **A**, rotates in the horizontal plane. A homogeneous bar of length L and mass m is connected to the arm at point **B** with a torsional spring of stiffness constant k. Frame $\mathcal{F}^A = [\mathbf{A}, \mathcal{A} = (\bar{a}_1, \bar{a}_2, \bar{a}_3)]$ is attached to the arm and frame $\mathcal{F}^B = [\mathbf{B}, \mathcal{B} = (\bar{b}_1, \bar{b}_2, \bar{b}_3)]$ is attached to the bar. A planar rotation of magnitude α about axis $\bar{i}_3$ brings basis $\mathcal{I}$ to $\mathcal{A}$. A planar rotation of magnitude β about axis $\bar{a}_2$ brings basis $\mathcal{A}$ to $\mathcal{B}$; the torsional spring is un-stretched when $\beta = \beta_0$. A torque Q is applied to the shaft at point **O**. *(1)* Use Lagrange's formulation to derive the equations of motion of the system. Use two generalized coordinates, α and β. *(2)* If the first planar rotation is constrained such that $\dot{\alpha} = \Omega$, *i.e.*, the shaft is rotating at a constant angular velocity Ω, find the applied torque Q. *(3)* In this latter case, find the equilibrium configuration of the system. *(4)* Find the natural frequency of the system for small amplitude oscillations about the equilibrium configuration. *(5)* If $\beta_0 = 0$, is the system stable at the equilibrium point? *(6)* If $\beta_0 = \pi/2$, is the system stable at the equilibrium point?

Problem 8.52. Inverted pendulum mounted on a track

Figure 6.37 depicts an inverted homogeneous pendulum of mass m and length ℓ. The pendulum is mounted on a cart of mass M free to translate along a horizontal track. A torsional spring of stiffness constant k restrains the pendulum at its attachment point. The spring is un-stretched when angle $\theta = \theta_0$. *(1)* Derive the two equations of motion of the system. *(2)* Linearize the equations of motions about an equilibrium point. *(3)* Let $\bar{g} = 0.1$. Find the equilibrium point(s) and study the characteristics of the motion about that point. *(4)* Let $\bar{g} = 20$. Find the equilibrium point(s) and study the characteristics of the motion about that point. Use the following data: $\mu = M/m = 1.5$ and $\bar{g} = g/(\ell\omega^2)$. Use non-dimensional time $\tau = \omega t$, where $\omega^2 = k/(m\ell^2)$ and $(\cdot)'$ denotes a derivative with respect to τ.

Part IV

Constrained dynamical systems

9

Constrained systems: preliminaries

In the previous chapter, variational and energy principles were derived for dynamical system. Various formulations were addressed including d'Alembert's principle, Hamilton's principle and Lagrange's formulation. In all cases, developments were limited to unconstrained dynamical systems. This means that the number of generalized coordinates used to represent the configuration of the system was equal the number of degrees of freedom of the system.

The idea of using more generalized coordinates than is strictly necessary to represent the configuration of a mechanical system is not appealing at first, because it seems to increase needlessly formulation complexity. It turns out, however, that increasing the number of generalized coordinates often simplifies the derivation of the governing equations of mechanical systems.

This chapter starts with an introductory problem dealing with a simple, single degree of freedom two-bar mechanism. In example 9.1, the single equation of motion of this problem is derived from Newton's formulation. An alternative formulation using a large number of generalized coordinates is then presented in a cursory manner in example 9.2. The advantages and drawbacks of the two approaches are contrasted.

The remainder of the chapter presents basic concepts associated with constrained dynamical systems. Section 9.1 introduces Lagrange's multiplier method, which is a theoretical underpinning for the analysis of constrained systems. Holonomic and nonholonomic constraints are presented and contrasted in section 9.2, which also introduces the constraint matrix.

The chapter concludes with the generalization of the principle of virtual work for constrained static problems. Arbitrary and kinematically admissible virtual displacements are contrasted. Finally, the combined use of the principle of virtual work and Lagrange's multipliers is investigated.

Example 9.1. Two-bar mechanism, Newtonian formulation

The two bar mechanism shown in fig. 9.1 comprises bar **OB** of length L_1 and mass m_1, and bar **BAT** of length L_2 and mass m_2. Bar **BAT** passes through a slider located at fixed point **A** but free to swivel about that point. A spring of stiffness constant k connects the tip of the bar at point **T** to the slider at point **A** and is of vanishing un-stretched length. A viscous friction force, $F_v = -c\dot{w}$, acts at the interface between the bar and the slider. Using Newton's formulation, derive the equation of motion of this single degree of freedom system represented by the single generalized coordinate θ_1.

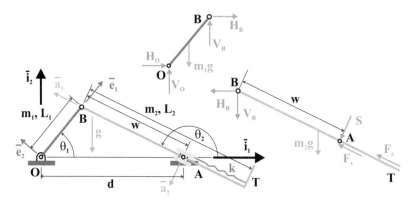

Fig. 9.1. Configuration of the two-bar mechanism.

The law of cosines applied to triangle **OAB** yields $w^2 = L_1^2 + d^2 - 2dL_1C_1$, where $C_1 = \cos\theta_1$. This leads to

$$\bar{w} = \frac{w}{L_1} = \sqrt{1 + \bar{d}^2 - 2\bar{d}C_1}, \tag{9.1}$$

where $\bar{d} = d/L_1$. The law of sines applied to the same triangle gives $L_1S_1 = wS_2$, where $S_1 = \sin\theta_1$, and similar notations are used for the trigonometric functions of angle θ_2. Projections of segments **OB** and **BAT** along the horizontal yield $L_1C_1 - wC_2 = d$. In non-dimensional form,

$$S_2 = \frac{S_1}{\bar{w}}, \quad C_2 = \frac{C_1 - \bar{d}}{\bar{w}}. \tag{9.2}$$

Let unit vectors $\bar{e}_1$ and $\bar{a}_1$ be aligned with bars **OA** and **BAT**, respectively, as shown in fig. 9.1. It then follows that $L_1\bar{e}_1 - w\bar{a}_1 = d\bar{\imath}_1$, and a time derivative of this expression yields $L_1\dot{\theta}_1\bar{e}_2 - \dot{w}\bar{a}_1 - w\dot{\theta}_2\bar{a}_2 = 0$. Projecting this equation along unit vectors $\bar{a}_2$ and $\bar{a}_1$ then leads to

$$\bar{w}\dot{\theta}_2 = \dot{\theta}_1C_{21}, \quad \dot{\bar{w}} = \dot{\theta}_1S_{21}, \tag{9.3}$$

where $C_{21} = \cos(\theta_2 - \theta_1)$ and $S_{21} = \sin(\theta_2 - \theta_1)$. The first equation expresses the angular velocity of bar **BAT** in terms of that of bar **OA**, and the second the velocity of bar **BAT** with respect to the slider. Finally, the angular acceleration of bar **BAT** is found as

$$\ddot{\theta}_2 = \ddot{\theta}_1\frac{C_{21}}{\bar{w}} + \dot{\theta}_1^2\frac{S_{21}}{\bar{w}}\left(1 - 2\frac{C_{21}}{\bar{w}}\right). \tag{9.4}$$

Figure 9.1 also shows a free body diagram of bar **OB**. Euler's laws applied to this planar rigid body problem yield

$$(H_O + H_B)\bar{\imath}_1 + (V_O + V_B)\bar{\imath}_2 - m_1 g \bar{\imath}_2 = \frac{m_1 L_1}{2}(-\dot{\theta}_1^2 \bar{e}_1 + \ddot{\theta}_1 \bar{e}_2), \qquad (9.5a)$$

$$L_1 C_1 V_B - L_1 S_1 H_B - \frac{m_1 L_1}{2} g C_1 = \frac{m_1 L_1^2}{3}\ddot{\theta}_1, \qquad (9.5b)$$

where the second equation corresponds to the pivot equation, eq. (6.38), written about point **O**. The components of the reaction force at point **O** are denoted H_O and V_O, in the horizontal and vertical directions, respectively. Similarly, the components of the internal force acting at point **B** are denoted H_B and V_B.

Figure 9.1 shows a free body diagram of bar **BAT** and Euler's laws yield

$$-H_B\bar{\imath}_1 - V_B\bar{\imath}_2 - m_2 g \bar{\imath}_2 + (F_v + F_s)\bar{a}_1 + S\bar{a}_2$$
$$= m_2 \left[L_1(-\dot{\theta}_1^2 \bar{e}_1 + \ddot{\theta}_1 \bar{e}_2) + \frac{L_2}{2}(-\dot{\theta}_1^2 \bar{e}_1 + \ddot{\theta}_1 \bar{e}_2) \right], \qquad (9.6a)$$

$$\frac{m_2 L_2}{2} g C_2 - wS = -m_2 \left[\frac{L_2}{2}\tilde{a}_1 L_1 (-\dot{\theta}_1^2 \bar{e}_1 + \ddot{\theta}_1 \bar{e}_2) \right]_3 + \frac{m_2 L_2^2}{3}\ddot{\theta}_2. \qquad (9.6b)$$

In this case, the second equation is written with respect to material point **B** of the bar, see eq. (6.39). The reaction force acting in the direction normal to the sliding direction at point **A** is denoted S. The elastic spring applies force $F_s = k(L_2 - w)$ to the tip of the bar and the viscous friction force acts at point **A**.

At this point, the six equations of dynamics have been written for this two-body planar problem. It involves six unknowns, five components of internal force, H_O, V_O, H_B, V_B, and S, and a single generalized coordinate, θ_1. The additional kinematic variables appearing in these equations, θ_2, w, and their time derivatives, should be expressed in terms of the generalized coordinate, θ_1, and its time derivatives, using the kinematic equations (9.1) to (9.4). To obtain the single equation of motion of this problem, the five internal force components must be eliminated from the six equations of dynamics through careful algebra.

The first step of the process is to use eq. (9.6b) to express the normal contact force at the slider

$$wS = \frac{m_2 L_2}{2} L_1 (\dot{\theta}_1^2 S_{21} + \ddot{\theta}_1 C_{21}) + \frac{m_2 L_2}{2} g C_2 - \frac{m_2 L_2^2}{3}\ddot{\theta}_2. \qquad (9.7)$$

Next, eq. (9.6a) is projected along unit vectors $\bar{\imath}_1$ and $\bar{\imath}_2$ to find the reaction force components, H_B and V_B, respectively. The following linear combination of these two components is then evaluated

$$L_1(C_1 V_B - S_1 H_B) = m_2 L_1 \left[-L_1\ddot{\theta}_1 - \frac{L_2}{2}(\dot{\theta}_2^2 S_{21} - \ddot{\theta}_2 C_{21}) \right] + SC_{21}L_1$$
$$+ (F_v + F_s)L_1 S_{21} - m_2 L_1 g C_1. \qquad (9.8)$$

The final step of the procedure is to introduce eq. (9.7) into eq. (9.8) to eliminate the internal force, S. The resulting expression is introduced into eq. (9.5b) to yield the desired equation of motion

$$
\left[\frac{m_1 L_1^2}{3} + m_2 L_1^2 + \frac{m_2 L_2^2}{3} \frac{C_{21}^2}{\bar{w}^2} - m_2 L_1 L_2 \frac{C_{21}^2}{\bar{w}} \right] \ddot{\theta}_1
$$

$$
+ \left[\frac{m_2 L_2^2}{3} S_{21} \frac{C_{21}}{\bar{w}^2} (1 - 2\frac{C_{21}}{\bar{w}}) - \frac{m_2 L_1 L_2}{2} S_{21} \frac{C_{21}}{\bar{w}} (2 - 3\frac{C_{21}}{\bar{w}}) \right] \dot{\theta}_1^2 \quad (9.9)
$$

$$
+ (\frac{m_1}{2} + m_2) L_1 g C_1 - \frac{m_2 L_2}{2} g C_2 \frac{C_{21}}{\bar{w}} - k(L_2 - w) S_{21} + c L_1^2 S_{21}^2 \dot{\theta}_1 = 0.
$$

In this equation, the following quantities must be expressed in terms of the generalized coordinate, θ_1,

$$
\bar{w} = \sqrt{1 + \bar{d}^2 - 2\bar{d}C_1}, \quad C_2 = \frac{C_1 - \bar{d}}{\bar{w}}, \quad C_{21} = \frac{1 - \bar{d}C_1}{\bar{w}}, \quad S_{21} = \frac{\bar{d}S_1}{\bar{w}}.
$$

This example illustrates one of the main problems associated with the derivation of the equations of motion of mechanical systems. While the system depicted in fig. 9.1 is a rather simple mechanical system that features a single degree of freedom, the procedure to derive the single equation of motion of the system is a complex analytical task. The use of Lagrange's formulation will streamline the process by eliminating the reaction and internal forces from the onset of the formulation. It is left to the reader to verify that the same equation of motion will be obtained, as expected.

From a mathematical viewpoint, the equation of motion is a second-order, ordinary differential equation in time. It is, however, a highly nonlinear differential equation, which cannot be solved in closed form. Fortunately, numerical procedures for the solution of this class of equations are widely available and hence, approximate solutions of eq. (9.9) are easily obtained.

Although numerical procedures ease the solution of the differential equation, its derivation remains an arduous, error-prone task. Symbolic manipulation software can be used to ease this task, but the equation of motion remains complex.

Example 9.2. Two-bar mechanism, alternative formulation

The two bar mechanism shown in fig. 9.2 comprises bar **OB**, of length L_1 and mass m_1, and bar **BAT**, of length L_2 and mass m_2. Bar **BAT** passes through a slider located at fixed point **A** but free to swivel about that point. A spring of stiffness constant k connects the tip of the bar at point **T** to the slider at point **A** and is of vanishing unstretched length. A viscous friction force, $F_v = -c\dot{w}$, acts at the interface between the bar and the slider. Derive the equation of motion of this single degree of freedom system using a highly redundant set of seven generalized coordinates: the components of the position vector of the center of mass of bar **OB** along axes $\bar{\imath}_1$ and $\bar{\imath}_2$, denoted x_1 and y_1, respectively, and its orientation, denoted θ_1, the components of the position vector of the center of mass of bar **BAT** along axes $\bar{\imath}_1$ and $\bar{\imath}_2$, denoted x_2 and y_2, respectively, and its orientation, denoted θ_2, and finally, the relative distance between points **B** and **A**, denoted w. The array of generalized coordinates is

$$q^T = \{x_1, y_1, \theta_1, x_2, y_2, \theta_2, w\}.$$

Figure 9.2 shows a free body diagram of bar **OB**. Euler's laws applied to this planar rigid body problem yield

$$(H_O + H_B)\bar{\imath}_1 + (V_O + V_B)\bar{\imath}_2 - m_1 g\bar{\imath}_2 = m_1(\ddot{x}_1\bar{\imath}_1 + \ddot{y}_1\bar{\imath}_2), \qquad (9.10a)$$

$$\frac{L_1}{2}C_1(V_B - V_O) - \frac{L_1}{2}S_1(H_B - H_O) = \frac{m_1 L_1^2}{12}\ddot{\theta}_1. \qquad (9.10b)$$

where the second equation is written with respect to the center of mass of bar 1. The reaction force at point **O**, denoted $\underline{R}_O$, has components in the horizontal and vertical directions denoted H_O and V_O, respectively. Similarly, the reaction force at point **B**, denoted $\underline{R}_B$, has components components denoted H_B and V_B.

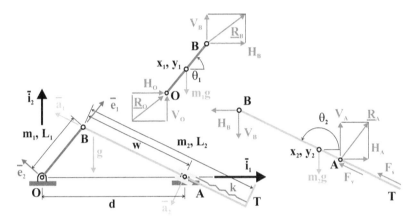

Fig. 9.2. Configuration of the two-bar mechanism.

Figure 9.2 also shows a free body diagram of bar **BAT** and Euler's laws yield

$$(H_A - H_B)\bar{\imath}_1 + (V_A - V_B)\bar{\imath}_2 - m_2 g\bar{\imath}_2 + (F_v + F_s)\bar{a}_1 = m_2(\ddot{x}_2\bar{\imath}_1 + \ddot{y}_2\bar{\imath}_2), \qquad (9.11a)$$

$$-\frac{L_2}{2}C_2 V_B + \frac{L_2}{2}S_2 H_B + (w - \frac{L_2}{2})(S_2 H_A - C_2 V_A) = \frac{m_2 L_2^2}{12}\ddot{\theta}_2. \qquad (9.11b)$$

Here again, the second equation is written with respect to the center of mass of bar 2. The reaction force acting at point **A**, denoted $\underline{R}_A$, has components components denoted H_A and V_A. The elastic spring applies a force $F_s = k(L_2 - w)$ to the tip of the bar and the viscous friction force acts at point **A**.

Because independent generalized coordinates have been used for bars **OB** (x_1, y_1, and θ_1) and **BAT** (x_2, y_2, and θ_2), the two bars are free to move independently. In particular, point **O** will not remain a fixed inertial point, the two bars will not remain connected at point **B**, nor will point **A** remain a fixed inertial point. All these conditions must be expressed as constraints on the generalized coordinates.

Consider point **O** at the tip of bar **OB**. Its horizontal and vertical position components are $x_1 - L_1 C_1/2$ and $y_1 - L_1 S_1/2$, respectively. Because point **O** must remain a fixed inertial point, the following two constraints must link the three generalized coordinates determining the configuration of bar **OB**,

$$\underline{C}_O = \begin{Bmatrix} x_1 - L_1 C_1/2 \\ y_1 - L_1 S_1/2 \end{Bmatrix} = \underline{0}. \tag{9.12}$$

Similarly, the horizontal and vertical position components of point **B** are $x_1 + L_1 C_1/2$ and $y_1 + L_1 S_1/2$, respectively, when computed based on the generalized coordinates defining the configuration of bar **OB**. The same horizontal and vertical position components of point **B**, evaluated based on the generalized coordinates defining the configuration of bar **BAT**, are $x_2 + L_2 C_2/2$ and $y_2 + L_2 S_2/2$, respectively. If the two bars are to remain connected at point **B** at all times, the following constraints must link the generalized coordinates of bars **OB** and **BAT**,

$$\underline{C}_B = \begin{Bmatrix} x_1 + L_1 C_1/2 - x_2 - L_2 C_2/2 \\ y_1 + L_1 S_1/2 - y_2 - L_2 S_2/2 \end{Bmatrix} = \underline{0}. \tag{9.13}$$

Finally, imposing the condition that point **A** must remain a fixed inertial point leads to the following constraints,

$$\underline{C}_A = \begin{Bmatrix} x_2 + (L_2/2 - w)C_2 - d \\ y_2 + (L_2/2 - w)S_2 \end{Bmatrix} = \underline{0}. \tag{9.14}$$

The formulation of the problem involves the seven generalized coordinates stored in array $\underline{q}$, and the six components of internal and reaction forces stored in arrays $\underline{R}_O$, $\underline{R}_B$, and $\underline{R}_A$, for a total of thirteen unknowns. Euler's laws applied to the two bars provide a total of six equations, eqs. (9.10) and (9.11). Six constraint equations must also be satisfied, eqs. (9.12), (9.13), and (9.14). Clearly, one additional equation is required to solve the problem. This equation is the static equilibrium equation at the slider: the reaction force at the slider should be normal to bar **BAT**, leading to

$$C_S = C_2 H_A + S_2 V_A = 0. \tag{9.15}$$

Note that the friction force at the slider is taken into account by vector $\underline{F}_v$, independently of the reaction force, $\underline{R}_A$.

At this point, the formulation of the problem is complete. It involves thirteen equations, six equations of dynamics, eqs. (9.10) and (9.11), one slider static equilibrium equation, eq. (9.15), and six constraint equations, eqs. (9.12) to (9.14), for a total of thirteen unknowns, seven generalized coordinates and six components of internal force. It is possible to eliminate the six components of internal force and six of the seven generalized coordinates to obtain a single equation of motion written in terms of a single generalized coordinate, say θ_1. As illustrated in example 9.1, this process is lengthy and error-prone. This elimination process will be even more arduous in the presence case, because a highly redundant set of generalized coordinates was selected from the onset of the formulation.

In this example, an alternative approach is followed. Instead of eliminating the redundant generalized coordinates and internal forces, all these unknowns are kept in the formulation. First, it is interesting to take a time derivative of the constraints; for instance, the first constraint expressed by eq. (9.12) is $x_1 - L_1 C_1/2 = 0$ and its time derivative is $\dot{x}_1 + L_1 S_1 \dot{\theta}_1/2 = 0$. Combining this expression with a similar treatment of the second constraint and recasting the results in a matrix form yields

$$\dot{\underline{C}}_O = \begin{bmatrix} 1 & 0 & L_1 S_1/2 & 0 & 0 & 0 & 0 \\ 0 & 1 & -L_1 C_1/2 & 0 & 0 & 0 & 0 \end{bmatrix} \dot{\underline{q}} = \underline{\underline{B}}_O \dot{\underline{q}} = \underline{0}, \tag{9.16}$$

where $\dot{\underline{q}}$ is the array of generalized velocities. Matrix $\underline{\underline{B}}_O(\underline{q})$ is called the constraint matrix.

Proceeding in the same manner with the constraints defined by eqs. (9.13) and (9.14), the following results are obtained

$$\dot{\underline{C}}_B = \begin{bmatrix} 1 & 0 & -L_1 S_1/2 & -1 & 0 & L_2 S_2/2 & 0 \\ 0 & 1 & L_1 C_1/2 & 0 & -1 & -L_2 C_2/2 & 0 \end{bmatrix} \dot{\underline{q}} = \underline{\underline{B}}_B \dot{\underline{q}} = \underline{0}, \tag{9.17}$$

$$\dot{\underline{C}}_A = \begin{bmatrix} 0 & 0 & 0 & 1 & 0 & -(L_2/2 - w)S_2 & -C_2 \\ 0 & 0 & 0 & 0 & 1 & (L_2/2 - w)C_2 & -S_2 \end{bmatrix} \dot{\underline{q}} = \underline{\underline{B}}_A \dot{\underline{q}} = \underline{0}, \tag{9.18}$$

where $\underline{\underline{B}}_B$ and $\underline{\underline{B}}_A$ are the constraint matrices associated with constraints $\underline{C}_B = \underline{0}$ and $\underline{C}_A = \underline{0}$, respectively.

It now becomes possible to write the dynamical equations of the problem in a compact manner as

$$\underline{\underline{M}} \ddot{\underline{q}} = \underline{\underline{B}}_O^T \underline{R}_O + \underline{\underline{B}}_B^T \underline{R}_B + \underline{\underline{B}}_A^T \underline{R}_A + \underline{F}_a, \tag{9.19}$$

where $\underline{\underline{M}} = \text{diag}(m_1, m_1, m_1 L_1^2/12, m_2, m_2, m_2 L_2^2/12, 0)$ is the mass matrix of the system. It is left to the reader to verify that the first three governing equations of system (9.19) are the dynamical equations for bar **OB**, eqs. (9.10). The contributions of the reaction forces acting on bar **OB** are written in terms of the constraint matrices. The next three equations of system (9.19) are the dynamical equations for bar **BAT**, eqs. (9.11). The last equation of system (9.19) is the equilibrium equation for the slider, eq. (9.15). The last term on the right-hand side of system (9.19), $\underline{F}_a^T = \{0, -m_1 g, 0, (F_v + F_s)C_2, -m_2 g + (F_v + F_s)S_2, 0, 0\}$, is the array of externally applied forces, where $F_v = -c\dot{w}$ and $F_s = k(L_2 - w)$. Combining system (9.19) with the six constraint equations, eqs. (9.12) to (9.14), yields a total of thirteen equations for the thirteen unknowns of the problem.

It is important to compare the characteristics of the present formulation to those of that presented in example 9.1. Instead of the single equation of motion in a single unknown obtained earlier, eq. (9.9), the present formulation leads to thirteen equations in thirteen unknowns. It must be noted, however, that the thirteen equations of the present formulation are far easier to derive and far less complex than the single equation of motion, eq. (9.9).

The present formulation requires writing the constraints equations, eqs. (9.12) to (9.14), and their time derivatives, eqs. (9.16) to (9.18). These kinematic tasks are far simpler than those associated with Newton's formulation, see eqs. (9.1) to (9.4), which requires the evaluation of the linear and angular accelerations of all bodies in terms of the generalized velocities and accelerations.

Newton's formulation leads to *ordinary differential equations* in time. Despite their complexity and high level of nonlinearity, numerical solution procedures for this type of equations are well developed and robust. In contrast, the present formulation leads to *differential-algebraic equations*; because this type of equation is less common than its ordinary differential counterpart, solution techniques for differential-algebraic equations are not as well developed or robust. This does not imply that the solution of differential-algebraic equations is computationally less efficient; indeed, although the number of equations and unknowns is typically higher, the simplicity and sparsity of the equations enables efficient solution procedures.

9.1 Lagrange's multiplier method

A fundamental tool used for the analysis of constrained dynamical systems is Lagrange's multiplier technique; a formal description of this method is presented here. Consider the problem of determining a stationary point of a function of several variables, $F = F(u_1, u_2, \ldots, u_n)$, as was discussed in section 7.1.1. In this case, however, the variables are not independent, rather, they are subjected to a constraint of the form

$$\mathcal{C}(u_1, u_2, \ldots, u_n) = 0. \tag{9.20}$$

Conceptually, this constraint could be used to express one variable, say u_n, in term of the others. Next, u_n would be eliminated from F to obtain a function of $n-1$ independent variables $F = F(u_1, u_2, \ldots, u_{n-1})$, a problem identical to that treated in section 7.1.1. In many practical situations, it might be cumbersome, undesirable, or even impossible, to completely eliminate one variable of the problem.

This elimination process can be avoided altogether by using an alternative approach. At a stationary point, the variation of function F vanishes

$$\delta F = \frac{\partial F}{\partial u_1} \delta u_1 + \frac{\partial F}{\partial u_2} \delta u_2 + \ldots + \frac{\partial F}{\partial u_n} \delta u_n = 0. \tag{9.21}$$

This statement, however, does not imply $\partial F/\partial u_i = 0$ for $i = 1, 2, \ldots, n$, because variations δu_i cannot be chosen arbitrarily. Indeed, they must satisfy the constraint expressed by eq. (9.20).

The relationship among the variations δu_i can be explicitly written by taking a variation of the constraint to find

$$\delta \mathcal{C} = \frac{\partial \mathcal{C}}{\partial u_1} \delta u_1 + \frac{\partial \mathcal{C}}{\partial u_2} \delta u_2 + \ldots + \frac{\partial \mathcal{C}}{\partial u_n} \delta u_n = 0. \tag{9.22}$$

This expression shows in an explicit manner that variations $\delta u_1, \delta u_2, \ldots, \delta u_n$, are not independent because a linear combination of these quantities must vanish. A linear combination of eqs. (9.21) and (9.22) is formed by multiplying eq. (9.22) by λ and summing the results with eq. (9.21) to find

$$\frac{\partial F}{\partial u_1} \delta u_1 + \ldots + \frac{\partial F}{\partial u_n} \delta u_n + \lambda \left[\frac{\partial \mathcal{C}}{\partial u_1} \delta u_1 + \ldots + \frac{\partial \mathcal{C}}{\partial u_n} \delta u_n \right] = 0.$$

Coefficient λ is an arbitrary function of the variables $u_1, u_2, \ldots, u_n$, called *Lagrange multiplier*.

Regrouping the various terms then leads to

$$\sum_{i=1}^{n} \left[\frac{\partial F}{\partial u_i} + \lambda \frac{\partial \mathcal{C}}{\partial u_i} \right] \delta u_i = 0. \tag{9.23}$$

Conceptually, variation δu_n could be expressed in terms of the other variations, δu_i, $i = 1, 2, \ldots, n-1$, using eq. (9.22), leaving $n-1$ independent, arbitrary variations. To avoid this cumbersome step, the arbitrary Lagrange multiplier is chosen such that

$$\frac{\partial F}{\partial u_n} + \lambda \frac{\partial \mathcal{C}}{\partial u_n} = 0.$$

With this choice, the last term of the sum in eq. (9.23) vanishes for all δu_n. Hence, there is no need to express this variation in terms of the $n-1$ others, which can now be treated as independent, arbitrary quantities, implying

$$\frac{\partial F}{\partial u_i} + \lambda \frac{\partial \mathcal{C}}{\partial u_i} = 0, \quad i = 1, 2, \ldots, n-1.$$

Combining the last two equations then leads to the condition that

$$\delta F + \lambda \delta \mathcal{C} = 0,$$

where all variations, δu_i, $i = 1, 2, \ldots, n$, are considered to be *independent*. Because the constraint expressed by eq. (9.20) must be satisfied, $\mathcal{C}\delta\lambda = 0$ for *any arbitrary* $\delta\lambda$, and the stationarity condition becomes

$$\delta F + \lambda \delta \mathcal{C} = \delta F + \lambda \delta \mathcal{C} + \mathcal{C}\delta\lambda = \delta(F + \lambda \mathcal{C}) = 0. \tag{9.24}$$

An augmented function, $F^+ = F + \lambda \mathcal{C}$, is now introduced; the above statement implies the vanishing of the variation in F^+ for *all arbitrary variations* δu_i, $i = 1, 2, \ldots, n$, *and* $\delta\lambda$.

In summary, the initial, *constrained problem* can be replaced by an *unconstrained problem*

$$\delta F^+ = 0, \quad F^+ = F + \lambda \mathcal{C}. \tag{9.25}$$

The augmented function, F^+, involves $n+1$ variables, u_i, $i = 1, 2, \ldots, n$ and λ. The vanishing of variations of the augmented function implies

$$\sum_{i=1}^{n} \left[\frac{\partial F}{\partial u_i} + \lambda \frac{\partial \mathcal{C}}{\partial u_i} \right] \delta u_i + \mathcal{C}\,\delta\lambda = 0. \tag{9.26}$$

Because δu_i, $i = 1, 2, \ldots, n$, and $\delta\lambda$ are all independent, arbitrary variations, it follows that

$$\frac{\partial F}{\partial u_i} + \lambda \frac{\partial \mathcal{C}}{\partial u_i} = 0, \quad i = 1, 2, \ldots, n, \quad \text{and} \quad \mathcal{C} = 0. \tag{9.27}$$

These form $n + 1$ equations to be solved for the $n + 1$ unknowns.

Lagrange's multiplier method results in an unconstrained problem, but *increases* the number of unknowns from n to $n + 1$; the additional unknown is Lagrange's multiplier. If the constraint is used to eliminate one of the unknowns, the resulting problem will feature $n - 1$ unconstrained unknowns.

Lagrange's multiplier method can be readily generalized to problems involving multiple constraints, $\mathcal{C}_i = 0$, $i = 1, 2, \ldots, m$. In the presence of m constraints, m Lagrange multipliers, λ_i, $i = 1, 2, \ldots, m$, are introduced. The augmented function then becomes $F^+ = F + \sum_{i=1}^{m} \lambda_i \mathcal{C}_i$.

Example 9.3. Minimum distance between a circle and a plane
Consider plane $\mathcal{P} = (\underline{x}_P, \bar{n})$ and circle $\mathcal{C} = (\underline{x}_C, \bar{k}, \rho)$, both shown in fig. 9.3. Find the shortest algebraic distance, d, between the disk and the plane.

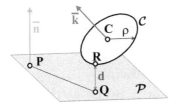

Fig. 9.3. The distance between a disk and a plane.

Let $\underline{x}_Q$ and $\underline{x}_R$ be the position vectors of points **Q** and **R**, respectively; **Q** is a point on the plane and **R** a point on the circle. Finding shortest distance can be cast as a minimization problem

$$d = \min_{\underline{x}_R, \underline{x}_Q} (\|\underline{x}_R - \underline{x}_Q\|), \tag{9.28}$$

for all point **R** and **Q**. This minimization problem, however, is subjected to several constraints: point **Q** must be on plane $\mathcal{P}$; in view of eq. (1.40), this implies

$$\bar{n}^T (\underline{x}_Q - \underline{x}_P) = 0, \tag{9.29}$$

and point **R** must be on circle $\mathcal{C}$,

$$\bar{k}^T (\underline{x}_R - \underline{x}_C) = 0, \quad \|\underline{x}_R - \underline{x}_C\| = \rho, \tag{9.30}$$

as required by eq. (1.42).

The constrained minimization problem defined by eq. (9.28) is now transformed into a an unconstrained minimization with the help of Lagrange's multiplier technique

$$d = \min_{\underline{x}_R, \underline{x}_Q, \lambda, \mu, \nu} \left[\|\underline{x}_R - \underline{x}_Q\| + \lambda \bar{n}^T(\underline{x}_Q - \underline{x}_P) + \mu \bar{k}^T(\underline{x}_R - \underline{x}_C) \right.$$
$$\left. + \nu(\|\underline{x}_R - \underline{x}_C\| - \rho)\right],$$

where λ, μ, and ν are three Lagrange multipliers used to enforce the three constraints defined by eqs. (9.29) to (9.30). Minimization of the augmented function with respect to $\underline{x}_R$ yields

$$\frac{\underline{x}_R - \underline{x}_Q}{\|\underline{x}_R - \underline{x}_Q\|} + \mu \bar{k} + \nu \frac{\underline{x}_R - \underline{x}_C}{\|\underline{x}_R - \underline{x}_C\|} = 0, \tag{9.31}$$

and minimization with respect to $\underline{x}_Q$ leads to

$$-\frac{\underline{x}_R - \underline{x}_Q}{\|\underline{x}_R - \underline{x}_Q\|} + \lambda \bar{n} = 0. \tag{9.32}$$

Of course, minimization with respect to λ, μ, and ν will yield the constraint equations, eqs. (9.29) to (9.30). Equations (9.29), (9.30), (9.31) and (9.32) form a set of nine equations for the nine unknowns of the problem, $\underline{x}_R$, $\underline{x}_Q$, λ, μ, and ν. The solution of this set of nonlinear algebraic equations will yield the desired minimum distance.

The scalar product of eq. (9.32) by $\bar{n}^T$ yields the first Lagrange multiplier as $\lambda = \bar{n}^T(\underline{x}_R - \underline{x}_Q)/\|\underline{x}_R - \underline{x}_Q\|$ and introducing this result back into eq. (9.32) then leads to $(\underline{I} - \bar{n}\bar{n}^T)(\underline{x}_R - \underline{x}_Q) = 0$. This implies the vanishing of the projection of vector $\underline{x}_R - \underline{x}_Q$ onto plane $\mathcal{P}$, and hence $\underline{x}_R - \underline{x}_Q = d\bar{n}$. This confirms the very intuitive fact that the minimum distance between the circle and the plane is achieved when vector $\underline{x}_R - \underline{x}_Q$ is perpendicular to plane $\mathcal{P}$.

Equation (9.31) now becomes $\bar{n} + \mu \bar{k} + \nu(\underline{x}_R - \underline{x}_C)/\rho = 0$ and a scalar product of this equation by $\bar{k}^T$ then yields the second Lagrange multiplier as $\mu = -\bar{n}^T\bar{k}$, where constraint eq. (9.32) was used. Introducing this result back into the equation yields $(1 - \bar{k}\bar{k}^T)\bar{n} = -\nu(\underline{x}_R - \underline{x}_C)/\rho$ and taking the norm of the relationship yields $\nu = \|\widetilde{k}\bar{n}\|$. It then follows that $\underline{x}_R - \underline{x}_C = -(\rho/\nu)(\underline{I} - \bar{k}\bar{k}^T)\bar{n}$ and finally, $\underline{x}_R - \underline{x}_C = (\rho/\|\widetilde{k}\bar{n}\|)\widetilde{k}\widetilde{k}\bar{n}$, where identity (1.33b) was used. The minimum distance between the plane and the circle is now $d = \bar{n}^T(\underline{x}_R - \underline{x}_Q) = \bar{n}^T(\underline{x}_R - \underline{x}_P +$ $\underline{x}_P - \underline{x}_Q) = \bar{n}^T(\underline{x}_R - \underline{x}_P)$, where constraint eq. (9.29) was used; it follows that $d = \bar{n}^T(\underline{x}_C - \underline{x}_P + \underline{x}_R - \underline{x}_C) = \bar{n}^T(\underline{x}_C - \underline{x}_P) + \rho\bar{n}^T\widetilde{k}\widetilde{k}\bar{n}/\|\widetilde{k}\bar{n}\|$ and finally

$$d = \bar{n}^T(\underline{x}_C - \underline{x}_P) - \rho\|\widetilde{k}\bar{n}\|.$$

Of course, the same result can be obtained in a simpler manner by using simple geometric arguments.

9.1.1 Problems

Problem 9.1. Minimum distance from point to line
Use Lagrange's multiplier technique to find the minimum distance between an arbitrary point **P** of coordinates $\underline{x}_P$ and a line $\mathcal{L} = (\underline{x}_Q, \bar{l})$.

Problem 9.2. Minimum distance from point to circle
Use Lagrange's multiplier technique to find the minimum distance between an arbitrary point **P** of coordinates $\underline{x}_P$ and a circle $\mathcal{C} = (\underline{x}_C, \bar{k}, \rho)$.

9.2 Constraints

The examples discussed in section 7.2 show the importance of constraints: the freedom of using a number of generalized coordinates that exceeds the number of degrees of freedom comes at the expense of adding kinematic constraints. Using more generalized coordinates than is necessary to represent the configuration of the system seems, at first, to be a poor idea because this increases the number of unknowns. The alternative, however, *i.e.*, the elimination of the redundant generalized coordinates, can lead to equations that are unduly complex and cumbersome to manipulate.

Kinematic constraints, also called holonomic constraints, are not the only type of constraints that are encountered in practice. Nonholonomic constraints are constraints that involve the generalized velocities of the system and cannot be integrated. Both types of constraints will be discussed in this section.

9.2.1 Holonomic constraints

Typical kinematic constraints take the form of nonlinear relationships among the generalized coordinates; in general, m such constraints might be imposed on the system

$$\mathcal{C}_i(q_1, q_2, \ldots q_n) = 0, \quad i = 1, 2, \ldots m. \tag{9.33}$$

To simplify the notation, an array of generalized coordinates is is defined, which stores the n generalized coordinates of the system, $\underline{q}^T = \{q_1, q_2, \ldots, q_n\}$. Next, an array of kinematic constraint is introduced that stores the m constraints applied on the system, $\underline{\mathcal{C}}^T = \{\mathcal{C}_1, \mathcal{C}_2, \ldots, \mathcal{C}_m\}$. The m constraints acting on the system are now expressed in a compact form as

$$\underline{\mathcal{C}}(\underline{q}) = 0. \tag{9.34}$$

Constraints of this form are called *kinematic constraints, configuration constraints*, or more generally, *holonomic constraints*. Because the constraints do not depend on time explicitly, they are said to be *scleronomic constraints*. Such constraints reduce the number of degrees of freedom of the system because each constraint could be used to eliminate one generalized coordinate of the system: a system described by n generalized coordinates and subjected to m holonomic constraints presents $d = n - m$ degrees of freedom and is called a *holonomic system*.

The differential of the ith constraint is written as

$$dC_i = \frac{\partial C_i}{\partial q_1}dq_1 + \frac{\partial C_i}{\partial q_2}dq_2 + \ldots + \frac{\partial C_i}{\partial q_n}dq_n = 0. \tag{9.35}$$

The differential of each constraint can be computed in a similar manner and expressed in a compact form as

$$d\underline{C}(\underline{q}) = \underline{\underline{B}}(\underline{q})d\underline{q} = 0, \tag{9.36}$$

where $d\underline{q}^T = \{dq_1, dq_2, \ldots, dq_n\}$ is the array of generalized coordinate differentials. Matrix $\underline{\underline{B}}$ is called the *constraint matrix* or *Jacobian matrix of the constraints*. Each line of this matrix stores the partial derivatives of one constraint with respect to the generalized coordinates,

$$\underline{\underline{B}}(\underline{q}) = \begin{bmatrix} \dfrac{\partial C_1}{\partial q_1} & \dfrac{\partial C_1}{\partial q_2} & \cdots & \dfrac{\partial C_1}{\partial q_n} \\ \dfrac{\partial C_2}{\partial q_1} & \dfrac{\partial C_2}{\partial q_2} & \cdots & \dfrac{\partial C_2}{\partial q_n} \\ \vdots & \vdots & \vdots & \vdots \\ \dfrac{\partial C_m}{\partial q_1} & \dfrac{\partial C_m}{\partial q_2} & \cdots & \dfrac{\partial C_m}{\partial q_n} \end{bmatrix}. \tag{9.37}$$

The kinematic constraints discussed thus far do not depend on time explicitly; such constraints are called *scleronomic constraints*. It is not uncommon for constraints to be explicit functions of time,

$$\underline{C}(\underline{q}, t) = 0; \tag{9.38}$$

such constraints are called *rheonomic constraints*.

The differential of rheonomic constraints involves partial derivatives with respect to time, together with the partial derivatives with respect to the generalized coordinates. The array of partial derivatives with respect to time is denoted

$$\underline{b}^T(\underline{q}, t) = \left\{ \frac{\partial C_1}{\partial t}, \frac{\partial C_2}{\partial t}, \ldots, \frac{\partial C_m}{\partial t} \right\}. \tag{9.39}$$

With the help of this notation, the differential of rheonomic constraints becomes

$$d\underline{C}(\underline{q}, t) = \underline{\underline{B}}(\underline{q}, t)d\underline{q} + \underline{b}(\underline{q}, t)dt = 0. \tag{9.40}$$

In the presence of a mixture of scleronomic and rheonomic constraints, the entries in array $\underline{b}$ corresponding to scleronomic constraints will vanish.

Kinematic constraints express relationships among the generalized coordinates that must hold at each instant in time. Consequently, time derivatives of the constraint must also vanish. Considering first a scleronomic constraint, see eq. (9.34), the time derivative is

$$\underline{\dot{C}}(\underline{q}) = \underline{\underline{B}}(\underline{q})\underline{\dot{q}} = 0. \tag{9.41}$$

The time derivative of rheonomic constraints, see eq. (9.38), is found in a similar manner as

$$\underline{\dot{C}}(\underline{q}, t) = \underline{\underline{B}}(\underline{q}, t)\underline{\dot{q}} + \underline{b}(\underline{q}, t) = 0. \tag{9.42}$$

These expressions are called *velocity level constraints*. Of course, higher-order derivatives could be computed: *acceleration level constraints*, obtained by taking second derivatives of the constraints, are commonly used in many computational schemes for constrained multibody systems. Expressions for the velocity level constraints are linear functions of the generalized velocities because they were obtained by taking time derivatives of the corresponding kinematic constraints. Similarly, acceleration level constraints are linear functions of the generalized accelerations.

9.2.2 Nonholonomic constraints

The constraints considered thus far are kinematic or configuration constraints of the form of eqs. (9.34) or (9.38). In some cases, however, the constraints imposed on a mechanical system are of a different nature from those discussed thus far; consider a differential relationship of the form of eq. (9.40),

$$\underline{\underline{B}}(\underline{q}, t) \, \mathrm{d}\underline{q} + \underline{b}(\underline{q}, t) \, \mathrm{d}t = 0, \tag{9.43}$$

such constraints are said to be in *Pfaffian form*.

For holonomic constraints, constraint matrix $\underline{\underline{B}}$ and array $\underline{b}$ store the partial derivatives of the constraints with respect to the generalized coordinates and time, respectively, see eqs. (9.37) and (9.39), respectively. For *nonholonomic constraints*, constraint matrix $\underline{\underline{B}}$ stores a set of arbitrary functions of the generalized coordinates and time

$$\underline{\underline{B}}(\underline{q}, t) = \begin{bmatrix} b_{11}(\underline{q}, t) & b_{12}(\underline{q}, t) & \cdots & b_{1n}(\underline{q}, t) \\ b_{21}(\underline{q}, t) & b_{22}(\underline{q}, t) & \cdots & b_{2n}(\underline{q}, t) \\ \vdots & \vdots & \vdots & \vdots \\ b_{m1}(\underline{q}, t) & b_{m2}(\underline{q}, t) & \cdots & b_{mn}(\underline{q}, t) \end{bmatrix}, \tag{9.44}$$

and array $\underline{b}$ stores arbitrary functions

$$\underline{b}^T(\underline{q}, t) = \{b_1(\underline{q}, t), b_2(\underline{q}, t), \ldots, b_m(\underline{q}, t)\}. \tag{9.45}$$

The fact that coefficient $b_{ij}(\underline{q}, t)$, $j = 1, 2, \ldots, n$, and $b_i(\underline{q}, t)$ are arbitrary implies that, in general, there exists no function, $C_i(\underline{q}, t)$, such that $b_{ij}(\underline{q}, t) = \partial C_i/\partial q_j$, $j = 1, 2, \ldots, n$ and $b_i = \partial C_i/\partial t$. If function $C_j(\underline{q}, t)$ does not exist, the constraint is *not integrable*, *i.e.*, it is nonholonomic.

When faced with a constraint written in the Pfaffian form, it is important to determine whether such constraint is holonomic or not, *i.e.*, whether it is integrable or not. Configuration constraints must be continuous, and hence, $\partial^2 C_i/\partial q_k \partial q_j = \partial^2 C_i/\partial q_j \partial q_k$, or $\partial b_{ik}/\partial q_j = \partial b_{ij}/\partial q_k$. Hence, a differential constraint can be integrated if and only if the following *integrability conditions* are met

$$\frac{\partial}{\partial q_j}(g_i b_{ik}) = \frac{\partial}{\partial q_k}(g_i b_{ij}), \quad j, k = 1, 2, \ldots, n, \ j \neq k, \tag{9.46a}$$

$$\frac{\partial}{\partial q_j}(g_i b_i) = \frac{\partial}{\partial t}(g_i b_{ij}), \quad j = 1, 2, \ldots, n, \tag{9.46b}$$

where $g_i(q,t)$ are integrating functions. If a constraint is not integrable, it can only be expressed in the Pfaffian form presented in eq. (9.43), or in the velocity form obtained by dividing this equation by dt to find

$$\underline{\mathcal{D}}(q,\dot{q},t) = \underline{\underline{B}}(q,t)\dot{q} + \underline{b}(q,t) = 0, \qquad (9.47)$$

These expressions are not completely general; indeed, the constraints are assumed to be linear functions of the generalized velocities. Constraints with arbitrary mathematical structures could be imagined, but it turns out that the nonholonomic constraints encountered in common mechanical system appear to all be linear functions of the generalized velocities. In the expressions above, time appears explicitly in the nonholonomic relationships; clearly, time-independent nonholonomic constraints could also occur.

If a constraint cannot be integrated, it is a nonholonomic constraint and hence, it cannot be used to eliminate a generalized coordinate: *nonholonomic constraints do not decrease the number of degrees of freedom of the system.* A system described by n generalized coordinates and subjected to a single nonholonomic constraint features n degrees of freedom.

Example 9.4. Two bar linkage tracking a curve
Figure 9.4 depicts a planar two bar linkage tracking a curve $\mathbb{C}$. The two bars are of length L_1 and L_2, respectively, and make angles θ_1 and θ_2 with the horizontal, respectively. The end point of the second bar tracks a fixed planar curve described by position vector $\underline{p}_0(s)$, where s defines the intrinsic parametrization of the curve.

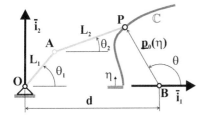

Fig. 9.4. Two bar linkage tracking a curve.

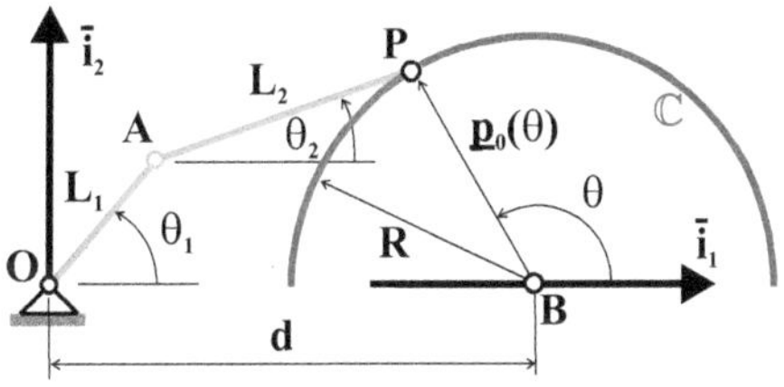

Fig. 9.5. Two bar linkage tracking a semi-circular curve.

This system clearly feature a single degree of freedom. The position vector of the end point $\mathbf{P}$ of the linkage can be expressed in two different manners

$$(L_1 \cos\theta_1 + L_2 \cos\theta_2)\,\bar{\imath}_1 + (L_1 \sin\theta_1 + L_2 \sin\theta_2)\,\bar{\imath}_2 = d\,\bar{\imath}_1 + \underline{p}_0(s),$$

where d is the distance from the origin at point $\mathbf{O}$ to point $\mathbf{B}$. The left-hand side of this equation corresponds to path $\mathbf{OAP}$, while the right-hand side corresponds to path $\mathbf{OBP}$. This vector equation expresses two kinematic constraints among three parameters, θ_1, θ_2, and s, confirming the fact that the system possess a single degree of freedom.

At first, curve $\mathbb{C}$ will be selected to be a circle of radius R and center **B**, as shown in fig. 9.5. The position vector of a point on the curve is then simply $\underline{p}_0(\theta) = R\cos\theta\,\bar{\imath}_1 + R\sin\theta\,\bar{\imath}_2$. The above kinematic constraints now become

$$L_1\cos\theta_1 + L_2\cos\theta_2 - d = R\cos\theta, \quad L_1\sin\theta_1 + L_2\sin\theta_2 = R\sin\theta. \quad (9.48)$$

Two generalized coordinates, $q_1 = \theta_1$ and $q_2 = \theta_2$, will be selected to represent the system. Kinematic constraints (9.48) are now used to eliminate θ, leading to a single constraint for the two generalized coordinates, $2L_1L_2\cos(q_1 - q_2) - 2dL_1\cos q_1 - 2dL_2\cos q_2 = R^2 - L_1^2 - L_2^2 - d^2$. This holonomic constraint could be used to eliminate either q_1 or q_2, but the trigonometric functions involved in the constraint will lead to complex expressions. The constraint matrix is

$$\underline{B}(\underline{q}) = \left[L_1 d\sin q_1 - L_1 L_2\sin(q_1 - q_2),\; L_2 d\sin q_2 + L_2 L_1\sin(q_1 - q_2)\right].$$

This is not the only way to proceed. It might be preferable to keep θ, the location along the circle, as a generalized coordinate. Hence, the generalized coordinates would be selected as $q_1 = \theta_1$ and $q_2 = \theta$. Kinematic constraints (9.48) are then used to eliminate θ_2, leading to a single constraint, $2RL_1\cos(q_1 - q_2) + 2dL_1\cos q_1 - 2dR\cos q_2 = L_1^2 - L_2^2 + R^2 + d^2$. The constraint matrix is

$$\underline{B}(\underline{q}) = \left[-L_1 d\sin q_1 - L_1 R\sin(q_1 - q_2),\; Rd\sin q_2 + RL_1\sin(q_1 - q_2)\right].$$

Finally, curve $\mathbb{C}$ is assumed to be arbitrary, as depicted in fig. 9.4. In this case, it might be very difficult to use the constraint conditions to eliminate any parameter. Consequently, it is convenient to use three generalized coordinates, $q_1 = \theta_1, q_2 = \theta_2$, and $q_3 = s$, linked by two kinematic constraints expressed as

$$(L_1\cos q_1 + L_2\cos q_2)\,\bar{\imath}_1 + (L_1\sin q_1 + L_2\sin q_2)\,\bar{\imath}_2 = d\,\bar{\imath}_1 + \underline{p}_0(q_3).$$

The constraint matrix becomes

$$\underline{B}(\underline{q}) = \begin{bmatrix} -L_1\sin q_1 & -L_2\sin q_2 & -\bar{\imath}_1^T\bar{t}(q_3) \\ L_1\cos q_1 & L_2\cos q_2 & -\bar{\imath}_2^T\bar{t}(q_3) \end{bmatrix},$$

where $\bar{t}(q_3) = d\underline{p}_0/dq_3$ is the unit tangent to the curve at location s.

Example 9.5. The rigid body/universal joint system

This example deals with a rigid body attached to the ground by means of a universal joint, as depicted in fig. 7.9 and discussed in example 7.5 on page 266. Component k of the universal joint, see fig. 7.10, is connected to the ground at point **O** by means of a bearing allowing rotation about axis $\bar{\imath}_3$. Component ℓ is connected to a rigid body at point **O'**.

The orientation of the rigid body will be defined by Euler angles, using the 3-1-2 sequence. A first planar rotation about axis $\bar{\imath}_3$, of magnitude ϕ, brings inertial basis $\mathcal{I} = (\bar{\imath}_1, \bar{\imath}_2, \bar{\imath}_3)$ to $\mathcal{A} = (\bar{a}_1, \bar{a}_2, \bar{a}_3)$, where $\bar{a}_1$ is aligned with unit vector $\bar{b}_1$ of the cruciform. This rotation is associated with a constant angular speed Ω, implying $\bar{a}_1(t) = \cos(\Omega t)\,\bar{\imath}_1 + \sin(\Omega t)\,\bar{\imath}_2$. A second planar rotation about axis $\bar{a}_1$, of magnitude θ, brings basis $\mathcal{A}$ to $\mathcal{B} = (\bar{b}_1, \bar{b}_2, \bar{b}_3)$, where $\bar{b}_2$ is the second unit vector aligned with the cruciform. Finally, a third planar rotation about axis $\bar{b}_2$, of magnitude ψ, bring basis $\mathcal{B}$ to $\mathcal{E} = (\bar{e}_1, \bar{e}_2, \bar{e}_3)$ that is attached to the rigid body. Points $\mathbf{O}$ and $\mathbf{O}'$ are coincident. The system present two degrees of freedom.

Rather than describing the configuration of the system with the sequence of three planar rotations discussed in the previous paragraph, it might be simpler to argue that the rigid body is free to rotate about fixed point $\mathbf{O}$ and hence, its orientation is determined by three generalized coordinates selected, for instance, as the three Euler angles associated with the rotation tensor $\underline{\underline{R}}$ that brings basis $\mathcal{I}$ to $\mathcal{E}$. In this scenario, the kinematic constraint associated with the universal joint imposes the following normality condition: $\mathcal{C} = \bar{e}_2^T \bar{a}_1(t) = 0$, a rheonomic kinematic constraint.

If the orientation of the rigid body is defined by Euler angles using the *3-1-3* sequence, the components of $\underline{\underline{R}}$, resolved in $\mathcal{I}$, are given by eq. (4.11), and the rheonomic kinematic constraint becomes

$$\mathcal{C}(\underline{q}, t) = \sin q_3 \cos(q_1 - \Omega t) + \cos q_2 \cos q_3 \sin(q_1 - \Omega t) = 0,$$

where $q_1 = \phi$, $q_2 = \theta$, and $q_3 = \psi$ are the three generalized coordinates of the problem. These angles are associated with the *3-1-3* sequence, rather than the *3-1-2* sequence that would more naturally describe the sequence of planar rotations inherent to the present mechanical system. The constraint matrix becomes

$$\underline{\underline{B}}^T(\underline{q}, t) = \begin{bmatrix} \cos q_2 \cos q_3 \cos(q_1 - \Omega t) - \sin q_3 \sin(q_1 - \Omega t) \\ -\sin q_2 \cos q_3 \sin(q_1 - \Omega t) \\ \cos q_3 \cos(q_1 - \Omega t) - \cos q_2 \sin q_3 \sin(q_1 - \Omega t) \end{bmatrix}.$$

In the presence of a single constraint, the constraint matrix reduces to a single line.

For this rheonomic constraint, the partial derivative of the constraint with respect to time is $\underline{b}(\underline{q}, t) = \Omega \left[\sin q_3 \sin(q_1 - \Omega t) - \cos q_2 \cos q_3 \cos(q_1 - \Omega t) \right]$. Array $\underline{b}$ features a single entry because the problem involves a single constraint.

Example 9.6. The skateboard

Figure 9.6 depicts the simplified configuration of a skateboard of mass m and moment of inertia I about its center of mass $\mathbf{G}$. The skateboard rolls without sliding on the horizontal plane by means of a wheel aligned with unit vector $\bar{e}_1$ of the skateboard and located at point $\mathbf{C}$, a distance ℓ from the center of mass. The position vector of the center of mass is written as $\underline{r}_G = x\,\bar{\imath}_1 + y\,\bar{\imath}_2$ and the axis of the skateboard makes an angle θ with the horizontal.

The equations of motion of this planar problem are readily obtained from Newton's second law as

$$F^C \sin \theta = m\ddot{x}, \quad -F^C \cos \theta = m\ddot{y}, \quad -\ell F^C = I\ddot{\theta}, \tag{9.49}$$

where $\underline{F}^C = -F^C \bar{e}_2$ is the contact force vector between the wheel and the ground.

The system is subjected to one constraint: because the wheel does not slip, the velocity vector of the contact point must be along unit vector $\bar{e}_1$. The velocity of point $\mathbf{C}$ is $\underline{v}_C = \dot{x}\bar{\imath}_1 + \dot{y}\bar{\imath}_2 + \ell\dot{\theta}\bar{e}_2$, and hence, the constraint becomes $\bar{e}_2^T \underline{v}_C = 0$, or

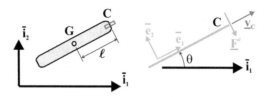

Fig. 9.6. Configuration of the skateboard.

$$\mathcal{D} = \ell\dot{\theta} - \dot{x}\sin\theta + \dot{y}\cos\theta = 0. \tag{9.50}$$

Let the generalized coordinates of the problem be $q_1 = x$, $q_2 = y$, and $q_3 = \theta$. The constraint equations now takes the form of eq. (9.47), where

$$\underline{B}(\underline{q}) = \left[-\sin\theta, \ \cos\theta, \ \ell\right], \quad \underline{b}(\underline{q}) = 0. \tag{9.51}$$

Clearly, the integrability conditions, eqs. (9.46a), are not satisfied, and hence, the constraint is nonholonomic. Time does not appear explicitly and the constraint is linear in terms of the generalized velocities.

Example 9.7. Contact between two rigid bodies

Consider a rigid body, denoted body k, whose outer shape is described by a surface, denoted $\mathbb{S}^k$, as indicated on fig. 9.7. The surface coordinates for this surface are denoted η_1^k and η_2^k, see section 2.4. A body attached frame, $\mathcal{F}^k = \left[\mathbf{B}^k, \mathcal{B}^k = (\bar{b}_1^k, \bar{b}_2^k, \bar{b}_3^k)\right]$, is defined by the position vector $\underline{u}^k$ of its origin with respect to point **O** and by the rotation tensor $\underline{R}^k$ that brings basis $\mathcal{I}$ to $\mathcal{B}^k$. Finally, the position vector of an arbitrary point on the surface is denoted $\underline{p}_0^k(\eta_1^k, \eta_2^k)$. Tensor components resolved in basis $\mathcal{B}^k$ are denoted with superscript $(\cdot)^*$.

Consider now a second body, denoted body ℓ, whose configuration is described in a manner identical to that used for body k, replacing the superscript $(\cdot)^k$ with $(\cdot)^\ell$, see fig. 9.7. Position vectors $\underline{x}^k$ and $\underline{x}^\ell$ of arbitrary points on bodies k and ℓ, respectively, with respect to point **O**, now become

$$\underline{x}^k = \underline{u}^k + \underline{R}^k \underline{p}_0^{*k}(\eta_1^k, \eta_2^k), \quad \underline{x}^\ell = \underline{u}^\ell + \underline{R}^\ell \underline{p}_0^{*\ell}(\eta_1^\ell, \eta_2^\ell).$$

The configuration of body k is represented by eight generalized coordinates: three displacement components for vector $\underline{u}^k$, three rotation components for the rotation tensor $\underline{R}^k$ and two surface coordinates η_1^k and η_2^k. To assess the effect of the contact force on the body, it is necessary to know the location the contact point, and hence, the last two generalized coordinates are an inherent part of the formulation.

At first, imagine that the bodies are at a short distance from each other: the points that are about to come in contact with each other, called *candidate contact points*, must satisfy a number of kinematic constraints. First, the tangent planes to bodies k and ℓ at the candidate contact points must be parallel. Second, the vector joining the two candidate contact points must be parallel to the common normal. The tangent planes are those spanned by the surface base vector $\underline{a}_1^k(\eta_1^k, \eta_2^k)$ and $\underline{a}_2^k(\eta_1^k, \eta_2^k)$ for

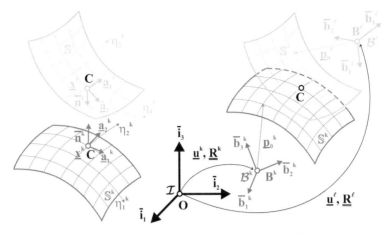

Fig. 9.7. Two rigid bodies contacting at point **C**.

surface $\mathbb{S}^k$, and $\underline{a}_1^\ell(\eta_1^\ell, \eta_2^\ell)$ and $\underline{a}_2^\ell(\eta_1^\ell, \eta_2^\ell)$ for surface $\mathbb{S}^\ell$; the normals to the same planes are denoted $\bar{n}^k(\eta_1^k, \eta_2^k)$ and $\bar{n}^\ell(\eta_1^\ell, \eta_2^\ell)$, respectively.

The kinematic conditions defining the location of the candidate contact points can now be expressed in the following manner,

$$\underline{a}_1^{kT}(\eta_1^k, \eta_2^k)\bar{n}^\ell(\eta_1^\ell, \eta_2^\ell) = 0, \tag{9.52a}$$

$$\underline{a}_2^{kT}(\eta_1^k, \eta_2^k)\bar{n}^\ell(\eta_1^\ell, \eta_2^\ell) = 0, \tag{9.52b}$$

$$\underline{a}_1^{kT}(\eta_1^k, \eta_2^k)\left[\underline{x}^\ell(\eta_1^\ell, \eta_2^\ell) - \underline{x}^k(\eta_1^k, \eta_2^k)\right] = 0, \tag{9.52c}$$

$$\underline{a}_2^{kT}(\eta_1^k, \eta_2^k)\left[\underline{x}^\ell(\eta_1^\ell, \eta_2^\ell) - \underline{x}^k(\eta_1^k, \eta_2^k)\right] = 0. \tag{9.52d}$$

The first two constraints express the parallelism between the two tangent planes, and the last two the parallelism between the normal and the vector joining the candidate contact points. These four holonomic constraints each involve the generalized coordinates η_1^k, η_2^k, η_1^ℓ, and η_2^ℓ, and could be used to solve for these four generalized coordinates, eliminating them from the formulation. For complex surfaces, however, the position, base, and normal vectors at a point are complex, nonlinear functions of two of the generalized coordinates that define the surface. Consequently, the elimination process will be arduous, if not outright impossible. This demonstrates the advantage working with redundant generalized coordinates.

Next, the two bodies are assumed to be in rolling contact with each other and point **C** is the instantaneous point of contact. Surface coordinates $(\bar{\eta}_1^k, \bar{\eta}_2^k)$ and $(\bar{\eta}_1^\ell, \bar{\eta}_2^\ell)$ denote the location of the contact point on surfaces $\mathbb{S}^k$ and $\mathbb{S}^\ell$, respectively, which satisfy the holonomic constraints expressed by eqs. (9.52). The velocities of this point of contact, computed using the configurations of bodies k and ℓ, now become $\underline{v}^k = \underline{\dot{u}}^k + \widetilde{\omega}^k \underline{R}^k \underline{p}_0^{*k}(\bar{\eta}_1^k, \bar{\eta}_2^k)$, and $\underline{v}^\ell = \underline{\dot{u}}^\ell + \widetilde{\omega}^\ell \underline{R}^\ell \underline{p}_0^{*\ell}(\bar{\eta}_1^\ell, \bar{\eta}_2^\ell)$, respectively. When taking a time derivative of the inertial position vectors to obtain velocities, the surface coordinates $(\bar{\eta}_1^k, \bar{\eta}_2^k)$ and $(\bar{\eta}_1^\ell, \bar{\eta}_2^\ell)$ *were held constant* because they are the fixed parameters that defined the material points on bodies k and l that are located at the instantaneous point of contact. The difference between the instantaneous point of contact between the two rigid bodies and the material points that are instantaneously located at this point of contact is discussed in section 5.4.

If the two bodies are in rolling contact with respect to each other, the tangential components of the instantaneous relative velocity of the points of contact on body k and ℓ must vanish, leading to the following two nonholonomic constraints,

$$\mathcal{D}_1 = \underline{a}_1^{kT}(\bar{\eta}_1^k, \bar{\eta}_2^k)\left[\underline{\dot{u}}^\ell - \underline{\dot{u}}^k + \tilde{\omega}^\ell \underline{\underline{R}}^\ell \underline{p}_0^{*\ell}(\bar{\eta}_1^\ell, \bar{\eta}_2^\ell) - \tilde{\omega}^k \underline{\underline{R}}^k \underline{p}_0^{*k}(\bar{\eta}_1^k, \bar{\eta}_2^k)\right] = 0,$$

$$\mathcal{D}_2 = \underline{a}_2^{kT}(\bar{\eta}_1^k, \bar{\eta}_2^k)\left[\underline{\dot{u}}^\ell - \underline{\dot{u}}^k + \tilde{\omega}^\ell \underline{\underline{R}}^\ell \underline{p}_0^{*\ell}(\bar{\eta}_1^\ell, \bar{\eta}_2^\ell) - \tilde{\omega}^k \underline{\underline{R}}^k \underline{p}_0^{*k}(\bar{\eta}_1^k, \bar{\eta}_2^k)\right] = 0.$$

These velocity level constraints are not integrable. Indeed, the surface coordinates of the instantaneous points of contact, $(\bar{\eta}_1^k, \bar{\eta}_2^k)$ and $(\bar{\eta}_1^\ell, \bar{\eta}_2^\ell)$, for body k and ℓ, respectively, are time varying functions that depend on the solution of the problem and must satisfy the holonomic constraints expressed by eqs. (9.52). As the bodies roll on each other, the material points of bodies k and ℓ that are instantaneously located at the point of contact describe complex curves embedded in surfaces $\mathbb{S}^k$ and $\mathbb{S}^\ell$, respectively.

These nonholonomic constraints do not reduce the number of degrees of freedom of the problem. The two rigid bodies are rolling against each other, but the trajectories of the instantaneous points of contact describe arbitrary curves on the two surfaces. The two surfaces could also be rotating with respect to each other about an axis passing through the point of contact and normal to the tangent plane at the contact point, in a manner such that the contact point coincides with fixed material points on either bodies.

9.2.3 Problems

Problem 9.3. Integrability conditions
Show that the integrability conditions are not satisfied for the nonholonomic constraint, eq. (9.50), associated with the skateboard system.

Problem 9.4. Spatial mechanism
The spatial mechanism depicted in fig. 5.6 consists of an arm of length L_a attached to the ground at point **S** and rotating about axis $\bar{\imath}_1$ of the inertial frame $\mathcal{F}^I = [\mathbf{O}, \mathcal{I} = (\bar{\imath}_1, \bar{\imath}_2, \bar{\imath}_3)]$; the time-dependent rotation angle of unit vector $\bar{s}_2$ with respect to axis $\bar{\imath}_2$ is denoted $\theta(t)$. A rigid link connects point **P**, at the tip of the arm, to point **Q** that is free to slide along axis $\bar{\imath}_1$. The link is of length L_b and the distance from point **O** to point **Q** is denoted x. (1) How many degrees of freedom does this mechanism present? (2) If θ and x are used as generalized coordinates, write the appropriate constraint (or constraints) applied on the system. (3) Derive the constraint matrix.

Problem 9.5. Crank-slider mechanism
The crank-slider mechanism depicted in fig. 9.8 consists of a uniform crank of length L_1 and mass m_1 connected to the ground at point **O**; let θ be the angle from the horizontal to the crank. At point **B**, the crank connects to a uniform linkage of length L_2 and mass m_2 that slides along point **P**, a fixed point in space, located at a distance d from point **O**. Let w

denote the distance from point **B** to point **P** and ϕ the angle from the horizontal to link **BP**. *(1)* How many degrees of freedom does this mechanism present? *(2)* If θ, ϕ, and w are used as generalized coordinates, write the appropriate constraint (or constraints) applied to the system. *(3)* Derive the constraint matrix.

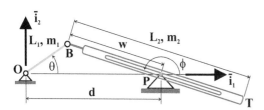

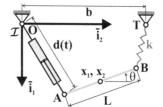

Fig. 9.8. Crank-slider mechanism rotating at a constant angular velocity.

Fig. 9.9. Configuration of the bar with an actuator.

Problem 9.6. Bar with actuator

Rigid homogeneous bar **AB** of length L is connected to the ground at point **T** by means of a spring of stiffness constant k as depicted in fig. 9.9. The other end of the bar is connected to the ground at point **O** through an actuator of prescribed length, $d(t)$. The configuration of bar **AB** is defined by three generalized coordinates, the coordinates of its center of mass, x_1 and x_2, and its orientation with respect to the horizontal, θ. *(1)* How many degrees of freedom does this mechanism present? *(2)* If x_1, x_2, and θ are used as generalized coordinates, write the appropriate constraint(s) applied to the system. *(3)* Derive the constraint matrix.

Problem 9.7. The two-bar linkage with slider system

The two-bar linkage with slider system shown in fig. 8.4 is a planar mechanism. It consists of a uniform crank of length L_1 and mass m_1 connected to the ground at point **O**; let θ be the angle from the horizontal to the crank. At point **B**, the crank slides over a uniform linkage of length L_2 and mass m_2 that is connected to the ground at point **A**. Let w denote the distance from point **B** to point **A** and ϕ the angle from the horizontal to link **BA**. *(1)* How many degrees of freedom does this mechanism present? *(2)* If θ, ϕ, and w are used as generalized coordinates, write the appropriate constraint (or constraints) applied to the system. *(3)* Derive the constraint matrix.

Problem 9.8. Particle in a circular slot with guiding arm

A particle of mass M slides along a circular slot of radius R, as shown in fig. 8.15. The particle also slides in a rectilinear slot in an arm of mass m and length L. The arm is pivoted to the ground at point **O** and is restrained by a torsional spring of stiffness constant k and a dashpot of constant c. The spring is un-stretched when angle $\phi = 0$. *(1)* How many degrees of freedom does this mechanism present? *(2)* If θ, ϕ, and w are used as generalized coordinates, write the appropriate constraint(s) applied to the system. *(3)* Derive the constraint matrix.

9.3 The principle of virtual work for constrained static problems

In section 7.5, the principle of virtual work was presented for a single particle and for systems of particles. It was pointed out that Newton's first law does not distinguish among various types of forces: "the sum of *all forces* must vanish." On the other hand, the nature of the applied forces profoundly affects the statement of the principle of virtual work: conservative forces can be derived from a potential, but non-conservative forces cannot, see section 3.2; this fundamental difference is reflected in statements of the principle.

Constraint forces form another important category of forces. In Newtonian mechanics, such forces are treated like any other applied force but in the principle of virtual, such forces are the object of special treatment, as explained in section 7.5.2. Indeed, it is possible to eliminate constraint forces from the formulation by choosing virtual displacements in a specific manner, as presented in the next sections.

9.3.1 The principle of virtual work for a constrained particle

Application of the principle of virtual work to constrained system will require a close scrutiny of the forces associated with the constraints because the virtual work done by these forces presents special properties. Consider the case of a particle constrained to move in a slot inclined at an angle ψ with respect to the horizontal, as depicted in fig. 9.10. The particle is connected to two springs of stiffness k_1 and k_2, respectively, that remain at all times horizontal and vertical, respectively. The particle is subjected to an externally applied force, $\underline{F}^a$. This simple problem will now be treated using various approaches.

Newtonian approach

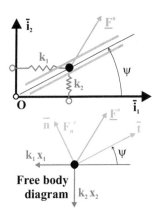

Fig. 9.10. Particle in a slot.

First, the problem depicted in fig. 9.10 will be solved using Newton's approach. If the particle is in static equilibrium, Newton's first law implies $\underline{F}^e + \underline{F}^c + \underline{F}^a = 0$, where $\underline{F}^e = -k_1 x_1 \bar{\imath}_1 - k_2 x_2 \bar{\imath}_2$ is the elastic force the two springs apply on the particle and $\underline{F}^c = F^c \bar{n}$ the constraint force applied by the slot. The tangent and normal to the slot, denoted $\bar{t} = \cos\psi\, \bar{\imath}_1 + \sin\psi\, \bar{\imath}_2$ and $\bar{n} = -\sin\psi\, \bar{\imath}_1 + \cos\psi\, \bar{\imath}_2$, respectively, are defined, and the position vector of the particle is $\underline{r} = x_1\, \bar{\imath}_2 + x_2\, \bar{\imath}_2$. The constraint force solely acts in the direction of the normal because the particle is free to move along the slot. The vector equilibrium equation, projected along unit vectors $\bar{\imath}_1$ and $\bar{\imath}_2$, yields two scalar equations,

$$-k_1 x_1 - \sin\psi F_n^c + F_1^a = 0, \quad \text{and} \quad -k_2 x_2 + \cos\psi F_n^c + F_2^a = 0, \quad (9.53)$$

respectively.

This problem involves three unknowns: the coordinates of the particle, x_1 and x_2, and the constraint force, F_n^c. The solution of the problem will require the above two equations of equilibrium complemented by the constraint equation, $\mathcal{C} = -x_1 \sin\psi +$

$x_2 \cos \psi = 0$. These three equations are easily solved by noticing that a solution of the form $x_1 = r \cos \psi$, $x_2 = r \sin \psi$, where r is the distance from point $\mathbf{O}$ to the particle, automatically satisfies the constraint equation. The equilibrium equations then become $-k_1 r \cos \psi - F_n^c \sin \psi + F_1^a = 0$ and $-k_2 r \sin \psi + F_n^c \cos \psi + F_2^a = 0$, respectively.

The solution is readily found to be $r = F_t^a / k$, where $F_t^a = F_1^a \cos \psi + F_2^a \sin \psi$ is the component of the applied force along tangent $\bar{t}$, and $k = k_1 \cos^2 \psi + k_2 \sin^2 \psi$ is the effective spring constant. The constraint force is $F_n^c = (k_2 - k_1) r \sin \psi - F_n^a$, where $F_n^a = -F_1^a \sin \psi + F_2^a \cos \psi$ is the component of the applied force acting along normal $\bar{n}$. The complete solution is then

$$x_1 = \frac{F_t^a}{k} \cos \psi, \quad x_2 = \frac{F_t^a}{k} \sin \psi, \quad F_n^c = \frac{k_2 - k_1}{k} F_t^a \sin \psi \cos \psi - F_n^a.$$

In conclusion, when treating this problem using Newtonian mechanics, constraint forces are an integral part of the problem; the equations of equilibrium, eqs. (9.53), cannot be written without explicitly taking these forces into account. The complete solution of the problem involves the determination of both displacements and constraint forces.

The principle of virtual work

Next, the same problem will be analyzed with the help of the principle of virtual work developed in section 7.5.1. The particle is in static equilibrium if and only if

$$\delta V^e = (\underline{F}^c + \underline{F}^a)^T \delta \underline{r}, \tag{9.54}$$

for all arbitrary virtual displacements, $\delta \underline{r} = \delta x_1 \bar{\imath}_1 + \delta x_2 \bar{\imath}_2$. The potential of the elastic forces in the springs is $V^e = k_1 x_1^2 / 2 + k_2 x_2^2 / 2$. Expanding the statement of the principle leads to $[k_1 x_1 + \sin \psi F_n^c - F_1^a] \delta x_1 + [k_2 x_2 - \cos \psi F_n^c - F_2^a] \delta x_2 = 0$; because the virtual displacement components are arbitrary, the bracketed terms must vanish, and equations of equilibrium (9.53) are recovered: as expected, the principle of virtual work is equivalent to Newton's first law. Here again, the force of constraint is an integral part of the formulation, and is treated like any other externally applied forces.

Virtual work done by the constraint force

The virtual work done by the constraint force is $\delta W^c = \underline{F}^{cT} \delta \underline{r} = F_n^c \bar{n}^T \delta \underline{r}$. Because virtual displacements are completely arbitrary, this virtual work does not necessarily vanish; indeed, virtual displacements are not required to satisfy the kinematic constraints of the problem.

The particle is confined to remain in the slot, it can move along the tangent vector to the slot only; this direction is called the *kinematically admissible direction*. The direction normal to the slot is called the *kinematically inadmissible direction*, or the *infeasible direction*, because the particle is not allowed to move in that direction.

It is interesting to contrast the *virtual work* done by the constraint forces with the corresponding *differential work*, $dW^c = F_n^c \, \bar{n}^T dr$. The differential displacement, dr, is the true, infinitesimal displacement of the particle along its path. Because it must satisfy the constraint, this differential displacement is along the slot tangential direction, $dr = \bar{t} \, dr$. It follows that the differential work done by the force of constraint does vanish, $dW^c = F_n^c \, \bar{n}^T dr = F_n^c \, \bar{n}^T \bar{t} \, dr = 0$.

Next, the constraint equation is written in a compact manner as $\mathcal{C} = \bar{n}^T r$ and a variation of this constraint is $\delta \mathcal{C} = \underline{B} \delta r = \bar{n}^T \delta r$: for this simple problem, the constraint matrix coincides with the normal to the slot. The virtual work done by the constraint force now becomes

$$\delta W^c = F_n^c \, \delta \mathcal{C}. \tag{9.55}$$

If the virtual displacements satisfy the constraint condition, *i.e.*, if they are along the tangent to the slot, $\delta \mathcal{C} = \bar{n}^T \delta r = 0$, and the virtual work done by the constraint forces vanishes.

Principle of virtual work with kinematically admissible virtual displacements

The concept of *kinematically admissible virtual displacements* was introduced in section 7.5.2; such virtual displacements are not completely arbitrary but are required to satisfy the kinematic constraints. For the case at hand, kinematically admissible virtual displacements are such that $\delta \mathcal{C} = \bar{n}^T \delta r = -\sin \psi \, \delta x_1 + \cos \psi \, \delta x_2 = 0$. This constraint will be automatically satisfied by selecting $\delta x_1 = \cos \psi \, \delta r$ and $\delta x_2 = \sin \psi \, \delta r$, which implies $\delta r = (\cos \psi \, \bar{\imath}_1 + \sin \psi \, \bar{\imath}_2) \delta r = \bar{t} \, \delta r$. As expected, kinematically admissible virtual displacements are virtual displacements along the sole direction compatible with the constraint condition, the direction tangent to the slot. Consequently, kinematically admissible virtual displacements are also called *virtual displacements compatible with the constraints*.

For kinematically admissible virtual displacements, $\delta \mathcal{C} = 0$, and eq. (9.55) implies the vanishing of the virtual work done by the constraint forces. The principle of virtual work now becomes: the particle is in equilibrium if and only if

$$\delta V^e = \underline{F}^{aT} \delta r, \tag{9.56}$$

for all kinematically admissible virtual displacements, $\delta r = \bar{t} \, \delta r$. This principle implies $k_1 x_1 \delta x_1 + k_2 x_2 \delta x_2 = \underline{F}^{aT} \bar{t} \, \delta r$ and finally $(k_1 \cos^2 \psi + k_2 \sin^2 \psi) r \delta r = F_t^a \delta r$. Because the kinematically admissible virtual displacement, δr, is arbitrary, it follows that $r = F_t^a / k$, the same solution as found above.

In summary, the principle of virtual work can be stated as follows.

Principle 17 (Principle of virtual work for a particle) *A particle is in static equilibrium if and only if the virtual work done by the externally applied forces vanishes for all kinematically admissible virtual displacements.*

Discussion

It is interesting to contrast the two statements of the principle of virtual work given by principle 8 on page 273 and principle 17 above. These two principles are nearly identical.

When the principle of virtual work is used with *arbitrary virtual displacements*, as principle 8, the virtual work done by the reactions forces must be included in the statement of the external virtual work because it does not vanish, see eq. (9.54). The principle of virtual work is only another form of Newton's first law stating that the sum of all externally applied forces must vanish for static equilibrium to occur. Newton's first law does not distinguish between various types of forces; "all externally applied forces" means all forces, including the constraint forces.

In contrast, when the principle of virtual work is used with *kinematically admissible virtual displacements*, as principle 17, the virtual work done by the externally applied forces does not include the reaction forces. Indeed, the virtual work they perform automatically vanishes because kinematically admissible virtual displacements are orthogonal to the reaction forces, see eq. (9.56).

These two principles are derived from Newton's law to which they are equivalent. When arbitrary virtual displacements are used, all equilibrium equations of the problem are recovered. If the virtual displacements are limited to those that are kinematically admissible, a subset of the equilibrium equations is recovered.

More generally, consider a system featuring n generalized coordinates and m kinematic constraints for a total of $d = n - m$ degrees of freedom. Application of the principle of virtual work with *arbitrary virtual displacements*, as principle 8, leads to n equations of equilibrium, identical to those obtained from Newton's first law. These equations will involve the n generalized coordinates as well as the m constraint forces associated with the m kinematic constraints. The n equations of equilibrium and m constraint equations are then solved to yield the n unknown generalized coordinates and m constraint forces.

In contrast, application of the principle of virtual with *kinematically admissible virtual displacements*, as principle 17, leads to d equations of equilibrium. These equations will involve the sole d degrees of freedom of the problem; the constraint forces vanish from the formulation.

It is often more convenient to use a formulation based on *kinematically admissible virtual displacements*; indeed, fewer equations are obtained, involving a smaller number of unknowns. This simplification, however, comes at the expense of eliminating the constraint forces from the formulation, thereby loosing all information about these important forces acting on the system.

The use of virtual displacements that violate the kinematic constraints is by no means incorrect. As mentioned earlier, virtual displacements can be interpreted as "mathematical experiments" or "what if?" scenarios. In contrast with real displacements that must indeed satisfy all kinematic constraints, virtual displacements are not constrained to satisfy these same conditions. Of course, if the analyst chooses to work with virtual displacements that violate kinematic constraints, the virtual work associated with the corresponding constraint forces must be taken into account.

Example 9.8. Particle sliding along a curve

The concepts discussed in the previous section will be illustrated by investigating the problem depicted in fig. 9.11. A particle of mass m is sliding along a track whose shape is defined by an arbitrary curve $\mathbb{C}$ described by its intrinsic parametrization; curvilinear variable s measures length along the curve. The particle is connected to point **O** by means of an elastic spring of stiffness constant k and vanishing unstretched length; it is subjected to an externally applied force $\underline{F}^a$.

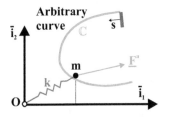
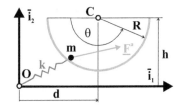

Fig. 9.11. Particle connected to a spring and sliding along a curve.

Fig. 9.12. Particle connected to a spring and sliding along a circle.

Newton's first law is used to derive the equilibrium equations of the system: $\underline{F}^e + \underline{F}^c + \underline{F}^a = 0$. The elastic force the spring applies on the particle is $\underline{F}^e = -kx_1\bar{\imath}_1 - kx_2\bar{\imath}_2$ and the constraint force the track applies on the particle is $\underline{F}^c = F_n^c\bar{n}$, where $\bar{n}$ is the unit vector normal to the track. Projecting Newton's first law along unit vectors $\bar{\imath}_1$ and $\bar{\imath}_2$ then leads to

$$-kx_1 + (\bar{\imath}_1^T\bar{n})F_n^c + F_1^a = 0, \quad \text{and} \quad -kx_2 + (\bar{\imath}_2^T\bar{n})F_n^c + F_2^a = 0, \qquad (9.57)$$

respectively.

These equations involve three unknowns: the coordinates of the particle, x_1 and x_2, and the magnitude of the normal contact force, F_n^c. A third equation is required to solve the problem: the definition of the curve that can be viewed as a kinematic constraint linking generalized coordinates, x_1 and x_2. In general, the shape of the curve will be defined through its intrinsic parametrization; in this case, the position of the particle is defined in terms of s, $x_1 = x_1(s)$ and $x_2 = x_2(s)$. The two equilibrium equations are then sufficient to solve for s and F_n^c.

Next, the principle of virtual work based on *arbitrary virtual displacements* is used to solve the problem. The position vector of the particle is $\underline{r} = x_1\bar{\imath}_1 + x_2\bar{\imath}_2$, and the virtual displacement vector is $\delta\underline{r} = \delta x_1\bar{\imath}_1 + \delta x_2\bar{\imath}_2$. The potential of the elastic forces in the spring is $V^e = k(x_1^2 + x_2^2)/2$. The principle of virtual work then leads to $[kx_1 - (\bar{\imath}_1^T\bar{n})F_n^c - F_1^a]\delta x_1 + [kx_2 - (\bar{\imath}_2^T\bar{n})F_n^c - F_2^a]\delta x_2 = 0$. Because the virtual displacement components are arbitrary, this statement involves the reaction force, F_n^c, and is identical to the equilibrium equations obtained from Newton's first law, eqs. (9.57).

Of course, it is also possible to use the principle of virtual work based on *kinematically admissible virtual displacements*. Kinematically admissible virtual displacements are readily obtained by considering the position of the particle to be an implicit function of the curvilinear coordinate, s, to find $\delta \underline{r} = (d\underline{r}/ds)\delta s = \bar{t}\,\delta s$, where $\bar{t}$ is the unit tangent vector to the curve. Arbitrary virtual changes in the curvilinear coordinate, δs, then generate virtual displacements, $\delta \underline{r} = \bar{t}\,\delta s$, that are compatible with the constraints; as expected, kinematically admissible virtual displacements are along the tangent to the curve.

As discussed in the previous section, the virtual work done by the constraint forces vanishes when using virtual displacements compatible with the constraints. Indeed, the constraint force acts along the normal to the curve and kinematically admissible displacements are along the tangent to the curve. Because the normal and tangent vectors are orthogonal to each other, the virtual work done by the constraint force vanishes, although the constraint force itself does not.

The potential of the elastic forces in the spring can be expressed in terms of curvilinear variable s as $V^e = V^e(s)$. The principle of virtual work now reduces to $(dV^e/ds)\,\delta s = \underline{F}^{aT}\bar{t}\,\delta s$. Because δs is arbitrary, it follows that

$$\frac{dV^e}{ds} = F_t^a, \tag{9.58}$$

where $F_t^a = \bar{t}^T \underline{F}^a$ is the tangential component of the externally applied force. This single equation can be solved to find the equilibrium position of the particle along the curve.

Example 9.9. Particle sliding along a circle

To illustrate the process described in example 9.8, let the curve be a circle of radius R, as shown in fig. 9.12. The coordinates of the particle then become $x_1 = d - R\cos\theta$ and $x_2 = h - R\sin\theta$, where d, h, and θ are defined in the figure. The unit tangent vector to the circle is easily found to be $\bar{t} = \sin\theta\,\bar{\imath}_1 - \cos\theta\,\bar{\imath}_2$, and the unit normal vector is $\bar{n} = -\cos\theta\,\bar{\imath}_1 - \sin\theta\,\bar{\imath}_2$. For a circle, it is more convenient to use an arbitrary parameterization of the curve, angle θ, rather than its intrinsic parameterization, $s = R\theta$.

The equilibrium equations obtained from Newton's first law, eqs. (9.57), are $-k(d - R\cos\theta) + \cos\theta F_n^c + F_1^a = 0$ and $-k(h - R\sin\theta) + \sin\theta F_n^c + F_2^a = 0$. These equations are expressed in terms of the angular position of the particle, angle θ, and the normal contact force, F_n^c. Eliminating the normal contact force yields the angular position of the particle as

$$\tan\theta = \frac{kh - F_2^a}{kd - F_1^a}. \tag{9.59}$$

Finally, the magnitude of the normal contact force becomes

$$F_n^c = \sqrt{(kd - F_1^a)^2 + (kh - F_2^a)^2} - kR. \tag{9.60}$$

To apply the principle of virtual work with kinematically admissible virtual displacements, the potential of the elastic spring is expressed in terms of the particle's angular position as $V^e = 1/2\,k[(d - R\cos\theta)^2 + (h - R\sin\theta)^2]$. It then follows that $dV^e/ds = (dV^e/d\theta)(d\theta/ds) = k(d\sin\theta - h\cos\theta)$. The principle of virtual work, expressed by eq. (9.58), then yields $k(d\sin\theta - h\cos\theta) = F_1^a\sin\theta - F_2^a\cos\theta$. Solving this single equation gives the angular position of the particle, eq. (9.59).

The use of the principle of virtual work based on the kinematically admissible virtual displacements is expeditious: it yields a single equation for the single degree of freedom of the problem. The constraint force does not appear in this equation, in contrast with the case of Newton's formulation.

9.3.2 The principle of virtual work and Lagrange multipliers

The problem discussed in the previous example and depicted in fig. 9.11 involves two scalar kinematic constraints: at all times, the particle must remain on curve $\mathbb{C}$. These constraints will be expressed as $\underline{\mathcal{C}} = \underline{r}(x_1, x_2) - \underline{p}_0(s) = 0$. The configuration of this single degree of freedom system is now represented by three generalized coordinates, the two Cartesian coordinates of the particle, x_1 and x_2, and the curvilinear variable, s, linked by two kinematic constraints. Variation of these constraint can be written as $\delta\underline{\mathcal{C}} = \delta\underline{r} - \bar{t}\,\delta s$, where $\bar{t}$ is the unit tangent vector to the curve.

In all previous examples, the forces of constraint were introduced at the onset of the problem to represent the effect of kinematic constraints. In this section, however, kinematic constraints will be enforced using Lagrange's multiplier method presented in section 9.1. In this approach, the constrained problem is transformed into an unconstrained problem based on an augmented potential, see eq. (9.25),

$$V^+ = V^e + \underline{\lambda}^T\underline{\mathcal{C}} = \frac{1}{2}k(x_1^2 + x_2^2) + \underline{\lambda}^T\left[\underline{r}(x_1, x_2) - \underline{p}_0(s)\right],$$

where the generalized coordinates, x_1, x_2, and s, and Lagrange's multipliers, $\underline{\lambda}$, are all unconstrained variables.

The principle of virtual work now implies $-\underline{F}^{eT}\delta\underline{r} + \underline{\lambda}^T(\delta\underline{r} - \bar{t}\,\delta s) + \delta\underline{\lambda}^T\underline{\mathcal{C}} = \underline{F}^{aT}\delta\underline{r}$. Since all variations $\delta\underline{r}$, δs, and $\delta\underline{\lambda}$ are arbitrary, the following equations are obtained

$$-\underline{F}^e + \underline{\lambda} = \underline{F}^a, \tag{9.61a}$$

$$\underline{\lambda}^T\bar{t} = 0, \tag{9.61b}$$

$$\underline{r} - \underline{p}_0(s) = 0. \tag{9.61c}$$

Equation (9.61a) is the equation of equilibrium for the particle, stating that the sum of the externally applied forces must vanish. Lagrange's multipliers can be interpreted as the constraint forces, $\underline{\lambda} = -\underline{F}^c$, and the equation then becomes $-\underline{F}^e - \underline{F}^c = \underline{F}^a$. Equation (9.61b) implies that Lagrange's multipliers are normal to the tangent vector, *i.e.*, are oriented along the normal to the curve. If Lagrange's multiplier are written as $\underline{\lambda} = \lambda_t\,\bar{t} + \lambda_n\,\bar{n}$, eq. (9.61b) implies $\lambda_t = 0$: the constraint force consists of a sole component acting along the normal direction. Finally, eq. (9.61c) is just the kinematic constraint. The principle of virtual work yields five equations to be solved for the three generalized coordinates, x_1, x_2, and s, and two Lagrange multipliers, λ_t and λ_n.

This simple example reveals the close relationship that exists between Lagrange's multipliers and constraint forces. In section 9.1, the multipliers were introduced as auxiliary mathematical variables devoid of any physical meaning. Within the framework of the principle of virtual work, Lagrange's multipliers are closely related to constraint forces.

It would be a mistake, however, to simply equate Lagrange multipliers and constraint forces. The term that augments the potential of the problem, $\underline{\lambda}^T \underline{C}$, could also be written as $(\underline{\lambda}^T/a)(a\underline{C})$, where a is an arbitrary constant. Lagrange's multipliers would then become proportional to the constraint forces, where a is the constant of proportionality. In the above example, $\underline{\lambda} = -\underline{F}^c$; had the constraint been written as $\underline{C} = -\underline{r} + \underline{p}_0(s)$, the corresponding result would have been $\underline{\lambda} = +\underline{F}^c$.

Finally, the term that augments the potential could be written as $\lambda_1 \mathcal{C}_1^2 + \lambda_2 \mathcal{C}_2^2$, where $\mathcal{C}_1$ and $\mathcal{C}_2$ are the components of constraint $\underline{C}$ resolved in basis $\mathcal{I}$. In this case, although related to the constraint forces, Lagrange's multipliers would not be proportional the constraint forces. Clearly, it is important to determine the precise physical meaning of Lagrange's multipliers to help explain the significance of the equations derived from the principle of virtual work.

Example 9.10. Particle sliding along a curve with friction

In this example, the problem of a particle sliding along a track whose shape is defined by an arbitrary curve $\mathbb{C}$, as depicted in fig. 9.11, will be investigated once again. This time, a friction force, F^f, acts between the particle and the track. The friction force will be assumed to obey Coulomb's law of static friction, *i.e.*, $|F^f| \leq \mu_s |F^n|$, where μ_s is the coefficient of static friction, and F^n the normal force at the frictional interface. The virtual work done by the friction force is $\delta W^f = F^f \delta s$, because the friction force acts in the direction tangent to the curve.

The principle of virtual work based on kinematically admissible virtual displacements is used first. This principle implies $(\mathrm{d}V^e/\mathrm{d}s)\,\delta s = \underline{F}^{aT}\bar{t}\,\delta s + F^f \delta s$, and the equilibrium equation becomes $\mathrm{d}V^e/\mathrm{d}s = F_t^a + F^f$. This equations includes the effect of the friction force and should be compared with its counterpart, eq. (9.58), that ignores this effect.

Unfortunately, this approach does not yield enough information to solve the problem: the single equation of equilibrium involves two unknowns, the curvilinear variable, s, and the friction force that depends on the unknown normal contact force, F^n. The main advantage of the principle of virtual work based on kinematically admissible virtual displacements is to eliminate the constraint forces from the formulation. This advantage turns out to be a drawback in the present situation: the constraint force is, in fact, the normal force at the frictional interface and is required to evaluate the friction force.

In contrast, the principle of virtual work in combination with Lagrange's multiplier technique yields an elegant solution to the problem. Using the notation defined in section 9.3.2, the principle of virtual work now implies $-\underline{F}^{eT}\delta\underline{r}+\underline{\lambda}^T(\delta\underline{r}-\bar{t}\delta s)+\delta\underline{\lambda}^T\underline{C} = \underline{F}^{aT}\delta\underline{r} + F^f \delta s$, and the governing equations of the problem become

$$-\underline{F}^e + \underline{\lambda} = \underline{F}^a, \tag{9.62a}$$

$$-\underline{\lambda}^T \underline{\bar{t}} = F^f, \tag{9.62b}$$

$$\underline{r} - \underline{p}_0(s) = 0. \tag{9.62c}$$

Here again, these equations should be compared with their counterpart, eq. (9.61), that ignore the effect of friction. The second equation only has been modified by the addition of friction: within a sign, the tangential component of Lagrange's multiplier can be interpreted as the friction force itself. Solving these equation with the condition $|F^f| \leq \mu_s |F^n|$ or $|\lambda_t| \leq \mu_s |\lambda_n|$ will yield a range of angular positions of the particle for which equilibrium is possible.

Of course, the principle of virtual work based on arbitrary virtual displacements would also provide enough information to solve the problem because it would bring to light the normal contact force required to quantify the friction force.

Example 9.11. Constrained system of particles, Lagrange multiplier approach
The two systems depicted in fig. 9.13, denoted *system 1* and *system 2*, respectively, are clearly different and feature different equilibrium configurations. Examples 7.14 and 7.15 have treated these two problems using the principle of virtual work, and the equations of equilibrium were found to be given by eq. (7.42a) and (7.42b), for the first and eq. (7.45) for the second.

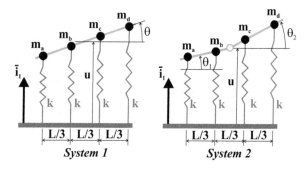

Fig. 9.13. Particle connected to a spring and sliding along a curve.

Imagine that moments of equal magnitudes and opposite signs are now applied at the mid-span hinge of *system 2*. Intuitively, if the magnitude of these moments is "just right," the two parts of the articulated bar will align, and $\theta_1 = \theta_2$. In this case, *systems 1* and *2* become equivalent, and the moment applied at the mid-span articulation of *system 2* is equal to the mid-span internal moment of *system 1*, given by eq. (7.43). Rather than finding the mid-span moment that is "just needed" to align the two articulated bar of *system 2*, it is possible to enforce the kinematic constraint, $\theta_1 = \theta_2$, by means of Lagrange's multiplier technique. In this case, Lagrange's multiplier associated with the constraint will be closely related to the moment applied at the hinge, and hence, to the mid-span internal moment of *system 1*.

Following this latter line of thought, the kinematic constraint $\mathcal{C} = \theta_1 - \theta_2 = 0$ is applied to *system 2* using Lagrange's multiplier technique. The potential of the problem, given by eq. (7.44), is augmented with the following term: $\lambda\mathcal{C} = \lambda(\theta_1 - \theta_2)$. All variables are considered to be unconstrained, leading to the following set of equations

$$
\begin{bmatrix}
4 & -2/3 & 2/3 & 0 \\
-2/3 & 5/18 & 0 & 1 \\
2/3 & 0 & 5/18 & -1 \\
0 & 1 & -1 & 0
\end{bmatrix}
\begin{Bmatrix}
\bar{u} \\ \theta_1 \\ \theta_2 \\ \bar{\lambda}
\end{Bmatrix}
=
\begin{bmatrix}
4\bar{\ell} - 4m\bar{g} \\
-(\bar{\ell}_a/2 + \bar{\ell}_b/6) + (m_a/2 + m_b/6)\bar{g} \\
(\bar{\ell}_d/2 + \bar{\ell}_c/6) - (m_d/2 + m_c/6)\bar{g} \\
0
\end{bmatrix},
$$

where $\bar{\lambda} = \lambda/(kL^2)$ is the non-dimensional multiplier. Note that the first 3×3 system of equations is identical to that derived for the unconstrained system, see eq. (7.45). As expected, the solutions for $\bar{u}$ and $\theta_1 = \theta_2$ are identical to those earlier, see eqs. (7.42a) and (7.42b), respectively. Lagrange's multiplier λ is the "force" that imposes the kinematic condition $\theta_1 = \theta_2$, *i.e.*, it is the mid-span bending moment in the now rigid bar. This can be verified by solving for Lagrange multiplier and observing that it is indeed equal to the mid-span bending moment given in eq. (7.43).

Example 9.12. System of particles with constraints, penalty method approach
Additional insight into Lagrange's multiplier technique can be gained by comparing it with the penalty method for enforcing constraints. Imagine that the two articulated bars shown in fig. 7.24 are connected by a torsional spring of stiffness constant p.

This problem can be solved by adding to the potential of the elastic forces acting in the linear springs, see eq. (7.44), a term for the mid-span torsional spring, $1/2\,p(\theta_1 - \theta_2)^2$, called the *penalty term*. The principle of virtual work based on kinematically admissible virtual displacements then yields the following equations

$$
\begin{bmatrix}
4 & -2/3 & 2/3 \\
-2/3 & 5/18 + \bar{p} & -\bar{p} \\
2/3 & -\bar{p} & 5/18 + \bar{p}
\end{bmatrix}
\begin{Bmatrix}
\bar{u} \\ \theta_1 \\ \theta_2
\end{Bmatrix}
=
\frac{1}{6}
\begin{bmatrix}
24\bar{\ell} - 24m\bar{g} \\
-(3\ell_a + \ell_b) + (3m_a + m_b)\bar{g} \\
(3\ell_d + \ell_c) - (3m_d + m_c)\bar{g}
\end{bmatrix},
$$

(9.63)

where $\bar{p} = p/(kL^2)$ is the non-dimensional stiffness of the mid-span torsional spring.

If the mid-span torsional spring is made increasingly stiffer, *i.e.*, as p increases, the relative rotation of the two rigid bars will become increasingly smaller. Indeed, as p increases, an increasing "penalty," $\bar{p}(\theta_1 - \theta_2)^2/2$, is payed for any violation of the constraint, $\theta_1 \neq \theta_2$. In fact, as $p \to \infty$, the relative rotation vanishes, $\theta_1 \to \theta_2$.

This technique is known as the "penalty method" for enforcing constraints. Figure 9.14 shows the solution of eqs. (9.63) as a function of the penalty factor, $\bar{p}$; note the logarithmic scale on the horizontal axis. The top figure shows the convergence of the mid-span displacement; as $\bar{p}$ increases, the prediction of the penalty method (in dashed lines) converges to the corresponding result for Lagrange's multiplier technique (in solid line). The middle figure shows the rotations of the two bars; both θ_1 and θ_2 converge to the same value, angle $\theta_1 = \theta_2$, predicted using the multiplier technique. Finally, the bottom figure compares the mid-span bending moments; for Lagrange's multiplier technique, the bending moment is simply Lagrange's multiplier, $\bar{M} = \bar{\lambda}$, whereas for the penalty method, $\bar{M} = \bar{p}(\theta_1 - \theta_2)$.

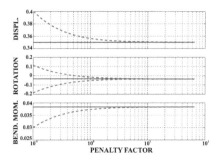

Fig. 9.14. Convergence of the mid-span displacement (top figure), rotations (middle figure), and mid-span bending moment (bottom figure) as the penalty factor, $\bar{p}$, increases. Lagrange multiplier method: solid line; penalty method: dashed line.

The penalty method is easy to use: it does not introduce additional variables, as is the case for Lagrange's multiplier technique, nor does it eliminate any variable. Unfortunately, it is not a robust method: the exact solution of the problem is only recovered when $p \to \infty$, but in practice, a finite value of $\bar{p}$ must be used to avoid the ill conditioning of the system equations, eqs. (9.63). Lagrange's multiplier technique is a rigorous approach to the enforcement of constraints; the penalty method is a convenient approach to obtaining approximate predictions.

The parallel between the two approaches underlines their common physical interpretation. In the penalty method, the penalty term, $V^c = 1/2 \, \bar{p}(\theta_1 - \theta_2)^2$, can be interpreted as the "potential of the constraint forces." Indeed, the constraint moment, $\bar{M}$, can be derived from this potential: $\bar{M} = \mathrm{d}V^c/\mathrm{d}\theta$, where $\theta = \theta_1 - \theta_2$ is the relative mid-span rotation. The constraint is exactly enforced in the limiting case where $\bar{p} \to \infty$, in such a manner that $\bar{p}(\theta_1 - \theta_2)$ converges to a finite value, $\bar{M}$. The potential of the constraint forces now becomes $V^c = 1/2 \, \bar{p}(\theta_1 - \theta_2)(\theta_1 - \theta_2) = 1/2 \, \bar{M}(\theta_1 - \theta_2) = \bar{\lambda}\mathcal{C}$. This means that the term $\bar{\lambda}\mathcal{C}$ introduced in Lagrange's multiplier technique is, in fact, the *potential of the constraint forces*. Because the constraint forces associated with kinematic constraints can be derived from a potential, such forces are conservative. Indeed, the work done by such constraint forces vanishes, leaving the total mechanical energy unchanged, the hallmark of conservative forces.

9.3.3 Problems

Problem 9.9. Particle on a circular track
Consider a particle connected to a fixed point **O** by an spring of stiffness constant k and sliding along a circular track, as depicted in fig. 9.15. Friction acts between the particle and the track. The friction force F^f is assumed to obey Coulomb's law of static friction, *i.e.*, $|F^f| \leq \mu_s |F^n|$, where μ_s is the coefficient of static friction, and F^n the normal force at the frictional interface. *(1)* Derive the governing equations of the problem using the principle of virtual work and Lagrange's multiplier technique. *(2)* Find the equilibrium position, θ_0, of

the particle in the absence of friction, *i.e.*, when $\mu_s = 0$. *(3)* Find the range of equilibrium positions, $\theta_\ell < \theta_e < \theta_u$, as a function of μ_s; θ_ℓ and θ_u are the lower and upper bounds, respectively, of the angular position of the particle for which equilibrium is possible. *(4)* Plot these bounds as a function of the static coefficient of friction μ_s, *i.e.*, plot $\theta_\ell = \theta_\ell(\mu_s)$ and $\theta_u = \theta_u(\mu_s)$. *(5)* Plot the normal contact forces, $\bar{F}_\ell^n(\mu_s)$ and $\bar{F}_u^n(\mu_s)$, acting on the particle when it is located at $\theta_\ell(\mu_s)$ and $\theta_u(\mu_s)$, respectively. *(6)* Plot the friction forces, $\bar{F}_\ell^f(\mu_s)$ and $\bar{F}_u^f(\mu_s)$, acting on the particle when it is located at $\theta_\ell(\mu_s)$ and $\theta_u(\mu_s)$, respectively. *(7)* Plot the total contact forces, $\bar{F}_\ell^c(\mu_s)$ and $\bar{F}_u^c(\mu_s)$, acting on the particle when it is located at $\theta_\ell(\mu_s)$ and $\theta_u(\mu_s)$, respectively. Use the following data: $\bar{d} = d/R = 1.5$; $\bar{h} = h/R = 2$. $\bar{F}^n = F^n/kR$.

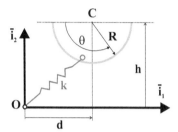
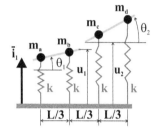

Fig. 9.15. Particle sliding along a curve with friction.

Fig. 9.16. Two bars supported by four springs.

Problem 9.10. Two bars supported by four springs

Figure 9.16 depicts a system consisting of two rigid bars connected to the ground by means of elastic springs of stiffness k. The un-stretched lengths of the four springs are ℓ_a, ℓ_b, ℓ_c, and ℓ_d, respectively, and the four masses are m_a, m_b, m_c, and m_d, respectively. The configuration of the left bar is represented by the displacement, u_1, of its right end and its orientation, θ_1. The configuration of the right bar is represented by the displacement, u_2, of its left end and its orientation, θ_2. *(1)* Use the principle of virtual work to find the solution of the problem. *(2)* Use Lagrange's multiplier technique to enforce two kinematic constraints: $\mathcal{C}_1 = u_1 - u_2 = 0$ and $\mathcal{C}_2 = \theta_1 - \theta_2 = 0$. *(3)* Solve the constrained problem and show that its solution is identical to that found in example 7.14. *(4)* What is the physical meaning of Lagrange's multipliers? *(5)* Validate your solution by comparing the value of Lagrange's multipliers with predictions based on statics arguments. *(6)* Enforce the constraints $\mathcal{C}_1$ and $\mathcal{C}_2$ using the penalty method. *(7)* Demonstrate the convergence of the solution of the penalty method to that of Lagrange's multiplier technique as the penalty factor increases. Plot the displacements, rotations, and loads as a function of the penalty factor for both solutions. Use the following data: $\bar{\ell}_a = \ell_a/L = 0.5$; $\bar{\ell}_b = \ell_b/L = 0.75$; $\bar{\ell}_c = \ell_c/L = 0.60$; $\bar{\ell}_d = \ell_d/L = 0.30$; $m_a = 1.2$ kg; $m_b = 1.50$ kg; $m_c = 0.60$ kg; $m_d = 0.45$ kg; $\bar{g} = g/kL = 0.2$.

Problem 9.11. Four springs supporting a rigid bar

Figure 7.23 depicts a system consisting of a rigid bar connected to the ground by means of elastic springs of stiffness k. The un-stretched lengths of the four springs are ℓ_a, ℓ_b, ℓ_c, and ℓ_d, respectively. The configuration of system is represented by the displacements, u_a, u_b, u_c, and u_d of the four masses, m_a, m_b, m_c, and m_d, respectively. *(1)* Use the principle of virtual work to find the solution of the unconstrained problem, *i.e.*, in the absence of the rigid bar. *(2)* Use Lagrange's multiplier technique to enforce the four kinematic constraints imposed by the rigid bar: $\mathcal{C}_1 = x_a - (x - L\theta/2) = 0$, $\mathcal{C}_2 = x_b - (x - L\theta/6) = 0$, $\mathcal{C}_3 = x_c - (x + L\theta/6) = 0$, $\mathcal{C}_4 = x_d - (x + L\theta/2) = 0$, where x is the mid-span displacement of the bar and θ its rotation. *(3)* Solve the constrained problem and show that its solution is identical to that found in example 7.14. *(4)* What is the physical meaning of Lagrange's multipliers? *(5)* Validate your solution by comparing the value of Lagrange's multipliers with predictions based on statics arguments. *(6)* Enforce the constraints $\mathcal{C}_1$, $\mathcal{C}_2$, $\mathcal{C}_3$, and $\mathcal{C}_4$ using the penalty method. *(7)* Demonstrate the convergence of the solution of the penalty method to that of Lagrange's multiplier technique as the penalty factor increases. Plot the displacements, rotations, and forces as a function of the penalty factor for both solutions. Use the following data: $\bar{\ell}_a = \ell_a/L = 0.5$; $\bar{\ell}_b = \ell_b/L = 0.75$; $\bar{\ell}_c = \ell_c/L = 0.60$; $\bar{\ell}_d = \ell_d/L = 0.30$; $m_a = 1.2$ kg; $m_b = 1.50$ kg; $m_c = 0.60$ kg; $m_d = 0.45$ kg; $\bar{g} = g/kL = 0.2$.

Problem 9.12. A rigid bar suspended by two spring

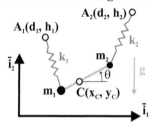

Figure 9.17 depicts a massless rigid bar of length ℓ with end masses, m_1 and m_2, suspended by two springs of stiffness constants k_1 and k_2. The springs, of un-stretched lengths ℓ_1 and ℓ_2, respectively, are connected to the ground at points $\mathbf{A}_1$ and $\mathbf{A}_2$, of coordinates (d_1, h_1) and (d_2, h_2), respectively. Gravity acts on the system in the direction indicated on the figure. Two different sets of generalized coordinates will be used to represent the system. A three generalized coordinate representation: x_c

Fig. 9.17. A rigid bar suspended by two spring.

and y_c, the position of the center of mass of the system, and θ, the orientation of the rigid bar. The second is a four generalized coordinate representation using x_1 and y_1, the coordinates of mass m_1, and x_2 and y_2, the coordinates of mass m_2. *(1)* Using the **first set** of generalized coordinates, find the equilibrium configuration of the system based on the principle of virtual work. *(2)* Find the internal force in the rigid bar. *(3)* For the **second set** of generalized coordinates, determine the kinematic constraint that links the four generalized coordinates. *(4)* Find the equilibrium configuration of the system based on the principle of virtual work and Lagrange's multiplier technique. *(5)* Provide a physical interpretation of Lagrange's multiplier. *(6)* Show that the solutions obtained with the two sets of generalized coordinates are identical. *(7)* Using the **second set** of generalized coordinates, find the equilibrium configuration of the system based on the principle of virtual work and the penalty method. *(8)* Study the convergence of this latter solution as the penalty coefficient increases. Use the following data: $\bar{d}_1 = d_1/\ell = 0.5$, $\bar{h}_1 = h_1/\ell = 1.5$, $\bar{d}_2 = d_2/\ell = 2$, $\bar{h}_2 = h_2/\ell = 0.8$; $\ell_1 = \ell_2 = 0$; $\bar{m}_1 = m_1 g/(k\ell) = 5$, $\bar{m}_2 = m_2 g/(k\ell) = 8$; $k = k_1 = k_2$.

10

Constrained systems: classical formulations

Chapter 8 presented variational and energy principles for unconstrained dynamical system. This chapter generalizes these formulations to enable the treatment of constrained systems. D'Alembert's principle is treated in section 10.1. The generalization of Hamilton's principle and Lagrange's formulation to systems with holonomic constraints is presented in section 10.2 and section 10.3 generalizes the same formulations to systems with nonholonomic constraints.

The second part of the chapter deals with constraints in multibody systems. While mechanical systems employ many types of joints, the lower pair joints are the most commonly used and section 10.4 describes their kinematic characteristics. Section 10.5 develops generic constraints that will be used for all lower pair joints, and the specific constraints associated with each of the six lower pair joints are detailed in section 10.6. The chapter concludes with a cursory look at a few additional joints.

10.1 D'Alembert's principle for constrained systems

D'Alembert's principle was derived in section 8.1 and is expressed by eq. (8.3). This principle involves the virtual work done by the various forces acting on the system: the inertial forces, the conservative forces, and the non-conservative forces. When using kinematically admissible virtual displacements, the virtual work done by the forces that impose kinematic constraints does vanish, as discussed in section 9.3.1; consequently, such forces do not appear in d'Alembert's formulation.

If redundant generalized coordinates are used, the system will be subjected to kinematic constraints, and virtual changes in these generalized coordinates will no longer be kinematically admissible. As discussed in section 9.3.1, it will then be required to take into account the virtual work done by the constraint forces, because this virtual work does not vanish for virtual displacements that are not kinematically admissible.

The virtual work done by the constraint forces is elegantly introduced into the formulation by means of Lagrange's multipliers; as presented in section 9.3.2, the potential of the conservative forces is augmented by the potential of the constraint forces expressed in terms of Lagrange's multipliers.

Example 10.1. The double pendulum

Consider the double pendulum system depicted in fig. 10.1. The first bar is of length L_1, mass m_1, and is connected by hinges to the ground at point $\mathbf{O}$ and to the second bar at point $\mathbf{A}$. The second bar is of length L_2 and mass m_2. The bars have orientation angles θ_1 and θ_2 with respect to the vertical, respectively. $\mathcal{I} = (\bar{\imath}_1, \bar{\imath}_2, \bar{\imath}_3)$ is an inertial basis; bases $\mathcal{E} = (\bar{e}_1, \bar{e}_2, \bar{e}_3)$ and $\mathcal{A} = (\bar{a}_1, \bar{a}_2, \bar{a}_3)$ are attached to the first and second bars, respectively. This problem was treated in example 8.2 on page 300 with d'Alembert's principle using kinematically admissible virtual displacements. Derive the equations of motion of the system using d'Alembert's principle with a redundant set of generalized coordinates.

It is assumed that the internal forces at point $\mathbf{A}$ must be evaluated as part of the solution process. The following generalized coordinates will be used: angles θ_1 and θ_2 and the position vector of point $\mathbf{A}$ with respect to point $\mathbf{O}$, denoted $\underline{r}_A$. Since the system features two degrees of freedom only, two constraints must exist between these four generalized coordinates: $\underline{\mathcal{C}} = \underline{r}_A - L_1\bar{e}_1 = 0$. These two constraints will be enforced using Lagrange's multiplier technique.

The configuration of the first bar is defined by the sole generalized coordinate θ_1, and hence, the difference between the variation of the potential energy and virtual work done by the inertial forces is identical to that given by eq. (8.12).

The configuration of the second bar now involves three generalized coordinates, the position of point $\mathbf{A}$, $\underline{r}_A$, and the orientation of the second bar, θ_2. The potential of the gravity forces acting of the bar is $V_2 = m_2 g(\bar{\imath}_2^T \underline{r}_A - L_2 C_2/2)$. The virtual work done by the inertial forces is obtained from eq. (8.8), using point $\mathbf{A}$ as the reference point now becomes

$$\delta W_2^I = -\delta \underline{r}_A^T m_2 \left[\underline{a}_A + \ddot{\theta}_2 \tilde{\imath}_3 \frac{L_2}{2} \bar{a}_1 - \dot{\theta}_2^2 \frac{L_2}{2} \bar{a}_1 \right] - \delta\theta_2 \left[m_2 \bar{\imath}_3^T \frac{L_2}{2} \tilde{a}_1 \underline{a}_A + I_2^A \ddot{\theta}_2 \right].$$

Although this expression is identical to that obtained earlier, see eq. (8.13), $\underline{r}_A$ is now an independent, unconstrained generalized coordinate, which implies that $\delta \underline{r}_A \neq L_1 \delta\theta_1 \bar{e}_2$.

According to Lagrange's multiplier technique, the potential of the constraint forces is $V^c = \underline{\mu}^T(\underline{r}_A - L_1\bar{e}_1)$, where $\underline{\mu}$ is the array of Lagrange's multipliers used to enforce the two constraints. Variation of this potential is

$$\delta V^c = \delta\underline{\mu}^T(\underline{r}_A - L_1\bar{e}_1) + \delta\underline{r}_A^T\underline{\mu} - \delta\theta_1 L_1\underline{\mu}^T\bar{e}_2.$$

D'Alembert's principle now implies $\delta(V_1 + V_2 + V^c) - \delta(W_1^I + W_2^I) = 0$ for all arbitrary variations $\delta\theta_1$, $\delta\theta_2$, and $\delta\underline{r}_A$, leading to the following equations of motion

$$\frac{m_1 L_1^2}{3}\ddot{\theta}_1 + m_1 g \frac{L_1}{2} S_1 - L_1 \bar{e}_2^T \mu = 0, \qquad (10.1a)$$

$$m_2 \frac{L_1 L_2}{2} C_{21}\ddot{\theta}_1 + \frac{m_2 L_2^2}{3}\ddot{\theta}_2 + m_2 g \frac{L_2}{2} S_2 + m_2 \frac{L_1 L_2}{2}\dot{\theta}_1^2 S_{21} = 0, \qquad (10.1b)$$

$$\mu + m_2 g \bar{\imath}_2 + m_2 L_1 (\ddot{\theta}_1 \bar{e}_2 - \dot{\theta}_1^2 \bar{e}_1) + \frac{m_2 L_2}{2}(\ddot{\theta}_2 \bar{a}_2 - \dot{\theta}_2^2 \bar{a}_1) = 0, \qquad (10.1c)$$

respectively.

Equation (10.1b), the coefficient of the arbitrary variation $\delta\theta_2$, is identical to the corresponding equation obtained by using kinematically admissible virtual displacements; it provides the first equation of motion for the problem. Equation (10.1a) is identical to that obtained from Newton's approach, eq. (8.10b), provided that Lagrange's multipliers are interpreted as the internal force at the joint, $\mu = H_A \bar{\imath}_1 + V_A \bar{\imath}_2$, where H_A and V_A are the horizontal and vertical components of the internal force transmitted at point **A**, respectively. Finally, eq. (10.1c) is identical to that obtained from the Newtonian approach, see eq. (8.11a), and provides an expression for Lagrange's multipliers, μ, which are the desired internal forces. Elimination of Lagrange's multipliers from the first and third equations yields the second equation of motion for the problem.

This example demonstrates the versatility of d'Alembert's principle coupled with Lagrange's multiplier technique. Through a judicious choice of generalized coordinates and constraints, a set of equations involving the desired unknowns of the problem is generated.

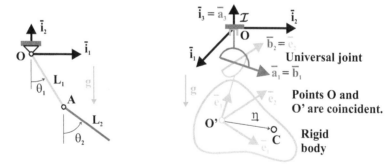

Fig. 10.1. Configuration of the double pendulum system.

Fig. 10.2. Configuration of the rigid body connected to a universal joint.

Example 10.2. The rigid body/universal joint system

This example deal with a rigid body attached to the ground by means of a universal joint, see fig. 10.2 and examples 7.5 and 8.3, on pages 266 and 302, respectively. Component k of the universal joint, see fig. 7.10, is connected to the ground at point **O** by means of a bearing allowing rotation about axis $\bar{\imath}_3$. Component ℓ is connected to a rigid body at point **O'**. The orientation of the rigid body will be defined by Euler angles, using the *3-1-2* sequence. A first planar rotation about axis $\bar{\imath}_3$, of magnitude ϕ, brings inertial basis $\mathcal{I} = (\bar{\imath}_1, \bar{\imath}_2, \bar{\imath}_3)$ to $\mathcal{A} = (\bar{a}_1, \bar{a}_2, \bar{a}_3)$, where $\bar{a}_1$ is aligned with unit vector $\bar{b}_1$ of the cruciform. This rotation is associated with a constant angular speed $\dot{\phi} = \Omega$, implying $\bar{a}_1(t) = \cos(\Omega t)\bar{\imath}_1 + \sin(\Omega t)\bar{\imath}_2$. A second planar rotation about axis $\bar{a}_1$, of magnitude θ, brings basis $\mathcal{A}$ to $\mathcal{B} = (\bar{b}_1, \bar{b}_2, \bar{b}_3)$, where $\bar{b}_2$ is the second unit vector aligned with the cruciform. Finally, a third planar rotation of magnitude ψ about axis $\bar{b}_2$ bring basis $\mathcal{B}$ to $\mathcal{E} = (\bar{e}_1, \bar{e}_2, \bar{e}_3)$ that is attached to the rigid body. The components of tensors in basis $\mathcal{E}$ will be denoted with the superscript $(\cdot)^*$. Points $\mathbf{O}$ and $\mathbf{O}'$ are coincident. Find the equations of motion of the system using d'Alembert's principle and the torque, $Q(t)$, required to drive the system at a constant angular velocity Ω.

This system features two degrees of freedom, but the configuration of the system will be represented by three generalized coordinates, the three Euler angles, ϕ, θ, and ψ, defined above. A rheonomic constraint, $\mathcal{C} = \phi - \Omega t$, will be enforced using Lagrange's multiplier technique by adding to the potential of the system the potential of the constraint force, $V^c = \lambda (\phi - \Omega t)$, where λ is Lagrange's multiplier. Variation of this potential can be written as

$$\delta V^c = \delta\lambda (\phi - \Omega t) + \delta\phi\lambda = \delta\lambda (\phi - \Omega t) + \{\delta\phi, \delta\theta, \delta\psi\} \lambda\bar{\imath}_1,$$

where $\bar{\imath}_1^T = \{1, 0, 0\}$. The term $\delta\phi\lambda$ corresponds to the virtual work done by a torque, λ, undergoing a virtual rotation, $\delta\phi$; hence, Lagrange's multiplier λ can be interpreted as the torque required to enforce the constraint $\phi - \Omega t = 0$, *i.e.*, λ is the driving torque.

D'Alembert's principle now states

$$\{\delta\phi, \delta\theta, \delta\psi\} \underline{\underline{H}}^{*T} \left[mg\widetilde{\underline{\eta}}^* \underline{\underline{R}}^T \bar{\imath}_3 + \underline{\underline{I}}^{O*} \underline{\dot{\omega}}^* + \widetilde{\underline{\omega}}^* \underline{\underline{I}}^{O*} \underline{\omega}^* \right]$$
$$= \delta\lambda (\phi - \Omega t) + \{\delta\phi, \delta\theta, \delta\psi\} \lambda\bar{\imath}_1,$$

where the tangent operator, $\underline{\underline{H}}^*$, is defined by eq. (4.80). Because Lagrange's multiplier technique is used, ϕ, θ, ψ, and λ all are unconstrained variables and their variations arbitrary, leading to the following equations of motion: $\phi - \Omega t = 0$, the constraint to be enforced, and

$$\underline{\underline{H}}^{*T} \left[mg\widetilde{\underline{\eta}}^* \underline{\underline{R}}^T \bar{\imath}_3 + \underline{\underline{I}}^{O*} \underline{\dot{\omega}}^* + \widetilde{\underline{\omega}}^* \underline{\underline{I}}^{O*} \underline{\omega}^* \right] = \lambda\bar{\imath}_1.$$

This matrix equation represents three independent scalar equations; the last two are identical to the equations of motion obtained above when using kinematically admissible virtual displacements, see eq. (8.15). The first equation yields the desired torque as

$$\lambda = \underline{h}^{*T} \left[mg\widetilde{\underline{\eta}}^* \underline{\underline{R}}^T \bar{\imath}_3 + \underline{\underline{I}}^{O*} \underline{\dot{\omega}}^* + \widetilde{\underline{\omega}}^* \underline{\underline{I}}^{O*} \underline{\omega}^* \right],$$

where $\underline{h}^*$ stores the first column of $\underline{\underline{H}}^*$. Through the proper selection of the constraint and associated Lagrange's multiplier, the system's equations of motion are obtained together with the additional equation defining the driving torque.

Example 10.3. The rigid body/universal joint system

The previous example focused on a rigid body attached to the ground by means of a universal joint, as depicted in fig. 10.2. The equations of motion of the system were obtained, and one additional equation determining the driving torque was derived based on Lagrange's formulation. For specific applications, the reaction forces acting at point $\mathbf{O}$ might also be important. Derive the equations of motion of the system and determine the reaction forces at point $\mathbf{O}$.

The reactions forces at point $\mathbf{O}$ do not appear in the formation presented in example 10.2 because the virtual displacement components at point $\mathbf{O}$ were selected to be kinematically admissible, *i.e.*, virtual displacements vanish at point $\mathbf{O}$, $\underline{r}_O = 0$, implying $\delta\underline{r}_O = 0$. Here again, Lagrange's multiplier technique will be used to generate the equations required to evaluate these forces: the displacement components at point $\mathbf{O}$, $\underline{r}_O$, are now considered to be generalized coordinates, and constraint $\mathcal{C} = \underline{r}_O = 0$ is imposed. The potential of the constraints now becomes $V^c = \lambda(\phi - \Omega t) + \underline{\mu}^T \underline{r}_O$, where $\underline{\mu}$ is a set of Lagrange's multipliers used to enforce the constraint $\underline{r}_O = 0$.

Variation of this potential can be written as

$$\delta V^c = \delta\lambda\,(\phi - \Omega t) + \{\delta\phi, \delta\theta, \delta\psi\}\,\lambda\bar{\imath}_1 + \delta\underline{\mu}^T \underline{r}_O + \delta\underline{r}_O^T \underline{\mu}.$$

The term $\delta\underline{r}_O^T \underline{\mu}$ corresponds to the virtual work done by a force, $\underline{\mu}$, undergoing a virtual displacement, $\delta\underline{r}_O$; hence, the array of Lagrange's multipliers, $\underline{\mu}$, can be interpreted as the force required to enforce the constraint $\underline{r}_O = 0$, *i.e.*, $\underline{\mu}$ is the reaction force vector at point $\mathbf{O}$.

The potential of the gravity forces must be updated to accommodate the new displacement field, $V = mg\bar{\imath}_3^T(\underline{r}_O + \underline{\eta})$, and variation of this new potential is $\delta V = \delta\underline{r}_O^T mg\,\bar{\imath}_3 + \delta\underline{\psi}^{*T} mg\underline{\widetilde{\eta}}^* \underline{R}^T \bar{\imath}_3$. D'Alembert's principle now states

$$\delta\underline{r}_O^T \underline{R}\left[mgR^T\bar{\imath}_3 + (\dot{\widetilde{\omega}}^* + \widetilde{\omega}^*\widetilde{\omega}^*)m\underline{\eta}^*\right] + \delta\underline{\psi}^{*T}\left[mg\underline{\widetilde{\eta}}^* \underline{R}^T\bar{\imath}_3 + \underline{I}^{O*}\dot{\underline{\omega}}^* + \widetilde{\omega}^*\underline{I}^{O*}\underline{\omega}^*\right]$$
$$= \delta\lambda\,(\phi - \Omega t) + \{\delta\phi, \delta\theta, \delta\psi\}\,\lambda\bar{\imath}_1 + \delta\underline{\mu}^T \underline{r}_O + \delta\underline{r}_O^T \underline{\mu}.$$

Because Lagrange's multiplier technique is used, the Euler angles, displacement $\underline{r}_O$, and multipliers λ and μ all are unconstrained variables and their variations arbitrary, leading to the following equations of motion: $\phi - \Omega t = 0$ and $\underline{r}_O = 0$, the constraints to be enforced, and

$$\underline{\mu} = \underline{R}\left[mgR^T\bar{\imath}_3 + (\dot{\widetilde{\omega}}^* + \widetilde{\omega}^*\widetilde{\omega}^*)m\underline{\eta}^*\right],$$
$$\underline{H}^{*T}\left[mg\underline{\widetilde{\eta}}^* \underline{R}^T\bar{\imath}_3 + \underline{I}^{O*}\dot{\underline{\omega}}^* + \widetilde{\omega}^*\underline{I}^{O*}\underline{\omega}^*\right] = \lambda\bar{\imath}_1.$$

As expected, the last three equations are identical to those obtained earlier and the first three equations yield the reaction force at point $\mathbf{O}$.

Clearly, the combination of d'Alembert's principle and Lagrange's multiplier technique provides a powerful approach to the analysis of constrained dynamical systems. Selecting various sets of generalized coordinates gives equations of motion involving the variables of interest. In particular, if the number of generalized coordinates equals the number of degrees of freedom of the system, a minimum set of equations is obtained from which reaction forces are completely eliminated. The equations required for the evaluation of the reaction forces can then be obtained by using Lagrange's multiplier technique.

10.1.1 Problems

Problem 10.1. The double pendulum

Consider the double pendulum system depicted in fig. 10.1. The first bar is of length L_1, mass m_1, and is connected by hinges to the ground at point **O** and to the second bar at point **A**. The second bar is of length L_2 and mass m_2. The bars have orientation angles θ_1 and θ_2 with respect to the vertical, respectively. $\mathcal{I} = (\bar{\imath}_1, \bar{\imath}_2, \bar{\imath}_3)$ is an inertial basis, and bases $\mathcal{E} = (\bar{e}_1, \bar{e}_2, \bar{e}_3)$ and $\mathcal{A} = (\bar{a}_1, \bar{a}_2, \bar{a}_3)$ are attached to the first and second bars, respectively. Use four generalized coordinates to represent the configuration of the system: the angles θ_1 and θ_2, and the position of point **O**, denoted $\underline{r}_O$. Enforce the two constraints $\underline{\mathcal{C}} = \underline{r}_O = 0$ using Lagrange's multiplier technique. *(1)* Derive the equations of motion of the system using d'Alembert's principle. *(2)* Prove that your equations are correct by comparing them to those obtained in example 10.1. *(3)* Give the physical interpretation of Lagrange's multipliers. *(4)* On one graph, plot the time history of the angles θ_1 and θ_2. *(5)* Plot the trajectories of the points at the tip of the first and second bars. *(6)* Plot the angular velocities of the two bars. *(7)* Plot the horizontal and vertical components of the internal force at point **A**. *(8)* Plot the horizontal and vertical components of the reaction force at point **O**. *(9)* Plot the kinetic, potential, and total mechanical energies of the system. Comment on your results. Use the following data: $m_1 = 1.2$ kg; $m_2 = 5$ kg; $L_1 = 0.4$ m; $L_2 = 0.6$ m; acceleration of gravity $g = 9.81$ m/s^2. Present all results for a period of 10 s. The initial conditions are: $\theta_1(t = 0) = \theta_2(t = 0) = \pi/2$; $\dot{\theta}_1(t = 0) = \dot{\theta}_2(t = 0) = 0$.

Problem 10.2. Crank-slider mechanism

The crank-slider mechanism depicted in fig. 9.8 consists of a uniform crank of length L_1 and mass m_1 connected to the ground at point **O**; let θ be the angle from the horizontal to the crank. At point **B**, the crank connects to a uniform linkage of length L_2 and mass m_2 that slides along point **P**, a fixed point in space, located at a distance d from point **O**. Let w denote the distance from point **B** to point **P** and ϕ the angle from the horizontal to link **BP**. The system is represented by three generalized coordinates: θ, ϕ, and w. *(1)* Derive the equations of motion of the system using d'Alembert's principle. *(2)* Give the physical interpretation of Lagrange's multipliers. *(3)* Find the single equation of motion of the system expressed in terms of a single degree of freedom, θ.

Problem 10.3. The two-bar linkage with slider system

The two-bar linkage with slider system shown in fig. 8.4 is a planar mechanism. It consists of a uniform crank of length L_1 and mass m_1 connected to the ground at point **O**; let θ be the angle from the horizontal to the crank. At point **B**, the crank slides over a uniform linkage of length L_2 and mass m_2 that is connected to the ground at point **A**. Let w denote the distance from point **B** to point **A** and ϕ the angle from the horizontal to link **BA**. *(1)* Using three generalized coordinates, θ, ϕ, and w, derive the equations of motion of the system using d'Alembert's principle. *(2)* Give the physical interpretation of Lagrange's multipliers. *(3)* Find the single equation of motion of the system expressed in terms of a generalized coordinate, θ.

Problem 10.4. Rigid body attached to universal joint

Figure 10.2 depicts a rigid body attached to the ground by means of a universal joint. Component k of the universal joint, see fig. 7.10, is connected to the ground at point $\mathbf{O}$ by means of a bearing allowing rotation about axis $\bar{\imath}_3$. Component ℓ is connected to a rigid body at point $\mathbf{O}'$. The orientation of the rigid body will be defined by Euler angles, using the *3-1-2* sequence. A first planar rotation about axis $\bar{\imath}_3$, of magnitude ϕ, brings inertial basis $\mathcal{I} = (\bar{\imath}_1, \bar{\imath}_2, \bar{\imath}_3)$ to $\mathcal{A} = (\bar{a}_1, \bar{a}_2, \bar{a}_3)$, where $\bar{a}_1$ is aligned with unit vector $\bar{b}_1$ of the cruciform. This rotation is associated with a constant angular speed $\dot{\phi} = \Omega$, implying $\bar{a}_1(t) = \cos(\Omega t)\bar{\imath}_1 - \sin(\Omega t)\bar{\imath}_3$. A second planar rotation about axis $\bar{a}_1$, of magnitude θ, brings basis $\mathcal{A}$ to $\mathcal{B} = (\bar{b}_1, \bar{b}_2, \bar{b}_3)$, where $\bar{b}_2$ is the second unit vector aligned with the cruciform. Finally, a third planar rotation of magnitude ψ about axis $\bar{b}_2$ bring basis $\mathcal{B}$ to $\mathcal{E} = (\bar{e}_1, \bar{e}_2, \bar{e}_3)$ that is attached to the rigid body. The components of tensors in basis $\mathcal{E}$ will be denoted with the superscript $(\cdot)^*$. Points $\mathbf{O}$ and $\mathbf{O}'$ are coincident. For all questions, use d'Alembert's principle and Lagrange's multiplier technique, when necessary. *(1)* Find the equations of motion of the system. *(2)* On one graph, plot the time history of angles θ and ψ. *(3)* On one graph, plot $\dot{\theta}$ and $\dot{\psi}$. *(4)* On one graph, plot the three components of the unit vector $\bar{e}_1$ in basis $\mathcal{I}$. *(5)* Same question for unit vectors $\bar{e}_2$ and $\bar{e}_3$. *(6)* Plot the trajectory of the center of mass of the rigid body in three-dimensional space. *(7)* On one graph, plot the three components of the angular velocity vector of the rigid body in the body attached basis $\mathcal{E}$. *(8)* Plot the components of the same vector in the inertial basis $\mathcal{I}$. *(9)* Plot the history of the driving torque required to maintain the constant angular velocity Ω. *(10)* Compute the cumulative work W done by the driving torque. *(11)* On one graph, plot the kinetic energy of the system, its potential energy and the cumulative work W. Will a combination of these quantities remain constant? *(12)* Plot the components of the reaction force at point $\mathbf{O}$ in basis $\mathcal{I}$. Use the following data: mass of the body $m = 2.8$ kg; principal mass moments of inertia about the center of mass $I_1^{*C} = 1.1$; $I_2^{*C} = 0.6$; $I_3^{*C} = 0.9$ kg·m^2; components of the relative position vector of the center of mass with respect to point $\mathbf{O}$, $\underline{\eta}^{*T} = \{0.1, -0.4, 0.3\}$ m; acceleration of gravity $g = 9.81$ m/s^2; angular velocity $\Omega = 2$ rad/s. Present all results in a non-dimensional manner; use the reference mass $m_r = m$, reference length $\ell_r = \|\underline{\eta}^*\|$ and reference time $t_r = 1/\Omega$. At the initial time, the principal axes of inertia are aligned with the inertial system and the body is at rest. Present the response of the system over a non-dimensional period of 8π.

Problem 10.5. Particle in a circular slot with guiding arm

A particle of mass M slides along a circular slot of radius R, as shown in fig. 8.15. The particle also slides in a rectilinear slot in an arm of mass m and length L. The arm is pivoted to the ground at point $\mathbf{O}$ and is restrained by a torsional spring of stiffness constant k and a dashpot of constant c. The spring is un-stretched when the arm is horizontal. A viscous friction force, $F^f = -\mu\dot{w}$ is acting at the interface between the particle and the arm. *(1)* Using three generalized coordinates, x, y, the coordinates of mass M, and w, the position of the particle along the arm, derive the equations of motion of the system using d'Alembert's principle. *(2)* Give the physical interpretation of Lagrange's multipliers.

Problem 10.6. The crank piston mechanism

The crank slider mechanism depicted in fig. 10.4 comprises a bar of length L_1 and mass m_1 connected to the ground at point $\mathbf{O}$ by means of a hinge. At point $\mathbf{A}$, a hinge connects the first bar to a second bar of length L_2 and mass m_2. A slider of mass M, that is constrained to move in the horizontal direction, is connected to this second bar. A spring of stiffness constant k connects the slider to the ground and is un-stretched when the two bars are aligned. This system will be represented with three generalized coordinates: x and y, the coordinates of point $\mathbf{A}$ and z, the horizontal position of point $\mathbf{B}$. *(1)* Write the constraint equations for this

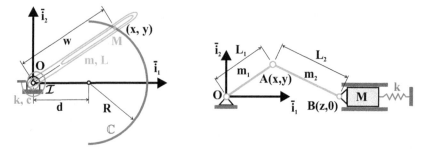

Fig. 10.3. Particle in a circular slot with guiding arm.

Fig. 10.4. Crank piston mechanism with spring.

problem. *(2)* Use d'Alembert's principle to derive the equation of motion of the system. *(3)* Give the physical interpretation of Lagrange's multipliers.

10.2 Hamilton's principle and Lagrange's formulation with holonomic constraints

As discussed in section 9.2, mechanical systems are often subjected to constraints which fall into two broad categories: holonomic and nonholonomic constraints. Furthermore, constraints can be scleronomic, if they are not an explicit function of time, or rheonomic, in the opposite case. Systems subjected to holonomic constraints are treated in this section, and those subjected to nonholonomic constraints are treated in section 10.3.

The most general type of constraints to be considered in this section are in the form of eq. (9.38), $\underline{C}(\underline{q}, t) = 0$, *i.e.*, rheonomic constraints. If time does not appear explicitly in the constraint, it is scleronomic, $\underline{C}(\underline{q}) = 0$. Constraints limit the allowable virtual displacement in such a way that

$$\delta \underline{C} = \underline{\underline{B}}(\underline{q}, t)\delta \underline{q} = 0. \tag{10.2}$$

Because a virtual displacement is an arbitrary change in displacement *at a given, fixed instant*, this expression does not involve the term $\underline{b}(\underline{q}, t)$ that appears in the differential or Pfaffian form of the constraints.

Hamilton's principle, eq. (8.20), will be written in the following form,

$$\int_{t_i}^{t_f} \left(\delta L + \delta W^{nc}\right) \, dt = 0, \tag{10.3}$$

for all arbitrary virtual displacements. The boundary terms at the initial and final times have been ignored. For constrained systems, virtual displacements are not arbitrary because they must satisfy the constraints as expressed by eq. (10.2).

A linear combination of eqs. (10.3) and (10.2) now yields

$$\int_{t_i}^{t_f} \left(\delta L + \underline{\lambda}^T \delta\underline{\mathcal{C}} + \delta\underline{\lambda}^T \underline{\mathcal{C}} + \delta W^{nc} \right) dt = 0, \tag{10.4}$$

where $\underline{\lambda}$ is an array of arbitrary Lagrange's multipliers. The term $\delta\underline{\lambda}^T \underline{\mathcal{C}}$ was added to the equation. Indeed, because the constraints must be satisfied, this term vanishes, and hence, can be added to the statement of Hamilton's principle. A reasoning similar to that developed in section 9.1 leads to the conclusion that the constrained problem expressed by eq. (10.3) is now replaced by an unconstrained problem, eq. (10.4), in which variations of the generalized coordinates and Lagrange's multipliers are unconstrained.

The second term in the integrand of eq. (10.4) affords an important physical interpretation. Indeed, in view of eq. (10.2), $\underline{\lambda}^T \delta\underline{\mathcal{C}} = \delta\underline{q}^T \underline{\underline{B}}^T(\underline{q},t)\underline{\lambda} = \delta\underline{q}^T \underline{F}^c$, where

$$\underline{F}^c = \underline{\underline{B}}^T(\underline{q},t)\underline{\lambda}, \tag{10.5}$$

are the *generalized forces of constraint*, i.e., the forces that must be applied on the system in the configuration space to guarantee the satisfaction of the constraints. It then follows that $\underline{\lambda}^T \delta\underline{\mathcal{C}} = \delta\underline{q}^T \underline{F}^c$ can be interpreted as the *virtual work done by the forces of constraint*.

The differential work done by these forces is expressed as $dW^c = d\underline{q}^T \underline{F}^c = d\underline{q}^T \underline{\underline{B}}^T(\underline{q},t)\underline{\lambda} = -\underline{\lambda}^T \underline{b}(\underline{q},t)dt$, where the last equality was obtained with the help of eq. (9.40). When dealing with a scleronomic constraint, $\underline{b}(\underline{q},t) = 0$ and the *differential work done by the forces constraint vanishes*. This contrasts with rheonomic constraints: in this case, the differential work does not necessarily vanish.

10.2.1 Hamilton's principle with holonomic constraints

Holonomic constraints are considered here. In this case, the constraints are integrable and can be written $\underline{\mathcal{C}}(\underline{q},t) = 0$. It then follows that $\underline{\lambda}^T \delta\underline{\mathcal{C}} + \delta\underline{\lambda}^T \underline{\mathcal{C}} = \delta(\underline{\lambda}^T \underline{\mathcal{C}})$ and $\delta L + \underline{\lambda}^T \delta\underline{\mathcal{C}} + \delta\underline{\lambda}^T \underline{\mathcal{C}} = \delta(K - V) + \delta(\underline{\lambda}^T \underline{\mathcal{C}}) = \delta K - \delta(V - \underline{\lambda}^T \underline{\mathcal{C}})$. The *potential of the constraint forces* is now defined as

$$V^c = -\underline{\lambda}^T \underline{\mathcal{C}}. \tag{10.6}$$

This quantity is indeed the "potential of the constraint forces" because constraint forces are derived from this potential, see eq. (7.21),

$$\frac{\partial V^c}{\partial \underline{q}} = -\frac{\partial \underline{\mathcal{C}}}{\partial \underline{q}}^T \underline{\lambda} = -\underline{\underline{B}}^T(\underline{q},t)\underline{\lambda} = -\underline{F}^c. \tag{10.7}$$

It is convenient to introduce the *augmented potential of the system*, defined as the sum of the potential of the conservative forces acting on the system and of the potential of the constraint forces

$$V^+ = V + V^c. \tag{10.8}$$

Finally, the augmented Lagrangian of the system is defined as

$$L^+ = L - V^c, \tag{10.9}$$

and clearly, $\delta L + \underline{\lambda}^T \delta \underline{C} + \delta \underline{\lambda}^T \underline{C} = \delta(L - V^c) = \delta L^+$.

In the presence of holonomic constraints, Hamilton's principle, eq. (10.4), now becomes

$$\int_{t_i}^{t_f} \left(\delta L^+ + \delta W^{nc} \right) \, dt = 0, \tag{10.10}$$

for all arbitrary variations in generalized coordinates and Lagrange's multipliers. Both *generalized coordinates and Lagrange's multipliers are unconstrained variables* in this principle. This principle is identical to that derived for unconstrained system, except that the Lagrangian has been replaced by the augmented Lagrangian and Lagrange's multipliers are additional, unconstrained variables.

10.2.2 Lagrange's formulation with holonomic constraints

As was done in section 8.3 for unconstrained systems, Lagrange's formulation for systems with holonomic constraints will be derived from Hamilton's principle. After the integration by parts expressed by eq. (8.55), Hamilton's principle, eq. (10.10), becomes

$$\int_{t_i}^{t_f} \delta \underline{q}^T \left[-\frac{d}{dt} \left(\frac{\partial L}{\partial \dot{\underline{q}}} \right) + \frac{\partial L}{\partial \underline{q}} + \underline{F}^c + \underline{Q}^{nc} \right] \, dt + \int_{t_i}^{t_f} \delta \underline{\lambda}^T \, [\underline{C}] \, dt = 0, \tag{10.11}$$

for all arbitrary variations $\delta \underline{q}$ and $\delta \underline{\lambda}$. Consequently, the bracketed terms must vanish, revealing *Lagrange's equations of motion for systems subjected to holonomic constraints*

$$\frac{d}{dt} \left(\frac{\partial L}{\partial \dot{\underline{q}}} \right) - \frac{\partial L}{\partial \underline{q}} = \underline{\underline{B}}^T (\underline{q}, t) \underline{\lambda} + \underline{Q}^{nc}, \quad \text{and} \quad \underline{C}(\underline{q}, t) = 0. \tag{10.12}$$

Here again, it is convenient to introduce the generalized momenta, see eq. (8.25), to simplify the writing of the equations of motion that become

$$\left(\dot{\underline{p}} - \frac{\partial K}{\partial \underline{q}} \right) + \frac{\partial V}{\partial \underline{q}} = \underline{\underline{B}}^T (\underline{q}, t) \underline{\lambda} + \underline{Q}^{nc}, \quad \text{and} \quad \underline{C}(\underline{q}, t) = 0. \tag{10.13}$$

The physical interpretation of Lagrange's equations for systems subjected to holonomic constraints is revealed by recasting eq. (10.13) as $\underline{Q}^I + \underline{F}^c + \underline{Q}^c + \underline{Q}^{nc} = 0$, where the generalized inertial forces, $\underline{Q}^I$, are given by eq. (8.58). Here again, Lagrange's equations are a *statement of dynamic equilibrium*: the sum of the inertial forces and all external forces applied on the system, *including the forces of constraint*, must vanish for dynamic equilibrium conditions to be satisfied.

Although Lagrange's equations of motion have the same physical meaning when dealing with both unconstrained and constrained systems, a marked difference is observed in the mathematical nature of the equations in these two cases. For un-constrained systems, Lagrange's equations of motions, eqs. (8.57), form a set of n *second-order, ordinary differential-equations in time*, or second-order ODE's. Indeed, second-order derivatives of the generalized coordinates are implied by the structure of Lagrange's equations.

If the system is subjected to holonomic constraints, second-order derivatives of the generalized coordinates will be present as well, but Lagrange's multipliers appear in the equations as algebraic variables, *i.e.*, as undifferentiated variables. Equation that feature this mixed differential-algebraic nature are called *differential-algebraic equations* or DAE's. Lagrange's equations form a set of $n + m$ DAE's. DAE's are typically more difficult to solve than ODE's; while methods for the numerical solution of ODE's are well developed, the solution of DAE's is still a challenging task.

Example 10.4. The simple pendulum

Figure 10.5 depicts a simple pendulum of length L featuring a bob of mass m. This single degree of freedom system will be represented using two generalized coordinates, the Cartesian coordinates of the bob, $\underline{r} = q_1 \bar{\imath}_1 + q_2 \bar{\imath}_2$. A single holonomic constraint must be enforced, $\mathcal{C} = 1/2 \left(\underline{r}^T \underline{r} - L^2 \right) = 0$. This constraint enforces the constant length condition for the pendulum; the constraint matrix is $\underline{\underline{B}}^T(\underline{q}) = \underline{r}$.

The Lagrangian of the system is easily evaluated as $L = m\, \underline{\dot{r}}^T \underline{\dot{r}}/2 - mg\, \bar{\imath}_1^T \underline{r}$. The potential of the constraint forces, eq. (10.6), is $V^c = \mathcal{C}\lambda = 1/2 \left(\underline{r}^T \underline{r} - L^2 \right) \lambda$; a single Lagrange multiplier is used here to enforce the single constraint. The generalized momenta of the system and partial derivatives of the Lagrangian with respect to the generalized coordinates are

$$\underline{p} = \frac{\partial L}{\partial \underline{\dot{r}}} = m\underline{\dot{r}}, \quad \text{and} \quad \frac{\partial L}{\partial \underline{r}} = -mg\, \bar{\imath}_1,$$

respectively.

Application of Lagrange's formulation for systems with holonomic constraints, eqs. (10.13), leads to $m\underline{\ddot{r}} + mg\bar{\imath}_1 - \lambda\underline{r} = 0$, and $1/2 \left(\underline{r}^T \underline{r} - L^2 \right) = 0$. These three equations can be used to solve for the three unknowns of the problem: q_1, q_2, and λ. Note the differential-algebraic nature of the equations: terms in $\ddot{q}_1$ and $\ddot{q}_2$ appear, but λ is not differentiated.

Of course, these equations are equivalent to the single equation of motion that would have been obtained had the single generalized coordinate θ been used to represent the configuration of the system. Indeed, the position vector can be written in polar coordinates as $\underline{r}^T = L \left\{ S_\theta, C_\theta \right\}$, where $S_\theta = \sin\theta$ and $C_\theta = \cos\theta$; the constraint is then automatically satisfied, as expected. The remaining equation of motion then yields two scalar equations: $\ddot{\theta} + g/L\, S_\theta = 0$ and $\lambda L = -mgC_\theta - mL\dot{\theta}^2$. Application of Newton's law to the free body diagram sketched on fig. 10.5 yields two equations $\ddot{\theta} + g/L\, S_\theta = 0$ and $T = mgC_\theta + mL\dot{\theta}^2$, where T is the tension in the string that can be interpreted as the constraint force, *i.e.*, the force that maintains the constant length of the pendulum.

The equations of motion are identical, and $T = -\lambda L$. As expected, Lagrange's multiplier is closely related to, although not identical, to the force of constraint; indeed, $\lambda = -T/L$. In the constrained formulation, the force of constraint is given by eq. (10.5) as $\underline{F}^c = \underline{\underline{B}}^T(\underline{q})\underline{\lambda} = \lambda\underline{r} = \lambda L\,\bar{e}_1$. This development explains the difference in sign: T, the tension in the string was chosen to act along axis $-\bar{e}_1$, but the constraint force $\underline{F}^c$ acts along axis $+\bar{e}_1$. This development also explains the presence of factor L: the tension in the string, T, has units of force, but Lagrange's multiplier has units of force over length because $\|\underline{F}^c\| = |\lambda L|$.

The physical meaning of Lagrange's multipliers must be clearly identified to ease the interpretation of the equations of motion generated by Lagrange's formulation. The definition of the constraint force, $\underline{F}^c = \underline{\underline{B}}^T(\underline{q}, t)\underline{\lambda}$, provides the physical interpretation of the multiplier. The constraint force, $\underline{F}^c$, is a generalized force of constraint, *i.e.*, a force acting in the configuration space. The meaning of the multiplier depends on the specific manner in which the constraint was written. For instance, writing the constraint as $\mathcal{C} = (L^2 - \underline{r}^T\underline{r}) = 0$ results in $\lambda = T/(2L)$.

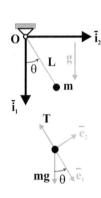

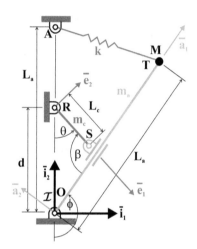

Fig. 10.5. Configuration of the simple pendulum.

Fig. 10.6. Configuration of the quick return mechanism.

Example 10.5. The quick return mechanism

The quick return mechanism shown in fig. 10.6 consists of a uniform crank of length L_c and mass m_c, and of a uniform arm of length L_a and mass m_a. The crank is pinned at point **R** and the arm at point **O**; the distance between these two points is denoted d. At point **S**, a slider allows the tip of the crank to slide along the arm. A mass M is attached at point **T**, the tip of the arm. A spring of stiffness constant k connects the tip of the arm, point **T**, to fixed point **A**; the spring is un-stretched when the arm is in the vertical position. This problem was treated using Lagrange's formulation in example 8.15 on page 329.

The generalized coordinates of the problem are selected to be the angular positions of the two bars, denoted θ and ϕ. This problem was treated in example 8.15 on page 329 using Lagrange's formulation with a single generalized coordinate, θ. Use

Lagrange's formulation for constrained systems to derive the equations of motion of the system.

Considering triangle **ORS**, it is clear that $\beta = \phi - \theta$, and the law of sines then yields $L_c \sin(\phi - \theta) = d \sin \phi$. This equation expresses the kinematic constraint between the two generalized coordinates of this single degree of freedom problem, $\mathcal{C} = d \sin \phi - L_c \sin(\phi - \theta) = 0$. The constraint matrix now becomes

$$\underline{\underline{B}} = \left[L_c C_{\phi-\theta} \ dC_\phi - L_c C_{\phi-\theta} \right], \tag{10.14}$$

where the following notation was used: $C_{\phi-\theta} = \cos(\phi - \theta)$, and $C_\phi = \cos \phi$.

The system's kinetic energy is $K = [m_c L_c^2 \dot{\theta}^2 / 3 + (M + m_a/3) L_a^2 \dot{\phi}^2]/2$, where the first term represents the kinetic energy of the crank and the second that of the arm. The potential energy of the spring is $V = 1/2 \, k \Delta^2$, where Δ is the stretch of the spring. The law of cosines applied to triangle **OMA** yields $\Delta^2 = L_a^2 + L_a^2 - 2 L_a^2 \cos(\pi - \phi) = 2 L_a^2 (1 + C_\phi)$, and the potential of the elastic spring is $V = k L_a^2 (1 + C_\phi)$. Finally, the potential of the single constraint of this problem is $V^c = [dS_\phi - L_c S_{\phi-\theta}] \lambda$, where λ is Lagrange's multiplier used to enforce the constraint, $S_{\phi-\theta} = \sin(\phi - \theta)$, and $S_\phi = \sin \phi$.

The Lagrangian is $L = K - V$, and the system's generalized momenta become

$$p_{\dot{\theta}} = \frac{\partial L}{\partial \dot{\theta}} = \frac{m_c L_c^2}{3} \dot{\theta}, \quad p_{\dot{\phi}} = \frac{\partial L}{\partial \dot{\phi}} = (M + \frac{m_a}{3}) L_a^2 \dot{\phi}.$$

The derivatives of the Lagrangian with respect to the generalized coordinates are

$$\frac{\partial L}{\partial \theta} = 0, \quad \frac{\partial L}{\partial \phi} = k L_a^2 S_\phi.$$

Lagrange's formulation for constrained systems then yields the equations of motion of the system,

$$\begin{bmatrix} m_c L_c^2/3 & 0 \\ 0 & (M + m_a/3) L_a^2 \end{bmatrix} \begin{Bmatrix} \ddot{\theta} \\ \ddot{\phi} \end{Bmatrix} - \begin{bmatrix} L_c C_{\phi-\theta} \\ dC_\phi - L_c C_{\phi-\theta} \end{bmatrix} \lambda = \begin{Bmatrix} 0 \\ k L_a^2 S_\phi \end{Bmatrix}. \tag{10.15}$$

The equations of motion for this single degree of freedom problem now take the form of three differential-algebraic equations. In the two equations given above, Lagrange's multiplier, λ, is an algebraic variable, but second time derivatives of the generalized coordinates, θ and ϕ, appear. The third equation is the holonomic constraint equation, $d \sin \phi - L_c \sin(\phi - \theta) = 0$, which is an algebraic equation.

It is interesting to compare the equation of motion obtained from Lagrange's formulation using a single generalized coordinate, eq. (8.61), to those obtained in the present development. Equation (8.61) is a single, ordinary differential equation for the single generalized coordinate, θ. In contrast, when using two generalized coordinates, the equations of motion take the form of three coupled differential-algebraic equations, because an additional variable, Lagrange's multiplier, was added to enforce the constraint.

In the absence constraints, Lagrange's formulation leads to a single equation of motion that is far more complex than those obtained in the constrained formulation, and involves a much higher level of nonlinearity. On the other hand, the constrained formulation leads to a higher number of equations, but these equations are easier to derive and present a lower level of nonlinearity. This ease of derivation of the equations of motion for constrained system is one of the major attractions of constrained formulations.

It is always instructive to provide a physical interpretation of Lagrange's multipliers. The first equation of system (10.15) reads $m_c L_c^2 \ddot{\theta}/3 = L_c C_{\phi-\theta} \lambda$, which corresponds to the pivot equation written for the crank about point $\mathbf{R}$. The term on the right-hand side of the equation corresponds to the moment of the normal force, λ, the crank applies on the arm. It is left to the reader to verify that the second equation of system (10.15) can be interpreted as the pivot equation written for the arm about point $\mathbf{O}$, and leads to an identical interpretation of the physical meaning of Lagrange's multiplier as the normal interaction force between the crank and the arm.

10.2.3 Problems

Problem 10.7. The 12 generalized coordinates rigid body
The configuration of a rigid body can be defined by 12 generalized coordinates: the position vector, $\underline{u}$ (3 coordinates), of its reference point $\mathbf{O}$, and the three vectors, $\underline{e}_1$, $\underline{e}_2$, and $\underline{e}_3$ (3 coordinates each), defining its orientation. Clearly, this set of coordinates is 6 times redundant and hence, 6 constraints must be added to the problem: three normality constraints $\underline{e}_1^T \underline{e}_1 = \underline{e}_2^T \underline{e}_2 = \underline{e}_3^T \underline{e}_3 = 1$, and three orthogonality constraints $\underline{e}_2^T \underline{e}_3 = \underline{e}_1^T \underline{e}_3 = \underline{e}_1^T \underline{e}_2 = 0$. Let array $\underline{q}$ store the generalized coordinates of the problem, $\underline{q}^T = \{\underline{u}^T, \underline{e}_1^T, \underline{e}_2^T, \underline{e}_3^T\}$. *(1)* Show that the kinetic energy of the rigid body can be written as $K = 1/2 \, \underline{\dot{q}}^T \underline{\underline{M}}^* \underline{\dot{q}}$, where $\underline{\underline{M}}^*$ is a 12 × 12 mass matrix. *(2)* If $\underline{f}$ and $\underline{m}$ are the force and moment vectors applied to the rigid body at point $\mathbf{O}$, show that the virtual work done by these forces is $\delta W = \delta \underline{q}^T \underline{F}$, where $\underline{F}$ is a 12 × 1 loading array. *(3)* Write the governing equations of motion for the rigid body.

Problem 10.8. The 9 generalized coordinates rigid body
Read the paper by García de Jalón *et al.* [28] describing the concept of *natural coordinates*. Consider a rigid body described by 9 generalized coordinates: the position vector, $\underline{u}$ (3 coordinates), of its reference point $\mathbf{O}$, and two unit vectors, $\underline{e}_1$ and $\underline{e}_2$ (3 coordinates each), defining its orientation. Let array $\underline{q}$ store the generalized coordinates of the problem, $\underline{q}^T = \{\underline{u}^T, \underline{e}_1^T, \underline{e}_2^T\}$. *(1)* Define the constraints associated with this representation. *(2)* Evaluate the kinetic energy of the rigid body based on these generalized coordinates. *(3)* If $\underline{f}$ and $\underline{m}$ are the force and moment vectors applied to the rigid body at point $\mathbf{O}$, find the associated generalized forces, $\underline{F}$, such that $\delta W = \delta \underline{q}^T \underline{F}$. *(4)* Write the governing equations of motion for the rigid body.

Problem 10.9. The crank piston mechanism
The crank slider mechanism depicted in fig. 10.7 comprises a bar of length L_1 and mass m_1 connected to the ground at point $\mathbf{O}$ by means of a hinge. The orientation of is bar with respect to the horizontal is denoted ϕ. At point $\mathbf{A}$, a hinge connects the first bar to a second bar of length L_2 and mass m_2. A slider of mass M, that is constrained to move in the horizontal direction, is connected to this second bar. A spring of stiffness constant k connects the slider

to the ground and is un-stretched when $\phi = 0$. Clearly, this system features a single degree of freedom. Two sets of generalized coordinates will be used to represent the system. For *representation 1*, a single generalized coordinate, ϕ, is used. For *representation 2*, seven generalized coordinates are used: u_1, v_1 and ϕ_1, respectively the horizontal and vertical position of the center of mass of the first bar, and its orientation with respect to the horizontal; u_2, v_2 and ϕ_2, the corresponding quantities for the second bar; and x, the displacement of the slider in the horizontal direction. The spring is un-stretched when $x = 0$. *(1)* With the help of Lagrange's formulation, derive the equation of motion of the system using *representation 1*. *(2)* Using *representation 2*, derive the equations of motion for the constrained system. *(3)* Discuss the relative merits of the two representations.

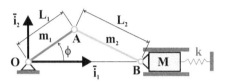

Fig. 10.7. Crank piston mechanism with spring.

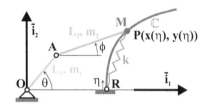

Fig. 10.8. Two bar linkage tracking a curve.

Problem 10.10. Two bar linkage tracking a curve

Figure 10.8 depicts a planar two bar linkage tracking curve $\mathbb{C}$. The first bar, of length L_1 and mass m_1, is connected to the ground at point **O**. The second bar, of length L_2 and mass m_2, connects to the first bar at point **A** and tracks curve $\mathbb{C}$ at point **P**. A concentrated mass, M, is located at point **P** and an elastic spring of stiffness constant constant k connects this mass to point **R**. Curve $\mathbb{C}$ is described by its coordinates $x(\eta)$ and $y(\eta)$, where η defines an arbitrary parametrization of the curve. This system will be represented by three generalized coordinates: angles θ and ϕ, as defined on the figure, and η, the parameter along curve $\mathbb{C}$. *(1)* Derive the constraints among the three generalized coordinates and the constraint matrix. *(2)* Use Lagrange's formulation for constrained systems to derive the equations of motion of the system. *(3)* Discuss the physical nature of Lagrange's multipliers.

Problem 10.11. Crank-slider mechanism

The crank-slider mechanism depicted in fig. 9.8 consists of a uniform crank of length L_1 and mass m_1 connected to the ground at point **O**; let θ be the angle from the horizontal to the crank. At point **B**, the crank connects to a uniform linkage of length L_2 and mass m_2 that slides along point **P**, a fixed point in space, located at a distance d from point **O**. Let w denote the distance from point **B** to point **P** and ϕ the angle from the horizontal to link **BP**. The system is represented by three generalized coordinates: θ, ϕ, and w. *(1)* Derive the constraints among the three generalized coordinates and the constraint matrix. *(2)* Use Lagrange's formulation for constrained systems to derive the equations of motion of the system. *(3)* Discuss the physical nature of Lagrange's multipliers.

Problem 10.12. The spatial mechanism

The spatial mechanism depicted in fig. 10.9 consists of a crank of length L_c and mass m_c attached to the ground at point **A** and rotating about axis $\bar{\imath}_1$; the crank moves in plane $(\bar{\imath}_2, \bar{\imath}_3)$. A rigid arm of length L_a and mass m_a connects point **P**, at the tip of the crank, to point **Q** that is free to slide along axis $\bar{\imath}_1$. The slider at point $\mathbf{Q}$ is of mass M. The generalized coordinates of the problem are y and z, defining the position of point $\mathbf{P}$ and x, defining the position of point $\mathbf{Q}$. (1) Derive the constraints among the three generalized coordinates and the constraint matrix. (2) Use Lagrange's formulation for constrained systems to derive the equations of motion of the system. (3) Discuss the physical nature of Lagrange's multipliers.

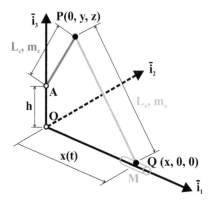

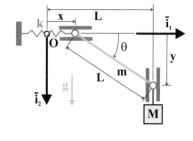

Fig. 10.9. Configuration of the spatial mechanism.

Fig. 10.10. Homogeneous bar sliding on guides at both ends.

Problem 10.13. Bar sliding on guides
Figure 10.10 depicts a homogeneous bar of length L and mass m sliding on two guides at its end points. At the left end, the bar is connected to a spring of stiffness constant k that is un-stretched when the bar is horizontal. At the right end, the bar is connected to a point mass M. Gravity acts along axis $\bar{\imath}_2$. This single degree of freedom system will be represented using three generalized coordinates: x, y, and θ. (1) Derive the constraints among the three generalized coordinates and the constraint matrix. (2) Use Lagrange's formulation for constrained systems to derive the equations of motion of the system. (3) Discuss the physical nature of Lagrange's multipliers.

Problem 10.14. Bar sliding on guides
Repeat the previous problem but use two generalized coordinates only, x and y.

Problem 10.15. The spatial mechanism
Read the paper by García de Jalón *et al.* [28] describing the concept of *natural coordinates*. The spatial mechanism depicted in fig. 10.9 consists of a crank of length L_c and mass m_c attached to the ground at point $\mathbf{A}$ and rotating about axis $\bar{\imath}_1$; the crank moves in plane $(\bar{\imath}_2, \bar{\imath}_3)$. A rigid arm of length L_a and mass m_a connects point $\mathbf{P}$, at the tip of the crank, to point $\mathbf{Q}$ that is free to slide along axis $\bar{\imath}_1$. The slider at point $\mathbf{Q}$ is of mass M. This mechanism will be described by the following 12 generalized coordinates: unit vector $\bar{n}_1$ along segment $\mathbf{AP}$, the position vector, $\underline{r}_P$, of point $\mathbf{P}$, the position vector, $\underline{r}_Q$, of point $\mathbf{Q}$, and unit vector $\bar{n}_2$ normal to segment $\mathbf{PQ}$. When bar $\mathbf{PQ}$ is in plane $(\bar{\imath}_1, \bar{\imath}_3)$, unit vector $\bar{n}_2$ lies in the same plane. (1) Derive the constraints among the 12 generalized coordinates and the constraint matrix. (2) Use Lagrange's formulation for constrained systems to derive the equations of motion of the system. (3) Discuss the physical nature of Lagrange's multipliers.

Problem 10.16. The two-bar linkage with slider system

The two-bar linkage with slider system shown in fig. 8.4 is a planar mechanism. It consists of a uniform crank of length L_1 and mass m_1 connected to the ground at point **O**; let θ be the angle from the horizontal to the crank. At point **B**, the crank slides over a uniform linkage of length L_2 and mass m_2 that is connected to the ground at point **A**. Let w denote the distance from point **B** to point **A** and ϕ the angle from the horizontal to link **BA**. *(1)* Derive the constraints among the three generalized coordinates and the constraint matrix. *(2)* Use Lagrange's formulation for constrained systems to derive the equations of motion of the system. *(3)* Discuss the physical nature of Lagrange's multipliers.

Problem 10.17. Pendulum mounted on a cart

Figure 8.5 shows a pendulum of length L and mass m mounted on a cart of mass M that is connected to the ground by means of a spring of stiffness constant k and of a dashpot of constant c. The displacement of the cart is denoted x which is also the stretch of the spring, and θ measures the angular deflection of the pendulum with respect to the vertical. Gravity acts on the system as indicated in fig. 8.5. *(1)* Based on Hamilton's principle, derive the equations of motion of the system using the following generalized coordinates: x, θ, and $\underline{r}_A$, the position of point **A**. Use Lagrange's multiplier technique to enforce the kinematic constraint $\underline{C} = \underline{r}_A - x\bar{\imath}_2 = \underline{0}$. *(2)* Interpret Lagrange's multipliers in physical terms. *(3)* Plot the time history of the cart displacement, x. *(4)* Plot the history of angle θ. *(5)* Plot the trajectory of the point at the tip of the pendulum. *(6)* Plot the cart velocity, $\dot{x}$. *(7)* Plot the angular velocity of the pendulum, $\dot{\theta}$. *(8)* Plot the system kinetic and potential energies and the energy dissipated in the damper. Check the energy closure equation. *(9)* Plot the components of the internal force at point **A**. Use the following data: $M = 5$ kg; $m = 2$ kg; $L = 0.4$ m; $k = 10$ N/m; acceleration of gravity $g = 9.81$ m/s^2; $c = 0.5$ N.s/m. Present all your results for a period of 10 s. Initial condition are at rest with $x(t = 0) = 0.2$ m and $\theta(t = 0) = \pi$.

Problem 10.18. Particle in a circular slot with guiding arm

A particle of mass M slides along a circular slot of radius R, as shown in fig. 8.15. The particle also slides in a rectilinear slot in an arm of mass m and length L. The arm is pivoted to the ground at point **O** and is restrained by a torsional spring of stiffness constant k and a dashpot of constant c. The spring is un-stretched when the arm is horizontal. A viscous friction force, $F^f = -\mu\dot{w}$ is acting at the interface between the particle and the arm. *(1)* Using three generalized coordinates, x, y, the coordinates of mass M, and w, the position of the particle along the arm, derive the equations of motion of the system using Lagrange's formulation for constrained systems. *(2)* Give the physical interpretation of Lagrange's multipliers.

Problem 10.19. The crank piston mechanism

The crank slider mechanism depicted in fig. 10.4 comprises a bar of length L_1 and mass m_1 connected to the ground at point **O** by means of a hinge. At point **A**, a hinge connects the first bar to a second bar of length L_2 and mass m_2. A slider of mass M, that is constrained to move in the horizontal direction, is connected to this second bar. A spring of stiffness constant k connects the slider to the ground and is un-stretched when the two bars are aligned. This system will be represented with three generalized coordinates: x and y, the coordinates of point **A** and z, the horizontal position of point **B**. *(1)* Write the constraint equations for this problem. *(2)* Use Lagrange's formulation for constrained systems to derive the equations of motion of the system. *(3)* Give the physical interpretation of Lagrange's multipliers.

Problem 10.20. Spinning arm

Figure 8.24 depicts a shaft of height h fixed at point **O** and free to rotate about axis $\bar{\imath}_3$. An arm of length d, rigidly attached to the shaft at point **A**, rotates in the horizontal plane.

A homogeneous bar of length L and mass m is connected to the arm at point **B** with a torsional spring of stiffness constant k. Gravity acts as indicated on the figure and the applied torque $Q = 0$. Frame $\mathcal{F}^A = [\mathbf{A}, \mathcal{A} = (\bar{a}_1, \bar{a}_2, \bar{a}_3)]$ is attached to the arm and frame $\mathcal{F}^B = [\mathbf{B}, \mathcal{B} = (\bar{b}_1, \bar{b}_2, \bar{b}_3)]$ is attached to the bar. A planar rotation of magnitude α about axis $\bar{\imath}_3$ brings basis $\mathcal{I}$ to $\mathcal{A}$. A planar rotation of magnitude β about axis $\bar{a}_2$ brings basis $\mathcal{A}$ to $\mathcal{B}$; the torsional spring is un-stretched when $\beta = \beta_0$. *(1)* Use Lagrange's formulation to derive the equations of motion of the system. Use two generalized coordinates, α and β. *(2)* Assume that the shaft is rotating at a constant angular velocity, $\dot{\alpha} = \Omega$ and impose this condition through a rheonomic constraint. *(3)* Use Lagrange's formulation for constrained systems to derive the equations of motion of the problem. *(4)* What is the physical meaning of Lagrange's multiplier. *(5)* Assume now instead that the bar is rotating at a constant angular velocity, $\dot{\beta} = \omega$ and impose this condition through a rheonomic constraint. *(6)* Use Lagrange's formulation for constrained systems to derive the equations of motion of the problem. *(7)* What is the physical meaning of Lagrange's multiplier.

10.3 Hamilton's principle and Lagrange's formulation with nonholonomic constraints

Systems subjected to holonomic constraints were studied in section 10.2. In the present section, attention turns to systems subjected to nonholonomic constraints. The most general type of constraint to be considered here are linear functions of the generalized velocities, as expressed by eq. (9.47), $\underline{\mathcal{D}} = \underline{\underline{B}}(\underline{q}, t)\underline{\dot{q}} + \underline{b}(\underline{q}, t) = 0$. If time appears explicitly in the expression of the constraint, it is rheonomic, otherwise it is scleronomic.

Nonholonomic constraints limit the allowable virtual displacement as expressed by eq. (10.2). Because a virtual displacement is an arbitrary change in displacement *at a given, fixed instant*, this expression does not involve the term $\underline{b}(\underline{q}, t)$ that appears in the differential or Pfaffian form of the constraints.

Hamilton's principle, eq. (8.20), will be written in the following form,

$$\int_{t_i}^{t_f} (\delta L + \delta W^{nc}) \, dt = 0, \tag{10.16}$$

for all arbitrary virtual displacements. The boundary terms at the initial and final times have been ignored. For constrained systems, virtual displacements are not arbitrary because they must satisfy the constraints as expressed by eq. (10.2).

A linear combination of eqs. (10.16) and (10.2) now yields

$$\int_{t_i}^{t_f} \left(\delta L + \underline{\lambda}^T \underline{\underline{B}}(\underline{q}, t)\delta\underline{q} + \delta\underline{\lambda}^T \underline{\mathcal{D}} + \delta W^{nc}\right) \, dt = 0, \tag{10.17}$$

where $\underline{\lambda}$ is an array of arbitrary Lagrange's multipliers. The term $\delta\underline{\lambda}^T\underline{\mathcal{D}}$ was added to the equation. Indeed, because the constraints must be satisfied, this term vanishes, and hence, can be added to the statement of Hamilton's principle. A reasoning similar to that developed in section 9.1 will lead to the conclusion that the constrained problem expressed by eq. (10.16) is now replaced by an unconstrained problem, eq. (10.17), in which variations of the generalized coordinates and Lagrange's multipliers are unconstrained.

10.3.1 Hamilton's principle with nonholonomic constraints

In the presence of nonholonomic constraints, Hamilton's principle takes the form of eq. (10.17). The second term in the integrand of this equation affords an important physical interpretation: $\delta \underline{q}^T \underline{\underline{B}}^T (\underline{q}, t) \underline{\lambda} = \delta \underline{q}^T \underline{F}^c$, where the generalized forces of constraint, $\underline{F}^c$, are defined by eq. (10.5). Clearly, this term can be interpreted as the virtual work done by the constraint forces, $\delta W^c = \delta \underline{q}^T \underline{F}^c = \delta \underline{q}^T \underline{\underline{B}}(\underline{q}, t) \underline{\lambda}$. This expression underlines the fundamental difference between holonomic and nonholonomic constraints: for holonomic constraints, *the constraint forces can be derived from the potential of the constraint forces*, see eq. (10.7), whereas for nonholonomic constraints *the virtual work done by the forces of constraint is non integrable, i.e.*, there exist no potential of the constraint forces.

10.3.2 Lagrange's formulation with nonholonomic constraints

Because both generalized coordinates and Lagrange's multipliers are unconstrained variables in eq. (10.17), Lagrange's equations of motion for systems subjected to nonholonomic constraints are obtained from this principle in a manner similar to that presented in section 8.3. After the integration by parts expressed by eq. (8.55), Hamilton's principle, eq. (10.17), becomes

$$\int_{t_i}^{t_f} \delta \underline{q}^T \left[-\frac{d}{dt} \left(\frac{\partial L}{\partial \underline{\dot{q}}} \right) + \frac{\partial L}{\partial \underline{q}} + \underline{F}^c + \underline{Q}^{nc} \right] dt + \int_{t_i}^{t_f} \delta \underline{\lambda}^T \left[\underline{\mathcal{D}} \right] dt = 0, \quad (10.18)$$

for all arbitrary variations in generalized coordinates and Lagrange's multipliers. Consequently, the bracketed terms must vanish, revealing *Lagrange's equations of motion for systems subjected to nonholonomic constraints*

$$\frac{d}{dt} \left(\frac{\partial L}{\partial \underline{\dot{q}}} \right) - \frac{\partial L}{\partial \underline{q}} = \underline{\underline{B}}^T (\underline{q}, t) \underline{\lambda} + \underline{Q}^{nc}, \quad \text{and} \quad \underline{\mathcal{D}}(\underline{q}, \underline{\dot{q}}, t) = 0. \quad (10.19)$$

Here again, it is convenient to introduce the generalized momenta, see eq. (8.25), to simplify the writing of the equations of motion that become

$$\left(\underline{\dot{p}} - \frac{\partial K}{\partial \underline{q}} \right) + \frac{\partial V}{\partial \underline{q}} = \underline{\underline{B}}^T (\underline{q}, t) \underline{\lambda} + \underline{Q}^{nc}, \quad \text{and} \quad \underline{\mathcal{D}}(\underline{q}, \underline{\dot{q}}, t) = 0. \quad (10.20)$$

These equations are identical to those obtained for systems subjected to holonomic constraints, eq. (10.12); the only difference is the form of the constraint to be enforced.

The derivation of Lagrange's equations of motion for constrained systems clearly underlines the differences between Newtonian and Lagrangian dynamics. In Newtonian mechanics, the focus is on forces and acceleration vectors for each particle of the system; both quantities are of a vectorial nature. No distinctions exist between various types of forces: the vector sum of all externally applied forces is parallel to the acceleration vector.

In Lagrangian dynamics, the focus is on two scalar quantities, the Lagrangian of the system and the virtual work done by the applied forces. These quantities are characteristics of the complete system, not of individual particles. The efficiency and elegance of Lagrange's approach stems from the fact that the equations of motion of complex mechanical systems can be derived from these two scalar quantities.

Various types of forces are treated differently in Lagrange's formulation. First, the virtual work done by the inertial forces is directly related to the kinetic energy of the system. Next, externally applied forces are divided into conservative and non-conservative forces. The virtual work done by the conservative forces equals the variation of the potential of these forces; the Lagrangian of the system is the difference between the kinetic and potential energies of the system. The virtual work done by the non-conservative forces gives rise to the generalized, non-conservative forces. This distinction is also present for constrained systems: for holonomic constraints, the constraint forces can be derived from a potential, whereas no such potential exists for nonholonomic constraints.

Example 10.6. The skateboard

Figure 10.11 depicts the simplified configuration of a skateboard of mass m and moment of inertia I about its center of mass **G**. The skateboard rolls without sliding on the horizontal plane by means of a wheel aligned with axis $\bar{e}_1$ of the skateboard and located at point **C**, a distance ℓ from the center of mass. The position vector of the center of mass is written as $\underline{r}_G = x\,\bar{\imath}_1 + y\,\bar{\imath}_2$, and the axis of the skateboard makes an angle θ with the horizontal. This system is subjected to a constraint: because the wheel does not slip, the velocity vector of the contact point must be along axis $\bar{e}_1$. This nonholonomic constraint and the corresponding constraint matrix are given by eqs. (9.50) and (9.51), respectively.

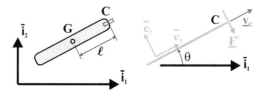

Fig. 10.11. Skateboard with front wheel.

The kinetic energy of the system is $K = m(\dot{x}^2 + \dot{y}^2)/2 + I\dot{\theta}^2/2$. The potential energy of the system vanishes, and hence, the Lagrangian of the system equals the kinetic energy. The generalized coordinates of the system are $q_1 = x$, $q_2 = y$, and $q_3 = \theta$. The generalized momenta of the system are $\underline{p}_{\dot{x}} = m\dot{x}$, $\underline{p}_{\dot{y}} = m\dot{y}$, and $\underline{p}_{\dot{\theta}} = I\dot{\theta}$. Lagrange's formulation then yields the equations of motion of the system

$$m\ddot{x} = -\lambda\sin\theta, \quad m\ddot{y} = \lambda\cos\theta, \quad I\ddot{\theta} = \lambda\ell,$$

where λ is Lagrange's multiplier associated with the nonholonomic constraint. These equations are identical to those obtained from the Newtonian approach, eqs. (9.49). Lagrange's multiplier, λ, is the contact force of the wheel on the ground, $\lambda = -F^C$, see fig. 10.11.

10.3.3 Problems

Problem 10.21. The skateboard

Figure 10.11 depicts the simplified configuration of a skateboard of mass m and moment of inertia I about axis $\bar{\imath}_3$, computed with respect to its center of mass **G**. The skateboard solely moves in the horizontal plane: the position vector of the center of mass is written as $\underline{r}_G = x\bar{\imath}_1 + y\bar{\imath}_2$, and the axis of the skateboard makes an angle θ with respect to $\bar{\imath}_1$. A wheel of mass M and radius R is connected to the front of the skateboard, at a distance ℓ from its center of mass. The wheel rolls on the ground without slipping. *(1)* Determine the number of degrees of freedom for this system. *(2)* Discuss the nature of the constraints. *(3)* Find the equations of motion of the system using Lagrange's approach. Use the following generalized coordinates: x, y, θ, and ψ, the rotation of the wheel.

10.4 The lower pair joints

A distinguishing feature of multibody systems is the presence of joints that impose constraints on the relative motion of the various bodies of the system. Most joints used in practical applications can be modeled in terms of the so called *lower pairs* [29]: the revolute, prismatic, screw, cylindrical, planar and spherical joints, depicted in fig. 10.12. In some cases, however, joints with specialized kinematic conditions must also be developed.

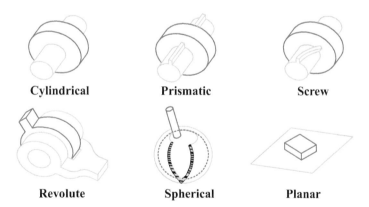

| Cylindrical | Prismatic | Screw |

| Revolute | Spherical | Planar |

Fig. 10.12. The six lower pairs.

10.4.1 Kinematics of a typical lower pair joint

Consider two bodies, denoted "body k" and "body ℓ," as shown in fig. 10.13. Quantities pertaining to body k and body ℓ will be indicated with superscripts $(\cdot)^k$ and $(\cdot)^\ell$, respectively. A lower pair joint, *i.e.*, anyone of the joints depicted in fig. 10.12, connects the two bodies at points $\mathbf{K}$ and $\mathbf{L}$, which are material points of bodies k and ℓ, respectively. In the reference configuration, two frames, $\mathcal{F}_0^k = \left[\mathbf{K}, \mathcal{B}_0^k = (\bar{e}_{01}^k, \bar{e}_{02}^k, \bar{e}_{03}^k)\right]$ and $\mathcal{F}_0^\ell = \left[\mathbf{L}, \mathcal{B}_0^\ell = (\bar{e}_{01}^\ell, \bar{e}_{02}^\ell, \bar{e}_{03}^\ell)\right]$, are attached to bodies k and ℓ, respectively. In the deformed configuration, the bodies are defined by two frames, $\mathcal{F}^k = \left[\mathbf{K}, \mathcal{B}^k = (\bar{e}_1^k, \bar{e}_2^k, \bar{e}_3^k)\right]$ and $\mathcal{F}^\ell = \left[\mathbf{L}, \mathcal{B}^\ell = (\bar{e}_1^\ell, \bar{e}_2^\ell, \bar{e}_3^\ell)\right]$, respectively.

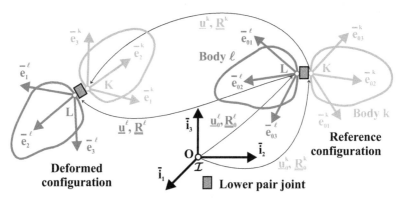

Fig. 10.13. Typical lower pair joint in the reference and deformed configurations.

10.4.2 Notational conventions

Unless otherwise indicated, the components of all tensors will be resolved in the inertial frame, denoted $\mathcal{F}^I = [\mathbf{O}, \mathcal{I} = (\bar{\imath}_1, \bar{\imath}_2, \bar{\imath}_3)]$. In the reference configuration, the position vectors of points $\mathbf{K}$ and $\mathbf{L}$ are denoted $\underline{u}_0^k$ and $\underline{u}_0^\ell$, respectively, and rotation tensors $\underline{\underline{R}}_0^k$ and $\underline{\underline{R}}_0^\ell$ describe the rotations from basis $\mathcal{I}$ to $\mathcal{B}_0^k$ and $\mathcal{I}$ to $\mathcal{B}_0^\ell$, respectively. The displacement vectors of these two points are denoted $\underline{u}^k$ and $\underline{u}^\ell$, respectively; tensors $\underline{\underline{R}}^k$ and $\underline{\underline{R}}^\ell$ describe the rotations from basis $\mathcal{B}_0^k$ to $\mathcal{B}^k$, and $\mathcal{B}_0^\ell$ to $\mathcal{B}^\ell$, respectively. The virtual rotations vectors are defined as $\delta\underline{\psi}^k = \text{axial}(\delta\underline{\underline{R}}^k \underline{\underline{R}}^{kT})$ and $\delta\underline{\psi}^\ell = \text{axial}(\delta\underline{\underline{R}}^\ell \underline{\underline{R}}^{\ell T})$. Finally, the relative displacement vector at the joint is defined as

$$\underline{u} = \underline{u}^\ell - \underline{u}^k. \tag{10.21}$$

With these notations, unit vectors of triads $\mathcal{B}^k$ and $\mathcal{B}^\ell$ can be expressed as

$$\bar{e}_\alpha^k = \underline{\underline{R}}^k \underline{\underline{R}}_0^k \bar{\imath}_\alpha \quad \text{and} \quad \bar{e}_\beta^\ell = \underline{\underline{R}}^\ell \underline{\underline{R}}_0^\ell \bar{\imath}_\beta, \tag{10.22}$$

respectively, where $\alpha = 1, 2,$ or 3 and $\beta = 1, 2,$ or 3. Variations of these unit vectors are readily found as

$$\delta \bar{e}_\alpha^k = \tilde{e}_\alpha^{kT} \underline{\delta \psi^k} \quad \text{and} \quad \delta \bar{e}_\beta^\ell = \tilde{e}_\beta^{\ell T} \underline{\delta \psi^\ell}, \tag{10.23}$$

respectively.

To simplify the expressions of the constraints associated with the various joints, the following notation is adopted for the scalar and vector products of two unit vectors of bases $\mathcal{B}^k$ and $\mathcal{B}^\ell$,

$$g_{\alpha\beta} = \bar{e}_\alpha^{kT} \bar{e}_\beta^\ell, \tag{10.24a}$$

$$\underline{h}_{\alpha\beta} = \tilde{e}_\alpha^k \bar{e}_\beta^\ell. \tag{10.24b}$$

Using eqs. (10.23), variations of these two quantities are easily obtained as

$$\delta g_{\alpha\beta} = (\underline{\delta\psi^k} - \underline{\delta\psi^\ell})^T \underline{h}_{\alpha\beta}, \tag{10.25a}$$

$$\delta \underline{h}_{\alpha\beta} = \underline{\delta\psi}^{kT} \underline{\underline{D}}_{\alpha\beta}^{k\ell} - \underline{\delta\psi}^{\ell T} \underline{\underline{D}}_{\beta\alpha}^{\ell k}. \tag{10.25b}$$

where the following matrices were defined

$$\underline{\underline{D}}_{\alpha\beta}^{k\ell} = \tilde{e}_\alpha^k \tilde{e}_\beta^\ell, \quad \underline{\underline{D}}_{\beta\alpha}^{\ell k} = \tilde{e}_\beta^\ell \tilde{e}_\alpha^k. \tag{10.26}$$

During the solution process, linearization of the forces of constraint will be required. In turn, this calls for the evaluation of increments in the kinematic quantities expressing the constraints. For instance, an increment in unit vector $\bar{e}_\alpha^k$ will be written as $\Delta \bar{e}_\alpha^k = \tilde{e}_\alpha^{kT} \underline{\Delta\psi^k} = \tilde{e}_\alpha^{kT} \underline{\underline{H}}^k \underline{\Delta r^k}$, where $\underline{\underline{H}}^k$ is the tangent operator. Relationship $\underline{\Delta\psi^k} = \underline{\underline{H}}^k \underline{\Delta r^k}$ expresses the differential rotation vector, $\underline{\Delta\psi^k}$, in terms of the increments of the rotation parameters, $\underline{\Delta r^k}$. If Euler angles with the *3-1-3* sequence are used to represent the finite rotation of body k that brings basis $\mathcal{B}_0^k$ to $\mathcal{B}^k$, $\underline{r}^{kT} = \{\phi, \theta, \psi\}$ and eqs. (4.68) define the tangent operator, $\underline{\underline{H}}^k$. Similar notation are used for the rotation of body ℓ; in summary,

$$\underline{\Delta\psi^k} = \underline{\underline{H}}^k \underline{\Delta r^k}, \quad \underline{\Delta\psi^\ell} = \underline{\underline{H}}^\ell \underline{\Delta r^\ell}. \tag{10.27}$$

10.4.3 Relative motions

In the reference configuration, the relative motions at the joint are assumed to vanish, *i.e.*, $\underline{u}_0^k = \underline{u}_0^\ell$ and $\underline{R}_0^k = \underline{R}_0^\ell$. In the deformed configuration, Δ_γ denote the relative displacement between the two bodies along unit vector $\bar{e}_\gamma^k$, and ϕ_γ the relative rotation about the same unit vector. More formally, the relative displacements and rotations of the two bodies are defined as

$$\bar{e}_\gamma^{kT} \underline{u} - \Delta_\gamma = 0, \tag{10.28a}$$

$$g_{\alpha\alpha} \sin \phi_\gamma + g_{\alpha\beta} \cos \phi_\gamma = 0, \tag{10.28b}$$

respectively. In the second equation, indices α, β, and γ, are such that $\epsilon_{\alpha\beta\gamma} = +1$, where the permutation symbol, $\epsilon_{\alpha\beta\gamma}$, is defined by eq. (1.25). For a planar rotation of magnitude θ about unit vector $\bar{e}_\gamma^k$, $g_{\alpha\alpha} = \cos\theta$, $g_{\alpha\beta} = -\sin\theta$, and eq. (10.28b) becomes $\sin(\phi_\gamma - \theta) = 0$ or $\phi_\gamma = \theta$, as expected.

Table 10.1 now formally defines the six lower pairs depicted in fig. 10.12 in terms of the relative displacement and/or rotation components that each joint allows or inhibits. If the two bodies are rigidly connected to each other, their six relative motions, three displacements and three rotations, must vanish at the connection point.

Table 10.1. Definition of the six lower pair joints. Symbols "✓" or "X" indicate that relative motion is allowed or inhibited, respectively. For the screw joint, p is the pitch of the screw.

Joint type	Relative displacements			Relative rotations		
	Δ_1	Δ_2	Δ_3	ϕ_1	ϕ_2	ϕ_3
Revolute	X	X	X	X	X	✓
Prismatic	X	X	✓	X	X	X
Screw	X	X	$p\phi_3$	X	X	✓
Cylindrical	X	X	✓	X	X	✓
Planar	✓	✓	X	X	X	✓
Spherical	X	X	X	✓	✓	✓

Setting $\Delta_\gamma = 0$ in eq. (10.28a) yields the constraint equation expressing the vanishing of the relative displacement along unit vector $\bar{e}_\gamma^k$. Similarly, setting $\phi_\gamma = 0$ in eq. (10.28b) expresses the vanishing of the relative rotation about unit vector $\bar{e}_\gamma^k$. On the other hand, if relative displacement along unit vector $\bar{e}_\gamma^k$ is allowed, eq. (10.28a) defines the magnitude of the relative displacement, Δ_γ, along that direction. Similarly, if relative rotation about unit vector $\bar{e}_\gamma^k$ is allowed, eq. (10.28b) defines the magnitude of the relative rotation, ϕ_γ, about that vector.

The explicit definition of the relative displacement and rotation components in lower pair joints as additional variables represents an important detail of the implementation. First, it allows the introduction of spring and/or damper elements in the joints, as usually required for modeling realistic configurations. Second, the time histories of joint relative motions can be driven according to suitably specified time functions or by actuators presenting their own physical characteristics.

10.5 Generic constraints for lower pair joints

Although the six lower pair joints depicted in fig. 10.12 are kinematically very different from each other, the constraints they impose on the bodies they are connected to are of two distinct types only. Lower pair joints inhibit one or more relative rotation components, and/or one or more relative displacement components. These two generic constraints are examined in details and the associated forces of constraint are derived from Lagrange's multiplier method in sections 10.5.1 and 10.5.2. In addition, the relative rotation and displacement components at a joint will be defined by means of constraint of two types, which are examined in sections 10.5.3 and 10.5.4. Once the formulation of these generic constraints has been developed, the formulations of the six lower pair joints will be presented in section 10.6.

10.5.1 First constraint: vanishing relative rotation

The constraint associated with the vanishing of the relative rotation at a lower pair joint connecting two bodies is readily obtained by imposing $\phi_\gamma = 0$ in eq. (10.28b) to find $g_{\alpha\beta} = 0$. In view of eq. (10.24a), this constraint imposes the orthogonality of two unit vectors, $\bar{e}_\alpha^{kT} \bar{e}_\beta^\ell = 0$, and is written in a generic manner as

$$C_I = g_{\alpha\beta} = \bar{e}_\alpha^{kT} \bar{e}_\beta^\ell = 0. \tag{10.29}$$

For various types of lower pair joints, indices α and β will take different values.

This holonomic constraint is enforced using Lagrange's multiplier method developed in section 10.2.1. The potential of the constraint forces is $V_I^c = \lambda_I C_I$, where λ_I is Lagrange's multiplier used to enforce this constraint; variation of this potential yields $\delta V_I^c = \delta\lambda_I C_I + \lambda_I \delta C_I$. The second term represents the virtual work done by the constraint force, $\delta W^c = \lambda_I \delta C_I = \delta \underline{q}^T \underline{F}_I^c$, where array $\delta \underline{q}$ stores variations of the generalized coordinates associated with this constraint and $\underline{F}_I^c$ the corresponding constraint forces. Variation of the constraint is expressed as $\delta C_I = \underline{B}_I \delta \underline{q}$, and it follows that $\underline{F}_I^c = \lambda_I \underline{B}_I^T$. Equation (10.25a) now yields

$$\delta\underline{q} = \begin{Bmatrix} \delta\underline{\psi}^k \\ \delta\underline{\psi}^\ell \end{Bmatrix}, \quad \underline{B}_I^T = \begin{Bmatrix} \underline{h}_{\alpha\beta} \\ -\underline{h}_{\alpha\beta} \end{Bmatrix}, \quad \underline{F}_I^c = \lambda_I \underline{B}_I^T. \tag{10.30}$$

Because the orthogonality constraint expressed by eq. (10.29) is nonlinear, numerical processes for the solution of constrained multibody systems will rely on successive linearizations of this constraint and associated forces. An increment in the constraint is expressed as

$$\Delta C_I = \frac{\partial C_I}{\partial \underline{q}} \Delta \underline{q} = \underline{Z}_I^T \Delta \underline{q}, \tag{10.31}$$

where array $\underline{Z}_I$ is easily found as

$$\Delta\underline{q} = \begin{Bmatrix} \Delta\underline{r}^k \\ \Delta\underline{r}^\ell \end{Bmatrix}, \quad \underline{Z}_I = \begin{Bmatrix} \underline{\underline{H}}^{kT} \underline{h}_{\alpha\beta} \\ -\underline{\underline{H}}^{\ell T} \underline{h}_{\alpha\beta} \end{Bmatrix}. \tag{10.32}$$

Arrays $\underline{r}^k$ and $\underline{r}^\ell$ store the rotation parameters representing the rotations of bodies k and ℓ, respectively, and $\underline{\underline{H}}^k$ and $\underline{\underline{H}}^\ell$ are the corresponding tangent operators defined by eqs. (10.27).

An increment in the forces of constraint is expressed as

$$\Delta\underline{F}_I^c = \frac{\partial \underline{F}_I^c}{\partial \underline{q}} \Delta \underline{q} = \underline{\underline{X}}_I \Delta \underline{q}, \tag{10.33}$$

where $\underline{\underline{X}}_I$ is the *equivalent stiffness matrix* for the constraint. Partial derivatives of the constraint forces yield the following expression for this matrix

$$\underline{\underline{X}}_I = \lambda_I \begin{bmatrix} \underline{\underline{D}}_{\beta\alpha}^{\ell k} \underline{\underline{H}}^k & -\underline{\underline{D}}_{\alpha\beta}^{k\ell} \underline{\underline{H}}^\ell \\ -\underline{\underline{D}}_{\beta\alpha}^{\ell k} \underline{\underline{H}}^k & \underline{\underline{D}}_{\alpha\beta}^{k\ell} \underline{\underline{H}}^\ell \end{bmatrix}, \tag{10.34}$$

where matrices $\underline{\underline{D}}_{\alpha\beta}^{k\ell}$ and $\underline{\underline{D}}_{\beta\alpha}^{\ell k}$ are defined by eqs. (10.26).

10.5.2 Second constraint: vanishing relative displacement

The constraint associated with the vanishing of the relative displacement at a lower pair joint connecting two bodies is readily obtained by imposing $\Delta_\alpha = 0$ in eq. (10.28a) to find $\bar{e}_\alpha^{kT} \underline{u} = 0$. This constraint imposes the orthogonality of the relative displacement vector defined by eq. (10.21) to unit vector $\bar{e}_\alpha^k$. This orthogonality constraint is written in a generic manner as

$$\mathcal{C}_{II} = \bar{e}_\alpha^{kT} \underline{u} = 0. \tag{10.35}$$

For various types of lower pair joints, index α will take different values.

This holonomic constraint is enforced using Lagrange's multiplier method, as discussed in section 10.5.1. The potential of the constraint forces is $V_{II}^c = \lambda_{II}\mathcal{C}_{II}$, and the virtual work done by the constraint force is $\delta W^c = \delta \underline{q}^T \underline{F}_{II}^c$. The variation of the constraint, $\delta\mathcal{C}_{II}$, is evaluated using eq. (10.23) to find

$$\delta\underline{q} = \left\{ \begin{array}{c} \delta\underline{u}^k \\ \delta\underline{\psi}^k \\ \delta\underline{u}^\ell \end{array} \right\}, \quad \underline{B}_{II}^T = \left\{ \begin{array}{c} -\bar{e}_\alpha^k \\ \tilde{e}_\alpha^k \underline{u} \\ \bar{e}_\alpha^k \end{array} \right\}, \quad \underline{F}_{II}^c = \lambda_{II}\underline{B}_{II}^T. \tag{10.36}$$

An increment in the constraint is expressed as $\Delta\mathcal{C}_{II} = \underline{Z}_{II}^T \Delta\underline{q}$, where array $\underline{Z}_{II}$ is easily found as

$$\Delta\underline{q} = \left\{ \begin{array}{c} \Delta\underline{u}^k \\ \Delta\underline{r}^k \\ \Delta\underline{u}^\ell \end{array} \right\}, \quad \underline{Z}_{II} = \left\{ \begin{array}{c} -\bar{e}_\alpha^k \\ \underline{\underline{H}}^{kT}\tilde{e}_\alpha^k \underline{u} \\ \bar{e}_\alpha^k \end{array} \right\}. \tag{10.37}$$

Array $\underline{r}^k$ stores the rotation parameters representing the rotations of body k, and $\underline{\underline{H}}^k$ is the corresponding tangent operator defined by eqs. (10.27).

The equivalent stiffness matrix for this constraint is

$$\underline{\underline{X}}_{II} = \lambda_{II} \begin{bmatrix} \underline{\underline{0}} & \tilde{e}_\alpha^k \underline{\underline{H}}^k & \underline{\underline{0}} \\ -\tilde{e}_\alpha^k \underline{\underline{H}}^k & \widetilde{u}\tilde{e}_\alpha^k \underline{\underline{H}}^k & \tilde{e}_\alpha^k \\ \underline{\underline{0}} & -\tilde{e}_\alpha^k \underline{\underline{H}}^k & \underline{\underline{0}} \end{bmatrix}. \tag{10.38}$$

10.5.3 Third constraint: definition of relative rotation

The constraint associated with the definition of the relative rotation at a lower pair joint connecting two bodies is given by eq. (10.28b). This constraint is written in a generic manner as

$$\mathcal{C}_{III} = g_{\alpha\alpha} \sin\phi_\gamma + g_{\alpha\beta} \cos\phi_\gamma = 0. \tag{10.39}$$

Indices α, β, and γ, are such that $\epsilon_{\alpha\beta\gamma} = +1$, where the permutation symbol, $\epsilon_{\alpha\beta\gamma}$, is defined by eq. (1.25). For various types of lower pair joints, index γ will take different values.

This holonomic constraint is enforced using Lagrange's multiplier method, as discussed in section 10.5.1. The potential of the constraint forces is $V^c_{III} = \lambda_{III}\mathcal{C}_{III}$, and the virtual work done by the constraint force is $\delta W^c = \delta\underline{q}^T \underline{F}^c_{III}$. The variation of the constraint, $\delta\mathcal{C}_{III}$, is evaluated using eq. (10.25a) to find

$$\delta\underline{q} = \left\{ \begin{array}{c} \delta\underline{\psi}^k \\ \delta\underline{\psi}^\ell \\ \delta\phi_\gamma \end{array} \right\}, \quad \underline{B}^T_{III} = \left\{ \begin{array}{c} \underline{w} \\ -\underline{w} \\ \tau \end{array} \right\}, \quad \underline{F}^c_{III} = \lambda_{III}\underline{B}^T_{III}, \tag{10.40}$$

where $\underline{w} = \underline{h}_{\alpha\alpha} \sin\phi_\gamma + \underline{h}_{\alpha\beta} \cos\phi_\gamma$ and $\tau = g_{\alpha\alpha} \cos\phi_\gamma - g_{\alpha\beta} \sin\phi_\gamma$.

An increment in the constraint is expressed as $\Delta\mathcal{C}_{III} = \underline{Z}^T_{III}\Delta\underline{q}$, where array $\underline{Z}_{III}$ is easily found as

$$\Delta\underline{q} = \left\{ \begin{array}{c} \Delta\underline{r}^k \\ \Delta\underline{r}^\ell \\ \Delta\phi_\gamma \end{array} \right\}, \quad \underline{Z}_{III} = \left\{ \begin{array}{c} \underline{H}^{kT}\underline{w} \\ -\underline{H}^{\ell T}\underline{w} \\ \tau \end{array} \right\}. \tag{10.41}$$

Arrays $\underline{r}^k$ and $\underline{r}^\ell$ store the rotation parameters representing the rotations of bodies k and ℓ, respectively, and $\underline{H}^k$ and $\underline{H}^\ell$ are the corresponding tangent operators defined by eqs. (10.27).

The equivalent stiffness matrix for this constraint is

$$\underline{\underline{X}}_{III} = \lambda_{III} \begin{bmatrix} \underline{\underline{E}}^T\underline{\underline{H}}^k & \underline{\underline{E}}\,\underline{\underline{H}}^\ell & \underline{z} \\ -\underline{\underline{E}}^T\underline{\underline{H}}^k & \underline{\underline{E}}\,\underline{\underline{H}}^\ell & -\underline{z} \\ \underline{z}^T\underline{\underline{H}}^k & -\underline{z}^T\underline{\underline{H}}^\ell & -\mathcal{C}_{III} \end{bmatrix}, \tag{10.42}$$

where $\underline{z} = \underline{h}_{\alpha\alpha} \cos\phi_\gamma - \underline{h}_{\alpha\beta} \sin\phi_\gamma$ and $\underline{\underline{E}} = \underline{\underline{D}}^{k\ell}_{\alpha\alpha} \sin\phi_\gamma + \underline{\underline{D}}^{k\ell}_{\alpha\beta} \cos\phi_\gamma$.

10.5.4 Fourth constraint: definition of relative displacement

The constraint associated with the definition of the relative displacement at a lower pair joint connecting two bodies is given by eq. (10.28a). This constraint is written in a generic manner as

$$\mathcal{C}_{IV} = \bar{e}^{kT}_\gamma\underline{u} - \Delta_\gamma = 0. \tag{10.43}$$

For various types of lower pair joints, index γ will take different values.

This holonomic constraint is enforced using Lagrange's multiplier method, as discussed in section 10.5.1. The potential of the constraint forces is $V_{IV}^c = \lambda_{IV} \mathcal{C}_{IV}$, and the virtual work done by the constraint force is $\delta W^c = \delta \underline{q}^T \underline{F}_{IV}^c$. The variation of the constraint, $\delta \mathcal{C}_{IV}$, is evaluated using eq. (10.23) to find

$$
\delta \underline{q} = \left\{ \begin{array}{c} \delta \underline{u}^k \\ \delta \underline{\psi}^k \\ \delta \underline{u}^\ell \\ \delta \Delta_\gamma \end{array} \right\}, \quad \underline{B}_{IV}^T = \left\{ \begin{array}{c} -\bar{e}_\alpha^k \\ \widetilde{e}_\alpha^k \underline{u} \\ \bar{e}_\alpha^k \\ -1 \end{array} \right\}, \quad \underline{F}_{IV}^c = \lambda_{IV} \underline{B}_{IV}^T. \tag{10.44}
$$

An increment in the constraint is expressed as $\Delta \mathcal{C}_{IV} = \underline{Z}_{IV}^T \Delta \underline{q}$, where array $\underline{Z}_{IV}$ is easily found as

$$
\Delta \underline{q} = \left\{ \begin{array}{c} \Delta \underline{u}^k \\ \Delta \underline{r}^k \\ \Delta \underline{u}^\ell \\ \Delta \Delta_\gamma \end{array} \right\}, \quad \underline{Z}_{IV} = \left\{ \begin{array}{c} -\bar{e}_\alpha^k \\ \underline{H}^{kT} \widetilde{e}_\alpha^k \underline{u} \\ \bar{e}_\alpha^k \\ -1 \end{array} \right\}. \tag{10.45}
$$

Array $\underline{r}^k$ stores the rotation parameters representing the rotations of body k, and $\underline{H}^k$ is the corresponding tangent operator defined by eqs. (10.27).

The equivalent stiffness matrix for this constraint is

$$
\underline{\underline{X}}_{IV} = \lambda_{IV} \begin{bmatrix} \underline{0} & \widetilde{e}_\alpha^k \underline{H}^k & \underline{0} & \underline{0} \\ -\widetilde{e}_\alpha^k & \widetilde{u}\widetilde{e}_\alpha^k \underline{\underline{H}}^k & \widetilde{e}_\alpha^k & \underline{0} \\ \underline{0} & -\widetilde{e}_\alpha^k \underline{\underline{H}}^k & \underline{0} & \underline{0} \\ \underline{0} & \underline{0} & \underline{0} & \underline{0} \end{bmatrix}. \tag{10.46}
$$

10.6 Constraints for the lower pair joints

In this section, the constraints associated with the six lower pair joints depicted in fig. 10.12 are detailed. The corresponding constraint forces are derived, and their physical nature is discussed.

10.6.1 Revolute joints

Figure 10.13 depicts two bodies linked together by a lower pair joint. The kinematics of the problem and the corresponding notational conventions are presented in sections 10.4.1 and 10.4.2, respectively. This section focuses on a specific type of joint, the *revolute joint*, depicted in fig. 10.14. For this joint, points $\mathbf{K}$ and $\mathbf{L}$ are coincident in both reference and deformed configurations. The revolute joint allows the two bodies it connects to rotate with respect to each other about a material axis, selected, by convention, to be $\bar{e}_3^k = \bar{e}_3^\ell$. This condition implies the orthogonality of $\bar{e}_3^k$ to both $\bar{e}_1^\ell$ and $\bar{e}_2^\ell$.

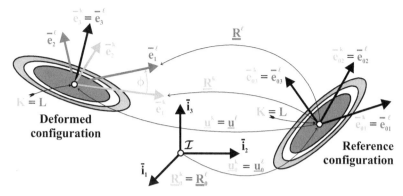

Fig. 10.14. Revolute joint in the reference and deformed configurations.

The revolute joint is characterized by the following kinematic constraints

$$\underline{\mathcal{C}}_1 = \underline{u}^\ell - \underline{u}^k = 0, \tag{10.47a}$$

$$\mathcal{C}_2 = \bar{e}_3^{kT} \bar{e}_1^\ell = g_{31} = 0, \tag{10.47b}$$

$$\mathcal{C}_3 = \bar{e}_3^{kT} \bar{e}_2^\ell = g_{32} = 0, \tag{10.47c}$$

$$\mathcal{C}_4 = g_{11} \sin\phi + g_{12} \cos\phi = 0. \tag{10.47d}$$

Constraint (10.47a) expresses the vanishing of the relative displacement at the joint; it is readily enforced by Boolean identification of the corresponding degrees of freedom. The second and third constraints are of type I, see section (10.5.1), with $\alpha = 3$, $\beta = 1$, and $\alpha = 3$, $\beta = 2$, for eqs. (10.47b) and (10.47c), respectively. Finally, the last constraint is of type III, see section 10.5.3, and defines the relative rotation about unit vector $\bar{e}_3^k = \bar{e}_3^\ell$, denoted ϕ in fig. 10.14.

The combination of eqs. (10.30) and (10.40) yields the forces associated with the revolute joint constraints,

$$\underline{F}^c = \left\{ \begin{array}{c} \underline{h}_{31} \\ -\underline{h}_{31} \\ 0 \end{array} \right\} \lambda_1 + \left\{ \begin{array}{c} \underline{h}_{32} \\ -\underline{h}_{32} \\ 0 \end{array} \right\} \lambda_2 + \left\{ \begin{array}{c} (\underline{h}_{11} \sin\phi + \underline{h}_{12} \cos\phi) \\ -(\underline{h}_{11} \sin\phi + \underline{h}_{12} \cos\phi) \\ (g_{11} \cos\phi - g_{12} \sin\phi) \end{array} \right\} \lambda_3, \tag{10.48}$$

where Lagrange's multipliers λ_1, λ_2, and λ_3 are associated with constraints (10.47b), (10.47c), and (10.47d), respectively.

When the constraints are satisfied, $\underline{h}_{31} = \bar{e}_2^\ell$ and $\underline{h}_{32} = -\bar{e}_1^\ell$. The forces of constraint associated with the first constraint correspond to two moments acting about unit vector $\bar{e}_2^\ell$ and of magnitudes $+\lambda_1$ and $-\lambda_1$, respectively, applied to bodies k and ℓ, respectively. The forces of constraint associated with the second constraint are readily interpreted in a similar manner. The moments associated with these first two constraints enforce the parallelism of unit vectors $\bar{e}_3^k$ and $\bar{e}_3^\ell$.

When the constraints are satisfied, $\underline{h}_{11} = \bar{e}_3^k \sin\phi$ and $\underline{h}_{12} = \bar{e}_3^k \cos\phi$, implying that $\underline{h}_{11} \sin\phi + \underline{h}_{12} \cos\phi = \bar{e}_3^k$; furthermore, $g_{11} \cos\phi - g_{12} \sin\phi = \cos\phi \cos\phi - (-\sin\phi) \sin\phi = 1$. To interpret the forces associated with the third constraint, it is assumed that a motor applies a torque Q at the revolute joint; the virtual work done by this torque is then $\delta W = Q\delta\phi$. Because Lagrange's multiplier technique is used to enforce the constraint, the relative rotation, ϕ, is now an unconstrained variable, and the corresponding equation of motion will be $\lambda_3 + Q = 0$: Lagrange's multiplier is of equal magnitude and opposite sign to the applied torque. The remaining components of the constraint forces correspond to two moments acting about unit vector $\bar{e}_3^k$ and of magnitude $-Q$ and $+Q$, respectively, transmitting the applied torque to bodies k and ℓ, respectively. If no torque is applied at the joint, Lagrange's multiplier vanishes, $\lambda_3 = 0$, and no forces are associated with this constraint, which simply defines variable ϕ but applies no forces to the system.

10.6.2 Prismatic joints

Figure 10.13 depicts two bodies linked together by a lower pair joint. The kinematics of the problem and the corresponding notational conventions are presented in sections 10.4.1 and 10.4.2, respectively. This section focuses on the *prismatic joint*, depicted in fig. 10.15. For this joint, the two bases coincide in the reference configuration, $\mathcal{B}_0^k = \mathcal{B}_0^\ell$, and in the deformed configuration, $\mathcal{B}^k = \mathcal{B}^\ell$. The prismatic joint allows the two bodies it connects to translate with respect to each other along a material axis, selected, by convention, to be $\bar{e}_3^k = \bar{e}_3^\ell$. This condition implies the orthogonality of unit vectors $\bar{e}_1^k$ and $\bar{e}_2^k$ to the relative displacement vector, $\underline{u} = \underline{u}^\ell - \underline{u}^k$.

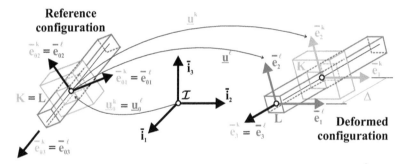

Fig. 10.15. Prismatic joint in the reference and deformed configurations.

The prismatic joint is characterized by the following kinematic constraints

$$\underline{\mathcal{C}}_1 = \underline{R}^\ell - \underline{R}^k = 0, \tag{10.49a}$$

$$\mathcal{C}_2 = \bar{e}_1^{kT}\underline{u} = 0, \tag{10.49b}$$

$$\mathcal{C}_3 = \bar{e}_2^{kT}\underline{u} = 0, \tag{10.49c}$$

$$\mathcal{C}_4 = \bar{e}_3^{kT}\underline{u} - \Delta = 0. \tag{10.49d}$$

Constraint (10.49a) expresses the vanishing of the relative rotation at the joint; it is readily enforced by Boolean identification of the corresponding degrees of freedom.

The second and third constraints are of type II, see section (10.5.2), with $\alpha = 1$ and 2, for eqs. (10.49b) and (10.49c), respectively. Finally, the last constraint is of type IV, see section 10.5.4, and defines the relative displacement along unit vector $\bar{e}_3^k = \bar{e}_3^\ell$, denoted Δ in fig. 10.15.

The combination of eqs. (10.36) and (10.44) yields the forces associated with the prismatic joint constraints,

$$
\underline{F}^c = \left\{ \begin{array}{c} -\bar{e}_1^k \\ \widetilde{e}_1^k \, \underline{u} \\ \bar{e}_1^k \\ 0 \end{array} \right\} \lambda_1 + \left\{ \begin{array}{c} -\bar{e}_2^k \\ \widetilde{e}_2^k \, \underline{u} \\ \bar{e}_2^k \\ 0 \end{array} \right\} \lambda_2 + \left\{ \begin{array}{c} -\bar{e}_3^k \\ \widetilde{e}_3^k \, \underline{u} \\ \bar{e}_3^k \\ -1 \end{array} \right\} \lambda_3. \tag{10.50}
$$

where Lagrange's multipliers λ_1, λ_2, and λ_3 are associated with constraints (10.49b), (10.49c), and (10.49d), respectively.

When the constraints are satisfied, $\widetilde{e}_1^k \, \underline{u} = -\|\underline{u}\| \bar{e}_2^k$, $\widetilde{e}_2^k \, \underline{u} = \|\underline{u}\| \bar{e}_1^k$, and $\widetilde{e}_3^k \, \underline{u} = 0$. The forces of constraint associated with the first constraint correspond to two forces acting along unit vector $\bar{e}_1^k$ and of magnitude $-\lambda_1$ and $+\lambda_1$, respectively, applied to bodies k and ℓ, respectively, and one moment acting about unit vector $\bar{e}_2^k$ and of magnitude $-\|\underline{u}\|\lambda_1$, applied to both bodies k and ℓ that share a common orientation.

The forces of constraint associated with the second constraint are readily interpreted in a similar manner. The forces associated with these first two constraints enforce the collinearity of unit vectors $\bar{e}_3^k$ and $\bar{e}_3^\ell$; the moments account for the fact that these aligning forces form couples with a moment arm $\|\underline{u}\|$.

To interpret the forces associated with the third constraint, it is assumed that an actuator applies a force F at the prismatic joint; the virtual work done by this force is then $\delta W = F \delta \Delta$. Because Lagrange's multiplier technique was used to enforce the constraint, the relative displacement, Δ, is now an unconstrained variable, and the corresponding equation of motion will be $\lambda_3 - F = 0$: Lagrange's multiplier equals the applied force. The remaining components of the constraint forces correspond to two forces along unit vector $\bar{e}_3^k$ and of magnitude $-F$ and $+F$, respectively, transmitting the applied force to bodies k and ℓ, respectively. If no force is applied at the joint, Lagrange's multiplier vanishes, $\lambda_3 = 0$, and no forces are associated with this constraint.

10.6.3 Cylindrical joints

Figure 10.16 depicts the *cylindrical joint*, which is one of the lower pair joints discussed in a generic manner in section 10.4.1. The cylindrical joint allows the two bodies it connects to rotate and translate with respect to each other about a material axis, implying the orthogonality of $\bar{e}_3^k$ to both $\bar{e}_1^\ell$ and $\bar{e}_2^\ell$ and the orthogonality of unit vectors $\bar{e}_1^k$ and $\bar{e}_2^k$ to the relative displacement vector, $\underline{u} = \underline{u}^\ell - \underline{u}^k$.

The cylindrical joint is characterized by the following kinematic constraints: constraints (10.47b) and (10.47c) expressing the orthogonality of unit vectors $\bar{e}_1^\ell$ and $\bar{e}_2^\ell$ to unit vector $\bar{e}_3^k$, and constraints (10.49b) and (10.49c) expressing the orthogonality of unit vectors $\bar{e}_1^k$ and $\bar{e}_2^k$ to the relative displacement vector, $\underline{u}$. The relative rotation

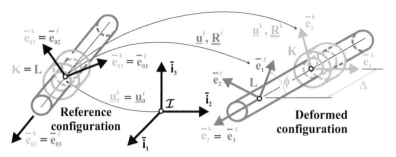

Fig. 10.16. Cylindrical joint in the reference and deformed configurations.

about unit vector $\bar{e}_3^k = \bar{e}_3^\ell$, denoted ϕ, and relative displacement along the same axis, denoted Δ, between the two bodies are defined by adding to the formulation constraints (10.47d) and (10.49d), respectively. Clearly, the cylindrical joint combines the constraints of the revolute and prismatic joints. The associated forces of constraints are identical to those developed in section 10.6.1 and 10.6.2 and will not be repeated here.

10.6.4 Screw joints

The kinematic constraints associated with the screw joint are identical to those of the cylindrical joint. An additional constraint imposes a linear relationship between the relative rotation, ϕ, and relative displacement, Δ,

$$C = \Delta - \frac{p}{2\pi}\phi = 0, \tag{10.51}$$

where p is the *pitch of the screw*.

10.6.5 Planar joints

Figure 10.13 depicts two bodies linked together by a lower pair joint. The kinematics of the problem and the corresponding notational conventions are presented in sections 10.4.1 and 10.4.2, respectively. This section focuses on the *planar joint*, depicted in fig. 10.17. The planar joint allows the two bodies it connects to translate with respect to each other within a material plane, selected by convention, to be normal to unit vector $\bar{e}_3^k = \bar{e}_3^\ell$. This condition implies the orthogonality of unit vector $\bar{e}_3^k$ to the relative displacement vector, $\underline{u} = \underline{u}^\ell - \underline{u}^k$. The planar joint further allows the two bodies to rotate with respect to each other about the axis perpendicular to the material plane, $\bar{e}_3^k = \bar{e}_3^\ell$. This condition implies the orthogonality of unit vector $\bar{e}_3^k$ to unit vectors $\bar{e}_1^\ell$ and $\bar{e}_2^\ell$.

The planar joint is characterized by the following kinematic constraints

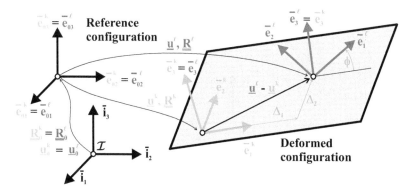

Fig. 10.17. Planar joint in the reference and deformed configurations.

$$\mathcal{C}_1 = \bar{e}_3^{kT} \underline{u} = 0, \tag{10.52a}$$

$$\mathcal{C}_2 = \bar{e}_3^{kT} \bar{e}_1^{\ell} = g_{31} = 0, \tag{10.52b}$$

$$\mathcal{C}_3 = \bar{e}_3^{kT} \bar{e}_2^{\ell} = g_{32} = 0, \tag{10.52c}$$

$$\mathcal{C}_4 = \bar{e}_1^{kT} \underline{u} - \Delta_1 = 0. \tag{10.52d}$$

$$\mathcal{C}_5 = \bar{e}_2^{kT} \underline{u} - \Delta_2 = 0. \tag{10.52e}$$

$$\mathcal{C}_6 = g_{11} \sin\phi + g_{12} \cos\phi = 0. \tag{10.52f}$$

Constraint (10.52a) is of type II, see section 10.5.2, with $\alpha = 3$. The second and third constraints are of type I, see section (10.5.1), with $\alpha = 3$, $\beta = 1$, and $\alpha = 3$, $\beta = 2$, for eqs. (10.52b) and (10.52c), respectively. The next two constraints are of type IV, see section 10.5.4, with $\alpha = 1$ and $\alpha = 2$ for eqs. (10.52d) and (10.52e), respectively. They define the relative displacements of the bodies along unit vectors $\bar{e}_1^k$ and $\bar{e}_2^k$, respectively, denoted Δ_1 and Δ_2, respectively. Finally, the last constraint is of type III, see section 10.5.3, and defines the relative rotation about unit vector $\bar{e}_3^k = \bar{e}_3^{\ell}$, denoted ϕ in fig. 10.17.

The planar joint combines the constraints of the revolute and prismatic joints. The associated forces of constraints are identical to those developed in sections 10.6.1 and 10.6.2.

10.6.6 Spherical joints

Figure 10.18 depicts the *spherical joint*, which is one of the lower pair joints discussed in a generic manner in section 10.4.1. The spherical joint allows the two bodies it connects to freely rotate with respect to each other about a material point, $\mathbf{K} = \mathbf{L}$, while preventing any relative displacement at this point, *i.e.*, $\underline{u}^k = \underline{u}^\ell$.

The spherical joint is characterized by constraints (10.47a), which prevents relative displacement between the bodies. This constraint is readily enforced by Boolean identification of the corresponding degrees of freedom.

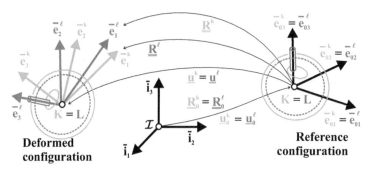

Fig. 10.18. Spherical joint in the reference and deformed configurations.

10.6.7 Problems

Problem 10.22. Relative rotation of a revolute joint
Consider the revolute joint discussed in section 10.6.1. Express the rotation of body ℓ, $\underline{\underline{R}}^\ell$, as a function of rotation of body k, $\underline{\underline{R}}^k$, and of the relative rotation angle ϕ.

Problem 10.23. Relative motion of a prismatic joint
Consider the prismatic joint discussed in section 10.6.2. Express the displacement of body ℓ, $\underline{u}^\ell$, as a function of displacement of body k, $\underline{u}^k$, and of the relative displacement Δ.

10.7 Other joints

Multibody systems often involve a variety of joints that impose constraints on the relative motion of the bodies of the system. The lower pairs described in the previous sections can be used to synthesize more complex joints: for instance, the universal joint depicted in fig. 10.19 can be viewed two revolute joints sharing a common axis of rotation along unit vector $\bar{e}_3^k$ and two more revolute joints sharing a common axis of rotation along unit vector $\bar{e}_3^\ell$. In many cases, however, joints with specialized kinematic conditions must be developed.

10.7.1 Universal joints

Although the universal joint depicted in fig. 10.19 is not a lower pair joint, the kinematic description and notational conventions presented in sections 10.4.1 and 10.4.2, respectively, will be used here again. At the heart of the universal joint is a cruciform, consisting of two rigidly connected bars assembled together at a 90 degree angle. Body k is allowed to rotate about unit vector $\bar{e}_3^k$, which is aligned with the first bar of the cruciform; body ℓ is allowed to rotate about unit vector $\bar{e}_3^\ell$, which is aligned with the second bar of the cruciform. It follows that unit vectors $\bar{e}_3^k$ and $\bar{e}_3^\ell$ are material axes of bodies k and ℓ, respectively, and point $\mathbf{K} = \mathbf{L}$ is a material point of both bodies.

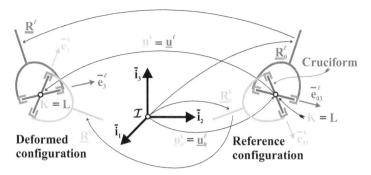

Fig. 10.19. Universal joint in the reference and deformed configurations.

The universal joint is characterized by the following kinematic constraints: constraints (10.47a), which prevents relative displacement between the bodies, and a single type I constraint, see eq. (10.29), with $\alpha = \beta = 3$. The universal joint uses a subset of the constraint developed for the revolute joint.

10.7.2 Curve sliding joints

The kinematic conditions associated with the sliding of a body along a flexible track have been presented by Li and Likins [30] within the framework of Kane's method. Cardona [31] derived a finite element based formulation for the sliding of a body along a prescribed curve. Finally, Bauchau [32] presented the formulation of a sliding joint that enforces the sliding of a body along a flexible beam. This formulation was later refined [33] to include constraints on the relative rotation between the sliding bodies. This section describes the curve sliding joint that enforces the sliding of a body on a rigid curve connected to another body.

Figure 10.20 depicts two bodies linked together by a curve sliding joint. Here again, the kinematic description and notational conventions presented in sections 10.4.1 and 10.4.2, respectively, will be used. Spatial curve $\mathbb{C}$, see section 2.2, is rigidly connected to body k.

A curve sliding joint involves displacement constraints requiring point **L**, a material point of body ℓ, to slide along curve $\mathbb{C}$, which is rigidly connected to body k. Let $\underline{p}^*(\eta)$ and $\underline{p}(\eta) = \underline{\underline{R}}^k \underline{\underline{R}}_0^k \underline{p}^*(\eta)$ be the components of the position vector of a point on curve $\mathbb{C}$ with respect to point **K**, resolved in bases $\mathcal{B}^k$ and $\mathcal{I}$, respectively. Moreover, let $\underline{P}^k$ be the components of the position vector of an arbitrary point on curve $\mathbb{C}$ with respect to point **O**, resolved in basis $\mathcal{I}$. It then follows that $\underline{P}^k = \underline{u}_0^k + \underline{u}^k + \underline{p}(\eta)$. Similarly, the components of the position vector of point **L** with respect to point **O**, resolved in basis $\mathcal{I}$, are denoted $\underline{P}^\ell = \underline{u}_0^\ell + \underline{u}^\ell$.

Because point **L** must be along curve $\mathbb{C}$, the following vector constraint must be satisfied

$$\underline{\mathcal{C}} = \underline{P}^k - \underline{P}^\ell = \underline{u}_0 + \underline{u} + \underline{p}(\eta) = \underline{0}, \qquad (10.53)$$

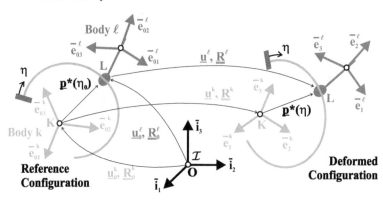

Fig. 10.20. Configuration of the curve sliding joint.

where $\underline{u}_0 = \underline{u}_0^k - \underline{u}_0^\ell$ and $\underline{u} = \underline{u}^k - \underline{u}^\ell$ are the relative displacement vectors of the two bodies, in the reference and deformed configurations, respectively.

This holonomic constraint is enforced using Lagrange's multiplier method. The potential of the constraint forces is $V^c = \underline{\lambda}^T \underline{C}$, and the virtual work done by the constraint force is $\delta W^c = \delta \underline{q}^T \underline{F}^c$. The variation of the constraint, $\delta \underline{C}$, is evaluated as

$$
\delta \underline{q} = \left\{ \begin{matrix} \delta \underline{u}^k \\ \delta \underline{\psi}^k \\ \delta \underline{u}^\ell \\ \delta \eta \end{matrix} \right\}, \quad \underline{\underline{B}}^T = \left\{ \begin{matrix} \underline{\underline{I}} \\ -\widetilde{p}(\eta) \\ -\underline{\underline{I}} \\ \underline{p}_1^T(\eta) \end{matrix} \right\}, \quad \underline{F}^c = \underline{\underline{B}}^T \underline{\lambda}. \tag{10.54}
$$

where $\underline{p}_1^* = \underline{p}^{*\prime}(\eta)$, $\underline{p}_1(\eta) = \underline{\underline{R}}^k \underline{\underline{R}}_0^k \underline{p}_1^*(\eta)$, and notation $(\cdot)'$ indicates a derivative with respect to η.

It is often necessary to know the curvilinear coordinate, s, along the curves. For instance, if a friction force of magnitude F^f is present between body ℓ and curve $\mathbb{C}$ at point $\mathbf{L}$, the formulation would require the evaluation of the virtual work done by this force, $\delta W = F^f \delta s$. It is often convenient to use the very versatile NURBS representation of curves [34, 35], but this approach is based on an arbitrary parameterization, $\eta \in [0, 1]$, as discussed in section 2.2.2. The intrinsic parameterization of the curve presented in section 2.2.1 directly uses curvilinear coordinate s, but is often very difficult to obtain.

To remedy the situation, an additional scalar constraint relating these two variables is necessary. Expressing the relationship between variables s and η is arduous, and more often that not, impossible. Equation (2.14), however, recast as $\dot{s} = p_1 \dot{\eta}$, provides a relationship between the corresponding generalized velocities,

$$
\mathcal{C} = \dot{s} - \|\underline{p}_1^*(\eta)\| \dot{\eta} = 0. \tag{10.55}
$$

In general, this constraint is not integrable and hence, must be treated as a nonlinear, nonholonomic constraint.

10.7.3 Sliding joints

The formulation of prismatic joints was presented in section 10.6.2. The prismatic joint is characterized by the following kinematic constraints: constraint (10.49a) that prevents relative rotation between the bodies, and constraints (10.49b) and (10.49c) that express the orthogonality of unit vectors $\bar{e}_1^k$ and $\bar{e}_2^k$ to the relative displacement vector $\underline{u}$. Note that although these constraints are expressed in terms of the kinematic variables at points $\mathbf{K}$ and $\mathbf{L}$, they imply the sliding of body ℓ on body k at point $\mathbf{K}$, *when body ℓ is rigid.*

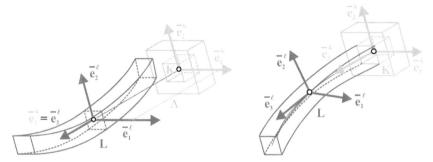

Fig. 10.21. Prismatic joint with flexible body.   **Fig. 10.22.** Sliding joint with flexible body.

The situation is sharply different when body ℓ is flexible, as shown in fig. 10.21. If conditions (10.49b), (10.49c) and (10.49a) are enforced, body ℓ is no longer sliding on body k at point $\mathbf{K}$, *i.e.*, contact between the bodies is no longer enforced. In actual systems, the piece of hardware corresponding to the prismatic joint implies the sliding of body ℓ on body k with contact at point $\mathbf{K}$ at all times, as depicted in fig. 10.22. In fact, in the presence of flexible bodies, such joint is more accurately described as a *sliding joint* [32, 33, 36].

Due to the flexibility of body ℓ, the kinematic variables at material points $\mathbf{K}$ and $\mathbf{L}$ are no longer related by conditions (10.49b), (10.49c) and (10.49a). Rather, constraint conditions must be enforced between the kinematic variables at point $\mathbf{K}$ of body k, and the kinematic variables at the material point of body ℓ which is in contact with body k at an instant. Clearly, kinematic constraints (10.49b), (10.49c) and (10.49a) associated with the classical formulation of prismatic joints, and the kinematic constraint associated with sliding in the presence of flexible bodies are fundamentally different and will lead to sharply different dynamic responses of the system. Although the above discussion has focused on prismatic joints, it is clear that identical remarks can be made concerning the classical formulation of cylindrical joints, and about their inadequacy to model sliding behavior in the presence of flexible bodies.

Figure 10.23 depicts two bodies linked together by a sliding joint. Body k is a flexible beam element whose displacement field is interpolated from nodal quantities, see sections 16.3 and 17.7. In the reference configuration, the coordinates of a point on the beam are

$$\underline{u}_0^k(\eta) = \underline{\underline{N}}(\eta)\hat{u}_0^k, \tag{10.56}$$

where $\hat{u}_0^k$ are the nodal positions in the reference configuration, $\underline{\underline{N}}(\eta)$ the displacement interpolation matrix defined by eq. (17.6), and $\eta \in [0, 1]$ a non-dimensional parameter indicating the location of a material particle along the beam axis in the reference configuration. Body ℓ can be a rigid or flexible element of the system. The position vector of a node point of this body is denoted $\underline{u}_0^\ell$ in the reference configuration.

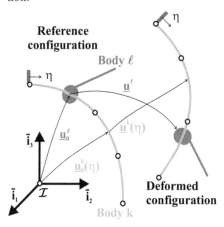

Fig. 10.23. Sliding joint in the reference and deformed configurations.

After deformation, the position vector of a point on the beam becomes $\underline{P}^k(\eta) = \underline{\underline{N}}(\eta)\,(\hat{u}_0^k + \hat{u}^k)$, where $\hat{u}^k$ are the nodal displacement vectors. Similarly, the position vector of the node on body ℓ is $\underline{P}^\ell = \hat{u}_0^\ell + \hat{u}^\ell$, where $\hat{u}^\ell$ is the nodal displacement vector.

The kinematic constraint associated with the condition of body ℓ freely sliding over the flexible beam is $\underline{\mathcal{C}} = \underline{P}^k(\eta) - \underline{P}^\ell = 0$.

Parameter η which determines the location of contact between bodies k and ℓ is, of course, a time varying unknown of the problem. This constraint will be enforced using the Lagrange multiplier method, see section 10.2.1. The virtual work done by the constraint force defined by eq. (10.5) becomes

$$\left\{\begin{matrix} \delta\hat{u}^k \\ \delta\eta \\ \delta\underline{u}^\ell \end{matrix}\right\}^T \underline{F}^c = \left\{\begin{matrix} \delta\hat{u}^k \\ \delta\eta \\ \delta\underline{u}^\ell \end{matrix}\right\}^T \left\{\begin{matrix} \underline{\underline{N}}^T \\ (\hat{u}_0^k + \hat{u}^k)^T \underline{\underline{N}}'^T \\ -1 \end{matrix}\right\} \underline{\lambda}, \tag{10.57}$$

where $(\cdot)'$ denotes a derivative with respect to η.

10.7.4 Problems

Problem 10.24. Two rigid bodies connected by a rigid link
Figure 10.24 shows two rigid bodies connected at point $\mathbf{K}'$ and $\mathbf{L}'$ by a rigid link. The kinematics of the two bodies is represented using the conventions described in section 10.4.1 and fig. 10.13. Points $\mathbf{K}'$ and $\mathbf{L}'$ are material points of bodies k and ℓ, respectively. In the reference configuration, the position of point $\mathbf{K}'$ with respect to point $\mathbf{K}$ is given by vector $\underline{s}^k$, with a similar definition for vector $\underline{s}^\ell$. In the deformed configuration, the corresponding position vectors are denoted $\underline{S}^k$ and $\underline{S}^\ell$, respectively. Due to the presence of the rigid link, the distance between points $\mathbf{K}'$ and $\mathbf{L}'$ must remain constant, leading to the following nonlinear holonomic constraint, $\mathcal{C} = (\|\underline{d}\|^2 - \ell^2)/2 = 0$, where $\ell^2 = \|\underline{d}_0\|^2$ is the constant length of the link. *(1)* Identify the array of generalized coordinates for this problem. *(2)* Determine the constraint

matrix. *(3)* If the constraint is enforced via Lagrange's multiplier method, derive the constraint forces. *(4)* Describe the physical interpretation of these forces of constraint. *(5)* Evaluate the equivalent stiffness matrix for the constraint.

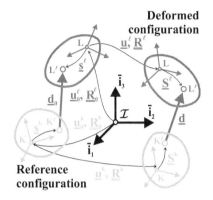

Fig. 10.24. Configuration of the rigid link.

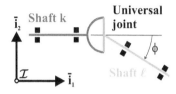

Fig. 10.25. Two rigid shafts connected by a universal joint.

Problem 10.25. Angular velocities of a universal joint

Consider two rigid shafts connected by a universal joint as depicted in fig. 10.25. The two shafts remain in the fixed plane defined by vectors $\bar{\imath}_1$ and $\bar{\imath}_2$, and the constant angle between the shafts is ϕ. Let ω^k denote the constant angular velocity of shaft k. *(1)* Find the angular velocity of shaft ℓ, denoted ω^ℓ. *(2)* Plot the angular velocity ratio, ω^ℓ/ω^k, over one period of rotation of shaft k. *(3)* Find the maximum value of this ratio as a function of the relative shaft angle ϕ.

Problem 10.26. Point associated with a rigid body

Figure 10.26 shows a rigid body in its reference and final configurations. The kinematics of the rigid body is represented using the conventions described in section 10.4.1 and fig. 10.13. Point **A** is a material point of the rigid body and its position vector with respect to reference point **B** of the rigid body is denoted $\underline{s}_A$ and $\underline{S}_A$ in the reference and final configurations, respectively. Let $\underline{u}_A$ denote the displacement vector of point **A**. Because $\underline{u} \neq \underline{u}_A$, it is often desirable to use $\underline{u}_A$ as an additional set of generalized coordinates, which are defined by the following constraints, $\underline{\mathcal{C}} = (\underline{u} + \underline{S}_A) - (\underline{u}_A + \underline{s}_A) = \underline{0}$. *(1)* Identify the array of generalized coordinates for this problem. *(2)* Determine the constraint matrix. *(3)* If the constraint is enforced via Lagrange's multiplier method, derive the constraint forces. *(4)* Describe the physical interpretation of these forces of constraint. *(5)* Evaluate the equivalent stiffness matrix for the constraint.

Problem 10.27. The curve sliding joint

Consider the curve sliding joint. *(1)* Compute the velocity of a material point on curve $\mathbb{C}$. This can be obtained by taking a time derivative of the position vector of an arbitrary point on curve $\mathbb{C}$, resolved in $\mathcal{I}$ considering η to be constant. *(2)* Compute the velocity of the material point on body ℓ. *(3)* Compute the relative velocity of body ℓ with respect to body k at the point of contact. *(4)* Use the constraint condition, eq. (10.53), to show that the component of relative velocity in the direction tangent to the curve is simply $\dot{s}$. *(5)* What are the components of relative velocity in the other directions?

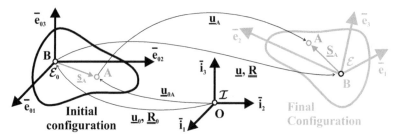

Fig. 10.26. Configuration of the rigid body with an associated point.

Problem 10.28. Relative rotation for a universal joint

The formulation of the universal joint was presented in section 10.7.1. In practice, a universal joint consists of a cruciform component connected to body k and ℓ by means of two pairs of revolute joints, as depicted in fig. 10.27. In the reference configuration, the following orthonormal bases are constructed: $\mathcal{B}_{0g}^k = (\bar{g}_{01}^k = \bar{e}_{03}^\ell, \bar{g}_{02}^k = \tilde{e}_{03}^k \bar{e}_{03}^\ell, \bar{g}_{03}^k = \bar{e}_{03}^k)$ and $\mathcal{B}_{0g}^\ell = (\bar{g}_{01}^\ell = \bar{e}_{03}^k, \bar{g}_{02}^\ell = \tilde{e}_{03}^k \bar{e}_{03}^\ell, \bar{g}_{03}^\ell = \bar{e}_{03}^\ell)$. Let $\underline{\underline{G}}^k$ and $\underline{\underline{G}}^\ell$ be the components of the rotation tensors defining the rotations from $\mathcal{I}$ to $\mathcal{B}_{0g}^k$ and $\mathcal{I}$ to $\mathcal{B}_{0g}^\ell$, respectively, resolved in $\mathcal{I}$. In the deformed configuration, the following orthonormal bases are constructed: $\mathcal{B}_g^k = (\bar{g}_1^k = \bar{e}_3^\ell, \bar{g}_2^k = \tilde{e}_3^k \bar{e}_3^\ell, \bar{g}_3^k = \bar{e}_3^k)$ and $\mathcal{B}_g^\ell = (\bar{g}_1^\ell = \bar{e}_3^k, \bar{g}_2^\ell = \tilde{e}_3^\ell \bar{e}_3^k, \bar{g}_3^\ell = \bar{e}_3^\ell)$. Let $\underline{\underline{G}}^k$ and $\underline{\underline{G}}^\ell$ be the components of the rotation tensors defining the rotations from $\mathcal{I}$ to $\mathcal{B}_g^k$ and $\mathcal{I}$ to $\mathcal{B}_g^\ell$, respectively, resolved in $\mathcal{I}$. Finally, two additional orthonormal bases, $\mathcal{B}_R^k = (\bar{g}_{R1}^k, \bar{g}_{R2}^k, \bar{g}_{R3}^k)$ and $\mathcal{B}_{Rg}^\ell = (\bar{g}_{R1}^\ell, \bar{g}_{R2}^\ell, \bar{g}_{R3}^\ell)$ are defined as $\bar{g}_{R\alpha}^k = \underline{\underline{R}}^k \underline{\underline{G}}_0^{k} \bar{\imath}_\alpha$ and $\bar{g}_{R\alpha}^\ell = \underline{\underline{R}}^\ell \underline{\underline{G}}_0^{\ell} \bar{\imath}_\alpha$, respectively. The rotation ϕ^k of the revolute joint between body k and the cruciform is defined as $(\bar{e}_3^{\ell T} \bar{g}_{R1}^k) \sin \phi^k - (\bar{e}_3^{\ell T} \bar{g}_{R2}^k) \cos \phi^k = 0$; and rotation ϕ^ℓ of the revolute joint between body ℓ and the cruciform is $(\bar{e}_3^{kT} \bar{g}_{R1}^\ell) \sin \phi^\ell - (\bar{e}_3^{kT} \bar{g}_{R2}^\ell) \cos \phi^\ell = 0$. *(1)* Show that $\underline{\underline{G}}^k = \underline{\underline{R}}^k \underline{\underline{R}}^*(\theta^k)$ and $\underline{\underline{G}}^\ell = \underline{\underline{R}}^\ell \underline{\underline{R}}^*(\theta^\ell)$, where θ^k and θ^ℓ are the angles defining the planar rotations from $\mathcal{B}_0^k$ to $\mathcal{B}_{0g}^k$ and $\mathcal{B}_0^\ell$ to $\mathcal{B}_{0g}^\ell$, respectively. *(2)* What are the values of angles ϕ^k and ϕ^ℓ in the reference configuration? *(3)* Find the relationship between the rotations of body ℓ, $\underline{\underline{R}}^\ell$, as a function of the rotation of body k, $\underline{\underline{R}}^k$ and of the rotations ϕ^k and ϕ^ℓ of the revolute joints.

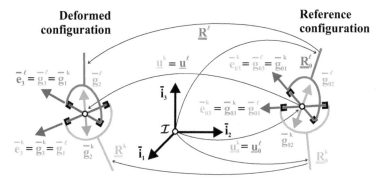

Fig. 10.27. Configuration of the universal joint.

Constrained systems: advanced formulations

Multibody systems are characterized by two distinguishing features: system components undergo finite relative rotations and these components are connected by mechanical joints that impose restrictions on their relative motion. Finite rotations introduce geometric nonlinearities, hence, multibody systems are inherently nonlinear. Mechanical joints, such as the lower pair joints presented in section 10.4, result in algebraic constraints leading to a set of governing equations that combines differential and algebraic equations.

Several textbooks are devoted to the description of the many formulations that have been developed to deal with these complex systems: see, for instance, Roberson and Schwertassek [37], Nikravesh [38], Amirouche [39], Schiehlen [40], García de Jalón and Bayo [41], or Shabana [42]. Computer implementations of a number of the proposed methods and a comparison of their salient features is given by Schiehlen [43]. Bauchau and Laulusa [44, 45] have presented a comprehensive review of the many formulations and numerical techniques that have been used to enforce constraints in multibody systems.

A survey paper by Schiehlen [46] summarizes different approaches to the derivation of the equations of motion for multibody systems. The choice of various frames of reference, system variables and mechanics principles are reviewed. While the dynamic behavior of the system is, of course, independent of the formalism used to describe it, the form of the equations of motion, the effort required to derive them, and the computational burden associated with their numerical solution are all affected by the choice of formalism. The same remarks apply to the methods used to enforce constraints: the effort involved in the derivation of the complete system of governing equations and associated constraints, the computational cost required for their solution, and the resulting accuracy all critically depend on the theoretical formalism and numerical methods used to solve the problem.

Chapter 10 generalizes the basic formulations of dynamics to constrained systems. Lagrange's multiplier technique, the key to this generalization, is shown to be both effective and elegant. Furthermore, because Lagrange's multipliers are closely related to the forces of constraint, these new variables are often physically meaningful. Unfortunately, the use of Lagrange's multipliers changes the mathematical nature of the equations of motion, which now become differential-algebraic equations rather than the ordinary differential equations that characterize unconstrained systems.

The systematic use of Lagrange's multipliers considerably simplifies the development of the equations of motion of complex mechanical systems, but results in large systems of nonlinear differential-algebraic equations. Consequently, appropriate numerical techniques must be developed to deal with this type of problems. In fact, the availability of computationally efficient and accurate numerical tools for the solution of systems of differential-algebraic equations enable the use of Lagrange's multiplier approach. This chapter surveys the numerical tools developed for this task and their theoretical underpinnings.

Typically, the equations of motion of constrained dynamical systems are cast in the form of Lagrange's equation of the first kind presented in section 11.1. Several approaches that eliminate these multipliers are presented in section 11.2: Maggi's, the index-1, the null space, and Udwadia and Kalaba's formulations are summarized in sections 11.2.1 to 11.2.7. A comparison of these various approaches appears in section 11.2.8. The geometric interpretation of the problem presented in section 11.3 presents valuable insight into the behavior of constrained systems. If projection operations are defined in a space endowed with a metric defined by the inverse of the mass matrix, the governing equations of motion of constrained system can be projected in the feasible and infeasible directions. Projections in the feasible direction yield the equations of motion of the system from which Lagrange's multipliers have been eliminated, and projections in the infeasible direction yield an expression for the forces of constraint.

Section 11.4 presents Gauss' principle that has also been used for the solution of constrained dynamical systems. Additional formulations of a more theoretical nature are summarized in section 11.5.

11.1 Lagrange's equations of the first kind

Consider a system represented by n generalized coordinates and subjected to m holonomic or nonholonomic constraints. Lagrange's formulation yields the equations of motion of the system in the form of eqs. (10.12) or (10.19), for holonomic or nonholonomic systems, respectively. These equations, often called *Lagrange's equations of the first kind*, take the following form

$$\underline{\underline{M}}_{(n \times n)} \underline{\ddot{q}}_{(n)} + \underline{\underline{B}}^T_{(n \times m)} \underline{\lambda}_{(m)} = \underline{F}_{(n)}, \tag{11.1}$$

where $\underline{\underline{M}} = \underline{\underline{M}}(\underline{q}, t)$ is the symmetric, positive-definite *mass matrix*, $\underline{F} = \underline{F}(\underline{q}, \underline{\dot{q}}, t)$ are the dynamic and externally applied forces, $\underline{\lambda}$ the array of m Lagrange's multipliers, and the subscripts indicate the sizes of the corresponding arrays. In the literature, Lagrange's equations of the first kind typically appear as in eqs. (11.1) instead (10.12): the constraint force term, $\underline{\underline{B}}^T \underline{\lambda}$, appears on the left- rather than right-hand side. This difference is unimportant because it corresponds to a change of sign of Lagrange's multipliers.

The constraints applied to the system are written as

$$\underline{\mathcal{D}}_{(m)} = \underline{\underline{B}}_{(m \times n)}(\underline{q}, t)\, \underline{\dot{q}}_{(n)} + \underline{b}_{(m)}(\underline{q}, t) = 0, \tag{11.2}$$

where $\underline{\underline{B}}(\underline{q}, t)$ is the constraint matrix. Constraints could be nonholonomic, in which case eq. (11.2) expresses relationships among the generalized velocities; it is assumed that these relationships depend on the generalized velocities in a linear manner. On the other hand, some of the constraints could be holonomic; this means that they can be integrated to the form $\underline{\mathcal{C}}(\underline{q}, t) = 0$, see section 9.2.

If all the constraint are holonomic, the n generalized coordinates are linked by m algebraic constraints. If these latter are independent, it is conceptually possible to partition the generalized coordinate array into *independent*, $\underline{q}^I_{(n-m)}$, and *dependent* coordinates, $\underline{q}^D_{(m)}$, such that $\underline{q}^T = \{\underline{q}^{IT}, \underline{q}^{DT}\}$ and $\underline{q}^D = \underline{f}(\underline{q}^I)$, leading to the elimination of the dependent variables. Unfortunately, this approach is fraught with difficulties: it is not clear which generalized coordinates should be selected to be independent. Furthermore, a poor selection of the independent set of coordinates might render $\underline{f}(\underline{q}^I)$ singular, a suitable set of independent coordinates might become unsuitable for different configurations of the system, and finally, function $\underline{f}(\underline{q}^I)$ might be so complex as to preclude any practical computations.

Lagrange's equations of the first kind form a set of $(n+m)$ Differential-Algebraic Equations (DAEs) for the $(n + m)$ unknowns, $\underline{q}$ and $\underline{\lambda}$; indeed, Lagrange's multipliers are *algebraic variables*, *i.e.*, no time derivatives of these variables appear in the equations, whereas first- and second-order derivatives of the generalized coordinates are present, as implied by Newton's second law. Gear, Petzold and co-workers [47, 48, 49], as well as Brennan [50], have given a formal definition of the *index of a system of DAEs*. The governing equations for mechanical systems with holonomic constraints are index-3 DAEs; typically, higher indices result in more arduous solution processes.

Constraints written in the form of eq. (11.2) are sometimes called *velocity level* constraints. A time derivative of these constraints then yields

$$\underline{\underline{B}}(\underline{q}, t)\, \underline{\ddot{q}} = -\underline{\dot{b}}(\underline{q}, t) - \underline{\dot{\underline{B}}}(\underline{q}, t)\, \underline{\dot{q}} = \underline{c}_{(m)}(\underline{q}, \underline{\dot{q}}, t), \tag{11.3}$$

the *acceleration level* constraints. The linear dependency of eq. (11.2) on the generalized velocities implies the linear dependency of eq. (11.3) on the generalized accelerations.

11.2 Algebraic elimination of Lagrange's multipliers

In this section, algebraic procedures for the solution of Lagrange's equations of the first kind are presented. In all cases, the approach eliminates Lagrange's multipliers to obtain a set of ordinary differential equations (ODEs). Maggi's formulation is presented in section 11.2.1 and introduces the important concepts of null space and orthogonal complements. Next, the index-1 and null space formulations are introduced in sections 11.2.3 and 11.2.5, respectively. Finally, Udwadia and Kalaba's formulation is presented in section 11.2.7. A comparison of the features of these various approaches is the focus of section 11.2.8.

11.2.1 Maggi's formulation

While the original derivation of Maggi's formulation appeared in 1896 [51], then again in 1901 [52], it is only in 1972 that it is presented in the English literature by Neimark and Fufaev [53]. More recently, Kurdila *et al.* [54] and Papastavridis [55] have shown the relevance of this formulation to computational methods for constrained multibody systems.

The formulation begins with the definition of a set of $(n-m)$ *kinematic characteristics*, denoted $\underline{e}$, that satisfy the following relationships

$$\left\{ \begin{matrix} \underline{0}_{(m)} \\ \underline{e}_{(n-m)} \end{matrix} \right\} = \left[\begin{matrix} \underline{\underline{B}}_{(m \times n)}(\underline{q},t) \\ \underline{\underline{\check{B}}}_{((n-m) \times n)}(\underline{q},t) \end{matrix} \right] \underline{\dot{q}} + \left\{ \begin{matrix} \underline{b}_{(m)}(\underline{q},t) \\ \underline{\check{b}}_{(n-m)}(\underline{q},t) \end{matrix} \right\}. \qquad (11.4)$$

The first m equations of this system represent the constraints, eq. (11.2), and the last $(n-m)$ equations define *independent* kinematic characteristics, which are sometimes called *kinematic parameters*, *generalized speeds* or *independent quasi-velocities*. In general, these quantities cannot be integrated, *i.e.*, no $\underline{p}$ exist such that $\underline{\dot{p}} = \underline{e}$.

The choice of the kinematic parameters, a crucial aspect of the procedure, is left to the analyst. The number of degrees of freedom of the system is $n-m$. The matrix formed by $\underline{\underline{B}}$ and $\underline{\underline{\check{B}}}$ defines a linear transformation that is assumed to be invertible; this implies a full rank constraint matrix and a judicious choice of the kinematic parameters.

The following notation is used

$$\underline{\underline{\mathcal{B}}}_{n \times n} = \left[\begin{matrix} \underline{\underline{B}}_{m \times n}(\underline{q},t) \\ \underline{\underline{\check{B}}}_{((n-m) \times n)}(\underline{q},t) \end{matrix} \right]. \qquad (11.5)$$

The inverse of this matrix is denoted

$$\underline{\underline{\mathcal{B}}}_{n \times n}^{-1} = \left[\underline{\underline{\check{\Gamma}}}_{(n \times m)}(\underline{q},t) \ \underline{\underline{\Gamma}}_{(n \times (n-m))}(\underline{q},t) \right]. \qquad (11.6)$$

The generalized velocities are now readily expressed in terms of the kinematic characteristics as

$$\underline{\dot{q}} = \left[\underline{\underline{\check{\Gamma}}}(\underline{q},t) \ \underline{\underline{\Gamma}}(\underline{q},t) \right] \left(\left\{ \begin{matrix} \underline{0} \\ \underline{e} \end{matrix} \right\} - \left\{ \begin{matrix} \underline{b} \\ \underline{\check{b}} \end{matrix} \right\} \right) = \underline{\underline{\Gamma}}\,\underline{e} - (\underline{\underline{\check{\Gamma}}}\,\underline{b} + \underline{\underline{\Gamma}}\,\underline{\check{b}}) = \underline{\underline{\Gamma}}\,\underline{e} - \underline{d}(\underline{q},t). \quad (11.7)$$

Because eqs. (11.4) and (11.7) are inverse of each other, the following relationships must be satisfied

$$\left[\begin{matrix} \underline{\underline{B}} \\ \underline{\underline{\check{B}}} \end{matrix} \right] \left[\underline{\underline{\check{\Gamma}}} \ \underline{\underline{\Gamma}} \right] = \left[\begin{matrix} \underline{\underline{B}}\,\underline{\underline{\check{\Gamma}}} & \underline{\underline{B}}\,\underline{\underline{\Gamma}} \\ \underline{\underline{\check{B}}}\,\underline{\underline{\check{\Gamma}}} & \underline{\underline{\check{B}}}\,\underline{\underline{\Gamma}} \end{matrix} \right] = \left[\begin{matrix} \underline{\underline{I}} & \underline{\underline{0}} \\ \underline{\underline{0}} & \underline{\underline{I}} \end{matrix} \right], \qquad (11.8)$$

and

$$\check{\underline{\Gamma}}\,\underline{B} + \underline{\Gamma}\,\check{\underline{B}} = \underline{I}. \tag{11.9}$$

Matrix $\underline{\Gamma}$ plays a central role in Maggi's formulation. The property $\underline{B}\,\underline{\Gamma} = \underline{0}$ implies that $\underline{\Gamma}$ spans the *null space* of the constraint matrix; $\underline{B}$ and $\underline{\Gamma}$ are *orthogonal complements*. Next, the generalized accelerations are expressed in terms of the kinematic characteristics by taking a time derivative of eq. (11.7), leading to

$$\ddot{\underline{q}} = \underline{\Gamma}\,\dot{\underline{e}} + \dot{\underline{\Gamma}}\,\underline{e} - \ddot{\underline{d}}. \tag{11.10}$$

The governing equations of the system can be expressed in terms of the kinematic characteristics by introducing eq. (11.10) into eq. (11.1), and pre-multiplying by $\underline{\Gamma}^T$ to find

$$\left(\underline{\Gamma}^T\underline{M}\,\underline{\Gamma}\right)\dot{\underline{e}} + \left(\underline{\Gamma}^T\underline{M}\,\dot{\underline{\Gamma}}\right)\underline{e} = \underline{\Gamma}^T\underline{F} + \underline{\Gamma}^T\underline{M}\,\ddot{\underline{d}}. \tag{11.11}$$

Lagrange's multipliers have been eliminated from Maggi's equations: in view of eq. (11.8), the term $\underline{\Gamma}^T\underline{B}^T\underline{\lambda}$ vanishes.

The choice of kinematic characteristics is not unique. Ignoring, for simplicity, arrays $\underline{b}$ and $\check{\underline{b}}$, eq. (11.7) becomes $\dot{\underline{q}} = \underline{\Gamma}\,\underline{e}$: the generalized velocities are uniquely defined, any set of linearly independent vectors can be selected to span the null space, $\underline{\Gamma}$, each leading to a new set of kinematic characteristics. Maggi's formulation consists of eqs. (11.11) and (11.7), which form a system of $(2n - m)$, first-order ODEs for the $(2n - m)$ unknown, $\underline{e}$ and $\underline{q}$.

Maggi's formulation eliminates Lagrange's multipliers from Lagrange's equations of the first kind. It is possible, however, to compute these multipliers once Maggi's equations have been solved. Properties (11.8) imply that $\underline{B}\,\check{\underline{\Gamma}} = \underline{I}$, hence, multiplying eq. (11.1) by $\check{\underline{\Gamma}}^T$ yields $\underline{\lambda} = -\check{\underline{\Gamma}}^T(\underline{M}\,\ddot{\underline{q}} - \underline{F})$. Introducing eq. (11.10) then leads to

$$\underline{\lambda} = -\check{\underline{\Gamma}}^T\left[\underline{M}\,\underline{\Gamma}\,\dot{\underline{e}} + \underline{M}\,\dot{\underline{\Gamma}}\,\underline{e} - \underline{M}\,\ddot{\underline{d}} - \underline{F}\right]. \tag{11.12}$$

Example 11.1. The simple pendulum

Figure 10.5 depicts a point of mass m and coordinates x and y, constrained to remain at a distance ℓ from an inertial point $\mathbf{O}$, and discussed in section 10.4. The generalized coordinates of the pendulum are $\underline{q}^T = \{x, y\}$. The constraint condition is $\mathcal{C} = (\underline{q}^T\underline{q} - \ell^2)/2\ell$ and the constraint matrix is now

$$\underline{B} = [x/\ell,\ y/\ell]. \tag{11.13}$$

A first form of the equations of motion for this constrained problem is Lagrange's equations of the first kind, eq. (11.1), or $\underline{M}\,\ddot{\underline{q}} - \underline{B}^T\underline{\lambda} = m\underline{g}$, where $\underline{g}^T = g\{1, 0\}$, g the acceleration of gravity and $\underline{M} = \mathrm{diag}(m, m)$ the mass matrix of the system.

To derive Maggi's equations for the pendulum, the kinematics of the problem are recast in the form of eq. (11.4), as

$$\begin{Bmatrix} 0 \\ e \end{Bmatrix} = \frac{1}{\ell}\begin{bmatrix} x & y \\ -y & x \end{bmatrix}\dot{\underline{q}},$$

where the second row of the matrix on the right-hand side defined matrix $\underline{\underline{\check{B}}}$.

The single kinematic characteristic of the system was defined as $e = (-y\dot{x} + x\dot{y})/\ell$. To interpret this quantity, the coordinates of the point mass are expressed in terms of angle θ, such that $x = \ell\cos\theta$ and $y = \ell\sin\theta$. It then follows that $e = \ell\dot{\theta}$, i.e., the kinematic characteristic is related to the angular velocity of the pendulum.

The relationship between the generalized velocities and the kinematic characteristics give rise to the matrices $\underline{\underline{\check{\varGamma}}}$ and $\underline{\underline{\varGamma}}$, defined in eq. (11.7),

$$\underline{\underline{\check{\varGamma}}} = \frac{1}{\ell}\begin{bmatrix} x \\ y \end{bmatrix}, \quad \underline{\underline{\varGamma}} = \frac{1}{\ell}\begin{bmatrix} -y \\ x \end{bmatrix}.$$

It is easily verified that the following relationship holds, as implied by eq. (11.8).

$$\begin{bmatrix} \underline{\underline{B}} \\ \underline{\underline{B}} \end{bmatrix}[\underline{\underline{\check{\varGamma}}}\ \underline{\underline{\varGamma}}] = \frac{1}{\ell}\begin{bmatrix} x & y \\ -y & x \end{bmatrix}\frac{1}{\ell}\begin{bmatrix} x & -y \\ y & x \end{bmatrix} = \begin{bmatrix} 1 & 0 \\ 0 & 1 \end{bmatrix}.$$

Simple algebraic manipulations show that $\underline{\underline{\varGamma}}^T\underline{\underline{M}}\,\underline{\underline{\varGamma}} = m$ and $\underline{\underline{\varGamma}}^T\underline{\underline{M}}\,\underline{\underline{\dot{\varGamma}}} = \underline{0}$. The equations of motion of the system then follow from eq. (11.11) as

$$m\dot{e} = \underline{\underline{\varGamma}}^T m\underline{g}. \tag{11.14}$$

It is easily verified that this equation is indeed the equation of motion of the pendulum: $\ddot{\theta} + g/\ell\,\sin\theta = 0$. Note that the right-hand side of Maggi's equation involves $\underline{\underline{\varGamma}}(\underline{q})$, a function of the generalized coordinates $\underline{q}$.

Example 11.2. Quick return mechanism

The quick return mechanism shown in fig. 10.6 consists of a uniform crank of length L_c and mass m_c, and of a uniform arm of length L_a and mass m_a. The crank is pinned at point **R** and the arm at point **O**; the distance between these two points is denoted d. At point **S**, a slider allows the tip of the crank to slide along the arm. Mass M is attached at point **T**, the tip of the arm. A spring of stiffness constant k connects the tip of the arm, point **T**, to fixed point **A**; the spring is un-stretched when the arm is in the vertical position.

This problem was treated in examples 8.15 and 10.5 on pages 329 and 396, respectively, using Lagrange's formulation with one and two generalized coordinates, respectively. Use Maggi's formulation to derive the equations of motion of the system with two generalized coordinates, angles θ and ϕ.

Considering triangle **ORS**, it is clear that $\beta = \phi - \theta$, and the law of sines then yields $L_c\sin(\phi - \theta) = d\sin\phi$. This equation expresses the kinematic constraint between the two generalized coordinates of this single degree of freedom problem, $\mathcal{C} = d\sin\phi - L_c\sin(\phi - \theta) = 0$, and the constraint matrix is given by eq. (10.14).

To start Maggi's formulation, the following linear transformation is constructed

$$\begin{Bmatrix} 0 \\ e \end{Bmatrix} = \underline{\underline{B}}\begin{Bmatrix} \dot{\theta} \\ \dot{\phi} \end{Bmatrix} = \begin{bmatrix} L_cC_{\phi-\theta}\ dC_\phi - L_cC_{\phi-\theta} \\ 1 & 0 \end{bmatrix}\begin{Bmatrix} \dot{\theta} \\ \dot{\phi} \end{Bmatrix}. \tag{11.15}$$

The first equation represents the velocity level constraint, $\dot{\mathcal{C}} = 0$, and the first line of matrix $\underline{\underline{B}}$ is the constraint matrix defined by eq. (10.14). The second line of matrix

$\underline{\underline{B}}$ defines the single kinematic characteristic of the problem, $e = \dot{\theta}$, chosen to be the angular velocity of the crank, $\dot{\theta}$. Of course, the choice of the kinematic characteristic is not unique. For instance, the kinematic characteristic could be chosen to be the angular velocity of the arm, $e = \dot{\phi}$.

Inverting eq. (11.15) yields

$$\begin{Bmatrix} \dot{\theta} \\ \dot{\phi} \end{Bmatrix} = \underline{\underline{B}}^{-1} \begin{Bmatrix} 0 \\ e \end{Bmatrix} = \frac{1}{L_c C_{\phi-\theta} - dC_\phi} \begin{bmatrix} 0 & L_c C_{\phi-\theta} - dC_\phi \\ -1 & L_c C_{\phi-\theta} \end{bmatrix} \begin{Bmatrix} 0 \\ e \end{Bmatrix}.$$

The first column of matrix $\underline{\underline{B}}^{-1}$ defines matrix $\underline{\underline{\check{\Gamma}}}$, and the second column defines the null space of the constraint matrix, $\underline{\underline{\Gamma}}$.

Maggi's formulation calls for the evaluation of the following terms

$$\underline{\Gamma}^T \underline{\underline{M}} \, \underline{\Gamma} = \frac{m_c L_c^2}{3} + (M + \frac{m_a}{3}) L_a^2 \bar{h}^2, \quad \underline{\Gamma}^T \underline{\underline{M}} \, \underline{\dot{\Gamma}} = -(M + \frac{m_a}{3}) L_a^2 \bar{h} \bar{g} e,$$

where $\bar{L}_c = L_c/d$, $\bar{h} = (\bar{L}_c C_{\phi-\theta})/(\bar{L}_c C_{\phi-\theta} - C_\phi)$, and $\bar{g} = \bar{L}_c (C_\phi S_{\phi-\theta} + S_\theta \bar{h})/(\bar{L}_c C_{\phi-\theta} - C_\phi)^2$.

In non-dimensional form, Maggi's equations become

$$\begin{Bmatrix} \bar{e} \\ \theta \\ \phi \end{Bmatrix}' = \begin{Bmatrix} (\alpha + \beta \bar{e}^2)/\gamma \\ \bar{e} \\ \bar{h} \bar{e} \end{Bmatrix}. \tag{11.16}$$

The following notations were defined: $\bar{L}_a = L_a/d$, $\alpha = \bar{h} S_\phi$, $\mu_a = m_a/M$, $\mu_c = m_c/M$, and

$$\beta = (1 + \frac{\mu_a}{3}) \bar{h} \bar{g}, \quad \gamma = \frac{\mu_c \bar{L}_c^2}{3\bar{L}_a^2} + (1 + \frac{\mu_a}{3}) \bar{h}^2, \quad \bar{e} = e/\omega.$$

The non-dimensional time $\tau = \omega t$, where $\omega^2 = k/M$, was introduced and notation $(\cdot)'$ indicates a derivative with respect to τ.

Maggi's equations, eqs. (11.16), take the form of three coupled differential equations for the kinematic characteristic and the two generalized coordinates of the problem. Because Lagrange's multiplier was eliminated, the equations are ordinary differential equations instead of the algebraic-differential equations that characterize Lagrange's equations of the first kind, eqs. (10.15).

Maggi's equations can be integrated using classical numerical tools developed for the solution of ordinary differential equations. It is left to the reader to verify that the dynamic response predicted by Maggi's formulation is identical to that obtained using Lagrange's formulation, see figs 8.10 to 8.12. Of course, slight discrepancies between the two predictions are expected due to the approximate nature of numerical solution techniques.

Maggi's formulation enforces velocity level constraints, $\dot{\mathcal{C}} = 0$, at each instant in time. Indeed, the velocity level constraint is the third equation of system (11.16). An exact solution of these governing equations would then imply $\mathcal{C} = 0$, provided

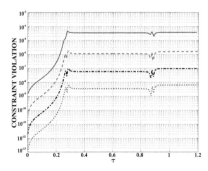

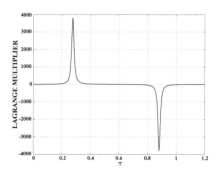

Fig. 11.1. Constraint violation versus non-dimensional time. Solid line, $N_s = 100$; dashed line, $N_s = 200$; dashed-dotted line, $N_s = 400$; dotted line, $N_s = 800$.

Fig. 11.2. Time history of Lagrange's multiplier.

that the constraint is satisfied at time $\tau = 0$. An approximate solution of the same equations results in an approximate satisfaction of the holonomic constraint equation, $\mathcal{C} \approx 0$.

Figure 11.1 shows the time history of the *constraint violation, i.e.*, $\mathcal{C}$ as a function of non-dimensional time. Maggi's equations were integrated for $\tau \in [0, 1.2]$ using a fourth-order Runge-Kutta algorithm [5] with $N_s = 100, 200, 400$, and 800 time steps of equal size. As the time step size is reduced, the constraint violation decreases; note the logarithmic scale along the vertical axis. The same data is also presented in table 11.1 on page 441: the constraint violations at time $\tau = 1$ are listed in the second column of the table for the four time step sizes.

For small time step sizes, accurate solutions are obtained that closely satisfy the constraint conditions. For long term simulations, the constraint violation will keep increasing as the simulation proceeds: this is known as the *drift phenomenon*. For short time simulations, the drift phenomenon can be overcome by using small time step sizes. For longer time simulations, the use of small time step might become unpractical, and constraint violation stabilization techniques should be used, see section 12.3.

Although Lagrange's multipliers have been eliminated from Maggi's formulation, they can be evaluated with the help of eq. (11.12). The following quantities are computed first,

$$\check{\underline{L}}^T \underline{M}\, \underline{\Gamma} = -(M + \frac{m_a}{3})\frac{L_a^2 \bar{h}^2}{L_c C_{\phi-\theta}}, \quad \check{\underline{L}}^T \underline{M}\, \dot{\underline{\Gamma}} = (M + \frac{m_a}{3})\frac{L_a^2 \bar{h}}{L_c C_{\phi-\theta}}\bar{g}e.$$

Equation (11.12) then yields the desired multipliers, expressed here in non-dimensional form as

$$\bar{\lambda} = \frac{\lambda}{kL_c} = \frac{\bar{L}_a^2}{\bar{L}_c^2 C_{\phi-\theta}}\left[-\alpha + (1 + \frac{\mu_a}{3})\bar{h}^2\bar{e}' - \beta\bar{e}^2\right].$$

Figure 11.2 shows the time history of the non-dimensional multiplier.

Example 11.3. The 9 degree of freedom rigid body

Consider a rigid body moving in space while one of its points, denoted **O**, remains fixed, as depicted in fig. 11.3. The orientation of the body is determined by an orthonormal basis $\mathcal{E} = (\bar{e}_1, \bar{e}_2, \bar{e}_3)$; nine generalized coordinates will be used to represent the configuration of the body, the nine components of unit vectors $\bar{e}_1$, $\bar{e}_2$, and $\bar{e}_3$. The inertial position vector of an arbitrary point **P** of the body is $\underline{r}_P = s_1^* \bar{e}_1 + s_2^* \bar{e}_2 + s_3^* \bar{e}_3$, where $\underline{s}^*$ are the components of the position vector of point **P** with respect to point **O**, resolved in basis $\mathcal{E}$.

The kinetic energy of the rigid body is readily found as $K = 1/2\, \dot{q}^T \underline{\underline{M}}^* \dot{q}$, where array q stores the generalized coordinates of the system $q^T = \left[\bar{e}_1^T, \, \bar{e}_2^T, \, \bar{e}_3^T \right]$, and the mass matrix is

$$\underline{\underline{M}}^* = \begin{bmatrix} M_{11}\underline{\underline{I}} & M_{12}\underline{\underline{I}} & M_{13}\underline{\underline{I}} \\ M_{12}\underline{\underline{I}} & M_{22}\underline{\underline{I}} & M_{23}\underline{\underline{I}} \\ M_{13}\underline{\underline{I}} & M_{23}\underline{\underline{I}} & M_{33}\underline{\underline{I}} \end{bmatrix}.$$

where the quantities $M_{ij} = \int_\mathcal{V} \rho s_i^* s_j^* \, \mathrm{d}\mathcal{V}$ are closely related to the components of the tensor of moments of inertia, ρ is the material density, and $\mathcal{V}$ the volume of the body.

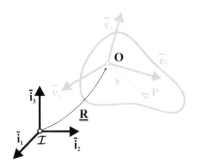

Fig. 11.3. Configuration of the rigid body.

The kinematics of the rigid body are defined by the nine generalized coordinates stored in array q. Clearly, six constraints must be imposed: three conditions on the normality of vectors $\bar{e}_1$, $\bar{e}_2$, and $\bar{e}_3$, and three additional constraints enforcing their orthogonality. The array of holonomic constraints and the constraint matrix then become

$$\underline{\mathcal{C}} = \begin{Bmatrix} (\bar{e}_1^T \bar{e}_1 - 1)/2 \\ (\bar{e}_2^T \bar{e}_2 - 1)/2 \\ (\bar{e}_3^T \bar{e}_3 - 1)/2 \\ \bar{e}_2^T \bar{e}_3 \\ \bar{e}_1^T \bar{e}_3 \\ \bar{e}_1^T \bar{e}_2 \end{Bmatrix}, \quad \text{and} \quad \underline{\underline{B}}(q) = \begin{bmatrix} \bar{e}_1^T & \underline{0}^T & \underline{0}^T \\ \underline{0}^T & \bar{e}_2^T & \underline{0}^T \\ \underline{0}^T & \underline{0}^T & \bar{e}_3^T \\ \underline{0}^T & \bar{e}_3^T & \bar{e}_2^T \\ \bar{e}_3^T & \underline{0}^T & \bar{e}_1^T \\ \bar{e}_2^T & \bar{e}_1^T & \underline{0}^T \end{bmatrix}, \qquad (11.17)$$

respectively.

The first form of the equations of motion for this constrained problem are Lagrange's equations of the first kind, eq. (11.1), $\underline{M}^*\underline{\ddot{q}} - \underline{B}^T\underline{\lambda} = \underline{F}^a$, where $\underline{F}^a$ are the externally applied forces.

To apply Maggi's formulation, the kinematic characteristics of the problem are selected as the components of the angular velocity resolved in the material frame. This means that $\underline{e} = \underline{\omega}^* = 1/2 \{\bar{e}_3^T\dot{\bar{e}}_2 - \bar{e}_2^T\dot{\bar{e}}_3, \ \bar{e}_1^T\dot{\bar{e}}_3 - \bar{e}_3^T\dot{\bar{e}}_1, \ \bar{e}_2^T\dot{\bar{e}}_1 - \bar{e}_1^T\dot{\bar{e}}_2\}$, see eq. (4.53), leading to the following expression for matrix $\underline{\check{B}}$,

$$\underline{\check{B}}(\underline{q}, t) = \frac{1}{2} \begin{bmatrix} 0 & \bar{e}_3^T & -\bar{e}_2^T \\ -\bar{e}_3^T & 0 & \bar{e}_1^T \\ \bar{e}_2^T & -\bar{e}_1^T & 0 \end{bmatrix},$$

and $\underline{b} = 0, \ \underline{\dot{b}} = 0$.

The relationship between the generalized velocities and the kinematic characteristics give rise to matrices $\underline{\check{\Gamma}}$ and $\underline{\Gamma}$, defined in eq. (11.7),

$$\underline{\check{\Gamma}} = \begin{bmatrix} \bar{e}_1 & \underline{0} & \underline{0} & \underline{0} & \bar{e}_3/2 & \bar{e}_2/2 \\ \underline{0} & \bar{e}_2 & \underline{0} & \bar{e}_3/2 & \underline{0} & \bar{e}_1/2 \\ \underline{0} & \underline{0} & \bar{e}_3 & \bar{e}_2/2 & \bar{e}_1/2 & \underline{0} \end{bmatrix}, \quad \underline{\Gamma} = \begin{bmatrix} 0 & -\bar{e}_3 & \bar{e}_2 \\ \bar{e}_3 & 0 & -\bar{e}_1 \\ -\bar{e}_2 & \bar{e}_1 & 0 \end{bmatrix}. \tag{11.18}$$

It is readily verified that the properties defined by eqs. (11.8) and (11.9) are satisfied; in particular, $\underline{B}\,\underline{\Gamma} = \underline{0}$: $\underline{\Gamma}$ is the null space of the constraint matrix.

The equations of motion of the system now follow from eq. (11.11); simple algebraic manipulations show that $\underline{\Gamma}^T\underline{M}^*\underline{\Gamma} = \underline{I}^*$, where $\underline{I}^*$ are the components of the mass moment of inertia tensor of the rigid body resolved in the material frame. Furthermore, $\underline{\Gamma}^T\underline{M}^*\underline{\dot{\Gamma}} = \tilde{d}$, where $\underline{d} = \left[1/2 \, \text{tr}(\underline{I}^*)\underline{I} - \underline{I}^*\right]\underline{e}$, leading to

$$\underline{I}^*\underline{\dot{e}} + \tilde{e}\underline{I}^*\underline{e} = \underline{M}_O^*,$$

where $\underline{M}_O^*$ are the components of the externally applied moment computed with respect to point **O**, expressed in the material frame. As expected, these equations are Euler's equations for the rigid body.

The choice of the kinematic characteristics is not unique. Indeed, instead of the components of the angular velocity vector resolved in the material frame, the components of the same vector in the inertial frame could have been selected as kinematic characteristics, $\underline{e} = \underline{\omega} = (\tilde{e}_1\dot{\tilde{e}}_1 + \tilde{e}_2\dot{\tilde{e}}_2 + \tilde{e}_3\dot{\tilde{e}}_3)/2$. This choice is associated with the following matrices

$$\underline{\check{B}}(\underline{q}, t) = \frac{1}{2} \begin{bmatrix} \tilde{e}_1 & \tilde{e}_2 & \tilde{e}_3 \end{bmatrix}, \quad \underline{\Gamma} = \begin{bmatrix} \tilde{e}_1^T \\ \tilde{e}_2^T \\ \tilde{e}_3^T \end{bmatrix}. \tag{11.19}$$

The derivation of Maggi's equations based on this set of kinematic characteristics is left to the reader as an exercise.

Clearly, the null space of the constraint matrix is unique; the choice of a set of linearly independent vectors spanning this null space, however, is not unique. Two specific choices were pointed out here: they correspond to the columns of matrix $\underline{\Gamma}$ as defined in eqs. (11.18) and (11.19), respectively.

Example 11.4. The skateboard

Figure 10.11 depicts the simplified configuration of a skateboard of mass m and moment of inertia I about its center of mass **G**, as discussed in examples 9.6 and 10.6. The skateboard rolls without sliding on the horizontal plane by means of a wheel aligned with axis $\bar{e}_1$ of the skateboard and located at point **C**, a distance ℓ from the center of mass. The position vector of the center of mass is written as $\underline{r}_G = x\,\bar{\imath}_1 + y\,\bar{\imath}_2$, and the axis of the skateboard makes an angle θ with the horizontal.

Let the generalized coordinates of the problem be $\underline{q}^T = \{x, y, \theta\}$. The system is subjected to a constraint: because the wheel does not slip, the velocity vector of the contact point must be along axis $\bar{e}_1$. The velocity of point **C** is $\underline{v}_C = \dot{x}\,\bar{\imath}_1 + \dot{y}\,\bar{\imath}_2 + \ell\dot{\theta}\,\bar{e}_2$, and hence, the constraint is $\bar{e}_2^T \underline{v}_C = 0$, leading to the constraint matrix

$$\underline{B} = [-\sin\theta, \cos\theta, \ell]. \tag{11.20}$$

This constraint is nonholonomic.

The first form of the equations of motion for this constrained problem are Lagrange's equations of the first kind, eq. (11.1), or $\underline{M}\,\ddot{\underline{q}} - \underline{B}^T\lambda = 0$, where the mass matrix is a diagonal matrix: $\underline{M} = \text{diag}(m, m, I)$.

To derive Maggi's equations for the skateboard, the kinematics of the problem are recast in the form of eq. (11.4), as

$$\begin{bmatrix} 0 \\ e_1 \\ e_2 \end{bmatrix} = \begin{bmatrix} -\sin\theta & \cos\theta & \ell \\ \cos\theta & \sin\theta & 0 \\ 0 & 0 & 1 \end{bmatrix} \dot{\underline{q}}, \tag{11.21}$$

where matrix $\check{\underline{B}}(q)$ is defined by the last two line of the matrix appearing on the right-hand side of this equation. The kinematic characteristics of the problem were selected as $e_1 = \dot{x}\cos\theta + \dot{y}\sin\theta$, the velocity of the wheel in the driving direction, and $e_2 = \dot{\theta}$, the angular velocity of the skateboard.

The relationship between the generalized velocities and the kinematic characteristics give rise to the matrices $\check{\underline{\Gamma}}$ and $\underline{\Gamma}$, defined by eq. (11.7),

$$\check{\underline{\Gamma}} = \begin{bmatrix} -\sin\theta \\ \cos\theta \\ 0 \end{bmatrix}, \quad \underline{\Gamma} = \begin{bmatrix} \cos\theta & \ell\sin\theta \\ \sin\theta & -\ell\cos\theta \\ 0 & 1 \end{bmatrix}. \tag{11.22}$$

Here again, matrix $\underline{\Gamma}$ defines the null space of the constraint matrix $\underline{B}$ since $\underline{B}\,\underline{\Gamma} = \underline{0}$. The equations of motion of the system now follow from eq. (11.11); simple algebraic manipulations yield Maggi's equations as

$$\begin{bmatrix} m & 0 \\ 0 & I + m\ell^2 \end{bmatrix} \dot{\underline{e}} + \begin{bmatrix} 0 & m\ell e_2 \\ -m\ell e_2 & 0 \end{bmatrix} \underline{e} = 0. \tag{11.23}$$

11.2.2 Problems

Problem 11.1. Bar sliding on guides

A homogeneous bar of length L and mass m slides on two guides at its end points, as shown in fig. 10.10. At the left end, the bar is connected to a spring of stiffness constant k that is unstretched when the bar is horizontal. At the right end, the bar is connected to a point mass M.

Gravity acts along axis $\bar{\imath}_2$. This single degree of freedom problem will be represented using three generalized coordinates: x, y, and θ. *(1)* Derive the constraints among the three generalized coordinates and the constraint matrix. *(2)* Use Lagrange's formulation for constrained systems to derive the equations of motion of the system. Interpret the physical meaning of the multipliers. *(3)* Derive and solve Maggi's equations. For this problem, the angular velocity of the bar is a good choice for the kinematic characteristic. *(4)* On one graph, plot the time history of x and y. *(5)* Plot the time history of angle θ. *(6)* On one graph, plot the time histories of $\dot{x}$ and $\dot{y}$. *(7)* Plot the time history of angular velocity of the bar. *(8)* On one graph, plot the time history of the constraint violations. Comment on your results. *(10)* Plot the time history of the total mechanical energy. *(11)* On one graph, plot the time history of Lagrange's multipliers. Use the following data: $L = 0.45$ m; $m = 5$ and $M = 5$ kg; $k = 150$ N/m; acceleration of gravity $g = 9.81$ m/s^2. At the initial time, $x = 0$, $\theta = 0$ and the system is at rest. Plot the results for $t \in [0, 2]$ s.

Problem 11.2. Bar sliding on guides
Repeat the previous problem using two generalized coordinates only, x and y.

Problem 11.3. Crank-slider mechanism
The crank-slider mechanism depicted in fig. 9.8 consists of a uniform crank of length L_1 and mass m_1 connected to the ground at point **O**; let θ be the angle from the horizontal to the crank. At point **B**, the crank connects to a uniform linkage of length L_2 and mass m_2 that slides along point **P**, a fixed point in space, located at a distance d from point **O**. Let w denote the distance from point **B** to point **P** and ϕ the angle from the horizontal to link **BP**. A linear spring, not shown on the figure, connects point **B** to the support at point **P**; the strain energy of this spring is $V = 1/2 \, kw^2$. The system is represented by three generalized coordinates: θ, ϕ, and w. *(1)* Derive the constraints among the three generalized coordinates and the constraint matrix. *(2)* Use Lagrange's formulation to derive the equations of motion of the constrained system. Interpret the physical meaning of the multipliers. *(3)* Derive and solve Maggi's equations. For this problem, the angular velocity of the crank is a good choice for the kinematic characteristic. *(4)* Plot the time history of the kinematic characteristic. *(5)* On one graph, plot the time histories of angles θ and ϕ. *(6)* On one graph, plot the time histories of angular velocities of the two bars. *(7)* Plot the time history of w. *(8)* Plot the time histories of $\dot{w}$. *(9)* On one graph, plot the time history of the constraint violations. Comment on your results. *(10)* Plot the time history of the total mechanical energy. *(11)* On one graph, plot the time history of Lagrange's multipliers. Use the following data: $L_1 = 0.25$, $L_2 = 0.75$ and $d = 0.35$ m; $m_1 = 1.5$ and $m_2 = 4$ kg; $k = 10$ kN/m. At the initial time, $x = 0$, $\theta = 0$ and the angular velocity of the first bar is $\omega_1 = 100$ rad/s. Plot the results for $t \in [0, 0.10]$ s.

Problem 11.4. The crank piston mechanism
The crank slider mechanism depicted in fig. 10.4 comprises a bar of length L_1 and mass m_1 connected to the ground at point **O** by means of a hinge. At point **A**, a hinge connects the first bar to a second bar of length L_2 and mass m_2. A slider of mass M, that is constrained to move in the horizontal direction, is connected to this second bar. A spring of stiffness constant k connects the slider to the ground and is un-stretched when the two bars are aligned. This system will be represented with three generalized coordinates: x and y, the coordinates of point **A** and z, the horizontal position of point **B**. *(1)* Write the constraint equations for this problem. *(2)* Use Lagrange's formulation to derive the equation of motion of the system. Interpret the physical meaning of the multipliers. *(3)* Derive and solve Maggi's equations for this problem. It will be convenient to use the angular velocity of the first bar as the kinematic characteristic. *(4)* Plot the kinematic characteristic as a function of time. *(5)* On one graph,

plot the three generalized coordinates as a function of time. *(6)* On one graph, plot the time history of the constraint violations. Comment on your results. *(7)* Plot the time history of the total mechanical energy. *(8)* On one graph, plot the time history of Lagrange's multipliers. Use the following data: $L_1 = 0.25$ and $L_2 = 0.45$ m; $m_1 = 1.5$, $m_2 = 2.5$, $M = 10$ kg; $k = 100$ N/m. At the initial time, $x = 0$, $y = L_1$ and the angular velocity of the first bar is $\omega_1 = 100$ rad/s. Plot the results for $t \in [0, 0.25]$ s.

Problem 11.5. Slider-arm mechanism

Consider the mechanism depicted in fig. 11.4 that consists of a slider of mass M free to move along axis $\bar{\imath}_1$ and connected to arm **AP** of length L_a and mass m_a. The arm is free to rotate in the plane normal to $\bar{\imath}_1$. A spring of stiffness constant k is attached to the slider and gravity acts in the direction indicated on the figure. This system will be represented with three generalized coordinates, the Cartesian coordinates of point **P**, x_1, x_2, and x_3. *(1)* Derive the constraints among the three generalized coordinates and the constraint matrix. *(2)* Use Lagrange's formulation to derive the equation of

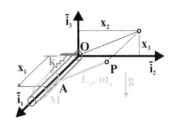

Fig. 11.4. Slider with arm mechanism.

motion of the system. Interpret the physical meaning of the multipliers. *(3)* Derive Maggi's equations for this problem.

Problem 11.6. The spatial mechanism

The spatial mechanism depicted in fig. 10.9 consists of a crank of length L_c and mass m_c attached to the ground at point **A** and rotating about axis $\bar{\imath}_1$. The crank moves in plane $(\bar{\imath}_2, \bar{\imath}_3)$. A rigid arm of length L_a and mass m_a connects point **P**, at the tip of the crank, to point **Q** that is free to slide along axis $\bar{\imath}_1$. The slider at point **Q** has a mass M. The generalized coordinates of the problem are y and z, defining the position of point **P** and x, defining the position of point **Q**. *(1)* Write the constraint equations for this problem. *(2)* Use Lagrange's formulation to derive the equation of motion of the system. Interpret the physical meaning of the multipliers. *(3)* Derive Maggi's equations for this problem. It will be convenient to use the angular velocity of the first bar as the kinematic characteristic. *(4)* Plot the kinematic characteristic as a function of time. *(5)* On one graph, plot the three generalized coordinates as a function of time. *(6)* On one graph, plot the three generalized velocities as a function of time. *(7)* On one graph, plot the time history of the constraint violations. Comment on your results. *(8)* Plot the time history of the total mechanical energy. Use the following data: $h = 0.15$, $L_c = 0.50$ and $L_a = 0.75$ m; $m_c = 1.4$, $m_a = 5$, $M = 125$ kg; $k = 500$ N/m. At the initial time, $y = 0$, $y = L_c + h$ and the angular velocity of the crank is $\omega_1 = 100$ rad/s. Plot the results for $t \in [0, 0.20]$ s.

Problem 11.7. The quick return mechanism

The quick return mechanism shown in fig. 10.6 consists of a uniform crank of length L_c and mass m_c, and of a uniform arm of length L_a and mass m_a. The crank is pinned at point **R** and the arm at point **O**; the distance between these two points is denoted d. At point **S**, a slider allows the tip of the crank to slide along the arm. A mass M is attached at point **T**, the tip of the arm. A spring of stiffness constant k connects the tip of the arm, point **T**, to fixed point **A**; the spring is un-stretched when the arm is in the vertical position. The generalized coordinates of the problem are the angles θ and ϕ as defined on the figure. *(1)* Derive the constraints among the two generalized coordinates and the constraint matrix. *(2)* Use Lagrange's formulation to derive the equation of motion of the system. Interpret the physical meaning of the multiplier.

(3) Derive Maggi's equations for this problem. Use the angular velocity of the arm as the kinematic characteristic. *(4)* Plot the kinematic characteristic as a function of time. *(5)* On one graph, plot the angles θ and ϕ as a function of time. *(6)* On one graph, plot the angular velocities of the two bars as a function of time. *(7)* On one graph, plot the time history of the constraint violations. Comment on your results. *(8)* Plot the time history of the normal force at point **S**. *(9)* Plot the time history of the total mechanical energy. Use the following data: $\bar{L}_c = L_c/d = 0.50$ and $\bar{L}_a = L_a/d = 3$ m; $\mu_c = m_c/M = 2.4$, $\mu_a = m_a/M = 1$. At the initial time, $\theta = 0$ and $\theta' = 2$, where $(\cdot)'$ indicates a derivative with respect to the non-dimensional time $\tau = \omega t$, $\omega^2 = k/M$. Plot the results for $\tau \in [0, 1.2]$.

Problem 11.8. Two bar linkage tracking a curve

Figure 10.8 shows a planar two bar linkage tracking curve $\mathbb{C}$. The first bar, of length L_1 and mass m_1, is connected to the ground at point **O**. The second bar, of length L_2 and mass m_2, connects to the first bar at point **A** and tracks curve $\mathbb{C}$ at point **P**. A concentrated mass, M, is located at point **P** and an elastic spring of stiffness constant constant k connects this mass to point **R**. Curve $\mathbb{C}$ is described by its coordinates $x(\eta)$ and $y(\eta)$, where η defines an arbitrary parametrization of the curve. This system will be represented by three generalized coordinates: angles θ and ϕ, as defined on the figure, and η, the parameter along curve $\mathbb{C}$. *(1)* Write the constraint equations for this problem. *(2)* Use Lagrange's formulation to derive the equation of motion of the system. Interpret the physical meaning of these multipliers. *(3)* Derive and solve Maggi's equations for this problem. It will be convenient to use the angular velocity of the first bar as the kinematic characteristic. *(4)* Plot the kinematic characteristic as a function of time. *(5)* On one graph, plot angles θ and ϕ as a function of time. *(6)* Plot η as a function of time. *(7)* On one graph, plot the angular velocities of the two bars. *(8)* Plot $\dot{\eta}$ as a function of time. *(9)* On one graph, plot the time history of the constraint violations. Comment on your results. *(10)* Plot the time history of the total mechanical energy. *(11)* On one graph, plot the time history of Lagrange's multipliers. Use the following data: $d = 1$, $L_1 = 0.20$ and $L_2 = 1.50$ m; $m_1 = 1.2$, $m_2 = 1.5$, $M = 25$ kg; $k = 500$ N/m. At the initial time, $\theta = 0$, $\eta > 0$ and the angular velocity of the first bar is $\omega_1 = 25$ rad/s. The curve is defined as $x(\eta) = d + a\eta$, $y(\eta) = b\eta^2$, where $a = 1$ and $b = 2$; note that η is not the intrinsic parametrization of the curve. Plot the results for $t \in [0, 1]$ s.

11.2.3 Index-1 formulation

Consider the system of equations consisting of the Lagrange's equations of the first kind, eqs. (11.1), and the acceleration level constraints, eq. (11.3), written in a matrix form as

$$\begin{bmatrix} \underline{\underline{M}} & \underline{\underline{B}}^T \\ \underline{\underline{B}} & \underline{\underline{0}} \end{bmatrix} \begin{bmatrix} \underline{\ddot{q}} \\ \underline{\lambda} \end{bmatrix} = \begin{bmatrix} \underline{F} \\ \underline{C} \end{bmatrix}. \tag{11.24}$$

This system is now an index-1 set of DAEs, which can be formally solved for the accelerations and Lagrange's multipliers. This system is equivalent to Lagrange's equations of the first kind, eq. (11.1), if and only if the initial conditions of the problem satisfy the constraint conditions, *i.e.*, $\underline{\mathcal{C}}(\underline{q}_0, t_0) = 0$ and $\underline{\mathcal{D}}(\underline{q}_0, \underline{\dot{q}}_0, t_0) = 0$ for holonomic and nonholonomic constraints, respectively, where $\underline{q}_0 = \underline{q}(t_0)$ and $\underline{\dot{q}}_0 = \underline{\dot{q}}(t_0)$ are the initial conditions of the problem.

It is easily shown that system (11.24) has a unique solution if and only if matrix $\underline{\underline{B}}$ is of full rank and the mass matrix is invertible, see *e.g.* Nikravesh [56]. To start, the first equations are pre-multiplied by $\underline{\underline{B}}\,\underline{\underline{M}}^{-1}$ and accelerations are eliminated with the help of the second equations, to find Lagrange's multipliers as

$$\underline{\lambda} = -(\underline{\underline{B}}\,\underline{\underline{M}}^{-1}\underline{\underline{B}}^T)^{-1}(\underline{c} - \underline{\underline{B}}\,\underline{\underline{M}}^{-1}\underline{F}). \tag{11.25}$$

Next, the multipliers are introduced in the first equations to yield the accelerations of the system as

$$\underline{\underline{M}}\,\underline{\ddot{q}} = \underline{F} + \underline{\underline{B}}^T\left(\underline{\underline{B}}\,\underline{\underline{M}}^{-1}\underline{\underline{B}}^T\right)^{-1}\left(\underline{c} - \underline{\underline{B}}\,\underline{\underline{M}}^{-1}\underline{F}\right), \tag{11.26}$$

and the forces of constraint, $\underline{F}^c = \underline{\underline{B}}^T\underline{\lambda}$, become

$$\underline{F}^c = -\underline{\underline{B}}^T\left(\underline{\underline{B}}\,\underline{\underline{M}}^{-1}\underline{\underline{B}}^T\right)^{-1}\left(\underline{c} - \underline{\underline{B}}\,\underline{\underline{M}}^{-1}\underline{F}\right). \tag{11.27}$$

System (11.26) is now a system of second-order ODEs, which can be solved to predict the dynamic behavior of the system. These equations have appeared in Hemami and Weimer [57], Lötstedt et al. [58, 48] and Gear et al. [59]

Example 11.5. Quick return mechanism
The quick return mechanism shown in fig. 10.6 consists of a uniform crank of length L_c and mass m_c, and of a uniform arm of length L_a and mass m_a. The crank is pinned at point **R** and the arm at point **O**; the distance between these two points is denoted d. At point **S**, a slider allows the tip of the crank to slide along the arm. A mass M is attached at point **T**, the tip of the arm. A spring of stiffness constant k connects the tip of the arm, point **T**, to fixed point **A**; the spring is un-stretched when the arm is in the vertical position. This problem was treated in examples 8.15 and 10.5 on pages 329 and 396, respectively, using Lagrange's formulation with one and two generalized coordinates, respectively. The same problem was treated with Maggi's formulation in example 11.2 on page 430. Use the index-1 formulation to derive the equations of motion of the system with two generalized coordinates, angles θ and ϕ.

The kinematics of the problem were presented in the previous examples dealing with this system and will not be repeated here. In the index-1 formulation, constraints are enforced at the acceleration level. For the quick return mechanism, the kinematic constraint is $\mathcal{C} = d\sin\phi - L_c\sin(\phi - \theta) = 0$; the velocity level constraint is $\dot{\mathcal{C}} = \underline{\underline{B}}\,\underline{\dot{q}} = 0$ where the constraint matrix is defined in eq. (10.14). Finally, using the notation of eq. (11.3), the acceleration level constraint is $\ddot{\mathcal{C}} = \underline{\underline{B}}\,\underline{\ddot{q}} - c = 0$, where $c = -\underline{\dot{\underline{\underline{B}}}}\,\underline{\dot{q}} = L_c S_{\phi-\theta}(\dot{\phi} - \dot{\theta})^2 - dS_\phi\dot{\phi}^2$.

The equations of motion written in index-1 form, eq. (11.24), are now

$$\begin{bmatrix} m_cL_c^2/3 & 0 & -L_cC_{\phi-\theta} \\ 0 & (M+m_a/3)L_a^2 & L_cC_{\phi-\theta}-dC_\phi \\ -L_cC_{\phi-\theta} & L_cC_{\phi-\theta}-dC_\phi & 0 \end{bmatrix}\begin{Bmatrix} \ddot{\theta} \\ \ddot{\phi} \\ \lambda \end{Bmatrix} = \begin{Bmatrix} 0 \\ kL_a^2S_\phi \\ c \end{Bmatrix}. \tag{11.28}$$

The first step of the index-1 formulation is to solve for Lagrange's multiplier using eq. (11.25) to find

$$\bar{L}_c(\bar{L}_cC_{\phi-\theta} - C_\phi)\bar{\lambda} = \frac{\mu_c\bar{L}_c^2}{3\gamma}\left[S_\phi - \frac{1 + \mu_a/3}{\bar{L}_cC_{\phi-\theta} - C_\phi}\bar{c}\right] \tag{11.29}$$

where the various non-dimensional quantities appearing in this expression were defined in example 11.2 and

$$\bar{c} = \frac{Mc}{kd} = \bar{L}_cS_{\phi-\theta}(\phi' - \theta')^2 - S_\phi\phi'^2. \tag{11.30}$$

The governing equations of the problem are given by eqs. (11.26), which become

$$\begin{bmatrix} \dfrac{\mu_c\bar{L}_c^2}{3} & 0 \\ 0 & (1 + \dfrac{\mu_a}{3})\bar{L}_a^2 \end{bmatrix}\begin{Bmatrix} \theta'' \\ \phi'' \end{Bmatrix} = \begin{Bmatrix} 0 \\ \bar{L}_a^2 S_\phi \end{Bmatrix} + \frac{\mu_c\bar{L}_c^2}{3\gamma}\left[S_\phi - \frac{(1 + \mu_a/3)\bar{c}}{\bar{L}_cC_{\phi-\theta} - C_\phi}\right]\begin{Bmatrix} \bar{h} \\ -1 \end{Bmatrix}.$$

These equations form a set of ordinary differential equations for the two generalized coordinates of this problem; Lagrange's multiplier is eliminated from the formulation.

The index-1 equations can be integrated using classical numerical tools developed for the solution of ordinary differential equations. The dynamic response predicted by the index-1 formulation is identical to that obtained by Lagrange's or Maggi's formulations, see figs 8.10 to 8.12. Of course, due to the approximations inherent to numerical solution techniques, slight discrepancies between the various predictions should be expected.

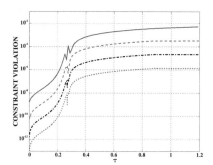

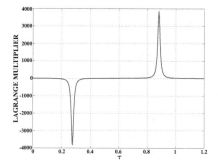

Fig. 11.5. Constraint violation versus non-dimensional time. Solid line, $N_s = 100$; dashed line, $N_s = 200$; dashed-dotted line, $N_s = 400$; dotted line, $N_s = 800$.

Fig. 11.6. Time history of Lagrange's multiplier.

The index-1 formulation enforces acceleration level constraints, $\ddot{\mathcal{C}} = 0$, at each instant in time. Indeed, the acceleration level constraint is the third equation of system (11.28). An exact solution of these governing equations would then imply $\mathcal{C} = 0$, provided that the displacement and velocity level constraints are satisfied at time $\tau = 0$. An approximate solution of the same equations results in an approximate satisfaction of the holonomic constraint equation, $\mathcal{C} \approx 0$.

Figure 11.5 shows the time history of the constraint violation, *i.e.*, $\mathcal{C}$ as a function of the non-dimensional time. The index-1 equations were integrated using a fourth-order Runge-Kutta algorithm [5] with $N_s =100$, 200, 400, and 800 time steps of equal size. As the time step size is reduced, the constraint violation decreases; note the logarithmic scale along the vertical axis. The same data is also presented in table 11.1: the constraint violations at time $\tau = 1$ are listed in the third column of the table for the four time step sizes.

Table 11.1. Constraint violations for Maggi's, index-1, and null space formulations.

Number of time steps	Maggi's formulation	Index-1 formulation	Null space formulation
100	3.5×10^{-05}	4.0×10^{-03}	4.0×10^{-03}
200	1.7×10^{-06}	3.1×10^{-04}	3.1×10^{-04}
400	1.0×10^{-07}	2.0×10^{-05}	2.0×10^{-05}
800	6.0×10^{-09}	1.2×10^{-06}	1.2×10^{-06}

It is important to compare columns two and three of table 11.1. These two columns list the constraint violation for Maggi's and index-1 formulations when the same integration technique and identical time step sizes are used. Much larger constraint violations are observed for the index-1 formulation, which enforces constraints at the acceleration level, than for Maggi's formulation, which enforces constraints at the velocity level. Clearly, larger constraint violation are expected for the formulations that enforce constraints at the acceleration level, resulting in a more pronounced drift phenomenon.

Although Lagrange's multipliers have been eliminated from the index-1 formulation, they can be evaluated with the help of eq. (11.29). Figure 11.6 shows the time history of the non-dimensional multiplier.

11.2.4 Problems

Problem 11.9. Bar sliding on guides
Treat problem 11.1 using the index-1 formulation.

Problem 11.10. Bar sliding on guides
Treat problem 11.2 using the index-1 formulation.

Problem 11.11. Crank-slider mechanism
Treat problem 11.3 using the index-1 formulation.

Problem 11.12. The crank piston mechanism
Treat problem 11.4 using the index-1 formulation.

Problem 11.13. Slider-arm mechanism
Treat problem 11.5 using the index-1 formulation.

Problem 11.14. The spatial mechanism
Treat problem 11.6 using the index-1 formulation.

Problem 11.15. The quick return mechanism
Treat problem 11.7 using the index-1 formulation.

Problem 11.16. Two bar linkage tracking a curve
Treat problem 11.8 using the index-1 formulation.

11.2.5 The null space formulation

It is also possible to solve system (11.24) in an expeditious manner with the help of the null space of the constraint matrix introduced in section 11.2.1. Pre-multiplying the first equations by $\underline{\underline{\varGamma}}^T$ eliminates Lagrange's multipliers and yields system accelerations as

$$\begin{bmatrix} \underline{\underline{\varGamma}}^T \underline{\underline{M}} \\ \underline{\underline{B}} \end{bmatrix} \underline{\ddot{q}} = \begin{bmatrix} \underline{\underline{\varGamma}}^T \underline{F} \\ \underline{c} \end{bmatrix}. \tag{11.31}$$

Here again, this system forms a set of second-order ODEs, the solution of which requires the constraint matrix to be of full rank and the mass matrix to be invertible.

A number of authors developed this formulation independently. Hemami and Weimer [57] introduced the orthogonal complement, $\underline{\underline{\varGamma}}$, of the constraint matrix to obtain eqs. (11.31) for small-scale systems, although no systematic procedure was proposed to determine the orthogonal complement. They also demonstrated the equivalence of this approach to Kane's equations [60, 61]. García de Jalón *et al.* [62, 63] also derived eqs. (11.31).

Borri *et al.* [64] proposed the *acceleration projection method*, which decomposes the generalized acceleration as $\underline{\ddot{q}} = \underline{\underline{\varGamma}} \, \underline{\zeta} + \underline{\underline{B}}^T \underline{\eta}$. Substituting this decomposition into eqs. (11.31) express $\underline{\zeta}$ and $\underline{\eta}$ in terms of $\underline{\underline{M}}, \underline{\underline{B}}, \underline{\underline{\varGamma}}^T, \underline{F}$ and $\underline{c}$, leading to second-order ODEs.

Example 11.6. Quick return mechanism
The quick return mechanism shown in fig. 10.6 consists of a uniform crank of length L_c and mass m_c, and of a uniform arm of length L_a and mass m_a. The crank is pinned at point **R** and the arm at point **O**; the distance between these two points is denoted d. At point **S**, a slider allows the tip of the crank to slide along the arm. A mass M is attached at point **T**, the tip of the arm. A spring of stiffness constant k connects the tip of the arm, point **T**, to fixed point **A**; the spring is un-stretched when the arm is in the vertical position. This problem was treated in examples 8.15 and 10.5 on pages 329 and 396, respectively, using Lagrange's formulation with one and two generalized coordinates, respectively. The same problem was treated with Maggi's and the index-1 formulations in example 11.2 and 11.5, respectively. Use the null space formulation to derive the equations of motion of the system with two generalized coordinates, angles θ and ϕ.

The kinematics of the problem were presented in the previous examples dealing with this system and will not be repeated here. In the null space formulation, constraints are enforced at the acceleration level. For the quick return mechanism, the kinematic constraint is $\mathcal{C} = d\sin\phi - L_c\sin(\phi - \theta) = 0$; the velocity level constraint is the $\dot{\mathcal{C}} = \underline{\underline{B}}\,\dot{q} = 0$ where the constraint matrix is defined in eq. (10.14). Finally, using the notation of eq. (11.3), the acceleration level constraint is $\ddot{\mathcal{C}} = \underline{\underline{B}}\,\ddot{q} - c = 0$, where $c = -\dot{\underline{\underline{B}}}\,\dot{q} = L_c S_{\phi-\theta}(\dot{\phi} - \dot{\theta})^2 - dS_\phi\dot{\phi}^2$.

The governing equations of the problem are given by eqs. (11.31), which become

$$\begin{bmatrix} \mu_c\bar{L}_c^2/3 & (1+\mu_a/3)\bar{L}_a^2\bar{h} \\ -\bar{L}_c C_{\phi-\theta} & \bar{L}_c C_{\phi-\theta} - C_\phi \end{bmatrix} \begin{Bmatrix} \theta'' \\ \phi'' \end{Bmatrix} = \begin{Bmatrix} \bar{h}\bar{L}_a^2 S_\phi \\ \bar{c} \end{Bmatrix}, \qquad (11.32)$$

where $\bar{c}$ is defined by eq. (11.30), and the various non-dimensional quantities appearing in this expression were defined in example 11.2. These equations form a set of ordinary differential equations for the two generalized coordinates of this problem; Lagrange's multiplier is eliminated from the formulation.

The null space equations can be integrated using classical numerical tools developed for the solution of ordinary differential equations. The dynamic response predicted by the index-1 formulation is identical to that obtained by Lagrange's, Maggi's, or index-1 formulations, see figs. 8.10 to 8.12. Of course, due to the approximations inherent to numerical solution techniques, slight discrepancies between the various predictions should be expected.

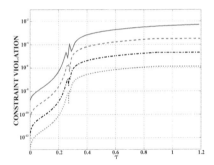

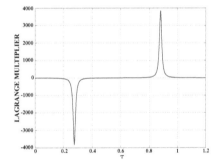

Fig. 11.7. Constraint violation versus non-dimensional time. Solid line, $N_s = 100$; dashed line, $N_s = 200$; dashed-dotted line, $N_s = 400$; dotted line, $N_s = 800$.

Fig. 11.8. Time history of Lagrange's multiplier.

The null space formulation enforces acceleration level constraints, $\ddot{\mathcal{C}} = 0$, at each instant in time. Indeed, the acceleration level constraint is the third equation of system (11.32). An exact solution of these governing equations would then imply $\mathcal{C} = 0$, provided that the displacement and velocity level constraints are satisfied at time $\tau = 0$. An approximate solution of the same equations results in an approximate satisfaction of the holonomic constraint equation, $\mathcal{C} \approx 0$.

Figure 11.7 shows the time history of the constraint violation, *i.e.*, $\mathcal{C}$ as a function of the non-dimensional time. The null space equations were integrated using a fourth-order Runge-Kutta algorithm [5] with $N_s = 100, 200, 400$, and 800 time steps of equal size. As the time step size is reduced, the constraint violation decreases; note the logarithmic scale along the vertical axis. The same data is also presented in table 11.1: the constraint violations at time $\tau = 1$ are listed in the fourth column of the table for the four time step sizes.

The behavior of the constraint violations are very similar for the index-1 and null space formulations, see figs. 11.5 and 11.7, respectively, or the third and fourth columns of table 11.1, respectively. This observation is consistent with the fact that both index-1 and null space formulations enforce acceleration level constraints.

11.2.6 Problems

The null space formulation requires the determination of the null space of the constraints. While the null space is uniquely defined, different sets linearly independent vectors spanning the null space can be selected, leading to different equations of motion. In the problems below, use Maggi's formulation to determine the null space; Problems 11.1 to 11.8 provide hints on how to select the kinematic parameters, thereby leading to a unique definition of the null space.

Problem 11.17. Bar sliding on guides
Treat problem 11.1 using the null space formulation.

Problem 11.18. Bar sliding on guides
Treat problem 11.2 using the null space formulation.

Problem 11.19. Crank-slider mechanism
Treat problem 11.3 using the null space formulation.

Problem 11.20. The crank piston mechanism
Treat problem 11.4 using the null space formulation.

Problem 11.21. Slider-arm mechanism
Treat problem 11.5 using the null space formulation.

Problem 11.22. The spatial mechanism
Treat problem 11.6 using the null space formulation.

Problem 11.23. The quick return mechanism
Treat problem 11.7 using the null space formulation.

Problem 11.24. Two bar linkage tracking a curve
Treat problem 11.8 using the null space formulation.

11.2.7 Udwadia and Kalaba's formulation

The results presented in the previous section can also be recast in a more compact and general form in terms of Moore-Penrose generalized inverses. The term featuring the matrix inverse in eq. (11.26) can be written as $\underline{\underline{B}}^T(\underline{\underline{B}}\,\underline{\underline{M}}^{-1}\underline{\underline{B}}^T)^{-1} =$

$\underline{\underline{M}}^{1/2}(\underline{\underline{B}}\,\underline{\underline{M}}^{-1/2})^T[(\underline{\underline{B}}\,\underline{\underline{M}}^{-1/2})(\underline{\underline{B}}\,\underline{\underline{M}}^{-1/2})^T]^{-1} = \underline{\underline{M}}^{1/2}(\underline{\underline{B}}\,\underline{\underline{M}}^{-1/2})^+$, where the last equality follows from eq. (18.16). Equations (11.26) and (11.27) now become

$$\underline{\underline{M}}\,\ddot{\underline{q}} = \underline{F} + \underline{\underline{M}}^{1/2}(\underline{\underline{B}}\,\underline{\underline{M}}^{-1/2})^+ \left(\underline{c} - \underline{\underline{B}}\,\underline{\underline{M}}^{-1}\underline{F}\right), \qquad (11.33)$$

and

$$\underline{F}^c = -\underline{\underline{M}}^{1/2}(\underline{\underline{B}}\,\underline{\underline{M}}^{-1/2})^+ \left(\underline{c} - \underline{\underline{B}}\,\underline{\underline{M}}^{-1}\underline{F}\right), \qquad (11.34)$$

respectively.

A process similar to that developed by Udwadia *et al.* [65] can be used to solve the system formed by eqs. (11.31) and $\underline{\underline{\Gamma}}^T\underline{F}^c = 0$, to recover eqs. (11.33) and (11.34). This underlines the close relationship between the present formulation and the null space formulation of section 11.2.5.

Equations (11.33) and (11.34) were first presented by Udwadia and Kalaba [66, 67, 68, 69], based on Gauss' principle: the explicit equations of motion were expressed as the solution of a quadratic minimization problem subjected to constraint conditions at the acceleration level.

In a later paper [70], the same equations were derived from d'Alembert's principle. This formulation is more general than those presented in the previous sections, which require the constraint matrix to be of full rank, whereas the Moore-Penrose generalized inverse is unique and always exists. The same authors [71] later presented a simpler derivation of their formulation that bypasses the concepts of generalized inverses. When $\text{rank}(B) = m$, they proved the existence of Lagrange's multipliers and expressed them in terms of the constraint forces; when $\text{rank}(B) < m$, Lagrange's multipliers are not unique, although the constraint forces are unique.

Udwadia *et al.* [65] presented an extended form of d'Alembert's principle that is able to deal with nonholonomic constraints, which might be nonlinear expressions of the generalized velocities. Furthermore, they showed that the previous formulation could be derived without invoking Moore-Penrose pseudo inverses. The geometric interpretation of the results in terms of projection operators, as presented in section 11.3, appeared in ref. [72] and the relationship of this formulation to Gibbs-Appell's equations was explored in ref. [73]. Finally, the same authors [74, 75] generalized the formulation to deal with mechanical system involving non-ideal constraints, *i.e.*, constraints associated with constraint forces whose virtual work might not vanish. The textbook authored by Udwadia and Kalaba [76] gives ample details on all aspects of the formulation.

11.2.8 Comparison of the ODE formulations

The formulations presented above all transform the $(2n + m)$ first-order DAEs associated with Lagrange's equations of the first kind into ODEs by eliminating Lagrange's multipliers. Maggi's formulation, eqs. (11.11) and (11.7), yields $(2n - m)$ first-order ODEs for the $(n - m)$ kinematic characteristics and n generalized coordinates. The index-1, eqs. (11.26), null space, eqs. (11.31), or Udwadia and Kalaba's, eqs. (11.33), formulations form sets of n second-order ODEs for the n generalized coordinates, which could alternatively be recast as sets of $2n$ first-order ODEs for the n generalized coordinates and n generalized velocities.

The first observation is that these methods decrease the size of the problem from $(2n + m)$ for Lagrange's equations of the first kind to $2n$ for the index-1, null space, and Udwadia and Kalaba's formulations, and $(2n - m)$ for Maggi's formulation. This dimensional reduction, however, comes at a price: the evaluation of the null space of the constraint matrix in Maggi's and null space formulations, or the evaluation of generalized inverses in Udwadia and Kalaba's formulation. Lagrange's equations of the first kind are typically formulated in terms of generalized coordinates that will render system matrices highly sparse, leading to efficient solution techniques, as discussed by Orlandea *et al.* [77, 78]. Hence, the main advantage of the above techniques is not so much the reduction in the number of equations, but rather the change in their mathematical nature, from DAEs to ODEs.

The second observation is that Maggi's formulation enforces *velocity* level constraints for holonomic constraints, whereas *acceleration* level constraints are enforced in the other formulations. This fact has important implications for numerical implementations of these approaches: the constraint drift phenomenon will be significantly more pronounced when using the latter formulations than when using Maggi's formulation. Typically, the constraint violation stabilization techniques described in section 12.3 will be required to compensate for this drift.

Although the enforcement of the constraints at the acceleration level is a widely used practice in multibody dynamics, it has the potential to adversely affect the numerical solution procedure. In fact, Campbell and Leimkuhler [79] studied the effects of differentiation of the constraints in DAEs; they concluded: "Thus, the differentiated system may be less well behaved numerically for a given method than either the original DAEs or an equivalent state-space form for that method. These numerical difficulties can take the form of increased stiffness, extraneous positive eigenvalues, and more stringent step-size restrictions."

Maggi's formulation requires the definition of a set of m kinematic parameters. Since the set of linearly independent vectors spanning the null space is not uniquely defined, the choice of kinematic parameters is not unique and the reduced equations can take a variety of forms. The same remark applies to the null space formulation. While the null space is uniquely defined, different sets linearly independent vectors spanning the null space can be selected, leading to different equations of motion.

In Maggi's formulation, the selected set of kinematic parameters appears explicitly in the equations of motion. In contrast, these additional variables do not appear in the index-1 and null space formulations, whose equations of motion are expressed in terms of the sole generalized coordinates originally used to describe the system.

Udwadia and Kalaba's formulation presents a number of advantages over the other formulations. The Moore-Penrose generalized inverse appearing in eq. (11.33) always exists, whereas the other formulations require a full rank constraint matrix. Hence, Udwadia and Kalaba's formulation is capable of dealing with problems featuring rank deficient constraint matrix, such as those involving redundant constraints. Furthermore, problems with variable number of degrees of freedom, such as intermittent contact problems, or problems involving rolling and slipping, are readily treated. In contrast, such situations will be problematic for minimum set approaches, since the number of kinematic parameters must change as the constraints change or become redundant.

11.2.9 Problems

Problem 11.25. Transformation of DAEs to ODEs
Consider a constrained dynamical system represented by n generalized coordinates and subjected to m holonomic constraints. The following questions deal with the solution of such problem using Maggi's, index-1, and null space formulations, which all three, transforms the equations of the problem from DAEs to ODEs. *(1)* Which of the three formulations is least sensitive to the drift phenomenon? Why? *(2)* For each of the three methods, what is the nature of the variables appearing in the final set of ODEs? *(3)* For each of the three methods, how many variables appear in the final set of ODEs? *(4)* For each of the three methods, in which form are the constraints enforced? *(5)* Are the variable appearing in Maggi's formulation uniquely defined? Why? *6)* Is the system of equations characterizing the null space formulation uniquely defined? Why?

11.3 The geometric interpretation of constraints

Many features of the dynamic response of constrained systems can be interpreted in a purely geometric manner. Consider the problem of a simple pendulum of mass m and length L, as depicted in fig. 11.9. This single degree of freedom system could be described by single generalized coordinate θ. Alternatively, two generalized coordinates, the Cartesian coordinates, q_1 and q_2, defining the position vector of the bob, $\underline{r} = q_1\bar{\imath}_1 + q_2\bar{\imath}_2$, could be used, subjected to a single holonomic constraint, $\mathcal{C} = (\underline{r}^T\underline{r}/L - L)/2 = 0$. This constraint enforces the constant length condition for the pendulum; the constraint matrix is $\underline{\underline{B}}^T(\underline{q}) = \underline{r}/L = C_\theta\bar{\imath}_1 + S_\theta\bar{\imath}_2$, where $C_\theta = \cos\theta$ and $S_\theta = \sin\theta$.

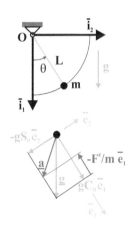

Fig. 11.9. Geometric interpretation of accelerations for a simple pendulum.

Figure 11.9 also shows the unit vectors, $\bar{e}_1$ and $\bar{e}_2$, of a polar coordinate system that can be used to conveniently compute the accelerations of the particle as $\underline{a} = -L\dot{\theta}^2\,\bar{e}_1 + L\ddot{\theta}\,\bar{e}_2$. Since the path of the particle is a circle of radius L, motion is allowed in the tangential direction, $\bar{e}_2$, but prohibited in the radial direction, $\bar{e}_1$. For this particular problem, the constraint matrix is a unit vector along $\bar{e}_1$ and defines the direction in which the motion is constrained.

In more general terms, the configuration space, defined here by the plane $(\bar{\imath}_1, \bar{\imath}_2)$, is divided into two mutually orthogonal subspaces: the subspace defined by unit vector $\bar{e}_2$ in which motion is allowed, and the subspace defined by unit vector $\bar{e}_1$, *i.e.*, the subspace defined by the constraint matrix, along which motion is prohibited. This discussion provides a geometric interpretation of the constraint matrix: it defines a subspace of the configuration space along which motion is not allowed.

Next, Newton's second law will be written for the particle. A free body diagram of the particle reveals that $gC_\theta - F^c/m = a_r$ and $-gS_\theta = a_\theta$, where $a_r = -L\dot{\theta}^2$, and $a_\theta = L\ddot{\theta}$ are the radial and tangential components of acceleration, respectively. The geometric interpretation of these results is depicted in the right portion of fig. 11.9. First, the constraint force is entirely contained in the subspace defined by the constraint matrix. Next, the externally applied load has a component, $-mgS_\theta$, in the feasible direction, $\bar{e}_2$; this component actually drives the motion of the particle. The externally applied load also has a component, mgC_θ, in the direction of prohibited motion, $\bar{e}_1$. Finally, due to this applied load, the particle experiences an acceleration featuring components in both feasible and infeasible directions. The projection of this acceleration along the feasible direction equals the component of load applied in this direction divided by the mass of the particle, $a_\theta = -gS_\theta$. The projection of the acceleration along the infeasible direction equals the component of load applied in this direction divided by the mass of the particle and corrected by the acceleration associated with the constraint force, F^c/m, $a_r = gC_\theta - F^c/m$.

This discussion outlines the geometric interpretation of all the quantities involved in this simple constrained dynamical problem. Of course, at this point, these observations are limited to the simple pendulum problem presented above. In the sections below, the above results will be shown to apply to all constrained dynamical systems. This geometric interpretation of the problem has been investigated by a number of authors using similar concepts: Brauchli *et al.* [80, 81] and Udwadia and Kalaba [72]. To generalize the above observations, however, the concept of projection must first be generalized. Instead of the orthogonal projections used in fig. 11.9, a more general type of projection, defined with respect to a certain metric of the configuration space, must be defined first.

11.3.1 The orthogonal projection operator

Consider a plane in a three-dimensional space; the plane is defined by its unit normal vector $\bar{n}$. This unit vector divides the three-dimensional space into two subspaces: the subspace spanned by $\bar{n}$, and the subspace orthogonal to $\bar{n}$, *i.e.*, the subspace spanned by two mutually orthogonal vectors, $\bar{u}$ and $\bar{v}$, spanning the plane normal to $\bar{n}$, as shown in fig. 11.10.

Intuitively, the projection operator, denoted $\underline{\underline{P}}_\parallel$, projects an arbitrary vector in the direction parallel to $\bar{n}$. It is readily verified that in the simple case considered here, the projection operator is $\underline{\underline{P}}_\parallel = \bar{n}\bar{n}^T$. Indeed

$$\underline{\underline{P}}_\parallel \underline{a} = \bar{n}\bar{n}^T\underline{a} = \|\underline{a}\| \cos\alpha\, \bar{n} = \underline{a}_\parallel. \tag{11.35}$$

As shown in fig. 11.10, $\underline{a}_\parallel$ is the projection of $\underline{a}$ along $\bar{n}$. Note that this is an orthogonal projection of $\underline{a}$ along unit vector $\bar{n}$.

The following terminology is now defined: $\bar{n}$ is the *image of the projector*, whereas $[\bar{u}, \bar{v}]$ is the *kernel* or *null space of the projector*. Note that while the kernel of the projector is uniquely defined as the subspace orthogonal to $\bar{n}$, vectors $[\bar{u}, \bar{v}]$ are not uniquely defined. Indeed, any two mutually orthogonal vectors normal to $\bar{n}$ could be selected. The meaning of this terminology is easily understood in view of the following results

$$\underline{\underline{P}}_{\parallel} \bar{n} = \bar{n}, \tag{11.36a}$$

$$\underline{\underline{P}}_{\parallel} [\bar{u}, \bar{v}] = 0. \tag{11.36b}$$

Equation (11.36a) implies that the projection of the image is the image itself, and eq. (11.36b) implies that the projection of the null space vanishes.

Another interpretation of these results is that the image of the projector spans its eigenvectors associated with unit eigenvalues, and the kernel of the projector spans its eigenvectors associated zero eigenvalues, hence, the kernel is also called the *null space*.

Finally, it is easily verified that

$$\underline{\underline{P}}_{\parallel}\underline{\underline{P}}_{\parallel} = \underline{\underline{P}}_{\parallel}. \tag{11.37}$$

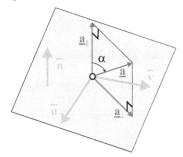

Fig. 11.10. Projection in three-dimensional space by a unit vector $\bar{n}$.

Geometrically, this corresponds to the fact that once an arbitrary vector has been projected along the image to find $\underline{a}_{\parallel}$, any further projection of $\underline{a}_{\parallel}$ will leave this vector unchanged.

The complementary projector, denoted $\underline{\underline{P}}_{\perp}$, projects an arbitrary vector in the direction perpendicular to $\bar{n}$: $\underline{a}_{\perp} = \underline{\underline{P}}_{\perp} \underline{a}$. Figure 11.10 reveals that $\underline{a} = \underline{a}_{\perp} + \underline{a}_{\parallel}$, and hence, $\underline{\underline{P}}_{\perp} \underline{a} = (\underline{\underline{I}} - \underline{\underline{P}}_{\parallel})\underline{a}$. Since this result must hold true for any arbitrary vector $\underline{a}$, it follows that

$$\underline{\underline{P}}_{\perp} = \underline{\underline{I}} - \underline{\underline{P}}_{\parallel} = \underline{\underline{I}} - \bar{n}\bar{n}^T. \tag{11.38}$$

The vectors $\bar{n}$, $\bar{u}$, and $\bar{v}$ form the column of an orthogonal matrix, and the following identity is readily verified $\bar{n}\bar{n}^T + \bar{u}\bar{u}^T + \bar{v}\bar{v}^T = \underline{\underline{I}}$. Hence, $\underline{\underline{P}}_{\perp} = \underline{\underline{I}} - \bar{n}\bar{n}^T = \bar{u}\bar{u}^T + \bar{v}\bar{v}^T$. This last result then implies

$$\underline{\underline{P}}_{\perp}\underline{\underline{P}}_{\parallel} = 0. \tag{11.39}$$

Finally, it is easily shown that the projections of an arbitrary vector in the two orthogonal subspaces are orthogonal to each other, as expected,

$$\underline{a}_{\perp}^T \underline{a}_{\parallel} = \underline{a}^T \underline{\underline{P}}_{\perp}\underline{\underline{P}}_{\parallel}\underline{a} = 0. \tag{11.40}$$

The developments presented in this section could have been started by considering an image subspace formed by the plane defined by units vector $\bar{u}$ and $\bar{v}$, the kernel or null space would then be unit vector $\bar{n}$. All results obtained above would be recovered, except for the fact that subscripts $(\cdot)_{\parallel}$ and $(\cdot)_{\perp}$ would be interchanged. The two projection operators are complementary of each other.

In this section, all projections were orthogonal projections: orthogonal vectors, $\underline{a}$ and $\underline{b}$, satisfy the condition $\underline{a}^T \underline{b} = 0$. It is possible to extend the concept of orthogonality: two vectors are said to be *orthogonal with respect to a metric*, $\underline{\underline{M}}$, of the space, if the following condition is satisfied

$$\underline{a}^T \underline{\underline{M}} \underline{b} = 0. \tag{11.41}$$

The *metric*, $\underline{\underline{M}}$, *of the space* is a symmetric, positive-definite matrix. In the present section, orthogonality was defined with respect to a metric equal to the identity matrix $\underline{\underline{M}} = \underline{\underline{I}}$. In the next section, the projection operator concept will be extended to a space where orthogonality is defined with respect to a non-identity metric.

11.3.2 The projection operator

Instead of following the rather intuitive, geometric development used in last section, a formal definition of the projection operator is the starting point of this section. A *projection operator*, $\underline{\underline{P}}_{\|}$, *is an $n \times n$ linear transformation that acts like the identity on its image*,

$$\underline{\underline{P}}_{\|} \underline{\underline{\mathcal{E}}} = \underline{\underline{\mathcal{E}}}, \tag{11.42}$$

where the *image of the projection*, $\underline{\underline{\mathcal{E}}}$, is an $n \times m$ matrix such that $\text{rank}(\underline{\underline{\mathcal{E}}}) = m \leq n$. The projection operation is assumed to act in a space where orthogonality is defined with respect to a metric, $\underline{\underline{M}}$, see eq. (11.41).

At first, the metric of the space is factorized as $\underline{\underline{M}} = \underline{\underline{S}}^T \underline{\underline{S}}$, using the Cholesky factorization [82], for instance. Equation (11.42), is rewritten as $(\underline{\underline{S}}\,\underline{\underline{P}}_{\|}\underline{\underline{S}}^{-1})(\underline{\underline{S}}\,\underline{\underline{\mathcal{E}}}) = (\underline{\underline{S}}\,\underline{\underline{\mathcal{E}}})$, or $\hat{\underline{\underline{P}}}_{\|}\hat{\underline{\underline{\mathcal{E}}}} = \hat{\underline{\underline{\mathcal{E}}}}$, where the following *scaled quantities* were defined: $\hat{\underline{\underline{\mathcal{E}}}} = \underline{\underline{S}}\,\underline{\underline{\mathcal{E}}}$ and $\hat{\underline{\underline{P}}}_{\|} = \underline{\underline{S}}\,\underline{\underline{P}}_{\|}\underline{\underline{S}}^{-1}$. Notation $(\hat{\cdot})$ indicates scaled quantities. The singular value decomposition presented in section 18.1 will be used extensively in this development. The scaled image, $\hat{\underline{\underline{\mathcal{E}}}}$, is factorized with the help of the singular value decomposition, eq. (18.10), to find

$$\hat{\underline{\underline{\mathcal{E}}}} = \check{\underline{\underline{U}}}\,\underline{\underline{\Sigma}}\,\underline{\underline{V}}^T. \tag{11.43}$$

The null space or kernel, $\underline{\underline{\Gamma}}$, of the projection is defined by eq. (18.8) as

$$\hat{\underline{\underline{\mathcal{E}}}}^T \underline{\underline{\Gamma}} = 0. \tag{11.44}$$

Introducing the singular value decomposition, eq. (11.43), into the definition of the projection operator yields $\hat{\underline{\underline{P}}}_{\|}\check{\underline{\underline{U}}}\,\underline{\underline{\Sigma}}\,\underline{\underline{V}}^T = \check{\underline{\underline{U}}}\,\underline{\underline{\Sigma}}\,\underline{\underline{V}}^T$. Since $\underline{\underline{V}}$ and $\underline{\underline{\Sigma}}$ are non singular matrices, it follows that

$$\hat{\underline{\underline{P}}}_{\|} = \check{\underline{\underline{U}}}\,\check{\underline{\underline{U}}}^T. \tag{11.45}$$

This result implies that $\hat{\underline{\underline{P}}}_{\|}$ is a symmetric matrix, and furthermore, in view of eq. (18.6b),

$$\hat{\underline{\underline{P}}}_{\|}\hat{\underline{\underline{P}}}_{\|} = \hat{\underline{\underline{P}}}_{\|}, \tag{11.46}$$

a characteristic property of projection operations.

An explicit expression of the projector can be obtained that does not rely on the singular value decomposition. In view of the orthogonality of matrix $\underline{\underline{V}}$, eq. (11.45) is recast as

$$\hat{\underline{\underline{P}}}_{\|} = \underline{\underline{\check{U}}}\left[\underline{\underline{\Sigma}}(\underline{\underline{V}}^T\underline{\underline{V}})\underline{\underline{\Sigma}}^{-2}(\underline{\underline{V}}^T\underline{\underline{V}})\underline{\underline{\Sigma}}\right]\underline{\underline{\check{U}}}^T = (\underline{\underline{\check{U}}}\,\underline{\underline{\Sigma}}\,\underline{\underline{V}}^T)(\underline{\underline{V}}\,\underline{\underline{\Sigma}}^{-2}\underline{\underline{V}}^T)(\underline{\underline{V}}\,\underline{\underline{\Sigma}}\,\underline{\underline{\check{U}}}^T). \tag{11.47}$$

Since $\hat{\underline{\underline{\mathcal{E}}}}^T\hat{\underline{\underline{\mathcal{E}}}} = \underline{\underline{V}}\,\underline{\underline{\Sigma}}^2\underline{\underline{V}}^T$, it follows that

$$\hat{\underline{\underline{P}}}_{\|} = \hat{\underline{\underline{\mathcal{E}}}}\left[\hat{\underline{\underline{\mathcal{E}}}}^T\hat{\underline{\underline{\mathcal{E}}}}\right]^{-1}\hat{\underline{\underline{\mathcal{E}}}}^T. \tag{11.48}$$

The *complementary projection operator*, $\underline{\underline{P}}_{\perp}$, is defined as

$$\underline{\underline{P}}_{\perp} = \underline{\underline{I}} - \underline{\underline{P}}_{\|}, \tag{11.49}$$

where $\underline{\underline{I}}$ is the identity matrix; this implies $\hat{\underline{\underline{P}}}_{\perp} = \underline{\underline{I}} - \hat{\underline{\underline{P}}}_{\|}$, where the scaled complementary projection operator is defined as $\hat{\underline{\underline{P}}}_{\perp} = \underline{\underline{S}}\,\underline{\underline{P}}_{\perp}\underline{\underline{S}}^{-1}$. It is readily verified that $\hat{\underline{\underline{P}}}_{\perp} = \underline{\underline{I}} - \hat{\underline{\underline{P}}}_{\|}$ is a symmetric matrix and that $\hat{\underline{\underline{P}}}_{\perp}\hat{\underline{\underline{P}}}_{\perp} = (\underline{\underline{I}} - \hat{\underline{\underline{P}}}_{\|})(\underline{\underline{I}} - \hat{\underline{\underline{P}}}_{\|}) = \hat{\underline{\underline{P}}}_{\perp}$, two characteristic properties of projection operators. In view of property (18.6a) of the singular value decomposition, $\underline{\underline{S}}^{-1}\underline{\underline{\check{U}}}\,\underline{\underline{\check{U}}}^T\underline{\underline{S}} + \underline{\underline{S}}^{-1}\underline{\underline{\Gamma}}\,\underline{\underline{\Gamma}}^T\underline{\underline{S}} = \underline{\underline{I}}$, or $\underline{\underline{S}}^{-1}\underline{\underline{\Gamma}}\,\underline{\underline{\Gamma}}^T\underline{\underline{S}} = \underline{\underline{I}} - \hat{\underline{\underline{P}}}_{\|}$, and hence, the explicit expression for the complementary projection becomes

$$\hat{\underline{\underline{P}}}_{\perp} = \underline{\underline{\Gamma}}\,\underline{\underline{\Gamma}}^T. \tag{11.50}$$

This shows that the image of the complementary projection is $\underline{\underline{\Gamma}}$, the null space or kernel of the projection. It is also clear that $\hat{\underline{\underline{P}}}_{\|}\hat{\underline{\underline{P}}}_{\perp} = \hat{\underline{\underline{P}}}_{\perp}\hat{\underline{\underline{P}}}_{\|} = 0$, where properties (18.7) were used.

Consider now an arbitrary vector, $\underline{a}$, and its scaled counterpart, $\hat{\underline{a}} = \underline{\underline{S}}\,\underline{a}$; the components, $\underline{c}$, of this vector along the orthogonal basis $\underline{\underline{U}}$ defined by the singular value decomposition of matrix $\hat{\underline{\underline{\mathcal{E}}}}$ are such that $\hat{\underline{a}} = \underline{\underline{U}}\,\underline{c}$. In view of the partition (18.3) of $\underline{\underline{U}}$, this becomes

$$\hat{\underline{a}} = \left[\underline{\underline{\check{U}}}\ \underline{\underline{\Gamma}}\right]\underline{c} = \underline{\underline{\check{U}}}\,\underline{c}_{\|} + \underline{\underline{\Gamma}}\,\underline{c}_{\perp} = \hat{\underline{a}}_{\|} + \hat{\underline{a}}_{\perp}. \tag{11.51}$$

The projections of vector $\hat{\underline{a}}$ are found to be

$$\hat{\underline{\underline{P}}}_{\|}\,\hat{\underline{a}} = \underline{\underline{\check{U}}}\,\underline{\underline{\check{U}}}^T\left[\underline{\underline{\check{U}}}\,\underline{c}_{\|} + \underline{\underline{\Gamma}}\,\underline{c}_{\perp}\right] = \underline{\underline{\check{U}}}\,\underline{c}_{\|} = \hat{\underline{a}}_{\|}. \tag{11.52a}$$

$$\hat{\underline{\underline{P}}}_{\perp}\,\hat{\underline{a}} = \underline{\underline{\Gamma}}\,\underline{\underline{\Gamma}}^T\left[\underline{\underline{\check{U}}}\,\underline{c}_{\|} + \underline{\underline{\Gamma}}\,\underline{c}_{\perp}\right] = \underline{\underline{\Gamma}}\,\underline{c}_{\perp} = \hat{\underline{a}}_{\perp}. \tag{11.52b}$$

Clearly, the projection and the complementary projection operators project an arbitrary vector onto the image of the projector and its null space, respectively.

These two complementary subspaces are orthogonal in metric $\underline{\underline{\mathcal{M}}}$; indeed, $\hat{\underline{a}}_{\|}^T \hat{\underline{a}}_\perp = \hat{\underline{a}}^T \hat{\underline{\underline{\mathcal{P}}}}_{\|}^T \hat{\underline{\underline{\mathcal{P}}}}_\perp \hat{\underline{a}} = 0$, and hence,

$$\underline{a}_\perp^T \underline{\underline{\mathcal{M}}} \, \underline{a}_{\|} = 0. \tag{11.53}$$

Projection operators are closely related to the Moore-Penrose generalized inverse defined in 18.2. Indeed, the Moore-Penrose inverse of the scaled image of the projector, $\hat{\underline{\underline{\mathcal{E}}}}^+$, satisfies condition (18.11), i.e., $\hat{\underline{\underline{\mathcal{E}}}} \hat{\underline{\underline{\mathcal{E}}}}^+ \hat{\underline{\underline{\mathcal{E}}}} = \hat{\underline{\underline{\mathcal{E}}}}$. Comparing this result to the scaled version of eq. (11.42) yields $\hat{\underline{\underline{\mathcal{P}}}}_{\|} = \hat{\underline{\underline{\mathcal{E}}}} \hat{\underline{\underline{\mathcal{E}}}}^+$, which, with the help of eq. (18.16), becomes $\hat{\underline{\underline{\mathcal{P}}}}_{\|} = \hat{\underline{\underline{\mathcal{E}}}} [\hat{\underline{\underline{\mathcal{E}}}}^T \hat{\underline{\underline{\mathcal{E}}}}]^{-1} \hat{\underline{\underline{\mathcal{E}}}}^T$. This simple manipulation re-establishes eq. (11.48) in an expeditious manner. The fact that the projector is a symmetric operator mirrors property (18.13) of the generalized inverse: $\hat{\underline{\underline{\mathcal{P}}}}_{\|} = \hat{\underline{\underline{\mathcal{E}}}} \hat{\underline{\underline{\mathcal{E}}}}^+ = \hat{\underline{\underline{\mathcal{E}}}}^{+T} \hat{\underline{\underline{\mathcal{E}}}}^T$. Transposing property (18.12) of the generalized inverse applied to the scaled image of the projector leads to $\hat{\underline{\underline{\mathcal{E}}}}^{+T} \hat{\underline{\underline{\mathcal{E}}}}^T \hat{\underline{\underline{\mathcal{E}}}}^{+T} = \hat{\underline{\underline{\mathcal{E}}}}^{+T}$, or

$$\hat{\underline{\underline{\mathcal{P}}}}_{\|} \hat{\underline{\underline{\mathcal{E}}}}^{+T} = \hat{\underline{\underline{\mathcal{E}}}}^{+T}. \tag{11.54}$$

This result implies that $\hat{\underline{\underline{\mathcal{E}}}}^{+T}$ is entirely contained in the subspace defined by the image, $\hat{\underline{\underline{\mathcal{E}}}}$, of the projector.

Example 11.7. Orthogonal projection in three-dimensional space

Consider a simple projection in three-dimensional space defined by the image $\underline{\underline{\mathcal{E}}} = \bar{n}$, where $\bar{n}$ is a unit vector, and a metric $\underline{\underline{\mathcal{M}}} = \underline{\underline{I}}$, where $\underline{\underline{I}}$ is the 3×3 identity, as depicted in fig. 11.10. It follows that the scaled quantities are identical to their unscaled counterparts. This problem was treated in an intuitive manner in section 11.3.1; in this example, the projection operator will be obtained from the more formal derivation presented above.

In view eq. (11.48), the projection operator becomes

$$\underline{\underline{\mathcal{P}}}_{\|} = \bar{n}(\bar{n}^T \bar{n})^{-1} \bar{n}^T = \bar{n}\bar{n}^T.$$

The kernel of the projector is $\underline{\underline{\Gamma}} = [\bar{u} \; \bar{v}]$, where $\bar{u}$ and $\bar{v}$ are two mutually orthogonal, unit vectors contained in the plane normal to $\bar{n}$. It satisfies eq. (11.44), $\bar{n}^T [\bar{u} \; \bar{v}] = 0$. Equation (11.50) now yields the complementary projection operator $\underline{\underline{\mathcal{P}}}_\perp = \underline{\underline{\Gamma}} \, \underline{\underline{\Gamma}}^T = \bar{u}\bar{u}^T + \bar{v}\bar{v}^T$. It is easily verified that $\underline{\underline{\mathcal{P}}}_{\|} + \underline{\underline{\mathcal{P}}}_\perp = \bar{n}\bar{n}^T + \bar{u}\bar{u}^T + \bar{v}\bar{v}^T = \underline{\underline{I}}$, where the last equality holds because $\bar{n}$, $\bar{u}$, and $\bar{v}$ form an orthogonal basis.

Consider now an arbitrary vector $\underline{a}$; application of the projection operators yields

$$\underline{\underline{\mathcal{P}}}_{\|} \, \underline{a} = \bar{n}\bar{n}^T \underline{a} = (\bar{n}^T \underline{a})\bar{n} = \|\underline{a}\| \cos \alpha \, \bar{n} = \underline{a}_{\|}.$$

and

$$\underline{\underline{\mathcal{P}}}_\perp \, \underline{a} = (\bar{u}^T \underline{a})\bar{u} + (\bar{v}^T \underline{a})\bar{v} = \underline{a}_\perp.$$

These results reproduce those obtained in a more intuitive manner in section 11.3.1.

Example 11.8. **Orthogonal projection in three-dimensional space**

Consider now a second example in three-dimensional space defined by the image $\underline{\mathcal{E}} = \underline{n}$, where $\underline{n}$ is a non-unit vector, and a metric $\underline{\underline{\mathcal{M}}} = \underline{\underline{S}}^T \underline{\underline{S}}$. The scaled image is $\hat{\underline{n}} = \underline{\underline{S}} \, \underline{n}$ and, in view eq. (11.48), the scaled projection operator becomes

$$\hat{\underline{\underline{P}}}_{\parallel} = \hat{\underline{n}} (\hat{\underline{n}}^T \hat{\underline{n}})^{-1} \hat{\underline{n}}^T = \frac{\hat{\underline{n}}}{\|\hat{\underline{n}}\|} \frac{\hat{\underline{n}}^T}{\|\hat{\underline{n}}\|}, \tag{11.55}$$

where $\hat{\underline{n}}^T/\|\hat{\underline{n}}\|$ represents the image of the projection normalized in the space of metric $\underline{\underline{\mathcal{M}}}$. Consider now an arbitrary vector $\underline{a}$, scaled as $\hat{\underline{a}} = \underline{\underline{S}} \, \underline{a}$; application of the projection operator, eq. (11.52a), yields

$$\hat{\underline{\underline{P}}}_{\parallel} \hat{\underline{a}} = \frac{\hat{\underline{n}}}{\|\hat{\underline{n}}\|} \frac{\hat{\underline{n}}^T}{\|\hat{\underline{n}}\|} \hat{\underline{a}} = \|\hat{\underline{a}}\| \cos\alpha \, \frac{\hat{\underline{n}}}{\|\hat{\underline{n}}\|} = \hat{\underline{a}}_{\parallel}, \tag{11.56}$$

where α is the angle between $\hat{\underline{a}}$ and $\hat{\underline{n}}$, *i.e.*, $\hat{\underline{a}}^T \hat{\underline{n}} = \|\hat{\underline{a}}\| \|\hat{\underline{n}}\| \cos\alpha$.

11.3.3 Projection of the equations of motion

The equations of motion of constrained dynamical systems have been cast in the form of eq. (11.1). Maggi's formulation, as presented in section 11.2.1, is a purely algebraic approach to the problem; this section focuses on a more geometric interpretation of constrained dynamical problems, which has been investigated by a number of authors: Brauchli *et al.* [80, 81], Udwadia and Kalaba [72], Blajer and his coworkers [83, 84, 85].

Because the mass matrix is a symmetric positive-definite matrix, it can be factorized as $\underline{\underline{M}} = \underline{\underline{S}} \, \underline{\underline{S}}^T$ using the Cholesky factorization [82], for instance. Multiplying the governing equations of motion by $\underline{\underline{S}}^{-1}$ then leads to

$$\ddot{\hat{\underline{q}}} - \hat{\underline{F}}^c = \hat{\underline{F}}, \tag{11.57}$$

where the scaled accelerations were defined as $\ddot{\hat{\underline{q}}} = \underline{\underline{S}}^T \ddot{\underline{q}}$, the scaled constraint forces as $\hat{\underline{F}}^c = \underline{\underline{S}}^{-1} \underline{F}^c$ and scaled forces as $\hat{\underline{F}} = \underline{\underline{S}}^{-1} \underline{F}$. Next, the scaled constraint matrix is defined as $\hat{\underline{\underline{E}}} = \underline{\underline{S}}^{-1} \underline{\underline{B}}^T$. Finally, the acceleration level constraint, eq. (11.3), is written as $\underline{\underline{B}} \, \underline{\underline{S}}^{-T} \underline{\underline{S}}^T \ddot{\underline{q}} = \underline{c}$, leading to the following scaled expression, $\hat{\underline{\underline{E}}}^T \ddot{\hat{\underline{q}}} = \underline{c}$.

A scaled projection operator is now introduced

$$\hat{\underline{\underline{P}}}_{\parallel} = \hat{\underline{\underline{E}}} \left[\hat{\underline{\underline{E}}}^T \hat{\underline{\underline{E}}} \right]^{-1} \hat{\underline{\underline{E}}}^T. \tag{11.58}$$

In view of eq. (11.48), the image of this projector is the scaled constraint matrix, $\hat{\underline{\underline{E}}}$, and it operates in a space of metric $\underline{\underline{\mathcal{M}}} = \underline{\underline{M}}^{-1}$, *i.e.*, the inverse of the mass matrix. By construction, the projection operator, $\hat{\underline{\underline{P}}}_{\parallel}$, projects an arbitrary vector into the subspace parallel to the constraint matrix; of course, the projection operation is performed in a space endowed with a metric defined by the inverse of the mass matrix. This basic property of the projector can be verified by projecting the constraint forces: since $\underline{\hat{F}}^c = \underline{\underline{S}}^{-1}\underline{\underline{B}}^T\underline{\lambda} = \underline{\hat{E}}\,\underline{\lambda}$, the constraint forces are entirely contained in the subspace defined by the constraint matrix, and hence, the projection operation has no effect

$$\underline{\underline{\hat{P}}}_{\parallel}\,\underline{\hat{F}}^c = \underline{\hat{F}}^c. \tag{11.59}$$

The projection of the scaled accelerations yields $\underline{\underline{\hat{P}}}_{\parallel}\,\underline{\ddot{\hat{q}}} = \underline{\hat{E}}[\underline{\hat{E}}^T\underline{\hat{E}}]^{-1}\underline{\hat{E}}^T\underline{\ddot{\hat{q}}} = \underline{\hat{E}}[\underline{\hat{E}}^T\underline{\hat{E}}]^{-1}\underline{c}$, and finally, in view of eq. (18.16),

$$\underline{\underline{\hat{P}}}_{\parallel}\,\underline{\ddot{\hat{q}}} = \underline{\hat{E}}^{+T}\underline{c} = \underline{\underline{\hat{P}}}_{\parallel}\underline{\hat{E}}^{+T}\underline{c}, \tag{11.60}$$

where the last equality follows from property (11.54) of the projection operator.

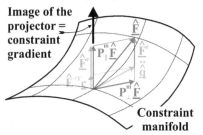

Image of the projector = constraint gradient

Constraint manifold

Fig. 11.11. Geometric representation of constraint dynamics with holonomic constraints. Although appearing as orthogonal projections in this illustration, projections are, in fact, operating in the metric of the inverse of the mass matrix.

The scaled equations of motion, eqs. (11.57), are multiplied by the projector to find

$$\underline{\underline{\hat{P}}}_{\parallel}\,\underline{\hat{F}}^c = \underline{\hat{F}}^c = \underline{\underline{\hat{P}}}_{\parallel}\,\underline{\ddot{\hat{q}}} - \underline{\underline{\hat{P}}}_{\parallel}\,\underline{\hat{F}}$$
$$= \underline{\underline{\hat{P}}}_{\parallel}\left(\underline{\hat{E}}^{+T}\underline{c} - \underline{\hat{F}}\right), \tag{11.61}$$

where the last equality was obtained with the help of eq. (11.60). This important relationship shows, once again, that the constraint forces are entirely contained within the image, $\underline{\hat{E}}$, of the projector, *i.e.*, the constraint forces belong to the space defined by the scaled constraint matrix.

Next, the accelerations of the system are computed from eq. (11.57) as

$$\underline{\ddot{\hat{q}}} = \underline{\hat{F}} + \underline{\hat{F}}^c = \underline{\hat{F}} + \underline{\underline{\hat{P}}}_{\parallel}\left(\underline{\hat{E}}^{+T}\underline{c} - \underline{\hat{F}}\right) = \underline{\underline{\hat{P}}}_{\perp}\,\underline{\hat{F}} + \underline{\hat{E}}^{+T}\underline{c}. \tag{11.62}$$

Clearly, system accelerations have a component in the image of the projector, $\underline{\underline{\hat{P}}}_{\parallel}\,\underline{\ddot{\hat{q}}} = \underline{\hat{E}}^{+T}\underline{c}$ and a component in the orthogonal subspace, $\underline{\underline{\hat{P}}}_{\perp}\,\underline{\ddot{\hat{q}}} = \underline{\underline{\hat{P}}}_{\perp}\,\underline{\hat{F}}$. The geometric interpretation of these results is illustrated in fig. 11.11.

According to eq. (11.57), the scaled unconstrained forces, $\underline{\hat{F}}$, are the sum of the scaled constraint forces, $\underline{\hat{F}}^c$, and the scaled accelerations of the system. The scaled constraint forces are the difference between vectors $\underline{\underline{\hat{P}}}_{\parallel}\,\underline{\hat{F}}$ and $\underline{\hat{E}}^{+T}\underline{c}$, both contained in the image of the projector, as implied by eq. (11.61). On the other hand, system accelerations are the sum of vector $\underline{\hat{E}}^{+T}\underline{c}$, contained in the image of the projector, and vector $\underline{\underline{\hat{P}}}_{\perp}\,\underline{\hat{F}}$, contained in the orthogonal subspace.

The various projection involved in this geometric interpretation of the equations of motion of constraint dynamical systems are not orthogonal projection. Indeed, the projection operator is defined in a space with a metric defined by the inverse of the mass matrix. Although appearing as orthogonal projections on this illustration, projections are, in fact, operating in the metric of the inverse of the mass matrix.

11.3.4 Elimination of Lagrange's multipliers

In the last section, geometric aspect of the problem have been underlined by using the geometric concept of projection operator. It is, however, easy to recast the main results of the last section in a form that does not make use of these geometric concepts. Indeed, eqs. (11.62) and (11.61) are rewritten as

$$\underline{\underline{M}}\,\underline{\ddot{q}} = \underline{F} + \underline{\underline{B}}^T \left(\underline{\underline{B}}\,\underline{\underline{M}}^{-1}\underline{\underline{B}}^T \right)^{-1} \left(\underline{c} - \underline{\underline{B}}\,\underline{\underline{M}}^{-1}\underline{F} \right), \tag{11.63a}$$

$$\underline{F}^c = \underline{\underline{B}}^T \left(\underline{\underline{B}}\,\underline{\underline{M}}^{-1}\underline{\underline{B}}^T \right)^{-1} \left(\underline{c} - \underline{\underline{B}}\,\underline{\underline{M}}^{-1}\underline{F} \right), \tag{11.63b}$$

respectively. These results are identical to those obtained by Udwadia and Kalaba [66, 69, 71, 65, 73].

Equations (11.63a) form a set of n second-order, ordinary differential for the n generalized coordinates $\underline{q}$. Clearly, Lagrange's multipliers present in the original equations, eqs. (11.1), have been eliminated; the governing equations of motion are now ordinary differential equations rather that differential-algebraic equations. Clearly, eqs. (11.63a) could be recast as a set of $2n$ first-order, ordinary differential for the n generalized coordinates $\underline{q}$ and n generalized velocities $\underline{\dot{q}}$. This contrasts with Maggi's equations, eqs. (11.11) and (11.7), that consist of $2n-m$ first-order, ordinary differential for the $n-m$ kinematic characteristics and n generalized coordinates. Although Lagrange's multipliers have been eliminated, an explicit expression for the forces of constraint is available as eq. (11.63b).

Example 11.9. The simple pendulum
Here again, the simple pendulum problem is considered as a first example. The constraint matrix of the problem is given by eq. (11.13), and hence $\left(\underline{\underline{B}}\,\underline{\underline{M}}^{-1}\underline{\underline{B}}^T \right)^{-1} =$ m, the mass of the particle. The acceleration level constraint, eq. (11.3), yields $c = -(\dot{x}^2 + \dot{y}^2)/\ell$. The system accelerations then follow from eq. (11.63a) as

$$m\underline{\ddot{q}} = mg\frac{y}{\ell}\frac{1}{\ell}\left\{\begin{matrix} y \\ -x \end{matrix}\right\} - m\frac{\dot{x}^2 + \dot{y}^2}{\ell}\frac{1}{\ell}\left\{\begin{matrix} x \\ y \end{matrix}\right\}.$$

The constraint forces are found with the help of eq. (11.63b)

$$\underline{F}^c = -mg\frac{x}{\ell}\frac{1}{\ell}\left\{\begin{matrix} x \\ y \end{matrix}\right\} - m\frac{\dot{x}^2 + \dot{y}^2}{\ell}\frac{1}{\ell}\left\{\begin{matrix} x \\ y \end{matrix}\right\}.$$

The geometric interpretation of this result, as discussed in section 11.3.4, is particularly striking for this simple pendulum example. For reference, the scaled projectors are easily found to be

$$\underline{\underline{\hat{P}}}_{\parallel} = \frac{1}{\ell^2}\begin{bmatrix} x^2 & xy \\ xy & y^2 \end{bmatrix}, \quad \text{and} \quad \underline{\underline{\hat{P}}}_{\perp} = \frac{1}{\ell^2}\begin{bmatrix} y^2 & -xy \\ -xy & x^2 \end{bmatrix}.$$

Example 11.10. The rigid body

Figure 11.3 depicts a rigid body moving in space while one of its points, denoted **O**, remains fixed. The orientation of the body is determined by an orthonormal basis $\mathcal{B} = (\bar{e}_1, \bar{e}_2, \bar{e}_3)$. The kinematics of the rigid body are defined by the nine generalized coordinates stored in array $\underline{q}^T = \{\bar{e}_1^T, \bar{e}_2^T, \bar{e}_3^T\}$. Clearly, six constraints must be imposed: three conditions on the normality of vectors $\bar{e}_1$, $\bar{e}_2$, and $\bar{e}_3$, and three additional constraints enforcing their orthogonality. The corresponding constraint matrix is given by eq. (11.17) and the projector defined by eq. (11.58) is found to be

$$
\underline{\hat{P}}_{\parallel} = \begin{bmatrix} \underline{I} - \frac{M_1 \bar{e}_2 \bar{e}_2^T}{M_1+M_2} - \frac{M_1 \bar{e}_3 \bar{e}_3^T}{M_1+M_3} & \frac{\sqrt{M_1 M_2}}{M_1+M_2} \bar{e}_2 \bar{e}_1^T & \frac{\sqrt{M_1 M_3}}{M_1+M_3} \bar{e}_3 \bar{e}_1^T \\ \frac{\sqrt{M_1 M_2}}{M_1+M_2} \bar{e}_1 \bar{e}_2^T & \underline{I} - \frac{M_2 \bar{e}_1 \bar{e}_1^T}{M_1+M_2} - \frac{M_2 \bar{e}_3 \bar{e}_3^T}{M_2+M_3} & \frac{\sqrt{M_2 M_3}}{M_2+M_3} \bar{e}_3 \bar{e}_2^T \\ \frac{\sqrt{M_1 M_3}}{M_1+M_3} \bar{e}_1 \bar{e}_3^T & \frac{\sqrt{M_2 M_3}}{M_2+M_3} \bar{e}_2 \bar{e}_3^T & \underline{I} - \frac{M_3 \bar{e}_1 \bar{e}_1^T}{M_3+M_1} - \frac{M_3 \bar{e}_2 \bar{e}_2^T}{M_3+M_2} \end{bmatrix},
$$

and the pseudo inverse of the image of the projector becomes

$$
\underline{\hat{E}}^{+T} = \begin{bmatrix} \sqrt{M_1}\bar{e}_1 & 0 & 0 & 0 & \frac{M_3\sqrt{M_1}}{M_1+M_3}\bar{e}_3 & \frac{M_2\sqrt{M_1}}{M_1+M_2}\bar{e}_2 \\ 0 & \sqrt{M_2}\bar{e}_2 & 0 & \frac{M_3\sqrt{M_2}}{M_2+M_3}\bar{e}_3 & 0 & \frac{M_1\sqrt{M_2}}{M_1+M_2}\bar{e}_1 \\ 0 & 0 & \sqrt{M_3}\bar{e}_3 & \frac{M_2\sqrt{M_3}}{M_2+M_3}\bar{e}_2 & \frac{M_1\sqrt{M_3}}{M_1+M_3}\bar{e}_1 & 0 \end{bmatrix},
$$

where, for simplicity, the mass matrix was assumed to be diagonal, $\underline{M}^* = \text{diag}(M_1 I, M_2 I, M_3 I)$. The scaled constraint forces and accelerations are then given by eqs. (11.61) and (11.62), respectively.

Example 11.11. The skateboard

Figure 9.6 depicts the simplified configuration of a skateboard of mass m and moment of inertia I about its center of mass **G**. The skateboard rolls without sliding on the horizontal plane by means of a wheel aligned with the axis $\bar{e}_1$ of the skateboard and located at point **C**, a distance ℓ from the center of mass. The position vector of the center of mass is written as $\underline{r}_G = x\,\bar{\imath}_1 + y\,\bar{\imath}_2$, and the axis of the skateboard makes an angle θ with the horizontal. Let the generalized coordinates of the problem be $\underline{q}^T = \{x, y, \theta\}$. Clearly, the system is subjected to a constraint: because the wheel does not slip, the velocity vector of the contact point must be along axis $\bar{e}_1$. The velocity of point **C** is $\underline{v}_C = \dot{x}\,\bar{\imath}_1 + \dot{y}\,\bar{\imath}_2 + \ell\dot{\theta}\,\bar{e}_2$, and hence, the constraint is $\bar{e}_2^T \underline{v}_C = 0$, leading to the constraint matrix given by eq. (11.20).

The procedure described in the previous section leads to the elimination of Lagrange's multiplier by constructing the projection operator. The accelerations of the system, given by eq. (11.63a), then become

$$
\underline{M}\,\ddot{\underline{q}} = \frac{mI}{I + m\ell^2}\dot{\theta}(\dot{x}\cos\theta + \dot{y}\sin\theta) \begin{Bmatrix} -\sin\theta \\ \cos\theta \\ \ell \end{Bmatrix},
$$

where the mass matrix is a diagonal matrix, $\underline{M} = \text{diag}(m, m, I)$. Since there are no externally applied forces, it follows that $\underline{F}^c = \underline{M}\,\ddot{\underline{q}}$.

11.4 Gauss' principle

Consider a dynamical system characterized by generalized coordinates denoted $\underline{q}$. By definition (8.1), the inertial forces acting on the system are $\underline{F}^I = \sum m_i \underline{a}_i$, where m_i are the masses of the particles and $\underline{a}_i$ their acceleration vectors. When the position vectors of all particles are expressed in terms of generalized coordinates, the inertial forces become $\underline{F}^I = -\underline{\underline{M}}\,\ddot{\underline{q}} - \underline{f}^I$, where $\underline{\underline{M}} = \underline{\underline{M}}(\underline{q}, t)$ is the symmetric, positive-definite mass matrix, and $\underline{f}^I = \underline{f}^I(\underline{q}, \dot{\underline{q}}, t)$ the dynamical forces. Note that inertial forces are linear functions of the generalized accelerations.

D'Alembert's principle, eq. (8.3), now implies that $\delta \underline{q}^T[-\underline{\underline{M}}\,\ddot{\underline{q}} - \underline{f}^I + \underline{f}^a] = 0$, for all kinematically admissible virtual changes in the generalized coordinates. In this statement of the principle, $\delta \underline{q}^T \underline{f}^a$ represents the virtual work done by all externally applied conservative and non-conservative forces. D'Alembert's principle is now recast in a compact manner as

$$\delta \underline{q}^T \left[\underline{\underline{M}}\,\ddot{\underline{q}} - \underline{F} \right] = 0, \tag{11.64}$$

where $\underline{F}(\underline{q}, \dot{\underline{q}}, t) = \underline{f}^a(\underline{q}, \dot{\underline{q}}, t) - \underline{f}^I(\underline{q}, \dot{\underline{q}}, t)$ is the sum of all dynamical and externally applied forces.

The generalized coordinates are functions of time and a Taylor series expansion yields $\underline{q}(t + dt) = \underline{q}(t) + \dot{\underline{q}}(t)dt + \ddot{\underline{q}}(t)dt^2/2 + \text{h.o.t.}$ The position and velocity vectors of all particles of the system are now assumed to be *given, fixed quantities* at time t, implying that $\delta \underline{q} = \underline{0}$ and $\delta \dot{\underline{q}} = \underline{0}$. Neglecting higher-order terms, variation of the series expansion now yields

$$\delta \underline{q}(t + dt) = \frac{1}{2}\delta \ddot{\underline{q}}(t)dt^2.$$

Introducing this result into d'Alembert's principle, eq. (11.64), leads to

$$\delta \ddot{\underline{q}}^T \left[\underline{\underline{M}}\,\ddot{\underline{q}} - \underline{F} \right] = 0, \tag{11.65}$$

for all kinematically admissible virtual changes in the generalized accelerations. Within the framework of the present development, variations of all quantities that are sole functions of the generalized coordinates and velocities vanish; for instance, $\delta \underline{\underline{M}}(\underline{q}, t) = 0$ or $\delta \underline{F}(\underline{q}, \dot{\underline{q}}, t) = 0$. This implies that $\delta \ddot{\underline{q}} = \delta[\underline{\underline{M}}^{-1}(\underline{\underline{M}}\,\ddot{\underline{q}} - \underline{F})]$ and eq. (11.65) becomes

$$\left[\underline{\underline{M}}\,\ddot{\underline{q}} - \underline{F} \right]^T \delta \left[\underline{\underline{M}}^{-1} \left(\underline{\underline{M}}\,\ddot{\underline{q}} - \underline{F} \right) \right] = 0,$$

and finally, $\delta G = 0$, where the *Gaussian* of the system is defined as

$$G = \frac{1}{2} \left[\underline{\underline{M}}\,\ddot{\underline{q}} - \underline{F} \right] \underline{\underline{M}}^{-1} \left[\underline{\underline{M}}\,\ddot{\underline{q}} - \underline{F} \right], \tag{11.66}$$

which is a quadratic function of the generalized accelerations.

Thus far, d'Alembert's principle has been used to prove that among all kinematically admissible generalized accelerations, the actual accelerations of the constrained system are a stationary point of its Gaussian, $\delta G = 0$.

If the system is constrained by a combination of holonomic and nonholonomic constraints, the accelerations level constraints are expressed by eqs. (11.3), which are linear functions of the generalized accelerations. Conceptually, generalized accelerations could be divided into two sets, the independent and dependent accelerations, denoted $\ddot{\underline{q}}^I$ and $\ddot{\underline{q}}^D$, respectively. The accelerations level constraints, eq. (11.3), would then yield a linear relationship between the two sets, $\ddot{\underline{q}}^D = \underline{\underline{T}}(\underline{q},t)\ddot{\underline{q}}^I + \underline{d}(\underline{q},\dot{\underline{q}},t)$. Introducing this expression into eq. (11.66), shows that the Gaussian now becomes a quadratic expression of the independent generalized accelerations, and hence, its stationary point corresponds to its absolute minimum.

This discussion establishes Gauss' principle [86, 87].

Principle 18 (Gauss' principle) *Among all kinematically admissible generalized accelerations, the actual acceleration of a constrained system minimizes its Gaussian.*

Because the Gaussian is a quadratic function of the generalized acceleration and because the acceleration level constraints are linear functions of the same variables, the stationary point of the Gaussian corresponds to it absolute minimum. Consequently, Gauss' principle is a true minimum condition, rather than the stationarity condition that characterizes d'Alembert's principle.

The use of Gauss' principle for the solution of constrained multibody systems was proposed by Lilov and Lorer [88] in 1982; their approach involves the Moore-Penrose inverse of the constraint matrix. The importance of Gauss' principle and its relationship to d'Alembert's principle was studied in a mathematical manner by Cardin and Zanzotto [89], within the framework of differential geometry. Possibly non-Riemannian mechanical systems with holonomic constraints were considered, generalizing Gauss' principle.

Example 11.12. The simple pendulum
Derive the equations of motion of the simple pendulum problem depicted in fig. 11.9 using Gauss' principle. Two generalized coordinates, the Cartesian coordinates, $\underline{q}^T = \{x, y\}$, defining the position vector of the bob are used, subjected to a single holonomic constraint: $\mathcal{C} = (\underline{q}^T\underline{q} - L^2)/2 = 0$. This constraint enforces the constant length condition for the pendulum; the constraint matrix of the problem is $\underline{\underline{B}}(\underline{q}) = [x, y]$.

The acceleration level constraint is $\ddot{\mathcal{C}} = \underline{\underline{B}}\,\ddot{\underline{q}} + \dot{x}^2 + \dot{y}^2 = 0$. Within the framework of Gauss' principle, variations of all quantities that are sole functions of the generalized coordinates and velocities vanish: $\delta\underline{\underline{B}} = 0$, $\delta\dot{x} = 0$, and $\delta\dot{y} = 0$. A variation of the acceleration level constraint then yields $x\delta\ddot{x} + y\delta\ddot{y} = 0$, which demonstrates that the variations in the generalized accelerations, $\delta\ddot{x}$ and $\delta\ddot{y}$, are not independent of each other. If $\ddot{y}$ is selected to be the independent acceleration component, $\delta\ddot{x} = -y\delta\ddot{y}/x$.

For this simple problem, the Gaussian is $G = m[(\ddot{x} - g)^2 + \ddot{y}^2]/2$, and the stationarity condition becomes $\delta G = m[(\ddot{x} - g)\delta\ddot{x} + \ddot{y}\delta\ddot{y}] = 0$. Because $\delta\ddot{x}$ and $\delta\ddot{y}$

are not independent quantities, the stationarity condition does not yield the equation of motion. Eliminating $\delta\ddot{x}$ leads to $\delta G = m[-y(\ddot{x} - g)/x + \ddot{y}]\delta\ddot{y} = 0$, and the equation of motion of the problem becomes

$$-y\ddot{x} + x\ddot{y} + gy = 0. \tag{11.67}$$

To verify that this result is correct, the position of the particle is written as $x = LC_\theta$, $y = LS_\theta$, where L is the length of the pendulum. Equation (11.67) can then be recast as $\ddot{\theta} + S_\theta g/L = 0$, as expected.

This example call for several remarks. First, the elimination process is fraught with difficulties. Indeed, equation $\delta\ddot{x} = -y\delta\ddot{y}/x$ becomes singular when $x = 0$. Selecting $\delta\ddot{x}$ to be the independent acceleration component would not circumvent the problem because equation $\delta\ddot{y} = -x\delta\ddot{x}/y$ would now becomes singular when $y = 0$.

Second, equation (11.67) is still expressed in terms of the two generalized coordinates selected to represent the configuration of the problem, x and y. Eliminating one of the two generalized coordinates leads to very complicated expressions for the single equation of motion and singularities will appear, whether x or y is selected as the independent variable.

Rather than eliminating one of the generalized coordinates, it is simple to append to equation (11.67) the acceleration level constraint, thereby creating a set of two ordinary differential equations for the two generalized coordinates. It is left to the reader to verify that this set of equations is identical to that generated by application of the null space formulation presented in section 11.2.5.

Example 11.13. Derivation of the index-1 formulation from Gauss' principle
Derive the index-1 formulation for constrained dynamical systems from Gauss' principle. As shown in the previous example, the elimination of the dependent accelerations is a perilous exercise. Furthermore, it is difficult, in general, to express dependent acceleration components in terms of their independent counterparts. To avoid this potentially difficult step, the constraints will be enforced using Lagrange's multiplier technique described in section 9.1.

The following augmented Gaussian, G^+, is introduced,

$$G^+ = G + \underline{\lambda}^T \left[\underline{B}(\underline{q}, t)\, \underline{\ddot{q}} - \underline{c}(\underline{q}, \underline{\dot{q}}, t)\right],$$

where the Gaussian of the system, G, is defined by eq. (11.66), and $\underline{\lambda}$ is the array of Lagrange multipliers used to enforce the constraints. The augmented Gaussian is now an unconstrained function of two sets of variables, the generalized accelerations, $\underline{\ddot{q}}$, and Lagrange's multipliers, $\underline{\lambda}$. Variation of the augmented Gaussian leads to $\delta G^+ = \delta\underline{\ddot{q}}^T \left[\underline{\underline{M}}\,\underline{\ddot{q}} - \underline{F} + \underline{\underline{B}}^T\underline{\lambda}\right] + \delta\underline{\lambda}^T \left[\underline{\underline{B}}\,\underline{\ddot{q}} - \underline{c}\right] = 0$, and because variations $\delta\underline{\ddot{q}}$ and $\delta\underline{\lambda}$ are arbitrary, the two bracketed terms must vanish, leading to the equations of motion of the problem. When recast in a matrix form, these two sets of equations are identical to those characterizing the index-1 formulation developed in section 11.2.3, see eq. (11.24).

Example 11.14. Independent quasi-accelerations

In general, it is difficult to express dependent velocity components in terms of their independent counterparts. To overcome this difficulty, kinematic parameters or independent quasi-velocities were introduced at the onset of the development of Maggi's formulation presented in section 11.2.1. Develop the corresponding concept of independent quasi-accelerations for Gauss' principle.

The relationship between generalized velocities and kinematic parameters expressed by eqs. (11.7) is at the heart of Maggi's formulation and introduces the null space, $\underline{\underline{\Gamma}}$. A time derivative of this expression yields eqs. (11.10), repeated here for convenience, $\ddot{\underline{q}} = \underline{\underline{\Gamma}}\,\dot{\underline{e}} + \dot{\underline{\underline{\Gamma}}}\,\underline{e} - \dot{\underline{d}}$.

Within the framework of Gauss' principle, variations of all quantities that are sole functions of the generalized coordinates and velocities vanish: $\delta\underline{\underline{\Gamma}}(q,t) = 0$, $\delta\underline{e} = 0$, $\delta\dot{\underline{\underline{\Gamma}}} = 0$, and $\delta\dot{\underline{d}} = 0$. Taking a variation of eqs. (11.10) then yields

$$\delta\ddot{\underline{q}} = \underline{\underline{\Gamma}}\,\delta\dot{\underline{e}}. \tag{11.68}$$

This important relationship expresses variations of the system's generalized accelerations in terms of variations of a set of independent quasi-accelerations, $\dot{\underline{e}}$, which are the time derivatives of the kinematic parameters introduced in Maggi's formulation. As mentioned earlier, the choice of the kinematic parameters is not unique and the selection of a specific set is left to the analyst. Similarly, the choice of specific quasi-accelerations is not unique

Example 11.15. Derivation Maggi's and null space formulations from Gauss' principle

Derive Maggi's and null space formulations from Gauss' principle. The condition of stationarity of the Gaussian implies $\delta\ddot{\underline{q}}^T \left[\underline{\underline{M}}\,\ddot{\underline{q}} - \underline{F}\right] = 0$. Because the generalized acceleration are not independent variables for a constrained dynamical system, this stationarity condition does not yield the equations of motion of the system.

To remedy this problem, the generalized accelerations are expressed in terms of independent quasi-accelerations using eq. (11.68), leading to $\delta\dot{\underline{e}}^T \underline{\underline{\Gamma}}^T \left[\underline{\underline{M}}\,\ddot{\underline{q}} - \underline{F}\right] = 0$. Because the quasi-accelerations are *independent variables*, the stationarity condition now yields the equations of motion of the problem as

$$\underline{\underline{\Gamma}}^T \left[\underline{\underline{M}}\,\ddot{\underline{q}} - \underline{F}\right] = 0. \tag{11.69}$$

Introducing the generalized accelerations from eqs. (11.10) into eqs. (11.69) then yields the governing equations, eqs. (11.11), of Maggi's formulation. On the other hand, appending the acceleration level constraints to eqs. (11.69) leads to the governing equations, eqs. (11.31), of the null space formulation.

The developments summarized in the last two examples show that the index-1, null space, and Maggi's formulations can all be derived from Gauss' principle. This should be expected because Gauss' principle is a fundamental principle of dynamics. It is indeed derived from d'Alembert's principle, which itself, was shown to be equivalent to Newton's second law.

11.5 Additional formulations

The theoretical developments presented in the above sections are well known, but additional formulations have also been presented in research papers, often of a more theoretical nature. In fact, many papers focus on explaining relationships, and often establishing equivalence, between various formulations rather than proposing practical numerical methods for the enforcement of constraints. For instance, Borri *et al.* [90] have shown the equivalence of Maggi's and Kane's equations [60, 61]. Angeles and Lee [91] independently derived Maggi's formulation for mechanical systems composed of rigid bodies coupled by holonomic constraints.

The formalism of Riemannian geometry was used by Maißer [92] to study holonomic multibody systems. This work focuses on rigid multibody systems with a tree topology and emphasizes the generation of the equations of motion within the Riemannian formalism with the help of Christoffel symbols. Coordinate partitioning was suggested as a solution method for the resulting equations. Jungnickel [93] further investigated the equations of motion for combined holonomic and general nonholonomic constraints, *i.e.*, constraints that might be nonlinear in the generalized velocities, within the framework of a Riemannian space endowed with a metric depending on the generalized mass matrix of the system and the constraints. The equations of motion were projected onto the tangent space resulting in index-1 DAEs from which Lagrange's multipliers were eliminated.

Essén [94] considered systems of particles and derived a minimal set of equations of motion for holonomic systems by projecting Newton's equations onto the space tangent to the constraint manifold. These projected Newton equations were then shown to be equivalent to Lagrange's equations. Generalizing to nonholonomic constraints in Pfaffian form, Essén obtained equations of motion in terms of quasi-velocities by projection of Newton's equations onto the null space of the constraint matrix. The resulting equations were shown to be general Boltzmann-Hamel equations. The relationship of this approach to Kane's method was also underlined.

Blajer [85, 95] summarized much of the work done within the framework of differential Riemannian geometry: index-1 formulations, null space formulations, and Maggi's formulation for combined holonomic and nonholonomic constraints have all been presented in this framework. For holonomic systems, the equivalence of Maggi's formulation and Boltzmann-Hamel equations was shown, as was the equivalence of the projective formulation and of the matrix setting of Gibbs-Appell equations. The author underlined the need to develop efficient methods for computing the time derivative of the null space, an indispensable ingredient for the application of Maggi's and projective formulations. He also proposed a technique for the elimination of constraint violations that affect the index-1, null space and Maggi's formulations. The Boltzmann-Hamel equations are immune from these violations because independent generalized coordinates are introduced.

12

Constrained systems: numerical methods

The classical and advanced formulations presented in chapter 10 and 11, respectively, provide the theoretical background for the analysis of constrained dynamical systems. In this chapter, practical numerical algorithm are described and compared.

Lagrange's equations of the first kind have been derived and form a set of index-3 differential-algebraic equations (DAEs). Gear [96] clearly underlined the difficulties associated with the solution of this type of equations. The same author and his coworkers [97, 47] have studied DAEs extensively and concluded in 1984: "If the index does not exceed 1, automatic codes [...] can solve the problem with no trouble." Furthermore, "If [...] the index is greater than one, the user should be encouraged to reduce it."

These observations prompted the multibody community to engage along two distinct avenues of research. First, the development of the ordinary differential equation techniques described in section 12.1, which eliminate Lagrange's multipliers all together, reducing the DAEs to ODEs. Methods developed for the solution of ordinary differential equations (ODEs) are then applicable to the reduced system of equations. Second, the index reduction techniques presented in section 12.2, which reduce the governing equations of motion to index-1 equations.

A survey paper by Haug [98] describes in a conceptual manner these two approaches to computational methods in constrained dynamics. Nikravesh [56] investigated two algorithms representative of those two approaches: the first algorithm reduces the problem to an index-1 system by enforcing the constraints at the acceleration level, the second used a coordinate partitioning method based on ref. [99].

Many of the methods proposed for the solution of constrained dynamical systems do not enforce constraints exactly, rather, small constraint violations are allowed that could grow over time. This phenomenon, called the drift phenomenon, was illustrated in examples 11.2, 11.5, and 11.6, when using Maggi's, the index-1, and the null space formulations, respectively. Figures 11.1, 11.5, and 11.7 show the time histories of the constraint violations and results are summarized in table 11.1. It was noted that the drift phenomenon is more pronounced for the index-1 and null space formulations than for Maggi's formulation because the two former approaches enforce the constraint at the acceleration level, but the latter enforces constraints at the velocity level.

Section 12.3 presents several constraint violation stabilization techniques that were developed to alleviate the drift phenomenon. Constraint violation elimination techniques that completely circumvent this problem are discussed in section 12.4.

Finally, in recent years, the finite element method has played an increasingly important role in multibody dynamics formulations and the tools and techniques used within this framework are the subject of section 12.5.

This chapter concludes with a detailed discussion of scaling methods presented in section 12.6. It is shown that with the proper scaling of the equations of motion, the index-3 DAEs stemming from the modeling of constrained dynamical systems are not more difficult to integrate than the ODEs characteristic of unconstrained systems.

12.1 Ordinary differential equation techniques

The challenges posed by the differential-algebraic nature of Lagrange's equations of the first kind can be dealt with by means of alternative formulations of the equations of motion. This section deals with methods that recast the governing equations of motion in terms of ODEs. A logical approach is to eliminate the redundant generalized coordinates to obtain a *minimum set of equations*, bypassing the need for constraints; this is the approach followed in Maggi's formulation presented in section 11.2.1. It is also possible, however, to obtain ODEs for all the generalized coordinates selected by the user to describe the system; this is the approach followed in the null space and Udwadia and Kalaba's formulations discussed in sections 11.2.5 and 11.2.7, respectively.

12.1.1 "Maggi-like" formulations

The essence of Maggi's formulation developed in section 11.2.1 is the construction of the null space, which enable the elimination of Lagrange's multipliers through the use of orthogonal complements, $\underline{\underline{B}}\,\underline{\underline{\Gamma}} = 0$. Because the constraint matrix is a function of time, the null space is itself a function of time, and in numerical implementations, it must be recomputed at each time step, a considerable computational burden. Hence, the vectors spanning the null space at two different steps could be different, resulting in a new set of kinematic characteristics at each time step.

To overcome these problems, many researchers have evaluated the null space at the beginning of the simulation, $\underline{\underline{\Gamma}}_0$, and kept it constant for the subsequent time steps of the analysis. When using this approach, Lagrange's multipliers are no longer eliminated, because $\underline{\underline{B}}(t)\underline{\underline{\Gamma}}_0 \neq 0$. At regular intervals, the null space is recomputed. Typically, a criterion is developed that identifies the appropriate time step when this expensive operation is to be performed; various criteria have been used by various researchers. It should be noted that these methods no longer represent a numerical implementation of Maggi's formulation; they might be better characterized as "Maggi-like" methods.

Kurdila *et al.* [54] and Papastavridis [55] first pointed out the unifying role of Maggi's formulation, which forms the basis for many of these coordinate reduction techniques that are equally applicable to holonomic and nonholonomic constraints. They point out that various methods only differ by the choice of the basis selected to span the null space of the constraint matrix, which in turns, determines the kinematic characteristics; the equations of motion are then projected onto this subspace. Clearly, matrix $\mathcal{B}$ and its inverse, as defined by eqs. (11.5) and (11.6), respectively, fully characterize Maggi's formulation.

A number of "Maggi-like" formulations only differ by the computational tool used to evaluate the null space of the constraint matrix. The following approaches have been used: the zero eigenvalue method [100], the coordinate partitioning method based on the LU factorization [99, 101], and the singular value decomposition method [102, 103]. Because these approaches were reviewed by Kurdila *et al.* [54], details are not repeated here. The following sections discuss methods that were developed after their review paper appeared.

The recursive Householder transformation method

Amirouche *et al.* [104] applied the Householder transformation technique to the transpose of the constraint matrix, assumed to be of full rank, to obtain a full rank, upper triangular matrix $\underline{\underline{B}}_{\mathrm{ut}(n\times m)} = \underline{\underline{H}}\,\underline{\underline{B}}^T$, where $\underline{\underline{H}}_{(n\times n)}$ is the product of successive Householder transformations. The Gram-Schmidt orthonormalization process was then employed to find an orthonormal basis, $\underline{\underline{D}}$, which was partitioned as $\underline{\underline{D}}_{(n\times n)} = \left[\underline{\underline{D}}_{1(n\times m)}\ \underline{\underline{D}}_{2(n\times(n-m))}\right]$. $\underline{\underline{D}}_1$ and $\underline{\underline{B}}_{\mathrm{ut}}$ span the same subspace, $\underline{\underline{B}}_{\mathrm{ut}}^T\underline{\underline{D}}_2 = \underline{\underline{B}}\,\underline{\underline{H}}^T\underline{\underline{D}}_2 = 0$, becuse $\underline{\underline{D}}_1^T\underline{\underline{D}}_2 = 0$. $\underline{\underline{D}}_2$ precisely spans the null space of $\underline{\underline{B}}\,\underline{\underline{H}}^T$, while $\underline{\underline{H}}^T\underline{\underline{D}}_2$ spans the null space of $\underline{\underline{B}}$. The fundamental matrices of Maggi's formulation are easily identified as $\underline{\mathcal{B}}^T = \left[\underline{\underline{B}}^T\ \underline{\underline{H}}^T\underline{\underline{D}}_2\right]$ and $\underline{\mathcal{B}}^{-1} = \left[\underline{\underline{H}}^T\underline{\underline{D}}_1(\underline{\underline{B}}\,\underline{\underline{H}}^T\underline{\underline{D}}_1)^{-1}\ \underline{\underline{H}}^T\underline{\underline{D}}_2\right]$. The authors pointed out that this approach is equivalent to the zero eigenvalue [100] and singular value decomposition methods [102, 103], while achieving higher computational efficiency.

The tangent coordinate method

Agrawal and Saigal [105] also used the Gram-Schmidt orthogonalization process to generate a basis of the null space of constraint matrix. For holonomic constraints, this null space is tangent to the constraint manifold, hence the name of the method. This approach is very similar to that presented by Liang and Lance [106], except that matrix $\underline{\underline{E}}$ is also constructed using the Gram-Schmidt process, a method that is faster and requires less computer memory. The process generates an orthogonal matrix $\underline{\underline{T}}$, which is partitioned as $\underline{\underline{T}}^T_{(n\times n)} = \left[\underline{\underline{T}}^T_{1(n\times m)}\ \underline{\underline{T}}^T_{2(n\times(n-m))}\right]$. $\underline{\underline{T}}^T_1$ and $\underline{\underline{B}}^T$ span the same subspace and $\underline{\underline{B}}\,\underline{\underline{T}}^T_2 = 0$, because $\underline{\underline{T}}_1\underline{\underline{T}}^T_2 = 0$, while $\underline{\underline{T}}^T_2$ precisely spans the tangent space. The fundamental matrices of Maggi's formulation are easily identified as $\underline{\underline{B}}^T = \begin{bmatrix} \underline{B}^T & \underline{\underline{T}}_2^T \end{bmatrix}$ and $\underline{\underline{B}}^{-1} = \begin{bmatrix} \underline{\underline{T}}_1^T (\underline{\underline{B}} \underline{\underline{T}}_1^T)^{-1} & \underline{\underline{T}}_2^T \end{bmatrix}$.

12.1.2 Maggi's formulations

Wampler *et al.* [107] devised a simple approach in which Maggi's kinematic characteristics are selected to be a subset of the generalized speed within the framework of Kane's method. They only presented analytical examples of their procedure. A few authors have developed approaches that update the null space at each time step. Kim and Vanderploeg [108] proposed an updating scheme which maintains the directional continuity of the null space. However, matrix Q of the underlying QR decomposition does not remain orthogonal, and hence, the full QR decomposition must be repeated at regular intervals, based on a criterion reflecting the condition number of a matrix involved in the null space update. This approach was reviewed by Kurdila *et al.* [54], details are not repeated here. The following sections discuss methods that were developed after their review paper appeared.

The Gram-Schmidt method

Liang and Lance [106] have used the Gram-Schmidt orthonormalization process to generate independent coordinates that are continuous and differentiable. At first, matrix $\underline{\underline{P}}_{(n \times n)} = \begin{bmatrix} \underline{\underline{B}}^T_{(n \times m)} & \underline{\underline{E}}^T_{(n \times (n-m))} \end{bmatrix}$ is constructed, where $\underline{\underline{E}}^T$ is an arbitrary matrix such that $\underline{\underline{P}}$ is nonsingular. Typically, $\underline{\underline{E}}$ is determined by singular value decomposition or by LU factorization. Matrix $\underline{\underline{P}}$ is then transformed into an orthogonal matrix $\underline{\underline{V}} = \begin{bmatrix} \underline{\underline{V}}_D & \underline{\underline{V}}_I \end{bmatrix}$ through the Gram-Schmidt process, where $\underline{\underline{V}}_D$ and $\underline{\underline{V}}_I$ are of the same dimensions as $\underline{\underline{B}}^T$ and $\underline{\underline{E}}^T$, respectively. $\underline{\underline{V}}_D$ and $\underline{\underline{B}}^T$ span the same subspace, hence, $\underline{\underline{B}} \underline{\underline{V}}_I = 0$ because $\underline{\underline{V}}_D^T \underline{\underline{V}}_I = 0$. $\underline{\underline{V}}_I$ precisely spans the null space of $\underline{\underline{B}}$. The fundamental matrices of Maggi's formulation are easily identified as $\underline{\underline{B}}^T = \begin{bmatrix} \underline{B}^T & \underline{\underline{V}}_I \end{bmatrix}$ and $\underline{\underline{B}}^{-1} = \begin{bmatrix} \underline{\underline{V}}_D (\underline{\underline{B}} \underline{\underline{V}}_D)^{-1} & \underline{\underline{V}}_I \end{bmatrix}$.

The extraction procedure approach

Constraints equations are intimately related to the choice of coordinates used to represent mechanical systems. García de Jalón *et al.* developed the concept of "basic coordinates" for systems composed rigid bodies; the approach was developed for the kinematic analysis of planar lower-pair mechanisms [109, 110] and later expanded to deal with spatial mechanisms [111, 112]. Serna *et al.* [113] used this framework to analyze the dynamic response of planar mechanisms. Maggi's and the null space formulations were both presented, together with an original approach to the determination of the null space.

The authors note that each column of the null space can be determined by means of the solution of an elementary velocity problem; this is a more physical approach that contrasts with the purely numerical procedures described in the previous sections. Similarly, the term $\dot{\underline{\Gamma}}\underline{e}$ that appears in Maggi's equations (11.11) can be evaluated as the solution of elementary acceleration problems. The same approach was used by García de Jalón *et al.* [114] who presented a formulation for both open- and closed-loop systems based on natural, or fully Cartesian coordinates. These coordinates have the advantages of leading to a constant mass matrix, and to relatively simple expression of the constraint matrix.

García de Jalón *et al.* [28] later showed how "natural coordinates" evolved from the earlier basic coordinates, and used this new concept to describe multibody systems. The null space is determined from eq. (11.7), written as $\dot{\underline{q}} = \underline{\Gamma}\,\underline{e}$, because all constraints are assumed to be holonomic and scleronomic. It is then possible to determine the null space corresponding to kinematic characteristics that are an "extraction" of components of the generalized velocity array. In view of eq. (11.4), this implies that each row of matrix $\underline{\underline{B}}$ has a single nonzero entry. A good choice of this extraction is initially determined by performing a Gaussian triangulation of the constraint matrix with full pivoting: the pivot locations indicate the generalized velocities to be extracted. This choice might become unsuitable during the simulation, when a previously selected pivot becomes very small; a new extraction is then selected. In a subsequent paper, García de Jalón *et al.* [62] also investigated the use of the singular value decomposition to identify the kinematic parameters. They concluded that while this approach might yield a set of kinematic parameters that are suitable over a longer period of the motion, it is also more expensive than the extraction approach.

Avello *et al.* [115] further elaborated on the extraction procedure by showing that it leads to a highly parallelizable algorithm. The columns of the null space, $\underline{\Gamma}$, are each computed in parallel as the solution of an elementary velocity problem, and furthermore, the triple product $\underline{\Gamma}^T \underline{\underline{M}} \, \underline{\Gamma}$ can also be evaluated in parallel. The terms of array $\dot{\underline{\Gamma}}\,\underline{e}$ appearing in Maggi's equations are also computed in parallel and correspond to solutions of elementary acceleration problems. For computational efficiency, the overall approach uses recursive techniques for open loop mechanisms; in the presence of closed loops, the augmented Lagrangian formulation is used, see section 12.3.2.

12.1.3 Discussion of the methods based on Maggi's formulation

While the null space of the constraint matrix is unique, individual vectors that span this subspace are not. The methods presented above all define the null space by different sets of vectors that are obtained by means of different computational processes. Two fundamental criteria can be used to assess the various approaches. First, is the subspace defined by a set of linearly independent vectors? Second, how robust and efficient is the numerical process used to generate the subspace? The first criterion is a necessary condition for the viability of the approach: if the vectors are not linearly independent, the null space is not properly defined. Kurdila *et al.* [54] pointed out that the approaches of Kane [60] and Wehage and Haug [99] are not robust because they sometimes lead to a poorly conditioned or even singular representations of the null space. To overcome this problem, most other approaches generate an orthogonal basis spanning the null space.

The second criterion deals with computational robustness and efficiency. Based on operation count, the computational cost of the singular value decomposition is known to be two to ten times higher than that of the QR algorithm, depending on the size of the constraint matrix. In turns, the QR algorithm is about two times more costly than the LU factorization. On the other hand, the singular value decomposition is many times more expensive than the Gram-Schmidt orthonormalization process. The singular value decomposition, however, is probably the most robust algorithm since it can be safely used even when the constraint matrix is not of full rank [102], as is the case in the presence of redundant constraints. Clearly, the singular value decomposition is the most robust and stable algorithm, but is also the most expensive.

An important feature of Maggi's formulation is that constraints are enforced at the velocity level. Hence, nonholonomic constraints will be satisfied to numerical accuracy, whereas holonomic constraints will drift due to the inherent errors associated with the integration process. This drift, however, is minimal, because the kinematic characteristics lie in the hyperplane tangent to the constraint manifold. In fact, Liang and Lance [106] mention that with their approach, "the numerical solution will be satisfactory without any positive constraint violation control or constraint violation stabilization." This is an important benefit of a rigorous application of Maggi's formulation. The situation, however, is different with the Maggi-like methods that do not update the null space, because the kinematic characteristics no longer exactly reside in the tangent hyperplane. To obtain accurate solutions, a Newton-Raphson iteration procedure that enforces the constraint is often added to the time integration process.

12.1.4 Null space formulations

This section discusses the approaches based on the null space formulation presented in section 11.2.5. Kamman and Huston [116, 117] developed an approach where the zero eigenvalue theorem was used to determine the null space of the constraint matrix. System dynamic response was then obtained based on the null space formulation. Borri *et al.* [64] pointed out that this approach is not much more computationally expensive than other null space methods because the most costly task is, by far, the determination of the null space.

Section 12.1.2 described the extraction procedure used by García de Jalón and his coworkers [62] to determine the null space of the constraint matrix. In these papers, the authors introduced the index-1, Maggi's and null space formulations for the modeling of rigid multibody systems within the framework of reference point and natural coordinates. Of particular interest is the second paper [63], which compares different approaches to the modeling of constrained mechanical systems. The salient conclusions of the work are as follows. First, the relative efficiency of all formulations depends on the number of generalized coordinates and degrees of freedom of the system. Second, the null space formulation tended to be more efficient than the index-1 approach. Finally, Maggi's formulation tended to outperform the null space formulation. Note that this study provides qualitative information for the specific framework described by the authors. For instance, it is unclear whether such conclusions would still hold when dealing with elastic multibody systems.

Chiou *et al.* [118] presented a numerical approach to the solution of the equations of motion expressed in terms of independent velocities. Based on a partitioning scheme that makes use of the velocity transformation relations, the null space of the constraint matrix was constructed. Maggi's equations and a system of ODEs were obtained for open- and closed-loop systems, respectively. The explicit-implicit staggered procedure devised by Park *et al.* [119] was employed to integrate the system of ODEs. A parallel implementation of the proposed approach was proposed but the authors underlined the need to increase the efficiency of the algorithm.

12.1.5 Udwadia and Kalaba's formulations

This section discusses the approaches based on Udwadia and Kalaba's formulation described in section 11.2.7. Arabyan and Wu [120] extended Udwadia and Kalaba's formulation, which was originally developed for systems of particles, to constrained rigid body problems. The main challenge to the use of this approach is that it calls for the computation of a generalized inverse at each time step, see eq. (11.33). The singular value decomposition is one tool to compute the generalized inverse, but it is very costly [82]. The authors proposed to use the Gram-Schmidt orthogonalization process [82] to this end; depending on the size of the constraint matrix, this process can be considerably cheaper than the singular value decomposition. Furthermore, the Gram-Schmidt algorithm is able to identify inconsistencies in the specification of constraints. These claims were substantiated by a number of examples comparing the performance of the index-1 approach to that based on the generalized inverse computed by both singular value decomposition and Gram-Schmidt algorithms.

12.1.6 The projective formulation

Blajer [83, 121] proposed a projection method for the analysis of constrained dynamical problems. Instead of introducing the concept of projectors, as discussed in section 11.3, Blajer uses differential Riemannian geometry: linear metric spaces in which vectors are resolved into their covariant and contravariant components, and the metric of the space is defined by the mass matrix. The effect of this metric is akin to the scaling of all quantities, as performed in section 11.3, and adds consistency to the formalism. The term "geometric projection" is used because the proposed method projects the index-1 equations onto the subspaces tangent and orthogonal to the admissible subspaces. Maggi equations (11.11) were then obtained by substitution of the independent variables into the equations projected on the tangent subspace. The independent variables form a set of independent quasi-velocities, in ref. [121], or independent quasi-accelerations, in ref. [83]. When applied to holonomic systems, the projective formulation is equivalent to Kane's form of Appell's equations [61]; for nonholonomic systems, it is equivalent to Maggi's formulation. Analytical examples were presented in these papers but numerical implementation and computational efficiency issues for complex multibody systems were not addressed. In a subsequent paper, Blajer et al. [84] used the projective formulation to devise a criterion for the optimal selection of independent coordinates, to be used in the coordinate partitioning method proposed by Wehage and Haug [99].

Blajer [122] also addressed the numerical implementation of the projective formulation. The Gram-Schmidt orthogonalization process was used to obtain a tangent subspace, as earlier suggested by other researchers [106, 105]. In this approach, however, orthogonality of the tangent and constraint subspaces is not achieved in a Cartesian space, as was the case for earlier methods, but rather in a space endowed with a metric defined by the mass matrix. The projective formulation requires the computation of the inverse of the mass matrix and of its time derivative, operations that are, in general, computationally expensive. Hence, Blajer recommends the use of this method in conjunction with absolute coordinates that lead to constant mass matrices; in such case, the inverse must be computed once only and its time derivative vanishes.

12.1.7 Modified phase space formulation

Borri et al. [123] derived governing DAEs that feature the following unknowns: the generalized coordinates, q, the modified momenta, p^*, which are related to the actual momenta, $p^* = p - \underline{B}^T \mu$, and the multipliers, μ, which are related to Lagrange's multipliers, $\mu = -\lambda$. Unlike the momenta, the modified momenta are unconstrained, i.e., the state vector (q, p^*) evolves in a modified, unconstrained phase space, and hence, significant reduction of the constraint violations can be expected. While the constraint forces, driven by Lagrange's multipliers, sometimes exhibit large amplitude oscillations, thus affecting the accuracy of the solution and imposing smaller time step sizes, the multipliers, μ, have a smoother behavior because they are integrals of Lagrange's multipliers, easing the integration process. The approach is robust in the presence of singular configurations. The DAEs are transformed into first-order ODEs in q and p^* for integration. Good numerical results were shown, particularly in terms of satisfaction of the constraint conditions. A penalty formulation of the approach, similar to that employed in Park and Chiou [124], was derived to render the method even more insensitive to singular configurations.

12.2 Index reduction techniques

Index reduction techniques are typically presented as mathematical processes that reduce the index of a set of DAEs. Numerical analysis techniques are then used to prove that the application of specific types of time integrators to the reduced order DAEs provides a reliable solution of the problem. Gear et al. [59] proposed a method, called the *stabilized index-2* or *GGL* method, that reduces the index from 3 to 2 and showed that variable-order, variable-step backward difference methods converge for the resulting index-2 problem. Later, Gear [125] developed an approach to further reduce the problem to index-1 DAEs. Of course, these approaches imply additional computational cost in the form of additional Lagrange multipliers to be solved for.

Gear [126] transformed first-order ODEs with equality and inequality invariants into index-2 DAEs, and provided, for equality invariants, a convergence analysis for variable-order, variable-step size, multi-step methods applied to the resulting DAEs. The proposed approach generalizes the one-step integrators proposed by Shampine [127], who minimally perturbed the solution of ODEs after each step to satisfy the invariants. This is equivalent to projecting the solution onto the invariant manifolds. Convergence was proved for one-step integration methods.

Lötstedt [58] studied rigid multibody systems subjected to unilateral holonomic and nonholonomic constraints. The equations of motion consisted of second-order ODEs and inequalities defining a linear complementarity problem. The occurrence of discontinuities in the displacement, velocity, and acceleration fields when constraint are activated or released was studied and bounds on the velocity vector were derived. Existence of solutions was discussed; particularly, displacements and constraint forces were proved to be unique in all configurations, whereas Lagrange multipliers are unique only when the constraint matrix has full rank. Because solutions of linear complementarity problems are also solutions of quadratic programming problems, Gauss' principle was shown to generalize to rigid body problems with unilateral constraints.

Lötstedt and Petzold [48] proved that k^{th}-order, constant step size, backward difference methods converge when applied to index-1, -2, or -3 DAEs; the numerical solution is accurate to order $\mathcal{O}(h^k)$, where h is the time step size. The same authors [49] further investigated the practical difficulties of implementing variables step size integration methods for the same types of problems. The difficulties associated with the solution of index-3 DAEs were underlined: the condition number of the Newton iteration matrix, *i.e.*, the tangent matrix used to solve the discretized nonlinear algebraic equations is $\mathcal{O}(h^{-3})$, resulting in increasingly ill conditioned problems for decreasing time step sizes. This conditioning problem can be completely eliminated for index-3 DAEs by using the scaling techniques presented in section 12.6.

Eich [128] provided a convergence analysis for a coordinate projection approach combined with backward difference methods to integrate index-1 DAEs. The approach projects the numerical solution of the underlying ODEs onto the position and velocity invariants to reduce constraint violations. The accuracy of the projected solution was shown to be identical to that of backward difference methods applied to the ODEs. For linear systems, it was shown that only the errors lying in the invariants were propagated, rendering the solution more accurate.

For holonomic systems, Yen *et al.* [129, 130] reduced the index-1 DAEs to ODEs by means of local parametrizations. The ODEs, which are similar to Boltzmann-Hamel equations, feature *local parameters* that implicitly define independent generalized coordinates and speeds. Then, using the local parametrization mapping and the constraint equations, the original generalized coordinates and velocities are recovered. A convergence analysis is presented that demonstrate an $\mathcal{O}(h^k)$ accuracy when k^{th}-order linear multistep methods are used. In numerical applications, local parameterizations were obtained using the generalized coordinate partitioning and tangent space methods. Local parameterizations using the tangent space determined by QR decomposition were also used by Potra and Yen [131]. A similar approach was used by Haug and Yen [132] who determined local parameters using the generalized coordinate partitioning technique. Based on backwards difference methods, the discretized Boltzmann-Hamel equations were shown to be equivalent to a set of discretized DAEs, involving the constraints at the displacement, velocity, and acceleration levels. For practical applications, these DAEs were used.

Yen *et al.* [133, 134] introduced the "coordinate-split formulation" for the solution of the index-2 DAEs characteristic of flexible multibody systems. A family of second-order α-methods with user controllable numerical dissipation was proposed, which extend the corresponding methods used for ODEs in structural dynamics, the HHT algorithm of Hilber *et al.* [135] and the generalized-α algorithm of Chung and Hulbert [136]. The coordinate-split method is a numerical implementation of the null space formulation applied to the stabilized index-2 approach of Gear *et al.* [59]; it eliminates the two sets of Lagrange multipliers associated with this approach. Projections in the space of the mass matrix were used to impose the constraints at both position and velocity levels. To deal with the highly oscillatory nature of the response of flexible multibody systems, the authors introduced a modification of the Newton iteration process, denoted "modified coordinate-split iteration." Improved convergence was proved mathematically and demonstrated by means of examples.

For holonomic systems, Tseng *et al.* [137] devised an algorithm, called "Maggi's equations with perturbation iteration," which further develops the modified coordinate-split iteration by perturbing the solution that is projected onto the constraint manifold to eliminate constraint violations. In this approach, the determination of the generalized accelerations, velocities, and displacements is separated from that of the Lagrange multipliers, which are recovered in a post-processing operation. Good numerical results were obtained, although the authors stressed the need for further validation of the approach, especially in the presence of flexible bodies. The authors considered the coordinate-split formulation to be a numerical implementation of Maggi's equations, but in the classification introduced herein, this approach belongs to null space formulations.

Another approach to index reduction is the embedded projection method developed by Borri *et al.* [138], which can be used to systematically reduce the index of the original DAEs system from 3 to 1. Furthermore, the method splits the original problem into its algebraic and differential parts, which can then be solved sequentially. While the accuracy and robustness of the procedure were demonstrated, its complexity is also apparent.

Parczewski and Blajer [139, 140] investigated systems subjected to *program constraints*, *i.e.*, systems forced to follow a prescribed path. The control forces that impose the prescribed motion might have components in directions both tangential and orthogonal to the constraint manifold. This feature of control problems contrasts with the classical theory of constrained dynamics, for which constraint forces are acting in the direction normal to the constraint manifold, see fig. 11.11. A classification of program constraint realizations was developed, which includes both orthogonal and tangent realizations, involving normal and tangent control forces, respectively. The authors provided several examples of program constraint realizations and determined the associated control forces; the difficulties inherent to non orthogonal realizations were underlined. Constraint forces that have components in directions both tangential and orthogonal to the constraint manifold are said to be "non-ideal" constraint forces. Udwadia and Kalaba [74] also studied such systems and gave explicit expressions for both ideal and non-ideal constraint forces in terms of Moore-Penrose generalized inverses.

12.3 Constraint violation stabilization techniques

A number of techniques impose the constraints at the acceleration level, as is the case for the index-1, null space, or Udwadia and Kalaba's formulations, see sections 11.2.3 and 11.2.5, respectively. Considering holonomic constraints, let $\underline{\mathcal{C}} = 0$, $\underline{\dot{\mathcal{C}}} = 0$, and $\underline{\ddot{\mathcal{C}}} = 0$ represent the displacement, velocity, and acceleration level constraints, respectively. The system consisting of the equations of motion and the acceleration level constraints then forms a set of index-1 DAEs *with invariants*. Indeed, for the exact solution, $\underline{\mathcal{C}} = 0$ and $\underline{\dot{\mathcal{C}}} = 0$ represent two invariants of the system.

Unfortunately, due to numerical approximations and round-off errors, numerical solutions will not evolve along the invariant manifolds, resulting in $\underline{\mathcal{C}} \neq 0$ and $\underline{\dot{\mathcal{C}}} \neq 0$. This phenomenon, called the drift phenomenon, was illustrated in examples 11.2, 11.5, and 11.6, when using Maggi's, the index-1, and the null space formulations, respectively. Figures 11.1, 11.5, and 11.7 show the time histories of the constraint violations and results are summarized in table 11.1. From a mathematical standpoint, equation $\underline{\ddot{\mathcal{C}}} = 0$ is not stable because its poles are located at the origin of the s-plane, where s is the variable of Laplace's transform; consequently, $\underline{\mathcal{C}}$ and $\underline{\dot{\mathcal{C}}}$ will not converge to zero if any deviation occurs. The constraint violation stabilization techniques presented in this section attempt to minimize or eliminate this drift; they are not, *per se*, solution methods for constrained dynamical problems, but rather, are used in conjunction with various solution techniques that are sensitive to the drift phenomenon.

If the mechanical system is conservative, the total mechanical energy, E, is an additional invariant of the system, $\dot{E} = 0$. If nonconservative forces are externally applied, the work they perform can be added to the total mechanical energy to form an invariant of the system. As was the case for the holonomic constraints considered above, due to numerical approximations and round-off errors, the solution will drift away from this manifold, *i.e.*, the total mechanical energy will not be preserved. In fact, the energy preservation constraint is a particular case of a nonholonomic constraint $\underline{\mathcal{D}}(\underline{q}, \underline{\dot{q}}) = 0$.

12.3.1 Control theory based stabilization techniques

The most popular stabilization technique is probably *Baumgarte's method*, which can be interpreted within the framework of control theory. Several researchers improved Baumgarte's original method and these efforts are described below.

Baumgarte's constraint violation stabilization

To compensate the observed drift of the solution, Baumgarte [141] introduced a stabilization method in which the original acceleration level constraint, $\ddot{\underline{C}} = 0$, is replaced by

$$\ddot{\underline{C}} + 2\alpha\dot{\underline{C}} + \beta^2\underline{C} = 0, \tag{12.1}$$

where α and β are user defined, positive parameters. In practical implementation, the choice $\alpha = \beta$ is often appropriate because critical damping is achieved. In the case of nonholonomic constraints, see eq. (11.2), the velocity level constraint, $\underline{D}(\dot{\underline{q}}, \underline{q}, t) = B(\underline{q}, t)\,\dot{\underline{q}} + \underline{b}(\underline{q}, t)$, is replaced by

$$\dot{\underline{D}} + \gamma\underline{D} = 0, \tag{12.2}$$

where γ is a user defined, positive parameter. The total energy constraint was treated in a similar manner. Baumgarte's stabilization method has been very widely used in multibody dynamics because it is easily implemented in conjunction with a variety of formulations of the equations of motion and time integration procedures. Ostermeyer [142] explained the effects of Baumgarte's stabilization method within the framework of control theory.

Unfortunately, parameters α and β are problem dependent, and no general procedure exists for their determination; hence, the approach tends to be unreliable and cannot be recommended for general purpose use in multibody dynamics because the constraints are never exactly satisfied. Eich and Hanke [143] mention that: "Choosing α and β too large results in stiff ODEs and a great amount of computing time." Nevertheless, some authors reported successful computations with Baumgarte's method. For instance, Nikravesh *et al.* [144] found an index-1 formulation in conjunction with Baumgarte's stabilization to be significantly more efficient computationally than the coordinate partitioning approach. They mention that: "Experience has shown that for most practical problems, positive values less than 5 for α and β are adequate. When $\alpha = \beta$, critical damping is achieved, which usually provides the fastest error reduction."

Park and Haug [145] have combined Baumgarte's stabilization method with the generalized coordinate partitioning method and shown that this hybrid approach outperforms both methods applied individually. They mention that: "Thus, the constraint stabilization method alone cannot handle every situation accurately and efficiently." Their rational for this conclusion is that the choice of α and β at each integration step is difficult and expensive, and erroneous solutions can appear when the constraint matrix is nearly singular.

Improvements of Baumgarte's stabilization method

Chang and Nikravesh [146] proposed an approach to adaptively determine the damping coefficient as the simulation proceeds. They assumed $\alpha = \beta$ and used adaptive control concepts to estimate optimal damping coefficients that are different for each

constraint. Numerical examples demonstrate that better control of constraint violations is achieved with the adaptive approach. Another improvement of Baumgarte's
method was proposed by Ostermeyer [142] who added to eq. (12.1) a term involving
the time integral of the constraint violation, based on optimum control theory.

Bae and Yang [147] also proposed an approach to the evaluation of the stabi-
lization parameters. First, they replaced eq. (12.1) by $\ddot{\underline{C}} + \alpha \dot{\underline{C}} + \alpha \underline{C} = 0$, where α
represents the magnitude of the penalty factor for both position and velocity con-
straint violations, arguing that both violations are equally undesirable. Larger values
of α will yield smaller constraint violations. If α is too large, however, the system becomes unstable. Hence, the value of α is limited by the stability characteristics of the
numerical procedure used to integrate the equations of motion; the Adams-Bashforth
integrator was used in their work. This condition yields a closed form expression for
α as a function of the time step size and stability boundaries of the integrator.

A similar study was undertaken by Yoon et al. [148] who showed that under
suitable assumptions, the constraint equation, written as $\ddot{\underline{C}} + \alpha \dot{\underline{C}} + \beta \underline{C} = \underline{d}$, where
$\underline{d}$ represent the disturbances due to truncation errors, is indirectly integrated with
the same numerical scheme as that used for the dynamics equations. This enables a
rigorous study of the accuracy and stability characteristics of Baumgarte's method
to be performed. In view of the complexity of the analysis, however, results were
only shown for one case, the simple pendulum. The authors also pointed out the
importance of stabilizing the energy preservation constraint.

Based on the input-output feedback linearization technique, Chiou and Wu [149]
transformed the nonlinear governing DAEs into a set of linear equations. Next, they
showed that a pole placement technique leads to Baumgarte's method and proposed
a new approach to stabilization based on the variable structure control technique.
While they demonstrated the superiority of their approach over Baumgarte's method
by means of examples, no guidelines were provided on how to select the constants
appearing in either approach.

Control theory concepts are also the basis for Lin and Hong's [150] stability analysis of Baumgarte's method using digital control theory. They notice that selecting
α and β to be positive numbers is not sufficient to guarantee convergence of $\underline{C}$ and $\dot{\underline{C}}$
to zero as implied by stability analysis applied to eq. (12.1). Hence, they performed
a stability analysis of the *discretized equations* using the Z-transform concept. They
defined two parameters, $\hat{\alpha} = \alpha/h$ and $\hat{\beta} = \beta/h^2$, and concluded that while $\hat{\alpha}$ and
$\hat{\beta}$ are independent of the problem and time step size, they do depend on the time
integration scheme used for the simulation. Desirable values of $\hat{\alpha}$ and $\hat{\beta}$ were given
for the Adams-Bashforth and Adams-Moulton predictor-corrector integrators.

12.3.2 Penalty based stabilization techniques

In penalty formulations, constraints are enforced by means of a *penalty term* added
to the Lagrangian of the system,

$$\frac{1}{2} \underline{C}^T \underline{\underline{P}} \, \underline{C}, \tag{12.3}$$

where $\underline{\underline{P}} = \mathrm{diag}(p_i^2)$, and p_i are the *penalty factors*. It is common practice to use the same penalty factor, $p = p_i$, for all constraints and hence, the penalty term is often written as $1/2\, p^2\, \underline{\underline{C}}^T \underline{\underline{C}}$. The idea behind this formulation is to choose large penalty factors so as to drive the constraints to zero, *i.e.*, $p \to \infty$ and $\underline{C} \to 0$. Taking a variation of the penalty term yields $\delta \underline{q}^T \underline{\underline{B}}^T p^2 \underline{C}$, revealing the equivalent externally applied generalized force, $\underline{\underline{B}}^T(p^2\underline{C})$, which at the limit, become $\underline{\underline{B}}^T \underline{\lambda}$, where $\underline{\lambda}$ are Lagrange's multipliers. Of course, in practical applications, a finite value of the penalty factor must be selected to avoid numerical ill conditioning and hence, the constraints are never exactly enforced and the quantities $\underline{\lambda} = p^2\underline{C}$ approximate Lagrange's multipliers.

The staggered stabilization technique

Park and Chiou [124] presented a stabilization technique based on a penalty formulation. The Lagrange multipliers associated with holonomic constraints were written as $\underline{\lambda} = \underline{C}(\underline{q},t)/\epsilon$, where $\epsilon = 1/p^2$ is the penalty factor; time differentiation of this expression then leads to

$$\underline{\dot{\lambda}} = \frac{1}{\epsilon}\,(\underline{\underline{B}}\,\underline{\dot{q}} + \frac{\partial \underline{C}}{\partial t}). \tag{12.4}$$

Taken together with eq. (11.1), these equations form a set of coupled ODEs. For nonholonomic constraints, a similar procedure can be followed by selecting Lagrange's multipliers as $\underline{\dot{\lambda}} = 1/\epsilon\,(\underline{\underline{B}}\,\underline{\ddot{q}} + \partial\underline{D}/\partial t)$. Introducing governing eq. (11.1) leads to $\epsilon\underline{\dot{\lambda}} + (\underline{\underline{B}}\,\underline{\underline{M}}^{-1}\underline{\underline{B}}^T)\underline{\lambda} = \underline{\underline{B}}\,\underline{\underline{M}}^{-1}\underline{F} + \partial\underline{D}/\partial t$. If Lagrange's multipliers are written as $\underline{\lambda} = \underline{\bar{\lambda}}\,\exp(\sigma t)$, the homogeneous part of this equations becomes $(\sigma + \underline{\underline{B}}\,\underline{\underline{M}}^{-1}\underline{\underline{B}}^T/\epsilon)\underline{\bar{\lambda}} = 0$. This implies that the constraint decay rates, σ_i, are the eigenvalues matrix $\underline{\underline{B}}\,\underline{\underline{M}}^{-1}\underline{\underline{B}}^T/\epsilon$; in other words, the decay rates are a function of the physical characteristics of the system, in contrast with Baumgarte's method that depends on abstract coefficients unrelated to system properties.

A single derivative of the constraints was taken, and hence, this approach will be less sensitive to the drift phenomenon than methods requiring two time derivatives. Furthermore, it depends on a single coefficient, the penalty factor. Examples treated by Park and Chiou [124] showed improved accuracy for displacement level constraint invariants as compared to the results of Baumgarte's method. This stabilized technique is robust as it can accommodate nearly rank deficient constraint matrices, while Baumgarte's technique cannot.

Park *et al.* [119] presented an *explicit-implicit, staggered* procedure to implement the stabilization procedure described in the previous paragraph. The approach calls for the developments of two distinct modules, one integrates the generalized coordinates knowing the constraint forces, the other integrates the Lagrange multipliers knowing the generalized coordinates. Calls to the two modules alternate, hence, the approach is called a "staggered procedure." Several application examples were given, demonstrating the accuracy and effectiveness of the procedure, which is robust but conditionally stable.

Augmented Lagrangian formulation

The augmented Lagrangian formulation developed by Bayo *et al.* [151] starts as a penalty formulation of the problem. Corresponding to the kth holonomic constraint, the following terms are added to the Lagrangian of the system: a penalty term, $1/2\ \alpha_k \omega_k \mathcal{C}_k^2$, a Rayleigh dissipative forces terms, $-2\ \alpha_k \omega_k \mu_k \dot{\mathcal{C}}_k$, and a fictitious kinetic energy term, $1/2\ \alpha_k \dot{\mathcal{C}}_k^2$. The governing equations of the system now become index-1 equations

$$\underline{\underline{M}}\,\ddot{\underline{q}} + \underline{\underline{B}}^T \underline{\underline{\alpha}}(\ddot{\underline{\mathcal{C}}} + 2\underline{\underline{\Omega}}\,\underline{\underline{\mu}}\,\dot{\underline{\mathcal{C}}} + \underline{\underline{\Omega}}^2 \underline{\mathcal{C}}) = \underline{F}, \tag{12.5}$$

where $\underline{\underline{\alpha}} = \text{diag}(\alpha_k)$, $\underline{\underline{\Omega}} = \text{diag}(\omega_k)$ and $\underline{\underline{\mu}} = \text{diag}(\mu_k)$. This penalty formulation will only yield accurate predictions for large penalty factors, $\alpha_k \to \infty$; the coefficients ω_k and μ_k play a stabilizing role similar to that of the corresponding coefficients of Baumgarte's method.

In the *augmented Lagrangian formulation*, a set of Lagrange multipliers is introduced together with the penalty terms, leading to

$$\underline{\underline{M}}\,\ddot{\underline{q}} + \underline{\underline{B}}^T \underline{\underline{\alpha}}(\ddot{\underline{\mathcal{C}}} + 2\underline{\underline{\Omega}}\,\underline{\underline{\mu}}\,\dot{\underline{\mathcal{C}}} + \underline{\underline{\Omega}}^2 \underline{\mathcal{C}}) = \underline{F} - \underline{\underline{B}}^T \underline{\lambda}^*. \tag{12.6}$$

Had the sole Lagrange multipliers been introduced, the governing equations would have been $\underline{\underline{M}}\,\ddot{\underline{q}} = \underline{F} - \underline{\underline{B}}^T \underline{\lambda}$, and hence, $\underline{\lambda} = \underline{\lambda}^* + \underline{\underline{\alpha}}(\ddot{\underline{\mathcal{C}}} + 2\underline{\underline{\Omega}}\,\underline{\underline{\mu}}\,\dot{\underline{\mathcal{C}}} + \underline{\underline{\Omega}}^2 \underline{\mathcal{C}})$. Because the Lagrange multipliers are sufficient, *per se*, to impose the constraints, the penalty coefficient is no longer required to be large; the formulation, however, now involves m additional unknowns. In the proposed approach, the Lagrange multipliers, $\underline{\lambda}^*$, are not treated as unknowns; rather, they are computed through an iterative process

$$\underline{\lambda}^{*(i+1)} = \underline{\lambda}^{*(i)} + \underline{\underline{\alpha}}(\ddot{\underline{\mathcal{C}}} + 2\underline{\underline{\Omega}}\,\underline{\underline{\mu}}\,\dot{\underline{\mathcal{C}}} + \underline{\underline{\Omega}}^2 \underline{\mathcal{C}})^{(i+1)}, \tag{12.7}$$

where the superscript $(\cdot)^{(i)}$ indicates the iteration number and $\underline{\lambda}^{*(0)} = 0$. Combining this iterative scheme with eq. (12.6) leads to

$$(\underline{\underline{M}} + \underline{\underline{B}}^T \underline{\underline{\alpha}}\,\underline{\underline{B}})\ddot{\underline{q}}^{(i+1)} = \underline{\underline{M}}\,\ddot{\underline{q}}^{(i)} - \underline{\underline{B}}^T \underline{\underline{\alpha}}(\dot{\underline{\underline{B}}}\,\dot{\underline{q}} + 2\underline{\underline{\Omega}}\,\underline{\underline{\mu}}\,\dot{\underline{\mathcal{C}}} + \underline{\underline{\Omega}}^2 \underline{\mathcal{C}}), \tag{12.8}$$

where $\underline{\underline{M}}\,\ddot{\underline{q}}^{(0)} = \underline{F}$.

The augmented Lagrangian formulation reduces the problem to a set of ODEs with no additional unknowns. The iterative solution of the Lagrange multipliers is inexpensive since iterations are already required for the solution of the nonlinear equations of motion. Numerical experimentation shows that accurate solutions can be obtained for a wide range of penalty factors, $\alpha_k \in [10^3, 10^9]$. The formulation can be generalized to accommodate nonholonomic constraints.

Bayo *et al.* [152] further elaborated the augmented Lagrangian formulation. The penalty term was simplified to keep two terms only, resulting in $\underline{\lambda} = \underline{\lambda}^* + \underline{\underline{\alpha}}(\underline{\mathcal{C}} + \underline{\underline{\mu}}\dot{\underline{\mathcal{C}}})$. They observed that the velocity level constraint in the penalty factor was necessary to prevent the appearance of high frequency numerical oscillations during the simulation. To integrate the equations of motion, the trapezoidal rule was used with accelerations or displacements as primary variables: the latter were shown to provide superior performance.

Using penalized potential and kinetic energies, and Raleigh dissipation functions, Kurdila *et al.* [153] formulated penalized equations of motion for holonomic systems. For conservative systems, convergence to the original, non penalized equations was shown, stability analysis was performed, and sufficient conditions for Lyapunov and asymptotic stability were given. Good numerical results were shown for singular configurations, due to the dissipative function. Systems with relatively large number of degrees of freedom and constraints could also be accurately simulated. The equations were developed for holonomic systems, and are thus somewhat limited in their applications.

12.4 Constraint violation elimination techniques

In contrast to the constraint violation stabilization techniques presented in the previous section, constraint violation elimination techniques are method which result in the exact satisfaction of the constraint, or at least to satisfaction of the constraints within machine accuracy.

12.4.1 Geometric projection approach to stabilization

Yoon *et al.* [154] developed an approach to constraint violation stabilization. Let $\bar{\underline{q}}_n$ and $\bar{\underline{v}}_n$ be the generalized coordinates and velocities, respectively, predicted by the integration of the equations of motion at the end of time step n. Due to numerical approximations, both holonomic and nonholonomic constraints will not be exactly satisfied, *i.e.*, $\underline{\mathcal{C}}(\bar{\underline{q}}_n, t_n) \neq 0$ and $\underline{\mathcal{D}}(\bar{\underline{q}}_n, \bar{\underline{v}}_n, t_n) \neq 0$, respectively. The approach consists in correcting or perturbing the generalized coordinates and velocities, $\underline{q}_n = \bar{\underline{q}}_n + \hat{\underline{q}}_n$, and $\underline{v}_n = \bar{\underline{v}}_n + \hat{\underline{v}}_n$, respectively, where $\hat{\underline{q}}_n$ and $\hat{\underline{v}}_n$ are the unknown coordinate and velocity corrections, respectively, both assumed to be small. The updated coordinates, $\underline{q}_n$, and velocities, $\underline{v}_n$, satisfy the holonomic and nonholonomic constraints, *i.e.*, $\underline{\mathcal{C}}(\underline{q}_n, t_n) = 0$ and $\underline{\mathcal{D}}(\underline{q}_n, \underline{v}_n, t_n) = 0$, respectively.

At first, the generalized coordinate corrections are evaluated by linearizing the holonomic constraints to find $\underline{\underline{B}}(\bar{\underline{q}}_n, t_n)\, \hat{\underline{q}}_n \approx -\underline{\mathcal{C}}(\bar{\underline{q}}_n, t_n)$. Since these equations are overdetermined, it is assumed that the corrections lie in the subspace defined by the constraint matrix, *i.e.*, $\hat{\underline{q}}_n = \underline{\underline{B}}^T \underline{\epsilon}_n$, where $\underline{\epsilon}_n$ is an unknown array. It then follows that $\underline{\epsilon}_n = -(\underline{\underline{B}}\,\underline{\underline{B}}^T)^{-1}\underline{\mathcal{C}}(\underline{q}_n, t_n)$, and finally

$$\hat{\underline{q}}_n = -\underline{\underline{B}}^T (\underline{\underline{B}}\,\underline{\underline{B}}^T)^{-1}\underline{\mathcal{C}}(\bar{\underline{q}}_n, t_n). \tag{12.9}$$

Next, the generalized velocity corrections are evaluated by linearizing the nonholonomic constraints to find $\underline{B}(\underline{q}_n, \bar{\underline{v}}_n, t_n)\, \hat{\underline{v}}_n \approx -\underline{\mathcal{D}}(\underline{q}_n, \bar{\underline{v}}_n, t_n)$. In this second phase, the generalized coordinates are kept constant, since they were corrected in the first phase of the procedure. Here again, these equations are overdetermined and it is assumed that $\hat{\underline{v}}_n = \underline{\underline{B}}^T \underline{\gamma}_n$. It then follows that $\underline{\gamma}_n = -(\underline{\underline{B}}\,\underline{\underline{B}}^T)^{-1}\underline{\mathcal{D}}(\underline{q}_n, \bar{\underline{v}}_n, t_n)$, and finally

$$\hat{\underline{v}}_n = -\underline{\underline{B}}^T(\underline{\underline{B}}\,\underline{\underline{B}}^T)^{-1}\underline{\mathcal{D}}(\underline{q}_n, \bar{\underline{v}}_n, t_n). \tag{12.10}$$

The procedure alleviates constraint violations without modifying the equations of motion, in contrast with Baumgarte's method. The approach is geometric in nature: $\underline{\underline{\Gamma}}^T(\underline{q}_n - \bar{\underline{q}}_n) = \underline{\underline{\Gamma}}^T \underline{\underline{B}}^T \underline{\epsilon}_n = 0$, the corrected solution is a projection of the approximate solution onto the constraint manifold. Clearly, the geometric procedure alleviates the constraint violations without eliminating them; complete elimination would require an iterative solution of the constraint equations. Yoon *et al.* [154] demonstrated the effectiveness of the procedure for holonomic, nonholonomic and energy constraints.

Blajer [85, 155] developed a similar approach to constraint elimination based on the geometric interpretation of constrained dynamics he presented with his coworkers in ref. [84]. Based on geometric arguments, the following correction schemes are found for the generalized coordinates

$$\hat{\underline{q}}_n = -\underline{\underline{M}}^{-1}\underline{\underline{B}}^T(\underline{\underline{B}}\,\underline{\underline{M}}^{-1}\underline{\underline{B}}^T)^{-1}\underline{\mathcal{C}}(\bar{\underline{q}}_n, t_n), \tag{12.11}$$

and velocities,

$$\hat{\underline{v}}_n = -\underline{\underline{M}}^{-1}\underline{\underline{B}}^T(\underline{\underline{B}}\,\underline{\underline{M}}^{-1}\underline{\underline{B}}^T)^{-1}\underline{\mathcal{D}}(\underline{q}_n, \bar{\underline{v}}_n, t_n), \tag{12.12}$$

respectively. Blajer's corrections, eqs. (12.11) and (12.12), are more physically consistent than Yoon's, eqs. (12.9) and (12.10), respectively, because matrix $(\underline{\underline{B}}\,\underline{\underline{M}}^{-1}\underline{\underline{B}}^T)$ involves terms that are of consistent units, in contrast with matrix $(\underline{\underline{B}}\,\underline{\underline{B}}^T)$ that does not. Indeed, when generalized coordinates have different units, such as displacements and rotations, matrix $(\underline{\underline{B}}\,\underline{\underline{B}}^T)$ weighs all components equally; in contrast, matrix $(\underline{\underline{B}}\,\underline{\underline{M}}^{-1}\underline{\underline{B}}^T)$ weighs each component by an appropriate inertial term. In numerical applications, the position corrections, eq. (12.11), are used first in an iterative manner until constraint violations are completely eliminated, *i.e.*, until the constraint equations are satisfied to machine accuracy. If nonholonomic constraints are present, the same process is applied to correct the velocities.

Baumgarte [156] developed a new stabilization method, which is derived from a modified statement of Hamilton's principle. The resulting equations of motion feature non classical Lagrangian multipliers and the holonomic constraints need to be differentiated only once with respect to time. Unfortunately, no applications were presented, making the assessment of the approach rather difficult.

Terze *et al.* [157] formulated a constraint elimination method within the framework of the null space approach. Using the projective criterion defined by Blajer *et al.* [84], they identified a set of independent variables. Displacement constraint violations were then iteratively eliminated by adjusting the sole dependent variables to satisfy the displacement level constraint equations. In a second step, the velocity constraint violations were eliminated using the velocity level constraint equations. During both correction steps, the independent displacements and velocities were kept unchanged.

12.4.2 The mass-orthogonal projection formulation

Bayo and Avello [158] proposed an augmented Lagrangian formulation based on the canonical equations of Hamilton. Compared to the index-1 based formulation derived earlier by Bayo *et al.* [151], the new formulation exhibits better accuracy and robustness in the presence of singular configurations. The improved performance was credited to the fact that a single differentiation of the holonomic constraints is required in canonical formulations, rather than the double differentiation associated with index-1 formulations. Effectiveness of the new approach was illustrated by numerical examples. A similar formulation was derived for nonholonomic constraints although no example was given.

While the augmented Lagrangian formulation presented in section 12.3.2 satisfies the weighted constraint, $\underline{\ddot{C}} + 2\underline{\Omega}\,\mu\,\underline{\dot{C}} + \underline{\Omega}^2\underline{C}$, see eq. (12.5), to machine accuracy, individual constraints at the position, $\underline{C} = 0$, velocity, $\underline{\dot{C}} = 0$, and acceleration levels, $\underline{\ddot{C}} = 0$, are not necessarily satisfied to the same level of accuracy. To improve this situation, Bayo and Ledesma [159] combined the augmented Lagrangian formulation with a mass orthogonal projection technique. To impose the position level constraint, they propose to minimize $V = 1/2\,(\underline{q} - \underline{q}^*)\underline{M}(\underline{q} - \underline{q}^*)$ subject to the constraint $\underline{C}(\underline{q}, t) = 0$, where $\underline{q}^*$ is the solution obtained at the end of a time step using the augmented Lagrangian formulation. This minimization problem is itself solved using an augmented Lagrangian formulation, transforming V into

$$V^* = \frac{1}{2}\,(\underline{q} - \underline{q}^*)^T \underline{M}(\underline{q} - \underline{q}^*) + \frac{1}{2}\,\underline{C}^T\underline{\alpha}\,\underline{C} + \underline{C}^T\underline{\lambda}. \tag{12.13}$$

Simple algebraic manipulations lead to the following iterative scheme to impose the position constraint

$$(\underline{M} + \underline{B}^T\underline{\alpha}\,\underline{B})\underline{\Delta}^{(i+1)} = -\underline{M}(\underline{q}^{(i)} - \underline{q}^*) - \underline{B}^T\underline{\lambda}^{(i)} \tag{12.14}$$

where $\underline{\Delta}^{(i+1)} = \underline{q}^{(i+1)} - \underline{q}^{(i)}$ and $\underline{\lambda}^{(i+1)} = \underline{\lambda}^{(i)} + \underline{\alpha}\,\underline{C}^{(i+1)}$. From a computational view point, this iterative procedure is not expensive because the system matrix, $(\underline{M} + \underline{B}^T\underline{\alpha}\,\underline{B})$, is identical to that of eq. (12.8). Hence, this matrix is factorized once only and the additional computational cost consists of the evaluation of the right-hand side of eq. (12.14) followed by forward reductions and backward substitutions. This contrasts with the approaches presented in section 12.4.1 that typically involve more computational effort. The constraints at the velocity and acceleration levels can be treated in a similar manner and are formulated in such a way that the resulting system matrix is identical to that of eq. (12.14), minimizing computational cost.

Bayo and Ledesma [159] illustrated their approach with several numerical examples. Application of the mass-orthogonal projection at each time step eliminates constraint violations to machine accuracy and dramatically increases the accuracy of the simulation. A mechanism presenting singular configurations was successfully simulated to demonstrate the robustness of the augmented Lagrangian formulation. The trapezoidal rule was used to integrate the equations of motion with accelerations or displacements as primary variables: the latter were shown to provide superior performance.

Comparative studies

For holonomic systems, Schiehlen [160] derived governing DAEs and ODEs, the Boltzmann-Hamel equations, as well as equations of motion based on a recursive approach. The recursive approach, suitable for chain topologies, can be much more competitive than the ODE formulation, although for small numbers of degrees of freedom, the latter is still competitive as the former is rather complex. In general, recursive approaches require $\mathcal{O}(n)$ operations, in contrast with ODE formulations that may need up to $\mathcal{O}(n^3)$ arithmetic operations. A comparative study showed that the ODE formulations are more efficient than their DAEs counterpart, although this conclusion was based on very simple, rigid multibody systems examples involving very few degrees of freedom.

Cuadrado *et al.* [161] compared four methods that are used to simulate multi-body dynamics with constraints. These methods are: the augmented Lagrangian formulation index-1 and index-3 with projections, a modified state-space formulation (equations of motion in independent coordinates) and a fully recursive formulation. Modifications were performed to the classical state-space formulation to improve its performance in the presence of stiff systems. The augmented Lagrangian index-1 and index-3 formulations used natural or fully Cartesian coordinates, as described in ref. [41]. These coordinates have the advantage of leading to a constant mass matrix.

A number of rigid multibody problems were solved with all four methods to compare their performance; none was found to be fully satisfactory. The index-3 formulation with projections failed to converge when using time step sizes smaller than 10^{-5} sec, while for time step sizes larger than 10^{-2} sec, the index-1 formulation failed to converge. The space-state and the fully recursive formulation lacked robustness as they failed to handle singular configurations. In addition, the fully recursive formulation behaved poorly in the presence of stiff systems or systems presenting redundant constraints. Nevertheless, for non-stiff problems of large size, this method became competitive. Of all the methods tested, the index-3 formulation with projections was the most efficient, while the index-1 formulation with projections was the most robust. The authors suggested that a combined index-1 and index-3 formulation would constitute a very good tool for solving multibody dynamics with constraints. Further evaluation of the methods was recommended, however, especially for large scale industrial problems. It is not clear how the various methods presented in this study would perform for elastic multibody systems.

12.5 Finite element based techniques

Multibody dynamics analysis was originally developed as a tool for modeling mechanisms with simple tree-like topologies composed of rigid bodies, but has considerably evolved to the point where it can handle nonlinear flexible systems with arbitrary topologies. The modeling of the elastic bodies is one of the most difficult aspects of multibody systems dynamics, and many different formulations have been presented in the literature. Comprehensive reviews of the state of the art in the field are given by Shabana [162], or Wasfy and Noor [163].

Traditionally, elasticity in multibody systems has been taken into account using the *floating frame of reference approach* [162], which is discussed in section 12.5.1. The displacement field of a flexible body is then decomposed into two additive parts, a rigid body and an elastic displacement field. The rigid body displacement field is represented by the arbitrarily large motion of a suitably selected frame of reference, which could be rigidly connected to a material point of the flexible body, or could be in motion with respect to the flexible body, hence the name "floating frame of reference." On the contrary, the elastic displacement field, resolved in the floating frame of reference, is assumed to remain small, and hence, is adequately represented using modal expansion techniques. If this assumption is satisfied, the elastic behavior of flexible bodies can be accurately captured using a small number of modal degrees of freedom. Component mode synthesis techniques, initially developed for finite element analysis, are now routinely used in flexible multibody dynamics and section 12.5.2 summarizes several commonly used approaches.

In *finite element based multibody dynamics approaches*, a given mechanism is modeled by an idealization process that represents each component of the flexible mechanism by an "element" chosen from an extensive library of elements implemented in the code. In fact, this approach is at the heart of the finite element method, which has enjoyed, for this very reason, an explosive growth in the last few decades. Each element provides a basic functional building block, for example a rigid or flexible member, a hinge, a motor, etc. Assembling the various elements, the construction of a mathematical description of the mechanism with the required level of accuracy becomes possible. In addition to the classical beam, plate, shell, and solid brick elements found in all finite element codes, kinematic constraints are also formulated as "finite elements," such as revolute joint or universal joint elements, to name but a few. A detailed description of the formulation is given by Géradin and Cardona [164].

12.5.1 Floating frame of reference approach

One of the most common approaches to the modeling of flexible multibody systems is based on the concept of floating frames [162]. The total motion of the flexible body is broken into two parts: rigid body motions represented by the motion of the floating frame, and superimposed "elastic motions." This decomposition allows the introduction of simplifying assumptions: although the total motion is always finite, the elastic motions may, in some cases, be assumed to give rise to infinitesimal deformations.

Many problems of great practical importance fall into this category. Consider, for instance, road or rail vehicles: the body of the vehicle undergoes large rigid body motions but the elastic deformations remain small. Of course, this assumption is no longer valid during a crash: in that case, large plastic deformations will be encountered. Other components of the vehicle, however, such as the suspension, wheels, and tires are inherently of a nonlinear nature. Another example is rotorcraft. Under normal operation, the fuselage undergoes large rigid body motions but small elastic deformations. During maneuvering flight, large rotations will be encountered. Here again, the other components of the rotorcraft, the main and tail rotor, and the landing gear, are inherently nonlinear.

When deformations remain small, it seems natural to use modal reduction techniques to represent the small elastic motions in an efficient manner. The system size will be considerably reduced, together with the resulting computational cost. In addition, since high frequency modes are eliminated, larger time step sizes can be used in the simulation.

Although the concept of floating frame seems rather intuitive, the implementation of a computational procedure based on this idea must deal with several thorny issues. First, the accuracy of the analysis will critically depend on the selection of a suitable modal basis. Second, a specific floating frame must be selected: it could be attached to a point of the elastic body or moving with respect to it. Third, the modal based elements should be easy to couple with the other components of the system modeled with multibody formulations. This points towards the use of component mode synthesis techniques that are well developed for structural dynamics problems. Fourth, in the absence of elastic deformations, the formulation should recover the exact equations of motion for a rigid body. Finally, the formulation should be independent of the finite element analysis package used to compute the modes of the elastic components. These various issues will be discussed in more detail in the following paragraphs.

The first critical step is the selection of a suitable modal basis, which is closely related to the choice of a specific floating frame of reference [165, 166, 167, 168]. Ideally, the selected modes should capture as accurately as possible the deformation patterns encountered during operation. Consequently, the analyst should be given the greatest possible freedom to select the type of modes he sees fit. The formulation should not put any restriction on the choice of the modal basis. Several authors have addressed the mode selection process [169, 170, 171, 172, 173, 174, 175, 176, 177].

Next, a specific floating frame must be selected. Since there exits no unique manner of defining the "rigid" and "elastic motions," the floating frame can be selected in a number of different ways and specific conditions must be selected to remove this indeterminacy. Several authors make use of body-attached frames, *i.e.*, the floating frame is attached to an arbitrary point in the body [178, 172, 174]. Other authors rely on floating frames moving with respect to the elastic body [167, 179, 177]. Cavanin and Likins [168] studied different options including frames attached to a material point of the flexible body, frames oriented along the principal axes of inertia, the Tisserand frame, the Buckens frame and the rigid body mode frame. They concluded that the Tisserand frame was the most advantageous choice and showed the equivalence of several of these choices when the body undergoes small deformations.

The moving frame approach seems to be more desirable than the body-attached approach because it eliminates the need to arbitrarily select a material point where to attach the floating frame. On the other hand, the moving frame approach also involves the analyst's insight since a specific condition must be selected to determine its location. Furthermore, this latter approach comes at the expense of additional computational complexity.

In some formulations, the choice of the floating frame is intimately linked to that of the modes used in the reduction technique [180, 181, 175, 176, 163, 182]. This connection hinders the selection of the most appropriate modes because the boundary conditions used to compute them do not necessarily match those of the flexible component once it is part of a multibody system.

In the classical application of modal analysis [183], the displacement field is represented as a linear combination of modes shapes. This type of representation has been used by some authors [184] in the context of multibody dynamics analysis, but it requires special techniques for coupling the modal based element with the other components of the system. Typically, this is done by formulating a constraint condition that equates the modal superposition to the physical displacement at a node of the model [185].

12.5.2 Component mode synthesis methods

These approaches do not take advantage of the component mode synthesis techniques that have been developed for structural dynamics problems over the past forty years. These techniques are aimed at computing the eigenmodes of very large structures in an efficient manner. The complete structure is broken into a number of substructures whose eigenmodes are easily computed. The substructures are then connected together to yield a lower-order model of the complete structure. Each substructure involves two types of degrees of freedom: physical degrees of freedom at a limited number of connection points (called "boundary nodes"), and modal degrees of freedom representing its internal flexibility.

Clearly, the need to interconnect the substructures is an integral part of the reduction technique and seems therefore ideally suited to the present problem. Among the most widely used component mode synthesis techniques are those of Craig and Bampton [186], MacNeal [187], Rubin [188]. Other efforts include those of Herting [189], Hintz [190], and refinements of the Craig-Bampton method [191].

Component mode synthesis methods have been used in the context of multibody dynamics by Shabana [178], Haug and coworkers [170, 171, 192, 193], and later by Cardona and coworkers [172, 174, 177] who used the Craig-Bampton method. Unfortunately, this method requires the use of modes associated with clamped conditions at the boundary nodes, thereby limiting the analyst's freedom to select the most appropriate modal basis. Consequently, the modal basis might poorly approximate the elastic behavior of the component.

This fundamental limitation of the approach was recognized by Craig and Bampton who suggested the use of "static correction modes" to alleviate the problem; Schwertassek and coworkers [175, 176] used this concept for flexible multibody systems. It also prompted the development of the MacNeal-Rubin method. In this case, however, free conditions must be used at all boundary nodes, limiting again the analyst's freedom. Furthermore, this method is more cumbersome to implement than the Craig-Bampton method.

Finally, Herting's method [189] offers a more general approach that enables the analyst to choose *any type of modes*. In fact, predictions based on the Craig-Bampton and MacNeal-Rubin methods were found to be in good agreement with those obtained with Herting's method [189]. Herting's method seems to be the most appropriate choice as it provides the analyst maximum flexibility in the choice of the modal basis. Furthermore, it allows independent choices to be made for the selection of the floating frame and of the modal basis [194, 195].

The formulation of modal based elements should be independent of the finite element analysis package used to compute the modes of the elastic components [170, 171, 172, 174, 177]. This means that the computation of the mass and stiffness coefficients used for the formulation of a modal based element should be solely based on the information readily provided by the finite element package. Some formulations have been proposed in which the finite element analysis tool is embedded in the multibody formulation [196]. Although higher accuracy can be achieved in that manner, this is clearly not a practical option if a large dimensional finite element model is required for the representation of the elastic components.

Yoo and Haug [170, 171] showed that by assuming a lumped mass representation of the elastic body, the modal based formulation could be fully decoupled from the finite element package. Unfortunately, the lumped mass approximation is rarely used in today's finite element models of complex structures. Cardona and Géradin [172, 174, 177] used corotational techniques to achieve the same decoupling without resorting to the lumped mass approximation.

Herting's transformation leads to an approximation that is fully independent of the finite element analysis package. The mass and stiffness coefficients of the modal based element are computed on the sole basis of the unconstrained mass and stiffness matrices of the elastic component and Herting's transformation, which also applies to the inertial velocities required to compute the kinetic energy of the elastic component, under the sole assumption of small displacements.

In summary, Herting's transformation is an attractive approach for the implementation of component mode synthesis techniques in flexible multibody systems. First, it allows the use of any modal basis the analyst sees fit. This contrasts with other approaches, such as those based on Craig-Bampton or Rubin-MacNeal transformations that require specific boundary conditions for the selected modes. Second, it can be used with both body-attached or moving frames of reference. Third, the modal based element is readily coupled to other components of the multibody system through the boundary nodes that retain physical degrees of freedom for this purpose. Fourth, the formulation recovers the exact equations of motion for a rigid body in the absence of elastic deformations. Finally, it is completely independent of the finite element package used to compute the modes of the elastic components.

12.5.3 Basic solution techniques for finite element models

Application of finite element concepts to multibody dynamics analysis has been the focus increased research in recent years. The textbook by Géradin and Cardona [164] describes such a procedure and presents numerical examples obtained with a commercial implementation the approach. As compared with rigid multibody dynamics or even flexible multibody dynamics using a modal approximation, a distinguishing feature of finite element methods is the much larger number of degrees of freedom used to model the system. While multibody systems involve tens or at the most a few hundreds of degrees of freedom, finite element element models often involve tens of even hundreds of thousands of degrees of freedom. Because solution costs grow as a power of the number of degrees of freedom, efficient solution techniques are an enabling technology for finite element formulations.

Because of the large number of degrees of freedom involved in finite element formulations and the likely presence of high frequencies associated with the discretization process, time integration relies almost exclusively on implicit schemes. For linear systems, the HHT integrator [135], the workhorse used in most commercial codes, is second-order accurate, unconditionally stable, and presents high frequency numerical damping; these three features are considered indispensable for the successful integration of large finite element systems, as discussed in textbooks such as Hughes [197] or Bathe [198]. This contrasts with multibody formulations that tend to use explicit, predictor multi-corrector algorithms such as the Adams-Bashforth integrator [199], for instance. Although of much higher-order accuracy, this integrator is conditionally stable.

Implicit integrators require the solution of a linear system at each time step. Typical solution procedures rely on the trifactorization of the sparse, banded dynamic stiffness matrix, $\underline{K}$, as $\underline{K} = \underline{L}\,\underline{D}\,\underline{L}^T$, where $\underline{L}$ is a lower triangular, $\underline{D}$ a diagonal, and $\underline{L}^T$ an upper triangular matrix, followed by a back-substitution phase. More details concerning this approach called the "skyline solver" or the "active column solver" are found in many textbooks, such as Bathe [198], for instance. The cost, C_b, of the trifactorization of a sparse, banded matrix can be roughly estimated as $C_b \propto nw^2$, where n is the number of freedom and w the average bandwidth of the dynamic stiffness matrix. If this matrix were to be fully populated, the factorization cost, C_{fp}, would become $C_{fp} \propto n^3$. For a finite element problem of modest size where $n = 10,000$ and $w = 100$, $C_b/C_{fp} = (m/n)^2 = 10^{-4}$; clearly, the advantage of the sparse solver is overwhelming and is an enabling technique of the finite element method.

An important implication of these observations is that any formulation that destroys the sparsity of the system matrix generated by the finite element method is unlikely to be effective. For instance, applications of Maggi's formulation presented in section 11.2.1 requires the computation of the null space of the constraint matrix. The various algorithms used to compute the null space, whether the LU factorization with pivoting, Gram-Schmidt orthogonalization algorithm, or singular value decomposition, all alter the band structure of the system matrix. The index-1 formulation requires the inverse of the mass matrix, another band destroying operation; of course, the null space formulation requires the evaluation of the null space; finally, the computation of the pseudo-inverse called for by Udwadia and Kalaba's formulation is once more an operation that does not preserve sparsity.

Clearly, far fewer methods are available for the effective enforcement of constraints when bandedness of the system matrix must be preserved. Hence, it should not come as a surprise that the sparsity based, index-3 DAEs formulation of Orlandea *et al.* [77, 78] discussed in section 12.6 has been used within the framework of finite element formulations. The penalty based stabilization techniques presented in section 12.3.2 have also been used in this framework. Finally, the staggered stabilization technique of Park and Chiou [124] and the augmented Lagrangian formulation of Bayo *et al.* [151] were originally developed for finite element formulations.

12.5.4 Numerically dissipative schemes

In view of the difficulties associated with the solution of index-3 DAEs, considerable effort was devoted to the development of time integration techniques suitable for large finite element systems. Cardona and Géradin [200, 31] showed that the classical Newmark [201] trapezoidal rule is unconditionally *unstable* for linear systems in the presence of constraints. The use of dissipative algorithms such as HHT [135] scheme, however, resulted stable behavior, even for nonlinear systems. Further work by Farhat *et al.* shows that both HHT and generalized-α [136] methods achieve stability for a class of constrained hybrid formulations. In these approaches, stabilization of the integration process is inherently associated with the dissipative nature of the algorithms. While stability is mathematically proven for linear systems, there is no guarantee when it comes to nonlinear systems [202]. A more detailed description of the generalized-α scheme is given in section 17.4.

12.5.5 Nonlinear unconditionally stable schemes

To remedy this situation, considerable work has been done in recent years with energy preserving schemes. In these schemes, unconditional nonlinear stability is achieved by proving a discrete energy preserving statement, $E_f = E_i$, where E denotes the total mechanical energy of the system, and the subscripts $(\cdot)_i$ and $(\cdot)_f$ denote the value of the corresponding quantity at the initial and final times of the time step, respectively, denoted t_i and t_f, respectively. This algorithmic preservation property is a direct consequence of the specific discretization used for the inertial and elastic forces acting on the system. In view of the positive-definite nature of the total mechanical energy, this discrete conservation law guarantees the stability of the computational scheme for nonlinear problems.

It is important to understand that while the exact solution of the equations of motion implies the exact preservation of the total mechanical energy, a numerical, *i.e.*, an inherently approximate, solution of the problem does not, in general, guarantee the preservation of energy at the discrete level. When using energy preserving schemes, the computed, approximate solution exactly satisfies the energy preservation condition. A number of researchers have developed energy preserving schemes for rigid bodies [203, 204, 205], beams [206, 207], and plates and shells [208, 209]. Section 17.5 describes energy preserving and decaying schemes in more details, but an exhaustive review of these schemes is beyond the scope of this book.

While nonlinear unconditional stability is the first step towards the development of robust algorithms, energy preserving schemes are not well suited for large finite element problems because high frequency oscillations, especially in the velocity and stress fields, can corrupt the computed system response, as observed by Bauchau *et al.* [210, 211]. Consequently, the presence of high frequency numerical dissipation is an indispensable feature of robust time integrators for multibody systems, a fact that was already observed for linear systems and prompted the development of numerically dissipative algorithms such as the HHT [135] or generalized-α [136] methods.

Numerically dissipative schemes that feature nonlinear unconditional stability can be developed by proving a discrete energy decay statement, $E_{i+1} = E_i - E_d$, where $E_d > 0$ is the energy dissipated within the time step. This approach was followed by a number of researchers who developed energy decaying schemes for beams [212], shells [213] and multibody systems [214, 215]. An exhaustive survey of energy decaying algorithms is beyond the scope of this book; Bottasso and Trainelli [216] attempted a classification of a number of such algorithms.

12.5.6 Enforcement of the constraints

The development of energy preserving and decaying algorithm has considerably increased the robustness of time integration schemes for multibody systems. The main idea behind these techniques is to develop discretizations of the equations of motion that imply algorithmic preservation of a known first integral of the motion, the total mechanical energy. When it comes to enforcement of the constraints, a similar path has been followed: the well known fact that the work done by the constraint forces must vanish is implemented at the algorithmic level [210, 217].

The work done by the constraint forces is $W^c = \int \underline{F}^{cT} \underline{\dot{q}} \, dt = \int \underline{\lambda}^T \underline{\underline{B}} \underline{\dot{q}} \, dt = \int \underline{\lambda}^T \underline{\dot{C}} \, dt$ and hence, the vanishing of this work is intimately linked to the vanishing of the constraint derivatives. This observation helps understand why it is important to enforce constraints at both displacement and velocity levels. Here again, it must be noted that an approximate solution of the constrained equations of motion will not necessarily imply the vanishing of the work done by the constraint forces at the algorithmic level. This provides a potential source of "numerical energy," which could destabilize the integration scheme.

Focusing on holonomic constraints, the following relationship is used to define the algorithmic constraint matrix, $\underline{\underline{B}}_m$, as

$$\underline{C}_f - \underline{C}_i = \underline{\underline{B}}_m (\underline{q}_f - \underline{q}_i), \tag{12.15}$$

where the subscript $(\cdot)_m$ indicates quantities evaluated at the midpoint of the time step. Note that the mean value theorem guarantees the existence of $\underline{\underline{B}}_m$. The discretized forces of constraint now become $\underline{F}^c_m = \underline{\underline{B}}^T_m \underline{\lambda}_m$, where $\underline{\lambda}_m$ are midpoint Lagrange's multipliers, and the work done by these discretized forces of constraint follows as $W^c = (\underline{q}_f - \underline{q}_i)^T \underline{F}^c_m = \underline{\lambda}^T_m \underline{\underline{B}}_m (\underline{q}_f - \underline{q}_i) = \underline{\lambda}^T_m (\underline{C}_f - \underline{C}_i)$. Clearly, the vanishing of the work done by the algorithmic forces of constraints implies $\Delta t \, \underline{\dot{C}}_m = \underline{C}_f - \underline{C}_i = 0$, which echoes, at the algorithmic level, the condition required for the exact solution, $\underline{\dot{C}} = 0$. Rather than imposing the condition $\underline{C}_f - \underline{C}_i = 0$, it is preferable to enforce $\underline{C}_f = 0$ at each time step, to avoid the drift phenomenon.

Discretizations of numerous constraints that satisfy eq. (12.15) can be found in the following references [218, 219, 220, 221]. Additional details about this approach are found in section 17.5.4.

Gonzalez [222] formulated an integration scheme for solving the equations of motion of Hamiltonian system expressed in the form of DAEs. Holonomic constraints were considered. The numerical scheme was based on the notion of *discrete derivative*, which satisfied properties such as directionality, consistency and orthogonality; eq. (12.15) is an example of discrete derivative. The proposed scheme satisfies the constraints, and leads to the conservation of the Hamiltonian and linear and angular momenta, but constraints are not satisfied at the velocity level. Bauchau [223] showed that the approach to modeling constraints characterized by eq. (12.15) is closely related to the stabilized index-2 method of Gear *et al.* [59], although no additional unknowns are required.

The approach summarized in the last two sections combines two algorithmic features: preservation/dissipation of energy and vanishing of the work done by the constraint forces. This provides a formal proof of numerical stability for the integration of nonlinear, flexible multibody systems, and constraints are enforced to machine accuracy, both at the displacement and velocity levels. The price to pay for these desirable features is that the discretization of inertial, elastic and constraint forces must be carefully crafted for each element type so that the preservation characteristics of the algorithms can be proved. This stands in sharp contrasts with the more traditional approach to multibody simulations that use a variety of formulations of the equations of motion, but rely on "black box" integration routine, which are designed for the solution of DAEs, but are otherwise unaware of the specific features and characteristics of the equations being solved.

12.5.7 The discrete null space approach

Betsch *et al.* [224, 225, 226] have recently proposed an original method for the time integration of constrained dynamical systems, based on Maggi's formulation. In this approach, the index-3 DAEs are first discretized with an energy/momentum preserving scheme based on the algorithms of Gonzalez [222], and Betsch and Steimann [227]. Next, the discrete Lagrange multipliers are eliminated using a *discrete null space*: using the notation of eq. (12.15), the discrete null space, $\underline{\underline{\Gamma}}_m$, is the orthogonal complement of the discrete constraint matrix, $\underline{\underline{B}}_m$, such that $\underline{\underline{\Gamma}}_m^T \underline{\underline{B}}_m^T = 0$. As discussed in section 12.5.6, analytical expressions of the discrete constraint matrix can be derived for a wide range of constraints; the originality of proposed approach is to show that analytical expressions of the discrete null space can also be obtained for numerous constraints. This approach bypasses the need for the numerical evaluation of the null space using the many techniques described in sections 12.1.1 and 12.1.2, and the associated numerical cost. Furthermore, the discrete null space is computed for each element of the system independently, and hence, can be used within the framework of finite element methods without harming the bandwidth of the system.

12.6 Scaling of Lagrange's equation of the first kind

Orlandea *et al.* [77, 78] have presented an approach to the dynamic analysis of mechanical systems based on the solution of Lagrange's equation of the first kind, a system of index-3 DAEs in the presence of holonomic constraints. While the number of generalized coordinates used in this approach is larger than the minimum set, they argue that numerical solutions of the resulting equations can be efficiently obtained by taking advantage of their sparsity through the use of appropriate algorithms.

To overcome the numerical problems associated with the solution of DAEs, numerically dissipative time integrators were used that are specifically designed for stiff problems, see Gear [228]. This early approach proposes a purely numerical solution to the challenges posed by Lagrange's equations of the first kind: stiff integrators are used to deal with DAEs.

Petzold and Lötstedt [49] have shown that index-3 DAEs are severely ill conditioned for small time step sizes when using backwards difference formulas: unless corrective actions are taken, the condition number of the iteration matrix is of $\mathcal{O}(h^{-3})$, where h denotes the integration time step size. Furthermore, errors in the displacement, velocity, and multiplier fields are shown to propagate at rates of $\mathcal{O}(h^{-1})$, $\mathcal{O}(h^{-2})$, and $\mathcal{O}(h^{-3})$, respectively. A perturbation analysis by Arnold [229] indicates that errors and constraint violations grow very rapidly as the time step size is reduced, preventing the practical use of time refinement procedures, and imposing tight tolerances on the solution of the nonlinear discrete equations.

Petzold and Lötstedt [49] presented a simple scaling transformation of the DAEs that yields a condition number of $\mathcal{O}(h^{-2})$ and an improvement of one order in the errors for all solution fields. Although the sensitivity to perturbations is reduced, numerical problems are still observed in practice. Their scaling, termed "left preconditioning," consists of dividing the constraint equations by the time step size, while the dynamic equilibrium equations are multiplied by the same quantity.

While the mathematical rational for preconditioning is recent, the technique has been used for a number of years by Cardona [31] or Bauchau *et al.* [230]. Clearly, scaling can and should be used in conjunction with other techniques for the solution of DAEs: it is easily implemented, does not require a reformulation of the equations of motion, and does not introduce additional unknowns.

In recent years the direct solution of index-3 DAEs has regained popularity, specially when finite element formulations are used to model flexible multibody systems, see section 12.5. Because of the large number of degrees of freedom involved in these formulations and the likely presence of high frequencies associated with the spatial discretization process, time integration relies almost exclusively on implicit schemes such as the HHT integrator [135], or more recently, the generalized-α scheme [136].

While dissipative time integration schemes seem to be indispensable to the successful integration of constrained dynamical systems modeled with index-3 DAEs, scaling of the governing equations and constraints seems to be an equally important technique, which is, in fact, hardly new.

In the framework of engineering optimization, scaling of constraint equations is a well-known practice that is recommended in numerous textbooks, such as Fox [231], 1971, or Reklaitis *et al.* [232], 1983. In his 1984 textbook, Vanderplaats [233] specifically mentions: "Often, numerical difficulties are encountered because one constraint function is of different magnitude or changes more rapidly than the others and therefore dominates the optimization process. [....,] we have normalized the constraints so they become of order of unity. This improves the conditioning of the optimization problem considerably, and *should always be done when formulating the problem.*" Although engineering optimization and multibody dynamics simulation are numerically similar problems that must both deal with constraints, it is disturbing to note that scaling of the constraint equations is rarely mentioned in multibody dynamics papers or textbooks.

Cardona and Géradin [234] showed that the condition number of the iteration matrix obtained from the HHT integrator is of $\mathcal{O}(h^{-4})$ and stated that "If we try to solve this problem without scaling, the Newton algorithm will not converge since round-off errors would become of the same order as the Newton correction itself." To remedy this problem, they proposed a symmetric scaling of the equations of motion that render the condition number of the system matrix independent of the time step size and of the mean value of the mass matrix.

A more systematic analysis of the scaling procedure was discussed by Bottasso *et al.* [235] who proposed a simple scaling transformation for the index-3 DAEs describing constrained multibody dynamical systems. The approach amounts to a left and right preconditioning of the iteration matrix, in an effort to decrease solution sensitivity to perturbation propagation. A remarkable result was obtained: both error propagation and iteration matrix conditioning are of $\mathcal{O}(h^0)$, and hence, the behavior of the numerical solution of index-3 DAEs is identical to that of regular ODEs. Bottasso *et al.* [236] later extended the same ideas to the Newmark family of integration schemes and provided a better theoretical foundation to explain how perturbations affect the solution process.

In section 12.6.1, physical arguments are used to derive a simple scaling procedure that is directly applied to the governing equations of motion, before the time discretization is performed, and an augmented Lagrangian term is added to the formulation, see section 12.6.2. Application of any time discretization scheme, such as that described in section 12.6.3, followed by a linearization of the resulting nonlinear algebraic equations then lead to a Jacobian matrix that is independent of the time step size; hence, the condition number of the Jacobian and error propagation are both of $\mathcal{O}(h^0)$: the numerical solution of index-3 DAEs behaves as in the case of regular ODEs. Since the scaling factor depends on the physical properties of the system, the proposed scaling decreases the dependency of this Jacobian on physical properties, further improving the numerical conditioning of the resulting linearized equations. Finally, the additional benefits stemming from the augmented Lagrangian term are discussed in section 12.6.5. Specifically, this term enables the use of sparse solvers that do not rely on pivoting for the stable and accurate solution of the linearized equations of motion.

12.6.1 Scaling of the equations of motion

In this section, simple physical arguments are used to scale Lagrange's equation of the first kind, eqs. (11.1), which form a set of index-3 DAEs,

$$\underline{\underline{M}}\frac{d^2\underline{q}}{dt^2} + \underline{\underline{B}}^T\underline{\lambda} = \underline{F}, \tag{12.16a}$$

$$\underline{C}(\underline{q}, t) = 0, \tag{12.16b}$$

where $\underline{\underline{M}} = \underline{\underline{M}}(\underline{q}, t)$ is the symmetric, semi positive-definite mass matrix, and $\underline{F} = \underline{F}(\underline{q}, \dot{\underline{q}}, t)$ the array of dynamic and externally applied forces. For simplicity of the exposition, the constraints are all assumed to be holonomic, but the derivation presented here equally applies to nonholonomic constraints, or a mixture thereof.

To ease the discussion, the damping and stiffness matrices will be explicitly shown in the equations of motion, and eqs. (12.16a) and (12.16b) are restated as

$$\underline{\underline{M}}\frac{d^2\underline{q}}{dt^2} + \underline{\underline{D}}\frac{d\underline{q}}{dt} + \underline{\underline{K}}\underline{q} + \underline{\underline{B}}^T\underline{\lambda} = \underline{G}, \tag{12.17a}$$

$$\underline{C}(\underline{q}, t) = 0, \tag{12.17b}$$

where $\underline{\underline{D}} = \underline{\underline{D}}(\underline{q})$ is the damping matrix, $\underline{\underline{K}} = \underline{\underline{K}}(\underline{q})$ the stiffness matrix, and $\underline{G} = \underline{G}(\underline{q}, \dot{\underline{q}}, t)$ the array of remaining dynamic and externally applied forces.

Following the advice of Vanderplaats [233] for optimization problems, constraints are normalized so as to become of the order of unity. This can be readily achieved by introducing normalized generalized coordinates, $\hat{\underline{q}}$, such that $\underline{q} = \ell_r\hat{\underline{q}}$, where ℓ_r is a reference or characteristic length of the system.

For dynamical systems, it is also important to introduce a normalized time variable, τ, such that $t = h\tau$, where h is the time step size. The equations of motion, eqs. (12.17a) and (12.17b), have not yet been discretized in time, but the time step size is anticipated to become an important characteristic time of the problem from a numerical standpoint.

The equations of motion now become

$$\underline{\underline{M}}\ddot{\hat{\underline{q}}} + h\underline{\underline{D}}\dot{\hat{\underline{q}}} + h^2\underline{\underline{K}}\hat{\underline{q}} + \underline{\underline{B}}^T h^2\underline{\lambda} = h^2\underline{G}, \tag{12.18a}$$

$$\underline{C}(\hat{\underline{q}}, \tau) = 0. \tag{12.18b}$$

Matrices $\underline{\underline{M}}, \underline{\underline{D}}, \underline{\underline{K}}$, and $\underline{\underline{B}}$ as well as arrays $\underline{G}$ and $\underline{C}$ are now expressed in terms of the normalized generalized coordinates. Matrices $\underline{\underline{M}}, \underline{\underline{D}}$, and $\underline{\underline{K}}$ have been multiplied by ℓ_r; for simplicity, the same notation is used from here on. Notation $(\dot{\ })$ is used to denote a derivative with respect to the non-dimensional time, τ. The equations of motion, eqs. (12.18a), were multiplied by h^2 to avoid division by a potentially small number, h^2.

A cursory examination of the normalized equations of motion, eqs. (12.18a) and (12.18b), reveals two obvious numerical problems. First, if the mass and/or damping and/or stiffness of the system become large, one or more of the first three terms of the equations of motion will become large, whereas the constraint equations remain unchanged. In other words, for systems with large mass, damping, or stiffness, the constraint equations become "invisible" to the numerical process. Second, the unknowns of the problem are of different orders of magnitude: displacements are typically very small quantities, but Lagrange's multipliers are force quantities, and hence, typically much larger, potentially by many orders of magnitude.

The first problem is easily solved by multiplying the constraint equations, eqs. (12.18b), by a scalar factor, called the *scaling factor*, s, to render the constraint equations and the equations of motion, eqs. (12.18a), of comparable magnitudes. Clearly, selecting $s = m_r + d_r h + k_r h^2$ accomplishes this goal. In this expression, m_r, d_r, and k_r represent characteristic mass, damping and stiffness coefficients of the system, which can be selected as $m_r = \|M\|_\infty$, $d_r = \|D\|_\infty$ and $k_r = \|K\|_\infty$; another convenient choice is to select m_r, d_r, and k_r as the average of the diagonal terms of the mass, damping and stiffness matrices, respectively.

The second problem can be solved by scaling Lagrange's multipliers by writing $h^2 \underline{\lambda} = s\hat{\underline{\lambda}}$. Clearly, in view of Newton's law, selecting $s = m_r + d_r h + k_r h^2$, makes $\hat{\underline{\lambda}}$ a quantity of magnitude comparable to that of displacement quantities. The equations of motion of the problem, eqs. (12.18a) and (12.18b), now become

$$\underline{\underline{M}}\,\ddot{\hat{q}} + h\underline{\underline{D}}\,\dot{\hat{q}} + h^2\underline{\underline{K}}\,\hat{q} + \underline{\underline{B}}^T s\hat{\underline{\lambda}} = h^2\underline{G}, \tag{12.19a}$$

$$s\underline{\mathcal{C}} = 0. \tag{12.19b}$$

The techniques used here are well-known scaling techniques for systems of equations, as discussed in textbooks on matrix computations. For instance, Golub and Van Loan [82] state: "The basic recommendation is that the scaling of the equations and unknowns must proceed on a problem-by-problem basis. General scaling strategies are unreliable. It is best to scale (if at all) on the basis of what the source problem proclaims about the significance of each a_{ij} [*i.e.*, each matrix entry]." In the proposed scaling strategy, the scaling factor was selected on the basis of physical arguments about the nature and order of magnitude of each term appearing in the equations of motion.

At this point, it is convenient to simplify the notation and write the scaled governing equations of index-3 multibody systems as

$$\underline{\underline{M}}\ddot{\hat{q}} + \underline{\underline{B}}^T s\hat{\underline{\lambda}} = h^2\underline{F}, \tag{12.20a}$$

$$s\underline{\mathcal{C}} = 0, \tag{12.20b}$$

where the scaling factor is defined as,

$$s = m_r + d_r h + k_r h^2. \tag{12.21}$$

Notation $(\dot{\cdot})$ indicates a derivative with respect to the non-dimensional time, $\tau = t/h$, and all generalized coordinates have been normalized by the reference length, ℓ_r.

12.6.2 The augmented Lagrangian term

An augmented Lagrangian term is now added to the scaled formulation of the equations of motion, as proposed by Bayo *et al.* [151, 152],

$$\underline{\underline{M}}\,\ddot{\hat{q}} + \underline{\underline{B}}^T s\dot{\underline{\lambda}} + \underline{\underline{B}}^T \rho s\underline{\mathcal{C}} = h^2 \underline{F}, \tag{12.22a}$$

$$s\underline{\mathcal{C}} = 0. \tag{12.22b}$$

The penalty factor, ρs, was defined as the product of the scaling factor defined in eq. (12.21) by ρ; for $\rho = 1$, the penalty factor is equal to the scaling factor. A set of modified Lagrange's multipliers,

$$\hat{\underline{\mu}} = \dot{\underline{\lambda}} + \rho\underline{\mathcal{C}}, \tag{12.23}$$

is introduced to simplify the above equations, leading to

$$\underline{\underline{M}}\,\ddot{\hat{q}} + \underline{\underline{B}}^T s\hat{\underline{\mu}} = h^2 \underline{F}, \tag{12.24a}$$

$$s\underline{\mathcal{C}} = 0. \tag{12.24b}$$

The equations of motion were scaled first, then the augmented Lagrangian term was added. Had this latter term be added from the onset of the formulation, the penalty factor would become $h^2 p$, *i.e.*, the penalty factor would vanish for small time step sizes, negating any advantage this term could have. It is possible to include the augmented Lagrangian term from the onset of the formulation by using a penalty factor written as $\bar{\rho}s = \rho s / h^2$, which yields results identical to those presented here.

12.6.3 Time discretization of the equations

To understand the implications of the scaling factor and augmented Lagrangian term presented above, the equations of motion will now be discretized in the time domain. A simple mid-point scheme is used for this task

$$\underline{\underline{M}}(\hat{\underline{v}}_f - \hat{\underline{v}}_i) + \underline{\underline{B}}_m^T s\hat{\underline{\mu}} = h^2 \underline{F}_m, \tag{12.25a}$$

$$\hat{\underline{q}}_f - \hat{\underline{q}}_i = (\hat{\underline{v}}_i + \hat{\underline{v}}_f)/2, \tag{12.25b}$$

$$s\underline{\mathcal{C}}_m = 0. \tag{12.25c}$$

Subscripts $(\cdot)_i$ and $(\cdot)_f$ indicate quantities at the beginning and end times of the time step, denoted t_i and t_f, respectively, $\underline{\underline{B}}_m = (\underline{\underline{B}}_i + \underline{\underline{B}}_f)/2$, $\underline{\mathcal{C}}_m = (\underline{\mathcal{C}}_i + \underline{\mathcal{C}}_f)/2$, $\underline{F}_m = (\underline{F}_i + \underline{F}_f)/2$, and $\hat{\underline{\mu}}_m$ are the mid-point, modified Lagrange multipliers. Equation (12.25b) is the discretized velocity-displacement relationship obtained from the mid-point rule; with the present notation, $\hat{\underline{v}} = \dot{\hat{q}} = d\hat{q}/d\tau = h\,d\hat{q}/dt$.

In view of the scaling of the time dimension performed in the previous section, the formulæ associated with time discretization are independent of the time step size, which is, in fact, taken to be unity; see eq. (12.25b), for example. This means that the time step size dependency of the various terms of the equations of motion indicated in eqs. (12.24a) and (12.24b) will not be affected by the time discretization, no matter what time integration scheme is used.

The unknown velocity, $\hat{\underline{v}}_f$, is easily eliminated from the discretized equations, leading to

$$2\underline{\underline{M}}(\hat{\underline{q}}_f - \hat{\underline{q}}_i - \hat{\underline{v}}_i) + \underline{\underline{B}}_m^T s\hat{\underline{\mu}} = h^2 \underline{\underline{F}}_m, \tag{12.26a}$$

$$s\underline{\underline{C}}_m = 0. \tag{12.26b}$$

Next, these nonlinear algebraic equations will be solved using a Newton-Ralphson iterative process based on the following set of linear algebraic equations

$$\underline{\hat{\underline{J}}}\Delta\hat{\underline{x}} = -\hat{\underline{b}}. \tag{12.27}$$

The Jacobian of the system, $\underline{\hat{\underline{J}}}$, is

$$\underline{\hat{\underline{J}}} = \begin{bmatrix} 2\underline{\underline{M}} + s(\underline{\underline{B}}^T \hat{\underline{\mu}})_{,\hat{\underline{q}}} - h^2 \underline{\underline{F}}_{,\hat{\underline{q}}} & s\underline{\underline{B}}^T \\ s\underline{\underline{C}}_{,\hat{\underline{q}}} & 0 \end{bmatrix}_m, \tag{12.28}$$

$$= \begin{bmatrix} \underline{\hat{\underline{J}}}_{11} & \underline{\hat{\underline{J}}}_{12} \\ \underline{\hat{\underline{J}}}_{21} & 0 \end{bmatrix}, \tag{12.29}$$

where the notation $(\cdot)_{,\hat{\underline{q}}}$ was used to indicate a derivative with respect to the generalized coordinates, and subscript $[\cdot]_m$ indicates that the Jacobian matrix is evaluated at the mid-point. The corrections to the unknowns of the problem are $\Delta\hat{\underline{x}}^T = \left\{ \Delta\hat{\underline{q}}_f^T, \Delta\hat{\underline{\lambda}}_m^T \right\}$, and the residual array is

$$\hat{\underline{b}} = \left\{ \begin{matrix} 2\underline{\underline{M}}(\hat{\underline{q}}_f - \hat{\underline{q}}_i - \hat{\underline{v}}_i) + \underline{\underline{B}}^T s\hat{\underline{\mu}} - h^2 \underline{\underline{F}} \\ s\underline{\underline{C}} \end{matrix} \right\}_m. \tag{12.30}$$

The asymptotic behavior of the Newton corrections, $\Delta\hat{\underline{x}}$, as the time step size tends to zero depends on the asymptotic behavior of both the Jacobian, $\underline{\hat{\underline{J}}}$, and the right-hand side, $\hat{\underline{b}}$. In fact,

$$\lim_{h \to 0}\left(\underline{\hat{\underline{J}}}\Delta\hat{\underline{x}} \right) = \lim_{h \to 0}\left(\underline{\hat{\underline{J}}} \right) \lim_{h \to 0}\left(\Delta\hat{\underline{x}} \right) = -\lim_{h \to 0} \hat{\underline{b}}, \tag{12.31}$$

and therefore, if $\lim_{h \to 0}(\underline{\hat{\underline{J}}}) = \underline{\underline{\mathcal{O}}}(h^0)$ and $\lim_{h \to 0}(\hat{\underline{b}}) = \underline{\mathcal{O}}(h^0)$, $\lim_{h \to 0}(\Delta\hat{\underline{x}}) = \underline{\mathcal{O}}(h^0)$.

The following results are easily obtained from examination of eqs. (12.28) and (12.30),

$$\underline{\hat{\underline{J}}} = \begin{bmatrix} \underline{\underline{\mathcal{O}}}(h^0) & \underline{\underline{\mathcal{O}}}(h^0) \\ \underline{\underline{\mathcal{O}}}(h^0) & \underline{0} \end{bmatrix}, \text{ and } \hat{\underline{b}} = \left\{ \begin{matrix} \underline{\mathcal{O}}(h^0) \\ \underline{\mathcal{O}}(h^0) \end{matrix} \right\}. \tag{12.32}$$

Furthermore, it is readily verified that the inverse Jacobian matrix is

$$\underline{\hat{\underline{J}}}^{-1} = \begin{bmatrix} \underline{\underline{\mathcal{O}}}(h^0) & \underline{\underline{\mathcal{O}}}(h^0) \\ \underline{\underline{\mathcal{O}}}(h^0) & \underline{\underline{\mathcal{O}}}(h^0) \end{bmatrix}. \tag{12.33}$$

It then follows that the condition number of the Jacobian matrix, $\kappa(\underline{\hat{\underline{J}}}) = \|\underline{\hat{\underline{J}}}\|_\infty \|\underline{\hat{\underline{J}}}^{-1}\|_\infty$, is clearly independent of the time step size, $\kappa(\underline{\hat{\underline{J}}}) = \mathcal{O}(h^0)$. In view of eqs. (12.27) and (12.31), it follows that

$$\Delta \hat{\underline{q}}_f = \mathcal{O}(h^0), \quad \Delta \hat{\lambda}_m = \mathcal{O}(h^0). \tag{12.34}$$

Of course, scaling of the variables has to be considered when the criterion for convergence of Newton iterations is evaluated.

This behavior is markedly different from what happens when scaling of the equations is not performed. Indeed, applying the mid-point time discretization to the unscaled, augmented equations of motion, eqs. (12.16a) and (12.16b), leads to

$$\frac{2M}{h^2}(\underline{q}_f - \underline{q}_i - h\frac{d\underline{q}_i}{dt}) + \underline{B}^T_m \underline{\mu}_m = \underline{F}_m, \tag{12.35a}$$

$$\underline{C}_f = 0, \tag{12.35b}$$

where the unscaled modified Lagrange multiplier is defined as $\mu = \lambda + \rho\underline{C}$. A Newton-Ralphson approach is taken again to solve this set of nonlinear algebraic equations; linearization leads to $\underline{J}\Delta\underline{x} = -\underline{b}$, where the Jacobian of the system, $\underline{J}$, is

$$\underline{J} = \begin{bmatrix} 2\underline{M}/h^2 + (\underline{B}^T\mu)_{,q} - \underline{F}_{,q} & \underline{B}^T \\ \underline{C}_{,q} & 0 \end{bmatrix}_m, \tag{12.36}$$

and the residual array is

$$\underline{b} = \left\{ \begin{array}{c} \frac{2\underline{M}}{h^2}(\underline{q}_f - \underline{q}_i - h\frac{d\underline{q}_i}{dt}) + \underline{B}^T\mu - \underline{F} \\ \underline{C} \end{array} \right\}_m. \tag{12.37}$$

The following results are easily obtained from examination of eqs. (12.36) and (12.37),

$$\underline{J} = \begin{bmatrix} \mathcal{O}(h^{-2}) & \mathcal{O}(h^0) \\ \mathcal{O}(h^0) & \underline{0} \end{bmatrix}, \quad \text{and } \underline{b} = \left\{ \begin{array}{c} \mathcal{O}(h^{-2}) \\ \mathcal{O}(h^0) \end{array} \right\}. \tag{12.38}$$

The inverse Jacobian matrix is

$$\underline{J}^{-1} = \begin{bmatrix} \mathcal{O}(h^2) & \mathcal{O}(h^0) \\ \mathcal{O}(h^0) & \mathcal{O}(h^{-2}) \end{bmatrix}. \tag{12.39}$$

It then follows that the condition number of the Jacobian matrix, $\kappa(\underline{J})$, exhibits a strong dependency on the time step size, $\kappa(\underline{J}) = \mathcal{O}(h^{-4})$, and

$$\Delta\underline{q}_f = \mathcal{O}(h^0), \quad \Delta\lambda_m = \mathcal{O}(h^{-2}). \tag{12.40}$$

Example 12.1. The simple pendulum

Figure 12.1 depicts a simple pendulum of length ℓ and bob of mass m. In this example, the root torsional spring is ignored. This problem will be treated with two generalized coordinates: the bob's horizontal and vertical Cartesian coordinates, denoted x_1 and x_2, respectively, $q^T = \{x_1, x_2\}$. Because the system features a single degree of freedom, a single constraint must be enforced: the pendulum arm must remain of constant length, ℓ. The governing equations of the problem are

$$M\frac{\mathrm{d}^2q}{\mathrm{d}t^2} + \underline{\underline{B}}^T\underline{\lambda} = 0, \tag{12.41a}$$

$$\underline{C} = 0, \tag{12.41b}$$

where $\underline{\underline{M}} = \mathrm{diag}(m, m)$, $\underline{\underline{B}} = 2\underline{q}^T$, $\underline{C} = \underline{q}^T\underline{q} - \ell^2$, and $\underline{\lambda} = \lambda_1$. The Jacobian of the unscaled system is readily obtained from eqs. (12.41a) and (12.41b) as

$$\underline{\underline{J}} = \begin{bmatrix} 2\underline{\underline{M}}/h^2 + (\underline{\underline{B}}^T\underline{\lambda})_{,\underline{q}} & \underline{\underline{B}}^T \\ \underline{C}_{,\underline{q}} & \underline{\underline{0}} \end{bmatrix}_m. \tag{12.42}$$

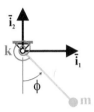

Fig. 12.1. Configuration of the pendulum with a root torsional spring.

These equations of motion can be scaled then augmented using the proposed approach, and with the help of the mid-point time discretization method, the Jacobian of the linearized system then becomes

$$\underline{\underline{\hat{J}}} = \begin{bmatrix} 2\underline{\underline{M}} + s(\underline{\underline{B}}^T\hat{\mu})_{,\hat{q}} & s\underline{\underline{B}}^T \\ s\underline{C}_{\hat{q}} & \underline{\underline{0}} \end{bmatrix}_m. \tag{12.43}$$

It is readily verified that all blocks of this Jacobian and of the corresponding right-hand side are $\mathcal{O}(h^0)$. For this simple problem, this is true even without the augmented Lagrangian term, *i.e.*, even if $\rho = 0$.

Example 12.2. Simple pendulum with torsional spring
The problem treated in example 12.1 will now be repeated with the addition of the root torsional spring of stiffness constant k, as depicted in fig. 12.1. Three generalized coordinates will be used here: the bob's horizontal and vertical Cartesian coordinates, and the root rotation angle, ϕ. Since the system features a single degree of freedom, two constraints must be enforced, the pendulum arm must remain of constant length, ℓ, and angle ϕ can be obtained from elementary trigonometric considerations. The governing equations of this problem are

$$M\frac{\mathrm{d}^2q}{\mathrm{d}t^2} + \underline{\underline{B}}^T\underline{\lambda} = 0, \tag{12.44a}$$

$$k\phi + C_{2,\phi}\lambda_2 = 0, \tag{12.44b}$$

$$\underline{C} = 0, \tag{12.44c}$$

where $C_\phi = \cos\phi$, $S_\phi = \sin\phi$, $\underline{\lambda}^T = \{\lambda_1, \lambda_2\}$, $\underline{C}^T = \{C_1, C_2\}$, $C_1 = \underline{q}^T\underline{q} - \ell^2$, $C_2 = q_1 C_\phi + q_2 S_\phi$, and

$$\underline{\underline{B}} = \begin{bmatrix} 2q_1 \ C_\phi \\ 2q_2 \ S_\phi \end{bmatrix}. \tag{12.45}$$

The relative rotation angle, ϕ, is an *algebraic variable*, which, in contrast with the Lagrange multipliers $\underline{\lambda}$, appears explicitly in the constraint equations, eq. (12.44b). This equation simply represents the static equilibrium of the spring and hence, involves no time derivative of this angle. The explicit definition of the relative rotations at the root readily allows for the introduction of root spring.

The Jacobian of the unscaled system is readily obtained from eqs. (12.44a) to (12.44c) as

$$\underline{\underline{J}} = \begin{bmatrix} 2\underline{\underline{M}}/h^2 + (\underline{\underline{B}}^T\underline{\lambda})_{,q} & (\underline{\underline{B}}^T\underline{\lambda})_{,\phi} & \underline{\underline{B}}^T \\ (C_{2,\phi}\lambda_2)_{,q} & k + (C_{2,\phi}\lambda_2)_{,\phi} & \underline{C}_{,\phi}^T \\ \underline{C}_{,q} & \underline{C}_{,\phi} & 0 \end{bmatrix}_m. \tag{12.46}$$

These equations of motion can be scaled and augmented using the proposed approach, and with the help of the mid-point time discretization method, the Jacobian of the linearized system then becomes

$$\hat{\underline{\underline{J}}} = \begin{bmatrix} 2\underline{\underline{M}} + s(\underline{\underline{B}}^T\hat{\mu})_{,\hat{q}} & s(\underline{\underline{B}}^T\hat{\mu})_{,\phi} & s\underline{\underline{B}}^T \\ s(C_{2,\phi}\hat{\mu}_2)_{,\hat{q}} & h^2 k + s(C_{2,\phi}\hat{\mu}_2)_{,\phi} & s\underline{C}_{,\phi}^T \\ s\underline{C}_{,\hat{q}} & s\underline{C}_{,\phi} & 0 \end{bmatrix}_m. \tag{12.47}$$

Here again, it is readily verified that all blocks of this Jacobian and of the corresponding right-hand side are of $\mathcal{O}(h^0)$. The key to this proof is in the fact that $s\hat{\mu} = s\hat{\lambda} + sp\underline{C} = h^2\underline{\lambda} + sp\underline{C} = \mathcal{O}(h^0)$. In contrast with example 12.1, the augmented Lagrangian term is indispensable to achieving this result; indeed, if $p = 0$, $s\hat{\mu} = s\hat{\lambda} = h^2\underline{\lambda} = \mathcal{O}(h^2)$.

Clearly, the proposed scaling of the unknowns and equations is sufficient to achieve time step size independent Jacobians when the sole algebraic variables of the problem are Lagrange's multipliers. When the problem involves additional algebraic variables, such as the relative rotation of this example, the scaling of the unknowns and of the equations must be used in conjunction with the augmented Lagrangian term to achieve time step size independent formulations.

12.6.4 Relationship to the preconditioning approach

A preconditioning approach for index-3 DAEs was proposed by Bottasso *et al.* [235, 236]. The starting point of their development is the Jacobian matrix resulting from the linearization of the governing equations (12.16a) and (12.16b). The Jacobian is multiplied by left and right preconditioning matrices, denoted $\underline{\underline{L}}$ and $\underline{\underline{R}}$, respectively, such that $\bar{\underline{\underline{J}}} = \underline{\underline{L}}\,\underline{\underline{J}}\,\underline{\underline{R}}$, where $\underline{\underline{L}} = \text{diag}(h^{\alpha_i})$ and $\underline{\underline{R}} = \text{diag}(h^{\beta_i})$. The powers of the time step size, *i.e.*, coefficients α_i and β_i, are selected to render the preconditioned

Jacobian, $\underline{\bar{J}}$, independent of h. To prevent confusion, it must be noted the scaling factor defined here, s, and that defined by Bottasso *et al.*, s' (but noted s in refs. [235, 236]), are different: $s' = s/h^2$.

For the problem presented in example 12.1, the preconditioning and scaling approaches yield identical Jacobian matrices if the preconditioning matrices are selected to be $\underline{L} = \text{diag}(h^2, s)$ and $\underline{R} = \text{diag}(1, s/h^2)$. For the problem presented in example 12.2, identical Jacobians are obtained by selecting $\underline{L} = \text{diag}(h^2, h^2, s)$ and $\underline{R} = \text{diag}(1, 1, s/h^2)$. Clearly, left and right preconditioning matrices can be found that will yield identical Jacobians for the two approaches.

For the problem presented in example 12.2, a time step size independent Jacobian is only obtained with the addition of an augmented Lagrangian term; indeed, without these terms, the Jacobian becomes

$$\underline{\bar{J}} = \begin{bmatrix} 2\underline{M} + (\underline{B}^T h^2 \underline{\lambda})_{,q} & (\underline{B}^T h^2 \underline{\lambda})_{,\phi} & \underline{B}^T \\ (\mathcal{C}_{2,\phi} h^2 \lambda_2)_{,q} & h^2 k + (\mathcal{C}_{2,\phi} h^2 \lambda_2)_{,\phi} & \underline{C}^T_{,\phi} \\ \underline{C}_{,q} & \underline{C}_{,\phi} & 0 \end{bmatrix}_m . \tag{12.48}$$

Clearly, not all blocks of this Jacobian are $\mathcal{O}(h^0)$. The reasons why this feature is desirable is discussed in the next section. While the use of the augmented Lagrangian term was not addressed in refs. [235, 236], it is clear that if such term is added to the equations of motion from the onset of the formulation, the two methods become equivalent.

12.6.5 Benefits of the augmented Lagrangian formulation

In practical implementations of the finite element method, the linearized set of governing equations is solved in two steps [198, 82]: first, the system Jacobian is factorized as $\underline{J} = \underline{L}\,\underline{D}\,\underline{L}^T$, where $\underline{L}$ is a lower triangular matrix and $\underline{D}$ a diagonal matrix, and second, the solution is found by back-substitution. The advantage of this approach is that it preserves the banded structure of the Jacobian, if its factorization is performed *without pivoting*. In general, factorization of the Jacobian without pivoting is numerically unstable, unless the Jacobian is symmetric and positive-definite [82]. This is always the case for the stiffness and mass matrices of structures because they can be derived from the minimization of quadratic energy functionals. Consequently, factorizations without pivoting, also called "skyline solvers," are used systematically in finite element codes.

The Jacobian matrices of constrained multibody systems, however, are not identical to the mass and stiffness matrices of structures. Consider the Jacobian obtained without the augmented Lagrangian term given by eq. (12.48), and note the presence of the factor h^2 along some columns of the matrix.

Consider next the very simple linear system, $\underline{J}\,\underline{x} = \underline{b}$, where

$$\underline{J} = \begin{bmatrix} 1 & 0 & 0 \\ 0 & h^2 & 1 \\ 0 & 1 & 0 \end{bmatrix}, \text{ and } \underline{b} = \begin{Bmatrix} 1 \\ 1 \\ 1 \end{Bmatrix}, \tag{12.49}$$

which shares the characteristics of eq. (12.48); although symmetric, the Jacobian is not positive-definite. It is easy to show that the condition number of this Jacobian is unity, and for $h = 0.001$, the exact solution is $x_1 = x_2 = 1$, and $x_3 = 0.999999$. Using finite precision arithmetic with five significant digits, the solution of the system with *full pivoting* yields $x_1 = x_2 = 1$, and $x_3 = 0.99999$, whereas solution *without pivoting* leads to an incorrect answer, $x_1 = 1$, $x_2 = 10$, and $x_3 = 0.99999$. Clearly, when using a skyline solver, *i.e.*, when factorization of the Jacobian is performed without pivoting, the condition number of the system matrix is not a good indicator of the accuracy of the solution.

While a low condition number is a necessary condition for obtaining accurate solutions of linear problems, it is not a sufficient condition when skyline solvers are used. Consider, for instance, the Jacobian matrices of example 12.2 defined in eqs. (12.47) and (12.48), obtained with and without the augmented Lagrangian term, respectively. Because of the presence of the multiplicative factor, h^2, across entire columns of the Jacobian in eq. (12.48), pivoting will be required to ensure accurate solutions. On the other hand, all the sub-matrices of the Jacobian obtained with the present scaling approach, see eq. (12.47), are independent of the time step size, enabling the safe use of skyline solvers.

The augmented Lagrangian term of the proposed formulation was shown above to be key to achieving time step size independent Jacobians, see eq. (12.28). The Hessian of the system, see eq. (12.29), can be expressed as $\hat{\underline{\underline{J}}}_{11} = 2\underline{\underline{M}} + s(\underline{\underline{B}}^T \hat{\underline{\lambda}})_{,\hat{q}} - h^2 \underline{\underline{F}}_{,\hat{q}} + s\rho \underline{\underline{B}}^T \underline{\underline{B}}$, where the last term represents the contribution of the penalty term and provide two further benefits.

First, consider the problem described in example 12.2 and assume that the system is at rest at $t = 0$. Because the first Lagrange multiplier represents the tension in the rod and the second the moment in the spring, it is clear that $\underline{\lambda} = 0$ at $t = 0$. In the absence of penalty term, *i.e.*, for $\rho = 0$, the Jacobian of the linearized system at that instant becomes

$$\underline{\underline{J}} = \begin{bmatrix} 2\underline{\underline{M}} & 0 & s\underline{\underline{B}}^T \\ 0 & 0 & s\underline{\underline{C}}^T_{,\phi} \\ s\underline{\underline{C}}_{\hat{q}} & s\underline{\underline{C}}_\phi & 0 \end{bmatrix}_m . \tag{12.50}$$

Although this Jacobian is not singular, a skyline solver will obviously fail if pivoting is not used. Clearly, if a skyline solver is used, the augmented Lagrangian term is indispensable to the success of the simulation's first time step.

Second, Gill *et al.* [237] showed that there always exists a ρ^* such that the Hessian of the augmented Lagrangian, $\hat{\underline{\underline{J}}}_{11}$, is positive-definite for all $\rho > \rho^*$. As mentioned earlier, positive-definiteness is key to the reliable use of skyline solvers: this implies that the sub-system $\hat{\underline{\underline{J}}}_{11} \Delta \hat{\underline{x}}^* = -\hat{\underline{b}}^*$, where $\hat{\underline{x}}^*$ and $\hat{\underline{b}}^*$ are vectors of appropriate dimensions, can be solved without pivoting. Experience shows that $\rho = 1$ is a good choice; this implies that the penalty factor is taken equal to the scaling factor.

Finally, now that it has been proved that the Hessian of the augmented Lagrangian, $\hat{\underline{\underline{J}}}_{11}$, can be factorized without pivoting, it must also be proved that the complete solution can be obtained without pivoting. At first, consider a system with a single constraint: $\hat{\underline{\underline{J}}}_{12}$ and $\hat{\underline{\underline{J}}}_{21}$ are then column and row vectors, respectively. Since the constraint matrix is assumed to be of full rank, its single column, $\hat{\underline{\underline{J}}}_{12}$, must contain a least one non-zero element, and hence, factorization without pivoting can proceed safely. Mathematical induction then implies that factorization without pivoting can proceed for systems with an arbitrary number of constraints, for as long as columns and rows of $\hat{\underline{\underline{J}}}_{12}$ and $\hat{\underline{\underline{J}}}_{21}$, respectively, are linearly independent, a property that is guaranteed by the fact that the constraint matrix is of full rank.

As a last note of interest, the proof presented above assumes that the degrees of freedom of the system are segregated: first, all the generalized coordinates of the system, then, Lagrange's multipliers. In practice, this ordering is not desirable because it does not minimize the bandwidth of the system of equations. It can be easily shown that generalized coordinates and Lagrange's multipliers can be interspersed, as desired for minimization of the bandwidth, while still using a skyline solver. The only requirement is that Lagrange's multipliers must be placed after the generalized coordinates that participate in the corresponding constraint equation, as was already observed by Cardona [31].

12.6.6 Using other time integration schemes

While the proposed scaling method has been presented so far within the framework of the mid-point time integration scheme, it is easily extended to the more advanced integration methods that are used for the analysis of realistic mechanical systems. Consider, for example, the generalized-α method [136] applied to the scaled general equations of motion of a multibody system, see eqs. (12.24a) and (12.24b). The resulting discretization is

$$\underline{\underline{M}}\,\breve{\underline{a}} + s\underline{\underline{B}}^T\breve{\underline{\ell}} = h^2\underline{F}, \tag{12.51a}$$

$$s\underline{\mathcal{C}} = 0. \tag{12.51b}$$

Here, the mass matrix, constraints, constraint matrix, and forces are evaluated using the algorithmic variables defined by eqs. (17.38) and (17.50). The corresponding variables at the end of the time step are related to their values at the beginning of the time step through eqs. (17.41).

Linearization of eqs. (12.51a) and (12.51b) with respect to these increments yields a system of algebraic equations identical to eq. (12.27) with a Jacobian matrix presenting the same structure as in eq. (12.29), where the sub-matrices are $\hat{\underline{\underline{J}}}_{11} = (1 - \alpha_M)/\beta\,\underline{\underline{M}} + h^2(1 - \alpha_F)\gamma/\beta\,\underline{F}_{,\hat{v}} + h^2(1 - \alpha_F)\underline{F}_{,\hat{q}} + s(\underline{\underline{B}}^T\hat{\mu})_{,\hat{q}}$, $\hat{\underline{\underline{J}}}_{12} = s(1 - \alpha_F)\underline{\underline{B}}^T$, and $\hat{\underline{\underline{J}}}_{21} = s(1 - \alpha_F)\underline{\mathcal{C}}_{,\hat{q}}$, respectively, and their asymptotic behavior is independent of the time step size as was observed for the simple mid-point scheme.

The developments presented above can be repeated for other integration schemes such as the well-known HHT scheme [135], implicit Runge-Kutta methods including the class of RADAU schemes [26], or backward difference formulæ (BDF) [96]. In all cases, the application of the time integration scheme to the proposed scaled equations, see eqs. (12.24a) and (12.24b), leads to a Jacobian matrix that is independent of the time step size.

Example 12.3. Scaling of a simple pendulum problem

Consider the simple pendulum problem described in example 12.2 and depicted in fig. 12.1, with $m = 1$ kg, $k = 10$ N·m/rad, and $\ell = 1$ m, simulated within the time range $t \in [0, 1]$ sec. Table 12.1 lists the condition numbers of iteration matrix, $\kappa(\underline{J})$, at convergence of the last time step, for time step size $h \in [10^{-1}, 10^{-5}]$ s.

Table 12.1. Condition numbers of the iteration matrix at convergence of the last time step.

h	No scaling	$s = 1$ in eq. (12.21)	s from eq. (12.21)
$1\ 10^{-1}$	$4\ 10^4$	10.	12.
$5\ 10^{-2}$	$6\ 10^5$	8.9	13.
$1\ 10^{-2}$	$3\ 10^8$	9.2	14.
$5\ 10^{-3}$	$5\ 10^9$	9.2	14.
$1\ 10^{-3}$	$3\ 10^{12}$	9.2	14.
$5\ 10^{-4}$	$5\ 10^{13}$	9.2	14.
$1\ 10^{-4}$	$3\ 10^{16}$	9.2	14.
$5\ 10^{-5}$	$5\ 10^{17}$	9.2	14.
$1\ 10^{-5}$	$3\ 10^{20}$	9.2	14.

Table 12.2. Condition numbers of the iteration matrix at convergence of the last time step.

Mass	No scaling	$s = 1$ in eq. (12.21)	s from eq. (12.21)
10^{-2}	$3\ 10^6$	$2\ 10^1$	13.
10^{-1}	$3\ 10^8$	$9\ 10^0$	14.
10^0	$3\ 10^{10}$	$4\ 10^2$	14.
10^1	$3\ 10^{12}$	$3\ 10^4$	14.
10^2	$3\ 10^{14}$	$3\ 10^6$	14.
10^3	$3\ 10^{16}$	$3\ 10^8$	14.
10^4	$3\ 10^{18}$	$3\ 10^{10}$	14.

The second column of table 12.1 lists the condition numbers in the absence of scaling. As predicted, $\kappa(\underline{J}) = \mathcal{O}(h^{-4})$, clearly demonstrating the need for scaling. The next two columns list the condition numbers with scaling factors $s = 1$ and s selected according to eq. (12.21), for the third and fourth columns, respectively.

Example 12.4. Scaling of a simple pendulum problem

Next, the same problem is solved with a fixed time step size, $h = 0.01$ s, and fixed spring stiffness constant, $k = 10$ N·m/rad, but for a range of mass values, $m \in [10^{-2}, 10^4]$ kg. Table 12.2 lists the condition numbers of iteration matrix, $\kappa(\underline{J})$, at convergence of the last time step.

The second column of table 12.2 lists the condition numbers in the absence of scaling. As the mass of the system increases, the condition number of the Jacobian matrix increases, demonstrating here again the need for scaling. The next two columns list the condition numbers with scaling factors $s = 1$ and s selected according to eq. (12.21), for the third and fourth columns, respectively. These results highlight the importance of scaling the problem with respect to its dependency on physical properties. Selecting the scaling factor according to eq. (12.21) renders the condition number of the Jacobian independent of the value of the mass. Of course, varying the spring stiffness constant would yield similar results.

Example 12.5. Flexible beam actuated by a crank

Figure 12.2 depicts a cantilevered beam of length $L = 1$ m actuated by a crank mechanism. The beam has a rectangular cross-section of depth $h = 0.1$ m and width $w = 2.5$ mm; it is made of aluminum of Young's modulus $E = 73$ GPa and Poisson's ratio $\nu = 0.3$. This beam is modeled by eight cubic beam elements; the geometrically exact beam element formulation is described in section 16.3. The tip of the beam is connected to a spherical joint at point **C** by means of a short connector modeled by two cubic elements and featuring physical properties identical to those of the beam. In turn, the spherical joint is connected to a flexible steel ($E = 210$ GPa and $\nu = 0.3$) link of length $L_\ell = 0.5$ m with a hollow circular cross-section of outer radius $R_o = 15$ mm and thickness $t = 8$ mm.

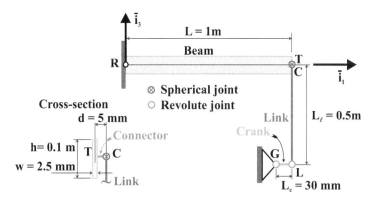

Fig. 12.2. Beam actuated by a tip crank.

Next, the link connects to a crank of length $L_c = 30$ mm by means of a revolute joint at point **L**; the crank's cross-section is identical to that of the link. Finally, a revolute joint connects the crank to the ground at point **G**. Points **G**, **L**, and **C** define the plane of the crank-link mechanism, which is offset by a distance $d = 5$ mm from plane $(\bar{\imath}_1, \bar{\imath}_3)$ of the cantilevered beam. The relative rotation of the revolute joint at point **G** is prescribed as $\phi = 1.6(1 - \cos 2\pi t/T)$ rad, where $T = 1.6$ s.

As the crank rotates up, the vertical transverse shear force in the beam increases, and the beam buckles laterally. Figure 12.3 shows the three displacement components at the beam's mid-point: at about 0.05 s in the simulation, the lateral displacement component, u_2, suddenly increases. Lateral buckling is accompanied by rotation of the beam's mid-section.

The following observations will be made concerning this simulation. First, in the absence of augmented Lagrangian terms, the simulation failed at the first iteration of the first time step. Indeed, at the first time step, the Jacobian of the system presents a structure similar to that presented in eq. (12.50), and the skyline solver fails to factorize the Jacobian.

Next, augmented Lagrangian terms were included in the simulation, but no scaling was used, *i.e.*, $s = 1$ was selected. In this case, the skyline solver was able to factorize the Jacobian at the first time step, but iterations failed to converge because of the poor conditioning of the system. Finally, when using the proposed scaling, the simulation ran smoothly to completion, as shown in fig. 12.3.

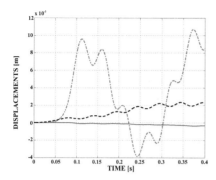

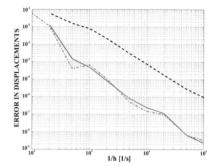

Fig. 12.3. Displacement components at the beam's mid-span. Component u_1: solid line; u_2: dashed-dotted line; and u_3: dashed line.

Fig. 12.4. Three integration schemes. Radau IIA: solid line; energy decaying scheme: dashed-dot line; HHT: dashed line.

The applicability of the proposed scaling to various time integration schemes will also be demonstrated here. Simulations were run with three integration schemes: the Radau IIA scheme [26], the energy decaying scheme [218], and the HHT scheme [135]. Figure 12.4 demonstrates the convergence characteristics of the three schemes by plotting the solution error as a function of the inverse of the time step size. Errors were computed with respect to a reference solution obtained by using the Radau IIA scheme with a very small time step size, $h = 5$ μsec. Note the good convergence of all three schemes, even for very small time step sizes.

12.7 Conclusions

This chapter has presented a comprehensive review of the numerical tools used for the enforcement of constraints in multibody systems. The classical formulation of Lagrange's equations of the first kind yields index-3, differential-algebraic equations. In view of the difficulties associated with the solution of these equations, several approaches have been used to algebraically eliminate Lagrange's multipliers.

The first approach is to use Maggi's formulation, which plays a pivotal role in constrained dynamics, although its importance was not initially recognized. The selection of a set kinematic characteristics is required, and for both holonomic and nonholonomic systems, constraints are enforced at the velocity level. The central ingredient of the approach is the null space of the constraint matrix.

The next two approaches are the index-1 formulation that is widely used in multibody codes, and the null space formulation that has also received considerable attention. Both methods enforce constraints at the acceleration level, and hence, significant drift of numerical solutions should be expected. Finally, Udwadia and Kalaba's formulation provides new insight to the behavior of constrained dynamical system. Here again, constraints are enforced at the acceleration level, but the use of Moore-Penrose inverses provides increased generality and robustness. The salient features of these four approaches were reviewed and compared.

Extensive mathematical studies of DAEs concluded that the best approach for the solution of DAEs is the reduce their index. Consequently, two distinct avenues of research were pursued: first, coordinate reduction techniques were developed to reduce the DAEs to ODEs, and second, index reduction techniques were proposed that bring the DAEs index from 3 to 2 or 1.

Maggi's method has been extensively used as a coordinate reduction technique that transform DAEs into ODEs. A distinction was made between Maggi-like methods that do not eliminate Lagrange's multipliers and true Maggi formulations for which these multipliers are completely eliminated. Many of these methods only differ by the numerical process used to compute the null space of the constraint matrix. This contrasts with the extraction procedure that evaluates the null space based on kinematic considerations. The null space formulation has also been used to obtain ODEs. Numerical implementations of Udwadia and Kalaba's formulation inherits the advantages of this powerful technique. Finally, more geometric arguments form the basis of the projective formulation, which uses the concepts of tangent and orthogonal subspaces to obtain ODEs.

Index reduction techniques are formal mathematical procedures that reduce the index of DAEs from 3 to 2 or 1. In many cases, properties of the proposed schemes are proved mathematically. While the order of accuracy of the solution is often provided, the problem of violation of the constraints was rarely addressed.

Due to approximations and round-off errors, many numerical solutions do not satisfy the constraints exactly, a phenomenon known as "drift." Numerous constraint violation stabilization techniques have been developed to remedy this problem, but Baumgarte's method is probably the most widely used. It presents two shortcomings: first, constraints are never exactly satisfied, and second, no general procedure exists to determine the problem dependent parameters appearing in the formulation. Consequently, this approach cannot be recommended for general use in multibody systems.

Penalty based formulations have also been used to control the drift phenomenon. The augmented Lagrangian formulation is probably the most robust and efficient method of that type. Next, constraint violation elimination techniques have been developed to enforce the exact satisfaction of the constraints, at least to machine accuracy. In those approaches, the solution obtained through the approximate time integration process is corrected or perturbed to satisfy the constraints. This correction is typically based on geometric concepts: the approximate solution is projected onto the constraint manifold and an iterative process is required to completely eliminate constraint violations.

Next, new algorithms have been developed for the enforcement of constraints within the framework of the finite element method, which has gained popularity for multibody dynamics applications. Based on the physical concepts of energy preservation and vanishing of the work done by the constraint forces, robust algorithms have been developed that present mathematical proofs of stability, but at the expense of more complex discretization schemes, moving away from the traditional "black box" approach to time integration schemes. Typically, constraints are satisfied to machine accuracy at both displacement and velocity levels.

For the several past decades, the numerical solution of DAEs has been known to be fraught with difficulties, mainly due to their undesirable behavior for vanishingly small time step sizes. The importance of scaling of both equations of motion and solution fields has been underlined and the following facts have been established.

1. Scaling can be performed at the level of the equations of motion, prior to time discretization. By curing problems a priori, benefits are reaped for all time integration schemes.
2. The proposed scaling factor depends on both time step size and system physical properties, further improving the numerical conditioning of the problem.
3. In multibody formulations, algebraic variables stem from the presence of Lagrange's multipliers, but also from the definition of additional algebraic variables such as relative motions. In such cases, scaling in conjunction with an augmented Lagrangian term was shown to yield time step size independent Jacobians.
4. The combined use of scaling and augmented Lagrangian term also enables the safe use of sparse linear equation solvers that do not rely on pivoting to ensure stable, accurate solutions. While finite element codes routinely rely on such skyline solvers, their safe use for DAEs has been justified and considerably improves the efficiency of the solution process.

Although further theoretical work is needed before more general conclusions can be drawn, the following facts are emerging from the discussion presented in section 12.6 and in refs. [234, 235, 236, 238].

1. High index DAEs, once properly scaled, are not more difficult to integrate than ODEs. Unless leading to computational savings, there is no reason to avoid Lagrange's multipliers, the main source of algebraic variables.
2. While numerous researchers have advocated the use of specific time integration schemes to overcome the ill-conditioning of the linearized index-3 equations, section 12.6 shows that these problems can be resolved a priori, for all stable integration schemes. Furthermore, scaling does not alter the basic properties of time integration schemes. If an integration scheme is energy preserving, its application to scaled equations of motion will still preserve energy.
3. Promoting index reduction techniques to avoid the perceived numerical problems associated with DAEs might be ill advised: section 12.6 results indicate that these techniques might not required. Furthermore, they might create difficulties that were not present in the original formulation based on DAEs; for instance, index reduction techniques often enforce constraints through their higher-order derivatives, leading to the drift phenomenon, which does not affect the direct solution of high-index DAEs. While the drift problem may be alleviated or completely eliminated by the use of projections onto the constraint manifold, the index-3 approach is conceptually simpler and possibly more efficient since it does not incur the extra costs of computing and applying projection operators.
4. The results presented in section 12.6 do not provide a general approach to the solution of DAEs. Indeed, the approach presented in that section only deals with the index-3 DAEs stemming from the modeling of mechanical systems, which

present a linear dependency on generalized accelerations, on Lagrange's multipliers, and on the generalized velocities appearing in nonholonomic constraints. The results presented in section 12.6 are limited to DAEs presenting these three characteristics.

Part V

Parameterization of rotation and motion

13

Parameterization of rotation

The effective description of rotations has led to the development of numerous parameterization techniques presenting various properties and advantages, as described in the following review papers [239, 240, 241, 242, 243, 244, 245]. Whether originating from geometric, algebraic, or matrix approaches, parameterization of rotation is most naturally categorized into two classes: *vectorial* and *non-vectorial* parameterizations. The former refers to parameterization in which a set of parameters, sometimes called rotational "quasi-coordinates," define a geometric vector, whereas the latter cannot be cast in the form of a vector. These two types of parameterizations are sometimes denoted as *invariant* and *non-invariant* parameterization, respectively.

The Cartesian rotation vector, the Euler-Rodrigues parameters, or the Wiener-Milenković parameters all are examples of vectorial parameterizations. These are all characterized by a minimal set of three parameters, which behave as the Cartesian components of a vector in three-dimensional space. Non-vectorial parameterizations, on the other hand, may be either minimal, as in the case of Euler angles, or "redundant," as for Euler parameters, Cayley-Klein parameters, and the matrix of direction cosines.

Redundancy arises when more than three parameters are employed: four in the case of Euler and Cayley-Klein parameters, nine in the case of direction cosines. In fact, rotation may be described as the motion of a point on a three-dimensional nonlinear manifold, the Lie group of special orthogonal transformations of the three-dimensional space. The various parameterizations of rotation are, in differential geometry terminology, different *charts* available for this particular manifold.

Stuelpnagel [246] provides a concise analysis of different parameterizations of rotations. He shows that the six parameter representation consisting of the first two columns of the rotation tensor yields a set of linear differential equations for the motion of a rigid body. He further proves that a minimum of five parameters is required to obtain a bijective mapping of the rotation group. This parameterization yields a set of nonlinear equations of motion for a rigid body and is not recommended for practical applications. Four parameter representations, such as the quaternion representation [247, 248, 249], are singularity free, in contrast with minimal set parameterizations, which he proves to always involve singularities.

The various parameterization techniques detailed in the literature present distinct advantages and drawbacks. Advantages can be of a theoretical nature, such as ease of geometric interpretation, or convenience in algebraic manipulations, for instance, or of a computational nature, low cost function evaluations, wide range of singularity-free behavior, etc. These features provide guidelines for selecting parameterizations that are best suited for specific applications. A survey of the literature reveals, however, that for both theoretical and numerical applications, the choice of parameterization is often based on personal taste and traditions rather than cost/benefit considerations.

Section 13.1 presents an algebraic description of rotation that contrasts with the geometric approach developed in chapter 4. Cayley's elegant formulation is introduced based on the fundamental property of the rotation operation: preservation the length of the rotated vector. Next, section 13.3 introduces the well-known Euler parameters [247, 246, 250, 249] that provide an elegant, purely algebraic representation of rotation. When using the quaternion algebra presented in section 13.2, all rotation operations become bi-linear expressions of quaternions. These advantages, however, come at a high cost: four parameters must be used instead of three, *i.e.*, Euler parameters do not form a minimum set.

Euler's theorem on rotations, see section 4.5, states that an arbitrary motion of a rigid body that leaves one of its point fixed can be represented by a single rotation of magnitude ϕ about unit vector $\bar{n}$. It is readily shown that the associated rotation tensor, $\underline{R}$, possesses a positive unit eigenvalue and the corresponding eigenvector is $\bar{n}$, see section 4.7.

The vectorial parameterization of rotation is introduced in section 13.4 and consists of minimal set of parameters defining the components of a *rotation parameter vector*, $\underline{p} = p(\phi)\bar{n}$, where $p(\phi)$ is the generating function. The vectorial parameterization of rotation presents two fundamental properties. First, it is tensorial in nature: the tensorial nature of the second-order rotation tensor implies and is implied by the tensorial nature of the rotation parameter vector, a first-order tensor. Second, rotation parameter vectors are parallel to the eigenvector of the rotation tensor corresponding to its unit eigenvalue. Because these two properties imply each other, either can be taken as the definition of the vectorial parameterization of rotation. A parameterization of rotation is tensorial if and only if the rotation parameter vector is parallel to the eigenvector of the rotation tensor associated with its unit eigenvalue.

The Cartesian rotation vector, the Cayley-Gibbs-Rodrigues parameters, or the Wiener-Milenković parameters all are special cases of the vectorial parameterization of rotation corresponding to specific choices of the generating function. Furthermore, these parameterizations are recovered as members of two different families: the sine and the tangent family. The occurrence of singularities in the proposed vectorial parameterization is the focus of section 13.6. Finally, section 13.7 details a number of useful parameterizations: the Cartesian rotation vector, the Euler-Rodrigues parameters, the Cayley-Gibbs-Rodrigues parameters, and the Wiener-Milenković parameters.

Euler parameters are closely related to the vectorial parameterization. On the other hand, minimal non-vectorial parameterizations such as Euler and Euler-type angles are not easily related to vectorial techniques. Rather, they may be investigated in terms of exponential coordinates of the second kind, in contrast with the exponential parameterization, which is an application of exponential coordinates of the first kind [251].

13.1 Cayley's rotation parameters

In chapter 4, rotation operations were described in geometric terms, based on the visualization of an orthonormal basis $\mathcal{B}$ rotating to a new basis $\mathcal{E}$. It is possible, however, to describe rotations without resorting to geometric concepts. Indeed, the fundamental property of rotation operations is to preserve length: the length of a vector is the same when computed from its components resolved in two arbitrary orthonormal bases.

Consider an arbitrary vector $\underline{b}$ and its components, $\underline{b}^{[\mathcal{B}]}$ and $\underline{b}^{[\mathcal{E}]}$, resolved in bases $\mathcal{B}$ and $\mathcal{E}$, respectively. Let $\underline{\underline{R}}$ be the rotation tensor that brings basis $\mathcal{B}$ to $\mathcal{E}$. The relationship between the components of $\underline{b}$ in these two bases is given by eq. (4.27) as $\underline{b}^{[\mathcal{E}]} = \underline{\underline{R}}^T \underline{b}^{[\mathcal{B}]}$. The basic property of rotation is to preserve the length, ℓ, of vector $\underline{b}$, i.e., $\ell^2 = \underline{b}^{[\mathcal{E}]T} \underline{b}^{[\mathcal{E}]} = \underline{b}^{[\mathcal{B}]T} \underline{b}^{[\mathcal{B}]}$, which implies

$$\left(\underline{b}^{[\mathcal{B}]T} + \underline{b}^{[\mathcal{E}]T}\right)\left(\underline{b}^{[\mathcal{B}]} - \underline{b}^{[\mathcal{E}]}\right) = 0. \tag{13.1}$$

Vectors $\underline{f}$ and $\underline{g}$ are now defined as $\underline{f} = \underline{b}^{[\mathcal{B}]} - \underline{b}^{[\mathcal{E}]} = (\underline{\underline{R}} - \underline{\underline{I}})\,\underline{b}^{[\mathcal{E}]}$, and $\underline{g} = \underline{b}^{[\mathcal{B}]} + \underline{b}^{[\mathcal{E}]} = (\underline{\underline{R}} + \underline{\underline{I}})\,\underline{b}^{[\mathcal{E}]}$, and the length preservation condition now simply states

$$\underline{g}^T \underline{f} = 0. \tag{13.2}$$

Eliminating $\underline{b}^{[\mathcal{E}]}$ from vectors $\underline{f}$ and $\underline{g}$ leads to $\underline{f} = \underline{\underline{C}}\,\underline{g}$, where

$$\underline{\underline{C}} = (\underline{\underline{R}} - \underline{\underline{I}})(\underline{\underline{R}} + \underline{\underline{I}})^{-1} = (\underline{\underline{R}} + \underline{\underline{I}})^{-1}(\underline{\underline{R}} - \underline{\underline{I}}). \tag{13.3}$$

Tensor $\underline{\underline{R}} + \underline{\underline{I}}$ is only singular when $\phi = \pm\pi$, because the rotation tensor then possesses an eigenvalue $\lambda = -1$. For all other rotations, $\underline{\underline{R}} + \underline{\underline{I}}$ can be inverted. The second equality of eq. (13.3) is readily obtained by noting that $(\underline{\underline{R}} - \underline{\underline{I}})(\underline{\underline{R}} + \underline{\underline{I}})^{-1} = (\underline{\underline{R}} + \underline{\underline{I}})^{-1}(\underline{\underline{R}} + \underline{\underline{I}})(\underline{\underline{R}} - \underline{\underline{I}})(\underline{\underline{R}} + \underline{\underline{I}})^{-1}$, where the second and third factors commute.

The length preservation condition, eq. (13.2), now becomes $\underline{g}^T \underline{\underline{C}}\,\underline{g} = 0$. Rotation operations preserve the length of any arbitrary vector. Consequently, scalar $\underline{g}^T \underline{\underline{C}}\,\underline{g}$ must vanish for any arbitrary vector $\underline{g}$, which implies that $\underline{\underline{C}}$ must be a skew-symmetric tensor, i.e., $\underline{\underline{C}} = \widetilde{a}$, where $\underline{a}$ are Cayley's rotation parameters.

The rotation tensor can be expressed in terms of this skew-symmetric tensor by solving eq. (13.3) for $\underline{\underline{R}}$ to find

$$\underline{\underline{R}} = (\underline{\underline{I}} - \underline{\underline{C}})^{-1}(\underline{\underline{I}} + \underline{\underline{C}}) = (\underline{\underline{I}} + \underline{\underline{C}})(\underline{\underline{I}} - \underline{\underline{C}})^{-1}. \tag{13.4}$$

The determinant of $\underline{\underline{I}} - \underline{\underline{C}}$ is $1 + \underline{a}^T \underline{a}$; hence, this matrix is always invertible. In summary, *there exists a one to one relationship between an orthogonal tensor $\underline{\underline{R}}$ and a skew-symmetric tensor $\underline{\underline{C}}$*. Equation (13.4) is known as *Cayley's formula*.

The structure of the rotation tensor is obtained from eq. (13.4)

$$\underline{\underline{R}} = \frac{1}{1 + \underline{a}^T \underline{a}} \left[(1 + \underline{a}^T \underline{a})\underline{\underline{I}} + 2\,\widetilde{\underline{a}} + 2\widetilde{\underline{a}}\widetilde{\underline{a}} \right]. \tag{13.5}$$

In expanded form, this becomes

$$\underline{\underline{R}} = \frac{1}{1 + \underline{a}^T \underline{a}} \begin{bmatrix} 1 + a_1^2 - a_2^2 - a_3^2 & 2(a_1 a_2 - a_3) & 2(a_1 a_3 + a_2) \\ 2(a_1 a_2 + a_3) & 1 - a_1^2 + a_2^2 - a_3^2 & 2(a_2 a_3 - a_1) \\ 2(a_1 a_3 - a_2) & 2(a_2 a_3 + a_1) & 1 - a_1^2 - a_2^2 + a_3^2 \end{bmatrix}. \tag{13.6}$$

The algebraic description of rotation is based on the length preservation property, eq. (13.1). This description implies the specific structure of the rotation tensor given by eq. (13.5), which explicitly shows the dependency of the rotation tensor on three parameters only. The geometric nature of these parameters, however, is not evident in this purely algebraic approach.

13.2 Quaternion algebra

In section 13.3, it will be shown that rotation operations are conveniently expressed in terms of quaternions. The present preparatory section focuses on the definition of quaternions, the derivation of a number of their properties, and the definition of the operators that ease quaternion algebra.

A *quaternion* [247] is defined as an array of four numbers

$$\hat{e} = \left\{ \begin{matrix} e_0 \\ \underline{e} \end{matrix} \right\}, \tag{13.7}$$

where e_0 is the *scalar part of the quaternion* and $\underline{e}$ the *vector part of the quaternion*. This four component array is not a vector, as it does not transform like a vector, see section 4.8.1. The norm of quaternion $\hat{p}$ is defined as

$$\|\hat{p}\| = \sqrt{\hat{p}^T \hat{p}} = \sqrt{p_0^2 + \underline{p}^T \underline{p}}. \tag{13.8}$$

Quaternions operators

Quaternion operations are conveniently performed using the following matrices or operators of size 4×4

$$\underline{\underline{A}}(\hat{e}) = \begin{bmatrix} e_0 & -\underline{e}^T \\ \underline{e} & e_0\underline{I} + \widetilde{e} \end{bmatrix} = \begin{bmatrix} e_0 & -e_1 & -e_2 & -e_3 \\ e_1 & e_0 & -e_3 & e_2 \\ e_2 & e_3 & e_0 & -e_1 \\ e_3 & -e_2 & e_1 & e_0 \end{bmatrix}, \qquad (13.9a)$$

$$\underline{\underline{B}}(\hat{e}) = \begin{bmatrix} e_0 & -\underline{e}^T \\ \underline{e} & e_0\underline{I} - \widetilde{e} \end{bmatrix} = \begin{bmatrix} e_0 & -e_1 & -e_2 & -e_3 \\ e_1 & e_0 & e_3 & -e_2 \\ e_2 & -e_3 & e_0 & e_1 \\ e_3 & e_2 & -e_1 & e_0 \end{bmatrix}, \qquad (13.9b)$$

$$\underline{\underline{C}}(\hat{e}) = \begin{bmatrix} e_0 & \underline{e}^T \\ \underline{e} & -e_0\underline{I} - \widetilde{e} \end{bmatrix} = \begin{bmatrix} e_0 & e_1 & e_2 & e_3 \\ e_1 & -e_0 & e_3 & -e_2 \\ e_2 & -e_3 & -e_0 & e_1 \\ e_3 & e_2 & -e_1 & -e_0 \end{bmatrix}. \qquad (13.9c)$$

If $\hat{p}$ is an arbitrary quaternion, these operators enjoy the following properties

$$\underline{\underline{A}}(\hat{p})\underline{\underline{A}}^T(\hat{p}) = \underline{\underline{B}}(\hat{p})\underline{\underline{B}}^T(\hat{p}) = \underline{\underline{C}}(\hat{p})\underline{\underline{C}}^T(\hat{p}) = \|\hat{p}\|^2\underline{\underline{I}}_4, \qquad (13.10)$$

where $\underline{\underline{I}}_4$ is the 4×4 identity matrix.

If $\hat{p}$ and $\hat{q}$ are arbitrary quaternions, the following matrix products commute

$$\underline{\underline{A}}(\hat{p})\underline{\underline{B}}^T(\hat{q}) = \underline{\underline{B}}^T(\hat{q})\underline{\underline{A}}(\hat{p}), \quad \underline{\underline{A}}^T(\hat{p})\underline{\underline{B}}(\hat{q}) = \underline{\underline{B}}(\hat{q})\underline{\underline{A}}^T(\hat{p}), \qquad (13.11a)$$

$$\underline{\underline{A}}(\hat{p})\underline{\underline{B}}(\hat{q}) = \underline{\underline{B}}(\hat{q})\underline{\underline{A}}(\hat{p}), \quad \underline{\underline{A}}^T(\hat{p})\underline{\underline{B}}^T(\hat{q}) = \underline{\underline{B}}^T(\hat{q})\underline{\underline{A}}^T(\hat{p}), \qquad (13.11b)$$

$$\underline{\underline{C}}(\hat{p})\underline{\underline{A}}^T(\hat{q}) = \underline{\underline{B}}(\hat{q})\underline{\underline{C}}(\hat{p}), \quad \underline{\underline{C}}^T(\hat{p})\underline{\underline{B}}(\hat{q}) = \underline{\underline{A}}^T(\hat{q})\underline{\underline{C}}^T(\hat{p}). \qquad (13.11c)$$

These identities then imply the following results

$$\underline{\underline{A}}(\hat{p})\hat{q} = \underline{\underline{B}}(\hat{q})\hat{p}, \quad \underline{\underline{A}}^T(\hat{p})\hat{q} = \underline{\underline{C}}^T(\hat{q})\hat{p}, \quad \underline{\underline{B}}^T(\hat{p})\hat{q} = \underline{\underline{C}}(\hat{q})\hat{p}. \qquad (13.12)$$

Next, the following results are easily checked

$$\underline{\underline{A}}(\hat{p})\underline{\underline{A}}(\hat{q}) = \underline{\underline{A}}(\hat{r}) \iff \hat{r} = \underline{\underline{A}}(\hat{p})\hat{q} = \underline{\underline{B}}(\hat{q})\hat{p}, \qquad (13.13a)$$

$$\underline{\underline{A}}(\hat{p})\underline{\underline{A}}^T(\hat{q}) = \underline{\underline{A}}(\hat{r}) \iff \hat{r} = \underline{\underline{C}}(\hat{p})\hat{q} = \underline{\underline{B}}^T(\hat{q})\hat{p}, \qquad (13.13b)$$

$$\underline{\underline{A}}^T(\hat{p})\underline{\underline{A}}(\hat{q}) = \underline{\underline{A}}(\hat{r}) \iff \hat{r} = \underline{\underline{A}}^T(\hat{p})\hat{q} = \underline{\underline{C}}^T(\hat{q})\hat{p}, \qquad (13.13c)$$

where the double-headed arrows indicate that the two equalities imply each other. Similarly

$$\underline{\underline{B}}(\hat{p})\underline{\underline{B}}(\hat{q}) = \underline{\underline{B}}(\hat{r}) \iff \hat{r} = \underline{\underline{B}}(\hat{p})\hat{q} = \underline{\underline{A}}(\hat{q})\hat{p}, \qquad (13.14a)$$

$$\underline{\underline{B}}(\hat{p})\underline{\underline{B}}^T(\hat{q}) = \underline{\underline{B}}(\hat{r}) \iff \hat{r} = \underline{\underline{C}}^T(\hat{p})\hat{q} = \underline{\underline{A}}^T(\hat{q})\hat{p}, \qquad (13.14b)$$

$$\underline{\underline{B}}^T(\hat{p})\underline{\underline{B}}(\hat{q}) = \underline{\underline{B}}(\hat{r}) \iff \hat{r} = \underline{\underline{B}}^T(\hat{p})\hat{q} = \underline{\underline{C}}(\hat{q})\hat{p}. \qquad (13.14c)$$

Finally, the skew-symmetric operator $\underline{\underline{S}}(\hat{p})$ is defined as

$$\underline{\underline{S}}(\hat{p}) = \begin{bmatrix} 0 & \underline{0}^T \\ \underline{0} & \widetilde{p} \end{bmatrix} = \frac{1}{2}[\underline{\underline{A}}(\hat{p}) - \underline{\underline{B}}(\hat{p})]. \qquad (13.15)$$

Unit quaternions

Quaternion $\hat{e}$ is said to be a *unit quaternion* if its norm, eq. (13.8), is unity, *i.e.*, $\|\hat{e}\| = 1$. In view of identity (13.10), operators $\underline{A}(\hat{e})$, $\underline{B}(\hat{e})$, and $\underline{C}(\hat{e})$ now become orthogonal matrices. A bi-linear operator is now defined

$$\underline{\underline{D}}(\hat{e}) = \underline{\underline{A}}(\hat{e})\underline{\underline{B}}^T(\hat{e}) = \underline{\underline{B}}^T(\hat{e})\underline{\underline{A}}(\hat{e}) = \underline{\underline{C}}(\hat{e})\,\underline{\underline{C}}(\hat{e}). \qquad (13.16)$$

It is now readily verified that for a unit quaternion $\hat{e}$,

$$\underline{\underline{D}}(\hat{e}) = \begin{bmatrix} 1 & \underline{0}^T \\ \underline{0} & \underline{\underline{R}}(\hat{e}) \end{bmatrix}, \qquad (13.17)$$

where $\underline{\underline{R}}(\hat{e}) = \underline{\underline{I}} + 2e_0\widetilde{e} + 2\widetilde{e}\widetilde{e}$.

Orthogonal quaternions

Two quaternions, $\hat{p}$ and $\hat{q}$, are said to be orthogonal if $\hat{p}^T\hat{q} = 0$. For such pair of quaternions, the following identities hold

$$\underline{\underline{A}}^T(\hat{p})\underline{\underline{A}}(\hat{q}) + \underline{\underline{A}}^T(\hat{q})\underline{\underline{A}}(\hat{p}) = \underline{\underline{0}}, \quad \underline{\underline{A}}(\hat{p})\underline{\underline{A}}^T(\hat{q}) + \underline{\underline{A}}(\hat{q})\underline{\underline{A}}^T(\hat{p}) = \underline{\underline{0}}. \qquad (13.18a)$$

$$\underline{\underline{B}}^T(\hat{p})\underline{\underline{B}}(\hat{q}) + \underline{\underline{B}}^T(\hat{q})\underline{\underline{B}}(\hat{p}) = \underline{\underline{0}}, \quad \underline{\underline{B}}(\hat{p})\underline{\underline{B}}^T(\hat{q}) + \underline{\underline{B}}(\hat{q})\underline{\underline{B}}^T(\hat{p}) = \underline{\underline{0}}. \qquad (13.18b)$$

13.3 Euler parameters

Euler parameters [248, 247, 246, 250, 249, 252] are presented in this section. These parameters lead to a very simple, purely algebraic representation of rotation. Four parameters, however, instead of three, are used in this representation. These four parameters are related by a normality condition, and are thus not independent.

The four *Euler parameters* are defined as follows

$$e_0 = \cos\frac{\phi}{2}, \quad \underline{e} = \sin\frac{\phi}{2}\bar{n}, \qquad (13.19)$$

where $\bar{n}$ is the unit vector about which the rotation of magnitude ϕ is taking place, according to Rodrigues' rotation formula, eq. (4.15). Consequently, Euler parameters define a rotation operation. Note the redundancy in this representation: four parameters are used instead of three. Of course, these four parameters are linked by the following constraint

$$e_0^2 + e_1^2 + e_2^2 + e_3^2 = 1, \qquad (13.20)$$

where $\underline{e}^T = \{e_1\ e_2\ e_3\}$. Euler parameters are conveniently interpreted as the components of a unit quaternion, $\hat{e}^T = \{e_0, \underline{e}\}$, because eq. (13.20) implies $\|\hat{e}\| = 1$.

13.3.1 The rotation tensor

The rotation tensor can be expressed in terms of the Euler parameters by introducing eqs. (13.19) into Rodrigues' rotation formula, eq. (4.15), to find

$$\underline{R}(\underline{e}) = \underline{I} + 2e_0\widetilde{e} + 2\widetilde{e}\widetilde{e}. \tag{13.21}$$

Rotation operations using Euler parameters are most easily expressed in terms of quaternions. Instead of working with the 3×3 rotation tensor, $\underline{R}$, it is easier to work with the 4×4 operator $\underline{D}$ defined by eq. (13.17), which is closely related to the rotation tensor. Identity (13.16) then gives various expressions for operator $\underline{D}$. The rotation tensor is now a purely algebraic function of Euler parameters.

13.3.2 The angular velocity vector

The components of the angular velocity vector resolved in the inertial basis, see section 4.10, are obtained from their definition, eq. (4.56). Here again, it is easier to work with the 4×4 operators defined in the previous section; indeed,

$$\underline{\dot{D}}(\hat{e})\underline{D}^T(\hat{e}) = \begin{bmatrix} 0 & \underline{0}^T \\ \underline{0} & \underline{\dot{R}}(\hat{e}) \end{bmatrix} \begin{bmatrix} 1 & \underline{0}^T \\ \underline{0} & \underline{R}^T(\hat{e}) \end{bmatrix} = \begin{bmatrix} 0 & \underline{0}^T \\ \underline{0} & \widetilde{\omega} \end{bmatrix} = \underline{S}(\hat{\omega}).$$

The following algebraic manipulations now relate the components of the angular velocity vector to Euler parameters

$$\underline{\dot{D}}(\hat{e})\underline{D}^T(\hat{e}) = \left[\underline{A}(\hat{e})B^T(\hat{e})\right]^{\cdot} \underline{B}(\hat{e})\underline{A}^T(\hat{e}) = A(\hat{e})\underline{A}^T(\hat{e}) + \underline{B}^T(\hat{e})\underline{B}(\hat{e}), \tag{13.22}$$

where identities (13.10) and (13.11a) were used. Note that $\hat{\dot{e}}^T = \{\dot{e}_0, \underline{\dot{e}}\}$ does not form a unit quaternion. Because $\underline{B}(\hat{e})$ is an orthogonal operator, $\underline{B}^T(\hat{\dot{e}})\underline{B}(\hat{e}) = -\underline{B}^T(\hat{e})\underline{B}(\hat{\dot{e}})$, and eq. (13.22) becomes

$$\underline{\dot{D}}(\hat{e})\underline{D}^T(\hat{e}) = \underline{A}(\hat{\dot{e}})\underline{A}^T(\hat{e}) - \underline{B}^T(\hat{e})\underline{B}(\hat{\dot{e}}) = \underline{A}(\frac{\hat{\omega}}{2}) - \underline{B}(\frac{\hat{\omega}}{2}) = \underline{S}(\hat{\omega}),$$

where identities (13.13) and (13.14) were used to find

$$\hat{\omega} = 2\underline{B}^T(\hat{e})\dot{\hat{e}} = 2\underline{C}(\hat{e})\hat{e}. \tag{13.23}$$

The vector part of quaternion $\hat{\omega}$ is the angular velocity vector. Its scalar part, ω_0, follows from the definition of operator $\underline{B}$, eq. (13.9b), as $\omega_0 = 2(e_0\dot{e}_0 + \underline{e}^T\underline{\dot{e}}) = 2\hat{e}^T\dot{\hat{e}} = 0$, because $\hat{e}$ is a unit quaternion.

The components of the angular velocity vector resolved in the rotating frame, see section 4.10, are obtained in a similar manner

$$\hat{\omega}^* = 2\underline{A}^T(\hat{e})\dot{\hat{e}} = 2\underline{C}^T(\hat{e})\hat{e}. \tag{13.24}$$

Time derivatives of Euler parameters can be related to the angular velocities by inverting eqs. (13.23) and (13.24) to find $\dot{\hat{e}} = 1/2\ \underline{B}(\hat{e})\hat{\omega}$ and $\dot{\hat{e}} = 1/2\ \underline{A}(\hat{e})\hat{\omega}^*$, respectively. Because operators $\underline{A}(\hat{e})$ and $\underline{B}(\hat{e})$ are orthogonal operators for unit quaternions, these relationships are free of singularities.

13.3.3 Composition of rotations

The concept of composition of rotations was discussed in section 4.9. Consider three unit quaternions $\hat{p}$, $\hat{q}$, and $\hat{r}$ such that

$$\underline{R}(\hat{r}) = \underline{R}(\hat{p})\underline{R}(\hat{q}). \tag{13.25}$$

The problem at hand is to determine unit quaternion $\hat{r}$ as a function of quaternions $\hat{p}$ and $\hat{q}$.

Here again, eq. (13.25) is expressed by means of 4×4 operators to ease the algebraic manipulations as $\underline{D}(\hat{r}) = \underline{D}(\hat{p})\underline{D}(\hat{q})$. With the help of identity (13.16), this expands to

$$\underline{A}(\hat{r})\underline{B}^T(\hat{r}) = \underline{A}(\hat{p})\underline{B}^T(\hat{p})\underline{A}(\hat{q})\underline{B}^T(\hat{q}) = \underline{A}(\hat{p})\underline{A}(\hat{q})\underline{B}^T(\hat{p})\underline{B}^T(\hat{q}),$$

where identity (13.11a) was used. Identities (13.13) and (13.14) then imply

$$\hat{r} = \underline{A}(\hat{p})\hat{q}. \tag{13.26}$$

It is readily shown that $\hat{r}$ is also a unit quaternion; indeed $\hat{r}^T\hat{r} = \hat{q}^T\underline{A}^T(\hat{p})\underline{A}(\hat{p})\hat{q} = \hat{q}^T\hat{q} = 1$, since $\hat{p}$ and $\hat{q}$ both are unit quaternions.

13.3.4 Determination of Euler parameters

Equation (13.21) expresses the rotation tensor in terms of Euler parameters. In this section, the inverse operation is developed, but unfortunately, it cannot be written in a simple manner. Indeed, any such expression will involve a division by a term that can vanish for certain specific rotation tensors. To overcome this problem, Klumpp [253] and Shepperd [254] introduced the procedure described in this section.

Consider the following symmetric matrix constructed from the components of the rotation tensor

$$\underline{T} = \begin{bmatrix} 1 + \mathrm{tr}(\underline{R}) & 2\,\mathrm{axial}^T(\underline{R}) \\ 2\,\mathrm{axial}(\underline{R}) & \left[1 - \mathrm{tr}(\underline{R})\right]\underline{I} + 2\,\mathrm{symm}(\underline{R}) \end{bmatrix}.$$

Introducing eq. (13.21) then yields

$$\underline{T} = 4 \begin{bmatrix} e_0^2 & e_0e_1 & e_0e_2 & e_0e_3 \\ e_0e_1 & e_1^2 & e_1e_2 & e_1e_3 \\ e_0e_2 & e_1e_2 & e_2^2 & e_2e_3 \\ e_0e_3 & e_1e_3 & e_2e_3 & e_3^2 \end{bmatrix} = 4\,\hat{e}\hat{e}^T. \tag{13.27}$$

Euler parameters can be computed from any column of this matrix, for instance, $e_i = T_{ik}/\Delta_k$, $i = 0, 1, 2, 3$, where $\Delta_k = 2\sqrt{T_{kk}}$. This expression shows the problem associated with the desired inverse relationships: the results become inaccurate when the denominator Δ_k becomes very small, or vanishes.

The most accurate results will be obtained by selecting the denominator of maximum magnitude. In other words, the best results will be obtained by extracting Euler parameters from the column of $\underline{\underline{T}}$ which presents the largest diagonal term. It can be readily shown that

$$\max (T_{00}, T_{11}, T_{22}, T_{33}) = \max \left(\text{tr}(\underline{\underline{R}}), R_{11}, R_{22}, R_{33} \right) . \tag{13.28}$$

If m is the index corresponding to the column with the maximum diagonal term, Euler parameters write

$$e_i = T_{im}/\Delta_m, \quad i = 0, 1, 2, 3. \tag{13.29}$$

The combination of eqs. (13.28) and (13.29) provides a singularity free algorithm for extracting Euler parameters from a given rotation tensor. In contrast, it is not possible to extract Euler angles from a given rotation tensor without encountering singularities, see eq. (4.13).

Example 13.1. *Kinetic energy of a rigid body*
The kinetic energy of a rigid body undergoing an arbitrary motion was developed in section 6.2. To illustrate the use of Euler parameters, the kinetic energy of a rigid body undergoing rotational motion about a fixed inertial point will be evaluated in this example.

The kinetic energy of the rigid body is given by eq. (6.16). To express this quantity in terms of Euler parameters, it is convenient to use quaternion $\hat{\omega}^{*T} = \{\omega_0 \ \underline{\omega}^{*T}\}$, where $\underline{\omega}^*$ are the components of the angular velocity vector resolved in the body attached basis. The scalar part, $\omega_0 = 2\hat{e}^T\dot{\hat{e}} = 0$, because $\hat{e}$ is a unit quaternion. Next, the following 4×4 matrix is introduced

$$\underline{\underline{M}}^{B*} = \begin{bmatrix} m^* & \underline{0}^T \\ \underline{0} & \underline{\underline{I}}^{B*} \end{bmatrix}, \tag{13.30}$$

where $\underline{\underline{I}}^{B*}$ is the mass moment of inertia tensor of the rigid body computed with respect to the fixed inertial point and m^* a representative mass moment of inertia component. Notation $(\cdot)^*$ indicates tensor components resolved in a body attached basis.

The kinetic energy now becomes

$$K = \frac{1}{2} \hat{\omega}^{*T} \underline{\underline{M}}^{B*} \hat{\omega}^*. \tag{13.31}$$

Because the scalar part of quaternion $\hat{\omega}^*$ vanishes, the specific value of coefficient m^* does not affect the value of the kinetic energy; hence, m^* is simply defined as a "representative mass moment of inertia component."

Quaternion $\hat{\omega}^*$ is now readily expressed in terms of Euler parameters using eq. (13.24), to find

$$K = 2 \, \dot{\hat{e}}^T \underline{\underline{A}}(\hat{e}) \underline{\underline{M}}^{B*} \underline{\underline{A}}^T(\hat{e}) \dot{\hat{e}} = 2 \, \hat{e}^T \underline{\underline{C}}(\dot{\hat{e}}) \underline{\underline{M}}^{B*} \underline{\underline{C}}^T(\dot{\hat{e}}) \hat{e}, \tag{13.32}$$

where the last equality follows from identity (13.12). Clearly, it is expeditious to express all quantities in the quaternion formalism before introducing Euler parameters.

Example 13.2. Lagrange's equations of the first kind for a rigid body
Express the equations of motion for a rigid body rotating about a fixed inertial point
in terms of Euler parameters. This representation involves four generalized coordinates, the four Euler parameters, which are linked by the kinematic constraint expressed by eq. (13.20). Lagrange's equations of the first kind will be derived for this
constrained system.

If the constraint equation is written as $C = (e_0^2 + e_1^2 + e_2^2 + e_3^2 - 1)/2 = 0$, the
constraint matrix becomes $\underline{\underline{B}} = \{e_0 \; e_1 \; e_2 \; e_3\} = \hat{e}^T$. The Lagrangian of the system
is simply $L = K - V = K$. The system's generalized momenta and the derivatives
of the Lagrangian with respect to the generalized coordinates are now

$$p = \frac{\partial L}{\partial \dot{\hat{e}}} = 4 \, \underline{\underline{A}}(\hat{e}) \underline{\underline{M}}^{B*} \underline{\underline{A}}^T(\hat{e}) \dot{\hat{e}}, \quad \frac{\partial L}{\partial \hat{e}} = 4 \, \underline{\underline{C}}(\dot{\hat{e}}) \underline{\underline{M}}^{B*} \underline{\underline{C}}^T(\dot{\hat{e}}) \hat{e}.$$

Let the rigid body be acted upon by an external moment, which components
resolved in the material basis are denoted $\underline{Q}^*$. The virtual work done by this mo-
ment is $\delta W^{nc} = \underline{Q}^{*T} \delta \underline{\psi}^*$, where $\delta \underline{\psi}^*$ are the components of the virtual rotation
vector resolved in the material basis. Here again, it is convenient to express the vir-
tual work in terms of quaternions as $\delta W^{nc} = \hat{Q}^{*T} \delta \hat{\psi}^*$, where $\hat{Q}^{*T} = \{0 \; \underline{Q}^{*T}\}$
and $\delta \hat{\psi}^{*T} = \{0 \; \delta \underline{\psi}^{*T}\}$. Both quaternions feature a vanishing scalar part. By anal-
ogy with eq. (13.24), $\delta \hat{\psi}^* = 2 \, \underline{\underline{A}}^T(\hat{e}) \delta \hat{e}$ and the virtual work becomes $\delta W^{nc} = \hat{Q}^{*T} 2 \, \underline{\underline{A}}^T(\hat{e}) \delta \hat{e}$.

Lagrange's equations of the first kind are readily found as

$$\underline{\underline{A}}(\hat{e}) \underline{\underline{M}}^{B*} \left[\underline{\underline{A}}^T(\hat{e}) \ddot{\hat{e}} + \underline{\underline{A}}^T(\dot{\hat{e}}) \dot{\hat{e}} \right] + \left[\underline{\underline{A}}(\dot{\hat{e}}) - \underline{\underline{C}}(\dot{\hat{e}}) \right] \underline{\underline{M}}^{B*} \underline{\underline{A}}^T(\hat{e}) \dot{\hat{e}} + \frac{\lambda}{4} \hat{e} = \frac{1}{2} \underline{\underline{A}}(\hat{e}) \hat{Q}^*,$$

(13.33)

where λ is Lagrange's multiplier used to enforce the normality of the Euler parameters. The equations of motion for the rigid body do not involve transcendental functions, only products of the generalized coordinates, $\hat{e}$.

Example 13.3. Maggi's formulation for a rigid body
Lagrange's equation of the first kind developed in the previous example are
differential-algebraic equations. Use Maggi's formulation presented in section 11.2.1
to derive ordinary equations of motion expressed in terms of Euler parameters.

Equation (13.24) is recast in the following form

$$\left\{ \begin{matrix} 0 \\ \underline{\omega}^* \end{matrix} \right\} = 2 \underline{\underline{A}}^T(\hat{e}) \dot{\hat{e}} = \begin{bmatrix} \underline{B} \\ \underline{\underline{B}} \end{bmatrix} \dot{\hat{e}} = \underline{\underline{B}} \dot{\hat{e}}.$$

The first row of matrix $\underline{\underline{B}}$ defines the constraint matrix, and the next three row define
the kinematic characteristics of the problem, selected to be the components of the
angular velocity vector resolved in the material basis. This linear transformation is
at the heart of Maggi's formulation, see eq. (11.4).

To eliminate Lagrange's multiplier from the formulation, Lagrange's equations
of the first kind, eqs. (13.33), are multiplied by $\underline{\underline{A}}^T(\hat{e})$ to yield

$$\underline{\underline{M}}^{B*}\dot{\hat{\omega}}^* + \underline{\underline{A}}^T(\hat{e}) \left[\underline{\underline{A}}(\dot{\hat{e}}) - \underline{\underline{C}}(\dot{\hat{e}}) \right] \underline{\underline{M}}^{B*}\hat{\omega}^* + \frac{\lambda}{2} \begin{Bmatrix} 1 \\ 0 \end{Bmatrix} = \hat{Q}^*. \tag{13.34}$$

The following identity is readily verified

$$\underline{\underline{A}}^T(\hat{e}) \left[\underline{\underline{A}}(\dot{\hat{e}}) - \underline{\underline{C}}(\dot{\hat{e}}) \right] = 2 \begin{bmatrix} 0 & \underline{r}^T \\ 0 & r_0\underline{I} - \tilde{r} \end{bmatrix},$$

where

$$\hat{r} = \underline{\underline{C}}^T(\hat{e})\dot{\hat{e}} = \frac{1}{2}\underline{\underline{C}}^T(\hat{e})\underline{\underline{A}}(\hat{e})\hat{\omega}^* = \frac{1}{2} \begin{Bmatrix} 0 \\ -\underline{\omega}^* \end{Bmatrix}.$$

These results indicate that expression $\underline{\underline{A}}^T(\hat{e})[\underline{\underline{A}}(\dot{\hat{e}}) - \underline{\underline{C}}(\dot{\hat{e}})]$ is closely related to the angular velocity vector.

Introducing these results into Maggi's equation, eqs. (13.34), leads to

$$\begin{bmatrix} m^* & \underline{0}^T \\ \underline{0} & \underline{\underline{I}}^{B*} \end{bmatrix} \begin{Bmatrix} 0 \\ \dot{\underline{\omega}}^* \end{Bmatrix} + \begin{bmatrix} 0 & -\underline{\omega}^{*T} \\ \underline{0} & \tilde{\underline{\omega}}^* \end{bmatrix} \begin{bmatrix} m^* & \underline{0}^T \\ \underline{0} & \underline{\underline{I}}^{B*} \end{bmatrix} \begin{Bmatrix} 0 \\ \underline{\omega}^* \end{Bmatrix} + \frac{\lambda}{2} \begin{Bmatrix} 1 \\ 0 \end{Bmatrix} = \begin{Bmatrix} 0 \\ \underline{Q}^* \end{Bmatrix}. \tag{13.35}$$

The first equation yields Lagrange's multiplier as $\lambda = 2\, \underline{\omega}^{*T}\underline{\underline{I}}^{B*}\underline{\omega}^* = 4K$: Lagrange's multiplier is a moment that enforces the normality condition for the Euler parameters and its magnitude equals four times the kinetic energy of the rigid body. The second equation is $\underline{\underline{I}}^{B*}\dot{\underline{\omega}}^* + \tilde{\underline{\omega}}^*\underline{\underline{I}}^{B*}\underline{\omega}^* = \underline{Q}^*$, which is the pivot equation, eq. (6.23), for the angular motion of a rigid body about a fixed inertial point. Of course, this equation could have been used from the onset of this development, bypassing the formal derivation of Maggi's formulation.

The complete formulation of the problem consists of seven ordinary differential equations combining Maggi's equations, eqs. (13.35), and the definition of the kinematic characteristics, eqs. (13.24),

$$\begin{Bmatrix} \underline{\omega}^* \\ \hat{e} \end{Bmatrix}' = \begin{Bmatrix} (\underline{\underline{I}}^{B*})^{-1}(\underline{Q}^* - \tilde{\underline{\omega}}^*\underline{\underline{I}}^{B*}\underline{\omega}^*) \\ \underline{\underline{A}}(\hat{e})\hat{\omega}^*/2 \end{Bmatrix}, \tag{13.36}$$

where the scalar part of quaternion $\hat{\omega}^*$ vanishes.

Example 13.4. Hamilton's principle for a rigid body

In example 8.16 on page 332, the equations of motion of a rigid body were derived from Lagrange's formulation. Because this formulation requires the explicit evaluation of the derivatives of the Lagrangian with respect to the generalized coordinates and generalized velocities, a specific parameterization of rotation must be specified at the onset of the formulation: for instance, eq. (13.32) shows the explicit expression of the kinetic energy in terms of Euler parameters and their time derivatives. Examples 8.16 and 13.2 both demonstrate the ensuing complexity of the analytical developments.

Application of Hamilton's principle to the rigid body problem leads to a compact form of the equations of motion, as was shown in example 8.6 on page 312. Derive the governing equations for the rotational motion of a rigid body in terms of Euler parameters based on Hamilton's principle.

The rigid body is acted upon by an external moment and the virtual work done by this moment is $\delta W^{nc} = \underline{Q}^T \delta \underline{\psi}$, where $\delta \underline{\psi}$ are the components of the virtual rotation vector resolved in the inertial basis. It is convenient to express the virtual work in terms of quaternions as $\delta W^{nc} = \hat{Q}^T \delta \hat{\psi}$, where $\hat{Q} = \{0 \ \underline{Q}\}$ and $\delta \hat{\psi}^T = \{\delta \psi_0 \ \delta \underline{\psi}^T\}$; $\delta \psi_0$ is the scalar part of the virtual rotation quaternion.

Similarly, the kinetic energy of the system can be expressed in terms of the angular velocity quaternion, see eq. (13.31), and variation of this quantity becomes $\delta K = \delta \hat{\omega}^{*T} \underline{\underline{M}}^{B*} \hat{\omega}^* = \delta \hat{\omega}^{*T} \hat{h}^*$, where the vector part of quaternion $\hat{h}^{*T} = \{0, \underline{h}^{*T}\}$ stores the components of the angular momentum vector resolved in the material basis. Because this quaternion has a vanishing scalar part, eq. (4.102b) yields $\delta K = \delta \underline{\omega}^{*T} \underline{h}^* = \delta \dot{\underline{\psi}}^T \underline{\underline{R}} \underline{h}^* = \delta \dot{\underline{\psi}}^T \underline{h}$, where array $\underline{h}$ stores the components of the angular momentum vector resolved in the inertial basis. Finally, $\delta K = \delta \dot{\hat{\psi}}^T \hat{h}$, where the angular momentum quaternion is defined as $\hat{h} = \{0, \underline{h}\}$.

The normality constraint to be imposed on Euler parameters is expressed as $\mathcal{C} = \hat{e}^T \hat{e} - 1 = 0$, and the potential of the constraint, eq. (10.6), becomes $V^c = -\lambda \mathcal{C}$, where λ is Lagrange's multiplier used to enforce the constraint. Variation of this potential is $\delta V^c = -\delta \lambda \ \mathcal{C} - 2\lambda \delta \hat{e}^T \hat{e}$. By analogy with eq. (13.23), the relationship between the virtual rotation quaternion and virtual changes in Euler parameters is $\delta \hat{\psi} = 2 \underline{\underline{B}}^T(\hat{e}) \delta \hat{e}$. Variation in the constraint potential becomes $\delta V^c = -\delta \lambda \ \mathcal{C} - \delta \hat{\psi}^T \lambda \hat{1}$, where $\hat{1} = \underline{\underline{B}}^T(\hat{e}) \hat{e} = \{1, \underline{0}^T\}$ is a quaternion with a unit scalar part and a vanishing vector part.

For this problem, $L^+ = K - V^c$ and Hamilton's principle for constrained system, eq. (10.10), now implies

$$\int_{t_i}^{t_f} \left(\delta \dot{\hat{\psi}} \hat{h} + \delta \lambda \ \mathcal{C} + \delta \hat{\psi}^T \lambda \hat{1} + \delta \hat{\psi}^T \hat{Q} \right) \, dt = 0.$$

All boundary terms are ignored here. After integration by parts of the first term, the equations of motion of the system are found to be $\mathcal{C} = 0$, the constraint equation, and $\dot{\hat{h}} - \lambda \hat{1} = \hat{Q}$. Lagrange's multiplier is readily eliminated from the last equation because quaternion $\hat{1}$ has a vanishing vector part, leading to $\dot{\underline{h}} = \underline{Q}$.

The last step of the procedure is to evaluate the time derivatives of Euler parameters with respect to time. First, eq. (13.24) yields $\dot{\hat{e}} = \underline{\underline{A}}(\hat{e}) \hat{\omega}^* / 2$, and finally, $\dot{\hat{e}} = \underline{\underline{A}}(\hat{e}) (\underline{\underline{M}}^{B*})^{-1} \underline{\underline{D}}^T(\hat{e}) \hat{h} / 2$, which leads to the following system of ordinary differential equations,

$$\left\{ \begin{matrix} \underline{h} \\ \hat{e} \end{matrix} \right\}' = \left\{ \begin{matrix} \underline{Q} \\ \underline{\underline{A}}(\hat{e}) \, (\underline{\underline{M}}^{B*})^{-1} \underline{\underline{D}}^T(\hat{e}) \hat{h} / 2 \end{matrix} \right\}, \tag{13.37}$$

where the scalar part of quaternion $\hat{h}$ vanishes.

It is left to the reader to show that eqs. (13.37) are identical eqs. (13.36) obtained from Lagrange's formulation from which Lagrange's multipliers have been eliminated using Maggi's formulation. The present procedure, based on Hamilton's principle, is far more expeditious.

13.3.5 Problems

Problem 13.1. Angular velocity with Euler parameters
Starting from eq. (4.55), prove that the components of the angular velocity vector expressed in the rotating system are given by eq. (13.24).

Problem 13.2. Euler parameters in terms of Euler angles
Consider the Euler angles with the *3-1-3* sequence described in section 4.11.1. Show that the Euler parameters defining this rotation are given in terms of Euler angles as

$$
\begin{Bmatrix} e_0 \\ e_1 \\ e_2 \\ e_3 \end{Bmatrix} = \begin{Bmatrix} \cos(\phi + \psi)/2 \ \cos\theta/2 \\ \cos(\phi - \psi)/2 \ \sin\theta/2 \\ \sin(\phi - \psi)/2 \ \sin\theta/2 \\ \sin(\phi + \psi)/2 \ \cos\theta/2 \end{Bmatrix}.
$$

Hint. Express the rotation as a succession of three planar rotations, see example 4.4. Find the Euler parameters of each planar rotation. Use the composition of rotation formula, eq. (13.26), to find the desired result.

Problem 13.3. Time dependent quaternions
Consider the following time dependent quaternion,

$$
\hat{e} = \begin{Bmatrix} \cos\phi/2 \\ \sin\phi/2 \ \sin\theta \ \cos\psi \\ \sin\phi/2 \ \sin\theta \ \sin\psi \\ \sin\phi/2 \ \cos\theta \end{Bmatrix},
$$

where $\phi(t) = 3t + 5t^2$, $\theta(t) = 2t$ and $\psi(t) = 7t - 3t^3$. *(1)* Show that $\hat{e}$ is a unit quaternion. *(2)* Compute the quaternion $\dot{\hat{e}}(t)$ and its norm. Is it a unit quaternion? *(3)* Plot the components of angular velocity vector in the fixed system for $t \in [0, 2]$ s. *(4)* Plot the components of angular velocity vector in the rotating system for $t \in [0, 2]$ s. *(5)* Consider the following approximation for the components of the angular velocity vector in the fixed system

$$
\tilde{\omega}(t + \Delta t/2) \approx \frac{\underline{\underline{R}}(t + \Delta t) - \underline{\underline{R}}(t)}{\Delta t} \frac{\underline{\underline{R}}^T(t + \Delta t) + \underline{\underline{R}}^T(t)}{2}.
$$

Prove that this can be written as

$$
\tilde{\omega}(t + \Delta t/2) \approx \frac{[\underline{\underline{R}}(t + \Delta t)\underline{\underline{R}}^T(t)] - [\underline{\underline{R}}(t + \Delta t)\underline{\underline{R}}^T(t)]^T}{2\Delta t}.
$$

(6) On one graph, plot the exact and approximate components of angular velocity vector in the fixed system for $t \in [0, 2]$ s. *(7)* Find the corresponding approximation for the components of the angular velocity vector in the rotating system. *(8)* On one graph, plot the exact and approximate components of angular velocity vector in the rotating system for $t \in [0, 2]$ s.

Problem 13.4. Composition of rotations with quaternions
Consider components of three rotation tensors $\underline{\underline{R}}_1$, $\underline{\underline{R}}_2$, and $\underline{\underline{R}} = \underline{\underline{R}}_1\underline{\underline{R}}_2$, all resolved in a single basis $\mathcal{B}$. The components of $\underline{\underline{R}}_1$ and $\underline{\underline{R}}_2$ are

$$
\underline{\underline{R}}_1 = \begin{bmatrix} 0.5996 & 0.7336 & 0.3199 \\ -0.4732 & 0.6473 & -0.5976 \\ -0.6455 & 0.2069 & 0.7352 \end{bmatrix}, \quad \underline{\underline{R}}_2 = \begin{bmatrix} -0.0282 & 0.2133 & 0.9766 \\ -0.4423 & 0.8735 & -0.2035 \\ -0.8964 & -0.4377 & 0.0697 \end{bmatrix}.
$$

(1) Extract Euler parameters $\hat{e}_1$ and $\hat{e}_2$ corresponding to rotation tensors $\underline{\underline{R}}_1$ and $\underline{\underline{R}}_2$, respectively, using the procedure described in section 13.3.4. *(2)* Compute Euler parameters $\hat{e}$ of $\underline{\underline{R}}$ using the composition formula, eq. (13.26). *(3)* Compute the components of the rotation tensor $\underline{\underline{R}} = \underline{\underline{R}}_1 \underline{\underline{R}}_2$. Extract Euler parameters $\hat{e}$ corresponding to rotation tensor $\underline{\underline{R}}$. Check that your answer is identical to that of question *(2)*.

Problem 13.5. Satellite dynamics with quaternions

Consider a satellite with a body attached frame $\mathcal{B} = (\bar{b}_1, \bar{b}_2, \bar{b}_3)$ that is aligned with the principal axes of inertia of the system. Let $\underline{\underline{R}}$ denote the components of the rotation tensor that brings the inertial frame $\mathcal{I}$ to the body attached frame $\mathcal{B}$. The components of the angular velocity vector of the satellite, resolved in $\mathcal{B}$, are denoted $\underline{\omega}^*$. The mass moments of inertia are $I_1^* = 12$, $I_2^* = 16$ and $I_3^* = 20$ kg·m². During a maneuver, thrusters apply a moment $\underline{M}(t)$ to the satellite $\underline{M}^*(t) = \underline{Q}^* \sin 2\pi t/T$ for $t \leq T$ and $\underline{M}^*(t) = 0$ for $t > T$, where $T = 5$ s. The initial angular velocity of the satellite is $\underline{\omega}^{*T}(t = 0) = \{0, 0.5, 0\}$ rad/s. The components of the moment vector $\underline{Q}^*$ in the body attached frame are $\underline{Q}^{*T} = \{5, 0, 0\}$ N·m. *(1)* Solve Euler's equation for the time history of the angular velocity of the satellite. *(2)* Simultaneously solve for the Euler parameters $\hat{e}$ parameterizing $\underline{\underline{R}}$. *(3)* On one graph, plot the three components of the angular velocity vector in the body attached frame as a function of time for $t \in [0, 30T]$. *(4)* Plot the Euler parameters e_1, e_2, and e_3. *(5)* The rotation $\underline{\underline{R}}(t)$ can be represented by a rotation of magnitude $\phi(t)$ about axis $\bar{n}(t)$. Plot the angle $\phi(t)$. *(6)* Plot the components of axis $\bar{n}(t)$ in frame $\mathcal{I}$. *(7)* Plot the components of axis $\bar{n}(t)$ in frame $\mathcal{B}$. Comment on your results. *(8)* Plot the direction cosines of axis $\bar{b}_1$ with respect to basis $\mathcal{I}$. *(9)* Same question for axis $\bar{b}_2$. *(10)* Same question for axis $\bar{b}_3$. *(11)* Plot the quantity $\hat{e}^T \hat{e} - 1$. Comment on your results.

13.4 The vectorial parameterization of rotation

The *vectorial parameterization of rotation* [255] consists of a minimum set of three parameters defining the components of a *rotation parameter vector*. The tensorial nature of this class of parameterization of rotation sets it apart from the parameterizations investigated in previous sections.

13.4.1 Fundamental properties

Consider three rotations of magnitudes ϕ_1, ϕ_2, and ϕ_3, about unit vectors $\bar{n}_1$, $\bar{n}_2$, and $\bar{n}_3$, respectively. The three rotations, denoted $(\phi_1, \bar{n}_1)$, $(\phi_2, \bar{n}_2)$, and $(\phi_3, \bar{n}_3)$, respectively, are associated with three rotation tensors, denoted $\underline{\underline{R}}_1$, $\underline{\underline{R}}_2$, and $\underline{\underline{R}}_3$, respectively, through Rodrigues' rotation formula, eq. (4.15).

Assume that the following triple product of rotation tensors relates these three quantities,

$$\underline{\underline{R}}_3 = \underline{\underline{R}}_2^T \underline{\underline{R}}_1 \underline{\underline{R}}_2. \tag{13.38}$$

As discussed in section 4.8.2, this operation corresponds to a change of basis for second-order tensors: $\underline{\underline{R}}_1$ and $\underline{\underline{R}}_3$ are the components of the same rotation tensor expressed in two bases related by rotation tensor $\underline{\underline{R}}_2$.

Using Rodrigues' rotation formula, eq. (4.15), eq. (13.38) becomes

$$\underline{\underline{R}}_3 = \underline{\underline{I}} + \sin \phi_1 \widetilde{\underline{\underline{R}}_2^T \bar{n}_1} + (1 - \cos \phi_1) \widetilde{\underline{\underline{R}}_2^T \bar{n}_1} \widetilde{\underline{\underline{R}}_2^T \bar{n}_1},$$

where eq. (4.30) was used. Comparing this result with Rodrigues' rotation formula implies that

$$\phi_3 = \phi_1, \tag{13.39a}$$

$$\bar{n}_3 = \underline{\underline{R}}_2^T \bar{n}_1. \tag{13.39b}$$

These equations express the two conditions required for the proper transformation of rotation tensors components under a change of basis.

Let $p(\phi)$ be an odd scalar function of angle ϕ; eq. (13.39a) then implies $p(\phi_3) = p(\phi_1)$. Multiplication of eq. (13.39b) by $p(\phi_3)$ on the left-hand side and $p(\phi_1) = p(\phi_3)$ on the right-hand side then yields

$$p(\phi_3)\bar{n}_3 = \underline{\underline{R}}_2^T p(\phi_1)\bar{n}_1. \tag{13.40}$$

This equation is equivalent to eqs. (13.39). Indeed, taking the norm of eq. (13.40) yields $p(\phi_3) = p(\phi_1)$, or $\phi_3 = \phi_1$, because $\bar{n}_1$ and $\bar{n}_1$ are unit vectors and $\underline{\underline{R}}_2$ an orthogonal tensor. Dividing eq. (13.40) by $p(\phi_3)$ then yields eq. (13.39b) because $p(\phi_3) = p(\phi_1)$.

The *vectorial parameterization of rotation* is defined as

$$\underline{p} = p(\phi)\bar{n}, \tag{13.41}$$

where $\underline{p}$ is the rotation parameter vector. Equation (13.40) can now be recast in a more compact manner as

$$\underline{p}_3 = \underline{\underline{R}}^T (\underline{p}_2) \, \underline{p}_1. \tag{13.42}$$

The discussion presented above establishes that the tensorial nature of the rotation tensor expressed by the transformation rule of its components, eq. (13.38), implies the tensorial nature of the rotation parameter vector expressed by the transformation rule of its components, eq. (13.42). It is easily shown that the process can be reversed, *i.e.*, tensorial nature of the rotation parameter vector implies that of the rotation tensor.

In summary, the vectorial parameterization of rotation presents two fundamental properties.

1. *The vectorial parameterization of rotation is tensorial in nature,* as expressed by the following equivalence,

$$\underline{\underline{R}}(\underline{p}_3) = \underline{\underline{R}}^T (\underline{p}_2)\underline{\underline{R}}(\underline{p}_1)\underline{\underline{R}}(\underline{p}_2) \iff \underline{p}_3 = \underline{\underline{R}}^T (\underline{p}_2)\underline{p}_1. \tag{13.43}$$

The tensorial nature of the second-order rotation tensor implies and is implied by the tensorial nature of the rotation parameter vector, a first-order tensor.
2. Rotation parameter vectors are parallel to the *eigenvector of the rotation tensor* corresponding to its unit eigenvalue. Because unit vector $\bar{n}$ is the eigenvector of the rotation tensor associated with its unit eigenvalue, eq. (4.24), the definition of the rotation parameter vector, eq. (13.41), implies its parallelism to $\bar{n}$.

Because these two properties imply each other, either can be taken as the definition of the vectorial parameterization of rotation. A parameterization of rotation is tensorial if and only if the rotation parameter vector is parallel to the eigenvector of the rotation tensor associated with its unit eigenvalue.

The rotation parameter vector is not yet fully defined because function $p(\phi)$, called the *generating function*, is still arbitrary. Generating functions must be odd functions of the rotation angle, ϕ, and present the following limit behavior

$$\lim_{\phi \to 0} p(\phi) = \phi, \qquad (13.44)$$

i.e., all rotation parameter vectors must approach the *infinitesimal rotation vector* when $\phi \to 0$. It will be shown that many widely used rotation parameterization belong to this class of vectorial parameterization.

13.4.2 The rotation tensor

The explicit expression of the rotation tensor in term of the vectorial parameterization is easily obtained from Rodrigues' rotation formula, eq. (4.15),

$$\underline{\underline{R}} = \underline{\underline{I}} + \zeta_1(\phi)\,\widetilde{p} + \zeta_2(\phi)\,\widetilde{p}\widetilde{p}, \qquad (13.45)$$

where $\zeta_1(\phi)$ and $\zeta_2(\phi)$ are even functions of the rotation angle, ϕ, defined as

$$\zeta_1(\phi) = \frac{\sin\phi}{p} = \nu\cos\frac{\phi}{2} = \frac{\nu^2}{\varepsilon}, \qquad (13.46a)$$

$$\zeta_2(\phi) = \frac{1 - \cos\phi}{p^2} = \frac{\nu^2}{2} = \frac{\varepsilon\zeta_1}{2}. \qquad (13.46b)$$

The following two even functions of the rotation angle play an important role in the vectorial parameterization of rotation,

$$\nu = \frac{2\sin\phi/2}{p}, \qquad (13.47a)$$

$$\varepsilon = \frac{2\tan\phi/2}{p} = \frac{\nu}{\cos\phi/2}. \qquad (13.47b)$$

In view of eq. (13.44), $\lim_{\phi \to 0}\nu = 1$, $\lim_{\phi \to 0}\varepsilon = 1$, $\lim_{\phi \to 0}\zeta_1 = 1$, $\lim_{\phi \to 0}\zeta_2 = 1/2$, and $\lim_{\phi \to 0}\underline{\underline{R}} = \underline{\underline{I}}$, as expected.

Functions $p(\phi)$, $\zeta_1(\phi)$, and $\zeta_2(\phi)$ solely depend on the magnitude, ϕ, of the rotation. Because this angle is invariant under a change of basis, these functions are also invariant under a change of basis, and hence, are zeroth order tensors. Since $\widetilde{p}$ is a second-order tensor, see eq. (4.30), eq. (13.45) proves that the rotation tensor is a second-order tensor because it is obtained through tensor operations from zeroth and second-order tensors. This proof of the tensorial nature of the rotation tensor mirrors that provided in section 13.4.1.

The components of the rotation tensor resolved in the canonical basis, $\mathcal{E}$, defined by eq. (4.32), are

$$\underline{\underline{R}}^{[\mathcal{E}]} = \begin{bmatrix} 1 & 0 & 0 \\ 0 & 1 - p^2\zeta_2 & -p\zeta_1 \\ 0 & p\zeta_1 & 1 - p^2\zeta_2 \end{bmatrix} = \begin{bmatrix} 1 & 0 & 0 \\ 0 & \cos\phi & -\sin\phi \\ 0 & \sin\phi & \cos\phi \end{bmatrix}. \tag{13.48}$$

Clearly, functions ζ_1 and ζ_2 are not independent because $(1 - p^2\zeta_2)^2 + (p\zeta_1)^2 = 1$ and hence, $\zeta_1^2 + p^2\zeta_2^2 = \nu^2$.

The two multiplicative decompositions of the rotation tensor, eqs. (4.20) and (4.22), are easily expressed in terms of the vectorial parameterization as

$$\underline{\underline{G}} = \underline{\underline{I}} + \frac{\nu}{2}\widetilde{p} + \frac{1 - \cos\phi/2}{p^2}\widetilde{p}\widetilde{p}, \tag{13.49}$$

leading to $\underline{\underline{R}} = \underline{\underline{G}}\,\underline{\underline{G}}$, and

$$\underline{\underline{R}} = (\underline{\underline{I}} + \frac{\varepsilon}{2}\widetilde{p})(\underline{\underline{I}} - \frac{\varepsilon}{2}\widetilde{p})^{-1} = (\underline{\underline{I}} - \frac{\varepsilon}{2}\widetilde{p})^{-1}(\underline{\underline{I}} + \frac{\varepsilon}{2}\widetilde{p}). \tag{13.50}$$

Here again,

$$(\underline{\underline{I}} - \frac{\varepsilon}{2}\widetilde{p})^{-1} = \frac{\underline{\underline{R}} + \underline{\underline{I}}}{2}. \tag{13.51}$$

This decomposition implies

$$\underline{\underline{R}} - \underline{\underline{I}} = \zeta_1\widetilde{p}(\underline{\underline{I}} + \frac{\varepsilon}{2}\widetilde{p}) = (\underline{\underline{I}} + \frac{\varepsilon}{2}\widetilde{p})\zeta_1\widetilde{p}, \tag{13.52a}$$

$$\underline{\underline{R}} - \underline{\underline{I}} = \varepsilon\widetilde{p}(\underline{\underline{I}} - \frac{\varepsilon}{2}\widetilde{p})^{-1} = (\underline{\underline{I}} - \frac{\varepsilon}{2}\widetilde{p})^{-1}\varepsilon\widetilde{p}. \tag{13.52b}$$

Tensors $\underline{\underline{R}}$ and $\underline{\underline{G}}$ are related by the following identities

$$\underline{\underline{R}} - \underline{\underline{I}} = \frac{\underline{\underline{R}} + \underline{\underline{I}}}{2}\varepsilon\widetilde{p} = \varepsilon\widetilde{p}\frac{\underline{\underline{R}} + \underline{\underline{I}}}{2} = \nu\underline{\underline{G}}\widetilde{p} = \widetilde{p}\nu\underline{\underline{G}}, \tag{13.53a}$$

$$\varepsilon\widetilde{p}\,(\frac{\underline{\underline{G}} + \underline{\underline{G}}^T}{2}) = (\frac{\underline{\underline{G}} + \underline{\underline{G}}^T}{2})\,\varepsilon\widetilde{p} = \nu\widetilde{p} = \underline{\underline{G}} - \underline{\underline{G}}^T, \tag{13.53b}$$

$$(\underline{\underline{I}} + \varepsilon\frac{\widetilde{p}}{2})^T\underline{\underline{G}} = (\underline{\underline{I}} + \varepsilon\frac{\widetilde{p}}{2})\underline{\underline{G}}^T = \left(\frac{\underline{\underline{G}} + \underline{\underline{G}}^T}{2}\right)^{-1}. \tag{13.53c}$$

13.4.3 The angular velocity vector

Taking a time derivative of the rotation parameter vector yields $\dot{p} = p'\dot{\phi}\bar{n} + p\dot{\bar{n}}$, where $p' = \mathrm{d}p/\mathrm{d}\phi$. Use of identity (1.33b) leads to $\widetilde{n}\widetilde{n}\dot{p} = p\widetilde{n}\dot{\bar{n}}\dot{\bar{n}} = -p\dot{\bar{n}} = p'\dot{\phi}\bar{n} - \dot{p}$, because $\bar{n}$ is a unit vector and hence, $\bar{n}^T\dot{\bar{n}} = 0$. Introducing these results into the expression for the angular velocity, eq. (4.58), then leads to

$$\underline{\omega} = \underline{\underline{H}}(\underline{p})\dot{\underline{p}}. \tag{13.54}$$

Operator $\underline{\underline{H}}(p)$ is given by

$$\underline{\underline{H}}(p) = \sigma_0(\phi)\,\underline{\underline{I}} + \sigma_1(\phi)\,\widetilde{p} + \sigma_2(\phi)\,\widetilde{p}\widetilde{p}, \tag{13.55}$$

where $\sigma_0(\phi)$, $\sigma_1(\phi)$, and $\sigma_2(\phi)$ are even functions of the rotation angle, ϕ, defined as

$$\sigma_0(\phi) = \frac{1}{p'}, \tag{13.56a}$$

$$\sigma_1(\phi) = \frac{1 - \cos\phi}{p^2} = \zeta_2, \tag{13.56b}$$

$$\sigma_2(\phi) = \frac{\sigma_0 - \zeta_1}{p^2}. \tag{13.56c}$$

These three functions are zeroth order tensors because they are functions of angle ϕ, which is invariant under a change of basis. Using eq. (13.44), $\lim_{\phi\to 0}\sigma_0 = 1$, $\lim_{\phi\to 0}\sigma_1 = 1/2$, $\lim_{\phi\to 0}\sigma_2 = 0$ and $\lim_{\phi\to 0}\underline{\underline{H}} = \underline{\underline{I}}$. Since $\underline{\underline{H}}\bar{n} = \sigma_0\bar{n}$, σ_0 is the eigenvalue of $\underline{\underline{H}}$ corresponding to the eigenvector $\bar{n}$.

Properties of the tangent tensor

The components of the tangent tensor, resolved in the canonical basis, $\mathcal{E}$, defined by eq. (4.32), are

$$\underline{\underline{H}}^{[\mathcal{E}]} = \begin{bmatrix} 1/p' & 0 & 0 \\ 0 & \zeta_1 & -p\sigma_1 \\ 0 & p\sigma_1 & \zeta_1 \end{bmatrix} = \nu \begin{bmatrix} 1/(\nu p') & 0 & 0 \\ 0 & \cos\phi/2 & -\sin\phi/2 \\ 0 & \sin\phi/2 & \cos\phi/2 \end{bmatrix}. \tag{13.57}$$

The eigenvalues, μ_k, of $\underline{\underline{H}}$ are $\mu_1 = \sigma_0$ and $\mu_{2,3} = \zeta_1 \pm ip\zeta_2 = \nu(\cos\phi/2 \pm i\sin\phi/2)$. The determinant of $\underline{\underline{H}}$ is readily obtained as

$$\det(\underline{\underline{H}}) = \frac{\nu^2}{p'}. \tag{13.58}$$

Time derivatives of the rotation parameter vector can be expressed in terms of angular velocity vector as

$$\dot{p} = \underline{\underline{H}}^{-1}(p)\underline{\omega}, \tag{13.59}$$

where

$$\underline{\underline{H}}^{-1}(p) = \chi_0\underline{\underline{I}} - \frac{1}{2}\widetilde{p} + \chi_2\widetilde{p}\widetilde{p}, \tag{13.60}$$

where $\chi_0(\phi)$ and $\chi_2(\phi)$ are even functions of the rotation angle, ϕ, defined as

$$\chi_0(\phi) = p', \tag{13.61a}$$

$$\chi_2(\phi) = \frac{1}{p^2}\left(p' - \frac{1}{\varepsilon}\right). \tag{13.61b}$$

These two functions are zeroth order tensors because they are functions of angle ϕ, which is invariant under a change of basis.

The components of tensor $\underline{\underline{H}}^{-1}$ resolved in the canonical base, $\mathcal{E}$, defined by eq. (4.32) are

$$\underline{\underline{H}}^{-1[\mathcal{E}]} = \frac{1}{\nu} \begin{bmatrix} \nu\chi_0 & 0 & 0 \\ 0 & \cos\phi/2 & \sin\phi/2 \\ 0 & -\sin\phi/2 & \cos\phi/2 \end{bmatrix}. \tag{13.62}$$

As expected, the eigenvalues, $\bar{\mu}_k$, of $\underline{\underline{H}}^{-1}$ are $\bar{\mu}_1 = \chi_0$ and $\bar{\mu}_{2,3} = (\cos\phi/2 \pm i\sin\phi/2)/\nu$.

Operator $\underline{\underline{H}}$ enjoys the following remarkable properties,

$$\underline{\underline{R}} = \underline{\underline{H}}\,\underline{\underline{H}}^{-T} = \underline{\underline{H}}^{-T}\underline{\underline{H}}, \tag{13.63a}$$

$$\underline{\underline{R}} - \underline{\underline{I}} = \widetilde{p}\underline{\underline{H}} = \underline{\underline{H}}\,\widetilde{p}, \tag{13.63b}$$

$$\underline{\underline{I}} - \underline{\underline{R}}^T = \nu^2\widetilde{p}\underline{\underline{H}}^{-1} = \nu^2\underline{\underline{H}}^{-1}\widetilde{p}, \tag{13.63c}$$

$$\widetilde{p} = \underline{\underline{H}}^{-T} - \underline{\underline{H}}^{-1}. \tag{13.63d}$$

Finally, the components of the angular velocity vector resolved in the rotating system are given by eq. (4.55) as $\underline{\omega}^* = \underline{\underline{R}}^T\underline{\omega}$. In view of eq. (13.63a), $\underline{\omega}^* = \underline{\underline{H}}^T(\underline{p})\dot{\underline{p}}$ and the inverse relationship is $\dot{\underline{p}} = \underline{\underline{H}}^{-T}(\underline{p})\underline{\omega}^*$.

As discussed in section 7.3, the virtual rotation vector, $\delta\underline{\psi}$, can be defined by analogy to the angular velocity vector as $\widetilde{\delta\psi} = \delta\underline{\underline{R}}\,\underline{\underline{R}}^T$. Hence, the relationship between the virtual rotation vector and virtual changes in the vectorial parameters is readily found to be $\delta\underline{\psi} = \underline{\underline{H}}(\underline{p})\delta\underline{p}$ and similarly, $\delta\underline{\psi}^* = \underline{\underline{H}}^T(\underline{p})\delta\underline{p}$.

Tensorial nature of the tangent operator

Operator $\underline{\underline{H}}$ is specific to a particular vectorial parameterization, *i.e.*, its expression depends on the choice of the generating function. It is, however, a second-order tensor. In equation (13.55), $\widetilde{p}$ is a second-order tensor, and scalars σ_0, σ_1, and σ_2 are zeroth order tensors. Consequently, $\underline{\underline{H}}$ must be a second-order tensor because it is obtained through tensor operations from zeroth and second-order tensors,

$$\underline{\underline{H}}(\underline{p}_3) = \underline{\underline{R}}^T(\underline{p}_2)\underline{\underline{H}}(\underline{p}_1)\underline{\underline{R}}(\underline{p}_2) \iff \underline{p}_3 = \underline{\underline{R}}^T(\underline{p}_2)\underline{p}_1. \tag{13.64}$$

Although tensor $\underline{\underline{H}}$ is not an intrinsic tensor because it depends on the choice of a specific generating function, it is a second-order tensor for all vectorial parameterizations of rotation.

13.4.4 Determination of the rotation parameter vector

The determination of the components of the vectorial parameterization from the rotation tensor is best accomplished through a two step procedure: first, extract Euler parameters from the rotation tensor using eq. (13.29), and second, express the vectorial parameterization in terms of Euler parameters. This second operation simply states that

$$\underline{p} = \frac{2}{\nu}\underline{e}. \tag{13.65}$$

13.4.5 Composition of rotations

The concept of composition of rotations was discussed in section 4.9 and is depicted in fig. 4.7. Let $\underline{p}_1$, $\underline{p}_2$, and $\underline{p}$ with rotation angles ϕ_1, ϕ_2, and ϕ, respectively, be the rotation parameter vectors of three rotation tensors such that $\underline{R}(\underline{p}) = \underline{R}(\underline{p}_1)\underline{R}(\underline{p}_2)$, the relationship between the various parameters then follows from eq. (13.26)

$$\cos\frac{\phi}{2} = \nu_1\nu_2\left(\frac{1}{\varepsilon_1\varepsilon_2} - \frac{1}{4}\underline{p}_1^T\underline{p}_2\right), \tag{13.66a}$$

$$\nu\underline{p} = \nu_1\nu_2\left(\frac{1}{\varepsilon_2}\underline{p}_1 + \frac{1}{\varepsilon_1}\underline{p}_2 + \frac{1}{2}\widetilde{p}_1\underline{p}_2\right). \tag{13.66b}$$

The first equation is used to compute ϕ and hence, ν. The second equation then yields the components of the rotation parameter vector.

13.4.6 Linearization of the tangent tensor

In many numerical procedures, linearization of the tangent tensor will be required. For example, an increment in the angular velocity vector defined by eq. (13.54) is $\Delta\underline{\omega} = \Delta[\underline{H}(\underline{p})\underline{\dot{p}}] = \underline{H}(\underline{p})\Delta\underline{\dot{p}} + \underline{M}(\underline{p}, \underline{\dot{p}})\Delta\underline{p}$. More generally, operator $\underline{M}$ is defined as $\underline{M}(\underline{p}, \underline{a}) = \partial(\underline{H}(\underline{p})\underline{a})/\partial\underline{p}$, where $\underline{a}$ is an arbitrary vector.

Tedious algebra shows that operator $\underline{M}$ is

$$\underline{M}(\underline{p}, \underline{a}) = (\hat{\sigma}_0 + \hat{\sigma}_1\widetilde{p} + \hat{\sigma}_2\widetilde{pp})\underline{a}\,\underline{p}^T - \sigma_1\widetilde{a} - \sigma_2(2\widetilde{pa} - \widetilde{ap}), \tag{13.67}$$

where $\hat{\sigma}_0 = \sigma_0'/(pp')$, $\hat{\sigma}_1 = \sigma_1'/(pp')$, and $\hat{\sigma}_2 = \sigma_2'/(pp')$. Notation $(\cdot)'$ indicates a derivative with respect to angle ϕ, and coefficients σ_0, σ_1, and σ_2 are given by eq. (13.56).

13.4.7 Problems

Problem 13.6. Angular velocity with vectorial parameterization
Prove relationships (13.55) and (13.60).

Problem 13.7. Properties of the vectorial parameterization
Prove relationships (13.63a) and (13.63b).

Problem 13.8. Angular velocity with vectorial parameterization
Prove that $\underline{\omega}^* = \underline{H}^T(\underline{p})\underline{\dot{p}}$ starting from the definition of the angular velocity vector expressed in the rotating system $\widetilde{\omega}^* = \underline{R}^T\underline{\dot{R}}$.

Problem 13.9. Tensorial nature of the tangent tensor
Prove that the tangent tensor is a tensor starting directly from its definition, eq. (13.55).

Problem 13.10. Composition of collinear rotations
Let $\underline{p}_1$, $\underline{p}_2$, and $\underline{p}$ be the rotation parameter vectors of three rotations with rotation angles ϕ_1, ϕ_2, and ϕ, respectively, such that $\underline{\underline{R}}(\underline{p}) = \underline{\underline{R}}(\underline{p}_1)\underline{\underline{R}}(\underline{p}_2)$. If $\underline{p}_1$ and $\underline{p}_2$ are two parallel vectors, prove that $\underline{p}$ is also parallel to them and that $\phi = \phi_1 + \phi_2$.

Problem 13.11. Relationship between tensors $\underline{\underline{G}}$ and $\underline{\underline{H}}$
Prove the following relationship

$$\underline{\underline{G}}^T \underline{\underline{H}} = \nu \underline{\underline{I}} + (\sigma_0 - \nu)\bar{n}\bar{n}^T = \sigma_0 \underline{\underline{I}} + (\sigma_0 - \nu)\widetilde{n}\widetilde{n}. \tag{13.68}$$

Problem 13.12. Linearization of the tangent tensor
Equation (13.67) defines operator $\underline{\underline{M}}$ involved in the linearization of the tangent operator. *(1)*
Determine operator $\underline{\underline{M}}^*(\underline{p}, \underline{a}) = \partial(\underline{\underline{H}}^T(\underline{p})\underline{a})/\partial\underline{p}$.

13.5 Specific choices of generating function

The formulation presented in the previous section is very general, but, in practice, a specific choice of the generating function, $p(\phi)$, must be selected. It seems natural to select a generating function that will simplify some of the operators involved in rotation manipulations.

Specific parameterizations

The simplest choice is to select the generating function as the rotation angle itself

$$p(\phi) = \phi. \tag{13.69}$$

This parameterization is called the *Cartesian rotation vector*, or the *exponential map* of rotation; details concerning this parameterization are given in section 13.7.1.

A second approach is to simplify the expression for the rotation tensor, $\underline{\underline{R}}$, given by eq. (13.45), by requiring $\zeta_1(\phi) = 1$. This yields

$$p(\phi) = \sin\phi. \tag{13.70}$$

This choice is called the *linear parameterization*. Note that requiring $\zeta_2(\phi) = 1$ yields $p(\phi) = \sqrt{2}\sin\phi/2$. Although this is a valid parameterization, it does not satisfy the limit condition, eq. (13.44).

An alternative approach is to require the last term of tensor $\underline{\underline{H}}$, eq. (13.55), to vanish, *i.e.*, $\sigma_2(\phi) = 0$. This leads to the nonlinear differential equation $p' \sin\phi = p$, the solution of which is $p(\phi) = c\tan(\phi/2)$, where c is an integration constant. The limit condition, eq. (13.44), implies $c = 2$, and hence,

$$p(\phi) = 2\tan\frac{\phi}{2}. \tag{13.71}$$

This parameterization is variously called after Cayley, Gibbs, Rodrigues, or some combinations of these names. It shall be referred to here as the *Cayley-Gibbs-Rodrigues parameterization*. Details concerning this parameterization are given in section 13.7.3.

Another approach is to require the last term of tensor $\underline{\underline{H}}^{-1}$, eq. (13.60), to vanish, i.e., $p' - 1/\varepsilon = 0$. This leads to the nonlinear differential equation $2p' \tan \phi/2 = p$, the solution of which is $p(\phi) = c \sin \phi/2$, where c is an integration constant. Here again, the limit condition yields the solution as

$$p(\phi) = 2 \sin \frac{\phi}{2}. \tag{13.72}$$

This parameterization is usually termed the *reduced Euler-Rodrigues parameterization*. It is closely connected to the parameterization technique employing unit quaternions: the parameter vector, $\underline{p}$, coincides with the vector part of the unit quaternion of the rotation. It shall be referred to here as the *Euler-Rodrigues parameterization*. Details concerning this parameterization are given in section 13.7.2.

In yet another approach, operators $\underline{\underline{H}}$ and $\underline{\underline{G}}$ are required to be multiples of each other, i.e., $\underline{\underline{H}} = \alpha(\phi)\underline{\underline{G}}$. This implies two conditions: $p' = 1/\alpha(\phi)$ and $p\alpha(\phi) = 2 \sin \phi/2$. Hence, the following differential equation must hold: $2p' \sin \phi/2 = p$. With the help of the limit condition, the solution becomes

$$p(\phi) = 4 \tan \frac{\phi}{4}. \tag{13.73}$$

This parameterization also bears various names in the literature: Wiener [256], Milenković [257], or modified Rodrigues parameterization [243, 258]. It is also known as the conformal rotation vector (CRV) parameterization. It shall be referred to here as the *Wiener-Milenković parameterization*. Details concerning this parameterization are given in section 13.7.4.

To avoid the appearance of singularities when manipulating tensor $\underline{\underline{H}}$, eq. (13.55), it might be desirable to work with a parameterization for which $\det(\underline{\underline{H}}) = c$, where c is a constant. From eq. (13.58), this requirement implies $cp' = \nu^2$. The solution of this nonlinear differential equation leads to

$$p(\phi) = \sqrt[3]{6(\phi - \sin \phi)}. \tag{13.74}$$

Constant c was evaluated with the help of the limit condition and found to be $c = 1$. Hence, this particular parameterization is such that $\det(\underline{\underline{H}}) = 1$ for all values of ϕ. Clearly, the complex expression for this parameterization makes it quite unpractical to use.

The sine and tangent families

This discussion indicates that two subclasses of vectorial parameterization enjoy interesting properties

$$p(\phi) = m \sin \frac{\phi}{m}, \quad \text{and} \quad p(\phi) = m \tan \frac{\phi}{m}. \tag{13.75}$$

To ease the manipulation of the trigonometric functions, m is typically selected to be an integer, but real values of m are equally valid.

As m increases, $p(\phi) \to \phi$ for $|\phi| < \pi$. This feature is illustrated in figs. 13.1 and 13.2 for the sine and tangent families, respectively. Note the convergence by lower and upper bound for the sine and tangent families, respectively.

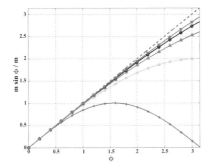

Fig. 13.1. Generating function versus angle ϕ for the sine family. m = 1 (+), 2 ($\square$), 3 ($\triangle$), 4 ($\circ$), 5 ($\triangledown$). The dotted line represents the generating function for the rotation vector, $p(\phi) = \phi$.

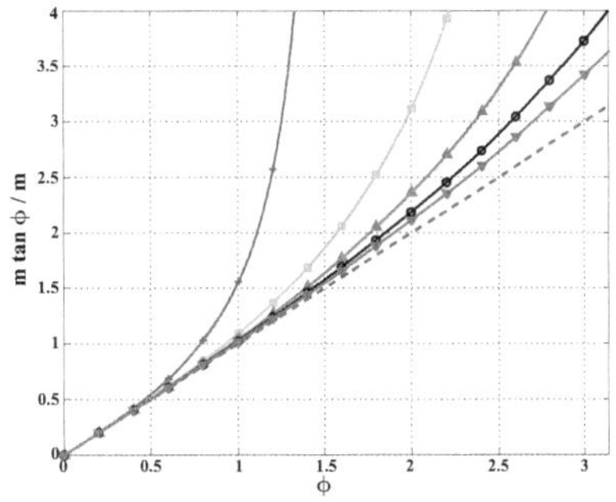

Fig. 13.2. Generating function versus angle ϕ for the tangent family. m = 1 (+), 2 ($\square$), 3 ($\triangle$), 4 ($\circ$), 5 ($\triangledown$). The dotted line represents the generating function for the rotation vector, $p(\phi) = \phi$.

13.6 The extended vectorial parameterization

The vectorial parameterization presented in the previous section exhibits desirable features, but also suffers serious drawbacks. In particular, for all generating function choices, singularities will occur for specific values of the rotation angle, as proved by Stuelpnagel [246].

13.6.1 Singularities of the vectorial parameterization

More specifically, singularities can first occur in the definition of the generating function when $p \to \infty$. For instance, the Cayley-Gibbs-Rodrigues parameterization is singular when $\phi = \pm\pi$. Because the representation of all arbitrary rotations requires a well defined parameterization for all $|\phi| \le \pi$, the Cayley-Gibbs-Rodrigues parameterization can not be used when dealing with rotations of arbitrary magnitude.

Next, problems can occur when determining the component of the vectorial parameterization from the rotation tensor. In view of eq. (13.65), singularities are encountered when $\nu \to 0$ or ∞. Linear parameters, for instance, experience such singularity when $\nu \to \infty$, $i.e.$, when $\phi = \pm\pi$.

534   13 Parameterization of rotation

Singularities also arise from the manipulation of the tangent tensor and of its inverse. Inspection of eqs. (13.55), (13.58), and (13.60), reveals that singularities will appear when $p' \to 0$ or ∞ and $\nu \to 0$ or ∞.

In summary, singularities will appear when $p \to \infty$, $\nu \to 0$ or ∞, and $p' \to 0$ or ∞. Table 13.1 lists the range of validity of various parameterizations. Figures 13.3 and 13.4 show the relevant function, $\nu(\phi)$, $p'(\phi)$, and $\det(\underline{H})$ for the sine and tangent families, respectively. Clearly, the parameterizations with larger values of m have an extended range of validity, although for $m = 4$ the range settles to $|\phi| < 2\pi$ for both the sine and tangent families, and does not increase with further increases in m.

Table 13.1. Various choices of the generating function.

Name	$p(\phi)$	p'	ν	ε	Validity range
Cartesian rotation vector	ϕ	1	$(\sin \frac{\phi}{2})/(\frac{\phi}{2})$	$(\tan \frac{\phi}{2})/(\frac{\phi}{2})$	$|\phi| < 2\pi$
Cayley-Gibbs-Rodrigues	$2 \tan \frac{\phi}{2}$	$1/\cos^2 \frac{\phi}{2}$	$\cos \frac{\phi}{2}$	1	$|\phi| < \pi$
Wiener-Milenković	$4 \tan \frac{\phi}{4}$	$1/\cos^2 \frac{\phi}{4}$	$\cos^2 \frac{\phi}{4}$	$1/(1 - \tan^2 \frac{\phi}{4})$	$|\phi| < 2\pi$
Linear Parameters	$\sin \phi$	$\cos \phi$	$1/(\cos \frac{\phi}{2})$	$1/(\cos^2 \frac{\phi}{2})$	$|\phi| < \pi$
Euler Rodrigues	$2 \sin \frac{\phi}{2}$	$\cos \frac{\phi}{2}$	1	$1/(\cos \frac{\phi}{2})$	$|\phi| < \pi$
	$4 \sin \frac{\phi}{4}$	$\cos \frac{\phi}{4}$	$\cos \frac{\phi}{4}$	$(\cos \frac{\phi}{4})/(\cos \frac{\phi}{2})$	$|\phi| < 2\pi$

All parameterizations with a validity range of $|\phi| > \pi$ are able to handle all rotations. Such parameterizations, however, are not necessarily "worry free." Indeed, rotation are often used in incremental procedures where a small incremental rotation is added to a rotation at each time step, for instance. In this case, rotation angles of arbitrary magnitude are routinely encountered; consider, for instance, a rotating shaft, or a satellite tumbling in space. In such cases, singularities will always appear as ϕ increases to large values. For the sine and tangent families, problems will be encountered when $|\phi|$ reaches 2π, for $m \geq 4$.

13.6.2 The rescaling operation

The range of validity of the sine and tangent parameterizations for $m = 4$ can be *extended* by using a *rescaling operation*. This operation is based on the observation that rotations of magnitudes ϕ and $\phi^\dagger = \phi \pm 2\pi$ about the same axis $\bar{n}$ correspond to the same final configuration.

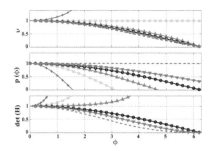

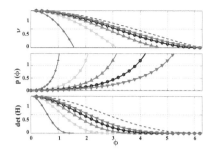

Fig. 13.3. Functions ν (top figure), p' (middle figure), and $\det(\underline{\underline{H}})$ (bottom figure), versus ϕ for the sine family. m = 1 (+), 2 (□), 3 (△), 4 (○), 5 (▽). The dotted line gives the corresponding quantities for the rotation vector, $p(\phi) = \phi$.

Fig. 13.4. Functions ν (top figure), p' (middle figure), and $\det(\underline{\underline{H}})$ (bottom figure), versus ϕ for the tangent family. m = 1 (+), 2 (□), 3 (△), 4 (○), 5 (▽). The dotted line gives the corresponding quantities for the rotation vector, $p(\phi) = \phi$.

Rescaling the Wiener-Milenković parameterization

The Wiener-Milenković parameterization characterized by the generating function $p = 4\bar{n}\tan\phi/4$ is considered first. The norm of the rotation parameter vector, $p = \|\underline{p}\|$, is such that $p \leq 4$ when $|\phi| \leq \pi$. Let $\underline{p}$ and $\underline{p}^{\dagger}$ be associated with the rotations ϕ and $\phi^{\dagger}$, respectively. The relationship between these two sets of parameters is

$$\underline{p}^{\dagger} = 4\bar{n}\tan\frac{\phi^{\dagger}}{4} = 4\bar{n}\tan\left(\frac{\phi}{4} \pm \frac{\pi}{2}\right) = -4\bar{n}\frac{1}{\tan\phi/4} = -\frac{\underline{p}}{\tan^2\phi/4}, \quad (13.76)$$

which implies

$$\underline{p}^{\dagger} = -\frac{\nu}{1-\nu}\underline{p}. \quad (13.77)$$

Taking the norm of eq. (13.76) yields $p^{\dagger} = p/\tan^2\phi/4$, and hence, $pp^{\dagger} = p^2/(\tan^2\phi/4)$, or $pp^{\dagger} = 16$. If $\pi < |\phi| < 2\pi$, $p > 4$, and hence $p^{\dagger} < 4$; in other words, the rescaling operation decreases the norm of the rotation parameter vector.

Let $\underline{p}_1$, $\underline{p}_2$, and $\underline{p}$ be the parameters of three rotation tensors such that $\underline{\underline{R}}(p) = \underline{\underline{R}}(p_1)\underline{\underline{R}}(p_2)$. Equation (13.66b) then provides the following rotation composition formula

$$\underline{p} = \frac{\nu_1\nu_2}{\nu}\left(\frac{1}{\varepsilon_2}\underline{p}_1 + \frac{1}{\varepsilon_1}\underline{p}_2 + \frac{1}{2}\widetilde{p}_1\underline{p}_2\right), \quad (13.78)$$

where, in view of eq. (13.66a), $2\nu - 1 = \cos\phi/2 = \nu_1\nu_2\left(1/\varepsilon_1\varepsilon_2 - \underline{p}_1^T\underline{p}_2/4\right)$. As incremental rotations are added to the initial orientation, p increases and when $|\phi|$ becomes larger than π, $p > 4$ and a rescaling operation, eq. (13.77), becomes necessary. The two operations, composition and rescaling, are conveniently combined into a single operation

$$
\underline{p} = \begin{cases}
\dfrac{\nu_1\nu_2}{\nu}\left(\dfrac{1}{\varepsilon_2}\underline{p}_1 + \dfrac{1}{\varepsilon_1}\underline{p}_2 + \dfrac{1}{2}\widetilde{p}_1\underline{p}_2\right), & \nu \geq \dfrac{1}{2}, \\[3mm]
-\dfrac{\nu_1\nu_2}{1-\nu}\left(\dfrac{1}{\varepsilon_2}\underline{p}_1 + \dfrac{1}{\varepsilon_1}\underline{p}_2 + \dfrac{1}{2}\widetilde{p}_1\underline{p}_2\right), & \nu \leq \dfrac{1}{2}.
\end{cases}
\tag{13.79}
$$

Rescaling the sine parameterization

Similar developments hold for the sine parameterization characterized by the generating function $p = 4\bar{n}\sin\phi/4$. The norm of the rotation parameter vector, $p = \|\underline{p}\|$, is such that $p^2 \leq 8$ when $|\phi| \leq \pi$. The rescaling operation now becomes

$$
\underline{p}^\dagger = \frac{\nu}{\sqrt{1-\nu^2}}\underline{p}.
\tag{13.80}
$$

and implies $p^2 + p^{\dagger 2} = 16$. Here again, the rescaling operation decreases the norm of the rotation parameter vector.

Using the same approach as for the Wiener-Milenković parameterization, the update and rescaling operations are conveniently combined into a single operation

$$
\underline{p} = \begin{cases}
\dfrac{\nu_1\nu_2}{\nu}\left(\dfrac{1}{\varepsilon_2}\underline{p}_1 + \dfrac{1}{\varepsilon_1}\underline{p}_2 + \dfrac{1}{2}\widetilde{p}_1\underline{p}_2\right), & \nu \geq \dfrac{1}{\sqrt{2}}, \\[3mm]
\dfrac{\nu_1\nu_2}{\sqrt{1-\nu^2}}\left(\dfrac{1}{\varepsilon_2}\underline{p}_1 + \dfrac{1}{\varepsilon_1}\underline{p}_2 + \dfrac{1}{2}\widetilde{p}_1\underline{p}_2\right), & \nu \leq \dfrac{1}{\sqrt{2}},
\end{cases}
\tag{13.81}
$$

where, in view of eq. (13.66a), $2\nu^2 - 1 = \cos\phi/2 = \nu_1\nu_2(1/\varepsilon_1\varepsilon_2 - \underline{p}_1^T\underline{p}_2/4)$.

The two parameterizations, $p = 4\bar{n}\sin\phi/4$ and $p = 4\bar{n}\tan\phi/4$, are now able to handle rotations of truly arbitrary magnitude provided that any update operation is combined with a possible rescale, as indicated in eqs. (13.79) and (13.81), respectively.

Compact notation

It is often convenient to indicate the composition of rotations combined with an optional rescaling by the following notation

$$
\underline{R}(\underline{p}) = \underline{R}(\underline{p}_1)\underline{R}(\underline{p}_2) \Leftrightarrow \underline{p} = \underline{p}_1 \oplus \underline{p}_2,
\tag{13.82}
$$

which implies that $\underline{p}$ is computed with the help of eq. (13.79) or (13.81) for the Wiener-Milenković or sine parameterization, respectively.

Composition operations such as $\underline{R}(\underline{p}) = \underline{R}^T(\underline{p}_1)\underline{R}(\underline{p}_2)$ are also commonly encountered. In view of eq. (13.45), $\underline{R}^T(\underline{p}_1) = \underline{R}(-\underline{p}_1)$ and hence, the following notation is used

$$
\underline{R}(\underline{p}) = \underline{R}^T(\underline{p}_1)\underline{R}(\underline{p}_2) \Leftrightarrow \underline{p} = \underline{p}_1^- \oplus \underline{p}_2,
\tag{13.83}
$$

where notation $\underline{p}_1^-$ indicates that the sign of the rotation parameter vector should be changed before using eqs. (13.79) or (13.81).

Note the simplicity of eqs. (13.79) or (13.81) as compared to the direct application of the composition equation. Indeed, given $\underline{p}_1$ and $\underline{p}_2$, equation $\underline{\underline{R}}(\underline{p}) = \underline{\underline{R}}(\underline{p}_1)\underline{\underline{R}}(\underline{p}_2)$ requires a four step procedure for the evaluation of $\underline{p}$: (1) evaluate $\underline{\underline{R}}(\underline{p}_1)$ using eq. (13.45), (2) evaluate $\underline{\underline{R}}(\underline{p}_2)$ using the same equation, (3) evaluate the matrix product $\underline{\underline{R}}(\underline{p}) = \underline{\underline{R}}(\underline{p}_1)\underline{\underline{R}}(\underline{p}_2)$, (4) extract parameters $\underline{p}$ from $\underline{\underline{R}}(\underline{p})$ using the procedure described in section 13.3.4 and eq (13.65).

13.7 Specific parameterizations of rotation

In this section, several parameterizations of rotation will be discussed that correspond to specific choices of the generating function. The *Cartesian rotation vector*, the *Euler-Rodrigues parameters*, *Cayley-Gibbs-Rodrigues parameters*, and the *Wiener-Milenković parameters* are discussed in the sections below.

13.7.1 The Cartesian rotation vector

The rotation vector is associated with the generating function $p(\phi) = \phi$, *i.e.*,

$$\underline{r} = \phi\,\bar{n}. \tag{13.84}$$

Important quantities associated with the rotation vector are $p' = 1$, $\nu = (\sin\phi/2)/(\phi/2)$, and $\varepsilon = (\tan\phi/2)/(\phi/2)$. Equation (13.45) yields the rotation tensor, using the following parameters: $\zeta_1 = (\sin\phi)/\phi$ and $\zeta_2 = (1 - \cos\phi)/\phi^2$. Tensor $\underline{\underline{H}}$ then follows from eq. (13.55) with $\sigma_0 = 1, \sigma_1 = \zeta_2$, and $\sigma_2 = (1-\zeta_1)/\phi^2$. Finally, tensor $\underline{\underline{H}}^{-1}$ is obtained from eq. (13.60) with $\chi_0 = 1, \chi_1 = -1/2$, and $\chi_2 = (1 - 1/\varepsilon)/\phi^2$.

An interesting expression can be found by expanding the trigonometric functions in infinite series and using identity (1.34b)

$$\underline{\underline{R}}(\underline{r}) = \underline{\underline{I}} + \widetilde{r} + \frac{1}{2!}\widetilde{r}\widetilde{r} + \frac{1}{3!}\widetilde{r}\widetilde{r}\widetilde{r} + \cdots = \exp(\widetilde{r}) \tag{13.85}$$

This form of the rotation tensor is called the *exponential map of rotation*.

This expansion of the rotation tensor provides a natural way of approximating rotations. For very small rotations, only the first-order term is kept, to yield

$$\underline{\underline{R}}(\underline{r}) \approx \underline{\underline{I}} + \widetilde{r} = \begin{bmatrix} 1 & -r_3 & r_2 \\ r_3 & 1 & -r_1 \\ -r_2 & r_1 & 1 \end{bmatrix}. \tag{13.86}$$

For moderate rotations, two terms of the expansion are kept, leading to

$$\underline{\underline{R}}(\underline{r}) \approx \underline{\underline{I}} + \widetilde{r} + \frac{1}{2}\widetilde{r}\widetilde{r} = \begin{bmatrix} 1 - \dfrac{r_2^2 + r_3^2}{2} & \dfrac{r_1 r_2}{2} - r_3 & \dfrac{r_1 r_3}{2} + r_2 \\ \dfrac{r_1 r_2}{2} + r_3 & 1 - \dfrac{r_1^2 + r_3^2}{2} & \dfrac{r_2 r_3}{2} - r_1 \\ \dfrac{r_1 r_3}{2} - r_2 & \dfrac{r_2 r_3}{2} + r_1 & 1 - \dfrac{r_1^2 + r_2^2}{2} \end{bmatrix}. \tag{13.87}$$

Composition of rotations

It is difficult to write the composition of rotation formulæ in terms of the rotation vector. The simplest way to proceed is to first transform the corresponding rotation vectors to an Euler parameter representation, compose the rotations with the help of eq. (13.26), then transform the result back to the rotation vector representation.

13.7.2 The Euler-Rodrigues parameters

The Euler-Rodrigues parameters are associated with the generating function $p(\phi) = 2\sin\phi/2$, i.e.,

$$\underline{v} = 2\bar{n}\sin\frac{\phi}{2}. \tag{13.88}$$

As its name suggests, this parameterization is closely related to Euler parameters. Indeed, $\underline{v} = 2\underline{e}$, and $\underline{e} = \underline{v}/2$.

The following parameter plays an important role in this parameterization,

$$v_0 = \sqrt{1 - \frac{\underline{v}^T\underline{v}}{4}} = \cos\frac{\phi}{2}. \tag{13.89}$$

This representation is limited to rotation angles $|\phi| < \pi$, so that $0 \le v_0 \le 1$. For $|\phi| > \pi$, this parameterization cannot distinguish between the distinct rotations of magnitudes ϕ and $\pi - \phi$.

Important quantities associated with the Euler-Rodrigues parameters are $p' = v_0$, $\nu = 1$, and $\varepsilon = 1/v_0$. Equation (13.45) yields the rotation tensor, using the following parameters: $\zeta_1 = v_0$ and $\zeta_2 = 1/2$. Tensor $\underline{\underline{H}}$ then follows from eq. (13.55) with $\sigma_0 = 1/v_0$, $\sigma_1 = 1/2$, and $\sigma_2 = 1/(4v_0)$. Finally, tensor $\underline{\underline{H}}^{-1}$ is obtained from eq. (13.60) with $\chi_0 = v_0$, $\chi_1 = -1/2$, and $\chi_2 = 0$.

Composition of rotations

Let $\underline{p}$ and $\underline{q}$ be the Euler-Rodrigues parameters of two successive rotations, and $\underline{r}$ the Euler-Rodrigues parameter of the composed rotation, such that $\underline{\underline{R}}(\underline{r}) = \underline{\underline{R}}(\underline{p})\underline{\underline{R}}(\underline{q})$. The composition formulæ, eqs. (13.66a) and (13.66b), then yield

$$r_0 = p_0 q_0 - \frac{1}{4}(\underline{p}^T\underline{q}), \tag{13.90a}$$

$$\underline{r} = q_0\underline{p} + p_0\underline{q} + \frac{1}{2}\widetilde{\underline{p}}\underline{q}. \tag{13.90b}$$

13.7.3 The Cayley-Gibbs-Rodrigues parameters

The Cayley-Gibbs-Rodrigues parameters are associated with the generating function $p(\phi) = 2\tan\phi/2$, i.e.,

$$\underline{r} = 2\bar{n} \tan \frac{\phi}{2}. \tag{13.91}$$

The following parameter plays an important role in this parameterization,

$$r_0 = \frac{1}{1 + (\underline{r}^T \underline{r})/4} = \cos^2 \frac{\phi}{2}. \tag{13.92}$$

This representation is limited to rotation angles of magnitude $|\phi| < \pi$, because it presents a singularity, $\underline{r} \to \infty$, when $|\phi| \to \pi$.

This parameterization is closely related to Euler parameters, $\underline{r} = 2\underline{e}/e_0$ and $\underline{e} = \sqrt{r_0}\underline{r}/2$. It is also closely related to Cayley's parameters presented in section 13.1, $\underline{r} = 2\underline{a}$.

Important quantities associated with the Cayley-Gibbs-Rodrigues parameters are $p' = 1/r_0$, $\nu = \sqrt{r_0}$, and $\varepsilon = 1$. Equation (13.45) yields the rotation tensor, using the following parameters: $\zeta_1 = r_0$ and $\zeta_2 = r_0/2$. Tensor $\underline{\underline{H}}$ then follows from eq. (13.55) with $\sigma_0 = r_0$, $\sigma_1 = r_0/2$, and $\sigma_2 = 0$. Finally, tensor $\underline{\underline{H}}^{-1}$ is obtained from eq. (13.60) with $\chi_0 = 1/r_0$, $\chi_1 = -1/2$, and $\chi_2 = 1/4$.

Composition of rotations

Let $\underline{p}$ and $\underline{q}$ be the Cayley-Gibbs-Rodrigues parameters of two successive rotations, and $\underline{r}$ the Cayley-Gibbs-Rodrigues parameters of the composed rotation, such that $\underline{\underline{R}}(\underline{r}) = \underline{\underline{R}}(\underline{p})\underline{\underline{R}}(\underline{q})$. The composition formulæ, eqs. (13.66a) and (13.66b), then yield

$$r_0 = p_0 q_0 \Delta_1^2, \tag{13.93a}$$

$$\underline{r} = \frac{1}{\Delta_1} \left(\underline{p} + \underline{q} + \frac{1}{2}\tilde{p}\underline{q} \right). \tag{13.93b}$$

where $\Delta_1 = 1 - (\underline{p}^T \underline{q})/4$.

13.7.4 The Wiener-Milenković parameters

The Wiener-Milenković parameters are associated with the generating function $p(\phi) = 4\tan \phi/4$, i.e.,

$$\underline{c} = 4\bar{n} \tan \frac{\phi}{4}. \tag{13.94}$$

The following parameter plays an important role in this parameterization,

$$c_0 = 2(1 - \tan^2 \frac{\phi}{4}) = 2 - \frac{\underline{c}^T \underline{c}}{8}. \tag{13.95}$$

This representation is limited to rotation angles of magnitude $|\phi| < 2\pi$, because it presents a singularity, $\underline{c} \to \infty$, when $|\phi| \to 2\pi$.

This parameterization is closely related to Euler parameters. Indeed, $\underline{c} = 4\underline{e}/(1 + e_0)$ and $c_0 = 4e_0/(1 + e_0)$. The inverse relationship is $\underline{e} = \underline{c}/(4 - c_0)$ and $e_0 =$

$c_0/(4 - c_0)$. Because the Wiener-Milenković parameters can be obtained from this conformal transformation, they are sometimes referred to as the *conformal rotation vector*. The following relationship ease the manipulations of this parameterization: $\cos \phi/2 = c_0/(4 - c_0)$.

Important quantities associated with the Wiener-Milenković parameters are $p' = 1/\nu$, $\nu = 2/(4 - c_0)$, and $\varepsilon = 2/c_0$. Equation (13.45) yields the rotation tensor, using the following parameters: $\zeta_1 = \nu^2 c_0/2$ and $\zeta_2 = \nu^2/2$. Tensor $\underline{\underline{H}}$ then follows from eq. (13.55) with $\sigma_0 = \nu$, $\sigma_1 = \nu^2/2$, and $\sigma_2 = \nu^2/8$. Finally, tensor $\underline{\underline{H}}^{-1}$ is obtained from eq. (13.60) with $\chi_0 = 1/\nu$, $\chi_1 = -1/2$, and $\chi_2 = 1/8$.

For the Wiener-Milenković parameterization, tensors $\underline{\underline{H}}$ and $\underline{\underline{G}}$ are closely related, $\underline{\underline{H}}(\underline{c}) = \nu \underline{\underline{G}}$ and $\underline{\underline{H}}^{-1}(\underline{c}) = \underline{\underline{G}}^T/\nu$.

Composition of rotations

Let $\underline{p}$ and $\underline{q}$ be the Wiener-Milenković parameters of two successive rotations, and $\underline{r}$ the conformal rotation parameters of the composed rotation, such that $\underline{\underline{R}}(\underline{r}) = \underline{\underline{R}}(\underline{p})\underline{\underline{R}}(\underline{q})$. The composition formulæ, eq. (13.66a) and (13.66b), then yield

$$r_0 = 4 \left(p_0 q_0 - \underline{p}^T \underline{q} \right)/(\Delta_1 + \Delta_2), \tag{13.96a}$$

$$\underline{r} = 4 \left(q_0 \underline{p} + p_0 \underline{q} + \widetilde{p}\underline{q} \right)/(\Delta_1 + \Delta_2), \tag{13.96b}$$

where $\Delta_1 = (4 - p_0)(4 - q_0)$ and $\Delta_2 = p_0 q_0 - \underline{p}^T \underline{q}$.

In most applications, it will be necessary to rescale the Wiener-Milenković parameters, as discussed in section 13.6.2. The two operations, composition and rescaling, are conveniently combined into a single operation, eq. (13.79), which, for Wiener-Milenković parameters, takes on a particularly simple form

$$\underline{r} = \begin{cases} 4 \left(q_0 \underline{p} + p_0 \underline{q} + \widetilde{p}\underline{q} \right)/(\Delta_1 + \Delta_2) & \text{if } \Delta_2 \geq 0, \\ -4 \left(q_0 \underline{p} + p_0 \underline{q} + \widetilde{p}\underline{q} \right)/(\Delta_1 - \Delta_2) & \text{if } \Delta_2 < 0. \end{cases} \tag{13.97}$$

The rescaling condition automatically selects the largest denominator, also guaranteeing the most accurate numerical evaluation of the composed rotation.

Time derivatives of the tangent tensor

When linearizing equations involving rotations expressed in terms of the Wiener-Milenković parameters, time derivatives of the tangent tensor are often required. First, this tensor is recast as $a\underline{\underline{H}} = c_0 + \widetilde{c} + \underline{c}\underline{c}^T/4$, where $a = (4 - c_0)^2/2$. Simple algebra then implies

$$\underline{\underline{H}} = \frac{c_0}{a} + \frac{\widetilde{c}}{a} + \frac{\underline{c}\,\underline{c}^T}{4\,a}, \tag{13.98a}$$

$$\dot{\underline{\underline{H}}} = \frac{\dot{c}_0}{a} + \frac{\dot{\widetilde{c}}}{a} + \frac{\dot{\underline{c}}\,\underline{c}^T}{a\,4} + \frac{\underline{c}\,\dot{\underline{c}}^T}{4\,a} - \frac{\dot{a}}{a}\underline{\underline{H}}, \tag{13.98b}$$

$$\ddot{\underline{\underline{H}}} = \frac{\ddot{c}_0}{a} + \frac{\ddot{\widetilde{c}}}{a} + \frac{\ddot{\underline{c}}\,\underline{c}^T}{a\,4} + \frac{\dot{\underline{c}}\,\dot{\underline{c}}^T}{2\,a} + \frac{\underline{c}\,\ddot{\underline{c}}^T}{4\,a} - \frac{2\dot{a}}{a}\dot{\underline{\underline{H}}} - \frac{\ddot{a}}{a}\underline{\underline{H}}. \tag{13.98c}$$

In these expressions, the following notation was used

$$c_0 = 2(1 - \frac{c^2}{16}), \quad \frac{\dot{c}_0}{a} = -\frac{\underline{c}^T \dot{\underline{c}}}{4\ a}, \qquad \frac{\ddot{c}_0}{a} = -\frac{a}{4} \frac{\dot{\underline{c}}^T \dot{\underline{c}}}{a\ a} - \frac{\underline{c}^T \ddot{\underline{c}}}{4\ a}, \tag{13.99a}$$

$$a = 2(1 + \frac{c^2}{16})^2, \quad \frac{\dot{a}}{a} = -2(1 + \frac{c^2}{16}) \frac{\dot{c}_0}{a}, \quad \frac{\ddot{a}}{a} = a(\frac{\dot{c}_0}{a})^2 - 2(1 + \frac{c^2}{16}) \frac{\ddot{c}_0}{a}. \tag{13.99b}$$

13.7.5 Problems

Problem 13.13. The exponential map of rotation
Prove that the rotation tensor expressed in terms of the rotation vector can be written as the exponential map, eq. (13.85).

Problem 13.14. Algebraic representations of rotation
(1) Show that for all vectorial parameterizations of rotation, tensors $\underline{\underline{R}}$, $\underline{\underline{H}}$ and $\underline{\underline{H}}^{-1}$, and the composition of rotation formulæ are expressed in terms of three parameters only, ν, ε, and p'. It then follows that if those three parameters can be expressed as algebraic functions of the rotation parameter vector, the corresponding vectorial parameterization of rotation enables the manipulation of rotation without using any trigonometric functions. *(2)* Show that the Wiener-Milenković parameters described in section 13.7.4 enables an algebraic representations of rotation. *(3)* Show that the following parameterization, $\underline{s} = 4 \sin \phi/4 \ \bar{n}$, also leads an algebraic representations of rotation.

Problem 13.15. Short questions
(1) Are the Euler angles using the *3-1-3* sequence a particular case of the vectorial parameterization of rotation? *(2)* Describe the singularities associated with the Euler-Rodrigues parameters. *(3)* Describe the procedure used to extract the Cayley-Gibbs-Rodrigues parameters from a given rotation tensor. *(4)* What is the main problem associated with the use of the Euler parameters? *(5)* Is it possible to find a vectorial parameterization of rotation for which the tangent tensor is orthogonal?

Problem 13.16. Relationship between tensors $\underline{\underline{G}}$ and $\underline{\underline{H}}$
Based on eq. (13.68), find the vectorial parameterization for which $\underline{\underline{H}} = \nu \underline{\underline{G}}$.

Problem 13.17. Study of the limit behavior
Prove the following results related to the limit behavior of the angular velocity and acceleration vectors as the rotation parameter vector vanishes. *(1)* $\lim_{\underline{p} \to 0} \underline{\underline{H}} = \underline{\underline{I}}$. *(2)* $\lim_{\underline{p} \to 0} \underline{\underline{M}} = -\tilde{a}/2$, see eq. (13.67). *(3)* $\lim_{\underline{p} \to 0} \dot{\underline{\underline{H}}} = \tilde{\dot{p}}/2$. *(4)* $\lim_{\underline{p} \to 0} \underline{\omega} = \dot{\underline{p}}$. *(5)* $\lim_{\underline{p} \to 0} \dot{\underline{\omega}} = \ddot{\underline{p}}$.

Problem 13.18. Composing rotations with the Wiener-Milenković parameters
Consider a sequence of rotation tensors $\underline{\underline{R}}_k = \underline{\underline{R}}(t_k)$, where $t_k = k\Delta t$ is a sequence of equally spaced discrete instants in time; $\Delta t = 0.01$ s is the time step size. Each tensor in the sequence is obtained from the previous one through an incremental rotation tensor $\hat{\underline{\underline{R}}}$ such that $\underline{\underline{R}}_{k+1} = \hat{\underline{\underline{R}}} \underline{\underline{R}}_k$. Let the Wiener-Milenković parameters $\underline{c}_k$ and $\hat{\underline{c}}$ be associated with $\underline{\underline{R}}_k$ and $\hat{\underline{\underline{R}}}$, respectively. The initial tensor is such that $\underline{c}_0^T = \{0, 0, 0\}$. The parameters of the incremental rotation are given as a function of time

$$\hat{\underline{c}} = 0.15 \begin{bmatrix} \sin\theta_k\,\cos\psi_k \\ \sin\theta_k\,\sin\psi_k \\ \cos\theta_k \end{bmatrix},$$

where $\theta_k = 3t_k$ and $\psi_k = 6t_k - 2t_k^3$. *(1)* Find the sequence of parameters $\underline{c}_k$, $k = 1, \ldots 200$. Use the formula for composition of rotation with rescaling. *(2)* On one graph, plot the three Wiener-Milenković parameters $\underline{c}_k$. *(3)* On one graph, plot the direction cosines R_{11}, R_{21}, R_{31} as a function of time. Comment on your results. *(4)* Show that the components of the angular velocity vector in the fixed system can be approximated in the following manner

$$\tilde{\omega}(t_k + \Delta t/2) \approx \frac{[\underline{\underline{R}}_{k+1}\underline{\underline{R}}_k^T] - [\underline{\underline{R}}_{k+1}\underline{\underline{R}}_k^T]^T}{2\Delta t}.$$

Relate the components of the angular velocity vector, $\underline{\omega}$, to the incremental rotation parameters $\hat{\underline{c}}$. *(5)* On one graph, plot the three components of the angular velocity vector in the fixed system as a function of time. *(6)* Derive a similar formula the the components of the angular velocity vector in the rotating system. *(7)* On one graph, plot the three components of the angular velocity vector in the rotating system as a function of time.

Problem 13.19. Rigid body tumbling in space

In example 8.7, the equations of motion for a rigid body with respect to an inertial point were derived. The dynamic equilibrium equations, eqs. (8.43), are $\dot{\underline{p}}_O = \underline{F}_O$ and $\dot{\underline{h}}_O = \underline{M}_O$. The relationship between the displacements and the momenta are in the form of eq. (8.44), recast here as

$$\left\{\begin{matrix} \underline{u} \\ \underline{c} \end{matrix}\right\}' = \begin{bmatrix} \underline{\underline{I}} & \underline{\underline{0}} \\ \underline{\underline{0}} & \underline{\underline{H}}^{-1}(\underline{c}) \end{bmatrix} \left(\underline{\underline{M}}_B\right)^{-1} \left\{\begin{matrix} \underline{p}_O \\ \underline{h}_O - (\tilde{r}_{B0} + \tilde{u})\underline{p}_O \end{matrix}\right\},$$

where $\underline{c}$ is the Wiener-Milenković rotation parameter vector for rotation tensor $\underline{\underline{R}}$ and mass matrix $\underline{\underline{M}}_B$ is defined by eq. (8.38). The mass properties of the body are $m = 6{,}900$ kg, $\eta^{*T} = \{1.5, 0.5, 0.7\}$ m, and $\underline{\underline{I}}^{B*} = \text{diag}(110, 9{,}000, 15{,}000)$ kg·m^2. At the initial time, the position and rotation parameter vectors are $\underline{u}(t = 0) = \underline{0}$ and $\underline{c}(t = 0) = \underline{0}$, respectively. The initial linear and angular momenta are $\underline{p}_O^T(t = 0) = \{45{,}000, 18{,}000, 25{,}000\}$ kg·m/s, and $\underline{h}_O^T(t = 0) = \{5{,}000, 4{,}000, 8{,}000\}$ kg·m^2·rad/s. The externally applied loads are $\underline{F}_O = \underline{0}$ and $\underline{M}_O = \underline{0}$. *(1)* Integrate the governing equations of motion of the rigid body to find its response for $t \in [0, 300]$ s. Plot the components of the position vector of the reference point, $\underline{u}$, and the orientation parameters of the rigid body, $\underline{c}$, as a function of time. *(2)* Plot the inertial components of the velocity vector, $\dot{\underline{u}}$, and those of the inertial angular velocity, $\underline{\omega}$, of the rigid body. *(3)* Plot the components of the linear and angular momenta, and the kinetic energy of the body as a function of time. Comment on your results. *(4)* Plot the motion and velocity of the center of mass of the rigid body.

14

Parameterization of motion

While the parameterization of rotation discussed in chapter 13 has received wide attention, much less emphasis has been placed on that of motion. This is probably due to the fact that the analysis of motion is often described in terms of rather abstract mathematical formulations.

For instance, Ball [259] and Angeles [260] used the concepts of twists and wrenches. A systematic, coordinate-free exposition of the different algebraic operations in the set of infinitesimal displacements (screws) and their relation with finite displacement was developed by Chevallier [261]. Euler motion parameters are closely related to the dual-number quaternion algebra techniques used in kinematics [262, 263, 264]. Dual numbers, vectors, and matrices are described in the textbook by Fisher [265] and have received considerable attention in kinematics [1, 23, 22], dynamics [266, 267], and elastodynamics [268]. Many of these studies have shown that the most efficient and elegant implementations of dual-number techniques are based on general screw theory with the screw expressed by means of Plücker coordinates, see section 5.1.

Borri *et al.*[269] addressed the problem of parameterization of motion by focusing on two representations, the exponential map of motion and Cayley's parameterization. Borri and Bottasso [270, 271] used these concepts to analyze curved beam in three-dimensional space, leading to the helicoidal approximation.

The present chapter focuses on parameterization techniques for motion. In contrast with other approaches, the exposition presented here is expressed in terms of linear algebra concepts, which are easily understood and implemented in computer software.

Cayley's formulation that led to an algebraic representation of rotation is generalized in section 14.1 to the problem of motion. Euler motion parameters are presented in section 14.3 and provide a purely algebraic representation of motion. When using the bi-quaternion algebra introduced in section 14.2, all motion operations become bilinear expressions of bi-quaternions. These advantages, however, come at a high cost: eight parameters must be used instead of six, *i.e.*, Euler motion parameters do not form a minimum set.

Mozzi-Chasles' theorem, presented in section 5.1, states that an arbitrary motion of a rigid body can be represented by a screw motion. The axis of the screw is called the *Mozzi-Chasles axis*, denoted $\underline{\mathcal{M}}$, and its Plücker coordinates have been evaluated in section 5.1. The Plücker coordinates of an arbitrary material line of a rigid body subjected to a screw motion are known to transform by the action of the *motion tensor*, and $\underline{\mathcal{M}}$ is an eigenvalue of this tensor associated with its positive unit eigenvalue.

The vectorial parameterization of motion is introduced in section 14.4 and consists of minimal set of parameters defining the components of a *motion parameter vector*. The vectorial parameterization of motion presents two fundamental properties. First, it is tensorial in nature: the tensorial nature of the second-order motion tensor implies and is implied by the tensorial nature of the motion parameter vector, a first-order tensor. Second, rotation parameter vectors are parallel to the eigenvector of the motion tensor corresponding to its unit eigenvalue. Because these two properties imply each other, either can be taken as the definition of the vectorial parameterization of motion. A parameterization of motion is vectorial if and only if the motion parameter vector is parallel an eigenvector of the motion tensor associated with its unit eigenvalue.

A complete description of motion is presented for a generic motion parameter vector. Relevant formulæ for specific parameterizations of this class are then easily obtained. Specific expressions are given for three parameterizations that present desirable properties: the exponential map of motion, the Cayley-Gibbs-Rodrigues motion parameters, and Wiener-Milenković motion parameters.

14.1 Cayley's motion parameters

In section 13.1, a purely algebraic description of rotation was obtained from the simple argument of length preservation. These developments lead to a multiplicative decomposition of the rotation tensor: $\underline{\underline{R}} = (\underline{\underline{I}} - \widetilde{a})^{-1}(\underline{\underline{I}} + \widetilde{a})$, see eq. (13.4). Cayley's parameters, $\underline{a}$, are a by-product of this decomposition. This multiplicative decomposition exists for all vectorial parameterizations, see eq. (13.50).

A similar decomposition is sought for the motion tensor, $\underline{\underline{C}}$, defined by eq. (5.35). Consider the following relationship

$$\begin{bmatrix} (\underline{\underline{I}} - \varepsilon\widetilde{p}/2) & -\varepsilon\widetilde{q}/2 \\ 0 & (\underline{\underline{I}} - \varepsilon\widetilde{p}/2) \end{bmatrix} \begin{bmatrix} \underline{\underline{R}} & \widetilde{u}\underline{\underline{R}} \\ 0 & \underline{\underline{R}} \end{bmatrix} = \begin{bmatrix} (\underline{\underline{I}} + \varepsilon\widetilde{p}/2) & \varepsilon\widetilde{q}/2 \\ 0 & (\underline{\underline{I}} + \varepsilon\widetilde{p}/2) \end{bmatrix}, \tag{14.1}$$

where q is an as yet unknown quantity. In view of eq. (13.50), three of the above submatrix equalities are readily satisfied. Using eq. (13.51), the last equality implies $\varepsilon\widetilde{q} = (\underline{\underline{I}} - \varepsilon\widetilde{p}/2)\widetilde{u}(\underline{\underline{I}} + \varepsilon\widetilde{p}/2) = \widetilde{u} + \varepsilon\widetilde{u}\widetilde{p}/2 + \varepsilon^2(p^T u)\widetilde{p}/4$. Clearly, q is related to the displacement vector, $\underline{u}$, $\varepsilon q = \left[\underline{\underline{I}} - \varepsilon\widetilde{p}/2 + \varepsilon^2 p p^T/4\right] \underline{u}$. Finally, simple vector identities and the definition of ε, eq. (13.47), lead to

$$\underline{q} = \frac{\underline{\underline{R}}^T + \underline{\underline{I}}}{2\zeta_1}\underline{u}. \tag{14.2}$$

The following notation is introduced

$$\underline{\mathcal{P}} = \left\{ \frac{q}{p} \right\}. \tag{14.3}$$

The two vectors, $\underline{q}$ and $\underline{p}$, form the *motion parameter vector*, $\underline{\mathcal{P}}$. The multiplicative decomposition, eq. (14.1), now becomes

$$\underline{\underline{C}} = \left(\underline{\underline{I}} + \frac{\varepsilon}{2} \widetilde{\mathcal{P}} \right) \left(\underline{\underline{I}} - \frac{\varepsilon}{2} \widetilde{\mathcal{P}} \right)^{-1} = \left(\underline{\underline{I}} - \frac{\varepsilon}{2} \widetilde{\mathcal{P}} \right)^{-1} \left(\underline{\underline{I}} + \frac{\varepsilon}{2} \widetilde{\mathcal{P}} \right), \tag{14.4}$$

where the generalized vector product tensor, $\widetilde{\mathcal{P}}$, is defined by eq. (5.52). It is readily verified that eq. (13.51) generalizes to

$$\left(\underline{\underline{I}} - \frac{\varepsilon}{2} \widetilde{\mathcal{P}} \right)^{-1} = \frac{1}{2} \left(\underline{\underline{C}} + \underline{\underline{I}} \right).$$

The motion tensor is now written in terms of the vectors $\underline{q}$ and $\underline{p}$ as

$$\underline{\underline{C}}(\underline{\mathcal{P}}) = \begin{bmatrix} \underline{\underline{R}}(p) & \underline{\underline{R}}(p) \dfrac{\underline{\underline{R}}(p) + \underline{\underline{I}}}{2} \, \varepsilon\widetilde{q} \, \dfrac{\underline{\underline{R}}(p) + \underline{\underline{I}}}{2} \\ 0 & \underline{\underline{R}}(p) \end{bmatrix}. \tag{14.5}$$

In summary, the motion tensor can be expressed in a purely algebraic form in terms of the six parameters, $\underline{\mathcal{P}}^T = \{\underline{q}^T, \underline{p}^T\}$. The rotation parameter vector, $\underline{p}$, determines the rotation tensor. Vector $\underline{q}$ is related to the displacement vector of the reference point through eq. (14.2) and determines the remaining entries of the motion tensor.

Similar developments lead to the following additional results

$$\underline{\underline{C}}^{-1} = \left(\underline{\underline{I}} - \frac{\varepsilon}{2} \widetilde{\mathcal{P}} \right) \left(\underline{\underline{I}} + \frac{\varepsilon}{2} \widetilde{\mathcal{P}} \right)^{-1} = \left(\underline{\underline{I}} + \frac{\varepsilon}{2} \widetilde{\mathcal{P}} \right)^{-1} \left(\underline{\underline{I}} - \frac{\varepsilon}{2} \widetilde{\mathcal{P}} \right),$$

$$\left(\underline{\underline{I}} + \frac{\varepsilon}{2} \widetilde{\mathcal{P}} \right)^{-1} = \frac{1}{2} \left(\underline{\underline{C}}^{-1} + \underline{\underline{I}} \right),$$

and

$$\underline{\underline{C}}^{-1}(\underline{\mathcal{P}}) = \begin{bmatrix} \underline{\underline{R}}^T & \underline{\underline{R}}^T \dfrac{\underline{\underline{R}}^T + \underline{\underline{I}}}{2} \, \varepsilon\widetilde{q}^T \, \dfrac{\underline{\underline{R}}^T + \underline{\underline{I}}}{2} \\ 0 & \underline{\underline{R}}^T \end{bmatrix}.$$

In section 13.1, Cayley's rotation parameters were shown to provide a purely algebraic description of rotation. Based on the length preservation property of rotation, Cayley's formula, eq. (13.4), was derived, which takes the form of a multiplicative decomposition of the rotation tensor. In this section, a multiplicative decomposition of the motion tensor was derived based on purely algebraic arguments. The resulting decomposition, eq. (14.4), mirrors Cayley's formula, eq. (13.4). The motion parameters defined by eq. (14.3) are a byproduct of the multiplicative decomposition and provide a purely algebraic description of the motion tensor, eq. (14.5), which explicitly shows the dependency of the motion tensor on six parameters only.

14.2 Bi-quaternion algebra

In section 14.3, it will be shown that motion operations are conveniently expressed in terms of bi-quaternions. The present preparatory section focuses on the definition of bi-quaternions, the derivation of a number of their properties, and the definition of the operators that ease bi-quaternion algebra.

A *bi-quaternion* is defined as an array of two quaternions

$$\breve{g} = \left\{ \begin{matrix} \hat{q} \\ \hat{e} \end{matrix} \right\}. \tag{14.6}$$

Bi-quaternion operators

Bi-quaternion operations are performed by using a number of matrices of size 8×8 defined as follows

$$\underline{\underline{A}}(\breve{g}) = \begin{bmatrix} \underline{A}(\hat{e}) & \underline{A}(\hat{q}) \\ \underline{0} & \underline{A}(\hat{e}) \end{bmatrix}, \quad \underline{\underline{\bar{A}}}(\breve{g}) = \begin{bmatrix} \underline{A}^T(\hat{e}) & \underline{A}^T(\hat{q}) \\ \underline{0} & \underline{A}^T(\hat{e}) \end{bmatrix}, \tag{14.7a}$$

$$\underline{\underline{B}}(\breve{g}) = \begin{bmatrix} \underline{B}(\hat{e}) & \underline{B}(\hat{q}) \\ \underline{0} & \underline{B}(\hat{e}) \end{bmatrix}, \quad \underline{\underline{\bar{B}}}(\breve{g}) = \begin{bmatrix} \underline{B}^T(\hat{e}) & \underline{B}^T(\hat{q}) \\ \underline{0} & \underline{B}^T(\hat{e}) \end{bmatrix}, \tag{14.7b}$$

$$\underline{\underline{C}}(\breve{g}) = \begin{bmatrix} \underline{C}(\hat{e}) & \underline{C}(\hat{q}) \\ \underline{0} & \underline{C}(\hat{e}) \end{bmatrix}, \quad \underline{\underline{\bar{C}}}(\breve{g}) = \begin{bmatrix} \underline{C}^T(\hat{e}) & \underline{C}^T(\hat{q}) \\ \underline{0} & \underline{C}^T(\hat{e}) \end{bmatrix}. \tag{14.7c}$$

Each operator is composed of four, 4×4 sub-matrices consisting of the quaternion operators defined by eqs. (13.9).

If $\breve{g}$ and $\breve{h}$ are two arbitrary bi-quaternions, the following matrix products commute

$$\underline{\underline{A}}(\breve{g})\underline{\underline{B}}(\breve{h}) = \underline{\underline{B}}(\breve{h})\underline{\underline{A}}(\breve{g}), \quad \underline{\underline{\bar{A}}}(\breve{g})\underline{\underline{B}}(\breve{h}) = \underline{\underline{B}}(\breve{h})\underline{\underline{\bar{A}}}(\breve{g}), \tag{14.8a}$$

$$\underline{\underline{A}}(\breve{g})\underline{\underline{\bar{B}}}(\breve{h}) = \underline{\underline{\bar{B}}}(\breve{h})\underline{\underline{A}}(\breve{g}), \quad \underline{\underline{\bar{A}}}(\breve{g})\underline{\underline{\bar{B}}}(\breve{h}) = \underline{\underline{\bar{B}}}(\breve{h})\underline{\underline{\bar{A}}}(\breve{g}), \tag{14.8b}$$

$$\underline{\underline{C}}(\breve{g})\underline{\underline{\bar{A}}}(\breve{h}) = \underline{\underline{B}}(\breve{h})\underline{\underline{C}}(\breve{g}), \quad \underline{\underline{\bar{C}}}(\breve{g})\underline{\underline{B}}(\breve{h}) = \underline{\underline{\bar{A}}}(\breve{h})\underline{\underline{C}}(\breve{g}). \tag{14.8c}$$

These commutativity properties mirror the corresponding properties of quaternion operators expressed by eqs. (13.11). These identities then imply the following results

$$\underline{\underline{A}}(\breve{g})\breve{h} = \underline{\underline{B}}(\breve{h})\breve{g}, \quad \underline{\underline{\bar{A}}}(\breve{g})\breve{h} = \underline{\underline{\bar{C}}}(\breve{h})\breve{g}, \quad \underline{\underline{B}}(\breve{g})\breve{h} = \underline{\underline{C}}(\breve{h})\breve{g}. \tag{14.9}$$

Here again, these properties mirror the corresponding properties of quaternion operators expressed by eqs. (13.12).

Next, the following results are easily checked

$$\underline{\underline{A}}(\breve{g})\underline{\underline{A}}(\breve{h}) = \underline{\underline{A}}(\breve{f}) \iff \breve{f} = \underline{\underline{A}}(\breve{g})\breve{h} = \underline{\underline{B}}(\breve{h})\breve{g}, \tag{14.10a}$$

$$\underline{\underline{A}}(\breve{g})\underline{\underline{\bar{A}}}(\breve{h}) = \underline{\underline{A}}(\breve{f}) \iff \breve{f} = \underline{\underline{C}}(\breve{g})\breve{h} = \underline{\underline{\bar{B}}}(\breve{h})\breve{g}, \tag{14.10b}$$

$$\underline{\underline{\bar{A}}}(\breve{g})\underline{\underline{A}}(\breve{h}) = \underline{\underline{A}}(\breve{f}) \iff \breve{f} = \underline{\underline{\bar{A}}}(\breve{g})\breve{h} = \underline{\underline{\bar{C}}}(\breve{h})\breve{g}, \tag{14.10c}$$

where the double-headed arrows indicate that the two equalities imply each other. These bi-quaternion identities are inherited from the quaternion counterparts, eqs. (13.13). Similarly

$$\underline{\mathbb{B}}(\breve{g})\underline{\mathbb{B}}(\breve{h}) = \underline{\mathbb{B}}(\breve{f}) \Longleftrightarrow \breve{f} = \underline{\mathbb{B}}(\breve{g})\breve{h} = \underline{\mathbb{A}}(\breve{h})\breve{g}, \tag{14.11a}$$

$$\underline{\mathbb{B}}(\breve{g})\underline{\mathbb{B}}(\breve{h}) = \underline{\mathbb{B}}(\breve{f}) \Longleftrightarrow \breve{f} = \underline{\bar{\mathbb{C}}}(\breve{g})\breve{h} = \underline{\bar{\mathbb{A}}}(\breve{h})\breve{g}, \tag{14.11b}$$

$$\underline{\bar{\mathbb{B}}}(\breve{g})\underline{\mathbb{B}}(\breve{h}) = \underline{\mathbb{B}}(\breve{f}) \Longleftrightarrow \breve{f} = \underline{\bar{\mathbb{B}}}(\breve{g})\breve{h} = \underline{\mathbb{C}}(\breve{h})\breve{g}. \tag{14.11c}$$

Finally, operator $\underline{\underline{\mathbb{W}}}$ is defined as

$$\underline{\underline{\mathbb{W}}}(\breve{g}) = \begin{bmatrix} \underline{\underline{S}}(\hat{e}) & \underline{\underline{S}}(\hat{q}) \\ \underline{\underline{0}} & \underline{\underline{S}}(\hat{e}) \end{bmatrix} = \frac{1}{2}[\underline{\underline{\mathbb{A}}}(\breve{g}) - \underline{\underline{\mathbb{B}}}(\breve{g})], \tag{14.12}$$

where quaternion operator $\underline{\underline{S}}$ is defined by eq. (13.15) which also implies the last equality.

Bi-quaternions composed of orthogonal quaternions

Two quaternions, $\hat{q}$ and $\hat{e}$, are said to be orthogonal if $\hat{q}^T \hat{e} = 0$. For bi-quaternions composed of such pair of quaternions, $\breve{g}^T = \{\hat{q}^T, \hat{e}^T\}$, the following identities can be shown to hold

$$\underline{\bar{\mathbb{A}}}(\breve{g})\underline{\mathbb{A}}(\breve{g}) = \underline{\mathbb{A}}(\breve{g})\underline{\bar{\mathbb{A}}}(\breve{g}) = \underline{\bar{\mathbb{B}}}(\breve{g})\underline{\mathbb{B}}(\breve{g}) = \underline{\mathbb{B}}(\breve{g})\underline{\bar{\mathbb{B}}}(\breve{g}) = |\hat{e}|^2\underline{\underline{\mathbb{I}}}, \tag{14.13}$$

where $\underline{\underline{\mathbb{I}}}$ is the 6×6 identity matrix. Identities (13.18) were used to prove the above relationships.

14.3 Euler motion parameters

The definition of the motion tensor, eq. (5.35), requires six parameters, three parameters to define the displacement vector, $\underline{u}$, and three parameters to define the rotation tensor, $\underline{\underline{R}}$. The goal of this section is to develop a parameterization of motion, *i.e.*, to find a set of parameters that ease the manipulation of motion operations.

In section 13.3, Euler parameters were introduced and quaternion algebra was shown to provide an effective tool to manipulate rotation operations in a purely algebraic manner. The vector part of Euler parameters is oriented along the eigenvector of the rotation tensor corresponding to a unit eigenvalue, see eq. (13.19).

To generalize Euler parameters to the problem of motion, a set of parameters are derived that are parallel to the eigenvector of the motion tensor corresponding to a unit eigenvalue. To ease the algebra, the motion tensor, eq. (5.35), is expanded to an 8×8 operator using the quaternion algebra operators defined in section 13.2,

$$\underline{\underline{\mathbb{C}}} = \begin{bmatrix} \underline{\underline{D}}(\hat{e}) & \underline{\underline{S}}(\hat{u})\underline{\underline{D}}(\hat{e}) \\ \underline{\underline{0}} & \underline{\underline{D}}(\hat{e}) \end{bmatrix}, \tag{14.14}$$

where $\hat{e}$ is the unit quaternion representing the rotation tensor $\underline{\underline{R}}$, and $\hat{u}^T = \{0, \underline{u}^T\}$ a non-unit quaternion with a vanishing scalar part. An eigenvector of this expanded motion tensor associated with the unit eigenvalue is

$$\underline{\mathbb{N}} = \left\{ \begin{matrix} \frac{1}{2}\underline{\underline{B}}(\hat{e})\hat{u} \\ \hat{e} \end{matrix} \right\} = \left\{ \begin{matrix} \hat{q} \\ \hat{e} \end{matrix} \right\}. \tag{14.15}$$

As discussed in section 5.5.2, several expressions for the eigenvector can be found, due to the multiplicity of two of the unit eigenvalue. The above expression, however, is convenient because a one to one correspondence exists between quaternions $\hat{u}$ and $\hat{q}$; indeed,

$$\hat{q} = \frac{1}{2}\underline{\underline{B}}(\hat{e})\hat{u} \iff \hat{u} = 2\underline{\underline{B}}^T(\hat{e})\hat{q}. \tag{14.16}$$

Because $\underline{\underline{B}}(\hat{e})$ is an orthogonal operator, this mapping presents no singularities.

The scalar part of $\hat{q}$ is $q_0 = -1/2\,\underline{e}^T\underline{u} = -d/2\,\sin\phi/2$, where d is the intrinsic displacement of the rigid body defined by eq. (5.7). On the other hand, the scalar part of $\hat{u}$ is $u_0 = 0 = 2\,\hat{e}^T\hat{q}$. This implies that quaternions $\hat{e}$ and $\hat{q}$ are orthogonal to each other

$$\hat{e}^T\hat{q} = 0. \tag{14.17}$$

Finally, it is readily verified that the norms of quaternions $\hat{u}$ and $\hat{q}$ are closely related, $\hat{u}^T\hat{u} = 4\,\hat{q}^T\hat{q}$.

The *Euler motion parameters* form a bi-quaternion and provide a convenient parameterization of the motion tensor,

$$\breve{g} = \left\{ \begin{matrix} \hat{q} \\ \hat{e} \end{matrix} \right\}. \tag{14.18}$$

Note the redundancy in this representation that requires eight parameters instead of the six forming a minimum set. The eight Euler motion parameters are subjected to two constraints: the normality condition for quaternion $\hat{e}$, see eq. (13.20), and the orthogonality of quaternions $\hat{e}$ and $\hat{q}$, see eq. (14.17).

14.3.1 The motion tensor

Equation (14.14) gives the expression for the motion tensor. The only term not expressed in terms of the Euler motion parameters is $\underline{\underline{S}}(\hat{u})\underline{\underline{D}}(\hat{e})$. With the help of identity (13.15), this term becomes

$$\underline{\underline{S}}(\hat{u})\underline{\underline{D}}(\hat{e}) = \frac{1}{2}\underline{\underline{A}}(\hat{u})\underline{\underline{A}}(\hat{e})\underline{\underline{B}}^T(\hat{e}) - \frac{1}{2}\underline{\underline{A}}(\hat{e})\underline{\underline{B}}(\hat{u})\underline{\underline{B}}^T(\hat{e}), \tag{14.19}$$

where operator $\underline{\underline{D}}(\hat{e})$ was expressed by eq. (13.16), and the commutativity property (13.11b) was used in the second term. Using eq. (13.13a), the first term of eq. (14.19) becomes $\underline{\underline{A}}(\hat{q})\underline{\underline{B}}^T(\hat{e})$, where $\hat{q}$ is given by eq. (14.16). Next, note that because quaternion $\hat{u}$ has a vanishing scalar part, $-\underline{B}(\hat{u}) = \underline{B}^T(\hat{u})$, and the second term of eq. (14.19) becomes $\underline{A}(\hat{e})\underline{B}^T(\hat{q})$, where $\hat{q}$ is given by eq. (14.16) once again.

The motion tensor written in terms of the Euler motion parameters becomes

$$\underline{\underline{C}}(\breve{g}) = \begin{bmatrix} \underline{D}(\hat{e}) & \underline{A}(\hat{q})\underline{B}^T(\hat{e}) + \underline{A}(\hat{e})\underline{B}^T(\hat{q}) \\ \underline{0} & \underline{D}(\hat{e}) \end{bmatrix}.$$

Introducing operators $\underline{\underline{A}}$ and $\underline{\underline{\bar{B}}}$ defined by eqs. (14.7a) and (14.7b), respectively, then leads to

$$\underline{\underline{C}}(\breve{g}) = \underline{\underline{A}}(\breve{g})\underline{\underline{\bar{B}}}(\breve{g}) = \underline{\underline{\bar{B}}}(\breve{g})\underline{\underline{A}}(\breve{g}). \tag{14.20}$$

The motion tensor becomes a bilinear function of Euler motion parameters. Note the parallel between this expression and that for the rotation tensor, eq. (13.16).

The inverse of the motion tensor is found using similar developments,

$$\underline{\underline{C}}^{-1}(\breve{g}) = \begin{bmatrix} \underline{D}^T(\hat{e}) & \underline{B}(\hat{q})\underline{A}^T(\hat{e}) + \underline{B}(\hat{e})\underline{A}^T(\hat{q}) \\ 0 & \underline{D}^T(\hat{e}) \end{bmatrix},$$

and finally,

$$\underline{\underline{C}}^{-1}(\breve{g}) = \underline{\underline{\bar{A}}}(\breve{g})\underline{\underline{B}}(\breve{g}) = \underline{\underline{B}}(\breve{g})\underline{\underline{\bar{A}}}(\breve{g}). \tag{14.21}$$

14.3.2 The velocity vector

The components of the velocity vector resolved in the fixed frame are obtained from their definition, eq. (5.69a). This definition is expressed by means of 8×8 operators to ease algebraic manipulations

$$\dot{\underline{\underline{C}}}(\breve{g})\underline{\underline{C}}^{-1}(\breve{g}) = \left[\underline{\underline{A}}(\breve{g})\underline{\underline{\bar{B}}}(\breve{g})\right]^{\cdot} \underline{\underline{B}}(\breve{g})\underline{\underline{\bar{A}}}(\breve{g}) = \underline{\underline{A}}(\dot{\breve{g}})\underline{\underline{\bar{A}}}(\breve{g}) + \underline{\underline{\bar{B}}}(\breve{g})\underline{\underline{B}}(\breve{g}),$$

where the commutativity relationships (14.8) and normality conditions (14.13) were used. Because $\hat{e}$ is a unit quaternion, eq. (14.13) implies $\underline{\underline{\bar{B}}}\,\underline{\underline{B}} = \underline{\underline{I}}$, and a time derivative yields $\dot{\underline{\underline{\bar{B}}}}\,\underline{\underline{B}} = -\underline{\underline{\bar{B}}}\,\dot{\underline{\underline{B}}}$. The above expression then becomes $\dot{\underline{\underline{C}}}(\breve{g})\underline{\underline{C}}^{-1}(\breve{g}) = \underline{\underline{A}}(\dot{\breve{g}})\underline{\underline{\bar{A}}}(\breve{g}) - \underline{\underline{\bar{B}}}(\breve{g})\underline{\underline{B}}(\dot{\breve{g}}) = \underline{\underline{A}}(\breve{v}/2) - \underline{\underline{B}}(\breve{v}/2)$, where, according to eq. (14.10b) and (14.11c), $\breve{v} = 2\underline{\underline{\bar{B}}}(\breve{g})\dot{\breve{g}}$.

In summary,

$$\dot{\underline{\underline{C}}}(\breve{g})\underline{\underline{C}}^{-1}(\breve{g}) = \underline{\underline{W}}(\breve{v}),$$

where $\breve{v}^T = \{\hat{v}^T, \hat{\omega}^T\}$ is the velocity bi-quaternion in the fixed frame given by

$$\breve{v} = 2\underline{\underline{\bar{B}}}(\breve{g})\dot{\breve{g}}, \tag{14.22}$$

The velocity bi-quaternion becomes a bilinear expression of the Euler motion parameters and their derivatives; note the parallel between eq. (14.22) for motion and its counterpart, eq. (13.23), for rotation.

The vector parts of quaternions $\hat{v}$ and $\hat{\omega}$ are the velocity and angular velocity vectors, respectively. The scalar part of the velocity quaternion is $v_0 = 2[\hat{e}^T\dot{\hat{q}} + \hat{q}^T\dot{\hat{e}}] =$

0, because $\hat{q}$ and $\hat{e}$ are orthogonal quaternions, and because $\hat{e}$ is a unit quaternion, the scalar part of the angular velocity quaternion vanishes, $\omega_0 = 2\hat{e}^T\dot{\hat{e}} = 0$.

The components of the velocity vector resolved in the material frame are obtained in a similar manner

$$\underline{\mathbb{C}}^{-1}(\check{g})\underline{\dot{\mathbb{C}}}(\check{g}) = \underline{\mathbb{W}}(\check{v}^*),$$

where $\check{v}^{*T} = \left\{\hat{v}^{*T}, \hat{\omega}^{*T}\right\}$ is the velocity bi-quaternion in the material frame. These results are written in a compact manner as

$$\check{v}^* = 2\underline{\bar{\mathbb{A}}}(\check{g})\dot{\check{g}}. \tag{14.23}$$

14.3.3 Composition of finite motions

Let $\check{g}^T = \left\{\hat{q}^T, \hat{e}^T\right\}$, $\check{g}_1^T = \left\{\hat{q}_1^T, \hat{e}_1^T\right\}$, and $\check{g}_2^T = \left\{\hat{q}_2^T, \hat{e}_2^T\right\}$ be the bi-quaternions of three motion tensors such that $\underline{\mathbb{C}}(\check{g}) = \underline{\mathbb{C}}(\check{g}_1)\underline{\mathbb{C}}(\check{g}_2)$. The problem at hand is to determine bi-quaternion $\check{g}$ as a function of the other two. With the help of eq. (14.20), this expands to $\underline{\mathbb{C}}(\check{g}) = \underline{\mathbb{A}}(\check{g}_1)\underline{\mathbb{B}}(\check{g}_1)\underline{\mathbb{A}}(\check{g}_2)\underline{\mathbb{B}}(\check{g}_2)$ and the commutativity property (14.8a) then implies $\underline{\mathbb{C}}(\check{g}) = \underline{\mathbb{A}}(\check{g}_1)\underline{\mathbb{A}}(\check{g}_2)\underline{\mathbb{B}}(\check{g}_1)\underline{\mathbb{B}}(\check{g}_2)$. Equations (14.10a) and (14.11a) then yield $\underline{\mathbb{C}}(\check{g}) = \underline{\mathbb{A}}(\check{g})\underline{\mathbb{B}}(\check{g})$, where $\check{g} = \underline{\mathbb{A}}(\check{g}_1)\check{g}_2$.

In summary, composition of motions expressed in terms of Euler motion parameters reduces to

$$\underline{\mathbb{C}}(\check{g}) = \underline{\mathbb{C}}(\check{g}_1)\underline{\mathbb{C}}(\check{g}_2) \iff \check{g} = \underline{\mathbb{A}}(\check{g}_1)\check{g}_2 = \underline{\mathbb{B}}(\check{g}_2)\check{g}_1. \tag{14.24}$$

This operation is bilinear in terms of the Euler motion parameters of the two motions; note the parallel between eq. (14.24) for motion and its counterpart, eq. (13.26), for rotation.

If bi-quaternions $\check{g}_1^T = \left\{\hat{e}_1^T, \hat{q}_1^T\right\}$ and $\check{g}_2^T = \left\{\hat{e}_2^T, \hat{q}_2^T\right\}$ are such that $\hat{e}_1$ and $\hat{e}_2$ are unit quaternions and $\hat{e}_1^T\hat{q}_1 = \hat{e}_2^T\hat{q}_2 = 0$, bi-quaternion $\check{g}^T = \left\{\hat{e}^T, \hat{q}^T\right\}$ enjoys the same properties. Indeed, $\hat{e}^T\hat{e} = \hat{e}_2^T A^T(\hat{e}_1)A(\hat{e}_1)\hat{e}_2 = \hat{e}_2^T\hat{e}_2 = 1$, because $\hat{e}_1$ and $\hat{e}_2$ both are unit quaternions. Furthermore, $\hat{e}^T\hat{q} = \hat{e}_2^T A^T(\hat{e}_1)A(\hat{e}_1)\hat{q}_2 + \hat{e}_2^T A^T(\hat{e}_1)A(\hat{q}_1)\hat{e}_2 = \hat{e}_2^T\hat{q}_2 + \hat{e}_1^T B^T(\hat{e}_2)B(\hat{e}_2)\hat{q}_1 = \hat{e}_2^T\hat{q}_2 + \hat{e}_1^T\hat{q}_1 = 0$.

14.3.4 Determination of Euler motion parameters

The last task is to determine Euler motion parameters given the components of the motion tensor. Unfortunately, this inverse relationship cannot be expressed in a simple manner. In view of eq. (14.20), the motion tensor written in the following form

$$\underline{\mathcal{C}}(\check{g}) = \begin{bmatrix} \underline{R}(\hat{e}) & \underline{\underline{Z}}(\hat{q}, \hat{e}) \\ \underline{0} & \underline{R}(\hat{e}) \end{bmatrix},$$

where $\underline{R}(\hat{e})$ is given by eq. (13.21) and

$$\underline{Z}(\hat{q}, \hat{e}) = 2\left[e_0\tilde{q} + q_0\tilde{e} + \tilde{e}\tilde{q} + \tilde{q}\tilde{e}\right]. \tag{14.25}$$

The determination of the bi-quaternion $\breve{g}^T = \{\hat{q}^T, \hat{e}^T\}$ proceeds in two steps. First, quaternion $\hat{e}$ is determined from the rotation tensor, $\underline{R}(\hat{e})$, by following the procedure described in section 13.3.4.

The second step is to determine quaternion $\hat{q}$ from operator $\underline{\underline{Z}}(\hat{q}, \hat{e})$. Consider the following symmetric matrix constructed from the components of $\underline{\underline{Z}}(\hat{q}, \hat{e})$,

$$
\underline{\underline{T}} = \begin{bmatrix} \text{tr}(\underline{\underline{Z}}) & Z_{32} - Z_{23} & Z_{13} - Z_{31} & Z_{21} - Z_{12} \\ Z_{32} - Z_{21} & & & \\ Z_{13} - Z_{31} & & \underline{\underline{Z}} + \underline{\underline{Z}}^T - \text{tr}(\underline{\underline{Z}})\underline{\underline{I}} & \\ Z_{21} - Z_{12} & & & \end{bmatrix}.
$$

Introducing the definition of matrix $\underline{\underline{Z}}$, eq. (14.25), then yields

$$
\underline{\underline{T}} = 4 \begin{bmatrix} e_0q_0 + q_0e_0 & e_1q_0 + q_1e_0 & e_2q_0 + q_2e_0 & e_3q_0 + q_3e_0 \\ e_0q_1 + q_0e_1 & e_1q_1 + q_1e_1 & e_2q_1 + q_2e_1 & e_3q_1 + q_3e_1 \\ e_0q_2 + q_0e_2 & e_1q_2 + q_1e_2 & e_2q_2 + q_2e_2 & e_3q_2 + q_3e_2 \\ e_0q_3 + q_0e_3 & e_1q_3 + q_1e_3 & e_2q_3 + q_2e_3 & e_3q_3 + q_3e_3 \end{bmatrix} = 4(\hat{e}\hat{q}^T + \hat{q}\hat{e}^T).
$$

Quaternion $\hat{q}$ can readily be computed from any column of this matrix. This determination, however, will involve a division by components of quaternion $\hat{e}$; hence, inaccurate results will be obtained when dividing by small, or zero values. The most accurate results will be obtained by selecting index m such that $|e_m| > |e_i|, i \neq m$. The components of $\hat{q}$ are then

$$
q_m = \frac{1}{e_m}\left[\frac{T_{mm}}{8}\right], \quad q_i = \frac{1}{e_m}\left[\frac{T_{mi}}{4} - e_iq_m\right], \quad i \neq m. \tag{14.26}
$$

Of course, the integrity of the data should be checked by verifying that $\hat{q}$ and $\hat{e}$ are orthogonal quaternions.

Example 14.1. Kinetic energy of a rigid body
The kinetic energy of a rigid body undergoing an arbitrary motion was developed in example 8.5, on page 311. Find the expression for the kinetic energy of a rigid body depicted in fig. 8.6 expressed in terms of Euler motion parameters.

Equation (8.31) gives the kinetic energy of the rigid body. To express this quantity in terms of Euler motion parameters, it is convenient to use the two quaternions, $\hat{v}^*$ and $\hat{\omega}^*$, introduced in section 14.3.2. These two quaternions with vanishing scalar parts are combined into a bi-quaternion, $\breve{v}^{*T} = \{\hat{v}^{*T}, \hat{\omega}^{*T}\}$.

Next, the following 8×8 mass matrix is introduced

$$
\underline{\underline{M}}^{B*} = \begin{bmatrix} m\underline{\underline{I}}_{4\times 4} & m\underline{\underline{S}}(\hat{\eta}^{*T}) \\ m\underline{\underline{S}}(\hat{\eta}^*) & \underline{\underline{M}}^{B*} \end{bmatrix}, \tag{14.27}
$$

where $\hat{\eta}^* = \{0, \underline{\eta}^*\}$ is a quaternion with a vanishing scalar part and whose vector part stores the components of the relative position vector of the body's center of mass with respect to the reference point, resolved in the body attached basis. Operator $\underline{\underline{S}}$

is defined by eq. (13.15), and matrix $\underline{\underline{M}}^{B*}$ by eq. (13.30). Notation $(\cdot)^*$ indicates tensor components resolved in the body attached basis.

The kinetic energy now becomes

$$K = \frac{1}{2} \check{v}^{*T} \underline{\underline{M}}^{B*} \check{v}^*. \tag{14.28}$$

Bi-quaternion $\check{v}^*$ is now readily expressed in terms of Euler motion parameters using eq. (14.23), to find

$$K = 2\, \dot{\check{g}}^T \underline{\underline{\bar{A}}}^T (\check{g}) \underline{\underline{M}}^{B*} \underline{\underline{\bar{A}}}(\check{g}) \dot{\check{g}} = 2\, \check{g}^T \underline{\underline{\bar{C}}}^T (\dot{\check{g}}) \underline{\underline{M}}^{B*} \underline{\underline{\bar{C}}}(\dot{\check{g}}) \check{g}, \tag{14.29}$$

where the last equality follows from identity (14.9). Clearly, it is expeditious to express all quantities in the bi-quaternion formalism before introducing Euler motion parameters.

Example 14.2. Hamilton's principle for a rigid body
Application of Hamilton's principle to the rigid body problem leads to a compact form of the equations of motion, as was shown in example 8.6 on page 312. Based on Hamilton's principle, derive the equations of motion of the rigid body depicted in fig. 8.6 in terms of Euler motion parameters.

As observed in the previous example, it is expeditious to express all quantities in terms of bi-quaternion before introducing Euler motion parameters. The virtual displacement and rotation quaternions are defined as $\hat{\delta u}^{*T} = \{\delta u_0^*, \underline{R}\,\delta\underline{u}^T\}$ and $\hat{\delta\psi}^{*T} = \{\delta\psi_0^*, \delta\underline{\psi}^{*T}\}$, respectively. These two quaternions are combined into the virtual motion bi-quaternion resolved in the material frame, $\check{\delta u}^{*T} = \{\hat{\delta u}^{*T}, \hat{\delta\psi}^{*T}\}$.

The force and moment quaternions are defined in a similar manner as $\hat{F}^{*T} = \{0, \underline{F}^{*T}\}$ and $\hat{Q}^{*T} = \{0, \underline{Q}^{*T}\}$, respectively. Both quaternions have a vanishing scalar part and are combined into the load bi-quaternion resolved in the material frame, $\check{a}^{*T} = \{\hat{F}^{*T}, \hat{Q}^{*T}\}$.

Finally, $\hat{p}^{*T} = \{0, \underline{p}^{*T}\}$ and $\hat{h}^{*T} = \{0, \underline{h}^{*T}\}$ are the linear and angular momentum quaternions, respectively. Both quaternions have a vanishing scalar part and are combined into the momentum bi-quaternion resolved in the material frame, $\check{p}^{*T} = \{\hat{p}^{*T}, \hat{h}^{*T}\}$.

With these notations at hand, the virtual work done by the applied loads defined by eq. (8.45) becomes $\delta W^{nc} = \check{a}^{*T} \check{\delta u}^* = \check{a}^T \check{\delta u}$. The load bi-quaternion, $\check{a} = \underline{\underline{C}}^{-1} \check{a}^*$, consists of two quaternions with vanishing scalar components, $\hat{F}^O$ and $\hat{Q}^O$, whose vector parts store the components of the applied force and moment, respectively, computed with respect to inertial point $\mathbf{O}$.

Virtual changes in the kinetic energy of the rigid body follow from eq. (14.28) as $\delta K = \delta\check{v}^{*T} \underline{\underline{M}}^{B*} \check{v}^* = \delta\check{v}^{*T} \check{p}^*$. By analogy with eq. (8.46), $\delta\check{v}^* = \underline{\underline{C}}^{-1} \dot{\check{\delta u}}$, and virtual changes in the kinetic energy then becomes $\delta K = \dot{\check{\delta u}}^T \check{p}$. The momentum bi-quaternion, $\check{p} = \underline{\underline{C}}^{-T} \check{p}^*$, consists of two quaternions with vanishing scalar components, $\hat{p}^O$ and $\hat{h}^O$, whose vector parts store the components of the linear and angular momenta, respectively, computed with respect to inertial point $\mathbf{O}$.

Euler motion parameters are subject to two constraints, the orthogonality constraint, $C_1 = 2\hat{e}^T\hat{q} = 0$, and the normality constraint, $C_2 = \hat{e}^T\hat{e} - 1 = 0$. The potential of these constraints, eq. (10.6), becomes $V^c = -\underline{\lambda}^T\underline{C}$, where $\underline{C}^T = \{C_1, C_2\}$ is the array of constraints and $\underline{\lambda}$ the array of Lagrange's multiplier used to enforce these constraints. Variation of the potential of the constraints yields

$$\delta V^c = -\delta\underline{\lambda}^T\underline{C} - 2\underline{\lambda}^T \begin{bmatrix} \hat{e}^T & \hat{q}^T \\ \hat{0}^T & \hat{e}^T \end{bmatrix} \delta\breve{g} = -\delta\underline{\lambda}^T\underline{C} - \underline{\lambda}^T \begin{bmatrix} \hat{e}^T & \hat{q}^T \\ \hat{0}^T & \hat{e}^T \end{bmatrix} \underline{\underline{B}}(\breve{g})\delta\breve{u}$$

$$= -\delta\underline{\lambda}^T\underline{C} - \delta\breve{u}^T \begin{bmatrix} \hat{1} & \hat{0} \\ \hat{0} & \hat{1} \end{bmatrix} \underline{\lambda},$$

where $\hat{0}$ denotes a vanishing quaternion and $\hat{1}^T = \{1, \underline{0}^T\}$ a quaternion with a unit scalar part and a vanishing vector part. By analogy with eq. (14.22), variations in Euler motion parameters are related to the virtual motion vector as $\delta\breve{g} = 1/2\,\underline{\underline{B}}(\breve{g})\delta\breve{u}$.

For this problem, $L^+ = K - V^c$ and Hamilton's principle for constrained system, eq. (10.10), now implies

$$\int_{t_i}^{t_f} \left(\dot{\delta u}^T\breve{p} + \delta\underline{\lambda}^T\underline{C} + \delta\breve{u}^T \begin{bmatrix} \hat{1} & \hat{0} \\ \hat{0} & \hat{1} \end{bmatrix} \underline{\lambda} + \delta\breve{u}^T\breve{a} \right)\, dt = 0.$$

All boundary terms are ignored here. After integration by parts of the first term, the equations of motion of the system are found to be $\underline{C} = \underline{0}$, the constraint equations, and

$$\left\{ \begin{matrix} \hat{p}_O \\ \hat{h}_O \end{matrix} \right\}^{\cdot} - \left\{ \begin{matrix} \lambda_1\hat{1} \\ \lambda_2\hat{1} \end{matrix} \right\} = \left\{ \begin{matrix} \hat{F}_O \\ \hat{Q}_O \end{matrix} \right\}.$$

Because bi-quaternions $\breve{p}$ and $\breve{a}$ are resolved in the inertial frame, their quaternion components are evaluated with respect to the origin of the inertial frame, point **O**.

Lagrange's multipliers are readily eliminated by only keeping the vector parts of the quaternion equations, leading to the following equations of motion for the rigid body, $\dot{\underline{p}}_O = \underline{F}_O$ and $\dot{\underline{h}}_O = \underline{Q}_O$. These equations are identical to those obtained earlier, see eqs. (8.48a).

The last step of the procedure is to evaluate the time derivatives of Euler motion parameters with respect to time. First, eq. (14.23) yields $\dot{\breve{g}} = \underline{\underline{A}}(\breve{g})\breve{v}^*/2$, and finally, $\dot{\breve{g}} = \underline{\underline{A}}(\breve{g})(\underline{\underline{M}}^{B*})^{-1}\underline{\underline{C}}^T(\breve{g})\breve{p}/2$, which leads to the following system of ordinary differential equations,

$$\left\{ \begin{matrix} \underline{p}_O \\ \underline{h}_O \\ \breve{g} \end{matrix} \right\}^{\cdot} = \left\{ \begin{matrix} \underline{F}_O \\ \underline{Q}_O \\ \underline{\underline{A}}(\breve{g})(\underline{\underline{M}}^{B*})^{-1}\underline{\underline{C}}^T(\breve{g})\breve{p}/2 \end{matrix} \right\}, \tag{14.30}$$

where the scalar parts of quaternions $\hat{p}_O$ and $\hat{h}_O$ vanish.

14.3.5 Problems

Problem 14.1. Inverse of the motion tensor
Prove eq. (14.21) starting from the definition of the motion tensor, eq. (5.59).

Problem 14.2. Angular velocity with Euler motion parameters

Prove eq. (14.23) by evaluating $\underline{\underline{C}}^{-1}(\check{g})\underline{\dot{\underline{C}}}(\check{g}) = \underline{\underline{W}}(\check{v}^*)$. Find the relationship between $\check{v}$, defined in eq. (14.22) and $\check{v}^*$, defined in eq. (14.23).

Problem 14.3. Eigenvectors of the motion tensor

Prove that two linearly independent eigenvectors of the motion tensor associated with its unit eigenvalue are $\mathbb{N}_1^T = \{\hat{e}^T, \hat{0}^T\}$ and $\mathbb{N}_2^T = \{\hat{u}^T \underline{\underline{B}}^T(\hat{e})/2, \hat{e}^T\}$.

14.4 The vectorial parameterization of motion

In the previous section, the Euler motion parameters have been shown to provide an elegant, purely algebraic representation of finite motion. In fact, when using bi-quaternions, all motion operations become bi-linear expressions of bi-quaternions. These advantages, however, come at a high cost: eight parameters must be used instead of six, *i.e.*, the Euler motion parameters do not form a minimal set. Furthermore, the normality and orthogonality conditions inherent to the representation must be enforced as constraints.

The *vectorial parameterization of motion* [272] consists of a minimal set of parameters defining the components of two vectors. The vectorial nature of this class of parameterization of motion sets it apart from the other parameterizations investigated earlier.

14.4.1 Fundamental properties

Consider three motions characterized by displacement vectors, $\underline{u}_1$, $\underline{u}_2$, and $\underline{u}_3$, and rotation tensors, $\underline{\underline{R}}_1$, $\underline{\underline{R}}_2$, and $\underline{\underline{R}}_3$, respectively. The three motions, denoted $(\underline{u}_1, \underline{\underline{R}}_1)$, $(\underline{u}_2, \underline{\underline{R}}_2)$, and $(\underline{u}_3, \underline{\underline{R}}_3)$, respectively, are associated with three motion tensors, $\underline{\underline{C}}_1$, $\underline{\underline{C}}_2$, and $\underline{\underline{C}}_3$, respectively, through the intrinsic expression of the motion tensor, eq. (5.53a).

Assume that the following triple product of motion tensors relates these three quantities,

$$\underline{\underline{C}}_3 = \underline{\underline{C}}_2^{-1}\underline{\underline{C}}_1\underline{\underline{C}}_2. \tag{14.31}$$

As discussed in section 5.6.3, this operation corresponds to a change of frame operation for motion tensors: $\underline{\underline{C}}_1$ and $\underline{\underline{C}}_3$ are the components of the same motion tensor expressed in two frames related by motion tensor $\underline{\underline{C}}_2$.

Using the intrinsic expression for the motion tensor, eq. (5.53a), eq. (14.31) now becomes

$$\underline{\underline{C}}_3 = \underline{\underline{I}} + \underline{\underline{Z}}(d_1 c_1, \sin\phi_1)\,\widetilde{\underline{\underline{C}}_2^{-1}\underline{\underline{N}}_1} + \underline{\underline{Z}}(d_1 c_2, 1 - \cos\phi_1)\,\widetilde{\underline{\underline{C}}_2^{-1}\underline{\underline{N}}_1}\,\widetilde{\underline{\underline{C}}_2^{-1}\underline{\underline{N}}_1},$$

where eq. (5.54) was used. Comparing this result with the intrinsic expression for the motion tensor implies

$$\phi_3 = \phi_1, \tag{14.32a}$$

$$\left\{ \frac{m_3}{\tilde{n}_3} \right\} = \underline{\mathcal{N}}_3 = \underline{\underline{C}}_2^{-1} \underline{\mathcal{N}}_1 = \left\{ \begin{matrix} \underline{\underline{R}}_2^T (\underline{m}_1 + \tilde{n}_1 \underline{u}_2) \\ \underline{\underline{R}}_2^T \tilde{n}_1 \end{matrix} \right\}. \tag{14.32b}$$

These equations express the two conditions required for the proper transformation of motion tensors components under a change of frame. Note that an additional condition is required, $d_3 = d_1$, but is implied by eqs. (14.32). Indeed, eq. (14.32b) yields $\tilde{n}_3^T m_3 = \tilde{n}_1^T \underline{\underline{R}}_2 \underline{\underline{R}}_2^T (\underline{m}_1 + \tilde{n}_1 \underline{u}_2) = \tilde{n}_1^T \underline{m}_1$, or $\lambda_3 = \lambda_1$, where λ is defined by eq. (5.41). In view of eq. (14.32a) and (5.41), $\lambda_3 = \lambda_1$ then yields $d_3 = d_1$.

Let $p(\phi)$ be an arbitrary scalar function of angle ϕ; eq. (14.32a) then implies $p(\phi_3) = p(\phi_1)$. Multiplication of eq. (14.32b) by $p_3 = p(\phi_3)$ on the left-hand side and $p_1 = p(\phi_1) = p(\phi_3)$ on the right-hand side then yields

$$p(\phi_3)\underline{\mathcal{N}}_3 = \left\{ \begin{matrix} p_3 \underline{m}_3 \\ p_3 \tilde{n}_3 \end{matrix} \right\} = \underline{\underline{C}}_2^{-1} \, p(\phi_1)\underline{\mathcal{N}}_1 = \left\{ \begin{matrix} \underline{\underline{R}}_2^T p_1 (\underline{m}_1 + \tilde{n}_1 \underline{u}_2) \\ \underline{\underline{R}}_2^T p_1 \tilde{n}_1 \end{matrix} \right\}. \tag{14.33}$$

This equation is equivalent to eqs. (14.32). Indeed, taking the norm of the last three of eqs. (14.33) yields $p_3 = p_1$, or $\phi_3 = \phi_1$, because $\bar{n}_1$ and $\bar{n}_3$ are unit vectors and $\underline{\underline{R}}_2$ an orthogonal tensor. Dividing eq. (14.33) by $p(\phi_3)$ then yields eq. (14.32b) because $p_3 = p_1$.

The *vectorial parameterization of motion* is defined as

$$\underline{\mathcal{P}} = p(\phi)\underline{\mathcal{N}}, \tag{14.34}$$

where $\underline{\mathcal{P}}$ is the motion parameter vector. Equation (14.33) can now be recast in a more compact manner as

$$\underline{\mathcal{P}}_3 = \underline{\underline{C}}_2^{-1} \underline{\mathcal{P}}_1. \tag{14.35}$$

The discussion presented above establishes that the tensorial nature of the motion tensor expressed by the transformation rule of its components, eq. (14.31), implies the tensorial nature of the rotation parameter vector expressed by the transformation rule of its components, eq. (14.35). It is easily shown that the process can be reversed, *i.e.*, tensorial nature of the rotation parameter vector implies that of the rotation tensor.

In summary, the vectorial parameterization of motion presents two fundamental properties.

1. *The vectorial parameterization of motion is tensorial in nature*, as expressed by the following equivalence,

$$\underline{\underline{C}}(\underline{\mathcal{P}}_3) = \underline{\underline{C}}^{-1}(\underline{\mathcal{P}}_2)\underline{\underline{C}}(\underline{\mathcal{P}}_1)\underline{\underline{C}}(\underline{\mathcal{P}}_2) \Longleftrightarrow \underline{\mathcal{P}}_3 = \underline{\underline{C}}^{-1}(\underline{\mathcal{P}}_2)\underline{\mathcal{P}}_1. \tag{14.36}$$

The tensorial nature of the second-order motion tensor implies and is implied by the tensorial nature of the motion parameter vector, a first-order tensor.

2. Motion parameter vectors are parallel to an *eigenvector of the motion tensor* associated with its unit eigenvalue. Equation (5.39) shows that vector $\underline{\mathcal{N}}$ is a linear

combination of two linearly independent eigenvectors of the motion tensor, both associated with its unit eigenvalue; equation (14.34) then implies that the motion parameter vector shares this property.

Because these two properties imply each other, either can be taken as the definition of the vectorial parameterization of motion. A parameterization of motion is vectorial if and only if the motion parameter vector is parallel an eigenvector of the motion tensor associated with its unit eigenvalue.

14.4.2 The motion parameter vector

A more explicit expression of the motion parameter vector is as follows

$$\underline{P} = \left\{ \begin{matrix} \underline{q} \\ \underline{p} \end{matrix} \right\} = p\underline{N} = \left\{ \begin{matrix} p\underline{m} \\ p\bar{n} \end{matrix} \right\} = \left\{ \begin{matrix} p\underline{\underline{E}}(\phi)\underline{u} \\ p\bar{n} \end{matrix} \right\} = \left\{ \begin{matrix} \underline{\underline{D}}(\underline{p})\underline{u} \\ \underline{p} \end{matrix} \right\}, \tag{14.37}$$

where $\underline{p}$ is the vectorial parameterization of rotation, see section 13.4, and tensor $\underline{\underline{E}}$ is defined by eq. (5.44). Using the notation developed for the vectorial parameterization of rotation, tensor $\underline{\underline{D}}$ becomes

$$\underline{\underline{D}}(\underline{p}) = \delta_0 - \frac{1}{2}\widetilde{p} + \delta_2\widetilde{p}\widetilde{p}, \tag{14.38}$$

where functions $\delta_0(\phi)$ and $\delta_2(\phi)$ are even functions of the rotation angle given by

$$\delta_0 = \frac{\alpha}{\nu}, \tag{14.39a}$$

$$\delta_2 = \frac{1}{p^2}\left(\delta_0 - \frac{1}{\varepsilon}\right). \tag{14.39b}$$

Tensor $\underline{\underline{D}}$ can be resolved in the canonical basis defined by eq. (4.32) to find

$$\underline{\underline{D}}^{[\mathcal{E}]} = \frac{1}{\nu}\begin{bmatrix} \alpha & 0 & 0 \\ 0 & \cos\phi/2 & \sin\phi/2 \\ 0 & -\sin\phi/2 & \cos\phi/2 \end{bmatrix}. \tag{14.40}$$

The determinant of this tensor is now simply evaluated as $\det(\underline{\underline{D}}) = \alpha/\nu^3$.

The inverse of this tensor is readily found as

$$\underline{\underline{F}}(\underline{p}) = \underline{\underline{D}}^{-1}(\underline{p}) = \varphi_0 + \varphi_1\widetilde{p} + \varphi_2\widetilde{p}\widetilde{p}, \tag{14.41}$$

where functions $\varphi_0(\phi)$, $\varphi_1(\phi)$, and $\varphi_2(\phi)$ are even functions of the rotation angle given by

$$\varphi_0 = \frac{\nu}{\alpha}, \tag{14.42a}$$

$$\varphi_1 = \zeta_2, \tag{14.42b}$$

$$\varphi_2 = \frac{1}{p^2}(\varphi_0 - \zeta_1). \tag{14.42c}$$

where coefficients ζ_1 and ζ_2 are given by eqs. (13.46).

Tensor $\underline{\underline{F}}$ enjoys the following remarkable properties

$$\underline{\underline{R}} = \underline{\underline{F}}\,\underline{\underline{F}}^{-T} = \underline{\underline{F}}^{-T}\underline{\underline{F}}, \tag{14.43a}$$

$$\underline{\underline{R}} - \underline{\underline{I}} = \underline{\underline{F}}\,\widetilde{p} = \widetilde{p}\underline{\underline{F}}, \tag{14.43b}$$

$$\widetilde{p} = \underline{\underline{F}}^{-T} - \underline{\underline{F}}^{-1}, \tag{14.43c}$$

which are similar to those of the tangent tensor, eqs. (13.63).

The motion parameter vector is not fully defined yet because it depends on the choice of the generating function, $p(\phi)$, of the vectorial parameterization of rotation and furthermore, parameter α can be selected arbitrarily. Generating functions must be odd functions of the rotation angle and present the limit behavior expressed by eq. (13.44), *i.e.*, all rotation parameter vectors must approach the infinitesimal rotation vector when $\phi \to 0$.

Similarly, the displacement related part of the motion parameter vector, q, should approach the infinitesimal displacement vector for vanishing motions. In view of eq. (14.37), this requirement implies $\lim_{\phi \to 0, d \to 0} \underline{\underline{D}}(p) = \underline{\underline{I}}$, or $\lim_{\phi \to 0, d \to 0} \alpha/\nu = 1$, and finally

$$\lim_{\phi \to 0, d \to 0} \alpha = 1. \tag{14.44}$$

Time derivative of the displacement

In the manipulation of the time derivatives of the motion tensor, it will be necessary to evaluate $\underline{\dot{u}} = (\underline{\underline{F}}\,\underline{q})^{\cdot} = \underline{\underline{\dot{F}}}\,\underline{q} + \underline{\underline{F}}\,\underline{\dot{q}}$, which can be written as $\underline{\dot{u}} = \underline{\underline{L}}(\underline{q}, V p)\underline{\dot{p}} + \underline{\underline{F}}\,\underline{\dot{q}}$, where operator $\underline{\underline{L}}$ is implicitly defined as follows

$$\underline{\underline{\dot{F}}}(p)\underline{q} = \underline{\underline{L}}(\underline{q}, p)\underline{\dot{p}}. \tag{14.45}$$

Using eq. (14.41), operator $\underline{\underline{L}}$ is easily found as

$$\underline{\underline{L}}(\underline{q}, p) = \frac{1}{p'}\left(\frac{\varphi_0'}{p} + \frac{\varphi_1'}{p}\widetilde{p} + \frac{\varphi_2'}{p}\widetilde{p}\widetilde{p}\right)\underline{q}\,p^T - \varphi_1\widetilde{q} - \varphi_2\left(2\widetilde{p}\widetilde{q} - \widetilde{q}\widetilde{p}\right), \tag{14.46}$$

where the notation $(\cdot)'$ indicates a derivative with respect to angle ϕ. Operator $\underline{\underline{L}}$ enjoys the following properties,

$$\underline{\underline{R}}_1^T \underline{\underline{L}}(\underline{q}, p)\underline{\underline{R}}_1 = \underline{\underline{L}}(\underline{\underline{R}}_1^T \underline{q}, \underline{\underline{R}}_1^T p), \tag{14.47a}$$

$$\underline{\underline{L}}(\underline{q}_1 + \underline{q}_2, p) = \underline{\underline{L}}(\underline{q}_1, p) + \underline{\underline{L}}(\underline{q}_2, p), \tag{14.47b}$$

$$\underline{\underline{L}}\widetilde{p} = \underline{\underline{\widetilde{F}q}} - \underline{\underline{F}}\,\widetilde{q}, \tag{14.47c}$$

$$\widetilde{p}\underline{\underline{L}} = \underline{\underline{\widetilde{F}q}} - \underline{\underline{R}}\widetilde{q}\underline{\underline{H}}^T, \tag{14.47d}$$

$$\underline{\underline{L}}(\widetilde{p}\underline{q}, p) = \widetilde{p}\underline{\underline{L}}(\underline{q}, p) - \underline{\underline{L}}(\underline{q}, p)\widetilde{p} = \underline{\underline{F}}\,\widetilde{q} - \underline{\underline{R}}\widetilde{q}\underline{\underline{H}}^T. \tag{14.47e}$$

The first property, eq. (14.47a), expresses the transformation of the components of operator $\underline{\underline{L}}$ under a change of basis of both of its arguments. The second property, eq. (14.47b), expresses the linearity of operator $\underline{L}$ with respect to its first argument. Property (14.47d) stems from the definition of operator $\underline{L}$, $\widetilde{p}\dot{\underline{F}}\,q = \widetilde{p}\underline{L}\,\dot{p}$, and noting that eq. (14.43b) implies $\widetilde{\dot{p}\underline{F}} = \dot{\underline{R}} - \dot{p}\underline{F}$.

14.4.3 The generalized vector product tensor

Skew-symmetric tensor $\widetilde{p}$ plays an important role in the vectorial parameterization of rotation as it appears in the explicit expression of all rotation related tensors, see section 13.4. The generalized vector product tensor, $\widetilde{P}$, plays an important role in the vectorial parameterization of motion.

The tensorial nature of the generalized vector product operator directly follows from eq. (5.54), leading to

$$\widetilde{P}_3 = \underline{C}^{-1}(\underline{P}_2)\widetilde{P}_1\underline{C}(\underline{P}_2) \iff \underline{P}_3 = \underline{C}^{-1}(\underline{P}_2)\underline{P}_1. \tag{14.48}$$

This statement generalizes eq. (4.30), which expresses the tensorial nature of the skew-symmetric operator, $\widetilde{p}$.

Identity (5.55) generalizes as

$$\widetilde{P}\widetilde{P}\widetilde{P} + \underline{Z}(2\varrho, p^2)\widetilde{P} = 0, \tag{14.49}$$

where tensor $\underline{Z}$ is defined by eq. (5.51) and scalar ϱ is closely related to the intrinsic displacement of the rigid body defined by eq. (5.7),

$$\varrho = \underline{p}^T\underline{q} = \frac{pd}{\varphi_0}. \tag{14.50}$$

14.4.4 The motion tensor

The motion tensor and its inverse are obtained from eqs. (5.53) as

$$\underline{C}(\underline{P}) = \underline{I} + \underline{Z}(\bar{\zeta}_1, \zeta_1)\widetilde{P} + \underline{Z}(\bar{\zeta}_2, \zeta_2)\widetilde{P}\widetilde{P}, \tag{14.51a}$$

$$\underline{C}^{-1}(\underline{P}) = \underline{I} - \underline{Z}(\bar{\zeta}_1, \zeta_1)\widetilde{P} + \underline{Z}(\bar{\zeta}_2, \zeta_2)\widetilde{P}\widetilde{P}. \tag{14.51b}$$

The parallel between the expressions for the rotation and motion tensors, eqs. (13.45) and (14.51), is now evident. Coefficients ζ_1 and ζ_2 are given by eqs. (13.46), and

$$\bar{\zeta}_1 = \varrho(\varphi_2 - \varphi_0\zeta_2), \tag{14.52a}$$

$$\bar{\zeta}_2 = \varrho(\varphi_2\zeta_1 - \zeta_2^2). \tag{14.52b}$$

The following multiplicative decomposition of the motion tensor generalizes the corresponding expression for the rotation tensor, eq. (13.50),

$$\begin{aligned}
\underline{C}(\underline{P}) &= \left[\underline{I} + \frac{1}{2}\underline{Z}(\bar{\varepsilon}, \varepsilon)\widetilde{P}\right]\left[\underline{I} - \frac{1}{2}\underline{Z}(\bar{\varepsilon}, \varepsilon)\widetilde{P}\right]^{-1} \\
&= \left[\underline{I} - \frac{1}{2}\underline{Z}(\bar{\varepsilon}, \varepsilon)\widetilde{P}\right]^{-1}\left[\underline{I} + \frac{1}{2}\underline{Z}(\bar{\varepsilon}, \varepsilon)\widetilde{P}\right].
\end{aligned} \tag{14.53}$$

Furthermore, eq. (13.51) also generalizes as

$$\left[\underline{\underline{I}} - \frac{1}{2}\underline{\underline{Z}}(\bar{\varepsilon}, \varepsilon)\widetilde{P}\right]^{-1} = \frac{\underline{\underline{C}} + \underline{\underline{I}}}{2}, \quad \left[\underline{\underline{I}} + \frac{1}{2}\underline{\underline{Z}}(\bar{\varepsilon}, \varepsilon)\widetilde{P}\right]^{-1} = \frac{\underline{\underline{C}}^{-1} + \underline{\underline{I}}}{2}. \tag{14.54}$$

In these last two equations, coefficient ε is given by eq. (13.47b) and

$$\zeta_1\bar{\varepsilon} = 2\bar{\zeta}_2 - \varepsilon\bar{\zeta}_1. \tag{14.55}$$

These equations also generalize Cayley's multiplicative decomposition presented in section 14.1.

14.4.5 The velocity vector

The velocity vector is obtained from a time derivative of the motion tensor, as indicated in eq. (5.64). For the vectorial parameterization of motion, this becomes

$$\underline{V} = \left\{\begin{array}{c} \underline{v} \\ \underline{\omega} \end{array}\right\} = \left\{\begin{array}{c} \dot{u} + \tilde{u}\underline{\omega} \\ \underline{\omega} \end{array}\right\} = \left\{\begin{array}{c} \underline{\underline{F}}(p)\underline{q} + \underline{\underline{F}}(p)\dot{\underline{q}} + \widetilde{\underline{\underline{F}}(p)q}\, \underline{\underline{H}}(p)\dot{p} \\ \underline{\underline{H}}(p)\dot{p} \end{array}\right\}.$$

The velocity vector is now related to the time derivative of the motion parameter vector, $\underline{V} = \underline{\underline{H}}\dot{\underline{P}}$, where tangent tensor $\underline{\underline{H}}$ is defined by eq. (14.57a). Using eq. (5.67), similar developments for the components of the velocity vector resolved in the material frame lead to $\underline{V}^* = \underline{\underline{H}}^*\dot{\underline{P}}$, where tensor $\underline{\underline{H}}^*$ is defined by eq. (14.57b).

In summary, the components of the velocity vector resolved in the inertial and material frames, denoted $\underline{V}$ and $\underline{V}^*$, respectively, are related to the time derivatives of the motion parameter vectors through the following relationships,

$$\underline{V} = \underline{\underline{H}}\dot{\underline{P}}, \tag{14.56a}$$

$$\underline{V}^* = \underline{\underline{H}}^*\dot{\underline{P}}. \tag{14.56b}$$

Explicit expressions for tensors $\underline{\underline{H}}$, $\underline{\underline{H}}^*$, and their inverses are

$$\underline{\underline{H}} = \begin{bmatrix} \underline{\underline{F}}\,\underline{\underline{L}} + \widetilde{\underline{\underline{F}}\,q}\,\underline{\underline{H}} \\ \underline{\underline{0}} \quad \underline{\underline{H}} \end{bmatrix}, \tag{14.57a}$$

$$\underline{\underline{H}}^* = \begin{bmatrix} \underline{\underline{F}}^T & \underline{\underline{R}}^T\underline{\underline{L}} \\ \underline{\underline{0}} & \underline{\underline{H}}^T \end{bmatrix}, \tag{14.57b}$$

$$\underline{\underline{H}}^{-1} = \begin{bmatrix} \underline{\underline{F}}^{-1} & -\underline{\underline{F}}^{-1}\left(\underline{\underline{L}}\,\underline{\underline{H}}^{-1} + \widetilde{\underline{\underline{F}}\,q}\right) \\ \underline{\underline{0}} & \underline{\underline{H}}^{-1} \end{bmatrix}, \tag{14.57c}$$

$$\underline{\underline{H}}^{*-1} = \begin{bmatrix} \underline{\underline{F}}^{-T} & -\underline{\underline{F}}^{-1}\underline{\underline{L}}\,\underline{\underline{H}}^{-T} \\ \underline{\underline{0}} & \underline{\underline{H}}^{-T} \end{bmatrix}, \tag{14.57d}$$

where operator $\underline{\underline{L}}(q, p)$ is defined by eq. (14.45).

Properties of the tangent tensor

Tangent tensors $\underline{\underline{\mathcal{H}}}$, $\underline{\underline{\mathcal{H}}}^*$, and their inverses enjoys the following remarkable properties,

$$\underline{\underline{\mathcal{C}}} = \underline{\underline{\mathcal{H}}}\,\underline{\underline{\mathcal{H}}}^{*-1}, \tag{14.58a}$$

$$\underline{\underline{\mathcal{C}}}^{-1} = \underline{\underline{\mathcal{H}}}^*\underline{\underline{\mathcal{H}}}^{-1}, \tag{14.58b}$$

$$\underline{\underline{\mathcal{C}}} - \underline{\underline{\mathcal{I}}} = \widetilde{\mathcal{P}}\,\underline{\underline{\mathcal{H}}} = \underline{\underline{\mathcal{H}}}\,\widetilde{\mathcal{P}}, \tag{14.58c}$$

$$\underline{\underline{\mathcal{C}}}^{-1} - \underline{\underline{\mathcal{I}}} = -\widetilde{\mathcal{P}}\,\underline{\underline{\mathcal{H}}}^* = -\underline{\underline{\mathcal{H}}}^*\,\widetilde{\mathcal{P}}, \tag{14.58d}$$

$$\widetilde{\mathcal{P}} = \underline{\underline{\mathcal{H}}}^{*-1} - \underline{\underline{\mathcal{H}}}^{-1}. \tag{14.58e}$$

These properties are established directly from the definition of the tangent tensor, eqs. (14.57), taking into account the properties of the tangent tensor for the vectorial parameterization of rotation, eqs. (13.63), and those of tensor $\underline{\underline{F}}$, eqs. (14.43). The properties of operator $\underline{L}$, eqs. (14.47), must also be used.

Tangent tensor $\underline{\underline{\mathcal{H}}}$ is specific to a particular vectorial parameterization of motion, *i.e.*, its expression depends on the choice of the generating function. It is, however, a second-order tensor because the following equivalence holds

$$\underline{\underline{\mathcal{H}}}(\underline{P}_3) = \underline{\underline{\mathcal{C}}}^{-1}(\underline{P}_2)\underline{\underline{\mathcal{H}}}(\underline{P}_1)\underline{\underline{\mathcal{C}}}(\underline{P}_2) \iff \underline{P}_3 = \underline{\underline{\mathcal{C}}}^{-1}(\underline{P}_2)\underline{P}_1. \tag{14.59}$$

Although tensor $\underline{\underline{\mathcal{H}}}$ is not an intrinsic tensor because it depends on the choice of a specific generating function, it is a second-order tensor for all vectorial parameterizations of motion. Equation (14.59) is established directly from the definition of tensor $\underline{\underline{\mathcal{H}}}$, eq. (14.57a), by using eqs. (13.63), (14.43), and eqs. (14.47).

Alternative expression of tangent tensor

Tangent tensors $\underline{\underline{\mathcal{H}}}$, $\underline{\underline{\mathcal{H}}}^*$, and their inverses are given by eqs. (14.57), which are valid for any choice of parameter α. If this parameter is selected to be $\alpha = \nu p'$, alternative expressions of tangent tensor can be obtained,

$$\underline{\underline{\mathcal{H}}}(\underline{P}) = \underline{\underline{\mathcal{Z}}}(\bar{\sigma}_0, \sigma_0) + \underline{\underline{\mathcal{Z}}}(\bar{\zeta}_2, \zeta_2)\widetilde{\mathcal{P}} + \underline{\underline{\mathcal{Z}}}(\bar{\sigma}_2, \sigma_2)\widetilde{\mathcal{P}}\widetilde{\mathcal{P}}, \tag{14.60a}$$

$$\underline{\underline{\mathcal{H}}}^*(\underline{P}) = \underline{\underline{\mathcal{Z}}}(\bar{\sigma}_0, \sigma_0) - \underline{\underline{\mathcal{Z}}}(\bar{\zeta}_2, \zeta_2)\widetilde{\mathcal{P}} + \underline{\underline{\mathcal{Z}}}(\bar{\sigma}_2, \sigma_2)\widetilde{\mathcal{P}}\widetilde{\mathcal{P}}, \tag{14.60b}$$

$$\underline{\underline{\mathcal{H}}}^{-1}(\underline{P}) = \underline{\underline{\mathcal{Z}}}(\bar{\chi}_0, \chi_0) - \widetilde{\mathcal{P}}/2 + \underline{\underline{\mathcal{Z}}}(\bar{\chi}_2, \chi_2)\widetilde{\mathcal{P}}\widetilde{\mathcal{P}}, \tag{14.60c}$$

$$\underline{\underline{\mathcal{H}}}^{*-1}(\underline{P}) = \underline{\underline{\mathcal{Z}}}(\bar{\chi}_0, \chi_0) + \widetilde{\mathcal{P}}/2 + \underline{\underline{\mathcal{Z}}}(\bar{\chi}_2, \chi_2)\widetilde{\mathcal{P}}\widetilde{\mathcal{P}}. \tag{14.60d}$$

The parallel between these expressions and those for the vectorial parameterization of rotation, eqs. (13.55) and (13.60), is evident.

Coefficients σ_0, σ_1, and σ_2 are given by eqs. (13.56), and

$$\bar{\sigma}_0 = \varrho\varphi_0'/(pp'), \tag{14.61a}$$

$$p^2\bar{\sigma}_2 = \bar{\sigma}_0 - 2\varrho\varphi_2 - \bar{\zeta}_1. \tag{14.61b}$$

Coefficient χ_0 and χ_2 are given by eqs. (13.61), and

$$\bar{\chi}_0 = -\chi_0 \delta_0 \bar{\sigma}_0, \tag{14.62a}$$

$$p^2 \zeta_2 \bar{\chi}_2 = \frac{\bar{\zeta}_2}{\varepsilon} - \frac{\bar{\zeta}_1}{2} + \bar{\chi}_0 \zeta_2 - 2\varrho \zeta_2 \delta_2. \tag{14.62b}$$

14.4.6 Determination of the motion parameter vector

The motion tensor can be written as

$$\underline{\underline{C}}(\mathcal{P}) = \begin{bmatrix} \underline{\underline{R}}(p) & \underline{\underline{Z}}(p,q) \\ \underline{\underline{0}} & \underline{\underline{R}}(p) \end{bmatrix},$$

where $\underline{\underline{Z}}(p,q) = \widetilde{\underline{\underline{F}}q}\underline{\underline{R}}$. To determine the components of the motion parameter vector from this motion tensor, the rotation parameter vector is first extracted from $\underline{\underline{R}}(p)$ using the procedure described in section 13.4.4. The displacement related vector, q, is then extracted from tensor $\underline{\underline{Z}}$

$$\underline{q} = \underline{\underline{F}}^{-1}(\underline{p})\text{axial}\left[\underline{\underline{Z}}(p,q)\underline{\underline{R}}^T(p)\right]. \tag{14.63}$$

14.4.7 Composition of finite motions

Let $\mathcal{P}$, $\mathcal{P}_1^T = \{\underline{q}_1^T, \underline{p}_1^T\}$, and $\mathcal{P}_2^T = \{\underline{q}_2^T, \underline{p}_2^T\}$ correspond to motion tensors $\underline{\underline{C}}(\mathcal{P})$, $\underline{\underline{C}}(\mathcal{P}_1)$, and $\underline{\underline{C}}(\mathcal{P}_2)$, respectively. If $\underline{\underline{C}}(\mathcal{P}) = \underline{\underline{C}}(\mathcal{P}_1)\underline{\underline{C}}(\mathcal{P}_2)$, the problem is to relate $\mathcal{P}$ to $\mathcal{P}_1$ and $\mathcal{P}_2$. The first step of the process is to note that $\underline{\underline{R}} = \underline{\underline{R}}_1\underline{\underline{R}}_2$, and hence, eqs. (13.66a) and (13.66b) yield $\underline{p}$ as a function of $\underline{p}_1$ and $\underline{p}_2$.

Next, the relationship between $\underline{q}$ and $\underline{q}_1$, $\underline{q}_2$ is obtained as $\underline{\underline{F}}(\underline{p})\underline{q} = \underline{\underline{F}}(\underline{p}_1)\underline{q}_1 + \underline{\underline{R}}(\underline{p}_1)\underline{\underline{F}}(\underline{p}_2)\underline{q}_2$, and finally

$$\underline{q} = \underline{\underline{F}}^{-1}(\underline{p})\underline{\underline{F}}(\underline{p}_1)\underline{q}_1 + \underline{\underline{F}}^{-T}(\underline{p})\underline{\underline{F}}^T(\underline{p}_2)\underline{q}_2. \tag{14.64}$$

14.5 Specific parameterizations of motion

The vectorial parameterization of motion presented in the previous sections consists of a set of displacement related parameters, $\underline{q} = \underline{\underline{D}}\underline{u}$, and of the rotation parameter vector, $\underline{p}$. The motion parameter vector is parallel to an eigenvector of the motion tensor associated with its unit eigenvalue. This leads to families of parameterizations that depend on two choices: the choice of the generating function, $p(\phi)$, and that of parameter α appearing in tensor $\underline{\underline{D}}$.

As discussed in section 13.5, the generating function can be selected to simplify some of the operators involved in manipulating rotations. But more importantly, judicious choices of this function can eliminate the singularities that occur in the various rotation operators.

The occurrence of singularities is also a major concern when dealing with the vectorial parameterization of motion. Two criteria guide the selection of function $\alpha(\phi)$. First, a one to one, singularity free relationship must exist between the displacement related part of motion parameter vector, $\underline{q}$, and the physical displacement vector, $\underline{u}$. Second, the limit behavior expressed by eq. (14.44) must be satisfied.

14.5.1 Alternative choices of the motion parameter vector

Arbitrary parameter α was introduced in section 5.5.2 to reflect the non-uniqueness of the eigenvector of the motion tensor associated with its unit eigenvalue. An additional constraint is required to evaluate this parameter. For instance, imposing the orthogonality condition, $\underline{q}^T\underline{p} = 0$, leads to $\alpha = 0$, and the resulting motion parameter vector then corresponds to the Plücker coordinates of the Mozzi-Chasles axis, as discussed in section 5.5.1. This choice, however, does satisfy the limit behavior condition, eq. (14.44), and furthermore, because $\det(\underline{\underline{D}}) = \alpha/\nu^3 = 0$, a one-to-one mapping between $\underline{q}$ and $\underline{u}$ ceases to exist for this choice.

Comparing eqs. (14.40) and (4.34), the close relationship between tensors $\underline{\underline{D}}^{[\mathcal{E}]}$ and $\underline{\underline{G}}^{[\mathcal{E}]}$ is apparent. Indeed, for $\alpha = 1$, tensors $\underline{\underline{D}}$ and $\underline{\underline{F}}$ become

$$\underline{\underline{D}} = \frac{1}{\nu}\underline{\underline{G}}^T, \quad \underline{\underline{F}} = \nu\underline{\underline{G}}, \tag{14.65}$$

respectively. The determinant of tensor $\underline{\underline{D}}$ is $\det(\underline{\underline{D}}) = 1/\nu^3$. This choice satisfies the limit behavior condition, eq. (14.44).

The close connection between tensors $\underline{\underline{D}}$ and $\underline{\underline{H}}^{-1}$ is evident when comparing eqs. (14.40) and (13.62). For $\alpha = \nu p'$, tensors $\underline{\underline{D}}$ and $\underline{\underline{F}}$ are identical to the inverse of the tangent tensor and to the tangent tensor, respectively, *i.e.*,

$$\underline{\underline{D}} = \underline{\underline{H}}^{-1}, \quad \underline{\underline{F}} = \underline{\underline{H}}, \tag{14.66}$$

respectively. The determinant of tensor $\underline{\underline{D}}$ is $\det(\underline{\underline{D}}) = p'/\nu^2$. Here again the limit behavior condition is satisfied.

It is possible to eliminate the quadratic term in $\underline{p}$ of tensor $\underline{\underline{D}}$ by choosing $\alpha = \nu/\varepsilon$. This leads to the following expressions for tensors $\underline{\underline{D}}$ and $\underline{\underline{F}}$

$$\underline{\underline{D}}(p) = \frac{1}{\varepsilon}\left(\underline{\underline{I}} - \frac{\varepsilon}{2}\widetilde{p}\right), \quad \underline{\underline{F}} = \varepsilon\frac{\underline{\underline{R}} + \underline{\underline{I}}}{2}, \tag{14.67}$$

respectively. The determinant of tensor $\underline{\underline{D}}$ is $\det(\underline{\underline{D}}) = 1/(\varepsilon\nu^2)$.

One final alternative is to select $\alpha = \varepsilon/\nu$, and comparing eqs. (14.40) and (13.48) then leads to

$$\underline{\underline{D}}(p) = \frac{\underline{\underline{I}} + \underline{\underline{R}}^T}{2\varsigma_1}, \quad \underline{\underline{F}} = \varsigma_1\left(\underline{\underline{I}} + \frac{\varepsilon}{2}\widetilde{p}\right), \tag{14.68}$$

and $\det(\underline{\underline{D}}) = 1/(\nu^2\varsigma_1)$. This choice leads to Cayley's motion parameters presented in section 14.1, see eq. (14.2).

Of all the choices presented in this section, $\alpha = \nu p'$ seems to the most desirable because it leads to $\underline{\underline{D}} = \underline{\underline{H}}^{-1}$. This choice satisfies the limit behavior expressed by eq. (14.44), and because the tangent tensor plays a critical role in manipulating rotations, eliminating singularities from this tensor is already a criterion for the selection of appropriate generating functions. A singularity free tangent tensor in the vectorial parameterization of rotation will then automatically lead to a one-to-one mapping between $\underline{q}$ and $\underline{u}$, avoiding the occurrence of singularities in the vectorial parameterization of motion.

14.5.2 The exponential map of motion

One of the simplest choices of the generating function is $p(\phi) = \phi$, which leads to the exponential map of motion [269]. This parameterization involves the evaluation of numerous trigonometric functions and will not be discussed here, although all the relevant formulae can be obtained by introducing the generating function, $p(\phi) = \phi$, into the expression given in the previous sections.

14.5.3 The Euler-Rodrigues motion parameters

The Euler-Rodrigues rotation parameters, $\underline{v}$, are associated with the generating function $p(\phi) = 2\sin\phi/2$ and the properties of this parameterization are detailed in section 13.7.2. The corresponding Euler-Rodrigues motion parameters are defined as

$$\underline{\mathcal{P}} = \left\{ \begin{array}{c} \underline{\underline{D}}\,\underline{u} \\ \underline{v} \end{array} \right\}. \tag{14.69}$$

The relevant coefficients for this parameterization, ζ_1, ζ_2, σ_0, σ_1, σ_2, χ_0, χ_1, and χ_2, are given in section 13.7.2. Parameters δ_0 and δ_2, and φ_0, φ_1, and φ_2, are given by eqs. (14.39) and (14.42), respectively.

Selecting, for instance, $\alpha = \nu p'$, eqs. (14.52) yield $\bar\zeta_1 = -\varrho/(4v_0)$ and $\bar\zeta_2 = 0$, where coefficient ϱ is given by eq. (14.50). Equations (14.51) then gives the motion tensor and its inverse. Next, eqs. (14.61) and (14.62) yield $\bar\sigma_0 = \varrho/(4v_0^3)$, $\bar\sigma_2 = -\varrho/(16v_0^3)$, $\bar\chi_0 = \varrho/(4v_0)$, and $\bar\chi_2 = 0$; tensors $\underline{\underline{\mathcal{H}}}$, $\underline{\underline{\mathcal{H}}}^*$ and their inverses then follow from eqs. (14.60). Finally, the multiplicative decomposition of the motion tensor given by eq. (14.53) is applicable to this parameterization using $\bar\varepsilon = \varrho/(4v_0^3)$.

14.5.4 The Cayley-Gibbs-Rodrigues motion parameters

The Cayley-Gibbs-Rodrigues rotation parameters, $\underline{r}$, are associated with the generating function $p(\phi) = 2\tan\phi/2$ and the properties of this parameterization are detailed in section 13.7.3. The corresponding Cayley-Gibbs-Rodrigues motion parameters are defined as

$$\underline{\mathcal{P}} = \left\{ \begin{array}{c} \underline{\underline{D}}\,\underline{u} \\ \underline{r} \end{array} \right\}. \tag{14.70}$$

The relevant coefficients for this parameterization, ζ_1, ζ_2, σ_0, σ_1, σ_2, χ_0, χ_1, and χ_2, are given in section 13.7.3. Parameters δ_0 and δ_2, and φ_0, φ_1, and φ_2, are given by eqs. (14.39) and (14.42), respectively.

Selecting, for instance, $\alpha = \nu p'$, eqs. (14.52) yield $\bar{\zeta}_1 = -\varrho r_0^2/2$ and $\bar{\zeta}_2 = -\varrho r_0^2/4$, where coefficient ϱ is given by eq. (14.50). Equations (14.51) then gives the motion tensor and its inverse. Next, eqs. (14.61) and (14.62) yield $\bar{\sigma}_0 = -\varrho r_0^2/2$, $\bar{\sigma}_2 = 0$, $\bar{\chi}_0 = \varrho/2$, and $\bar{\chi}_2 = 0$; tensors $\underline{\mathcal{H}}$, $\underline{\mathcal{H}}^*$ and their inverses then follow from eqs. (14.60). Finally, the multiplicative decomposition of the motion tensor given by eq. (14.53) is applicable to this parameterization using $\bar{\varepsilon} = 0$.

14.5.5 The Wiener-Milenković motion parameters

The Wiener-Milenković rotation parameters, $\underline{c}$, are associated with the generating function $p(\phi) = 4 \tan \phi/4$ and the properties of this parameterization are detailed in section 13.7.4. The corresponding Cayley-Gibbs-Rodrigues motion parameters are defined as

$$\underline{\mathcal{P}} = \left\{ \frac{\underline{D}\, u}{\underline{c}} \right\}, \tag{14.71}$$

The Wiener-Milenković parameterization of rotation is singularity free for rotations of arbitrary magnitude when using the rescaling technique. Consequently, the Wiener-Milenković motion parameterization is singularity free for displacements and rotations of arbitrary magnitude provided that the rescaling operation is applied to the rotation parameter vector. The relevant coefficients for this parameterization, ζ_1, ζ_2, σ_0, σ_1, σ_2, χ_0, χ_1, and χ_2, are given in section 13.7.4. Parameters δ_0 and δ_2, and φ_0, φ_1, and φ_2, are given by eqs. (14.39) and (14.42), respectively.

Selecting, for instance, $\alpha = \nu p'$, eqs. (14.52) yield $\bar{\zeta}_1 = \varrho\nu^2(1-4\nu)/8$ and $\bar{\zeta}_2 = -\varrho\nu^3/8$, where coefficient ϱ is given by eq. (14.50). Equations (14.51) then gives the motion tensor and its inverse. Next, eqs. (14.61) and (14.62) yield $\bar{\sigma}_0 = -\varrho\nu^2/8$, $\bar{\sigma}_2 = -\varrho\nu^3/32$, $\bar{\chi}_0 = \varrho/8$, and $\bar{\chi}_2 = 0$; tensors $\underline{\mathcal{H}}$, $\underline{\mathcal{H}}^*$ and their inverses then follow from eqs. (14.60). Finally, the multiplicative decomposition of the motion tensor given by eq. (14.53) is applicable to this parameterization using $\bar{\varepsilon} = \varrho\varepsilon^2/8$.

Using eq. (5.35), the motion tensor can also be expressed as

$$\underline{\underline{C}} = \begin{bmatrix} \underline{\underline{R}} & \underline{\underline{G}}\nu\widetilde{q}\underline{\underline{G}} \\ \underline{\underline{0}} & \underline{\underline{R}} \end{bmatrix} = \begin{bmatrix} \underline{\underline{G}} & \underline{\underline{0}} \\ \underline{\underline{0}} & \underline{\underline{G}} \end{bmatrix} \begin{bmatrix} \underline{\underline{I}} & \nu\widetilde{q} \\ \underline{\underline{0}} & \underline{\underline{I}} \end{bmatrix} \begin{bmatrix} \underline{\underline{G}} & \underline{\underline{0}} \\ \underline{\underline{0}} & \underline{\underline{G}} \end{bmatrix}. \tag{14.72}$$

where the second equality follows from eq. (4.19). This factorization affords the following geometric interpretation: the motion is decomposed into the half-angle rotation characterized by the rotation tensor $\underline{\underline{G}}$, followed by a translation of magnitude νq, and finally a half-angle rotation.

14.5.6 Problems

Problem 14.4. Motion tensor for the Cayley-Gibbs-Rodrigues parameters
Show that the motion tensor and its inverse, expressed in terms of Cayley-Gibbs-Rodrigues motion parameters, can be written as

$$\underline{\underline{C}} = \begin{bmatrix} \underline{\underline{R}} & \dfrac{\underline{\underline{R}}+\underline{\underline{I}}}{2}\widetilde{\underline{q}}\dfrac{\underline{\underline{R}}+\underline{\underline{I}}}{2} \\ \underline{0} & \underline{\underline{R}} \end{bmatrix}, \quad \underline{\underline{C}}^{-1} = \begin{bmatrix} \underline{\underline{R}}^T & \dfrac{\underline{\underline{R}}^T+\underline{\underline{I}}}{2}\widetilde{\underline{q}}^T\dfrac{\underline{\underline{R}}^T+\underline{\underline{I}}}{2} \\ \underline{0} & \underline{\underline{R}}^T \end{bmatrix}.$$

Problem 14.5. Prove relationship
Prove the following relationship

$$\frac{1}{2}\underline{\underline{Z}}(\bar{\varepsilon},\varepsilon)\widetilde{\underline{P}} = (\underline{\underline{C}}+\underline{\underline{I}})^{-1}(\underline{\underline{C}}-\underline{\underline{I}}) = (\underline{\underline{C}}-\underline{\underline{I}})(\underline{\underline{C}}+\underline{\underline{I}})^{-1}.$$

Problem 14.6. Prove the properties of the tangent tensor
Prove properties (14.58c) of the tangent tensor.

Problem 14.7. Prove the properties of the tangent tensor
Prove properties (14.58d) of the tangent tensor.

Problem 14.8. Prove the tensorial nature of the tangent tensor
Prove property (14.59) of the tangent tensor.

Problem 14.9. The half-motion tensor
The half-rotation tensor $\underline{\underline{G}}$ defined by eq. (13.49), play an important role in the vectorial parameterization of rotation. Find the half-motion tensor, $\underline{\underline{\mathcal{G}}}$, such that $\underline{\underline{C}} = \underline{\underline{\mathcal{G}}}\,\underline{\underline{\mathcal{G}}}$.

Part VI

Flexible multibody dynamics

15

Flexible multibody systems: preliminaries

Multibody systems are characterized by two distinguishing features: system components undergo finite relative rotations and these components are connected by mechanical joints that impose restrictions on their relative motion. Broadly speaking, multibody systems can be divided into three categories, rigid multibody systems, linearly elastic multibody systems, and nonlinearly elastic multibody systems. This classification and its implication on modeling techniques for multibody systems are discussed in section 15.1.

Section 15.2 presents a review of the basic equations of three-dimensional, linear elastodynamics. Geometrically nonlinear problems are characterized by nonlinear strain-displacement relationships, which are the subject of section 15.3. In section 15.5, special attention is devoted to the formulation of problems where structures undergo arbitrarily large displacements and rotations although strain components are assumed to remain small at all points of the structure.

15.1 Classification of multibody systems

Multibody systems can be divided into three categories, rigid multibody systems, linearly elastic multibody systems, and nonlinearly elastic multibody systems. Systems of the first category involves rigid bodies only, but those of the latter two categories comprise both rigid and flexible bodies. Section 12.5.1 introduced the concept of floating frame of reference in which the total motion of flexible bodies is broken into two parts: rigid body motions represented by the motion of the floating frame of reference and superimposed elastic motions. By definition, rigid body motions generate no strains. The elastic motions typically consist of displacement and rotation fields, which generate an associated strain field, denoted $\underline{\epsilon}$. For rigid multibody system, the strain field vanishes in all bodies, *i.e.*, $\underline{\epsilon} = 0$ in each body. The distinction between linearly and nonlinearly elastic multibody systems stems from the characteristics of the strain field.

The characteristics of the three types of multibody systems are as follows.

1. *Rigid multibody systems* consist of an assemblage of rigid bodies connected together through mechanical joints and in arbitrary motion with respect to each other. Although all bodies are rigid, *i.e.*, $\underline{\epsilon} = 0$ in each body, lumped elastic components, also called *flexible joints*, *bushing elements* or *force elements*, could be placed between two components of the system to represent localized elasticity. These flexible joints exhibit arbitrary constitutive behavior.
2. *Linearly elastic multibody systems* consist of an assemblage of both elastic and rigid bodies connected together through mechanical joints and in arbitrary motion with respect to each other. For linearly elastic multibody systems, it is assumed that the strain-displacement relationships remain linear and that strains components remain very small at all times, *i.e.*, $\underline{\epsilon} \ll 1$ for all elastic bodies. Efficient analysis techniques for this type of problems typically rely on modal expansions of the elastic displacement field.
3. *Nonlinearly elastic multibody systems* consist of an assemblage of both elastic and rigid bodies connected together through mechanical joints and in arbitrary motion with respect to each other. For the elastic bodies, the strain-displacement relationships become nonlinear, or the strain components become large, or both. Nonlinear strain-displacement relationships characterize *geometrically nonlinear problems*, *i.e.*, problems involving large elastic displacements, or rotations, or both. When strain components become large, nonlinear material constitutive laws must be used, a characteristic of *materially nonlinear problems*. For nonlinearly elastic multibody systems the accuracy and reliability of modal expansion of the elastic displacement field become questionable.

Because the overall motions of all bodies of a multibody system are large and because the relative motions between the system's various components are also large, multibody system dynamics is an inherently nonlinear problem. The qualifiers "linearly" and "nonlinearly elastic" used in the classification above specifically refer to the elastic behavior of the bodies. The modeling of linearly elastic multibody systems leads to nonlinear dynamical equations of motion, although the representation of the elastic behavior of the bodies could be largely linearized.

15.1.1 Linearly and nonlinearly elastic multibody systems

The demarcation between linearly and nonlinearly elastic multibody systems is sometimes blurry. Consider, for instance, the problem of a helicopter rotor blade. As the blade rotates, elastic displacements and rotations remain very small, and the blade is designed to undergo small strains at all time to ensure safety of flight and guarantee structural fatigue life. This problem seems to fall into the category of linearly elastic multibody systems.

Due to the high angular speed of the rotor, however, large centrifugal forces appear in the blade, leading to considerable centrifugal stiffening of the blade and nonlinear coupling between its two bending and torsional deformations. To accurately capture these effects, nonlinear strain-displacement relationships must be used, although linear constitutive laws adequately represent material behavior. These geometric nonlinearities squarely put the helicopter rotor blade problem into the category of nonlinearly elastic multibody systems.

Because wind turbine blades rotate at a much lower angular speed and are far stiffer than helicopter rotor blades, assuming wind turbines to be linearly elastic multibody systems often leads to reliable predictions. On the other hand, efforts to design ever increasingly efficient turbines call for ever increasing rotor diameters. It is possible that future generation turbines will become large enough to operate in the geometrically nonlinear regime, requiring the use of formulations capable of dealing with nonlinearly elastic multibody systems to predict accurately the dynamic response of such highly flexible machines.

The distinction between linearly and nonlinearly elastic systems is further complicated by the fact that both linearly and nonlinearly elastic components could appear simultaneously in a given multibody system. For instance, the modeling of a complete helicopter in flight calls for the coupled simulation of the rotor and fuselage. As explained above, the rotor problem is inherently nonlinear, but it is reasonable to assume that the fuselage behaves in a linearly elastic manner, even during large angle maneuvers. For such problems, different formulations could be used to model the rotor and fuselage that reflect the distinctly different behavior of these two components. Similar remarks could be made concerning wind turbines. Whereas the rotor blades could be treated as linearly or nonlinearly elastic bodies depending on the magnitude of the elastic motions, it seems reasonable to assume the supporting tower to behave in a linearly elastic manner.

The remaining chapters of this book focus on nonlinearly elastic multibody systems. As discussed in section 12.5.1, one of the most common approaches to the modeling of linearly elastic multibody systems is based on the concept of floating frames of reference [162] and the component mode synthesis based methods described in section 12.5.2 are then used to approximate the elastic displacement field. For nonlinearly elastic multibody systems the accuracy and reliability of these approaches become questionable for the reasons detailed in the next section.

15.1.2 Shortcomings of modal analysis applied to nonlinear systems

The natural vibration modes of a structure are the eigenfunctions of the system's equations of motion linearized about one of its equilibrium configurations. Therefore, these modes characterize the dynamic behavior of small perturbations about an equilibrium configuration of the system. Clearly, natural vibration modes are inherently linearized quantities that provide no information about the nonlinear behavior of the system.

Consider a simple cantilevered beam under transverse, time-dependent loading. The beam's mode shapes computed with respect to its unloaded configuration include the familiar bending, axial, and torsional modes of vibration, which are derived in numerous structural vibration textbooks such as those of Timoshenko and Young [273] or Meirovitch [183]. If the beam is subjected to a tip transverse load that generates small transverse deflections of the beam, a modal expansion in terms of transverse bending modes will accurately predict the dynamic behavior of the system, even when using a small number of modes. Of course, resonance conditions should be avoided as large deflections would result. Furthermore, all the modes associated with natural frequencies of magnitude comparable to that of the excitation frequency should be included in the modal expansion.

Modal analysis is a natural approach to reducing the number of degrees of freedom involved in structural dynamics problems, and has the added advantage of involving degrees of freedom that have a direct physical meaning. Modal analysis can be viewed as a two step process.

In the first step of the process, a change of variables is performed by projecting the equations of motion expressed in term of physical variables to the modal domain. For a simple cantilevered beam problem, the physical variables would typically consist of the displacement and rotation components at discrete nodes along the beam. The modal variables are the amplitudes of excitation of the beam's eigenmodes. If all the modes of the structure are used, this change of variable is a purely mathematical operation that involves no approximations.

In the second step of the process, modal truncation is performed. Based on physical arguments, a small subset of all the eigenmodes of the structure is retained. For linear systems, modal truncation is a simple operation: modes associated with frequencies far higher than the excitation frequencies are simply eliminated from the modal basis because their contribution to structural response is negligible.

This assumption yields the major advantages of modal analysis. First, a dramatic reduction in the number of degrees of freedom is achieved, leading to considerable computational savings. For complex structures, thousands or even hundreds of thousands of degrees of freedom might be involved in a detailed finite element model, whereas ten or fifteen modes only could be sufficient to capture accurately the overall dynamic behavior of the structure.

Second, because high frequency modes have been eliminated, larger time step sizes can be used to integrate the system's equations of motion, leading to additional computational savings. Finally, the modal degrees of freedom are easily interpreted in a physical manner. For instance, if a structure responds "mainly in its second bending mode," or "primarily in its first torsion mode," it is easy to visualize the overall deformation of even very complex structure.

Projection of the equations of motion into the modal domain is a purely mathematical step that does not involve any approximation. Modal truncation is an assumption. Indeed, when applied to linear systems, the projection of the equations of motion to the modal space decouples the governing equations. In the modal space, even the most complex structures can be viewed as a superposition of linear, single degree of freedom oscillators. The elimination of specific modes to obtain a reduced modal basis is based on the well known properties of these oscillators.

When applied to nonlinear structures, the first step of the modal analysis process is also the projection of the equations of motion onto the modal domain. As was the case for linear structures, no assumption is involved here. Modal truncation, however is fraught with difficulties. For nonlinear problems, the equations of motion projected in the modal space are still nonlinear and in general, do not decouple. The nonlinear system cannot be viewed as the superposition of linear, single degree of freedom oscillators, nor can it be viewed as the superposition of nonlinear oscillators. Consequently, the physical arguments invoked to eliminate specific modes from the modal basis no longer apply, or become more tenuous.

To illustrate the problems encountered by modal analysis applied to nonlinear structures, consider the simply supported beam depicted in fig. 15.1. The beam is subjected to time-dependent loading, $p_2(x_1,t)$, and features a large axial stiffness. For case (a), the end points of the beam cannot move axially. In contrast, axial displacements are allowed at point **T** for case (b). For case (a), large transverse displacements of the beam

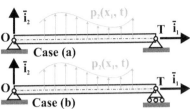

Fig. 15.1. Simply supported beam with different end conditions.

generate axial deformations, which in turn, cause large axial forces to appear due to the beam's high axial stiffness. As the magnitude of the transverse displacements increases, a considerable stiffening of the system is observed, leading to pronounced nonlinear behavior. This phenomenon is much less severe for case (b) because the beam is free to move axially at point **T**.

To simplify the discussion, the problems depicted in fig. 15.1 are limited to the planar case where all displacements take place in plane $(\bar{\imath}_1, \bar{\imath}_2)$. The beam's natural vibration modes about its unloaded configuration are easily obtained and consist of transverse bending and axial modes. Because they are linearized quantities, the bending modes involve transverse displacement components only. Note that the beam's bending modes are identical for cases (a) and (b), although its axial modes differ.

If small, time-dependent transverse loads are applied to the beam, modal analysis using a modal basis consisting of a few bending modes yields accurate predictions of the dynamic response of the system. In the linear range, axial modes are not excited and need not be present in the modal basis. The solutions for cases (a) and (b) are identical in the linear range, a feature that is correctly reproduced by the modal solution because the axial displacement boundary condition is not reflected in the modal basis. The modal basis does not "feel" the difference between cases (a) and (b).

Next, larger transverse loads are applied to the beam, which now responds in the nonlinear range. If bending modes only are used in the modal approximation, the beam's dynamic response is no longer predicted accurately. The situation is somewhat improved by adding axial vibration modes, but a large number of these modes is required to obtained a good solution.

The reason for this behavior is twofold. First, because the beam's axial displacement field is not captured accurately by the modal approximation, errors are to be expect in the estimation of the axial strain field. Due to the beam's large axial stiffness, small errors in the axial strain field lead to a grossly erroneous axial force field and the nonlinear stiffening effects it induces are poorly captured by the modal analysis. Clearly, the foreshortening of the blade, an inherent part of its nonlinear response, must be modeled precisely to predict accurately the beam's nonlinear response.

Second, the blade's axial displacement field is primarily due to foreshortening (a purely kinematic, nonlinear phenomenon), whereas axial vibration modes characterize true axial vibration (a purely vibratory, linear phenomenon). In other words, modal analysis attempts to "synthesize" a nonlinear kinematic mode shape as a superposition of linear vibratory modes. Because these two phenomena are not physically related, accurate predictions should hardly be expect from this superposition.

Thus far, the discussion has focussed on nonlinear foreshortening effects in a simple planar problem. The above arguments, however, also apply to other kinematic nonlinearities found in three-dimensional problems. For instance, transverse deflections due to transverse loads applied to the beam in two orthogonal directions create a torsional moment, thereby coupling bending and torsion responses.

When applied to nonlinear problems, convergence and accuracy of modal analysis are not guaranteed. To improve the situation, it seems natural to investigate the selection of alternative mode shapes that contain information about the nonlinear behavior of the structure. Several concepts have been proposed to improve the quality of the modal basis when dealing with nonlinear problems.

The conceptually simplest method it to recalculate a new set of natural vibration modes once the deformations of the blade become significant [274]. Due to the large relative motions between the components of multibody systems, the equilibrium configuration of the system is time-dependent, and hence, the modal basis is itself time-dependent. Although this approach might give good results, it does so at a tremendous computational costs, because the modal basis must now be updated during the response calculation, and the modal reduction scheme must be repeated at each update.

Another approach is to include in the modal basis natural vibration modes about different equilibrium configurations of the structure. This method is attractive because the evaluation of the various equilibrium configurations and associated modal bases only require a modest increase in computational costs. In some cases, this method appears to give accurate results, as documented by Nickell [274].

The concept of perturbation modes was introduced by Thompson and Walker [275] to study the nonlinear behavior of beam structures. The authors demonstrated improved accuracy compared with classical modal analysis based on natural vibration modes. The same concept was later refined by Noor *et al.* [276, 277] for the nonlinear static analysis of beam and shell structures in conjunction with the finite element method. Bauchau and Guernsey [278] also used these concepts for the analysis of helicopter rotors.

Mixed finite element formulations have also been developed to improve the accuracy and reliability of modal methods. Ruzicka and Hodges [279, 280] have demonstrated applications of this method to rotorcraft problems.

While the various approaches described in the previous paragraphs are capable of extending the range of validity of modal methods, they also present many drawbacks. First, they require extensive numerical developments and more often that not, the simplicity of modal analysis is lost. Second, they are based of formulations that are not widely available. For instance, commercial finite element codes are rarely based on mixed formulations. Finally, different approaches are required to deal with different problem, *i.e.*, these methods are not general purpose methods that can be used reliably for general multibody systems.

15.1.3 Finite element based modeling of flexible multibody systems

Multibody dynamics analysis was originally developed as a tool for modeling rigid multibody systems with simple tree-like topologies, but has considerably evolved to the point where it can handle linearly and nonlinearly elastic multibody systems with arbitrary topologies. It is now used widely as a fundamental design tool in many areas of engineering.

In the automotive industry, for instance, multibody dynamics analysis is used routinely for optimizing vehicle ride qualities, a complex multidisciplinary task that involves the simulation of many different sub-components. Modern multibody codes can deal with complex mechanisms of arbitrary topologies including sensors, actuators, and controls, are interfaced with CAD solid modeling programs that allow to import directly the problem geometry, and have sophisticated graphics, animation, and post-processing features [38, 39].

The success of multibody dynamics analysis tools stems from their versatility: a given mechanism can be modeled by an idealization process that identifies the mechanical components from within a large library of elements implemented in the code. Each element provides a basic functional building block, for example, a rigid or flexible member, a revolute joint, or a motor. Assembling the various elements, it is then possible to construct a mathematical description of the mechanism with the required level of accuracy.

The modeling of linearly elastic multibody systems relies predominantly on the floating frame of reference approach discussed in section 12.5.1, in which the elastic displacement field is approximated using modal expansion techniques. In the last two decades, new approaches have emerged that bypass the need for the floating frame of reference and eliminate modal expansions. These approaches are closely related to the finite element method which they effectively generalize to enable the treatment of multibody systems.

In the finite element method, the solution domain is first divided into a finite number of sub-domains called *finite elements*. Within each element, the solution is then approximated by a small number of continuous functions, based on the value of these functions at discrete points, often called *nodes*, associated with the element. The main advantage of this two-step approximation process is that many aspects of the solution procedure can be carried out at the element level, *i.e.*, by considering one single element at a time, independently of all others.

The continuity of the solution across elements is guaranteed by the fact that neighboring elements share common nodes, *i.e.*, share common degrees of freedom. This aspect of the formulation is key to ensuring the continuity of the displacement field over the entire system, an indispensable requirement for displacement based finite element formulations.

Consider a beam connected at one of its end points to a revolute joint. Complete definition of the beam requires geometric data: typically, a local frame is used to define the beam's cross-sectional plane and its physical mass and stiffness properties are given with respect to this local frame. Similarly, definition of the revolute joint also requires geometric data: typically, a local frame is used to define the plane of the joint and the unit vector about which relative rotation is allowed, see fig. 10.14. These local frames are independent of each other and are used solely by the elements for which they are defined.

In contrast, the components of the displacement and rotation vectors at the connection point between the beam and revolute joint must be uniquely defined. If the displacement components are resolved in the local frame of the beam element, the revolute joint will not be able to interpret these components properly because they are resolved in a local frame whose orientation it does not know. Vice versa, if the displacement components are resolved in the local frame of the revolute joint, the beam will not be able to interpret these components adequately because they are resolved in a local frame whose orientation it does not know.

To resolve this conflict, all nodal displacement and rotation components must be defined in a common frame, which is conveniently selected to be the *inertial frame of reference*. Because the components of multibody systems typically undergo large displacements and rotations, these inertial, or absolute, displacement and rotation components must be treated rigorously as large displacement and rotation components. Consequently, all the elements of the multibody system must be able to handle arbitrarily large displacements and rotations exactly. This is why these elements are sometimes called "geometrically exact elements," although the term "kinematically exact elements" would be more appropriate.

Finite element based modeling of flexible multibody systems makes use of a library of elements consisting of *structural elements* and *joint elements*. Structural elements, such as cables, membranes, beams, plates, shells, or three-dimensional elements are similar to the corresponding elements found in all finite element packages. The geometrically exact formulations of cables, beams, and plates and shells are presented in sections 16.2, 16.3, and 16.4, respectively. The dynamic equations of equilibrium of these elements are written in an absolute Cartesian frame.

Joint elements characterize multibody systems and are absent from most finite element codes. In typical multibody formulations, joints are modeled as idealized components, *i.e.*, joints are not modeled *per se*. Rather, the effects of joints are represented by the kinematic constraints they impose on the components they are connected to. For instance, section 10.6 details the constraints associated with the lower pair joints, which are enforced via Lagrange's multiplier technique.

The assembly of the equations of motion of both structural and joint elements leads to systems of equations that are highly sparse, although not of minimal size, a characteristic of the approach pioneered by Orlandea *et al.* [77]. Because it is an extension of the finite element method to multibody systems, algorithms such as sparse solvers, assembly procedures, and data structures developed for finite element analysis are directly applicable to finite element based modeling of flexible multibody systems.

This approach can readily treat configurations of arbitrarily complex topologies through the assembly of basic components chosen from an extensive library of structural and joint elements. In fact, this concept is at the heart of the finite element method which has enjoyed, for this very reason, an explosive growth in the last few decades. This analysis approach leads to new comprehensive simulation tools that are modular and expandable. Modularity implies that all the basic building blocks can be validated independently, easing the more challenging task of validating complete simulation procedures. Because they are applicable to configurations with arbitrary topologies, including those not yet foreseen, such simulation tools will enjoy a longer life span, a critical requirement for any complex software tool.

Example 15.1. *Modeling helicopter rotors with flexible multibody dynamics*
Historically, the classical approach to rotor dynamics modeling has relied on modal reduction approaches, as pioneered by Houbolt and Brooks [281]. Typical models were limited to single articulated blades connected to an inertial point and the control chain was ignored. The blade's equations of motion were written in the rotating system, and ordering schemes were used to decrease the number of nonlinear terms appearing in the modal expansion [282].

In time, more detailed rotor models were developed to improve accuracy and to account for various design complexities such as gimbal mounts, swashplates, or bearingless root retention beams, among many others. The relevant equations of motion were derived for the specific configurations at hand. In fact, the various codes developed in-house by rotorcraft manufacturers are geared towards the modeling of the specific configuration they produce. This approach severely limits the generality and flexibility of the resulting codes.

In recent years, a number of new rotorcraft configurations have been proposed: bearingless rotors with redundant load paths, tilt rotors, co-axial rotors, or variable diameter tilt rotors, to name just a few. Developing a new simulation tool for each novel configuration is a daunting task, and software validation is an even more difficult issue. Furthermore, the requirement for ever more accurate predictions calls for increasingly detailed and comprehensive models. For instance, modeling the interaction of the rotor with a flexible fuselage or with the control chain must be considered to capture specific phenomena or instabilities.

The finite element based flexible multibody dynamics formulation outlined above appears to be readily applicable to the rotorcraft dynamics analysis, because a rotorcraft system can be viewed as a complex flexible mechanism. It is now becoming the industry norm for this complex, nonlinear problem.

Figure 15.2 depicts the conceptual representation of a rotorcraft system as a flexible multibody system. The various mechanical components of the system are associated with elements found in the library of typical multibody analysis tools. The figure shows a classical configuration for the control chain, consisting of a swashplate with rotating and non-rotating components. The lower swashplate motion is controlled by actuators that provide the vertical and angular control inputs. The upper swashplate is connected to the rotor shaft through a scissors-like mechanism and controls the blade pitching motions through pitch-links.

This control linkage configuration can be modeled using the following elements: rigid bodies, used to model the non-rotating and rotating swashplate components and

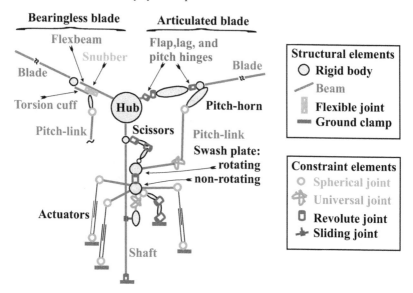

Fig. 15.2. Detailed multibody representation of a rotor system.

scissors links, and beams for modeling the flexible shaft and pitch-link. These bodies are connected through mechanical joints. For instance, a revolute joint, described in section 10.6.1, connects the rotating and non-rotating swashplates, allowing the former to rotate at the shaft angular velocity while the latter is non-rotating. Revolute joints also connect the scissors links to each other and to the upper swashplate, thereby synchronizing the angular speeds of the shaft and upper swashplate.

Other types of joints are required for the model. For instance, the non-rotating swashplate is allowed to tilt with respect to an element that slides along the shaft, but does not rotate about the shaft direction. The universal joint, described in section 10.7.1, serves this purpose. Similarly, the pitch-link is connected to the pitch-horn by means of a spherical joint, see section 10.6.6, that allows the connected components to be at an arbitrary orientation with respect to each other. The other end of the pitch-link is attached to the swashplate by means of a universal joint.

Figure 15.2 also shows two different rotor configurations: a classical, fully articulated design on the right portion of the figure and a bearingless design on the left. The articulated blade is connected to the hub through three consecutive revolute joints, that allow out-of-plane, in-plane, and torsional motions of the blade. For rotorcraft, these joints are called the flap, lag, and pitch hinges, respectively. In some designs, these joints are collocated, while other designs call for offset distances between these joints. In the latter case, rigid or flexible bodies of finite dimensions would be used to connect the three joints.

For bearingless designs, the blade connects to the hub through a flexible component, called the flexbeam. The control input coming from the pitch-link is transmitted to the blade via the torsion cuff, represented by a flexible beam. To ensure a purely rotational motion of the torsion cuff, it is also connected to the flexbeam through a snubber, which is typically modeled as a flexible joint, see section 16.1. The bearing-less design is a multiple load path configuration: the flexbeam and torsion cuff are assembled in parallel and connected by a snubber.

Because they eliminates the flap, lag, and pitch hinges characteristic of fully articulated designs, bearingless designs are mechanically simpler and more robust. On the other hand, the blade's control motions are accommodated through flexible elements, which could be subjected to higher stresses than those observed in fully articulated design.

When using the finite element based flexible multibody dynamics formulation outlined above, the two designs, fully articulated and bearingless, can be modeled by assembling different sets of elements from the multibody library of elements. There is no need to derive and validate two different sets of equations for the two configurations.

The blade itself is modeled by an appropriate beam element that should account for shearing deformations and for all elastic couplings that arise from the use of composite materials [283]. Furthermore, the center of mass, center of tension, and shear center of the blade are at distinct geometric locations of the blade's cross-section, further complicating the modeling task.

Of course, the level of detail presented in fig. 15.2 is not always needed: some or all of the control chain components could be omitted, and the blade could be represented by a rigid body rather than beam elements, if a crude model is desired.

15.2 The elastodynamics problem

Figure 15.3 depicts an elastic body of arbitrary shape subjected to time-dependent surface tractions and body forces. Geometric boundary conditions consist of time-dependent prescribed displacements at a point or over a portion of the body's outer surface. The volume of the body is denoted $\mathcal{V}$ and its outer surface $\mathcal{S}$. Unit vector $\bar{n}$ is the normal to its outer surface. The dynamic response of the system is studied between initial and a final times, denoted t_i and t_f, respectively. The displacement field at a point of the body is denoted $\underline{u}(x_1, x_2, x_3, t)$, where t denotes time and x_1, x_2, and x_3 the Cartesian coordinates of a point of the body resolved in inertial frame $\mathcal{F}^I = [\mathbf{O}, \mathcal{I} = (\bar{\imath}_1, \bar{\imath}_2, \bar{\imath}_3)]$.

Over the outer surface of the body, displacements and surface tractions are de-noted $\underline{\hat{u}}(t)$ and $\underline{\hat{t}}(t)$, respectively. Over portion $\mathcal{S}_1$ of the body's outer surface, the surface tractions are given, prescribed quantities; this includes the portion of the outer surface that is traction free, because vanishing surface tractions, $\underline{\hat{t}}(t) = 0$, are prescribed over that portion of the outer surface. Over portion $\mathcal{S}_2$ of the body's outer surface, the displacements are given, prescribed quantities.

Surfaces $\mathcal{S}_1$ and $\mathcal{S}_2$ share no common point because *displacements and tractions cannot be prescribed simultaneously* at the same point, and hence, $\mathcal{S} = \mathcal{S}_1 + \mathcal{S}_2$. Over $\mathcal{S}_1$, $\underline{\hat{t}}(t)$ represents the prescribed surface tractions, and $\underline{\hat{u}}(t)$ the resulting displace-ments. Over $\mathcal{S}_2$, $\underline{\hat{u}}(t)$ represents the prescribed displacements, and $\underline{\hat{t}}(t)$ the resulting traction, also called *reaction forces*. Reaction forces are those forces arising from the enforcement of the prescribed displacements.

At the initial and final times, the momenta and displacements are denoted $\hat{p}$ and $\hat{u}$, respectively. At these times, displacements could be given, prescribed quantities, and simultaneously, momenta could also be given, prescribed quantities. Note that at the initial and final times, *both displacements and momenta can be prescribed simultaneously*. For instance, in an initial value problem, both initial displacements and momenta are prescribed at the initial time, and the values of both quantities at the final time will result from the analysis.

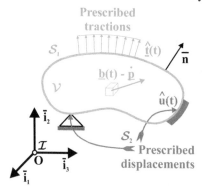

Fig. 15.3. General elastodynamics problem.

Time-dependent body forces, $\underline{b}(t)$, might also be applied over the entire volume of the body. Gravity forces are a typical example of body forces, but such forces can also arise from electric or magnetic fields. In dynamic problems, inertial forces can be considered to be externally applied body forces, in accordance with d'Alembert's principle. The momentum vector for a differential element of the body is $\underline{p} = \rho\underline{v}$, where ρ is the material mass density and $\underline{v}$ the element's inertial velocity vector. The inertial forces are then $-\dot{\underline{p}}$.

The basic equations of elastodynamics form a set of first-order partial differential equations in space and time that can be solved to find the displacement, velocity, strain, stress, and momentum fields at all points of the body and all instants in time. These equations will be reviewed in section 15.2.1 where several important definitions are also introduced. In the subsequent sections, a number of variational and energy principles are presented that provide an alternative formalism for the solution of elasticity problems. This formalism is the basis for powerful numerical techniques, such as the *finite element method*, that are routinely used to obtain approximate solutions to complex elastodynamics problems.

15.2.1 Review of the equations of linear elastodynamics

As discussed in section 8.1, d'Alembert's principle reduces dynamics problems to statics problems, provided that inertial forces are treated as externally applied forces. This implies that elastodynamics problems reduce to elasticity problems, provided that inertial forces are treated as externally applied forces. The equations presented in this section are, in fact, the general equations of elasticity [284, 285], in which the inertial forces taken into account as externally applied body forces. The equations of elastodynamics can be broken into three groups: *(1)* the equations of dynamic equilibrium, *(2)* the strain-displacement and velocity-displacement equations, and *(3)* the constitutive laws.

The equations of dynamic equilibrium

The *equations of dynamic equilibrium* are the most fundamental equations of elasticity. They are derived from Newton's first law, see section 3.1.2, which states the conditions for static equilibrium of a differential element of the body. D'Alembert's principle 12 then states the conditions for dynamic equilibrium of the differential element of the body,

$$\frac{\partial \sigma_1}{\partial x_1} + \frac{\partial \tau_{21}}{\partial x_2} + \frac{\partial \tau_{31}}{\partial x_3} + b_1 - \dot{p}_1 = 0, \tag{15.1a}$$

$$\frac{\partial \tau_{12}}{\partial x_1} + \frac{\partial \sigma_2}{\partial x_2} + \frac{\partial \tau_{32}}{\partial x_3} + b_2 - \dot{p}_2 = 0, \tag{15.1b}$$

$$\frac{\partial \tau_{13}}{\partial x_1} + \frac{\partial \tau_{23}}{\partial x_2} + \frac{\partial \sigma_3}{\partial x_3} + b_3 - \dot{p}_3 = 0, \tag{15.1c}$$

where σ_1, σ_2, and σ_3 are the *direct stress* components and τ_{23}, τ_{13}, and τ_{12} the *shear stress* components. These are the components of the stress tensor [284, 285] resolved in basis $\mathcal{I}$. The components of the body force and momentum vectors were resolved in the inertial basis as $\underline{b}^T = \{b_1, b_2, b_3\}$ and $\underline{p}^T = \{p_1, p_2, p_3\}$, respectively. Equations (15.1) are first-order, partial differential equations in space and time must be satisfied at all points of the body and all instants in time.

The *surface equilibrium equations* state that at all points on $\mathcal{S}$ and all instants in time,

$$t_1 = n_1 \sigma_1 + n_2 \tau_{21} + n_3 \tau_{31} = \hat{t}_1, \tag{15.2a}$$

$$t_2 = n_1 \tau_{12} + n_2 \sigma_2 + n_3 \tau_{32} = \hat{t}_2, \tag{15.2b}$$

$$t_3 = n_1 \tau_{13} + n_2 \tau_{23} + n_3 \sigma_3 = \hat{t}_3, \tag{15.2c}$$

where the components of the unit vector normal to the outer surface of the body, the surface traction vector, and the prescribed surface traction vector, all resolved in basis $\mathcal{I}$, are denoted $\underline{n}^T = \{n_1, n_2, n_3\}$, $\underline{t}^T = \{t_1, t_2, t_3\}$, and $\hat{\underline{t}}^T = \{\hat{t}_1, \hat{t}_2, \hat{t}_3\}$, respectively. Over $\mathcal{S}_1$, these conditions are also called the *force*, or *natural boundary conditions*. The stress array

$$\underline{\sigma}^T(t) = \{\sigma_1, \sigma_2, \sigma_3, \tau_{23}, \tau_{13}, \tau_{12}\}, \tag{15.3}$$

will be used whenever that notation is convenient to represent the stress field.

Finally, additional conditions are imposed on the momentum vectors at the initial and final times

$$\underline{p}(t_i) = \hat{\underline{p}}_i, \quad \underline{p}(t_f) = \hat{\underline{p}}_f. \tag{15.4}$$

These conditions are called the *boundary conditions in time*.

Definition 15.1 (Admissible stress and momentum fields). *Stress fields, $\underline{\sigma}(x_1, x_2, x_3, t)$, and momentum fields, $\underline{p}(x_1, x_2, x_3, t)$, are said to be admissible if, at all times, they satisfies the dynamic equilibrium equations, eqs. (15.1), at all points in $\mathcal{V}$, the surface equilibrium equations, eqs. (15.2), at all points on $\mathcal{S}$, and the time boundary conditions, eqs. (15.4), at times t_i and t_f.*

The strain-displacement and velocity-displacement relationships

The *strain-displacement equations* merely define the strain components that are used for the characterization of the deformation of the body at a point. The strain-displacement relationships are derived from purely geometric considerations. Similarly, the *velocity-displacement equations* simply define the velocity components at a point of the body.

When the displacements are small, deformations at a point of the body are conveniently measured by the engineering strain components [284, 285], defined as,

$$\varepsilon_1 = \frac{\partial u_1}{\partial x_1}, \qquad \varepsilon_2 = \frac{\partial u_2}{\partial x_2}, \qquad \varepsilon_3 = \frac{\partial u_3}{\partial x_3}, \tag{15.5a}$$

$$\gamma_{23} = \frac{\partial u_2}{\partial x_3} + \frac{\partial u_3}{\partial x_2}, \quad \gamma_{13} = \frac{\partial u_1}{\partial x_3} + \frac{\partial u_3}{\partial x_1}, \quad \gamma_{12} = \frac{\partial u_1}{\partial x_2} + \frac{\partial u_2}{\partial x_1}, \tag{15.5b}$$

where $\varepsilon_1, \varepsilon_2$, and ε_3 are the *relative elongations* or *direct strain* components of a material line and γ_{23}, γ_{13}, and γ_{12} the *angular distortions* or *shear strain* components of two material lines.

To compute strain components, the displacement field must be continuous and differentiable. Furthermore, over S and at the initial and final times, the following *displacement boundary conditions* must be met

$$\underline{u}(t) = \hat{\underline{u}}(t). \tag{15.6}$$

Over S_2, these conditions are called the *geometric boundary conditions*. The strain array

$$\underline{\varepsilon}^T(t) = \left\{ \varepsilon_1, \ \varepsilon_2, \ \varepsilon_3, \ \gamma_{23}, \ \gamma_{13}, \ \gamma_{12} \right\}, \tag{15.7}$$

will be used whenever that notation is convenient to represent the strain field.

The components of the velocity vector $\underline{v}(t)$ are the time derivatives of the displacement vector,

$$\underline{v}(t) = \dot{\underline{u}}. \tag{15.8}$$

Definition 15.2 (Kinematically admissible displacement field). *A displacement field, $\underline{u}(x_1, x_2, x_3, t)$, is said to be kinematically admissible if, at all time, it is continuous and differentiable at all points in V and satisfies the displacement boundary conditions, eqs. (15.6), at all points on S and the initial and final times.*

Definition 15.3 (Compatible strain field). *A strain field, $\underline{\varepsilon}(x_1, x_2, x_3, t)$, is said to be compatible if, at all times, it is derived from a kinematically admissible displacement field through the strain-displacement relationships, eqs. (15.5).*

Definition 15.4 (Compatible velocity field). *A velocity field, $\underline{v}(x_1, x_2, x_3, t)$, is said to be compatible if, at all times, it is derived from a kinematically admissible displacement field through the velocity-displacement relationships, eqs. (15.8).*

The constitutive laws

The *constitutive laws* relate the stress and strain fields. They consist of a mathematical idealization of the experimentally observed behavior of materials. Hooke's law is commonly used to model the behavior of homogeneous, isotropic, linearly elastic materials operating in the small strain regime. Many materials may present one or more of the following features: anisotropy, plasticity, visco-elasticity, or creep, to name just a few commonly observed behaviors of engineering materials. A second set of constitutive laws relates the momentum vector to the velocity vector.

The stress and strain fields must satisfy the constitutive laws at all points in $\mathcal{V}$. For small strains, Hooke's law [284, 285] represents the behavior of homogeneous, isotropic, linearly elastic materials in a approximate manner by the following linear relationship

$$\underline{\sigma} = \underline{\underline{C}}\,\underline{\epsilon}, \quad \underline{\epsilon} = \underline{\underline{S}}\,\underline{\sigma}, \tag{15.9}$$

where $\underline{\underline{C}}$ is a symmetric, positive-definite *material stiffness matrix*, and $\underline{\underline{S}} = \underline{\underline{C}}^{-1}$ a symmetric, positive-definite *material compliance matrix*.

A constitutive law is also required for the momentum field. This law is, in fact, the definition of the momentum vector

$$\underline{p} = \rho\,\underline{v}, \tag{15.10}$$

where ρ is the material mass density.

Summary

Complete solutions of elastodynamics problems involves the following fields.

1. Admissible stress and momentum fields, see definition 15.1.
2. Kinematically admissible displacement fields, see definition 15.2, and associated compatible strain and velocity fields, see definition 15.3 and 15.4, respectively.
3. Stress and momentum fields that satisfy the constitutive laws, eqs. (15.9) and (15.10), respectively, at all points in $\mathcal{V}$.

All these equations must be satisfied at all instants in time.

15.2.2 The principle of virtual work

Consider an elastic body that is in dynamic equilibrium under applied body forces and surface tractions. This implies that the stress and momentum fields are admissible, see definition 15.1. The following statement is now constructed

$$\int_{t_i}^{t_f} \int_{\mathcal{V}} \left[\left(\frac{\partial \sigma_1}{\partial x_1} + \frac{\partial \tau_{21}}{\partial x_2} + \frac{\partial \tau_{31}}{\partial x_3} + b_1 - \dot{p}_1 \right) \delta u_1 \right.$$

$$+ \left(\frac{\partial \tau_{12}}{\partial x_1} + \frac{\partial \sigma_2}{\partial x_2} + \frac{\partial \tau_{32}}{\partial x_3} + b_2 - \dot{p}_2 \right) \delta u_2$$

$$\left. + \left(\frac{\partial \tau_{13}}{\partial x_1} + \frac{\partial \tau_{23}}{\partial x_2} + \frac{\partial \sigma_3}{\partial x_3} + b_3 - \dot{p}_3 \right) \delta u_3 \right] \, \mathrm{d}\mathcal{V} \mathrm{d}t \tag{15.11}$$

$$- \int_{t_i}^{t_f} \int_{\mathcal{S}} (\underline{t} - \hat{\underline{t}})^T \delta \underline{u} \, \mathrm{d}\mathcal{S} \mathrm{d}t + \left[\int_{\mathcal{V}} (\underline{p} - \hat{\underline{p}})^T \delta \underline{u} \, \mathrm{d}\mathcal{V} \right]_{t_i}^{t_f} = 0.$$

This statement was constructed in the following manner. Each of the three dynamic equilibrium equations, eqs. (15.1), was multiplied by an arbitrary, virtual change in displacement, then integrated over the range of validity of the equation. Similarly, each of the three surface equilibrium equations, eqs. (15.2), was multiplied by an arbitrary, virtual change in displacement, then integrated over the range of validity of the equation. Finally, each of the three boundary conditions in time, eqs. (15.4), was multiplied by an arbitrary, virtual change in displacement, then integrated over the range of validity of the equation.

Because the stress and momentum fields are admissible, each term in parenthesis is zero, and multiplication by an arbitrary quantity results in a zero product. Each of the three integrals then vanishes, as does their sum.

Next, integration by parts is performed. Using Green's theorem [2], the first term of the volume integral becomes

$$\int_{\mathcal{V}} \frac{\partial \sigma_1}{\partial x_1} \delta u_1 \, \mathrm{d}\mathcal{V} = - \int_{\mathcal{V}} \sigma_1 \frac{\partial \delta u_1}{\partial x_1} \, \mathrm{d}\mathcal{V} + \int_{\mathcal{S}} n_1 \sigma_1 \delta u_1 \, \mathrm{d}\mathcal{S},$$

where n_1 is the component of the outward unit normal along $\bar{\imath}_1$, see fig. 15.3. A similar operation is performed on each stress derivative terms appearing in eq. (15.11). For the momentum terms, integration by parts yields

$$- \int_{t_i}^{t_f} \dot{\underline{p}}^T \delta \underline{u} \, \mathrm{d}t = \int_{t_i}^{t_f} \underline{p}^T \delta \dot{\underline{u}} \, \mathrm{d}t - \left[\underline{p}^T \delta \underline{u} \right]_{t_i}^{t_f} = \int_{t_i}^{t_f} \underline{p}^T \delta \underline{v} \, \mathrm{d}t - \left[\underline{p}^T \delta \underline{u} \right]_{t_i}^{t_f}.$$

Introducing the results of these integrations by parts into eq. (15.11) then yields

$$- \int_{t_i}^{t_f} \left\{ \int_{\mathcal{V}} \left[\underline{\sigma}^T \delta \underline{\varepsilon} + \underline{p}^T \delta \underline{v} \right] \, \mathrm{d}\mathcal{V} + \int_{\mathcal{V}} \underline{b}^T \delta \underline{u} \, \mathrm{d}\mathcal{V} + \int_{\mathcal{S}} \hat{\underline{t}}^T \delta \underline{u} \, \mathrm{d}\mathcal{S} \right\} \mathrm{d}t$$

$$- \left[\int_{\mathcal{V}} \hat{\underline{p}}^T \delta \underline{u} \, \mathrm{d}\mathcal{V} \right]_{t_i}^{t_f} = 0, \tag{15.12}$$

where the *virtual, compatible strain field* was defined as

$$\delta \varepsilon_1 = \frac{\partial \delta u_1}{\partial x_1}, \qquad \delta \varepsilon_2 = \frac{\partial \delta u_2}{\partial x_2}, \qquad \delta \varepsilon_3 = \frac{\partial \delta u_3}{\partial x_3}, \tag{15.13a}$$

$$\delta \gamma_{23} = \frac{\partial \delta u_2}{\partial x_3} + \frac{\partial \delta u_3}{\partial x_2}, \; \delta \gamma_{13} = \frac{\partial \delta u_1}{\partial x_3} + \frac{\partial \delta u_3}{\partial x_1}, \; \delta \gamma_{12} = \frac{\partial \delta u_1}{\partial x_2} + \frac{\partial \delta u_2}{\partial x_1}, \tag{15.13b}$$

and the *virtual, compatible velocity field* as

$$\delta \underline{v} = \delta \underline{\dot{u}}. \tag{15.14}$$

Equation (15.12) can be restated as

$$-\int_{t_i}^{t_f} \int_{\mathcal{V}} \left[\underline{\sigma}^T \delta \underline{\varepsilon} - \underline{p}^T \delta \underline{v}\right] \, d\mathcal{V} dt$$
$$+\int_{t_i}^{t_f} \left\{ \int_{\mathcal{V}} \underline{b}^T \delta \underline{u} \, d\mathcal{V} + \int_{S} \hat{\underline{t}}^T \delta \underline{u} \, dS \right\} dt - \left[\int_{\mathcal{V}} \hat{\underline{p}}^T \delta \underline{u} \, d\mathcal{V}\right]_{t_i}^{t_f} = 0. \tag{15.15}$$

The first term represents the virtual work done by the internal stresses and momenta and the remaining correspond to the virtual work done by the externally applied loads and momenta.

It has been shown thus far that if the stress and momenta fields are admissible, eq. (15.15) must hold. It can also be shown that if this equation holds, the stress and momenta fields must be admissible. Indeed, eq. (15.15) implies eq. (15.12), which in turns implies eq. (15.11) by reversing the integration by parts process. Finally, the volume and surface equilibrium equations are recovered because eq. (15.11) must hold for all arbitrary, kinematically admissible virtual displacements fields. Statement (15.15) can thus be interpreted as follows.

Principle 19 (Principle of virtual work) *A body is in dynamic equilibrium if the sum of the internal and external virtual work vanishes for all arbitrary kinematically admissible virtual displacement fields and corresponding compatible strain and velocity fields.*

This principle is illustrated in fig. 15.4. The surface tractions act at the spatial boundaries of the problem and play a role similar to that of the momenta at the temporal boundaries of the problem.

In summary, the equations of dynamic equilibrium, eqs. (15.1), (15.2), and (15.4), and the principle of virtual work are two equivalent statements. Furthermore, because the equations of dynamic equilibrium are a statement of

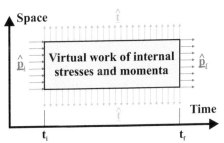

Fig. 15.4. Illustration of the principle of virtual work.

Newton's second law, the principle of virtual work and Newton's second law are two equivalent statements.

Because the principle of virtual work is solely a statement of equilibrium, it is always true. However, for the solution of specific elastodynamics problems, it must be complemented with stress-strain and momentum-velocity relationships, and constitutive laws. More details about the principle of virtual work and its application to structures can be found in numerous textbooks such as [284, 286, 287, 285].

15.2.3 Hamilton's principle

If the internal forces in the solid are assumed to be conservative, they can be derived from a potential, as discussed in section 3.2. In this case, the internal forces are the components of stress, and the potential is the *strain energy density function*. If the stresses in the solid can be derived from a strain energy density function, $a(\underline{\varepsilon})$,

$$\underline{\sigma} = \frac{\partial a(\underline{\varepsilon})}{\partial \underline{\varepsilon}}, \qquad (15.16)$$

the material is said to be an *elastic material*. Assuming the material to be elastic or assuming the existence of a strain energy density function are two equivalent assumptions. Linearly elastic materials are elastic materials for which the stress-strain relationship is linear.

If the material is elastic, the work done by the internal stresses when the system is brought from one state of deformation to another depends only on the two states of deformations, but not on the specific path that the system followed from one deformation state to the other. This restricts the types of material constitutive laws that can be expressed in terms of a strain energy density function. For instance, if a material is deformed in the plastic range, the work of deformation will depend on the specific deformation history; hence, there exists no strain energy density function that describes material behavior when plastic deformations are involved.

For instance, the strain energy density function of a linearly elastic material is

$$a(\underline{\varepsilon}) = \frac{1}{2}\underline{\varepsilon}^T \underline{\underline{C}}\,\underline{\varepsilon}. \qquad (15.17)$$

Introducing this function into eq. (15.16), yields $\underline{\sigma} = \underline{\underline{C}}\,\underline{\varepsilon}$, the constitutive law for a linearly elastic material.

Consider a general elastic body that is in equilibrium under applied body forces and surface tractions, and therefore, the principle of virtual work, eq. (15.15), must apply. It is now assumed that the constitutive law for the material can be expressed in terms of a strain energy density function, eq. (15.16). The virtual work done by the internal stresses appears in the first term of eq. (15.15), and it is readily evaluated as

$$\int_{\mathcal{V}} \delta\underline{\varepsilon}^T \underline{\sigma}\, \mathrm{d}\mathcal{V} = \int_{\mathcal{V}} \delta\underline{\varepsilon}^T \frac{\partial a(\underline{\varepsilon})}{\partial \underline{\varepsilon}}\, \mathrm{d}\mathcal{V} = \int_{\mathcal{V}} \delta a(\underline{u})\, \mathrm{d}\mathcal{V} = \delta \int_{\mathcal{V}} a(\underline{u})\, \mathrm{d}\mathcal{V} = \delta A(\underline{u}),$$

where the chain rule for derivatives is used to obtain the second equality.

The strain energy density and the *total strain energy* of the body, $A = \int_{\mathcal{V}} a\, \mathrm{d}\mathcal{V}$, must be expressed in terms of the displacement field using the strain-displacement relationships because the principle of virtual work requires a compatible strain field.

In a similar manner, the use of the dynamic constitutive law, eq. (15.10), leads to

$$\int_{\mathcal{V}} \underline{p}^T \delta\underline{v}\, \mathrm{d}\mathcal{V} = \int_{\mathcal{V}} \rho\underline{v}^T \delta\underline{v}\, \mathrm{d}\mathcal{V} = \int_{\mathcal{V}} \delta k(\underline{v})\, \mathrm{d}\mathcal{V} = \delta K(\underline{v}),$$

where $k(\underline{v}) = 1/2\, \rho\underline{v}^T \underline{v}$ is the *kinetic energy density function*, and $K(\underline{v})$ the *total kinetic energy* of the body.

The principle of virtual work, eq. (15.15), now becomes

$$
-\delta \int_{t_i}^{t_f} (A - K)\, dt
$$
$$
+ \int_{t_i}^{t_f} \left\{ \int_{\mathcal{V}} \underline{b}^T \delta \underline{u}\, d\mathcal{V} + \int_{\mathcal{S}} \underline{\hat{t}}^T \delta \underline{u}\, d\mathcal{S} \right\} dt - \left[\int_{\mathcal{V}} \underline{\hat{p}}^T \delta \underline{u}\, d\mathcal{V} \right]_{t_i}^{t_f} = 0. \tag{15.18}
$$

The first term on the second line of this statement represents the *virtual work done by the externally applied loads*,

$$
\delta W_{\text{ext}} = \int_{\mathcal{V}} \underline{b}^T \delta \underline{u}\, d\mathcal{V} + \int_{\mathcal{S}} \underline{\hat{t}}^T \delta \underline{u}\, d\mathcal{S}. \tag{15.19}
$$

With this definition, the principle of virtual work, eq. (15.18), becomes

$$
-\delta \int_{t_i}^{t_f} (A - K)\, dt + \int_{t_i}^{t_f} \delta W_{\text{ext}}\, dt - \left[\int_{\mathcal{V}} \underline{\hat{p}}^T \delta \underline{u}\, d\mathcal{V} \right]_{t_i}^{t_f} = 0. \tag{15.20}
$$

Next, the body forces and surface tractions are assumed to be conservative, *i.e.*, they can be derived from a potential,

$$
\underline{b} = -\frac{\partial \phi}{\partial \underline{u}}; \quad \underline{\hat{t}} = -\frac{\partial \psi}{\partial \underline{u}},
$$

where ϕ is the *potential of the body forces*, and ψ the *potential of the surface tractions*. For instance, the potential of fixed surface tractions is simply $\psi = -\underline{\hat{t}}^T \underline{u}$, or the potential of the body forces associated with a gravity field, $\underline{g}$, is $\phi = -\rho\, \underline{g}^T \underline{u}$.

The two terms of the virtual work done by the external forces, eq. (15.19), now become

$$
\int_{\mathcal{V}} \underline{b}^T \delta \underline{u}\, d\mathcal{V} = -\int_{\mathcal{V}} \frac{\partial \phi}{\partial \underline{u}}^T \delta \underline{u}\, d\mathcal{V} = -\delta \int_{\mathcal{V}} \phi(\underline{u})\, d\mathcal{V},
$$
$$
\int_{\mathcal{S}} \underline{\hat{t}}^T \delta \underline{u}\, d\mathcal{S} = -\int_{\mathcal{S}} \frac{\partial \psi}{\partial \underline{u}}^T \delta \underline{u}\, d\mathcal{S} = -\delta \int_{\mathcal{S}} \psi(\underline{u})\, d\mathcal{S}.
$$

Combining these two loading terms then yields

$$
\delta W_{\text{ext}} = -\delta \int_{\mathcal{V}} \phi(\underline{u})\, d\mathcal{V} - \delta \int_{\mathcal{S}} \psi(\underline{u})\, d\mathcal{S} = -\delta \Phi(\underline{u}),
$$

where $\Phi(\underline{u})$ is the *total potential the externally applied loads*. Introducing this result into eq. (15.20) leads to

$$
\delta \int_{t_i}^{t_f} (A - K + \Phi)\, dt + \left[\int_{\mathcal{V}} \underline{\hat{p}}^T \delta \underline{u}\, d\mathcal{V} \right]_{t_i}^{t_f} = 0. \tag{15.21}
$$

The *Lagrangian* of the system is now defined as

$$
L = K(\underline{v}) - A(\underline{u}) - \Phi(\underline{u}), \tag{15.22}
$$

and it follows that

$$
\int_{t_i}^{t_f} \delta L \, dt = \left[\int_{\mathcal{V}} \underline{\hat{p}}^T \delta \underline{u} \, d\mathcal{V} \right]_{t_i}^{t_f}
\tag{15.23}
$$

Hamilton's principle can be stated as

Principle 20 (Hamilton's principle) *An elastic system is in dynamic equilibrium if and only if equation (15.23) holds for all arbitrary virtual displacements.*

Clearly, this statement generalizes the version of Hamilton's principle derived in section 8.2 for systems of particles. If the momenta vanish at the initial and final times, it simply becomes $\delta L = 0$.

Many other variational principles exist in elasticity. Particularly noteworthy are the principle of complementary virtual work and the principle of minimum complementary energy [286, 287, 285]. Two or three field principles can also be developed, such as Hellinger-Reissner's and Hu-Washizu's principles, respectively. All these principle can be extended to elastodynamics problem by invoking d'Alembert's principle, inertial forces are included as externally applied forces.

15.3 Finite displacements kinematics for flexible bodies

While flexible multibody systems are characterized by large relative motions at the joints, it is often the case that individual flexible bodies undergo small deformations. Conceptually, the displacement field at a point of the flexible body can be decomposed into rigid body and elastic displacement field [167], where the latter field is responsible for the straining of the body whereas the former, by definition, generates no deformation.

It is not uncommon for structures such as slender beams or thin plates and shells, to undergo large rigid body displacements and rotations while the strains remain small at all points of the structure. This behavior is often the consequence of careful planning: to avoid premature failure, the structure is designed to operate in the small strain regime.

When a structure operates in the small strain regime, the three groups of equations of elastodynamics described in section 15.2.1 still apply, but must be updated to account for the large displacements. Of course, Newton's law still applies, but in this case, the equilibrium conditions must be enforced on the deformed configuration of the structure. The strain-displacement relationships now become nonlinear equations, rather than the linear relationships that characterize small displacement problems, eqs. (15.5). Finally, the constitutive laws remain unchanged, although the stress and strain components are now those resolved in the *convected* or *material basis*.

This section presents a brief discussion of the state of deformation in the neighborhood of a material point in a flexible body. Two configuration of this body will be defined: a *reference configuration*, and a *deformed configuration*. The following notational convention will be used: lower-case symbols refer to quantities defined in the reference configuration, and upper-case symbols refer to the corresponding quantities in the deformed configuration.

Material coordinates

Figure 15.5 depicts a body in its reference and deformed configurations and an inertial frame, $\mathcal{F}^I = [\mathbf{O}, \mathcal{I} = (\bar{\imath}_1, \bar{\imath}_2, \bar{\imath}_3)]$. This section focuses on the relationships between the deformed and reference configuration of the solid without any consideration for the loads that create the deformation.

Let point $\mathbf{P}$ be a material point of the body, and the position vectors of this material point are denoted $\underline{x}$ and $\underline{X}$, in the reference and deformed configuration, respectively. Each material particle of the body will be identified by a label consisting of a triplet of real numbers. This label will remain attached to the material particle throughout the deformation process. This label is called the *material coordinates* of material point $\mathbf{P}$, and is denoted $(\alpha_1, \alpha_2, \alpha_3)$.

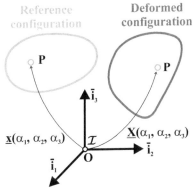

Fig. 15.5. The reference and deformed configurations of a body.

The position vectors of point $\mathbf{P}$ in the reference and deformed configurations are

$$\underline{x} = \underline{x}(\alpha_1, \alpha_2, \alpha_3), \tag{15.24a}$$
$$\underline{X} = \underline{X}(\alpha_1, \alpha_2, \alpha_3), \tag{15.24b}$$

respectively. Because the material coordinates are an identifying label for a material particle, they can be chosen arbitrarily. A convenient choice for the material coordinates consists of the components of the position vector resolved in basis $\mathcal{I}$

$$\underline{x}(\alpha_1, \alpha_2, \alpha_3) = \alpha_1\,\bar{\imath}_1 + \alpha_2\,\bar{\imath}_2 + \alpha_3\,\bar{\imath}_3. \tag{15.25}$$

This particular choice of the material coordinates is called the *Lagrangian representation*.

A *material line* is an ensemble of material particles forming a straight line in the reference configuration of the body. For instance, fig. 15.6 shows segments $\mathbf{PR}$, $\mathbf{PS}$, and $\mathbf{PT}$ of the reference configuration, which are are material lines intersecting at point $\mathbf{P}$. Due to the deformation of the body, all the material particles forming material line $\mathbf{PR}$ will move to segment $\mathbf{PR}$ in the deformed configuration. Because segment $\mathbf{PR}$ is of differential length, it can be assumed to remain straight in the deformed configuration.

Base vectors and metric tensor

The *base vectors* are vectors tangent to these material line

$$\bar{g}_i = \frac{\partial \underline{x}}{\partial \alpha_i} = \bar{\imath}_i. \qquad (15.26)$$

In the reference configuration, the base vectors are mutually orthogonal unit vectors.

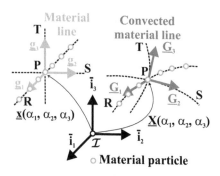
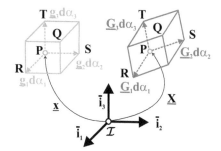

Fig. 15.6. The base vectors in the reference and deformed configurations.

Fig. 15.7. Volume elements in the reference and deformed configurations.

As the deformation takes place, the material lines are convected with the body. The convected material lines now describe curves in space intersecting at the new location of the particle. The base vectors in the deformed configuration are defined in a manner similar to those of the reference configuration

$$\underline{G}_i = \frac{\partial \underline{X}}{\partial \alpha_i}. \qquad (15.27)$$

These base vectors are not mutually orthogonal, nor are they unit vectors.

To visualize this deformation, fig. 15.7 shows the small rectangular parallelepiped **PQRST** of differential size $d\alpha_1$ by $d\alpha_2$ by $d\alpha_3$ cut in the neighborhood of point **P**. The reference configuration is the configuration of the solid in its undeformed state, and rectangular parallelepiped **PQRST** is spanned by vectors $\bar{g}_1 d\alpha_1$, $\bar{g}_2 d\alpha_2$, and $\bar{g}_3 d\alpha_3$. Under the action of applied loads, the body deforms and assumes a new configuration, called the deformed configuration. All the material particles that formed the rectangular parallelepiped **PQRST** in the reference configuration now form parallelepiped **PQRST**, which is spanned by vectors $\underline{G}_1 d\alpha_1$, $\underline{G}_2 d\alpha_2$, and $\underline{G}_3 d\alpha_3$ in the deformed configuration. The state of strain at a point characterizes the deformation of the parallelepiped without any consideration for the loads that created the deformation.

Increments in position vector are denoted $d\underline{x}$ and $d\underline{X}$ in the reference and deformed configurations, respectively, and are expressed as

$$d\underline{x} = \frac{\partial \underline{x}}{\partial \alpha_i} \, d\alpha_i = \bar{g}_i \, d\alpha_i,$$

$$d\underline{X} = \frac{\partial \underline{X}}{\partial \alpha_i} \, d\alpha_i = \underline{G}_i \, d\alpha_i,$$

where summation is implied by the repeated indices.

The lengths of these increments, denoted $\mathrm{d}s$ and $\mathrm{d}S$ in the reference and deformed configurations, respectively, are readily found as

$$\mathrm{d}s^2 = \mathrm{d}\underline{x}^T \mathrm{d}\underline{x} = \bar{g}_i^T \bar{g}_j\, \mathrm{d}\alpha_i \mathrm{d}\alpha_j = g_{ij}\, \mathrm{d}\alpha_i \mathrm{d}\alpha_j, \tag{15.28a}$$

$$\mathrm{d}S^2 = \mathrm{d}\underline{X}^T \mathrm{d}\underline{X} = \underline{G}_i^T \underline{G}_j\, \mathrm{d}\alpha_i \mathrm{d}\alpha_j = G_{ij}\, \mathrm{d}\alpha_i \mathrm{d}\alpha_j. \tag{15.28b}$$

These relationships define the components of the *metric tensors* in the reference and deformed configurations, denoted g_{ij} and G_{ij}, respectively, as

$$g_{ij} = \bar{g}_i^T \bar{g}_j, \tag{15.29a}$$

$$G_{ij} = \underline{G}_i^T \underline{G}_j. \tag{15.29b}$$

The symmetry of both tensors is apparent from these definitions.

Displacement field

The difference between the positions of a material point in the deformed and reference configurations defines the *displacement vector* as

$$\underline{u}(\alpha_1, \alpha_2, \alpha_3) = \underline{X} - \underline{x}. \tag{15.30}$$

The displacement and position vectors in the deformed configuration are now resolved along the base vectors of the reference configuration as

$$\underline{u} = u_i\, \bar{g}_i, \tag{15.31a}$$

$$\underline{X} = X_i\, \bar{g}_i. \tag{15.31b}$$

With these definitions, eq. (15.30) now becomes

$$X_i = \alpha_i + u_i. \tag{15.32}$$

The base vector in the deformed configuration is related to the displacement field

$$\underline{G}_i = \frac{\partial \underline{X}}{\partial \alpha_i} = (\delta_{ij} + u_{j,i})\, \bar{g}_j, \tag{15.33}$$

where δ_{ij} is Kronecker's symbol defined by eq. (1.14) and notation $(\cdot)_{,i}$ indicates a derivative with respect to material coordinate α_i.

Many different measures can be used to characterize the state of deformation at a point. Some measures are directly related to the physical concept of strain, *i.e.*, a relative change in length, but are not necessarily of a tensorial nature. Some other measures, clearly related to the physical concept of strain can be shown to be tensors.

15.3.1 The engineering strain components

The motion of segment **PR** from its reference to deformed configuration depicted in fig. 15.7 consists of two parts: a change in orientation and a change in length. Clearly, the change in length is a deformation or stretching of the material line. Similarly, segments **PR** and **PS** form a rectangle in the reference configuration, but form a parallelogram in the deformed configuration. The angular distortion of the rectangle into a parallelogram represents a deformation of the body. Stretching of a material line and angular distortion between two material lines will be selected as measures of the state of strain at a point.

The stretching or *relative elongations* of material lines **PR**, **PS**, and **PT** will be denoted as ε_1, ε_2, and ε_3, respectively. The *angular distortions* between segments **PS** and **PT**, **PR** and **PT**, and **PR** and **PS** will be denoted γ_{23}, γ_{13}, and γ_{12}, respectively.

The relative elongation, ε_1, of material line **PR**, see fig. 15.7, is defined as

$$
\varepsilon_1 = \frac{\|\mathbf{PR}\|_{\text{def}} - \|\mathbf{PR}\|_{\text{ref}}}{\|\mathbf{PR}\|_{\text{ref}}}, \tag{15.34}
$$

where the subscripts $(\cdot)_{\text{ref}}$ and $(\cdot)_{\text{def}}$ are used to indicate the reference and deformed configurations, respectively. The relative elongation is a non-dimensional quantity. Similar definitions hold for ε_2 and ε_3, the relative elongation of material lines **PQ** and **PT**.

The angular distortion, γ_{23}, between two material lines **PT** and **PS** is defined as the change of the initially right angle, $\gamma_{23} = \pi/2 - \angle\mathbf{TPS}_{\text{def}}$, where notation $\angle\mathbf{TPS}$ is used to indicate the angle between segments **PT** and **PS**. This can also written as

$$
\gamma_{23} = \arcsin \frac{G_2^T G_3}{\|G_2\| \, \|G_3\|}. \tag{15.35}
$$

Angular distortion are non-dimensional quantities. Similar definitions hold for the angular distortion γ_{13} and γ_{12} of the angles between material lines **PR** and **PT**, and **PS** and **PT**, respectively. The engineering strain components do not form a second-order tensor. They are often called *physical strain components*.

15.3.2 The deformation gradient tensor

A widely used strain measure is the *deformation gradient tensor* defined as

$$
F_{ij} = \frac{\partial X_i}{\partial \alpha_j}. \tag{15.36}
$$

In the following sections, the index notation will be used to represent second-order tensors. For instance, the deformation gradient tensor is denoted F_{ij} rather than the less explicit $\underline{\underline{F}} = \partial \underline{X}/\partial\underline{\alpha}$.

Resolving the base vector in the deformed configuration, eq. (15.27), along the reference frame yields

$$\underline{G}_i = \frac{\partial \underline{X}}{\partial \alpha_i} = F_{ji}\,\bar{g}_j. \tag{15.37}$$

A scalar product of this relationship by $\bar{g}_l$ yields an alternative definition of the deformation gradient tensor

$$F_{ij} = \bar{g}_i^T\,\underline{G}_j. \tag{15.38}$$

With the help of the chain rule for derivatives, an explicit expression of the inverse of the deformation gradient tensor can be obtained

$$F_{ij}^{-1} = \frac{\partial \alpha_i}{\partial X_j}. \tag{15.39}$$

Introducing the displacement field, eq. (15.32) into eq. (15.36) yields the deformation gradient tensor in terms of the displacement vector components

$$F_{ij} = \delta_{ij} + u_{i,j}. \tag{15.40}$$

15.3.3 The metric tensor

Relationship (15.28b) shows that the metric tensor, G_{ij}, is, in fact, a measure of the deformation. When used as strain measure, the metric tensor is also called the *Green deformation tensor*, or the *Cauchy-Green deformation tensor*. The metric tensor is clearly related to the engineering strain components. Indeed, eq. (15.34) implies

$$\varepsilon_1 = \frac{\|\underline{G}_1 d\alpha_1\| - \|\bar{g}_1 d\alpha_1\|}{\|\bar{g}_1 d\alpha_1\|} = \sqrt{G_{11}} - 1. \tag{15.41}$$

Similar relations hold for ε_2 and ε_3. The angular distortion, eq. (15.35) becomes

$$\gamma_{23} = \arcsin \frac{G_{23}}{\sqrt{G_{22}G_{33}}}. \tag{15.42}$$

The inverse relationships are readily obtained as

$$G_{11} = (1 + \varepsilon_1)^2, \tag{15.43a}$$
$$G_{23} = (1 + \varepsilon_2)(1 + \varepsilon_3)\sin\gamma_{23}. \tag{15.43b}$$

The metric tensor in the deformed configuration is closely related to the deformation gradient tensor. Introducing eq. (15.37) into the definition of the metric tensor, eq. (15.29b), yields

$$G_{ij} = F_{ki}F_{kj}. \tag{15.44}$$

15.3.4 The Green-Lagrange strain tensor

A widely used strain measure is the *Green-Lagrange strain tensor*, defined as

$$e_{ij} = \frac{1}{2}(G_{ij} - g_{ij}). \tag{15.45}$$

It is also called the *Lagrangian strain tensor*, or the *Green-Saint Venant strain tensor*. The Green-Lagrange strain tensor is closely related to the metric tensor, and eqs. (15.43) reveal its connection to the engineering strain components,

$$e_{11} = \frac{1}{2}(G_{11} - 1) = \varepsilon_1 + \frac{1}{2}\varepsilon_1^2, \tag{15.46a}$$

$$e_{23} = \frac{1}{2}(G_{23} - 0) = \frac{1}{2}(1 + \varepsilon_2)(1 + \varepsilon_3)\sin\gamma_{23}. \tag{15.46b}$$

If the deformation of the body is such that the strain components remain far smaller than unity, the above relations simplify to

$$e_{11} \approx \varepsilon_1, \quad e_{23} \approx \frac{1}{2}\gamma_{23}. \tag{15.47}$$

The Green-Lagrange strain tensor is closely related to the deformation gradient tensor. Indeed, introducing eq. (15.44) into eq. (15.45) yields

$$e_{ij} = \frac{1}{2}(F_{ki}F_{kj} - g_{ij}). \tag{15.48}$$

The Green-Lagrange strain tensor is also closely related to the change in length of the increment of the position vector. Indeed, eqs. (15.28a) and (15.28b) yield

$$\frac{1}{2}(dS^2 - ds^2) = \frac{1}{2}(G_{ij} - g_{ij})\, d\alpha_i d\alpha_j = e_{ij}\, d\alpha_i d\alpha_j. \tag{15.49}$$

Finally, the Green-Lagrange strain tensor is readily expressed in terms of the displacement components by introducing eq. (15.40) into eq. (15.48) to find

$$e_{ij} = \frac{1}{2}(u_{i,j} + u_{j,i} + u_{k,i}u_{k,j}). \tag{15.50}$$

15.4 Strain measures for various differential elements

The previous section has focused on the state of strain at a point of a three-dimensional solid. It is often useful, however, to characterize the straining of a differential line, surface, or volume element of the body. These issues are addressed in the following sections.

15.4.1 Stretch of a material line

In the reference configuration, the orientation of material line **PQ** is defined by a unit vector, denoted $\bar{n}$, defined as

$$\bar{n} = \frac{\mathbf{PQ}}{\|\mathbf{PQ}\|} = \frac{d\alpha_i}{ds}\bar{g}_i = n_i\bar{g}_i. \tag{15.51}$$

The stretch, λ, of this material line is defined as the ratio of the length of the differential elements in the reference and deformed configurations, given by eqs. (15.28a) and (15.28b), respectively, to find $\lambda^2 = dS^2/ds^2 = G_{ij}(d\alpha_i/ds)(d\alpha_j/ds) = G_{ij}n_in_j$. The stretch of the line element is now

$$\lambda = \sqrt{G_{ij}n_in_j} = \sqrt{F_{ki}F_{kj}n_in_j}. \tag{15.52}$$

15.4.2 Angle between two material lines

Consider two material lines defined by unit vectors $\bar{n}$ and $\bar{n}'$ with stretches λ and λ', respectively. The scalar product of the position increments corresponding to these material lines is $d\underline{X}^T d\underline{X}' = \|d\underline{X}\| \, \|d\underline{X}'\| \cos\theta$, where θ is the angle between the two material lines in the deformed configuration. Solving for this angle yields $\cos\theta = (d\underline{X}/dS)^T (d\underline{X}'/dS')$. Using the chain rule for derivatives, and introducing the definition, eq. (15.36), of the deformation gradient tensor yields

$$\cos\theta = F_{ik}F_{il} \frac{d\alpha_k}{ds} \frac{ds}{dS} \frac{d\alpha_l'}{ds'} \frac{ds'}{dS'},$$

and finally

$$\cos\theta = \frac{F_{ik}F_{il}\, n_k n_l'}{\lambda\lambda'} = \frac{G_{ij}\, n_i n_j'}{\lambda\lambda'}. \tag{15.53}$$

15.4.3 Surface dilatation

Consider now the area of the rectangle defined by vectors $\bar{g}_2 \, \alpha_2$ and $\bar{g}_3 \, \alpha_3$. The material particles forming that surface before deformation are located in the surface defined by vectors $\underline{G}_2 \, d\alpha_2$ and $\underline{G}_3 \, d\alpha_3$ after deformation. The initial area da_1, is found from fig. 15.8 as

$$da_1 = \|\tilde{g}_2 \, \bar{g}_3 \, d\alpha_2 d\alpha_3\| = d\alpha_2 d\alpha_3. \tag{15.54}$$

The area in the deformed configuration, dA_1, is similarly found

$$dA_1 = \|\widetilde{\underline{G}}_2 \, \underline{G}_3 d\alpha_2 d\alpha_3\| = \sqrt{\underline{G}_3^T \widetilde{G}_2^T \, \widetilde{\underline{G}}_2 \, \underline{G}_3}\, d\alpha_2 d\alpha_3 = \sqrt{G_{22}G_{33} - G_{23}^2}\, d\alpha_2 d\alpha_3.$$

Clearly, quantity $G_{22}G_{33} - G_{23}^2 = GG_{11}^{-1}$, where G_{ij}^{-1} is the inverse of the metric tensor and G its determinant. Hence,

$$dA_1 = \sqrt{GG_{11}^{-1}}\, da_1, \tag{15.55}$$

where

$$G = \det(G_{ij}). \tag{15.56}$$

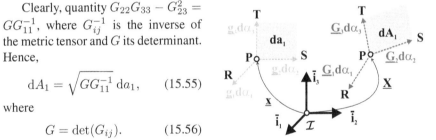

Fig. 15.8. Dilatation of a differential element of area.

Similar developments yield expressions for areas da_2 and da_3, and dA_2 and dA_3 for the reference and deformed configurations, respectively. Combining all results then yields

$$dA_i = \sqrt{GG_{ii}^{-1}}\, da_i, \quad \text{no sum on i.} \tag{15.57}$$

The *surface dilatation*, Σ_i, is defined as the relative change in area of a differential element in the deformed and reference configurations and is readily found from eq. (15.57) as

$$\Sigma_i = \frac{dA_i - da_i}{da_i} = \sqrt{GG_{ii}^{-1}} - 1, \quad \text{no sum on i.} \tag{15.58}$$

15.4.4 Volume dilatation

Figure 15.7 depicts the volume spanned by vectors $\bar{g}_1 \, \mathrm{d}\alpha_1$, $\bar{g}_2 \, \mathrm{d}\alpha_2$, and $\bar{g}_3 \, \mathrm{d}\alpha_3$. The particles contained in that volume before deformation are located in the volume defined by vectors $\underline{G}_1 \, \mathrm{d}\alpha_1$, $\underline{G}_2 \, \mathrm{d}\alpha_2$, and $\underline{G}_3 \, \mathrm{d}\alpha_3$ after deformation. The volume, $\mathrm{d}v$, in the reference configuration is found from eq. (1.32) as

$$\mathrm{d}v = \bar{g}_1^T \tilde{g}_2 \, \bar{g}_3 \, \mathrm{d}\alpha_1 \mathrm{d}\alpha_2 \mathrm{d}\alpha_3 = \mathrm{d}\alpha_1 \mathrm{d}\alpha_2 \mathrm{d}\alpha_3. \tag{15.59}$$

The volume, $\mathrm{d}V$, in the deformed configuration is

$$\mathrm{d}V = \underline{G}_1^T \tilde{G}_2 \, \underline{G}_3 \, \mathrm{d}\alpha_1 \mathrm{d}\alpha_2 \mathrm{d}\alpha_3 = \det(F_{ij}) \, \mathrm{d}\alpha_1 \mathrm{d}\alpha_2 \mathrm{d}\alpha_3. \tag{15.60}$$

In view of eq. 15.44, $\det(G_{ij}) = \det(F_{ki} F_{kj})$, and hence $\det(F_{ij}) = \sqrt{\det(G_{ij})} = \sqrt{G}$. The *volumetric strain*, or relative change in volume is now defined as

$$\varDelta = \frac{\mathrm{d}V - \mathrm{d}v}{\mathrm{d}v} = \sqrt{G} - 1. \tag{15.61}$$

15.4.5 Problems

Problem 15.1. Deformed elastic body
Figure 15.9 depicts an elastic body in its reference and deformed configurations. The displacement field is given as $u_1 = \alpha_1 \alpha_2 / 4$ and $u_2 = -\alpha_1 \alpha_2 / 8$. *(1)* Evaluate the base vectors in the reference and deformed configurations. *(2)* Find the deformation gradient tensor. *(3)* Determine the metric tensors in the reference and deformed configurations. *(4)* Evaluate the Green-Lagrange strain tensor, and *(5)* the physical strain components. Consider point **A** (located at $\alpha_1 = \alpha_2 = 1$) and two material lines $\bar{n}_1$ and $\bar{n}_2$ parallel to $\bar{\imath}_1$ and $\bar{\imath}_2$, respectively. *(1)* Find the stretch of the material lines $\bar{n}_1$ and $\bar{n}_2$. *(2)* Determine the angle between the two material lines in the deformed configuration, *(3)* the surface dilatation, Σ_3, at point **A**, and *(4)* the volumetric strain, $\varDelta$, at point **A**.

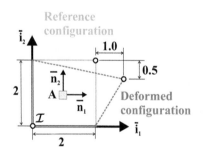

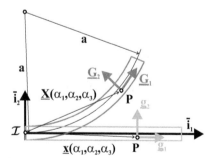

Fig. 15.9. Configuration of the elastic body.

Fig. 15.10. Configuration of the cantilevered beam.

Problem 15.2. Deformed cantilevered beam

Figure 15.10 shows a cantilevered beam of length a and depth h in its reference and deformed configurations. The position vector in the reference configuration is $\underline{x} = \alpha_1 \bar{\imath}_1 + \alpha_2 \bar{\imath}_2 + \alpha_3 \bar{\imath}_3$, and in the deformed configuration

$$\underline{X} = \left\{ (a - \alpha_2) \sin \left[(1 + \varDelta) \frac{\alpha_1}{a} \right] \right\} \bar{\imath}_1 + \left\{ a - (a - \alpha_2) \cos \left[(1 + \varDelta) \frac{\alpha_1}{a} \right] \right\} \bar{\imath}_2 + \alpha_3 \bar{\imath}_3,$$

where $\varDelta$ is a strain measure. (1) Find the base vectors in the reference and deformed configurations. (2) Determine the deformation gradient tensor, (3) the metric tensor, (4) the Green-Lagrange strain tensor, (5) the physical strain components, and (6) the volume dilatation at points **A** (located at $\alpha_1 = a/2, \alpha_2 = h/2, \alpha_3 = 0$) and **B** (located at $\alpha_1 = a/2, \alpha_2 = -h/2, \alpha_3 = 0$). (7) Evaluate the surface dilatations Σ_1, Σ_2, and Σ_3 at points **A** and **B**.

15.5 The formulation of small strain problems

At the heart of the formulation of problems involving small strain is the decomposition of the deformation gradient tensor into rigid body motion and deformation. In section 15.5.1, the decomposition of the deformation gradient tensor is described. This leads to a modified principle of virtual work. The implications of the small strain assumption are then discussed in detail in section 15.5.2.

15.5.1 Decomposition of the deformation gradient tensor

Fig. 15.11 depicts the base vectors at a material point of a deformable body in the reference and deformed configurations. The analysis of the metric tensor presented in section 15.3.3 demonstrates that the base vectors in the deformed configuration do not form an orthogonal basis. Indeed, eqs. (15.43) show that these base vectors are not unit vectors, nor are they mutually orthogonal.

Consider an orthonormal basis of arbitrary orientation denoted $\mathcal{J} = (\bar{\jmath}_1, \bar{\jmath}_2, \bar{\jmath}_3)$. The position vector of a material point in the deformed configuration is now resolved in this basis as

$$\underline{X} = X_i^* \bar{\jmath}_i. \tag{15.62}$$

The base vectors in the deformed configuration then become

$$\underline{G}_i = \frac{\partial \underline{X}}{\partial \alpha_i} = \frac{\partial X_j^*}{\partial \alpha_i} \bar{\jmath}_j = \hat{F}_{ji} \bar{\jmath}_j, \tag{15.63}$$

where the following modified deformation gradient tensor was defined

$$\hat{F}_{ij} = \frac{\partial X_i^*}{\partial \alpha_j}. \tag{15.64}$$

A scalar product of eq. (15.63) by $\bar{\jmath}_l$ yields an alternative definition of this tensor

$$\hat{F}_{ij} = \bar{\jmath}_i^T \underline{G}_j. \tag{15.65}$$

The two deformation gradient tensors defined in eqs. (15.38) and (15.65) can be related by equating the two expressions for the base vectors in the deformed configuration, eqs. (15.37) and (15.63), to find $\underline{G}_i = \hat{F}_{ji}\,\bar{\jmath}_j = F_{ji}\,\bar{g}_j$. A scalar product of this result by $\bar{g}_l$ yields the desired relationship, $F_{ij} = (\bar{g}_i^T\bar{\jmath}_k)\,\hat{F}_{kj}$. Because the base vectors in the reference configuration form the orthonormal basis $\mathcal{I}$, see eq. (15.26), rotation tensor $\underline{R}$ brings basis $\mathcal{I}$ to basis $\mathcal{J}$, and $R_{ij} = \bar{g}_i^T\bar{\jmath}_j$. The relationship between the two deformation gradient tensors now simply becomes

$$F_{ij} = R_{ik}\hat{F}_{kj}. \tag{15.66}$$

This decomposition expresses the deformation gradient tensor as the product of rotation tensor $\underline{R}$, defining a rigid body rotation, by deformation gradient tensor $\hat{\underline{F}}$, defining the deformation of the body at that point.

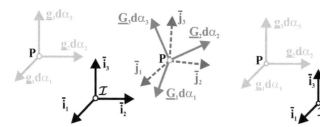

Fig. 15.11. The base vectors in the reference and deformed configurations.

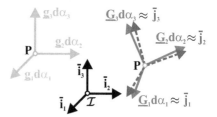

Fig. 15.12. Deformation of a differential element.

15.5.2 The small strain assumption

Consider the deformation of a differential element of the body depicted in fig. 15.12. The norm of base vector $\underline{G}_1$ in the deformed configuration was found in eq. (15.43a) to be closely related to the engineering strain component, ε_1,

$$\|\underline{G}_1\|^2 = G_{11} = (1 + \varepsilon_1)^2 = 1 + 2\varepsilon_1 + \varepsilon_1^2. \tag{15.67}$$

Similarly, the angular distortion between $\underline{G}_2$ and $\underline{G}_3$ is closely related to the engineering strain component, γ_{23}, see eq. (15.43b),

$$\frac{G_2^T G_3}{\|\underline{G}_2\|\|\underline{G}_3\|} = \sin\gamma_{23}. \tag{15.68}$$

In numerous applications, thin structures such as cables, membranes, beam, plates, and shells undergo finite displacements and rotations while strain components remain very small. The small strain assumption states that *relative elongations and angular distortions are negligible compared to unity*, i.e.,

$$|\varepsilon_1|, |\varepsilon_2|, |\varepsilon_3| \ll 1, \quad |\gamma_{23}|, |\gamma_{13}|, |\gamma_{12}| \ll 1. \tag{15.69}$$

Introducing this approximation in eq. (15.67) and (15.68) yields

$$\|\underline{G}_1\|^2 = G_{11} \approx 1, \qquad \frac{\underline{G}_2^T \underline{G}_3}{\|\underline{G}_2\|\|\underline{G}_3\|} \approx 0.$$

In other words, the base vectors in the deformed configuration approximately define an orthonormal basis because each base vector is approximately of unit length, and they are nearly orthogonal to each other. For small strain problems, orthonormal basis $\mathcal{J}$ defined in section 15.5.1 will be selected to be nearly coincident with the base vectors in the deformed configuration, $i.e.$,

$$\underline{G}_i \approx \bar{\jmath}_i. \tag{15.70}$$

Consequently, basis $\mathcal{J}$ is called the $convected$ or $material$ frame. Introducing this approximation in eq. (15.63) leads to $\underline{G}_i = \hat{F}_{ji}\bar{\jmath}_j \approx \bar{\jmath}_i$. A scalar product of this result by $\bar{\jmath}_p$ then yields $\hat{F}_{ij} \approx \delta_{ij}$. Finally, using eq. (15.66)

$$F_{ij} \approx R_{ij}. \tag{15.71}$$

In other words, when the strain components are very small, the deformation gradient tensor is approximately equal to the finite rotation tensor that brings basis $\mathcal{I}$ to $\mathcal{J}$.

Under the assumption of small strains, it can be shown that the principle of virtual work becomes

$$\int_{t_i}^{t_f} \int_{\mathcal{V}} \left[\tau^{*ij}\delta\gamma_{ij} - \underline{p}^T\delta\underline{v} \right] \, d\mathcal{V}dt = \int_{t_i}^{t_f} \left\{ \int_{\mathcal{V}} \underline{b}^T\delta\underline{u} \, d\mathcal{V} + \int_{\mathcal{S}} \hat{\underline{t}}^T\delta\underline{u} \, d\mathcal{S} \right\} dt$$
$$- \left[\int_{\mathcal{V}} \hat{\underline{p}}^T\delta\underline{u} \, d\mathcal{V} \right]_{t_i}^{t_f}.$$
$$\tag{15.72}$$

In this principle, the strain measures are defined as

$$\gamma_{ij} = \frac{1}{2}(\hat{F}_{ij} + \hat{F}_{ji}) - \delta_{ij}, \tag{15.73}$$

and the stress measures, τ^{*ij}, form the convected Cauchy stress tensor, $i.e.$, the components of the true stress tensor in basis $\mathcal{J}$. This statement of the principle of virtual work will be used in subsequent sections to derive the governing equations of structures undergoing large displacements and rotations but small strains.

Formulation of flexible elements

This chapter deals with the formulations of flexible elements such as flexible joints, cables, beams, and plates and shells, which are presented in sections 16.1, 16.2, 16.3, and 16.4, respectively. In all cases, geometrically exact formulations are derived, *i.e.*, the displacements and rotations of the elements are arbitrarily large, although strain components are assumed remain small, a feature that significantly simplifies the governing equations of motion of these structural components.

16.1 Formulation of flexible joints

Flexible joints, sometimes called bushing elements or force elements, are found in all multibody dynamics codes. In their simplest form, flexible joints consist of sets of three linear and three torsional springs placed between two nodes of a multibody system. For infinitesimal deformations, the selection of the lumped spring constants is an easy task, which can be based on a numerical simulation of the joint or on experimental measurements.

If the joint undergoes finite deformations, identification of its stiffness characteristics is not so simple, specially if the joint is itself a complex system. When finite deformations occur, the definition of deformation measures becomes a critical issue. Indeed, for finite deformation, the observed nonlinear behavior of materials is partly due to material characteristics, and partly due to kinematics.

This section focuses on the determination of the proper finite deformation measures for elastic bodies of *finite dimension*. In contrast, classical strain measures, such as the Green-Lagrange strains presented in section 15.3.4, among many others, characterize finite deformations of *infinitesimal elements of a body*. It is argued that proper finite deformation measures must be of a tensorial nature, *i.e.*, must present specific invariance characteristics. This requirement is satisfied if and only if deformation measures are parallel to the eigenvector of the motion tensor.

Anand [288, 289] has shown that the classical strain energy function for infinitesimal isotropic elasticity is in good agreement with experiment for a wide class of materials for moderately large deformations, provided the infinitesimal strain measure used in the strain energy function is replaced by the Hencky or logarithmic measure of finite strain. This means that the behavior of materials for moderate deformations can be captured accurately using linear constitutive laws, but replacing the infinitesimal strain measures by finite deformation measures that are nonlinear functions of the displacements.

These nonlinear deformation measures capture the observed nonlinear behavior associated with the nonlinear kinematics of the problem. Degener *et al.* [290] also reported similar findings for the torsional behavior of beams subjected to large axial elongation.

Much attention has been devoted to the problem of synthesizing accurate constitutive properties for the modeling of flexible bushings presenting complex, time-dependent rheological behavior [291, 292]. It is worth stressing, however, that the literature seldom addresses three-dimensional joint deformations.

Much like multibody codes, most finite element codes also support the modeling of lumped structural elements. While linear analysis is easily implemented, problems are encountered when dealing with finite displacements and rotations, as pointed out by Masarati and Morandini [293]. Structural analysis codes, either specifically intended for multibody dynamics analysis, like MSC/ADAMS, or for nonlinear finite element codes with multibody capabilities, like Abaqus/Standard, allow arbitrarily large absolute displacements and rotations of the nodes and correctly describe their rigid-body motion. When lumped deformable joints are used, relative displacements and rotations are often required to remain moderate, although not necessarily infinitesimal.

Such restrictions occur when using the FIELD element of MSC/ADAMS, a linear element that implements an orthotropic torsional spring based on a constant, orthotropic constitutive matrix [294]. Similarly, the JOINTC element implemented in Abaqus/Standard, describes the interaction between two nodes when the second node can "displace and rotate slightly with respect to the first node [295]," because its formulation is based on an approximate relative rotation measure.

The formulations and implementations of flexible joints available in research and commercial codes do not appear to allow arbitrarily large relative displacements and rotations. Moreover, in many cases, the ordering sequence of the nodes connected to the joint matters, because the behavior of the flexible joint is biased towards one of the nodes. This problem is known to experienced analysts using these codes. To the authors' knowledge, these facts are rarely acknowledged in the literature. It appears that little effort has been devoted to the elimination of these shortcomings from the formulations found in research and commercially available codes, although the predictions of these codes might be unexpected.

This section presents families of finite deformation measures that can be used to characterize the deformation of flexible joints. These deformation measures are closely related to the tensorial parameterization motion developed in chapter 14. Because they are of a tensorial nature, these deformation measures are intrinsic and invariant. Furthermore, it will be shown that using these strain measures in combination with the linear constitutive laws of the joint enable the accurate prediction of joint behavior under moderate deformation.

16.1.1 Flexible joint configuration

Figure 16.1 shows a flexible joint in its reference and deformed configurations. It consists of a three-dimensional elastic body of finite dimension and of two rigid bodies, called handle k and handle ℓ, that are rigidly connected to the elastic body. In the reference configuration, the configuration of the handles is defined by frame $\mathcal{F}_0 = \left[\mathbf{K} = \mathbf{L}, \mathcal{B}_0 = (\bar{b}_{01}, \bar{b}_{02}, \bar{b}_{03}) \right]$, where $\mathcal{B}_0$ forms an orthonormal basis. Points $\mathbf{K}$ and $\mathbf{L}$ are material points of handles k and ℓ, respectively, with coincident geometric locations.

In the deformed configuration, the two handles move to new positions and the elastic body deforms. Points $\mathbf{K}$ and $\mathbf{L}$ are now at distinct locations; the relative displacement vector of point $\mathbf{L}$ with respect to point $\mathbf{K}$ is denoted $\underline{u}$. The configurations of the two handles are now distinct and two distinct frames, $\mathcal{F}^k = \left[\mathbf{K}, \mathcal{B}^k = (\bar{b}_1^k, \bar{b}_2^k, \bar{b}_3^k) \right]$ and $\mathcal{F}^\ell = \left[\mathbf{L}, \mathcal{B}^\ell = (\bar{b}_1^\ell, \bar{b}_2^\ell, \bar{b}_3^\ell) \right]$, define the configurations of handle k and ℓ, respectively. The relative rotation tensor of basis $\mathcal{B}^\ell$ with respect to basis $\mathcal{B}^k$ is denoted $\underline{\underline{R}}$.

The deformation of the flexible joint stems from applied forces and moments. At point $\mathbf{K}$, the applied force and moment vectors are denoted $\underline{F}_k$ and $\underline{M}_k$, respectively; the corresponding quantities applied at point $\mathbf{L}$ are denoted $\underline{F}_\ell$ and $\underline{M}_\ell$, respectively. The loading applied to the flexible joint is defined in the following manner

$$
\underline{A}_k = \left\{ \begin{matrix} F_k \\ M_k \end{matrix} \right\}, \quad \underline{A}_\ell = \left\{ \begin{matrix} F_\ell \\ M_\ell \end{matrix} \right\}, \quad (16.1)
$$

where $\underline{A}_k$ and $\underline{A}_\ell$ denote the loads applied at points $\mathbf{K}$ and $\mathbf{L}$, respectively. According to Newton's third law, these loads must be in equilibrium, i.e.,

$$
\underline{A}_k = - \left[\begin{matrix} \underline{\underline{I}} & \underline{\underline{0}} \\ \underline{\underline{\tilde{u}}} & \underline{\underline{I}} \end{matrix} \right] \underline{A}_\ell. \quad (16.2)
$$

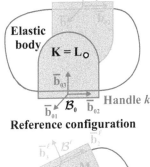

Reference configuration

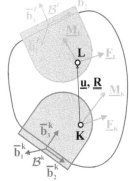

Deformed configuration

Fig. 16.1. Configuration of the flexible joint.

The joint is assumed to be massless, i.e., inertial forces associated with its motion are neglected.

The state of deformation of the elastic body depends on the relative displacement and rotation of the two handles and is unaffected by rigid body motions. Consequently, it is possible to assume that handle k does not move, and the relative displacement and rotation of handle ℓ with respect to handle k then simply becomes its absolute motion, as illustrated in fig. 16.2. This configuration is denoted scenario ℓ. Of course, scenario k could also be defined in a similar manner if the location of handle ℓ is assumed to remain fixed in space.

Consider the differential displacement of point **L** shown in fig 16.2. The components of this differential displacement vector in bases $\mathcal{B}^k$ and $\mathcal{B}^\ell$ are $d\underline{u}^+$ and $\underline{\underline{R}}^T d\underline{u}^+$, respectively. The components of the differential rotation vector of handle ℓ are denoted $d\underline{\psi}^+ = \text{axial}(d\underline{\underline{R}}\,\underline{\underline{R}}^T)$ and $d\underline{\psi}^* = \text{axial}(\underline{\underline{R}}^T d\underline{\underline{R}}) = \underline{\underline{R}}^T d\underline{\psi}^+$ when resolved in the same bases, respectively. The differential motion vector of point **L** is now defined as

$$d\underline{\mathcal{U}}_\ell^* = \left\{ \begin{matrix} \underline{\underline{R}}^T d\underline{u}^+ \\ \underline{\underline{R}}^T d\underline{\psi}^+ \end{matrix} \right\}. \tag{16.3}$$

Superscripts $(\cdot)^+$ and $(\cdot)^*$ indicate tensor components resolved in basis $\mathcal{B}^k$ and $\mathcal{B}^\ell$, respectively.

The differential motion of the point of handle ℓ that instantaneously coincides with the origin of reference frame $\mathcal{F}_0$, denoted $d\underline{\mathcal{U}}_\ell^+$, is found from the following frame change operation

$$d\underline{\mathcal{U}}_\ell^+ = \left\{ \begin{matrix} d\underline{u}^+ + \widetilde{u}^+ d\underline{\psi}^+ \\ d\underline{\psi}^+ \end{matrix} \right\} \tag{16.4}$$
$$= \underline{\underline{C}}(\underline{u}^+, \underline{\underline{R}}^+) d\underline{\mathcal{U}}_\ell^*,$$

where $\underline{\underline{R}}^+ = \underline{\underline{R}}$.

The load externally applied at point **L**, denoted $\underline{\mathcal{A}}_\ell$, was defined in eq. (16.1). These applied force and moment vectors are now resolved in basis $\mathcal{B}^\ell$ to form $\underline{\mathcal{A}}_\ell^{*T} = \{\underline{F}_\ell^{*T}, \underline{M}_\ell^{*T}\}$. The

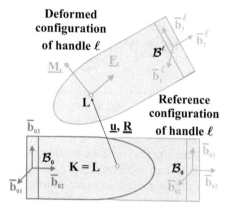

Fig. 16.2. Configuration of the flexible joint for scenario ℓ. For clarity of the figure, the elastic body is not shown.

following change of frame operation is now considered

$$\underline{\mathcal{A}}_\ell^+ = \left\{ \begin{matrix} \underline{F}_\ell^+ \\ \underline{M}_\ell^+ + \widetilde{u}^+ \underline{F}_\ell^+ \end{matrix} \right\} = \underline{\underline{C}}^{-T}(\underline{u}^+, \underline{\underline{R}}^+) \underline{\mathcal{A}}_\ell^*. \tag{16.5}$$

Note the parallel between vector $\underline{\mathcal{A}}_\ell^+$ and the second Piola-Kirchhoff stress tensor [296]. Indeed, $\underline{\mathcal{A}}_\ell^+$ represents the true loads applied to handle ℓ in its deformed configuration, but transferred to the original location of their application point in the reference configuration. Loads $\underline{\mathcal{A}}_\ell^+$ and $\underline{\mathcal{A}}_\ell^*$ form a set of equipollent loads applied to handle ℓ. The change of frame operation described by eq. (16.5), expresses, in fact, a condition of equipollence.

16.1.2 Flexible joint differential work

The differential work, dW, done by the forces applied to the joint is

$$dW = \underline{F}_\ell^{+T} d\underline{u}^+ + \underline{M}_\ell^{+T} d\underline{\psi}^+ = \underline{\mathcal{A}}_\ell^{*T} d\underline{\mathcal{U}}_\ell^* = \underline{\mathcal{A}}_\ell^{+T} d\underline{\mathcal{U}}_\ell^+, \tag{16.6}$$

where the last two equalities follow from eqs. (5.57) and (5.61), respectively. Because handle k does not move, the forces and moments applied at point $\mathbf{K}$ do not work.

Let $\underline{\mathcal{E}}_\ell^+$ be a set of six generalized coordinates that uniquely define the configuration of handle ℓ, *i.e.*, a one-to-one mapping is assumed to exist between these generalized coordinates and the configuration of handle ℓ. It then follows that a one-to-one mapping must exist between the handle's differential motion and differentials of the generalized coordinates

$$\underline{\mathrm{d}\mathcal{U}_\ell^*} = \underline{\underline{\mathcal{H}}}^*(\underline{\mathcal{E}}_\ell^+)\mathrm{d}\underline{\mathcal{E}}_\ell^+, \quad \underline{\mathrm{d}\mathcal{U}_\ell^+} = \underline{\underline{\mathcal{H}}}(\underline{\mathcal{E}}_\ell^+)\mathrm{d}\underline{\mathcal{E}}_\ell^+. \tag{16.7}$$

Matrix $\underline{\underline{\mathcal{H}}}(\underline{\mathcal{E}}_\ell^+)$ is the Jacobian matrix or tangent operator of the coordinate transformation.

The differential work done by the forces applied to the joint, eq. (16.6), now becomes

$$\mathrm{d}W = \underline{\mathcal{A}}_\ell^{*T}\underline{\underline{\mathcal{H}}}^*(\underline{\mathcal{E}}_\ell^+)\mathrm{d}\underline{\mathcal{E}}_\ell^+ = \underline{\mathcal{A}}_\ell^{+T}\underline{\underline{\mathcal{H}}}(\underline{\mathcal{E}}_\ell^+)\mathrm{d}\underline{\mathcal{E}}_\ell^+ = \underline{\mathcal{L}}_\ell^{+T}\mathrm{d}\underline{\mathcal{E}}_\ell^+, \tag{16.8}$$

where the generalized forces associated with the generalized coordinates are defined as

$$\underline{\mathcal{L}}_\ell^+ = \underline{\underline{\mathcal{H}}}^{*T}(\underline{\mathcal{E}}_\ell^+)\underline{\mathcal{A}}_\ell^* = \underline{\underline{\mathcal{H}}}^T(\underline{\mathcal{E}}_\ell^+)\underline{\mathcal{A}}_\ell^+. \tag{16.9}$$

It is now assumed that the flexible joint is made of an elastic material [285], which implies that the generalized forces can be derived from a potential, the strain energy of the joint, denoted A,

$$\underline{\mathcal{L}}_\ell^+ = \frac{\partial A(\underline{\mathcal{E}}_\ell^+)}{\partial \underline{\mathcal{E}}_\ell^+}. \tag{16.10}$$

The differential work now becomes

$$\mathrm{d}W = \mathrm{d}\underline{\mathcal{E}}_\ell^{+T} \frac{\partial A(\underline{\mathcal{E}}_\ell^+)}{\partial \underline{\mathcal{E}}_\ell^+} = \mathrm{d}(A), \tag{16.11}$$

and can be expressed as the differential of a scalar function, the strain energy.

The reasoning presented in this section could be repeated for scenario k. Because scenarios k and ℓ only differ by a rigid body motion, identical results should be obtained. In particular, the differential work for the two scenarios should be identical, leading to $\mathrm{d}W = \underline{\mathcal{A}}_\ell^{+T}\underline{\mathrm{d}\mathcal{U}_\ell^+} = \underline{\mathcal{A}}_k^{+T}\underline{\mathrm{d}\mathcal{U}_k^+}$. Loading $\underline{\mathcal{A}}_\ell^+$ and $\underline{\mathcal{A}}_k^+$ are referred to the same point, the origin of frame $\mathcal{F}_0$, and expressed in the same basis, $\mathcal{B}_0$; Newton's first law then implies $\underline{\mathcal{A}}_\ell^+ + \underline{\mathcal{A}}_k^+ = 0$, leading to the intuitive result that

$$\underline{\mathrm{d}\mathcal{U}_k^+} = -\underline{\mathrm{d}\mathcal{U}_\ell^+}. \tag{16.12}$$

16.1.3 The deformation measures

In the previous section, quantities $\underline{\mathcal{E}}_\ell^+$ were defined as "a set of generalized coordinates that uniquely define the configuration of handle ℓ," but were otherwise left undefined. For scenario ℓ, the configuration of handle ℓ defines the deformation of the elastic body, and hence, these generalized coordinates are, in fact, deformation measures for the flexible joint. The following notation is introduced

$$\underline{\mathcal{E}}_\ell^+ = \left\{ \begin{matrix} \underline{\epsilon}^+ \\ \underline{\kappa}^+ \end{matrix} \right\}. \tag{16.13}$$

The first three components of this array form the stretch vector, denoted $\underline{\epsilon}$, and the last three the wryness vector, denoted $\underline{\kappa}$. Both quantities are assumed to form first-order tensors.

Because the deformation measures uniquely define the configuration of handle ℓ relative to handle k, the motion tensor, $\underline{\underline{C}}(\underline{u}, \underline{\underline{R}})$, can be expressed as $\underline{\underline{C}} = \underline{\underline{C}}(\underline{\mathcal{E}}_\ell^+)$. It follows that the deformation measures form a parameterization of the motion tensor. In general, the deformation measures are nonlinear functions of six quantities, the three components of the relative displacement vector, $\underline{u}$, and the three parameters that define the relative rotation tensor, $\underline{\underline{R}}$.

For instance, the stretch vector could be selected as the position vector of point $\mathbf{L}$, $\underline{\epsilon}^+ = \underline{u}^+$; note that $\underline{\epsilon}^* = \underline{\underline{R}}^T \underline{\epsilon}^+$, as expected from the tensorial nature of the stretch vector. The Euler angles associated with rotation tensor $\underline{\underline{R}}$ form a valid set of generalized coordinates to characterize the angular motion of handle ℓ, but cannot be the components of the wryness vector because Euler angles do not form the components of a vector. Any vectorial parameterization of rotation, see section 13.4, is a suitable choice for the wryness vector.

16.1.4 Change of reference frame

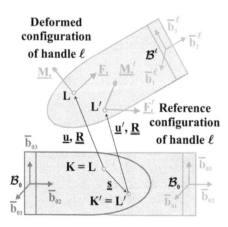

Fig. 16.3. Configuration of the flexible joint for scenario ℓ. For clarity of the figure, the elastic body is not shown.

The results derived in section 16.1.2 were based on the selection of an arbitrary reference frame, $\mathcal{F}_0$, defined by coincident reference points, $\mathbf{K} = \mathbf{L}$, which are material points of handles k and ℓ, respectively, and an orthonormal basis, $\mathcal{B}_0$. Another reference frame could have been selected, $\mathcal{F}_0' = [\mathbf{K}' = \mathbf{L}', \mathcal{B}_0' = (\bar{b}_{01}', \bar{b}_{02}', \bar{b}_{03}')]$; figure 16.3 shows the new reference points, $\mathbf{K}' = \mathbf{L}'$, which are material points of handles k and ℓ, respectively. For clarity, the new basis, $\mathcal{B}_0'$, is not shown on the figure. The position vector of point $\mathbf{K}'$ with respect to point $\mathbf{K}$ is denoted $\underline{s}$ and the relative rotation tensor of basis $\mathcal{B}_0'$ with respect to basis $\mathcal{B}_0$ is denoted $\underline{\underline{S}}$. The motion tensor that brings frame to $\mathcal{F}_0$ to $\mathcal{F}_0'$ is denoted $\underline{\underline{C}}'(\underline{s}, \underline{\underline{S}})$.

The development presented in section 16.1.2 could now be repeated for this new choice of basis and reference points, leading to a new set of applied loads, $\underline{A}_\ell'^+$, deformation measures, $\underline{\mathcal{E}}_\ell'^+$, tangent operator, $\underline{\underline{\mathcal{H}}}(\underline{\mathcal{E}}_\ell'^+)$, and associated generalized forces, $\underline{\mathcal{L}}_\ell'^+$.

If the same problem is treated with scenario ℓ in the two different frames, the sets of loads applied at points $\mathbf{L}$ and $\mathbf{L}'$ must be *equipollent*, which implies

$$\underline{A}_\ell^+ = \underline{\underline{C}}'^{-T}(\underline{s}, \underline{S})\underline{A}_\ell'^+. \tag{16.14}$$

This equation expresses the relationship between the components of the loading vector in the two frames, $\mathcal{F}_0$ to $\mathcal{F}_0'$, *i.e.*, the equipollence condition implies that the loading vector is a first-order tensor, see eq. (16.5).

To be physically meaningful, the corresponding deformation measures must also be invariant with respect to a change of frame, *i.e.*, they must also be first-order tensors, and their components in two frames, $\mathcal{F}_0$ and $\mathcal{F}_0'$, denoted $\underline{\mathcal{E}}_\ell^+$ and $\underline{\mathcal{E}}_\ell'^+$, respectively, must transform according to the rules of transformation for kinematic quantities given by eq. (16.4), *i.e.*,

$$\underline{\mathcal{E}}_\ell^+ = \underline{\underline{C}}'(\underline{s}, \underline{S})\underline{\mathcal{E}}_\ell'^+. \tag{16.15}$$

This equation expresses the desired invariance of the deformation measure.

The motion tensor is a second-order tensor and the deformation measure is a parameterization of this motion tensor. According to eq. (16.15), this parameterization must be a first-order tensor. This implies that the deformation measure must be a tensorial parameterization of motion.

16.1.5 Deformation measure invariance

Equation (16.6) expresses the invariance of the differential work with respect to a change of frame. The equipollence condition of the applied load is expressed by eq. (16.14) and introducing this condition into eq. (16.6) yields

$$\mathrm{d}\underline{\mathcal{U}}_\ell^+ = \underline{\underline{C}}'(\underline{s}, \underline{S})\mathrm{d}\underline{\mathcal{U}}_\ell'^+. \tag{16.16}$$

The equipollence of the applied load and invariance of the differential work imply that the components of the loading vector transform according to the first-order tensor transformation rule expressed by eq. (16.5) for loading quantities and the components of the differential displacement vector according to that expressed by eq. (16.4) for kinematic quantities.

Introducing eq. (16.16) into eq. (16.7) and pre-multiplying by $\underline{\underline{C}}'^{-1}$ yields $\mathrm{d}\underline{\mathcal{U}}_\ell'^+ = \underline{\underline{C}}'^{-1}\underline{\underline{\mathcal{H}}}(\underline{\mathcal{E}}_\ell^+)\underline{\underline{C}}'\underline{\underline{C}}'^{-1}\mathrm{d}\underline{\mathcal{E}}_\ell^+$, where $\underline{\underline{C}}'\underline{\underline{C}}'^{-1} = \underline{\underline{I}}$. Introducing eq. (14.59) then yields $\mathrm{d}\underline{\mathcal{U}}_\ell'^+ = \underline{\underline{\mathcal{H}}}(\underline{\mathcal{E}}_\ell'^+)\underline{\underline{C}}'^{-1}\mathrm{d}\underline{\mathcal{E}}_\ell^+$, which leads to the expected transformation rule for the components of the differential deformation measure

$$\mathrm{d}\underline{\mathcal{E}}_\ell^+ = \underline{\underline{C}}'\mathrm{d}\underline{\mathcal{E}}_\ell'^+. \tag{16.17}$$

The invariance of the differential work written in the form of eq. (16.8) requires $\mathrm{d}W = \underline{\mathcal{L}}_\ell^{+T}\mathrm{d}\underline{\mathcal{E}}_\ell^+ = \underline{\mathcal{L}}_\ell'^{+T}\mathrm{d}\underline{\mathcal{E}}_\ell'^+$. Introducing eq. (16.17) then yields

$$\underline{\mathcal{L}}_\ell^+ = \underline{\underline{C}}'^{-T} \underline{\mathcal{L}}_\ell'^+. \tag{16.18}$$

In summary, the formulation developed in section 16.1.2, is frame invariant. Under a change of frame, the components of the applied and generalized loads transform according to eqs. (16.14) and (16.18), respectively. The components of the deformation measure, differential displacement, and differential deformation measure transform according to eqs. (16.15), (16.16), and (16.17), respectively. These energetically conjugate first-order tensors present different transformation rules under a change of frame to guarantee the required invariance of the differential work.

The invariance of the various quantities involved in the formulation stems from the tensorial nature of the deformation measure. Because this measure is selected to be the tensorial parameterization of motion, it must be an eigenvector of the motion tensor, i.e., $\underline{\mathcal{E}}_\ell^+ = \underline{\underline{C}}\,\underline{\mathcal{E}}_\ell^+$. Since the deformation measure is a kinematic quantity, $\underline{\mathcal{E}}_\ell^+ = \underline{\underline{C}}\,\underline{\mathcal{E}}_\ell^*$, and it follows that $\underline{\mathcal{E}}_\ell^+ = \underline{\mathcal{E}}_\ell^*$, i.e., the components of the deformation measure are identical in frames $\mathcal{F}_0$ and $\mathcal{F}^\ell$. This implies that the deformation measure is identical when viewed by observers in frames $\mathcal{F}_0$ or $\mathcal{F}^\ell$. Consequently, the deformation measure is not biased towards one of the nodes of the joint, a shortcoming of many of the formulation presently implemented in research and commercial codes.

Equation (16.12) implies $\underline{\mathcal{E}}_k^+ = -\underline{\mathcal{E}}_\ell^+$, which simply corresponds to a sign convention. Henceforth, notation $\underline{\mathcal{E}} = \underline{\mathcal{E}}_\ell^+ = \underline{\mathcal{E}}_\ell^*$ is used, which emphasizes the intrinsic nature of the deformation measure; of course, a change of sign is required for scenario k. Finally, eq. (16.10) implies $\underline{\mathcal{L}} = \underline{\mathcal{L}}_\ell^+ = \underline{\mathcal{L}}_\ell^*$, which shows the intrinsic nature of the generalized forces; here again, a change of sign is required for scenario k.

The proposed deformation measures are parallel to the eigenvector of the motion tensor associated with its unit eigenvalue. Because this eigenvalue has a multiplicity of two, two linearly independent eigenvectors exist, and the deformation measure is a linear combination of these two eigenvectors. An explicit expression of the deformation measure, see eq. (14.37), is

$$\underline{\mathcal{E}} = \left\{ \begin{matrix} \underline{\epsilon} \\ \underline{\kappa} \end{matrix} \right\} = \left\{ \begin{matrix} \underline{\underline{D}}(\underline{\kappa})\underline{u} \\ \underline{\kappa} \end{matrix} \right\}, \tag{16.19}$$

where the stretch vector, $\underline{\epsilon}$, is related to the displacement vector, $\underline{u}$, of the handle, the wryness vector, $\underline{\kappa}$, is the vectorial parameterization of rotation, and tensor $\underline{\underline{D}}$ is defined by eq. (14.38).

16.1.6 Flexible joint constitutive laws

The strain energy of the flexible joint is assumed to be a quadratic function of the deformation measures, $A = 1/2\,\underline{\mathcal{E}}^T \underline{\underline{K}}\,\underline{\mathcal{E}}$, where $\underline{\underline{K}}$ is the joint's stiffness matrix for infinitesimal deformations. The generalized forces now become $\underline{\mathcal{L}} = \underline{\underline{K}}\,\underline{\mathcal{E}}$, and eq. (16.9) then yields

$$\underline{A}_\ell^+ = \underline{\underline{H}}^{-T}(\underline{\mathcal{E}})\underline{\underline{K}}\,\underline{\mathcal{E}}, \tag{16.20a}$$

$$\underline{A}_\ell^* = \underline{\underline{H}}^{*-T}(\underline{\mathcal{E}})\underline{\underline{K}}\,\underline{\mathcal{E}}. \tag{16.20b}$$

Due to the presence of the tangent tensor, the load-deformation relationships are non-linear, and the deformation-displacement relationships, eqs. (16.19), are also nonlinear.

The loads applied to handle ℓ resolved in basis $\mathcal{B}_0$, denoted $\underline{\mathcal{A}}$, are $\underline{\mathcal{A}} = \underline{\underline{T}}^T \underline{\mathcal{A}}_\ell^+ = \underline{\underline{R}}\,\underline{\mathcal{A}}_\ell^*$. The joint's constitutive laws now become

$$
\underline{\mathcal{A}} = \left[\begin{array}{cc} \underline{\underline{F}}^{-T}(\underline{\kappa}) & \underline{\underline{0}} \\ -\underline{\underline{H}}^{-T}(\underline{\kappa})\underline{\underline{L}}^T(\underline{\epsilon},\underline{\kappa})\underline{\underline{F}}^{-T}(\underline{\kappa}) & \underline{\underline{H}}^{-T}(\underline{\kappa}) \end{array} \right] \underline{\underline{K}}\,\underline{\mathcal{E}}. \tag{16.21}
$$

where tensors $\underline{\underline{F}}$, $\underline{\underline{H}}$, and $\underline{\underline{L}}$ are defined by eqs. (14.41), (13.55), and (14.46), respectively.

Finally, inversion of this equation gives the constitutive laws in compliance form as

$$
\underline{\mathcal{E}} = \underline{\underline{S}} \left[\begin{array}{cc} \underline{\underline{F}}^T(\underline{\kappa}) & \underline{\underline{0}} \\ \underline{\underline{L}}^T(\underline{\epsilon},\underline{\kappa}) & \underline{\underline{H}}^T(\underline{\kappa}) \end{array} \right] \underline{\mathcal{A}}, \tag{16.22}
$$

where $\underline{\underline{S}} = \underline{\underline{K}}^{-1}$ is the compliance matrix for infinitesimal deformations. Given the externally applied loads, $\underline{\mathcal{A}}$, this nonlinear equation yield the joints deformations, in terms of the stretch vector, $\underline{\epsilon}$, and the wryness vector, $\underline{\kappa}$.

This section has focused on the definition of appropriate deformation measures for elastic bodies of finite dimension, in contrast with classical strain measures that are defined for infinitesimal elements of an elastic body. It was first argued that to be physically meaningful, these deformation measures must be of a tensorial nature. Next, it was proved that this requirement is satisfied if and only if the deformation measures are parallel to the eigenvector of the motion tensor associated with its unit eigenvalue.

Equipped with these deformation measures, constitutive laws for the flexible joint were derived by assuming the existence of a strain energy function that is a quadratic form of these deformation measures. Because all the quantities involved in the formulation are of a tensorial nature, the behavior of the joint presents the required invariance with respect to changes of basis or reference point. Furthermore, the proposed strain measures are unbiased. Flexible joint formulations described in the literature up to date do not appear to present these desirable characteristics.

Example 16.1. Simple beam treated as a flexible joint
The load-deformation and deformation-configuration relationships developed above will be tested on a number of simple examples involving a flexible beam. Figure 16.4 shows the beam of length L along unit vector $\bar{b}_{01}$, width b along $\bar{b}_{02}$, and height h along $\bar{b}_{03}$. The beam is made of a homogeneous material of Young's modulus E and shear modulus G. The examples presented below use the following data: $L = 0.6$ m, $b = 5$ mm, $h = 15$ mm, $E = 73$ GPa, and $G = E/(2(1+\nu))$, where $\nu = 0.3$.

Handles k and ℓ are rigidly attached to the root and tip of the beam, respectively. Elementary structural analysis [285] yields the compliance matrix of the joint

$$\underline{\underline{S}} = \begin{bmatrix} L/S & 0 & 0 & 0 & 0 & 0 \\ 0 & L^3/3H_{33} & 0 & 0 & 0 & L^2/2H_{33} \\ 0 & 0 & L^3/3H_{22} & 0 & -L^2/2H_{22} & 0 \\ 0 & 0 & 0 & L/H_{11} & 0 & 0 \\ 0 & 0 & -L^2/2H_{22} & 0 & L/H_{22} & 0 \\ 0 & L^2/2H_{33} & 0 & 0 & 0 & L/H_{33} \end{bmatrix}, \quad (16.23)$$

where $S = Ebh$, $H_{22} = Ebh^3/12$, $H_{33} = Ehb^3/12$, and $H_{11} = Ghb^3/3$ are the beam's axial stiffness, bending stiffness with respect the unit vector $\bar{\imath}_2$, bending stiffness with respect to the unit vector $\bar{\imath}_3$, and torsional stiffness, respectively.

Various combinations of forces and moments are applied to handle k, and the resulting displacements and rotations are then evaluated using the joint's constitutive laws, eqs. (16.22). These predictions are compared with those of a finite element solution for the geometrically exact beam model presented in section 16.3, which provide an exact treatment of the kinematics of the system, but assume the strains to remain small at all time. This latter assumption is equivalent to assuming a constant compliance matrix, as done here. All the numerical solutions shown below are obtained by modeling the beam with 12 cubic elements, corresponding to a total 216 degrees of freedom.

In the first example, the joint is subjected to a single bending moment about unit vector $\bar{\imath}_3$, denoted M_3. For this simple case, eqs. (16.22) can be solved analytically to yield $\kappa_3(\phi) = \sigma_0(\phi)\bar{M}_3$, where ϕ is the rotation angle of handle ℓ about unit vector $\bar{\imath}_3$ and $\bar{M}_3 = LM_3/H_{33}$. The displacement components of handle ℓ along unit vectors $\bar{\imath}_1$ and $\bar{\imath}_2$ are then $\bar{u}_1 = u_1/L = -(1-\cos\phi)/2$ and $\bar{u}_2 = u_2/L = 1/2 \sin\phi$.

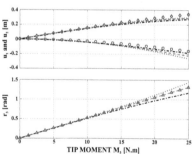

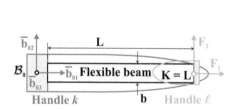

Fig. 16.4. Reference configuration of the flexible beam.

Fig. 16.5. Joint deformation under a single moment. Top figure: displacement components u_1 (o) and u_2 (◇); bottom figure: rotation r_3 (△). Exact solution: symbols. Present solution: $\kappa(\phi) = \phi$, dashed line; $\kappa(\phi) = 4\sin\phi/4$, dotted line; $\kappa(\phi) = 4\tan\phi/4$, dashed-dotted line.

The exact solution of this problem is easily found because the beam deforms into an arc of circle under the single applied moment, leading to $\phi = \bar{M}_3$, $\bar{u}_1 = -(1 - \sin\phi/\phi)/2$, and $\bar{u}_2 = (1 - \cos\phi)/\phi$, see fig. 16.5. Three approximate solutions obtained from the proposed approach for three different generating functions, $\kappa(\phi) = \phi$, $\kappa(\phi) = 4\sin\phi/4$, and $\kappa(\phi) = 4\tan\phi/4$, are also depicted in this figure. For $\kappa(\phi) = \phi$, corresponding to the exponential map of rotation, the proposed approach gives the exact solution of the joint's relative rotation. The transverse displacement of the joint is well captured up to very large displacement magnitudes, $u_2 \approx 0.3$ m, for a beam of length $L = 0.6$ m. The beam's foreshortening, a higher-order nonlinear effect, is also well predicted up to large transverse displacements.

If the joint were made of a nonlinear material, the curvature-relative rotation relationship would become nonlinear, and the generating function could be selected to approximate this numerically or experimentally observed behavior as closely as possible. This will enable the present approach to deal with nonlinear elastic manner in an approximate manner. This effect is apparent in fig 16.5 that depicts the curvature-relative rotation relationship for generating functions $\kappa(\phi) = 4\sin\phi/4$ and $\kappa(\phi) = 4\tan\phi/4$, which give rise to softening or stiffening material behaviors, respectively.

The second example involves the same flexible joint now subjected to two moment components, $M_2 = 3\lambda$ N·m and $M_3 = \lambda$ N·m, acting about unit vectors $\bar{\imath}_2$ and $\bar{\imath}_3$, respectively, where $\lambda \in [0, 12]$ is the loading factor.

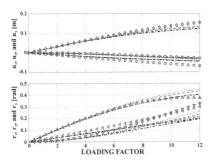 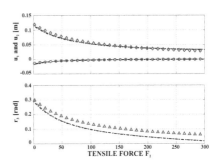

Fig. 16.6. Joint deformation under two moments. Top figure: displacement components u_1 (○), u_2 (◇), and u_3 (□); bottom figure: exponential map components r_1 (▽), r_2 (◁), and r_3 (△). Finite element solution: symbols. Present solution: $\kappa(\phi) = \phi$, dashed line; $\kappa(\phi) = 2\sin\phi/2$, dotted line; $\kappa(\phi) = 2\tan\phi/2$, dashed-dotted line.

Fig. 16.7. Joint deformation under two forces. Top figure: displacement components u_1 (○) and u_2 (◇); bottom figure: exponential map component r_3 (△). Finite element solution: symbols. Present solution: $\kappa(\phi) = \phi$, dashed line; $\kappa(\phi) = 2\sin\phi/2$, dotted line; $\kappa(\phi) = 2\tan\phi/2$, dashed-dotted line.

Figure 16.6 illustrate the ability of the proposed approach to capture the coupled, three-dimensional response of the joint up to large relative displacements and rotations.

In the next example, the joint is subjected to two forces: a constant force $F_2 = 20$ N and a linearly increasing tensile force $F_1 \in [0, 300]$ N, acting along unit vectors $\bar{\imath}_2$ and $\bar{\imath}_1$, respectively. Under the effect of the tensile force, the joint stiffens and the displacement component u_2 resulting from the constant force component F_2 decreases, as shown in fig. 16.7. Here again, the predictions of the proposed approach are found to be in qualitative agreement with the finite element solution.

The stiffening of the joint under a tensile force is a nonlinear effect that is captured by the proposed approach because the equilibrium equations of the joint are expressed in the deformed configuration of the system. This prompts the following question: is the proposed formulation able to predict the instability of the joint under compressive load?

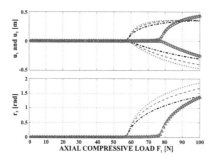

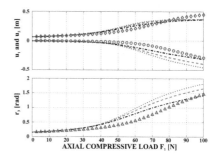

Fig. 16.8. Joint deformation under compressive force. Top figure: displacement components u_1 (o) and u_2 (◇); bottom figure: exponential map component r_3 (△). Finite element solution: symbols. Present solution: $\kappa(\phi) = \phi$, dashed line; $\kappa(\phi) = 4 \sin \phi/4$, dotted line; $\kappa(\phi) = 4 \tan \phi/4$, dashed-dotted line.

Fig. 16.9. Joint deformation under two forces. Top figure: displacement components u_1 (o) and u_2 (◇); bottom figure: exponential map component r_3 (△). Finite element solution: symbols. Present solution: $\kappa(\phi) = \phi$, dashed line; $\kappa(\phi) = 2 \sin \phi/2$, dotted line; $\kappa(\phi) = 2 \tan \phi/2$, dashed-dotted line.

Figure 16.8 shows the response of the system subjected to a small, constant load $F_2 = 0.1$ N and a linearly increasing compressive load, $F_1 \in [0, 100]$ N. The Euler buckling load of the beam [285] is $P_{\text{Euler}} = \pi^2 H_{33}/(4L^2) = 78$ N, which is accurately predicted by the finite element model. The present model also predicts the buckling phenomenon, although for a lower compressive load of about 60 N. The inaccurate prediction of the present model is due to the fact that it uses 6 degrees of freedom only, in contrast with the 216 degrees of freedom used in the reference solution. Modeling the problem with a single two-node beam element also results in an inaccurate prediction of the buckling load, which is over-predicted by about 50%.

It is also possible to trace the post-buckling path of the system. If a constant load $F_2 = 10$ N and a compressive load $F_1 \in [0, 100]$ N are applied to the joint, it quickly enters the post-buckling regime, as depicted in fig. 16.9. The proposed model traces the post-buckling path for up to very large displacements and rotations: for a compressive load of 100 N, the relative rotation of the joint is of about 180 degrees.

All the predictions presented in this example are in good qualitative agreement with exact solutions for geometrically exact beams obtained from nonlinear finite element simulations, up to very large relative displacements and rotations of the flexible joint. For small to moderate displacements and rotations, the agreement between the predictions of the proposed formulation and exact solutions is accurate.

It must be emphasized that the present formulation only "knows" the linearized compliance matrix of the joint. The nonlinear governing equations of geometrically exact beams are not derived. Yet, the proposed deformation measures used in conjunction with the linearized compliance matrix provide constitutive laws for the flexible joint that qualitatively describe its behavior up to large relative displacements and rotations. Instabilities, such as buckling under large compressive load or lateral buckling under large transverse loads (not shown here for brevity sake) are also predicted by the proposed formulation. For small displacements and rotations, accurate predictions are obtained.

While the proposed deformation measures remain tensorial for deformations of arbitrary magnitude, nonlinear constitutive laws should be used if the joint undergoes large deformations. The numerical examples presented in this example use linear constitutive laws to model a joint consisting of a simple flexible beam. The behavior joint is accurately predicted for small and moderate deformations and the correct qualitative behavior for up to very large displacements and rotations is observed.

16.2 Formulation of cable equations

Cables are one-dimensional, flexible structures that can only carry axial forces, *i.e.*, forces acting in the direction tangent to the cable. In contrast with beams, described in section 16.3, cables present no bending, torsional, or shearing stiffness. The kinematic description of cable structures in presented in section 16.2.1 and leads to the definition of the strain components in section 16.2.2. The governing equations for the static behavior of elastic cables are derived in section 16.2.3 and section 16.2.4 extends the formulation to dynamics problems.

16.2.1 The kinematics of the problem

Figure 16.10 shows a flexible cable idealized as a curve in space. The reference and deformed configurations of the cable will be described with respect to an inertial reference frame, $\mathcal{F}^I = [\mathbf{O}, \mathcal{I} = (\bar{\imath}_1, \bar{\imath}_2, \bar{\imath}_3)]$. Material point $\mathbf{P}$ of the cable is defined by its curvilinear coordinate, α_1, which measures length along the reference configuration of the cable.

The position vector of point $\mathbf{P}$ is

$$\underline{x} = \underline{x}(\alpha_1). \tag{16.24}$$

Using eq. (15.26), base vector $\bar{g}_1$ becomes

$$\bar{g}_1 = \frac{\partial \underline{x}}{\partial \alpha_1}.$$
(16.25)

The base vector is the unit tangent to the curve that defines the geometry of the cable in its reference configuration; indeed, as discussed in section 2.2.1, curvilinear variable α_1 represents an intrinsic parameterization of the curve.

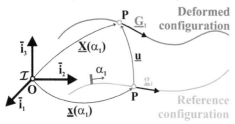

Fig. 16.10. Cable in the reference and deformed configurations.

In the deformed configuration, the position vector of point **P**, denoted $\underline{X}(\alpha_1)$, becomes

$$\underline{X}(\alpha_1) = \underline{x}(\alpha_1) + \underline{u}(\alpha_1),$$
(16.26)

where $\underline{u}$ is the displacement vector of point **P**. The base vector in the deformed configuration becomes

$$\underline{G}_1 = \frac{\partial \underline{X}}{\partial \alpha_1} = \bar{g}_1 + \underline{u}'$$
(16.27)

where notation $(\cdot)'$ indicates a derivative with respect to α_1. Because the cable undergoes axial deformations, material coordinate α_1 no longer measures length along the deformed configuration of the cable; hence, as discussed in section 2.2.2, it represents an arbitrary parameterization of the curve defining the geometry of the cable in its deformed configuration. Base vector $\underline{G}_1$ is tangent to the deformed configuration of the cable, but it is not a unit vector.

Let unit vector $\bar{\jmath}_1$ be parallel to base vector $\underline{G}_1$,

$$\underline{G}_1 = (1 + \bar{e}_{11})\,\bar{\jmath}_1,$$
(16.28)

where $\bar{e}_{11}$ is a strain related parameter which can be expressed in terms of displacements with the help of eqs. (16.27) and (16.28)

$$(1 + \bar{e}_{11})^2 = (\bar{g}_1 + \underline{u}')^T (\bar{g}_1 + \underline{u}').$$
(16.29)

Because the cable is a one dimensional structure, the metric tensor reduces to a single component, $G_{11} = (1 + \bar{e}_{11})^2$. The only non vanishing component of the Green-Lagrange strain tensor, eq. (15.45), is

$$e_{11} = \frac{1}{2}\left[(1 + \bar{e}_{11})^2 - 1\right] = \bar{e}_{11} + \frac{1}{2}\,\bar{e}_{11}^2 = \bar{g}_1^T \underline{u}' + \frac{1}{2}\,\underline{u}'^T \underline{u}',$$
(16.30)

where the strain parameter was expressed in terms of displacement using eq. (16.29).

16.2.2 The small strain assumption

The strain-displacement relation, eq. (16.30), is valid for arbitrarily large displacements and strains. If the strain component can be assumed to remain much smaller that unity, a simplified strain-displacement relationship can be obtained. The modified deformation gradient tensor defined in section 15.5.1 reduces to a single component, $\hat{F}_{11}$, obtained from eqs. (15.65) and (16.28) as

$$\hat{F}_{11} = 1 + \bar{e}_{11}. \tag{16.31}$$

The small strain measure then follows from eq. (15.73)

$$\gamma_{11} = \bar{e}_{11} \approx \bar{g}_1^T \underline{u}' + \frac{1}{2} \underline{u}'^T \underline{u}'. \tag{16.32}$$

The small strain assumption was used to approximate eq. (16.30), $e_{11} = \bar{e}_{11} + \bar{e}_{11}^2/2 \approx \bar{e}_{11}$, leading to the second equality of eq. (16.32). When the strains are small, it is clear that the strain parameter, $\bar{e}_{11}$, is equal to the axial strain in the cable, γ_{11}. Variation of the small strain measure is

$$\delta\gamma_{11} = \delta\underline{u}'^T \left(\bar{g}_1 + \underline{u}' \right) = \delta\underline{u}'^T \underline{G}_1. \tag{16.33}$$

16.2.3 Governing equations

The governing equations of the static problem are readily obtained from the principle of virtual work, eq. (15.72), which states

$$\int_0^L \int_{\mathcal{A}} \tau^{*11} \delta\gamma_{11} \, d\mathcal{A} d\alpha_1 = \delta W_{\text{ext}}, \tag{16.34}$$

for all arbitrary virtual displacements. The length of the cable in the reference configuration is denoted L, $\mathcal{A}$ is its cross-section area, and δW_{ext} the virtual work done by the externally applied loads. Integrating the left-hand side over the cross-sectional area of the cable yields

$$\int_0^L F^* \delta\gamma_{11} \, d\alpha_1 = \delta W_{\text{ext}}, \tag{16.35}$$

where $F^* = \int_{\mathcal{A}} \tau^{*11} \, d\mathcal{A}$ is the total axial force in the cable along material axis $\bar{\jmath}_1$.

The virtual work done by the forces externally applied to the cable is expressed as $\delta W_{\text{ext}} = \int_0^L \delta\underline{u}^T \underline{f} \, d\alpha_1$, where $\underline{f}$ is the externally applied load per unit length of the cable's reference configuration. Introducing the strain variation, eq. (16.33), into eq. (16.35) then leads to

$$\int_0^L \delta\underline{u}'^T F^* \underline{G}_1 \, d\alpha_1 = \int_0^L \delta\underline{u}^T \underline{f} \, d\alpha_1. \tag{16.36}$$

Integration by parts then yields the governing equations of the problem,

$$[F^* \underline{G}_1]' = -\underline{f}. \tag{16.37}$$

If the cable is assumed to present a linear elastic behavior, the constitutive law simply states the proportionality of the axial force to the axial strain,

$$F^* = S\gamma_{11}, \tag{16.38}$$

where S is the axial stiffness of the cable. Introducing the constitutive law into the governing equation, eq. (16.37),

$$\left[S\underline{u}'^T (\bar{g}_1 + \underline{u}'/2)\underline{G}_1 \right]' = -\underline{f}. \tag{16.39}$$

When written in this form, the high level of nonlinearity of the equation governing the cable's displacement field is apparent.

16.2.4 Extension to dynamic problems

The formulation presented thus far has focused on static problems. If the cable's configuration changes in time, the inertial velocity of a material point of the cable is $\underline{v} = \underline{\dot{u}}$. The cable's total kinetic energy is then

$$K = \frac{1}{2} \int_0^L \int_{\mathcal{A}} \rho \underline{\dot{u}}^T \underline{\dot{u}} \, \mathrm{d}\mathcal{A} \mathrm{d}\alpha_1 = \frac{1}{2} \int_0^L m \underline{\dot{u}}^T \underline{\dot{u}} \, \mathrm{d}\alpha_1, \tag{16.40}$$

where ρ is the cable's mass density, and $m = \int_{\mathcal{A}} \rho \, \mathrm{d}\mathcal{A}$ its mass per unit span in the reference configuration.

Variation of the kinetic energy is

$$\delta K = \int_0^L \delta \underline{\dot{u}}^T m \underline{\dot{u}} \, \mathrm{d}\alpha_1 = \int_0^L \delta \underline{\dot{u}}^T \underline{p} \, \mathrm{d}\alpha_1, \tag{16.41}$$

where $\underline{p} = m\underline{\dot{u}}$ is the momentum vector. Hamilton's principle now yields the equations of motion of the problem

$$m\underline{\ddot{u}} - [F^* \underline{G}_1]' = \underline{f}. \tag{16.42}$$

These equations of motion are valid for arbitrarily large displacements of the cable when the strain components are assumed to remain small.

16.2.5 Problems

Problem 16.1. Linear elastic cable
Consider a cable with a linear elastic constitutive law: $F^* = S\gamma_{11}$, where S is the axial stiffness of the cable. The cable is unloaded. Prove: (1) the preservation the total linear momentum of the cable; (2) the preservation the total angular momentum of the cable; (3) the preservation the total mechanical energy of the cable. If the cable is subjected to distributed external loads $\underline{f}(\alpha_1, t)$ and end forces $F_1(t)$ and $F_2(t)$ at $\alpha_1 = 0$ and L, respectively, what happens to the above three preservation laws?

Problem 16.2. Pre-stretched cable
Consider a straight, pre-stretched cable of length L. The constitutive law for the cable is $F^* = S(\bar{e} + \gamma_{11})$, where $\bar{e}$ is the pre-stretch, S the axial stiffness, and hence, $T = S\bar{e}$ the pre-tension in the cable. Linearize the governing equations by assuming displacement field to remain small. Find the equilibrium configuration of the cable under a uniform transverse loading f_0. For the unloaded cable under pre-tension find the natural frequencies and mode shapes of the system.

16.3 Formulation of beam equations

A beam is defined as a structure having one of its dimensions much larger than the other two. The axis of the beam is defined along that longer dimension and its cross-section is normal to this axis. The cross-section's geometric and physical properties are assumed to vary smoothly along the beam's span. Civil engineering structures often consist of assemblies or grids of beams with cross-sections having shapes such as T's or I's. A large number of machine parts also are beam-like structures: linkages, transmission shafts, robotic arms, etc. Aeronautical structures such as aircraft wings or helicopter rotor blades are often treated as thin-walled beams. Finally, both tower and rotor blades of wind turbines also fall within the category of beams structures.

The solid mechanics theory of beams, more commonly referred to simply as "beam theory," plays an important role in structural analysis because it provides designers with simple tools to analyze numerous structures [285]. Within the framework of multibody dynamics, the governing equations for beam structures are nonlinear partial differential equations, and the finite element method is often used to obtain approximate numerical solutions of these equations. Of course, the same finite element approach could also be used to model the same structures based on plate and shell, or even three-dimensional elasticity models, but at a much higher computation cost. Beam models are often used at a pre-design stage because they provide valuable insight into the behavior of structures.

Several beam theories have been developed based on various assumptions, and lead to different levels of accuracy. One of the simplest and most useful of these theories is due to Euler who analyzed the elastic deformation of a slender beam, a problem known as Euler's Elastica [297]. Euler-Bernoulli beam theory [285] is now commonly used in many civil, mechanical and aerospace applications, although shear deformable beam theories [298, 299], often called "Timoshenko beams," have also found wide acceptance. Reissner investigated beam theory for large strains [300] and large displacements of spatially curved members [301, 302].

In this section, the *geometrically exact beam theory* will be presented. The kinematic description of the problem developed in section 16.3.1 accounts for arbitrarily large displacements and rotation, hence the term "geometrically exact," although the strain components are assumed to remain small, see section 15.5.2. The kinematics of geometrically beams was first presented by Simo *et al.* [303, 304], but similar developments were proposed by Borri and Merlini [305] or Danielson and Hodges [306, 307].

In many applications, however, beams are, in fact, complex build-up structures with solid or thin-walled cross-sections. In aeronautical constructions, for instance, the increasing use of laminated composite materials leads to heterogeneous, highly anisotropic structures. The analysis of complex cross-sections featuring composite materials and the determination of the associated sectional properties was first presented by Giavotto *et al.* [308, 309]. Their approach, based on linear elasticity theory, leads to a two-dimensional analysis of the beam's cross-section using finite elements, which yields the sectional stiffness characteristics in the form of a 6×6 stiffness matrix relating the six sectional deformations, three strains and three curvatures, to the sectional loads, three forces and three moments. Furthermore, the three-dimensional strain field at all points of the cross-section can be recovered once the sectional strains are known.

For nonlinear problems, the decomposition of the beam problem into a linear, two-dimensional analysis over the cross-section, and a nonlinear, one-dimensional analysis along its span was first proposed by Berdichevsky [310]. Hodges [311] has reviewed many approaches to beam modeling; he points out that although the two-dimensional finite element analysis of the cross-section seems to be computationally expensive, it is, in fact, a preprocessing step that is performed once only.

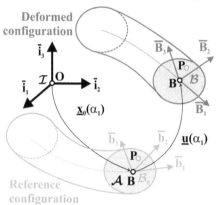

Fig. 16.11. Curved beam in the reference and deformed configurations.

A unified theory presenting both linear, two-dimensional analysis over the cross-section, and a nonlinear, one-dimensional analysis along the beam's span was further refined by Hodges and his co-workers [312, 313]. The nonlinear, one-dimensional analysis along the beam's span corresponds the geometrically exact beam theory developed earlier based on simplified kinematic assumptions. More sophisticated beam theories have been developed that account for Vlasov effects [314] or the trapeze effect [315]. Detailed developments of nonlinear composite beam theory developed by Hodges and his coworkers are found in his textbook [316] and applications to multibody systems in ref. [283].

16.3.1 Kinematics of the problem

Figure 16.11 depicts an initially curved and twisted beam of length L, with a cross-section of arbitrary shape and area $\mathcal{A}$. The volume of the beam is generated by sliding the cross-section along the reference line of the beam, which is defined by an arbitrary curve in space. Curvilinear coordinate α_1 defines the intrinsic parameterization of this curve, section 2.2.1, *i.e.*, it measures length along the beam's reference line. Point **B** is located at the intersection of the reference line with the plane of the cross-section.

In the reference configuration, an orthonormal basis, $\mathcal{B}_0(\alpha_1) = (\bar{b}_1, \bar{b}_2, \bar{b}_3)$, is defined at point **B**. Vector $\bar{b}_1$ is the unit tangent vector to the reference curve at that point, and unit vectors $\bar{b}_2$ and $\bar{b}_3$ define the plane to the cross-section. An inertial reference frame, $\mathcal{F}^I = [\mathbf{O}, \mathcal{I} = (\bar{\imath}_1, \bar{\imath}_2, \bar{\imath}_3)]$, is defined, and the components of the rotation tensor that brings basis $\mathcal{I}$ to $\mathcal{B}_0$, resolved in basis $\mathcal{I}$, are denoted $\underline{\underline{R}}_0(\alpha_1)$.

The position vector of point **B** along the beam's reference line is denoted $\underline{x}_0(\alpha_1)$. The position vector of material point **P** of the beam then becomes $\underline{x}(\alpha_1, \alpha_2, \alpha_3) = \underline{x}_0(\alpha_1) + \alpha_2\,\bar{b}_2 + \alpha_3\,\bar{b}_3$, where α_2 and α_3 are the material coordinates along unit vectors $\bar{b}_2$ and $\bar{b}_3$, respectively. Coordinates α_1, α_2, and α_3 form a natural choice of coordinates to represent the configuration of the beam.

The displacement field

In the deformed configuration, all the material points located on a cross-section of the beam move to new positions. This motion is decomposed into two parts, a rigid body motion and a warping displacement field. The rigid body motion consists of a translation of the cross-section, characterized by displacement vector $\underline{u}(\alpha_1)$ of reference point **B**, and of a rotation of the cross-section, which brings basis $\mathcal{B}_0$ to $\mathcal{B}(\alpha_1) = (\bar{B}_1, \bar{B}_2, \bar{B}_3)$, see fig. 16.11. The components of the rotation tensor that brings basis $\mathcal{B}_0$ to $\mathcal{B}$, resolved in basis $\mathcal{I}$, are denoted $\underline{\underline{R}}(\alpha_1)$.

The warping displacement field is defined as $\underline{w}(\alpha_1, \alpha_2, \alpha_3) = w_1\bar{B}_1 + w_2\bar{B}_2 + w_3\bar{B}_3$. This displacement field represents a warping that includes both in-plane and out-of-plane deformations of the cross-section. To be uniquely defined, the warping field should be orthogonal to the rigid body motion [308, 316]. Consequently, unit vectors $\bar{B}_2$ and $\bar{B}_3$ define the average plane of the cross-section and vector $\bar{B}_1$ is orthogonal to that plane.

The position vector of point **P** in the deformed configuration now becomes

$$\underline{X}(\alpha_1, \alpha_2, \alpha_3) = \underline{X}_0 + w_1\,\bar{B}_1 + (w_2 + \alpha_2)\bar{B}_2 + (w_3 + \alpha_3)\bar{B}_3. \tag{16.43}$$

The position of point **B** is expressed as $\underline{X}_0(\alpha_1) = \underline{x}_0 + \underline{u}$. Because $\bar{B}_i = \underline{\underline{R}}\,\bar{b}_i = (\underline{\underline{R}}\,\underline{\underline{R}}_0)\,\bar{\imath}_i$, eq. (16.43) becomes

$$\underline{X}(\alpha_1, \alpha_2, \alpha_3) = \underline{x}_0 + \underline{u} + (\underline{\underline{R}}\,\underline{\underline{R}}_0)(\underline{w} + \alpha_2\,\bar{\imath}_2 + \alpha_3\,\bar{\imath}_3). \tag{16.44}$$

The warping displacement field is computed from the geometric and stiffness properties of the cross-section, typically by solving a two-dimensional finite element problem over the cross-section, as described in refs. [308, 316].

The sectional strain measures

The sectional strain measures for beams with shallow curvature are defined as

$$\underline{e} = \left\{ \begin{matrix} \underline{\epsilon} \\ \underline{\kappa} \end{matrix} \right\} = \left\{ \begin{matrix} \underline{x}_0' + \underline{u}' - (\underline{\underline{R}}\,\underline{\underline{R}}_0)\,\bar{\imath}_1, \\ \underline{k} + \underline{\underline{R}}\,\underline{k}_i \end{matrix} \right\}, \tag{16.45}$$

where $\underline{k} = \text{axial}(\underline{\underline{R}}'\underline{\underline{R}}^T)$ are the components of the sectional curvature vector resolved in the inertial basis and $\underline{k}_i$ the components of the corresponding curvature vector in the reference configuration. Notation $(\cdot)'$ indicates a derivative with respect to α_1. The strain components resolved in the convected material basis, $\mathcal{B}$, are denoted $\underline{\epsilon}^* = (\underline{\underline{R}}\,\underline{\underline{R}}_0)^T\underline{\epsilon}$ and consist of the sectional axial and shear strains. The curvature components resolved in the same material basis are denoted $\underline{\kappa}^* = (\underline{\underline{R}}\,\underline{\underline{R}}_0)^T\underline{\kappa}$

and consist of the sectional twisting and bending curvatures. Notation $(\cdot)^*$ indicates the components of vectors and tensors resolved in the material basis.

By definition, a rigid body motion is a motion that generates no strains. This implies that the following rigid body motion, $\underline{u}(\alpha_1) = \underline{u}^R + (\underline{\underline{R}}^R - \underline{\underline{I}})\underline{x}_0(\alpha_1)$, $\underline{\underline{R}}(\alpha_1) = \underline{\underline{R}}^R$, consisting of a translation, $\underline{u}^R$, and a rotation about the origin characterized by a rotation matrix, $\underline{\underline{R}}^R$, should generate no straining of the beam. It can be readily verified with the help of eqs. (16.45) that such rigid body motion results in $\underline{\epsilon} = 0$ and $\underline{\kappa} = 0$, as expected.

16.3.2 Governing equations

For the problem at hand, the principle of virtual work states

$$\int_0^L (\delta\underline{\epsilon}^{*T}\underline{N}^* + \delta\underline{\kappa}^{*T}\underline{M}^*)\,\mathrm{d}\alpha_1 = \delta W_{\text{ext}}, \tag{16.46}$$

where $\underline{N}^*$ and $\underline{M}^*$ are the beam's sectional forces and moments, respectively. The sectional constitutive law relates the sectional strain measures to the sectional loads,

$$\left\{\begin{matrix}\underline{N}^* \\ \underline{M}^*\end{matrix}\right\} = \underline{\underline{C}}^*\left\{\begin{matrix}\underline{\epsilon}^* \\ \underline{\kappa}^*\end{matrix}\right\}, \tag{16.47}$$

where $\underline{\underline{C}}^*$ is the beam's 6×6 sectional stiffness matrix. This matrix is a byproduct of a two-dimensional finite element analysis over the beam's cross-section, as discussed in refs. [308, 316]. For homogeneous sections of simple geometry, exact or approximate analytical expressions are available for the stiffness matrix.

Variations in strain components are expressed using eq. (16.45) to find

$$\delta\underline{\epsilon}^* = (\underline{\underline{R}}\,\underline{\underline{R}}_0)^T\left[\delta\underline{u}' + (\tilde{x}_0' + \tilde{u}')\delta\underline{\psi}\right], \tag{16.48a}$$

$$\delta\underline{\kappa}^* = (\underline{\underline{R}}\,\underline{\underline{R}}_0)^T\delta\underline{\psi}'. \tag{16.48b}$$

where $\delta\underline{\psi} = \text{axial}(\delta\underline{\underline{R}}\,\underline{\underline{R}}^T)$ is the virtual rotation vector. The principle of virtual work, eq. (16.46), now becomes

$$\int_0^L \left\{\left[\delta\underline{u}'^T + \delta\underline{\psi}^T(\tilde{x}_0' + \tilde{u}')^T\right]\underline{N} + \delta\underline{\psi}'^T\underline{M}\right\}\,\mathrm{d}\alpha_1 = \delta W_{\text{ext}}, \tag{16.49}$$

where $\underline{N} = (\underline{\underline{R}}\,\underline{\underline{R}}_0)\underline{N}^*$ and $\underline{M} = (\underline{\underline{R}}\,\underline{\underline{R}}_0)\underline{M}^*$ are the beam's internal forces and moments, respectively, resolved in the inertial basis.

The virtual work done by the externally applied forces is expressed as $\delta W_{\text{ext}} = \int_0^L [\delta\underline{u}^T\underline{f} + \delta\underline{\psi}^T\underline{m}]\,\mathrm{d}\alpha_1$, where $\underline{f}$ and $\underline{m}$ denote the externally applied forces and moments per unit span of the beam, respectively.

The governing equations of the static problem then follow as

$$\underline{N}' = -\underline{f}, \tag{16.50a}$$

$$\underline{M}' + (\tilde{x}_0' + \tilde{u}')\underline{N} = -\underline{m}. \tag{16.50b}$$

Example 16.2. The cantilevered beam under tip loading
Consider a cantilevered beam of length L with a rectangular cross-section of width b and height h. The beam is made of a homogeneous material of Young's modulus E and shear modulus G and is subjected to a tip axial load, N_T, tip transverse load, P_T, and tip moment, M_T. The beam is of bending stiffness $H_{33} = Ebh^3/12$, axial stiffness $S = Ebh$, and shearing stiffness $K_{22} = 5Gbh/6$.

The loading is acting in plane $(\bar{\imath}_1, \bar{\imath}_2)$, and due to the symmetry of the problem, the beam deforms in that plane only. The rotation tensor, $\underline{\underline{R}}$, then corresponds to a planar rotation, eq. (4.6), and the displacement vector, $\underline{u}$, is two-dimensional

$$\underline{\underline{R}} = \begin{bmatrix} C_\theta & -S_\theta \\ S_\theta & C_\theta \end{bmatrix}, \quad \underline{u} = \begin{Bmatrix} u_1 \\ u_2 \end{Bmatrix},$$

where $C_\theta = \cos\theta$, $S_\theta = \sin\theta$, angle θ is the average rotation of the cross-section, and u_1 and u_2 the displacement components along unit vectors $\bar{\imath}_1$ and $\bar{\imath}_2$, respectively.

Because the beam is not subjected to distributed transverse loads, the first equation of equilibrium, eq. (16.50a), reduces to $\underline{N}' = 0$. Consequently, the sectional force, $\underline{N}^T = \{N_1, V_2\}$, remain constant, where N_1 and V_2 are the sectional forces along unit vectors $\bar{\imath}_1$ and $\bar{\imath}_2$, respectively. Since equilibrium must be satisfied at the tip of the beam, $N_1(\alpha_1) = N_T$ and $V_2(\alpha_1) = P_T$.

The second equation of equilibrium, eq. (16.50b), now becomes $\underline{M}' = -(\tilde{x}'_0 + \tilde{u}')\underline{N}$, and since the sectional forces are constant, this equation integrates to $M_3(\alpha_1) = u_2 N_T - (\alpha_1 + u_1)P_T + c$, where c is an integration constant. Because the problem is two-dimensional, the other two moment components, M_1 and M_2, vanish. Imposing the moment equilibrium condition at the tip of the beam yields the integration constant and finally,

$$M_3(\alpha_1) = M_T + (L - \alpha_1 + u_1^T - u_1)P_T - (u_2^T - u_2)N_T,$$

where u_1^T and u_2^T are the beam's tip displacements along unit vectors $\bar{\imath}_1$ and $\bar{\imath}_2$, respectively.

The constitutive law for the bending moment is simply $M_3 = H_{33}\theta'$; indeed, for this two-dimensional problem, the curvature vector reduces to a single non vanishing component, $\kappa_3 = \theta'$. The constitutive laws for the sectional forces becomes

$$\begin{Bmatrix} N_T \\ P_T \end{Bmatrix} = \begin{Bmatrix} N_1 \\ V_2 \end{Bmatrix} = \underline{\underline{R}} \begin{Bmatrix} N_1^* \\ V_2^* \end{Bmatrix} = \underline{\underline{R}}\,\underline{\underline{C}}^*\,\underline{\underline{R}}^T \left(\begin{Bmatrix} 1 + u'_1 \\ u'_2 \end{Bmatrix} - \begin{Bmatrix} C_\theta \\ S_\theta \end{Bmatrix} \right),$$

where N_1^* and V_2^* are the sectional axial and shear forces, respectively, resolved in the material system, and $\underline{\underline{C}}^* = \mathrm{diag}(S, K_{22})$. Combining all the relationships obtained above yields the governing equations of the problem,

$$\begin{Bmatrix} u_1 \\ u_2 \\ \theta \end{Bmatrix}' = \begin{Bmatrix} C_\theta - 1 + \dfrac{N_T}{S}C_\theta^2 + \dfrac{N_T}{K_{22}}S_\theta^2 + (\dfrac{P_T}{S} - \dfrac{P_T}{K_{22}})S_\theta C_\theta \\[2mm] S_\theta + (\dfrac{N_T}{S} - \dfrac{N_T}{K_{22}})S_\theta C_\theta + \dfrac{P_T}{S}S_\theta^2 + \dfrac{P_T}{K_{22}}C_\theta^2 \\[2mm] \dfrac{M_T}{H_{33}} + (L - \alpha_1 + u_1^T - u_1)\dfrac{P_T}{H_{33}} - (u_2^T - u_2)\dfrac{N_T}{H_{33}} \end{Bmatrix}.$$

To better understand these equations, it is convenient to normalize all quantities. First, the material coordinate, α_1, is normalized by the length of the beam, $\eta = \alpha_1/L$, and notation $(\cdot)^+$ denotes a derivative with respect to η. The displacement components are also normalized by the beam's length, $\bar{u}_1 = u_1/L$ and $\bar{u}_2 = u_2/L$. The non-dimensional loading parameters are $\bar{N} = N_T L^2/H_{33}$, $\bar{P} = P_T L^2/H_{33}$, and $\bar{M} = M_T L/H_{33}$, and the governing equations now become

$$
\left\{ \begin{array}{c} \bar{u}_1 \\ \bar{u}_2 \\ \theta \end{array} \right\}^+ = \left\{ \begin{array}{c} C_\theta - 1 + \bar{N}(\bar{a}^2 C_\theta^2 + \bar{s}^2 S_\theta^2) - \bar{P}(\bar{s}^2 - \bar{a}^2)S_\theta C_\theta \\ S_\theta + \bar{P}(\bar{a}^2 S_\theta^2 + \bar{s}^2 C_\theta^2) - \bar{N}(\bar{s}^2 - \bar{a}^2)S_\theta C_\theta \\ \bar{M} + \bar{P}(1 - \eta + \bar{u}_1^T - \bar{u}_1) - \bar{N}(\bar{u}_2^T - \bar{u}_2) \end{array} \right\}, \qquad (16.51)
$$

where the non-dimensional stiffness properties of the beam are defined as

$$
\bar{a}^2 = \frac{H_{33}}{SL^2} = \frac{1}{12}\left(\frac{h}{L}\right)^2, \quad \bar{s}^2 = \frac{H_{33}}{K_{22}L^2} = \frac{1}{10}\left(\frac{E}{G}\right)\left(\frac{h}{L}\right)^2.
$$

The axial stiffness coefficient, $\bar{a}^2$, is the ratio of the bending to the axial stiffness of the beam, and the shear stiffness coefficient, $\bar{s}^2$, is the ratio of the bending to the shear stiffness of the beam. For long, slender beams, both coefficients are very small as $(h/L)^2 \to 0$ and can be assumed to vanish without noticeably affecting the predictions.

The governing equations of the problem, eqs. (16.51), take the form of three coupled first-order differential equations for the three variables of the problem, $\bar{u}_1$, $\bar{u}_2$, and θ. These equations are nonlinear due to the presence of trigonometric functions, but also because the beam's unknown tip deflections, $\bar{u}_1^T$ and $\bar{u}_2^T$, appear on the right-hand side of the equations. A convenient solution technique is to assume $\bar{u}_1^T = \bar{u}_2^T = 0$ and integrate eqs. (16.51) numerically. The solution yields an estimate of the beam's tip deflections, which are then used to obtain a refined solution by integrating eqs. (16.51) once again. An iterative procedure then yields the desired solution. This crude solution process will become unstable for large deflections of the beam. Using a relaxation factor when updating the tip deflections is often sufficient to stabilize the computation.

The deflected shape of the beam under a tip loads N_T and P_T was computed using the procedure described above. The following parameters were used: $E/G = 2.6$ and $L/h = 10$. Simulations were performed first for $\bar{N} = 0$ and $\bar{P} = 0.5$, 1, 2, and 4. Figure 16.12 shows the predictions of the simulations. The tip deflection of the beam is not proportional to the applied load, as expected for this nonlinear problem.

A second set of simulation was performed for $\bar{P} = 4$ and $\bar{N} = 0, 4, 8$, and 12. As the axial tip force, N_T, increases, the effective stiffness of the beam increases and the tip deflection under the constant tip transverse force decreases.

16.3.3 Extension to dynamic problems

The developments presented thus far have focused on static problems. The inertial velocity vector, $\underline{v}$, of a material point is found by taking a time derivative of its inertial position vector, eq. (16.44), to find

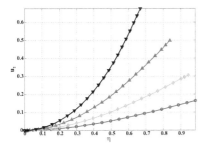

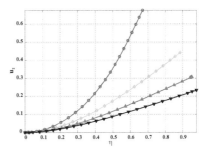

Fig. 16.12. Deflected shape of the beam under a tip transverse load for $\bar{N} = 0$. $\bar{P} = 0.5$, 1, 2, and 4, indicated with symbols $\circ$, $\diamond$, $\triangle$, and $\triangledown$, respectively.

Fig. 16.13. Deflected shape of the beam under a tip transverse load for $\bar{P} = 4$. $\bar{N} = 0$, 4, 8, and 12, indicated with symbols $\circ$, $\diamond$, $\triangle$, and $\triangledown$, respectively.

$$\underline{v} = \dot{\underline{u}} + \dot{\underline{R}}\,\underline{R}_0\,\underline{s}^* = \dot{\underline{u}} + (\underline{R}\,\underline{R}_0)\widetilde{\omega}^*\underline{s}^* = \dot{\underline{u}} + (\underline{R}\,\underline{R}_0)\widetilde{s}^{*T}\underline{\omega}^*, \tag{16.52}$$

where contributions of warping of the cross-section have been ignored and $\underline{s}^{*T} = \{0, \alpha_2, \alpha_3\}$. Notation $(\dot{\cdot})$ indicates a derivative with respect to time and $\underline{\omega}^*$ are the components of the angular velocity vector in the material system, $\underline{\omega}^* = (\underline{R}\,\underline{R}_0)^T\underline{\omega}$, where $\underline{\omega} = \mathrm{axial}(\dot{\underline{R}}\,\underline{R})$.

The components of the inertial velocity vector of a material point resolved in the material frame now become

$$\underline{v}^* = (\underline{R}\,\underline{R}_0)^T\underline{v} = (\underline{R}\,\underline{R}_0)^T\dot{\underline{u}} + \widetilde{s}^{*T}\underline{\omega}^*. \tag{16.53}$$

The total inertial velocity of a material point has two components: the first term, $(\underline{R}\,\underline{R}_0)^T\dot{\underline{u}}$, due to the translation of the cross-section, and the second term, $\widetilde{s}^{*T}\underline{\omega}^*$, due to its rotation.

The kinetic energy

The kinetic energy, K, of the beam is

$$K = \frac{1}{2}\int_0^L \int_{\mathcal{A}} \rho\,\underline{v}^{*T}\underline{v}^*\,\mathrm{d}\mathcal{A}\mathrm{d}\alpha_1, \tag{16.54}$$

where ρ is the mass density of the material per unit volume of the reference configuration. Introducing eq. (16.53) for the inertial velocity yields

$$K = \frac{1}{2}\int_0^L \int_{\mathcal{A}} \rho\left[\dot{\underline{u}}^T(\underline{R}\,\underline{R}_0) + \underline{\omega}^{*T}\widetilde{s}^*\right]\left[(\underline{R}\,\underline{R}_0)^T\dot{\underline{u}} + \widetilde{s}^{*T}\underline{\omega}^*\right]\mathrm{d}\mathcal{A}\mathrm{d}\alpha_1. \tag{16.55}$$

The following sectional mass constants are defined

$$m = \int_{\mathcal{A}} \rho\,\mathrm{d}\mathcal{A}, \quad \underline{\eta}^* = \frac{1}{m}\int_{\mathcal{A}} \rho\,\underline{s}^*\,\mathrm{d}\mathcal{A}, \quad \underline{\varrho}^* = \int_{\mathcal{A}} \rho\widetilde{s}^*\widetilde{s}^{*T}\,\mathrm{d}\mathcal{A}, \tag{16.56}$$

where m is the mass of the beam per unit span, η^* the components of the position vector of the sectional center of mass with respect to point **B**, see fig. 16.11, and ϱ^* the components of the sectional tensor of inertia per unit span, all resolved in the material basis.

After integration over the beam's cross-section, the kinetic energy, eq. (16.55), becomes

$$K = \frac{1}{2} \int_0^L \left[m\underline{\dot{u}}^T \underline{\dot{u}} + 2m\underline{\dot{u}}^T (\underline{\underline{R}}\,\underline{\underline{R}}_0)\,\widetilde{\eta}^{*T} \underline{\omega}^* + \underline{\omega}^{*T} \underline{\underline{\varrho}}^* \underline{\omega}^* \right] \, d\alpha_1$$

$$= \frac{1}{2} \int_0^L \underline{\mathcal{V}}^{*T} \underline{\underline{\mathcal{M}}}^* \underline{\mathcal{V}}^* \, d\alpha_1.$$

To obtain the compact form expressed by the second equality, the sectional mass matrix of the cross-section, resolved in the material basis, is defined as

$$\underline{\underline{\mathcal{M}}}^* = \begin{bmatrix} m\underline{\underline{I}} & m\widetilde{\eta}^{*T} \\ m\widetilde{\eta}^* & \underline{\underline{\varrho}}^* \end{bmatrix}, \tag{16.57}$$

and the sectional velocities, also resolved in the material basis, are given by

$$\underline{\mathcal{V}}^* = \left\{ \begin{matrix} (\underline{\underline{R}}\,\underline{\underline{R}}_0)^T \underline{\dot{u}} \\ \underline{\omega}^* \end{matrix} \right\} = \begin{bmatrix} (\underline{\underline{R}}\,\underline{\underline{R}}_0)^T & \underline{\underline{0}} \\ \underline{\underline{0}} & (\underline{\underline{R}}\,\underline{\underline{R}}_0)^T \end{bmatrix} \left\{ \begin{matrix} \underline{\dot{u}} \\ \underline{\omega} \end{matrix} \right\} = (\underline{\underline{\mathcal{R}}}\,\underline{\underline{\mathcal{R}}}_0)^T \underline{\mathcal{V}}. \tag{16.58}$$

In this expression, the sectional velocities resolved in the inertial system were defined as $\underline{\mathcal{V}}^T = \{\underline{\dot{u}}^T, \underline{\omega}^T\}$ and the following notation was introduced

$$\underline{\underline{\mathcal{R}}}\,\underline{\underline{\mathcal{R}}}_0 = \begin{bmatrix} (\underline{\underline{R}}\,\underline{\underline{R}}_0) & \underline{\underline{0}} \\ \underline{\underline{0}} & (\underline{\underline{R}}\,\underline{\underline{R}}_0) \end{bmatrix}. \tag{16.59}$$

The components of the sectional linear and angular momenta resolved in the material system, denoted $\underline{h}^*$ and $\underline{g}^*$, respectively, are

$$\underline{\mathcal{P}}^* = \left\{ \begin{matrix} \underline{h}^* \\ \underline{g}^* \end{matrix} \right\} = \underline{\underline{\mathcal{M}}}^* \underline{\mathcal{V}}^*. \tag{16.60}$$

The governing equations

Variation of the kinetic energy is $\delta K = \int_0^L \delta \underline{\mathcal{V}}^{*T} \underline{\underline{\mathcal{M}}}^* \underline{\mathcal{V}}^* \, d\alpha_1$, where the variations in velocities are $\delta[\underline{\dot{u}}^T (\underline{\underline{R}}\,\underline{\underline{R}}_0)] = (\delta\underline{\dot{u}}^T + \delta\underline{\psi}^T \underline{\dot{\widetilde{u}}}^T)(\underline{\underline{R}}\,\underline{\underline{R}}_0)$ and $\delta\underline{\omega}^{*T} = \delta\underline{\dot{\psi}}^T (\underline{\underline{R}}\,\underline{\underline{R}}_0)$. Introducing these variations in the expression for the kinetic energy yields

$$\delta K = \int_0^L \left[(\delta\underline{\dot{u}}^T + \delta\underline{\psi}^T \underline{\dot{\widetilde{u}}}^T)(\underline{\underline{R}}\,\underline{\underline{R}}_0)\,\underline{h}^* + \delta\underline{\dot{\psi}}^T (\underline{\underline{R}}\,\underline{\underline{R}}_0)\,\underline{g}^* \right] \, d\alpha_1,$$

The components of the sectional linear and angular momenta, denoted $\underline{h}$ and $\underline{g}$, respectively, resolved in the inertial system are

$$\underline{P} = \left\{ \begin{matrix} h \\ g \end{matrix} \right\} = (\underline{\underline{R}}\,\underline{\underline{R}}_0)\underline{P}^*, \tag{16.61}$$

where $\underline{P}^*$ are the corresponding quantities resolved in the material frame, see eq. (16.60). The variation in kinetic energy finally can be written as

$$\delta K = \int_0^L (\delta \underline{\dot{u}}^T\,\underline{h} + \delta \underline{\psi}^T\,\underline{\dot{\tilde{u}}}^T\,\underline{h} + \delta \underline{\dot{\psi}}^T\,\underline{g})\,d\alpha_1. \tag{16.62}$$

With the help of eqs. (16.49) and (16.62), the governing equations of motion of the problem are obtained from Hamilton's principle, which states that

$$\int_{t_i}^{t_f} \int_0^L \left\{ (\delta \underline{\dot{u}}^T + \delta \underline{\psi}^T\,\underline{\dot{\tilde{u}}}^T)\underline{h} + \delta \underline{\dot{\psi}}\,\underline{g} - (\delta \underline{u}'^T + \delta \underline{\psi}^T\,\tilde{E}_1^T)\underline{N} \right.$$
$$\left. - \delta \underline{\psi}'^T\,\underline{M} + \delta \underline{u}^T\,\underline{f} + \delta \underline{\psi}^T\,\underline{m} \right\} d\alpha_1\,dt = 0.$$

Integration by parts yields the equations of motion of the problem

$$\underline{\dot{h}} - \underline{N}' = \underline{f}, \tag{16.63a}$$

$$\underline{\dot{g}} + \underline{\dot{\tilde{u}}}\,\underline{h} - \underline{M}' - (\tilde{x}_0' + \tilde{u}')\underline{N} = \underline{m}. \tag{16.63b}$$

Example 16.3. The four-bar mechanism

Figure 16.14 depicts a flexible four bar mechanism. Bar 1 is of length 0.12 m and is connected to the ground at point **A** by means of a revolute joint. Bar 2 is of length 0.24 m and is connected to bar 1 at point **B** with a revolute joint. Finally, bar 3 is of length 0.12 m and is connected to bar 2 and the ground at points **C** and **D**, respectively, by means of two revolute joints.

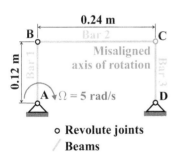

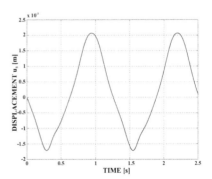

Fig. 16.14. Configuration of the four bar mechanism.

Fig. 16.15. Out-of-plane displacement u_3 at point **C**. ($\rho_\infty = 0$)

In the reference configuration, the bars of this planar mechanism intersect each other at 90 degree angles and the axes of rotation of the revolute joints at points **A**, **B**, and **D** are normal to the plane of the mechanism. The axis of rotation of the revolute joint at point **C** is at a 5 degree angle with respect to this normal to simulate an initial defect in the mechanism. The angular velocity at point **A** of bar 1 is prescribed to be $\Omega = 5$ rad/s.

If the bars were infinitely rigid, no motion would be possible because the mechanism locks. For elastic bars, motion becomes possible, but generates large, rapidly varying internal forces. Bar 1 has the following physical characteristics: axial stiffness, $EA = 40$ MN, bending stiffnesses, $EI_{22} = EI_{33} = 2.4$ MN·m^2, torsional stiffness, $GJ = 0.28$ MN·m^2, shearing stiffnesses, $K_{22} = K_{33} = 2$ MN, mass per unit span, $m = 3.2$ kg/m, and mass moments of inertia, $m_{22} = m_{33} = 0.012$ kg·m. Bars 2 and 3 have the following physical characteristics: axial stiffness, $EA = 4$ MN, bending stiffnesses, $EI_{22} = EI_{33} = 0.24$ MN·m^2, torsional stiffness, $GJ = 0.028$ MN·m^2, shearing stiffnesses, $K_{22} = K_{33} = 0.2$ MN, mass per unit span, $m = 1.6$ kg/m, and mass moments of inertia, $m_{22} = m_{33} = 0.06$ kg·m.

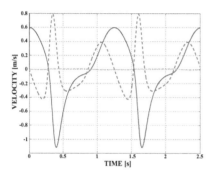

Fig. 16.16. Velocity components at point **C**. Solid line: v_1; dashed line: v_2; dashed-dotted line: v_3. ($\rho_\infty = 0$)

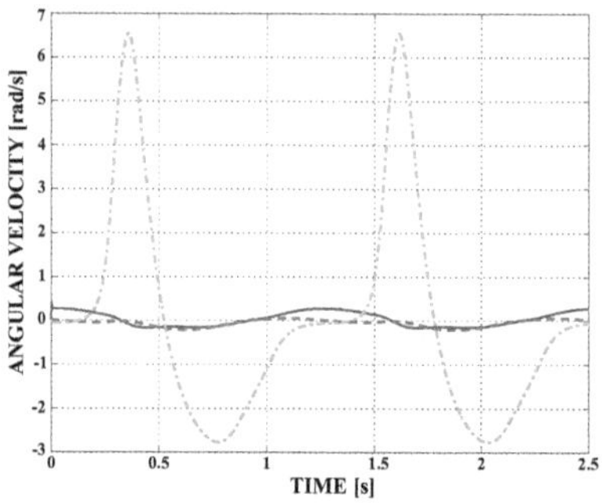

Fig. 16.17. Angular velocity components at point **C**. Solid line: ω_1; dashed line: ω_2; dashed-dotted line: ω_3. ($\rho_\infty = 0$)

This problem was simulated for a total of 2.5 s using the generalized-α scheme described in section 17.4 with $\rho_\infty = 0$; a time step of constant size $\Delta t = 2$ ms was used. If the four revolute joints had their axes of rotation orthogonal to the plane of the mechanism, the response of the system would be purely planar, and bars 1 and 3 would rotate at constant angular velocities around points **A** and **D**, respectively. The initial defect in the mechanism causes a markedly different response. Bar 1 rotates at the constant prescribed angular velocity, but bar 3 now oscillates back and forth, never completing an entire turn.

When the direction of rotation of bar 3 reverses, bar 2 undergoes large rotations, instead of near translation, and sharp increases in velocities are observed, as depicted in figs. 16.16 and 16.17, which show the three components of velocity and angular velocity at point **C**, respectively. Furthermore, fig. 16.15 depicts the time history of out-of-plane displacements at point **C**; clearly, the response of the system is three-dimensional: the out-of-plane displacement at point **C** has a magnitude of up to about 2 mm.

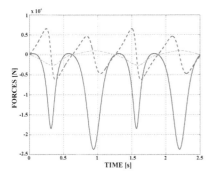
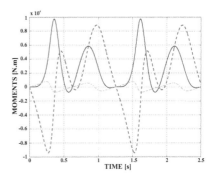

Fig. 16.18. Bar 1 force components at point **A**. Solid line: F_1; dashed line: F_2; dashed-dotted line: F_3. ($\rho_\infty = 0$)

Fig. 16.19. Bar 1 moment components at point **A**. Solid line: M_1; dashed line: M_2; dashed-dotted line: M_3. ($\rho_\infty = 0$)

The time history of the three components of internal forces and bending moments in bar 1 at point **A** are shown in fig. 16.18 and 16.19, respectively. These large internal forces and moments are all caused by the initial imperfection of the mechanism.

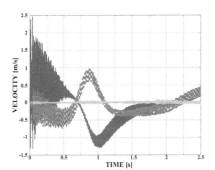
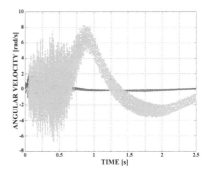

Fig. 16.20. Bar 1 force components at point **A**. Solid line: F_1; dashed line: F_2; dashed-dotted line: F_3. ($\rho_\infty = 0$)

Fig. 16.21. Bar 1 moment components at point **A**. Solid line: M_1; dashed line: M_2; dashed-dotted line: M_3. ($\rho_\infty = 0$)

Next, the same simulation was run using the generalized-α scheme with $\rho_\infty = 0.85$, see eq. (17.39). In the previous simulation, the spectral radius at infinity $\rho_\infty = 0$ achieves asymptotic annihilation, see fig. 17.19; in contrast, the present simulation uses $\rho_\infty = 0.85$, which generates very little algorithmic damping, even at high frequencies. Figures 16.20 and 16.21 show the three components of velocity and angular velocity at point **C**, respectively, for $\rho_\infty = 0.85$, and should be compared with their counterparts, figs. 16.16 and 16.17, respectively, obtained for $\rho_\infty = 0$.

Using an initial time step size of $\Delta t = 2$ ms, the simulation with $\rho_\infty = 0.85$ failed to converge at time steps 10 and 14. In both cases, the time step size was halved to allow the simulation to continue. Note that very high frequency oscillations of a purely numerical origin are predicted. The asymptotic annihilation achieved for $\rho_\infty = 0$ effectively eliminates this undesirable numerical noise.

16.3.4 Problems

Problem 16.3. Conservation properties for beams
Consider an unloaded beam with linearly elastic constitutive laws. *(1)* Prove the preservation the total linear momentum of the beam. *(2)* Prove the preservation the total angular momentum of the beam. *(3)* Prove the preservation the total mechanical energy of the beam. If the beam is subjected to distributed external loads and concentrated end forces what happens to the above three preservation laws?

16.4 Formulation of plate and shell equations

Section 16.3 presents the formulation of beams, which are structures possessing one dimension that is much larger than the other two. The present section focuses on another type of structural components, plates, for which one dimension is far smaller than the other two. The mid-plane of the plate lies along the two long dimensions of the plate, and the normal to the plate extends along the shorter dimension. The term "plate" is usually reserved for flat structures, while the term "shell" refers to a curved plate.

Solid mechanics theories describing plates, more commonly referred to as *plate theories*, play an important role in structural analysis because they provide tools for the analysis of these commonly used structural components. Although more sophisticated formulations, such as three-dimensional elasticity theory, could be used for the analysis of plates and shells, the associated computational burden is often too heavy, and furthermore, plate and shell models provide valuable insight into the behavior of these structures at a much reduced computational cost. It is beyond the scope of this text to review the numerous formulations that have been developed for the analysis of plate and shell structures; comprehensive reviews of the topic are given by Noor *et al.* [317, 318].

Beam theories reduce the analysis of complex, three-dimensional structures to one-dimensional problems. Indeed, the governing equations for geometrically exact beams, eqs. (16.50), are ordinary differential equations expressed in terms of a single variable along the axis of the beam. In contrast, plate theories reduce the analysis of three-dimensional structures to two-dimensional problems. The equations of plate theory are partial differential equations in the two dimensions defining the mid-plane of the plate.

16.4.1 Kinematics of the shell problem

Figure 16.22 depicts a shell of thickness h and mid-plane surface $\mathcal{S}_m$. Let $\underline{x}_0(\alpha_1, \alpha_2)$ be the position vector of an arbitrary point $\mathbf{B}$ on the shell's mid-surface and let α_1 and α_2 be two coordinates that parameterize the mid-surface, see section 2.4.

If the mid-surface of the shell is represented by an arbitrary set of coordinates, the expressions for the first and second metric tensors of the surface, given by eqs. (2.37) and (2.47), respectively, will be complex. Consequently, it is natural to use the concept of lines of curvature introduced in section 2.4.5. In fact, shell theories are developed almost exclusively with the help of lines of curvature.

In the reference configuration frame $\mathcal{F}_0 = \left[\mathbf{B}, \mathcal{B}_0(\alpha_1, \alpha_2) = (\bar{b}_1, \bar{b}_2, \bar{b}_3) \right]$ is defined at point $\mathbf{B}$. Vector $\bar{b}_1 = \underline{x}_{0,1}/\|\underline{x}_{0,1}\|$ and $\bar{b}_2 = \underline{x}_{0,2}/\|\underline{x}_{0,2}\|$ are unit vectors defining the plane tangent to the shell's mid-surface and unit vector $\bar{b}_3$ is the unit normal to this tangent plane. Notations $(\cdot)_{,1}$ and $(\cdot)_{,2}$ indicate derivatives with respect to α_1 and α_2, respectively. An inertial reference frame, $\mathcal{F}^I = [\mathbf{O}, \mathcal{I} = (\bar{\imath}_1, \bar{\imath}_2, \bar{\imath}_3)]$, is defined and the components of the rotation tensor that brings basis $\mathcal{I}$ to $\mathcal{B}_0$, resolved in basis $\mathcal{I}$, are denoted $\underline{\underline{R}}_0(\alpha_1)$.

The position vector of point $\mathbf{B}$ on the shell's mid-surface is denoted $\underline{x}_0(\alpha_1, \alpha_2)$. The position vector of material point $\mathbf{P}$ of the shell then becomes $\underline{x}(\alpha_1, \alpha_2, \zeta) = \underline{x}_0 + \zeta \bar{b}_3$, where ζ is the material coordinate measuring length along the normal to the mid-surface. Unit vector $\bar{b}_3$ defines a material line, *i.e.*, a set of material particles that are normal to the shell's mid-surface in the reference configuration. Coordinates α_1, α_2, and ζ form a set of curvilinear coordinates that is a natural choice of coordinates to represent the shell.

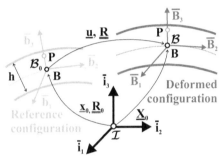

Fig. 16.22. Shell in the reference and deformed configurations.

The displacement field

In the deformed configuration, all the material points located on a normal material line of the shell move to new positions. This motion is decomposed into two parts, a rigid body motion and a warping displacement field. The rigid body motion consists of a translation of the normal material line, characterized by displacement vector $\underline{u}(\alpha_1, \alpha_2)$ of reference point $\mathbf{B}$, and a rotation of the material line, which brings basis $\mathcal{B}_0$ to $\mathcal{B}(\alpha_1) = (\bar{B}_1, \bar{B}_2, \bar{B}_3)$, see fig. 16.22. Unit vectors $\bar{B}_1$ and $\bar{B}_2$ define the plane tangent to the deformed mid-surface of the shell and unit vector $\bar{B}_3$ is normal to this plane. The components of the rotation tensor that brings basis $\mathcal{B}_0$ to $\mathcal{B}$, resolved in basis $\mathcal{I}$, are denoted $\underline{\underline{R}}(\alpha_1, \alpha_2)$.

The warping displacement field is defined as $\underline{w}(\alpha_1, \alpha_2, \zeta) = w_1 \bar{B}_1 + w_2 \bar{B}_2 + w_3 \bar{B}_3$. This displacement field represents a warping that includes all possible deformations of the normal material line. To be uniquely defined, the warping field should be orthogonal to the rigid body motion [319, 320].

The position vector of point $\mathbf{P}$ in the deformed configuration now becomes

$$\underline{X}(\alpha_1, \alpha_2, \zeta) = \underline{X}_0 + w_1 \, \bar{B}_1 + w_2 \, \bar{B}_2 + (w_3 + \zeta) \, \bar{B}_3. \tag{16.64}$$

The position of point $\mathbf{B}$ is expressed as $\underline{X}_0(\alpha_1, \alpha_2) = \underline{x}_0 + \underline{u}$. To uniquely define the orientations of unit vectors $\bar{B}_1$ and $\bar{B}_2$, the following condition is imposed, $\bar{B}_1^T \underline{X}_{0,2} = \bar{B}_2^T \underline{X}_{0,1}$. Because $\bar{B}_i = \underline{\underline{R}} \, \bar{b}_i = (\underline{\underline{R}}\,\underline{\underline{R}}_0) \, \bar{\imath}_i$, eq. (16.64) becomes

$$\underline{X}(\alpha_1, \alpha_2, \zeta) = \underline{x}_0 + \underline{u} + (\underline{\underline{R}}\,\underline{\underline{R}}_0)(\underline{w} + \zeta \, \bar{\imath}_3), \tag{16.65}$$

The warping displacement field is computed from the geometric and stiffness properties of the normal material line, typically by solving a one-dimensional finite element problem over the material line, as described in refs. [319, 320].

The sectional strain measures

The two-dimensional generalized strain measures for shallow shells are now defined. They are conveniently divided into three groups, the mid-surface in-plane strain components, the transverse shear strain components, and the curvature components. The mid-surface in-plane strain components are

$$e_{11} = \left(\hat{\underline{E}}_1^T \hat{\underline{E}}_1 - 1 \right) / 2, \tag{16.66a}$$

$$e_{22} = \left(\hat{\underline{E}}_2^T \hat{\underline{E}}_2 - 1 \right) / 2, \tag{16.66b}$$

$$2e_{12} = \hat{\underline{E}}_1^T \hat{\underline{E}}_2. \tag{16.66c}$$

The transverse shearing strain components are

$$2e_{13} = \hat{\underline{E}}_1^T \hat{\underline{E}}_3, \tag{16.67a}$$

$$2e_{23} = \hat{\underline{E}}_2^T \hat{\underline{E}}_3. \tag{16.67b}$$

Finally, the curvature components are

$$\kappa_{11} = \hat{\underline{E}}_1^T \frac{\hat{\underline{E}}_{3,1}}{\sqrt{a_{11}}} + \frac{1}{R_1}, \tag{16.68a}$$

$$\kappa_{22} = \hat{\underline{E}}_2^T \frac{\hat{\underline{E}}_{3,2}}{\sqrt{a_{22}}} + \frac{1}{R_2}, \tag{16.68b}$$

$$\kappa_{12} = \hat{\underline{E}}_1^T \frac{\hat{\underline{E}}_{3,2}}{\sqrt{a_{22}}} + \hat{\underline{E}}_2^T \frac{\hat{\underline{E}}_{3,1}}{\sqrt{a_{11}}}, \tag{16.68c}$$

where R_1 and R_2 are the principal radii of curvature of the shell's reference configuration as defined by eqs. (2.54).

The shell's deformation measures are defined in terms three vectors,

$$\hat{\underline{E}}_1 = \bar{b}_1 + \underline{u}_{,1}/\sqrt{a_{11}}, \tag{16.69a}$$

$$\hat{\underline{E}}_2 = \bar{b}_2 + \underline{u}_{,2}/\sqrt{a_{22}}, \tag{16.69b}$$

$$\hat{\underline{E}}_3 = \bar{B}_3, \tag{16.69c}$$

where $a_{11} = \left\|\underline{x}_{0,1}\right\|^2$ and $a_{22} = \left\|\underline{x}_{0,2}\right\|^2$ are the diagonal terms of the shell's first metric tensor in its reference configuration, see section 2.4.1 and eq. (2.37). Vector $\hat{\underline{E}}_3$ is the unit vector normal to the deformed mid-surface of the shell.

The generalized strain measures are expressed in terms of five parameters: the three components of the displacement vector, $\underline{u}$, appearing in the definition of vectors $\hat{\underline{E}}_1$ and $\hat{\underline{E}}_2$, eqs. (16.69a) and (16.69b), respectively, and the two parameters defining the orientation of the unit normal vector, $\hat{\underline{E}}_3$.

16.4.2 Governing equations

The governing equations of the problem are obtained from the principle of virtual work, which states that $\delta W_{\text{int}} + \delta W_{\text{ext}} = 0$, where δW_{int} and δW_{ext} are the virtual works done by the internal forces and externally applied loads, respectively.

Virtual work done by internal forces

For simplicity, the shell's two-dimensional generalized strain measures are collected into a single array, $\underline{e}^*$, defined as

$$\underline{e}^{*T} = \left\{e_{11}, e_{22}, e_{12}, e_{13}, e_{23}, \kappa_{11}, \kappa_{22}, \kappa_{12}\right\}.$$

The first three entries are the mid-surface in-plane strain components defined by eqs. (16.66), the next two entries the transverse shearing strain components defined by eqs. (16.67), and the last three entries the curvature components defined by eqs. (16.68).

The corresponding stress resultants are also collected in a single array, $\underline{F}^*$, defined as

$$\underline{F}^{*T} = \left\{N_{11}^*, N_{22}^*, N_{12}^*, N_{13}^*, N_{23}^*, M_{11}^*, M_{22}^*, M_{12}^*\right\}.$$

The first three entries are the in-plane forces; N_{11}^* and N_{22}^* are the stress resultants along unit vectors $\bar{B}_1$ and $\bar{B}_2$, respectively, and N_{12}^* is the in-plane shear force. The next two entries are the transverse shear forces; N_{13}^* and N_{23}^* act on faces normal to unit vectors $\bar{B}_1$ and $\bar{B}_2$, respectively. Finally, the last three entries are the bending and twisting moments. Both forces and moments are measured per unit length of the shell, and resolved in material basis $\mathcal{B}$.

Evaluating the variation of the strain components given in eqs. (16.66), (16.67), and (16.68), the virtual work done by the internal forces becomes

$$\delta W_{\text{int}} = -\int_{\mathcal{S}_m} \delta \underline{e}^{*T} \underline{F}^* \, \mathrm{d}\mathcal{S}_m = -\int_{\mathcal{S}_m} \left\{ \delta \underline{u}_{,1}^T \underline{N}_1 + \delta \underline{u}_{,2}^T \underline{N}_2 \right.$$
$$\left. + \delta \hat{\underline{E}}_{3,1}^T \underline{M}_1 + \delta \hat{\underline{E}}_{3,2}^T \underline{M}_2 + \delta \hat{\underline{E}}_3^T \underline{N}_3 \right\} \mathrm{d}\mathcal{S}_m. \tag{16.70}$$

To simplify this expression, the following quantities were introduced

$$\underline{N}_1 = \frac{1}{\sqrt{a_{11}}} \left[N_{11}^* \hat{\underline{E}}_1 + N_{12}^* \hat{\underline{E}}_2 + N_{13}^* \hat{\underline{E}}_3 + M_{11}^* \frac{\hat{\underline{E}}_{3,1}}{\sqrt{a_{11}}} + M_{12}^* \frac{\hat{\underline{E}}_{3,2}}{\sqrt{a_{22}}} \right], \tag{16.71a}$$

$$\underline{N}_2 = \frac{1}{\sqrt{a_{22}}} \left[N_{12}^* \hat{\underline{E}}_1 + N_{22}^* \hat{\underline{E}}_2 + N_{23}^* \hat{\underline{E}}_3 + M_{12}^* \frac{\hat{\underline{E}}_{3,1}}{\sqrt{a_{11}}} + M_{22}^* \frac{\hat{\underline{E}}_{3,2}}{\sqrt{a_{22}}} \right], \tag{16.71b}$$

$$\underline{N}_3 = N_{13}^* \hat{\underline{E}}_1 + N_{23}^* \hat{\underline{E}}_2, \tag{16.71c}$$

$$\underline{M}_1 = \frac{1}{\sqrt{a_{11}}} \left[M_{11}^* \hat{\underline{E}}_1 + M_{12}^* \hat{\underline{E}}_2 \right], \tag{16.71d}$$

$$\underline{M}_2 = \frac{1}{\sqrt{a_{22}}} \left[M_{12}^* \hat{\underline{E}}_1 + M_{22}^* \hat{\underline{E}}_2 \right]. \tag{16.71e}$$

Constitutive laws

The stress resultants are related to the strain measures through the constitutive law

$$\underline{F}^* = \underline{\underline{C}}^* \underline{e}^*. \tag{16.72}$$

where $\underline{\underline{C}}^*$ is the shells's 8×8 sectional stiffness matrix. This matrix is a byproduct of a one-dimensional finite element analysis through the shell's thickness, as discussed in refs. [319, 320].

Virtual work done by externally applied loads

Let $\underline{f}$ and $\underline{m}$ denote the force and moment vectors applied to the shell's mid-surface per unit area, respectively. The virtual work done by these externally applied loads is expressed as

$$\delta W_{\text{ext}} = \int_{\mathcal{S}_m} \left(\delta \underline{u}^T \underline{f} + \delta \underline{\psi}^T \underline{m} \right) \mathrm{d}\mathcal{S}_m, \tag{16.73}$$

where $\delta \underline{u}$ is the virtual displacement vector of the point of application of the force and $\delta \underline{\psi}$ the virtual rotation vector of the same point.

Unit vector $\hat{\underline{E}}_3$ is a director, as defined in section 4.15, and can be expressed as $\hat{\underline{E}}_3 = (\underline{\underline{R}}\, \underline{\underline{R}}_0)\, \bar{\imath}_3$. A virtual change in this director's orientation then becomes $\delta \hat{\underline{E}}_3 = (\underline{\underline{R}}\, \underline{\underline{R}}_0)\, \bar{\imath}_3^T \underline{\underline{b}}\, \delta \underline{\alpha}^*$, see eq. (4.113), where $\delta \underline{\alpha}^*$ is a two-parameter virtual rotation vector resolved in material basis $\mathcal{B}$ and matrix $\underline{\underline{b}}$ is defined by eq. (4.112).

The virtual work done by the applied moment becomes $\delta \underline{\psi}^T \underline{m} = \delta \underline{\psi}^{*T} \underline{m}^* = \delta \underline{\alpha}^{*T} \underline{\underline{b}}^T \underline{m}^*$, where $\underline{m}^*$ denotes the components of the applied moment vector, resolved in material basis $\mathcal{B}$, $\underline{m}^* = (\underline{\underline{R}}\, \underline{\underline{R}}_0)^T \underline{m}$. Because the last row of matrix $\underline{\underline{b}}$ stores two vanishing entries, see eq. (4.112), the product $\underline{b}^T \underline{m}^*$, ignores the last component of vector $\underline{m}^*$. This last component, called the *drilling moment*, is the component of the externally applied moment acting about the normal to the shell's mid-surface. Because the shell presents no stiffness about this axis, it cannot carry a drilling moment. The virtual work done by the externally applied moment, $\delta\underline{\alpha}^{*T}\underline{b}^T\underline{m}^*$, automatically filters out the contribution of the drilling moment.

Equations of motion

Now that the virtual work done by both internal force and externally applied loads have been evaluated, the principle of virtual work states that

$$
\int_{\mathcal{S}_m} \left\{ \delta\underline{u}^T\underline{f} + \delta\underline{\alpha}^{*T}\underline{b}^T\underline{m}^* - \delta\underline{u}^T \left[-\underline{N}_{1,1} - \underline{N}_{2,2} \right] \right.
$$
$$
\left. -\delta\underline{\alpha}^{*T}\underline{b}^T\tilde{\imath}_3(\underline{R}\,\underline{R}_0)^T \left[\underline{N}_3 - \underline{M}_{1,1} - \underline{M}_{2,2} \right] \right\} \mathrm{d}\mathcal{S}_m = 0.
\tag{16.74}
$$

The governing equations finally become

$$
\underline{N}_{1,1} + \underline{N}_{2,2} = -\underline{f}, \tag{16.75a}
$$
$$
\underline{b}^T\tilde{\imath}_3(\underline{R}\,\underline{R}_0)^T \left(\underline{M}_{1,1} + \underline{M}_{2,2} - \underline{N}_3 \right) = -\underline{b}^T\underline{m}^*. \tag{16.75b}
$$

16.4.3 Extension to dynamic problems

The velocity of a material of the shell is computed as a time derivative of the position vector, eq. (16.65), to find $\dot{\underline{X}} = \dot{\underline{u}} + \zeta\dot{\underline{B}}_3$, where velocity components associated with the warping field have been ignored.

The kinetic energy of the shell then becomes

$$
K = \frac{1}{2} \int_{\mathcal{S}_m} \int_h \rho\,(\dot{\underline{u}}^T + \zeta\dot{\underline{E}}_3^T)(\dot{\underline{u}} + \zeta\dot{\underline{E}}_3)\,\mathrm{d}\zeta\mathrm{d}\mathcal{S}_m,
$$

where ρ is the material density. Integration through the shell thickness then yields

$$
K = \frac{1}{2} \int_{\mathcal{S}_m} \underline{\mathcal{V}}^{*T}\underline{\underline{\mathcal{M}}}^*\underline{\mathcal{V}}^*\,\mathrm{d}\mathcal{S}_m,
$$

where $\underline{\mathcal{V}}^{*T} = \{\dot{\underline{u}}, \dot{\underline{B}}_3\}$ is the velocity vector and the 6×6 mass matrix, $\underline{\underline{\mathcal{M}}}^*$, is defined as

$$
\underline{\underline{\mathcal{M}}}^* = \begin{bmatrix} m\underline{I} & m^*\underline{I} \\ m^*\underline{I} & M^*\underline{I} \end{bmatrix}.
$$

The following mass coefficients were defined

$$
m = \int_h \rho\,\mathrm{d}\zeta, \quad m^* = \int_h \rho\zeta\,\mathrm{d}\zeta, \quad M^* = \int_h \rho\zeta^2\,\mathrm{d}\zeta,
$$

where m is the mass of the shell per unit mid-surface area, m^*/m the location of the center mass, and M^*/m the square of the radius of gyration.

Virtual changes in the kinetic energy become

$$\delta K = \int_{\mathcal{S}_m} (\delta \underline{\dot{u}}^T \underline{h} + \delta \underline{\dot{E}}_3^T \underline{g}) \, d\mathcal{S}_m, \tag{16.76}$$

where $\underline{h} = m\underline{\dot{u}} + m^* \underline{\dot{E}}_3$ and $\underline{g} = m^* \underline{\dot{u}} + M^* \underline{\dot{E}}_3$ are the linear and angular momentum vectors, respectively.

The governing equations of motion are then obtained from Hamilton's principle that becomes

$$\int_t \int_{\mathcal{S}_m} \left\{ \delta \underline{u}^T \left[-\underline{\dot{h}} + \underline{N}_{1,1} + \underline{N}_{2,2} \right] + \delta \underline{\alpha}^{*T} \underline{b}^T \tilde{i}_3 (\underline{R}\,\underline{R}_0)^T \right.$$
$$\left. \left[-\underline{\dot{g}} - \underline{N}_3 + \underline{M}_{1,1} + \underline{M}_{2,2} \right] + \delta \underline{u}^T \underline{f} + \delta \underline{\alpha}^{*T} \underline{b}^T \underline{m}^* \right\} \, d\mathcal{S}_m dt = 0.$$

The governing equations of motion finally become

$$\underline{\dot{h}} - \underline{N}_{1,1} - \underline{N}_{2,2} = \underline{f}, \tag{16.77a}$$
$$\underline{b}^T \tilde{i}_3 (\underline{R}\,\underline{R}_0)^T \left[\underline{\dot{g}} + \underline{N}_3 - \underline{M}_{1,1} - \underline{M}_{2,2} \right] = \underline{b}^T \underline{m}^*. \tag{16.77b}$$

16.4.4 Mixed interpolation of tensorial components

Several recently developed shell elements have distinguished themselves from other shell formulations because of their versatility, accuracy and robustness. One of these is the mixed interpolation of tensorial components (MITC) element developed by Bathe and his co-workers [321, 322, 323]. The MITC approach is based on the interpolation of strains at chosen sampling points (so-called "tying points"). The key issue of this approach is the selection of the tying points and corresponding interpolation functions. In case of the nine-noded MITC9 element, the interpolated strain components are defined as

$$e_{11} = \sum_{\alpha} g_{rr}^{\alpha} e_{11}^{\alpha}, \quad e_{22} = \sum_{\alpha} g_{ss}^{\alpha} e_{22}^{\alpha}, \quad e_{12} = \sum_{\alpha} g_{rs}^{\alpha} e_{12}^{\alpha}; \tag{16.78a}$$

$$e_{13} = \sum_{\alpha} g_{rr}^{\alpha} e_{13}^{\alpha}, \quad e_{23} = \sum_{\alpha} g_{ss}^{\alpha} e_{23}^{\alpha}. \tag{16.78b}$$

where g_{rr}^{α}, g_{ss}^{α}, and g_{rs}^{α} are the strain interpolation functions and e_{ij} the strain components at the α tying point, which are obtained by direct interpolation using the finite element displacement assumptions. The location of the tying points and corresponding strain interpolation functions can be found, for example, in [322, 323] for each strain component. For the MITC9 element, the strain components e_{11} and e_{13} are interpolated based on six tying points, using the shape functions g_{rr}^{α}. The strain components e_{22} and e_{23} are interpolated based on six tying points, using the shape functions g_{ss}^{α}. Finally, the in-plane shearing strain component e_{12} is interpolated based on four tying points, using the shape functions g_{rs}^{α}. This approach takes care of both membrane and transverse shearing strain locking problems. The stiffness matrix of the element is then formed based on these interpolated strain components and full integration is used. The element does not present any spurious mechanism. In view of the more complicated strain interpolation and full integration scheme, the MITC9 element is a more computationally expensive element, but it is accurate and fairly insensitive to element deformations.

Example 16.4. Lateral buckling of a thin plate

Figure 16.23 depicts a thin cantilevered plate acted upon by a crank and link mechanism. The plate is of length $L = 1$m, height $h = 80$ mm, thickness $t = 2$ mm, and is made of steel with the following properties: Young's modulus $E = 210$ GPa, Poisson ratio $\nu = 0.25$ and density $\rho = 7870$ kg/m³. It is clamped along edge **AB** and a reinforcing beam is located along edge **CD**.

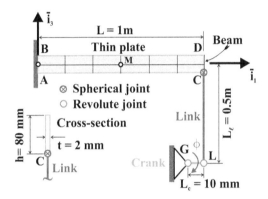

Fig. 16.23. Thin plate actuated by a crank.

At point **C**, the reinforcing beam connects to a crank and link mechanism through a spherical joint. The crank of length $L_c = 10$ mm is attached to the ground at point **G** and the link is of length $L_\ell = 0.5$ m. The ground, crank, and link are connected together by means of revolute joints. The crank is modeled as a rigid body and its rotation is prescribed as $\phi = \pi(1 - \cos 2\pi t/T)/4$ for $t \le T/2$ s and $\phi = \pi/2$ for $t > T/2$, where $T = 1.6$ s.

The reinforcing beam has the following physical characteristics: axial stiffness, $EA = 3.36$ MN, bending stiffnesses, $EI_{22} = EI_{33} = 4.48$ N·m², torsional stiffness, $GJ = 3.02$ N·m², shearing stiffnesses, $K_{22} = K_{33} = 1.12$ MN, mass per unit span, $m = 0.126$ kg/m, and mass moments of inertia, $m_{22} = m_{33} = 0.168$ mg·m. The link has the following physical characteristics: axial stiffness, $EA = 44$ MN, bending stiffnesses, $EI_{22} = EI_{33} = 0.3$ MN·m², torsional stiffness, $GJ = 28$ kN·m², shearing stiffnesses, $K_{22} = K_{33} = 2.4$ MN, mass per unit span, $m = 1.6$ kg/m, and mass moments of inertia, $m_{22} = m_{33} = 0.011$ kg·m.

The link is modeled as a geometrically exact beam, see section 16.3. The thin plate is modeled with a 2×6 mesh of quadratic elements. The system is simulated for 1.4 s using a constant time step $\Delta t = 0.5$ ms using the generalized-α scheme with $\rho_\infty = 0$. As the crank rotates, the plate deflects downwards then snaps laterally when its buckling load is reached. In the post-buckling regime, the plate becomes significantly softer in bending due to its large twisting allowed by the spherical joint.

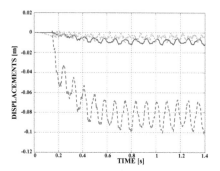

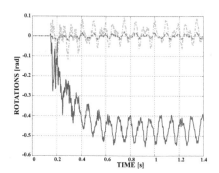

Fig. 16.24. Displacement components at point **M**. Solid line: u_1; dashed line: u_2; dashed-dotted line: u_3.

Fig. 16.25. Rotation components at point **M**. Solid line: r_1; dashed line: r_2; dashed-dotted line: r_3.

The plate's displacement components at point **M** are shown in fig. 16.24. At time $t = 0.145$ s, the plate buckles laterally and the transverse displacement, which was vanishingly small up to that time, suddenly becomes very large. For time $t > 0.8$ s, the crank angle remains constant at $\phi = \pi/2$, but the plate continues to vibrate because no dissipative mechanism is present in the system. The components of the Wiener-Milenković vectorial parameterization of rotation at point **M** are shown in fig. 16.25.

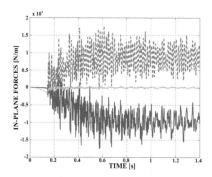

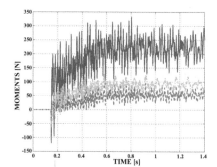

Fig. 16.26. In-plane force components at point **M**. Solid line: N_{11}^*; dashed line: N_{22}^*; dashed-dotted line: N_{12}^*.

Fig. 16.27. Moment components at point **M**. Solid line: M_{11}^*; dashed line: M_{22}^*; dashed-dotted line: M_{12}^*.

The force and moments components in the plate at point **M** are depicted in figs. 16.26 and 16.27, respectively. Prior to buckling, the plate resists the bending loads applied by the driving mechanism with very little deformations. The in-plane shear force component, N_{12}^*, reflects the tip shear force applied by the crank and link mechanism, but all other force and moment components vanish. Once buckling has occurred, twisting of the plate renders it much softer in the vertical direction, offering little resistance to crank motion. Because the lateral buckling occurs so suddenly, high frequency vibrations are observed.

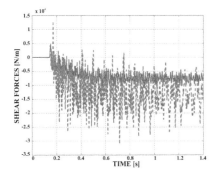

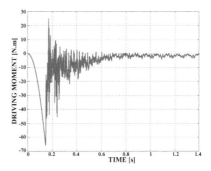

Fig. 16.28. Transverse shear force components at point **M**. Solid line: N_{13}^*; dashed line: N_{23}^*.

Fig. 16.29. Rotation components at mid-point. Solid line: r_1; dashed line: r_2; dashed-dotted line: r_3.

Figure 16.28 shows the corresponding transverse shear force components. Finally, the driving torque, *i.e.*, to torque applied to the crank at point **G** to achieve the prescribed schedule of angle ϕ is depicted in fig. 16.29. This quantity is, in fact, Lagrange's multiplier used to enforce the prescribed rotation holonomic constraint. As soon as the plate buckles, the magnitude if this moment decreases suddenly because of the plate's apparent softening when it buckles laterally.

Finally, the same problem was simulated using the generalized-α scheme with $\rho_\infty = 1$. In this case, due to the lack of numerical dissipation, high frequency oscillations with amplitudes an order of magnitude larger than those predicted for $\rho_\infty = 0$ are observed. This numerical noise completely obscures the results of the simulation demonstrating here again the need for integration schemes presenting numerical dissipation.

17

Finite element tools

Numerous textbooks [324, 197, 198] present detailed development of the theoretical and numerical concepts underpinning the finite element method. Similar developments are clearly beyond the scope of this text. The present chapter focuses on specific details of the finite element method that are relevant to its application to flexible multibody systems. Techniques for interpolation of displacement and specially rotation fields are presented in sections 17.1 and 17.2, respectively.

Next, general processes for the linearization of the governing equations are presented in section 17.3. Both statics and dynamics problems are addressed, with special emphasis of the equations characterizing flexible multibody systems subjected to both holonomic and nonholonomic constraints.

Time integrations schemes are a crucial part of the solution process for multibody dynamics codes. Because finite element based formulations of multibody dynamics inherit many of the characteristics of the finite element methods, it should not come as a surprise if the time integration schemes used in finite element implementations are also used for multibody dynamics. In particular, the HHT scheme and its generalization, the generalized-α scheme, both workhorses used in most commercial codes, are reviewed in section 17.4. Section 17.5 discusses energy preserving and decaying schemes that have been developed in recent years for application to flexible multibody systems.

The chapter closes with a detailed presentation of the implementation of two elements: the cable and the beam element are presented in sections 17.6 and 17.7, respectively.

17.1 Interpolation of displacement fields

The present section discusses simple numerical tools used to interpolate displacement fields within one-dimensional finite elements, such as the cable or beam elements discussed in sections 16.2 or 16.3, respectively. Interpolation is a linear operation that has been used for decades to interpolate displacement fields, which form a linear space. The interpolation of rotation fields is addressed in section 17.2.

In the *finite element method*, the solution domain is first divided into a finite number of sub-domains called *finite elements*. Within each element, the solution is then approximated by a finite number of continuous functions, based on the value of these functions at discrete points, called *nodes*, associated with the element. The main advantage of this two-step approximation process is that many aspects of the solution procedure can be carried out at the element level, *i.e.*, by considering one single element at a time, independently of all others. The continuity of the solution across element boundaries can be guaranteed by the fact that neighboring elements share common nodes.

Consider an element of length ℓ, described by material coordinate α_1. To illustrate the process, the geometry and displacement field of the element are assumed to be defined at three nodes along the element. The first two nodes, denoted nodes 1 and 2, are located at the end points of the element, and one additional node, denoted node 3, is inside the element. Let $\underline{r}_1$, $\underline{r}_2$, and $\underline{r}_3$ be the position vectors of nodes 1, 2, and 3, respectively; similarly, let $\underline{u}_1$, $\underline{u}_2$, and $\underline{u}_3$ be the displacement vectors of nodes 1, 2, and 3, respectively. The geometry and displacement field of the element are now interpolated based on the values of the position and displacement vectors at the nodes using *shape functions* denoted $h_1(s)$, $h_2(s)$, and $h_3(s)$,

$$\underline{r}(s) = h_1(s)\underline{r}_1 + h_2(s)\underline{r}_2 + h_3(s)\underline{r}_3, \tag{17.1a}$$
$$\underline{u}(s) = h_1(s)\underline{u}_1 + h_2(s)\underline{u}_2 + h_3(s)\underline{u}_3, \tag{17.1b}$$

where variable s is a non-dimensional quantity defined along the span of the element. Node 1, 2, and 3 are located at $s = -1, +1$, and 0, respectively.

The shape functions are as yet undetermined, but at $s = -1, +1$, and 0, the approximation must recover nodal values exactly. For instance, at $s = -1$, eq. (17.1a) yields $\underline{r}(-1) = h_1(-1)\underline{r}_1 + h_2(-1)\underline{r}_2 + h_3(-1)\underline{r}_3 = \underline{r}_1$, which implies $h_1(-1) = 1, h_2(-1) = h_3(-1) = 0$. Proceeding similarly at $s = +1$ and 0 leads to the following conditions that must be satisfied by the shape functions,

$$
\begin{array}{lll}
h_1(-1) = 1, & h_1(+1) = 0, & h_1(0) = 0, \\
h_2(-1) = 0, & h_2(+1) = 1, & h_2(0) = 0, \\
h_3(-1) = 0, & h_3(+1) = 0, & h_3(0) = 1.
\end{array}
\tag{17.2}
$$

Conditions (17.2) alone do not uniquely define the shape functions. It is convenient, however, to select the shape functions in the form of quadratic polynomials because each shape function is then uniquely defined by conditions (17.2). It is easily verified that the desired shape functions are

$$h_1(s) = -\frac{1}{2}s(1-s), \quad h_2(s) = \frac{1}{2}s(1+s), \quad h_3(s) = 1 - s^2. \tag{17.3}$$

The reasoning developed in the previous paragraphs can be repeated for elements presenting two, three, or four nodes, leading to linear, quadratic, or cubic polynomial shape functions, respectively. For elements featuring two nodes located at their end points, the two linear shape functions are

$$h_1(s) = \frac{1}{2}(1 - s), \quad h_2(s) = \frac{1}{2}(1 + s). \tag{17.4}$$

For elements with four nodes, two at their end points and two internal nodes located at $s = \mp 1/3$, the four cubic shape functions are

$$h_1(s) = \frac{9}{16}\left(s^2 - \frac{1}{9}\right)(1 - s), \quad h_3(s) = -\frac{27}{16}(1 - s^2)\left(s - \frac{1}{3}\right),$$
$$h_2(s) = \frac{9}{16}\left(s^2 - \frac{1}{9}\right)(1 + s), \quad h_4(s) = \frac{27}{16}(1 - s^2)\left(s + \frac{1}{3}\right). \tag{17.5}$$

The shape functions defined by eqs. (17.4), (17.3), and (17.5) are depicted in the top, middle, and bottom portions of fig. 17.1, respectively. Derivatives of the shape functions with respect to variable s will also be necessary and are readily computed from eqs. (17.4), (17.3), and (17.5). Figure 17.2 depicts these derivatives.

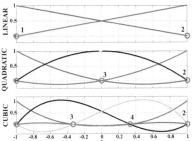

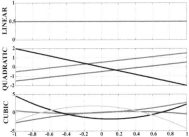

Fig. 17.1. Shape function. Linear: top figure. Quadratic: middle figure. Cubic: bottom figure. Red circles indicate the node locations.

Fig. 17.2. Derivatives of the shape function. Linear: top figure. Quadratic: middle figure. Cubic: bottom figure.

It will be convenient to introduce a compact notation for the interpolation operation expressed by eq. (17.1b). First, the *displacement interpolation matrix*, $\underline{\underline{N}}(s)$, is defined as

$$\underline{\underline{N}}(s) = \left[h_1(s)\underline{\underline{I}}, \; h_2(s)\underline{\underline{I}}, \; h_3(s)\underline{\underline{I}}\right]. \tag{17.6}$$

Next, the nodal displacements are stored in a single array, denoted $\hat{\underline{u}}$, and defined as

$$\hat{\underline{u}}^T = \left\{\underline{u}_1^T, \underline{u}_2^T, \underline{u}_3^T\right\}. \tag{17.7}$$

With this notation, eq. (17.1b) simply becomes

$$\underline{u}(s) = \underline{\underline{N}}(s)\hat{\underline{u}}. \tag{17.8}$$

The sizes of the displacement interpolation matrix and nodal displacement array will vary according to the number of nodes used for the interpolation.

In many applications, the derivatives of the displacement field will also be required and are easily found as

$$\underline{u}^+(s) = \underline{\underline{N}}^+(s)\hat{\underline{u}}, \qquad (17.9)$$

where notation $(\cdot)^+$ indicates a derivative with respect to s, and the displacement derivative interpolation matrix, $\underline{\underline{N}}^+(s)$, simply stores the derivatives of the shape functions,

$$\underline{\underline{N}}^+(s) = \left[h_1^+(s)\underline{\underline{I}},\, h_2^+(s)\underline{\underline{I}},\, h_3^+(s)\underline{\underline{I}} \right]. \qquad (17.10)$$

Example 17.1. Interpolation of a displacement field

Consider a displacement field with the following nodal values, $u_1 = 0.9$, $u_2 = 0.1$, $u_3 = 0.3$, and $u_4 = 0.5$. For this simple example, a single component of the displacement vector is considered. Find the interpolated displacement field over the element.

Consider first the case of linear interpolation. In this case, the two end nodes only will be used, and the displacement interpolation matrix defined by eq. (17.6) reduces to

$$\underline{\underline{N}}(s) = \left[(1-s)/2,\, (1+s)/2 \right].$$

The single line corresponds to the single displacement component and the two columns correspond to the two nodes of the element. The linear shape functions are those of eq. (17.4). Equation (17.8) now yields the interpolated displacement field as $\underline{u}(s) = 1/2\,(1-s)u_1 + 1/2\,(1+s)u_2 = 1/2\,(1-s)\,0.9 + 1/2\,(1+s)\,0.1$. This corresponds to the straight line interpolation depicted in the top portion of fig. 17.3.

Quadratic interpolation is considered next. The two end nodes are used, together with the third node, where the displacement is $u_3 = 0.3$. The displacement interpolation matrix is now of size 1×3, and the quadratic shape functions defined by eq. (17.3) appear along the single line of this matrix. Equation (17.8) now yields the interpolated displacement field as $\underline{u}(s) = -1/2\,s(1-s)\,0.9 + 1/2\,s(1+s)\,0.1 + (1-s^2)\,0.3$. This corresponds to the parabolic interpolation depicted in the middle portion of fig. 17.3.

Finally, cubic interpolation is used. The cubic shape functions defined in eq. (17.5) are now stored in the displacement interpolation matrix of size 1×4. Equation (17.8) now yields the interpolated displacement field as $\underline{u}(s) = 9/16\,(s^2 - 1/9)(1-s)\,0.9 + 9/16\,(s^2 - 1/9)(1+s)\,0.1 - 27/16\,(1-s^2)(s - 1/3)\,0.3 + 27/16\,(1-s^2)(s+1/3)\,0.5$, which corresponds to the cubic interpolation depicted in the bottom portion of fig. 17.3.

The derivative of the displacement field is readily obtained from eq. (17.9) and is depicted in fig. 17.4.

Example 17.2. Evaluation of the strain field

The shape functions developed in section 17.1 are defined in terms of the non-dimensional variable, s, and the derivative of the displacement field with respect to this non-dimensional variable is easily evaluated. The strain field, however, is defined as the derivative of the displacement field with respect to material coordinate α_1. Find the strain field.

Consider the axial displacement field, $u(\alpha_1)$, of a beam, for instance. The axial strain field is

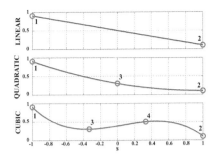

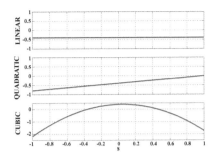

Fig. 17.3. Shape function. Linear: top figure. Quadratic: middle figure. Cubic: bottom figure. Red circles indicate the node locations.

Fig. 17.4. Derivatives of the shape function. Linear: top figure. Quadratic: middle figure. Cubic: bottom figure.

$$\epsilon = \frac{du}{d\alpha_1} = u' = \frac{du}{ds}\frac{ds}{d\alpha_1} = \frac{1}{J}\frac{du}{ds}, \qquad (17.11)$$

where the second equality follows from the chain rule for derivatives and $J = d\alpha_1/ds$ is the Jacobian of the coordinate transformation. Notation $(\cdot)'$ indicates a derivative with respect to material coordinate α_1.

To evaluate the Jacobian, an increment of the position vector is evaluated based on eq. (17.1a), $d\underline{r} = \underline{N}^{+}\hat{r}\,ds$, where $\underline{N}(s)$ is the displacement interpolation matrix defined by eq. (17.6) and $\hat{r}$ the array of nodal position vectors. It then follows that $d\alpha_1^2 = d\underline{r}^T d\underline{r} = \hat{r}^T \underline{N}^{+T}\underline{N}^{+}\hat{r}\,ds^2$, and the Jacobian becomes

$$J = \frac{d\alpha_1}{ds} = \sqrt{\hat{r}^T\underline{N}^{+T}\underline{N}^{+}\hat{r}}. \qquad (17.12)$$

The strain field, eq. (17.11), is now

$$\epsilon = u' = \frac{1}{J}u^{+} = \frac{1}{J}\underline{N}^{+}(s)\hat{u}, \qquad (17.13)$$

Example 17.3. Evaluation of the strain energy
The axial strain energy stored in a beam element is

$$A = \frac{1}{2}\int_0^{\ell} S(\alpha_1)\epsilon^2(\alpha_1)\,d\alpha_1, \qquad (17.14)$$

where ℓ is the length of the beam element, S its axial stiffness, and $\epsilon(\alpha_1)$ the axial strain field. Express the strain energy in terms of the elements nodal displacements.

Equation (17.13) gives the axial strain field in terms of the nodal displacements. Introducing this expression into eq. (17.14) leads to

$$A = \frac{1}{2}\int_0^{\ell} S(\alpha_1)\frac{1}{J^2}\hat{u}^T\underline{N}^{+T}(s)\underline{N}^{+}(s)\hat{u}\,\frac{d\alpha_1}{ds}\,ds$$
$$= \frac{1}{2}\hat{u}^T\left[\int_0^{\ell} S(\alpha_1)\frac{1}{J^2}\underline{N}^{+T}(s)\underline{N}^{+}(s)\,Jds\right]\hat{u} = \frac{1}{2}\hat{u}^T\underline{k}\,\hat{u}.$$

In the second equality, the nodal displacement array was placed outside of the integral because nodal quantities are independent of the integration variable, s. The bracketed quantity is the stiffness matrix of the element. The above equation gives the desired result, the strain energy expressed in terms of the nodal displacements.

The element's stiffness matrix is defined as

$$\underline{\underline{k}} = \int_0^\ell \frac{S(\alpha_1)}{J(\alpha_1)} \underline{\underline{N}}^{+T}(s) \underline{\underline{N}}^+(s) \, ds.$$

Typically, numerical integration is used to evaluate this integral, leading to the following expression for the stiffness matrix

$$\underline{\underline{k}} = \sum_{i=1}^{N_G} \frac{w_i S(s_i)}{J(s_i)} \underline{\underline{N}}^{+T}(s_i) \underline{\underline{N}}^+(s_i).$$

where s_i and w_i are Gauss' points and weights, respectively, as discussed in section 18.3.

17.2 Interpolation of rotation fields

The exact treatment of finite rotations is particularly important in multibody dynamics because finite rotations associated with the finite relative motions of the system's components are combined with the finite elastic motions of the flexible components. Consider, for instance, the motion of a helicopter rotor blade in which elastic deformation of the blade are superimposed onto the rigid body rotation of the entire rotor.

Interpolation of displacement fields is at the heart of the finite element method and basic interpolation techniques based on the definition of appropriate shape functions are reviewed in section 17.1. Application of the same, linear interpolation technique to finite rotation fields has been the subject of controversy, because finite rotation fields do not form a linear space.

Crisfield and Jelenić [325] were the first to point out a major deficiency of the classical interpolation techniques applied to rotation fields: its lack of objectivity. By definition, a rigid body motion generates no strains; hence, the strain field associated with a given displacement field must remain unaffected by the addition of a rigid body motion to the displacement field. If a computational scheme satisfies this condition, it is said to be "objective." Crisfield and Jelenić [325, 326] showed that classical interpolation formulæ applied to finite rotation fields violate this objectivity criterion. They prove the non-objectivity of the direct interpolation of total rotations [327], incremental rotations [328] and iterative rotations [304].

Crisfield and Jelenić argue that "all of these formulations can be regarded as stemming from the same family, for which the following is valid: the interpolation is applied to the rotation between a particular reference configuration and the current configuration. With hindsight, the nature of this interpolation is bound to make all of these formulations non-objective. The rotations interpolated in this way in general include rigid body rotations, so that the error, introduced by the interpolation, makes the resulting strain measures dependent on the rigid body rotation."

They also point out, however, that while the errors in the computed strain field are small and decrease with mesh p- or h-refinement, lack of objectivity persists if rotation increments or Newton-Raphson updates are interpolated. Crisfield and Jelenić proposed a novel interpolation technique that guarantees objectivity by splitting rotations into rigid and elastic components: the sole elastic component is interpolated. This approach is akin to the co-rotational formulation [329], but retains the fully nonlinear strain-configuration equations, rather than their linearized counterparts.

Betsch and Steinmann [330] proposed an alternative approach to achieving objectivity: instead of interpolating finite rotation parameters, they interpolate the unit vectors forming the columns of the finite rotation tensor and proved that this approach also satisfies the objectivity criterion. Linear interpolation of unit vectors, however, does not yield unit vectors, nor does it preserve their orthogonality. Special procedures were developed to guarantee that the interpolated results lead to orthogonal rotation tensors. Numerical examples were shown that demonstrate the accuracy of numerical predictions.

Romero *et al.* [331, 332] presented a comparison of different interpolation methods including the direct interpolation of finite rotations, the interpolation method proposed by Crisfield and Jelenić [325], and two new approaches, based on 1) the non-orthogonal interpolation of rotations with modification of geometrically exact beam theory and 2) the isoparametric interpolation of rotations followed by orthogonalization using polar decomposition. Numerical tests of all four methods showed that with the exception of the direct interpolation of finite rotations, the other methods are objective, path-independent and preserve the orthogonality of the rotation tensor. The proposed interpolation approaches, however, were shown to soften structural response, and could converge to erroneous solutions. They recommend the use of the interpolation approach of Crisfield and Jelenić.

Finally, Ibrahimbegović and Taylor [333] also proposed interpolation techniques that satisfy the objectivity criterion for geometrically exact structural models. Update formulæ are based on an incremental approach and rely on the representation of finite rotations based on quaternion quantities, which must be stored at each node of the model. Special attention was paid to the implementation details for applied support rotations and the corresponding modifications of the residual vector and tangent matrix introduced by the follower forces and moments.

Because of the percieved difficulties associated with the treatment of finite rotations, "rotationless formulations" have appeared in recent years. For instance, in the absolute nodal coordinate formulation [334], absolute displacements and global slopes are used as nodal coordinates, bypassing the need for finite rotations. Betsch and Steinmann [227] have advocated the use of the direction cosine matrix to represent finite rotations. It should be noted, however, that these rotationless formulations use more coordinates than the minimum set required to represent finite rotations, and hence, typically require more computational resources than their counterparts based on minimum set representations.

In this section, the problem of interpolation of finite rotations within the framework of geometrically exact structural elements is examined. For computational efficiency, it is desirable to use a minimal set representation of finite rotations, *i.e.*, three parameters only. While quaternions have been used in multibody dynamics simulations [250, 252], the computational costs of dealing with four parameters and the enforcement of the associated normality condition, eq. (13.20), have limited their use. The rescaling operation presented in section 13.6.2 is used to systematically eliminate singularities associated with such minimal set representations.

The rescaling operation is based on the observation that addition of a rotation of magnitude $\phi = \pm 2\pi$ to a finite rotation leaves the associated rotation tensor unchanged. While the concept of objectivity is based on the invariance of the strain field with respect to the addition of a rigid body motion to the rotation field, the concept of rescaling is based on the invariance of the rotation tensor with respect to the addition of a rotation of magnitude $\phi = \pm 2\pi$, *i.e.*, $\underline{\underline{R}}(\phi, \bar{n}) = \underline{\underline{R}}(\phi \pm 2\pi, \bar{n})$. In turn, this raises the question of invariance on the interpolation of finite rotation with respect to rescaling. It is shown that the basic interpolation algorithm proposed by Crisfield and Jelenić [325] to achieve objectivity, is also invariant with respect to rescaling operations.

The simple numerical tools used to interpolate displacement fields within one-dimensional finite elements, such as the cable or beam elements were presented in section 17.1. The challenges associated with the interpolation of rotation fields are addressed in section 17.2.1, with special attention devoted to the impact of rescaling operations. Rescaling also impacts the choice of unknowns, as discussed in section 17.2.2, and a new algorithm is proposed for the interpolation of incremental quantities. Finally, numerical examples are discussed that demonstrated the simplicity and efficiency of the proposed approach when applied to complex, flexible multibody systems.

17.2.1 Finite element discretization

Interpolation of the displacement field within an element is at the heart of the finite element discretization procedure and is summarized in section 17.1. Interpolation is a linear operation, acting on the displacement field which forms a linear space.

Interpolation of the displacement field

Let arrays $\hat{\underline{u}}_i$, $\hat{\underline{u}}$ and $\hat{\underline{u}}_f$ store the nodal displacements of the element at the beginning of a time step, the incremental nodal displacements, and the displacements at the end of a time step, respectively. Furthermore, let the displacement update at the nodes be written as $\hat{\underline{u}}_f = \hat{\underline{u}}_i + \hat{\underline{u}}$. It then follows that

$$\underline{u}_i(s) + \underline{u}(s) = \underline{\underline{N}}(s)(\hat{\underline{u}}_i + \hat{\underline{u}}) = \underline{\underline{N}}(s)\hat{\underline{u}}_f = \underline{u}_f(s), \qquad (17.15)$$

where $\underline{\underline{N}}(s)$ is the displacement interpolation matrix defined by eq. (17.6) and variable s a non-dimensional quantity defined along the span of the element. This important relationship implies that initial, final, and incremental fields can all three be interpolated with the same shape functions, and a simple update of the nodal values then guarantees compatibility of the interpolated displacement field over the entire element.

Interpolation of the rotation field

When formulating beam and shell elements, the kinematic description of the problem also requires interpolation of the rotation field and its derivative, written as

$$\underline{c}(s) = \underline{\underline{N}}(s)\hat{\underline{c}}, \quad \text{and} \quad \underline{c}'(s) = \underline{\underline{N}}'(s)\hat{\underline{c}} = \frac{1}{J}\underline{\underline{N}}^{+}(s)\hat{\underline{c}}, \qquad (17.16)$$

respectively, where $\hat{\underline{c}}$ is the array that stores the rotation parameter vectors at the nodes of the element. This interpolation simply provides an approximation to the rotation field within the element.

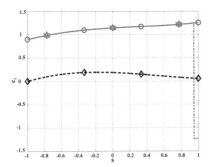

Fig. 17.5. Wiener-Milenković parameter, c_1, for the given rotation field; no rescaling is used at node 4. Nodal rotations: ($\circ$). Interpolation using eq. (17.16): solid line, corresponding Gauss point values: ($\triangle$). Relative nodal rotations: ($\diamond$). Interpolation of relative rotations: dashed line. Interpolation computed by algorithm 1: dashed-dotted line, corresponding Gauss point values: ($\triangledown$).

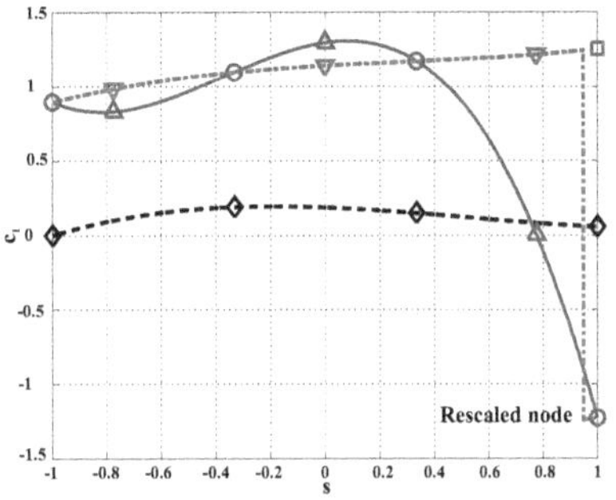

Fig. 17.6. Wiener-Milenković parameter, c_1, for the given rotation field; node 4 has been rescaled; for reference, the unscaled note 4 is indicated by ($\square$). Nodal rotations: ($\circ$). Interpolation using eq. (17.16): solid line, corresponding Gauss point values: ($\triangle$). Relative nodal rotations: ($\diamond$). Interpolation of relative rotations: dashed line. Interpolation computed by algorithm 1: dashed-dotted line, corresponding Gauss point values: ($\triangledown$).

Figure 17.5 shows the interpolated rotation field for a four-noded beam element using the cubic interpolation functions given by eq. (17.5). The rotations at the four nodes are defined by four rotation angles, $\phi_1 = 145°$, $\phi_2 = 160°$, $\phi_3 = 170°$, $\phi_4 = 181°$, and associated unit vectors,

$$[\bar{n}_1, \bar{n}_2, \bar{n}_3, \bar{n}_4] = \begin{bmatrix} 0.3049 & 0.3262 & 0.3193 & 0.3105 \\ 0.6097 & 0.6461 & 0.6095 & 0.5485 \\ 0.7316 & 0.6900 & 0.7256 & 0.7763 \end{bmatrix}.$$

The interpolated rotation field was computed using eq. (17.16), and the first component, c_1, of the Wiener-Milenković rotation parameter vector is shown in fig. 17.5; similar results are obtained for the other two components, c_2 and c_3. The curvature can be computed in a similar manner as $\underline{\kappa}(s) = \underline{\underline{H}}\,\underline{c}'(s)$, where the tangent tensor, $\underline{\underline{H}}(\underline{c})$, is defined by eq. (13.55) and $\underline{c}(s)$ and $\underline{c}'(s)$ by eq. (17.16). Figure 17.7 shows the first component, $\hat{\kappa}_1$, of the curvature vector.

Although the interpolation procedure of eq. (17.16) looks reasonable considering the results shown in fig. 17.5, it suffers several serious drawbacks. First, let $\underline{\hat{c}}_i$, $\underline{\hat{c}}$ and $\underline{\hat{c}}_f$ be the nodal rotation parameter vectors at the beginning of a time step, for the incremental rotation, and at the end of a time step, respectively. Proceeding as was done above for the displacement field implies that $\underline{c}_f(s) = \underline{c}_i(s) + \underline{c}(s)$ if the nodal updates are selected as $\underline{\hat{c}}_f = \underline{\hat{c}}_i + \underline{\hat{c}}$.

Unfortunately, as discussed in section 4.9, rotations do not form a linear space; they must be composed, not added. At the nodes, rotations should be updated using the following composition formula, $\underline{\hat{c}}_f = \underline{\hat{c}} \oplus \underline{\hat{c}}_i$, where the notation $\oplus$ is used to indicate the composition operation, as defined in eq. (13.82). Furthermore, the nonlinear character of composition operations and the linear character of the interpolation operation imply that $\underline{c}_f(s) \neq \underline{c}(s) \oplus \underline{c}_i(s)$ if $\underline{\hat{c}}_f = \underline{\hat{c}} \oplus \underline{\hat{c}}_i$. Consequently, if the nodal rotations are updated using the composition formula, the interpolated rotation field does not satisfy the same composition formula at all points along the element.

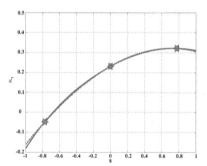

Fig. 17.7. First component of the curvature vector, κ_1, based on interpolation using eq. (17.16): solid line, corresponding Gauss point values: ($\triangleleft$). Curvatures computed by algorithm 1: dashed-dotted line, corresponding Gauss point values: ($\triangleright$).

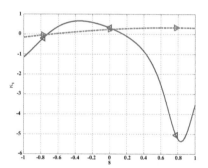

Fig. 17.8. First component of the curvature vector, κ_1, based on interpolation using eq. (17.16): solid line, corresponding Gauss point values: ($\triangleleft$). Curvatures computed by algorithm 1: dashed-dotted line, corresponding Gauss point values: ($\triangleright$).

The second drawback becomes obvious once the rescaling operation presented in section 13.6.2 is taken into account. The rotation at the fourth node of the element is of magnitude $\phi_4 = 181° > 180°$, and hence, should be rescaled to avoid singularities. The Wiener-Milenković rotation parameter vector at this node is $\underline{c}_4^T = \{1.253, 2.214, 3.132\}$, $\|\underline{c}_4\| = 4.035 > 4$, and the rescaled parameter vector is $\underline{c}_4^{\dagger T} = \{-1.231, -2.175, -3.078\}$, $\|\underline{c}_4^{\dagger}\| = 3.965 < 4$, as expected from the results presented in section 13.6.2.

Figure 17.6 shows the rotation field interpolated using eq. (17.16) in the presence of rescaling. Note that the results presented in this figure should be identical to those shown in fig. 17.5 because they correspond to the interpolation of identical configurations: indeed, the rotation tensor at node 4 is uniquely defined, but represented by different rotation parameters, $\underline{c}_4$ and $\underline{c}_4^{\dagger}$, due to rescaling. Clearly, the linear interpolation operation of eq. 17.16 is not invariant under the rescaling operation.

The curvature field is shown in fig. 17.8 and clearly, in the presence of rescaling, the results are erroneous: without rescaling, the three Gauss point values of the first curvature component are $\kappa_1 = -0.048, 0.230$, and 0.322, respectively, as compared to $\kappa_1 = -0.208, 0.308$, and -5.0521, respectively, in the presence of rescaling.

The Crisfield and Jelenić interpolation algorithm

Clearly, a more robust interpolation approach is necessary to deal with rotations in the presence of rescaling; the following algorithm was proposed by Crisfield and Jelenić [325].

Algorithm 1 (Rotation interpolation) *Interpolation of a rotation field defined by its rotation parameter vectors, $\hat{\underline{c}}$, at the nodes of a finite element.*

Step 1. Compute the nodal relative rotations, $\hat{\underline{r}}$, by removing the rigid body rotation, $\hat{\underline{c}}_1$, from the rotation at each node, $\hat{\underline{r}} = \hat{\underline{c}}^{1-} \oplus \hat{\underline{c}}$.
Step 2. Interpolate the relative rotation field, $\underline{r}(s) = \underline{N}(s)\hat{\underline{r}}$, and its derivative, $\underline{r}'(s) = \underline{N}'(s)\hat{\underline{r}}$. Find the curvature field, $\underline{\kappa} = \underline{R}(\underline{c}_1)\underline{H}(\underline{r})\,\underline{r}'$.
Step 3. Restore the rigid body rotation removed in step 1, $\underline{c}(s) = \underline{c}_1 \oplus \underline{r}(s)$.

Algorithm 1 removes the possible effects of rescaling from the interpolation procedure. In step 1, the relative rotations of the nodes with respect to node 1 are computed using the composition formula; note that the relative rotation field could be computed with respect to any of the nodes of the element, node 1 is simply a convenient choice. It is assumed here that the relative rotations within one single element are small enough that no rescaling is needed within the element, *i.e.*, within the element, $|\phi_r| < \pi$. If this condition were not to be satisfied, a finer mesh would be required to limit the relative rotation within each element. Next, these relative rotations are interpolated using standard procedures. Finally, the interpolated relative rotation is composed with the rotation of node 1 to find the interpolated rotation field.

The interpolated rotation field computed by algorithm 1 is also shown in figs. 17.5 and 17.6. Because the nodal rotations presented in these figures only differ by the rescaling of node 4, the relative rotation fields are identical, the corresponding curvature fields are identical, as are the interpolated rotation fields. The interpolated rotation field seems to present a discontinuity at $s = 0.973$ in both figures: this is due to the rescaling operation in step 3 of algorithm 1, but does not affect the quality of the interpolation. In fact, the interpolation procedure of algorithm 1 is able to deal with the discontinuities inherent to the required rescaling operations. The presence of these discontinuities, however, has implications on the linearization of the equations of motion, as discussed in section 17.2.2.

The third drawback of interpolation based on eq. (17.16) is its lack of objectivity when computing strain components. As shown in section 16.3.1, the strain measures of geometrically exact beam theory are invariant with respect to the addition of a rigid body motion. Because algorithm 1 is based on the interpolation of relative rotation, the addition of a rigid body motion is automatically filtered out from the interpolation step, ensuring the objectivity of the process. Jelenić and Crisfield [326] studied the lack of objectivity of interpolation schemes based on eq. (17.16) and concluded that "The non-invariance and path-dependence in these formulations decrease with both p-refinement and h-refinement and in practical applications cannot always be easily spotted."

These conclusions are supported by the data presented here: in fig. 17.7, the curvatures computed based on eq. (17.16) (non-objective) are nearly identical to those computed with algorithm 1 (objective). In fact, at the Gauss points, the curvature component, κ_1, computed by the two approaches only differ by 0.16, -0.085 and 0.16%, respectively. These discrepancies are minute compared to the gross disparities observed in fig. 17.8 in the presence of rescaling. The objectivity of the strain interpolation resulting from the use of algorithm 1 typically provides modest improvements in the quality of the interpolated strain field, but is indispensable when dealing with rotation fields involving potential rescaling.

17.2.2 Total versus incremental unknowns

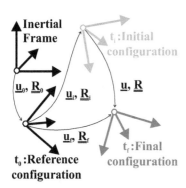

Fig. 17.9. Configuration of the system at various instants in time.

Multibody simulations typically proceed in discrete time steps. Figure 17.9 shows the inertial frame of reference, the reference, *i.e.*, unstressed, configuration of the beam at time $t = 0$, and its configurations at the beginning and end times of a typical time step, denoted t_i and t_f, respectively. Each frame is related to its parent frame by a finite motion characterized by a displacement vector and a rotation tensor, all resolved in the inertial frame. It is assumed that the dynamic simulation has successfully proceeded up to time t_i, *i.e.*, the corresponding displacement and rotation fields, denoted $\underline{u}_i$ and $\underline{\underline{R}}_i$, respectively, are known. Let $\underline{c}_i$ be a vectorial parameterization of the rotation tensor $\underline{\underline{R}}_i$.

To advance the solution from the initial to the final time of the time step, two sets of unknowns can be selected: the incremental displacements and rotations, denoted $\underline{u}$ and $\underline{\underline{R}}$, respectively, or the total displacements and rotations, denoted $\underline{u}_f$ and $\underline{\underline{R}}_f$, respectively, see fig. 17.9. Let $\underline{c}$ and $\underline{c}_f$ be parameterizations of the rotation tensors $\underline{\underline{R}}$ and $\underline{\underline{R}}_f$, respectively. From a kinematic viewpoint, both sets of unknowns are equivalent.

In typical dynamic simulations, however, small time steps must be selected to achieve convergence and guarantee the accuracy of the solution. Consequently, it can be assumed that incremental rotations will be of magnitude $|\phi| < \pi$; in fact, for most practical cases, $|\phi| \ll \pi$; indeed, $|\phi| = \pi$ implies that a component of the system rotates by 180° within a single time step. It cannot be assumed, however, that $|\phi_f|$, the rotation associated with rotation tensor $\underline{\underline{R}}_f$, is small, in fact, $|\phi_f| > \pi$ is likely to occur.

The implication of these observations is clear: if total rotations are used as unknowns, some of the rotation parameters, $\underline{c}_f$, will be rescaled, as required, whereas if incremental rotations are used as unknowns, none of the unknown parameters, $\underline{c}$, will be rescaled. Interpolation algorithm 1 was shown to seamlessly handle rescaling, however, when dealing with dynamic simulations, additional considerations must be taken into account.

Spatial and time discretization algorithms typically transform the governing partial differential equations of complex multibody systems into a set of nonlinear algebraic equations, which are solved in an iterative manner using the Newton-Raphson method. Inherent to this approach is a linearization process that transforms the nonlinear algebraic equations into their linearized counterparts. Consider, for instance, the linearization of the curvature vector, $\underline{\kappa} = \underline{\underline{H}}(\underline{c})\underline{c}'$, which will appear in the expression for the elastic forces of a beam element. Application of the linearization procedure leads to $\Delta\underline{\kappa} = \underline{\underline{H}}(\underline{c})\Delta\underline{c}' + \underline{\underline{M}}(\underline{c}, \underline{c}')\Delta\underline{c}$, where operator $\underline{\underline{M}}(\underline{c}, \underline{c}')$ is defined by eq. (13.67), and hence, operators $\underline{\underline{H}}(\underline{c})$ and $\underline{\underline{M}}(\underline{c}, \underline{c}')$ appear in the expression of the tangent stiffness matrix of the element.

Let $\underline{c}$ and $\underline{c}^\dagger$ denote a rotation parameter vector and its rescaled counterpart, respectively, as discussed in section 13.6.2. By construction of the rescaling operation, $\underline{\underline{R}}(\underline{c}) = \underline{\underline{R}}(\underline{c}^\dagger)$, but it is easily verified that $\underline{\underline{H}}(\underline{c}) \neq \underline{\underline{H}}(\underline{c}^\dagger)$ and $\underline{\underline{M}}(\underline{c}, \underline{c}') \neq \underline{\underline{M}}(\underline{c}^\dagger, \underline{c}^{\dagger\prime})$. Clearly, while intrinsic quantities such as the rotation tensor, the curvature vector, or elemental elastic forces are invariant to rescaling, and while the interpolation operation can be made invariant to the same rescaling through the use of algorithm 1, the tangent stiffness matrix is not invariant to rescaling.

The implications of this lack of invariance are easily understood by considering the situation depicted in fig. 17.6. Evaluation of the tangent stiffness matrix at the Gauss points uses the interpolated rotation field and its derivative, but ignores the fact node 4 was rescaled. The tangent stiffness matrix will be evaluated as if the rescaling of node 4 never took place, i.e., the equations are linearized about the wrong point. Hence, the search direction in the Newton-Raphson iteration process will be erroneous, which can ultimately cause failure of the iteration process.

17.2.3 Interpolation of incremental rotations

In view of the above discussion, it is desirable to work with incremental rotations because they remain small and do not require rescaling. The tangent stiffness matrix then always corresponds to the correct linearization of the problem. This contrasts with the choice of total rotations as unknowns for which these desirable features cannot be guaranteed.

The choice of incremental nodal rotations as unknowns requires interpolation of the incremental rotation field to compute the elemental elastic forces and tangent stiffness matrix. This task cannot be performed with the help of eq. (17.16): as already pointed out in section 13.6.2, the nonlinear nature of the composition operation is incompatible with the linear interpolation operation. An alternative approach is proposed for this operation.

Algorithm 2 (Incremental rotation interpolation) *Interpolation of the incremental rotation field between two configurations defined by nodal rotation parameter vectors, $\hat{\underline{c}}_i$ and $\hat{\underline{c}}_f$, of a finite element.*

Step 1. Use algorithm 1 to compute the interpolated rotation field, $\underline{c}_i(s)$, based on nodal values $\hat{\underline{c}}_i$.

Step 2. Use algorithm 1 to compute the interpolated rotation field, $\underline{c}_f(s)$, based on nodal values $\hat{\underline{c}}_f = \hat{\underline{c}} \oplus \hat{\underline{c}}_i$.

Step 3. Compute the incremental rotation field by composition: $\underline{c}(s) = \underline{c}_f(s) \oplus \underline{c}_i^-(s)$.

This approach is different from that proposed by Cardona and Géradin, who directly interpolated incremental rotations using eq. 17.16. It is also different from the algorithm proposed by Crisfield and Jelenić [325].

Example 17.4. The use of total versus incremental unknowns

In this example, the use of total versus incremental unknowns will be contrasted, to underline the difficulties associated with the use of total rotations in the formulation of dynamic problems. Consider a free-free beam featuring the following physical properties: axial stiffness $S = 9.28$ kN, shearing stiffness $K_{22} = K_{33} = 3.57$ kN, torsional stiffness $J = 65.2$ N·m², bending stiffness $I_{22} = I_{33} = 32.6$ N·m², and mass per unit length $m = 0.35$ kg/m. The beam is modeled using a single cubic element and is subjected to two mutually orthogonal end bending moments Q_2 and Q_3, both acting in directions normal to the axis of the beam. Both bending moments have a triangular time history: starting from zero value at time $t = 0$, growing linearly to a maximum value of 0.3 N·m at $t = 0.5$ s, linearly decreasing to a zero value at time $t = 1$ s, and remaining zero at all subsequent times.

The dynamic response of the beam was computed using time step sizes $\Delta t = 1$ and 0.1 ms, with formulations using both total and incremental unknowns. Algorithms 1 and 2 were used to interpolate the total and incremental rotations, respectively. Figure 17.10 shows the third component of the Wiener-Milenković rotation parameter vector at the beam's end opposite to the applied bending moments, for $\Delta t = 1$ ms; the formulations using total and incremental unknowns lead to nearly identical predictions.

The rescaling operation that occurs at time $t = 0.929$ s is evident in fig. 17.10. All four nodes of the element, however, are rescaled simultaneously and the rotation interpolation procedure performs well with both total and incremental unknowns.

Next, the time step size was reduced to $\Delta t = 0.1$ ms. Due to this smaller time step size, the node at the unloaded end of the beam was rescaled at time $t = 0.9284$

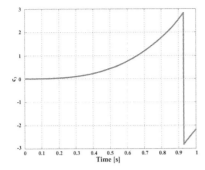

Fig. 17.10. Time histories of third component of the conformal rotation vector at the end node: incremental formulation: solid line; total formulation: dashed line.

Fig. 17.11. Rotating cantilevered beam subjected to transverse tip force.

s, while the other three nodes of the element were not. As expected, the formulation using total unknowns failed to converge at that time step, in contrast with that using incremental unknowns that proceeded uneventfully.

This example call for the following observations. If the rotation field is interpolated with eq. (17.16) without ever rescaling the rotation parameters, the computation will proceed smoothly at first; although the interpolated strain field is not objective, errors remain small, particularly if higher-order elements are used with fine meshes.

During the simulation, rotation magnitudes grow; no matter what parameterization is used to represent rotations, a singularity will eventually be reached and the simulation will fail at that point. On the other hand, if the rotation field is interpolated with eq. (17.16) with rescaling of the rotation parameters, the computation will proceed smoothly at first, although the interpolated strain field is not objective.

When the first node of the model is rescaled, the strain field computed in the elements connected to this node will be grossly erroneous, see fig. 17.8, and typically, convergence will not be reached for that time step at which rescaling occurs.

Finally, if algorithm 1 is used for the interpolation of the strain field, the simulation is not affected by rescaling of the rotation parameters that takes place whenever required, and the computed strain field is objective. The rescaling operation becomes transparent to the computation process.

However, evaluations of the tangent stiffness matrix based on interpolations of total unknowns computed with algorithm 1 can yield erroneous search directions in the Newton-Raphson process used to solve the nonlinear equations, which are inherent to time-stepping procedures. This can destabilize simulations.

Therefore, the use incremental unknowns in conjunction with algorithm 2 is recommended. This method preserves the objectivity of geometrically exact formulations, yields tangent stiffness matrices and residual vectors that are invariant to the rescaling of rotations, and enables the use of geometrically exact structural models in multibody simulations.

Example 17.5. Assessing the accuracy of algorithm 2

Figure 17.11 depicts a cantilevered beam rotating about an axis normal to its axis and passing through its root. The beam's physical properties are identical to those used in example 17.4 and it is subjected to a transverse tip load, F, linearly increasing from 0 to 50 N in one second. The beam rotates at an angular speed, Ω, linearly increasing from 0 to 4 rad/s in the same time. The system was simulated for 1.5 s with a time step size $h = 0.01$ s.

Given the results of example 17.4, the simulations presented here only use incremental unknowns. The direct interpolation of rotation increments through eq. (17.16) will be contrasted with the interpolation technique described in algorithm 2. Figure 17.12 shows the error in the beam's root shear force as a function of the number of linear elements used to mesh the beam. Figure 17.13 shows the corresponding results for quadratic elements. The reference solution for the error analysis was obtained using a 250 cubic element mesh for which convergence was established.

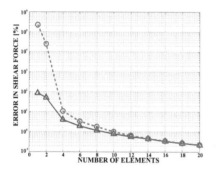

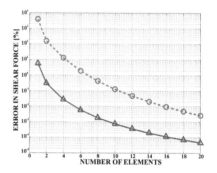

Fig. 17.12. Beam root shear force error versus number of linear elements. Interpolation using algorithm 2: solid line; direct interpolation using eq. 17.16: dashed line.

Fig. 17.13. Beam root shear force error versus number of quadratic elements. Interpolation using algorithm 2: solid line; direct interpolation using eq. 17.16: dashed line.

For both linear and quadratic elements, direct rotation interpolation using eq. (17.16) leads to large errors when coarse meshes are used, but these errors decrease rapidly for both h- and p-refinements. Indeed, the errors observed for the quadratic element mesh are far smaller than those for the linear element mesh. When algorithm 2 is used to interpolate rotation increments, errors are further reduced, although this reduction is less pronounced for finer meshes.

Since the computational cost associated with the use of algorithm 2 is nearly identical to that of using eq. (17.16), the use of the former is advisable. Indeed, achieving a 0.01% error in root shear force with quadratic elements requires 5 elements with algorithm 2, but 16 elements for eq. (17.16); this will result in a nearly threefold gain in computational cost when using algorithm 2.

Example 17.6. Rotorcraft tail rotor transmission

This example presents the modeling of a helicopter supercritical tail rotor transmission. Figure 17.14 shows the configuration of the system. The aft part of the helicopter is modeled and consists of a 6 m fuselage section that connects at a 45 degree angle to a 1.2 m projected length tail section. This structure supports the transmission to which it is connected at points **M** and **T** by means of 0.25 m support brackets. The transmission is broken into two shafts, each connected to flexible couplings at either end. The flexible couplings are represented by flexible joints, consisting of concentrated springs.

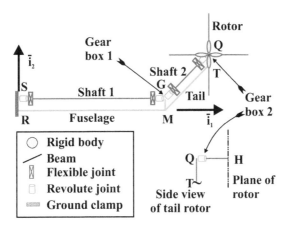

Fig. 17.14. Configuration of a tail rotor transmission.

Shaft 1 is connected to a revolute joint at point **S**, and gear box 1 at point **G**. Shaft 2 is connected to gear box 1 and gear box 2 which in turn, transmits power to the tail rotor. The plane of the tail rotor is at a 0.3 m offset with respect to the plane defined by the fuselage and tail, and its hub is connected to gear box 2 by means of a short shaft. Each tail rotor blade has a length of 0.8 m and is connected to the rotor hub at point **H** through a rigid root attachment of length 0.2 m. The gear ratios for gear boxes 1 and 2 are 1:1 and 2:1, respectively.

The fuselage has the following physical characteristics: axial stiffness $S = 687$ MN, bending stiffnesses $I_{22} = 19.2$ and $I_{33} = 26.9$ MN·m², torsional stiffness $J = 8.77$ MN·m², and mass per unit span $m = 15.65$ kg/m. The properties of the tail are one third of those of the fuselage.

Shafts 1 and 2 have the following physical characteristics: axial stiffness $S = 22.9$ MN, bending stiffnesses $I_{22} = 26.7$ and $I_{33} = 27.7$ kN·m², torsional stiffness $J = 22.1$ kN·m², and mass per unit span $m = 0.848$ kg/m. The center of mass of the shaft has a 1 mm offset with respect to the shaft reference line. The small difference in bending stiffnesses together with the center of mass offset are meant to represent an initial manufacturing imperfection or an unbalance in the shaft.

The stiffness properties of the flexible couplings are as follows: axial stiffness 5 kN/m and damping 0.5 N·sec/m, transverse stiffnesses 1 MN/m, torsional stiffness 0.1 MN·m/rad, and bending stiffnesses 0.1 kN·m/rad. Finally, gear boxes 1 and 2 have a concentrated mass of 5 kg each, and the tail rotor a 15 kg mass with a polar moment of inertia of 3 kg·m^2.

At first, a static analysis of the system was performed for various constant angular velocities of the drive train. The natural frequencies of the system were computed about each equilibrium configuration. When shaft 1 does not rotate, its two lowest natural frequencies of shaft 1 were found to be $\omega_1 = 46.9$ and $\omega_2 = 49.1$ rad/s. According to linear theory, the system is stable when the shaft angular velocity is below ω_1 or above ω_2, but unstable between theses two speeds.

The system was loaded by a torque acting at the root of shaft 1, featuring the following time history: $Q(t) = 50 \ (1 - \cos 2\pi t)$ for $0 < t < 1$ s, $Q(t) = 0$ for $1 < t < 2$ s, $Q(t) = 6 \ (1 - \cos 2\pi t)$ for $2 < t < 3$ s, and $Q(t) = 0$ for $3 < t < 6$ s. After 1 s, the angular velocity of shaft 1 stabilizes at about 45 rad/s, below the critical speed. The torque applied for $2 < t < 3$ s then accelerates the transmission through the critical zone to reach an angular velocity of 50.5 rad/s. A constant time step size $h = 0.5$ ms was used for the entire simulation.

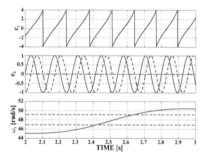

Fig. 17.15. Top figure: first component of the Wiener-Milenković rotation parameter vector. Middle figure: components of unit vector $\bar{e}_2$. Bottom figure: shaft 1 mid-span angular speed.

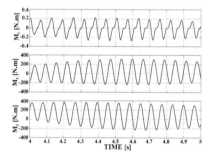

Fig. 17.16. Moments at the mid-span location of shaft 1. Top figure: torque, M_1. Middle figure: bending moments M_2. Bottom figure: bending moments M_3.

Figure 17.15 shows the dynamic response at shaft 1 mid-span position for $2 < t < 3$ s. The top portion of the figure shows the first component of the Wiener-Milenković rotation parameter vector: a rescaling operation occurs for each complete revolution of the shaft. The middle portion of the figure shows the components of the unit vector $\bar{e}_2$, *i.e.*, the second column of the rotation tensor. As expected, these quantities are continuous, as they do not "see" the rescaling operations. Finally, the bottom portion of the figure shows the angular velocity of the shaft. The horizontal dashed line indicate the unstable zone for the shaft.

The angular velocity of shaft 1 passes through this critical zone fast enough to avoid the build up of lateral vibrations. Here again, the angular velocity is continuous, unaffected by the rotation rescaling operations. Figure 17.16 shows the torque, M_1, and the two bending moments, M_2 and M_3, at shaft 1's mid-span, for $4 < t < 5$ s. Since the shaft has just passed through the critical zone, fairly large bending moments are observed. Here again, all quantities are continuous, despite the multiple rescaling operations. This example demonstrates the ability of the algorithm 2 to handle rotations of arbitrary magnitudes in complex, flexible multibody systems. The rescaling operations are applied at those nodes where they are required to avoid singularities in rotation representations. All other quantities, such as the rotation tensor, angular velocities, or bending moments are continuous and unaffected by the rescaling operations.

17.3 Governing equations and linearization process

The governing equations for multibody systems can take many different forms. Prior to performing a dynamic analysis, it is often informative to carry out a static analysis, for which all inertial forces are assumed to vanish. Both linear and nonlinear static problems arise both with and without kinematic constraints, as discussed in section 17.3.1. Next, the equations governing linear structural dynamics problems will be reviewed in section 17.3.3, leading to the nonlinear problems discussed in section 17.3.4. Finally, the governing equations for typical multibody systems are presented, for holonomic and nonholonomic systems in sections 17.3.5 and 17.3.6, respectively.

17.3.1 Statics problems

Consider first a simple linear, unconstrained static problem characterized by a system of linear equations,

$$\underline{\underline{K}}\,\underline{q} = \underline{f}, \tag{17.17}$$

where array $\underline{q}$ stores the n generalized coordinates of the system, $\underline{\underline{K}}$ is the constant stiffness matrix, and $\underline{f}$ the externally applied forces.

For complex elastic structures, static problems are typically formulated using finite element techniques [324, 197, 198]. A large number generalized coordinates, corresponding to the displacement components at all nodes, will be present and a two step procedure is generally used to solve linear system (17.17). First, the symmetric stiffness matrix is factorized as $\underline{\underline{K}} = \underline{\underline{L}}\,\underline{\underline{D}}\,\underline{\underline{L}}^T$, where $\underline{\underline{L}}$ is a lower triangular matrix and $\underline{\underline{D}}$ a diagonal matrix. Next, the solution is found through backsubstitution. This algorithm, often referred to as the *active column solver*, is well documented numerous textbooks [324, 197, 198].

Nonlinear static problems

More often than not, the elastic forces are nonlinear functions of the generalized coordinates, and the governing equations express the equilibrium of the system as $\underline{f}^E(\underline{q}) = \underline{f}$, where $\underline{f}^E$ are the elastic forces. This nonlinear static problem will be solved through successive linearization of the governing equations.

Given an approximate solution, $\underline{q}$, a more accurate solution, $\underline{q} + \Delta\underline{q}$, is sought, where $\Delta\underline{q}$ are the unknown increments in generalized coordinates. The new solution is assumed to satisfy the governing equations of the problem, i.e., $\underline{f}^E(\underline{q} + \Delta\underline{q}) = \underline{f}$. A Taylor series expansion is performed about the known, approximate equilibrium solution, to find

$$\underline{f}^E(\underline{q}) + \frac{\partial \underline{f}^E}{\partial \underline{q}} \Delta\underline{q} + \text{h.o.t.} = \underline{f}. \tag{17.18}$$

The externally applied load array, $\underline{f}$, is assumed here to be known and independent of the generalized coordinates. If this is not the case, this array would also be approximated using a Taylor series expansion and treated in a manner similar to the elastic forces.

Next, the solution increments, $\Delta\underline{q}$, are assumed to be small quantities and the higher-order terms in the Taylor series expansion are neglected, leading to a linearized problem for these increments,

$$\underline{\underline{K}} \Delta\underline{q} = \underline{f} - \underline{f}^E(\underline{q}), \tag{17.19}$$

where array $\underline{f} - \underline{f}^E(\underline{q})$ is called the *out-of-balance force array*. The stiffness matrix, $\underline{\underline{K}}(\underline{q})$, corresponds to the derivatives of the elastic forces with respected to the generalized coordinates,

$$\underline{\underline{K}}(\underline{q}) = \frac{\partial \underline{f}^E(\underline{q})}{\partial \underline{q}}. \tag{17.20}$$

This solution procedure, known as the *Newton-Raphson method*, is of an iterative nature. Starting from an approximate solution, $\underline{q}^{(1)}$, the stiffness matrix, $\underline{\underline{K}}(\underline{q}^{(1)})$, is evaluated first and increments are obtained from the solution of system (17.19) as $\Delta\underline{q}^{(1)} = \underline{\underline{K}}^{-1}(\underline{q}^{(1)})[\underline{f} - \underline{f}^E(\underline{q}^{(1)})]$. The new solution is then $\underline{q}^{(2)} = \underline{q}^{(1)} + \Delta\underline{q}^{(1)}$.

At the kth iteration, the approximate solution is denoted $\underline{q}^{(k)}$ and the next approximate solution, denoted $\underline{q}^{(k+1)}$, is obtained as $\underline{q}^{(k+1)} = \underline{q}^{(k)} + \underline{\underline{K}}^{-1}(\underline{q}^{(k)})[\underline{f} - \underline{f}^E(\underline{q}^{(k)})]$. At convergence, the norm of the out-of-balance force array becomes small, $\|\underline{f} - \underline{f}^E(\underline{q}^{(k)})\| < \epsilon$, where ϵ is a small positive number, which implies that $\underline{q}^{(k)}$ is a good approximation to the exact solution of the nonlinear problem.

Because the most computationally expensive step of the procedure is the factorization of the stiffness matrix, it often efficient to keep the stiffness matrix unchanged for several iterations. For instance, at the second iteration, the new approximation is obtained as $\underline{q}^{(3)} = \underline{q}^{(2)} + \underline{\underline{K}}^{-1}(\underline{q}^{(1)})\{\underline{f} - \underline{f}^E(\underline{q}^{(2)})\}$. The sole elastic forces, $\underline{f}^E(\underline{q}^{(2)})$, are evaluated at the second iteration; the stiffness matrix and its factorization are kept unchanged, resulting in considerable computational savings. This approach is known as the *modified Newton-Raphson method*.

Constrained statics problems

Consider now a static problem subjected to m holonomic constraints, denoted $\underline{C}(\underline{q}) = 0$. The nonlinear equilibrium equations of the problem are stated as

$$\underline{f}^E(\underline{q}) + \underline{\underline{B}}^T(\underline{q})\underline{\lambda} = \underline{f}, \tag{17.21a}$$

$$\underline{C}(\underline{q}) = \underline{0}. \tag{17.21b}$$

The constraint forces, $\underline{\underline{B}}^T\underline{\lambda}$, appear in the equilibrium equation of the system, eq. (17.21a), and eq. (17.21b) states the constraints imposed on the system. The constraints were enforced via the Lagrange multiplier technique; $\underline{\lambda}$ denotes Lagrange's multipliers and $\underline{B}(q)$ the constraint matrix defined by eq. (9.37).

Here again, system (17.21) will be solved using a linearization technique. The two sets of equations are expanded using Taylor series about a known, approximate solution, $(\underline{q}, \underline{\lambda})$, leading to

$$\underline{f}^E(\underline{q}) + \underline{\underline{K}}\,\Delta\underline{q} + \underline{\underline{B}}^T(\underline{q})\underline{\lambda} + \underline{\underline{B}}^T(\underline{q})\Delta\underline{\lambda} + \underline{\underline{K}}^b(\underline{q},\underline{\lambda})\Delta\underline{q} + \text{h.o.t.} = \underline{f}, \tag{17.22a}$$

$$\underline{C}(\underline{q}) + \underline{\underline{K}}^c(\underline{q})\Delta\underline{q} + \text{h.o.t.} = \underline{0}. \tag{17.22b}$$

The linearization of the elastic forces involves the stiffness matrix defined by eq. (17.20). The linearization of the constraint forces involves two matrices, the constraint matrix $\underline{\underline{B}}$ defined by eq. (9.37) and the constraint related stiffness matrix,

$$\underline{\underline{K}}^b(\underline{q},\underline{\lambda}) = \frac{\partial(\underline{\underline{B}}^T\underline{\lambda})}{\partial\underline{q}}. \tag{17.23}$$

Finally, the linearization of the constraint gives rise to matrix $\underline{\underline{K}}^c$ defined as

$$\underline{\underline{K}}^c(\underline{q},t) = \frac{\partial\underline{C}}{\partial\underline{q}}. \tag{17.24}$$

Here again, solution increments are assumed to be small quantities and higher-order terms in the Taylor series expansion are neglected, leading to a linearized problem for these increments,

$$\begin{bmatrix} \underline{\underline{K}} + \underline{\underline{K}}^b & \underline{\underline{B}}^T \\ \underline{\underline{K}}^c & \underline{\underline{0}} \end{bmatrix} \begin{Bmatrix} \Delta\underline{q} \\ \Delta\underline{\lambda} \end{Bmatrix} = \begin{Bmatrix} \underline{f} - \underline{f}^E(\underline{q}) \\ -\underline{C}(\underline{q}) \end{Bmatrix}. \tag{17.25}$$

An iterative approach based on the Newton-Raphson or modified Newton-Raphson method then yields the solution of the nonlinear constrained system.

17.3.2 Problems

Problem 17.1. Rigid body with a root spring
Consider the rigid bar of length L with a root spring, as depicted in fig. 17.17. The bar is subjected to a tip vertical force P. The root spring is nonlinear such that $M = k_1\theta + k_3\theta^3$,

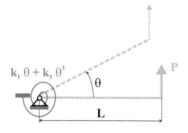

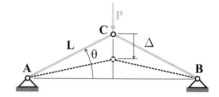

Fig. 17.17. Rigid body with a root spring.   **Fig. 17.18.** Shallow arch under center load.

where M is the root moment and θ the root rotation. Let $k_1/L = 50$ N/rad and $k_3/L = 15$ N/rad^3. (1) Write the nonlinear governing equations of the system. (2) Linearize the governing equations. (3) Use an iterative technique to find the response of the system θ as a function of the applied load P. Plot θ as a function of $P \in [0, 600]$ N.

Problem 17.2. Snap-through behavior of a shallow arch
The shallow arch depicted in fig. 17.18 is subjected to a center vertical load, P. The shallow arch is modeled by two articulated bars of length L, pinned at points **A**, **B**, and **C**. At point **C**, a vertical load P is applied and the deflection of point **C** under the load is denoted Δ. The constitutive law for the two identical bars is $F = ke$, where k is the axial stiffness of the bar, F the applied axial load, and $e = (L' - L)/L$ the resulting axial strain. The initial length of the bar is L and its length under load is denoted L'. (1) Find the strain in the bar in terms of the non-dimensional vertical displacement, $\bar{\Delta} = \Delta/L$. (2) Find the non-dimensional applied load, $\bar{P} = P/(2k)$ versus $\bar{\Delta}$. (3) On one graph, plot $\bar{P}$ versus $\bar{\Delta}$ for $\theta = 10, 20,$ and 30 degrees. (4) The shallow arch snaps through when $dP/d\Delta = 0$. Find the strain at snap-through, e_s and plot e_s versus angle θ. (5) Find the vertical deflection at snap-through, $\bar{\Delta}_s$ and plot $\bar{\Delta}_s$ versus θ. (6) Find the applied load at snap-through, $\bar{P}_s$ and plot $\bar{P}_s$ versus θ.

17.3.3 Linear structural dynamics problems

Consider next a linear structural dynamics problem characterized by the following equations of motion

$$\underline{\underline{M}}\,\ddot{\underline{q}} + \underline{\underline{C}}\,\dot{\underline{q}} + \underline{\underline{K}}\,\underline{q} = \underline{f}(t), \tag{17.26}$$

where array $\underline{q}$ stores the n generalized coordinates, $\underline{\underline{M}}$, $\underline{\underline{C}}$, and $\underline{\underline{K}}$ are the constant mass, damping, and stiffness matrices of the system, respectively, and $\underline{f}(t)$ the externally applied, time-dependent force array. These equations of motion form a set of linear, second-order, coupled ordinary differential equations.

For complex elastic structures, dynamics problems are typically formulated using the finite element method. Procedures for formulating the mass and stiffness matrices of such problems are well documented in textbooks [324, 197, 198]. Because energy dissipating mechanisms are difficult to model rigorously, the damping matrix is often approximated. The *Rayleigh damping* assumption is often used and corresponds to the choice of a damping matrix written as a linear combination of the mass and stiffness matrices, $\underline{\underline{C}} = \alpha\underline{\underline{M}} + \beta\underline{\underline{K}}$, where parameters α and β are selected based on experimental observations.

The initial conditions of the problem are the initial displacements and velocities of the system, $\underline{q}(t_i) = \underline{q}_i$, and $\underline{\dot{q}}(t_i) = \underline{v}_i$, where t_i is the initial time of the simulation. The initial accelerations can be obtained by expressing the dynamic equilibrium conditions, eq. (17.26), at time t_i, to find $\underline{\ddot{q}}(t_i) = \underline{M}^{-1}\left[\underline{f}(t_i) - \underline{C}\,\underline{v}_i - \underline{K}\,\underline{q}_i\right]$.

17.3.4 Nonlinear structural dynamics problems

Many practical engineering problems involve dynamical systems presenting large displacements and rotations, *i.e.*, *geometric nonlinearities*, or large deformations resulting in nonlinear material behavior, *i.e.*, *material nonlinearities*. Such nonlinear structural dynamics problems are described by the following dynamic equilibrium equations

$$\underline{M}(\underline{q}, t)\underline{a} + \underline{f}(\underline{q}, \underline{v}, t) = \underline{0}, \tag{17.27}$$

where arrays $\underline{q}$, $\underline{v} = \underline{\dot{q}}$, and $\underline{a} = \underline{\ddot{q}}$ store the n generalized displacement, velocity, and acceleration components of the system, respectively. The mass matrix is symmetric, positive-definite, and array $\underline{f}$ stores the dynamic and externally applied forces. These equations of motion form a set of second-order, coupled, nonlinear, ordinary differential equations, and exhibit a linear dependency on generalized accelerations because they are derived from Newton's second law.

For multibody systems, the generalized coordinates are likely to involve both displacements and rotations. For instance, six generalized coordinates, three displacements and three rotations, are used to represent the configuration of rigid bodies and the formulation of geometrically exact beams presented in section 16.3.1 calls for both displacements and rotation fields. Rotations could be represented using the vectorial parameterization of rotation discussed in section 13.4.

Because the problem is nonlinear, it is necessary to linearize the equations of motion following a procedure similar to that developed in section 17.3.1 for nonlinear static problems. Given an approximate solution characterized by generalized displacement, velocity, and acceleration arrays, denoted $\underline{q}$, $\underline{v}$, and $\underline{a}$, respectively, a more accurate solution characterized by arrays $\underline{q}+\Delta\underline{q}$, $\underline{v}+\Delta\underline{v}$, and $\underline{a}+\Delta\underline{a}$, is sought, where $\Delta\underline{q}$, $\Delta\underline{v}$, and $\Delta\underline{a}$ are the unknown increments in generalized displacements, velocities, and accelerations, respectively.

The new solution is assumed to satisfy the governing equations of the problem, *i.e.*, $\underline{M}(\underline{q} + \Delta\underline{q}, t)(\underline{a} + \Delta\underline{a}) + \underline{f}(\underline{q} + \Delta\underline{q}, \underline{v} + \Delta\underline{v}, t) = \underline{0}$. A Taylor series expansion is performed about the known, approximate solution, to find

$$\underline{M}(\underline{q}, t)\underline{a} + \underline{f}(\underline{q}, \underline{v}, t) + \underline{K}(\underline{q}, \underline{v}, \underline{a}, t)\Delta\underline{q} + \underline{G}(\underline{q}, \underline{v}, t)\Delta\underline{v} + \underline{M}(\underline{q}, t)\Delta\underline{a} + \text{h.o.t} = \underline{0}.$$

The stiffness, gyroscopic, and mass matrices, denoted $\underline{K}$, $\underline{G}$, and $\underline{M}$, respectively, are defined as the derivative of all forces with respect to the displacement, velocity, and acceleration components, respectively, *i.e.*,

$$K(q, v, a, t) = \frac{\partial(M a + f)}{\partial q}, \tag{17.28a}$$

$$G(q, v, t) = \frac{\partial(M a + f)}{\partial v}, \tag{17.28b}$$

$$M(q, t) = \frac{\partial(M a + f)}{\partial a}, \tag{17.28c}$$

Next, the solution increments, Δq, Δv, and Δa, are assumed to be small quantities and the higher-order terms in the Taylor series expansion are neglected, leading to a linearized problem for these increments,

$$K(q, v, a, t)\Delta q + G(q, v, t)\Delta v + M(q, t)\Delta a = -M(q, t)a - f(q, v, t). \tag{17.29}$$

In their linearized form, the governing equations of the system now resemble their counterparts for linear systems, eqs. (17.26). The mass, gyroscopic, and stiffness matrices, however, are now functions of the states of the system. As discussed in section 12.6.1, it is desirable to scale the equations of motion equations to find

$$h^2 K \Delta q + h G h \Delta v + M h^2 \Delta a = - \left(M h^2 a + h^2 f \right), \tag{17.30}$$

where h is the time step size that will be use in the time integration procedure.

17.3.5 Multibody dynamics problems with holonomic constraints

Multibody systems are typically subjected to constraints; the present section deals with nonlinear multibody systems featuring n generalized coordinates and m holonomic. Problems involving nonholonomic constraints will be addressed in section 17.3.6.

Nonlinear multibody systems subjected to holonomic constraints are described by Lagrange's equations of the first kind developed in section 11.1 and repeated here for convenience,

$$M(q, t)a + f(q, v, t) + B^T(q, t)\lambda = 0, \tag{17.31a}$$

$$C(q, t) = 0. \tag{17.31b}$$

Array f stores the dynamic and externally applied forces. The constraint forces associated with the holonomic constraints are given by the term $B^T \lambda$, where B is the constraint matrix and λ the array of Lagrange's multipliers used to enforce the constraint. Although equations (17.31) describe fully nonlinear multibody systems, their dependency on Lagrange's multipliers is linear.

Because the governing equations are nonlinear, their solution calls once more for a linearization process. The linearization of the dynamical terms, $M(q, t)a + f(q, v, t)$, gives rise to the stiffness, gyroscopic, and mass matrices defined in eqs. (17.28). Linearization of the constraint forces gives rise to matrix K^b defined by eq. (17.23), and similarly, linearization of the constraint gives rise to matrix $\underline{\underline{K}}^c$ defined by eq. (17.24).

Following a procedure identical to that developed in section 17.3.4 for the equations of nonlinear structural dynamics, the linearized equations for the present problem are obtained. As discussed in section 12.6.1, it is desirable to scale the equations of motion equations to find

$$(h^2 \underline{\underline{K}} + s\underline{\underline{K}}^b)\Delta\underline{q} + h\underline{\underline{G}}h\Delta\underline{v} + \underline{\underline{M}}h^2\Delta\underline{a} + s\underline{\underline{B}}^T\Delta\hat{\underline{\mu}} = -\underline{\underline{M}}h^2\underline{a} - h^2\underline{f} - s\underline{\underline{B}}^T\hat{\underline{\mu}},$$
(17.32a)

$$s\underline{\underline{K}}^c\Delta\underline{q} = -s\underline{\mathcal{C}}(\underline{q},t).$$
(17.32b)

where $\hat{\underline{\mu}}$ is the array of modified Lagrange's multipliers defined by eq. (12.23).

17.3.6 Multibody dynamics problems with nonholonomic constraints

If the multibody system is subjected to nonholonomic constraints of the form given by eq. (11.2), Lagrange's equations of the first kind developed in section 11.1 are still applicable,

$$\underline{\underline{M}}(\underline{q},t)\underline{a} + \underline{f}(\underline{q},\underline{v},t) + \underline{\underline{B}}^T(\underline{q},t)\underline{\lambda} = \underline{0},$$
(17.33a)

$$\underline{\mathcal{D}} = \underline{0}.$$
(17.33b)

The nonholonomic constraints are expressed by eq. (11.2) as $\underline{\mathcal{D}} = \underline{\underline{B}}(\underline{q},t)\underline{v} + \underline{b}(\underline{q},t) = \underline{0}$ and are assumed to present a linear dependency on the generalized velocities. The constraint forces associated with the nonholonomic constraints are given by the term $\underline{\underline{B}}^T\underline{\lambda}$, where $\underline{\underline{B}}$ is the constraint matrix and $\underline{\lambda}$ the array of Lagrange's multipliers used to enforce the constraint. Although equations (17.33) describe fully nonlinear multibody systems, their dependency on Lagrange's multipliers is linear.

Because the governing equations are nonlinear, their solution calls once more for a linearization process. The linearization of the dynamical terms, $\underline{\underline{M}}(\underline{q},t)\underline{a} + \underline{f}(\underline{q},\underline{v},t)$, gives rise to the stiffness, gyroscopic, and mass matrices defined in eqs. (17.28). Linearization of the constraint forces gives rise to matrix $\underline{\underline{K}}^b$ defined by eq. (17.23). Finally, linearization of the nonholonomic constraints introduces the following matrix

$$\underline{\underline{K}}^d(\underline{q},\underline{v},t) = \frac{\partial\underline{\mathcal{D}}}{\partial\underline{q}}.$$
(17.34)

Following, once again, a procedure identical to that developed in section 17.3.4 for the equations of nonlinear structural dynamics, the linearized equations of present problem are obtained. As discussed in section 12.6.1, it is desirable to scale the equations of motion equations to find

$$(h^2\underline{\underline{K}} + s\underline{\underline{K}}^b)\Delta\underline{q} + h\underline{\underline{G}}h\Delta\underline{v} + \underline{\underline{M}}h^2\Delta\underline{a} + s\underline{\underline{B}}^T\Delta\hat{\underline{\mu}} = -\underline{\underline{M}}h^2\underline{a} - h^2\underline{f} - s\underline{\underline{B}}^T\hat{\underline{\mu}},$$
(17.35a)

$$sh\underline{\underline{K}}^d\Delta\underline{q} + s\underline{\underline{B}}h\Delta\underline{v} = -s\underline{\underline{B}}(\underline{q},t)\bar{\underline{v}} - sh\underline{b}(\underline{q},t),$$
(17.35b)

where $\hat{\underline{\mu}}$ is the array of modified Lagrange's multipliers defined by eq. (12.23).

17.4 The generalized-α time integration scheme

Typical equations for static, structural dynamic, and multibody dynamic problems have been presented in the previous sections. For nonlinear statics problems, the Newton-Raphson procedure outlined in section 17.3.1 is used and transforms the solution of nonlinear algebraic problems into the solution of a sequence of linear, algebraic problems. The situation is different for structural and multibody dynamic problems.

Linear structural dynamics problems were presented in section 17.3.3 and are characterized by eq. (17.26), which forms a set of ordinary differential equations in time. Time integration schemes transform these ordinary differential equations into a set of linear algebraic equations. For nonlinear structural and multibody dynamic problems, a similar path is followed. The Newton-Raphson procedure outlined in section 17.3.4 for structural dynamics or in section 17.3.5 or 17.3.6 for multibody dynamics problems with holonomic or nonholonomic constraints, respectively, is first used to transform the nonlinear, ordinary differential equations of motion into a sequence of linear, ordinary differential equations. Time integration schemes then finally lead to sets of linear algebraic equations.

Numerous time integration schemes have been used in multibody dynamics. For systems presenting a small number of degrees of freedom, explicit, predictor multi-corrector algorithms such as the Adams-Bashforth integrator [199] are often used. Hairer and Wanner [26] present an exhaustive review of this field.

Because numerous degrees of freedom are generated by the discretization process inherent to finite element formulations, the resulting equations of motion typically involve many high frequency modes that are an artifact of the discretization process. Consequently, implicit schemes are used almost exclusively when dealing with finite element discretizations. The Hilber-Hughes-Taylor (HHT) integrator [135], the workhorse used in most commercial codes, is described in textbooks such as Hughes [197] or Bathe [198].

The HHT scheme was originally developed for linear structural dynamics problems [135]. Chung and Hulbert [136] later generalized this scheme as the generalized-α scheme. Because the HHT scheme is a particular case of the latter, the presentation focuses on the generalized-α scheme. Applications of this scheme to nonlinear structural and multibody problems are presented next.

The presentation of the generalized-α scheme given in the sections below is limited to a description of the scheme and of its basic properties. Chung and Hulbert [136] proved that for linear structural dynamics problems, the scheme is second-order accurate, unconditionally stable, and presents high frequency numerical damping; these three features are considered indispensable for the successful integration of large finite element systems [197, 198].

The need for numerical dissipation in time integration of large systems of linear equations was identified very early. Indeed, the average acceleration scheme proposed by Newmark [201] in 1959 is an energy preserving scheme for linear systems; the strict preservation of energy at each time step of the integration process precludes the presence of any numerical dissipation. Undesirable characteristics of this scheme were reported by Hughes [335]: in large systems, numerical round-off errors are sufficient to provide excitation of the high frequency modes of the system. This energy does not dissipate, due to the strict energy preservation characteristic of the algorithm. This prompted the development of algorithms presenting high frequency numerical dissipation in linear systems, such as the HHT and generalized-α schemes.

Because the generalized-α scheme was shown to be a powerful tool for the time integration of large structural dynamics problems, extending its use to the simulation of constrained dynamical systems seems natural. Cardona and Géradin [200] have shown that numerical damping is critical to avoid numerical oscillations in Lagrange's multipliers. Although their analysis is restricted to linear problems, stable predictions are presented for nonlinear test cases. A more formal study of the spectral behavior of the HHT schemes in constrained linear system is given by Farhat *et al.* [336].

The generalized-α scheme has also been applied to the solution of the equations of constrained dynamical systems after index reduction, see section 12.2. Lunk and Simeon [337] and Jay and Negrut [338] have proved second-order accuracy when applied to the stabilized index-2 or GGL method [59]. Formal results concerning the application of the generalized-α scheme to index-3 constrained dynamic systems are presented by Negrut *et al.* [339] and Arnold and Brüls [340]. Of course, in all cases, the equations of motion should be properly scaled, as discussed in section 12.6.

17.4.1 Linear structural dynamics problems

The generalized-α scheme [136] was introduced for linear structural dynamics problems of the form described in section 17.3.3. The equations of motion are in the form given by eq. (17.26). A typical time step starts and ends at times t_i and t_f, respectively, and $h = t_f - t_i$ is the time step size. Subscripts $(\cdot)_i$ and $(\cdot)_f$ are used to identify quantities evaluated at times t_i and t_f, respectively. The generalized displacement, velocity, and acceleration arrays at time t_i are denoted $\underline{q}_i$, $\underline{v}_i$, and $\underline{a}_i$, respectively. Similar notations are defined at the end of the time step using subscript $(\cdot)_f$.

In the generalized-α scheme, the solution at the end of the time step is written as

$$\underline{q}_f = \underline{q}_i + h\underline{v}_i + \left[\left(\tfrac{1}{2} - \beta\right) h^2\underline{a}_i + \beta h^2\underline{a}_f \right], \tag{17.36a}$$

$$h\underline{v}_f = h\underline{v}_i + \left[(1 - \gamma) h^2\underline{a}_i + \gamma h^2\underline{a}_f \right], \tag{17.36b}$$

where β and γ are two parameters that will be selected later to optimize the performance of the scheme.

Instead of being satisfied at each instant in time, the equations of motion, eqs. (17.26), are satisfied at discrete instants only. The discrete statement of dynamic equilibrium is stated as

$$\underline{\underline{M}}h^2\underline{\breve{a}} + h\underline{\underline{C}}\,h\underline{\breve{v}} + h^2\underline{\underline{K}}\,\underline{\breve{q}} = h^2\underline{f}(\breve{t}), \tag{17.37}$$

where the displacement, velocity, and acceleration stages are defined as

$$\check{q} = \hat{\alpha}_F \underline{q}_f + \alpha_F \underline{q}_i, \tag{17.38a}$$

$$h\check{\underline{v}} = \hat{\alpha}_F h\underline{v}_f + \alpha_F h\underline{v}_i, \tag{17.38b}$$

$$h^2\check{\underline{a}} = \hat{\alpha}_M h^2\underline{a}_f + \alpha_M h^2\underline{a}_i, \tag{17.38c}$$

$$\check{t} = \hat{\alpha}_F t_f + \alpha_F t_i. \tag{17.38d}$$

Coefficients α_M and α_F are two additional quantities that characterize the generalized-α family of integration schemes. Coefficients β, γ, α_M, and α_F will be selected to optimize the accuracy and stability characteristics of the algorithm. The following simplifying notation was adopted, $\hat{\alpha}_F = 1 - \alpha_F$ and $\hat{\alpha}_M = 1 - \alpha_M$.

For the generalized-α scheme [136], the four coefficients are expressed in terms of the spectral radius at infinity, denoted ρ_∞. At first, α_M and α_F are chosen as

$$\alpha_M = \frac{2\rho_\infty - 1}{\rho_\infty + 1}, \quad \alpha_F = \frac{\rho_\infty}{\rho_\infty + 1}, \tag{17.39}$$

with $\rho_\infty \in [0, 1]$. The two remaining coefficients are then computed as

$$\gamma = \frac{1}{2} - \alpha_M + \alpha_F, \quad \beta = \frac{1}{4}(1 - \alpha_M + \alpha_F)^2. \tag{17.40}$$

The HHT-α scheme [135] is a subset of the generalized-α scheme for which the first two coefficients are selected as $\alpha_M = 0$ and $\alpha_F = -\alpha$, with $\alpha \in [-0.3, 0]$. The two remaining coefficients are then computed using eq. (17.40)

To facilitate the solution process, the solution at the end of the time step given by eqs. (17.36) is recast as

$$\underline{q}_f = \underline{q}_i + h\underline{v}_i + \frac{h^2}{2}\underline{a}_i + \beta h^2(\underline{a}_f - \underline{a}_i) = \underline{q}_i + h\underline{v}_i + \frac{h^2}{2}\underline{a}_i + \quad \Delta\underline{q}, \tag{17.41a}$$

$$h\underline{v}_f = \quad h\underline{v}_i + h^2\underline{a}_i + \gamma h^2(\underline{a}_f - \underline{a}_i) = \quad h\underline{v}_i + h^2\underline{a}_i + \frac{\gamma}{\beta}\Delta\underline{q}, \tag{17.41b}$$

$$h^2\underline{a}_f = \quad h^2\underline{a}_i + h^2(\underline{a}_f - \underline{a}_i) = \quad h^2\underline{a}_i + \frac{1}{\beta}\Delta\underline{q}. \tag{17.41c}$$

Introducing these expressions into eq. (17.37) then leads to

$$\left[\frac{\hat{\alpha}_M}{\beta}\underline{\underline{M}} + \frac{\gamma\hat{\alpha}_F}{\beta}h\underline{\underline{C}} + \hat{\alpha}_F h^2\underline{\underline{K}}\right]\Delta q = h^2\underline{f}(\check{t})$$

$$- \underline{\underline{M}}h^2\underline{a}_i - h\underline{\underline{C}}\left[\hat{\alpha}_F h^2\underline{a}_i + h\underline{v}_i\right] - h^2\underline{\underline{K}}\left[\frac{\hat{\alpha}_F}{2}h^2\underline{a}_i + \hat{\alpha}_F h\underline{v}_i + \underline{q}_i\right]. \tag{17.42}$$

This linear set of algebraic equations is solved for the increments in the generalized coordinates, Δq. Equations (17.41) then yield the generalized displacements, velocities, and accelerations at the end of the time step.

Example 17.7. Single degree of freedom problem

Consider a simple, single degree of freedom spring, mass, dashpot system. For this system, it is now easily shown that

$$\frac{\hat{\alpha}_M}{\beta}\underline{\underline{M}} + \frac{\gamma\hat{\alpha}_F}{\beta}h\underline{\underline{C}} + \hat{\alpha}_F h^2 \underline{\underline{K}} = m\left[\frac{\hat{\alpha}_M}{\beta} + 2\frac{\gamma\hat{\alpha}_F}{\beta}\zeta\mu + \hat{\alpha}_F\mu^2\right] = m\mathcal{G},$$

where $\mu = \omega h = 2\pi h/T$ and ζ is the damping of the system, expressed as a fraction of the critical damping rate.

Equation (17.42) is now restated as

$$\underline{\underline{G}}\Delta q = \left[\frac{h^2}{m}f - h^2 a_i - 2\zeta\mu\left(\hat{\alpha}_F h^2 a_i + hv_i\right) - \mu^2\left(\frac{\hat{\alpha}_F}{2}h^2 a_i + \hat{\alpha}_F hv_i + q_i\right)\right].$$

Finally, the displacements, velocities, and accelerations at the end of the time step are now expressed in terms of their counterparts at the beginning of the time step with the help of eqs. (17.41) as

$$\begin{Bmatrix} q_f \\ hv_f \\ h^2 a_f \end{Bmatrix} = \frac{h^2}{m\mathcal{G}}f\begin{Bmatrix} 1 \\ \gamma/\beta \\ 1/\beta \end{Bmatrix} + \underline{\underline{A}}\begin{Bmatrix} q_i \\ hv_i \\ h^2 a_i \end{Bmatrix}. \tag{17.43}$$

In the absence of external excitation, the first term on the right-hand side vanishes, and matrix $\underline{\underline{A}}$, called the *amplification matrix*, then relates the displacements, velocities, and accelerations of the system at the beginning of the time step to the corresponding quantities at the end of the time step. The amplification matrix is defined as $\underline{\underline{A}} = \underline{\underline{A}}_1 - \underline{\underline{A}}_2\underline{\underline{A}}_3^T/\mathcal{G}$, where

$$\underline{\underline{A}}_1 = \begin{bmatrix} 1 & 1 & 1/2 \\ 0 & 1 & 1 \\ 0 & 0 & 1 \end{bmatrix}, \quad \underline{\underline{A}}_2 = \begin{Bmatrix} 1 \\ \gamma/\beta \\ 1/\beta \end{Bmatrix}, \quad \underline{\underline{A}}_3 = \begin{Bmatrix} \mu^2 \\ 2\zeta\mu + \mu^2\hat{\alpha}_F \\ 1 + 2\zeta\mu\hat{\alpha}_F + \mu^2\hat{\alpha}_F/2 \end{Bmatrix}.$$

The largest eigenvalue of the amplification matrix is called its *spectral radius*. Figure 17.19 shows this spectral radius as a function of h/T for $\zeta = 0$ and several values of ρ_∞. The corresponding results for the HHT scheme are shown in fig. 17.20, when $\zeta = 0$ and coefficient α takes different values. For this single degree of freedom linear oscillator, the generalized-α scheme can be viewed as a low-pass filter. For small time step sizes, $h/T \ll 1$, the integrator yields accurate predictions. For large time step sizes, $h/T > 1$, the response of the system is dramatically attenuated, and for $\rho_\infty = 0$, asymptotic annihilation is achieved, *i.e.*, the numerical prediction of the system's response vanishes after a single time step. Example 16.3 shows the implications of the choice of ρ_∞ on the performance of the generalized-α scheme.

17.4.2 Nonlinear structural dynamics problems

Nonlinear structural dynamics problems were investigated in section 17.3.4, with equations of motion cast in the form of eq. (17.27). The linearization process described in this section leads to the linearized equations of motion given by eq. (17.30). Since the generalized-α scheme was introduced for linear structural dynamics problems, it seems logical to extend its application to nonlinear structural

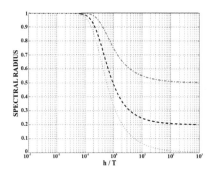

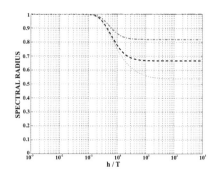

Fig. 17.19. Spectral radius of the generalized-α scheme versus h/T for $\rho_\infty = 1$: solid line; $\rho_\infty = 0.5$: dashed-dotted line; $\rho_\infty = 0.2$: dashed line; $\rho_\infty = 0$: dotted line.

Fig. 17.20. Spectral radius of the HHT scheme versus h/T for $\alpha = 0$: solid line; $\alpha = -0.1$: dashed-dotted line; $\alpha = -0.2$: dashed line; $\alpha = -0.3$: dotted line.

dynamics problems by applying the scheme to the linearized equations of motion. The scaled, linearized equations are recast here as

$$h^2 \underline{\underline{K}} \Delta \breve{q} + h \underline{\underline{G}} h \Delta \breve{v} + \underline{\underline{M}} h^2 \Delta \breve{a} = - \left(\underline{\underline{M}} h^2 \breve{a} + h^2 \underline{F} \right), \tag{17.44}$$

where $\breve{q}$, $\breve{v}$, and $\breve{a}$ are the stages defined in eqs. (17.38), and the following notations were defined for the mass, gyroscopic and stiffness matrices,

$$\underline{\underline{M}} = \underline{\underline{M}}(\breve{q}, \breve{t}), \quad \underline{\underline{G}} = \underline{\underline{G}}(\breve{q}, \breve{v}, \breve{t}), \quad \underline{\underline{K}} = \underline{\underline{K}}(\breve{q}, \breve{v}, \breve{a}, \breve{t}), \tag{17.45}$$

respectively, and the dynamic load vector,

$$\underline{F} = \underline{f}(\breve{q}, \breve{v}, \breve{t}). \tag{17.46}$$

Increments in the stages are readily obtained from eq. (17.38) as

$$\Delta \breve{q} = \beta \hat{\alpha}_F h^2 \Delta \underline{a}_f = \qquad \Delta \breve{q} = \quad \hat{\alpha}_F \Delta \underline{q}, \tag{17.47a}$$

$$h \Delta \breve{v} = \gamma \hat{\alpha}_F h^2 \Delta \underline{a}_f = \quad \frac{\gamma}{\beta} \Delta \breve{q} = \frac{\gamma \hat{\alpha}_F}{\beta} \Delta \underline{q}, \tag{17.47b}$$

$$h^2 \Delta \breve{a} = \hat{\alpha}_M h^2 \Delta \underline{a}_f = \frac{\hat{\alpha}_M}{\beta \hat{\alpha}_F} \Delta \breve{q} = \frac{\hat{\alpha}_M}{\beta} \Delta \underline{q}, \tag{17.47c}$$

where the second set of equalities were obtained from eq. (17.41). Introducing these results into eq. (17.44) and multiplying by β leads to

$$\left[\hat{\alpha}_M \underline{\underline{M}} + \gamma \hat{\alpha}_F h \underline{\underline{G}} + \beta \hat{\alpha}_F h^2 \underline{\underline{K}} \right] \Delta \underline{q} = -\beta \left(\underline{\underline{M}} h^2 \breve{a} + h^2 \underline{F} \right). \tag{17.48}$$

These linearized equations are solved sequentially as part of an iterative procedure up to convergence. Increments in the displacement, velocity, and acceleration stages are then obtained from eq. (17.47).

17.4.3 Multibody dynamics problems with holonomic constraints

Multibody dynamics problems with holonomic constraints were investigated in section 17.3.5, with equations of motion cast in the form of eqs. (17.31). The linearization process described in that section leads to the linearized equations of motion given by eqs. (17.32). Since the generalized-α scheme was introduced for linear structural dynamics problems, it seems logical to extend its application to multibody dynamics problems with holonomic constraints by applying the scheme to the linearized equations of motion. The scaled, linearized equations are recast here as

$$(h^2\underline{\underline{K}} + s\underline{\underline{K}}^b)\Delta\breve{q} + h\underline{\underline{G}}h\Delta\breve{v} + \underline{\underline{M}}h^2\Delta\breve{a} + s\underline{\underline{B}}^T\Delta\breve{\ell} \tag{17.49a}$$
$$= -\left(\underline{\underline{M}}h^2\breve{a} + h^2\underline{F} + s\underline{\underline{B}}^T\breve{\ell}\right),$$
$$s\underline{\underline{K}}^c\Delta\breve{q} = -s\underline{C}, \tag{17.49b}$$

where the stiffness, gyroscopic, and mass matrices were defined in eq. (17.45), the dynamic load vector by eq. (17.46), and

$$\breve{\ell} = \hat{\mu} \tag{17.50}$$

are the Lagrange multiplier stages. Additionally, the following notations were introduced

$$\underline{\underline{B}} = \underline{\underline{B}}(\breve{q},\breve{t}), \quad \underline{\underline{K}}^b = \underline{\underline{K}}^b(\breve{q},\breve{\ell},\breve{t}), \quad \underline{\underline{K}}^c = \underline{\underline{K}}^c(\breve{q},\breve{t}). \tag{17.51}$$

Introducing the increments in the stages as defined in eqs. (17.47) into eqs. (17.49) yields the following discrete equations

$$\begin{bmatrix} \hat{\alpha}_M\underline{\underline{M}} + \gamma\hat{\alpha}_F h\underline{\underline{G}} + \beta\hat{\alpha}_F(h^2\underline{\underline{K}} + s\underline{\underline{K}}^b) & \beta\hat{\alpha}_F s\underline{\underline{B}}^T \\ \hat{\alpha}_F s\underline{\underline{K}}^c & 0 \end{bmatrix} \begin{Bmatrix} \Delta q \\ \Delta\hat{\mu} \end{Bmatrix}$$
$$= \begin{Bmatrix} -\beta\left(\underline{\underline{M}}h^2\breve{a} + h^2\underline{F} + s\underline{\underline{B}}^T\breve{\ell}\right) \\ -s\underline{C} \end{Bmatrix}. \tag{17.52}$$

These linearized equations are solved sequentially as part of an iterative procedure up to convergence.

17.4.4 Multibody dynamics problems with nonholonomic constraints

Multibody dynamics problems with nonholonomic constraints were investigated in section 17.3.6, with equations of motion cast in the form of eqs. (17.33). The linearization process described in that section leads to the linearized equations of motion given by eqs. (17.35). The scaled, linearized equations are recast here as

$$(h^2\underline{\underline{K}} + s\underline{\underline{K}}^b)\Delta\breve{q} + h\underline{\underline{G}}h\Delta\breve{v} + \underline{\underline{M}}h^2\Delta\breve{a} + s\underline{\underline{B}}^T\Delta\breve{\ell} \tag{17.53a}$$
$$= -(\underline{\underline{M}}h^2\breve{a} + h^2\underline{F} + s\underline{\underline{B}}^T\breve{\ell}),$$
$$sh\underline{\underline{K}}^d\Delta\breve{q} + s\underline{\underline{B}}h\Delta\breve{v} = -(s\underline{\underline{B}}h\breve{v} + sh\underline{b}). \tag{17.53b}$$

where the stiffness, gyroscopic, and mass matrices were defined in eq. (17.45), the dynamic load vector by eq. (17.46), the constraint related matrices by eq. (17.51), $\check{\ell} = \hat{\mu}$ are the Lagrange multiplier stages, and the following notation was used $\underline{b} = \underline{b}(\check{q}, \check{t})$, and $\underline{\underline{K}}^d = \underline{\underline{K}}^d(\check{q}, \check{v}, \check{t})$.

Introducing the increments in the stages as defined in eqs. (17.47) into eqs. (17.53) yields the following discrete equations

$$
\begin{bmatrix}
\hat{\alpha}_M \underline{\underline{M}} + \gamma \hat{\alpha}_F h \underline{\underline{G}} + \beta \hat{\alpha}_F (h^2 \underline{\underline{K}} + s \underline{\underline{K}}^b) & \beta \hat{\alpha}_F s \underline{\underline{B}}^T \\
\beta \hat{\alpha}_F s h \underline{\underline{K}}^d + \gamma \hat{\alpha}_F s \underline{\underline{B}} & \underline{0}
\end{bmatrix}
\begin{Bmatrix} \Delta \underline{q} \\ \Delta \underline{\hat{\mu}} \end{Bmatrix}
$$
$$
=
\begin{bmatrix}
-\beta \left(\underline{\underline{M}} h^2 \check{a} + h^2 \underline{F} + s \underline{\underline{B}}^T \check{\ell} \right) \\
-\beta (s \underline{\underline{B}} \, h \check{v} + s h \underline{b})
\end{bmatrix}.
\tag{17.54}
$$

These linearized equations are solved sequentially as part of an iterative procedure up to convergence.

17.5 Energy preserving and decaying schemes

The equations of motion resulting from finite element based modeling of nonlinearly elastic multibody systems present distinguishing features: they are stiff, nonlinear, differential-algebraic equations. The stiffness of the system stems from the presence of high frequencies in the elastic members, but also from the infinite frequencies associated with the kinematic constraints.

The main focus of this section is the derivation of algorithms presenting high frequency numerical dissipation and for which unconditional stability can be proven in the nonlinear case. An energy decay argument will be used to establish stability [202].

The Newmark algorithm [201] is widely used in structural dynamics. In particular, the average acceleration scheme, also known as the trapezoidal rule, is an unconditionally stable, second-order accurate scheme when applied to linear problems. The classical stability analysis of this scheme is readily found in textbooks [197] and shows that the spectral radius of the amplification matrix equals unity at all frequencies. An alternative way of proving stability is based on an energy argument. Indeed, it is easily shown that the average acceleration scheme exactly preserves the total energy of the system [202].

Finite element discretizations of complex structures involve numerous degrees of freedom. Consequently, high frequency modes are present in the models and high frequency numerical dissipation is a desirable, if not indispensable feature for robust time integration schemes. Numerical dissipation cannot be introduced in the Newmark method without degrading its accuracy. The HHT and generalized-α schemes presented in section 17.4 remedy this situation by achieving high frequency dissipation while minimizing unwanted low frequency dissipation. Both methods have been successfully used for both linear and nonlinear problems, although unconditional stability is proved for linear systems only.

Simo and his coworkers presented energy preserving schemes for rigid body dynamics [203], elastodynamics [204], beams [206], and plates and shells [208]. These schemes were presented as second-order accurate, finite difference schemes based on a mid-point approximation. Finite rotations were parameterized with the rotation vector, then using Cayley's algebraic form of finite rotations. The unconditional stability of these schemes stems from a proof of preservation of the total energy of the system. It is important to understand that while the exact solution of the equations of motion implies the exact preservation of the total mechanical energy, a numerical, *i.e.*, an inherently approximate solution of the problem does not, in general, guarantee the preservation of energy at the discrete level. When using energy preserving schemes, the computed, approximate solution exactly satisfies the energy preservation condition.

An energy preserving scheme for nonlinear elastic multibody systems was proposed by Bauchau [210]. In this scheme, the discretization of the equations of motion implies the conservation of the total energy for the elastic components of the system, and that of the forces of constraint associated with the kinematic constraints implies the vanishing of work they perform. The combination of these two features of the discretization guarantees the unconditional stability of the numerical integration process for nonlinearly elastic multibody systems.

When rotationless formulations of dynamics are used, see example 11.3, the governing equations of motion for rigid bodies, beams, and plates and shells involve algebraic nonlinearities only, of the second degree at most. It is remarkable, that for these problems, the mid-point time integration scheme naturally leads to discrete equations that satisfy energy and momentum conservation conditions. Betsch and his coworkers used this approach to develop energy preserving schemes for rigid body dynamics [205], beams [207], and plates and shells [209].

Although energy preserving schemes perform well for simple problems, their lack of high frequency numerical dissipation can cause problems [210]. First, the time histories of internal forces and velocities often present considerable high frequency content of a purely numerical origin. Second, these high frequency oscillations can hinder the convergence process for the solution of the nonlinear equations of motion. The selection of a smaller time step size does not necessarily help the convergence process, because smaller time step sizes allow even higher frequency oscillations to be present in the response. Finally, it seems that the presence of high frequency oscillations also renders strict energy preservation difficult to obtain. This could prove to be a real limitation of energy preserving schemes when applied to more and more complex problems, for which the use of integration schemes presenting high frequency numerical dissipation become desirable, if not indispensable.

It appears that the development of "energy decaying schemes," *i.e.*, schemes eliminating the energy associated with vibratory motions at high frequency, is desirable. This is particularly important when dealing with problems presenting a complex dynamic response such as nonlinearly elastic multibody problems.

The key to the development of an energy decaying scheme is the derivation of an *energy decay inequality* [202] rather that the *discrete energy conservation law* which is central to energy preserving schemes. A methodology that can systematically lead to an energy decay inequality is the *time discontinuous Galerkin method* [341, 342, 343] which was initially developed for hyperbolic equations.

Hughes and Hulbert [344, 345] have investigated the use of the time discontinuous Galerkin methodology for linear elastodynamics. They point out that "classical elastodynamics can be converted to first-order symmetric hyperbolic form, which has proved useful in theoretical studies. Finite element methods for first-order symmetric hyperbolic system are thus immediately applicable. However, there seems to be several disadvantages: in symmetric hyperbolic form the state vector consists of displacements, velocities, and stresses which is computationally uneconomical; and the generalization to nonlinear elastodynamics seems possible only in special circumstances." Indeed, writing the nonlinear equations of motion of geometrically exact beams in this symmetric hyperbolic form does not appear to be possible.

In this section, an alternative route is taken. Practical time integration schemes that do not rely on the symmetric hyperbolic form of the equations of motion are developed. These schemes are of a finite difference nature, and imply an energy balance condition that is obtained by a direct computation of the work done by the discretized inertial and elastic forces over a time step. The mean value theorem guarantees the existence of discretizations leading to these energy preservation, or energy decay statements, leading to a rigorous proof of unconditional stability for the scheme.

Energy decaying schemes were presented by Bauchau and his co-workers for beams [346], elastodynamics [347], and multibody systems [217, 218, 211], and plates and shells [348, 213]. These schemes originate from Galerkin and time discontinuous Galerkin approximations of the equations of motion written in the symmetric hyperbolic form. Finite rotations were parameterized using the conformal rotation parameters.

Bottasso and Borri proposed both energy preserving and decaying schemes for beams [212, 349] and multibody systems [214, 215]. Their schemes were cast within the framework of finite elements in time at first, then as 2-stage FSAL Runge-Kutta methods. The rotation vector was used to represent finite rotations. Some of the proposed schemes also imply the conservation of momenta, or are geometrically invariant [350, 214, 215, 219]. These additional features are easily obtained by recasting the field equations in *fixed pole* form [269], see example 8.9.

17.5.1 The symmetric hyperbolic form

Consider a dynamical system described by a kinetic energy $K = K(\dot{q}, q)$, and a strain energy $V = V(q)$, where q are the system's generalized coordinates.

Classical forms of the equations of motion

The Lagrangian of the system is defined as $L(\dot{q}, q) = K - V$, and the equations of motion of the system in Lagrangian form are then

$$\frac{\mathrm{d}}{\mathrm{d}t} \left(L_{,\dot{q}} \right) - L_{,q} = 0. \tag{17.55}$$

Notation $(\cdot)_{,q}$ is used here to indicate a derivative with respect to q. Hamilton's formulation is obtained with the help of Legendre's transformation [87]. First, the momenta are defined as $\underline{p}(\underline{\dot{q}}, \underline{q}) = L_{,\underline{\dot{q}}}$, and these relationships can be inverted to yield $\underline{\dot{q}} = \underline{\dot{q}}(\underline{q}, \underline{p})$. The Hamiltonian of the system is now defined $H(\underline{q}, \underline{p}) = \underline{p}^T \underline{\dot{q}}(\underline{q}, \underline{p}) - L(\underline{q}, \underline{p})$. The equations of motion of the system in Hamiltonian form are then

$$\underline{\dot{q}} = \ H_{,p}, \tag{17.56a}$$

$$\underline{\dot{p}} = -H_{,q}. \tag{17.56b}$$

The symmetric hyperbolic form

The symmetric hyperbolic form stems from a second Legendre transformation. The following variables are defined first $\underline{f}(\underline{q}, \underline{p}) = H_{,q}$ and $\underline{v}(\underline{q}, \underline{p}) = H_{,p}$. These relations can be inverted to yield $\underline{q} = \underline{q}(\underline{f}, \underline{v})$ and $\underline{p} = \underline{p}(\underline{f}, \underline{v})$. A new function is now defined

$$G(\underline{f}, \underline{v}) = \underline{f}^T \underline{q}(\underline{f}, \underline{v}) + \underline{v}^T \underline{p}(\underline{f}, \underline{v}) - H(\underline{f}, \underline{v}), \tag{17.57}$$

implying $\underline{q} = G_{,f}$ and $\underline{p} = G_{,v}$. It can be readily shown that the Hessians of H and G are the inverse of each other. Hence, if H is a positive-definite function, so is G. Hamilton's equations, eqs. (17.56), can be expressed in terms of the new variables, $\underline{f}$ and $\underline{v}$, to find the symmetric hyperbolic form of the equations of motion $G_{,ff} \, \underline{\dot{f}} + G_{,fv} \, \underline{\dot{v}} - \underline{v} = \underline{0}$ and $G_{,vf} \, \underline{\dot{f}} + G_{,vv} \, \underline{\dot{v}} + \underline{f} = \underline{0}$. To simplify the notation, an implicit form of the equations is preferred

$$\underline{\dot{q}}(\underline{f}, \underline{v}) - \underline{v} = \underline{0}, \tag{17.58a}$$

$$\underline{\dot{p}}(\underline{f}, \underline{v}) + \underline{f} = \underline{0}. \tag{17.58b}$$

The Galerkin approximation

In the Galerkin approximation, the equations of motion are enforced in a weak, integral manner. Fig. 17.21 shows a time interval from t_i to t_f, and an approximate solution over that interval. Subscripts $(\cdot)_i$ and $(\cdot)_f$ will be used to indicate the value of a quantity at times t_i and t_f, respectively. The Galerkin approximation of the equations of motion in implicit symmetric hyperbolic form (17.58) writes

$$\int_{t_i}^{t_f} \left\{ \underline{w}_1^T \left[\underline{\dot{q}}(\underline{f}, \underline{v}) - \underline{v} \right] + \underline{w}_2^T \left[\underline{\dot{p}}(\underline{f}, \underline{v}) + \underline{f} \right] \right\} \mathrm{d}t = 0, \tag{17.59}$$

where $\underline{w}_1$ and $\underline{w}_2$ are arbitrary test functions. Integration by parts yields

$$\int_{t_i}^{t_f} \left[-\underline{\dot{w}}_1^T \underline{q} - \underline{\dot{w}}_2^T \underline{p} - \underline{w}_1^T \underline{v} + \underline{w}_2^T \underline{f} \right] \mathrm{d}t + \underline{w}_{1f}^T \underline{q}_f + \underline{w}_{2f}^T \underline{p}_f - \underline{w}_{1i}^T \underline{q}_i - \underline{w}_{2i}^T \underline{p}_i = 0.$$

This approximation of the equations of motion enjoys remarkable properties. Indeed, selecting the test functions as $\underline{w}_1 = \underline{f}$ and $\underline{w}_2 = \underline{v}$ yields $\int_{t_i}^{t_f} [-\underline{\dot{f}}^T G_{,f} - \underline{\dot{v}}^T G_{,v} -$

$\underline{f}^T \underline{v} + \underline{v}^T \underline{f}] \, dt + \underline{f}_f^T \underline{q}_f + \underline{v}_f^T \underline{p}_f - \underline{f}_i^T \underline{q}_i - \underline{v}_i^T \underline{p}_i = 0$. The time integral clearly has a closed form solution, leading to $G_i - G_f + \underline{f}_f^T \underline{q}_f + \underline{v}_f^T \underline{p}_f - \underline{f}_i^T \underline{q}_i - \underline{v}_i^T \underline{p}_i = 0$. Finally, function G is expressed in terms of the Hamiltonian, H, with the help of eq. (17.57) to find

$$H_f = H_i. \tag{17.60}$$

In summary, the Galerkin approximation, eq. (17.59), of the equations of motion written in symmetric hyperbolic form implies a discrete Hamiltonian preservation statement (17.60). If the Hamiltonian is a positive-definite function, this statement implies the *unconditional stability* of integration schemes based on eq. (17.59).

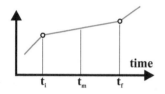

Fig. 17.21. The time continuous Galerkin approximation.

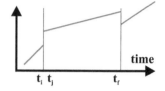

Fig. 17.22. The time discontinuous Galerkin approximation.

The time discontinuous Galerkin approximation

In the time discontinuous Galerkin approximation, the solution is allowed to present discontinuities in the displacement and velocity fields at discrete times. Figure 17.22 shows a time interval from t_i to t_f and the approximate solution over that interval. At the initial instant, the solution presents a jump. Subscripts $(\cdot)_i$ will be used to denote the value of a discontinuous quantity to the left side of the jump, whereas a subscript $(.)_j$ indicates the value of that quantity to the right side of the jump. The equations of motion and initial conditions are enforced in a weak, integral manner. The time discontinuous Galerkin approximation of the equations of motion in implicit symmetric hyperbolic form (17.58) writes

$$\int_{t_j}^{t_f} \left\{ \underline{w}_1^T \left[\underline{\dot{q}}(\underline{f}, \underline{v}) - \underline{v} \right] + \underline{w}_2^T \left[\underline{\dot{p}}(\underline{f}, \underline{v}) + \underline{f} \right] \right\} dt + \underline{w}_{1j}^T \langle \underline{q} \rangle + \underline{w}_{2j}^T \langle \underline{p} \rangle = 0, \tag{17.61}$$

where notation $\langle \cdot \rangle$ indicates the jump in a quantity at the initial time, *i.e.*, $\langle \underline{q} \rangle = \underline{q}_j - \underline{q}_i$ and $\langle \underline{p} \rangle = \underline{p}_j - \underline{p}_i$.

This approximation of the equations of motion also enjoys remarkable properties. Indeed, integrating by parts and selecting the test functions as $\underline{w}_1 = \underline{f}$ and $\underline{w}_2 = \underline{v}$ yields $\int_{t_j}^{t_f} [-\underline{\dot{f}} G_{,\underline{f}} - \underline{\dot{v}} G_{,\underline{v}} - \underline{f}^T \underline{v} + \underline{v}^T \underline{f}] \, dt + \underline{f}_f^T \underline{q}_f + \underline{v}_f^T \underline{p}_f - \underline{f}_j^T \underline{q}_i - \underline{v}_j^T \underline{p}_i = 0$. The time integral clearly has a closed form solution, leading to $G_j - G_f + \underline{f}_f^T \underline{q}_f +$

$\underline{v}_f^T \underline{p}_f - \underline{f}_j^T \underline{q}_i - \underline{v}_j^T \underline{p}_i = 0$. Finally, we express G in terms of the Hamiltonian H with the help of eq. (17.57) to find

$$H_f - H_j + \underline{f}_j^T \langle \underline{q} \rangle + \underline{v}_j^T \langle \underline{p} \rangle = 0. \tag{17.62}$$

Because the Hamiltonian is a continuous function of $\underline{q}$ and $\underline{p}$, the mean value theorem implies

$$H_j = H_i + \underline{f}_j^T \langle \underline{q} \rangle + \underline{v}_j^T \langle \underline{p} \rangle - \frac{1}{2} \left[\langle \underline{q} \rangle^T H_{,\underline{q}\underline{q}} \langle \underline{q} \rangle + \langle \underline{q} \rangle^T H_{,\underline{q}\underline{p}} \langle \underline{p} \rangle + \\ \langle \underline{p} \rangle^T H_{,\underline{p}\underline{q}} \langle \underline{q} \rangle + \langle \underline{p} \rangle^T H_{,\underline{p}\underline{p}} \langle \underline{p} \rangle \right]_h = H_i + \underline{f}_j^T \langle \underline{q} \rangle + \underline{v}_j^T \langle \underline{p} \rangle - c^2, \tag{17.63}$$

where the last equality holds if the Hamiltonian is a positive-definite function. Combining eqs. (17.62) and (17.63) then yields

$$H_f = H_i - c^2, \Rightarrow H_f \leq H_i. \tag{17.64}$$

In summary, the time discontinuous Galerkin approximation (17.61) of the equations of motion written in symmetric hyperbolic form implies a Hamiltonian decay inequality, eq. (17.64), if the Hamiltonian is a positive-definite quantity. This inequality implies the *unconditional stability* of time integration schemes based on eq. (17.61).

Example 17.8. Linear spring-mass system

To illustrate the procedures described in the previous sections, a very simple example will be treated here. Consider a linear spring-mass system with a kinetic energy $K = 1/2\, m\, \dot{u}^2$, a strain energy $V = 1/2\, k u^2$, and subjected to an external force $F^a(t)$. In this simple case, $f = k\,u$ and $v = p/m$, and the symmetric hyperbolic form of the equations of motion becomes: $\dot{p} + ku = F^a$; $\dot{u} - p/m = 0$. The Galerkin approximation (17.59) for this problem writes

$$\int_{t_i}^{t_f} \left\{ w_1 [\dot{u} - \frac{p}{m}] + w_2 [\dot{p} + ku - F^a] \right\} \, dt = 0.$$

Using a linear in time approximation for the displacement and momentum, and a constant in time approximation for the test functions, the following discrete equations are obtained

$$F_m^I + F_m^E = F_m^a, \tag{17.65}$$

where subscript $(\cdot)_m$ denotes a quantity at the mid-point t_m, see fig. 17.21. The discretized inertial forces are $F_m^I = m(\dot{u}_f - \dot{u}_i)/\Delta t$, and the following velocity-displacement and force-displacement relationships are used

$$\frac{u_f - u_i}{\Delta t} = \frac{\dot{u}_f + \dot{u}_i}{2}, \quad F_m^E = k \frac{u_f + u_i}{2}, \tag{17.66}$$

where Δt indicates the time step size. Finally, the discretized applied forces are

$$F_m^a = \frac{1}{2} \int_{-1}^{1} F^a(\tau)\, \mathrm{d}\tau, \qquad (17.67)$$

where τ is a non-dimensional time variable such that $\tau = -1$ or $+1$ at times t_i and t_f, respectively. The properties of this integration scheme can be investigated using the classical techniques for the analysis of linear schemes. The spectral radius of the amplification matrix is always equal to unity, implying unconditional stability. This scheme is identical to the Newmark scheme [201] with $\gamma = 1/2$ and $\beta = 1/4$. It can be readily shown that the discrete equations of motion (17.65) imply a discrete energy preservation statement $E_f = E_i$, where $E = K + V$ is the total mechanical energy, as expected from the theoretical developments presented above.

The same problem can be treated with the time discontinuous Galerkin approximation, eq. (17.61), which writes

$$\int_{t_j}^{t_f} \left\{ w_1[\dot{u} - \frac{p}{m}] + w_2[\dot{p} + ku - F^a] \right\} \mathrm{d}t + w_{1j}\langle u \rangle + w_{2j}\langle p \rangle = 0. \qquad (17.68)$$

Using a linear in time approximation for the displacement, momentum, and test functions, the following discrete equations are obtained

$$F_m^I + F_g^E = F_g^a, \quad \text{and} \quad F_h^I - [F_g^E - f_j]/3 = F_h^a. \qquad (17.69)$$

Subscript $(\cdot)_g$ denotes a quantity at the midpoint between times t_j and t_f, and $(\cdot)_h$ denotes a quantity at the midpoint between times t_i and t_j, see fig. 17.22. The discretized inertial forces are $F_m^I = m(\dot{u}_f - \dot{u}_i)/\Delta t$ and $F_h^I = m(\dot{u}_j - \dot{u}_i)/\Delta t$, and the following velocity-displacement and force-displacement relationships are used

$$\frac{u_f - u_i}{\Delta t} = \frac{\dot{u}_f + \dot{u}_j}{2}, \quad 3\frac{u_j - u_i}{\Delta t} = -\frac{\dot{u}_f - \dot{u}_j}{2}, \quad F_g^E = k\frac{u_f + u_j}{2}. \qquad (17.70)$$

Finally, the discretized applied forces are

$$F_g^a = \frac{1}{2} \int_{-1}^{1} F^a\, \mathrm{d}\tau, \quad \text{and} \quad F_h^a = -\frac{1}{2} \int_{-1}^{1} F^a\, \tau \mathrm{d}\tau. \qquad (17.71)$$

It can be readily shown that the discrete equations of motion (17.69) imply a discrete energy decay inequality $E_f \leq E_i$. This is a direct consequence of (17.64), since the Hamiltonian is equal to the total energy of the system for this simple problem.

This can be confirmed by a conventional analysis of the scheme based on the characteristics of the amplification matrix. The period elongation is $\Delta T/T = \omega^4 \Delta t^4/270 + O(\omega^6 \Delta t^6)$, while the algorithmic damping is $\zeta = \omega^3 \Delta t^3/72 + O(\omega^5 \Delta t^5)$, where $\omega^2 = k/m$. Hence, the scheme is third-order accurate. The spectral radius, shown in fig. 17.23 as function of $\Delta t/T = \omega \Delta t/(2\pi)$, is compared with that of the generalized-α scheme for three different values of spectral radius at infinity, $\rho_\infty = 0.9, 0.5$, and 0. Asymptotic annihilation is obtained with the time discontinuous Galerkin scheme. The scheme is unconditionally stable since the spectral radius is always smaller than unity.

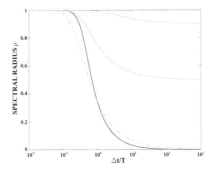

Fig. 17.23. Comparison of spectral radii of various time integration schemes. Solid line: time discontinuous Galerkin scheme. Dotted, dash-doted, and dash-double dotted lines: generalized-α scheme with $\rho_\infty = 0.9, 0.5$, and 0, respectively.

17.5.2 Discussion

Both Galerkin (17.59) and time discontinuous Galerkin (17.61) approximations applied to the equations of motion written in the symmetric hyperbolic form (17.58) have been shown to provide a systematic way of deriving unconditionally stable time integration schemes, provided the Hamiltonian is a positive-definite function. The energy decay inequality associated with the time discontinuous Galerkin approximation implies the presence of numerical dissipation in the resulting time integration schemes, whereas such dissipation is ruled out by the strict energy preservation associated with the Galerkin approximation. Since the presence of numerical dissipation is highly desirable, the time discontinuous Galerkin approach appears to be the most promising method.

However, both of these approaches present a major drawback: it is not always possible to recast the equations of motion of general systems into the symmetric hyperbolic form. In particular, it does not seem possible to cast the governing equations of constrained multibody systems in the symmetric hyperbolic form. Furthermore, the time discontinuous Galerkin approach require two level of unknowns (at t_j and t_f). In elastodynamics, three fields are required for the symmetric hyperbolic form: displacements, stresses and momenta. Hence, the final discrete equations will involve $6N$ unknowns, resulting in unacceptably high computational cost [344].

17.5.3 Practical time integration schemes

In this section, time integration schemes applicable to nonlinear elastic multibody systems will be developed, without resorting to the symmetric hyperbolic form of the equations of motion. The investigation will focus on dynamical system defined by a kinetic energy, $K = 1/2\, \underline{v}^T \underline{\underline{M}}\, \underline{v}$, and a strain energy, $V = 1/2\, \underline{\varepsilon}^T \underline{\underline{C}}\, \underline{\varepsilon}$. The mass matrix $\underline{\underline{M}}$ and stiffness matrix $\underline{\underline{C}}$ are symmetric and positive-definite; the velocities and strains are given as $\underline{v} = \overline{\underline{\underline{T}}}(\underline{q})\dot{\underline{q}}$, and $\underline{\varepsilon} = \underline{\varepsilon}(\underline{q})$, respectively. Velocities are assumed to be linear functions of the generalized velocities, resulting in a kinetic energy that is a quadratic form of the same quantities. Under these conditions the total mechanical energy of the system is preserved [87].

The equations of motion of such systems simply write $\underline{F}^I + \underline{F}^E = \underline{F}^a(t)$, where $\underline{F}^a(t)$ are the time dependent external forces. The inertial and elastic forces, $\underline{F}^I$ and $\underline{F}^e$, respectively, are

$$\underline{F}^I = \frac{\mathrm{d}}{\mathrm{d}t}(\underline{\underline{T}}^T \underline{p}) - \underline{v}_{,q}^T \underline{p}, \tag{17.72}$$

$$\underline{F}^E = \underline{\varepsilon}_{,q}^T \underline{f}, \tag{17.73}$$

where $\underline{p} = \underline{\underline{M}}\,\underline{v}$, $\underline{f} = \underline{\underline{C}}\,\underline{\varepsilon}$ and notation $(\cdot)_{,q}$ indicates a derivative with respect to $\underline{q}$.

The energy preservation statement can be obtained by evaluating the work done by the inertial, elastic, and applied forces. The work done by the inertial forces is computed first $W^I = \int_{t_i}^{t_f} \underline{\dot{q}}^T \underline{F}^I \, \mathrm{d}t = K_f - K_i$. Next, the work done by the elastic forces is evaluated $W^E = \int_{t_i}^{t_f} \underline{\dot{q}}^T \underline{F}^E \, \mathrm{d}t = V_f - V_i$. Finally, the work done by the applied forces is $W^a = \int_{t_i}^{t_f} \underline{\dot{q}}^T \underline{F}^a \, \mathrm{d}t$. Hence, the equations of motion imply the following work balance equation

$$K_f - K_i + V_f - V_i = W^a, \quad \Rightarrow E_f - E_i = W^a, \tag{17.74}$$

where the total mechanical energy $E = K + V$. In the absence of externally applied forces $W^a = 0$ and the total energy is preserved.

The goal is to obtain *discretized equations of motion* that will imply an exact energy preservation condition (17.74), or an energy decay inequality. At first, discretizations of the inertial and elastic forces will proposed, then energy preserving and energy decaying schemes will be derived.

Discretization of inertial and elastic forces

Consider a time interval $t_i\ t_f$, and an approximate solution over this interval, as shown in fig. 17.21. The following discretizations of the inertial (17.72) and elastic (17.73) forces are proposed:

$$\underline{F}_m^I = \frac{\underline{\underline{T}}_f^T \underline{p}_f - \underline{\underline{T}}_i^T \underline{p}_i}{\Delta t} - \left(\underline{v}_{,q}^T\right)_m \frac{\underline{p}_f + \underline{p}_i}{2}; \tag{17.75}$$

$$\underline{F}_m^E = \left(\underline{\varepsilon}_{,q}^T\right)_m \underline{f}_m, \tag{17.76}$$

where the quantities $(\underline{v}_{,q})_m$, $(\underline{\varepsilon}_{,q})_m$ and $\underline{f}_m$ are as yet undetermined. The work done by the discretized inertial forces is $W^I = (\underline{q}_f - \underline{q}_i)\underline{F}_m^I$, and regrouping the term yields:

$$W^I = \frac{\underline{q}_f^T - \underline{q}_i^T}{\Delta t} \left\{ \left[\underline{\underline{T}}_f^T - \frac{\Delta t}{2} \left(\underline{v}_{,q}^T\right)_m \right] \underline{p}_f - \left[\underline{\underline{T}}_i^T + \frac{\Delta t}{2} \left(\underline{v}_{,q}^T\right)_m \right] \underline{p}_i \right\}. \tag{17.77}$$

The following condition is now imposed

$$\underline{v}_m = \left[\underline{\underline{T}}_f - \frac{\Delta t}{2} \left(\underline{v}_{,q} \right)_m \right] \frac{\underline{q}_f - \underline{q}_i}{\Delta t} = \left[\underline{\underline{T}}_i + \frac{\Delta t}{2} \left(\underline{v}_{,q} \right)_m \right] \frac{\underline{q}_f - \underline{q}_i}{\Delta t}. \qquad (17.78)$$

These relationships define both $(\underline{v}_{,q})_m$ and $\underline{v}_m$. Note that the existence of $(\underline{v}_{,q})_m$ satisfying eq. (17.78) is guaranteed by the mean value theorem which states that $\underline{v}_f = \underline{v}_i + (\underline{v}_{,q})_m(\underline{q}_f - \underline{q}_i)$. The work done by the discretized inertial forces now becomes $W^I = (\underline{v}_f^T - \underline{v}_i^T)\underline{\underline{M}}\,\underline{v}_m$.

Next, the work done by the discretized elastic forces is evaluated $W^E = (\underline{q}_f^T - \underline{q}_i^T)(\underline{\varepsilon}_{,q})_m\underline{f}_m$. The following condition is now imposed

$$\underline{\varepsilon}_f - \underline{\varepsilon}_i = \left(\underline{\varepsilon}_{,q}^T \right)_m (\underline{q}_f - \underline{q}_i). \qquad (17.79)$$

Here again, the existence of $(\underline{\varepsilon}_{,q})_m$ satisfying this condition is guaranteed by the mean value theorem. The work done by the discretized elastic forces now becomes $W^E = (\underline{\varepsilon}_f^T - \underline{\varepsilon}_i^T)\underline{f}_m$.

Energy preserving scheme

The discretized equations of motion for the energy preserving scheme mimic those obtained in example 17.8 for the Galerkin approximation of the linear spring-mass problem (17.65)

$$\underline{F}_m^I + \underline{F}_m^E = \underline{F}_m^a, \qquad (17.80)$$

where $\underline{F}_m^I$ and $\underline{F}_m^E$ are now given by (17.75) and (17.76), respectively; and $\underline{F}_m^a = 1/2 \int_{-1}^{1} \underline{F}^a(\tau)\, d\tau$, as in eq. (17.67). The work done by these discretized forces can be evaluated, as was done earlier. With the help of conditions (17.78) and (17.79), equations of motion (17.80) imply a work balance statement $(\underline{v}_f^T - \underline{v}_i^T)\underline{\underline{M}}\,\underline{v}_m + (\underline{\varepsilon}_f^T - \underline{\varepsilon}_i^T)\underline{f}_m = W_m^a$. The following algorithmic velocity-displacement and force-strain relationship are now selected, see eq. (17.66),

$$\underline{v}_m = \frac{\underline{\dot{q}}_f + \underline{\dot{q}}_i}{2}, \quad \underline{f}_m = \underline{\underline{C}}\frac{\underline{\varepsilon}_f + \underline{\varepsilon}_i}{2}. \qquad (17.81)$$

The work balance equation then becomes $K_f - K_i + V_f - V_i = W^a$, and finally, the discrete energy preservation statement is obtained, $E_f - E_i = W^a$.

In summary, discretization (17.80) implies the discrete energy preservation statement provided that relationships (17.78) and (17.79) are satisfied, and that the algorithmic velocity-displacement and force-strain relationships (17.81) are used.

Energy Decaying Scheme

The discretized equations of motion for the energy decaying scheme mimic those obtained in example 17.8 for the time discontinuous Galerkin approximation of a linear spring-mass system (17.69)

$$\underline{F}_m^I + \underline{F}_g^E = \underline{F}_g^a, \quad \text{and} \quad \underline{F}_h^I - \frac{1}{3}\left[\underline{F}_g^E - \left(\underline{v}_q\right)_h \underline{f}_j\right] = \underline{F}_h^a \tag{17.82}$$

where $\underline{F}_m^I$, and $\underline{F}_h^I$ are given by (17.75) using subscripts $(\cdot)_f$, $(\cdot)_i$ and $(\cdot)_j$, $(\cdot)_i$, respectively; $\underline{F}_g^E$ is given by eq. (17.76) using subscripts $(\cdot)_f$, $(\cdot)_j$; and $\underline{F}_g^a = 1/2 \int_{-1}^1 \underline{F}^a \, d\tau$ and $\underline{F}_h^a = 1/2 \int_{-1}^1 \underline{F}^a \, \tau d\tau$, as in eq. 17.71.

The work done by the discretized inertial forces is $W^I = (\underline{q}_f^T - \underline{q}_i^T)\underline{F}_m^I + 3\langle \underline{q}\rangle^T \underline{F}_h^I$. With the help of condition (17.78) this becomes

$$W^I = \underline{v}_m^T \underline{M}(\underline{v}_f - \underline{v}_i) + 3\underline{v}_h^T \underline{M}\langle\underline{v}\rangle. \tag{17.83}$$

The work done by the discretized elastic forces is $W^E = (\underline{q}_f^T - \underline{q}_i^T)\underline{F}_g^E - \langle\underline{q}\rangle^T[\underline{F}_g^E - (\underline{\varepsilon}_{,q})_h \underline{f}_j]$. With the help of condition (17.79) this becomes $W^E = (\underline{\varepsilon}_f^T - \underline{\varepsilon}_j^T) \underline{C} \underline{f}_g + \langle\underline{\varepsilon}\rangle^T \underline{C} \underline{f}_j$. The following velocity-displacement and force-strain relationship are now selected, see eq. (17.70),

$$\underline{v}_m = \frac{\dot{\underline{q}}_f + \dot{\underline{q}}_j}{2}, \quad 3\underline{v}_h = -\frac{\dot{\underline{q}}_f - \dot{\underline{q}}_j}{2}, \quad \underline{f}_g = \underline{C}\frac{\underline{\varepsilon}_f + \underline{\varepsilon}_j}{2}. \tag{17.84}$$

The work balance equation now writes

$$E_f - E_j + \underline{v}_j^T \underline{M} \langle\underline{v}\rangle + \underline{f}_j^T \underline{C} \langle\underline{\varepsilon}\rangle = W^a, \tag{17.85}$$

which mirrors eq. (17.62). Since the total mechanical energy is a positive-definite function of the velocities and strains, the mean value theorem implies

$$E_j = E_i + \underline{v}_j^T \underline{M}\langle\underline{v}\rangle + \underline{f}_j^T \underline{C}\langle\underline{\varepsilon}\rangle - \frac{1}{2}\left[\langle\underline{v}\rangle^T E_{,vv}^h \langle\underline{v}\rangle + \langle\underline{v}\rangle^T E_{,v\varepsilon}^h \langle\underline{\varepsilon}\rangle\right.$$
$$\left. + \langle\underline{\varepsilon}\rangle^T E_{,\varepsilon v}^h \langle\underline{v}\rangle + \langle\underline{\varepsilon}\rangle^T E_{,\varepsilon\varepsilon}^h \langle\underline{\varepsilon}\rangle\right] = E_i + \underline{v}_j^T \underline{M} \langle\underline{v}\rangle + \underline{f}_j^T \underline{C} \langle\underline{\varepsilon}\rangle - c^2, \tag{17.86}$$

which is equivalent to eq. (17.63). Combining eqs. (17.85) and (17.86) yields $E_f = E_i - c^2 + W^a$ and finally, the energy decay statement $E_f \leq E_i + W^a$.

In summary, discretization (17.82) implies the energy decay statement provided that relationships (17.78) and (17.79) are satisfied, and that the algorithmic velocity-displacement and force-strain relationships (17.84) are used.

Example 17.9. Nonlinear spring-mass system

Consider a nonlinear spring mass oscillator defined by kinetic energy $K = 1/2\, m\dot{u}^2$, strain energy $V = 1/2\, k\varepsilon^2$, and strain $\varepsilon = u^2$. For this example $m = k = 1$. It is clear that condition (17.79) implies $(\varepsilon_{,u})_m = u_f + u_i$ in this case, and $f_m = k(\varepsilon_f + \varepsilon_i)/2$.

For the trapezoidal rule scheme, the discretized equation of motion is $(m\dot{u}_f - m\dot{u}_i)/\Delta t + 2\, k[(u_f + u_i)/2]^3 = 0$. Although this scheme is unconditionally stable for linear system, there is no guarantee of stability when applied to nonlinear systems. Figure 17.24 shows the response of the system for initial conditions $u_0 = 1$ and $\dot{u}_0 =$

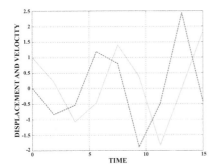

Fig. 17.24. Displacement response for the trapezoidal rule (Initial conditions: $u_0 = 1$ and $\dot{u}_0 = 0$). Solid line: displacement; dashed line: velocity.

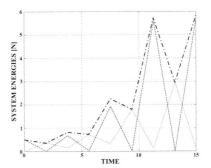

Fig. 17.25. Energy response for the trapezoidal rule. Solid line: kinetic energy; dashed line: strain energy; dashed-dotted line: total mechanical energy.

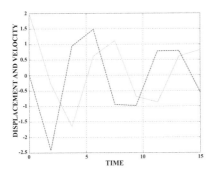

Fig. 17.26. System response for the trapezoidal rule (Initial conditions: $u_0 = 2$ and $\dot{u}_0 = 0$). Solid line: displacement; dashed line: velocity.

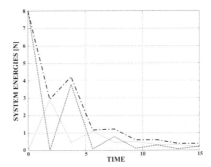

Fig. 17.27. Energy response for the trapezoidal rule. Solid line: kinetic energy; dashed line: strain energy; dashed-dotted line: total mechanical energy.

0. The total mechanical energy rapidly increases as shown by fig. 17.25, although not monotonously. For initial conditions $u_0 = 2$ and $\dot{u}_0 = 0$, the corresponding results are shown in figs. 17.26 and 17.27, which now show a rapid decrease in the total mechanical energy, although here again, not monotonously.

For the energy preserving scheme, the discretized equation of motion is $(m\dot{u}_f - m\dot{u}_i)/\Delta t + (\varepsilon_{,u})_m f_m = 0$. Figure 17.28 shows the response of the system for initial conditions $u_0 = 1$ and $\dot{u}_0 = 0$. The total mechanical energy of the system, shown in fig. 17.29, is preserved exactly, as expected from the energy preservation condition that characterizes this scheme.

Finally, for the energy decaying scheme, the discretized equations of motion are $(m\dot{u}_f - m\dot{u}_i)/\Delta t + (\varepsilon_{,u})_g f_g = 0$ and $(m\dot{u}_j - m\dot{u}_i)/\Delta t - [(\varepsilon_{,u})_g f_g - (\varepsilon_{,u})_h k\varepsilon_j]/3 = 0$. Figure 17.30 shows the response of the system for initial conditions $u_0 = 1$ and $\dot{u}_0 = 0$. The total mechanical energy of the system, shown in fig. 17.31, decays

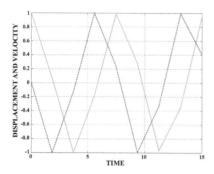

Fig. 17.28. System response for the energy preserving scheme (Initial conditions: $u_0 = 1$ and $\dot{u}_0 = 0$). Solid line: displacement; dashed line: velocity.

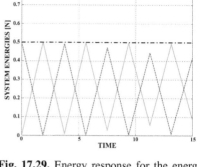

Fig. 17.29. Energy response for the energy preserving scheme. Solid line: kinetic energy; dashed line: strain energy; dashed-dotted line: total mechanical energy.

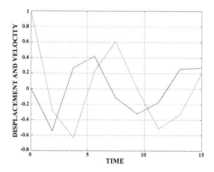

Fig. 17.30. System response for the energy decaying scheme (Initial conditions: $u_0 = 1$ and $\dot{u}_0 = 0$). Solid line: displacement; dashed line: velocity.

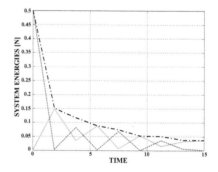

Fig. 17.31. Energy response for the energy decaying scheme. Solid line: kinetic energy; dashed line: strain energy; dashed-dotted line: total mechanical energy.

monotonously, as expected from the energy decay inequality that characterize this scheme.

17.5.4 Enforcement of the constraints

Consider a multibody system subjected to holonomic constraints $\underline{\mathcal{C}}(\underline{q}) = 0$. If the Lagrange multiplier method is used to enforce these constraints, a constraint potential $V^c = \underline{\lambda}^T \underline{\mathcal{C}}$ is added to the strain energy of the system, where $\underline{\lambda}$ are the Lagrange multipliers. The corresponding forces of constraint are $\underline{F}^c = \underline{\underline{B}}^T \underline{\lambda}$, where $\underline{\underline{B}}$ is the constraint matrix. The work done by these forces is $W^c = \int_{t_i}^{t_f} \dot{\underline{q}}^T \underline{F}^c \, dt = \int_{t_i}^{t_f} \underline{\lambda}^T \dot{\underline{\mathcal{C}}} \, dt$. Since $\underline{\mathcal{C}}$ must vanish at all times, $\dot{\underline{\mathcal{C}}} = 0$, and $W^c = 0$, i.e., the work done by the forces of constraint vanishes exactly.

Energy preserving scheme

Lagrange's multiplier approach will be used to enforce the constraint. The discretized forces of constraint are selected in the following manner

$$\underline{F}_m^c = \left(\underline{\mathcal{C}}_{,q}^T\right)_m \underline{\lambda}_m, \tag{17.87}$$

where Lagrange's multipliers, $\underline{\lambda}_m$, are additional unknowns of the problem and the mean value theorem, $\mathcal{C}_f - \mathcal{C}_i = (\underline{\mathcal{C}}_{,q})_m (\underline{q}_f - \underline{q}_i)$, defines quantity $(\underline{\mathcal{C}}_{,q})_m$. The work done by these forces of constraint then becomes $W^c = (\mathcal{C}_f - \mathcal{C}_i) \underline{\lambda}_m$. The additional equations required to solve the problem are obtained by enforcing the exact vanishing for the work done by the discretized forces of constraint. Since $\underline{\lambda}_m \neq 0$, this implies $\mathcal{C}_f - \mathcal{C}_i = 0$. To avoid the drift phenomenon, it is preferable to enforce the condition $\mathcal{C}_f = 0$ at each time step.

In summary, the discretization of the forces of constraint, eq. (17.87) together with the definition of $(\underline{\mathcal{C}}_{,q})_m$ and the discrete constraint $\mathcal{C}_f = 0$ imply the exact vanishing of the work done by the discretized forces of constraint.

Energy decaying scheme

The Lagrange multiplier method for the energy decaying scheme is obtained in a similar manner. The discretized forces of constraint are selected as follows

$$\underline{F}_g^c = \left(\underline{\mathcal{C}}_{,q}^T\right)_g \underline{\lambda}_g, \quad \underline{F}_h^c = -\frac{1}{3}\left[\underline{F}_g^c - \left(\underline{\mathcal{C}}_{,q}^T\right)_h \underline{\lambda}_j\right]. \tag{17.88}$$

The work done by these forces become $W^c = (\mathcal{C}_f - \mathcal{C}_j)\underline{\lambda}_g + (\mathcal{C}_j - \mathcal{C}_i)\underline{\lambda}_g$, and vanishes only if $\mathcal{C}_f - \mathcal{C}_j = 0$ and $\mathcal{C}_j - \mathcal{C}_i = 0$. Here again, it is preferable to enforce the constraints as $\mathcal{C}_f = \mathcal{C}_j = 0$, to avoid the drift phenomenon. In summary, the discretization of the forces of constraint (17.88) together with the discrete constraints $\mathcal{C}_f = \mathcal{C}_j = 0$ imply the exact vanishing of the work done by the discretized forces of constraint.

Example 17.10. The pendulum problem
Consider a pendulum problem defined by kinetic energy $K = 1/2\, m\dot{\underline{q}}^T \dot{\underline{q}}$, potential energy $V = -m\, g q_2$, and constraint $\mathcal{C} = (\underline{q}^T \underline{q} - \ell^2)/2 = 0$, where ℓ is the length of the pendulum. The generalized coordinates of the problem, $\underline{q}^T = \{q_1, q_2\}^T$, correspond to the horizonal and vertical displacements of the bob, respectively. For this example, $m = 1$ kg, $\ell = 0.5$ m, $v_0 = 1.695$ m/s, and $g = 9.81$ m/s^2.

It is clear that $(\mathcal{C}_{,q})_m = \underline{q}_m$, where $\underline{q}_m = (\underline{q}_f + \underline{q}_i)/2$. The governing equations for the trapezoidal rule and energy preserving schemes are $(m\dot{\underline{q}}_f - m\dot{\underline{q}}_i)/\Delta t + \underline{q}_m \lambda_m = m\underline{g}$. For the trapezoidal rule the constraint condition is $\mathcal{C}_m = (\underline{q}_m^T \underline{q}_m - \ell^2)/2 = 0$, whereas for the energy preserving it is $\mathcal{C}_f = (\underline{q}_f^T \underline{q}_f - \ell^2)/2 = 0$.

Finally, the governing equations for the energy decaying scheme are $(m\dot{\underline{q}}_f - m\dot{\underline{q}}_i)/\Delta t + \underline{q}_g \lambda_g = m\underline{g}$ and $(m\dot{\underline{q}}_j - m\dot{\underline{q}}_i)/\Delta t - (\underline{q}_g \lambda_g - \underline{q}_h \lambda_j)/3 = 0$, subjected to two constraint conditions $\mathcal{C}_f = 0$ and $\mathcal{C}_j = (\underline{q}_j^T \underline{q}_j - \ell^2)/2 = 0$.

Figures 17.32, 17.33 and 17.34 show the time history of the pendulum displacements, velocities, and Lagrange multiplier, respectively, for the trapezoidal rule. Although the displacement history is accurately predicted, the velocities and Lagrange multipliers present violent oscillations of a purely numerical origin. The sharp rise in total energy shown in fig. 17.35 clearly indicates the unstable nature of this scheme.

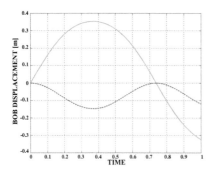

Fig. 17.32. Response for the trapezoidal rule. Solid line: q_1; dashed line: q_2.

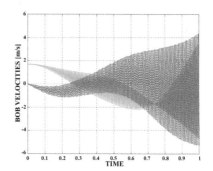

Fig. 17.33. Response for the trapezoidal rule. Solid line: $\dot{q}_1$; dashed line: $\dot{q}_2$.

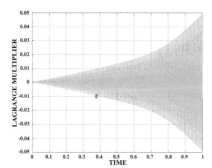

Fig. 17.34. Response for the trapezoidal rule. Solid line: Lagrange multiplier.

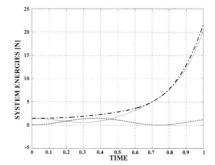

Fig. 17.35. Energy response for the trapezoidal rule. Solid line: kinetic energy; dashed line: strain energy; dashed-dotted line: total mechanical energy.

Figures 17.36 to 17.39 show the corresponding results for the energy preserving and decaying schemes which are in very close agreement. All predicted histories are smooth. The total energy is exactly preserved for the energy preserving scheme, and for the energy decaying scheme, the amount of dissipated energy is very small for this simple problem.

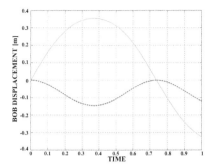

Fig. 17.36. Response for the energy preserving and decaying schemes. Solid line: q_1; dashed line: q_2.

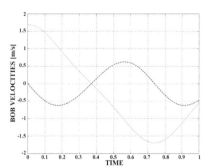

Fig. 17.37. Response for the energy preserving and decaying schemes. Solid line: $\dot{q}_1$; dashed line: $\dot{q}_2$.

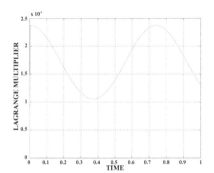

Fig. 17.38. Response for the energy preserving and decaying schemes. Solid line: Lagrange multiplier.

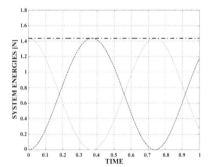

Fig. 17.39. Energy response for the energy preserving and decaying schemes. Solid line: kinetic energy; dashed line: strain energy; dashed-dotted line: total mechanical energy.

17.6 Implementation of cable elements

The formulation of the governing equations for cables is presented in section 16.2. Due to the high level of nonlinearity of these equations, analytical solutions cannot be obtained for all but the simplest problems. This section develops the implementation of the cable governing equations within the finite element framework.

Sections 17.6.1 and 17.6.2 detail the definition and linearization of the inertial and elastic forces, respectively. It is sometimes necessary to include dissipative forces in the formulation, as explained in section 17.6.3. The last section presents the discretization of all forces, leading to the discretized force array and associated mass, gyroscopic, and stiffness matrices.

17.6.1 Inertial forces

The inertial forces acting on the cable are obtained from the governing equations, eqs. (16.42),

$$\underline{\mathcal{F}}^I = m\underline{\ddot{u}}. \tag{17.89}$$

Linearization of inertial forces

Since the expression for the inertial forces is already linear, the increment of inertial forces is simply

$$\Delta\underline{\mathcal{F}}^I = \underline{\underline{M}}\Delta\underline{\ddot{u}}, \tag{17.90}$$

where $\underline{\underline{M}}$ is the mass matrix associated with the inertial forces, defined as

$$\underline{\underline{M}} = m\underline{\underline{I}}. \tag{17.91}$$

17.6.2 Elastic forces

The elastic forces acting on the cable are also obtained from eqs. (16.42),

$$\underline{\mathcal{F}}^C = F^*\underline{G}_1, \tag{17.92}$$

where $\underline{G}_1 = \bar{g}_1 + \underline{u}'$, and the axial elastic force, F^*, is related to the axial strain through the constitutive law, eq. (16.38).

Linearization of elastic forces

Since the expression for the elastic forces is nonlinear, the computational process will require a linearization. At first, the increment in strain is evaluated to find $\Delta\gamma_{11} = \underline{G}_1^T \Delta\underline{u}'$. Next, the increment in the elastic force is computed

$$\Delta F^* = S\underline{G}_1^T \Delta\underline{u}', \tag{17.93}$$

where S is the axial stiffness of the cable.

Taking variations of eq. (17.92) yields the following expression for increments in the elastic forces

$$\Delta\underline{\mathcal{F}}^C = \underline{\underline{S}}\Delta\underline{u}', \tag{17.94}$$

where the stiffness matrix, $\underline{\underline{S}}$, is defined as

$$\underline{\underline{S}} = F^*\underline{\underline{I}} + S\underline{G}_1\underline{G}_1^T. \tag{17.95}$$

17.6.3 Dissipative forces

The cable model discussed in the previous section is a purely conservative model, because the elastic forces are proportional the strain measures. It is often desirable to also introduce dissipative forces in the cable model. By an analogy to eq. (16.38), the dissipative force will be written as

$$F_d = \mu S \dot{\gamma}_{11}. \tag{17.96}$$

where μ is the damping coefficient of units 1/s, and $\dot{\gamma}_{11}$ the time rate of change of the strain. Because the dissipative mechanisms in the cable are not well understood, it is postulated that the damping coefficient is proportional to the stiffness coefficient. The time rate of change of the sectional strain is readily obtained from eq. (16.32) as $\dot{\gamma}_{11} = \underline{G}_1^T \underline{\dot{u}}'$. The dissipative forces, $\underline{F}^{dC}$, associated with the cable element then become

$$\underline{F}^{dC} = F_d \underline{G}_1. \tag{17.97}$$

Linearization of dissipative forces

Because the expression for the dissipative forces is nonlinear, linearization is required here again. At first, the increment in the time rate of change of the strain is evaluated to find

$$\Delta \dot{\gamma}_{11} = \underline{G}_1^T \Delta \underline{\dot{u}}' + \underline{\dot{u}}'^T \Delta \underline{u}'. \tag{17.98}$$

Next, the increment in the dissipative forces is computed

$$\Delta F_d = \mu S \left(\underline{\dot{u}}'^T \Delta \underline{u}' + \underline{G}_1^T \Delta \underline{\dot{u}}' \right). \tag{17.99}$$

Taking variations of eq. (17.97) yields the following expression for increments in the dissipative forces

$$\Delta \underline{F}^{dC} = \underline{\underline{S}}^d \Delta \underline{u}' + \underline{\underline{E}}^d \Delta \underline{\dot{u}}', \tag{17.100}$$

where the dissipative matrices, $\underline{\underline{S}}^d$ and $\underline{\underline{E}}^d$, are defined as $\underline{\underline{S}}^d = F_d \underline{\underline{I}} + \mu S \underline{G}_1 \underline{\dot{u}}'^T$ and $\underline{\underline{E}}^d = \mu S \underline{G}_1 \underline{G}_1^T$, respectively.

17.6.4 Gravity forces for cables

Gravity forces will be applied on cable due to their mass distribution. The potential of the gravity forces is written as $V = -m\underline{g}^T (\underline{x} + \underline{u})$, where $\underline{g}$ is the gravity vector. A variation of this potential is $\delta V = -m\underline{g}^T \delta \underline{u}$. The gravity forces acting on a material point of cable then becomes

$$\underline{F}^G = m\underline{g}. \tag{17.101}$$

17.6.5 Finite element formulation of cables

With the notation defined in eqs. (17.89), (17.92), (17.97), and (17.101), the equations of motion of cable, eqs. (16.42), can be recast in the following form

$$
\underline{\mathcal{F}}^I - \left(\underline{\mathcal{F}}^C + \underline{\mathcal{F}}^{dC} \right)' = \underline{\mathcal{F}}^G + \underline{\mathcal{F}}^{\text{ext}},
$$

where $\underline{\mathcal{F}}^{\text{ext}}$ are the external forces applied to the cable. A weighted residual formulation will be used here to enforce these dynamic equilibrium equations

$$
\int_0^\ell \underline{\underline{N}}^T \left[\underline{\mathcal{F}}^I - \left(\underline{\mathcal{F}}^C + \underline{\mathcal{F}}^{dC} \right)' - \underline{\mathcal{F}}^G - \underline{\mathcal{F}}^{\text{ext}} \right] \mathrm{d}\alpha_1 = 0, \tag{17.102}
$$

where ℓ is the length of the cable element and $\underline{\underline{N}}(\alpha_1)$ a matrix storing the selected test functions, see eq. (17.6). An integration by parts is performed on the second term of this equation, leading to

$$
\int_0^\ell \left[\underline{\underline{N}}^T \underline{\mathcal{F}}^I + \underline{\underline{N}}'^T \left(\underline{\mathcal{F}}^C + \underline{\mathcal{F}}^{dC} \right) \right] \mathrm{d}\alpha_1 = \int_0^\ell \underline{\underline{N}}^T (\underline{\mathcal{F}}^G + \underline{\mathcal{F}}^{\text{ext}}) \, \mathrm{d}\alpha_1.
$$

Because this set of algebraic equations is nonlinear, a linearization process is required to solve it. Equations (17.90), (17.93), and (17.100) are introduced to find

$$
\int_0^\ell \left[\underline{\underline{N}}^T \left(\underline{\mathcal{F}}^I + \underline{\underline{M}}\Delta\underline{\ddot{u}} \right) + \underline{\underline{N}}'^T \left(\underline{\mathcal{F}}^C + \underline{\underline{S}}\Delta\underline{u}' + \underline{\mathcal{F}}^{dC} + \underline{\underline{S}}^d\Delta\underline{u}' + \underline{\underline{\mathcal{E}}}^d\Delta\underline{\dot{u}}' \right) \right] \mathrm{d}\alpha_1
$$
$$
= \int_0^\ell \underline{\underline{N}}^T (\underline{\mathcal{F}}^G + \underline{\mathcal{F}}^{\text{ext}}) \, \mathrm{d}\alpha_1.
$$

Next, the displacement, velocity, and acceleration fields of the element are expressed in terms of their nodal values using the assumed shape functions to find $\underline{u}(\alpha_1) = \underline{\underline{N}}(\alpha_1)\underline{\hat{u}}$, $\underline{\dot{u}}(\alpha_1) = \underline{\underline{N}}(\alpha_1)\underline{\hat{\dot{u}}}$, and $\underline{\ddot{u}}(\alpha_1) = \underline{\underline{N}}(\alpha_1)\underline{\hat{\ddot{u}}}$, where $\underline{\hat{u}}$, $\underline{\hat{\dot{u}}}$, and $\underline{\hat{\ddot{u}}}$ are the nodal displacements, velocities, and accelerations, respectively. With the help of these interpolations, the weak statement of dynamic equilibrium, eq. (17.102), becomes

$$
\underline{\underline{\hat{M}}}\Delta\underline{\hat{\ddot{u}}} + \underline{\underline{\hat{G}}}\Delta\underline{\hat{\dot{u}}} + \underline{\underline{\hat{K}}}\Delta\underline{\hat{u}} = \underline{\hat{F}}^G + \underline{\hat{F}}^{\text{ext}} - \underline{\hat{F}}.
$$

The mass, gyroscopic, and stiffness matrices of the cable element are

$$
\underline{\underline{\hat{M}}} = \int_0^\ell \underline{\underline{N}}^T \underline{\underline{M}} \, \underline{\underline{N}} \, \mathrm{d}\alpha_1, \tag{17.103a}
$$

$$
\underline{\underline{\hat{G}}} = \int_0^\ell \underline{\underline{N}}'^T \underline{\underline{\mathcal{E}}}^d \underline{\underline{N}}' \, \mathrm{d}\alpha_1, \tag{17.103b}
$$

$$
\underline{\underline{\hat{K}}} = \int_0^\ell \underline{\underline{N}}'^T \left(\underline{\underline{S}} + \underline{\underline{S}}^d \right) \underline{\underline{N}}' \, \mathrm{d}\alpha_1, \tag{17.103c}
$$

respectively, and the internal, gravity, and externally applied loads are

$$\hat{\underline{F}} = \int_0^\ell \left[\underline{N}^T \underline{\mathcal{F}}^I + \underline{N}'^T \left(\underline{\mathcal{F}}^C + \underline{\mathcal{F}}^{dC} \right) \right] \, \mathrm{d}\alpha_1, \tag{17.104a}$$

$$\hat{\underline{F}}^G = \int_0^\ell \underline{N}^T \underline{\mathcal{F}}^G \, \mathrm{d}\alpha_1, \quad \hat{\underline{F}}^{\text{ext}} = \int_0^\ell \underline{N}^T \underline{\mathcal{F}}^{\text{ext}} \, \mathrm{d}\alpha_1, \tag{17.104b}$$

respectively.

17.7 Finite element implementation of beam elements

The formulation of the governing equations for beams is presented in section 16.3. This section develops the implementation of the beam governing equations within the finite element framework. Because the expressions for the various forces present in the beam are far more complex and nonlinear than those characterizing cables, the linearization process is more arduous, although the final discretized equations are formally identical. Inertial, elastic and dissipative forces and their linearizations are presented in sections 17.7.1, 17.7.2, and 17.7.3, respectively. The last section presents the discretization of all forces, leading to the discretized force array and associated mass, gyroscopic, and stiffness matrices.

17.7.1 Inertial forces

The inertial forces actin in the beam are obtained from the governing equations of motion, eqs. (16.63),

$$\underline{\mathcal{F}}^I = \dot{\underline{P}} + \begin{bmatrix} \underline{0} & \underline{0} \\ \widetilde{u} & \underline{0} \end{bmatrix} \underline{P}, \tag{17.105}$$

where $\underline{P}$ is the momentum array resolved in the inertial system, defined by eq. (16.61). In view of eq. (16.60), this momentum array can be expressed as

$$\underline{P} = \underline{\mathcal{M}} \underline{V}, \tag{17.106}$$

where the sectional mass matrix resolved in the inertial system is

$$\underline{\mathcal{M}} = (\underline{\underline{R}}\,\underline{\underline{R}}_0)\underline{\mathcal{M}}^*(\underline{\underline{R}}\,\underline{\underline{R}}_0)^T = \begin{bmatrix} m\underline{I} & m\widetilde{\eta}^T \\ m\widetilde{\eta} & \underline{\varrho} \end{bmatrix}. \tag{17.107}$$

The location of the sectional center of mass and its moment of inertia tensor, both resolved in the inertial frame, are defined as $\underline{\eta} = (\underline{\underline{R}}\,\underline{\underline{R}}_0)\underline{\eta}^*$, $\underline{\varrho} = (\underline{\underline{R}}\,\underline{\underline{R}}_0)\underline{\varrho}^*(\underline{\underline{R}}\,\underline{\underline{R}}_0)^T$, respectively.

Expanding eq. (17.106) now leads to

$$\underline{P} = \begin{Bmatrix} m\dot{\underline{u}} + m\widetilde{\eta}^T \underline{\omega} \\ m\widetilde{\eta}\dot{\underline{u}} + \underline{\varrho}\,\underline{\omega} \end{Bmatrix}. \tag{17.108}$$

The time derivatives of the location of the sectional mass center and its moment of inertia tensor are $m\dot{\eta} = \widetilde{\omega}m\eta$, and $\dot{\varrho} = \widetilde{\omega}\varrho + \varrho\widetilde{\omega}^T$, respectively. The time derivative of this momentum array, eq. (17.106), then becomes

$$\dot{\underline{P}} = \left\{ \begin{array}{c} m\ddot{u} + (\dot{\widetilde{\omega}} + \widetilde{\omega}\widetilde{\omega})m\eta \\ m\widetilde{\eta}\ddot{u} + \dot{\widetilde{u}}^T\widetilde{\omega}m\eta + \widetilde{\omega}\varrho\,\omega + \varrho\,\dot{\omega} \end{array} \right\}. \tag{17.109}$$

Finally, the inertial forces, eq. (17.105), can be written in a compact form as

$$\underline{F}^I = \left\{ \begin{array}{c} m\ddot{u} + (\dot{\widetilde{\omega}} + \widetilde{\omega}\widetilde{\omega})m\eta \\ m\widetilde{\eta}\ddot{u} + \varrho\,\dot{\omega} + \widetilde{\omega}\varrho\,\omega \end{array} \right\}. \tag{17.110}$$

The inertial forces are expressed in terms of physical quantities, the angular velocity and acceleration. In practical implementations of the finite element method, nodal rotations must parameterized using any of the techniques described in chapter 13. When using the vectorial parameterization of rotation, the angular velocity and acceleration vectors are expressed as $\underline{\omega} = \underline{\underline{H}}\,\dot{p}$, eq. (13.54), and $\dot{\underline{\omega}} = \underline{\underline{H}}\,\dot{p} + \underline{\underline{H}}\,\ddot{p}$, respectively, where $\underline{p}$ is rotation parameter vector and $\underline{\underline{H}}(\underline{p})$ the tangent tensor, eq. (13.55). Of course, the interpolation the rotation field must be performed carefully, according to the algorithm presented in section 17.2.

Linearization of inertial forces

The expression for the inertial forces given above is nonlinear, and the solution process will require linearization of these forces. First, it will be necessary to compute increments of the sectional mass center location and of the sectional moment of inertia tensor, which are found to be $m\Delta\eta = m\widetilde{\eta}^T\underline{\Delta\psi}$ and $(\Delta\varrho)\underline{b} = (\varrho\widetilde{b} - \widetilde{\varrho b})\underline{\Delta\psi}$, respectively, where $\underline{b}$ is an arbitrary vector. Linearization of the inertial forces then yields

$$\Delta\underline{F}^I = \underline{\underline{K}} \left\{ \begin{array}{c} \Delta u \\ \Delta\psi \end{array} \right\} + \underline{\underline{G}} \left\{ \begin{array}{c} \Delta\dot{u} \\ \Delta\omega \end{array} \right\} + \underline{\underline{M}} \left\{ \begin{array}{c} \Delta\ddot{u} \\ \Delta\dot{\omega} \end{array} \right\}, \tag{17.111}$$

where $\underline{\underline{K}}$, $\underline{\underline{G}}$, and $\underline{\underline{M}}$ are the stiffness, gyroscopic, and mass matrices associated with the inertial forces, respectively. Simple algebra yields

$$\underline{\underline{K}} = \left[\begin{array}{cc} \underline{\underline{0}} & (\dot{\widetilde{\omega}} + \widetilde{\omega}\widetilde{\omega})m\widetilde{\eta}^T \\ \underline{\underline{0}} & \ddot{u}m\widetilde{\eta} + (\varrho\dot{\widetilde{\omega}} - \widetilde{\varrho\dot{\omega}}) + \widetilde{\omega}(\varrho\widetilde{\omega} - \widetilde{\varrho\omega}) \end{array} \right], \tag{17.112a}$$

$$\underline{\underline{G}} = \left[\begin{array}{cc} \underline{\underline{0}} & \widetilde{\widetilde{\omega}m\eta}^T + \widetilde{\omega}m\widetilde{\eta}^T \\ \underline{\underline{0}} & \widetilde{\omega}\varrho - \widetilde{\varrho\omega} \end{array} \right], \tag{17.112b}$$

$$\underline{\underline{M}} = \left[\begin{array}{cc} m\underline{\underline{I}} & m\widetilde{\eta}^T \\ m\widetilde{\eta} & \varrho \end{array} \right]. \tag{17.112c}$$

Here again, these matrices are expressed in terms of physical quantities, the angular velocity and acceleration. Increments of these quantities are now related to increments in the rotation parameters using the chain rule for derivatives

$$\Delta\underline{\omega} = \frac{\partial\underline{\omega}}{\partial\underline{p}}\Delta\underline{p} + \frac{\partial\underline{\omega}}{\partial\underline{\dot{p}}}\Delta\underline{\dot{p}} = (\underline{\dot{H}} - \widetilde{\omega}\underline{H})\Delta\underline{p} + \underline{H}\Delta\underline{\dot{p}}, \tag{17.113}$$

where $\underline{p}$ is the rotation parameter vector and eq. (4.84), recast as $\partial\underline{\omega}/\partial\underline{p} = \underline{\dot{H}} - \widetilde{\omega}\underline{H}$, was used.

Next, increments in the angular acceleration vector are evaluated by taking a time derivative of the above equation

$$\Delta\underline{\dot{\omega}} = (\underline{\ddot{H}} - \widetilde{\dot{\omega}}\underline{\dot{H}} - \widetilde{\dot{\omega}}\underline{H})\Delta\underline{p} + (2\underline{\dot{H}} - \widetilde{\omega}\underline{H})\Delta\underline{\dot{p}} + \underline{H}\Delta\underline{\ddot{p}}. \tag{17.114}$$

Introducing these results into eq. (17.111) yields the following expression for increments in the inertial forces

$$\Delta\underline{\mathcal{F}}^I = \underline{\underline{K}}^I\Delta\underline{q} + \underline{\underline{G}}^I\Delta\underline{\dot{q}} + \underline{\underline{M}}\Delta\underline{\ddot{q}}, \tag{17.115}$$

where $\underline{\underline{K}}^I$, $\underline{\underline{G}}^I$, and $\underline{\underline{M}}$ are the stiffness, gyroscopic, and mass matrices associated with the inertial forces, respectively. The incremental arrays of displacement, velocity, and acceleration arrays are defined as $\Delta\underline{q}^T = \{\Delta\underline{u}^T, \Delta\underline{p}^T\}$, $\Delta\underline{\dot{q}}^T = \{\Delta\underline{\dot{u}}^T, \Delta\underline{\dot{p}}^T\}$, and $\Delta\underline{\ddot{q}}^T = \{\Delta\underline{\ddot{u}}^T, \Delta\underline{\ddot{p}}^T\}$, respectively.

In summary, the inertial forces can be written in the following form

$$\underline{\mathcal{F}}^I = \begin{Bmatrix} m\underline{\ddot{u}} + \underline{\alpha} \\ m\widetilde{\eta}\underline{\ddot{u}} + \underline{\gamma} \end{Bmatrix}, \tag{17.116}$$

and the stiffness, gyroscopic, and mass matrices are

$$\underline{\underline{K}}^I = \begin{bmatrix} \underline{0} & m\widetilde{\eta}^T\underline{\ddot{H}} - 2\widetilde{\beta}\underline{\dot{H}} - \widetilde{\alpha}\underline{H} \\ \underline{0} & \underline{\varrho}\underline{\ddot{H}} + \underline{\varepsilon}\underline{\dot{H}} + (\widetilde{u}m\widetilde{\eta} - \widetilde{\gamma})\underline{H} \end{bmatrix}, \tag{17.117a}$$

$$\underline{\underline{G}}^I = \begin{bmatrix} \underline{0} & 2m\widetilde{\eta}^T\underline{\dot{H}} - 2\widetilde{\beta}\underline{H} \\ \underline{0} & 2\underline{\varrho}\underline{\dot{H}} + \underline{\varepsilon}\underline{H} \end{bmatrix}, \tag{17.117b}$$

$$\underline{\underline{M}} = \begin{bmatrix} m\underline{I} & m\widetilde{\eta}^T\underline{H} \\ m\widetilde{\eta} & \underline{\varrho}\underline{H} \end{bmatrix}. \tag{17.117c}$$

The following notations were introduced to simply the writing of the above expressions $\underline{\alpha} = (\widetilde{\dot{\omega}} + \widetilde{\omega}\widetilde{\omega})m\underline{\eta}$, $\underline{\beta} = \widetilde{\omega}m\underline{\eta}$, $\underline{\gamma} = \underline{\varrho}\underline{\dot{\omega}} + \widetilde{\omega}\underline{\varrho}\underline{\omega}$, and $\underline{\varepsilon} = \widetilde{\omega}\underline{\varrho} + (\widetilde{\omega}\underline{\varrho})^T - \widetilde{\varrho\omega}$.

17.7.2 Elastic forces

The elastic forces acting in the beam element are obtained from eqs. (16.63) and will be treated in two separate components, denoted $\underline{\mathcal{F}}^C$ and $\underline{\mathcal{F}}^D$, defined as

$$\underline{\mathcal{F}}^C = \underline{f} = \begin{Bmatrix} \underline{N} \\ \underline{M} \end{Bmatrix}, \text{ and } \underline{\mathcal{F}}^D = \begin{bmatrix} \underline{0} & \underline{0} \\ (\widetilde{u}'_0 + \widetilde{u}')^T & \underline{0} \end{bmatrix}\underline{f} = \begin{Bmatrix} \underline{0} \\ (\widetilde{u}'_0 + \widetilde{u}')^T\underline{N} \end{Bmatrix}, \tag{17.118}$$

respectively. The components of the beam's sectional force and moment vectors resolved in the inertial basis are denoted $\underline{N}$ and $\underline{M}$, respectively.

The sectional strains and curvatures defined in eq. (16.45) are recast in the following compact notation

$$
\underline{e} = \left\{ \begin{array}{c} \underline{u}'_0 + \underline{u}' - (\underline{\underline{R}}\,\underline{\underline{R}}_0)\bar{\imath}_1 \\ \underline{k} + \underline{\underline{R}}\,\underline{k}_i \end{array} \right\}, \tag{17.119}
$$

where $\underline{k} = \mathrm{axial}(\underline{\underline{R}}'\underline{\underline{R}}^T)$ are the components of the sectional curvature vector resolved in the inertial basis and $\underline{k}_i$ the components of the corresponding curvature vector in the beam's reference configuration. The corresponding strain components resolved in the material basis are

$$
\underline{e}^* = \left\{ \begin{array}{c} (\underline{\underline{R}}\,\underline{\underline{R}}_0)^T \underline{E}_1 - \bar{\imath}_1 \\ (\underline{\underline{R}}\,\underline{\underline{R}}_0)^T (\underline{k} + \underline{\underline{R}}\,\underline{k}_i) \end{array} \right\}. \tag{17.120}
$$

The elastic forces in the beam are then $\underline{f} = \underline{\underline{C}}\,\underline{e}$, where $\underline{\underline{C}} = (\underline{\underline{R}}\,\underline{\underline{R}}_0)\underline{\underline{C}}^*(\underline{\underline{R}}\,\underline{\underline{R}}_0)^T$ is the sectional stiffness matrix resolved in the inertial basis, and the corresponding stiffness matrix resolved in the material basis, $\underline{\underline{C}}^*$, is defined by eq. (16.47).

The elastic forces are expressed in terms of physical quantities, the sectional strain and curvatures. In practical implementations of the finite element method, nodal rotations must parameterized using any of the techniques described in chapter 13. When using the vectorial parameterization of rotation, the curvature vector will be expressed as $\underline{k} = \underline{\underline{H}}\,\underline{p}'$, eq. (13.54), where $\underline{p}$ is rotation parameter vector and $\underline{\underline{H}}(\underline{p})$ the tangent tensor, eq. (13.55).

Linearization of elastic forces

The expressions for the elastic forces given above are nonlinear, and the finite element process will require a linearization of these forces. First, increments in the curvature vector are computed using the chain rule for partial derivatives

$$
\Delta\underline{k} = \frac{\partial\underline{k}}{\partial\underline{p}}\Delta\underline{p} + \frac{\partial\underline{k}}{\partial\underline{p}'}\Delta\underline{p}' = (\underline{\underline{H}}' - \widetilde{\underline{k}}\underline{\underline{H}})\Delta\underline{p} + \underline{\underline{H}}\Delta\underline{p}', \tag{17.121}
$$

where $\underline{p}$ is the rotation parameter vector and eq. (4.84), recast as $\partial\underline{k}/\partial\underline{p} = \underline{\underline{H}}' - \widetilde{\underline{k}}\underline{\underline{H}}$, was used.

Increments in the strain components, eq. (17.119), are now easily evaluated to find

$$
\Delta\underline{e} = \left\{ \begin{array}{c} \Delta\underline{u}' + \widetilde{(\underline{\underline{R}}\,\underline{\underline{R}}_0)\bar{\imath}_1}\,\underline{\underline{H}}\Delta\underline{p} \\ \underline{\underline{H}}\Delta\underline{p}' + \underline{\underline{H}}'\Delta\underline{p} - \widetilde{(\underline{k} + \underline{\underline{R}}\,\underline{k}_i)}\underline{\underline{H}}\Delta\underline{p} \end{array} \right\}.
$$

This leads to the following expression for increments in the elastic forces

$$
\Delta\underline{f} = \left\{ \begin{array}{c} \widetilde{\underline{N}}^T\underline{\underline{H}}\Delta\underline{p} \\ \widetilde{\underline{M}}^T\underline{\underline{H}}\Delta\underline{p} \end{array} \right\} + \underline{\underline{C}} \left\{ \begin{array}{c} (\widetilde{\underline{u}}'_0 + \widetilde{\underline{u}}')\underline{\underline{H}}\Delta\underline{p} + \Delta\underline{u}' \\ \underline{\underline{H}}\Delta\underline{p}' + \underline{\underline{H}}'\Delta\underline{p} \end{array} \right\}.
$$

Taking variations of eq. (17.118) yields the following expression for increments in the elastic forces

$$\Delta \underline{\mathcal{F}}^C = \underline{\underline{S}} \Delta \underline{q}' + \underline{\underline{Q}} \Delta \underline{q}, \quad \Delta \underline{\mathcal{F}}^D = \underline{\underline{P}} \Delta \underline{q}' + \underline{\underline{Q}} \Delta \underline{q}. \tag{17.122}$$

The incremental arrays of displacement and displacement rates are defined as $\Delta \underline{q}^T = \{\Delta \underline{u}^T, \Delta \underline{p}^T\}$, and $\Delta \underline{q}'^T = \{\Delta \underline{u}'^T, \Delta \underline{p}'^T\}$, respectively.

In summary, the elastic forces can be written in the following form

$$\underline{\mathcal{F}}^C = \underline{\underline{C}} \, \underline{e}, \quad \underline{\mathcal{F}}^D = \underline{\underline{\varUpsilon}} \, \underline{\mathcal{F}}^C, \tag{17.123}$$

and the stiffness matrices are

$$\underline{\underline{S}} = \underline{\underline{C}} \begin{bmatrix} \underline{\underline{I}} & \underline{\underline{0}} \\ \underline{\underline{0}} & \underline{\underline{H}} \end{bmatrix}, \qquad \underline{\underline{Q}} = \begin{bmatrix} \underline{\underline{0}} & \widetilde{N}^T \underline{\underline{H}} \\ \underline{\underline{0}} & \widetilde{M}^T \underline{\underline{H}} \end{bmatrix} + \underline{\underline{C}} \begin{bmatrix} \underline{\underline{0}} & (\widetilde{u}_0' + \widetilde{u}')\underline{\underline{H}} \\ \underline{\underline{0}} & \underline{\underline{H}}' \end{bmatrix}, \tag{17.124a}$$

$$\underline{\underline{P}} = \begin{bmatrix} \underline{\underline{0}} & \underline{\underline{0}} \\ \widetilde{N} & \underline{\underline{0}} \end{bmatrix} + \underline{\underline{\varUpsilon}} \, \underline{\underline{S}}, \quad \underline{\underline{Q}} = \underline{\underline{\varUpsilon}} \, \underline{\underline{Q}}. \tag{17.124b}$$

The following notation was introduced

$$\underline{\underline{\varUpsilon}} = \begin{bmatrix} \underline{\underline{0}} & \underline{\underline{0}} \\ (\widetilde{u}_0' + \widetilde{u}')^T & \underline{\underline{0}} \end{bmatrix}. \tag{17.125}$$

17.7.3 Dissipative forces

The beam model discussed in the previous section is a purely conservative model, because the elastic force are proportional the strain measures. It is often desirable to also introduce dissipative forces in the beam model. By analogy to eq. (16.47), the dissipative forces in the material frame, $\underline{f}^{d*}$, will be written as

$$\underline{f}^{d*} = \left\{ \begin{matrix} \underline{N}^{d*} \\ \underline{M}^{d*} \end{matrix} \right\} = \mu \underline{\underline{C}}^* \underline{\dot{e}}^*, \tag{17.126}$$

where μ is the damping coefficient of units 1/s, and $\underline{\dot{e}}^*$ the time rate of change of the strains measured in the material frame. Since the dissipative mechanisms in the beam are not well understood, it is postulated that the damping matrix is proportional to the stiffness matrix. The time rate of change of the sectional strains in the material frame are readily obtained from eq. (17.120) as

$$\underline{\dot{e}}^* = \left\{ \begin{matrix} (\underline{\underline{R}} \, \underline{\underline{R}}_0)^T (\underline{\dot{u}} + \widetilde{E}_1 \underline{\omega}) \\ (\underline{\underline{R}} \, \underline{\underline{R}}_0)^T \underline{\omega}' \end{matrix} \right\}. \tag{17.127}$$

The dissipative forces in the inertial frame now become

$$\underline{f}^d = \mu \underline{\underline{C}} \, \underline{\dot{e}}, \tag{17.128}$$

where $\underline{f}^{dT} = \{\underline{N}^{dT}, \underline{M}^{dT}\}$, and $\underline{N}^d = (\underline{\underline{R}} \, \underline{\underline{R}}_0) \underline{N}^{d*}$ and $\underline{M}^d = (\underline{\underline{R}} \, \underline{\underline{R}}_0) \underline{M}^{d*}$ are the are the sectional dissipative force and moment vector components in the inertial frame, respectively, $\underline{\underline{C}} = (\underline{\underline{R}}\,\underline{\underline{R}}_0)\underline{\underline{C}}^*(\underline{\underline{R}}\,\underline{\underline{R}}_0)^T$ is the sectional stiffness matrix resolved in the inertial basis, and $\underline{\dot{e}}$ is defined as

$$\underline{\dot{e}} = \left\{ \begin{array}{c} \dot{u} + \tilde{E}_1\underline{\omega} \\ \underline{\omega}' \end{array} \right\}. \tag{17.129}$$

The dissipative forces will be treated in two separate components, denoted $\mathcal{F}^{dC}$ and $\mathcal{F}^{dD}$,

$$\mathcal{F}^{dC} = \underline{f}^d = \left\{ \begin{array}{c} \underline{N}^d \\ \underline{M}^d \end{array} \right\}, \quad \mathcal{F}^{dD} = \begin{bmatrix} \underline{\underline{0}} & \underline{\underline{0}} \\ \tilde{E}_1^T & \underline{\underline{0}} \end{bmatrix} \underline{f}^d = \left\{ \begin{array}{c} \underline{0} \\ \tilde{E}_1^T \underline{N}^d \end{array} \right\}. \tag{17.130}$$

Linearization of dissipative forces

Because the expression for the dissipative forces is nonlinear, the solution process will require a linearization. Increments in the strain array are evaluated to find

$$\Delta\underline{\dot{e}} = \left\{ \begin{array}{c} \Delta\dot{u} + \tilde{E}_1(\underline{\dot{H}} - \tilde{\omega}\underline{H})\Delta p + \tilde{E}_1\underline{H}\Delta\dot{p} + \tilde{\omega}^T\Delta u' \\ (\underline{\dot{H}}' - \tilde{\omega}'\underline{H} - \tilde{\omega}\underline{H}')\Delta p + (\underline{\dot{H}} - \tilde{\omega}\underline{H})\Delta p' + \underline{H}'\Delta p + \underline{H}\Delta p' \end{array} \right\}. \tag{17.131}$$

Next, increments in the dissipative forces are computed

$$\Delta\underline{f}^d = \left\{ \begin{array}{c} \tilde{N}^{dT}\underline{H}\Delta p \\ \tilde{M}^{dT}\underline{H}\Delta p \end{array} \right\} + \mu\underline{\underline{C}} \left\{ \begin{array}{c} (\dot{\tilde{u}} + \tilde{E}_1\tilde{\omega} - \tilde{\omega}\tilde{E}_1)\underline{H}\Delta p \\ \tilde{\omega}'\underline{H}\Delta p \end{array} \right\} + \mu\underline{\underline{C}}\Delta\underline{\dot{e}}. \tag{17.132}$$

Taking variations of eq. (17.130) yields the following expression for increments in the dissipative forces

$$\Delta\mathcal{F}^{dC} = \underline{\underline{S}}^d\Delta\underline{q}' + \underline{\underline{O}}^d\Delta\underline{q} + \underline{\underline{G}}^d\Delta\underline{\dot{q}} + \underline{\underline{E}}^d\Delta\underline{\dot{q}}', \tag{17.133a}$$

$$\Delta\mathcal{F}^{dD} = \underline{\underline{P}}^d\Delta\underline{q}' + \underline{\underline{Q}}^d\Delta\underline{q} + \underline{\underline{X}}^d\Delta\underline{\dot{q}} + \underline{\underline{Y}}^d\Delta\underline{\dot{q}}'. \tag{17.133b}$$

In summary, the dissipative forces can be written in the following form

$$\mathcal{F}^{dC} = \mu\underline{\underline{C}}\,\underline{\dot{e}}, \quad \mathcal{F}^{dD} = \underline{\underline{\Upsilon}}\,\mathcal{F}^{dC}, \tag{17.134}$$

where the gyroscopic and stiffness matrices are

$$\underline{\underline{S}}^d = \mu\underline{\underline{C}} \begin{bmatrix} \tilde{\omega}^T & \underline{0} \\ \underline{0} & (\underline{\dot{H}} - \tilde{\omega}\underline{H}) \end{bmatrix}, \quad \underline{\underline{O}}^d = \begin{bmatrix} \underline{0} & \tilde{N}^{dT}\underline{H} \\ \underline{0} & \tilde{M}^{dT}\underline{H} \end{bmatrix} + \mu\underline{\underline{C}} \begin{bmatrix} \underline{0} & (\dot{\tilde{u}} - \tilde{\omega}\tilde{E}_1)\underline{H} + \tilde{E}_1\underline{\dot{H}} \\ \underline{0} & \underline{\dot{H}}' - \tilde{\omega}\underline{H}' \end{bmatrix}, \tag{17.135}$$

$$\underline{\underline{G}}^d = \mu\underline{\underline{C}} \begin{bmatrix} \underline{I} & \tilde{E}_1\underline{H} \\ \underline{0} & \underline{H}' \end{bmatrix}, \quad \underline{\underline{E}}^d = \mu\underline{\underline{C}} \begin{bmatrix} \underline{0} & \underline{0} \\ \underline{0} & \underline{H} \end{bmatrix}, \tag{17.136}$$

and

$$\underline{\underline{P}}^d = \begin{bmatrix} \underline{0} & \underline{0} \\ \tilde{N}^d & \underline{0} \end{bmatrix} + \underline{\underline{\Upsilon}}\,\underline{\underline{S}}^d, \quad \underline{\underline{Q}}^d = \underline{\underline{\Upsilon}}\,\underline{\underline{O}}^d, \quad \underline{\underline{X}}^d = \underline{\underline{\Upsilon}}\,\underline{\underline{G}}^d, \quad \underline{\underline{Y}}^d = \underline{\underline{\Upsilon}}\,\underline{\underline{E}}^d. \tag{17.137}$$

Matrix $\underline{\underline{\Upsilon}}$ is defined in eq. (17.125).

17.7.4 Gravity forces for beams

For many applications, the gravity forces associated with the beam's distributed mass must be taken into account. The potential of these gravity forces is $V = m\underline{g}^T(\underline{u}_0 + \underline{u} + \underline{\eta})$, where vector $\underline{\eta}$ defines the location of the sectional mass center, see eq. (16.56).

A variation of this potential is easily found to be $\delta V = \underline{g}^T(m\delta\underline{u} + m\widetilde{\underline{\eta}}^T\delta\underline{\psi})$ and the gravity forces acting on the cross-section are readily found as

$$\underline{\mathcal{F}}^G = \left\{\begin{matrix} m\underline{g} \\ m\widetilde{\underline{\eta}}\underline{g} \end{matrix}\right\}. \tag{17.138}$$

17.7.5 Finite element formulation of beams

With the notation defined in eqs. (17.105) and (17.118), the equations of motion of curved beams, eqs. (16.63), can be recast in the following compact form, $\underline{\mathcal{F}}^I - \underline{\mathcal{F}}^{C\prime} + \underline{\mathcal{F}}^D = \underline{\mathcal{F}}^G + \underline{\mathcal{F}}^{\text{ext}}$, where $\underline{\mathcal{F}}^{\text{ext}}$ are the external forces applied to the beam.

A weighted residual formulation will be used here to enforce these dynamic equilibrium conditions

$$\int_0^\ell \underline{N}^T\left(\underline{\mathcal{F}}^I - \underline{\mathcal{F}}^{C\prime} + \underline{\mathcal{F}}^D - \underline{\mathcal{F}}^G - \underline{\mathcal{F}}^{\text{ext}}\right)\mathrm{d}\alpha_1 = 0,$$

where ℓ is the length of the beam element and $\underline{N}$ a matrix storing the selected test functions, see eq. (17.6). An integration by parts is performed on the second term of this equation, leading to

$$\int_0^\ell \left(\underline{\underline{N}}^T\underline{\mathcal{F}}^I + \underline{\underline{N}}^{\prime T}\underline{\mathcal{F}}^C + \underline{\underline{N}}^T\underline{\mathcal{F}}^D\right)\mathrm{d}\alpha_1 = \int_0^\ell \underline{\underline{N}}^T(\underline{\mathcal{F}}^G + \underline{\mathcal{F}}^{\text{ext}})\,\mathrm{d}x_1.$$

Since this set of algebraic equations is nonlinear, a linearization process is required to solve it. Equations (17.115) and (17.122) are introduced to find

$$\int_0^\ell \left[\underline{N}^T\left(\underline{\mathcal{F}}^I + \underline{\underline{K}}^I\Delta\underline{q} + \underline{\underline{G}}^I\Delta\underline{v} + \underline{\underline{M}}\Delta\underline{a} + \underline{\mathcal{F}}^D + \underline{\underline{P}}\Delta\underline{q}' + \underline{\underline{Q}}\Delta\underline{q}\right)\right.$$
$$\left. + \underline{N}^{\prime T}\left(\underline{\mathcal{F}}^C + \underline{\underline{S}}\Delta\underline{q}' + \underline{\underline{O}}\Delta\underline{q}\right)\right]\mathrm{d}\alpha_1 = \int_0^\ell \underline{\underline{N}}^T(\underline{\mathcal{F}}^G + \underline{\mathcal{F}}^{\text{ext}})\,\mathrm{d}\alpha_1.$$

Next, the elemental displacement, velocity, and acceleration fields are expressed in terms of their nodal values using the assumed shape functions, $\underline{q}(x_1) = \underline{N}\,\hat{\underline{q}}$, $\underline{q}'(x_1) = \underline{N}'\hat{\underline{q}}$, $\underline{v}(x_1) = \underline{N}\,\hat{\underline{v}}$, $\underline{a}(x_1) = \underline{N}\,\hat{\underline{a}}$, where $\hat{\underline{q}}$, $\hat{\underline{v}}$, and $\hat{\underline{a}}$ are the nodal values of the displacements, velocities, and accelerations, respectively. With the help of these interpolations of elemental fields, the weak statement of dynamic equilibrium becomes

$$\hat{\underline{M}}\Delta\hat{\underline{a}} + \hat{\underline{G}}\Delta\hat{\underline{v}} + \hat{\underline{K}}\Delta\hat{\underline{q}} = \hat{\underline{F}}^G + \hat{\underline{F}}^{\text{ext}} - \hat{\underline{F}}. \tag{17.139}$$

The mass, gyroscopic, and stiffness matrices of the beam element are

$$\underline{\underline{\hat{M}}} = \int_0^\ell \underline{\underline{N}}^T \underline{\underline{M}} \underline{\underline{N}} \, \mathrm{d}\alpha_1, \tag{17.140a}$$

$$\underline{\underline{\hat{G}}} = \int_0^\ell \underline{\underline{N}}^T \underline{\underline{G}}^I \underline{\underline{N}} \, \mathrm{d}\alpha_1, \tag{17.140b}$$

$$\underline{\underline{\hat{K}}} = \int_0^\ell \left[\underline{\underline{N}}^T (\underline{\underline{K}}^I + \underline{\underline{Q}})\underline{\underline{N}} + \underline{\underline{N}}^T \underline{\underline{P}} \, \underline{\underline{N}}' + \underline{\underline{N}}'^T \underline{\underline{S}} \, \underline{\underline{N}}' + \underline{\underline{N}}'^T \underline{\underline{O}} \, \underline{\underline{N}} \right] \, \mathrm{d}\alpha_1, \tag{17.140c}$$

respectively, whereas the elemental forces, gravity loads, and externally applied loads are

$$\underline{\hat{F}} = \int_0^\ell \left(\underline{\underline{N}}^T \underline{\mathcal{F}}^I + \underline{\underline{N}}^T \underline{\mathcal{F}}^D + \underline{\underline{N}}'^T \underline{\mathcal{F}}^C \right) \, \mathrm{d}\alpha_1, \tag{17.141a}$$

$$\underline{\hat{F}}^G = \int_0^\ell \underline{\underline{N}}^T \underline{\mathcal{F}}^G \, \mathrm{d}\alpha_1, \quad \underline{\hat{F}}^{\mathrm{ext}} = \int_0^\ell \underline{\underline{N}}^T \underline{\mathcal{F}}^{\mathrm{ext}} \, \mathrm{d}\alpha_1, \tag{17.141b}$$

respectively.

18

Mathematical tools

18.1 The singular value decomposition

The *singular value decomposition theorem* [82] states that an arbitrary, $n \times m$ matrix $\underline{\underline{A}}$ $(n > m)$, of rank r, $r \leq m$ can be decomposed into the following matrix product

$$\underline{\underline{A}}_{(n \times m)} = \underline{\underline{U}}_{(n \times n)} \begin{bmatrix} \underline{\underline{\Sigma}}_{(r \times r)} & \underline{\underline{0}}_{(r \times (m-r))} \\ \underline{\underline{0}}_{((n-r) \times r)} & \underline{\underline{0}}_{((n-r) \times (m-r))} \end{bmatrix} \underline{\underline{V}}^T_{(m \times m)}, \tag{18.1}$$

where $n > m$, $r \leq m$, $\underline{\underline{U}}$ and $\underline{\underline{V}}$ are orthogonal matrices, and $\underline{\underline{\Sigma}} = \mathrm{diag}(\sigma_i)$ a unique diagonal matrix with real, non-negative elements. The other matrices in eq. (18.1) are zero matrices with the corresponding size indicated by their subscript. The elements of $\underline{\underline{\Sigma}}$ are arranged in descending order as

$$\sigma_1 \geq \sigma_2 \geq \sigma_3 \geq \ldots \geq \sigma_r > \sigma_{r+1} = \ldots = \sigma_m = 0, \tag{18.2}$$

where the σ_i are called the *singular values* of $\underline{\underline{A}}$, and, again, $r = \mathrm{rank}(\underline{\underline{A}})$; if $\underline{\underline{A}}$ has *full rank*, $r = m$. Matrices $\underline{\underline{U}}$ and $\underline{\underline{V}}$ can be partitioned as

$$\underline{\underline{U}}_{(n \times n)} = \begin{bmatrix} \underline{\underline{\breve{U}}}_{(n \times r)} & \underline{\underline{\Gamma}}_{(n \times (n-r))} \end{bmatrix}, \text{ and } \underline{\underline{V}}_{(m \times m)} = \begin{bmatrix} \underline{\underline{V}}_{1\ (m \times r)} & \underline{\underline{V}}_{2\ (m \times (m-r))} \end{bmatrix}, \tag{18.3}$$

respectively, and hence, eq. (18.1) can be recast as

$$\underline{\underline{A}}_{(n \times m)} = \begin{bmatrix} \underline{\underline{\breve{U}}}_{(n \times r)} & \underline{\underline{\Gamma}}_{(n \times (n-r))} \end{bmatrix} \begin{bmatrix} \underline{\underline{\Sigma}}_{(r \times r)} & \underline{\underline{0}}_{(r \times (m-r))} \\ \underline{\underline{0}}_{((n-r) \times r)} & \underline{\underline{0}}_{((n-r) \times (m-r))} \end{bmatrix} \begin{bmatrix} \underline{\underline{V}}^T_{1\ (r \times m)} \\ \underline{\underline{V}}^T_{2\ ((m-r) \times m)} \end{bmatrix}, \tag{18.4}$$

where the size of the matrices $\underline{\underline{\breve{U}}}$, $\underline{\underline{\Gamma}}$, $\underline{\underline{V}}^T_1$ and $\underline{\underline{V}}^T_2$ are indicated by their subscript. Thus, eq. (18.4) can be simplified to be

$$\underline{\underline{A}} = \underline{\underline{\breve{U}}}\,\underline{\underline{\Sigma}}\,\underline{\underline{V}}^T_1, \tag{18.5}$$

i.e., $\underline{\underline{\breve{U}}}$ and $\underline{\underline{V}}_1$ are the left and right *singular vectors* of $\underline{\underline{A}}$, respectively. The orthogonality of $\underline{\underline{U}}$ implies the following relationships.

$$\check{\underline{U}}\,\check{\underline{U}}^T + \underline{\Gamma}\,\underline{\Gamma}^T = \underline{I}, \qquad (18.6a)$$

$$\check{\underline{U}}^T\check{\underline{U}} = \underline{I}, \qquad (18.6b)$$

$$\underline{\Gamma}^T\underline{\Gamma}^T = \underline{I}, \qquad (18.6c)$$

and finally

$$\check{\underline{U}}^T\underline{\Gamma} = 0, \quad \underline{\Gamma}^T\check{\underline{U}} = 0. \qquad (18.7)$$

Transposing eq. (18.5) and post multiplying by $\underline{\Gamma}$ leads to

$$\underline{A}^T\underline{\Gamma} = 0, \qquad (18.8)$$

where property (18.7) was used; clearly, $\underline{\Gamma}$ forms the *null space* of $\underline{A}^T$.

When matrix $\underline{A}$ has full rank, *i.e.* $r = m$, eq. (18.1) reduces to

$$\underline{A} = \underline{U}\left[\frac{\underline{\Sigma}}{\underline{0}}\right]\underline{V}^T, \qquad (18.9)$$

i.e. the partition of V is itself, and eq. (18.9) simplifies to

$$\underline{A} = \check{\underline{U}}\,\underline{\Sigma}\,\underline{V}^T \qquad (18.10)$$

18.2 The Moore-Penrose generalized inverse

The *Moore-Penrose generalized inverse* of matrix $\underline{A}_{(n \times m)}$ with $n \geq m$ is the unique matrix, denoted $\underline{A}^+$, that features the following properties

$$\underline{A}\,\underline{A}^+\underline{A} = \underline{A}, \qquad (18.11)$$

$$\underline{A}^+\underline{A}\,\underline{A}^+ = \underline{A}^+, \qquad (18.12)$$

$$(\underline{A}\,\underline{A}^+) = (\underline{A}\,\underline{A}^+)^T, \qquad (18.13)$$

$$(\underline{A}^+\underline{A}) = (\underline{A}^+\underline{A})^T. \qquad (18.14)$$

The Moore-Penrose inverse is most elegantly computed using the singular value decomposition. If $\underline{A}$ is of full rank, eq. (18.10) implies $\underline{A} = \check{\underline{U}}\,\underline{\Sigma}\,\underline{V}^T$, and its Moore-Penrose inverse is then

$$\underline{A}^+ = \underline{V}\,\underline{\Sigma}^{-1}\check{\underline{U}}^T. \qquad (18.15)$$

It is readily verified that this expression satisfies the four conditions for a Moore-Penrose inverse, eqs. (18.11) to (18.14). It is also possible to express the Moore-Penrose inverse without resorting to the singular value decomposition; indeed, it is readily verified that

$$\underline{A}^+ = \left[\underline{A}^T\underline{A}\right]^{-1}\underline{A}^T, \qquad (18.16)$$

verifies, once again, the four conditions (18.11) to (18.14).

18.2.1 Problems

Problem 18.1. Properties of Moore-Penrose inverse
 Verify that the Moore-Penrose inverse given by eq. (18.15) satisfies the four conditions (18.11) to (18.14).

Problem 18.2. Properties of Moore-Penrose inverse
 Verify that the Moore-Penrose inverse given by eq. (18.16) satisfies the four conditions (18.11) to (18.14).

Problem 18.3. Properties of Moore-Penrose inverse
 Consider matrix $\underline{\underline{A}}_{(m \times n)} = \underline{\underline{B}}_{(m \times m)} \underline{\underline{C}}_{(m \times n)}$, $m > n$, where $\text{rank}(\underline{\underline{A}}) = \text{rank}(\underline{\underline{C}}) = n$ and $\text{rank}(\underline{\underline{B}}) = m$. Is the following statement true: $\underline{\underline{A}}^+ = \underline{\underline{C}}^+ \underline{\underline{B}}^{-1}$?

Problem 18.4. Properties of Moore-Penrose inverse
 Consider matrix $\underline{\underline{A}}_{(m \times n)} = \underline{\underline{B}}_{(m \times r)} \underline{\underline{C}}_{(r \times n)}$, $m > n$, where $\text{rank}(\underline{\underline{A}}) = \text{rank}(\underline{\underline{B}}) = \text{rank}(\underline{\underline{C}}) = r$. Is the following statement true: $\underline{\underline{A}}^+ = \underline{\underline{C}}^T (\underline{\underline{C}}\,\underline{\underline{C}}^T)^{-1} (\underline{\underline{B}}^T \underline{\underline{B}})^1 \underline{\underline{B}}^T$?

18.3 Gauss-Legendre quadrature

When applying energy methods, the computation of the stiffness matrix and load array involves integrations of the product of the shape functions by the stiffness properties of the structure. As the number of assumed shape functions increases, it becomes increasingly cumbersome to perform all these integrations in closed form, specially when the expression for the shape functions becomes complex.

 To circumvent this problem, numerical integration can be used. A very powerful tool for numerical integration is the Gauss-Legendre quadrature scheme. In its simplest form [5], this scheme approximately evaluates an integral by the following sum

$$\int_{-1}^{+1} f(s)\,\mathrm{d}s \approx \sum_{i=1}^{N_G} w_i f(s_i), \tag{18.17}$$

where s_i, $i = 1, 2, \ldots N_G$ are the *Gauss-Legendre quadrature points*, and w_i the associated *weights*. The Gauss-Legendre quadrature points are often called *sampling points*, because the integral is evaluated by sampling the value of the integrand at these points. Table 18.1 lists the Gauss-Legendre quadrature points and associated weights for $N_G = 2$, 3, and 4. The fundamental property of the N_G point Gauss-Legendre quadrature scheme is that it exactly integrates a polynomial of degree $2N_G - 1$.

 To illustrate the application of the Gauss-Legendre quadrature scheme, consider the following integral

$$I = \int_{-1}^{+1} \left[x^4 - 5x^3 + 3x^2 + 5x \right]\,\mathrm{d}x = 2.4.$$

At first, the 2-point quadrature formula is used to find

Table 18.1. Gauss points and associated weights for $N_G = 2$, 3, and 4.

N_G	s_i	w_i	N_G	s_i	w_i
2	$\pm\sqrt{1/3}$	1	4	$\pm\sqrt{(3-2\sqrt{6/5})/7}$	$(18+\sqrt{30})/36$
3	0	8/9		$\pm\sqrt{(3+2\sqrt{6/5})/7}$	$(18-\sqrt{30})/36$
	$\pm\sqrt{3/5}$	5/9			

$$I \approx \left[\left(\frac{1}{3}\right)^2 + 5\left(\frac{1}{3}\right)^{3/2} + 3\frac{1}{3} - 5\left(\frac{1}{3}\right)^{1/2}\right]$$
$$+ \left[\left(\frac{1}{3}\right)^2 - 5\left(\frac{1}{3}\right)^{3/2} + 3\frac{1}{3} + 5\left(\frac{1}{3}\right)^{1/2}\right] = \frac{20}{9} = 2.22.$$

This 2-point formula exactly integrates a polynomial of degree $2 \times 2 - 1 = 3$; hence, an approximate answer is expected for this integral involving a polynomial of degree four. The approximate answer only incurs a 7.4% error. Next, the 3-point quadrature formula is used, leading to

$$I \approx \frac{5}{9}\left[\left(\frac{3}{5}\right)^2 + 5\left(\frac{3}{5}\right)^{3/2} + 3\frac{3}{5} - 5\left(\frac{3}{5}\right)^{1/2}\right]$$
$$+ \frac{5}{9}\left[\left(\frac{3}{5}\right)^2 - 5\left(\frac{3}{5}\right)^{3/2} + 3\frac{3}{5} + 5\left(\frac{3}{5}\right)^{1/2}\right] = \frac{60}{25} = 2.4.$$

This 3-point formula exactly integrates a polynomial of degree $3 \times 2 - 1 = 5$; hence, the exact solution is recovered.

Next, consider the following integral involving transcendental function

$$I = \int_1^5 \frac{1}{x}\, dx = [\ln x]_1^5 = \ln 5 = 1.609.$$

To recast the problem in the standard form, a change of variable, $x = 2s + 3$, is first performed. The Jacobian of the coordinate transformation is readily evaluated, $dx/ds = 2$. The 2-point quadrature formula then yields a first approximation of the integral

$$I = \int_{-1}^{+1} \frac{1}{2s+3}\frac{dx}{ds}\, ds \approx 2\left[\frac{1}{-2\sqrt{1/3}+3} + \frac{1}{2\sqrt{1/3}+3}\right] = \frac{36}{23} = 1.565,$$

which only involves a 2.75% error. To improve the approximation, the 3-point quadrature formula is used, leading to

$$I \approx \frac{2}{9}\left[\frac{5}{-2\sqrt{3/5}+3} + \frac{8}{3} + \frac{5}{2\sqrt{3/5}+3}\right] = \frac{476}{297} = 1.603.$$

The error is now reduced to about 0.42%. Higher order Gauss-Legendre quadrature scheme can be derived that involve an increasing number of sampling points and associated weights. This data have been tabulated, see Abramowitz and Stegun [351], or can be readily calculated [5].

For integration over a rectangular domain, the basic Gauss-Legendre quadrature scheme of eq. (18.17) is generalized as

$$\int_{-1}^{+1} \int_{-1}^{+1} f(s,t) \, \mathrm{d}s\mathrm{d}t \approx \sum_{i=1}^{N_G} \sum_{j=1}^{M_G} w_i w_j f(s_i, t_j), \qquad (18.18)$$

where the sampling points, s_i and t_j, and associated weights, w_i and w_j, respectively, are those listed in table 18.1.

References

1. O. Bottema and B. Roth. *Theoretical Kinematics*. Dover Publications, Inc., New York, 1979.
2. F.B. Hildebrand. *Advanced Calculus for Applications*. Prentice Hall, Inc., Englewood Cliffs, New Jersey, second edition, 1976.
3. W. Flügge. *Tensor Analysis and Continuum Mechanics*. Springer-Verlag, New York, Heidelberg, Berlin, 1972.
4. G. Dahlquist and Å. Björck. *Numerical Methods*. Prentice Hall, Inc., Englewood Cliffs, New Jersey, 1974.
5. W.H. Press, B.P. Flannery, S.A. Teutolsky, and W.T. Vetterling. *Numerical Recipes. The Art of Scientific Computing*. Cambridge University Press, Cambridge, 1990.
6. F. Pfeiffer and C. Glocker. *Multi-Body Dynamics with Unilateral Contacts*. John Wiley & Sons, Inc, New York, 1996.
7. P.R. Dahl. Solid friction damping of mechanical vibrations. *AIAA Journal*, 14:1675–1682, 1976.
8. T. Baumeister, E.A. Avallone, and T. Baumeister III (eds.). *Marks' Mechanical Engineers Handbook*. McGraw-Hill Book Company, New-York, 1978.
9. J.E. Shigley and C.R. Mischke. *Mechanical Engineering Design*. McGraw-Hill Book Company, New York, 1989.
10. A.K. Banerjee and T.R. Kane. Modeling and simulation of rotor bearing friction. *Journal of Guidance, Control and Dynamics*, 17:1137–1151, 1994.
11. J. Srnik and F. Pfeiffer. Dynamics of CVT chain drives: Mechanical model and verification. In *Proceedings of the 16th Biennial Conference on Mechanical Vibration and Noise, Sacramento, CA, Sept. 14-17*, 1997.
12. E. Rabinowicz. *Friction and Wear of Materials*. John Wiley & Sons, New York, second edition, 1995.
13. J.C. Oden and J.A.C. Martins. Models and computational methods for dynamic friction phenomena. *Computer Methods in Applied Mechanics and Engineering*, 52:527–634, 1985.
14. L. Euler. Découverte d'un nouveau principe de mécanique. *Mémoires de l'Académie des Sciences de Berlin*, 6(1752):185–217, 1750.
15. L. Euler. Nova methodus motum corporum rigidorum determinandi. *Novi Commentari Academiae Scientiarum Imperialis Petropolitanae*, 20:208–238, 1775.
16. L. Euler. De motu corporum circa pumctum fixum mobilium. *Opera Mechanica et Astronomica*, 9(Series Secunda):413–441, 1776. Leonhardi Euleri Opera Omnia.

17. L. Euler. Formulae generales pro translatione quacunque corporum rigidorum. *Novi Commentari Academiae Scientiarum Imperialis Petropolitanae*, 20:189–207, 1775.

18. O. Rodrigues. Des lois géometriques qui régissent les déplacements d'un système solide dans l'espace, et de la variation des coordonnées provenant de ces déplacements considérés indépendamment des causes qui peuvent les produire. *Journal de Mathématiques Pures et Appliquées*, 5:380–440, 1840.

19. J. Angeles. *Fundamentals of Robotic Mechanical Systems. Theory, Methods, and Algorithms*. Springer-Verlag, New York, 1997.

20. G. Mozzi. *Discorso Matematico Sopra il Rotamento Momentaneo dei Corpi*. Stamperia di Donato Campo, Napoli, Italy, 1763.

21. M. Chasles. Note sur les propriétés générales du système de deux corps semblables entre eux et placés d'une manière quelconque dans l'espace; et sur le déplacement fini, ou infiniment petit d'un corps solide libre. *Bulletin des Sciences Mathématiques de Férussac*, 14:321–326, 1830.

22. J. Angeles. The application of dual algebra to kinematic analysis. In J. Angeles and E. Zakhariev, editors, *Computational Methods in Mechanical Systems*, volume 161, pages 3–31. Springer-Verlag, Heidelberg, 1998.

23. A.K. Pradeep, P.J. Yoder, and R. Mukundan. On the use of dual matrix exponentials in robot kinematics. *International Journal of Robotics Research*, 8(5):57–66, 1989.

24. D.J. Ewins. *Modal testing: theory and practice*. Wiley, New York, 1984.

25. R. Weinstock. *Calculus of Variations with Applications to Physics and Engineering*. Dover Publications, Inc., New York, 1974.

26. E. Hairer and G. Wanner. *Solving Ordinary Differential Equations II : Stiff and Differential-Algebraic Problems*. Springer, Berlin, 1996.

27. H. Hochstadt. *Differential Equations*. Dover Publications, Inc., New York, 1964.

28. J. García de Jalón, J. Unda, and A. Avello. Natural coordinates for the computer analysis of multibody systems. *Computer Methods in Applied Mechanics and Engineering*, 56:309–327, 1986.

29. J. Angeles. *Spatial Kinematic Chains*. Springer-Verlag, Berlin, 1982.

30. D. Li and P.W. Likins. Dynamics of a multibody system with relative translation on curved, flexible tracks. *Journal of Guidance, Control and Dynamics*, 10:299–306, 1987.

31. A. Cardona. *An Integrated Approach to Mechanism Analysis*. PhD thesis, Université de Liège, Belgium, 1989.

32. O.A. Bauchau. On the modeling of prismatic joints in flexible multi-body systems. *Computer Methods in Applied Mechanics and Engineering*, 181:87–105, 2000.

33. O.A. Bauchau and C.L. Bottasso. Contact conditions for cylindrical, prismatic, and screw joints in flexible multi-body systems. *Multibody System Dynamics*, 5:251–278, 2001.

34. L. Piegl and W. Tiller. *The Nurbs Book*. Springer-Verlag, Berlin, New Jersey, second edition, 1997.

35. G.E. Farin. *Curves and Surfaces for Computer Aided Geometric Design*. Academic Press, Inc., Boston, third edition, 1992.

36. A. Cardona and M. Géradin. Finite element modeling of flexible tracks. In C. L. Kirk and J.L. Junkins, editors, *Dynamics of Flexible Structures in Space*, pages 411–424. Springer-Verlag, 1990.

37. R.E. Roberson and R. Schwertassek. *Dynamics of Multibody Systems*. Springer-Verlag, Berlin, 1988.

38. P.E. Nikravesh. *Computer-Aided Analysis of Mechanical Systems*. Prentice-Hall, Englewood Cliffs, New Jersey, 1988.

39. F.M.L. Amirouche. *Computational Methods in Multibody Dynamics*. Prentice-Hall, Englewood Cliffs, New Jersey, 1992.
40. W.O. Schiehlen. *Advanced Multibody System Dynamics*. Kluwer Academic Publishers, Doordrecht, The Netherlands, 1993.
41. J. García de Jalón and E. Bayo. *Kinematic and Dynamic Simulation of Multibody Systems. The Real-Time Challenge*. Springer-Verlag, New York, 1994.
42. A.A. Shabana. *Dynamics of Multibody Systems*. Cambridge University Press, third edition, 2005.
43. W.O. Schiehlen, editor. *Multibody System Handbook*. Springer-Verlag, Berlin, 1990.
44. A. Laulusa and O.A. Bauchau. Review of classical approaches for constraint enforcement in multibody systems. *Journal of Computational and Nonlinear Dynamics*, 3(1):011004 1–8, January 2008.
45. O.A. Bauchau and A. Laulusa. Review of contemporary approaches for constraint enforcement in multibody systems. *Journal of Computational and Nonlinear Dynamics*, 3(1):011005 1–8, January 2008.
46. W.O. Schiehlen. Dynamics of complex multibody systems. *SM Archives*, 9:159–195, 1984.
47. C.W. Gear and L.R. Petzold. ODE methods for the solution of differential/algebraic systems. *SIAM Journal on Numerical Analysis*, 21(4):716–728, 1984.
48. P. Lötstedt and L.R. Petzold. Numerical solution of nonlinear differential equations with algebraic constraints I: Convergence results for backward differentiation formulas. *Mathematics of Computation*, 46(174):491–516, April 1986.
49. L.R. Petzold and P. Lötstedt. Numerical solution of nonlinear differential equations with algebraic constraints. II: Practical implications. *SIAM Journal on Scientific and Statistical Computing*, 7(3):720–733, July 1986.
50. K.E. Brenan, S.L. Campbell, and L.R. Petzold. *Numerical Solution of Initial-Value Problems in Differential-Algebraic Problems*. North-Holland, New York, 1989.
51. G.A. Maggi. *Principii della Teoria Matematica del Movimento dei Corpi: Corso di Meccanica Razionale*. Ulrico Hoepli, Milano, 1896.
52. G.A. Maggi. Di alcune nuove forme delle equazioni della dinamica applicabili ai systemi anolonomi. *Rendiconti della Regia Accademia dei Lincei, Serie V*, X:287–291, 1901.
53. J.I. Neimark and N.A. Fufaev. *Dynamics of Nonholonomic Systems*. American Mathematical Society, Providence, Rhode Island, 1972.
54. A. Kurdila, J.G. Papastavridis, and M.P. Kamat. Role of Maggi's equations in computational methods for constrained multibody systems. *Journal of Guidance, Control, and Dynamics*, 13(1):113–120, 1990.
55. J.G. Papastavridis. Maggi's equations of motion and the determination of constraint reactions. *Journal of Guidance, Control, and Dynamics*, 13(2):213–220, 1990.
56. P.E. Nikravesh. Some methods for dynamic analysis of constrained mechanical systems: A survey. In E.J. Haug, editor, *Computer Aided Analysis and Optimization of Mechanical Systems Dynamics*, pages 351–367. Springer-Verlag, Berlin, Heidelberg, 1984.
57. H. Hemami and F.C. Weimer. Modeling of nonholonomic dynamic systems with applications. *Journal of Applied Mechanics*, 48:177–182, March 1981.
58. P. Lötstedt. Mechanical systems of rigid bodies subjected to unilateral constraints. *SIAM Journal of Applied Mathematics*, 42(2):281–296, April 1982.
59. C.W Gear, B. Leimkuhler, and G.K. Gupta. Automatic integration of Euler-Lagrange equations with constraints. *Journal of Computational and Applied Mathematics*, 12 & 13:77–90, 1985.
60. T.R. Kane and C.F. Wang. On the derivation of equations of motion. *Journal of the Society for Industrial and Applied Mathematics*, 13(2):487–492, June 1965.

61. T.R. Kane and D.A. Levinson. *Dynamics: Theory and Applications*. McGraw-Hill Book Company, New York, 1985.

62. J. García de Jalón, J. Unda, A. Avello, and J.M. Jiménez. Dynamic analysis of three-dimensional mechanisms in "natural" coordinates. *Journal of Mechanisms, Transmissions, and Automation in Design*, 109:460–465, December 1987.

63. J. Unda, J. García de Jalón, F. Losantos, and R. Enparantza. A comparative study on some different formulations of the dynamic equations of constrained mechanical systems. *Journal of Mechanisms, Transmissions, and Automation in Design*, 109:466–474, December 1987.

64. M. Borri, C.L. Bottasso, and P. Mantegazza. Acceleration projection method in multibody dynamics. *European Journal of Mechanics, A/Solids*, 11(3):403–418, 1992.

65. F.E. Udwadia, R.E. Kalaba, and H.C. Eun. Equations of motion for constrained mechanical systems and the extended d'Alembert's principle. *Quarterly of Applied Mathematics*, LV(2):321–331, 1997.

66. F.E. Udwadia and R.E. Kalaba. A new perspective on constrained motion. *Proceedings of the Royal Society London, Series A*, 439:407–410, 1992.

67. R.E. Kalaba and F.E. Udwadia. On constrained motion. *Applied Mathematics and Computation*, 51:85–86, 1992.

68. R.E. Kalaba and F.E. Udwadia. Equations of motion for nonholonomic, constrained dynamical systems via Gauss's principle. *ASME Journal of Applied Mechanics*, 60:662–668, 1993.

69. R.E. Kalaba and F.E. Udwadia. Lagrangian mechanics, Gauss's principle, quadratic programming, and generalized inverses: New equations for nonholonomically constrained discrete mechanical systems. *Quarterly of Applied Mathematics*, LII(2):229–241, 1994.

70. F.E. Udwadia and R.E. Kalaba. On motion. *Journal of the Franklin Institute*, 330(3):571–577, 1993.

71. F.E. Udwadia and R.E. Kalaba. Equations of motion for mechnical systems. *Journal of Aerospace Engineering*, 9(3):64–69, July 1996.

72. F.E. Udwadia and R.E. Kalaba. The geometry of constrained motion. *Zeitschrift für angewandte Mathematik und Mechanik*, 75(8):637–640, 1995.

73. F.E. Udwadia and R.E. Kalaba. The explicit Gibbs-Appell equation and generalized inverse forms. *Quarterly of Applied Mathematics*, LVI(2):277–288, 1998.

74. F.E. Udwadia and R.E. Kalaba. What is the general form of the explicit equations of motion for constrained mechanical system. *Journal of Applied Mechanics*, 69:335–339, May 2002.

75. F.E. Udwadia and R.E. Kalaba. On the foundations of analytical dynamics. *International Journal of Non-Linear Mechanics*, 37:1079–1090, 2002.

76. F.E. Udwadia and R.E. Kalaba. *Analytical Dynamics: A New Approach*. Cambridge University Press, Cambridge, 1996.

77. N. Orlandea, M.A. Chace, and D.A. Calahan. A sparsity-oriented approach to the dynamic analysis and design of mechanical systems. Part I. *ASME Journal of Engineering for Industry*, 99(3):773–779, 1977.

78. N. Orlandea, D.A. Calahan, and M.A. Chace. A sparsity-oriented approach to the dynamic analysis and design of mechanical systems. Part II. *ASME Journal of Engineering for Industry*, 99(3):780–784, 1977.

79. S.L. Campbell and B. Leimkuhler. Differentiation of constraints in differential-algebraic equations. *Mechanics of Structures and Machines*, 19(1):19–39, 1991.

80. H. Brauchli. Mass-orthogonal formulation of equations of motion for multibody systems. *Journal of Applied Mathematics and Physics (ZAMP)*, 42:169–182, 1991.

81. H. Brauchli and R. Weber. Dynamical equations in natural coordinates. *Computer Methods in Applied Mechanics and Engineering*, 91:1403–1414, 1991.

82. G.H. Golub and C.F. van Loan. *Matrix Computations*. The Johns Hopkins University Press, Baltimore, second edition, 1989.

83. W. Blajer. A projection method approach to constrained dynamic analysis. *Journal of Applied Mechanics*, 59:643–649, September 1992.

84. W. Blajer, W. Schiehlen, and W. Schirm. A projective criterion to the coordinate partitioning method for multibody dynamics. *Archive of Applied Mechanics*, 64:86–98, 1994.

85. W. Blajer. A geometric unification of constrained system dynamics. *Multibody System Dynamics*, 1:3–21, 1997.

86. C.F. Gauss. Über ein neues algemeines Grundgesetz der Mechanik. *Zeitschrift für die reine und angewandte Mathematik*, 4:232–235, 1829.

87. C. Lanczos. *The Variational Principles of Mechanics*. Dover Publications, Inc., New York, 1970.

88. L. Lilov and M. Lorer. Dynamic analysis of multirigid-body systems based on Gauss principle. *Zeitschrift für angewandte Mathematik und Mechanik*, 62:539–545, 1982.

89. F. Cardin and G. Zanzotto. On constrained mechnical systems: D'Alembert's and Gauss' principles. *Journal of Mathematical Physics*, 30(7):1473–1479, July 1989.

90. M. Borri, C.L. Bottasso, and P. Mantegazza. Equivalence of Kane's and Maggi's equations. *Meccanica*, 25:272–274, 1990.

91. J. Angeles and S.K. Lee. The formulation of dynamical equations of holonomic mechanical systems using a natural orthogonal complement. *ASME Journal of Applied Mechanics*, 55:243–244, 1988.

92. P. Maißer. Analytical dynamics of multibody systems. *Computer Methods in Applied Mechanics and Engineering*, 91:1391–1396, 1991.

93. U. Jungnickel. Differential-algebraic equations in Riemannian spaces and applications to multibody system dynamics. *Zeitschrift für angewandte Mathematik und Mechanik*, 74(9):409–415, 1994.

94. H. Essén. On the geometry of nonholonomic dynamics. *Journal of Applied Mechanics*, 61:689–694, September 1994.

95. W. Blajer. A geometrical interpretation and uniform matrix formulation of multibody system dynamics. *Zeitschrift für angewandte Mathematik und Mechanik*, 81(4):247–259, 2001.

96. C.W. Gear. Simultaneous numerical solution of differential-algebraic equations. *IEEE Transactions on Circuit Theory*, CT-18(1):89–95, January 1971.

97. C.W. Gear. Differential-algebraic equations. In E.J. Haug, editor, *Computer Aided Analysis and Optimization of Mechanical Systems Dynamics*, pages 323–334. Springer-Verlag, Berlin, Heidelberg, 1984.

98. E.J. Haug. Elements and methods of computational dynamics. In E.J. Haug, editor, *Computer Aided Analysis and Optimization of Mechanical Systems Dynamics*, pages 3–38. Springer-Verlag, Berlin, Heidelberg, 1984.

99. R.A. Wehage and E.J. Haug. Generalized coordinate partitioning for dimension reduction in analysis of constrained dynamic systems. *ASME Journal of Mechanical Design*, 104(1):247–255, January 1982.

100. W.C. Walton and E.C. Steeves. A new matrix theorem and its application for establishing independent coordinates for complex dynamical systems with constraints. Technical Report NASA TR R-326, NASA, 1969.

101. P.E. Nikravesh and I.S. Chung. Application of Euler parameters to the dynamic analysis of three-dimensional constrained mechanical systems. *Journal of Mechanical Design*, 104:785–791, October 1982.

102. R.P. Singh and P.W. Likins. Singular value decomposition for constrained dynamical systems. *Journal of Applied Mechanics*, 52:943–948, December 1985.

103. N.K. Mani, E.J. Haug, and K.E. Atkinson. Application of singular value decomposition for analysis of mechanical system dynamics. *Journal of Mechanisms, Transmissions, and Automation in Design*, 107:82–87, March 1985.

104. F.M.L. Amirouche, T. Jia, and S.K. Ider. A recursive Householder transformation for complex dynamical systems with constraints. *Journal of Applied Mechanics*, 55:729–734, September 1988.

105. O.P. Agrawal and S. Saigal. Dynamic analysis of multi-body systems using tangent coordinates. *Computers & Structures*, 31(3):349–355, 1989.

106. C.G. Liang and G.M. Lance. A differentiable null space method for constrained dynamic analysis. *Journal of Mechanisms, Transmissions and Automation in Design*, 109:405–411, September 1987.

107. C. Wampler, K. Buffinton, and J. Shu-hui. Formulation of equations of motion for systems subject to constraints. *Journal of Applied Mechanics*, 52:465–470, June 1985.

108. S.S. Kim and M.J. Vanderploeg. QR decomposition for state space representation of constrained mechanical dynamic systems. *Journal of Mechanisms, Transmissions and Automation in Design*, 108:183–188, June 1986.

109. J. García de Jalón, M.A. Serna, and R. Avilés. Computer method for kinematic analysis of lower-pair mechanisms - I Velocities and accelerations. *Mechanism and Machine Theory*, 16(5):543–556, 1981.

110. J. García de Jalón, M.A. Serna, and R. Avilés. Computer method for kinematic analysis of lower-pair mechanisms - II Position problems. *Mechanism and Machine Theory*, 16(5):557–566, 1981.

111. J. García de Jalón, M.A. Serna, F. Viadero, and J. Flaquer. A simple numerical method for the kinematic analysis of spatial mechanisms. *Journal of Mechanical Design*, 104:78–82, January 1982.

112. J.A. Tárrago, M.A. Serna, C. Bastero, and J. García de Jalón. A computer method for the finite displacement problem in spatial mechanisms. *Journal of Mechanical Design*, 104:869–874, October 1982.

113. M.A. Serna, R. Avilés, and J. García de Jalón. Dynamic analysis of plane mechanisms with lower pairs in basic coordinates. *Mechanism and Machine Theory*, 17(6):397–403, 1982.

114. J. García de Jalón, J.M. Jiménez, A. Avello, F. Martín, and J. Cuadrado. Real time simulation of complex 3-D multibody systems with realistic graphics. In E.J. Haug and R.C. Deyo, editors, *Real-Time Integration Methods for Mechanical System Simulation*, pages 265–292. Springer-Verlag, Berlin, Heidelberg, 1990.

115. A. Avello, J.M. Jiménez, E. Bayo, and J. García de Jalón. A simple and highly parallelizable method for real-time dynamic simulation based on velocity transformations. *Computer Methods in Applied Mechanics and Engineering*, 107:313–339, 1993.

116. J.W. Kamman and R.L. Huston. Constrained multibody system dynamics - An automated approach. *Computers & Structures*, 18(6):999–1003, 1984.

117. J.W. Kamman and R.L. Huston. Dynamics of constrained multibody systems. *ASME Journal of Applied Mechanics*, 51:899–903, December 1984.

118. J.C. Chiou, K.C. Park, and C. Farhat. A natural partitioning scheme for parallel simulation of multibody systems. *International Journal for Numerical Methods in Engineering*, 36:945–967, 1993.

119. K.C. Park, J.C. Chiou, and Downer J.D. Explicit-implicit staggered procedure for multi-body dynamics analysis. *Journal of Guidance, Control, and Dynamics*, 13(3):562–570, May-June 1990.

120. A. Arabyan and F. Wu. An improved formulation for constrained mechanical systems. *Multibody System Dynamics*, 2:49–69, 1998.

121. W. Blajer. Projective formulation of Maggi's method for nonholonomic system analysis. *Journal of Guidance, Control, and Dynamics*, 15(2):522–525, 1992.

122. W. Blajer. An orthonormal tangent space method for constrained multibody systems. *Computer Methods in Applied Mechanics and Engineering*, 121:45–57, 1995.

123. M. Borri, C.L. Bottasso, and P. Mantegazza. A modified phase space formulation for constrained mechanical systems - Differential approach. *European Journal of Mechanics, A/Solids*, 11(5):701–727, 1992.

124. K.C. Park and J.C. Chiou. Stabilization of computational procedures for constrained dynamical systems. *Journal of Guidance, Control, and Dynamics*, 11(4):365–370, July-August 1988.

125. C.W. Gear. Differential-algebraic equation index transformations. *SIAM Journal on Scientific and Statistical Computing*, 9(1):40–47, January 1988.

126. C.W. Gear. Maintaining solution invariants in the numerical solution of ODEs. *SIAM Journal on Scientific and Statistical Computing*, 7(3):734–743, July 1986.

127. L.F. Shampine. Conservation laws and the numerical solution of ODEs. *Computers and Mathematics with Applications*, 12B(5/6):1287–1296, 1986.

128. E. Eich. Convergence results for a coordinate projection method applied to mechanical systems with algebraic constraints. *SIAM Journal on Numerical Analysis*, 30(5):1467–1482, October 1993.

129. J. Yen, E.J. Haug, and T.O. Tak. Numerical methods for constrained equations of motion in mechanical system dynamics. *Mechanics of Structures and Machines*, 19(1):41–76, 1991.

130. J. Yen. Constrained equations of motion in multibody dynamics as ODEs on manifolds. *SIAM Journal on Numerical Analysis*, 30(2):553–568, 1993.

131. F.A. Potra and J. Yen. Implicit numerical integration for euler-lagrange equations via tangent space parameterization. *Mechanics of Structures and Machines*, 19(1):77–98, 1991.

132. E.J. Haug and J. Yen. Implicit numerical integration of constrained equations of motion via generalized coordinate partitioning. *Journal of Mechanical Design*, 114:296–304, June 1992.

133. J. Yen and L.R. Petzold. An efficient Newton-type iteration for the numerical solution of highly oscillatory constrained multibody dynamic systems. *SIAM Journal on Scientific Computing*, 19(5):1513–1534, September 1998.

134. J. Yen, L.R. Petzold, and S. Raha. A time integration algorithm for flexible mechanism dynamics: The DAE α-method. *Computer Methods in Applied Mechanics and Engineering*, 158:341–355, 1998.

135. H.M. Hilber, T.J.R. Hughes, and R.L. Taylor. Improved numerical dissipation for time integration algorithms in structural dynamics. *Earthquake Engineering and Structural Dynamics*, 5:283–292, 1977.

136. J. Chung and G.M. Hulbert. A time integration algorithm for structural dynamics with improved numerical dissipation: The generalized-α method. *Journal of Applied Mechanics*, 60:371–375, 1993.

137. F.C. Tseng, Z.D. Ma, and G.M. Hulbert. Efficient numerical solution of constrained multibody dynamics systems. *Computer Methods in Applied Mechanics and Engineering*, 192:439–472, 2003.

138. M. Borri, L. Trainelli, and A. Croce. The embedded projection method: A general index reduction procedure for constrained system dynamics. *Computer Methods in Applied Mechanics and Engineering*, 195(50-51):6974–6992, 2006.

139. J. Parczewski and W. Blajer. On realization of program constraints: Part I - Theory. *Journal of Applied Mechanics*, 56:676–679, September 1989.

140. W. Blajer and J. Parczewski. On realization of program constraints. Part II - Practical implications. *Journal of Applied Mechanics*, 56:680–684, September 1989.

141. J.W. Baumgarte. Stabilization of constraints and integrals of motion in dynamic systems. *Computer Methods in Applied Mechanics and Engineering*, 1:1–16, 1972.

142. G.P. Ostermeyer. On Baumgarte stabilization for differential algebraic equations. In E.J. Haug and R.C. Deyo, editors, *Real-Time Integration Methods for Mechanical System Simulation*, pages 193–207. Springer-Verlag, Berlin, Heidelberg, 1990.

143. E. Eich and M. Hanke. Regularization methods for constrained mechanical multibody systems. *Zeitschrift für angewandte Mathematik und Mechanik*, 75(10):761–773, 1995.

144. P.E. Nikravesh, R.A. Wehage, and O.K. Kwon. Euler parameters in computational dynamics and kinematics. Part I and Part II. *Journal of Mechanisms, Transmissions, and Automation in Design*, 107(3):358–369, September 1985.

145. T.W. Park and E.J. Haug. A hybrid numerical integration method for machine dynamic simulation. *Journal of Mechanisms, Transmissions, and Automation in Design*, 108:211–216, June 1986.

146. C.O. Chang and P.E. Nikravesh. An adaptive constraint violation stabilization method for dynamic analysis of mechanical systems. *Journal of Mechanisms, Transmissions, and Automation in Design*, 107:488–492, December 1985.

147. D.S. Bae and S.M. Yang. A stabilization method for kinematic and kinetic constraint equations. In E.J. Haug and R.C. Deyo, editors, *Real-Time Integration Methods for Mechanical System Simulation*, pages 209–232. Springer-Verlag, Berlin, Heidelberg, 1990.

148. S. Yoon, R.M. Howe, and D.T. Greenwood. Stability and accuracy analysis of Baumgarte's constraint violation stabilization method. *Journal of Mechanical Design*, 117:446–453, September 1995.

149. J.C. Chiou and S.D. Wu. Constraint violation stabilization using input-output feedback linearization in multibody dynamic analysis. *Journal of Guidance, Control, and Dynamics*, 21(2):222–228, March-April 1998.

150. S.T. Lin and M.C. Hong. Stabilization method for numerical integration of multibody mechanical systems. *Journal of Mechanical Design*, 120:565–572, December 1998.

151. E. Bayo, J. García de Jalón, and M.A. Serna. A modified Lagrangian formulation for the dynamic analysis of constrained mechanical systems. *Computer Methods in Applied Mechanics and Engineering*, 71:183–195, November 1988.

152. E. Bayo, J. García de Jalón, A. Avello, and J. Cuadrado. An efficient computational method for real time multibody dynamic simulation in fully Cartesian coordinates. *Computer Methods in Applied Mechanics and Engineering*, 92:377–395, 1991.

153. A.J. Kurdila, J.L. Junkins, and S. Hsu. Lyapunov stable penalty methods for imposing nonholonomic constraints in multibody system dynamics. *Nonlinear Dynamics*, 4:51–82, 1993.

154. S. Yoon, R.M. Howe, and D.T. Greenwood. Geometric elimination of constraint violations in numerical simulation of lagrangian equations. *Journal of Mechanical Design*, 116:1058–1064, December 1994.

155. W. Blajer. Elimination of constraint violation and accuracy aspects in numerical simulation of multibody systems. *Multibody System Dynamics*, 7:265–284, 2002.

156. J.W. Baumgarte. A new method of stabilization for holonomic constraints. *ASME Journal of Applied Mechanics*, 50:869–870, December 1983.

157. Z. Terze, D. Lefeber, and O. Muftić. Null space integration method for constrained multibody system simulation with no constraint violation. *Multibody System Dynamics*, 6:229–243, 2001.

158. E. Bayo and A. Avello. Singularity-free augmented Lagrangian algorithms for constrained multibody dynamics. *Nonlinear Dynamics*, 5:209–231, 1994.

159. E. Bayo and R. Ledesma. Augmented Lagrangian and mass-orthogonal projection methods for constrained multibody dynamics. *Nonlinear Dynamics*, 9:113–130, 1996.

160. W.O. Schiehlen. Computational aspects in multibody system dynamics. *Computer Methods in Applied Mechanics and Engineering*, 90:569–582, 1991.

161. J. Cuadrado, J. Cardenal, and Bayo. E. Modeling and solution methods for efficient real-time simulation of multibody dynamics. *Multibody System Dynamics*, 1:259–280, 1997.

162. A.A. Shabana. Flexible multibody dynamics: Review of past and recent developments. *Multibody System Dynamics*, 1(2):189–222, June 1997.

163. T.M. Wasfy and A.K. Noor. Computational strategies for flexible multibody systems. *ASME Applied Mechanics Reviews*, 56(2):553–613, 2003.

164. M. Géradin and A. Cardona. *Flexible Multibody System: A Finite Element Approach.* John Wiley & Sons, New York, 2001.

165. P.W. Likins. Modal method for analysis of free rotations spacecraft. *AIAA Journal*, 5(7):1304–1308, July 1967.

166. R.D. Milne. Some remarks on the dynamics of deformable bodies. *AIAA Journal*, 6(3):556–558, March 1968.

167. B. Fraeijs de Veubeke. The dynamics of flexible bodies. *International Journal of Engineering Science*, 14(10):895–913, 1976.

168. J.R. Cavanin and P.W. Likins. Floating reference frames for flexible spacecraft. *Journal of Spacecraft*, 14(12):724–732, 1977.

169. O.P. Agrawal and A.A. Shabana. Dynamic analysis of multibody systems using component modes. *Computers & Structures*, 21(6):1303–1312, 1985.

170. W.S. Yoo and E.J. Haug. Dynamics of articulated structures. Part I. Theory. *Journal of Structural Mechanics*, 14:105–126, 1986.

171. W.S. Yoo and E.J. Haug. Dynamics of articulated structures. Part II. Computer implementation and applications. *Journal of Structural Mechanics*, 14:177–189, 1986.

172. A. Cardona and M. Géradin. Modelling of superelements in mechanism analysis. *International Journal for Numerical Methods in Engineering*, 32:1565–1593, 1991.

173. O. Friberg. A method for selecting deformation modes in flexible multibody dynamics. *International Journal for Numerical Methods in Engineering*, 32:1637–1655, 1991.

174. A. Cardona and M. Géradin. A superelement formulation for mechanism analysis. *Computer Methods in Applied Mechanics and Engineering*, 100:1–29, 1992.

175. R. Schwertassek, O. Wallrapp, and A.A. Shabana. Flexible multibody simulation and choice of shape functions. *Nonlinear Dynamics*, 20(4):361–380, 1999.

176. R. Schwertassek, S.V. Dombrowski, and O. Wallrapp. Modal representation of stress in flexible multibody simulation. *Nonlinear Dynamics*, 20(4):381–399, 1999.

177. A. Cardona. Superelements modelling in flexible multibody dynamics. *Multibody System Dynamics*, 4:245–266, 2000.

178. A.A. Shabana. Substructure synthesis methods for dynamic analysis of multi-body systems. *Computers & Structures*, 20:737–744, 1985.

179. O.P. Agrawal and A.A. Shabana. Application of deformable-body mean axis to flexible multibody system dynamics. *Computer Methods in Applied Mechanics and Engineering*, 56(2):217–245, 1986.

180. A.A. Shabana. Resonance conditions and deformable body co-ordinate systems. *Journal of Sound and Vibration*, 192:389–398, 1996.

181. P. Ravn. *Analysis and Synthesis of Planar Mechanical Systems Including Flexibility, Contact and Joint Clearance*. PhD thesis, Technical University of Denmark, 1998.

182. C.B. Drab, J.R. Haslinger, R.U. Pfau, and G. Offner. Comparison of the classical formulation with the reference conditions formulation for dynamic flexible multibody systems. *Journal of Computational and Nonlinear Dynamics*, 2(4):337–344, 2007.

183. L. Meirovitch. *Elements of Vibration Analysis*. McGraw-Hill Book Company, New York, 1975.

184. J.A.C. Ambrósio and J.P.C. Gonçalves. Complex flexible multibody systems with application to vehicle dynamics. In Jorge A.C. Ambrósio and Werner O. Schiehlen, editors, *Advances in Computational Multibody Dynamics, IDMEC/IST, Lisbon, Portugal, Sept. 20-23, 1999*, pages 241–258, 1999.

185. J.A.C. Ambrósio. Geometric and material nonlinear deformations in flexible multibody systems. In Jorge Ambrósio and Michal Kleiber, editors, *Proceedings of Computational Aspects of Nonlinear Structural Systems with Large Rigid Body Motion, NATO Advanced Research Workshop, Pultusk, Poland, July 2-7*, pages 91–115, 2000.

186. R.R. Craig and M.C. Bampton. Coupling of substructures for dynamic analyses. *AIAA Journal*, 6:1313–1319, 1968.

187. R.H. MacNeal. A hybrid method of component mode synthesis. *Computers & Structures*, 1(4):581–601, 1971.

188. S. Rubin. Improved component-mode representation for structural dynamic analysis. *AIAA Journal*, 13:995–1006, 1975.

189. D.N. Herting. A general purpose, multi-stage, component modal synthesis method. *Finite Elements in Analysis and Design*, 1:153–164, 1985.

190. R.M. Hintz. Analytical methods in component modal synthesis. *AIAA Journal*, 13:1007–1016, 1975.

191. R.R. Craig and C. Chang. Free-interface methods of substructure coupling for dynamic analysis. *AIAA Journal*, 14:1633–1635, 1976.

192. W.S. Yoo and E.J. Haug. Dynamics of flexible mechanical systems using vibration and static correction modes. *Journal of Mechanisms, Transmissions, and Automation in Design*, 108:315–322, 1986.

193. E.J. Wu, S. Haug. Geometric non-linear substructuring for dynamics of flexible mechanical elements. *International Journal for Numerical Methods in Engineering*, 26:2211–2226, 1988.

194. O.A. Bauchau and J. Rodriguez. Formulation of modal based elements in nonlinear, flexible multibody dynamics. *Journal of Multiscale Computational Engineering*, 1(2):161–180, 2003.

195. O.A. Bauchau, J. Rodriguez, and S.Y. Chen. Coupled rotor-fuselage analysis with finite motions using component mode synthesis. *Journal of the American Helicopter Society*, 49(2):201–211, 2004.

196. A.A. Shabana and R.A. Wehage. A coordinate reduction technique for dynamic analysis of spatial substructures with large angular rotations. *Journal of Structural Mechanics*, 11(3):401–431, March 1983.

197. T.J.R. Hughes. *The Finite Element Method*. Prentice Hall, Inc., Englewood Cliffs, New Jersey, 1992.

198. K.J. Bathe. *Finite Element Procedures*. Prentice Hall, Inc., Englewood Cliffs, New Jersey, 1996.

199. L.F. Shampine and M.K. Gordon. *Computer Solution of Ordinary Differential Equations: The Initial Value Problem*. W.J. Freeman, San Francisco, CA, 1975.

200. A. Cardona and M. Géradin. Time integration of the equations of motion in mechanism analysis. *Computers & Structures*, 33(3):801–820, 1989.

201. N.M. Newmark. A method of computation for structural dynamics. *Journal of the Engineering Mechanics Division*, 85:67–94, 1959.

202. T.J.R. Hughes. Analysis of transient algorithms with particular reference to stability behavior. In T. Belytschko and T.J.R. Hughes, editors, *Computational Methods for Transient Analysis*, pages 67–155. North-Holland, Amsterdam, 1983.

203. J.C. Simo and K. Wong. Unconditionally stable algorithms for rigid body dynamics that exactly preserve energy and momentum. *International Journal for Numerical Methods in Engineering*, 31:19–52, 1991.

204. J.C. Simo and N. Tarnow. The discrete energy-momentum method. Conserving algorithms for nonlinear dynamics. *ZAMP*, 43:757–792, 1992.

205. P. Betsch and P. Steinmann. Constrained integration of rigid body dynamics. *Computer Methods in Applied Mechanics and Engineering*, 191:467–488, 2001.

206. J.C. Simo, N. Tarnow, and M. Doblare. Non-linear dynamics of three-dimensional rods: Exact energy and momentum conserving algorithms. *International Journal for Numerical Methods in Engineering*, 38:1431–1473, 1995.

207. S. Leyendecker, P. Betsch, and P. Steinmann. Objective energy-momentum conserving integration for the constrained dynamics of geometrically exact beams. *Computer Methods in Applied Mechanics and Engineering*, 195:2313–2333, 2006.

208. J.C. Simo and N. Tarnow. A new energy and momentum conserving algorithm for the nonlinear dynamics of shells. *International Journal for Numerical Methods in Engineering*, 37:2527–2549, 1994.

209. P. Betsch and N. Sänger. On the use of geometrically exact shells in a conserving framework for flexible multibody dynamics. *Computer Methods in Applied Mechanics and Engineering*, 198:1609–1630, 2009.

210. O.A. Bauchau, G. Damilano, and N.J. Theron. Numerical integration of nonlinear elastic multi-body systems. *International Journal for Numerical Methods in Engineering*, 38:2727–2751, 1995.

211. O.A. Bauchau, C.L. Bottasso, and L. Trainelli. Robust integration schemes for flexible multibody systems. *Computer Methods in Applied Mechanics and Engineering*, 192(3-4):395–420, 2003.

212. C.L. Bottasso and M. Borri. Energy preserving/decaying schemes for non-linear beam dynamics using the helicoidal approximation. *Computer Methods in Applied Mechanics and Engineering*, 143:393–415, 1997.

213. O.A. Bauchau, J.Y. Choi, and C.L. Bottasso. Time integrators for shells in multibody dynamics. *International Journal of Computers and Structures*, 80:871–889, 2002.

214. C.L. Bottasso, M. Borri, and L. Trainelli. Integration of elastic multibody systems by invariant conserving/dissipating algorithms. Part I: formulation. *Computer Methods in Applied Mechanics and Engineering*, 190:3669–3699, 2001.

215. C.L. Bottasso, M. Borri, and L. Trainelli. Integration of elastic multibody systems by invariant conserving/dissipating algorithms. Part II: numerical schemes and applications. *Computer Methods in Applied Mechanics and Engineering*, 190:3701–3733, 2001.

216. C.L. Bottasso and L. Trainelli. An attempt at the classification of energy decaying schemes for structural and multibody dynamics. *Multibody System Dynamics*, 12:173–185, 2004.

217. O.A. Bauchau and N.J. Theron. Energy decaying schemes for nonlinear elastic multibody systems. *Computers & Structures*, 59(2):317–331, 1996.

218. O.A. Bauchau. Computational schemes for flexible, nonlinear multi-body systems. *Multibody System Dynamics*, 2(2):169–225, 1998.

219. O.A. Bauchau and C.L. Bottasso. On the design of energy preserving and decaying schemes for flexible, nonlinear multi-body systems. *Computer Methods in Applied Mechanics and Engineering*, 169(1-2):61–79, 1999.

220. M. Borri, C.L. Bottasso, and L. Trainelli. Integration of elastic multibody systems by invariant conserving/dissipating algorithms. Part I: Formulation. *Computer Methods in Applied Mechanics and Engineering*, 190:3669–3699, 2001.

221. M. Borri, C.L. Bottasso, and L. Trainelli. Integration of elastic multibody systems by invariant conserving/dissipating algorithms. Part II: Numerical schemes and applications. *Computer Methods in Applied Mechanics and Engineering*, 190:3701–3733, 2001.

222. O. Gonzalez. Mechanical systems subject to holonomic constraints: Differential-algebraic formulations and conservative integration. *Physica D*, 132:165–174, 1999.

223. O.A. Bauchau. A self-stabilized algorithm for enforcing constraints in multibody systems. *International Journal of Solids and Structures*, 40(13-14):3253–3271, 2003.

224. P. Betsch. The discrete null space method for the energy consistent integration of constrained mechanical systems. Part I: Holonomic constraints. *Computer Methods in Applied Mechanics and Engineering*, 194(50-52):5159–5190, 2005.

225. P. Betsch and S. Leyendecker. The discrete null space method for the energy consistent integration of constrained mechanical systems. Part II: Multibody dynamics. *International Journal for Numerical Methods in Engineering*, 67:499–552, 2006.

226. S. Leyendecker, P. Betsch, and P. Steinmann. The discrete null space method for the energy consistent integration of constrained mechanical systems. Part III: Flexible multibody dynamics. *Multibody System Dynamics*, 19(1-2):45–72, 2008.

227. P. Betsch and P. Steinmann. A DAE approach to flexible multibody dynamics. *Multibody System Dynamics*, 8:367–391, 2002.

228. C.W. Gear. *Numerical Initial Value Problems in Ordinary Differential Equations*. Prentice-Hall, Englewood Cliff, N.J., 1971.

229. M. Arnold. A perturbation analysis for the dynamical simulation of mechanical multibody systems. *Applied Numerical Mathematics*, 18:37–56, 1995.

230. O.A. Bauchau, C.L. Bottasso, and Y.G. Nikishkov. Modeling rotorcraft dynamics with finite element multibody procedures. *Mathematical and Computer Modeling*, 33(10-11):1113–1137, 2001.

231. R.L. Fox. *Optimization Methods for Engineering Design*. Addison-Wesley Publishing Company, Reading, Massachusetts, 1971.

232. G.V. Reklaitis, A. Ravindran, and K.M. Ragsdell. *Engineering Optimization. Methods and Applications*. John Wiley & Sons, New York, 1983.

233. G.N. Vanderplaats. *Numerical Optimization Techniques for Engineering: With Applications*. McGraw-Hill Book Company, New-York, 1984.

234. A. Cardona and M. Géradin. Numerical integration of second order differential-algebraic systems in flexible mechanism dynamics. In J. Ambrosio and M. Seabra Pereira, editors, *Computer-Aided Analysis Of Rigid And Flexible Mechanical Systems*, pages 501–529. NATO ASI Series, Kluwer Academic Publishers, 1993.

235. C.L. Bottasso, O.A. Bauchau, and A. Cardona. Time-step-size-independent conditioning and sensitivity to perturbations in the numerical solution of index three differential algebraic equations. *SIAM Journal on Scientific Computing*, 29(1):397–414, 2007.

236. C.L. Bottasso, D. Dopico, and L. Trainelli. On the optimal scaling of index-three DAEs in multibody dynamics. *Multibody System Dynamics*, 19:3–20, 2008.

237. P.E. Gill, W. Murray, M.A. Saunders, and M.H. Wright. Sequential quadratic programming methods for nonlinear programming. In E.J. Haug, editor, *Computer-Aided Analysis and Optimization of Mechanical System Dynamics*, pages 679–697. Springer-Verlag, Berlin, Heidelberg, 1984.

238. O.A. Bauchau, A. Epple, and C.L. Bottasso. Scaling of constraints and augmented La-grangian formulations in multibody dynamics simulations. *Journal of Computational and Nonlinear Dynamics*, 4(2):021007 1–9, April 2009.

239. T.R. Kane. *Dynamics*. Holt, Rinehart and Winston, Inc, New York, 1968.

240. J. Argyris. An excursion into large rotations. *Computer Methods in Applied Mechanics and Engineering*, 32(1–3):85–155, 1982.

241. P.C. Hughes. *Spacecraft Attitude Dynamics*. John Wiley & Sons, New York, 1986.

242. H. Cheng and K.C. Gupta. A historical note on finite rotations. *Journal of Applied Mechanics*, 56:139–145, 1989.

243. M.D. Shuster. A survey of attitude representations. *Journal of the Astronautical Sciences*, 41(4):439–517, 1993.

244. A. Ibrahimbegović. On the choice of finite rotation parameters. *Computer Methods in Applied Mechanics and Engineering*, 149:49–71, 1997.

245. P. Betsch, A. Menzel, and Stein. E. On the parameterization of finite rotations in com-putational mechanics. A classification of concepts with application to smooth shells. *Computer Methods in Applied Mechanics and Engineering*, 155(3-4):273–305, 1998.

246. J. Stuelpnagel. On the parameterization of the three-dimensional rotation group. *SIAM Review*, 6(4):422–430, 1964.

247. W.R. Hamilton. *Elements of Quaternions*. Cambridge University Press, Cambridge, 1899.

248. A. Cayley. On certain results relating to quaternions. *Philosophical Magazine*, 26:141–145, 1845.

249. K.W. Spring. Euler parameters and the use of quaternion algebra in the manipulation of finite rotations: A review. *Mechanism and Machine Theory*, 21:365–373, 1986.

250. R.A. Wehage. Quaternions and Euler parameters. A brief exposition. In E.J. Haug, editor, *Computer Aided Analysis and Optimization of Mechanical Systems Dynamics*, pages 147–180. Springer-Verlag, Berlin, Heidelberg, 1984.

251. R.M. Murray, Z. Li, and S.S. Sastry. *A Mathematical Introduction to Robotic Manipu-lation*. CRC Press, 1994.

252. M. Géradin and A. Cardona. Kinematics and dynamics of rigid and flexible mechanisms using finite elements and quaternion algebra. *Computational Mechanics*, 4:115–135, 1989.

253. A.R. Klumpp. Singularity-free extraction of a quaternion from a direction-cosine matrix. *Journal of Spacecraft and Rockets*, 13:754–755, December 1976.

254. S.W. Shepperd. Quaternion from rotation matrix. *Journal of Guidance and Control*, 1:223–224, May-June 1978.

255. O.A. Bauchau and L. Trainelli. The vectorial parameterization of rotation. *Nonlinear Dynamics*, 32(1):71–92, 2003.

256. T.F. Wiener. *Theoretical Analysis of Gimballess Inertial Reference Equipment Using Delta-Modulated Instruments*. PhD thesis, Massachusetts Institute of Technology, Cam-bridge, Massachusetts, 1962. Department of Aeronautical and Astronautical Engineer-ing.

257. V. Milenković. Coordinates suitable for angular motion synthesis in robots. In *Proceed-ings of the Robot VI Conference, Detroit MI, March 2-4, 1982*, 1982. Paper MS82-217.

258. S.R. Marandi and V.J. Modi. A preferred coordinate system and the associated orienta-tion representation in attitude dynamics. *Acta Astronautica*, 15:833–843, 1987.

259. R.S. Ball. *A Treatise on the Theory of Screws*. Cambridge University Press, Cambridge, 1998.

260. J. Angeles. On twist and wrench generators and annihilators. In M.F.O. Seabra Pereira and J.A.C. Ambrosio, editors, *Computer-Aided Analysis Of Rigid And Flexible Mechanical Systems*, pages 379–411, Dordrecht, 1993. NATO ASI Series, Kluwer Academic Publishers.

261. D.P. Chevallier. Lie algebra, modules, dual quaternions and algebraic methods in kinematics. *Mechanism and Machine Theory*, 26(6):613–627, 1991.

262. A.T. Yang and F. Freudenstein. Application of dual-number quaternion algebra to the analysis of spatial mechanims. *ASME Journal of Applied Mechanics*, 86:300–308, 1964.

263. O.P. Agrawal. Hamilton operators and dual-number quaternions in spatial kinematics. *Mechanism and Machine Theory*, 22(6):569–575, 1987.

264. J.M. McCarthy. *An Introduction to Theoretical Kinematics*. The MIT Press, Cambridge, MA, 1990.

265. I.S. Fischer. *Dual Number Methods in Kinematics, Statics and Dynamics*. CRC Press, Boca Raton, 1999.

266. V. Brodsky and M. Shoham. The dual inertia operator and its application to robot dynamics. *Journal of Mechanical Design*, 116:1089–1095, 1994.

267. E. Pennestrì and R. Stefanelli. Linear algebra and numerical algorithms using dual numbers. *Multibody System Dynamics*, 18:323–344, 2007.

268. T. Merlini and M. Morandini. The helicoidal modeling in computational finite elasticity. Part I: Variational formulation. *International Journal of Solids and Structures*, 41(18-19):5351–5381, 2004.

269. M. Borri, L. Trainelli, and C.L. Bottasso. On representations and parameterizations of motion. *Multibody Systems Dynamics*, 4:129–193, 2000.

270. M. Borri and C.L. Bottasso. An intrinsic beam model based on a helicoidal approximation. Part I: Formulation. *International Journal for Numerical Methods in Engineering*, 37:2267–2289, 1994.

271. M. Borri and C.L. Bottasso. An intrinsic beam model based on a helicoidal approximation. Part II: Linearization and finite element implementation. *International Journal for Numerical Methods in Engineering*, 37:2291–2309, 1994.

272. O.A. Bauchau and J.Y. Choi. The vector parameterization of motion. *Nonlinear Dynamics*, 33(2):165–188, 2003.

273. S.P. Timoshenko and D.H. Young. *Vibration Problems in Engineering*. John Wiley & Sons, New York, 1974.

274. R. E. Nickell. Nonlinear dynamics by mode superposition. *Computer Methods in Applied Mechanics and Engineering*, 7(1):107–129, 1976.

275. J.M.T. Thompson and A.C. Walker. The non-linear perturbation analysis of discrete structural systems. *International Journal of Solids and Structures*, 4(8):281–288, 1968.

276. A.K. Noor and J.M. Peters. Reduced basis technique for nonlinear analysis of structures. *AIAA Journal*, 18(4):455–462, 1980.

277. A.K. Noor. Recent advances in reduction methods for nonlinear problems. *Computers and Structures*, 13(1-3):31–44, 1981.

278. O.A. Bauchau and D. Guernsey. On the choice of appropriate bases for nonlinear dynamic modal analysis. *Journal of the American Helicopter Society*, 38(4):28–36, 1993.

279. G.C. Ruzicka and D.H. Hodges. Application of the mixed finite element method to rotor blade modal reduction. *Mathematical and Computer Modelling*, 33(10-11):1177–1202, 2001.

280. G.C. Ruzicka and D.H. Hodges. A mixed finite element treatment of rotor blade elongation. *Journal of the American Helicopter Society*, 48(3):167–175, 2003.

281. J.C. Houbolt and G.W. Brooks. Differential equations of motion for combined flapwise bending, chordwise bending, and torsion of twisted nonuniform rotor blades. Technical Report 1348, NACA Report, 1958.

282. D.H. Hodges and E.H. Dowell. Nonlinear equations of motion for the elastic bending and torsion of twisted nonuniform rotor blades. Technical report, NASA TN D-7818, 1974.

283. O.A. Bauchau and D.H. Hodges. Analysis of nonlinear multi-body systems with elastic couplings. *Multibody System Dynamics*, 3(2):168–188, May 1999.

284. S.P. Timoshenko and Goodier J.N. *Theory of Elasticity*. McGraw-Hill Book Company, New York, third edition, 1970.

285. O.A. Bauchau and J.I. Craig. *Structural Analysis with Application to Aerospace Structures*. Springer, Dordrecht, Heidelberg, London, New-York, 2009.

286. K. Washizu. *Variational Methods in Elasticity and Plasticity*. Pergamon Press, Oxford, U.K., 1975.

287. I.H. Shames and C.L. Dym. *Energy and Finite Element Methods in Structural Mechanics*. Hemisphere Publishing Corp., 1985.

288. L. Anand. On H. Hencky's approximate strain-energy function for moderate deformations. *Journal of Applied Mechanics*, 46:78–82, March 1979.

289. L. Anand. Moderate deformations in extension-torsion of incompressible isotropic elastic materials. *Journal of the Mechanics and Physics of Solids*, 34(3):293–304, 1986.

290. M. Degener, D.H. Hodges, and D. Petersen. Analytical and experimental study of beam torsional stiffness with large axial elongation. *Journal of Applied Mechanics*, 55:171–178, March 1988.

291. R. Ledesma, Z.-D. Ma, G. Hulbert, and A. Wineman. A nonlinear viscoelastic bushing element in multibody dynamics. *Computational Mechanics*, 17(5):287–296, September 1996.

292. J. Kadlowec, A. Wineman, and G. Hulbert. Elastomer bushing response: Experiments and finite element modeling. *Acta Mechanica*, 163(5):25–38, 2003.

293. P. Masarati and M. Morandini. Intrinsic deformable joints. *Multibody System Dynamics*, 23(4):361–386, 2010.

294. *MSC/ADAMS User's Manual*, 2007.

295. *Abaqus Theory Manual, Abaqus Version 6.7 edition*.

296. L.E. Malvern. *Introduction to the Mechanics of a Continuous Medium*. Prentice Hall, Inc., Englewood Cliffs, New Jersey, 1969.

297. L. Euler. *Methodus Inveniendi Lineas Curvas Maximi Minimive Proprietate Gaudentes*. Bousquet, Lausanne and Geneva, 1744. Appendix I: De Curvis Elasticis.

298. S.P. Timoshenko. On the correction factor for shear of the differential equation for transverse vibrations of bars of uniform cross-section. *Philosophical Magazine*, 41:744–746, 1921.

299. S.P. Timoshenko. On the transverse vibrations of bars of uniform cross-section. *Philosophical Magazine*, 43:125–131, 1921.

300. E. Reissner. On one-dimensional finite-strain beam theory: the plane problem. *Zeitschrift für angewandte Mathematik und Physik*, 23:795–804, 1972.

301. E. Reissner. On one-dimensional large-displacement finite-strain beam theory. *Studies in Applied Mathematics*, 52:87–95, 1973.

302. E. Reissner. On finite deformations of space-curved beams. *Zeitschrift für angewandte Mathematik und Physik*, 32:734–744, 1981.

303. J.C. Simo. A finite strain beam formulation. the three-dimensional dynamic problem. Part I. *Computer Methods in Applied Mechanics and Engineering*, 49:55–70, 1985.

304. J.C. Simo and L. Vu-Quoc. A three dimensional finite strain rod model. Part II: Computational aspects. *Computer Methods in Applied Mechanics and Engineering*, 58:79–116, 1986.

305. M. Borri and T. Merlini. A large displacement formulation for anisotropic beam analysis. *Meccanica*, 21:30–37, 1986.

306. D.A Danielson and D.H. Hodges. Nonlinear beam kinematics by decomposition of the rotation tensor. *Journal of Applied Mechanics*, 54(2):258–262, 1987.

307. D.A Danielson and D.H. Hodges. A beam theory for large global rotation, moderate local rotation, and small strain. *Journal of Applied Mechanics*, 55(1):179–184, 1988.

308. V. Giavotto, M. Borri, P. Mantegazza, G. Ghiringhelli, V. Carmaschi, G.C. Maffioli, and F. Mussi. Anisotropic beam theory and applications. *Computers & Structures*, 16(1-4):403–413, 1983.

309. M. Borri, G.L. Ghiringhelli, and T. Merlini. Composite beam analysis: Linear analysis of naturally curved and twisted anisotropic beams. Technical report, Politecnico di Milano, 1992.

310. V.L. Berdichevsky. On the energy of an elastic rod. *Prikladnaya Matematika y Mekanika*, 45(4):518–529, 1982.

311. D.H. Hodges. A review of composite rotor blade modeling. *AIAA Journal*, 28(3):561–565, March 1990.

312. A.R. Atilgan, D.H. Hodges, and M.V. Fulton. Nonlinear deformation of composite beams: Unification of cross-sectional and elastica analyses. *Applied Mechanics Reviews*, 44(11):S9–S15, November 1991.

313. A.R. Atilgan and D.H. Hodges. Unified nonlinear analysis for nonhomogeneous anisotropic beams with closed cross sections. *AIAA Journal*, 29(11):1990 – 1999, November 1991.

314. W.B. Yu, D.H. Volovoi, V.V. Hodges, and X. Hong. Validation of the variational asymptotic beam sectional (VABS) analysis. *AIAA Journal*, 40(10):2105–2112, October 2002.

315. B. Popescu and Hodges D.H. Asymptotic treatment of the trapeze effect in finite element cross-sectional analysis of composite beams. *International Journal of Non-Linear Mechanics*, 34(4):709–721, 1999.

316. D.H. Hodges. *Nonlinear Composite Beam Theory*. AIAA, Reston, Virginia, 2006.

317. A.K. Noor, T. Belytschko, and J.C. Simo, editors. *Analytical and Computational Models of Shells*, volume 3. ASME, CED, 1989.

318. A.K. Noor. Bibliography of monographs and surveys on shells. *Applied Mechanics Review*, 43(9):223–234, 1990.

319. W.B. Yu, D.H. Hodges, and V.V. Volovoi. Asymptotic generalization of Reissner-Mindlin theory: Accurate three-dimensional recovery for composite shells. *Computer Methods in Applied Mechanics and Engineering*, 191(44):5087–5109, 2002.

320. W.B. Yu and D.H. Hodges. A geometrically nonlinear shear deformation theory for composite shells. *Journal of Applied Mechanics*, 71(1):1–9, 2004.

321. K.J. Bathe and E.N. Dvorkin. A four-node plate bending element based on Mindlin/Reissner plate theory and a mixed interpolation. *International Journal for Numerical Methods in Engineering*, 21:367–383, 1985.

322. K.J. Bathe and E.N. Dvorkin. A formulation of general shell elements - The use mixed interpolation of tensorial components. *International Journal for Numerical Methods in Engineering*, 22:697–722, 1986.

323. M.L. Bucalem and K.J. Bathe. Higher-order MITC general shell elements. *International Journal for Numerical Methods in Engineering*, 36:3729–3754, 1993.

324. R.D. Cook, Malkus D.S., and M.E. Plesha. *Concept and Applications of the Finite Elements Method*. John Wiley & Sons, New York, 1989.

325. M.A. Crisfield and G. Jelenić. Objectivity of strain measures in the geometrically exact three-dimensional beam theory and its finite-element implementation. *Proceedings of the Royal Society, London: Mathematical, Physical and Engineering Sciences*, 455(1983):1125–1147, 1999.

326. G. Jelenić and M.A. Crisfield. Geometrically exact 3D beam theory: Implementation of a strain-invariant finite element for static and dynamics. *Computer Methods in Applied Mechanics and Engineering*, 171:141–171, 1999.

327. A. Ibrahimbegović, F. Frey, and I. Kozar. Computational aspects of vector-like parameterization of three-dimensional finite rotations. *International Journal for Numerical Methods in Engineering*, 38(21):3653–3673, 1995.

328. A. Cardona and M. Géradin. A beam finite element non-linear theory with finite rotation. *International Journal for Numerical Methods in Engineering*, 26:2403–2438, 1988.

329. M.A. Crisfield. A consistent co-rotational formulation for non-linear, three-dimensional beam-elements. *Computer Methods in Applied Mechanics and Engineering*, 81:131–150, 1990.

330. P. Betsch and P. Steinmann. Frame-indifferent beam element based upon the geometrically exact beam theory. *International Journal for Numerical Methods in Engineering*, 54:1775–1788, 2002.

331. I. Romero and F. Armero. An objective finite element approximation of the kinematics of geometrically exact rods and its use in the formulation of an energymomentum conserving scheme in dynamics. *International Journal for Numerical Methods in Engineering*, 54:1683–1716, 2002.

332. I. Romero. The interpolation of rotations and its application to finite element models of geometrically exact rods. *Computational Mechanics*, 34(2):121–133, 2004.

333. A. Ibrahimbegović and R.L. Taylor. On the role of frame-invariance in structural mechanics models at finite rotations. *Computer Methods in Applied Mechanics and Engineering*, 191:5159–5176, 2002.

334. A.A. Shabana. A computer implementation of the absolute nodal coordinate formulation for flexible multibody dynamics. *Nonlinear Dynamics*, 16(3):293–306, 1998.

335. T.J.R. Hughes. Stability, convergence, and growth and decay of energy of the average acceleration method in nonlinear structural dynamics. *Computers & Structures*, 6:313–324, 1976.

336. C. Farhat, L. Crivelli, and M. Géradin. Implicit time integration of a class of constrained hybrid formulations - Part I: Spectral stability theory. *Computer Methods in Applied Mechanics and Engineering*, 125:71–107, 1995.

337. C. Lunk and B. Simeon. Solving constrained mechanical systems by the family of Newmark and α-methods. *Zeitschrift für angewandte Mathematik und Mechanik*, 86(10):772–784, 2006.

338. L.O. Jay and D. Negrut. Extensions of the HHT-α method to differential algebraic equations in mechanics. *Electronic Transactions on Numerical Analysis*, 26(1):190–208, 2007.

339. D. Negrut, R. Rampalli, G. Ottarsson, and A. Sajdak. On an implementation of the HHT method in the context of index 3 differential-algebraic equations of multibody dynamics. *Journal of Computational and Nonlinear Dynamics*, 2(1):73–85, January 2007.

340. M. Arnold and O. Brüls. Convergence of the generalized-α scheme for constrained mechanical systems. *Multibody System Dynamics*, 18(2):185–202, 2007.

341. C. Johnson. *Numerical Solutions of Partial Differential Equations by the Finite Element Method*. Cambridge University Press, Cambridge, 1987.

720   References

342. C. Johnson, U. Nävert, and V. Pitkäranta. Finite element methods for linear hyperbolic problems. *Computer Methods in Applied Mechanics and Engineering*, 45:285–312, 1984.

343. C. Johnson and V. Szepessy. Convergence of a finite element method for a nonlinear hyperbolic conservation law. Technical Report 1985-25, Mathematics Department, Chalmers University of Technology, Göteborg, Sweden, 1985.

344. T.R.J. Hughes and M. Hulbert. Space-time finite element formulations for elastodynamics: Formulation and error estimates. *Computer Methods in Applied Mechanics and Engineering*, 66:339–363, 1988.

345. M. Hulbert. *Space-Time Finite Element Methods for Second-Order Hyperbolic Equations*. PhD thesis, Stanford University, 1989.

346. O.A. Bauchau and N.J. Theron. Energy decaying scheme for non-linear beam models. *Computer Methods in Applied Mechanics and Engineering*, 134:37–56, 1996.

347. O.A. Bauchau and T. Joo. Computational schemes for nonlinear elasto-dynamics. *International Journal for Numerical Methods in Engineering*, 45:693–719, 1999.

348. O.A. Bauchau, J.Y. Choi, and C.L. Bottasso. On the modeling of shells in multibody dynamics. *Multibody System Dynamics*, 8(4):459–489, 2002.

349. C.L. Bottasso and M. Borri. Integrating finite rotations. *Computer Methods in Applied Mechanics and Engineering*, 164:307–331, 1998.

350. M. Borri, C.L. Bottasso, and L. Trainelli. A novel momentum-preserving energy-decaying algorithm for finite element multibody procedures. In Jorge Ambrósio and Michal Kleiber, editors, *Proceedings of Computational Aspects of Nonlinear Structural Systems with Large Rigid Body Motion, NATO Advanced Research Workshop, Pultusk, Poland, July 2-7*, pages 549–568, 2000.

351. M. Abramowitz and I.A. Stegun. *Handbook of Mathematical Functions*. Dover Publications, Inc., New York, 1964.

Index